Mathematics for the Managerial, Life, and Social Sciences

S. T. TAN

Stonehill College

PWS Publishing Company

I(T)P **An International Thomson Publishing Company**

Boston · Albany · Bonn · Cincinnati · Detroit · London · Madrid · Melbourne · Mexico City
New York · Paris · San Francisco · Singapore · Tokyo · Toronto · Washington

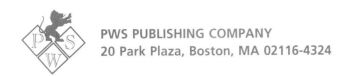

PWS PUBLISHING COMPANY
20 Park Plaza, Boston, MA 02116-4324

I(T)P™
International Thomson Publishing
The trademark ITP is used under license.

For more information contact:

PWS Publishing Company
20 Park Plaza
Boston, MA 02116

International Thomson Publishing
  Europe
Berkshire House 168-173
High Holborn
London WC1V 7AA
England

Thomas Nelson Australia
102 Dodds Street
South Melbourne, 3205
Victoria, Australia

Nelson Canada
1120 Birchmount Road
Scarborough, Ontario
Canada M1K 5G4

International Thomson Editores
Campos Eliseos 385, Piso 7
Col. Polanco
11560 Mexico D.F., Mexico

International Thomson Publishing
  GmbH
Königswinterer Strasse 418
53227 Bonn, Germany

International Thomson Publishing
  Asia
221 Henderson Road
#05-10 Henderson Building
Singapore 0315

International Thomson Publishing
  Japan
Hirakawacho Kyowa Building, 31
2-2-1 Hirakawacho
Chiyoda-ku, Tokyo 102
Japan

Library of Congress Cataloging-in-Publication Data

Tan, Soo Tang.
    Mathematics for the managerial, life, and social sciences / Soo
  Tang Tan.
      p.   cm.
    Includes index.
    ISBN 0-534-95184-8
    1. Mathematics.   I. Title.
  QA37.2.T364   1996
  510—dc20   95-42398
                                                                              CIP

Developmental Editor: Barbara Lovenvirth
Editorial Assistant: Anna Aleksandrowicz
Production Coordinator: Patricia Adams
Production: The Cooper Company
Market Development Manager: Marianne C.P. Rutter
Manufacturing Coordinator: Wendy Kilborn

Interior and Cover Design: Julie Gecha
Interior Illustrators: Kristi Mendola, Deborah Ivanoff,
    and ST Associates, Inc.
Compositor: Bi-Comp Incorporated
Cover Printer: Coral Graphic Services
Text Printer and Binder: Quebecor/Hawkins

This book is printed on recycled, acid-free paper.

Printed and bound in the United States of America.
        98 99— 10 9 8 7 6 5 4 3

To Pat, Bill, and Michael

CONTENTS

PREFACE

Mathematics for the Managerial, Life, and Social Sciences is designed for students in business, management, economics, and the social and life sciences who have completed at least one year of high school algebra. The text is appropriate for use in mathematics courses that include topics from algebra, finite mathematics, and calculus. The author's objective in writing this book is to provide a textbook that is both readable by students and useful as a teaching tool for instructors. The text's approach is both problem-solving and applications-oriented. Wherever possible, mathematical concepts are presented intuitively and in the context of real-life settings to make them more meaningful to the students. Numerous examples and solved problems are used to amplify each new concept or result, in order to facilitate students' comprehension of the material. Graphs and pictures are used extensively to help students visualize the concepts and ideas presented.

Text Features

Coverage of topics As the book contains more material than needed for a traditional two-semester or three-quarter course, the instructor may be flexible in choosing the topics most suitable for his or her course. The chart on page xiv on chapter dependency is provided to help the instructor design the most appropriate course.

Use of technology is optional. Discussion is placed at the end of appropriate sections, which are highlighted in the table of contents for easy reference. A technology index is also provided, written in a user-friendly "How Do I" format. Throughout the text, examples employing the use of technology are highlighted by an icon ▦. Appendix A provides step-by-step instructions for the TI-85 graphing calculator for students who may need assistance. Although the TI-85 is used as a model, the technology is not limited to this calculator.

The presentation in this text is unique in that the use of technology is encouraged after the students have gained a firm grasp of concepts, much as grade-school students learn arithmetic before using calculators to do mathematical computations. Rather than simply teaching students to push buttons, the graphing calculator is presented as a tool for enhancing students' understanding of concepts and to facilitate computing. Whenever a choice of methods is available to solve a problem using a graphics calculator, the graphical method is emphasized. This approach gives the student an opportunity to "see" what he or she is doing.

The technology sections are designed to allow the instructor the flexibility of either using them in the classroom or allowing the students to work through them independently. These sections are written in the traditional example-exercise format (with answers provided at the back of the book),

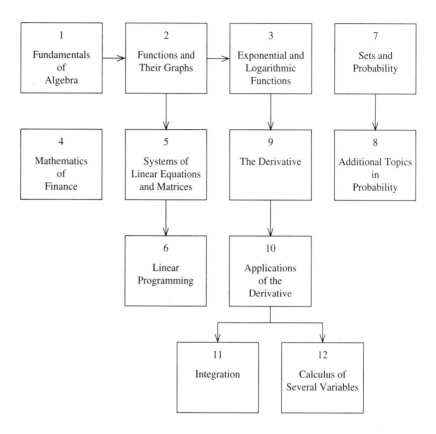

allowing students to immediately test themselves to see if they are using the calculator correctly. The technology sections also contain real-life applications, often referenced by sources. The students can readily see how problems of a more complex nature can be solved using a graphing calculator. Freed of cumbersome computations, the students can easily interpret their results, further enhancing their understanding of the concepts involved.

Real-life applications and examples   are drawn from fields of business, economics, social and behavioral sciences, life sciences, physical sciences, and other fields of general interest to students. These applications and examples are drawn from newspapers, magazines, and other weekly periodicals, and sources are noted. For ease in locating applications relevant to specific fields, they have been titled.

Self-tests   are placed throughout the text to help students to monitor their own progress.

Remarks   help clarify discussion and point out certain subtleties in the material being presented.

Caution symbols ⚠ alert students to common errors and pitfalls.

Portfolios   contain accounts of professionals who relate their real-world experiences and how they use mathematics in their professions.

Extended problems and projects are found at the end of each chapter. Here students are presented with questions that require more in-depth thinking and in many instances the discovery of mathematical results related to the concepts presented in the chapter. Careful effort has been taken to keep these problems and projects at a level that is both doable by students and at the same time challenging to them.

Exercise sets contain an ample set of routine problems, computational in nature, to help students master newly learned techniques. These are followed by an extensive array of application-oriented problems that will test their mastery of the topics. Critical-thinking questions requiring the interpretation of results and graphs are also found in each section. The answers to selected exercises that require written answers are included in the answer section as a further aid to the student, as experience has taught the author that many students need help getting started with this type of question. Solutions to the self-check exercises are found at the end of each exercise set. Answers to both even and odd exercises in the review sections are included in the answer section.

Supplements

- *Student's Solutions Manual,* available to both students and instructors, includes the solutions to odd-numbered exercises.

- *Instructor's Complete Solutions Manual* includes solutions to all exercises.

- *Test Bank with Chapter Tests,* free to adopters of the book, contains sample tests for each chapter.

- *EXPTest 6.0* and *Chariot MicroTest III.* With these computerized testing systems for IBM PCs and Macintosh, respectively, instructors can select or modify questions from prepared test banks or add their own test items to create any number of tests. These tests can be viewed on screen and printed with typeset-quality mathematical symbols, notation, graphs, and diagrams.

- *Graphing Calculator Supplement,* by Robert E. Seaver of Lorain County Community College, is available to both students and instructors. The manual develops selected examples and exercises and also includes additional problems for reinforcement. It is specifically written for use with the TI line of programmable graphics calculators.

Acknowledgments

I wish to express my personal appreciation to each of the following reviewers, whose many suggestions have helped make a much improved book.

J. Curtis Chipman
Oakland University

Joyce Longman
Villanova University

C. D. Luning
Sam Houston State University

Donald K. Mason
Elmhurst College

Wayne B. Powell
Oklahoma State University

Laurence Small
L. A. Pierce College

Ray Stanton
Fresno City College

I also wish to express my thanks to the editorial and production staffs of PWS Publishing Company, including Barbara Lovenvirth and Patty Adams, for their thoughtful contributions and patient assistance and cooperation during the development and production of this book. Finally, I wish to thank Cecile Joyner of The Cooper Company and Carol Dondrea for doing an excellent job ensuring the accuracy and readability of this book.

S. T. Tan

Fundamentals of Algebra

1

*T*his chapter contains a brief review of the algebra you will need for this course. In solving many practical problems, it is often necessary to simplify algebraic expressions before moving on to the next step. Thus, it is important that you reestablish familiarity with the process. This chapter also contains a short review of inequalities and absolute value, the uses of which range from describing the domains of functions to the formulation of practical problems.

1.1   Real Numbers

The Set of Real Numbers

We use *real numbers* everyday to describe quantities such as temperature, salary, annual percentage rate, shoe size, grade point average, and so on. Some of the symbols we use to represent real numbers are

$$3, \quad -17, \quad \sqrt{2}, \quad 0.666\ldots, \quad 113, \quad 3.9, \quad 0.12875$$

To construct the set of real numbers, we start with the set of **natural numbers** (also called counting numbers)

$$N = \{1, 2, 3, \ldots\}$$

and adjoin other numbers to it. The set

$$W = \{0, 1, 2, 3, \ldots\}$$

of **whole numbers** is obtained by adjoining the single number 0 to N. By adjoining the negatives of the natural numbers to the set W, we obtain the set of **integers**

$$I = \{\ldots -3, -2, -1, 0, 1, 2, 3, \ldots\}$$

Next, we consider the set Q of **rational numbers,** numbers of the form a/b, where a and b are integers, with $b \neq 0$. Using set notation, we write

$$Q = \{a/b \mid a \text{ and } b \text{ are integers}, \quad b \neq 0\}$$

Observe that I is contained in Q since each integer may be written in the form a/b, with $b = 1$. For example, the integer 6 may be written in the form 6/1. Symbolically, we express the fact that I is contained in Q by writing

$$I \subset Q$$

However, Q is not contained in I since fractions such as 1/2 and 23/25 are not integers. To show the relationships of the sets N, W, I, and Q, we write

$$N \subset W \subset I \subset Q$$

which says that N is a proper subset of W, W is a proper subset of I, and so on.* (See Figure 1.)

Finally, we obtain the set of real numbers by adjoining the set of rational numbers to the set of **irrational numbers** (Ir)—numbers that cannot be expressed in the form a/b, where a, b are integers ($b \neq 0$). Examples

FIGURE 1

The set of all real numbers consists of the set of rational numbers plus the set of irrational numbers.

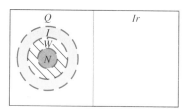

Q = Rationals
I = Integers
W = Whole numbers
N = Natural numbers
Ir = Irrationals

---

*A set A is a proper subset of a set B if every element of a set A is also an element of a set B and there exists at least one element in B that is not in A.

of irrational numbers are $\sqrt{2}$, $\sqrt{3}$, π, and so on. Thus, the set

$$R = Q \cup Ir$$

comprising all rational numbers and irrational numbers is called the set of **real numbers.**

Representing Real Numbers as Decimals

Every real number can be written as a decimal. A rational number can be represented as either a repeating or terminating decimal. For example, 2/3 is represented by the repeating decimal

0.66666666 . . .   Repeating decimal—note that the integer 6 repeats.

which may also be written $0.\overline{6}$, where the bar above the 6 indicates that the 6 repeats indefinitely. The number 1/2 is represented by the terminating decimal

0.5   Terminating decimal

When an irrational number is represented as a decimal, it neither terminates nor repeats. For example,

$$\sqrt{2} = 1.41421\ldots \qquad \text{and} \qquad \pi = 3.14159\ldots$$

Table 1 summarizes this classification of real numbers.

TABLE 1
The set of real numbers

Set	Description	Examples	Decimal Representation
Natural numbers	Counting numbers	1, 2, 3, . . .	Terminating decimals
Whole numbers	Counting numbers and 0	0, 1, 2, 3, . . .	Terminating decimals
Integers	Natural numbers, their negatives, and 0	. . . −3, −2, −1, 0, 1, 2, 3, . . .	Terminating decimals
Rational numbers	Numbers that can be written in the form a/b, where a and b are integers and $b \neq 0$	-3, $-3/4$, $-0.22\overline{2}$, 0, 5/6, 2, $4.311\overline{1}$	Terminating or repeating decimals
Irrational numbers	Numbers that cannot be written in the form a/b, where a and b are integers and $b \neq 0$	$\sqrt{2}$, $\sqrt{3}$, π 1.414213 . . . , 1.732050 . . . ,	Nonterminating, nonrepeating decimals
Real numbers	Rational and irrational numbers	All of the above	All types of decimals

The Real Number Line

Real numbers may be represented geometrically by points on a line. This *real number,* or *coordinate, line* is constructed as follows: Arbitrarily select a point on a straight line to represent the number zero. This point is called the *origin.* Choose a point at a convenient distance to the right of the origin to represent the number 1. This determines the scale for the number line.

The point representing each positive real number x lies x units to the right of zero, and the point representing each negative real number x lies $-x$ units to the left of zero. Thus, real numbers may be represented by points on a line in such a way that corresponding to each real number there is exactly one point on a line, and vice versa. In this way, a *one-to-one correspondence* is set up between the set of real numbers and the set of points on the number line, with all the positive numbers lying to the right of the origin and all the negative numbers lying to the left of the origin (Figure 2).

FIGURE 2

The real number line

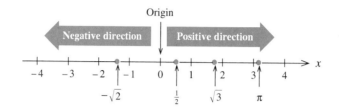

Operations with Real Numbers

Two real numbers may be combined to obtain a real number. The operation of addition, written $+$, enables us to combine any two numbers a and b to obtain their *sum,* denoted by $a + b$. Another operation, multiplication, written $\cdot$, enables us to combine any two real numbers a and b to form their product, the number $a \cdot b$ (more simply written ab). These two operations are subject to the rules of operation given in Table 2.

TABLE 2

Rules of operation for real numbers

Rule		Illustration
I. Under addition		
1. $a + b = b + a$	Commutative law of addition	$2 + 3 = 3 + 2$
2. $a + (b + c) = (a + b) + c$	Associative law of addition	$4 + (2 + 3) = (4 + 2) + 3$
3. $a + 0 = a$	Identity law of addition	$6 + 0 = 6$
4. $a + (-a) = 0$	Inverse law of addition	$5 + (-5) = 0$
II. Under multiplication		
1. $ab = ba$	Commutative law of multiplication	$3 \cdot 2 = 2 \cdot 3$
2. $a(bc) = (ab)c$	Associative law of multiplication	$4(3 \cdot 2) = (4 \cdot 3)2$
3. $a \cdot 1 = 1 \cdot a$	Identity law of multiplication	$4 \cdot 1 = 1 \cdot 4$
4. $a(1/a) = 1 \quad (a \neq 0)$	Inverse law of multiplication	$3(\frac{1}{3}) = 1$
III. Under addition and multiplication		
1. $a(b + c) = ab + ac$	Distributive law for multiplication with respect to addition	$3(4 + 5) = 3 \cdot 4 + 3 \cdot 5$

The operation of subtraction is defined in terms of addition. Thus,

$$a + (-b)$$

where $-b$ is the additive inverse of b, may be written in the more familiar form $a - b$, and we say that b is subtracted from a. Similarly, the operation of division is defined in terms of multiplication. Recall that the multiplicative inverse of a nonzero real number b is $1/b$, also written b^{-1}. Then,

$$a(1/b)$$

is written a/b, and we say that a is divided by b. Thus, $4(1/3) = 4/3$. Remember, 0 does not have a multiplicative inverse since division by zero is not defined.

Do the operations of associativity and commutativity hold for subtraction and division? Looking first at associativity, we see that the answer is no since

$$a - (b - c) \neq (a - b) - c \qquad 7 - (4 - 2) \neq (7 - 4) - 2, \text{ or } 5 \neq 1$$

and

$$a \div (b \div c) \neq (a \div b) \div c \qquad 8 \div (4 \div 2) \neq (8 \div 4) \div 2, \text{ or } 4 \neq 1$$

Similarly, commutativity does not hold because

$$a - b \neq b - a \qquad 7 - 4 \neq 4 - 7, \text{ or } 3 \neq -3$$

and

$$a \div b \neq b \div a \qquad 8 \div 4 \neq 4 \div 8, \text{ or } 2 \neq \tfrac{1}{2}$$

EXAMPLE 1   State the real number property that justifies each statement:

Statement	Property
a. $4 + (x - 2) = 4 + (-2 + x)$	The commutative law of addition
b. $(a + 2b) + c = a + (2b + c)$	The associative law of addition
c. $x(y - z + 2) = (y - z + 2)x$	The commutative law of multiplication
d. $4(xy^2) = (4x)y^2$	The associative law of multiplication
e. $x(y - 2) = xy - 2x$	The distributive law for multiplication under addition ∎

Using the properties of real numbers listed earlier, we can derive all the other algebraic properties of real numbers. Some of the more important properties are given in Tables 3–5.

TABLE 3
Properties of negatives

Property	Illustration
1. $-(-a) = a$	$-(-6) = 6$
2. $(-a)b = -(ab) = a(-b)$	$(-3)4 = -(3 \cdot 4) = 3(-4)$
3. $(-a)(-b) = ab$	$(-3)(-4) = 3 \cdot 4$
4. $(-1)a = -a$	$(-1)5 = -5$

TABLE 4
Properties involving zero

Property
1. $a \cdot 0 = 0$
2. If $ab = 0$, then $a = 0$, $b = 0$, or both.

TABLE 5
Properties of quotients

Property		Illustration
1. $\dfrac{a}{b} = \dfrac{c}{d}$ if $ad = bc$	$b, d \neq 0$	$\dfrac{3}{4} = \dfrac{9}{12}$ because $3 \cdot 12 = 9 \cdot 4$
2. $\dfrac{ca}{cb} = \dfrac{a}{b}$	$b, c \neq 0$	$\dfrac{4 \cdot 3}{4 \cdot 8} = \dfrac{3}{8}$
3. $\dfrac{a}{-b} = \dfrac{-a}{b} = -\dfrac{a}{b}$	$b \neq 0$	$\dfrac{4}{-3} = \dfrac{-4}{3} = -\dfrac{4}{3}$
4. $\dfrac{a}{b} \cdot \dfrac{c}{d} = \dfrac{ac}{bd}$	$b, d \neq 0$	$\dfrac{3}{4} \cdot \dfrac{5}{2} = \dfrac{15}{8}$
5. $\dfrac{a}{b} \div \dfrac{c}{d} = \dfrac{a}{b} \cdot \dfrac{d}{c} = \dfrac{ad}{bc}$	$b, c, d \neq 0$	$\dfrac{3}{4} \div \dfrac{5}{2} = \dfrac{3}{4} \cdot \dfrac{2}{5} = \dfrac{3}{10}$
6. $\dfrac{a}{b} + \dfrac{c}{d} = \dfrac{ad + bc}{bd}$	$b, d \neq 0$	$\dfrac{3}{4} + \dfrac{5}{2} = \dfrac{3 \cdot 2 + 4 \cdot 5}{8} = \dfrac{13}{4}$
7. $\dfrac{a}{b} - \dfrac{c}{d} = \dfrac{ad - bc}{bd}$	$b, d \neq 0$	$\dfrac{3}{4} - \dfrac{5}{2} = \dfrac{3 \cdot 2 - 4 \cdot 5}{8} = -\dfrac{7}{4}$

REMARK   In the rest of this book, we will assume that all variables are restricted so that division by zero is excluded.   □

EXAMPLE 2   State the real number property that justifies each statement:

Statement	Property
a. $-(-4) = 4$	Property 1 of negatives
b. If $(4x - 1)(x + 3) = 0$, then $x = \frac{1}{4}$ or $x = -3$.	Property 2 of zero properties
c. $\dfrac{(x - 1)(x + 1)}{(x - 1)(x - 3)} = \dfrac{x + 1}{x - 3}$	Property 2 of quotients
d. $\dfrac{x - 1}{y} \div \dfrac{y + 1}{x} = \dfrac{x - 1}{y} \cdot \dfrac{x}{y + 1}$	Property 5 of quotients
$= \dfrac{x(x - 1)}{y(y + 1)}$	
e. $\dfrac{x}{y} + \dfrac{x}{y + 1} = \dfrac{x(y + 1) + xy}{y(y + 1)}$	Property 6 of quotients
$= \dfrac{xy + x + xy}{y(y + 1)}$	Distributive law
$= \dfrac{2xy + x}{y(y + 1)}$	

■

SELF-CHECK EXERCISES 1.1

State the property (or properties) that justify each statement.

a. $(3v + 2) - w = 3v + (2 - w)$

b. $(3s)(4t) = 3[s(4t)]$

c. $-(-s + t) = s - t$

d. $\dfrac{2}{-(u - v)} = -\dfrac{2}{u - v}$

Solutions to Self-Check Exercises 1.1 can be found on page 8.

1.1 EXERCISES

In exercises 1–10, classify the number as to type. (For example, 1/2 is rational and real, whereas $\sqrt{5}$ is irrational and real.)

1. -3 **2.** -420 **3.** $\dfrac{3}{8}$ **4.** $-\dfrac{4}{125}$

5. $\sqrt{11}$ **6.** $-\sqrt{5}$ **7.** $\dfrac{\pi}{2}$ **8.** $\dfrac{2}{\pi}$

9. $2.\overline{421}$ **10.** $2.71828\ldots$

In exercises 11–16, indicate whether the statement is true or false.

11. Every integer is a whole number.

12. Every integer is a rational number.

13. Every natural number is an integer.

14. Every rational number is a real number.

15. Every natural number is an irrational number.

16. Every irrational number is a real number.

In exercises 17–36, state the real number property that justifies the statement.

17. $(2x + y) + z = z + (2x + y)$

18. $3x + (2y + z) = (3x + 2y) + z$

19. $u(3v + w) = (3v + w)u$

20. $a^2(b^2c) = (a^2b^2)c$

21. $u(2v + w) = 2uv + uw$

22. $(2u + v)w = 2uw + vw$

23. $(2x + 3y) + (x + 4y) = 2x + [3y + (x + 4y)]$

24. $(a + 2b)(a - 3b) = a(a - 3b) + 2b(a - 3b)$

25. $a - [-(c + d)] = a + (c + d)$

26. $-(2x + y)[-(3x + 2y)] = (2x + y)(3x + 2y)$

27. $0(2a + 3b) = 0$

28. If $(x - y)(x + y) = 0$, then $x = y$ or $x = -y$.

29. If $(x - 2)(2x + 5) = 0$, then $x = 2$ or $x = -\frac{5}{2}$.

30. If $x(2x - 9) = 0$, then $x = 0$ or $x = \frac{9}{2}$.

31. $\dfrac{(x + 1)(x - 3)}{(2x + 1)(x - 3)} = \dfrac{x + 1}{2x + 1}$

32. $\dfrac{(2x + 1)(x + 3)}{(2x - 1)(x + 3)} = \dfrac{2x + 1}{2x - 1}$

33. $\dfrac{a + b}{b} \div \dfrac{a - b}{ab} = \dfrac{a(a + b)}{a - b}$

34. $\dfrac{x + 2y}{3x + y} \div \dfrac{x}{6x + 2y} = \dfrac{x + 2y}{3x + y} \cdot \dfrac{2(3x + y)}{x} = \dfrac{2(x + 2y)}{x}$

35. $\dfrac{a}{b + c} + \dfrac{c}{b} = \dfrac{ab + bc + c^2}{b(b + c)}$

36. $\dfrac{x + y}{x + 1} - \dfrac{y}{x} = \dfrac{x^2 - y}{x(x + 1)}$

In exercises 37–42, indicate whether the statement is true or false.

37. If $ab = 1$, then $a = 1$ or $b = 1$.

38. If $ab = 0$ and $a \neq 0$, then $b = 0$.

39. $a - b = b - a$

40. $a \div b = b \div a$

41. $(a - b) - c = a - (b - c)$

42. $a \div (b \div c) = (a \div b) \div c$

SOLUTIONS TO SELF-CHECK EXERCISES 1.1

a. Associative law of addition: $a + (b + c) = (a + b) + c$

b. Associative law of multiplication: $a(bc) = (ab)c$

c. Distributive law for multiplication: $a(b + c) = ab + ac$

Properties 1 and 4 of negatives: $-(-a) = a; (-1)a = -a$

d. Property 3 of quotients: $\dfrac{a}{-b} = \dfrac{-a}{b} = -\dfrac{a}{b}$

1.2   Polynomials

Exponents

Expressions such as 2^5, $(-3)^2$, and $(1/4)^4$ are exponential expressions. More generally, if n is a natural number and a is a real number, then a^n represents the product of the real number a and itself n times.

EXPONENTIAL NOTATION

If a is a real number and n is a natural number, then

$$a^n = \underbrace{a \cdot a \cdot a \cdots \cdot a}_{n \text{ factors}} \qquad 3^4 = \underbrace{3 \cdot 3 \cdot 3 \cdot 3}_{4 \text{ factors}}$$

The natural number n is called the **exponent** and the real number a is called the **base**.

EXAMPLE 1

a. $4^4 = (4)(4)(4)(4) = 256$

b. $(-5)^3 = (-5)(-5)(-5) = -125$

c. $\left(\frac{1}{2}\right)^3 = \left(\frac{1}{2}\right)\left(\frac{1}{2}\right)\left(\frac{1}{2}\right) = \frac{1}{8}$

d. $\left(-\frac{1}{3}\right)^2 = \left(-\frac{1}{3}\right)\left(-\frac{1}{3}\right) = \frac{1}{9}$

When we evaluate expressions such as $3^2 \cdot 3^3$, we use the following property of exponents to write the product in exponential form.

PROPERTY 1

If m and n are natural numbers and a is any real number, then

$$a^m \cdot a^n = a^{m+n} \qquad 3^2 \cdot 3^3 = 3^{2+3} = 3^5$$

To verify that property 1 follows from the definition of an exponential expression, we note that the total number of factors in the exponential expression

$$a^m \cdot a^n = \underbrace{a \cdot a \cdot a \cdots \cdot a}_{m \text{ factors}} \cdot \underbrace{a \cdot a \cdot a \cdots \cdot a}_{n \text{ factors}}$$

is $m + n$.

EXAMPLE 2

a. $3^2 \cdot 3^3 = 3^{2+3} = 3^5 = 243$

b. $(-2)^2 \cdot (-2)^5 = (-2)^{2+5} = (-2)^7 = -128$

c. $(3x) \cdot (3x)^3 = (3x)^{1+3} = (3x)^4 = 81x^4$

Be careful to apply the exponent to the indicated base only. Note that

$$\underset{\uparrow \text{ The exponent applies only to the base } x.}{\overset{\ulcorner \text{ The exponent applies to } 4x.}{4 \cdot x^2 = 4x^2 \neq (4x)^2 = 4^2 \cdot x^2 = 16x^2}}$$

and

$$\underset{\uparrow \text{ The exponent applies only to the base } 3.}{\overset{\ulcorner \text{ The exponent applies to } -3.}{-3^2 = -9 \neq (-3)^2 = 9}}$$

Polynomials

We begin our discussion of *polynomials* by recalling that a *variable* is a letter that is used to represent any element of a given set. However, unless specified otherwise, variables in this text will represent real numbers. Sometimes physical considerations impose restrictions on the values a variable may assume. For example, if the variable x denotes the number of television sets sold daily in an appliance store, then x must be a nonnegative integer. At other times, restrictions must be imposed on x in order for an expression to make sense. For example, in the expression $\dfrac{1}{x+2}$, x may not take on the value -2 since division by zero is not permitted. We call the set of all real numbers that a variable is allowed to assume the *domain of the variable*.

In contrast to a variable, a *constant* is a fixed number or letter whose value remains fixed throughout a particular discussion. For example, in the expression $\frac{1}{2} gt^2$, which gives the distance in feet covered by a free-falling body near the surface of the earth, t seconds from rest, the letter g represents the constant of acceleration due to gravity (approximately 32 ft/sec^2), whereas the letter t is a variable with domain consisting of nonnegative real numbers.

By combining constants and variables through the use of addition, subtraction, multiplication, division, exponentiation, and root extraction, we obtain *algebraic expressions*. Examples of algebraic expressions are

$$3x - 4y, \quad 2x^2 - y + \frac{1}{xy}, \quad \frac{ax - b}{1 - x^2}, \quad \frac{3xy^{-2} + \pi}{x^2 + y^2 + z^2}$$

where a and b are constants, and x, y, and z are variables. Intimidating as some of these expressions might be, remember that they are just real numbers. For example, if $x = 1$ and $y = 4$, then the second expression represents the number

$$2(1)^2 - 4 + \frac{1}{(1)(4)}$$

or $-7/4$, obtained by replacing x and y in the expression by the appropriate values.

Polynomials are an important class of algebraic expressions. The simplest polynomials are those involving *one* variable.

POLYNOMIAL IN ONE VARIABLE

A **polynomial** in x is an expression of the form

$$a_n x^n + a_{n-1} x^{n-1} + \cdots + a_1 x + a_0$$

where n is a nonnegative integer and $a_0, a_1, \ldots, a_n$ are real numbers, with $a_n \neq 0$.

The expressions $a_k x^k$ in the sum are called the *terms* of the polynomial. The numbers $a_0, a_1, \ldots, a_n$ are called the *coefficients* of $1, x, x^2, \ldots, x^n$, respectively. The coefficient a_n of x^n (the highest power in x) is called the *leading coefficient* of the polynomial. The nonnegative integer n gives the *degree* of the polynomial. For example, consider the polynomial

$$-2x^5 + 8x^3 - 6x^2 + 3x + 1$$

Then

1. The terms of the polynomial are $-2x^5, 8x^3, -6x^2, 3x$, and 1.
2. The coefficients of $1, x, x^2, x^3, x^4$, and x^5 are $1, 3, -6, 8, 0$, and -2, respectively.
3. The leading coefficient of the polynomial is -2.

and

4. The degree of the polynomial is 5.

A polynomial having just one term (such as $2x^3$) is called a *monomial*; a polynomial having exactly two terms (such as $x^3 + x$) is called a *binomial*; and a polynomial having only three terms (such as $-2x^3 + x - 8$) is called a *trinomial*. Also, a polynomial consisting of one (constant) term a_0 (such as the monomial -8) is called a *constant polynomial*. Observe that the degree of a constant polynomial a_0, with $a_0 \neq 0$, is zero because we can write $a_0 = a_0 x^0$ and see that $n = 0$ in this situation. If all the coefficients of a polynomial are zero, it is called the *zero polynomial* and is denoted by 0. The zero polynomial is not assigned a degree.

Most of the terminology introduced in conjunction with the case of a polynomial in one variable carries over to the discussion of polynomials in several variables. But the *degree of a term* in a polynomial in several variables is obtained by adding the powers of all the variables in the term, and the *degree of the polynomial* is given by the highest degree of all its terms. For example, the polynomial

$$2x^2 y^5 - 3xy^3 + 8xy^2 - 3y + 4$$

is a polynomial in the two variables x and y. It has five terms with degrees $7, 4, 3, 1$, and 0, respectively. Accordingly, the degree of the polynomial is 7.

Adding and Subtracting Polynomials

Constant terms and terms having the same variable and exponent are called *like* or *similar* terms. Like terms may be combined by adding or subtracting their numerical coefficients. For example,

$$3x + 7x = (3 + 7)x = 10x \qquad \text{Like terms are added.}$$

and

$$\frac{1}{2}m^2 - 3m^2 = \left(\frac{1}{2} - 3\right)m^2 = -\frac{5}{2}m^2 \qquad \text{Like terms are subtracted.}$$

The distributive property of the real number system,

$$ab + ac = a(b + c)$$

is used to justify this procedure.

To add or subtract two or more polynomials, first remove the parentheses and then combine like terms. The resulting expression is then written in order of decreasing degree from left to right.

EXAMPLE 3

a. $(3x^3 + 2x^2 - 4x + 5) + (-2x^3 - 2x^2 - 2)$

$$\begin{aligned} &= 3x^3 + 2x^2 - 4x + 5 - 2x^3 - 2x^2 - 2 \qquad &&\text{Remove parentheses.} \\ &= 3x^3 - 2x^3 + 2x^2 - 2x^2 - 4x + 5 - 2 \qquad &&\text{Group like terms together.} \\ &= x^3 - 4x + 3 \qquad &&\text{Combine like terms.} \end{aligned}$$

b. $(2x^4 + 3x^3 + 4x + 6) - (3x^4 + 9x^3 + 3x^2)$

$$\begin{aligned} &= 2x^4 + 3x^3 + 4x + 6 - 3x^4 - 9x^3 - 3x^2 \qquad &&\text{Remove parentheses. Note that the minus sign preceding the second polynomial changes the sign of each term of that polynomial.} \\ &= 2x^4 - 3x^4 + 3x^3 - 9x^3 - 3x^2 + 4x + 6 \qquad &&\text{Group like terms.} \\ &= -x^4 - 6x^3 - 3x^2 + 4x + 6 \qquad &&\text{Combine like terms.} \end{aligned}$$

Multiplying Polynomials

To find the product of two polynomials, we again use the distributive property for real numbers. For example, to compute the product $3x(4x - 2)$, we use the distributive law to obtain

$$3x(4x - 2) = (3x)(4x) + (3x)(-2) \qquad a(b + c) = ab + ac$$

$$= 12x^2 - 6x$$

Observe that each term of one polynomial is multiplied by each term of the other. The resulting expression is then simplified by combining like terms. In general, an algebraic expression is *simplified* if none of its terms are similar.

EXAMPLE 4   Find the product $(3x + 5)(2x - 3)$.

Solution $(3x + 5)(2x - 3)$

$$\begin{aligned} &= 3x(2x - 3) + 5(2x - 3) \qquad &&\text{Distributive property} \\ &= (3x)(2x) + (3x)(-3) \qquad &&\text{Distributive property} \\ &\quad + (5)(2x) + (5)(-3) \\ &= 6x^2 - 9x + 10x - 15 \qquad &&\text{Multiply terms.} \\ &= 6x^2 + x - 15 \qquad &&\text{Combine like terms.} \end{aligned}$$

EXAMPLE 5   Find the product $(2t^2 - t + 3)(2t^2 - 1)$.

Solution $(2t^2 - t + 3)(2t^2 - 1)$

$\quad\quad = 2t^2(2t^2 - 1) - t(2t^2 - 1) + 3(2t^2 - 1)$ Distributive property

$\quad\quad = (2t^2)(2t^2) + (2t^2)(-1) + (-t)(2t^2)$ Distributive property
$\quad\quad\quad + (-t)(-1) + (3)(2t^2) + (3)(-1)$

$\quad\quad = 4t^4 - 2t^2 - 2t^3 + t + 6t^2 - 3$ Multiply terms.

$\quad\quad = 4t^4 - 2t^3 + 4t^2 + t - 3$ Combine terms.

Alternative Solution   We can also find the product by arranging the polynomials vertically and multiplying:

$$2t^2 - t + 3$$

$$\underline{2t^2 - 1}$$
$$4t^4 - 2t^3 + 6t^2$$

$$\underline{\quad\quad\quad -2t^2 + t - 3}$$
$$4t^4 - 2t^3 + 4t^2 + t - 3$$

     The polynomials in Examples 4 and 5 are polynomials in one variable. The operations of addition, subtraction, and multiplication are performed on polynomials of more than one variable in the same way as they are for polynomials in one variable.

EXAMPLE 6   Multiply: $(3x - y)(4x^2 - 2y)$.

Solution $(3x - y)(4x^2 - 2y)$

$\quad\quad = 3x(4x^2 - 2y) - y(4x^2 - 2y)$ Distributive law

$\quad\quad = 12x^3 - 6xy - 4x^2y + 2y^2$ Distributive property

$\quad\quad = 12x^3 - 4x^2y - 6xy + 2y^2$ Arrange terms in order of descending powers of x.

     Several commonly used products of polynomials are summarized in Table 6. Since products of this type occur so frequently, these formulas should be memorized.

TABLE 6
Special products

Formula	Illustration
1. $(a + b)^2 = a^2 + 2ab + b^2$	$(2x + 3y)^2 = (2x)^2 + 2(2x)(3y) + (3y)^2$ $= 4x^2 + 12xy + 9y^2$
2. $(a - b)^2 = a^2 - 2ab + b^2$	$(4x - 2y)^2 = (4x)^2 - 2(4x)(2y) + (2y)^2$ $= 16x^2 - 16xy + 4y^2$
3. $(a + b)(a - b) = a^2 - b^2$	$(2x + y)(2x - y) = (2x)^2 - (y)^2$ $= 4x^2 - y^2$

EXAMPLE 7   Use the special product formulas to compute

a. $(2x + y)^2$ **b.** $(3a - 4b)^2$ **c.** $(\frac{1}{2}x - 1)(\frac{1}{2}x + 1)$

Solution

a. $(2x + y)^2 = (2x)^2 + 2(2x)(y) + y^2$ Formula 1
$= 4x^2 + 4xy + y^2$

b. $(3a - 4b)^2 = (3a)^2 - 2(3a)(4b) + (4b)^2$ Formula 2
$= 9a^2 - 24ab + 16b^2$

c. $(\frac{1}{2}x - 1)(\frac{1}{2}x + 1) = (\frac{1}{2}x)^2 - 1 = \frac{1}{4}x^2 - 1$ Formula 3

Order of Operations

The common steps in Examples 1–7 have been to remove parentheses and combine like terms. If more than one grouping symbol is present, the innermost symbols are removed first. As you work through Example 8, note the order in which the grouping symbols are removed: parentheses ( ) first, brackets [ ] second, and finally braces { }. Also, note that the operations of multiplication and division take precedence over addition and subtraction.

EXAMPLE 8   Perform the indicated operations:

$$2t^3 - \{t^2 - [t - (2t - 1)] + 4\}$$

Solution $2t^3 - \{t^2 - [t - (2t - 1)] + 4\}$
$= 2t^3 - \{t^2 - [t - 2t + 1] + 4\}$ Remove parentheses.
$= 2t^3 - \{t^2 - [-t + 1] + 4\}$ Combine like terms within the brackets.
$= 2t^3 - \{t^2 + t - 1 + 4\}$ Remove brackets.
$= 2t^3 - \{t^2 + t + 3\}$ Combine like terms within the braces.
$= 2t^3 - t^2 - t - 3$ Remove braces.

EXAMPLE 9   Simplify $2\{3 - 2[x - 2x(3 - x)]\}$.

Solution $2\{3 - 2[x - 2x(3 - x)]\}$
$= 2\{3 - 2[x - 6x + 2x^2]\}$ Remove parentheses.
$= 2\{3 - 2[-5x + 2x^2]\}$ Combine like terms.
$= 2\{3 + 10x - 4x^2\}$ Remove brackets.
$= 6 + 20x - 8x^2$ Remove braces.
$= -8x^2 + 20x + 6$ Write answer in order of descending powers of x.

**SELF-CHECK
EXERCISES 1.2**

1. Find the product $(2x + 3y)(3x - 2y)$.

2. Simplify $3x - 2\{2x - [x - 2(x - 2)] + 1\}$.

Solutions to Self-Check Exercises 1.2 can be found on page 15.

1.2 EXERCISES

In exercises 1–12, evaluate the expression.

1. 3^4

2. $(-2)^5$

3. $\left(\frac{2}{3}\right)^3$

4. $\left(-\frac{3}{4}\right)^2$

5. -4^3

6. $-\left(-\frac{4}{5}\right)^3$

7. $-2\left(\frac{2}{5}\right)^3$

8. $\left(-\frac{2}{3}\right)^2\left(-\frac{3}{4}\right)^3$

9. $2^3 \cdot 2^5$

10. $(-3)^2 \cdot (-3)^3$

11. $(3y)^2(3y)^3$

12. $(-2x)^3(-2x)^2$

In exercises 13–48, perform the indicated operations and simplify.

13. $(2x + 3) + (4x - 6)$

14. $(-3x + 2) - (4x - 3)$

15. $(7x^2 - 2x + 5) + (2x^2 + 5x - 4)$

16. $(3x^2 + 5xy + 2y) + (4 - 3xy - 2x^2)$

17. $(5y^2 - 2y + 1) - (y^2 - 3y - 7)$

18. $(2x^2 - 3x + 4) - (-x^2 + 2x - 6)$

19. $(2.4x^3 - 3x^2 + 1.7x - 6.2)$
$- (1.2x^3 + 1.2x^2 - 0.8x + 2)$

20. $(1.4x^3 - 1.2x^2 + 3.2) - (-0.8x^3 - 2.1x - 1.8)$

21. $2m(3m - 4) + m(m - 1)$

22. $-3x(2x^2 + 3x - 5) + 2x(x^2 - 3)$

23. $3(2a - b) - 4(b - 2a)$

24. $2(3m - 1) - 3(-4m + 2n)$

25. $(2x + 3)(3x - 2)$

26. $(3r - 1)(2r + 5)$

27. $(2x - 3y)(3x + 2y)$

28. $(5m - 2n)(5m + 3n)$

29. $(3r + 2s)(4r - 3s)$

30. $(2m + 3n)(3m - 2n)$

31. $(0.2x + 1.2y)(0.3x - 2.1y)$

32. $(3.2m - 1.7n)(4.2m + 1.3n)$

33. $(2x - y)(3x^2 + 2y)$

34. $(3m - 2n^2)(2m^2 + 3n)$

35. $(2x + 3y)^2$

36. $(3m - 2n)^2$

37. $(2u - v)(2u + v)$

38. $(3r + 4s)(3r - 4s)$

39. $(t^2 - 2t + 4)(2t^2 + 1)$

40. $(3m^2 - 1)(2m^2 + 3m - 4)$

41. $2x - \{3x - [x - (2x - 1)]\}$

42. $3m - 2\{m - 3[2m - (m - 5)] + 4\}$

43. $x - \{2x - [-x - (1 - x)]\}$

44. $3x^2 - \{x^2 + 1 - x[x - (2x - 1)]\} + 2$

45. $(2x - 3)^2 - 3(x + 4)(x - 4) + 2(x - 4) + 1$

46. $(x - 2y)^2 + 2(x + y)(x - 3y) + x(2x + 3y + 2)$

47. $2x\{3x[2x - (3 - x)] + (x + 1)(2x - 3)\}$

48. $-3[(x + 2y)^2 - (3x - 2y)^2 + (2x - y)(2x + y)]$

49. *True or False.* The degree of the product of a polynomial of degree m and a polynomial of degree n is mn.

50. *True or False.* Suppose p and q are polynomials of degree n. Then $p + q$ is a polynomial of degree n.

51. Suppose p is a polynomial of degree m and q is a polynomial of degree n, where $m > n$. What is the degree of $p - q$?

52. Manufacturing Profits The total revenue realized in the sale of x units of the Lectro-Copy photocopying machine is

$$-0.04x^2 + 2000x$$

dollars per week, and the total cost incurred in manufacturing x units of the machines is

$$0.000002x^3 - 0.02x^2 + 1000x + 120{,}000$$

dollars per week ($0 \le x \le 50{,}000$). Find an expression giving the total weekly profit of the company. [*Hint:* The profit is revenue minus cost.]

53. Manufacturing Profits A manufacturer of tennis rackets finds that the total cost of manufacturing x rackets per day is given by

$$0.0001x^2 + 4x + 400$$

dollars. Each racket can be sold at a price of p dollars, where

$$p = -0.0004x + 10$$

Find an expression giving the daily profit for the manufacturer, assuming that all the rackets manufactured can be sold. [*Hint:* The total revenue is given by the total number of rackets sold multiplied by the price of the racket. The profit is given by revenue minus cost.]

54. Prison Overcrowding The 1980s saw a trend toward old-fashioned punitive deterrence as opposed to the more liberal penal policies and community-based corrections popular in the 1960s and early 1970s. As a result, prisons became more crowded, and the gap between the number of people in prison and prison capacity widened. Based on figures from the U.S. Department of Justice, the number of prisoners (in thousands) in federal and state prisons is approximately

$$3.5t^2 + 26.7t + 436.2 \qquad 0 \le t \le 10$$

and the number of inmates (in thousands) for which prisons were designed is given by

$$24.3t + 365 \qquad 0 \le t \le 10$$

where t is measured in years and $t = 0$ corresponds to 1984. Find an expression giving the gap between the number of prisoners and the number for which the prisons were designed at any time t.
Source: U.S. Dept. of Justice

55. Health-care Spending Health-care spending per person (in dollars) by the private sector comprises payments by individuals, corporations, and their insurance companies, and is approximated by

$$2.5t^2 + 31.3t + 406 \qquad 0 \le t \le 20$$

where t is measured in years and $t = 0$ corresponds to the beginning of 1975. The corresponding government spending (in dollars), comprising expenditures for Medicaid and other federal, state, and local government public health care, is

$$1.4t^2 + 29.6t + 251 \qquad 0 \le t \le 20$$

where t has the same meaning as before. Find an expression for the difference between private and government expenditures per person at any time t. What was the difference between private and government expenditures per person at the beginning of 1985? At the beginning of 1993?
Source: Health Care Financing Administration and U.S. Census Bureau

---

SOLUTIONS TO SELF-CHECK EXERCISES 1.2

1. $(2x + 3y)(3x - 2y) = 2x(3x - 2y) + 3y(3x - 2y)$
$$= 6x^2 - 4xy + 9xy - 6y^2$$
$$= 6x^2 + 5xy - 6y^2$$

2. $3x - 2\{2x - [x - 2(x - 2)] + 1\} = 3x - 2\{2x - [x - 2x + 4] + 1\}$
$$= 3x - 2\{2x - [-x + 4] + 1\}$$
$$= 3x - 2\{2x + x - 4 + 1\}$$
$$= 3x - 2\{3x - 3\}$$
$$= 3x - 6x + 6$$
$$= -3x + 6$$

1.3 # Factoring Polynomials

Factoring

Factoring a polynomial is the process of expressing it as a product of two or more polynomials. For example, by applying the distributive property we may write

$$3x^2 - x = x(3x - 1)$$

and we say that x and $3x - 1$ are factors of $3x^2 - x$.

How do we know if a polynomial is completely factored? Recall that an integer greater than 1 is *prime* if its only positive integer factors are itself and 1. For example, the number 3 is prime because its only factors are 3 and 1. In the same way, a polynomial is said to be prime over the set of integer coefficients if it cannot be expressed as a product of two or more polynomials of positive degree with integer coefficients. For example,

$x^2 + 2x + 2$ is a prime polynomial relative to the set of integers, whereas $x^2 - 9$ is not a prime polynomial, since $x^2 - 9 = (x + 3)(x - 3)$. Finally, a polynomial is said to be *completely factored* over the set of integers if it is expressed as a product of prime polynomials with integral coefficients.

REMARK   Unless otherwise mentioned, we will only consider factorization over the set of integers in this text. Hence, when the term *factor* is used, it will be understood that the factorization is to be completed over the set of integers.   □

Common Factors

The first step in factoring a polynomial is to see if it contains any common factors. If it does, the common factor of highest degree is then factored out. For example, the greatest common factor of $2a^2x + 4ax + 6a$ is $2a$ because

$$2a^2x + 4ax + 6a = \mathbf{2a} \cdot ax + \mathbf{2a} \cdot 2x + \mathbf{2a} \cdot 3$$
$$= 2a(ax + 2x + 3)$$

EXAMPLE 1   Factor out the greatest common factor:

a. $-3t^2 + 3t$ **b.** $6a^4b^4c - 9a^2b^2$

Solution

a. Since $3t$ is a common factor of each term, we have

$$-3t^2 + 3t = 3t(-t + 1) = 3t(1 - t)$$

b. Since $3a^2b^2$ is the common factor of highest degree, we have

$$6a^4b^4c - 9a^2b^2 = 3a^2b^2(2a^2b^2c - 3)$$ ■

Some Important Formulas

Having checked for common factors, the next step in factoring a polynomial is to express the polynomial as the product of a constant and/or one or more prime polynomials. The formulas given in Table 7 (at the top of page 17) are useful in factoring polynomials and should be committed to memory.

REMARK   Observe that a formula is given for factoring the sum of two cubes, but none is given for factoring the sum of two squares since $x^2 + a^2$ is prime over the set of integers.   □

EXAMPLE 2   Factor:

a. $x^2 - 9$ **b.** $16x^4 - 81y^4$ **c.** $(a - b)^2 - (a^2 + b)^2$

Solution   Observe that each of the polynomials in parts a–c is the difference of two squares. Using the formula given in Table 7, we have

a. $x^2 - 9 = x^2 - 3^2 = (x + 3)(x - 3)$
b. $16x^2 - 81y^4 = (4x)^2 - (9y^2)^2 = (4x + 9y^2)(4x - 9y^2)$

TABLE 7
Factoring formulas

Formula	Illustration
Difference of two squares: $a^2 - b^2 = (a + b)(a - b)$	$x^2 - 36 = (x + 6)(x - 6)$ $8x^2 - 2y^2 = 2(4x^2 - y^2) = 2[(2x)^2 - y^2]$ $\qquad = 2(2x + y)(2x - y)$ $9 - a^6 = 3^2 - (a^3)^2 = (3 + a^3)(3 - a^3)$
Perfect square trinomial: $a^2 + 2ab + b^2 = (a + b)^2$ $a^2 - 2ab + b^2 = (a - b)^2$	$x^2 + 8x + 16 = (x + 4)^2$ $4x^2 - 4xy + y^2 = (2x)^2 - 2(2x)(y) + y^2$ $\qquad = (2x - y)^2$
Sum of two cubes: $a^3 + b^3 = (a + b)(a^2 - ab + b^2)$	$z^3 + 27 = z^3 + (3)^3$ $\qquad = (z + 3)(z^2 - 3z + 9)$
Difference of two cubes: $a^3 - b^3 = (a - b)(a^2 + ab + b^2)$	$8x^3 - y^6 = (2x)^3 - (y^2)^3$ $\qquad = (2x - y^2)(4x^2 + 2xy^2 + y^4)$

c. $(a - b)^2 - (a^2 + b)^2 = [(a - b) + (a^2 + b)][(a - b) - (a^2 + b)]$

$\qquad = [a - b + a^2 + b][a - b - a^2 - b]$ Remove parentheses.

$\qquad = (a + a^2)(-a^2 + a - 2b)$ Combine like terms.

$\qquad = a(1 + a)(-a^2 + a - 2b)$ ■

EXAMPLE 3   Factor:

a. $x^2 + 4xy + 4y^2$ **b.** $4a^2 - 12ab + 9b^2$

Solution   Recognizing each of these polynomials as a perfect square trinomial, we use the formula given in Table 7 to factor each polynomial. Thus,

a. $x^2 + 4xy + 4y^2 = x^2 + 2x(2y) + (2y)^2 = (x + 2y)(x + 2y) = (x + 2y)^2$

b. $4a^2 - 12ab + 9b^2 = (2a)^2 - 2(2a)(3b) + (3b)^2$

$\qquad = (2a - 3b)(2a - 3b) = (2a - 3b)^2$ ■

EXAMPLE 4   Factor:

a. $x^3 + 8y^3$ **b.** $27a^3 - 64b^3$

Solution

a. This polynomial is the sum of two cubes. Using the formula given in Table 7, we have

$$x^3 + 8y^3 = x^3 + (2y)^3 = (x + 2y)[x^2 - x(2y) + (2y)^2]$$

$$= (x + 2y)(x^2 - 2xy + 4y^2)$$

b. Using the formula for the difference of two cubes, we have

$$27a^3 - 64b^3 = [(3a)^3 - (4b)^3]$$
$$= (3a - 4b)[(3a)^2 + (3a)(4b) + (4b)^2]$$
$$= (3a - 4b)(9a^2 + 12ab + 16b^2) \qquad \blacksquare$$

Trial-and-Error Factorization

The factors of the second-degree polynomial $px^2 + qx + r$, where p, q, and r are integers, have the form

$$(ax + b)(cx + d)$$

where $ac = p$, $ad + bc = q$, and $bd = r$. Since only a limited number of choices are possible, we use a trial-and-error method to factor polynomials having this form.

For example, to factor $x^2 - 2x - 3$, we first observe that the only possible first-degree terms are

$$(x \quad)(x \quad) \qquad \text{Since the coefficient of } x^2 \text{ is 1}$$

Next, we observe that the product of the constant terms is (-3). This gives us the following possible factors:

$$(x - 1)(x + 3)$$
$$(x + 1)(x - 3)$$

Looking once again at the polynomial $x^2 - 2x - 3$, we see that the coefficient of x is -2. Checking to see which set of factors yields -2 for the coefficient of x, we find that

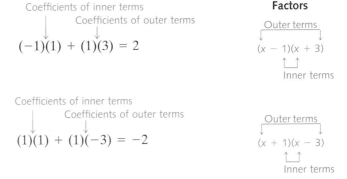

and we conclude that the correct factorization is

$$x^2 - 2x - 3 = (x + 1)(x - 3)$$

With practice, you will soon find that you can perform many of these steps mentally, and the need to write out each step will be eliminated.

EXAMPLE 5   Factor:

a. $3x^2 + 4x - 4$ **b.** $3x^2 - 6x - 24$

Solution

a. Using trial and error, we find that the correct factorization is

$$3x^2 + 4x - 4 = (3x - 2)(x + 2)$$

b. Since each term has the common factor 3, we have

$$3x^2 - 6x - 24 = 3(x^2 - 2x - 8)$$

Using the trial-and-error method of factorization, we find that

$$x^2 - 2x - 8 = (x - 4)(x + 2)$$

Thus, we have

$$3x^2 - 6x - 24 = 3(x - 4)(x + 2)$$ ■

Factoring by Regrouping

Sometimes a polynomial may be factored by regrouping and rearranging terms so that a common term can be factored out. This technique is illustrated in Example 6.

EXAMPLE 6 Factor:

a. $x^3 + x + x^2 + 1$ **b.** $2ax + 2ay + bx + by$

Solution

a. We begin by rearranging the terms in order of descending powers of x. Thus,

$$\begin{aligned}
x^3 + x + x^2 + 1 &= x^3 + x^2 + x + 1 \\
&= x^2(x + 1) + x + 1 &\text{Factoring the first two terms} \\
&= (x + 1)(x^2 + 1) &\text{Factoring the common term } x + 1
\end{aligned}$$

b. First factor the common term $2a$ from the first two terms and the common term b from the last two terms. Thus,

$$2ax + 2ay + bx + by = 2a(x + y) + b(x + y)$$

Since $(x + y)$ is common to both terms of the polynomial on the right, we may factor it out. Hence,

$$2a(x + y) + b(x + y) = (x + y)(2a + b)$$ ■

More Examples on Factoring

EXAMPLE 7 Factor:

a. $4x^6 - 4x^2$ **b.** $18x^4 - 3x^3 - 6x^2$

Solution

a. $\begin{aligned}[t]
4x^6 - 4x^2 &= 4x^2(x^4 - 1) &\text{Common factor} \\
&= 4x^2(x^2 - 1)(x^2 + 1) &\text{Difference of two squares} \\
&= 4x^2(x - 1)(x + 1)(x^2 + 1) &\text{Difference of two squares}
\end{aligned}$

b. $18x^4 - 3x^3 - 6x^2 = 3x^2(6x^2 - x - 2)$ Common factor
$$= 3x^2(3x - 2)(2x + 1)$$ Trial-and-error factorization

EXAMPLE 8 Factor:

a. $3x^2y + 9x^2 - 12y - 36$ **b.** $(x - at)^3 - (x + at)^3$

Solution

a. $3x^2y + 9x^2 - 12y - 36 = 3(x^2y + 3x^2 - 4y - 12)$ Common factor
$$= 3[x^2(y + 3) - 4(y + 3)]$$ Regrouping
$$= 3(y + 3)(x^2 - 4)$$ Common factor
$$= 3(y + 3)(x - 2)(x + 2)$$ Difference of two squares

b. $(x - at)^3 - (x + at)^3$
$$= [(x - at) - (x + at)]$$
$$\cdot [(x - at)^2 + (x - at)(x + at) + (x + at)^2]$$ Difference of two cubes
$$= -2at(x^2 - 2atx + a^2t^2 + x^2 - a^2t^2 + x^2 + 2atx + a^2t^2)$$
$$= -2at(3x^2 + a^2t^2)$$

Be sure you become familiar with all the factorization processes discussed in this chapter, as we will be using them throughout the text. As with many other algebraic techniques, you will find that proficiency in the technique of factoring will be gained from working through the exercises.

SELF-CHECK EXERCISES 1.3

1. Factor: **a.** $4x^3 - 2x^2$ **b.** $3(a^2 + 2b^2) + 4(a^2 + 2b^2)^2$

2. Factor: **a.** $6x^2 - x - 12$ **b.** $4x^2 + 10x - 6$

Solutions to Self-Check Exercises 1.3 can be found on page 21.

1.3 EXERCISES

In exercises 1–10, factor out the greatest common factor.

1. $6m^2 - 2m$ **2.** $4t^4 - 12t^3$

3. $9ab^2 - 6a^2b$ **4.** $12x^3y^5 + 16x^2y^3$

5. $10m^2n - 15mn^2 + 20mn$

6. $6x^4y - 4x^2y^2 + 2x^2y^3$

7. $3x(2x + 1) - 5(2x + 1)$

8. $2u(3v^2 + w) + 5v(3v^2 + w)$

9. $(3a + b)(2c - d) + 2a(2c - d)^2$

10. $4uv^2(2u - v) + 6u^2v(v - 2u)$

In exercises 11–54, factor the polynomial. If the polynomial is prime, state it.

11. $2m^2 - 11m - 6$ **12.** $6x^2 - x - 1$

13. $x^2 - xy - 6y^2$ **14.** $2u^2 + 5uv - 12v^2$

15. $x^2 - 3x - 1$ **16.** $m^2 + 2m + 3$

17. $4a^2 - b^2$ **18.** $12x^2 - 3y^2$

19. $u^2v^2 - w^2$ **20.** $4a^2b^2 - 25c^2$

21. $z^2 + 4$ **22.** $u^2 + 25v^2$

23. $x^2 + 6xy + y^2$ **24.** $4u^2 - 12uv + 9v^2$

25. $x^2 + 3x - 4$ **26.** $3m^3 + 3m^2 - 18m$ **41.** $u^2v^6 - 8u^2$ **42.** $r^6s^6 + 8s^3$

27. $12x^2y - 10xy - 12y$ **28.** $12x^2y - 2xy - 24y$ **43.** $2x^3 + 6x + x^2 + 3$ **44.** $2u^4 - 4u^2 + 2u^2 - 4$

29. $35r^2 + r - 12$ **30.** $6uv^2 + 9uv - 6v$ **45.** $3ax + 6ay + bx + 2by$

31. $9x^3y - xy^3$ **32.** $4u^4v - 9u^2v^3$ **46.** $6ux - 4uy + 3vx - 2vy$

33. $x^4 - 16y^2$ **34.** $16u^4v - 9v^3$ **47.** $u^4 - v^4$ **48.** $u^4 - u^2v^2 - 6v^4$

35. $(a - 2b)^2 - (a + 2b)^2$ **49.** $4x^3 - 9xy^2 + 4x^2y - 9y^3$

36. $2x(x + y)^2 - 8x(x + y^2)^2$ **50.** $4u^4 + 11u^2v^2 - 3v^4$ **51.** $x^4 + 3x^3 - 2x - 6$

37. $8m^3 + 1$ **38.** $27m^3 - 8$ **52.** $a^2 - b^2 + a + b$ **53.** $au^2 + (a + c)u + c$

39. $8r^3 - 27s^3$ **40.** $x^3 + 64y^3$ **54.** $ax^2 - (1 + ab)xy + by^2$

SOLUTIONS TO SELF-CHECK EXERCISES 1.3

1. a. The common factor is $2x^2$. Therefore,

$$4x^3 - 2x^2 = 2x^2(2x - 1)$$

b. The common factor is $a^2 + 2b^2$. Therefore,

$$3(a^2 + 2b^2) + 4(a^2 + 2b^2)^2 = (a^2 + 2b^2)[3 + 4(a^2 + 2b^2)]$$
$$= (a^2 + 2b^2)(3 + 4a^2 + 8b^2)$$

2. a. Using the trial-and-error method of factorization, we find that

$$6x^2 - x - 12 = (3x + 4)(2x - 3)$$

b. We first factor out the common factor 2. Thus,

$$4x^2 + 10x - 6 = 2(2x^2 + 5x - 3)$$

Using the trial-and-error method of factorization, we find that

$$2x^2 + 5x - 3 = (2x - 1)(x + 3)$$

and, consequently,

$$4x^2 + 10x - 6 = 2(2x - 1)(x + 3)$$

1.4

Rational Expressions

Quotients of polynomials are called **rational expressions.** Examples of rational expressions are

$$\frac{6x - 1}{2x + 3} \quad \text{and} \quad \frac{3x^2y^3 - 2xy}{4x - y}$$

Because division by zero is not allowed, the denominator of a rational expression must not be equal to zero. Thus, in the first example, $x \neq -3/2$, and in the second example, $y \neq 4x$.

Since rational expressions are quotients in which the variables represent real numbers, the properties of real numbers apply to rational expressions as well. For this reason, operations with rational fractions are performed in the same way as operations with arithmetic fractions.

Simplifying Rational Expressions

A rational expression is *simplified*, or *reduced to lowest terms*, if the numerator and the denominator have no common factors other than 1 and -1. If a rational expression does contain common factors, we use the properties of the real number system to write

$$\frac{ac}{bc} = \frac{a}{b} \cdot \frac{c}{c} = \frac{a}{b} \cdot 1 = \frac{a}{b} \qquad a, b, c, \text{ real numbers and } bc \neq 0$$

This process is often called "canceling common factors." To indicate this process, we often write

$$\frac{a\,\cancel{c}}{b\,\cancel{c}} = \frac{a}{b}$$

where a slash is shown through the common factors. As another example, the rational expression

$$\frac{(x + 2)(x - 3)}{(x - 2)(x - 3)} \qquad x \neq 2, 3$$

is simplified by canceling the common factors $(x - 3)$ and writing

$$\frac{(x + 2)(\cancel{x - 3})}{(x - 2)(\cancel{x - 3})} = \frac{x + 2}{x - 2}$$

 $\dfrac{\cancel{3} + 4x}{\cancel{3}} = 1 + 4x$ is an example of incorrect cancellation. Instead we write

$$\frac{3 + 4x}{3} = \frac{3}{3} + \frac{4x}{3} = 1 + \frac{4x}{3}.$$

EXAMPLE 1 Simplify:

a. $\dfrac{x^2 + 2x - 3}{x^2 + 4x + 3}$ **b.** $\dfrac{3 - 4x - 4x^2}{2x - 1}$ **c.** $\dfrac{(k + 4)^2(k - 1)}{k^2 - 16}$

Solution

a. $\dfrac{x^2 + 2x - 3}{x^2 + 4x + 3} = \dfrac{(x + 3)(x - 1)}{(x + 3)(x + 1)} = \dfrac{x - 1}{x + 1}$ Factor numerator and denominator and cancel common factors.

b. $\dfrac{3 - 4x - 4x^2}{2x - 1} = \dfrac{(1 - 2x)(3 + 2x)}{2x - 1}$

$\qquad\qquad = -\dfrac{(2x - 1)(2x + 3)}{2x - 1}$ Rewrite the term $1 - 2x$ in the equivalent form $-(2x - 1)$.

$\qquad\qquad = -(2x + 3)$ Cancel common factors.

c. $\dfrac{(k + 4)^2(k - 1)}{k^2 - 16} = \dfrac{(k + 4)^2(k - 1)}{(k + 4)(k - 4)} = \dfrac{(k + 4)(k - 1)}{k - 4}$ ■

Multiplication and Division

The operations of multiplication and division are performed with rational expressions in the same way they are with arithmetic fractions (Table 8).

TABLE 8
Multiplication and division of rational expressions

Operation	Illustration
If P, Q, R, and S are polynomials, then	
Multiplication:	
$\dfrac{P}{Q} \cdot \dfrac{R}{S} = \dfrac{PR}{QS} \qquad (Q, S \neq 0)$	$\dfrac{2x}{y} \cdot \dfrac{(x+1)}{(y-1)} = \dfrac{2x(x+1)}{y(y-1)}$
Division:	
$\dfrac{P}{Q} \div \dfrac{R}{S} = \dfrac{P}{Q} \cdot \dfrac{S}{R} = \dfrac{PS}{QR} \qquad (Q, R, S \neq 0)$	$\dfrac{x^2+3}{y} \div \dfrac{y^2+1}{x} = \dfrac{x^2+3}{y} \cdot \dfrac{x}{y^2+1}$
	$= \dfrac{x(x^2+3)}{y(y^2+1)}$

When rational expressions are multiplied and divided, the resulting expressions should be simplified.

EXAMPLE 2 Perform the indicated operations and simplify:

a. $\dfrac{2x-8}{x+2} \cdot \dfrac{x^2+4x+4}{x^2-16}$ **b.** $\dfrac{x^2-6x+9}{3x+12} \div \dfrac{x^2-9}{6x^2+18x}$

Solution

a. $\dfrac{2x-8}{x+2} \cdot \dfrac{x^2+4x+4}{x^2-16}$

$\qquad = \dfrac{2(x-4)}{x+2} \cdot \dfrac{(x+2)^2}{(x+4)(x-4)}$ Factor numerators and denominators.

$\qquad = \dfrac{2(x-4)(x+2)(x+2)}{(x+2)(x+4)(x-4)}$

$\qquad = \dfrac{2(x+2)}{x+4}$ Cancel the common factors $(x+2)(x-4)$.

b. $\dfrac{x^2-6x+9}{3x+12} \div \dfrac{x^2-9}{6x^2+18x}$

$\qquad = \dfrac{x^2-6x+9}{3x+12} \cdot \dfrac{6x^2+18x}{x^2-9}$

$\qquad = \dfrac{(x-3)^2}{3(x+4)} \cdot \dfrac{6x(x+3)}{(x+3)(x-3)}$

$\qquad = \dfrac{(x-3)(x-3)(6x)(x+3)}{3(x+4)(x+3)(x-3)}$

$\qquad = \dfrac{2x(x-3)}{x+4}$

Addition and Subtraction

For rational expressions, the operations of addition and subtraction are performed by finding a common denominator for the fractions and then adding or subtracting the fractions. Table 9 shows the rules for fractions with common denominators.

TABLE 9
Adding and subtracting fractions with common denominators

Operation	Illustration

If P, Q, and R are polynomials, then

Addition:

$$\frac{P}{R} + \frac{Q}{R} = \frac{P + Q}{R} \qquad (R \neq 0) \qquad\qquad \frac{2x}{x + 2} + \frac{6x}{x + 2} = \frac{2x + 6x}{x + 2} = \frac{8x}{x + 2}$$

Subtraction:

$$\frac{P}{R} - \frac{Q}{R} = \frac{P - Q}{R} \qquad (R \neq 0) \qquad\qquad \frac{3y}{y - x} - \frac{y}{y - x} = \frac{3y - y}{y - x} = \frac{2y}{y - x}$$

To add or subtract fractions that have different denominators, first find a common denominator, preferably the least common denominator (LCD). To find the least common denominator (LCD) of two or more rational expressions, follow these steps:

1. *Find the prime factors* of each denominator.

2. *Form the product of the different prime factors* that occur in the denominators. Each prime factor in this product should be raised to the highest power of that factor appearing in the denominators.

After finding the LCD, carry out the indicated operations following the procedure for adding and subtracting fractions with common denominators.

EXAMPLE 3   Simplify:

a. $\dfrac{3x + 4}{4x} + \dfrac{4y - 2}{3y}$ **b.** $\dfrac{2x}{x^2 - 1} + \dfrac{3x + 1}{2x^2 - x - 1}$ **c.** $\dfrac{1}{x + h} - \dfrac{1}{x}$

Solution

a. $\dfrac{3x + 4}{4x} + \dfrac{4y - 2}{3y} = \dfrac{3x + 4}{4x} \cdot \dfrac{3y}{3y} + \dfrac{4y - 2}{3y} \cdot \dfrac{4x}{4x}$ $\qquad$ LCD $= (4x)(3y) = 12xy$

$$= \frac{9xy + 12y}{12xy} + \frac{16xy - 8x}{12xy}$$

$$= \frac{25xy - 8x + 12y}{12xy}$$

b. $\dfrac{2x}{x^2 - 1} + \dfrac{3x + 1}{2x^2 - x - 1} = \dfrac{2x}{(x + 1)(x - 1)} + \dfrac{3x + 1}{(2x + 1)(x - 1)}$

$$= \frac{2x(2x + 1) + (3x + 1)(x + 1)}{(x + 1)(x - 1)(2x + 1)} \quad \text{LCD} = (2x + 1)$$
$$\cdot (x + 1)(x - 1)$$

$$= \frac{4x^2 + 2x + 3x^2 + 3x + x + 1}{(x+1)(x-1)(2x+1)}$$

$$= \frac{7x^2 + 6x + 1}{(x+1)(x-1)(2x+1)}$$

c. $\dfrac{1}{x+h} - \dfrac{1}{x} = \dfrac{1}{x+h} \cdot \dfrac{x}{x} - \dfrac{1}{x} \cdot \dfrac{x+h}{x+h}$ LCD $= x(x+h)$

$$= \frac{x}{x(x+h)} - \frac{x+h}{x(x+h)}$$

$$= \frac{x-x-h}{x(x+h)}$$

$$= -\frac{h}{x(x+h)}$$

Compound Fractions

A fractional expression that contains fractions in its numerator or denominator is called a **compound fraction.** The techniques used to simplify rational expressions may also be used to simplify these fractions.

EXAMPLE 4   Simplify:

a. $\dfrac{1 + \dfrac{1}{x+1}}{x - \dfrac{4}{x}}$ **b.** $\dfrac{\dfrac{1}{x} + \dfrac{1}{y}}{\dfrac{1}{x^2} - \dfrac{1}{y^2}}$

Solution

a. We first express the numerator and denominator of the given expression as a single quotient. Thus

$$\frac{1 + \dfrac{1}{x+1}}{x - \dfrac{4}{x}} = \frac{1 \cdot \dfrac{x+1}{x+1} + \dfrac{1}{x+1}}{x \cdot \dfrac{x}{x} - \dfrac{4}{x}}$$

The LCD for the fraction in the numerator is $x + 1$ and the LCD for the fraction in the denominator is x.

$$= \frac{\dfrac{x+1+1}{x+1}}{\dfrac{x^2-4}{x}}$$

$$= \frac{\dfrac{x+2}{x+1}}{\dfrac{x^2-4}{x}}$$

We then invert the denominator and multiply, obtaining

$$\frac{x+2}{x+1} \cdot \frac{x}{x^2-4} = \frac{x+2}{x+1} \cdot \frac{x}{(x-2)(x+2)}$$ Factor the denominator of the second fraction.

$$= \frac{x}{(x+1)(x-2)}$$ Cancel the common factors.

b. As before, we first write the numerator and denominator of the given expression as a single quotient and then simplify the resulting fraction.

$$\frac{\dfrac{1}{x} + \dfrac{1}{y}}{\dfrac{1}{x^2} - \dfrac{1}{y^2}} = \frac{\dfrac{y+x}{xy}}{\dfrac{y^2 - x^2}{x^2 y^2}}$$

The LCD for the fraction in the numerator is xy and the LCD for the fraction in the denominator is $x^2 y^2$.

$$= \frac{y+x}{xy} \cdot \frac{x^2 y^2}{y^2 - x^2}$$

Invert the fraction in the denominator and multiply.

$$= \frac{y+x}{xy} \cdot \frac{x^2 y^2}{(y+x)(y-x)}$$

$$= \frac{xy}{y-x}$$

Cancel common factors.

**SELF-CHECK
EXERCISES 1.4**

1. Simplify: $\dfrac{3a^2 b^3}{2ab^2 + 4ab} \cdot \dfrac{b^2 + 4b + 4}{6a^2 b^5}$

2. Simplify: $\dfrac{\dfrac{x}{y} - \dfrac{y}{x}}{\dfrac{x^2 + 2xy + y^2}{x^2 - y^2}}$

Solutions to Self-Check Exercises 1.4 can be found on page 27.

1.4   EXERCISES

In exercises 1–10, simplify the expression.

1. $\dfrac{4x + 12}{3x + 9}$

2. $\dfrac{12m - 6}{18m - 9}$

3. $\dfrac{6x^2 - 3x}{6x^2}$

4. $\dfrac{8y^2}{4y^3 - 4y^2 + 8y}$

5. $\dfrac{x^2 + x - 2}{x^2 + 3x + 2}$

6. $\dfrac{2y^2 - y - 3}{2y^2 + y - 1}$

7. $\dfrac{x^2 - 4}{2x^2 - x - 6}$

8. $\dfrac{6y^2 + 11y + 3}{4y^2 - 9}$

9. $\dfrac{x^3 + y^3}{x^2 - xy + y^2}$

10. $\dfrac{8r^3 - s^3}{2r^2 + rs - s^2}$

In exercises 11–40, perform the indicated operations and simplify.

11. $\dfrac{3x}{x + 2y} \cdot \dfrac{5x + 10y}{6}$

12. $\dfrac{4y + 12}{y + 2} \cdot \dfrac{3y + 6}{2y - 1}$

13. $\dfrac{2m + 6}{3} \div \dfrac{3m + 9}{6}$

14. $\dfrac{3y - 6}{4y + 6} \div \dfrac{6y + 24}{8y + 12}$

15. $\dfrac{6r^2 - r - 2}{2r + 4} \cdot \dfrac{6r + 12}{4r + 2}$

16. $\dfrac{x^2 - x - 6}{2x^2 + 7x + 6} \cdot \dfrac{2x^2 - x - 6}{x^2 + x - 6}$

17. $\dfrac{k^2 - 2k - 3}{k^2 - k - 6} \div \dfrac{k^2 - 6k + 8}{k^2 - 2k - 8}$

18. $\dfrac{6y^2 - 5y - 6}{6y^2 + 13y + 6} \div \dfrac{6y^2 - 13y + 6}{9y^2 - 12y + 4}$

19. $\dfrac{2}{2x + 3} \cdot + \dfrac{3}{2x - 1}$

20. $\dfrac{2x - 1}{x + 2} - \dfrac{x + 3}{x - 1}$

21. $\dfrac{3}{x^2 - x - 6} + \dfrac{2}{x^2 + x - 2}$

22. $\dfrac{4}{x^2 - 9} - \dfrac{5}{x^2 - 6x + 9}$

23. $\dfrac{2m}{2m^2 - 2m - 1} + \dfrac{3}{2m^2 - 3m + 3}$

24. $\dfrac{t}{t^2 + t - 2} - \dfrac{2t - 1}{2t^2 + 3t - 2}$

25. $\dfrac{x}{1 - x} + \dfrac{2x + 3}{x^2 - 1}$

26. $2 + \dfrac{1}{a + 2} - \dfrac{2a}{a - 2}$

27. $x - \dfrac{x^2}{x + 2} + \dfrac{2}{x - 2}$

28. $\dfrac{y}{y^2 - 1} + \dfrac{y - 1}{y + 1} - \dfrac{2y}{1 - y}$

29. $\dfrac{x}{x^2 + 5x + 6} + \dfrac{2}{x^2 - 4} - \dfrac{3}{x^2 + 3x + 2}$

30. $\dfrac{2x + 1}{2x^2 - x - 1} - \dfrac{x + 1}{2x^2 + 3x + 1} + \dfrac{4}{x^2 + 2x - 3}$

31. $\dfrac{x}{ax - ay} + \dfrac{y}{by - bx}$

32. $\dfrac{ax + by}{ax - bx} + \dfrac{ay - bx}{by - ay}$

33. $\dfrac{1 + \dfrac{1}{x}}{1 - \dfrac{1}{x}}$

34. $\dfrac{2 + \dfrac{2}{x}}{x - \dfrac{2}{x}}$

35. $\dfrac{\dfrac{1}{x} + \dfrac{1}{y}}{1 - \dfrac{1}{xy}}$

36. $\dfrac{1 + \dfrac{x}{y}}{1 - \dfrac{x^2}{y^2}}$

37. $\dfrac{\dfrac{1}{x^2} - \dfrac{1}{y^2}}{x + y}$

38. $\dfrac{\dfrac{1}{x^3} - \dfrac{1}{y^3}}{\dfrac{1}{x} - \dfrac{1}{y}}$

39. $\dfrac{\dfrac{1}{2(x + h)} - \dfrac{1}{2x}}{h}$

40. $\dfrac{\dfrac{1}{(x + h)^2} - \dfrac{1}{x^2}}{h}$

SOLUTIONS TO SELF-CHECK EXERCISES 1.4

1. Factoring the numerator and denominator of each expression, we have

$$\frac{3a^2 b^3}{2ab^2 + 4ab} \cdot \frac{b^2 + 4b + 4}{6a^2 b^5} = \frac{3a^2 b^3}{2ab(b + 2)} \cdot \frac{(b + 2)^2}{(3a^2 b^3)(2b^2)}$$

$$= \frac{b + 2}{2ab(2b^2)} \qquad \text{Canceling common factors}$$

$$= \frac{b + 2}{4ab^3}$$

2. Writing the numerator of the given expression as a single quotient, we have

$$\frac{\dfrac{x}{y} - \dfrac{y}{x}}{\dfrac{x^2 + 2xy + y^2}{x^2 - y^2}} = \frac{\dfrac{x^2 - y^2}{xy}}{\dfrac{x^2 + 2xy + y^2}{x^2 - y^2}} \qquad \text{The LCD for the fraction in the numerator is } xy.$$

$$= \frac{\dfrac{(x + y)(x - y)}{xy}}{\dfrac{(x + y)(x + y)}{(x + y)(x - y)}} \qquad \text{Factoring}$$

$$= \frac{(x + y)(x - y)}{xy} \cdot \frac{(x + y)(x - y)}{(x + y)(x + y)} \qquad \text{Inverting the fraction in the denominator and multiplying}$$

$$= \frac{(x - y)^2}{xy} \qquad \text{Canceling common factors}$$

1.5 Integer Exponents

Exponents

We begin by recalling the definition of the exponential expression a^n, where a is a real number and n is a positive integer.

**EXPONENTIAL
EXPRESSIONS**

If a is any real number and n is a natural number, then the expression a^n (read "a to the power n") is defined as the number

$$a^n = \underbrace{a \cdot a \cdot a \cdot \cdots \cdot a}_{n \text{ factors}}$$

Recall that the number a is the *base*, and the superscript n is the *exponent*, or *power*, to which the base is raised.

EXAMPLE 1   Write each of the following numbers without using exponents:

a. 2^5 **b.** $\left(\dfrac{2}{3}\right)^3$

Solution

a. $2^5 = 2 \cdot 2 \cdot 2 \cdot 2 \cdot 2 = 32$

b. $\left(\dfrac{2}{3}\right)^3 = \left(\dfrac{2}{3}\right)\left(\dfrac{2}{3}\right)\left(\dfrac{2}{3}\right) = \dfrac{8}{27}$

   Next we extend our definition of a^n to include $n = 0$; that is, we define the expression a^0. Observe that if a is any real number and m and n are positive integers, then we have the rule

$$a^m a^n = \underbrace{(a \cdot a \cdot \cdots \cdot a)}_{m \text{ factors}}\underbrace{(a \cdot a \cdot \cdots \cdot a)}_{n \text{ factors}} = \underbrace{a \cdot a \cdot \cdots \cdot a}_{(m+n) \text{ factors}} = a^{m+n}$$

Now, if we require that this rule hold for the zero exponent as well, then we must have, upon setting $m = 0$,

$$a^0 a^n = a^{0+n} = a^n \qquad \text{or} \qquad a^0 a^n = a^n$$

Therefore, if $a \neq 0$, we can divide both sides of this last equation by a^n to obtain $a^0 = 1$. This motivates the following definition.

**ZERO
EXPONENT**

For any nonzero real number a,

$$a^0 = 1$$

The expression 0^0 is not defined.

EXAMPLE 2

a. $2^0 = 1$ **b.** $(-2)^0 = 1$ **c.** $(\pi)^0 = 1$ **d.** $\left(\tfrac{1}{3}\right)^0 = 1$

   Next, we extend our definition to include expressions of the form a^n, where the exponent is a negative integer. Once again we use the rule

$$a^m a^n = a^{m+n}$$

where n is a positive integer. Now, if we require that this rule hold for negative integral exponents as well, upon setting $m = -n$, we have

$$a^{-n} a^n = a^{-n+n} = a^0 = 1 \qquad \text{or} \qquad a^{-n} a^n = 1$$

Therefore, if $a \neq 0$, we can divide both sides of this last equation by a^n to obtain $a^{-n} = 1/a^n$. This motivates the following definition.

EXPONENTIAL EXPRESSIONS WITH NEGATIVE EXPONENTS

If a is any nonzero real number and n is a positive integer, then

$$a^{-n} = \frac{1}{a^n}$$

EXAMPLE 3   Write each of the following numbers without using exponents:

a. 4^{-2} **b.** 3^{-1} **c.** -2^{-3} **d.** $\left(\frac{2}{3}\right)^{-1}$ **e.** $\left(\frac{3}{2}\right)^{-3}$

Solution

a. $4^{-2} = \dfrac{1}{4^2} = \dfrac{1}{16}$ **b.** $3^{-1} = \dfrac{1}{3^1} = \dfrac{1}{3}$ **c.** $-2^{-3} = -\dfrac{1}{2^3} = -\dfrac{1}{8}$

d. $\left(\dfrac{2}{3}\right)^{-1} = \dfrac{1}{\left(\dfrac{2}{3}\right)^1} = \dfrac{1}{\dfrac{2}{3}} = \dfrac{3}{2}$ **e.** $\left(\dfrac{3}{2}\right)^{-3} = \dfrac{1}{\left(\dfrac{3}{2}\right)^3} = \dfrac{2^3}{3^3} = \dfrac{8}{27}$

REMARK   In Example 3, parts d and e, the intermediate steps may be omitted by observing that

$$\left(\frac{a}{b}\right)^{-n} = \left(\frac{b}{a}\right)^{n} \qquad \text{For example, } \left(\frac{2}{3}\right)^{-1} = \left(\frac{3}{2}\right)^{1}$$

since

$$\left(\frac{a}{b}\right)^{-n} = \frac{1}{\left(\frac{a}{b}\right)^n} = \frac{1}{\dfrac{a^n}{b^n}} = 1 \cdot \frac{b^n}{a^n} = \left(\frac{b}{a}\right)^n$$

Five basic properties of exponents are given in Table 10.

TABLE 10
Properties of exponents

Property	Illustration
1. $a^m \cdot a^n = a^{m+n}$	$x^2 \cdot x^3 = x^{2+3} = x^5$
2. $\dfrac{a^m}{a^n} = a^{m-n}$	$\dfrac{x^7}{x^4} = x^{7-4} = x^3$
3. $(a^m)^n = a^{mn}$	$(x^4)^3 = x^{4 \cdot 3} = x^{12}$
4. $(ab)^n = a^n \cdot b^n$	$(2x)^4 = 2^4 \cdot x^4 = 16x^4$
5. $\left(\dfrac{a}{b}\right)^n = \dfrac{a^n}{b^n} \qquad (b \neq 0)$	$\left(\dfrac{x}{2}\right)^3 = \dfrac{x^3}{2^3} = \dfrac{x^3}{8}$

It can be shown that these laws are valid for any real numbers a and b and any integers m and n.

Simplifying Exponential Expressions

The next two examples illustrate the use of the laws of exponents.

EXAMPLE 4   Simplify the expression, writing your answer using positive exponents only.

a. $(2x^3)(3x^5)$ **b.** $\dfrac{2x^5}{3x^4}$ **c.** $(x^{-2})^{-3}$ **d.** $(2u^{-1}v^3)^3$ **e.** $\left(\dfrac{2m^3n^4}{m^5n^3}\right)^{-1}$

Solution

a. $(2x^3)(3x^5) = 6x^{3+5} = 6x^8$ Property 1

b. $\dfrac{2x^5}{3x^4} = \dfrac{2}{3}x^{5-4} = \dfrac{2}{3}x$ Property 2

c. $(x^{-2})^{-3} = x^{(-2)(-3)} = x^6$ Property 3

d. $(2u^{-1}v^3)^3 = 2^3u^{(-1)(3)}v^{3(3)} = 8u^{-3}v^9 = \dfrac{8v^9}{u^3}$ Property 4

e. $\left(\dfrac{2m^3n^4}{m^5n^3}\right)^{-1} = (2m^{3-5}n^{4-3})^{-1}$ Property 2

$= (2m^{-2}n)^{-1}$ Property 1

$= \dfrac{1}{2m^{-2}n}$

$= \dfrac{m^2}{2n}$

EXAMPLE 5   Simplify the expression, writing your result using positive exponents only.

a. $(2^2)^3 - (3^2)^2$ **b.** $(x^{-1} + y^{-1})^{-1}$ **c.** $\dfrac{2^{-4} \cdot (2^{-1})^2}{(2^0 + 1)^{-1}}$

Solution

a. $(2^2)^3 - (3^2)^2 = 2^6 - 3^4 = 64 - 81 = -17$

b. $(x^{-1} + y^{-1})^{-1} = \left(\dfrac{1}{x} + \dfrac{1}{y}\right)^{-1} = \left(\dfrac{y+x}{xy}\right)^{-1} = \dfrac{xy}{y+x}$

c. $\dfrac{2^{-4} \cdot (2^{-1})^2}{(2^0 + 1)^{-1}} = \dfrac{2^{-4} \cdot 2^{-2}}{(2)^{-1}} = 2^{-4-2+1} = 2^{-5} = \dfrac{1}{2^5} = \dfrac{1}{32}$

SELF-CHECK EXERCISES 1.5

1. Simplify the expression, writing your result using positive exponents only.

a. $(3a^4)(4a^3)$ **b.** $\left(\dfrac{u^{-3}}{u^{-5}}\right)^{-2}$

2. Simplify the expression, writing your result using positive exponents only.

a. $(x^2y^3)^3(x^5y)^{-2}$ **b.** $\left(\dfrac{a^2b^{-1}c^3}{a^3b^{-2}}\right)^2$

Solutions to Self-Check Exercises 1.5 can be found on page 31.

1.5   EXERCISES

In exercises 1–20, rewrite the number without using exponents.

1. $(-2)^3$

2. $\left(-\dfrac{2}{3}\right)^4$

3. 7^{-2}

4. $\left(\dfrac{3}{4}\right)^{-2}$

5. $-\left(-\dfrac{1}{4}\right)^{-2}$

6. -4^2

7. $2^{-2} + 3^{-1}$

8. $-3^{-2} - \left(-\dfrac{2}{3}\right)^2$

9. $(0.02)^2$

10. $(-0.3)^{-2}$

11. 1996^0

12. $(18 + 25)^0$

13. $(ab^2)^0$, where $a, b \neq 0$

14. $(3x^2y^3)^0$, where $x, y \neq 0$

15. $\dfrac{2^3 \cdot 2^5}{2^4 \cdot 2^9}$

16. $\dfrac{6 \cdot 10^4}{3 \cdot 10^2}$

17. $\dfrac{2^{-3} \cdot 2^{-4}}{2^{-5} \cdot 2^{-2}}$

18. $\dfrac{4 \cdot 2^{-3}}{2 \cdot 4^{-2}}$

19. $\left(\dfrac{3^4 \cdot 3^{-3}}{3^{-2}}\right)^{-1}$

20. $\left(\dfrac{5^{-2} \cdot 5^{-2}}{5^{-5}}\right)^{-2}$

In exercises 21–54, simplify the expression, writing your answer using positive exponents only.

21. $(2x^3)\left(\dfrac{1}{4}x^2\right)$

22. $(-2x^2)(3x^{-4})$

23. $\dfrac{3x^3}{2x^4}$

24. $\dfrac{(3x^2)(4x^3)}{2x^4}$

25. $(a^{-2})^3$

26. $(-a^2)^{-3}$

27. $(2x^{-2}y^2)^3$

28. $(3u^{-1}v^{-2})^{-3}$

29. $(4x^2y^{-3})(2x^{-3}y^2)$

30. $\left(\dfrac{1}{2}u^{-2}v^3\right)(4v^3)$

31. $(-x^2y)^3\left(\dfrac{2y^2}{x^4}\right)$

32. $\left(-\dfrac{1}{2}x^2y\right)^{-2}$

33. $\left(\dfrac{2u^2v^3}{3uv}\right)^{-1}$

34. $\left(\dfrac{a^{-2}}{2b^2}\right)^{-3}$

35. $(3x^{-2})^3(2x^2)^5$

36. $(2^{-1}r^3)^{-2}(3s^{-1})^2$

37. $\dfrac{3^0 \cdot 4x^{-2}}{16 \cdot (x^2)^3}$

38. $\dfrac{5x^2(3x^{-2})}{(4x^{-1})(x^3)^{-2}}$

39. $\dfrac{2^2u^{-2}(v^{-1})^3}{3^2(u^{-3}v)^2}$

40. $\dfrac{(3a^{-1}b^2)^{-2}}{(2a^2b^{-1})^{-3}}$

41. $(-2x)^{-2}(3y)^{-3}(4z)^{-2}$

42. $(3x^{-1})^2(4y^{-1})^3(2z)^{-2}$

43. $(a^2b^{-3})^2(a^{-2}b^2)^{-3}$

44. $(5u^2v^{-3})^{-1} \cdot 3(2u^2v^2)^{-2}$

45. $\left[\left(\dfrac{a^{-2}b^{-2}}{3a^{-1}b^2}\right)^2\right]^{-1}$

46. $\left[\left(\dfrac{x^2y^{-3}z^{-4}}{x^{-2}y^{-1}z^2}\right)^{-2}\right]^3$

47. $\left(\dfrac{3^2u^{-2}v^2}{2^2u^3v^{-3}}\right)^{-2}\left(\dfrac{3^2v^5}{4^2u}\right)^2$

48. $\left[\left(-\dfrac{2^2x^{-2}y^0}{3^2x^3y^{-2}}\right)^{-2}\right]^{-2}$

49. $\dfrac{x^{-1} - 1}{x^{-1} + 1}$

50. $\dfrac{x^{-1} - y^{-1}}{x^{-1} + y^{-1}}$

51. $\dfrac{u^{-1} - v^{-1}}{v - u}$

52. $\dfrac{(uv)^{-1}}{u^{-1} + v^{-1}}$

53. $\left(\dfrac{a^{-1} - b^{-1}}{a^{-1} + b^{-1}}\right)^{-1}$

54. $[(a^{-1} + b^{-1})(a^{-1} - b^{-1})]^{-2}$

55. *True or False.* If a and b are real numbers and m and n are natural numbers, then $a^m b^n = (ab)^{mn}$.

56. *True or False.* If a and b are real numbers ($b \neq 0$) and m and n are natural numbers, then

$$\dfrac{a^m}{b^n} = \left(\dfrac{a}{b}\right)^{m-n}$$

57. *True or False.* If a and b are real numbers and n is a natural number, then $(a + b)^n = a^n + b^n$.

SOLUTIONS TO SELF-CHECK EXERCISES 1.5

1. a. $(3a^4)(4a^3) = 3 \cdot a^4 \cdot 4 \cdot a^3 = 12a^{4+3} = 12a^7$

b. $\left(\dfrac{u^{-3}}{u^{-5}}\right)^{-2} = \dfrac{u^{(-3)(-2)}}{u^{(-5)(-2)}} = \dfrac{u^6}{u^{10}} = u^{6-10} = u^{-4} = \dfrac{1}{u^4}$

2. a. $(x^2y^3)^3(x^5y)^{-2} = x^{2\cdot3}y^{3\cdot3}x^{5(-2)}y^{-2} = x^6y^9x^{-10}y^{-2}$

$$= x^{6-10}y^{9-2} = x^{-4}y^7$$

$$= \frac{y^7}{x^4}$$

b. $\left(\dfrac{a^2b^{-1}c^3}{a^3b^{-2}}\right)^2 = \dfrac{a^{2\cdot2}b^{(-1)(2)}c^{3\cdot2}}{a^{3\cdot2}b^{(-2)(2)}} = \dfrac{a^4b^{-2}c^6}{a^6b^{-4}}$

$$= a^{4-6}b^{-2+4}c^6 = a^{-2}b^2c^6$$

$$= \frac{b^2c^6}{a^2}$$

1.6 Rational Exponents and Radicals

nth Roots of Real Numbers

Thus far, we have described the expression a^n, where a is a real number and n is an integer. We now direct our attention to a closely related topic: roots of real numbers. As we will soon see, expressions of the form a^n for fractional (rational) powers of n may be defined in terms of the roots of a.

nTH ROOT OF A REAL NUMBER

If n is a natural number and a and b are real numbers such that

$$a^n = b$$

then we say that a is the **nth root** of b.

For $n = 2$ and $n = 3$, the roots are commonly referred to as the **square roots** and **cube roots**, respectively. Some examples of roots follow:

-2 and 2 are square roots of 4 because $(-2)^2 = 4$ and $2^2 = 4$

-3 and 3 are fourth roots of 81 because $(-3)^4 = 81$ and $3^4 = 81$

-4 is a cube root of -64 because $(-4)^3 = -64$

$\dfrac{1}{2}$ is a fifth root of $\dfrac{1}{32}$ because $\left(\dfrac{1}{2}\right)^5 = \dfrac{1}{32}$

How many real roots does a real number b have?

1. *When* n *is even, the real* n*th roots of a positive real number* b *must come in pairs—one positive and the other negative.* For example, the real fourth roots of 81 include -3 and 3.

2. *When* n *is even and* b *is a negative real number, there are no real* n*th roots of* b. For example, if $b = -9$ and the real number a is a square root of b, then by definition $a^2 = -9$. But this is a contradiction since the square of a real number cannot be negative, and we conclude that b has no real roots in this case.

3. *When* n *is odd, then there is only one real* n*th root of* b. For example, the cube root of -64 is -4.

As you can see from (1), given a number b, there might be more than one real root. So, to avoid ambiguity, we define the *principal* nth root of a positive real number, when n is even, to be the positive root. Thus, the principal square root of a real number, when n is even, is the positive root. For example, the principal square root of 4 is 2 and the principal fourth root of 81 is 3. Of course, the principal nth root of any real number b, when n is odd, is given by the (unique) nth root of b. For example, the principal cube root of -64 is -4 and the principal fifth root of $1/32$ is $1/2$.

A summary of the number of roots of a real number b is given in Table 11.

TABLE 11
Number of roots of a real number b

Index	b	Number of Roots
n even	$b > 0$	Two real roots (One principal root)
	$b < 0$	No real roots
	$b = 0$	One real root
n odd	$b > 0$	One real root
	$b < 0$	One real root
	$b = 0$	One real root

We use the notation $\sqrt[n]{b}$, called a **radical,** to denote the principal nth root of b. The symbol $\sqrt{}$ is called a **radical sign,** and the number b within the radical sign is called the **radicand.** The positive integer n is called the **index** of the radical. For square roots ($n = 2$), we write $\sqrt{b}$ instead of $\sqrt[2]{b}$.

 A common mistake is to write $\sqrt[4]{16} = \pm 2$. This is wrong because $\sqrt[4]{16}$ denotes the principal fourth root of 16, which is the positive root 2. Of course, the negative of the fourth root of 16 is $-\sqrt[4]{16} = -(2) = -2$.

EXAMPLE 1 Determine the number of roots of each real number.

a. $\sqrt{25}$ **b.** $\sqrt[5]{0}$ **c.** $\sqrt[3]{-27}$ **d.** $\sqrt{-27}$

Solution

a. Here $b > 0$, n is even, and there is one principal root. Thus, $\sqrt{25} = 5$.
b. Here $b = 0$, n is odd, and there is one root. Thus, $\sqrt[5]{0} = 0$.
c. Here $b < 0$, n is odd, and there is one root. Thus, $\sqrt[3]{-27} = -3$.
d. Here $b < 0$, n is even, and no real root exists. Thus, $\sqrt{-27}$ is not defined.   ■

 Note that $(-81)^{1/4}$ does not exist because n is even and $b < 0$, but $-81^{1/4} = -(81)^{1/4} = -3$. The first expression is "the fourth root of -81," whereas the second expression is "the negative of the fourth root of 81."

EXAMPLE 2 Evaluate the following radicals:

a. $\sqrt[6]{64}$ **b.** $\sqrt[5]{-32}$ **c.** $\sqrt[3]{\dfrac{8}{27}}$ **d.** $-\sqrt{\dfrac{4}{25}}$

Solution

a. $\sqrt[6]{64} = 2$ because $2^6 = 64$.

b. $\sqrt[5]{-32} = -2$ because $(-2)^5 = -32$.

c. $\sqrt[3]{\dfrac{8}{27}} = \dfrac{2}{3}$ because $\left(\dfrac{2}{3}\right)^3 = \dfrac{8}{27}$.

d. $-\sqrt{\dfrac{4}{25}} = -\dfrac{2}{5}$ because $\sqrt{\dfrac{4}{25}} = \dfrac{2}{5}$ and so $-\sqrt{\dfrac{4}{25}} = -\dfrac{2}{5}$.   ∎

Rational Exponents and Radicals

In Section 1.5, we defined expressions such as 2^{-3}, $(1/2)^2$, and $1/\pi^3$ involving integer exponents. But how do we evaluate expressions such as $8^{1/3}$, where the exponent is a rational number? From the definition of the *n*th root of a real number, we know that $\sqrt[3]{8} = 2$. Using rational exponents, we write this same result in the form $8^{1/3} = 2$. More generally, we have the following definitions.

Rational Exponents	Illustration
a. If n is a natural number and b is a real number, then $$b^{1/n} = \sqrt[n]{b}$$ (If $b < 0$ and n is even, $b^{1/n}$ is not defined.)	$9^{1/2} = \sqrt{9} = 3$ $(-8)^{1/3} = \sqrt[3]{-8} = -2$
b. If m/n is a rational number reduced to lowest terms (m, n natural numbers) then $$b^{m/n} = (b^{1/n})^m$$ or, equivalently, $$b^{m/n} = \sqrt[n]{b^m}$$ whenever it exists.	$(27)^{2/3} = (27^{1/3})^2 = 3^2 = 9$ or $(27)^{2/3} = [(27^2)]^{1/3} = (729)^{1/3} = 9$ $(-27)^{2/3} = (-27^{1/3})^2$ $= (-3)^2 = 9$

EXAMPLE 3

a. $(64)^{1/3} = \sqrt[3]{64} = 4$ **b.** $(81)^{3/4} = (81^{1/4})^3 = 3^3 = 27$

c. $(-8)^{5/3} = (-8^{1/3})^5 = (-2)^5 = -32$

d. $\left(\dfrac{1}{27}\right)^{2/3} = \left[\left(\dfrac{1}{27}\right)^{1/3}\right]^2 = \left(\dfrac{1}{3}\right)^2 = \dfrac{1}{9}$ ∎

Expressions involving *negative* rational exponents are taken care of by the following definition.

NEGATIVE EXPONENTS

$$a^{-m/n} = \frac{1}{a^{m/n}} \qquad (a \neq 0)$$

EXAMPLE 4

a. $4^{-5/2} = \dfrac{1}{4^{5/2}} = \dfrac{1}{(4^{1/2})^5} = \dfrac{1}{2^5} = \dfrac{1}{32}$

b. $(-8)^{-1/3} = \dfrac{1}{(-8)^{1/3}} = \dfrac{1}{-2} = -\dfrac{1}{2}$ ∎

All the properties of integral exponents listed on page 29 hold for rational exponents. Examples 5 and 6 illustrate the use of these properties.

EXAMPLE 5

a. $\dfrac{16^{5/4}}{16^{1/2}} = 16^{5/4-1/2} = 16^{5/4-2/4} = 16^{3/4} = (16^{1/4})^3 = 2^3 = 8$

b. $(6^{2/3})^3 = 6^{(2/3)\cdot 3} = 6^{6/3} = 6^2 = 36$

c. $\left(\dfrac{16}{81}\right)^{3/4} = \left[\left(\dfrac{16}{81}\right)^{1/4}\right]^3 = \left(\dfrac{16^{1/4}}{81^{1/4}}\right)^3 = \left(\dfrac{2}{3}\right)^3 = \dfrac{8}{27}$ ∎

EXAMPLE 6   Evaluate:

a. $2x^{1/2}(x^{2/3} - x^{1/4})$ **b.** $(x^{1/3} - y^{2/3})^2$

Solution

a. $2x^{1/2}(x^{2/3} - x^{1/4}) = 2x^{1/2}(x^{2/3}) - 2x^{1/2}(x^{1/4})$

$$= 2x^{1/2+2/3} - 2x^{1/2+1/4} = 2x^{3/6+4/6} - 2x^{2/4+1/4}$$

$$= 2x^{7/6} - 2x^{3/4}$$

b. $(x^{1/3} - y^{2/3})^2 = (x^{1/3})^2 - 2x^{1/3}y^{2/3} + (y^{2/3})^2$

$$= x^{2/3} - 2x^{1/3}y^{2/3} + y^{4/3}$$ ∎

Simplifying Radicals

The properties of radicals given in Table 12 follow directly from the properties of exponents discussed earlier (page 29).

TABLE 12
Properties of radicals

If m and n are natural numbers and a and b are real numbers for which the indicated roots exist, then

Property	Illustration
1. $(\sqrt[n]{a})^n = a$	$(\sqrt[3]{2})^3 = (2^{1/3})^3 = 2^1 = 2$
2. $\sqrt[n]{ab} = \sqrt[n]{a} \cdot \sqrt[n]{b}$	$\sqrt[3]{216} = \sqrt[3]{27 \cdot 8} = \sqrt[3]{27} \cdot \sqrt[3]{8} = 3 \cdot 2 = 6$
3. $\sqrt[n]{\dfrac{a}{b}} = \dfrac{\sqrt[n]{a}}{\sqrt[n]{b}} \quad (b \neq 0)$	$\sqrt[3]{\dfrac{8}{64}} = \dfrac{\sqrt[3]{8}}{\sqrt[3]{64}} = \dfrac{2}{4} = \dfrac{1}{2}$
4. $\sqrt[m]{\sqrt[n]{a}} = \sqrt[mn]{a}$	$\sqrt[3]{\sqrt{64}} = \sqrt[2\cdot3]{64} = \sqrt[6]{64} = 2$
5. If n is even: $\sqrt[n]{x^n} = |x|$	$\sqrt{(-3)^2} = |-3| = 3$
If n is odd: $\sqrt[n]{x^n} = x$	$\sqrt[3]{-8} = -2$

A common error is to write $\sqrt[n]{x^n} = x$. This is not true if x is negative (see illustration 5 in Table 12). Thus, unless the variable x is known to be nonnegative, the correct equation is given by property 5.

When we work with algebraic expressions involving radicals, we usually express the radical in simplified form.

SIMPLIFYING RADICALS

An expression involving radicals is simplified if the following conditions are satisfied:

1. The powers of all factors under the radical sign are less than the index of the radical.
2. The index of the radical has been reduced as far as possible.
3. No radical appears in a denominator.
4. No fraction appears within a radical.

EXAMPLE 7   Determine whether each radical is in simplified form. If not, state which condition is violated.

a. $\dfrac{1}{\sqrt[4]{x^5}}$ **b.** $\sqrt[6]{y^2}$ **c.** $\sqrt{\dfrac{5}{4}}$

Solution   None of the three radicals are in simplified form. The radical in part a violates conditions 1 and 3; that is, the power of x is 5, which is greater than 4, the index of the radical, and a radical appears in the denominator. The radical in part b violates condition 2 since the index of the radical can be reduced; that is,

$$\sqrt[6]{y^2} = \sqrt[3\cdot2]{y^2} = \sqrt[3]{y}$$

The radical in part c violates condition 4 since there is a fraction within the radical. Rewriting the radical, we see that

$$\sqrt{\frac{5}{4}} = \frac{\sqrt{5}}{\sqrt{4}} = \frac{\sqrt{5}}{2} = \frac{1}{2}\sqrt{5}$$

EXAMPLE 8   Simplify:

a. $\sqrt[3]{375}$ **b.** $\sqrt[3]{8x^3y^6z^9}$ **c.** $\sqrt[6]{81x^4y^2}$

Solution

a. $\sqrt[3]{375} = \sqrt[3]{3 \cdot 125} = \sqrt[3]{3} \cdot \sqrt[3]{5^3} = 5\sqrt[3]{3}$ $\sqrt[3]{5^3} = 5$

b. $\sqrt[3]{8x^3y^6z^9} = \sqrt[3]{2^3 \cdot (xy^2z^3)^3}$
$\qquad\qquad\quad = \sqrt[3]{2^3} \cdot \sqrt[3]{(xy^2z^3)^3}$
$\qquad\qquad\quad = 2xy^2z^3$ $\sqrt[3]{2^3} = 2, \quad \sqrt[3]{(xy^2z^3)^3} = xy^2z^3$

c. $\sqrt[6]{81x^4y^2} = \sqrt[6]{9^2(x^2y)^2}$

$\qquad\qquad = \sqrt[3\cdot2]{9^2} \cdot \sqrt[3\cdot2]{(x^2y)^2}$

$\qquad\qquad = \sqrt[3]{9} \cdot \sqrt[3]{x^2y} \qquad\qquad \sqrt[3\cdot2]{9^2} = \sqrt[3]{9}, \quad \sqrt[3\cdot2]{(x^2y)^2} = \sqrt[3]{x^2y}$

$\qquad\qquad = \sqrt[3]{9x^2y}$

$\blacksquare$

As mentioned earlier, a simplified rational expression should have a denominator that does not involve radicals. For example, $3/\sqrt{5}$ is *not* in simplified form, whereas $3\sqrt{5}/5$ is since the denominator of the latter is free of radicals. How do we get rid of the radical $\sqrt{5}$ in the fraction $3/\sqrt{5}$? Obviously, multiplying by $\sqrt{5}$ does the job since $(\sqrt{5})(\sqrt{5}) = \sqrt{25} = 5$! But we cannot multiply the denominator of a fraction by any number other than 1 without changing the fraction. So, the solution is to multiply *both* the numerator and the denominator of $3/\sqrt{5}$ by $\sqrt{5}$. Equivalently, we multiply $3/\sqrt{5}$ by $\sqrt{5}/\sqrt{5}$ (which is equal to 1). Thus,

$$\frac{3}{\sqrt{5}} = \frac{3}{\sqrt{5}} \cdot \frac{\sqrt{5}}{\sqrt{5}} = \frac{3\sqrt{5}}{\sqrt{25}} = \frac{3\sqrt{5}}{5}$$

This procedure of eliminating a radical from the denominator of an algebraic expression is referred to as *rationalizing the denominator* and is illustrated in Examples 9 and 10.

EXAMPLE 9   Rationalize the denominator:

a. $\dfrac{1}{\sqrt{2}}$ **b.** $\dfrac{3x}{2\sqrt{x}}$ **c.** $\dfrac{x}{\sqrt[3]{y}}$

Solution

a. $\dfrac{1}{\sqrt{2}} \cdot \dfrac{\sqrt{2}}{\sqrt{2}} = \dfrac{\sqrt{2}}{2} = \dfrac{1}{2}\sqrt{2}$

b. $\dfrac{3x}{2\sqrt{x}} \cdot \dfrac{\sqrt{x}}{\sqrt{x}} = \dfrac{3x\sqrt{x}}{2x} = \dfrac{3}{2}\sqrt{x}$

c. $\dfrac{x}{\sqrt[3]{y}} \cdot \dfrac{\sqrt[3]{y^2}}{\sqrt[3]{y^2}} = \dfrac{x\sqrt[3]{y^2}}{\sqrt[3]{y^3}} = \dfrac{x\sqrt[3]{y^2}}{y}$

$\blacksquare$

Note that in Example 9c, we multiplied the denominator by $\sqrt[3]{y^2}$ so that we would obtain $\sqrt[3]{y^3}$, which is equal to y. In general, to rationalize a denominator involving an nth root, we multiply the numerator and the denominator by a factor that will yield a product in the denominator involving an nth power. For example,

$$\frac{1}{\sqrt[5]{x^3}} = \frac{1}{\sqrt[5]{x^3}} \cdot \frac{\sqrt[5]{x^2}}{\sqrt[5]{x^2}} = \frac{\sqrt[5]{x^2}}{\sqrt[5]{x^5}} = \frac{\sqrt[5]{x^2}}{x} \qquad \text{Since } \sqrt[5]{x^3} \cdot \sqrt[5]{x^2} = x^{3/5} \cdot x^{2/5} = x^{5/5} = x$$

EXAMPLE 10   Rationalize the denominator:

a. $\sqrt[3]{\dfrac{8}{3}}$ **b.** $\sqrt[3]{\dfrac{x}{y^2}}$

Solution

a. $\sqrt[3]{\dfrac{8}{3}} = \dfrac{\sqrt[3]{8}}{\sqrt[3]{3}} = \dfrac{2}{\sqrt[3]{3}} \cdot \dfrac{\sqrt[3]{3^2}}{\sqrt[3]{3^2}} = \dfrac{2\sqrt[3]{3^2}}{3} = \dfrac{2}{3}\sqrt[3]{9}$

b. $\sqrt[3]{\dfrac{x}{y^2}} = \dfrac{\sqrt[3]{x}}{\sqrt[3]{y^2}} = \dfrac{\sqrt[3]{x}}{\sqrt[3]{y^2}} \cdot \dfrac{\sqrt[3]{y}}{\sqrt[3]{y}} = \dfrac{\sqrt[3]{xy}}{\sqrt[3]{y^3}} = \dfrac{\sqrt[3]{xy}}{y}$ ∎

How do we rationalize the denominator of a fraction like $\dfrac{1}{1 - \sqrt{3}}$? Rather than multiplying by $\dfrac{\sqrt{3}}{\sqrt{3}}$ (which does not eliminate the radical in the denominator), we multiply by $\dfrac{1 + \sqrt{3}}{1 + \sqrt{3}}$, obtaining

$$\frac{1}{1 - \sqrt{3}} \cdot \frac{1 + \sqrt{3}}{1 + \sqrt{3}} = \frac{1 + \sqrt{3}}{1 - (\sqrt{3})^2} \qquad (a - b)(a + b) = a^2 - b^2$$

$$= \frac{1 + \sqrt{3}}{1 - 3} = \frac{1 + \sqrt{3}}{-2}$$

$$= -\frac{1 + \sqrt{3}}{2}$$

In general, to rationalize a denominator of the form $a + \sqrt{b}$, we multiply by $\dfrac{a - \sqrt{b}}{a - \sqrt{b}}$. Similarly, to rationalize a denominator of the form $a - \sqrt{b}$, we multiply by $\dfrac{a + \sqrt{b}}{a + \sqrt{b}}$. We refer to the quantities $a + \sqrt{b}$ and $a - \sqrt{b}$ as **conjugates** of each other.

EXAMPLE 11   Rationalize the denominator in the expression $\dfrac{3}{2 - \sqrt{5}}$.

Solution   The conjugate of $2 - \sqrt{5}$ is $2 + \sqrt{5}$. Therefore, we multiply the given fraction by $\dfrac{2 + \sqrt{5}}{2 + \sqrt{5}}$, obtaining

$$\frac{3}{2 - \sqrt{5}} \cdot \frac{2 + \sqrt{5}}{2 + \sqrt{5}} = \frac{3(2 + \sqrt{5})}{(2 - \sqrt{5})(2 + \sqrt{5})} = \frac{3(2 + \sqrt{5})}{2^2 - (\sqrt{5})^2}$$

$$= \frac{3(2 + \sqrt{5})}{-1}$$

$$= -3(2 + \sqrt{5}) \qquad ∎$$

Sometimes you may find it easier to convert an expression containing a radical to one involving rational exponents before evaluating the expression.

EXAMPLE 12   Evaluate:

a. $\sqrt[3]{3^2} \cdot \sqrt[4]{9^3}$ **b.** $\sqrt[3]{x^2 y} \cdot \sqrt{xy}$

Solution

a. $\sqrt[3]{3^2} \cdot \sqrt[4]{9^3} = 3^{2/3} \cdot 9^{3/4} = 3^{2/3} \cdot 3^{6/4} = 3^{2/3+3/2} = 3^{4/6+9/6} = 3^{13/6}$

b. $\sqrt[3]{x^2 y} \cdot \sqrt{xy} = x^{2/3} y^{1/3} \cdot x^{1/2} y^{1/2} = x^{2/3+1/2} y^{1/3+1/2} = x^{4/6+3/6} y^{2/6+3/6} = x^{7/6} y^{5/6}$

Note how much easier it is to work with rational exponents in this case.

SELF-CHECK EXERCISES 1.6

1. Simplify:

   a. $\dfrac{8^{2/3} \cdot 8^{4/3}}{8^{3/2}}$ **b.** $\left(\dfrac{3y^{-4} y^4}{z^{-2}}\right)^3$ **c.** $\sqrt{5} \cdot \sqrt{45}$

2. Rationalize the denominator:

   a. $\dfrac{1}{\sqrt[3]{xy}}$ **b.** $\dfrac{4}{3 + \sqrt{8}}$

Solutions to Self-Check Exercises 1.6 can be found on page 40.

1.6 EXERCISES

In exercises 1–20, rewrite the number without radicals or exponents.

1. $\sqrt{81}$

2. $\sqrt[3]{-27}$

3. $\sqrt[4]{256}$

4. $\sqrt[5]{-32}$

5. $9^{1/2}$

6. $625^{1/4}$

7. $8^{2/3}$

8. $32^{2/5}$

9. $-25^{1/2}$

10. $-16^{3/2}$

11. $(-8)^{2/3}$

12. $(-32)^{3/5}$

13. $\left(\dfrac{4}{9}\right)^{1/2}$

14. $\left(\dfrac{9}{25}\right)^{3/2}$

15. $\left(\dfrac{27}{8}\right)^{2/3}$

16. $\left(-\dfrac{8}{125}\right)^{1/3}$

17. $8^{-2/3}$

18. $81^{-1/4}$

19. $-\left(\dfrac{27}{8}\right)^{-1/3}$

20. $-\left(-\dfrac{8}{27}\right)^{-2/3}$

In exercises 21–40, carry out the indicated operation and write your answer using positive exponents only.

21. $3^{1/3} \cdot 3^{5/3}$

22. $2^{6/5} \cdot 2^{-1/5}$

23. $\dfrac{4^{1/2}}{4^{5/2}}$

24. $\dfrac{3^{-5/4}}{3^{-1/4}}$

25. $\dfrac{2^{-1/2} \cdot 3^{2/3}}{2^{3/2} \cdot 3^{-1/3}}$

26. $\dfrac{4^{1/3} \cdot 4^{-2/5}}{4^{2/3}}$

27. $(2^{3/2})^4$

28. $[(-3)^{1/3}]^2$

29. $x^{2/5} \cdot x^{-1/5}$

30. $y^{-3/8} \cdot y^{1/4}$

31. $\dfrac{x^{3/4}}{x^{-1/4}}$

32. $\dfrac{x^{7/3}}{x^{-2}}$

33. $\left(\dfrac{x^3}{-27x^{-6}}\right)^{-2/3}$

34. $\left(\dfrac{27x^{-3} y^2}{8x^{-2} y^{-5}}\right)^{1/3}$

35. $\left(\dfrac{x^{-3}}{y^{-2}}\right)^{1/2} \left(\dfrac{y}{x}\right)^{3/2}$

36. $\left(\dfrac{r^n}{r^{5-2n}}\right)^4$

37. $x^{2/5}(x^2 - 2x^3)$

38. $s^{1/3}(2s - s^{1/4})$

39. $2p^{3/2}(2p^{1/2} - p^{-1/2})$

40. $3y^{1/3}(y^{2/3} - 1)^2$

In exercises 41–52, write the expression in simplest radical form.

41. $\sqrt{32}$

42. $\sqrt{45}$

43. $\sqrt[3]{-54}$

44. $-\sqrt[4]{48}$

45. $\sqrt{16x^2 y^3}$

46. $\sqrt{40a^3 b^4}$

47. $\sqrt[3]{m^6 n^3 p^{12}}$

48. $\sqrt[3]{-27p^2 q^3 r^4}$

49. $\sqrt[3]{\sqrt{9}}$

50. $\sqrt[5]{\sqrt[3]{9}}$

51. $\sqrt[3]{\sqrt{x}}$

52. $\sqrt[3]{-\sqrt[4]{x^3}}$

In exercises 53–68, rationalize the denominator of the expression.

53. $\dfrac{2}{\sqrt{3}}$

54. $\dfrac{3}{\sqrt{5}}$

55. $\dfrac{3}{2\sqrt{x}}$

56. $\dfrac{3}{\sqrt{xy}}$

57. $\dfrac{2y}{\sqrt{3y}}$

58. $\dfrac{5x^2}{\sqrt{3x}}$

59. $\dfrac{1}{\sqrt[3]{x}}$

60. $\sqrt{\dfrac{2x}{y}}$

61. $\dfrac{2}{1+\sqrt{3}}$

62. $\dfrac{3}{1-\sqrt{2}}$

63. $\dfrac{1+\sqrt{2}}{1-\sqrt{2}}$

64. $\dfrac{9+\sqrt{2}}{3-\sqrt{2}}$

65. $\dfrac{q}{\sqrt{q}-1}$

66. $\dfrac{xy}{\sqrt{x}+\sqrt{y}}$

67. $\dfrac{y}{\sqrt[3]{x^2z}}$

68. $\dfrac{2x}{\sqrt[3]{xy^2}}$

In exercises 69–76, write the expression in simplest radical form.

69. $\sqrt{\dfrac{16}{3}}$

70. $-\sqrt{\dfrac{8}{3}}$

71. $\sqrt[3]{\dfrac{2}{3}}$

72. $\sqrt[4]{\dfrac{81}{4}}$

73. $\sqrt{\dfrac{3}{2x^2}}$

74. $\sqrt{\dfrac{x^3y^5}{4}}$

75. $\sqrt[3]{\dfrac{2y^2}{3}}$

76. $\sqrt[4]{\dfrac{3a^3}{b^2}}$

In exercises 77–84, simplify the expression.

77. $\dfrac{1}{\sqrt{a}}+\sqrt{a}$

78. $\dfrac{x}{\sqrt{x-y}}-\sqrt{x-y}$

79. $\dfrac{\sqrt{x}}{\sqrt{x}+\sqrt{y}}+\dfrac{\sqrt{y}}{\sqrt{x}-\sqrt{y}}$

80. $\dfrac{a}{\sqrt{a^2-b^2}}-\dfrac{\sqrt{a^2-b^2}}{a}$

81. $(x+1)^{1/2}+\dfrac{1}{2}x(x+1)^{-1/2}$

82. $\dfrac{1}{2}x^{-1/2}(x+y)^{1/3}+\dfrac{1}{3}x^{1/2}(x+y)^{-2/3}$

83. $\dfrac{\frac{1}{2}(1+x^{1/3})x^{-1/2}-\frac{1}{3}x^{1/2}\cdot x^{-2/3}}{(1+x^{1/3})^2}$

84. $\dfrac{\frac{1}{2}x^{-1/2}(x+y)^{1/2}-\frac{1}{2}x^{1/2}(x+y)^{-1/2}}{x+y}$

SOLUTIONS TO SELF-CHECK EXERCISES 1.6

1. a. $\dfrac{8^{2/3}\cdot 8^{4/3}}{8^{3/2}}=8^{2/3+4/3-3/2}=8^{6/3-3/2}=8^{1/2}=\sqrt{8}=2\sqrt{2}$

b. $\left(\dfrac{3y^{-4}y^4}{z^{-2}}\right)^3=\left(\dfrac{3y^{-4+4}}{z^{-2}}\right)^3=\left(\dfrac{3y^0}{z^{-2}}\right)^3=(3z^2)^3=3^3z^{2\cdot3}=27z^6$

c. $\sqrt{5}\cdot\sqrt{45}=\sqrt{5\cdot45}=\sqrt{225}=15$

2. a. $\dfrac{1}{\sqrt[3]{xy}}\cdot\dfrac{\sqrt[3]{x^2y^2}}{\sqrt[3]{x^2y^2}}=\dfrac{\sqrt[3]{x^2y^2}}{\sqrt[3]{x^3y^3}}=\dfrac{\sqrt[3]{x^2y^2}}{xy}$

b. $\dfrac{4}{3+\sqrt{8}}\cdot\dfrac{3-\sqrt{8}}{3-\sqrt{8}}=\dfrac{4(3-\sqrt{8})}{3^2-(\sqrt{8})^2}$

$=\dfrac{4(3-\sqrt{8})}{9-8}=4(3-\sqrt{8})$

1.7   Quadratic Equations

The equation

$$2x^2+3x+1=0$$

is an example of a quadratic equation. In general, a **quadratic equation** in the variable x is any equation that can be written in the form

$$ax^2+bx+c=0$$

where a, b, and c are constants and $a \neq 0$. We refer to this form as the *standard form*. Equations such as

$$3x^2 + 4x = 1 \qquad \text{and} \qquad 2t^2 = t + 1$$

are quadratic equations in nonstandard form, but they can be easily transformed into standard form. For example, adding -1 to both sides of the first equation leads to $3x^2 + 4x - 1 = 0$, which is in standard form. Similarly, by subtracting $(t + 1)$ from both sides of the second equation, we obtain $2t^2 - t - 1 = 0$, which is also in standard form.

Solving by Factoring

We solve a quadratic equation in x by finding its roots. The *roots* of a quadratic equation in x are precisely the values of x that satisfy the equation. The method of solving quadratic equations by *factoring* relies on the following zero property of real numbers, which we restate here.

ZERO-PRODUCT PROPERTY OF REAL NUMBERS

If a and b are real numbers and $ab = 0$, then $a = 0$, or $b = 0$, or both a, $b = 0$.

Simply stated, this property says that the product of two real numbers is equal to zero if and only if one (or both) of the factors is equal to zero.

EXAMPLE 1 Solve $x^2 - 3x + 2 = 0$ by factoring.

Solution Factoring the given equation, we find that

$$x^2 - 3x + 2 = (x - 2)(x - 1) = 0$$

By the zero-product property of real numbers, we have

$$x - 2 = 0 \quad \text{or} \quad x - 1 = 0$$

from which we see that $x = 2$ or $x = 1$ are the roots of the equation.

If a quadratic equation is not in standard form, we first rewrite it in standard form and then factor the equation to find its roots.

EXAMPLE 2 Solve by factoring:

a. $2x^2 - 7x = -6$ **b.** $4x^2 = 3x$ **c.** $2x^2 = 6x - 4$

Solution

a. Rewriting the equation in standard form, we have

$$2x^2 - 7x + 6 = 0$$

Factoring this equation, we obtain

$$(2x - 3)(x - 2) = 0$$

Therefore,

$$2x - 3 = 0 \quad \text{or} \quad x - 2 = 0$$

and

$$x = \frac{3}{2} \quad \text{or} \quad x = 2$$

b. First we write the equation in standard form:

$$4x^2 - 3x = 0$$

Factoring this equation, we have

$$x(4x - 3) = 0$$

Therefore,

$$x = 0 \quad \text{or} \quad 4x - 3 = 0$$

and

$$x = 0 \quad \text{or} \quad x = \frac{3}{4}$$

(Observe that if we had divided $4x^2 - 3x = 0$ by x before factoring the original equation, we would have lost the solution $x = 0$.)

c. Rewriting the equation in standard form, we have

$$2x^2 - 6x + 4 = 0$$

Factoring, we have

$$2(x^2 - 3x + 2) = 0 \qquad \text{2 is a common factor.}$$

$$2(x - 2)(x - 1) = 0$$

and

$$x = 2 \quad \text{or} \quad x = 1$$

Solving by Completing the Square

The method of factoring works well for equations that are easily factored. But what about equations such as $x^2 - 2x - 2 = 0$ that are not easily factored? Equations of this type may be solved by using the method of *completing the square.*

The Method of Completing the Square	**Illustration**
1. Write the equation in the form $$x^2 + \frac{b}{a}x = \frac{c}{a}$$ Coefficient of x ⌐ ⌐ Constant term where the coefficient of x^2 is 1 and the constant term is on the right side of the equation.	$$x^2 - 2x = 2$$ Coefficient⌐ ⌐Constant of x term
2. Square half of the coefficient of x.	$$\left(-\frac{2}{2}\right)^2 = 1$$

3. Add the number obtained in step 2 to both sides of the equation, factor, and solve for x.

$$x^2 - 2x + 1 = 2 + 1$$
$$(x - 1)^2 = 3$$
$$x - 1 = \pm\sqrt{3}$$
$$x = 1 \pm \sqrt{3}$$

EXAMPLE 3   Solve by completing the square:

a. $4x^2 - 3x - 2 = 0$ **b.** $6x^2 - 27 = 0$

Solution

a. Step 1   First write

$$x^2 - \frac{3}{4}x - \frac{1}{2} = 0$$
Divide the original equation by 4, the coefficient of x^2.

and

$$x^2 - \frac{3}{4}x = \frac{1}{2}$$
Add $\frac{1}{2}$ to both sides so that the constant term is on the right side.

Step 2   Square half of the coefficient of x, obtaining

$$\left(\frac{-\dfrac{3}{4}}{2}\right)^2 = \left(-\frac{3}{8}\right)^2 = \frac{9}{64}$$

Step 3   Add $\dfrac{9}{64}$ to both sides of the equation:

$$x^2 - \frac{3}{4}x + \frac{9}{64} = \frac{1}{2} + \frac{9}{64}$$
$$= \frac{41}{64}$$

Factoring, we have

$$\left(x - \frac{3}{8}\right)^2 = \frac{41}{64}$$
$$x - \frac{3}{8} = \pm\frac{\sqrt{41}}{8}$$
$$x = \frac{3}{8} \pm \frac{\sqrt{41}}{8} = \frac{1}{8}(3 \pm \sqrt{41})$$

b. This equation is much easier to solve because the coefficient of x is 0. As before, we write the equation in the form

$$6x^2 = 27 \quad \text{or} \quad x^2 = \frac{9}{2}$$

Taking the square root of both sides, we have

$$x = \pm\sqrt{\frac{9}{2}} = \pm\frac{3}{\sqrt{2}} = \pm\frac{3\sqrt{2}}{2}$$

Using the Quadratic Formula

By using the method of completing the square on the general quadratic equation

$$ax^2 + bx + c = 0 \qquad (a \neq 0)$$

we obtain the quadratic formula. (For the derivation of this formula, see Project 1, page 58). This formula can be used to solve any quadratic equation.

THE QUADRATIC FORMULA

The solutions of $ax^2 + bx + c = 0$ $(a \neq 0)$ are given by

$$x = \frac{-b \pm \sqrt{b^2 - 4ac}}{2a}$$

EXAMPLE 4   Use the quadratic formula to solve

a. $2x^2 + 5x - 12 = 0$ **b.** $x^2 + 165x + 6624 = 0$ **c.** $x^2 = -3x + 8$

Solution

a. The equation is in standard form, with $a = 2$, $b = 5$, and $c = -12$. Using the quadratic formula, we find

$$x = \frac{-b \pm \sqrt{b^2 - 4ac}}{2a} = \frac{-5 \pm \sqrt{5^2 - 4(2)(-12)}}{2(2)}$$

$$= \frac{-5 \pm \sqrt{121}}{4} = \frac{-5 \pm 11}{4}$$

$$= -4 \text{ or } \frac{3}{2}$$

Observe that this equation can also be solved by factoring. Thus

$$2x^2 + 5x - 12 = (2x - 3)(x + 4) = 0$$

from which we see that the desired roots are $x = 3/2$ or $x = -4$, as obtained earlier.

b. The equation is in standard form, with $a = 1$, $b = 165$, and $c = 6624$. Using the quadratic formula, we find

$$x = \frac{-b \pm \sqrt{b^2 - 4ac}}{2a} = \frac{-165 \pm \sqrt{165^2 - 4(1)(6624)}}{2(1)}$$

$$= \frac{-165 \pm \sqrt{729}}{2}$$

$$= \frac{-165 \pm 27}{2} = -96 \quad \text{or} \quad -69$$

In this case, using the quadratic formula is preferable to factoring the quadratic equation.

c. We first rewrite the given equation in the standard form $x^2 + 3x - 8 = 0$, from which we see that $a = 1$, $b = 3$, and $c = -8$. Using the quadratic formula, we find

$$x = \frac{-b \pm \sqrt{b^2 - 4ac}}{2a} = \frac{-3 \pm \sqrt{3^2 - 4(1)(-8)}}{2(1)}$$

$$= \frac{-3 \pm \sqrt{41}}{2}$$

That is, the solutions are $\dfrac{-3 + \sqrt{41}}{2} \approx 1.7$ or $\dfrac{-3 - \sqrt{41}}{2} \approx -4.7$. In this case, the quadratic formula proves quite handy! ■

EXAMPLE 5 Use the quadratic formula to solve $9x^2 - 12x + 4 = 0$.

Solution Using the quadratic formula with $a = 9$, $b = -12$, and $c = 4$, we find that

$$x = \frac{-b \pm \sqrt{b^2 - 4ac}}{2a} = \frac{-(-12) \pm \sqrt{(-12)^2 - 4(9)(4)}}{2(9)}$$

$$= \frac{12 \pm \sqrt{144 - 144}}{18} = \frac{12}{18} = \frac{2}{3}$$

Here the only solution is $x = 2/3$. We refer to 2/3 as a *double root*. (This equation could also be solved by factoring. Try it!) ■

The quantity $b^2 - 4ac$, which appears under the radical sign in the quadratic formula, is called the **discriminant.** We use the discriminant to determine the number of solutions of a quadratic equation. If the discriminant is positive, the equation has two real and distinct roots (Example 4a–c). If the discriminant is equal to zero, the equation has one double root. Finally, if the discriminant is negative, the equation has no real roots. These results are summarized in Table 13.

TABLE 13
Solutions to a quadratic equation

Discriminant ($b^2 - 4ac$)	Number of Solutions
Positive	Two real and distinct solutions
Equal to zero	One real solution
Negative	No real solution

EXAMPLE 6 Use the discriminant to determine the number of real solutions of each equation: **a.** $x^2 - 7x + 4 = 0$ **b.** $2x^2 - 3x + 4 = 0$

Solution

a. Here $a = 1$, $b = -7$, and $c = 4$. Therefore,

$$b^2 - 4ac = (-7)^2 - 4(1)(4) = 49 - 16$$

$$= 33$$

and we conclude that the equation has two real and distinct solutions.

b. Here $a = 2$, $b = -3$, and $c = 4$. Therefore,

$$b^2 - 4ac = (-3)^2 - 4(2)(4) = 9 - 32$$
$$= -23$$

Since the discriminant is negative, we conclude that the equation has no real solution.

SELF-CHECK EXERCISES 1.7

1. Solve by factoring: **a.** $x^2 - 5x + 6 = 0$ **b.** $4t^2 - 4t = 3$

2. Solve the equation $2x^2 - 4x - 8 = 0$

    a. by completing the square   **b.** by using the quadratic formula

Solutions to Self-Check Exercises 1.7 can be found on page 47.

1.7 EXERCISES

In exercises 1–16, solve the equation by factoring, if required.

1. $(x + 3)(x - 2) = 0$ **2.** $(y - 3)(y - 4) = 0$

3. $x^2 - 4 = 0$ **4.** $2m^2 - 32 = 0$

5. $x^2 + x - 12 = 0$ **6.** $3x^2 - x - 4 = 0$

7. $4t^2 + 2t - 2 = 0$ **8.** $-6x^2 + x + 12 = 0$

9. $\frac{1}{4}x^2 - x + 1 = 0$ **10.** $\frac{1}{2}a^2 + a - 12 = 0$

11. $2m^2 + m = 6$ **12.** $6x^2 = -5x + 6$

13. $4x^2 - 9 = 0$ **14.** $8m^2 + 64m = 0$

15. $z(2z + 1) = 6$ **16.** $13m = -5 - 6m^2$

In exercises 17–26, solve the equation by completing the square.

17. $x^2 + 2x - 8 = 0$ **18.** $x^2 - x - 6 = 0$

19. $6x^2 - 12x = 3$ **20.** $2x^2 - 6x = 20$

21. $m^2 - 3 = -m$ **22.** $p^2 - 4 = -2p$

23. $3x - 4 = -2x^2$ **24.** $10x - 5 = 4x^2$

25. $4x^2 - 13 = 0$ **26.** $7p^2 - 20 = 0$

In exercises 27–36, solve the equation by using the quadratic formula.

27. $2x^2 - x - 6 = 0$ **28.** $6x^2 - 7x - 3 = 0$

29. $m^2 = 4m - 1$ **30.** $2x^2 = 8x - 3$

31. $8x + 3 = 8x^2$ **32.** $6p - 6 = p^2$

33. $4x = -2x^2 + 3$ **34.** $15 - 2y^2 = 7y$

35. $2.1x^2 - 4.7x - 6.2 = 0$

36. $0.2m^2 + 1.6m + 1.2 = 0$

In exercises 37–44, solve the equation.

37. $x^4 - 5x^2 + 6 = 0$ [*Hint:* Let $m = x^2$. Then solve the quadratic equation in m.]

38. $m^4 - 13m^2 + 36 = 0$ [*Hint:* Let $x = m^2$. Then solve the quadratic equation in x.]

39. $y^4 - 5y^2 + 6 = 0$ [*Hint:* Let $x = y^2$.]

40. $4x^4 - 21x^2 + 5 = 0$ [*Hint:* Let $y = x^2$.]

41. $6(x + 2)^2 + 7(x + 2) - 3 = 0$ [*Hint:* Let $y = x + 2$.]

42. $8(2m + 3)^2 + 14(2m + 3) - 15 = 0$ [*Hint:* Let $x = 2m + 3$.]

43. $1 - \frac{5}{2y} - \frac{6}{y^2} = 0$ [*Hint:* Multiply by y^2.]

44. $6 + \frac{1}{k} - \frac{2}{k^2} = 0$ [*Hint:* Multiply by k^2.]

In exercises 45–52, use the discriminant to determine the number of real solutions of the equation.

45. $x^2 - 6x + 5 = 0$ **46.** $2m^2 + 5m + 3 = 0$

47. $3y^2 - 4y + 5 = 0$ **48.** $2p^2 + 5p + 6 = 0$

49. $4x^2 + 12x + 9 = 0$ **50.** $25x^2 - 80x + 64 = 0$

51. $\frac{6}{k^2} + \frac{1}{k} - 2 = 0$

52. $(2p + 1)^2 - 3(2p + 1) + 4 = 0$

**SOLUTIONS TO
SELF-CHECK
EXERCISES 1.7**

1. a. Factoring the given equation, we have

$$x^2 - 5x + 6 = (x - 3)(x - 2)$$

and $x = 3$ or $x = 2$.

b. Rewriting the given equation, we have

$$4t^2 - 4t - 3 = 0 \qquad \text{Add } -3 \text{ to both sides of the equation.}$$

Factoring this equation gives

$$(2t - 3)(2t + 1) = 0$$

and $t = 3/2$ or $t = -1/2$.

2. a. Step 1   First write

$$x^2 - 2x - 4 = 0 \qquad \text{Divide the original equation by 2,} \\ \text{the coefficient of } x^2.$$

and

$$x^2 - 2x = 4 \qquad \text{Add 4 to both sides so that the} \\ \text{constant term is on the right side.}$$

Step 2   Square half of the coefficient of x, obtaining

$$\left(\frac{-2}{2}\right)^2 = 1$$

Step 3   Add 1 to both sides of the equation:

$$x^2 - 2x + 1 = 5$$

Factoring, we have

$$(x - 1)^2 = 5$$

$$x - 1 = \pm\sqrt{5}$$

and

$$x = 1 \pm \sqrt{5}$$

b. Using the quadratic formula, with $a = 2$, $b = -4$, and $c = -8$, we obtain

$$x = \frac{-b \pm \sqrt{b^2 - 4ac}}{2a} = \frac{-(-4) \pm \sqrt{(-4)^2 - 4(2)(-8)}}{2(2)}$$

$$= \frac{4 \pm \sqrt{80}}{4} = \frac{4 \pm 4\sqrt{5}}{4}$$

$$= 1 \pm \sqrt{5}$$

1.8   Inequalities and Absolute Value

Intervals

We described the system of real numbers and its properties in Section 1.1. Often we will restrict our attention to certain subsets of the set of real numbers. For example, if x denotes the number of cars rolling off an assembly line each day in an automobile assembly plant, then x must be nonnegative; that is, $x \geq 0$. Taking this example one step further, suppose management decides that the daily production must not exceed 200 cars. Then x must satisfy the inequality $0 \leq x \leq 200$.

More generally, we will be interested in certain subsets of real numbers called *finite intervals* and *infinite intervals*. **Finite intervals** are open, closed, or half-open. The set of all real numbers that lie *strictly* between two fixed numbers a and b is called an **open interval** (a, b). It consists of all real numbers x that satisfy the inequalities $a < x < b$; it is called "open" because neither of its endpoints is included in the interval. A **closed interval** contains both of its endpoints. Thus, the set of all real numbers x that satisfy the inequalities $a \leq x \leq b$ is the closed interval $[a, b]$. Notice that square brackets are used to indicate that the endpoints are included in this interval. **Half-open intervals** (also called *half-closed intervals*) contain only *one* of their endpoints. The interval $[a, b)$ is the set of all real numbers x that satisfy $a \leq x < b$, whereas the interval $(a, b]$ is described by the inequalities $a < x \leq b$. Examples of these finite intervals are illustrated in Table 14.

TABLE 14
Finite intervals

Interval		Graph	Example	
Open	(a, b)		$(-2, 1)$	
Closed	$[a, b]$		$[-1, 2]$	
Half-open	$(a, b]$		$(\frac{1}{2}, 3]$	
Half-open	$[a, b)$		$[-\frac{1}{2}, 3)$	

Infinite intervals include the half-lines (a, ∞), $[a, \infty)$, $(-\infty, a)$, and $(-\infty, a]$, defined by the set of all real numbers that satisfy $x > a$, $x \geq a$, $x < a$, and $x \leq a$, respectively. The symbol ∞, called *infinity*, is not a real number. It is used here only for notational purposes in conjunction with the definition of infinite intervals. The notation $(-\infty, \infty)$ is used for the set of real numbers x, since, by definition, the inequalities $-\infty < x < \infty$ hold for any real number x. These infinite intervals are illustrated in Table 15.

TABLE 15
Infinite intervals

Interval	Graph	Example	
(a, ∞)		$(2, \infty)$	
$[a, \infty)$		$[-1, \infty)$	
$(-\infty, a)$		$(-\infty, 1)$	
$(-\infty, a]$		$(-\infty, -\frac{1}{2}]$	

Inequalities

In practical applications, intervals are often found by solving one or more inequalities involving a variable. To solve these inequalities, we use the properties listed in Table 16.

TABLE 16
Properties of inequalities

Property	Illustration
If a, b, and c are any real numbers, then	
1. If $a < b$ and $b < c$, then $a < c$.	$2 < 3$ and $3 < 8$, so $2 < 8$.
2. If $a < b$, then $a + c < b + c$.	$-5 < -3$ so $-5 + 2 < -3 + 2$; that is, $-3 < -1$.
3. If $a < b$ and $c > 0$, then $ac < bc$.	$-5 < -3$ and $2 > 0$, so $(-5)(2) < (-3)(2)$; that is, $-10 < -6$.
4. If $a < b$ and $c < 0$, then $ac > bc$.	$-2 < 4$ and $-3 < 0$, so $(-2)(-3) > (4)(-3)$; that is, $6 > -12$.

Similar properties hold if each inequality sign, $<$, between a and b is replaced by $\geq$, $>$, or $\leq$.

A real number is a *solution of an inequality* involving a variable if a true statement is obtained when the variable is replaced by that number. The set of all real numbers satisfying the inequality is called the *solution set*.

EXAMPLE 1 Solve $3x - 2 < 7$.

Solution Add 2 to each side of the inequality, obtaining

$$3x - 2 + 2 < 7 + 2$$

$$3x < 9$$

Next, multiply each side of the inequality by 1/3, obtaining

$$\frac{1}{3}(3x) < \frac{1}{3}(9)$$

$$x < 3$$

The solution is the set of all values of x in the interval $(-\infty, 3)$. ■

EXAMPLE 2 Solve $-1 \leq 2x - 5 < 7$ and graph the solution set.

Solution Add 5 to each member of the double inequality, obtaining

$$4 \leq 2x < 12$$

Next, multiply each member of the resulting double inequality by 1/2, yielding

$$2 \leq x < 6$$

FIGURE 3
$-1 \leq 2x - 5 < 7$

Thus, the solution is the set of all values of x lying in the interval $[2, 6)$. The graph of the solution set is shown in Figure 3. ■

Solving Inequalities by Factoring

One method of solving inequalities that involve polynomials of degree 2 or higher is the method of factoring. This method relies on the principle that a polynomial changes sign only at a point where its value is zero. To find the values of x where the polynomial is equal to zero, we set the polynomial equal to zero and then solve for x. The values obtained can then be used to help us solve the given inequality. In Example 3 detailed steps are provided for this technique.

EXAMPLE 3   Solve $x^2 - 5x + 6 > 0$.

Solution

1. *Set the polynomial in the given inequality equal to zero:*
$$x^2 - 5x + 6 = 0$$

2. *Factor the polynomial:*
$$(x - 3)(x - 2) = 0$$

3. *Construct a sign diagram for the factors of the polynomial.* We use a + to indicate that a factor is positive for a given value of x, a − to indicate that it is negative, and a 0 to indicate that it is equal to 0. Now, $x - 3 < 0$ if $x < 3$, $x - 3 > 0$ if $x > 3$, and $x - 3 = 0$ if $x = 3$. Similarly, $x - 2 < 0$ if $x < 2$, $x - 2 > 0$ if $x > 2$, and $x - 2 = 0$ if $x = 2$. Using this information we construct the sign diagram shown in Figure 4.

FIGURE 4

Sign of
$(x - 3)$ – – – – – – – – – – – 0 + + + + +
$(x - 2)$ – – – – – – – – – 0 + + + + + + +

4. *Determine the intervals that satisfy the given inequality.* Since $x^2 - 5x + 6 > 0$, we require that the product of the two factors be positive—that is, that both factors have the same sign. From the sign diagram, we see that the two factors have the same sign when $x < 2$ or $x > 3$. Thus, the solution set is $(-\infty, 2) \cup (3, \infty)$. ∎

EXAMPLE 4   Solve $x^2 + 2x - 8 < 0$.

Solution

Step 1 $x^2 + 2x - 8 = 0$

Step 2 $(x + 4)(x - 2) = 0$, so
$$x = -4 \quad \text{or} \quad x = 2$$

Step 3 $x + 4 > 0$ when $x > -4$, $x + 4 < 0$ when $x < -4$, and $x + 4 = 0$ when $x = -4$. Similarly, $x - 2 > 0$ when $x > 2$, $x - 2 < 0$ when $x < 2$, and $x - 2 = 0$ when $x = 2$. Using these results, we construct the sign diagram for the factors of $x^2 + 2x - 8$ (Figure 5).

Step 4   Since $x^2 + 2x - 8 < 0$, the product of the two factors must be negative; that is, the signs of the two factors must differ. From the

FIGURE 5

Sign of
$(x + 4)$ $- - 0 + + + + + + + + + + + + + + + + +$
$(x - 2)$ $- - - - - - - - - - - - - - 0 + + + + + +$

sign diagram, we see that the two factors $x + 4$ and $x - 2$ have opposite signs when x lies strictly between -4 and 2. Therefore, the required solution is the interval $(-4, 2)$. ■

Solving Inequalities Involving a Quotient

The next two examples show how an inequality involving the quotient of two algebraic expressions is solved.

EXAMPLE 5   Solve $\dfrac{x + 1}{x - 1} \geq 0$.

Solution   The quotient $(x + 1)/(x - 1)$ is positive (greater than zero) when the numerator and denominator have the *same* sign. The signs of $x + 1$ and $x - 1$ are shown in Figure 6.

FIGURE 6

Sign of
$(x + 1)$ $- - - - - - - 0 + + + + + + + + + + +$
$(x - 1)$ $- - - - - - - - - - 0 + + + + + +$

From the sign diagram, we see that $x + 1$ and $x - 1$ have the same sign when $x < -1$ or $x > 1$. The quotient $(x + 1)/(x - 1)$ is equal to zero when $x = -1$. It is undefined at $x = 1$ since the denominator is zero at that point. Therefore, the required solution is the set of all x in the intervals $(-\infty, -1]$ and $(1, \infty)$. ■

EXAMPLE 6   Solve $\dfrac{2x - 1}{x - 2} \geq 1$.

Solution   We rewrite the given inequality so that the right side is equal to zero:

$$\frac{2x - 1}{x - 2} - 1 \geq 0$$

$$\frac{2x - 1 - (x - 2)}{x - 2} \geq 0$$

$$\frac{2x - 1 - x + 2}{x - 2} \geq 0$$

$$\frac{x + 1}{x - 2} \geq 0.$$

Next we construct the sign diagram for the factors in the numerator and the denominator (Figure 7).

FIGURE 7

Sign of

(x + 1)   – – – – 0 + + + + + + + + + +

(x – 2)   – – – – – – – – – – 0 + + + +

$$\xrightarrow{\hspace{4cm}} x$$

    –1   0   2

Since the quotient of these two factors must be positive or equal to zero, we require that the sign of each factor be the same or that the quotient of the two factors be equal to zero. From the sign diagram, we see that the solution set is given by $(-\infty, -1]$ and $(2, \infty)$. Note that $x = 2$ is not included in the second interval since division by zero is not allowed. ■

Application

EXAMPLE 7   The gross domestic product (GDP) of a certain country is projected to be $t^2 + 2t + 50$ billion dollars t years from now. Find the time when the GDP of the country will first exceed \$58 billion.

Solution   The GDP of the country will exceed \$58 billion when

$$t^2 + 2t + 50 \geq 58$$

In order to solve this inequality for t, we first write it in the form

$$t^2 + 2t - 8 \geq 0$$

or

$$(t + 4)(t - 2) \geq 0$$

The sign diagram for the factors of $t^2 + 2t - 8$ is shown in Figure 8.

FIGURE 8

Sign of

(t + 4)   – – – 0 + + + + + + + + + + + +

(t – 2)   – – – – – – – – – – 0 + + + + +

$$\xrightarrow{\hspace{4cm}} t$$

    –4   0   2

From the sign diagram, we see that the required solution set is $(-\infty, -4] \cup [2, \infty)$. Since t must be nonnegative for the problem to be meaningful, we see that the GDP of the country is greater than or equal to \$58 billion when $t \geq 2$; that is, the GDP will first exceed \$58 billion when $t = 2$, or two years from now. ■

Absolute Value

ABSOLUTE VALUE

The absolute value of a number a is denoted by $|a|$ and is defined by

$$|a| = \begin{cases} a \text{ if } a \geq 0 \\ -a \text{ if } a < 0 \end{cases}$$

Since $-a$ is a positive number when a is negative, it follows that the absolute value of a number is always nonnegative. For example, $|5| = 5$ and $|-5| = -(-5) = 5$. Geometrically, $|a|$ is the distance between the origin and the point on the number line that represents the number a (Figures 9a and b).

FIGURE 9

(a)   (b)

The absolute value properties are given in Table 17. Property 4 is called the **triangle inequality**.

TABLE 17
Absolute value properties

Property	Illustration
If a and b are any real numbers, then	
1. $\lvert -a \rvert = \lvert a \rvert$	$\lvert -3 \rvert = -(-3) = 3 = \lvert 3 \rvert$
2. $\lvert ab \rvert = \lvert a \rvert \lvert b \rvert$	$\lvert (2)(-3) \rvert = \lvert -6 \rvert = 6$
	$= \lvert 2 \rvert \lvert -3 \rvert$
3. $\left\lvert \dfrac{a}{b} \right\rvert = \dfrac{\lvert a \rvert}{\lvert b \rvert}$	$\left\lvert \dfrac{(-3)}{(-4)} \right\rvert = \left\lvert \dfrac{3}{4} \right\rvert = \dfrac{3}{4} = \dfrac{\lvert -3 \rvert}{\lvert -4 \rvert}$
4. $\lvert a + b \rvert \le \lvert a \rvert + \lvert b \rvert$	$\lvert 8 + (-5) \rvert = \lvert 3 \rvert = 3$
	$\le \lvert 8 \rvert + \lvert -5 \rvert$
	$= 13$

EXAMPLE 8   Evaluate:

a. $\lvert \pi - 5 \rvert + 3$ **b.** $\lvert \sqrt{3} - 2 \rvert + \lvert 2 - \sqrt{3} \rvert$

Solution

a. Since $\pi - 5 < 0$, we see that $\lvert \pi - 5 \rvert = -(\pi - 5)$. Therefore,

$$\lvert \pi - 5 \rvert + 3 = -(\pi - 5) + 3 = -\pi + 5 + 3 = 8 - \pi$$

b. Since $\sqrt{3} - 2 < 0$, we see that $\lvert \sqrt{3} - 2 \rvert = -(\sqrt{3} - 2)$. Next, observe that $2 - \sqrt{3} > 0$, so $\lvert 2 - \sqrt{3} \rvert = 2 - \sqrt{3}$. Therefore,

$$\lvert \sqrt{3} - 2 \rvert + \lvert 2 - \sqrt{3} \rvert = -(\sqrt{3} - 2) + (2 - \sqrt{3}) = -\sqrt{3} + 2 + 2 - \sqrt{3}$$
$$= 4 - 2\sqrt{3} = 2(2 - \sqrt{3}) \qquad \blacksquare$$

EXAMPLE 9   Solve the inequalities $\lvert x \rvert \le 5$ and $\lvert x \rvert \ge 5$.

Solution   First, we consider the inequality $\lvert x \rvert \le 5$. If $x > 0$, then $\lvert x \rvert = x$, so $\lvert x \rvert \le 5$ implies $x \le 5$ in this case. However, if $x < 0$, then $\lvert x \rvert = -x$, so $\lvert x \rvert \le 5$ implies $-x \le 5$, or $x \ge -5$. Thus, $\lvert x \rvert \le 5$ means $-5 \le x \le 5$ (Figure 10a). To obtain an alternative solution observe that $\lvert x \rvert$ is the distance from the point x to 0, so the inequality $\lvert x \rvert \le 5$ implies immediately that $-5 \le x \le 5$.

Next, the inequality $\lvert x \rvert \ge 5$ states that the distance from x to 0 is greater than or equal to 5. This observation yields the result $x \ge 5$ or $x \le -5$ (Figure 10b). $\blacksquare$

FIGURE 10

(a) $\lvert x \rvert \le 5$

(b) $\lvert x \rvert \ge 5$

The results of Example 9 may be generalized. Thus, if $k > 0$, then $|x| \leq k$ is equivalent to $-k \leq x \leq k$ and $|x| \geq k$ is equivalent to $x \geq k$ or $x \leq -k$.

EXAMPLE 10   Solve the inequality $|2x - 3| \leq 1$.

Solution   The inequality $|2x - 3| \leq 1$ is equivalent to the inequalities $-1 \leq 2x - 3 \leq 1$ (see Example 9). Then

$$2 \leq 2x \leq 4 \qquad \text{Add 3 to each member of the inequality.}$$

and

$$1 \leq x \leq 2 \qquad \text{Multiply each member of the inequality by } \tfrac{1}{2}.$$

Therefore, the solution is given by the set of all x in the interval $[1, 2]$ (Figure 11).

FIGURE 11
$|2x - 3| \leq 1$

EXAMPLE 11   Solve the inequality $|5x + 7| \geq 18$.

Solution   Referring to Example 9 once again, we see that $|5x + 7| \geq 18$ is equivalent to

$$5x + 7 \leq -18 \quad \text{or} \quad 5x + 7 \geq 18$$

that is,

$$5x \leq -25 \quad \text{or} \qquad 5x \geq 11$$

$$x \leq -5 \quad \text{or} \qquad x \geq \frac{11}{5}$$

Therefore, the solution is given by the set of all x in the interval $(-\infty, -5]$ or the interval $[11/5, \infty)$ [Figure 12].

FIGURE 12
$|5x + 7| \geq 18$

SELF-CHECK EXERCISES 1.8

1. Solve $-1 < 2x - 1 \leq 5$ and graph the solution set.

2. Solve $6x^2 - 5x - 4 \leq 0$.

Solutions to Self-Check Exercises 1.8 can be found on page 56.

1.8 EXERCISES

In exercises 1–4, determine whether the statement is true or false.

1. $-3 < -20$

2. $-5 \leq -5$

3. $\dfrac{2}{3} > \dfrac{5}{6}$

4. $-\dfrac{5}{6} < -\dfrac{11}{12}$

In exercises 5–10, show the interval on a number line.

5. $(3, 6)$

6. $(-2, 5]$

7. $[-1, 4)$

8. $\left[-\dfrac{6}{5}, -\dfrac{1}{2}\right]$

9. $(0, \infty)$

10. $(-\infty, 5]$

In exercises 11–28, find the values of x that satisfy the inequalities.

11. $2x + 4 < 8$

12. $-6 > 4 + 5x$

13. $-4x \geq 20$

14. $-12 \leq -3x$

15. $-6 < x - 2 < 4$

16. $0 \leq x + 1 \leq 4$

17. $x + 1 > 4 \quad \text{or} \quad x + 2 < -1$

18. $x + 1 > 2 \quad \text{or} \quad x - 1 < -2$

19. $x + 3 > 1 \quad \text{and} \quad x - 2 < 1$

20. $x - 4 \leq 1 \quad \text{and} \quad x + 3 > 2$

21. $(x + 3)(x - 5) \leq 0$

22. $(2x - 4)(x + 2) \geq 0$

23. $(2x - 3)(x - 1) \geq 0$

24. $(3x - 4)(2x + 2) \leq 0$

25. $\dfrac{x + 3}{x - 2} \geq 0$

26. $\dfrac{2x - 3}{x + 1} \geq 4$

27. $\dfrac{x - 2}{x - 1} \leq 2$

28. $\dfrac{2x - 1}{x + 2} \leq 4$

In exercises 29–38, evaluate the expression.

29. $|-6 + 2|$

30. $4 + |-4|$

31. $\dfrac{|-12 + 4|}{|16 - 12|}$

32. $\left|\dfrac{0.2 - 1.4}{1.6 - 2.4}\right|$

33. $\sqrt{3}|-2| + 3|-\sqrt{3}|$

34. $|-1| + \sqrt{2}|-2|$

35. $|\pi - 1| + 2$

36. $|\pi - 6| - 3$

37. $|\sqrt{2} - 1| + |3 - \sqrt{2}|$

38. $|2\sqrt{3} - 3| - |\sqrt{3} - 4|$

In exercises 39–44, suppose that a and b are real numbers other than zero and that $a > b$. State whether the inequality is true or false.

39. $b - a > 0$

40. $\dfrac{a}{b} > 1$

41. $a^2 > b^2$

42. $\dfrac{1}{a} > \dfrac{1}{b}$

43. $a^3 > b^3$

44. $-a < -b$

45. Write the inequality $|x - a| < b$ without using absolute values.

46. Write the inequality $|x - a| \geq b$ without using absolute values.

In exercises 47–52, determine whether the statement is true for all real numbers a and b.

47. $|-a| = a$

48. $|b^2| = b^2$

49. $|a - 4| = |4 - a|$

50. $|a + 1| = |a| + 1$

51. $|a + b| = |a| + |b|$

52. $|a - b| = |a| - |b|$

53. Find the minimum cost C (in dollars) given that
$$5(C - 25) \geq 1.75 + 2.5C$$

54. Find the maximum profit P (in dollars) given that
$$6(P - 2500) \leq 4(P + 2400)$$

55. Driving Range of a Car An advertisement for a certain car states that the EPA fuel economy is 20 mpg city and 27 mpg highway and that the car's fuel-tank capacity is 18.1 gallons. Assuming ideal driving conditions, determine the driving range for the car from the foregoing data.

56. Celsius and Fahrenheit Temperatures The relationship between Celsius (°C) and Fahrenheit (°F) temperatures is given by the formula
$$C = \frac{5}{9}(F - 32)$$

a. If the temperature range for Montreal during the month of January is $-15° < C° < -5°$, find the range in degrees Fahrenheit in Montreal for the same period.

b. If the temperature range for New York City during the month of June is $63° < F° < 80°$, find the range in degrees Celsius in New York City for the same period.

57. Meeting Sales Targets A salesman's monthly commission is 15% on all sales over $12,000. If his goal is to make a commission of at least $1,000 per month, what minimum monthly sales figures must he attain?

58. Markup on a Car The markup on a used car was at least 30% of its current wholesale price. If the car was sold for $2800, what was the maximum wholesale price?

59. Meeting Profit Goals A manufacturer of a certain commodity has estimated that her profit in thousands of dollars is given by the expression
$$-6x^2 + 30x - 10$$
where x (in thousands) is the number of units produced. What production range will enable the manufacturer to realize a profit of at least $14,000 on the commodity?

60. Effect of Bactericide The number of bacteria in a certain culture t minutes after an experimental bactericide is introduced is given by
$$\frac{10,000}{t^2 + 1} + 2000$$
Find the time when the number of bacteria will have dropped below 4000.

61. A ball is thrown straight up so that its height after t seconds is
$$128t - 16t^2 + 4$$
feet. Determine the length of time the ball stays above a height of 196 ft.

62. Distribution of Income The distribution of income in a certain city can be described by the exponential model $y = (2.8 \cdot 10^{11})(x)^{-1.5}$, where y is the number of families with an income of x or more dollars.

a. How many families in this city have an income of $20,000 or more?
b. How many families have an income of $40,000 or more?
c. How many families have an income of $100,000 or more?

63. **Quality Control** The P.A.R. Manufacturing Company manufactures steel rods. Suppose the rods ordered by a customer are manufactured to a specification of 0.5 inch and are acceptable only if they are within the *tolerance limits* of 0.49 inch and 0.51 inch.

Letting x denote the diameter of a rod, write an inequality using absolute value to express a criterion involving x that must be satisfied in order for a rod to be acceptable.

64. **Quality Control** The diameter x (in inches) of a batch of ball bearings manufactured by the P.A.R. Manufacturing Company satisfies the inequality

$$|x - 0.1| \le 0.01$$

What is the smallest diameter a ball bearing in the batch can have? The largest diameter?

SOLUTIONS TO SELF-CHECK EXERCISES 1.8

1. $-1 < 2x - 1 \le 5$

   $-1 + 1 < 2x - 1 + 1 \le 5 + 1$ Add 1 to each member of the inequality.

   $0 < 2x \le 6$ Combine like terms.

   $0 < x \le 3$ Multiply each member of the inequality by $\frac{1}{2}$.

   We conclude that the solution set is (0, 3]. The graph of the solution set is shown in the following figure:

2. **Step 1** $6x^2 - 5x - 4 \le 0$

   Step 2 $(3x - 4)(2x + 1) \le 0$

   Step 3 $3x - 4 > 0$ when $x > 4/3$, $3x - 4 = 0$ when $x = 4/3$, and $3x - 4 < 0$ when $x < 4/3$. Similarly, $2x + 1 > 0$ when $x > -1/2$, $2x + 1 = 0$ when $x = -1/2$, and $2x + 1 < 0$ when $x < -1/2$. Using these results, we construct the following sign diagram for the factors of $6x^2 - 5x - 4$:

   Sign of
   $(3x-4)$ $- - - - - - - - - - 0 + + + +$
   $(2x+1)$ $- - - - 0 + + + + + + + + + +$

   $$-\frac{1}{2} \quad 0 \quad \frac{4}{3}$$

   Step 4 Since $6x^2 - 5x - 4 \le 0$, the signs of the two factors must differ or be equal to zero. From the sign diagram, we see that x must lie between $-1/2$ and $4/3$, inclusive. Therefore, the required solution is $\left[-\frac{1}{2}, \frac{4}{3}\right]$.

CHAPTER ONE SUMMARY OF PRINCIPAL FORMULAS AND TERMS

Formulas

1. Product formula

   $$(a + b)^2 = a^2 + 2ab + b^2$$
   $$(a - b)^2 = a^2 - 2ab + b^2$$
   $$(a + b)(a - b) = a^2 - b^2$$

2. Quadratic formula

   $$x = \frac{-b \pm \sqrt{b^2 - 4ac}}{2a}$$

Terms

Natural number	Radical
Whole number	Radical sign
Integer	Radicand
Rational number	Index
Irrational number	Conjugate
Real number	Quadratic equation
Exponent	Discriminant
Base	Finite interval
Polynomial	Open interval
Rational expression	Closed interval
Compound fraction	Half-open interval
nth root	Infinite interval
Square root	Absolute value
Cube root	Triangle inequality

CHAPTER 1 REVIEW EXERCISES

In exercises 1–6, classify the number as to type.

1. $\dfrac{7}{8}$ **2.** $\sqrt{13}$ **3.** -2π

4. 0 **5.** $2.\overline{71}$ **6.** $3.14159\ldots$

In exercises 7–12, evaluate the expression.

7. $\left(\dfrac{9}{4}\right)^{3/2}$ **8.** $\dfrac{5^6}{5^4}$

9. $(3\cdot4)^{-2}$ **10.** $(-8)^{5/3}$

11. $\dfrac{(3\cdot2^{-3})(4\cdot3^5)}{2\cdot9^3}$ **12.** $\dfrac{3\sqrt[3]{54}}{\sqrt[3]{18}}$

In exercises 13–20, simplify the expression.

13. $\dfrac{4(x^2+y)^3}{x^2+y}$ **14.** $\dfrac{a^6b^{-5}}{(a^3b^{-2})^{-3}}$

15. $\dfrac{\sqrt[4]{16x^5yz}}{\sqrt[4]{81xyz^5}}$ **16.** $(2x^3)(-3x^{-2})\left(\dfrac{1}{6}x^{-1/2}\right)$

17. $\left(\dfrac{3xy^2}{4x^3y}\right)^{-2}\left(\dfrac{3xy^3}{2x^2}\right)^3$ **18.** $(-3a^2b^3)^2(2a^{-1}b^{-2})^{-1}$

19. $\sqrt[4]{81x^5y^{10}}\sqrt[4]{9xy^2}$ **20.** $\left(\dfrac{-x^{1/2}y^{2/3}}{x^{1/3}y^{3/4}}\right)^6$

In exercises 21–26, perform the indicated operations and simplify the expression.

21. $(2x+3y)^2-(3x+1)(2x-3)$

22. $2(3a+b)-3[(2a+3b)-(a+2b)]$

23. $\dfrac{(t+6)(60)-(60t+180)}{(t+6)^2}$

24. $\dfrac{6x}{2(3x^2+2)}+\dfrac{1}{4(x+2)}$

25. $\dfrac{2}{3}\left(\dfrac{4x}{2x^2-1}\right)+3\left(\dfrac{3}{3x-1}\right)$

26. $\dfrac{-2x}{\sqrt{x+1}}+4\sqrt{x+1}$

In exercises 27–34, factor the expression.

27. $-2\pi^2r^3+100\pi r^2$ **28.** $2v^3w+2vw^3+2u^2vw$

29. $16-x^2$ **30.** $12t^3-6t^2-18t$

31. $-2x^2-4x+6$ **32.** $12x^2-92x+120$

33. $9a^2-25b^2$ **34.** $8u^6v^3+27u^3$

In exercises 35–42, solve the equation.

35. $8x^2 + 2x - 3 = 0$ **36.** $-6x^2 - 10x + 4 = 0$

37. $2x^2 - 3x - 4 = 0$ **38.** $x^2 + 5x + 3 = 0$

39. $2y^2 - 3y + 1 = 0$

40. $0.3m^2 - 2.1m - 3.2 = 0$

41. $-x^3 - 2x^2 + 3x = 0$ **42.** $2x^4 + x^2 = 1$

In exercises 43–46, find the values of x that satisfy the inequalities.

43. $-x + 3 \le 2x + 9$

44. $-2 \le 3x + 1 \le 7$

45. $x - 3 > 2$ or $x + 3 < -1$

46. $2x^2 > 50$

In exercises 47–50, evaluate the expression.

47. $|-5 + 7| + |-2|$ **48.** $\left| \dfrac{5 - 12}{-4 - 3} \right|$

49. $|2\pi - 6| - \pi$ **50.** $|\sqrt{3} - 4| + |4 - 2\sqrt{3}|$

In exercises 51–54, find the value(s) of x that satisfy the expression.

51. $2x^2 + 3x - 2 \le 0$ **52.** $\dfrac{1}{x + 2} > 2$

53. $|2x - 3| < 5$ **54.** $\left| \dfrac{x + 1}{x - 1} \right| = 5$

55. Rationalize the numerator.
$$\frac{\sqrt{x} - 1}{x - 1}$$

56. Rationalize the denominator.
$$\frac{\sqrt{x} - 1}{2\sqrt{x}}$$

In exercises 57 and 58, use the quadratic formula to solve the quadratic equation.

57. $x^2 - 2x - 5 = 0$ **58.** $2x^2 + 8x + 7 = 0$

59. Find the minimum cost C (in dollars) given that
$$2(1.5C + 80) \le 2(2.5C - 20)$$

60. Find the maximum revenue R (in dollars) given that
$$12(2R - 320) \le 4(3R + 240)$$

CHAPTER 1   EXTENDED PROBLEMS AND PROJECTS

1. Derive the quadratic formula for solving the equation $ax^2 + bx + c = 0$ ($a \ne 0$) using the following steps:
   a. Complete the square in x to show that the equation is equivalent to
$$a\left(x + \frac{b}{2a} \right)^2 + \frac{4ac - b^2}{4a} = 0$$
   b. Solve the equation in part a for x to obtain
$$x = \frac{-b \pm \sqrt{b^2 - 4ac}}{2a}$$

2. Let $T = \dfrac{1}{\sqrt{2x^2 - 4x + 3}}$.
   a. Complete the square in x to show that
$$2x^2 - 4x + 3 = 2(x - 1)^2 + 1$$
   b. Use the result from part a to show that $\sqrt{2x^2 - 4x - 3}$ attains its smallest value if $x = 1$.
   c. Use the result from part b to find the largest value of T.

2

Functions and
Their Graphs

*T*his chapter introduces the Cartesian coordinate system, a system that allows us to represent points in the plane in terms of ordered pairs of real numbers. This, in turn, enables us to study geometry using algebraic methods. We will also study straight lines. *Linear functions,* whose graphs are straight lines, can be used to describe many relationships between two quantities. These relationships can be found in fields of study as diverse as business, economics, the social sciences, physics, and medicine. We will also see how some practical problems can be solved by finding the point(s) of intersection of two straight lines.

2.1   The Cartesian Coordinate System and Straight Lines

The Cartesian Coordinate System

In Section 1.1 we saw how a one-to-one correspondence between the set of real numbers and the points on a straight line leads to a coordinate system on a line (a one-dimensional space).

A similar representation for points in a plane (a two-dimensional space) is realized through the **Cartesian (rectangular) coordinate system,** which is constructed as follows: Take two perpendicular lines, one of which is usually chosen to be horizontal. These lines intersect at a point O, called the **origin** (Figure 1). The horizontal line is called the *x-axis,* and the vertical line is called the *y-axis.* A number scale is set up along the *x*-axis, with the positive numbers lying to the right of the origin and the negative numbers lying to the left of it. Similarly, a number scale is set up along the *y*-axis, with the positive numbers lying above the origin and the negative numbers lying below it.

The number scales on the two axes need not be the same. Indeed, in many applications different quantities are represented by *x* and *y*. For example, *x* may represent the number of typewriters sold and *y* the total revenue resulting from the sales. In such cases it is often desirable to choose different number scales to represent the different quantities. Note, however, that the zeros of both number scales coincide at the origin of the two-dimensional coordinate system.

A point in the plane can now be represented uniquely in this coordinate system by an **ordered pair** of numbers—that is, a pair (*x*, *y*) where *x* is the first number and *y* the second. To see this, let *P* be any point in the plane (Figure 2). Draw perpendiculars from *P* to the *x*-axis and *y*-axis, respectively. Then the number *x* is precisely the number that corresponds to the point on the *x*-axis at which the perpendicular through *P* hits the *x*-axis. Similarly, *y* is the number that corresponds to the point on the *y*-axis at which the perpendicular through *P* crosses the *y*-axis.

Conversely, given an ordered pair (*x*, *y*) with *x* as the first number and *y* the second, a point *P* in the plane is uniquely determined as follows: Locate the point on the *x*-axis represented by the number *x* and draw a line through that point parallel to the *y*-axis. Next, locate the point on the *y*-axis represented by the number *y* and draw a line through that point parallel to the *x*-axis. The point of intersection of these two lines is the point *P* (Figure 2).

In the ordered pair (*x*, *y*), *x* is called the *abscissa,* or *x-coordinate,* *y* is called the *ordinate,* or *y-coordinate,* and *x* and *y* together are referred to as the **coordinates** of the point *P*.

Letting (*a*, *b*) denote the point *P* with *x*-coordinate *a* and *y*-coordinate *b*, the points $A = (2, 3)$, $B = (-2, 3)$, $C = (-2, -3)$, $D = (2, -3)$, $E = (3, 2)$, $F = (9/2, 0)$, and $G = (0, -5)$ are plotted in Figure 3. The fact that, in general, $(x, y) \neq (y, x)$ is clearly illustrated by points *A* and *E*.

The axes divide the plane into four quadrants. Quadrant I consists of the points (*x*, *y*) that satisfy $x > 0$ and $y > 0$; Quadrant II, the points

FIGURE 1

The Cartesian coordinate system

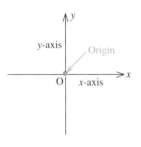

FIGURE 2

An ordered pair (x, y)

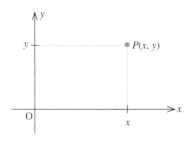

FIGURE 3

Several points in the Cartesian plane

FIGURE 3

Several points in the Cartesian plane

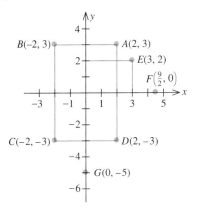

FIGURE 4

The four quadrants in the Cartesian plane

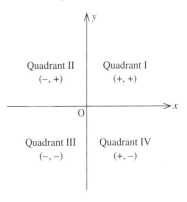

FIGURE 5

Linear depreciation of an asset

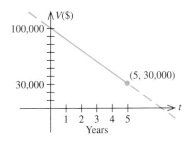

FIGURE 6

The slope is undefined.

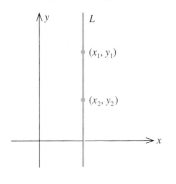

(x, y) where $x < 0$ and $y > 0$; Quadrant III, the points (x, y) where $x < 0$ and $y < 0$; and Quadrant IV, the points (x, y) where $x > 0$ and $y < 0$ (Figure 4).

Straight Lines

In computing income tax, business firms are allowed by law to depreciate certain assets such as buildings, machines, furniture, and automobiles over a period of time. Linear depreciation, or the straight-line method, is often used for this purpose. The graph of the straight line shown in Figure 5 describes the book value V of a computer that has an initial value of $100,000 and that is being depreciated linearly over five years with a scrap value of $30,000. Note that only the solid portion of the straight line is of interest here.

The book value of the computer at the end of year t, where t lies between 0 and 5, can be read directly from the graph. But there is one shortcoming in this approach: The result depends on how accurately you draw and read the graph. A better and more accurate method is based on finding an *algebraic* representation of the depreciation line.

In order to see how a straight line in the xy-plane may be described algebraically, we need to first recall certain properties of straight lines.

Slope of a Line

Let L denote the unique straight line that passes through the two distinct points (x_1, y_1) and (x_2, y_2). If $x_1 = x_2$, then L is a vertical line and the slope is undefined (Figure 6).

If $x_1 \neq x_2$, we define the slope of L as follows:

**SLOPE OF A
NONVERTICAL LINE**

If (x_1, y_1) and (x_2, y_2) are any two distinct points on a nonvertical line L (Figure 7), then the slope m of L is given by

$$m = \frac{\Delta y}{\Delta x} = \frac{y_2 - y_1}{x_2 - x_1} \qquad (1)$$

FIGURE 7

FIGURE 8

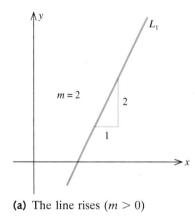

(a) The line rises ($m > 0$)   **(b)** The line falls ($m < 0$)

FIGURE 9

A family of straight lines

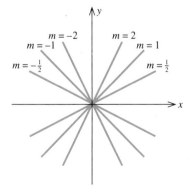

The slope of a straight line is a constant, whenever it is defined. The number $\Delta y = y_2 - y_1$ (Δy is read "delta y") is a measure of the vertical change in y, and $\Delta x = x_2 - x_1$ is a measure of the horizontal change in x as shown in Figure 7. From this figure we can see that the slope m of a straight line L is a measure of the *rate of change of y with respect to x.*

Figure 8a shows a straight line L_1 with slope 2. Observe that L_1 has the property that a unit increase in x results in a 2-unit increase in y. To see this, let $\Delta x = 1$ in (1) so that $m = \Delta y$. Since $m = 2$, we conclude that $\Delta y = 2$. Similarly, Figure 8b shows a line L_2 with slope -1. Observe that a straight line with positive slope slants upward from left to right (y increases as x increases), whereas a line with negative slope slants downward from left to right (y decreases as x increases). Finally, Figure 9 shows a family of straight lines passing through the origin with indicated slopes.

EXAMPLE 1   Sketch the straight line that passes through the point $(-2, 5)$ and has slope $-4/3$.

Solution   First plot the point $(-2, 5)$ (Figure 10). Next, recall that a slope of $-4/3$ indicates that an increase of 1 unit in the x-direction produces a *decrease* of $4/3$ units in the y-direction, or equivalently, a 3-unit increase in the x-direction produces a $3(4/3)$, or 4-unit, decrease in the y-direction. Using this information, we plot the point $(1, 1)$ and draw the line through the two points.

FIGURE 10
L has slope $-4/3$ and passes through $(-2, 5)$.

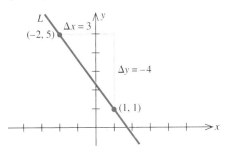

EXAMPLE 2   Find the slope m of the line that passes through the points $(-1, 1)$ and $(5, 3)$.

Solution   Choose (x_1, y_1) to be the point $(-1, 1)$ and (x_2, y_2) to be the point $(5, 3)$. Then, with $x_1 = -1$, $y_1 = 1$, $x_2 = 5$, and $y_2 = 3$, we find

$$m = \frac{y_2 - y_1}{x_2 - x_1} = \frac{3 - 1}{5 - (-1)} = \frac{1}{3} \qquad \text{Using (1)}$$

(Figure 11). Verify that the result obtained would have been the same had we chosen the point $(-1, 1)$ to be (x_2, y_2) and the point $(5, 3)$ to be (x_1, y_1).

FIGURE 11
L passes through $(5, 3)$ *and* $(-1, 1)$.

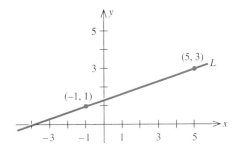

EXAMPLE 3   Find the slope of the line that passes through the points $(-2, 5)$ and $(3, 5)$.

FIGURE 12
The slope of the horizontal line L is 0.

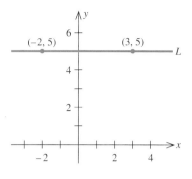

Solution The slope of the required line is given by

$$m = \frac{5 - 5}{3 - (-2)} = \frac{0}{5} = 0$$

(Figure 12).   ■

REMARK   In general, the slope of a horizontal line is zero.   □

We can use the slope of a straight line to determine whether a line is parallel to another line.

PARALLEL LINES

Two distinct lines are **parallel** if and only if their slopes are equal or their slopes are undefined.

EXAMPLE 4   Let L_1 be a line that passes through the points $(-2, 9)$ and $(1, 3)$, and let L_2 be the line that passes through the points $(-4, 10)$ and $(3, -4)$. Determine whether L_1 and L_2 are parallel.

Solution The slope m_1 of L_1 is given by

$$m_1 = \frac{3 - 9}{1 - (-2)} = -2$$

The slope m_2 of L_2 is given by

$$m_2 = \frac{-4 - 10}{3 - (-4)} = -2$$

Since $m_1 = m_2$, the lines L_1 and L_2 are in fact parallel (Figure 13).

FIGURE 13
L_1 and L_2 have the same slope and hence are parallel.

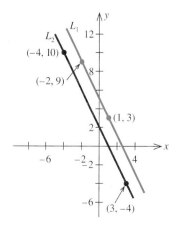

SELF-CHECK EXERCISE 2.1	Determine the number a so that the line passing through the points $(a, 2)$ and $(3, 6)$ is parallel to a line with slope 4.

Solution to Self-Check Exercise 2.1 can be found on page 66.

2.1   EXERCISES

In exercises 1–6, refer to the accompanying figure and determine the coordinates of the point and the quadrant in which it is located.

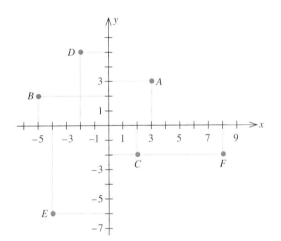

1. *A*

3. *C*

5. *E*

2. *B*

4. *D*

6. *F*

In exercises 7–12, refer to the accompanying figure.

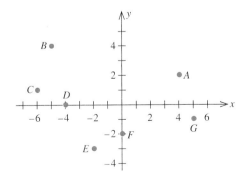

7. Which point has coordinates (4, 2)?

8. What are the coordinates of point *B*?

9. Which points have negative *y*-coordinates?

10. Which point has a negative *x*-coordinate and a negative *y*-coordinate?

11. Which point has an *x*-coordinate that is equal to zero?

12. Which point has a *y*-coordinate that is equal to zero?

In exercises 13–20, sketch a set of coordinate axes and plot the point.

13. $(-2, 5)$

15. $(3, -1)$

17. $(8, -\frac{7}{2})$

19. $(4.5, -4.5)$

14. $(1, 3)$

16. $(3, -4)$

18. $(-\frac{5}{2}, \frac{3}{2})$

20. $(1.2, -3.4)$

In exercises 21–24, find the slope of the line shown in each figure.

21.

22.

23.

24.

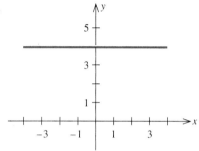

In exercises 25–30, find the slope of the line that passes through the pair of points.

25. $(4, 3)$ and $(5, 8)$ **26.** $(4, 5)$ and $(3, 8)$

27. $(-2, 3)$ and $(4, 8)$ **28.** $(-2, -2)$ and $(4, -4)$

29. (a, b) and (c, d)

30. $(-a + 1, b - 1)$ and $(a + 1, -b)$

31. Given the equation $y = 4x - 3$, answer the following questions:
   a. If x increases by 1 unit, what is the corresponding change in y?
   b. If x decreases by 2 units, what is the corresponding change in y?

32. Given the equation $2x + 3y = 4$, answer the following questions:
   a. Is the slope of the line described by this equation positive or negative?
   b. As x increases in value, does y increase or decrease?
   c. If x decreases by 2 units, what is the corresponding change in y?

In exercises 33 and 34, determine whether the lines through the pairs of points are parallel.

33. $A(1, -2)$, $B(-3, -10)$ and $C(1, 5)$, $D(-1, 1)$

34. $A(2, 3)$, $B(2, -2)$ and $C(-2, 4)$, $D(-2, 5)$

35. If the line passing through the points $(1, a)$ and $(4, -2)$ is parallel to the line passing through the points $(2, 8)$ and $(-7, a + 4)$, what is the value of a?

36. If the line passing through the points $(a, 1)$ and $(5, 8)$ is parallel to the line passing through the points $(4, 9)$ and $(a + 2, 1)$, what is the value of a?

37. Show that two distinct lines with equations $a_1x + b_1y + c_1 = 0$ and $a_2x + b_2y + c_2 = 0$, respectively, are parallel if and only if $a_1b_2 - b_1a_2 = 0$. [*Hint:* Write each equation in the slope-intercept form and compare.]

SOLUTION TO SELF-CHECK EXERCISE 2.1

The slope of the line that passes through the points $(a, 2)$ and $(3, 6)$ is

$$m = \frac{6 - 2}{3 - a}$$

$$= \frac{4}{3 - a}$$

Because this line is parallel to a line with slope 4, m must be equal to 4; that is,

$$\frac{4}{3 - a} = 4$$

or, upon multiplying both sides of the equation by $3 - a$,

$$4 = 4(3 - a)$$

$$4 = 12 - 4a$$

$$4a = 8$$

and

$$a = 2$$

2.2 Equations of Lines

Equations of Lines

We will now show that every straight line lying in the xy-plane may be represented by an equation involving the variables x and y. One immediate benefit of this is that problems involving straight lines may be solved algebraically.

Let L be a straight line parallel to the y-axis (perpendicular to the x-axis) (Figure 14). Then L crosses the x-axis at some point $(a, 0)$, with the x-coordinate given by $x = a$, where a is some real number. Any other point on L has the form $(a, \bar{y})$, where $\bar{y}$ is an appropriate number. Therefore, the vertical line L is described by the sole condition

$$x = a$$

and this is accordingly the equation of L. For example, the equation $x = -2$ represents a vertical line 2 units to the left of the y-axis, and the equation $x = 3$ represents a vertical line 3 units to the right of the y-axis (Figure 15).

FIGURE 14

The vertical line $x = a$

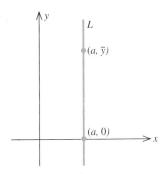

FIGURE 15

The vertical lines $x = -2$ and $x = 3$

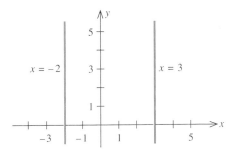

FIGURE 16

L passes through (x_1, y_1) and has slope m.

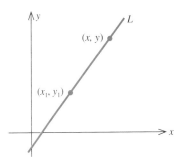

Next, suppose that L is a nonvertical line so that it has a well-defined slope m. Suppose (x_1, y_1) is a fixed point lying on L and (x, y) is a variable point on L distinct from (x_1, y_1) (Figure 16).

Using (1) with the point $(x_2, y_2) = (x, y)$, we find that the slope of L is given by

$$m = \frac{y - y_1}{x - x_1}$$

Upon multiplying both sides of the equation by $x - x_1$, we obtain (2).

POINT-SLOPE FORM

An equation of the line that has slope m and passes through the point (x_1, y_1) is given by

$$y - y_1 = m(x - x_1) \qquad \textbf{\textit{(2)}}$$

Equation (2) is called the **point-slope form** of the equation of a line, since it utilizes a given point (x_1, y_1) on a line and the slope m of the line.

EXAMPLE 1   Find an equation of the line that passes through the point (1, 3) and has slope 2.

Solution   Using the point-slope form of the equation of a line with the point (1, 3) and $m = 2$, we obtain

$$y - 3 = 2(x - 1) \qquad y - y_1 = m(x - x_1)$$
$$= 2x - 2$$

or $2x - y + 1 = 0$

(Figure 17). ◼

EXAMPLE 2   Find an equation of the line that passes through the points (−3, 2) and (4, −1).

Solution   The slope of the line is given by

$$m = \frac{-1 - 2}{4 - (-3)} = -\frac{3}{7}$$

Using the point-slope form of the equation of a line with the point (4, −1) and the slope $m = -3/7$, we have

$$y + 1 = -\frac{3}{7}(x - 4) \qquad y - y_1 = m(x - x_1)$$
$$7y + 7 = -3x + 12$$

or $3x + 7y - 5 = 0$

(Figure 18).

FIGURE 17

L passes through (1, 3) *and has slope* 2.

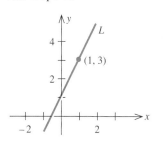

FIGURE 18

L passes through (−3, 2) *and (4, −1).*

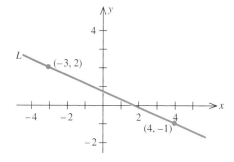

Perpendicular Lines

We can use the slope of a straight line to determine if a line is perpendicular to another line.

PERPENDICULAR LINES

If L_1 and L_2 are two distinct nonvertical lines that have slopes m_1 and m_2, respectively, then L_1 is **perpendicular** to L_2 (written $L_1 \perp L_2$) if and only if

$$m_1 = -\frac{1}{m_2}$$

FIGURE 19

L_2 is perpendicular to L_1 and passes through (3, 1).

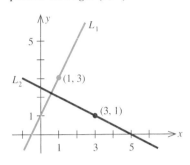

If the line L_1 is vertical (so that its slope is undefined), then L_1 is perpendicular to another line, L_2, if and only if L_2 is horizontal (so that its slope is zero). For a proof of these results, see exercise 55 page 75.

EXAMPLE 3 Find an equation of the line that passes through the point (3, 1) and is perpendicular to the line of Example 1.

Solution Since the slope of the line in Example 1 is 2, the slope of the required line is given by $m = -1/2$, the negative reciprocal of 2. Using the point-slope form of the equation of a line, we obtain

$$y - 1 = -\tfrac{1}{2}(x - 3) \qquad y - y_1 = m(x - x_1)$$
$$2y - 2 = -x + 3$$
or
$$x + 2y - 5 = 0$$

FIGURE 20

The line L has x-intercept a and y-intercept b.

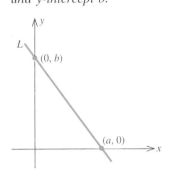

(Figure 19).

Slope-Intercept Form

A straight line L that is neither horizontal nor vertical cuts the x-axis and the y-axis at, say, points $(a, 0)$ and $(0, b)$, respectively (Figure 20). The numbers a and b are called the **x-intercept** and **y-intercept,** respectively, of L.

Now, let L be a line with slope m and y-intercept b. Using (2), the point-slope form of the equation of a line, with the point $(0, b)$ and slope m, we have

$$y - b = m(x - 0)$$
or
$$y = mx + b$$

This is called the **slope-intercept form** of an equation of a line.

SLOPE-INTERCEPT FORM

The equation of a line that has slope m and intersects the y-axis at the point $(0, b)$ is given by

$$y = mx + b \qquad \qquad (3)$$

EXAMPLE 4 Find an equation of the line that has slope 3 and y-intercept -4.

Solution Using (3) with $m = 3$ and $b = -4$, we obtain the required equation

$$y = 3x - 4$$

A pleasant consequence of writing an equation of a line in the slope-intercept form is that the slope and y-intercept of the line can be easily read off, as the next example shows.

EXAMPLE 5   Determine the slope and y-intercept of the line whose equation is $3x - 4y = 8$.

Solution   Rewrite the given equation in the slope-intercept form and obtain

$$y = \frac{3}{4}x - 2$$

Comparing this result with (3), we find $m = 3/4$ and $b = -2$, and we conclude that the slope and y-intercept of the given line are $3/4$ and -2, respectively. ▪

Applications

EXAMPLE 6   The sales manager of a local sporting goods store plotted sales versus time for the last five years and found the points to lie approximately along a straight line (Figure 21). By using the points corresponding to the first and fifth years, find an equation of the trend line. What sales figure can be predicted for the sixth year?

Solution   Using equation (1) with the points (1, 20) and (5, 60), we find that the slope of the required line is given by

$$m = \frac{60 - 20}{5 - 1} = 10$$

Next, using the point-slope form of the equation of a line with the point (1, 20) and $m = 10$, we obtain

$$y - 20 = 10(x - 1) \qquad {\scriptstyle y - y_1 = m(x - x_1)}$$

or

$$y = 10x + 10$$

as the required equation.

The sales figure for the sixth year is obtained by letting $x = 6$ in the last equation, giving

$$y = 70$$

or $70,000. ▪

FIGURE 21

Sales of a sporting goods store

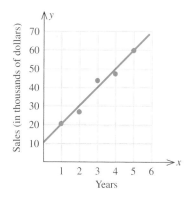

EXAMPLE 7   Suppose that an art object purchased for $50,000 is expected to appreciate in value at a constant rate of $5,000 per year for the next five years. Use (3) to write an equation predicting the value of the art object in the next several years. What will its value be three years from the date of purchase?

Solution   Let x denote the time (in years) that has elapsed since the date the object was purchased and let y denote the object's value (in dollars). Then $y = 50,000$ when $x = 0$. Furthermore, the slope of the required equation is given by $m = 5,000$, since each unit increase in x (one year) implies an increase of 5,000 units (dollars) in y. Using (3) with $m = 5,000$ and $b = 50,000$, we obtain

$$y = 5,000x + 50,000 \qquad {\scriptstyle y = mx + b}$$

Three years from the date of purchase, the value of the object will be given by

$$y = 5{,}000(3) + 50{,}000$$

or $65,000. ◼

General Equation of a Line

We have considered several forms of the equation of a straight line in the plane. These different forms of the equation are equivalent to each other. In fact, each is a special case of the following equation:

GENERAL FORM OF A LINEAR EQUATION

The equation

$$Ax + By + C = 0, \qquad \textbf{(4)}$$

where A, B, and C are constants and A and B are not both zero, is called the general form of a linear equation in the variables x and y.

We will now state (without proof) an important result concerning the algebraic representation of straight lines in the plane.

LINEAR EQUATIONS AND STRAIGHT LINES

An equation of a straight line is a linear equation; conversely, every linear equation represents a straight line.

This result justifies the use of the adjective *linear* in describing equation (4).

EXAMPLE 8 Sketch the straight line represented by the equation

$$3x - 4y - 12 = 0$$

Solution Since every straight line is uniquely determined by two distinct points, we need find only two such points through which the line passes in order to sketch it. For convenience, let us compute the x- and y-intercepts. Setting $y = 0$, we find $x = 4$; thus the x-intercept is 4. Setting $x = 0$ gives $y = -3$, and the y-intercept is -3. A sketch of the line appears in Figure 22. ◼

FIGURE 22
To sketch $3x - 4y - 12 = 0$, first find the x-intercept, 4, and the y-intercept, -3.

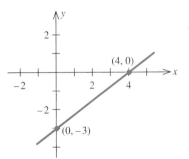

Following is a summary of the common forms of the equations of straight lines discussed in this section.

EQUATIONS OF STRAIGHT LINES

Vertical line:	$x = a$
Horizontal line:	$y = b$
Point-slope form:	$y - y_1 = m(x - x_1)$
Slope-intercept form:	$y = mx + b$
General form:	$Ax + By + C = 0$

**SELF-CHECK
EXERCISES 2.2**

1. Find an equation of the line that passes through the point $(3, -1)$ and is perpendicular to a line with slope $-\frac{1}{2}$.

2. Does the point $(3, -3)$ lie on the line with equation $2x - 3y - 12 = 0$?

3. The percentage of people over 65 who have high school diplomas is summarized in the following table:

Year (x)	1960	1965	1970	1975	1980	1985	1990
Percent with diplomas (y)	20	25	30	36	42	47	52

   a. Plot the percentage of people over 65 who have high school diplomas (y) versus the year (x).
   b. Draw the straight line L through the points (1960, 20) and (1990, 52).
   c. Find an equation of the line L.
   d. If the trend continues, estimate the percentage of people over 65 who will have high school diplomas by the year 2000.

   Source: The World Almanac

Solutions to Self-Check Exercises 2.2 can be found on page 75.

2.2   EXERCISES

1. Match each of the graphs i–vi with one of the statements a–f.
   a. The slope of the line is zero.
   b. The slope of the line is undefined.
   c. The slope of the line is positive and its y-intercept is positive.
   d. The slope of the line is positive and its y-intercept is negative.
   e. The slope of the line is negative and its x-intercept is negative.
   f. The slope of the line is negative and its x-intercept is positive.

ii.

i.

iii.

iv.

v.

vi.

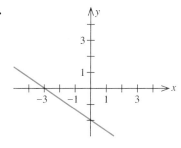

In exercises 2 and 3, determine if the lines through the given pairs of points are perpendicular.

2. $A(-2, 5)$, $B(4, 2)$ and $C(-1, -2)$, $D(3, 6)$

3. $A(2, 0)$, $B(1, -2)$ and $C(4, 2)$, $D(-8, 4)$

4. Find an equation of the horizontal line that passes through $(-4, -3)$.

5. Find an equation of the vertical line that passes through $(0, 5)$.

In exercises 6–9, find an equation of the line that passes through the point and has the indicated slope m.

6. $(3, -4)$; $m = 2$ **7.** $(2, 4)$; $m = -1$

8. $(-3, 2)$; $m = 0$ **9.** $(1, 2)$; $m = -\frac{1}{2}$

In exercises 10–13, find an equation of the line that passes through the points.

10. $(2, 4)$ and $(3, 7)$ **11.** $(2, 1)$ and $(2, 5)$

12. $(1, 2)$ and $(-3, -2)$ **13.** $(-1, -2)$ and $(3, -4)$

In exercises 14–17, find an equation of the line that has slope m and y-intercept b.

14. $m = 3$; $b = 4$ **15.** $m = -2$; $b = -1$

16. $m = 0$; $b = 5$ **17.** $m = -\frac{1}{2}$; $b = \frac{3}{4}$

In exercises 18–23, write the equation in the slope-intercept form and then find the slope and y-intercept of the corresponding line.

18. $x - 2y = 0$ **19.** $y - 2 = 0$

20. $2x - 3y - 9 = 0$ **21.** $3x - 4y + 8 = 0$

22. $2x + 4y = 14$ **23.** $5x + 8y - 24 = 0$

24. Find an equation of the line that passes through the point $(-2, 2)$ and is parallel to the line $2x - 4y - 8 = 0$.

25. Find an equation of the line that passes through the point $(2, 4)$ and is perpendicular to the line $3x + 4y - 22 = 0$.

In exercises 26–31, find an equation of the line that satisfies the given condition.

26. The line parallel to the x-axis and 6 units below it.

27. The line passing through the origin and parallel to the line joining the points $(2, 4)$ and $(4, 7)$.

28. The line passing through the point (a, b) with slope equal to zero.

29. The line passing through $(-3, 4)$ and parallel to the x-axis.

30. The line passing through $(-5, -4)$ and parallel to the line joining $(-3, 2)$ and $(6, 8)$.

31. The line passing through (a, b) with undefined slope.

32. Given that the point $P(-3, 5)$ lies on the line $kx + 3y + 9 = 0$, find k.

33. Given that the point $P(2, -3)$ lies on the line $-2x + ky + 10 = 0$, find k.

In exercises 34–39, sketch the straight line defined by the linear equation by finding the x- and y-intercepts. [Hint: See Example 8.]

34. $3x - 2y + 6 = 0$ **35.** $2x - 5y + 10 = 0$

36. $x + 2y - 4 = 0$ **37.** $2x + 3y - 15 = 0$

38. $y + 5 = 0$ **39.** $-2x - 8y + 24 = 0$

40. Show that an equation of a line through the points $(a, 0)$ and $(0, b)$ with $a \neq 0$ and $b \neq 0$ can be written in the form

$$\frac{x}{a} + \frac{y}{b} = 1$$

(Recall that the numbers a and b are the x- and y-intercepts, respectively, of the line. This form of an equation of a line is called the *intercept form*.)

In exercises 41–44, use the results of exercise 40 to find an equation of a line with the given x- and y-intercepts.

41. x-intercept 3; y-intercept 4

42. x-intercept -2; y-intercept -4

43. x-intercept $-\frac{1}{2}$; y-intercept $\frac{3}{4}$

44. x-intercept 4; y-intercept $-\frac{1}{2}$

In exercises 45 and 46, determine whether the given points lie on a straight line.

45. $A(-1, 7)$, $B(2, -2)$, and $C(5, -9)$

46. $A(-2, 1)$, $B(1, 7)$, and $C(4, 13)$

47. Social Security Contributions   For wages less than the maximum taxable wage base, social security contributions by employees are 7.65% of the employee's wages.
   a. Find an equation that expresses the relationship between the wages earned (x) and the social security taxes paid (y) by an employee who earns less than the maximum taxable wage base.
   b. For each additional dollar that an employee earns, by how much is his or her social security contribution increased? (Assume that the employee's wages are less than the maximum taxable wage base.)
   c. What social security contributions will an employee who earns \$35,000 (which is less than the maximum taxable wage base) be required to make?

48. College Admissions   Using data compiled by the Admissions Office at Faber University, college admissions officers estimate that 55% of the students who are offered admission to the freshman class at the university will actually enroll.
   a. Find an equation that expresses the relationship between the number of students who actually enroll (y) and the number of students who are offered admission to the university (x).
   b. If the desired freshman class size for the upcoming academic year is 1100 students, how many students should be offered admission?

49. Weight of Whales   The equation $W = 3.51L - 192$, expressing the relationship between the length L (in

feet) and the expected weight W (in British tons) of adult blue whales, was adopted in the late 1960s by the International Whaling Commission.
   a. What is the expected weight of an 80-foot blue whale?
   b. Sketch the straight line that represents the equation.

50. Cost of a Commodity   A manufacturer obtained the following data relating the cost y (in dollars) to the number of units (x) of a commodity produced:

Number of units produced (x)	0	20	40	60	80	100
Cost in dollars (y)	200	208	222	230	242	250

   a. Plot the cost (y) versus the quantity produced (x).
   b. Draw the straight line through the points $(0, 200)$ and $(100, 250)$.
   c. Derive an equation of the straight line of (b).
   d. Taking this equation to be an approximation of the relationship between the cost and the level of production, estimate the cost of producing 54 units of the commodity.

51. Ideal Heights and Weights for Women   The Venus Health Club for Women provides its members with the following table, which gives the average desirable weight for women of a certain height:

Height x (in inches)	60	63	66	69	72
Weight y (in pounds)	108	118	129	140	152

   a. Plot the weight (y) versus the height (x).
   b. Draw a straight line L through the points corresponding to heights of 5 feet and 6 feet.
   c. Derive an equation of the line L.
   d. Using the equation of (c), estimate the average desirable weight for a woman who is 5 feet 5 inches tall.

52. Sales Growth   Metro Department Store's annual sales (in millions of dollars) during the past five years were:

Annual sales (y)	5.8	6.2	7.2	8.4	9.0
Year (x)	1	2	3	4	5

   a. Plot the annual sales (y) versus the year (x).
   b. Draw a straight line L through the points corresponding to the first and fifth years.

c. Derive an equation of the line L.

d. Using the equation found in (c), estimate Metro's annual sales four years from now ($x = 9$).

53. **Welfare Costs** According to the Massachusetts Department of Welfare, the spending (in billions of dollars) by Medicaid, the national health-care plan for the poor, over the 5-year period from 1988 to 1992 is summarized in the following table. (Here $x = 0$ represents the beginning of the year 1988.)

Year (x)	0	1	2	3	4
Expenditure (y)	1.550	1.662	1.786	1.888	2.009

a. Plot the points (x, y) in a rectangular coordinate system.

b. It can be shown that the straight line that fits these data "best" (in some sense) is given by the line passing through the points (0, 1.55) and (5, 2.12). (See Project 1, page 112.) Sketch this "trend line" (called the regression line).

c. Use the trend line to predict the Medicaid spending for the year 1994, assuming this trend continued.

Source: Massachusetts Department of Welfare

54. **Health-Care Spending** The following data, compiled by the Organization for Economic Cooperation and Development (OECD) in 1990, give the per capita gross domestic product (GDP) and the corresponding per capita spending on health care for selected countries.

Country	Turkey	Spain	Netherlands	Sweden	Switzerland	Canada
GDP (per capita in thousands of dollars)	4.25	10	14	15.5	17.8	19.5
Health-care spending (per capita in dollars)	178	667	1194	1500	1388	1640

a. Letting x denote a country's GDP (in thousands of dollars per capita) and y denote the per capita health-care spending (in dollars), plot the points (x, y) in a rectangular coordinate system.

b. It can be shown that the straight line that fits these data "best" (in some sense) is given by the line passing through the points (5, 259.2) and (10, 750.1). (See Project 1, page 112.) Sketch this "trend line" (called the regression line).

c. The per capital GDP of the United States is $20,000. If the health-care spending of the United States were in line with that of these sample OECD countries, what would it be? (*Note:* The actual per capita health-care spending of the United States in 1990 was $2,444.)

Source: Organization for Economic Cooperation and Development

55. Prove that if a line L_1 with slope m_1 is perpendicular to a line L_2 with a slope m_2, then $m_1 m_2 = -1$. [*Hint:* Refer to the accompanying figure. Show that $m_1 = b$ and $m_2 = c$. Next, apply the Pythagorean Theorem and the distance formula to the triangles OAC, OCB, and OBA to show that $1 = -bc$.]

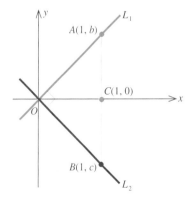

1. Since the required line L is perpendicular to a line with slope $-\frac{1}{2}$, the slope of L is

$$-\frac{1}{-\frac{1}{2}} = 2$$

Next, using the point-slope form of the equation of a line, we have

$$y - (-1) = 2(x - 3)$$

$$y + 1 = 2x - 6$$

or

$$y = 2x - 7$$

2. Substituting $x = 3$ and $y = -3$ into the left-hand side of the given equation, we find

$$2(3) - 3(-3) - 12 = 3$$

which is not equal to zero (the right-hand side). Therefore, $(3, -3)$ does not lie on the line with equation $2x - 3y - 12 = 0$.

3. **a.** and **b.** See the accompanying figure.

c. The slope of L is

$$m = \frac{52 - 20}{1990 - 1960} = \frac{32}{30} = \frac{16}{15}$$

When we use the point-slope form of the equation of a line with the point $(1960, 20)$, we find

$$y - 20 = \frac{16}{15}(x - 1960) = \frac{16}{15}x - \frac{(16)(1960)}{15}$$

or

$$y = \frac{16}{15}x - \frac{6272}{3} + 20$$

$$= \frac{16}{15}x - \frac{6212}{3}$$

d. To estimate the percentage of people over 65 who will have high school diplomas by the year 2000, let $x = 2000$ in the equation obtained in (c). Thus, the required estimate is

$$y = \frac{16}{15}(2000) - \frac{6212}{3} \approx 62.67$$

or approximately 63%.

2.3   # Mathematical Models and Functions

Mathematical Models

Before we can apply mathematics to solving real-world problems, we need to formulate these problems in the language of mathematics. This process is referred to as *mathematical modeling.* A *mathematical model* may give a precise description of the problem under consideration, or, more likely, it may provide only an acceptable approximation of the problem. For example, the accumulated amount A at the end of t years when a sum of P dollars is deposited in a bank account and earns interest at the rate of r percent per year compounded m times a year is given *exactly* by the formula (model)

$$A = P \left(1 + \frac{r}{m} \right)^{mt}$$

On the other hand, the size of a cancer tumor may be approximated by the volume of a sphere,

$$V = \frac{4}{3} \pi r^3$$

where r is the radius of the tumor in centimeters.

The many techniques used in constructing mathematical models of practical problems range from theoretical considerations of a problem on the one extreme to an interpretation of data associated with the problem on the other. The model giving the accumulated amount for a fixed bank account mentioned earlier may be derived theoretically (see Chapter 4). In Section 2.2 we saw how linear equations (models) can be constructed from a given set of data points. Also, in the ensuing chapters we will see how other mathematical models, including statistical and probability models, are used to describe and analyze real-world situations.

Functions

Many mathematical models are described using the notion of a *function.* A manufacturer would like to know how his company's profit is related to its production level; a biologist would like to know how the size of the population of a certain culture of bacteria will change over time; a psychologist would like to know the relationship between the learning time of an individual and the length of a vocabulary list; and a chemist would like to know how the initial speed of a chemical reaction is related to the amount of substrate used. In each instance we are concerned with the same question: How does one quantity depend upon another? This dependency between two quantities is conveniently described in mathematics using a function.

FUNCTION

A **function** is a rule that assigns to each element in a set A one and only one element in a set B.

The set A is called the *domain* of the function. It is customary to denote a function by a letter of the alphabet, such as the letter f. If x is an element in the domain of a function f, then the element in B that f associates with x is written $f(x)$ (read "f of x") and is called the *value of f at x*.

We can think of a function f as a machine. The domain is the set of inputs (raw material) for the machine, the rule describes how the input is to be processed, and the value(s) of the function are the outputs of the machine (Figure 23).

FIGURE 23
A function machine

FIGURE 24
The function f viewed as a mapping

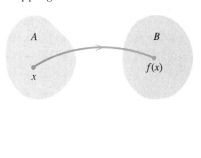

We can also think of a function f as a mapping in which an element x in the domain of f is mapped onto a unique element $f(x)$ in B (Figure 24).

REMARKS

1. It is important to understand that the output $f(x)$ associated with an input x is unique. To appreciate the importance of this uniqueness property, consider a rule that associates with each item x in a department store its selling price y. Then, each x must correspond to *one and only one y*. Notice, however, that different x's may be associated with the same y. In the context of the present example, this says that different items may have the same price.

2. Although the sets A and B that appear in the definition of a function may be quite arbitrary, in this book they will denote sets of real numbers, unless stated otherwise.   □

An example of a function may be taken from the familiar relationship between the area of a circle and its radius. Letting x and y denote the radius and area of a circle, respectively, we have, from elementary geometry,

$$y = \pi x^2 \tag{5}$$

Equation (5) defines y as a function of x, since for each admissible value of x (that is, for each nonnegative number representing the radius of a certain circle) there corresponds precisely one number $y = \pi x^2$ that gives the area of the circle. The rule defining this "area function" may be written as

$$f(x) = \pi x^2 \tag{6}$$

To compute the area of a circle of radius 5 inches, we simply replace x in (6) with the number 5. Thus, the area of the circle is

$$f(5) = \pi 5^2 = 25\pi$$

or 25π square inches.

In general, to evaluate a function at a specific value of x, we replace x with that value, as illustrated in Examples 1 and 2.

EXAMPLE 1   Let the function f be defined by the rule $f(x) = 2x^2 - x + 1$. Compute:

a. $f(1)$ **b.** $f(-2)$ **c.** $f(a)$ **d.** $f(a + h)$

Solution

a. $f(1) = 2(1)^2 - (1) + 1 = 2 - 1 + 1 = 2$
b. $f(-2) = 2(-2)^2 - (-2) + 1 = 8 + 2 + 1 = 11$
c. $f(a) = 2(a)^2 - (a) + 1 = 2a^2 - a + 1$
d. $f(a + h) = 2(a + h)^2 - (a + h) + 1 = 2a^2 + 4ah + 2h^2 - a - h + 1$ ∎

EXAMPLE 2   The Thermo-Master Company manufactures an indoor-outdoor thermometer at its Mexico subsidiary. Management estimates that the profit (in dollars) realizable by Thermo-Master in the manufacture and sale of x thermometers per week is

$$P(x) = -0.001x^2 + 8x - 5000$$

Find Thermo-Master's weekly profit if its level of production is (a) 1000 thermometers per week and (b) 2000 thermometers per week.

Solution

a. The weekly profit relizable by Thermo-Master when the level of production is 1000 units per week is found by evaluating the profit function P at $x = 1000$. Thus,

$$P(1000) = -0.001(1000)^2 + 8(1000) - 5000 = 2000$$

or $2000.

b. When the level of production is 2000 units per week, the weekly profit is given by

$$P(2000) = -0.001(2000)^2 + 8(2000) - 5000 = 7000$$

or $7000. ∎

Determining the Domain of a Function

Suppose we are given the function $y = f(x)$.* Then the variable x is called the **independent variable.** The variable y, whose value depends on x, is called the **dependent variable.**

The set of all values that may be assumed by x is called the **domain** of the function f, and the set comprising all the values assumed by $y = f(x)$ as x takes on all possible values in its domain is called the **range** of the function f. For the area function (6), the domain of f is the set of all nonnegative numbers x, and the range of f is the set of all nonnegative numbers y.

If a function is defined by a rule relating x to $f(x)$ without specific mention of its domain, it is understood that the domain will consist of all

---

*It is customary to refer to a function f as $f(x)$.

values of x for which $f(x)$ is a real number. In this connection, you should keep in mind that (1) division by zero is not permitted and (2) the square root of a negative number is not defined.

EXAMPLE 3   Find the domain of each of the functions defined by the following equations:

a. $f(x) = \sqrt{x - 1}$ **b.** $f(x) = \dfrac{1}{x^2 - 4}$ **c.** $f(x) = x^2 + 3$

Solution

a. Since the square root of a negative number is undefined, it is necessary that $x - 1 \geq 0$. The inequality is satisfied by the set of real numbers $x \geq 1$. Thus, the domain of f is the interval $[1, \infty)$.

b. The only restriction on x is that $x^2 - 4$ be different from zero, since division by zero is not allowed. But, $(x^2 - 4) = (x + 2)(x - 2) = 0$ if $x = -2$ or $x = 2$. Thus, the domain of f in this case consists of the intervals $(-\infty, -2)$, $(-2, 2)$, and $(2, \infty)$.

c. Here, any real number satisfies the equation, so the domain of f is the set of all real numbers.   ■

Graphs of Functions

If f is a function with domain A, then corresponding to each real number x in A there is precisely one real number $f(x)$. We can also express this fact by using *ordered pairs* of real numbers. Write each number x in A as the first member of an ordered pair and each number $f(x)$ corresponding to x as the second member of the ordered pair. This gives exactly one ordered pair $(x, f(x))$ for each x in A. This observation leads to an alternative definition of a function f:

A **function** f with domain A is the set of all ordered pairs $(x, f(x))$ where x belongs to A.

Observe that the condition that there be one and only one number $f(x)$ corresponding to each number x in A translates into the requirement that *no two ordered pairs have the same first number.*

Since ordered pairs of real numbers correspond to points in the plane, we have found a way to exhibit a function graphically.

GRAPH OF A FUNCTION OF ONE VARIABLE

The *graph of a function f* is the set of all points (x, y) in the *xy*-plane such that x is in the domain of f and $y = f(x)$.

Much information about the graph of a function can be gained by plotting a few points on its graph. Later on we will develop more systematic and sophisticated techniques for graphing functions.

FIGURE 25
The graph of $y = x^2 + 1$ is a parabola.

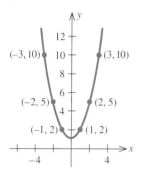

EXAMPLE 4 Sketch the graph of the function defined by the equation $y = x^2 + 1$.

Solution The domain of the function is the set of all real numbers. By assigning several values to the variable x and computing the corresponding values for y, we obtain the following solutions to the equation $y = x^2 + 1$:

x	-3	-2	-1	0	1	2	3
y	10	5	2	1	2	5	10

By plotting these points and then connecting them with a smooth curve, we obtain the graph of $y = f(x)$, which is a parabola (Figure 25). ■

Some functions are defined in a piecewise fashion, as Examples 5 and 6 show.

EXAMPLE 5 The Madison Finance Company plans to open two branch offices two years from now in two separate locations: an industrial complex and a newly developed commercial center in the city. As a result of these expansion plans, Madison's total deposits during the next five years are expected to grow in accordance with the rule

$$f(x) = \begin{cases} \sqrt{2x} + 20 & 0 \le x \le 2 \\ \frac{1}{2}x^2 + 20 & 2 < x \le 5 \end{cases}$$

where $y = f(x)$ gives the total amount of money (in millions of dollars) on deposit with Madison in year x ($x = 0$ corresponds to the present). Sketch the graph of the function f.

Solution The function f is defined in a piecewise fashion on the interval $[0, 5]$. In the subdomain $[0, 2]$, the rule for f is given by $f(x) = \sqrt{2x} + 20$. The values of $f(x)$ corresponding to $x = 0$, 1, and 2 may be tabulated as follows:

x	0	1	2
$f(x)$	20	21.4	22

Next, in the subdomain $(2, 5]$, the rule for f is given by $f(x) = (1/2)x^2 + 20$. The values of $f(x)$ corresponding to $x = 3$, 4, and 5 are shown in the following table:

x	3	4	5
$f(x)$	24.5	28	32.5

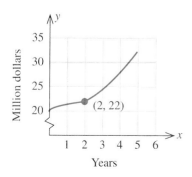

Using the values of $f(x)$ in this table, we sketch the graph of the function f as shown in Figure 26. ■

EXAMPLE 6   Sketch the graph of the function f defined by

$$f(x) = \begin{cases} -x & \text{if } x < 0 \\ \sqrt{x} & \text{if } x \geq 0 \end{cases}$$

Solution   The function f is defined in a piecewise fashion on the set of all real numbers. In the subdomain $(-\infty, 0)$, the rule for f is given by $f(x) = -x$. The equation $y = -x$ is a linear equation in the slope-intercept form (with slope -1 and intercept 0). Therefore the graph of f corresponding to the subdomain $(-\infty, 0)$ is the half-line shown in Figure 27. Next, in the subdomain $[0, \infty)$, the rule for f is given by $f(x) = \sqrt{x}$. The values of $f(x)$ corresponding to $x = 0, 1, 2, 3, 4, 9,$ and 16 are shown in the following table:

x	0	1	2	3	4	9	16
$f(x)$	0	1	$\sqrt{2}$	$\sqrt{3}$	2	3	4

Using these values, we sketch the graph of the function f as shown in Figure 27.

FIGURE 27
The graph of $y = f(x)$ is obtained by graphing $y = -x$ over $(-\infty, 0)$ and $y = \sqrt{x}$ over $[0, \infty)$.

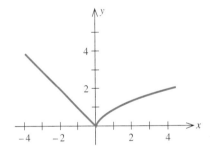

The Vertical-Line Test

Although it is true that every function f of a variable x has a graph in the xy-plane, it is important to realize that not every curve in the xy-plane is the graph of a function. For example, consider the curve depicted in Figure 28. This is the graph of the equation $x = y^2$. In general, the *graph of an*

FIGURE 28
Since a vertical line passes through the curve at more than one point, we deduce that it is not the graph of a function.

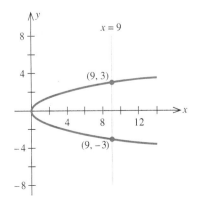

equation is the set of all ordered pairs (x, y) that satisfy the given equation. Observe that the points $(9, -3)$ and $(9, 3)$ both lie on the curve. This implies that the number $x = 9$ is associated with *two* numbers: $y = -3$ and $y = 3$. But this clearly violates the uniqueness property of a function. Thus, we conclude that the curve under consideration cannot be the graph of a function.

This example suggests the following test for determining when a curve is the graph of a function:

VERTICAL-LINE TEST   A curve in the xy-plane is the graph of a function $y = f(x)$ if and only if each vertical line intersects it in at most one point.

EXAMPLE 7   Determine which of the curves shown in Figure 29 are the graphs of functions of x.

FIGURE 29

The vertical-line test can be used to determine which of these curves are graphs of functions.

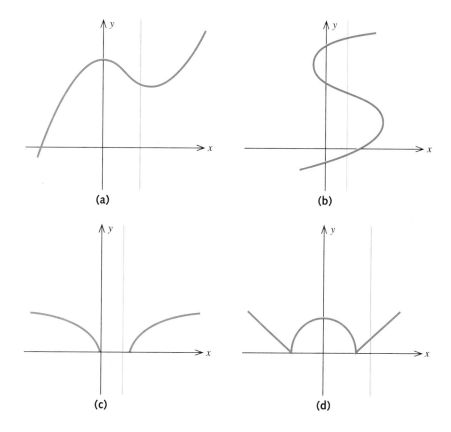

(a)

(b)

(c)

(d)

Solution   The curves depicted in Figures 29a, 29c, and 29d are graphs of functions because each curve satisfies the requirement that each vertical line intersect the curve in at most one point. Note that the vertical line shown in Figure 29c does not intersect the graph, because the point on the x-axis through which this line passes does not lie in the domain of the function. The curve depicted in Figure 29b is *not* the graph of a function, because the vertical line shown there intersects the graph at three points.

SELF-CHECK
EXERCISES 2.3

1. Let f be the function defined by

$$f(x) = \frac{\sqrt{x+1}}{x}$$

    a. Find the domain of f.

    b. Compute $f(3)$.

    c. Compute $f(a + h)$.

c 2. Statistics obtained by Amoco Corporation show that more and more motorists are pumping their own gas. The following function gives self-serve sales as a percentage of all U.S. gas sales:

$$f(t) = \begin{cases} 6t + 17 & 0 \le t \le 6 \\ 15.98(t-6)^{1/4} + 53 & 6 < t \le 20 \end{cases}$$

Here t is measured in years, with $t = 0$ corresponding to the beginning of 1974.

    a. Sketch the graph of the function f.

    b. What percentage of all gas sales at the beginning of 1978 were self-serve? At the beginning of 1994?

    Source: Amoco Corporation

3. Let $f(x) = \sqrt{2x+1} + 2$. Determine whether the point $(4, 6)$ lies on the graph of f.

Solutions to Self-Check Exercises 2.3 can be found on page 87.

2.3   EXERCISES

1. Let f be the function defined by $f(x) = 5x + 6$. Find $f(3)$, $f(-3)$, $f(a)$, $f(-a)$, and $f(a + 3)$.

2. Let f be the function defined by $f(x) = 4x - 3$. Find $f(4)$, $f(\frac{1}{4})$, $f(0)$, $f(a)$, and $f(a + 1)$.

3. Let g be the function defined by $g(x) = 3x^2 - 6x - 3$. Find $g(0)$, $g(-1)$, $g(a)$, $g(-a)$, and $g(x + 1)$.

4. Let h be the function defined by $h(x) = x^3 - x^2 + x + 1$. Find $h(-5)$, $h(0)$, $h(a)$, and $h(-a)$.

5. Let s be the function defined by $s(t) = 2t/(t^2 - 1)$. Find $s(4)$, $s(0)$, $s(a)$, $s(2 + a)$, and $s(t + 1)$.

6. Let g be the function defined by $g(u) = (3u - 2)^{3/2}$. Find $g(1)$, $g(6)$, $g(\frac{11}{3})$, and $g(u + 1)$.

7. Let f be the function defined by $f(t) = 2t^2/\sqrt{t-1}$. Find $f(2)$, $f(a)$, $f(x + 1)$, and $f(x - 1)$.

8. Let f be the function defined by $f(x) = 2 + 2\sqrt{5 - x}$. Find $f(-4)$, $f(1)$, $f(\frac{11}{4})$, and $f(x + 5)$.

9. Let f be the function defined by

$$f(x) = \begin{cases} x^2 + 1 & \text{if } x \le 0 \\ \sqrt{x} & \text{if } x > 0 \end{cases}$$

Find $f(-2)$, $f(0)$, and $f(1)$.

10. Let g be the function defined by

$$g(x) = \begin{cases} -\frac{1}{2}x + 1 & \text{if } x < 2 \\ \sqrt{x - 2} & \text{if } x \ge 2 \end{cases}$$

Find $g(-2)$, $g(0)$, $g(2)$, and $g(4)$.

11. Let f be the function defined by

$$f(x) = \begin{cases} -\frac{1}{2}x^2 + 3 & \text{if } x < 1 \\ 2x^2 + 1 & \text{if } x \ge 1 \end{cases}$$

Find $f(-1)$, $f(0)$, $f(1)$, and $f(2)$.

12. Let f be the function defined by

$$f(x) = \begin{cases} 2 + \sqrt{1 - x} & \text{if } x \le 1 \\ \dfrac{1}{1 - x} & \text{if } x > 1 \end{cases}$$

Find $f(0)$, $f(1)$, and $f(2)$.

In exercises 13–16, determine whether the point lies on the graph of the function.

13. $(2, \sqrt{3})$; $g(x) = \sqrt{x^2 - 1}$

14. $(3, 3)$; $f(x) = \dfrac{x + 1}{\sqrt{x^2 + 7}} + 2$

15. $(-2, -3)$; $f(t) = \dfrac{|t - 1|}{t + 1}$

16. $\left(-3, -\dfrac{1}{13}\right)$; $h(t) = \dfrac{|t + 1|}{t^3 + 1}$

In exercises 17–30, find the domain of the function.

17. $f(x) = x^2 + 3$

18. $f(x) = 7 - x^2$

19. $f(x) = \dfrac{3x + 1}{x^2}$

20. $g(x) = \dfrac{2x + 1}{x - 1}$

21. $f(x) = \sqrt{x^2 + 1}$

22. $f(x) = \sqrt{x - 5}$

23. $f(x) = \sqrt{5 - x}$

24. $g(x) = \sqrt{2x^2 + 3}$

25. $f(x) = \dfrac{x}{x^2 - 1}$

26. $f(x) = \dfrac{1}{x^2 + x - 2}$

27. $f(x) = (x + 3)^{3/2}$

28. $g(x) = 2(x - 1)^{5/2}$

29. $f(x) = \dfrac{\sqrt{1 - x}}{x^2 - 4}$

30. $f(x) = \dfrac{\sqrt{x - 1}}{(x + 2)(x - 3)}$

31. Let f be a function defined by the rule $f(x) = x^2 - x - 6$.
   a. Find the domain of f.
   b. Compute $f(x)$ for $x = -3, -2, -1, 0, \frac{1}{2}, 1, 2, 3$.
   c. Use the results obtained in (a) and (b) to sketch the graph of f.

32. Let f be a function defined by the rule $f(x) = 2x^2 + x - 3$.
   a. Find the domain of f.
   b. Compute $f(x)$ for $x = -3, -2, -1, -\frac{1}{2}, 0, 1, 2, 3$.
   c. Use the results obtained in (a) and (b) to sketch the graph of f.

In exercises 33–44, sketch the graph of the function f with the given rule.

33. $f(x) = 2x^2 + 1$

34. $f(x) = 9 - x^2$

35. $f(x) = 2 + \sqrt{x}$

36. $g(x) = 4 - \sqrt{x}$

37. $f(x) = \sqrt{1 - x}$

38. $f(x) = \sqrt{x - 1}$

39. $f(x) = |x| - 1$

40. $f(x) = |x| + 1$

41. $f(x) = \begin{cases} x & \text{if } x < 0 \\ 2x + 1 & \text{if } x \ge 0 \end{cases}$

42. $f(x) = \begin{cases} 4 - x & \text{if } x < 2 \\ 2x - 2 & \text{if } x \ge 2 \end{cases}$

43. $f(x) = \begin{cases} -x + 1 & \text{if } x \le 1 \\ x^2 - 1 & \text{if } x > 1 \end{cases}$

44. $f(x) = \begin{cases} -x - 1 & \text{if } x < -1 \\ 0 & \text{if } -1 \le x \le 1 \\ x + 1 & \text{if } x > 1 \end{cases}$

In exercises 45–52, use the vertical-line test to determine whether the graph represents y as a function of x.

45.

46.

47.

48.

49.

50.

51. **52.**

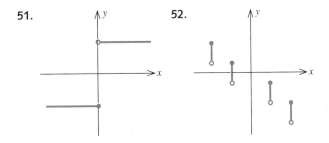

53. The circumference of a circle is given by $C(r) = 2\pi r$, where r is the radius of the circle. What is the circumference of a circle with a 5-inch radius?

54. The volume of a sphere of radius r is given by $V(r) = \frac{4}{3}\pi r^3$. Compute $V(2.1)$ and $V(2)$. What does the quantity $V(2.1) - V(2)$ measure?

55. **Consumption Function** The consumption function in a certain economy is given by the equation

$$C(y) = 0.75y + 6$$

where $C(y)$ is the personal consumption expenditure, y is the disposable personal income, and both $C(y)$ and y are measured in billions of dollars. Find $C(0)$, $C(50)$, and $C(100)$.

56. **Sales Taxes** In a certain state, the sales tax T on the amount of taxable goods is 6% of the value of the goods purchased (x), where both T and x are measured in dollars.
 a. Express T as a function of x.
 b. Find $T(200)$ and $T(5.65)$.

57. **Surface Area of a Single-Celled Organism** The surface area S of a single-celled organism may be found by multiplying 4π times the square of the radius r of the cell. Express S as a function of r.

58. **Friend's Rule** Friend's rule, a method for calculating pediatric drug dosages, is based on a child's age. If a denotes the adult dosage (in mg) and if t is the age of the child (in years), then the child's dosage is given by

$$D(t) = \frac{2}{25}ta$$

If the adult dose of a substance is 500 mg, how much should a four-year-old child receive?

59. **COLAs** Social security recipients receive an automatic cost-of-living adjustment (COLA) once each year. Their monthly benefit is increased by the amount that consumer prices increased during the preceding year. Suppose that consumer prices increased by 5.3% during the preceding year.

 a. Express the adjusted monthly benefit of a social security recipient as a function of his or her current monthly benefit.
 b. If Mr. Harrington's monthly social security benefit is now $620, what will his adjusted monthly benefit be?

60. **Cost of Renting a Truck** The Ace Truck Leasing Company leases a certain size truck at $30 a day and 15 cents a mile, whereas the Acme Truck Leasing Company leases the same size truck at $25 a day and 20 cents a mile.
 a. Find the daily cost of leasing from each company as a function of the number of miles driven.
 b. Sketch the graphs of the two functions on the same set of axes.
 c. Which company should a customer rent a truck from for one day if she plans to drive at most 70 miles and wishes to minimize her cost?

61. **Boyle's Law** As a consequence of Boyle's law, the pressure P of a fixed sample of gas held at a constant temperature is related to the volume V of the gas by the rule

$$P = f(V) = \frac{k}{V}$$

where k is a constant. What is the domain of the function f? Sketch the graph of the function f.

62. **Poiseuille's Law** According to a law discovered by the nineteenth-century physician Poiseuille, the velocity (in centimeters per second) of blood r centimeters from the central axis of an artery is given by

$$v(r) = k(R^2 - r^2)$$

where k is a constant and R is the radius of the artery. Suppose that for a certain artery, $k = 1000$ and $R = 0.2$ so that $v(r) = 1000(0.04 - r^2)$.
 a. What is the domain of the function $v(r)$?
 b. Compute $v(0)$, $v(0.1)$, and $v(0.2)$ and interpret your results.

63. **Population Growth** A study prepared for a certain Sunbelt town's Chamber of Commerce projected that the population of the town in the next three years will grow according to the rule

$$P(x) = 50{,}000 + 30x^{3/2} + 20x$$

where $P(x)$ denotes the population x months from now. By how much will the population increase during the next 9 months? During the next 16 months?

64. **Worker Efficiency** An efficiency study conducted for the Elektra Electronics Company showed that the number of "Space Commander" walkie-talkies assembled by the average worker t hours after starting work at 8:00 A.M. is given by

$$N(t) = -t^3 + 6t^2 + 15t \qquad 0 \le t \le 4$$

How many walkie-talkies can an average worker be expected to assemble between 8:00 and 9:00 A.M.? Between 9:00 and 10:00 A.M.?

65. Learning Curves The Emory Secretarial School finds from experience that the average student taking Advanced Typing will progress according to the rule

$$N(t) = \frac{60t + 180}{t + 6}$$

where $N(t)$ measures the number of words per minute the student can type after t weeks in the course. How fast can the average student be expected to type after two weeks in the course? After four weeks in the course?

66. Politics Political scientists have discovered the following empirical rule, known as the "cube rule,"

which gives the relationship between the proportion of seats in the House of Representatives won by Democratic candidates $s(x)$ and the proportion of popular votes x received by the Democratic presidential candidate:

$$s(x) = \frac{x^3}{x^3 + (1 - x)^3} \qquad 0 \le x \le 1$$

Compute $s(0.6)$ and interpret your result.

67. Home Shopping Industry According to industry sources, revenue from the home shopping industry for the years since its inception may be approximated by the function

$$R(t) = \begin{cases} -0.03t^3 + 0.25t^2 - 0.12t & 0 \le t \le 3 \\ 0.57t - 0.63 & 3 < t \le 11 \end{cases}$$

where $R(t)$ measures the revenue in billions of dollars and t is measured in years, with $t = 0$ corresponding to the beginning of 1984. What was the revenue at the beginning of 1985? At the beginning of 1993?
Source: Paul Kagen Associates

SOLUTIONS TO SELF-CHECK EXERCISES 2.3

1. a. The expression under the radical sign must be nonnegative, so $x + 1 \ge 0$ or $x \ge -1$. Also, $x \ne 0$, because division by zero is not permitted. Therefore, the domain of f is $[-1, 0) \cup (0, \infty)$.

b. $f(3) = \dfrac{\sqrt{3 + 1}}{3} = \dfrac{\sqrt{4}}{3} = \dfrac{2}{3}$

c. $f(a + h) = \dfrac{\sqrt{(a + h) + 1}}{a + h} = \dfrac{\sqrt{a + h + 1}}{a + h}$

2. a. For t in the subdomain $[0, 6]$, the rule for f is given by $f(t) = 6t + 17$. The equation $y = 6t + 17$ is a linear equation, so that a portion of the graph of f is the line segment joining the points $(0, 17)$ and $(6, 53)$. Next, in the subdomain $(6, 20]$, the rule for f is given by $f(t) = 15.98(t - 6)^{1/4} + 53$. Using a calculator, we construct the following table of values of $f(t)$ for selected values of t:

t	6	8	10	12	14	16	18	20
$f(t)$	53	72	75.6	78	79.9	81.4	82.7	83.9

We included $t = 6$ in the table, although it does not lie in the subdomain of the function under consideration, in order to help us obtain a better sketch of that portion of the graph of f in the subdomain $(6, 20]$. The graph of f is as follows:

b. The percentage of all self-serve gas sales at the beginning of 1978 is found by evaluating f at $t = 4$. Since this point lies in the interval $[0, 6]$, we use the rule $f(t) = 6t + 17$ and find

$$f(4) = 6(4) + 17$$

giving 41% as the required figure. The percentage of all self-serve gas sales at the beginning of 1994 is given by

$$f(20) = 15.98(20 - 6)^{1/4} + 53$$

or approximately 83.9%.

3. A point (x, y) lies on the graph of the function f if and only if the coordinates satisfy the equation $y = f(x)$. Now

$$f(4) = \sqrt{2(4) + 1} + 2 = \sqrt{9} + 2 = 5 \neq 6$$

and we conclude that the given point does not lie on the graph of f.

Using Technology

Graphing Utilities

Most of the graphs of functions in this book can be plotted with the help of a graphing utility. Furthermore, a graphing utility can be used to analyze the nature of a function. However, the amount and accuracy of the information obtained using a graphing utility depends on the experience and sophistication of the user. As you progress through this book, you will see that the more knowledge of calculus you gain, the more effective the graphing utility will prove as a tool in problem solving.

In this book we use Texas Instruments' TI-85 Graphics Calculator as our reference, although any other graphing calculator or computer software program with similar capabilities can be used. The terminology used here is that associated with the TI-85. Step-by-step guidelines for each procedure, including the necessary keystrokes, are given in Appendix A. A technology index at the back of the book contains a list of all the procedures described in the technology sections and the appendix.

Graphing a Function

The first step in plotting the graph of a function with a graphing utility is to select a suitable viewing rectangle. We usually do this by experimenting. For example, you might first plot the graph using the *standard viewing rectangle* $[-10, 10]$ by $[-10, 10]$. If necessary, you then might adjust the viewing rectangle by enlarging it or reducing it to obtain a sufficiently complete view of the graph or at least the portion of the graph that is of interest. (See Section A.1, Graphing a Function, in Appendix A.)

EXAMPLE 1   Plot the graph of $f(x) = 2x^2 - 4x - 5$ in the standard viewing rectangle.

Solution   The graph of f, shown in Figure 1, is a parabola. From our previous work (Section 2.3), we know that the figure does give a good view of the graph.

FIGURE 1
The graph of
$y = 2x^2 - 4x - 5$
in the viewing rectangle
$[-10, 10] \times [-10, 10]$

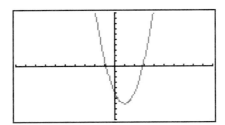

EXAMPLE 2   Let $f(x) = x^3(x - 3)^4$.

a. Plot the graph of f in the standard viewing rectangle.

b. Plot the graph of f in the rectangle $[-1, 5] \times [-40, 40]$.

Solution

a. The graph of f in the standard viewing rectangle is shown in Figure 2(a). Since the graph does not appear to be complete, we need to adjust the viewing rectangle.

FIGURE 2
An incomplete sketch of
$f(x) = x^3(x - 3)^4$ *is shown in*
(a) and a complete sketch is
shown in (b).

(a)   (b)

b. The graph of f in the rectangle $[-1, 5] \times [-40, 40]$, shown in Figure 2(b), is an improvement over the previous graph. (Later we will be able to show that the figure does, in fact, give a rather complete view of the graph of f.) ■

Evaluating a Function

A graphing utility can be used to find the value of a function with minimal effort, as the next example shows. (See Section A.2, Evaluating a Function, in Appendix A.)

EXAMPLE 3   Let $f(x) = x^3 - 4x^2 + 4x + 2$.

a. Plot the graph of f in the standard viewing rectangle.

b. Use **EVAL** to find $f(3)$ and verify your result by direct computation.

c. Use **EVAL** to find $f(4.215)$.

Solution

a. The graph of f is shown in Figure 3.

FIGURE 3
The graph of
$f(x) = x^3 - 4x^2 + 4x + 2$
in the standard
viewing rectangle

b. Using **EVAL** and the value of 3 for x, we find $y = 5$. This result is verified by computing

$$f(3) = 3^3 - 4(3^2) + 4(3) + 2 = 27 - 36 + 12 + 2 = 5$$

c. Using **EVAL** and the value 4.215 for x, we find $y = 22.679738375$. Thus, $f(4.215) = 22.679738375$. The efficacy of the graphing utility is clearly demonstrated here! ■

 Remember that when evaluating $f(x)$ at $x = a$, the number a must lie between xMin and xMax.

   We can also use the **evalF** function of a graphing utility to find the value of a function at a given point without graphing the function.

EXAMPLE 4   Use the **evalF** function to find $f(4.215)$, where

$$f(x) = x^3 - 4x^2 + 4x + 2$$

Solution   We find

$$f(4.215) = \text{evalF}(x^3 - 4x^2 + 4x + 2, x, 4.215) = 22.679738375$$

■

EXERCISES

In exercises 1–8, plot the graph of the function f in the standard viewing window.

1. $f(x) = 2x^2 - 16x + 29$ **2.** $f(x) = -x^2 - 10x - 20$

3. $f(x) = x^3 - 2x^2 + \dot{x} - 2$

4. $f(x) = -2.01x^3 + 1.21x^2 - 0.78x + 1$

5. $f(x) = 0.2x^4 - 2.1x^2 + 1$

6. $f(x) = -0.4x^4 + 1.2x - 1.2$

7. $f(x) = 2x\sqrt{x^2 + 1}$ **8.** $f(x) = \dfrac{\sqrt{x} + 1}{\sqrt{x} - 1}$

In exercises 9–20, plot the graph of the function f in (a) the standard viewing window and (b) the indicated window.

9. $f(x) = 2x^2 - 32x + 125; [5, 15] \times [-5, 10]$

10. $f(x) = x^2 + 20x + 95; [-20, 10] \times [-15, -5]$

11. $f(x) = x^3 - 20x^2 + 8x - 10; [-20, 20] \times [-1200, 100]$

12. $f(x) = -2x^3 + 10x^2 - 15x - 5; [-10, 10] \times [-100, 100]$

13. $f(x) = x^4 - 2x^2 + 8; [-2, 2] \times [6, 10]$

14. $f(x) = x^4 - 2x^3; [-1, 3] \times [-2, 2]$

15. $f(x) = x + \dfrac{1}{x}; [-1, 3] \times [-5, 5]$

16. $f(x) = \dfrac{4}{x^2 - 8}; [-5, 5] \times [-5, 5]$

17. $f(x) = 2 - \dfrac{1}{x^2 + 1}; [-3, 3] \times [0, 3]$

18. $f(x) = x - 2\sqrt{x}; [0, 20] \times [-2, 10]$

19. $f(x) = x\sqrt{4 - x^2}; [-3, 3] \times [-2, 2]$

20. $f(x) = \dfrac{\sqrt{x} - 1}{x}; [0, 50] \times [-0.25, 0.25]$

In exercises 21–30, plot the graph of the function f in an appropriate viewing window. Note: The answer is not unique.

21. $f(x) = x^2 - 4x + 16$ **22.** $f(x) = -x^2 + 2x - 11$

23. $f(x) = 2x^3 - 10x^2 + 5x - 10$

24. $f(x) = -x^3 + 5x^2 - 14x + 20$

25. $f(x) = 2x^4 - 3x^3 + 5x^2 - 20x + 40$

26. $f(x) = -2x^4 + 5x^2 - 4$

27. $f(x) = \dfrac{x^3}{x^3 + 1}$ **28.** $f(x) = \dfrac{2x^4 - 3x}{x^2 - 1}$

29. $f(x) = 0.2\sqrt{x} - 0.3x^3$ **30.** $f(x) = \sqrt{x}(2x - 1)^3$

*In exercises 31–34, use **evalF** to find the value of f at the given value of x and verify your result by direct computation.*

31. $f(x) = -3x^3 + 5x^2 - 2x + 8; x = -1$

32. $f(x) = 2x^4 - 3x^3 + 2x^2 + x - 5; x = 2$

33. $f(x) = \dfrac{x^4 - 3x^2}{x - 2}; x = 1$ **34.** $f(x) = \dfrac{\sqrt{x^2 - 1}}{3x + 4}; x = 2$

*In exercises 35–42, use **EVAL** to find the value of f at the indicated value of x. Express your answer accurate to four decimal places.*

35. $f(x) = 3x^3 - 2x^2 + x - 4; x = 2.145$

36. $f(x) = 2x^3 + 5x^2 + 3x + 1; x = -0.27$

37. $f(x) = 5x^4 - 2x^2 + 8x - 3; x = 1.28$

38. $f(x) = 4x^4 - 3x^3 + 1; x = -2.42$

39. $f(x) = \dfrac{2x^3 - 3x + 1}{3x - 2}; x = 2.41$

40. $f(x) = \dfrac{2x + 5}{3x^2 - 4x + 1}; x = -1.72$

41. $f(x) = \sqrt{2x^2 + 1} + \sqrt{3x^2 - 1}; x = 0.62$

42. $f(x) = 2x(3x^3 + 5)^{1/3}; x = -6.24$

43. Manufacturing Capacity   Data obtained from the Federal Reserve show that the annual increase in manufacturing capacity between 1988 and 1994 is given by

$$f(t) = 0.0388889t^3 - 0.283333t^2 + 0.477778t$$
$$+ 2.04286 \qquad 0 \le t \le 6$$

where $f(t)$ is a percentage and t is measured in years with $t = 0$ corresponding to the beginning of 1988.
a. Use a graphing utility to plot the graph of f in the viewing rectangle $[0, 8] \times [0, 4]$.
b. What is the annual increase in manufacturing capacity at the beginning of 1990 ($t = 2$)? At the beginning of 1992 ($t = 4$)?

Source: Federal Reserve

2.4   # Linear Functions

We now focus our attention on an important class of functions known as linear functions. Recall that a linear equation in x and y has the form

$Ax + By + C = 0$, where A, B, and C are constants and A and B are not both zero. If $B \neq 0$, the equation can always be solved for y in terms of x; in fact, as we saw in Section 2.2, the equation may be cast in the slope-intercept form:

$$y = mx + b \qquad m, b \text{ constants} \tag{7}$$

Equation (7) defines y as a function of x. The domain and range of this function is the set of all real numbers. Furthermore, the graph of the function, as we saw in Section 2.2, is a straight line in the plane. For this reason, the function $f(x) = mx + b$ is called a *linear function*.

LINEAR FUNCTION

The function f defined by

$$f(x) = mx + b$$

where m and b are constants, is called a **linear function.**

Linear functions play an important role in the quantitative analysis of business and economic problems. First, many problems arising in these and other fields are *linear* in nature or are *linear* in the intervals of interest and thus can be formulated in terms of linear functions. Second, because linear functions are relatively easy to work with, assumptions involving linearity are often made in the formulation of problems. In many cases these assumptions are justified, and acceptable mathematical models are obtained that approximate real-life situations.

In the rest of this section we look at several applications that can be modeled using linear functions.

Simple Depreciation

We first discussed linear depreciation in Section 2.1 as a real-world application of straight lines. The following example illustrates how to derive an equation describing the book value of an asset being depreciated linearly.

EXAMPLE 1   A printing machine has an original value of $100,000 and is to be depreciated linearly over five years with a $30,000 scrap value. Find an expression giving the book value at time t ($0 \leq t \leq 5$). What will the book value of the machine be at time $t = 2$? What is the rate of depreciation of the printing machine?

Solution   Let V denote the printing machine's book value at the end of the tth year. Since the depreciation is linear, V is a linear function of t. Equivalently, the graph of the function is a straight line. Now, in order to find an equation of the straight line, observe that $V = 100,000$ when $t = 0$; this tells us that the line passes through the point $(0, 100,000)$. Similarly, the condition that $V = 30,000$ when $t = 5$ says that the line also passes through the point $(5, 30,000)$. The slope of the line is given by

$$m = \frac{100,000 - 30,000}{0 - 5} = -\frac{70,000}{5} = -14,000$$

Using the point-slope form of the equation of a line with the point $(0, 100{,}000)$ and the slope $m = -14{,}000$, we have

$$V - 100{,}000 = -14{,}000(t - 0)$$

or

$$V = -14{,}000t + 100{,}000$$

the required expression. The book value at the end of the second year is given by

$$V = -14{,}000(2) + 100{,}000 = 72{,}000$$

or $72,000. The rate of depreciation of the machine is given by the negative of the slope of the depreciation line. Since the slope of the line is $a = -14{,}000$, the rate of depreciation is $14,000 per year. The graph of $V = -14{,}000t + 100{,}000$ is shown in Figure 30. ■

FIGURE 30

Linear depreciation of an asset

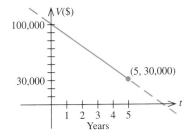

Linear Cost, Revenue, and Profit Functions

Whether a business is a sole proprietorship or a large corporation, the owner or chief executive must constantly keep track of operating costs, revenue resulting from the sale of products or services, and perhaps most important, the profits realized. Three functions provide management with a measure of these quantities: the total cost function, the revenue function, and the profit function.

COST, REVENUE, AND PROFIT FUNCTIONS

Let x denote the number of units of a product manufactured or sold. Then, the **total cost function** is

$C(x) = $ Total cost of manufacturing x units of the product

The **revenue function** is

$R(x) = $ Total revenue realized from the sale of x units of the product

The **profit function** is

$P(x) = $ Total profit realized from manufacturing and selling x units of the product

Generally speaking, the total cost, revenue, and profit functions associated with a company are more likely than not to be nonlinear (these functions are best studied using the tools of calculus). But *linear* cost, revenue, and profit functions do arise in practice, and we consider such functions in this section. Before deriving explicit forms of these functions, however, we need to recall some common terminology.

The costs incurred in operating a business are usually classified into two categories. Costs that remain more or less constant regardless of the firm's activity level are called *fixed costs.* Examples of fixed costs are rental fees and executive salaries. Costs that vary with production or sales are called *variable costs.* Examples of variable costs are wages and costs for raw materials.

Suppose a firm has a fixed cost of F dollars, a production cost of c dollars per unit, and a selling price of s dollars per unit. Then the *cost function* $C(x)$, the *revenue function* $R(x)$, and the *profit function* $P(x)$ for the firm are given by

$$C(x) = cx + F$$

$$R(x) = sx$$

and
$$P(x) = R(x) - C(x) \qquad \text{Revenue} - \text{Cost}$$

$$= (s - c)x - F$$

where x denotes the number of units of the commodity produced and sold. The functions C, R, and P are linear functions of x.

EXAMPLE 2   Puritron, a manufacturer of water filters, has a monthly fixed cost of \$20,000, a production cost of \$20 per unit, and a selling price of \$30 per unit. Find the cost function, the revenue function, and the profit function for Puritron.

Solution   Let x denote the number of units produced and sold. Then

$$C(x) = 20x + 20{,}000$$

$$R(x) = 30x$$

and
$$P(x) = R(x) - C(x)$$

$$= 30x - (20x + 20{,}000)$$

$$= 10x - 20{,}000 \qquad \blacksquare$$

Intersection of Straight Lines

The solution of certain practical problems involves finding the point of intersection of two straight lines. To see how such a problem may be solved algebraically, suppose we are given two straight lines L_1 and L_2 with equations

$$y = m_1 x + b_1 \qquad \text{and} \qquad y = m_2 x + b_2$$

(where m_1, b_1, m_2, and b_2 are constants) that intersect at the point $P(x, y)$ (Figure 31).

The point $P(x, y)$ lies on the line L_1 and so it satisfies the equation $y = m_1 x + b_1$. It also lies on the line L_2 and so it satisfies the equation $y = m_2 x + b_2$. Therefore, to find the point of intersection $P(x, y)$ of the lines L_1 and L_2, we solve the system comprising the two equations

$$y = m_1 x + b_1 \qquad \text{and} \qquad y = m_2 x + b_2$$

for x and y.

EXAMPLE 3   Find the point of intersection of the straight lines that have equations $y = x + 1$ and $y = -2x + 4$.

Solution   We solve the given simultaneous equations. Substituting the value y as given in the first equation into the second, we obtain

$$x + 1 = -2x + 4$$

$$3x = 3$$

or
$$x = 1$$

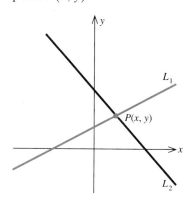

FIGURE 31
L_1 *and* L_2 *intersect at the point* $P(x, y)$.

Substituting this value of x into either one of the given equations gives $y = 2$. Therefore, the required point of intersection is $(1, 2)$ (Figure 32).

FIGURE 32

The point of intersection of L_1 and L_2 is $(1, 2)$.

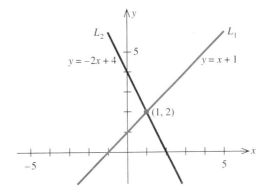

We now turn to some applications involving the intersections of pairs of straight lines.

Break-Even Analysis

Consider a firm with (linear) cost function $C(x)$, revenue function $R(x)$, and profit function $P(x)$ given by

$$C(x) = cx + F$$

$$R(x) = sx$$

and

$$P(x) = R(x) - C(x) = (s - c)x - F$$

where c denotes the unit cost of production, s denotes the selling price per unit, F denotes the fixed cost incurred by the firm, and x denotes the level of production and sales. The level of production at which the firm neither makes a profit nor sustains a loss is called the *break-even level of operation* and may be determined by solving the equations $p = C(x)$ and $p = R(x)$ simultaneously. At the level of production, x_0, the profit is zero, so

$$P(x_0) = R(x_0) - C(x_0) = 0$$

and

$$R(x_0) = C(x_0)$$

The point $P_0(x_0, p_0)$, the solution of the simultaneous equations $p = R(x)$ and $p = C(x)$, is referred to as the *break-even point;* the number x_0 and the number p_0 are called the *break-even quantity* and the *break-even revenue,* respectively.

    Geometrically, the break-even point $P_0(x_0, p_0)$ is just the point of intersection of the straight lines representing the cost and revenue functions, respectively. This follows because $P_0(x_0, p_0)$, being the solution of the simultaneous equations $p = R(x)$ and $p = C(x)$, must lie on both these lines simultaneously (Figure 33).

    Note that if $x < x_0$, then $R(x) < C(x)$ so that $P(x) = R(x) - C(x) < 0$ and thus the firm sustains a loss at this level of production. However, if $x > x_0$, then $P(x) > 0$ and the firm operates at a profitable level.

FIGURE 33

P_0 is the break-even point.

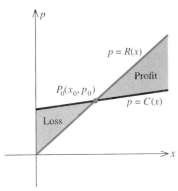

FIGURE 34

The point at which $R(x) = C(x)$ is the break-even point.

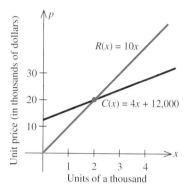

EXAMPLE 4   Prescott, Inc., manufactures its products at a cost of $4 per unit and sells them for $10 per unit. If the firm's fixed cost is $12,000 per month, determine the firm's break-even point.

Solution   The cost function C and the revenue function R are given by $C(x) = 4x + 12,000$ and $R(x) = 10x$, respectively (Figure 34).
Setting $R(x) = C(x)$, we obtain

$$10x = 4x + 12,000$$
$$6x = 12,000$$
$$x = 2,000$$

Substituting this value of x into $R(x) = 10x$ gives

$$R(2,000) = (10)(2,000) = 20,000$$

So, for a break-even operation, the firm should manufacture 2,000 units of its product, resulting in a break-even revenue of $20,000.   ■

EXAMPLE 5   Using the data given in Example 4, answer the following questions:

a. What is the loss sustained by the firm if only 1,500 units are produced and sold per month?
b. What is the profit if 3,000 units are produced and sold per month?
c. How many units should the firm produce in order to realize a minimum monthly profit of $9,000?

Solution   The profit function P is given by the rule

$$P(x) = R(x) - C(x)$$
$$= 10x - (4x + 12,000)$$
$$= 6x - 12,000$$

a. If 1,500 units are produced and sold per month, we have

$$P(1,500) = 6(1,500) - 12,000 = -3,000$$

so the firm will sustain a loss of $3,000 per month.

b. If 3,000 units are produced and sold per month, we have

$$P(3,000) = 6(3,000) - 12,000 = 6,000$$

or a monthly profit of $6,000.

c. Substituting 9,000 for $P(x)$ in the equation $P(x) = 6x - 12,000$, we obtain

$$9,000 = 6x - 12,000$$
$$6x = 21,000$$
$$x = 3,500$$

Thus, the firm should produce at least 3,500 units in order to realize a $9,000 minimum monthly profit.   ■

EXAMPLE 6   The management of the Robertson Controls Company must decide between two manufacturing processes for its Model C electronic thermostat. The monthly cost of the first process is given by $C_1(x) =$

$20x + 10,000$ dollars, where x is the number of thermostats produced, and the monthly cost of the second process is given by $C_2(x) = 10x + 30,000$ dollars. If the projected sales are 800 thermostats at a unit price of $40, which process should management choose in order to maximize the company's profit?

Solution The break-even level of operation using the first process is obtained by solving the equation

$$40x = 20x + 10,000$$

or

$$20x = 10,000$$

Thus

$$x = 500$$

giving an output of 500 units. Next, we solve the equation

$$40x = 10x + 30,000$$

or

$$30x = 30,000$$

Thus

$$x = 1,000$$

giving an output of 1,000 units for a break-even operation using the second process. Since the projected sales are 800 units, we conclude that management should choose the first process, which would give the firm a profit. ∎

EXAMPLE 7 Referring to Example 6, which process should Robertson's management choose if the projected sales are (a) 1,500 units and (b) 3,000 units?

Solution In both cases, the production is past the break-even level. Since the revenue is the same regardless of which process is employed, the decision will be based on how much each process costs.

a. If $x = 1,500$, then

$$C_1(x) = (20)(1,500) + 10,000 = 40,000$$
$$C_2(x) = (10)(1,500) + 30,000 = 45,000$$

Hence, management should choose the first process.

b. If $x = 3,000$, then

$$C_1(x) = (20)(3,000) + 10,000 = 70,000$$
$$C_2(x) = (10)(3,000) + 30,000 = 60,000$$

In this case, management should choose the second process. ∎

**SELF-CHECK
EXERCISES 2.4**

1. A manufacturer has a monthly fixed cost of $60,000 and a production cost of $10 for each unit produced. The product sells for $15 per unit.
   a. What is the cost function?
   b. What is the revenue function?
   c. What is the profit function?
   d. Compute the profit (loss) corresponding to production levels of 10,000 and 14,000 units.

2. Find the point of intersection of the straight lines with equations $2x + 3y = 6$ and $x - 3y = 4$.

Solutions to Self-Check Exercises 2.4 can be found on page 99.

2.4 EXERCISES

In exercises 1–10, determine whether the equation defines y as a linear function of x. If so, write it in the form y = mx + b.

1. $2x + 3y = 6$ **2.** $-2x + 4y = 7$

3. $x = 2y - 4$ **4.** $2x = 3y + 8$

5. $2x - 4y + 9 = 0$ **6.** $3x - 6y + 7 = 0$

7. $2x^2 - 8y + 4 = 0$ **8.** $3\sqrt{x} + 4y = 0$

9. $2x - 3y^2 + 8 = 0$ **10.** $3x + \sqrt{y} - 4 = 0$

11. A manufacturer has a monthly fixed cost of $40,000 and a production cost of $8 for each unit produced. The product sells for $12 per unit.
a. What is the cost function?
b. What is the revenue function?
c. What is the profit function?
d. Compute the profit (loss) corresponding to production levels of 8,000 and 12,000 units.

12. A manufacturer has a monthly fixed cost of $100,000 and a production cost of $14 for each unit produced. The product sells for $20 per unit.
a. What is the cost function?
b. What is the revenue function?
c. What is the profit function?
d. Compute the profit (loss) corresponding to production levels of 12,000 and 20,000 units.

In exercises 13–18, find the point of intersection of each pair of straight lines.

13. $y = 3x + 4$
$y = -2x + 14$

14. $y = -4x - 7$
$-y = 5x + 10$

15. $2x - 3y = 6$
$3x + 6y = 16$

16. $2x + 4y = 11$
$-5x + 3y = 5$

17. $y = \frac{1}{4}x - 5$
$2x - \frac{3}{2}y = 1$

18. $y = \frac{2}{3}x - 4$
$x + 3y + 3 = 0$

In exercises 19–22, find the break-even point for the firm whose cost function C and revenue function R are given.

19. $C(x) = 5x + 10,000$, $R(x) = 15x$

20. $C(x) = 15x + 12,000$, $R(x) = 21x$

21. $C(x) = 0.2x + 120$, $R(x) = 0.4x$

22. $C(x) = 150x + 20,000$, $R(x) = 270x$

23. Linear Depreciation   An office building worth $1 million when completed in 1995 is being depreci-

ated linearly over 50 years. What will the book value of the building be in 2000? In 2005? (Assume the scrap value is $0.)

24. Linear Depreciation   An automobile purchased for use by the manager of a firm at a price of $14,000 is to be depreciated using the straight-line method over five years. What will the book value of the automobile be at the end of three years? (Assume the scrap value is $0.)

25. Profit Functions   Auto-Time, a manufacturer of 24-hour variable timers, has a monthly fixed cost of $48,000 and a production cost of $8 for each timer manufactured. The timers sell for $14 each.
a. What is the cost function?
b. What is the revenue function?
c. What is the profit function?
d. Compute the profit (loss) corresponding to production levels of 4,000, 6,000, and 10,000 timers, respectively.

26. Profit Functions   The management of T.M.I., Inc., finds that the monthly fixed costs attributable to the company's blank tape division amount to $12,100. If the cost for producing each reel of tape is $.60 and each reel of tape sells for $1.15, find the company's cost function, revenue function, and profit function.

27. Linear Depreciation   In 1992, the National Textile Company installed a new machine in one of its factories at a cost of $250,000. The machine is depreciated linearly over ten years with a scrap value of $10,000.
a. Find an expression for the machine's book value in the tth year of use ($0 \le t \le 10$).
b. Sketch the graph of the function of (a).
c. Find the machine's book value in 1996.
d. Find the rate at which the machine is being depreciated.

28. Linear Depreciation   A minicomputer purchased at a cost of $60,000 in 1994 has a scrap value of $12,000 at the end of four years. Assume the straight-line method of depreciation is used.
a. Find the rate of depreciation.
b. Find the linear equation expressing the minicomputer's book value at the end of t years.
c. Sketch the graph of the function of (b).
d. Find the minicomputer's book value at the end of the third year.

29. Linear Depreciation   Suppose that an asset has an original value of $$C$ and is depreciated linearly over N years with a scrap value of $$S$. Show that the asset's book value at the end of the tth year is described by the function

$$V(t) = C - \left(\frac{C - S}{N}\right) t$$

[*Hint:* Find an equation of the straight line passing through the points $(0, C)$ and (N, S). (Why?)]

30. Rework exercise 23 using the formula derived in exercise 29.

31. Rework exercise 24 using the formula derived in exercise 29.

32. Drug Dosages   A method sometimes used by pediatricians to calculate the dosage of medicine for children is based on the child's surface area. If a denotes the adult dosage (in mg), and if S is the child's surface area (in square meters), then the child's dosage is given by

$$D(S) = \frac{Sa}{1.7}$$

   a. Show that D is a linear function of S. [*Hint:* Think of D as having the form $D(S) = mS + b$. What is the slope m and the y-intercept b?]
   b. If the adult dose of a drug is 500 mg, how much should a child whose surface area is 0.4 square meter receive?

33. Drug Dosages   Cowling's Rule is a method for calculating pediatric drug dosages. If a denotes the adult dosage (in mg) and if t is the child's age (in years), then the child's dosage is given by

$$D(t) = \left(\frac{t + 1}{24}\right) a$$

   a. Show that D is a linear function of t. [*Hint:* Think of $D(t)$ as having the form $D(t) = mt + b$. What is the slope m and the y-intercept b?]
   b. If the adult dose of a drug is 500 mg, how much should a four-year-old child receive?

34. Break-Even Analysis   Auto-Time, a manufacturer of 24-hour variable timers, has a monthly fixed cost of $48,000 and a production cost of $8 for each timer manufactured. The units sell for $14 each.
   a. Sketch the graphs of the cost function and the revenue function, and hence find the break-even point graphically.
   b. Find the break-even point algebraically.
   c. Sketch the graph of the profit function.
   d. At what point does the graph of the profit function cross the x-axis? Interpret your result.

c 35. Break-Even Analysis   A division of Carter Enterprises produces "Personal Income Tax" diaries. Each diary sells for $8. The monthly fixed costs incurred by the division are $25,000, and the variable cost of producing each diary is $3.
   a. Find the break-even point for the division.
   b. What should the level of sales be in order for the division to realize a 15% profit over the cost of making the diaries?

c 36. Break-Even Analysis   A division of the Gibson Corporation manufactures bicycle pumps. Each pump sells for $9, and the variable cost of producing each unit is 40% of the selling price. The monthly fixed costs incurred by the division are $50,000. What is the break-even point for the division?

c 37. Decision Analysis   A product may be made using machine I or machine II. The manufacturer estimates that the monthly fixed costs of using machine I are $18,000, whereas the monthly fixed costs of using machine II are $15,000. The variable costs of manufacturing one unit of the product using machine I and machine II are $15 and $20, respectively. The product sells for $50 each.
   a. Find the cost functions associated with using each machine.
   b. Sketch the graphs of the cost functions of (a) and the revenue functions on the same set of axes.
   c. Which machine should management choose to maximize their profit if the projected sales are 450 units? 550 units? 650 units?
   d. What is the profit for each case in (c)?

SOLUTIONS TO SELF-CHECK EXERCISES 2.4

1. Let x denote the number of units produced and sold. Then
   a. $C(x) = 10x + 60,000$
   b. $R(x) = 15x$
   c. $P(x) = R(x) - C(x) = 15x - (10x + 60,000)$
   $$= 5x - 60,000$$
   d. $P(10,000) = 5(10,000) - 60,000$
   $$= -10,000$$
   or a loss of $10,000.
   $P(14,000) = 5(14,000) - 60,000$
   $$= 10,000$$
   or a profit of $10,000.

2. The point of intersection of the two straight lines is found by solving the system of linear equations

$$2x + 3y = 6$$

$$x - 3y = 4$$

Adding the second equation to the first, we obtain

$$3x = 10$$

or $x = 10/3$. Substituting this value of x into the second equation, we find

$$\tfrac{10}{3} - 3y = 4$$

$$3y = -\tfrac{2}{3}$$

or $\qquad\qquad\qquad y = -\tfrac{2}{9}$

Therefore, the point of intersection is $(10/3, -2/9)$.

Using Technology

The **ISECT** operation of a graphing utility can be used to find the point(s) of intersection of the graphs of two functions. (See Section A.3, Finding the Points of Intersection of Two Graphs, in Appendix A.)

EXAMPLE 1   Use **ISECT** to find the points of intersection of the graphs of

$$f(x) = 0.3x^2 - 1.4x - 3 \qquad \text{and} \qquad g(x) = -0.4x^2 + 0.8x + 6.4$$

Solution   The graphs of both f and g in the standard viewing rectangle are shown in Figure 1. Moving the cursor near each point of intersection in succession and pressing ENTER twice, we find the points of intersection, accurate to four decimal places, to be $(-2.4158, 2.1329)$ and $(5.5587, -1.5125)$.

FIGURE 1

The graphs of f and g in the standard viewing window

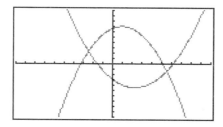

EXERCISES

In exercises 1–6, find the points of intersection of the graphs of the functions. Express your answer accurate to four decimal places.

1. $f(x) = 1.2x + 3.8; g(x) = -0.4x^2 + 1.2x + 7.5$

2. $f(x) = 0.2x^2 - 1.3x - 3; g(x) = -1.3x + 2.8$

3. $f(x) = 0.3x^2 - 1.7x - 3.2; g(x) = -0.4x^2 + 0.9x + 6.7$

4. $f(x) = -0.3x^2 + 0.6x + 3.2; g(x) = 0.2x^2 - 1.2x - 4.8$

5. $f(x) = 0.3x^3 - 1.8x^2 + 2.1x - 2; g(x) = 2.1x - 4.2$

6. $f(x) = -0.2x^3 + 1.2x^2 - 1.2x + 2; g(x) = -0.2x^2 + 0.8x + 2.1$

2.5 Quadratic Functions

Quadratic Functions

A **quadratic function** is one of the form

$$f(x) = ax^2 + bx + c$$

where a, b, and c are constants and $a \neq 0$. For example, the function $f(x) = 2x^2 - 3x + 4$ is quadratic (here $a = 2$, $b = -3$, and $c = 4$). Also, the function $f(x) = x^2 + 1$ of Example 4, Section 2.3, is quadratic (here $a = 1$, $b = 0$, and $c = 1$). Its graph illustrates the general shape of the graph of a quadratic function, which is called a *parabola* (Figure 35).

In general, the graph of a quadratic function is a parabola that opens upward or downward (Figure 36). Furthermore, the parabola is symmetric with respect to a vertical line called the *axis of symmetry* (shown dotted in Figure 36). This line also passes through the lowest point or the highest point of the parabola. This point of intersection of the parabola with its axis of symmetry is called the *vertex* of the parabola. We can use these properties to help us sketch the graph of a quadratic function. For example, suppose we want to sketch the graph of

$$f(x) = 2x^2 - 4x + 1$$

If we complete the square in x, we obtain

$$f(x) = 2(x^2 - 2x \quad) + 1$$

Factor out the coefficient of x^2 from the first two terms.

Adding and subtracting 2
$$= 2[x^2 - 2x + (-1)^2] + 1 - 2$$

Because of the 2 outside the brackets, we have added 2(1) and must therefore subtract 2.

$$= 2(x - 1)^2 - 1$$

Factor the terms within the brackets.

FIGURE 35

The graph of $f(x) = x^2 + 1$ is a parabola.

(−3, 10) 10
(3, 10)
12
8
(−2, 5) (2, 5)
6
(−1, 2) (1, 2)
4
−4   4

FIGURE 36

Graphs of quadratic functions are parabolas.

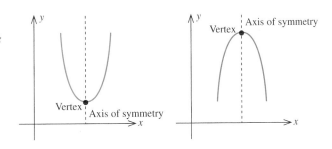

Observe that the first term, $2(x - 1)^2$, is nonnegative. In fact, it is equal to zero when $x = 1$ and is greater than zero if $x \neq 1$. Consequently, we see that $f(x) \geq -1$ for all values of x. This tells us that the vertex (in this case, the lowest point) of the parabola is the point $(1, -1)$. The axis of symmetry of the parabola is the vertical line $x = 1$. Finally, plotting the vertex and a few additional points on either side of the axes of symmetry of the parabola, we obtain the graph shown in Figure 37.

FIGURE 37

The graph of
$f(x) = 2x^2 - 4x + 1$

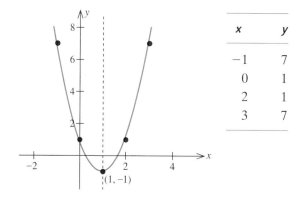

x	y
−1	7
0	1
2	1
3	7

The x-intercepts of f, the x-coordinates of the points at which the parabola intersects the x-axis, can be found by solving the equation $f(x) = 0$. Here we use the quadratic formula to find that

$$x = \frac{4 \pm \sqrt{16 - 8}}{4} = \frac{4 \pm \sqrt{8}}{4} = \frac{4 \pm 2\sqrt{2}}{4} = 1 \pm \frac{\sqrt{2}}{2}$$

Therefore, the x-intercepts are $1 + \sqrt{2}/2 \approx 1.71$ and $1 - \sqrt{2}/2 \approx 0.29$. The y-intercept of f (obtained by setting $x = 0$) is $f(0) = 1$.

The technique that we used to analyze $f(x) = 2x^2 - 4x + 1$ can be used to study the general quadratic function

$$f(x) = ax^2 + bx + c \qquad a \neq 0$$

If we complete the square in x, we find

$$f(x) = a\left(x + \frac{b}{2a}\right)^2 + \frac{4ac - b^2}{4a} \qquad \text{See Exercise 42.}$$

From this equation, we obtain the following properties of the quadratic function f.

PROPERTIES OF THE QUADRATIC FUNCTION

$$f(x) = ax^2 + bx + c \qquad a \neq 0$$

1. The domain of f is the set of all real numbers.

2. If $a > 0$, the parabola opens upward and if $a < 0$, it opens downward.

3. The vertex of the parabola is $\left(-\dfrac{b}{2a}, f\left(-\dfrac{b}{2a}\right)\right)$.

4. The axis of symmetry of the parabola is $x = -\dfrac{b}{2a}$.

5. The x-intercepts (if any) are found by solving $f(x) = 0$. The y-intercept is $f(0) = c$.

EXAMPLE 1   Given the quadratic function

$$f(x) = -2x^2 + 5x - 2$$

a. Find the vertex of the parabola.

b. Find the x-intercepts (if any) of the parabola.

c. Sketch the parabola.

Solution

a. Comparing $f(x) = -2x^2 + 5x - 2$ with the general form of the quadratic equation, we find that $a = -2$, $b = 5$, and $c = -2$. Therefore, the x-coordinate of the vertex of the parabola is

$$-\frac{b}{2a} = -\frac{5}{2(-2)} = \frac{5}{4}$$

Next, to find the y-coordinate of the vertex, we evaluate f at $x = 5/4$, obtaining

$$f\left(\frac{5}{4}\right) = -2\left(\frac{5}{4}\right)^2 + 5\left(\frac{5}{4}\right) - 2$$

$$= -\frac{25}{8} + \frac{25}{4} - 2 \qquad -2\left(\frac{25}{16}\right) = -\frac{25}{8}$$

$$= \frac{9}{8} \qquad\qquad -\frac{25}{8} + \frac{50}{8} - \frac{16}{8} = \frac{9}{8}$$

b. To find the x-intercepts of the parabola, we solve the equation

$$-2x^2 + 5x - 2 = 0$$

using the quadratic formula with $a = -2$, $b = 5$, and $c = -2$. We find

$$x = \frac{-5 \pm \sqrt{25 - 4(-2)(-2)}}{2(-2)}$$

$$= \frac{-5 \pm \sqrt{9}}{-4}$$

$$= \frac{-5 \pm 3}{-4}$$

$$= \frac{1}{2} \quad \text{or} \quad 2$$

Thus, the x-intercepts of the parabola are 1/2 and 2.

c. Since $a = -2 < 0$, the parabola opens downward. The vertex of the parabola (5/4, 9/8) is therefore the highest point on the curve. The

FIGURE 38

The graph of
$f(x) = -2x^2 + 5x - 2$

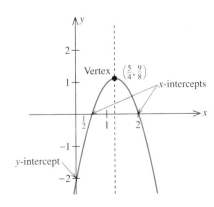

parabola crosses the x-axis at the points $(1/2, 0)$ and $(2, 0)$. Setting $x = 0$ gives -2 as the y-intercept of the curve. Finally, using this information, we sketch the parabola shown in Figure 38. ∎

EXAMPLE 2   The quarterly profit of Cunningham Realty (in thousands of dollars) is given by

$$P(x) = -\frac{1}{3}x^2 + 7x + 30 \qquad 0 \le x \le 50$$

where x (in thousands of dollars) is the amount of money Cunningham spends on advertising per quarter. Find the amount of money Cunningham should spend on advertising in order to realize a maximum quarterly profit. What is the maximum quarterly profit realizable by Cunningham?

Solution   The profit function P is a quadratic function and so its graph is a parabola. Furthermore, the coefficient of x^2 is $a = -1/3 < 0$, and so the parabola opens downward. The x-coordinate of the vertex of the parabola is

$$-\frac{b}{2a} = -\frac{7}{2\left(-\frac{1}{3}\right)} = \frac{21}{2} = 10.5$$

The corresponding y-coordinate is

$$f\left(\frac{21}{2}\right) = -\frac{1}{3}\left(\frac{21}{2}\right)^2 + 7\left(\frac{21}{2}\right) + 30 = \frac{267}{4} = 66.75$$

Therefore, the vertex of the parabola is $(21/2, 267/4)$. Since the parabola opens downward, the vertex of the parabola is the highest point on the parabola. Accordingly, the y-coordinate of the vertex gives the maximum value of P. This implies that the maximum quarterly profit of \$66,750 (remember that $P(x)$ is measured in thousands of dollars) is realized if Cunningham spends \$10,500 per quarter on advertising. The graph of P is shown in Figure 39. ∎

FIGURE 39
The graph of the profit func-
tion $P(x) = -\dfrac{1}{3}x^2 + 7x + 30$

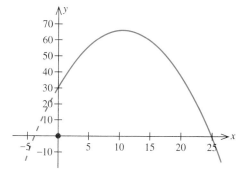

Demand and Supply Curves

In a free market economy, consumer demand for a particular commodity depends on the commodity's unit price. A *demand equation* expresses the relationship between the unit price and the quantity demanded. The graph of the demand equation is called a *demand curve*. In general, the quantity demanded of a commodity decreases as the commodity's unit price increases, and vice versa. Accordingly, a **demand function,** defined by $p = f(x)$, where p measures the unit price and x measures the number of units of the commodity in question, is generally characterized as a decreasing function of x; that is, $p = f(x)$ decreases as x increases. Since both x and p assume only nonnegative values, the demand curve is that part of the graph of $f(x)$ that lies in the first quadrant (Figure 40).

FIGURE 40

A demand curve

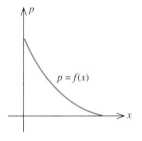

EXAMPLE 3   The demand function for a certain brand of videocassette is given by

$$p = d(x) = -0.01x^2 - 0.2x + 8$$

where p is the wholesale unit price in dollars and x is the quantity demanded each week, measured in units of a thousand. Sketch the corresponding demand curve. Above what price will there be no demand? What is the maximum quantity demanded per week?

Solution   The given function is quadratic, and its graph, which appears in Figure 41, may be sketched using the methods just developed. The p-intercept, 8, gives the wholesale unit price above which there will be no demand. To obtain the maximum quantity demanded, set $p = 0$, which gives

$$-0.01x^2 - 0.2x + 8 = 0$$

$$x^2 + 20x - 800 = 0 \qquad \text{Upon multiplying both sides of the equation by } -100$$

or $\qquad (x + 40)(x - 20) = 0$

FIGURE 41

A demand curve

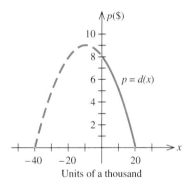

Units of a thousand

that is, $x = -40$ or $x = 20$. Since x must be nonnegative, we reject the root $x = -40$. Thus, the maximum number of videocassettes demanded per week is 20,000. ∎

In a competitive market, a relationship also exists between the unit price of a commodity and the commodity's availability in the market. In general, an increase in the commodity's unit price induces the producer to increase the supply of the commodity. Conversely, a decrease in the unit price generally leads to a drop in the supply. The equation that expresses the relation between the unit price and the quantity supplied is called a *supply equation,* and its graph is called a *supply curve.* A **supply function,** defined by $p = f(x)$, is generally characterized as an increasing function of x; that is, $p = f(x)$ increases as x increases. Since both x and p assume only nonnegative values, the supply curve is that part of the graph of $f(x)$ that lies in the first quadrant (Figure 42).

FIGURE 42

A supply curve

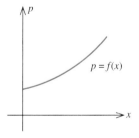

EXAMPLE 4   The supply function for a certain brand of videocassette is given by

$$p = s(x) = 0.01x^2 + 0.1x + 3$$

where p is the unit wholesale price in dollars and x stands for the quantity that will be made available in the market by the supplier (measured in units of a thousand). Sketch the corresponding supply curve. What is the lowest price at which the supplier will make the videocassettes available in the market?

Solution   A sketch of the supply curve appears in Figure 43. The p-intercept, 3, gives the lowest price at which the supplier will make the videocassettes available in the market.

FIGURE 43

A supply curve

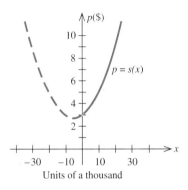

Units of a thousand

Market Equilibrium

FIGURE 44

Market equilibrium corresponds to (x_0, p_0), the point at which the supply and demand curves intersect.

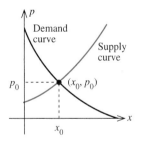

Under pure competition, the price of a commodity will eventually settle at a level dictated by the following condition: the supply of the commodity will be equal to the demand for it. If the price is too high, the consumer will not buy, and if the price is too low, the supplier will not produce. **Market equilibrum** prevails when the quantity produced is equal to the quantity demanded. The quantity produced at market equilibrium is called the *equilibrium quantity,* and the corresponding price is called the *equilibrium price.*

Market equilibrium corresponds to the point at which the demand curve and the supply curve intersect. In Figure 44, x_0 represents the equilibrium quantity and p_0 the equilibrium price. The point (x_0, p_0) lies on the supply curve and therefore satisfies the supply equation. At the same time, it also lies on the demand curve and therefore satisfies the demand equation. Thus, to find the point (x_0, p_0), and hence the equilibrium quantity and price, we solve the demand and supply equations simultaneously for x and p. For meaningful solutions, x and p must both be positive.

EXAMPLE 5   Refer to Examples 3 and 4. The demand function for a certain brand of videocassette is given by

$$p = d(x) = -0.01x^2 - 0.2x + 8$$

and the corresponding supply function is given by

$$p = s(x) = 0.01x^2 + 0.1x + 3$$

where p is expressed in dollars and x is measured in units of a thousand. Find the equilibrium quantity and price.

Solution   We solve the following system of equations:

$$p = -0.01x^2 - 0.2x + 8$$

and $$p = 0.01x^2 + 0.1x + 3$$

Substituting the first equation into the second yields

$$-0.01x^2 - 0.2x + 8 = 0.01x^2 + 0.1x + 3$$

FIGURE 45
The supply curve and the demand curve intersect at the point (10, 5).

Units of a thousand

which is equivalent to

$$0.02x^2 + 0.3x - 5 = 0 \quad \text{Transpose terms.}$$
$$2x^2 + 30x - 500 = 0 \quad \text{Multiply by 100.}$$
$$x^2 + 15x - 250 = 0 \quad \text{Divide by 2.}$$

or $$(x + 25)(x - 10) = 0$$

Thus, $x = -25$ or $x = 10$. Since x must be nonnegative, the root $x = -25$ is rejected. Therefore, the equilibrium quantity is 10,000 videocassettes. (Remember x is measured in units of 1,000.) The equilibrium price is given by

$$p = 0.01(10)^2 + 0.1(10) + 3 = 5$$

or $5 per videocassette (Figure 45). ■

SELF-CHECK EXERCISE 2.5

Given the quadratic function

$$f(x) = 2x^2 - 3x - 3$$

a. Find the vertex of the parabola.

b. Find the x-intercepts (if any) of the parabola.

c. Sketch the parabola.

Solution to Self-Check Exercise 2.5 can be found on page 109.

2.5 EXERCISES

In exercises 1–18, find the vertex, the x-intercepts (if any), and sketch the parabola.

1. $f(x) = x^2 + x - 6$

2. $f(x) = 3x^2 - 5x - 2$

3. $f(x) = x^2 - 4x + 4$

4. $f(x) = x^2 + 6x + 9$

5. $f(x) = -x^2 + 5x - 6$

6. $f(x) = -4x^2 + 4x + 3$

7. $f(x) = 3x^2 - 5x + 1$

8. $f(x) = -2x^2 + 6x - 3$

9. $f(x) = 2x^2 - 3x + 3$

10. $f(x) = 3x^2 - 4x + 2$

11. $f(x) = x^2 - 4$

12. $f(x) = 2x^2 + 3$

13. $f(x) = 16 - x^2$

14. $f(x) = 5 - x^2$

15. $f(x) = \frac{3}{8}x^2 - 2x + 2$

16. $f(x) = \frac{3}{4}x^2 - \frac{1}{2}x + 1$

17. $f(x) = 1.2x^2 + 3.2x - 1.2$

18. $f(x) = 2.3x^2 - 4.1x + 3$

In exercises 19–24, find the points of intersection of the graphs of the functions.

19. $f(x) = -x^2 + 4; g(x) = x + 2$

20. $f(x) = x^2 - 5x + 6; g(x) = \frac{1}{2}x + \frac{3}{2}$

21. $f(x) = -x^2 + 2x + 6; g(x) = x^2 - 6$

22. $f(x) = x^2 - 2x - 2; g(x) = -x^2 - x + 1$

23. $f(x) = 2x^2 - 5x - 8; g(x) = -3x^2 + x + 5$

24. $f(x) = 0.2x^2 - 1.2x - 4; g(x) = -0.3x^2 + 0.7x + 8.2$

For the demand equation in exercises 25 and 26, where x represents the quantity demanded in units of a thousand and p is the unit price in dollars, (a) sketch the demand curve and (b) determine the quantity demanded when the unit price is set at $p.

25. $p = -x^2 + 36; p = 11$ **26.** $p = -x^2 + 16; p = 7$

For the supply equation in exercises 27 and 28, where x is the quantity supplied in units of a thousand and p is the unit price in dollars, (a) sketch the supply curve and (b) determine the price at which the supplier will make 2000 units of the commodity available in the market.

27. $p = 2x^2 + 18$ **28.** $p = x^2 + 16x + 40$

For each pair of supply and demand equations in exercises 29–32, where x represents the quantity demanded in units of a thousand and p the unit price in dollars, find the equilibrium quantity and the equilibrium price.

29. $p = -2x^2 + 80$ and $p = 15x + 30$

30. $p = -x^2 - 2x + 100$ and $p = 8x + 25$

31. $11p + 3x - 66 = 0$ and $2p^2 + p - x = 10$

32. $p = 60 - 2x^2$ and $p = x^2 + 9x + 30$

33. A stone is thrown straight up from the roof of an 80-foot building. The distance of the stone from the ground at any time t (in seconds) is given by

$$h(t) = -16t^2 + 64t + 80$$

a. Sketch the graph of h.
b. At what time does the stone reach the highest point? What is the stone's maximum height from the ground?

34. Maximizing Profits   Lynbrook West, an apartment complex, has 100 two-bedroom units. The monthly profit realized from renting out x apartments is given by

$$P(x) = -10x^2 + 1760x - 50,000$$

dollars. How many units should be rented out in order to maximize the monthly rental profit? What is the maximum monthly profit realizable?

35. Maximizing Profits   The estimated monthly profit realizable by the Cannon Precision Instruments Corporation for manufacturing and selling x units of its model M1 cameras is

$$P(x) = -0.04x^2 + 240x - 10,000$$

dollars. Determine how many cameras Cannon should produce per month in order to maximize its profits.

36. Effect of Advertising on Profits   The relationship between Cunningham Realty's quarterly profits, $P(x)$, and the amount of money x spent on advertising per quarter is described by the function

$$P(x) = -\tfrac{1}{8}x^2 + 7x + 30 \qquad 0 \le x \le 50$$

where both $P(x)$ and x are measured in thousands of dollars.
a. Sketch the graph of P.
b. Find the amount of money the company should spend on advertising per quarter in order to maximize its quarterly profits.

37. Maximizing Revenue   The monthly revenue R (in hundreds of dollars) realized in the sale of Royal electric shavers is related to the unit price p (in dollars) by the equation

$$R(p) = -\tfrac{1}{2}p^2 + 30p$$

a. Sketch the graph of R.
b. At what unit price is the monthly revenue maximized?

38. Supply Functions   The supply function for the Luminar desk lamp is given by

$$p = -0.1x^2 + 0.5x + 15$$

where x is the quantity supplied (in thousands) and p is the unit price in dollars. Sketch the graph of the supply function. What unit price will induce the supplier to make 5000 lamps available in the marketplace?

39. Market Equilibrium   The weekly demand and supply functions for Sportsman 5×7 tents are given by

$$p = -0.1x^2 - x + 40$$

and

$$p = 0.1x^2 + 2x + 20$$

respectively, where p is measured in dollars and x is measured in units of a hundred. Find the equilibrium quantity and price.

40. Market Equilibrium   The management of the Titan Tire Company has determined that the weekly demand and supply functions for their Super Titan tires are given by

$$p = 144 - x^2$$

and

$$p = 48 + \tfrac{1}{2}x^2$$

respectively, where p is measured in dollars and x is measured in units of a thousand. Find the equilibrium quantity and price.

41. Poiseuille's Law   According to a law discovered by the nineteenth-century physician Poiseuille, the velocity (in centimeters per second) of blood r centimeters from the central axis of an artery is given by

$$v(r) = k(R^2 - r^2)$$

where k is a constant and R is the radius of the artery. Suppose that for a certain artery, $k = 1000$ and $R = 0.2$ so that $v(r) = 1000(0.04 - r^2)$.
a. Sketch the graph of v.
b. For what value of r is $v(r)$ largest? Smallest? Interpret your results.

42. Let $f(x) = ax^2 + bx + c$ $(a \neq 0)$. By completing the square in x, show that

$$f(x) = a\left(x + \frac{b}{2a}\right)^2 + \frac{4ac - b^2}{4a}$$

SOLUTION TO SELF-CHECK EXERCISE 2.5

a. Here $a = 2$, $b = -3$, and $c = -3$. The x-coordinate of the vertex is

$$-\frac{b}{2a} = -\frac{(-3)}{2(2)} = \frac{3}{4}$$

The corresponding y-coordinate is

$$f\left(\frac{3}{4}\right) = 2\left(\frac{3}{4}\right)^2 - 3\left(\frac{3}{4}\right) - 3 = \frac{9}{8} - \frac{9}{4} - 3 = -\frac{33}{8}$$

Therefore, the vertex of the parabola is $(3/4, -33/8)$.

b. Solving the equation $2x^2 - 3x - 3 = 0$, we find

$$x = \frac{3 \pm \sqrt{(-3)^2 - 4(2)(-3)}}{2(2)} = \frac{3 \pm \sqrt{33}}{4}$$

So the x-intercepts are $3/4 - \sqrt{33}/4 \approx -0.7$ and $3/4 + \sqrt{33}/4 \approx 2.2$.

c. Since $a = 2 > 0$, the parabola opens upward. The y-intercept is -3. The graph of the parabola is shown in the accompanying figure.

The graph of
$$f(x) = 2x^2 - 3x - 3$$

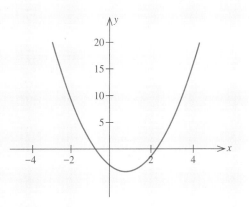

CHAPTER TWO SUMMARY OF PRINCIPAL FORMULAS AND TERMS

Formulas

1. Slope of a line		$m = \dfrac{y_2 - y_1}{x_2 - x_1}$
2. Equation of a vertical line		$x = a$
3. Equation of a horizontal line		$y = b$
4. Point-slope form of the equation of a line		$y - y_1 = m(x - x_1)$
5. Slope-intercept form of the equation of a line		$y = mx + b$
6. General equation of a line		$Ax + By + C = 0$

Terms

Cartesian coordinate system	Ordered pair
Origin	Coordinates

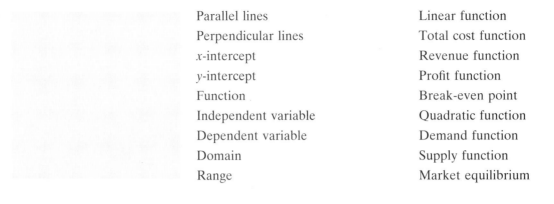

Parallel lines

Perpendicular lines

x-intercept

y-intercept

Function

Independent variable

Dependent variable

Domain

Range

Linear function

Total cost function

Revenue function

Profit function

Break-even point

Quadratic function

Demand function

Supply function

Market equilibrium

CHAPTER 2 REVIEW EXERCISES

In exercises 1–6, find an equation of the line L that passes through the point $(-2, 4)$ *and satisfies the given condition.*

1. L is a vertical line.

2. L is a horizontal line.

3. L passes through the point $(3, 7/2)$.

4. The x-intercept of L is 3.

5. L is parallel to the line $5x - 2y = 6$.

6. L is perpendicular to the line $4x + 3y = 6$.

7. Find an equation of the line with slope $-1/2$ and y-intercept -3.

8. Find the slope and y-intercept of the line with equation $3x - 5y = 6$.

9. Find an equation of the line passing through the point $(2, 3)$ and parallel to the line with equation $3x + 4y - 8 = 0$.

10. Find an equation of the line passing through the point $(-1, 3)$ and parallel to the line joining the points $(-3, 4)$ and $(2, 1)$.

11. Find an equation of the line passing through the point $(-2, -4)$ that is perpendicular to the line with equation $2x - 3y - 24 = 0$.

In exercises 12 and 13, sketch the graph of the equation.

12. $3x - 4y = 24$ **13.** $-2x + 5y = 15$

In exercises 14 and 15, find the domain of the function.

14. $f(x) = \sqrt{9 - x}$ **15.** $f(x) = \dfrac{x + 3}{2x^2 - x - 3}$

16. Let $f(x) = 3x^2 + 5x - 2$. Find
  a. $f(-2)$ **b.** $f(a + 2)$
  c. $f(2a)$ **d.** $f(a + h)$

17. Let $y^2 = 2x + 1$.
  a. Sketch the graph of this equation.
  b. Is y a function of x? Why?
  c. Is x a function of y? Why?

18. Sketch the graph of the function defined by

$$f(x) = \begin{cases} x + 1 & \text{if } x < 1 \\ -x^2 + 4x - 1 & \text{if } x \geq 1 \end{cases}$$

In exercises 19 and 20, find the vertex, the x-intercepts, and sketch the parabola.

19. $6x^2 - 11x - 10 = 0$ **20.** $-4x^2 + 4x + 3 = 0$

In exercises 21 and 22, find the point of intersection of the lines with the given equations.

21. $3x + 4y = -6$ and $2x + 5y = -11$

22. $y = \frac{3}{4}x + 6$ and $3x - 2y + 3 = 0$

23. Sales of a certain clock/radio are approximated by the relationship

$$S(t) = 6000t + 30{,}000 \qquad 0 \leq t \leq 5$$

where $S(t)$ denotes the number of clock/radios sold in year t ($t = 0$ corresponds to the year 1995). Find the number of clock/radios expected to be sold in the year 2000.

24. A company's total sales (in millions of dollars) are approximately linear as a function of time (in years). Sales in 1990 were \$2.4 million, whereas sales in 1995 amounted to \$7.4 million.

a. Find an equation giving the company's sales as a function of time.
b. What were the sales in 1993?

25. Clark's Rule is a method for calculating pediatric drug dosages based on a child's weight. If a denotes the adult dosage (in mg) and if w is the child's weight (in lb), then the child's dosage is given by

$$D(w) = \frac{aw}{150}$$

a. Show that D is a linear function of w.
b. If the adult dose of a substance is 500 mg, how much should a child who weighs 35 lb receive?

26. An office building worth $6 million when it was completed in 1990 is being depreciated linearly over 30 years.
a. What is the rate of depreciation?
b. What will the book value of the building be in 2000?

27. In 1995, a manufacturer installed a new machine in her factory at a cost of $300,000. The machine is depreciated linearly over 12 years with a scrap value of $30,000.
a. What is the rate of depreciation of the machine per year?
b. Find an expression for the book value of the machine in year t ($0 \le t \le 12$).

28. A company has a fixed cost of $30,000 and a production cost of $6 for each unit it manufactures. A unit sells for $10.
a. What is the cost function?
b. What is the revenue function?
c. What is the profit function?
d. Compute the profit (loss) corresponding to production levels of 6,000, 8,000, and 12,000 units, respectively.

29. There is no demand for a certain commodity when the unit price is $200 or more, but for each $10 decrease in price below $200, the quantity demanded increases by 200 units. Find the demand equation and sketch its graph.

30. Bicycle suppliers will make 200 bicycles available in the market per month when the unit price is $50 and 2000 bicycles available per month when the unit price is $100. Find the supply equation if it is known to be linear.

31. The cost function and the revenue function for a certain firm are given by $C(x) = 12x + 20,000$ and $R(x) = 20x$, respectively. Find the break-even point for the company.

32. Given the demand equation $3x + p - 40 = 0$ and the supply equation $2x - p + 10 = 0$, where p is the unit price in dollars and x represents the quantity in units of a thousand, determine the equilibrium quantity and the equilibrium price.

33. The accompanying data were compiled by the Admissions Office of Carter College during the past five years. The data relate the number of college brochures and follow-up letters (x) sent to a preselected list of high school juniors who took the PSAT and the number of completed applications (y) received from these students (both measured in thousands).

Number of brochures sent (x)	1.8	2	3.2
Number of completed applications (y)	0.4	0.5	0.7

Number of brochures sent (x)	4	4.8
Number of completed applications (y)	1	1.3

a. Plot these points and draw a straight line through the points (2, 0.5) and (4, 1).
b. Find an equation of the trend line passing through the points (2, 0.5) and (4, 1).
c. Use this equation to predict the number of completed applications that might be expected if 6400 brochures and follow-up letters are sent out during the next year.

34. Linear Depreciation   A new machine was purchased by the National Textile Company for $120,000. For income tax purposes, the machine is depreciated linearly over ten years; that is, the book value of the machine decreases at a constant rate, so that at the end of ten years the book value is zero.
a. Express the book value of the machine (V) as a function of the age (in years) of the machine (n).
b. Sketch the graph of the function in (a).
c. Find the book value of the machine at the end of the sixth year.
d. Find the rate at which the machine is being depreciated each year.

35. Linear Depreciation   Refer to exercise 34. An office building worth $1 million when completed in 1980 was depreciated linearly over 50 years. What will the book value of the building be in 1999? In 2003? (Assume that the book value of the building will be zero at the end of the 50th year.)

CHAPTER 2   EXTENDED PROBLEMS AND PROJECTS

1. The Method of Least Squares can be used to find a straight line that, in some sense, best fits a set of data points in a plane that are scattered about a straight line (Figure 46).

FIGURE 46

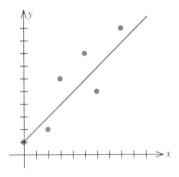

Suppose we are given n data points

$$P_1(x_1, y_1), P_2(x_2, y_2), P_3(x_3, y_3), \ldots, P_n(x_n, y_n)$$

Then the least-squares (regression) line for the data is given by the linear equation (function)

$$y = f(x) = mx + b$$

where the constants m and b satisfy the normal equations

$$nb + (x_1 + x_2 + \cdots + x_n)m = y_1 + y_2 + \cdots + y_n$$

and

$$(x_1 + x_2 + \cdots + x_n)b + (x_1^2 + x_2^2 + \cdots + x_n^2)m$$
$$= x_1y_1 + x_2y_2 + \cdots + x_ny_n$$

simultaneously. Use these formulas to find the least-squares line for the data $P_1(1, 1)$, $P_2(2, 3)$, $P_3(3, 4)$, $P_4(4, 3)$, and $P_5(5, 6)$.

2. **Waste Generation**   According to data from the Council on Environmental Quality, the amount of waste (in millions of tons per year) generated in the United States from 1960 to 1990 was:

Year	1960	1965	1970	1975
Amount (y)	81	100	120	124

Year	1980	1985	1990
Amount (y)	140	152	164

a. Find an equation of the least-squares line for these data. (Let t be in units of 5, and let $t = 1$ represent 1960.)

b. Use the result of (a) to estimate the amount of waste generated in the year 2000, assuming the trend continues.

Source: The Environmental Almanac

3. **Worldwide PC Shipments**   The following data describe the worldwide PC shipments from 1992 through 1996:

Year	1992	1993	1994	1995	1996
Number (y) (in millions)	31	39	48	52	61

a. Find an equation of the least-squares line for these data. (Let t denote time, with $t = 0$ corresponding to 1992.)

b. Use the result of (a) to estimate the worldwide shipment of PCs in 1997, assuming the trend continues.

Source: Dataquest Inc.

3

Exponential and Logarithmic Functions

3.1 Exponential Functions

3.2 Logarithmic Functions

3.3 Exponential Functions as Mathematical Models

*T*he exponential function is, without doubt, the most important function in mathematics and its applications. After a brief introduction to exponential and logarithmic functions, we will explore some of the many applications involving exponential functions, such as the growth of a bacteria population in the laboratory, the way radioactive matter decays, the rate at which a factory worker learns a certain process, and the rate at which a communicable disease is spread over time. Exponential functions also play an important role in computing interest earned in a bank account, a topic to be discussed in Chapter 4.

3.1   Exponential Functions

Exponential Functions and Their Graphs

Suppose that you deposit a sum of $1000 in an account earning interest at the rate of 10% per year *compounded continuously* (the way most financial institutions compute interest). Then the accumulated amount at the end of t years ($0 \le t \le 20$) is described by the function f, whose graph appears in Figure 1.* Such a function is called an *exponential function.* Observe that the graph of f rises rather slowly at first but very rapidly as time goes by. For purposes of comparison, we have also shown the graph of the function $y = g(t) = 1000(1 + 0.10t)$ giving the accumulated amount for the same principal ($1000) but earning *simple* interest at the rate of 10% per year. Later we will be able to show that the accumulated amount after 20 years under continuous compounding is $7389.06, whereas the accumulated amount over the same period earning simple interest is $3000 (see Chapter 4).

FIGURE 1
Under continuous compounding, a sum of money grows exponentially.

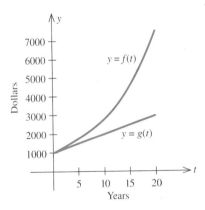

Exponential functions play an important role in many real-world applications, as you will see throughout this chapter.

In order to define an exponential function, observe that whenever b is a positive number and n is any real number, the expression b^n is a real number. This enables us to define an *exponential function* as follows:

EXPONENTIAL FUNCTION

The function defined by

$$f(x) = b^x \qquad b > 0, b \ne 1$$

is called an **exponential function with base b and exponent x.** The domain of f is the set of all real numbers.

For example, the exponential function with base 2 is the function

$$f(x) = 2^x$$

---

*We will derive the rule for f in Section 4.1.

with domain $(-\infty, \infty)$. The values of $f(x)$ for selected values of x follow:

$$f(3) = 2^3 = 8, \qquad f\left(\frac{3}{2}\right) = 2^{3/2} = 2 \cdot 2^{1/2} = 2\sqrt{2}, \qquad f(0) = 2^0 = 1,$$

$$f(-1) = 2^{-1} = \frac{1}{2}, \qquad \text{and} \qquad f\left(-\frac{2}{3}\right) = 2^{-2/3} = \frac{1}{2^{2/3}} = \frac{1}{\sqrt[3]{4}}$$

Computations involving exponentials are facilitated by the laws of exponents. These laws were stated in Section 1.5, and you might want to review the material there. For convenience, however, we will restate these laws.

THE LAWS OF EXPONENTS

Let a and b be positive numbers and let x and y be real numbers. Then,

1. $a^x \cdot a^y = a^{x+y}$ **4.** $(ab)^x = a^x b^x$

2. $\dfrac{a^x}{a^y} = a^{x-y}$ **5.** $\left(\dfrac{a}{b}\right)^x = \dfrac{a^x}{b^x}$

3. $(a^x)^y = a^{xy}$

The use of the laws of exponents is illustrated in Example 1.

EXAMPLE 1

a. $16^{7/4} \cdot 16^{-1/2} = 16^{7/4 - 1/2} = 16^{5/4} = 2^5 = 32$ Law 1

b. $\dfrac{8^{5/3}}{8^{-1/3}} = 8^{5/3 - (-1/3)} = 8^2 = 64$ Law 2

c. $(64^{4/3})^{-1/2} = 64^{(4/3)(-1/2)} = 64^{-2/3}$

$$= \frac{1}{64^{2/3}} = \frac{1}{(64^{1/3})^2} = \frac{1}{4^2} = \frac{1}{16} \qquad \text{Law 3}$$

d. $(16 \cdot 81)^{-1/4} = 16^{-1/4} \cdot 81^{-1/4} = \dfrac{1}{16^{1/4}} \cdot \dfrac{1}{81^{1/4}} = \dfrac{1}{2} \cdot \dfrac{1}{3} = \dfrac{1}{6}$ Law 4

e. $\left(\dfrac{3^{1/2}}{2^{1/3}}\right)^4 = \dfrac{3^{4/2}}{2^{4/3}} = \dfrac{9}{2^{4/3}}$ Law 5

EXAMPLE 2   Let $f(x) = 2^{2x-1}$. Find the value of x for which $f(x) = 16$.

Solution   We want to solve the equation

$$2^{2x-1} = 16 = 2^4$$

But this equation holds if and only if

$$2x - 1 = 4, \qquad b^m = b^n \Rightarrow m = n$$

or $\qquad\qquad\qquad\qquad 2x = 5$

giving $x = \frac{5}{2}$.

Exponential functions play an important role in mathematical analysis. Because of their special characteristics, they are some of the most useful functions and are found in virtually every field where mathematics is applied. Under ideal conditions, for example, the number of bacteria present at any time t in a culture may be described by an exponential function of t; radioactive substances decay over time in accordance with an "exponential" law of decay; money left on fixed deposit and earning compound interest grows exponentially; and some of the most important distribution functions encountered in statistics are exponential.

Let us begin our investigation into the properties of exponential functions by studying their graphs.

EXAMPLE 3   Sketch the graph of the exponential function $y = 2^x$.

Solution   First, as discussed earlier, the domain of the exponential function $y = f(x) = 2^x$ is the set of real numbers. Next, putting $x = 0$ gives $y = 2^0 = 1$, the y-intercept of f. There is no x-intercept since there is no value of x for which $y = 0$. To find the range of f, consider the following table of values:

x	-5	-4	-3	-2	-1	0	1	2	3	4	5
y	$\frac{1}{32}$	$\frac{1}{16}$	$\frac{1}{8}$	$\frac{1}{4}$	$\frac{1}{2}$	1	2	4	8	16	32

We see from these computations that 2^x decreases and approaches zero as x decreases without bound and that 2^x increases without bound as x increases without bound. Thus, the range of f is the interval $(0, \infty)$—that is, the set of positive real numbers. Finally, we sketch the graph of $y = f(x) = 2^x$ in Figure 2. ∎

EXAMPLE 4   Sketch the graph of the exponential function $y = (1/2)^x$.

Solution   The domain of the exponential function $y = (1/2)^x$ is the set of all real numbers. The y-intercept is $(1/2)^0 = 1$; there is no x-intercept since there is no value of x for which $y = 0$. From the following table of values, we deduce that $(1/2)^x = 1/2^x$ increases without bound as x decreases without bound and that $(1/2)^x$ decreases and approaches zero as x increases without bound.

x	-5	-4	-3	-2	-1	0	1	2	3	4	5
y	32	16	8	4	2	1	$\frac{1}{2}$	$\frac{1}{4}$	$\frac{1}{8}$	$\frac{1}{16}$	$\frac{1}{32}$

Thus, the range of f is the interval $(0, \infty)$. The graph of $y = f(x) = (1/2)^x$ is sketched in Figure 3. ∎

The functions $y = 2^x$ and $y = (1/2)^x$, whose graphs you studied in Examples 3 and 4, are special cases of the exponential function $y = f(x) = b^x$, obtained by setting $b = 2$ and $b = 1/2$, respectively. In general, the exponential function $y = b^x$ with $b > 1$ has a graph similar to $y = 2^x$, while the graph of $y = b^x$ for $0 < b < 1$ is similar to that of $y = (1/2)^x$ (see exercises 21 and 22). When $b = 1$, the function $y = b^x$ reduces to the constant function $y = 1$. For comparison, the graphs of all three functions are sketched in Figure 4.

FIGURE 2
The graph of $y = 2^x$

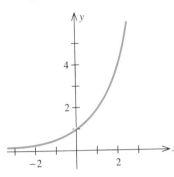

FIGURE 3
The graph of $y = (1/2)^x$

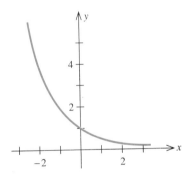

FIGURE 4
$y = b^x$ *is an increasing function of x if $b > 1$, a constant function if $b = 1$, and a decreasing function if $0 < b < 1$.*

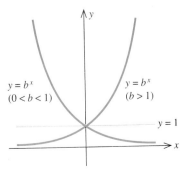

PROPERTIES OF THE EXPONENTIAL FUNCTION

The exponential function $y = b^x$ $(b > 0, b \neq 1)$ has the following properties:

1. Its domain is $(-\infty, \infty)$.
2. Its range is $(0, \infty)$.
3. Its graph passes through the point $(0, 1)$.
4. Its graph is an unbroken curve devoid of holes or jumps.
5. Its graph rises from left to right if $b > 1$ and falls from left to right if $b < 1$.

The Base e

It can be shown, although we will not do so here, that as m increases without bound, the value of the expression

$$\left(1 + \frac{1}{m}\right)^m$$

approaches the irrational number 2.7182818, which we denote by e. You may convince yourself of the plausibility of this definition of the number e by examining Table 1, which may be constructed with the help of a calculator. (Also, see Using Technology, exercise 11, page 120.)

TABLE 1

m	10	100	1000	10,000	100,000	1,000,000
$\left(1 + \dfrac{1}{m}\right)^m$	2.59374	2.70481	2.71692	2.71815	2.71827	2.71828

Exponential functions to the base e play an important role in both theoretical and applied problems.

FIGURE 5
The graph of $y = e^x$

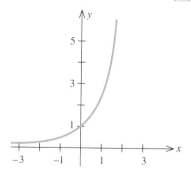

EXAMPLE 5 Sketch the graph of the function $y = e^x$.

Solution Since $e > 1$, it follows from our previous discussion that the graph of $y = e^x$ is similar to the graph of $y = 2^x$ (see Figure 2). With the aid of a calculator, we obtain the following table:

x	-3	-2	-1	0	1	2	3
y	0.05	0.14	0.37	1	2.72	7.39	20.09

The graph of $y = e^x$ is sketched in Figure 5. ■

Next, we consider another exponential function to the base e that is closely related to the previous function and is particularly useful in constructing models that describe "exponential decay."

FIGURE 6
The graph of $y = e^{-x}$

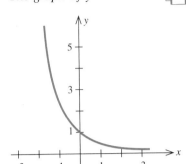

EXAMPLE 6   Using a calculator, sketch the graph of the function $y = e^{-x}$.

Solution   Since $e > 1$, it follows that $0 < 1/e < 1$, so $f(x) = e^{-x} = 1/e^x = (1/e)^x$ is an exponential function with base less than 1. Therefore, it has a graph similar to that of the exponential function $y = (1/2)^x$. We construct the following table of values of $y = e^{-x}$ for selected values of x:

x	-3	-2	-1	0	1	2	3
y	20.09	7.39	2.72	1	0.37	0.14	0.05

Using this table, we sketch the graph of $y = e^{-x}$ in Figure 6.   ∎

**SELF-CHECK
EXERCISES 3.1**

1. Solve the equation $2^{2x+1} \cdot 2^{-3} = 2^{x-1}$.

2. Sketch the graph of $y = e^{0.4x}$.

Solutions to Self-Check Exercises 3.1 can be found on page 119.

3.1 EXERCISES

In exercises 1–6, evaluate the expression.

1. a. $4^{-3} \cdot 4^5$ **b.** $3^{-3} \cdot 3^6$

2. a. $(2^{-1})^3$ **b.** $(3^{-2})^3$

3. a. $9(9)^{-1/2}$ **b.** $5(5)^{-1/2}$

4. a. $[(-\frac{1}{2})^3]^{-2}$ **b.** $[(-\frac{1}{3})^2]^{-3}$

5. a. $\dfrac{(-3)^4(-3)^5}{(-3)^8}$ **b.** $\dfrac{(2^{-4})(2^6)}{2^{-1}}$

6. a. $3^{1/4} \cdot (9)^{-5/8}$ **b.** $2^{3/4} \cdot (4)^{-3/2}$

In exercises 7–12, simplify the expression.

7. a. $(64x^9)^{1/3}$ **b.** $(25x^3y^4)^{1/2}$

8. a. $(2x^3)(-4x^{-2})$ **b.** $(4x^{-2})(-3x^5)$

9. a. $\dfrac{6a^{-5}}{3a^{-3}}$ **b.** $\dfrac{4b^{-4}}{12b^{-6}}$

10. a. $y^{-3/2}y^{5/3}$ **b.** $x^{-3/5}x^{8/3}$

11. a. $(2x^3y^2)^3$ **b.** $(4x^2y^2z^3)^2$

12. a. $\dfrac{5^0}{(2^{-3}x^{-3}y^2)^2}$ **b.** $\dfrac{(x+y)(x-y)}{(x-y)^0}$

In exercises 13–20, solve the equation for x.

13. $6^{2x} = 6^4$ **14.** $5^{-x} = 5^3$

15. $3^{3x-4} = 3^5$ **16.** $10^{2x-1} = 10^{x+3}$

17. $(2.1)^{x+2} = (2.1)^5$ **18.** $(-1.3)^{x-2} = (-1.3)^{2x+1}$

19. $8^x = \left(\dfrac{1}{32}\right)^{x-2}$ **20.** $3^{x-x^2} = \dfrac{1}{9^x}$

c *In exercises 21–29, sketch the graphs of the given functions on the same axes. A calculator is recommended for these exercises.*

21. $y = 2^x$, $y = 3^x$, and $y = 4^x$

22. $y = (\frac{1}{2})^x$, $y = (\frac{1}{3})^x$, and $y = (\frac{1}{4})^x$

23. $y = 2^{-x}$, $y = 3^{-x}$, and $y = 4^{-x}$

24. $y = 4^{0.5x}$ and $y = 4^{-0.5x}$

25. $y = 4^{0.5x}$, $y = 4^x$, and $y = 4^{2x}$

26. $y = e^x$, $y = 2e^x$, and $y = 3e^x$

27. $y = e^{0.5x}$, $y = e^x$, and $y = e^{1.5x}$

28. $y = e^{-0.5x}$, $y = e^{-x}$, and $y = e^{-1.5x}$

29. $y = 0.5e^{-x}$, $y = e^{-x}$, and $y = 2e^{-x}$

SOLUTIONS TO SELF-CHECK EXERCISES 3.1

1. $2^{2x+1} \cdot 2^{-3} = 2^{x-1}$

$$\frac{2^{2x+1}}{2^{x-1}} \cdot 2^{-3} = 1 \qquad \text{Dividing both sides by } 2^{x-1}$$

$$2^{(2x+1)-(x-1)-3} = 1$$

$$2^{x-1} = 1$$

This is true if and only if $x - 1 = 0$ or $x = 1$.

2. We first construct the following table of values:

x	-3	-2	-1	0	1	2	3	4
$y = e^{0.4x}$	0.3	0.5	0.7	1	1.5	2.2	3.3	5

Next, we plot these points and join them by a smooth curve to obtain the graph of f shown in the accompanying figure.

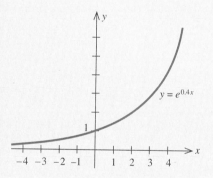

$$y = e^{0.4x}$$

U s i n g   T e c h n o l o g y

Although the proof is outside the scope of this book, it can be proved that an exponential function of the form $f(x) = b^x$, where $b > 1$, will ultimately grow faster than the power function $g(x) = x^n$ for *any* positive real number n. To give a visual demonstration of this result for the special case of the exponential function $f(x) = e^x$, we can use a graphing calculator to plot the graphs of both f and g (for selected values of n) on the same set of axes in an appropriate viewing rectangle and observe that the graph of f ultimately lies above that of g.

EXAMPLE 1   Use a graphing utility to plot the graphs of (a) $f(x) = e^x$ and $g(x) = x^3$ on the same set of axes in the viewing rectangle $[0, 6] \times [0, 250]$ and (b) $f(x) = e^x$ and $g(x) = x^5$ in the viewing rectangle $[0, 20] \times [0, 1{,}000{,}000]$.

Solution

a. The graphs of $f(x) = e^x$ and $g(x) = x^3$ in the viewing rectangle $[0, 6] \times [0, 250]$ are shown in Figure 1a.

b. The graphs of $f(x) = e^x$ and $g(x) = x^5$ in the viewing rectangle $[0, 20] \times [0, 1{,}000{,}000]$ are shown in Figure 1b.

FIGURE 1

(a) The graphs of $f(x) = e^x$ and $g(x) = x^3$ in the viewing rectangle $[0, 6] \times [0, 250]$

(b) The graphs of $f(x) = e^x$ and $g(x) = x^5$ in the viewing rectangle $[0, 20] \times [0, 1,000,000]$

In the exercises that follow, you will be asked to use a graphing utility to reveal the properties of exponential functions.

EXERCISES

In exercises 1 and 2, use a graphing utility to plot the graphs of the functions f and g on the same set of axes in the specified viewing rectangle.

1. $f(x) = e^x$ and $g(x) = x^2$; $[0, 4] \times [0, 30]$

2. $f(x) = e^x$ and $g(x) = x^4$; $[0, 15] \times [0, 20,000]$

In exercises 3 and 4, use a graphing utility to plot the graphs of the functions f and g on the same set of axes in an appropriate viewing rectangle to demonstrate that f ultimately grows faster than g. [Note: Your answer will not be unique.]

3. $f(x) = 2^x$ and $g(x) = x^{2.5}$

4. $f(x) = 3^x$ and $g(x) = x^3$

5. Use a graphing utility to plot the graphs of $f(x) = 2^x$, $g(x) = 3^x$, and $h(x) = 4^x$ on the same set of axes in the viewing rectangle $[0, 5] \times [0, 100]$. Comment on the relationship between the base b and the growth of the function $f(x) = b^x$.

6. Use a graphing utility to plot the graphs of $f(x) = (1/2)^x$, $g(x) = (1/3)^x$, and $h(x) = (1/4)^x$ on the same set of axes in the viewing rectangle $[0, 4] \times [0, 1]$. Comment on the relationship between the base b and the growth of the function $f(x) = b^x$.

7. Use a graphing utility to plot the graphs of $f(x) = e^x$, $g(x) = 2e^x$, and $h(x) = 3e^x$ on the same set of

axes in the viewing rectangle $[-3, 3] \times [0, 10]$. Comment on the role played by the constant k in the graph of $f(x) = ke^x$.

8. Use a graphing utility to plot the graphs of $f(x) = -e^x$, $g(x) = -2e^x$, and $h(x) = -3e^x$ on the same set of axes in the viewing rectangle $[-3, 3] \times [-10, 0]$. Comment on the role played by the constant k in the graph of $f(x) = ke^x$.

9. Use a graphing utility to plot the graphs of $f(x) = e^{0.5x}$, $g(x) = e^x$, and $h(x) = e^{1.5x}$ on the same set of axes in the viewing rectangle $[-2, 2] \times [0, 4]$. Comment on the role played by the constant k in the graph of $f(x) = e^{kx}$.

10. Use a graphing utility to plot the graphs of $f(x) = e^{-0.5x}$, $g(x) = e^{-x}$, and $h(x) = e^{-1.5x}$ on the same set of axes in the viewing rectangle $[-2, 2] \times [0, 4]$. Comment on the role played by the constant k in the graph of $f(x) = e^{kx}$.

11. In order to obtain a visual confirmation of the fact that the expression $(1 + 1/m)^m$ approaches the number $e = 2.71828 \ldots$ as m increases without bound, plot the graph of $f(x) = (1 + 1/x)^x$ in a suitable viewing rectangle and observe that $f(x)$ approaches $2.71828 \ldots$ as x increases without bound. Use **TRACE** to find the value of $f(x)$ for large values of x. (See Section A.12, Using the TRACE Function, in Appendix A.)

3.2   Logarithmic Functions

Logarithms

You are already familiar with exponential equations of the form

$$b^y = x \qquad b > 0, b \neq 1$$

where the variable x is expressed in terms of a real number b and a variable y. But what about solving this same equation for y? You may recall from your study of algebra that the number y is called the *logarithm of x to the base b* and is denoted by $\log_b x$. It is the *exponent* to which the base b must be raised in order to obtain the number x.

**THE LOGARITHM OF
x TO THE BASE b**

$$y = \log_b x \quad \text{if and only if} \quad x = b^y \qquad x > 0$$

 Observe that the logarithm $\log_b x$ is defined only for positive values of x.

EXAMPLE 1

a. $\log_{10} 100 = 2$, since $100 = 10^2$

b. $\log_5 125 = 3$, since $125 = 5^3$

c. $\log_3 \dfrac{1}{27} = -3$, since $\dfrac{1}{27} = \dfrac{1}{3^3} = 3^{-3}$

d. $\log_{20} 20 = 1$, since $20 = 20^1$

EXAMPLE 2   Solve each of the following equations for x:

a. $\log_3 x = 4$ **b.** $\log_{16} 4 = x$ **c.** $\log_x 8 = 3$

Solution

a. By definition, $\log_3 x = 4$ implies $x = 3^4 = 81$.

b. $\log_{16} 4 = x$ is equivalent to $4 = 16^x = (4^2)^x = 4^{2x}$, or $4^1 = 4^{2x}$ from which we deduce that

$$2x = 1 \qquad \scriptstyle b^m = b^n \Rightarrow m = n$$

or $\qquad\qquad\qquad x = \tfrac{1}{2}$

c. Referring once again to the definition, we see that the equation $\log_x 8 = 3$ is equivalent to

$$8 = (2^3) = x^3$$

or $\qquad\qquad\qquad x = 2 \qquad \scriptstyle a^m = b^m \Rightarrow a = b$

The two widely used systems of logarithms are the system of **common logarithms,** which uses the number 10 as the base, and the system of **natural logarithms,** which uses the irrational number $e = 2.71828 \ldots$ as the base. Also, it is standard practice to write **log** for $\log_{10}$ and **ln** for $\log_e$.

**LOGARITHMIC
NOTATION**

$$\log x = \log_{10} x \qquad \text{Common logarithm}$$

$$\ln x = \log_e x \qquad \text{Natural logarithm}$$

The system of natural logarithms is widely used in theoretical work. Using natural logarithms rather than logarithms to other bases often leads to simpler expressions.

Laws of Logarithms

Computations involving logarithms are facilitated by the following *laws of logarithms.*

LAWS OF LOGARITHMS

If m and n are positive numbers, then

1. $\log_b mn = \log_b m + \log_b n$

2. $\log_b \dfrac{m}{n} = \log_b m - \log_b n$

3. $\log_b m^n = n \log_b m$

4. $\log_b 1 = 0$

5. $\log_b b = 1$

Do not confuse the expression $\log m/n$ (law 2) with the expression $\log m/\log n$. For example,

$$\log \frac{100}{10} = \log 100 - \log 10 = 2 - 1 = 1 \neq \frac{\log 100}{\log 10} = \frac{2}{1} = 2$$

You will be asked to prove these laws in exercises 63–65 at the end of this section. Their derivations are based on the definition of a logarithm and the corresponding laws of exponents. The following examples illustrate the properties of logarithms.

EXAMPLE 3

a. $\log(2 \cdot 3) = \log 2 + \log 3$

b. $\ln \dfrac{5}{3} = \ln 5 - \ln 3$　　　　　　**c.** $\log \sqrt{7} = \log 7^{1/2} = \dfrac{1}{2} \log 7$

d. $\log_5 1 = 0$　　　　　　　　　　**e.** $\log_{45} 45 = 1$　　■

EXAMPLE 4　Given that $\log 2 \approx 0.3010$, $\log 3 \approx 0.4771$, and $\log 5 \approx 0.6990$, use the laws of logarithms to find

a. $\log 15$　　**b.** $\log 7.5$　　**c.** $\log 81$　　**d.** $\log 50$

Solution

a. Note that $15 = 3 \cdot 5$, so by law 1 for logarithms,

$$\begin{aligned}
\log 15 &= \log 3 \cdot 5 \\
&= \log 3 + \log 5 \\
&\approx 0.4771 + 0.6990 \\
&= 1.1761
\end{aligned}$$

b. Observing that $7.5 = 15/2 = (3 \cdot 5)/2$, we apply laws 1 and 2 to obtain

$$\begin{aligned}
\log 7.5 &= \log \frac{(3)(5)}{2} \\
&= \log 3 + \log 5 - \log 2 \\
&\approx 0.4771 + 0.6990 - 0.3010 \\
&= 0.8751
\end{aligned}$$

c. Since $81 = 3^4$, we apply law 3 to obtain

$$\log 81 = \log 3^4$$
$$= 4 \log 3$$
$$\approx 4(0.4771)$$
$$= 1.9084$$

d. We write $50 = 5 \cdot 10$ and find

$$\log 50 = \log(5)(10)$$
$$= \log 5 + \log 10$$
$$\approx 0.6990 + 1 \qquad \text{Using law 5}$$
$$= 1.6990$$

EXAMPLE 5 Expand and simplify the following expressions:

a. $\log_3 x^2 y^3$ **b.** $\log_2 \dfrac{x^2 + 1}{2^x}$ **c.** $\ln \dfrac{x^2 \sqrt{x^2 - 1}}{e^x}$

Solution

a. $\log_3 x^2 y^3 = \log_3 x^2 + \log_3 y^3 \qquad$ Law 1
$= 2 \log_3 x + 3 \log_3 y \qquad$ Law 3

b. $\log_2 \dfrac{x^2 + 1}{2^x} = \log_2(x^2 + 1) - \log_2 2^x \qquad$ Law 2

$= \log_2(x^2 + 1) - x \log_2 2 \qquad$ Law 3
$= \log_2(x^2 + 1) - x \qquad$ Law 5

c. $\ln \dfrac{x^2 \sqrt{x^2 - 1}}{e^x} = \ln \dfrac{x^2 (x^2 - 1)^{1/2}}{e^x} \qquad$ Rewriting

$= \ln x^2 + \ln (x^2 - 1)^{1/2} - \ln e^x \qquad$ Laws 1 and 2
$= 2 \ln x + \tfrac{1}{2} \ln (x^2 - 1) - x \ln e \qquad$ Law 3
$= 2 \ln x + \tfrac{1}{2} \ln (x^2 - 1) - x \qquad$ Law 5

Examples 6 and 7 illustrate how the properties of logarithms are used to solve equations.

EXAMPLE 6 Solve $\log_3(x + 1) - \log_3(x - 1) = 1$ for x.

Solution Using the properties of logarithms, we obtain

$$\log_3(x + 1) - \log_3(x - 1) = 1$$

$$\log_3 \frac{x + 1}{x - 1} = 1 \qquad \text{Law 2}$$

$$\frac{x + 1}{x - 1} = 3^1 = 3 \qquad \text{Definition of logarithms}$$

and so

$$x + 1 = 3(x - 1)$$
$$x + 1 = 3x - 3$$
$$4 = 2x$$

or

$$x = 2$$

EXAMPLE 7   Solve $\log x + \log(2x - 1) = \log 6$.

Solution   We have

$$\log x + \log(2x - 1) = \log 6$$

$$\log x + \log(2x - 1) - \log 6 = 0$$

$$\log\left(\frac{x(2x - 1)}{6}\right) = 0 \qquad\qquad \text{Laws 1 and 2}$$

$$\frac{x(2x - 1)}{6} = 10^0 = 1 \qquad\qquad \text{Definition of logarithms}$$

So, $x(2x - 1) = 6$

$$2x^2 - x - 6 = 0$$

$$(2x + 3)(x - 2) = 0$$

and $x = -\dfrac{3}{2} \quad \text{or} \quad 2$

Since the domain of $\log(2x - 1)$ is the interval $(1/2, \infty)$ (because $2x - 1$ must be positive), we reject the root $-3/2$ of the quadratic equation and conclude that the solution of the given equation is $x = 2$. ■

REMARK   Using the fact that $\log a = \log b$ if and only if $a = b$, we can also solve the equation of Example 7 as follows:

$$\log x + \log(2x - 1) = \log 6$$

$$\log x(2x - 1) = \log 6$$

$$x(2x - 1) = 6$$

The rest of the solution is the same as that in Example 7. □

Logarithmic Functions and Their Graphs

The definition of a logarithm implies that if b and n are positive numbers and b is different from 1, then the expression $\log_b n$ is a real number. This enables us to define a *logarithmic function* as follows:

LOGARITHMIC FUNCTION

The function defined by

$$f(x) = \log_b x \qquad b > 0, b \neq 1$$

is called the **logarithmic function with base b.** The domain of f is the set of all positive numbers.

One way to obtain the graph of the logarithmic function $y = \log_b x$ is to construct a table of values of the logarithm (base b). However, another method—and a more instructive one—is based on exploiting the intimate relationship between logarithmic and exponential functions.

If a point (u, v) lies on the graph of $y = \log_b x$, then

$$v = \log_b u$$

But we can also write this equation in exponential form as

$$u = b^v$$

So the point (v, u) also lies on the graph of the function $y = b^x$. Let us look at the relationship between the points (u, v) and (v, u) and the line $y = x$ (Figure 7). If we think of the line $y = x$ as a mirror, then the point (v, u) is the mirror reflection of the point (u, v). Similarly, the point (u, v) is a mirror reflection of the point (v, u). We can take advantage of this relationship to help us draw the graph of logarithmic functions. For example, if we wish to draw the graph of $y = \log_b x$, where $b > 1$, then we need only draw the mirror reflection of the graph of $y = b^x$ with respect to the line $y = x$ (Figure 8).

FIGURE 7

The points (u, v) and (v, u) are mirror reflections of each other.

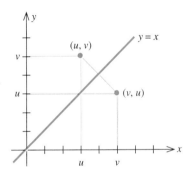

FIGURE 8

The graphs of $y = b^x$ and $y = \log_b x$ are mirror reflections of each other.

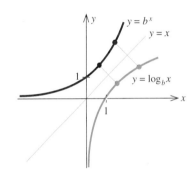

You may discover the following properties of the logarithmic function by taking the reflection of the graph of an appropriate exponential function (see exercises 45 and 46).

PROPERTIES OF THE LOGARITHMIC FUNCTION

The logarithmic function $y = \log_b x$ $(b > 0, b \neq 1)$ has the following properties:

1. Its domain is $(0, \infty)$.
2. Its range is $(-\infty, \infty)$.
3. Its graph passes through the point $(1, 0)$.
4. Its graph is an unbroken curve devoid of holes or jumps.
5. Its graph rises from left to right if $b > 1$ and falls from left to right if $b < 1$.

EXAMPLE 8   Sketch the graph of the function $y = \ln x$.

Solution   We first sketch the graph of $y = e^x$. Then the required graph is obtained by tracing the mirror reflection of the graph of $y = e^x$ with respect to the line $y = x$ (Figure 9).

FIGURE 9

The graph of $y = \ln x$ is the mirror reflection of the graph of $y = e^x$.

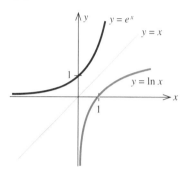

Properties Relating the Exponential and Logarithmic Functions

We made use of the relationship that exists between the exponential function $f(x) = e^x$ and the logarithmic function $g(x) = \ln x$ when we sketched the graph of g in Example 8. This relationship is further described by the following properties, which are an immediate consequence of the definition of the logarithm of a number:

PROPERTIES RELATING e^x and $\ln x$

$$e^{\ln x} = x \qquad x > 0 \tag{1}$$

$$\ln e^x = x \qquad \text{for any real number } x \tag{2}$$

(Try to verify these properties.)

The relationships expressed in (1) and (2) are useful in solving equations that involve exponentials and logarithms.

EXAMPLE 9   Solve the equation $2e^{x+2} = 5$.

Solution   We first divide both sides of the equation by 2 to obtain

$$e^{x+2} = \frac{5}{2} = 2.5$$

Next, taking the natural logarithm of each side of the equation and using (2), we have

$$\ln e^{x+2} = \ln 2.5$$

$$x + 2 = \ln 2.5$$

$$x = -2 + \ln 2.5$$

$$\approx -1.08$$

EXAMPLE 10   Solve the equation $5 \ln x + 3 = 0$.

Solution   Adding -3 to both sides of the equation leads to

$$5 \ln x = -3$$

$$\ln x = -\frac{3}{5} = -0.6$$

and so $e^{\ln x} = e^{-0.6}$

Using (1), we conclude that

$$x = e^{-0.6}$$

$$\approx 0.55$$

SELF-CHECK EXERCISES 3.2

1. Sketch the graph of $y = 3^x$ and $y = \log_3 x$ on the same set of axes.

2. Solve the equation $3e^{x+1} - 2 = 4$.

Solutions to Self-Check Exercises 3.2 can be found on page 129.

3.2   EXERCISES

In exercises 1–10, express the equation in logarithmic form.

1. $2^6 = 64$

2. $3^5 = 243$

3. $3^{-2} = \frac{1}{9}$

4. $5^{-3} = \frac{1}{125}$

5. $\left(\frac{1}{3}\right)^1 = \frac{1}{3}$

6. $\left(\frac{1}{2}\right)^{-4} = 16$

7. $32^{3/5} = 8$

8. $81^{3/4} = 27$

9. $10^{-3} = 0.001$

10. $16^{-1/4} = 0.5$

In exercises 11–16, use the facts that $\log 3 = 0.4771$ and $\log 4 = 0.6021$ to find the value of the logarithm.

11. $\log 12$

12. $\log \frac{3}{4}$

13. $\log 16$

14. $\log \sqrt{3}$

15. $\log 48$

16. $\log \frac{1}{300}$

In exercises 17–26, use the laws of logarithms to expand and simplify the expression.

17. $\log x(x + 1)^4$

18. $\log x(x^2 + 1)^{-1/2}$

19. $\log \dfrac{\sqrt{x + 1}}{x^2 + 1}$

20. $\ln \dfrac{e^x}{1 + e^x}$

21. $\ln xe^{-x^2}$

22. $\ln x(x + 1)(x + 2)$

23. $\ln \dfrac{x^{1/2}}{x^2\sqrt{1 + x^2}}$

24. $\ln \dfrac{x^2}{\sqrt{x}(1 + x)^2}$

25. $\ln x^x$

26. $\ln x^{x^2 + 1}$

In exercises 27–40, use the laws of logarithms to solve the equation.

27. $\log_2 x = 3$

28. $\log_3 x = 2$

29. $\log_2 8 = x$

30. $\log_3 27 = 2x$

31. $\log_x 10^3 = 3$

32. $\log_x \frac{1}{16} = -2$

33. $\log_2(2x + 5) = 3$

34. $\log_4(5x - 4) = 2$

35. $\log_2 x - \log_2(x - 2) = 3$

36. $\log x - \log(x + 6) = -1$

37. $\log_5(2x + 1) - \log_5(x - 2) = 1$

38. $\log(x + 7) - \log(x - 2) = 1$

39. $\log x + \log(2x - 5) = \log 3$

40. $\log_3(x + 1) + \log_3(2x - 3) = 1$

In exercises 41–44, sketch the graph of the equation.

41. $y = \log_3 x$

42. $y = \log_{1/3} x$

43. $y = \ln 2x$

44. $y = \ln \frac{1}{2} x$

In exercises 45 and 46, sketch the graphs of the given equations on the same coordinate axes.

45. $y = 2^x$ and $y = \log_2 x$

46. $y = e^{3x}$ and $y = \ln 3x$

In exercises 47–56, use logarithms to solve the equation for t.

47. $e^{0.4t} = 8$

48. $\frac{1}{3} e^{-3t} = 0.9$

49. $5e^{-2t} = 6$

50. $4e^{t-1} = 4$

51. $2e^{-0.2t} - 4 = 6$

52. $12 - e^{0.4t} = 3$

53. $\dfrac{50}{1 + 4e^{0.2t}} = 20$

54. $\dfrac{200}{1 + 3e^{-0.3t}} = 100$

55. $A = Be^{-t/2}$

56. $\dfrac{A}{1 + Be^{t/2}} = C$

57. Blood Pressure A normal child's systolic blood pressure may be approximated by the function

$$p(x) = m(\ln x) + b$$

where $p(x)$ is measured in millimeters of mercury, x is measured in pounds, and m and b are constants. Given that $m = 19.4$ and $b = 18$, determine the systolic blood pressure of a child who weighs 92 pounds.

58. Magnitude of Earthquakes On the Richter scale, the magnitude R of an earthquake is given by the formula

$$R = \log \frac{I}{I_0}$$

where I is the intensity of the earthquake being measured and I_0 is the standard reference intensity.
a. Express the intensity I of an earthquake of magnitude $R = 5$ in terms of the standard intensity I_0.
b. Express the intensity I of an earthquake of magnitude $R = 8$ in terms of the standard intensity I_0. How many times greater is the intensity of an earthquake of magnitude 8 than one of magnitude 5?
c. In modern times, the greatest loss of life attributable to an earthquake occurred in eastern China in 1976. Known as the Tangshan earthquake, it registered 8.2 on the Richter scale. How does the intensity of this earthquake compare with the intensity of an earthquake of magnitude $R = 5$?

59. Sound Intensity The relative loudness of a sound D of intensity I is measured in decibels, where

$$D = 10 \log \frac{I}{I_0}$$

and I_0 is the standard threshold of audibility.
a. Express the intensity I of a 30-decibel sound (the sound level of normal conversation) in terms of I_0.
b. Determine how many times greater the intensity of an 80-decibel sound (rock music) is than that of a 30-decibel sound.
c. Prolonged noise above 150 decibels causes immediate and permanent deafness. How does the intensity of a 150-decibel sound compare with the intensity of an 80-decibel sound?

60. Barometric Pressure Halley's Law states that the barometric pressure (in inches of mercury) at an altitude of x miles above sea level is approximately given by the equation

$$p(x) = 29.92 e^{-0.2x} \qquad x \geq 0$$

If the barometric pressure as measured by a hot air balloonist is 20 inches of mercury, what is the balloonist's altitude?

61. Forensic Science Forensic scientists use the following law to determine the time of death of accident or murder victims. If T denotes the temperature of a body t hours after death, then

$$T = T_0 + (T_1 - T_0)(0.97)^t$$

where T_0 is the air temperature and T_1 is the body temperature at the time of death. John Doe was found murdered at midnight in his house, when the room temperature was 70°F and his body temperature was 80°F. When was he killed? Assume that the normal body temperature is 98.6°F.

62. a. Given that $2^x = e^{kx}$, find k.
b. Show that, in general, if b is a nonnegative real number, then any equation of the form $y = b^x$ may be written in the form $y = e^{kx}$, for some real number k.

63. Use the definition of a logarithm to prove
a. $\log_b mn = \log_b m + \log_b n$

b. $\log_b \dfrac{m}{n} = \log_b m - \log_b n$

[*Hint:* Let $\log_b m = p$ and $\log_b n = q$. Then $b^p = m$ and $b^q = n$.]

64. Use the definition of a logarithm to prove

$$\log_b m^n = n \log_b m$$

65. Use the definition of a logarithm to prove
a. $\log_b 1 = 0$ **b.** $\log_b b = 1$

SOLUTIONS TO
SELF-CHECK
EXERCISES 3.2

1. First, sketch the graph of $y = 3^x$ with the help of the following table of values.

x	-3	-2	-1	0	1	2	3
$y = 3^x$	$\frac{1}{27}$	$\frac{1}{9}$	$\frac{1}{3}$	0	3	9	27

Next, take the mirror reflection of this graph with respect to the line $y = x$ to obtain the graph of $y = \log_3 x$.

2.
$$3e^{x+1} - 2 = 4$$
$$3e^{x+1} = 6$$
$$e^{x+1} = 2$$
$$\ln e^{x+1} = \ln 2$$
$$(x + 1)\ln e = \ln 2 \quad \text{Law 3}$$
$$x + 1 = \ln 2 \quad \text{Law 5}$$
$$x = \ln 2 - 1$$
$$\approx -0.3069$$

3.3 Exponential Functions as Mathematical Models

Exponential Growth

FIGURE 10
Exponential growth

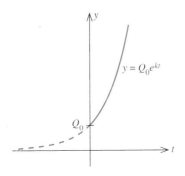

Many problems arising from practical situations can be described mathematically in terms of exponential functions or functions closely related to the exponential function. In this section, we will look at some applications involving exponential functions from the fields of the life and social sciences.

In Section 3.1 we saw that the exponential function $f(x) = b^x$ is an increasing function when $b > 1$. In particular, the function $f(x) = e^x$ shares this property. From this result one may deduce that the function $Q(t) = Q_0 e^{kt}$, where Q_0 and k are positive constants, has the following properties:

1. $Q(0) = Q_0$

2. $Q(t)$ increases "rapidly" without bound as t increases without bound (Figure 10).

The exponential function

$$Q(t) = Q_0 e^{kt} \qquad 0 \le t < \infty \qquad \textbf{(3)}$$

provides us with a mathematical model of a quantity $Q(t)$ that is initially present in the amount of $Q(0) = Q_0$ and whose rate of growth at any time t is directly proportional to the amount of the quantity present at time t (see Example 5, Section 9.7). Such a quantity is said to exhibit **exponential growth,** and the constant k is called the **growth constant.** Interest earned on a fixed deposit when compounded continuously exhibits exponential growth. Other examples of exponential growth follow.

 EXAMPLE 1   Under ideal laboratory conditions, the number of bacteria in a culture grows in accordance with the law $Q(t) = Q_0 e^{kt}$, where Q_0 denotes the number of bacteria initially present in the culture, k is some constant determined by the strain of bacteria under consideration, and t is the elapsed time measured in hours. Suppose that there are 10,000 bacteria present initially in the culture and 60,000 present two hours later. How many bacteria will there be in the culture at the end of four hours?

Solution   We are given that $Q(0) = Q_0 = 10,000$, so $Q(t) = 10,000 e^{kt}$. Next, the fact that 60,000 bacteria are present two hours later translates into $Q(2) = 60,000$. Thus

$$60,000 = 10,000 e^{2k}$$

$$e^{2k} = 6$$

Taking the natural logarithm on both sides of the equation, we obtain

$$\ln e^{2k} = \ln 6$$

$$2k = \ln 6, \qquad \text{Since } \ln e = 1$$

or $$k \approx 0.8959$$

Thus, the number of bacteria present at any time t is given by

$$Q(t) = 10,000 e^{0.8959t}$$

In particular, the number of bacteria present in the culture at the end of four hours is given by

$$Q(4) = 10,000 e^{0.8959(4)}$$

$$= 360,029 \qquad \blacksquare$$

Exponential Decay

In contrast to exponential growth, a quantity exhibits **exponential decay** if it decreases at a rate that is directly proportional to its size. Such a quantity may be described by the exponential function

$$Q(t) = Q_0 e^{-kt} \qquad t \in [0, \infty) \qquad \textbf{(4)}$$

where the positive constant Q_0 measures the amount present initially ($t = 0$) and k is some suitable positive number, called the **decay constant.** The choice of this number is determined by the nature of the substance under consideration. The graph of this function is sketched in Figure 11.

FIGURE 11
Exponential decay

EXAMPLE 2   Radioactive substances decay exponentially. For example, the amount of radium present at any time t obeys the law $Q(t) = Q_0 e^{-kt}$, where Q_0 is the initial amount present and k is a suitable positive constant. The **half-life** of a radioactive substance is the time required for a given amount to be reduced by one-half. Now, it is known that the half-life of radium is approximately 1600 years. Suppose that initially there are 200 milligrams of pure radium. Find the amount left after t years. What is the amount left after 800 years?

Solution   The initial amount of radium present is 200 milligrams, so $Q(0) = Q_0 = 200$. Thus $Q(t) = 200 e^{-kt}$. Next, the datum concerning the half-life of radium implies that $Q(1600) = 100$, and this gives

$$100 = 200 e^{-1600k}$$

$$e^{-1600k} = \frac{1}{2} \qquad \text{Dividing both sides of the equation by 200}$$

Taking the natural logarithm on both sides of this equation yields

$$-1600k \ln e = \ln \frac{1}{2}$$

$$-1600k = \ln \frac{1}{2} \qquad \text{ln } e = 1$$

$$k = -\frac{1}{1600} \ln \left(\frac{1}{2} \right) = 0.0004332$$

Therefore, the amount of radium left after t years is

$$Q(t) = 200 e^{-0.0004332t}$$

In particular, the amount of radium left after 800 years is

$$Q(800) = 200 e^{-0.0004332(800)} \approx 141.42$$

or approximately 141 milligrams.

EXAMPLE 3   Carbon 14, a radioactive isotope of carbon, has a half-life of 5770 years. What is its decay constant?

Solution   We have $Q(t) = Q_0 e^{-kt}$. Since the half-life of the element is 5770 years, half of the substance is left at the end of that period. That is,

$$Q(5770) = Q_0 e^{-5770k} = \frac{1}{2} Q_0$$

or
$$e^{-5770k} = \frac{1}{2}$$

Taking the natural logarithm on both sides of this equation, we have

$$\ln e^{-5770k} = \ln \frac{1}{2}$$

$$-5770k = -0.693147$$

$$\approx 0.00012$$

Carbon 14 dating is a well-known method used by anthropologists to establish the age of animal and plant fossils. This method assumes that the proportion of carbon 14 (C-14) present in the atmosphere has remained constant over the past 50,000 years. Professor Willard Libby, recipient of the Nobel Prize in chemistry in 1960, proposed this theory.

The amount of C-14 in the tissues of a living plant or animal is constant. However, when an organism dies it stops absorbing new quantities of C-14 and the amount of C-14 in the remains diminishes because of the natural decay of the radioactive substance. Thus, the approximate age of a plant or animal fossil can be determined by measuring the amount of C-14 present in the remains.

EXAMPLE 4   A skull from an archeological site has one-tenth the amount of C-14 that it originally contained. Determine the approximate age of the skull.

Solution   Here

$$Q(t) = Q_0 e^{-kt}$$

$$= Q_0 e^{-0.00012t}$$

where Q_0 is the amount of C-14 present originally and k, the decay constant, is equal to 0.00012 (see Example 3). Since $Q(t) = (1/10)Q_0$, we have

$$\frac{1}{10} Q_0 = Q_0 e^{-0.00012t}$$

$$\ln \frac{1}{10} = -0.00012t \qquad \text{Taking the natural logarithm on both sides}$$

$$t = \frac{\ln \dfrac{1}{10}}{-0.00012}$$

$$\approx 19{,}200$$

or approximately 19,200 years.   ■

Learning Curves

The next example shows how the exponential function may be applied to describe certain types of learning processes. Consider the function

$$Q(t) = C - Ae^{-kt}$$

where C, A, and k are positive constants. The graph of the function Q is shown in Figure 12, where that part of the graph corresponding to the negative values of t is drawn with a dashed line since, in practice, one normally restricts the domain of the function to the interval $[0, \infty)$. Observe that $Q(t)$ $(t > 0)$ increases rather rapidly initially but that the rate of increase slows down considerably after a while.

This behavior of the graph of the function Q closely resembles the learning pattern experienced by workers engaged in highly repetitive work. For example, the productivity of an assembly line worker increases very rapidly in the early stages of the training period. This productivity increase is a direct result of the worker's training and accumulated experience. But

FIGURE 12

A learning curve

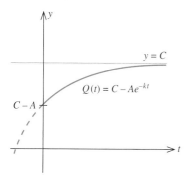

the rate of increase of productivity slows down as time goes by and the worker's productivity level approaches some fixed level due to the limitations of the worker and the machine. Because of this characteristic, the graph of the function $Q(t) = C - Ae^{-kt}$ is often called a *learning curve.*

 EXAMPLE 5 The Camera Division of the Eastman Optical Company produces a 35-mm single-lens reflex camera, model F. Eastman's training department determines that after completing the basic training program, a new, previously inexperienced employee will be able to assemble

$$Q(t) = 50 - 30e^{-0.5t}$$

model F cameras per day, t months after the employee starts work on the assembly line.

a. How many model F cameras can a new employee assemble per day after basic training?

b. How many model F cameras can an employee with one month of experience assemble per day? An employee with two months of experience? An employee with six months of experience?

c. How many model F cameras can the average experienced employee assemble per day?

Solution

a. The number of model F cameras a new employee can assemble is given by

$$Q(0) = 50 - 30 = 20 \qquad e^0 = 1$$

b. The number of model F cameras that an employee with one month of experience, two months of experience, and six months of experience can assemble per day is given by

$$Q(1) = 50 - 30e^{-0.5} \approx 31.80$$
$$Q(2) = 50 - 30e^{-1} \approx 38.96$$
$$Q(6) = 50 - 30e^{-3} \approx 48.51$$

or approximately 32, 39, and 49, respectively.

c. As t increases without bound, $Q(t)$ approaches 50. Hence, the average experienced employee can ultimately be expected to assemble 50 model F cameras per day. ∎

Other applications of the learning curve are found in models that describe the dissemination of information about a product or the velocity of an object dropped into a viscous medium.

Logistic Growth Functions

Our last example of an application of exponential functions to the description of natural phenomena involves the **logistic** (also called the *S-shaped,* or *sigmoidal*) **growth curve,** which is the graph of the function

$$Q(t) = \frac{A}{1 + Be^{-kt}}$$

FIGURE 13

A logistic curve

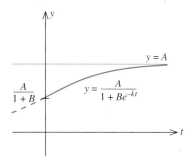

where A, B, and k are positive constants. The graph of the function Q is sketched in Figure 13.

Observe that $Q(t)$ increases rather rapidly for small values of t. In fact, for small values of t, the logistic curve resembles an exponential growth curve. However, the *rate of growth* of $Q(t)$ decreases quite rapidly as t increases and $Q(t)$ approaches the number A as x increases without bound.

Thus, the logistic curve exhibits both the property of rapid growth of the exponential growth curve as well as the "saturation" property of the learning curve. Because of these characteristics, the logistic curve serves as a suitable mathematical model for describing many natural phenomena. For example, if a small number of rabbits were introduced to a tiny island in the South Pacific, the rabbit population might be expected to grow very rapidly at first, but the growth rate would decrease quickly as overcrowding, scarcity of food, and other environmental factors affected it. The population would eventually stabilize at a level compatible with the life-support capacity of the environment. Models describing the spread of rumors and epidemics are other examples of the application of the logistic curve.

 EXAMPLE 6   The number of men at Fort MacArthur who contracted influenza after t days during a flu epidemic is approximated by the exponential model

$$Q(t) = \frac{5000}{1 + 1249e^{-kt}}$$

If 40 men contracted the flu by the seventh day, find how many men contracted the flu by the fifteenth day.

Solution   The given information implies that

$$Q(7) = 40$$

Thus,
$$Q(7) = \frac{5000}{1 + 1249e^{-7k}}$$

$$= 40$$

or $40(1 + 1249e^{-7k}) = 5000$

$$1 + 1249e^{-7k} = \frac{5000}{40} = 125$$

$$e^{-7k} = \frac{124}{1249}$$

$$-7k = \ln\frac{124}{1249} \qquad \text{Taking the natural logarithm on both sides}$$

$$k = -\frac{\ln\dfrac{124}{1249}}{7} \approx 0.33.$$

Therefore, the number of men who contracted the flu after t days is given by

$$Q(t) = \frac{5000}{1 + 1249e^{-0.33t}}$$

In particular, the number of men who contracted the flu by the fifteenth day is given by

$$Q(15) = \frac{5000}{1 + 1249e^{-15(0.33)}}$$

$$\approx 508$$

or approximately 508 men.

SELF-CHECK EXERCISE 3.3

Suppose that the population of a country (in millions) at any time t grows in accordance with the rule

$$P = \left(P_0 + \frac{I}{k}\right)e^{kt} - \frac{I}{k}$$

where P denotes the population at any time t, k is a constant reflecting the natural growth rate of the population, I is a constant giving the (constant) rate of immigration into the country, and P_0 is the total population of the country at time $t = 0$. The population of the United States in the year 1980 ($t = 0$) was 226.5 million. If the natural growth rate is 0.8 percent annually ($k = 0.008$) and net immigration is allowed at the rate of half a million people per year ($I = 0.5$) until the end of the century, what will the population of the United States be in the year 2000?

Solutions to Self-Check Exercise 3.3 can be found on page 137.

3.3 EXERCISES

c *A calculator is recommended for this exercise set.*

1. **Exponential Growth** Given that a quantity $Q(t)$ is described by the exponential growth function

$$Q(t) = 400e^{0.05t}$$

where t is measured in minutes, answer the following questions:
a. What is the growth constant?
b. What quantity is present initially?
c. Using a calculator, complete the following table of values:

t	0	10	20	100	1000
Q					

2. **Exponential Decay** Given that a quantity $Q(t)$ exhibiting exponential decay is described by the function

$$Q(t) = 2000e^{-0.06t}$$

where t is measured in years, answer the following questions:
a. What is the decay constant?
b. What quantity is present initially?
c. Using a calculator, complete the following table of values:

t	0	5	10	20	100
Q					

3. **Growth of Bacteria** The growth rate of bacterium *Escherichia coli*, a common bacterium found in the human intestine, is proportional to its size. Under ideal laboratory conditions, when this bacterium is grown in a nutrient broth medium, the number of cells in a culture doubles approximately every 20 minutes.
a. If the initial cell population is 100, determine the function $Q(t)$ that expresses the exponential

growth of the number of cells of this bacterium as a function of time t (in minutes).

b. How long will it take for a colony of 100 cells to increase to a population of one million?

c. If the initial cell population were 1000, how would this alter our model?

4. World Population   The world population at the beginning of 1990 was 5.3 billion. Assuming that the population continues to grow at its present rate of approximately 2% per year, find the function $Q(t)$ that expresses the world population (in billions) as a function of time t (in years).

a. Using this function, complete the following table of values and sketch the graph of the function Q:

Year	1990	1995	2000	2005
World Population				

Year	2010	2015	2020	2025
World Population				

b. Find the estimated rate of growth in the year 2000.

5. World Population   Refer to exercise 4.

a. If the world population continues to grow at its present rate of approximately 2% per year, find the length of time t_0 required for the world population to triple in size.

b. Using the time t_0 found in (a), what would the world population be if the growth rate were reduced to 1.8%?

6. Resale Value   A certain piece of machinery was purchased three years ago by the Garland Mills Company for $500,000. Its present resale value is $320,000. Assuming that the machine's resale value decreases exponentially, what will it be four years from now?

7. Radioactive Decay   If the temperature is constant, then the atmospheric pressure P (in pounds per square inch) varies with the altitude above sea level h in accordance with the law

$$P = p_0 e^{-kh}$$

where p_0 is the atmospheric pressure at sea level and k is a constant. If the atmospheric pressure is 15 pounds per square inch at sea level and 12.5 pounds per square inch at 4,000 feet, find the atmospheric pressure at an altitude of 12,000 feet.

8. Radioactive Decay   The radioactive element polonium decays according to the law

$$Q(t) = Q_0 \cdot 2^{-(t/140)}$$

where Q_0 is the initial amount and the time t is measured in days. If the amount of polonium left after 280 days is 20 milligrams, what was the initial amount present?

9. Radioactive Decay   Phosphorus 32 has a half-life of 14.2 days. If 100 grams of this substance are present initially, find the amount present after t days. What amount will be left after 7.1 days?

10. Nuclear Fallout   Strontium 90, a radioactive isotope of strontium, is present in the fallout resulting from nuclear explosions. It is especially hazardous to animal life, including humans, because upon ingestion of contaminated food it is absorbed into the bone structure. Its half-life is 27 years. If the amount of strontium 90 in a certain area is found to be four times the "safe" level, find how much time must elapse before an "acceptable level" is reached.

11. Carbon 14 Dating   Wood deposits recovered from an archeological site contain 20% of the carbon 14 they originally contained. How long ago did the tree from which the wood was obtained die?

12. Carbon 14 Dating   Skeletal remains of the so-called Pittsburgh Man, unearthed in Pennsylvania, had lost 82% of the carbon 14 they originally contained. Determine the approximate age of the bones.

13. Learning Curves   The American Stenographic Institute finds that the average student taking Advanced Shorthand, an intensive 20-week course, progresses according to the function

$$Q(t) = 120(1 - e^{-0.05t}) + 60 \qquad 0 \le t \le 20$$

where $Q(t)$ measures the number of words (per minute) of dictation that the student can take in shorthand after t weeks in the course. Sketch the graph of the function Q and answer the following questions.

a. What is the beginning shorthand speed for the average student in this course?

b. What shorthand speed does the average student attain halfway through the course?

c. How many words per minute can the average student take after completing this course?

14. Effect of Advertising on Sales   The Metro Department Store found that t weeks after the end of a sales promotion the volume of sales was given by a function of the form

$$S(t) = B + Ae^{-kt} \qquad 0 \le t \le 4$$

where $B = 50,000$ and is equal to the average weekly volume of sales before the promotion. The sales volumes at the end of the first and third weeks were

$83,515 and $65,055, respectively. Assume that the sales volume is decreasing exponentially and find
a. the decay constant k
b. the sales volume at the end of the fourth week

15. **Demand for Computers** The Universal Instruments Company found that the monthly demand for its new line of Galaxy Home Computers t months after placing the line on the market was given by

$$D(t) = 2000 - 1500e^{-0.05t} \qquad t > 0$$

Graph this function and answer the following questions.
a. What is the demand after one month? One year? Two years? Five years?
b. At what level is the demand expected to stabilize?

16. **Newton's Law of Cooling** Newton's law of cooling states that the rate at which the temperature of an object changes is proportional to the difference in temperature between the object and that of the surrounding medium. Thus, the temperature $F(t)$ of an object that is greater than the temperature of its surrounding medium is given by

$$F(t) = T + Ae^{-kt}$$

where t is the time expressed in minutes, T is the temperature of the surrounding medium, and A and k are constants. Suppose that a cup of instant coffee is prepared with boiling water (212°F) and left to cool on the counter in a room where the temperature is 72°F. If $k = 0.1865$, determine when the coffee will be cool enough to drink (say, 110°F).

17. **Spread of an Epidemic** During a flu epidemic, the number of children in the Woodbridge Community School System who contracted influenza after t days was given by

$$Q(t) = \frac{1000}{1 + 199e^{-0.8t}}$$

a. How many children were stricken by the flu after the first day?
b. How many children had the flu after ten days?

c. How many children eventually contracted the disease?

18. **Growth of a Fruit Fly Population** On the basis of data collected during an experiment, a biologist found that the growth of the fruit fly (*Drosophila*) with a limited food supply could be approximated by the exponential model

$$N(t) = \frac{400}{1 + 39e^{-0.16t}}$$

where t denotes the number of days since the beginning of the experiment.
a. What was the initial fruit fly population in the experiment?
b. What was the maximum fruit fly population that could be expected under this laboratory condition?
c. What was the population of the fruit fly colony on the twentieth day?

19. **Percentage of Households with VCRs** According to estimates by Paul Kroger Associates, the percentage of households that own videocassette recorders (VCRs) is given by

$$P(t) = \frac{68}{1 + 21.67e^{-0.62t}} \qquad 0 \le t \le 12$$

where t is measured in years, with $t = 0$ corresponding to the beginning of 1985. What percentage of households owned VCRs at the beginning of 1985? At the beginning of 1995?
Source: Paul Kroger Associates

20. **Spread of a Rumor** Three hundred students attended the dedication ceremony of a new building on a college campus. The president of the traditionally female college announced a new expansion program, which included plans to make the college coeducational. The number of students who learned of the new program t hours later is given by the function

$$f(t) = \frac{3000}{1 + Be^{-kt}}$$

If 600 students on campus had heard about the new program two hours after the ceremony, how many students had heard about the policy after four hours?

SOLUTION TO SELF-CHECK EXERCISE 3.3

We are given that $P_0 = 226.5$, $k = 0.008$, and $I = 0.5$. So

$$P = \left(226.5 + \frac{0.5}{0.008}\right)e^{0.008t} - \frac{0.5}{0.008}$$

$$= 289e^{0.008t} - 62.5$$

Therefore, the population in the year 2000 will be given by

$$P(20) = 289e^{0.16} - 62.5$$

$$\approx 276.6$$

or approximately 276.6 million.

CHAPTER THREE SUMMARY OF PRINCIPAL FORMULAS AND TERMS

Formulas

1.	Exponential function with base b	$y = b^x$
2.	Exponential function with base e	$y = e^x$
3.	Logarithmic function with base b	$y = \log_b x$
4.	Logarithmic function with base e	$y = \ln x$
5.	Exponential-logarithmic relationships	$\ln e^x = x$ and $e^{\ln x} = x$

Terms

Common logarithm   Exponential decay

Natural logarithm   Decay constant

Exponential growth   Half-life

Growth constant   Logistic growth curve

CHAPTER 3 REVIEW EXERCISES

In exercises 1–4, sketch the graph of the function.

1. $f(x) = 5^x$

2. $y = \left(\tfrac{1}{5}\right)^x$

3. $f(x) = \log_4 x$

4. $y = \log_{1/4} x$

In exercises 5–8, express each equation in logarithmic form.

5. $3^4 = 81$

6. $9^{1/2} = 3$

7. $\left(\tfrac{2}{3}\right)^{-3} = \tfrac{27}{8}$

8. $16^{-3/4} = 0.125$

In exercises 9–12, given that $\ln 2 \approx 0.6931$, $\ln 3 \approx 1.0986$, and $\ln 5 \approx 1.6094$, find the value of the expression.

9. $\ln 30$

10. $\ln 9$

11. $\ln 3.6$

12. $\ln 75$

In exercises 13–18, solve for x without using a calculator.

13. $2^{2x-3} = 8$

14. $e^{x^2+x} = e^2$

15. $3^{x-1} = 9^{x+2}$

16. $2^{x^2+x} = 4^{x^2-3}$

17. $\log_4(2x + 1) = 2$

18. $\ln(x - 1) + \ln 4 = \ln(2x + 4) - \ln 2$

In exercises 19–32, solve for x, giving your answer accurate to four decimal places.

19. $4^x = 5$

20. $3^{-2x} = 8$

21. $3 \cdot 2^{-x} = 17$

22. $2e^{-x} = 7$

23. $0.2e^x = 3.4$

24. $e^{2x-1} = 14$

25. $5^{3x+1} = 16$

26. $2^{3x+1} = 3^{2x-3}$

27. $2^{x^2} = 12$ **28.** $3e^{\sqrt{x}} = 15$

29. $4e^{-0.1x} - 2 = 8$ **30.** $8 - e^{0.2x} = 2$

31. $\dfrac{20}{1 + 2e^{0.2x}} = 4$ **32.** $\dfrac{30}{1 + 2e^{-0.1x}} = 5$

33. Sketch the graph of the function $y = \log_2(x + 3)$.

34. Sketch the graph of the function $y = \log_3(x + 1)$.

35. A culture of bacteria that initially contained 2,000 bacteria has a count of 18,000 bacteria after two hours.
   a. Determine the function $Q(t)$ that expresses the exponential growth of the number of cells of this bacterium as a function of time t (in minutes).
   b. Find the number of bacteria present after four hours.

36. The radioactive element radium has a half-life of 1600 years. What is its decay constant?

37. The V.C.A. Television Company found that the monthly demand for its new line of videodisc players t months after placing the players on the market is given by

$$D(t) = 4000 - 3000e^{-0.06t} \qquad t \geq 0$$

Graph this function and answer the following questions.
   a. What was the demand after one month? After one year? After two years?
   b. At what level is the demand expected to stabilize?

38. During a flu epidemic, it was found that the number of students at a certain university who contracted influenza after t days could be approximated by the exponential model

$$Q(t) = \frac{3000}{1 + 499e^{-kt}}$$

If 90 students contracted the flu by the tenth day, find how many students contracted the flu by the twentieth day.

CHAPTER 3   EXTENDED PROBLEMS AND PROJECTS

1. One method for determining the half-life of a radioactive substance (see Example 2, Section 3.3) follows: Measure the amount of the substance present at time t_1—call it Q_1. At some time t_2 later, measure the amount of the substance present again—call it Q_2. Show that the half-life T of the radioactive substance is

$$T = \frac{(t_2 - t_1)\ln 2}{\ln(Q_1/Q_2)}$$

2. Refer to exercise 1. The amount of a radioactive substance present at $t_1 = 1$ is 8.77 grams, and the amount present at $t_2 = 3$ is 6.74 grams, where t is measured in years. Determine the half-life of the substance.

Mathematics of Finance

*I*nterest that is periodically added to the principal and thereafter itself earns interest is called *compound interest.* We begin this chapter by deriving the *compound interest formula,* which gives the amount of money accumulated when an initial amount of money is invested in an account for a fixed term and earns compound interest.

An *annuity* is a sequence of payments made at regular intervals. We will derive formulas giving the *future value of an annuity* (what you end up with) and the *present value of an annuity* (the lump sum that, when invested now, will yield the same future value of the annuity). Then, using these formulas, we will answer questions involving the amortization of certain types of installment loans and questions involving *sinking funds* (funds that are set up to be used for a specific purpose at a future date).

Compound Interest

Simple Interest

FIGURE 1
The accumulated amount is a linear function of t.

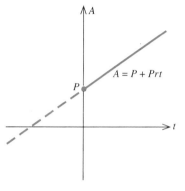

A natural application of linear functions to the business world is found in the computation of **simple interest**—interest that is computed on the original principal only. Thus, if I denotes the interest on a principal P (in dollars) at an interest rate of r per year for t years, then we have

$$I = Prt$$

The **accumulated amount** A, the sum of the principal and interest after t years, is given by

$$A = P + I = P + Prt$$

or

$$A = P(1 + rt)$$

and is a linear function of t (see exercise 32 at the end of this section). In business applications we are normally interested only in the case where t is positive, so only that part of the line that lies in Quadrant I is of interest to us (Figure 1).

**SIMPLE INTEREST
FORMULAS**

Interest:	$I = Prt$	*(1a)*
Accumulated amount:	$A = P(1 + rt)$	*(1b)*

EXAMPLE 1   A bank pays simple interest at the rate of 8% per year for certain deposits. If a customer deposits \$1000 and makes no withdrawals for three years, what is the total amount on deposit at the end of three years? What is the interest earned in that period of time?

Solution   Using (1b) with $P = 1000$, $r = 0.08$, and $t = 3$, we see that the total amount on deposit at the end of three years is given by

$$A = P(1 + rt)$$
$$= 1000[1 + (0.08)(3)] = 1240$$

or \$1240.

The interest earned over the 3-year period is given by

$$I = Prt \qquad \text{Using 1a}$$
$$= 1000(0.08)(3) = 240$$

or \$240.

EXAMPLE 2   An amount of \$2000 is invested in a 10-year trust fund that pays 6% annual simple interest. What is the total amount of the trust fund at the end of ten years?

Solution   The total amount of the trust fund at the end of ten years is given by

$$A = P(1 + rt)$$
$$= 2000[1 + (0.06)(10)] = 3200$$

or \$3200.

Earned interest that is periodically added to the principal and thereafter itself earns interest at the same rate is called **compound interest.** In order to find a formula for the accumulated amount, let us consider a numerical example. Suppose that $1000 (the principal) is deposited in a bank for a term of three years, earning interest at the rate of 8% per year (called the **nominal, or stated, rate**) compounded annually. Then, using (1b) with $P = 1000$, $r = 0.08$, and $t = 1$, we see that the accumulated amount at the end of the first year is

$$A_1 = P(1 + rt)$$
$$= 1000[1 + 0.08(1)] = 1000(1.08) = 1080$$

or $1080.

To find the accumulated amount A_2 at the end of the second year, we use (1b) once again, this time with $P = A_1$. (Remember, the principal *and* interest now earn interest over the second year.) We obtain

$$A_2 = P(1 + rt) = A_1(1 + rt)$$
$$= 1000[1 + 0.08(1)][1 + 0.08(1)]$$
$$= 1000(1 + 0.08)^2 = 1000(1.08)^2 = 1166.40$$

or $1166.40.

Finally, the accumulated amount A_3 at the end of the third year is found using (1b) with $P = A_2$, giving

$$A_3 = P(1 + rt) = A_2(1 + rt)$$
$$= 1000[1 + 0.08(1)]^2[1 + 0.08(1)]$$
$$= 1000[1 + 0.08]^3 = 1000(1.08)^3 \approx 1259.71$$

or approximately $1259.71.

If we reexamine our calculations, we see that the accumulated amounts at the end of each year have the following form:

First year: $A_1 = 1000(1 + 0.08)$, or $A_1 = P(1 + r)$
Second year: $A_2 = 1000(1 + 0.08)^2$, or $A_2 = P(1 + r)^2$
Third year: $A_3 = 1000(1 + 0.08)^3$, or $A_3 = P(1 + r)^3$

These observations suggest the following general result: If P dollars is invested over a term of t years earning interest at the rate of r per year compounded annually, then the accumulated amount is

$$A = P(1 + r)^t \qquad (2)$$

Formula (2) was derived under the assumption that interest was compounded *annually*. In practice, however, interest is usually compounded more than once a year. The interval of time between successive interest calculations is called the *conversion period.*

If interest at a nominal rate of r per year is compounded m times a year on a principal of P dollars, then the simple interest rate per conversion period is

$$i = \frac{r}{m} \qquad \frac{\text{Annual interest rate}}{\text{Periods per year}}$$

For example, if the nominal interest rate is 8% per year ($r = 0.08$) and interest is compounded quarterly ($m = 4$), then

$$i = \frac{r}{m} = \frac{0.08}{4} = 0.02$$

or 2% per period.

In order to find a general formula for the accumulated amount when a principal of P dollars is deposited in a bank for a term of t years and earns interest at the (nominal) rate of r per year compounded m times per year, we proceed as before, using (1b) repeatedly with the interest rate $i = r/m$. We see that the accumulated amount at the end of each period is

First period: $A_1 = P(1 + i)$
Second period: $A_2 = A_1(1 + i)$ $= [P(1 + i)](1 + i) = P(1 + i)^2$
Third period: $A_3 = A_2(1 + i)$ $= [P(1 + i)^2](1 + i) = P(1 + i)^3$
$$\vdots \qquad\qquad\qquad\qquad \vdots$$
nth period: $A_n = A_{n-1}(1 + i) = [P(1 + i)^{n-1}](1 + i) = P(1 + i)^n$

But there are $n = mt$ periods in t years (number of conversion periods times the term). Therefore, the accumulated amount at the end of t years is given by

$$A = P(1 + i)^n$$

**COMPOUND
INTEREST
FORMULA**

$$A = P(1 + i)^n \qquad\qquad (3)$$

where $i = r/m$, $n = mt$, and

A = Accumulated amount at the end of n conversion periods

P = Principal

r = Nominal interest rate per year

m = Number of conversion periods per year

t = Term (number of years)

EXAMPLE 3   Find the accumulated amount after three years if $1000 is invested at 8% per year compounded (a) annually, (b) semiannually, (c) quarterly, and (d) monthly.

Solution

a. Here $P = 1000$, $r = 0.08$, and $m = 1$. Thus, $i = r = 0.08$ and $n = 3$, so formula (3) gives

$$A = 1000(1.08)^3$$

$$= 1259.71$$

or $1259.71.

b. Here $P = 1000$, $r = 0.08$, and $m = 2$. Thus, $i = 0.08/2 = 0.04$ and $n = (3)(2) = 6$, so (3) gives

$$A = 1000(1.04)^6$$
$$= 1265.32$$

or \$1265.32.

c. In this case, $P = 1000$, $r = 0.08$, and $m = 4$. Thus, $i = 0.08/4 = 0.02$ and $n = (3)(4) = 12$, so (3) gives

$$A = 1000(1.02)^{12}$$
$$= 1268.24$$

or \$1268.24.

d. Here $P = 1000$, $r = 0.08$, and $m = 12$. Thus, $i = 0.08/12$ and $n = (3)(12) = 36$, so (3) gives

$$A = 1000\left(1 + \frac{0.08}{12}\right)^{36}$$
$$= 1270.24$$

or \$1270.24. These results are summarized in Table 1.

TABLE 1

Nominal Rate (r)	Conversion Period	Interest Rate/ Conversion Period	Initial Investment	Accumulated Amount
8%	Annual ($m = 1$)	8%	\$1000	\$1259.71
8%	Semiannual ($m = 2$)	4%	\$1000	\$1265.32
8%	Quarterly ($m = 4$)	2%	\$1000	\$1268.24
8%	Monthly ($m = 12$)	2/3%	\$1000	\$1270.24

Continuous Compounding of Interest

One question that arises naturally in the study of compound interest is: What happens to the accumulated amount over a fixed period of time if the interest is computed more and more frequently?

Intuition suggests that the more often interest is compounded, the larger the accumulated amount will be. This is confirmed by the results of Example 3, where we found that the accumulated amounts did, in fact, increase when we increased the number of conversion periods per year.

This leads us to another question: Does the accumulated amount approach a limit (that is, does it approach a fixed number) when the interest is computed more and more frequently over a fixed period of time?

To answer this question, let us look again at the compound interest

formula

$$A = P\left(1 + \frac{r}{m}\right)^{mt} \qquad (4)$$

Recall that m is the number of conversion periods per year. So to find an answer to our question, we should let m get larger and larger in (4). If we let $u = m/r$ so that $m = ru$, then (4) becomes

$$A = P\left(1 + \frac{1}{u}\right)^{urt} \qquad \frac{r}{m} = \frac{1}{u}$$

$$= P\left[\left(1 + \frac{1}{u}\right)^{u}\right]^{rt} \qquad \text{Since } b^{xy} = (b^x)^y$$

Now observe that u increases without bound as m increases without bound. But, from our work in Section 3.1, we know that $(1 + 1/u)^u$ approaches e as u increases without bound. Using this result, we see that, as m increases without bound A approaches $P(e)^{rt} = Pe^{rt}$. In this situation, we say that interest is *compounded continuously.*

Let us summarize this important result.

CONTINUOUS COMPOUND INTEREST FORMULA

$$A = Pe^{rt} \qquad (5)$$

where

$P = $ Principal

$r = $ Annual interest rate compounded continuously

$t = $ Time in years

$A = $ Accumulated amount at the end of t years

EXAMPLE 4   Find the accumulated amount after three years if $1000 is invested at 8% per year compounded (a) daily (take the number of days in a year to be 365) and (b) continuously.

Solution

a. Using formula (3) with $P = 1000$, $r = 0.08$, $m = 365$, and $n = (365)(3) = 1095$, we find

$$A = 1000\left(1 + \frac{0.08}{365}\right)^{1095} \approx 1271.22$$

or $1271.22.

b. Here we use formula (5), with $P = 1000$, $r = 0.08$, and $t = 3$, obtaining

$$A = 1000e^{(0.08)(3)}$$

$$\approx 1271.25 \qquad \text{Using the } e^x \text{ key}$$

or $1271.25.

Observe that the accumulated amounts corresponding to interest compounded daily and interest compounded continuously differ by very little. The continuous compound interest formula is a very important tool in theoretical work in financial analysis.

Effective Rate of Interest

As we saw in Examples 3 and 4, the interest actually earned on an investment depends on the frequency with which the interest is compounded. Thus, the stated, or nominal, rate of 8% per year does not reflect the actual rate at which interest is earned. This suggests that we need to find a common basis for comparing interest rates. One such way of comparing interest rates is provided by use of the *effective rate*. The **effective rate** is the *simple* interest rate that would produce the same accumulated amount in one year as the nominal rate compounded m times a year. The effective rate is also called the *effective annual yield*.

To derive a relationship between the nominal interest rate, r per year compounded m times, and its corresponding effective rate, R per year, let us assume an initial investment of P dollars. Then the accumulated amount after one year at a simple interest rate of R per year is

$$A = P(1 + R)$$

Also, the accumulated amount after one year at an interest rate of r per year compounded m times a year is

$$A = P(1 + i)^n = P\left(1 + \frac{r}{m}\right)^m \qquad \text{Since } i = \frac{r}{m}$$

Equating the two expressions gives

$$P(1 + R) = P\left(1 + \frac{r}{m}\right)^m$$

$$1 + R = \left(1 + \frac{r}{m}\right)^m, \qquad \text{Dividing both sides by } P$$

or, upon solving for R, we obtain the following formula for computing the effective rate of interest.

EFFECTIVE RATE OF INTEREST FORMULA

$$r_{\text{eff}} = \left(1 + \frac{r}{m}\right)^m - 1 \qquad \qquad \textbf{(6)}$$

where r_{eff} = Effective rate of interest

r = Nominal interest rate per year

m = Number of conversion periods per year

EXAMPLE 5   Find the effective rate of interest corresponding to a nominal rate of 8% per year compounded (a) annually, (b) semiannually, (c) quarterly, and (d) monthly.

Solution

a. The effective rate of interest corresponding to a nominal rate of 8% per year compounded annually is, of course, given by 8% per year. This result is also confirmed by using formula (6) with $r = 0.08$ and $m = 1$.

   Thus, $\qquad\qquad r_{\text{eff}} = (1 + 0.08) - 1 = 0.08$

b. Let $r = 0.08$ and $m = 2$. Then (6) yields

$$r_{\text{eff}} = \left(1 + \frac{0.08}{2}\right)^2 - 1$$

$$= (1.04)^2 - 1$$

$$= 0.0816$$

   so the required effective rate is 8.16% per year.

c. Let $r = 0.08$ and $m = 4$. Then (6) yields

$$r_{\text{eff}} = \left(1 + \frac{0.08}{4}\right)^4 - 1$$

$$= (1.02)^4 - 1$$

$$= 0.08243$$

   so the corresponding effective rate in this case is 8.243% per year.

d. Let $r = 0.08$ and $m = 12$. Then (6) yields

$$r_{\text{eff}} = \left(1 + \frac{0.08}{12}\right)^{12} - 1$$

$$= 0.08300$$

   so the corresponding effective rate in this case is 8.300% per year.

        Now if the effective rate of interest r_{eff} is known, the accumulated amount after t years on an investment of P dollars may be more readily computed by using the formula

$$A = P(1 + r_{\text{eff}})^t$$

        The Truth in Lending Act passed by Congress in 1968 requires that the effective rate of interest be disclosed in all contracts involving interest charges. The passage of this act has benefited consumers because they now have a common basis for comparing the various nominal rates quoted by different financial institutions. Furthermore, knowing the effective rate enables consumers to compute the actual charges involved in a transaction. Thus, if the effective rates of interest found in Example 3 were known, the accumulated values of Example 5, shown in Table 2, could have been readily found.

TABLE 2

Nominal Rate	Frequency of Interest Payment	Effective Rate	Initial Investment	Accumulated Amount after 3 Years	
8%	Annually	8%	$1000	$1000(1 + 0.08)^3$	$= \$1259.71$
8%	Semiannually	8.16%	$1000	$1000(1 + 0.0816)^3$	$= \$1265.32$
8%	Quarterly	8.243%	$1000	$1000(1 + 0.08243)^3$	$= \$1268.23$
8%	Monthly	8.300%	$1000	$1000(1 + 0.08300)^3$	$= \$1270.24$

Present Value

Let us return to the compound interest formula (3), which expresses the accumulated amount at the end of n periods when interest at the rate of r is compounded m times a year. The principal P in (3) is often referred to as the *present value,* and the accumulated value A is called the *future value,* since it is realized at a future date. In certain instances, an investor may wish to determine how much money he should invest now, at a fixed rate of interest, so that he will realize a certain sum at some future date. This problem may be solved by expressing P in terms of A. Thus, from formula (3) we find

$$P = A(1 + i)^{-n}$$

Here, as before, $i = r/m$, where m is the number of conversion periods per year.

PRESENT VALUE FORMULA FOR COMPOUND INTEREST

$$P = A(1 + i)^{-n} \qquad\qquad (7)$$

 EXAMPLE 6   Find how much money should be deposited in a bank paying interest at a rate of 6% per year compounded monthly so that at the end of three years the accumulated amount will be $20,000.

Solution   Here, $r = 0.06$ and $m = 12$, so $i = 0.06/12 = 0.005$ and $n = (3)(12) = 36$. Thus, the problem is to determine P given that $A = 20,000$. Using formula (7), we obtain

$$P = 20,000(1.005)^{-36} \approx 16,713$$

or $16,713.

 EXAMPLE 7   Find the present value of $49,158.60 due in five years at an interest rate of 10% per year compounded quarterly.

Solution   Using formula (7) with $r = 0.1$ and $m = 4$, so that $i = 0.1/4 = 0.025$, $n = (4)(5) = 20$, and $A = 49,158.6$, we obtain

$$P = (49,158.6)(1.025)^{-20} = 30,000.07$$

or approximately $30,000.

Applications

The returns on certain investments such as zero coupon certificates of deposit (CDs) and zero coupon bonds are compared by quoting the time it takes for each investment to triple, or even quadruple. These calculations use the compound interest formula (3).

EXAMPLE 8 Jane has narrowed her investment options down to two:

a. Purchase a CD that matures in 12 years and pays interest upon maturity at the rate of 10% per year compounded daily (assume 365 days in a year)

b. Purchase a zero coupon CD that will triple her investment in the same period

Which option will optimize her investment?

Solution Let us compute the accumulated amount under option (a). Here

$$r = 0.10, \quad m = 365, \quad \text{and} \quad t = 12$$

so $n = 12(365) = 4380$ and $i = 0.10/365 \approx 0.0002740$. The accumulated amount at the end of 12 years (after 4380 conversion periods) is

$$A = P(1.0002740)^{4380} \approx 3.32P$$

or $3.32. If Jane chooses option (b), the accumulated amount of her investment after 12 years will be $3P. Therefore, she should choose option (a).

◼

EXAMPLE 9 Diane has an Individual Retirement Account (IRA) with a brokerage firm. Her money is invested in a money market mutual fund that pays interest on a daily basis. Over a 2-year period in which no deposits or withdrawals were made, her account grew from $4500 to $5268.24. Find the effective rate at which Diane's account was earning interest over that period (assume 365 days in a year).

Solution Let r_{eff} denote the required effective rate of interest. We have

$$5268.24 = 4500(1 + r_{\text{eff}})^2$$

$$(1 + r_{\text{eff}})^2 = 1.17072$$

$$1 + r_{\text{eff}} = 1.081998, \qquad \text{Taking the square root on both sides}$$

or $r_{\text{eff}} = 0.081998$. Therefore, the required effective rate is 8.20% per year.

◼

Present Value Under Continuous Compounding

If we solve formula (7) for P, we obtain

$$P = Ae^{-rt} \tag{8}$$

which gives the present value in terms of the future (accumulated) value for the case of continuous compounding.

EXAMPLE 10 The Blakely Investment Company owns an office building located in the commercial district of a city. As a result of the continued success of an urban renewal program, local business is enjoying a miniboom. The market value of Blakely's property is

$$V(t) = 300,000e^{\sqrt{t}/2}$$

where $V(t)$ is measured in dollars and t is the time in years from the present. If the expected rate of inflation is 9% compounded continuously for the next ten years, find an expression for the present value $P(t)$ of the market price of the property valid for the next ten years. Compute $P(7)$, $P(8)$, and $P(9)$ and interpret your results.

Solution   Using formula (8) with $A(t) = V(t)$ and $r = 0.09$, we find that the present value of the market price of the property t years from now is

$$P(t) = V(t)e^{-0.09t}$$

$$= 300{,}000e^{-0.09t+\sqrt{t}/2} \qquad 0 \le t \le 10$$

Letting $t = 7$, 8, and 9, respectively, we find that

$$P(7) = 300{,}000e^{-0.09(7)+\sqrt{7}/2} = 599{,}837 \quad \text{or} \quad \$599{,}837$$

$$P(8) = 300{,}000e^{-0.09(8)+\sqrt{8}/2} = 600{,}640 \quad \text{or} \quad \$600{,}640$$

$$P(9) = 300{,}000e^{-0.09(9)+\sqrt{9}/2} = 598{,}115 \quad \text{or} \quad \$598{,}115$$

From the results of these computations, we see that the present value of the market price of the property seems to decrease after a certain period of growth. This suggests that there is an optimal time for the owners to sell the property. Later we will show that the highest present value of the market price of the property is $600,779, which occurs at time $t = 7.72$ years.

**SELF-CHECK
EXERCISES 4.1**

1. Find the present value of $20,000 due in three years at an interest rate of 12% per year compounded monthly.

2. Mr. Baker is a retiree living on social security and the income from his investment. Currently, his $100,000 investment in a 1-year CD is yielding 10.6% interest compounded daily. If he reinvests the principal ($100,000) on the due date of the CD in another 1-year CD paying 9.2% interest compounded daily, find the net decrease in his yearly income from his investment.

Solutions to Self-Check Exercises 4.1 can be found on page 153.

4.1   EXERCISES

A calculator is recommended for this exercise set.

1. Find the simple interest on a $500 investment made for two years at an interest rate of 8% per year. What is the accumulated amount?

2. Find the simple interest on a $1000 investment made for three years at an interest rate of 5% per year. What is the accumulated amount?

3. Find the accumulated amount at the end of nine months on an $800 deposit in a bank paying simple interest at a rate of 6% per year.

4. Find the accumulated amount at the end of eight months on a $1200 bank deposit paying simple interest at a rate of 7% per year.

5. If the accumulated amount is $1160 at the end of two years and the simple rate of interest is 8% per year, what is the principal?

6. A bank deposit paying simple interest at the rate of 5% per year grew to a sum of $3100 in ten months. Find the principal.

7. How many days will it take for a sum of $1000 to earn $20 interest if it is deposited in a bank paying

ordinary simple interest at the rate of 5% per year? (Use a 365-day year.)

8. How many days will it take for a sum of $1500 to earn $25 interest if it is deposited in a bank paying 5% per year? (Use a 365-day year.)

9. A bank deposit paying simple interest grew from an initial sum of $1000 to a sum of $1075 in nine months. Find the interest rate.

10. Determine the simple interest rate at which $1200 will grow to $1250 in eight months.

In exercises 11–20, find the accumulated amount A if the principal P is invested at the interest rate of r per year for t years.

11. $P = \$1000$, $r = 7\%$, $t = 8$, compounded annually

12. $P = \$1000$, $r = 8\frac{1}{2}\%$, $t = 6$, compounded annually

13. $P = \$2500$, $r = 7\%$, $t = 10$, compounded semiannually

14. $P = \$2500$, $r = 9\%$, $t = 10.5$, compounded semiannually

15. $P = \$12,000$, $r = 8\%$, $t = 10.5$, compounded quarterly

16. $P = \$42,000$, $r = 7\frac{3}{4}\%$, $t = 8$, compounded quarterly

17. $P = \$150,000$, $r = 14\%$, $t = 4$, compounded monthly

18. $P = \$180,000$, $r = 9\%$, $t = 6\frac{1}{4}$, compounded monthly

19. $P = \$150,000$, $r = 12\%$, $t = 3$, compounded daily

20. $P = \$200,000$, $r = 8\%$, $t = 4$, compounded daily

In exercises 21–24, find the effective rate corresponding to the given nominal rate.

21. 10% compounded semiannually

22. 9% compounded quarterly

23. 8% compounded monthly

24. 8% compounded daily

In exercises 25–28, find the present value of $40,000 due in four years at the given rate of interest.

25. 8% compounded semiannually

26. 8% compounded quarterly

27. 7% compounded monthly

28. 9% compounded daily

29. Find the accumulated amount after four years if $5000 is invested at 8% per year compounded continuously.

30. Find the accumulated amount after six years if $6500 is invested at 7% per year compounded continuously.

31. Consumer Decisions   Mr. Mitchell has been given the option of either paying his $300 bill now or settling it for $306 after one month. If he chooses to pay after one month, find the simple interest rate at which he would be charged.

32. Write equation (1b) in the slope-intercept form and interpret the meaning of the slope and the A-intercept in terms of r and P. [*Hint:* Refer to Figure 1.]

33. Hospital Costs   If the cost of a semiprivate room in a hospital was $280 per day five years ago, and hospital costs have risen at the rate of 12% per year since that time, what rate would you expect to pay for a semiprivate room today?

34. Family Food Expenditure   Today a typical family of four spends $560 monthly for food. If inflation occurs at the rate of 5% per year over the next six years, how much should the typical family of four expect to spend for food six years from now?

35. Housing Appreciation   The Brennans are planning to buy a house four years from now. Housing experts in their area have estimated that the cost of a home will increase at a rate of 7% per year during that period. If this economic prediction holds true, how much can the Brennans expect to pay for a house that currently costs $100,000?

36. Electricity Consumption   A utility company in a western city of the United States expects the consumption of electricity to increase by 8% per year during the next decade, due mainly to the expected increase in population. If consumption does increase at this rate, find the amount by which the utility company will have to increase its generating capacity to meet the needs of the area at the end of the decade.

37. Pension Funds   The managers of a pension fund have invested $1.5 million in U.S. government certificates of deposit that pay interest at the rate of 9.5% per year compounded semiannually over a period of ten years. At the end of this period, how much will the investment be worth?

38. Trust Funds   A young man is the beneficiary of a trust fund established for him 21 years ago at his birth. If the original amount placed in trust was $10,000, how much will he receive if the money has earned interest at the rate of 8% per year (a) compounded annually? (b) compounded quarterly? (c) compounded monthly?

39. Investment Planning   Find how much money should be deposited in a bank paying interest at the rate of

8.5% per year compounded quarterly so that at the end of five years the accumulated amount will be $40,000.

40. Promissory Notes An individual purchased a 4-year, $10,000 promissory note with an interest rate of 8.5% per year compounded semiannually. How much did she purchase the note for?

41. Financing a College Education The parents of a child have just come into a large inheritance and wish to establish a trust fund for the child's college education. If they estimate that they will need $60,000 in 13 years, how much should they set aside in the trust now if they can invest the money at $8\frac{1}{2}\%$ per year (a) compounded annually? (b) compounded semiannually? (c) compounded quarterly?

42. Investments Mr. Kaplan invested a sum of money five years ago in a savings account that has since paid interest at the rate of 8% per year compounded quarterly. His investment is now worth $22,289.22. How much did he originally invest?

43. Loan Consolidation The proprietors of The Coachmen Inn secured two loans from the Union Bank: one for $8,000 due in three years and one for $15,000 due in six years, both at an interest rate of 10% per year compounded semiannually. The bank has agreed to allow the two loans to be consolidated into one loan payable in five years at the same interest rate. What amount will the proprietors of the inn be required to pay the bank at the end of five years?

44. Effective Rate of Interest Find the effective rate of interest corresponding to a nominal rate of 9% per year compounded (a) annually, (b) semiannually, (c) quarterly, (d) monthly.

45. Housing Appreciation Ms. Collins purchased a house in 1989 for $80,000. In 1995 she sold the house and made a net profit of $28,000. Find the effective annual rate of return on her investment over the 6-year period.

46. Common Stock Transaction Mr. Stevens purchased 1000 shares of a certain stock for $25,250 (including commissions). He sold the shares two years later and received $32,100 after deducting commissions. Find the effective annual rate of return on his investment over the 2-year period.

47. Zero Coupon Bonds Carol purchased a zero coupon bond for $6,595.37. The bond matures in five years and has a face value of $10,000. Find the effective rate of interest for the bond if interest is compounded semiannually. [*Hint:* Assume that the purchase price

of the bond is the initial investment and that the face value of the bond is the accumulated amount.]

48. Money Market Mutual Funds John invested $5000 in a money market mutual fund that pays interest on a daily basis. The balance in his account at the end of eight months (245 days) was $5347.09. Find the effective rate at which John's account earned interest over this period (assume 365 days in a year).

49. Investment Options Investment A offers a 10% return compounded semiannually, and investment B offers a 9.75% return compounded continuously. Which investment has a higher rate of return over a 4-year period?

50. Present Value Find the present value of $59,673 due in five years at an interest rate of 8% per year compounded continuously.

51. Saving for College Having received a large inheritance, a child's parents wish to establish a trust for the child's college education. If they need an estimated $70,000 seven years from now, how much should they set aside in trust now, if they invest the money at 10.5% compounded (a) quarterly? (b) continuously?

52. Effect of Inflation on Salaries Mr. Lyons's current annual salary is $25,000. Ten years from now, how much will he need to earn in order to retain his present purchasing power if the rate of inflation over that period is 6% per year? Assume that inflation is continuously compounded.

53. Pensions Ms. Lindstrom, who is now 50 years old, is employed by a firm that guarantees her a pension of $40,000 per year at age 65. What is the present value of her first year's pension if inflation over the next 15 years is (a) 6%? (b) 8%? (c) 12%? Assume that inflation is continuously compounded.

54. Real Estate Investments An investor purchased a piece of waterfront property. Because of the development of a marina in the vicinity, the market value of the property is expected to increase according to the rule

$$V(t) = 80,000e^{\sqrt{t}/2}$$

where $V(t)$ is measured in dollars and t is the time in years from the present. If the rate of inflation is expected to be 9% compounded continuously for the next eight years, find an expression for the present value $P(t)$ of the market price of the property valid for the next eight years. What is the market value of the property expected to be in four years?

SOLUTIONS TO SELF-CHECK EXERCISES 4.1

1. Using formula (7) with $r = 0.12$ and $m = 12$, so that

$$i = \frac{0.12}{12} = 0.01, \quad n = (12)(3) = 36, \quad \text{and} \quad A = 20{,}000$$

we find the required present value to be

$$P = 20{,}000(1.01)^{-36} = 13{,}978.50$$

or $13,978.50.

2. The accumulated amount of Mr. Baker's current investment is found by using formula (3) with $P = 100{,}000$, $r = 0.106$, and $m = 365$. Thus,

$$i = \frac{0.106}{365} \quad \text{and} \quad n = 365$$

so the required accumulated amount is

$$A = 100{,}000\left(1 + \frac{0.106}{365}\right)^{365} = 111{,}180.48$$

or $111,180.48. Next, we compute the accumulated amount of Mr. Baker's reinvestment. Once again, using formula (3) with $P = 100{,}000$, $r = 0.092$, and $m = 365$ so that

$$i = \frac{0.092}{365} \quad \text{and} \quad n = 365$$

we find the required accumulated amount in this case to be

$$\overline{A} = 100{,}000\left(1 + \frac{0.092}{365}\right)^{365}$$

or $109,635.21. Therefore, Mr. Baker can expect to experience a net decrease in yearly income of

$$111{,}180.48 - 109{,}635.21$$

or $1,545.27.

Using Technology

EXERCISES

1. In the opening paragraph of Section 3.1, we pointed out that the accumulated amount of an account earning interest *compounded continuously* will eventually outgrow by far the accumulated amount of an account earning interest at the same nominal rate but earning simple interest. Illustrate this fact using the following example.

Suppose you deposit $1000 in account I earning interest at the rate of 10% per year compounded continuously so that the accumulated amount at the end of t years is $A_1(t) = 1000e^{0.1t}$. Suppose you also deposit $1000 in account II earning simple interest at the rate of 10% per year so that the accumulated amount at the end of t years is $A_2(t) = 1000(1 + 0.1t)$. Use a graphing utility to sketch the graphs of the functions A_1 and A_2 in the viewing rectangle $[0, 20] \times [0, 10{,}000]$ to see the accumulated amounts $A_1(t)$ and $A_2(t)$ over a 20-year period.

2. The effective rate of interest is given by

$$r_{\text{eff}} = \left(1 + \frac{r}{m}\right)^m - 1$$

where the number of conversion periods per year is m [see formula (6), page 146]. In calculus (see Exercises 9.1, exercise 82) you will be asked to show that the effective rate of interest r_{eff} corresponding to a nominal interest rate r per year compound continuously is given by

$$\hat{r}_{\text{eff}} = e^r - 1$$

In order to obtain a visual confirmation of this result, consider the special case where $r = 0.1$ (10% per annum).

a. Use a graphing utility to plot the graph of both

$$y_1 = \left(1 + \frac{0.1}{x}\right)^x - 1 \quad \text{and} \quad y_2 = e^{0.1} - 1$$

in the viewing rectangle $[0, 3] \times [0, 0.12]$.

b. Does your result seem to imply that

$$\left(1 + \frac{r}{m}\right)^m - 1$$

approaches

$$\hat{r}_{\text{eff}} = e^r - 1$$

as m increases without bound for the special case $r = 0.1$?

4.2   Annuities

Future Value of an Annuity

An **annuity** is a sequence of payments made at regular time intervals. The time period in which these payments are made is called the *term* of the annuity. Depending on whether the term is given by a *fixed time interval,* a time interval that begins at a definite date but extends indefinitely, or one that is not fixed in advance, an annuity is called an **annuity certain,** a *perpetuity,* or a *contingent annuity,* respectively. In general, the payments in an annuity need not be equal, but in many important applications they are equal. In this section, we assume that annuity payments are equal. Examples of annuities are regular deposits to a savings account, monthly home mortgage payments, and monthly insurance payments.

Annuities are also classified by payment dates. An annuity in which the payments are made at the *end* of each payment period is called an **ordinary annuity,** whereas an annuity in which the payments are made at the beginning of each period is called an *annuity due.* Furthermore, an annuity in which the payment period coincides with the interest conversion period is called a **simple annuity,** whereas an annuity in which the payment period differs from the interest conversion period is called a *complex annuity.*

In this section we consider ordinary annuities that are certain and simple, with periodic payments that are equal in size. In other words, we study annuities that are subject to the following conditions:

1. The terms are given by fixed time intervals.

2. The periodic payments are equal in size.

3. The payments are made at the *end* of the payment periods.

4. The payment periods coincide with the interest conversion periods.

In order to find a formula for the accumulated amount S of an annuity, suppose a sum of $100 is paid into an account at the end of each quarter over a period of three years. Furthermore, suppose that the account earns interest on the deposit at the rate of 8% per year, compounded quarterly. Then the first payment of $100 made at the end of the first quarter earns interest at the rate of 8% per year compounded four times a year (or $8/4 = 2$ percent per quarter) over the remaining 11 quarters and therefore, by the compound interest formula, has an accumulated amount of

$$100\left(1 + \frac{0.08}{4}\right)^{11} \quad \text{or} \quad 100(1 + 0.02)^{11}$$

dollars at the end of the term of the annuity (Figure 2).

FIGURE 2

The sum of the accumulated amounts is the amount of the annuity.

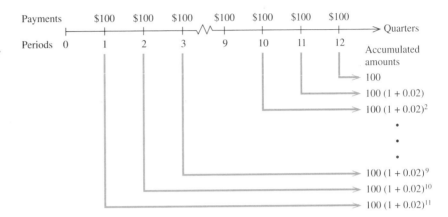

The second payment of $100 made at the end of the second quarter earns interest at the same rate over the remaining 10 quarters and therefore has an accumulated amount of

$$100(1 + 0.02)^{10}$$

dollars at the end of the term of the annuity, and so on. The last payment earns no interest, since it is due at the end of the term. The amount of the annuity is obtained by adding all the terms in Figure 2. Thus,

$$S = 100 + 100(1 + 0.02) + 100(1 + 0.02)^2 + \cdots + 100(1 + 0.02)^{11}$$

The sum on the right is the sum of the first n terms of a *geometric progression* with first term R and common ratio $(1 + i)$. We will show in Section 4.4 that the sum S can be written in the more compact form

$$S = 100\left[\frac{(1 + 0.02)^{12} - 1}{0.02}\right]$$

$$\approx 1341.21$$

or approximately $1341.21.

To find a general formula for the accumulated amount S of an annuity, suppose a sum of R is paid into an account at the end of each period for n periods, and suppose the account earns interest at the rate of i per period. Then, proceeding as we did with the numerical example, we see that

$$S = R + R(1 + i) + R(1 + i)^2 + \cdots + R(1 + i)^{n-1}$$

$$= R\left[\frac{(1 + i)^n - 1}{i}\right] \tag{9}$$

The expression inside the brackets is commonly denoted by $s_{\overline{n}|i}$ (read "s angle n at i") and is called the *compound amount factor*. Extensive tables have been constructed that give values of $s_{\overline{n}|i}$ for different values of i and n (such as Table I, Appendix B). In terms of the compound amount factor,

$$S = Rs_{\overline{n}|i} \qquad\qquad (10)$$

The quantity S in formulas (9) and (10) is realizable at some future date and is accordingly called the *future value of an annuity*.

FUTURE VALUE OF AN ANNUITY

The **future value S of an annuity** of n payments of R dollars each, paid at the end of each investment period into an account that earns interest at the rate of i per period, is

$$S = R\left[\frac{(1 + i)^n - 1}{i}\right]$$

 EXAMPLE 1   Find the amount of an ordinary annuity of 12 monthly payments of $100 that earn interest at 12% per year compounded monthly.

Solution   Since i is the interest rate per *period* and interest is compounded monthly in this case, we have $i = 0.12/12 = 0.01$. Using formula (9) with $R = 100$, $n = 12$, and $i = 0.01$, we have

$$S = \frac{100[(1.01)^{12} - 1]}{0.01}$$

$$= 1268.25 \qquad\qquad \text{Using a calculator}$$

or $1268.25. The same result is obtained by observing that

$$S = 100s_{\overline{12}|0.01}$$

$$= 100(12.6825)$$

$$= 1268.25 \qquad\qquad \text{Using Table I, Appendix B}$$

Present Value of an Annuity

Suppose you want to establish a retirement account by making a sequence of equal periodic payments into the account over a certain period of time. After each payment is made, the new balance continues to earn interest at some nominal rate. What is the current value P of this sequence of payments together with the interest earned? The amount P is referred to as the *present value of an annuity*.

To derive a formula for determining the present value P of an annuity, we may argue as follows. The amount P invested now and earning interest at the rate of i per period will have an accumulated value of $P(1 + i)^n$ at the end of n periods. But this must be equal to the future value of the annuity S given by formula (9). Therefore, equating the two expressions, we have

$$P(1 + i)^n = R\left[\frac{(1 + i)^n - 1}{i}\right]$$

Multiplying both sides of this equation by $(1 + i)^{-n}$ gives

$$P = R(1 + i)^{-n}\left[\frac{(1 + i)^n - 1}{i}\right]$$

$$= R\left[\frac{(1 + i)^n(1 + i)^{-n} - (1 + i)^{-n}}{i}\right] \qquad (1 + i)^n(1 + i)^{-n} = 1$$

$$= R\left[\frac{1 - (1 + i)^{-n}}{i}\right]$$

$$= Ra_{\overline{n}|i}$$

where the factor $a_{\overline{n}|i}$ (read "*a* angle *n* at *i*") represents the expression inside the brackets. Extensive tables have also been constructed giving values of $a_{\overline{n}|i}$ for different values of i and n (see Table I, Appendix B).

PRESENT VALUE OF AN ANNUITY

The **present value** P **of an annuity** of n payments of R dollars each, paid at the end of each investment period into an account that earns interest at the rate of i per period, is

$$P = R\left[\frac{1 - (1 + i)^{-n}}{i}\right] \qquad\qquad (11)$$

EXAMPLE 2   Find the present value of an ordinary annuity of 24 payments of $100 each made monthly and earning interest at 9% per year compounded monthly.

Solution   Here $R = 100$, $i = r/m = 0.09/12 = 0.0075$, and $n = 24$, so by (11)

$$P = \frac{100[1 - (1.0075)^{-24}]}{0.0075}$$

$$= 2188.92$$

or $2188.92. The same result may be obtained by using Table I, Appendix B. Thus,

$$P = 100a_{\overline{24}|0.0075}$$

$$= 100(21.8892)$$

$$= 2188.92$$

Applications

EXAMPLE 3   As a savings program toward their child's college education, parents decide to deposit $100 at the end of every month into a bank account paying interest at the rate of 6% per year compounded monthly. If the savings program began when the child was 6 years old, how much money would have accumulated by the time the child turns 18?

Solution   By the time the child turns 18, the parents would have made 144 deposits into the account. Thus, $n = 144$. Furthermore, we have $R = 100$, $r = 0.06$, and $m = 12$, so $i = 0.06/12 = 0.005$. Using formula (9), we find that the amount of money that would have accumulated is given by

$$S = \frac{100[(1.005)^{144} - 1]}{0.005}$$

$$= 21,015$$

or $21,015.

 EXAMPLE 4   After making a down payment of $2000 for an automobile, Mr. McMurphy paid $200 per month for 36 months with interest charged at 12% per year compounded monthly on the unpaid balance. What was the original cost of the car? What portion of Mr. McMurphy's total car payments went toward interest charges?

Solution   The loan taken up by Mr. McMurphy is given by the present value of the annuity

$$P = \frac{200[1 - (1.01)^{-36}]}{0.01} = 200a_{\overline{36}|0.01}$$

$$= 6021.50$$

or $6021.50. Therefore, the original cost of the automobile is $8021.50 ($6021.50 plus the $2000 down payment). The interest charges paid by Mr. McMurphy are given by $(36)(200) - 6021.50$, or $1178.50.

One important application of annuities is in the area of tax planning. During the 1980s, Congress created many tax-sheltered retirement savings plans, such as Individual Retirement Accounts (IRAs), Keogh plans, and Simplified Employee Pension (SEP) plans. These plans are examples of annuities in which the individual is allowed to make contributions (which are often tax deductible) to an investment account. The amount of the contribution is limited by congressional legislation. The taxes on the contributions and/or the interest accumulated in these accounts are deferred until the money is withdrawn, ideally during retirement, when tax brackets should be lower. In the interim period, the individual has the benefit of tax-free growth on his or her investment.

Suppose, for example, that you are eligible to make a fully deductible contribution to an IRA and that you are in a marginal tax bracket of 28%. Additionally, suppose that you receive a year-end bonus of $2000 from your employer and have the option of depositing the $2000 into either an IRA or a regular savings account, both accounts earning interest at an effective annual rate of 8% per year. If you choose to invest your bonus in a regular savings account, you will first have to pay taxes on the $2000, leaving $1440 to invest. At the end of one year, you will also have to pay taxes on the interest earned, leaving you with

$$
\begin{array}{ccccc}
\text{Accumulated} & - & \text{Tax on} & = & \text{Net} \\
\text{amount} & & \text{interest} & & \text{amount}
\end{array}
$$

$$1555.20 - 32.26 = 1522.94$$

or $1522.94.

However, if you put the money into the IRA account, the entire sum will earn interest, and at the end of one year you will have (1.08)($2000), or $2160, in your account. Of course, you will still have to pay taxes on this money when you withdraw it, but you will have gained the advantage of tax-free growth of the larger principal over the years. The disadvantage of this option is that if you withdraw the money before you reach the age of $59\frac{1}{2}$, you will be liable for taxes on both your contributions and the interest earned and you will also have to pay a 10% penalty.

REMARK In practice, the size of the contributions an individual might make to the various retirement plans might vary from year to year. Also, he or she might make the contributions at different payment periods. To simplify our discussion, we will consider examples in which fixed payments are made at regular intervals. □

EXAMPLE 5 Ms. Coughlin is planning to make the maximum contribution of $2000 on January 31 of each year into an IRA earning interest at an effective rate of 9% per year. After she makes her 25th payment on January 31 of the year following her retirement, how much will she have in her IRA?

Solution The amount of money Ms. Coughlin will have after her 25th payment into her account is found by using formula (9) with $R = 2000$, $r = 0.09$, and $m = 1$, so $i = r/m = 0.09$ and $n = 25$. The required amount is given by

$$S = \frac{2000[(1.09)^{25} - 1]}{0.09}$$

$$= 169,401.79$$

or $169,401.79. ■

Tax-deferred annuities are another type of investment vehicle that allow an individual to build assets for retirement, college funds, or other future needs. The advantage gained in this type of investment is that the tax on the accumulated interest is deferred to a later date. Note that in this type of investment the contributions themselves are not tax deductible. At first glance the advantage thus gained may seem to be relatively inconsequential, but its true effect is illustrated by the next example.

EXAMPLE 6 Both Mr. Clark and Mr. Colby are salaried individuals, 45 years of age, who are saving for their retirement 20 years from now. Both Clark and Colby are also in the 28% marginal tax bracket. Mr. Clark makes a $1000 contribution annually on December 31 into a savings account earning an effective rate of 8% per year. At the same time, Mr. Colby makes a $1000 annual payment to an insurance company for a tax-deferred annuity. The annuity also earns interest at an effective rate of 8% per year. (Assume that both men remain in the same tax bracket throughout this period, and disregard state income taxes.)

a. Calculate how much each man will have in his investment account at the end of 20 years.

b. Compute the interest earned on each account.

c. Show that even if the interest on Mr. Colby's investment were subjected to a tax of 28% upon withdrawal of his investment at the end of 20 years, the net accumulated amount of his investment would still be greater than that of Mr. Clark's.

Solution

a. Because Mr. Clark is in the 28% marginal tax bracket, the net yield for his investment is $(0.72)(8)$, or 5.76%, per year.

Using formula (9) with $R = 1000$, $r = 0.0576$, and $m = 1$, so that $i = 0.0576$ and $n = 20$, we see that Clark's investment will be worth

$$S = \frac{1000[(1 + 0.0576)^{20} - 1]}{0.0576}$$

$$= 35{,}850.49$$

or \$35,850.49 at his retirement.

Mr. Colby has a tax-sheltered investment with an effective yield of 8% per year. Using (9) with $R = 1000$, $r = 0.08$, and $m = 1$, so that $i = 0.08$ and $n = 20$, we see that Colby's investment will be worth

$$S = \frac{1000[(1 + 0.08)^{20} - 1]}{0.08}$$

$$= 45{,}761.96$$

or \$45,761.96 at his retirement.

b. Each man will have paid $20(1{,}000)$, or 20,000 dollars, into his account. Therefore, the total interest earned in Mr. Clark's account will be $(35{,}850.49 - 20{,}000)$, or \$15,850.49, whereas the total interest earned in Mr. Colby's account would be $(45{,}761.96 - 20{,}000)$, or \$25,761.96.

c. From (b) we see that the total interest earned in Mr. Colby's account would be \$25,761.96. If it were taxed at 28%, he would still end up with $(0.72)(25{,}761.96)$, or \$18,548.61. This is larger than the total interest of \$15,850.49 earned by Mr. Clark.   ■

SELF-CHECK EXERCISES 4.2

1. Mr. and Mrs. Fletcher opened a spousal IRA on January 31, 1985, with the maximum permissible contribution of \$2250. They plan to make a contribution of \$2250 thereafter on January 31 of each year until their retirement in the year 2004 (20 payments). If the account earns interest at the rate of 8% per year compounded yearly, how much will the Fletchers have in their account when they retire?

2. The Denver Wildcatting Company has an immediate need for a loan. In an agreement worked out with its banker, Denver assigns its royalty income of \$4800 per month for the next three years from certain oil properties to the bank, with the first payment due at the end of the first month. If the bank charges interest at the rate of 9% per year compounded monthly, what is the amount of the loan negotiated between the parties?

Solutions to Self-Check Exercises 4.2 can be found on page 162.

c *A calculator is recommended for this exercise set.*

In exercises 1–6, find the amount (future value) of each ordinary annuity.

1. $1000 a year for 10 years at 10% per year compounded annually

2. $1500 per semiannual period for 8 years at 9% per year compounded semiannually

3. $1800 per quarter for 6 years at 8% per year compounded quarterly

4. $500 per semiannual period for 12 years at 11% per year compounded semiannually

5. $600 per quarter for 9 years at 12% per year compounded quarterly

6. $150 per month for 15 years at 10% per year compounded monthly

In exercises 7–12, find the present value of each ordinary annuity.

7. $5000 a year for eight years at 8% per year compounded annually

8. $1200 per semiannual period for six years at 10% per year compounded semiannually

9. $4000 a year for five years at 9% per year compounded yearly

10. $3000 per semiannual period for six years at 11% per year compounded semiannually

11. $800 per quarter for seven years at 12% per year compounded quarterly

12. $150 per month for ten years at 8% per year compounded monthly

13. **IRAs** If a merchant deposits $1500 annually at the end of each tax year in an IRA account paying interest at the rate of 8% per year compounded annually, how much will she have in her account at the end of 25 years?

14. **Savings Accounts** If Mr. Pruitt deposits $100 at the end of each month in a savings account earning interest at the rate of 8% per year compounded monthly, how much will he have on deposit in his savings account at the end of six years, assuming that he makes no withdrawals during that period?

15. **Savings Accounts** Mrs. Lynde has joined a "Christmas Fund Club" at her bank. At the end of every month, December through October inclusive, she will make a deposit of $40 in her fund. If the money earns interest at the rate of 7% per year compounded monthly, how much will she have in her account on December 1 of the following year?

16. **Keogh Accounts** Mr. Robinson, who is self-employed, contributes $5000 a year into a Keogh account. How much will he have in the account after 25 years if the account earns interest at the rate of 8.5% per year compounded yearly?

17. **Retirement Planning** As a fringe benefit for the past 12 years, Mr. Collins's employer has contributed $100 at the end of each month into an employee retirement account for Collins that pays interest at the rate of 7% per year compounded monthly. Mr. Collins has also contributed $2000 at the end of each of the last eight years into an IRA that pays interest at the rate of 9% per year compounded yearly. How much does Mr. Collins have in his retirement fund at this time?

18. **Savings Account** The Pirerras are planning to go to Europe three years from now and have agreed to set aside $150 each month for their trip. If they deposit this money at the end of each month into a savings account paying interest at the rate of 8% per year compounded monthly, how much money will be in their travel fund at the end of the third year?

19. **Auto Leasing** Mr. and Mrs. Betz have leased an auto for two years at $450 per month. If money is worth 9% per year compounded monthly, what is the equivalent cash payment (present value) of this annuity?

20. **Auto Financing** Ms. Newburg made a down payment of $2000 toward the purchase of a new car. In order to pay the balance of the purchase price, she has secured a loan from her bank at the rate of 12% per year compounded monthly. Under the terms of her finance agreement, she is required to make payments of $210 per month for 36 months. What is the cash price of the car?

21. **Installment Plans** R. G. Pierce Publishing Company sells encyclopedias under two payment plans: cash or installment. Under the installment plan, the customer

pays $22 per month over three years with interest charged on the balance at a rate of 18% per year compounded monthly. Find the cash price for a set of encyclopedias if it is equivalent to the price paid by a customer using the installment plan.

22. **Lottery Payouts** A state lottery commission pays the winner of the "Million Dollar" lottery 20 installments of $50,000 per year. The commission makes the first payment of $50,000 immediately and the other $n = 19$ payments at the end of each of the next 19 years. Determine how much money the commission should have in the bank initially to guarantee the payments assuming that the balance on deposit with the bank earns interest at the rate of 8% per year compounded yearly. [*Hint:* Find the present value of an annuity.]

23. **Purchasing a Home** The Johnsons have accumulated a nest egg of $25,000 that they intend to use as a down payment toward the purchase of a new house.

Because their present gross income has placed them in a relatively high tax bracket, they have decided to invest a minimum of $800 per month in monthly payments (to take advantage of their tax deductions) toward the purchase of their house. However, because of other financial obligations their monthly payments should not exceed $1000. If local mortgage rates are 9.5% per year compounded monthly for a conventional 30-year mortgage, what is the price range of houses that they should consider?

24. **Purchasing a Home** Refer to exercise 23. If local mortgage rates were increased to 10%, how would this affect the price range of houses the Johnsons should consider?

25. **Purchasing a Home** Refer to exercise 23. If the Johnsons decide to secure a 15-year mortgage instead of a 30-year mortgage, what is the price range of houses they should consider when the local mortgage rate for this type of loan is 9%?

SOLUTIONS TO SELF-CHECK EXERCISES 4.2

1. The amount the Fletchers will have in their account when they retire may be found by using (9) with $R = 2250$, $r = 0.08$, $m = 1$, so that $i = r = 0.08$ and $n = 20$. Thus

$$S = \frac{2250[(1.08)^{20} - 1]}{0.08}$$

$$= 102,964.42$$

or $102,964.42.

2. We want to find the present value of an ordinary annuity of 36 payments of $4800 each made monthly and earning interest at 9% per year compounded monthly. Using formula (11) with $R = 4800$, $m = 12$, so that $i = r/m = 0.09/12 = 0.0075$ and $n = (12)(3) = 36$, we find

$$P = \frac{4800[1 - (1.0075)^{-36}]}{0.0075} = 150,944.67$$

or $150,944.67, the amount of the loan negotiated.

4.3   Amortization and Sinking Funds

Amortization of Loans

The annuity formulas derived in Section 4.2 may be used to answer questions involving the amortization of certain types of installment loans. For example, in a typical housing loan, the mortgagor makes periodic payments toward reducing his indebtedness to the lender, who charges interest at a fixed rate on the unpaid portion of the debt. In practice, the borrower is required to repay the lender in periodic installments, usually of the same

John Decker

Decker's job is simple: to loan money to people who want to buy a home. The hard part is deciding whether applicants qualify for one of Shawmut's 40 different mortgage plans. Sifting through income figures, current indebtedness, and credit history helps Decker determine which individuals make the best mortgage candidates.

Title: Mortgage Counselor   **Institution:** Shawmut Bank

Decker stresses that he and his colleagues "strive to grant loans. That's our job." But before someone files a formal application, Decker takes the person over several "pre-qualification hurdles."

To start, Decker relies on a two-tiered, debt-to-income ratio to see whether a person has sufficient income to make payments. Under the first tier, the proposed monthly payment (principal and interest, property tax, homeowner's insurance, and, when applicable, a condo fee) cannot exceed 28% of an individual's gross monthly income. If these calculations are positive, Decker then determines the individual's ability to meet monthly mortgage payments while also repaying other debts, such as car and student loans, credit cards, alimony, and so on. These combined payments cannot exceed 36% of gross monthly income. A typical person with an $800 mortgage obligation and $550 in other payments would have to earn $3750 per month.

Contrary to popular belief, bankers *want* to lend money. "The idea is to grant mortgages," says Decker, because they contribute substantially to a bank's profitability. But lending money means making sensible decisions about how much to lend as well as about a person's ability to repay the loan.

Using a loan-to-value formula, banks might lend 80% of a property's value, with the applicant putting 20% down. Or Shawmut might lend up to 95% or as little as 75% of the appraised value.

Understandably, banks don't like to see bankruptcies, late payments, or liens on a personal credit history. Decker notes, however, that even these don't necessarily mean failure in securing a mortgage.

Once Decker has put together a successful mortgage application, it is reviewed internally. When the mortgage is granted, it might come through at a slightly higher interest rate—10.4% instead of the expected 10% rate, for instance. Calculus is used to compute the change in monthly payments, for even a small change on a large mortgage could affect the customer's ability to make monthly payments.

Using the debt-to-income ratio, Decker plugs the new variable into his formulas to determine whether a problem exists. The formulas used to compute the monthly payments involve the use of exponential functions. On a $100,000, 30-year, fixed-rate mortgage at 10.4% instead of 10%, payments will increase about $30 per month. Such a small increase doesn't usually pose a problem, but if the customer can't make the new payment, Decker works to find a solution. He says, "If there is any possible way to give a loan, we're going to make it work."

size and over a fixed term, so that the loan (principal plus interest charges) is amortized at the end of the term.

By thinking of the monthly loan repayments R as the payments in an annuity, we see that the original amount of the loan is given by P, the present value of the annuity. From equation (11), Section 4.2, we have

$$P = R\left[\frac{1 - (1 + i)^{-n}}{i}\right] = Ra_{\overline{n}|i} \qquad (12)$$

A question a financier might ask is: How much should the monthly install-ment be so that the loan will be amortized at the end of the term of the loan? To answer this question, we simply solve (12) for R in terms of P, obtaining

$$R = \frac{Pi}{1 - (1 + i)^{-n}} = \frac{P}{a_{\overline{n}|i}}$$

AMORTIZATION FORMULA

The periodic payment R on a loan of P dollars to be amortized over n periods with interest charged at the rate of i per period is

$$R = \frac{Pi}{1 - (1 + i)^{-n}} \qquad (13)$$

 EXAMPLE 1   A sum of $50,000 is to be repaid over a 5-year period through equal installments made at the end of each year. If an interest rate of 8% per year is charged on the unpaid balance, and interest calculations are made at the end of each year, determine the size of each installment so that the loan (principal plus interest charges) is amortized at the end of five years. Verify the result by displaying the amortization schedule.

Solution   Substituting $P = 50,000$, $i = r = 0.08$ (here $m = 1$), and $n = 5$ into (13), we obtain

$$R = \frac{(50,000)(0.08)}{1 - (1.08)^{-5}} = 12,522.82$$

giving the required yearly installment as $12,522.82.

The amortization schedule is presented in Table 3.

TABLE 3

End of Period	Interest Charged	Repayment Made	Payment Toward Principal	Outstanding Principal
0				50,000.00
1	4,000.00	12,522.82	8,522.82	41,477.18
2	3,318.17	12,522.82	9,204.65	32,272.53
3	2,581.80	12,522.82	9,941.02	22,331.51
4	1,786.52	12,522.82	10,736.30	11,595.21
5	927.62	12,522.82	11,595.20	0.01

The outstanding principal at the end of five years is, of course, zero. (The figure of $.01 in Table 3 is the result of round-off errors.) Observe that initially the larger portion of the repayment goes toward payment of interest charges, but as time goes by the larger portion of the installment goes toward repayment of the principal.   ◼

Financing a Home

EXAMPLE 2   The Blakelys borrowed $80,000 from a bank to help finance the purchase of a house. The bank charges interest at a rate of 9% per year on the unpaid balance, with interest computations made at the end of each month. The Blakelys have agreed to repay the loan in equal monthly installments over 30 years. How much should each payment be if the loan is to be amortized at the end of the term?

Solution   Here $P = 80{,}000$, $i = r/m = 0.09/12 = 0.0075$, and $n = (30)(12) = 360$. Using formula (13), we find that the size of each monthly installment required is given by

$$R = \frac{(80{,}000)(0.0075)}{1 - (1.0075)^{-360}}$$

$$= 643.70$$

or $643.70.   ◼

EXAMPLE 3   Refer to Example 2 and find the amount of equity the Blakelys will have in their home (disregarding the down payment and appreciation) after the

a. 60th payment (5 years)

b. 240th payment (20 years)

Solution

a. After 60 payments have been made, the outstanding principal is given by the sum of the present values of the remaining installments (that is, $360 - 60 = 300$ installments). But this sum is just the present value of an annuity with $n = 300$, $R = 643.70$, and $i = 0.0075$ (see Example 2). Using (11), we find that

$$P = 643.70 \left[\frac{1 - (1 + 0.0075)^{-300}}{0.0075} \right]$$

$$= 76{,}704.34$$

or $76,704.34. Therefore, we may find how much equity the Blakelys have in their home by subtracting the outstanding principal from the original principal:

$$80{,}000 - 76{,}704.34 = 3{,}295.66$$

or $3,295.66.

b. As in (a), we compute the present value of the annuity with $n = 360 - 240 = 120$. Thus,

$$P = 643.70 \left[\frac{1 - (1 + 0.0075)^{-120}}{0.0075} \right]$$

$$= 50,814.77$$

or \$50,814.77. Therefore, the Blakelys' equity after the 240th payment is

$$80,000 - 50,814.77 = 29,185.23$$

or \$29,185.23.   ◼

EXAMPLE 4   The Jacksons have determined that after making a down payment they could afford at most \$1000 for a monthly house payment. The bank charges interest at the rate of 9.6% per year on the unpaid balance, with interest computations made at the end of each month. If the loan is to be amortized in equal monthly installments over 30 years, what is the maximum amount that the Jacksons could borrow from the bank?

Solution   Here $i = r/m = 0.096/12 = 0.008$, $n = (30)(12) = 360$, $R = 1000$, and we are required to find P. From (11), we have

$$P = \frac{R[1 - (1 + i)^{-n}]}{i}$$

Substituting the numerical values for R, n, and i into this expression for P, we obtain

$$P = \frac{1000[1 - (1.008)^{-360}]}{0.008} \approx 117,902$$

Therefore, the Jacksons could borrow at most \$117,902.   ◼

Sinking Funds

Sinking funds are another important application of the annuity formulas. Simply stated, a **sinking fund** is a fund that is set up for a specific purpose at some future date. For example, an individual might establish a sinking fund for the purpose of discharging a debt at a future date. A corporation might establish a sinking fund in order to accumulate sufficient capital to replace equipment that is expected to be obsolete at some future date.

By thinking of the amount to be accumulated by a specific date in the future as the future value of an annuity (equation (9), Section 4.2), we can answer questions about a large class of sinking fund problems.

EXAMPLE 5   The proprietor of Carson Hardware has decided to set up a sinking fund for the purpose of purchasing a computer in two years' time. It is expected that the computer will cost \$30,000. If the fund earns 10% interest per year compounded quarterly, determine the size of each (equal) quarterly installment the proprietor should pay into the fund. Verify the result by displaying the schedule.

Solution   The problem at hand is to find the size of each quarterly payment R of an annuity given that its future value is $S = 30,000$, the interest earned per conversion period is $i = r/m = 0.1/4 = 0.025$, and the number of payments is $n = (2)(4) = 8$. The formula for the annuity,

$$S = R\left[\frac{(1 + i)^n - 1}{i}\right]$$

when solved for R yields

$$R = \frac{iS}{(1 + i)^n - 1} \qquad (14)$$

or, equivalently,

$$R = \frac{S}{s_{\overline{n}|i}}$$

Substituting the appropriate numerical values for i, S, and n into (14), we obtain the desired quarterly payment

$$R = \frac{(0.025)(30,000)}{(1.025)^8 - 1} = 3434.02$$

or $3434.02. Table 4 shows the required schedule.

TABLE 4

End of Period	Deposit Made	Interest Earned	Addition to Fund	Accumulated Amount in Fund
1	3,434.02	0	3,434.02	3,434.02
2	3,434.02	85.85	3,519.87	6,953.89
3	3,434.02	173.85	3,607.87	10,561.76
4	3,434.02	264.04	3,698.06	14,259.82
5	3,434.02	356.50	3,790.52	18,050.34
6	3,434.02	451.26	3,885.28	21,935.62
7	3,434.02	548.39	3,982.41	25,918.03
8	3,434.02	647.95	4,081.97	30,000.00

The formula derived in this last example is restated below.

SINKING FUND PAYMENT

The periodic payment R required to accumulate a sum of S dollars over n periods with interest charged at the rate of i per period is

$$R = \frac{iS}{(1 + i)^n - 1}$$

SELF-CHECK EXERCISES 4.3

1. The Moores wish to borrow $100,000 from a bank to help finance the purchase of a house. Their banker has offered the following plans for their consideration. In plan I, the Moores have 30 years to repay the loan in monthly installments with interest on the unpaid balance charged at 10.5% per year

compounded monthly. In plan II, the loan is to be repaid in monthly install-ments over 15 years with interest on the unpaid balance charged at 9.75% per year compounded monthly.

a. Find the monthly repayment for each plan.

b. What is the difference in total payments made under each plan?

2. Mr. Harris, a self-employed individual who is 46 years old, is setting up a Defined-Benefit Keogh Plan for his retirement. If he wishes to have $250,000 in this Keogh account by age 65, what is the size of each yearly installment he will be required to make into a savings account earning interest at $8\frac{1}{4}\%$ per year? (Assume that Mr. Harris is eligible to make each of the 20 re-quired contributions.)

Solutions to Self-Check Exercises 4.3 can be found on page 170.

4.3   EXERCISES

c *A calculator is recommended for this exercise set.*

In exercises 1–8, find the periodic payment R required to amortize a loan of P dollars over n periods with interest earned at the rate of i per period.

1. $P = \$100,000$, $i = 0.08$, $n = 10$

2. $P = \$40,000$, $i = 0.015$, $n = 30$

3. $P = \$5,000$, $i = 0.01$, $n = 12$

4. $P = \$16,000$, $i = 0.0075$, $n = 48$

5. $P = \$25,000$, $i = 0.0075$, $n = 48$

6. $P = \$80,000$, $i = 0.00875$, $n = 180$

7. $P = \$80,000$, $i = 0.00875$, $n = 360$

8. $P = \$100,000$, $i = 0.00875$, $n = 300$

In exercises 9–14, find the periodic payment R required to accumulate a sum of S dollars over n periods with interest earned at the rate of i per period.

9. $S = \$20,000$, $i = 0.02$, $n = 12$

10. $S = \$40,000$, $i = 0.01$, $n = 36$

11. $S = \$100,000$, $i = 0.0075$, $n = 120$

12. $S = \$120,000$, $i = 0.0075$, $n = 180$

13. $S = \$250,000$, $i = 0.00875$, $n = 300$

14. $S = \$350,000$, $i = 0.00625$, $n = 120$

15. Loan Amortization   A sum of $100,000 is to be re-paid over a 10-year period through equal installments

made at the end of each year. If an interest rate of 10% per year is charged on the unpaid balance, and interest calculations are made at the end of each year, determine the size of each installment so that the loan (principal plus interest charges) is amortized at the end of ten years.

16. Loan Amortization   What monthly payment is re-quired to amortize a loan of $30,000 over ten years if interest at the rate of 12% per year is charged on the unpaid balance, and interest calculations are made at the end of each month?

17. Home Mortgages   Complete the following table, which shows the monthly payments on a $100,000, 30-year mortgage at the interest rates shown. Use this information to answer the questions that follow.

Amount of Mortgage	Interest Rate	Monthly Payment
$100,000	7%	$665.30
$100,000	8%	. . .
$100,000	9%	. . .
$100,000	10%	. . .
$100,000	11%	. . .
$100,000	12%	$1,028.61

a. What is the difference in monthly payments be-tween a $100,000, 30-year mortgage secured at 7% per year and one secured at 10% per year?

b. Use the table to calculate the monthly mortgage payments on a $150,000 mortgage at 10% per year over 30 years and a $50,000 mortgage at 10% per year over 30 years.

18. **Financing a Home**   The Flemings secured a bank loan of $96,000 to help finance the purchase of a house. The bank charges interest at a rate of 9% per year on the unpaid balance, and interest computations are made at the end of each month. The Flemings have agreed to repay the loan in equal monthly installments over 25 years. What should the size of each repayment be if the loan is to be amortized at the end of the term?

19. **Financing a Car**   The price of a new car is $16,000. Assume an individual makes a down payment of 25% toward the purchase of the car and secures financing for the balance at the rate of 10% per year compounded monthly.
   a. What monthly payment will she be required to make if the car is financed over a period of 36 months? Over a period of 48 months?
   b. What will the interest charges be if she elects the 36-month plan? The 48-month plan?

20. **Financial Analysis**   A group of private investors purchased a condominium complex for $2 million. They made an initial down payment of 10% and obtained financing for the balance. If the loan is to be amortized over 15 years at an interest rate of 12% per year compounded quarterly, find the required quarterly payment.

21. **Financing a Home**   The Taylors have purchased a $100,000 house. They made an initial down payment of $20,000 and secured a mortgage with interest charged at the rate of 8% per year on the unpaid balance. Interest computations are made at the end of each month. If the loan is to be amortized over 30 years, what monthly payment will the Taylors be required to make? What is their equity (disregarding appreciation) after 5 years? After 10 years? After 20 years?

22. **Sinking Funds**   A city has $2.5 million worth of school bonds that are due in 20 years and has established a sinking fund to retire this debt. If the fund earns interest at the rate of 7% per year compounded annually, what amount must be deposited annually in this fund?

23. **Sinking Funds**   The Lowell Corporation wishes to establish a sinking fund to retire a $200,000 debt that is due in ten years. If the investment will earn interest at the rate of 9% per year compounded quarterly, find the amount of the quarterly deposit that must be made in order to accumulate the required sum.

24. **Trust Funds**   John Carlson is the beneficiary of a $20,000 trust fund set up for him by his grandparents.

Under the terms of the trust, he is to receive the money over a 5-year period in equal installments at the end of each year. If the fund earns interest at the rate of 9% per year compounded annually, what amount will he receive each year?

25. **Sinking Funds**   The management of the Gibraltar Brokerage Services, Inc., anticipates a capital expenditure of $20,000 in three years' time for the purpose of purchasing new fax machines and has decided to set up a sinking fund to finance this purchase. If the fund earns interest at the rate of 10% per year compounded quarterly, determine the size of each (equal) quarterly installment that should be deposited in the fund.

26. **Keogh Accounts**   Ms. Anderson, a self-employed individual, wishes to accumulate a retirement fund of $250,000. How much should she deposit each month into her Keogh account, which pays interest at the rate of 8.5% per year compounded monthly, to reach her goal upon retirement 25 years from now?

27. **IRAs**   Mr. Berger has deposited $375 in his IRA at the end of each quarter for the past 20 years. His investment has earned interest at the rate of 8% per year compounded quarterly over this period. Now, at age 60, he is considering retirement. What quarterly payment will he receive over the next 15 years? (Assume that the money is earning interest at the same rate and payments are made at the end of each quarter.) If he continues working and makes quarterly payments of the same amount in his IRA account until age 65, what quarterly payment will he receive from his fund upon retirement over the following 10 years?

28. **Financing a Car**   Ms. Dwyer purchased a new car during a special sales promotion by the manufacturer. She secured a loan from the manufacturer in the amount of $10,000 at a rate of 7.9% per year compounded monthly. Her bank is now charging 11.5% per year compounded monthly for new car loans. Assuming that each loan would be amortized by 36 equal monthly installments, determine the amount of interest she would have paid at the end of three years for each loan. How much less will she have paid in interest payments over the life of the loan by borrowing from the manufacturer instead of her bank?

29. **Financing a Home**   The Sandersons are planning to refinance their home. The outstanding principal on their original loan is $100,000 and was to be amortized in 240 equal monthly installments at an interest rate of 10% per year compounded monthly. The new loan they expect to secure is to be amortized over the same period at an interest rate of 7.8% per year compounded monthly. How much less can they expect to pay over the life of the loan in interest payments by refinancing the loan at this time?

30. **Financing a Home**   After making a down payment of $25,000, the Meyers need to secure a loan of $90,000 to purchase a certain house. Their bank's current rate for 25-year home loans is 11% compounded monthly. The owner has offered to finance the loan at 9.8% compounded monthly. Assuming that both loans would be amortized over a 25-year period by 300 equal monthly installments, determine the difference in the amount of interest the Meyers would pay by choosing the seller's financing rather than their bank's.

SOLUTIONS TO SELF-CHECK EXERCISES 4.3

1. a. We use formula (13) in each instance. Under plan I,

$$P = 100{,}000, \quad i = \frac{r}{m} = \frac{0.105}{12} = 0.00875$$

and

$$n = (30)(12) = 360$$

Therefore, the size of each monthly repayment under plan I is

$$R = \frac{100{,}000(0.00875)}{1 - (1.00875)^{-360}}$$

$$= 914.74$$

or $914.74.
   Under plan II,

$$P = 100{,}000, \quad i = \frac{r}{m} = \frac{0.0975}{12} = 0.008125$$

and

$$n = (15)(12) = 180$$

Therefore, the size of each monthly repayment under plan II is

$$R = \frac{100{,}000(0.008125)}{1 - (1.008125)^{-180}}$$

$$= 1059.36$$

or $1059.36.

b. Under plan I, the total amount of repayments would have been

$$(360)(914.74) = 329{,}306.40 \qquad \begin{array}{l}\text{Number of payments}\\ \text{times the size}\\ \text{of each installment}\end{array}$$

or $329,306.40. Under plan II, the total amount of repayments would have been

$$(180)(1059.36) = 190{,}684.80$$

or $190,684.80. Therefore, the difference in payments is

$$329{,}306.40 - 190{,}684.80 = 138{,}621.60$$

or $138,621.60.

2. We use (14) with

$$S = 250{,}000,$$

$$i = r = 0.0825 \text{ (because } m = 1) \quad \text{and} \quad n = 20$$

giving the required size of each installment as

$$R = \frac{(0.0825)(250{,}000)}{(1.0825)^{20} - 1}$$

$$= 5313.59$$

or $5313.59.

4.4 Arithmetic and Geometric Progressions (Optional)

Arithmetic Progressions

An **arithmetic progression** is a sequence of numbers in which each term after the first is obtained by adding a constant d to the preceding term. The constant d is called the **common difference.** For example, the sequence

$$3, 6, 9, 12, \ldots$$

is an arithmetic progression with common difference equal to 3.

Observe that an arithmetic progression is completely determined if the first term and the common difference are known. In fact, if

$$a_1, a_2, a_3, \ldots, a_n, \ldots$$

is an arithmetic progression with the first term given by a and common difference given by d, then by definition,

$$a_1 = a$$

$$a_2 = a_1 + d = a + d$$

$$a_3 = a_2 + d = (a + d) + d = a + 2d$$

$$a_4 = a_3 + d = (a + 2d) + d = a + 3d$$

$$\vdots$$

$$a_n = a_{n-1} + d = a + (n - 2)d + d = a + (n - 1)d$$

Thus we see that the nth term of an arithmetic progression with first term a and common difference d is given by

$$a_n = a + (n - 1)d$$

nth TERM OF AN ARITHMETIC PROGRESSION

The nth term of an arithmetic progression with first term a and common difference d is given by

$$a_n = a + (n - 1)d \tag{15}$$

EXAMPLE 1   Find the twelfth term of the arithmetic progression

$$2, 7, 12, 17, 22, \ldots$$

Solution   The first term of the arithmetic progression is $a_1 = a = 2$, and the common difference is $d = 5$, so upon setting $n = 12$ in (15), we find

$$a_{12} = 2 + (12 - 1)5 = 57 \qquad \blacksquare$$

EXAMPLE 2   Write the first five terms of an arithmetic progression whose third and eleventh terms are 21 and 85, respectively.

Solution   Using (15), we obtain

$$a_3 = a + 2d = 21$$

$$a_{11} = a + 10d = 85$$

Subtracting the first equation from the second gives $8d = 64$, or $d = 8$. Substituting this value of d into the first equation yields $a + 16 = 21$, or $a = 5$. Thus, the required arithmetic progression is given by the sequence

$$5, 13, 21, 29, 37, \ldots \qquad \blacksquare$$

Let S_n denote the sum of the first n terms of an arithmetic progression with first term $a_1 = a$ and common difference d. Then

$$S_n = a + (a + d) + (a + 2d) + \cdots + [a + (n - 1)d] \qquad \textbf{(16)}$$

Rewriting the expression for S_n with the terms in reverse order gives

$$S_n = [a + (n - 1)d] + [a + (n - 2)d] + \cdots + (a + d) + a \qquad \textbf{(17)}$$

Adding (16) and (17), we obtain

$$2S_n = [2a + (n - 1)d] + [2a + (n - 1)d]$$

$$+ \cdots + [2a + (n - 1)d]$$

$$= n[2a + (n - 1)d],$$

or
$$S_n = \frac{n}{2}[2a + (n - 1)d]$$

SUM OF TERMS IN AN ARITHMETIC PROGRESSION

The sum of the first n terms of an arithmetic progression with first term a and common difference d is given by

$$S_n = \frac{n}{2}[2a + (n - 1)d] \qquad \textbf{(18)}$$

EXAMPLE 3   Find the sum of the first 20 terms of the arithmetic progression of Example 2.

Solution   Letting $a = 2$, $d = 5$, and $n = 20$ in (18), we obtain

$$S_{20} = \frac{20}{2}[2 \cdot 2 + 19 \cdot 5] = 990 \qquad \blacksquare$$

EXAMPLE 4   The Madison Electric Company had sales of $200,000 in its first year of operation. If the sales increased by $30,000 per year thereafter, find Madison's sales in the fifth year and its total sales over the first five years of operation.

Solution   Madison's yearly sales follow an arithmetic progression with the first term given by $a = 200,000$ and the common difference given by $d = 30,000$. The sales in the fifth year are found by using (15) with $n = 5$. Thus,

$$a_5 = 200,000 + (5 - 1)30,000 = 320,000$$

or $320,000.

Madison's total sales over the first five years of operation are found by using (18) with $n = 5$. Thus,

$$S_5 = \frac{5}{2}[2(200,000) + (5 - 1)30,000]$$

$$= 1,300,000$$

or $1,300,000. ∎

Geometric Progressions

A **geometric progression** is a sequence of numbers in which each term after the first is obtained by multiplying the preceding term by a constant r. The constant r is called the **common ratio.**

A geometric progression is completely determined if the first term and the common ratio are known. Thus, if

$$a_1, a_2, a_3, \ldots, a_n, \ldots$$

is a geometric progression with the first term given by a and common ratio given by r, then by definition,

$$a_1 = a$$

$$a_2 = a_1 r = ar$$

$$a_3 = a_2 r = ar^2$$

$$a_4 = a_3 r = ar^3$$

$$\vdots$$

$$a_n = a_{n-1} r = ar^{n-1}$$

Thus we see that the nth term of a geometric progression with first term a and common ratio r is given by

$$a_n = ar^{n-1}$$

nth TERM OF A GEOMETRIC PROGRESSION

The nth term of a geometric progression with first term a and common ratio r is given by

$$a_n = ar^{n-1} \tag{19}$$

EXAMPLE 5   Find the eighth term of a geometric progression whose first five terms are 162, 54, 18, 6, 2.

Solution   The common ratio is found by taking the ratio of any term other than the first to the preceding term. Taking the ratio of the fourth term to the third term, for example, gives $r = 6/18 = 1/3$. To find the eighth term of the geometric progression, use (19) with $a = 162$, $r = 1/3$, and $n = 8$, obtaining

$$a_8 = 162 \left(\frac{1}{3}\right)^7$$

$$= \frac{2}{27}$$

EXAMPLE 6   Find the tenth term of a geometric progression whose third term is 16 and whose seventh term is 1.

Solution   Using (19) with $n = 3$ and $n = 7$, respectively, yields

$$a_3 = ar^2 = 16$$

and

$$a_7 = ar^6 = 1$$

Dividing a_7 by a_3 gives

$$\frac{ar^6}{ar^2} = \frac{1}{16}$$

from which we obtain $r^4 = 1/16$, or $r = 1/2$. Substituting this value of r into the expression for a_3, we obtain

$$a\left(\frac{1}{2}\right)^2 = 16 \qquad \text{or} \qquad a = 64$$

Finally, using (19) once again with $a = 64$, $r = 1/2$, and $n = 10$ gives

$$a_{10} = 64 \left(\frac{1}{2}\right)^9 = \frac{1}{8}$$

To find the sum of the first n terms of a geometric progression with the first term $a_1 = a$ and common ratio r, denote the required sum by S_n. Then

$$S_n = a + ar + ar^2 + \cdots + ar^{n-2} + ar^{n-1} \tag{20}$$

Upon multiplying (20) by r, we obtain

$$rS_n = ar + ar^2 + ar^3 + \cdots + ar^{n-1} + ar^n \tag{21}$$

Subtracting (21) from (20) gives

$$S_n - rS_n = a - ar^n$$

or

$$(1 - r)S_n = a(1 - r^n)$$

If $r \neq 1$, we may divide both sides of the last equation by $(1 - r)$, obtaining

$$S_n = \frac{a(1 - r^n)}{(1 - r)}$$

If $r = 1$, then (20) gives

$$S_n = a + a + a + \cdots + a \qquad (n \text{ terms})$$

$$= na$$

Thus, $\qquad\qquad\qquad S_n = \begin{cases} \dfrac{a(1 - r^n)}{1 - r} & \text{if } r \neq 1 \\[2ex] na & \text{if } r = 1 \end{cases}$

SUM OF TERMS IN A GEOMETRIC PROGRESSION

The sum of the first n terms of a geometric progression with first term a and common ratio r is given by

$$S_n = \begin{cases} \dfrac{a(1 - r^n)}{1 - r} & \text{if } r \neq 1 \\[2ex] na & \text{if } r = 1 \end{cases} \qquad\qquad \textbf{(22)}$$

EXAMPLE 7   Find the sum of the first six terms of the geometric progression

$$3, 6, 12, 24, \ldots$$

Solution   Here $a = 3$ and $r = 6/3 = 2$, so formula (22) gives

$$S_6 = \frac{3(1 - 2^6)}{1 - 2} = 189$$

EXAMPLE 8   The Michaelson Land Development Company had sales of $1 million in its first year of operation. If the sales increased by 10% per year thereafter, find Michaelson's sales in the fifth year and its total sales over the first five years of operation.

Solution   Michaelson's yearly sales follow a geometric progression with the first term given by $a = 1{,}000{,}000$ and the common ratio given by $r = 1.1$. The sales in the fifth year are found by using (19) with $n = 5$. Thus,

$$a_5 = 1{,}000{,}000(1.1)^4 = 1{,}464{,}100$$

or $1,464,100.

Michaelson's total sales over the first five years of operation are found by using (22) with $n = 5$. Thus,

$$S_5 = \frac{1{,}000{,}000[1 - (1.1)^5]}{1 - 1.1}$$

$$= 6{,}105{,}100$$

or $6,105,100.

Double Declining-Balance Method of Depreciation

In Section 2.4 we discussed the straight-line, or linear, method of depreciating an asset. Linear depreciation assumes that the asset depreciates at a constant rate. For certain assets such as machines, whose market values drop rapidly in the early years of usage and thereafter less rapidly, another method of depreciation called the *double declining-balance method* is often used. In practice, a business firm normally employs the double declining-balance method for depreciating such assets for a certain number of years and then switches over to the linear method.

To derive an expression for the book value of an asset being depreciated by the double declining-balance method, let C (in dollars) denote the original cost of the asset and suppose the asset is to be depreciated over N years. Using this method, the amount depreciated each year is $2/N$ times the value of the asset at the beginning of that year. Thus, the amount by which the asset is depreciated in its first year of use is given by $2C/N$, so if $V(1)$ denotes the book value of the asset at the end of the first year, then

$$V(1) = C - \frac{2C}{N} = C\left(1 - \frac{2}{N}\right)$$

Next, if $V(2)$ denotes the book value of the asset at the end of the second year, then a similar argument leads to

$$V(2) = C\left(1 - \frac{2}{N}\right) - C\left(1 - \frac{2}{N}\right)\frac{2}{N}$$

$$= C\left(1 - \frac{2}{N}\right)\left(1 - \frac{2}{N}\right)$$

$$= C\left(1 - \frac{2}{N}\right)^2$$

Continuing, we find that if $V(n)$ denotes the book value of the asset at the end of n years, then the terms $C, V(1), V(2), \ldots, V(n)$ form a geometric progression with first term C and common ratio $(1 - 2/N)$. Consequently, the nth term, $V(n)$, is given by

$$V(n) = C\left(1 - \frac{2}{N}\right)^n \qquad 1 \le n \le N \tag{23}$$

Also, if $D(n)$ denotes the amount by which the asset has been depreciated by the end of the nth year, then

$$D(n) = C - C\left(1 - \frac{2}{N}\right)^n$$

$$= C\left[1 - \left(1 - \frac{2}{N}\right)^n\right] \tag{24}$$

 EXAMPLE 9   A tractor purchased at a cost of $60,000 is to be depreciated by the double declining-balance method over ten years. What is the book value of the tractor at the end of five years? By what amount has the tractor been depreciated by the end of the fifth year?

Solution   We have $C = 60,000$ and $N = 10$. Thus, using (23) with $n = 5$ gives the book value of the tractor at the end of five years as

$$V(5) = 60,000\left(1 - \frac{2}{10}\right)^5$$

$$= 60,000\left(\frac{4}{5}\right)^5 = 19,660.80$$

or $19,660.80.

The amount by which the tractor has been depreciated by the end of the fifth year is given by

$$60,000 - 19,660.80 = 40,339.20$$

or \$40,339.20. You may verify the last result by using (24) directly.   ◼

SELF-CHECK EXERCISES 4.4

1. Find the sum of the first five terms of the geometric progression with first term -24 and common ratio $-1/2$.

2. Office equipment purchased for \$75,000 is to be depreciated by the double declining-balance method over five years. Find the book value at the end of three years.

3. Derive the formula for the future value of an annuity (equation (9), Section 4.2).

Solutions to Self-Check Exercises 4.4 can be found on page 179.

4.4   EXERCISES

In exercises 1–4, find the nth term of the arithmetic progression that has the given values of a, d, and n.

1. $a = 6$, $d = 3$, $n = 9$

2. $a = -5$, $d = 3$, $n = 7$

3. $a = -15$, $d = 3/2$, $n = 8$

4. $a = 1.2$, $d = 0.4$, $n = 98$

5. Find the first five terms of the arithmetic progression whose fourth and eleventh terms are 30 and 107, respectively.

6. Find the first five terms of the arithmetic progression whose 7th and 23rd terms are -5 and -29, respectively.

7. Find the seventh term of the arithmetic progression: $x, x + y, x + 2y, \ldots$.

8. Find the eleventh term of the arithmetic progression: $a + b, 2a, 3a - b, \ldots$.

9. Find the sum of the first 15 terms of the arithmetic progression: 4, 11, 18, ...

10. Find the sum of the first 20 terms of the arithmetic progression: 5, -1, -7, ...

11. Find the sum of all the odd integers between 14 and 58.

12. Find the sum of the even integers between 21 and 99.

13. Find $f(1) + f(2) + f(3) + \cdots + f(20)$, given that $f(x) = 3x - 4$.

14. Find $g(1) + g(2) + g(3) + \cdots + g(50)$, given that $g(x) = 12 - 4x$.

15. Show that equation (18) can be written as $S_n = \dfrac{n}{2}(a + a_n)$, where a_n represents the last term of an arithmetic progression. Use this formula to find:
    a. The sum of the first 11 terms of the arithmetic progression whose 1st and 11th terms are 3 and 47, respectively
    b. The sum of the first 20 terms of the arithmetic progression whose 1st and 20th terms are 5 and -33, respectively

16. **Sales Growth**   The Moderne Furniture Company had sales of \$150,000 during its first year of operation. If the sales increased by \$16,000 per year thereafter, find Moderne's sales in the fifth year and its total sales over the first five years of operation.

17. **Exercise Program**   As part of her fitness program, Karen has taken up jogging. If she jogs 1 mile the first day and increases her daily run by 1/4 mile every week, how long will it take her to reach her goal of 10 miles per day?

18. **Cost of Drilling**   A 100-foot oil well is to be drilled. The cost of drilling the first foot is \$6 and the cost of drilling each additional foot is \$2.50 more than

that of the preceding foot. Find the cost of drilling the entire 100 feet.

19. **Consumer Decisions**   A tourist wishes to go from the airport to his hotel, which is 25 miles away. The taxi rate is $1 for the first mile and 60 cents for each additional mile. The airport limousine also goes to his hotel and charges a flat rate of $7.50. How much money will the tourist save by taking the airport limousine?

20. **Salary Comparisons**   A recent college graduate received two job offers. Company A offered her an initial salary of $30,800 with guaranteed annual increases of $1,250 per year for the first five years. Company B offered an initial salary of $32,000 with guaranteed annual increases of $900 per year for the first five years.
    a. Which company is offering a higher salary for the fifth year of employment?
    b. Which company is offering more money for the first five years of employment?

21. **Sum-of-the-Years-Digits Method of Depreciation**   One of the methods that the Internal Revenue Service allows for computing depreciation of certain business property is the *sum-of-the-years-digits method*. If a property valued at C dollars has an estimated useful life of N years and a salvage value of S dollars, then the amount of depreciation D_n allowed during the nth year is given by

$$D_n = (C - S)\frac{N - (n - 1)}{S_N}   \qquad 0 \le n \le N$$

where S_N is the sum of the first N positive integers representing the estimated useful life of the property. Thus,

$$S_N = 1 + 2 + \cdots + N = \frac{N(N + 1)}{2}$$

    a. Verify that the sum of the arithmetic progression $S_N = 1 + 2 + \cdots + N$ is given by

$$\frac{N(N + 1)}{2}$$

    b. If office furniture worth $6000 is to be depreciated by this method over $N = 10$ years, and the salvage value of the furniture is $500, find the depreciation for the third year by computing D_3.

22. **Sum-of-the-Years-Digits Method of Depreciation**   Refer to Example 1, Section 2.4. The amount of depreciation allowed for a printing machine that has an estimated useful life of five years and an initial value of $100,000 (with no salvage value) was $20,000 per year using the straight-line method of deprecia-

tion. Determine the amount of depreciation that would be allowed for the first year if the printing machine were depreciated using the sum-of-the-years-digits method described in exercise 21. Which method would result in a larger depreciation of the asset in its first year of use?

In exercises 23–28, determine which of the given sequences are geometric progressions. For each geometric progression, find the seventh term and the sum of the first seven terms.

23. $4, 8, 16, 32, \ldots$

24. $1, -\frac{1}{2}, \frac{1}{4}, -\frac{1}{8}, \ldots$

25. $\frac{1}{2}, -\frac{3}{8}, \frac{1}{4}, -\frac{9}{64}, \ldots$

26. $0.004, 0.04, 0.4, 4, \ldots$

27. $243, 81, 27, 9, \ldots$

28. $-1, 1, 3, 5, \ldots$

29. Find the 20th term and sum of the first 20 terms of the geometric progression $-3, 3, -3, 3, \ldots$

30. Find the 23rd term in a geometric progression having the first term $a = 0.1$ and ratio $r = 2$.

31. **Population Growth**   It has been projected that the population of a certain city in the Southwest will increase by 8% during each of the next five years. If the current population is 200,000, what is the expected population in five years?

32. **Sales Growth**   The Metro Cable T.V. Company had sales of $2,500,000 in its first year of operation. If thereafter the sales increased by 12% of the previous year, find the sales of the company in the fifth year and the total sales over the first five years of operation.

33. **COLAs**   Suppose that the cost-of-living index had increased by 9% during each of the past six years and that a member of the E.U.W. Union had been guaranteed an annual increase equal to 2% above the cost-of-living index over the period. What would the present salary of a union member be whose salary was $18,000 six years ago?

34. **Savings Plans**   The parents of a nine-year-old boy have agreed to deposit $10 in their son's bank account on his tenth birthday and to double the size of their deposit every year thereafter until his eighteenth birthday.
    a. How much will they have to deposit on his eighteenth birthday?
    b. How much will they have deposited by his eighteenth birthday?

35. Salary Comparisons Suppose that an employee of the Stenton Printing Company whose current annual salary is $28,000 has the option of taking an annual raise of 8% per year for the next four years or a fixed annual raise of $1,500 per year. Which option would be more profitable to him considering his total earnings over the 4-year period?

36. Bacteria Growth A culture of a certain bacteria is known to double in number every three hours. If the culture has an initial count of 20, what will the population of the culture be at the end of 24 hours?

37. Trust Funds Ms. Simpson is the recipient of a trust fund that she will receive over a period of six years. Under the terms of the trust, she is to receive $10,000 the first year and each succeeding annual payment is to be increased by 15%.

  a. How much will she receive during the sixth year?
  b. What is the total amount of the six payments she will receive?

In exercises 38–40, find the book value of office equipment purchased at a cost C at the end of the nth year if it is to be depreciated by the double declining-balance method over ten years. Assume a salvage value of $0.

38. $C = \$20,000, n = 4$

39. $C = \$150,000, n = 8$

40. $C = \$80,000, n = 7$

41. Double Declining-Balance Method of Depreciation Restaurant equipment purchased at a cost of $150,000 is to be depreciated by the double declining-balance method over ten years. What is the book value of the equipment at the end of six years? By what amount has the equipment been depreciated at the end of the sixth year?

42. Double Declining-Balance Method of Depreciation Refer to exercise 22. Recall that a printing machine that had an estimated useful life of five years and an initial value of $100,000 (with no salvage value) was to be depreciated. At the end of the first year using the straight-line method of depreciation, the amount of depreciation allowed was $20,000, and when the sum-of-the-years-digits method was used, the depreciation was $33,333. Determine the amount of depreciation that would be allowed for the first year if the printing machine were depreciated by the double declining-balance method. Which of these three methods would result in the largest depreciation of the printing machine at the end of its first year of use?

SOLUTIONS TO SELF-CHECK EXERCISES 4.4

1. Use (22) with $a = -24$ and $r = -1/2$, obtaining

$$S_5 = \frac{-24\left[1 - \left(-\frac{1}{2}\right)^5\right]}{1 - \left(-\frac{1}{2}\right)}$$

$$= \frac{-24\left(1 + \frac{1}{32}\right)}{\frac{3}{2}} = -\frac{33}{2}$$

2. Use (23) with $C = 75,000$, $N = 5$, and $n = 3$, giving the book value of the office equipment at the end of three years as

$$V(3) = 75,000\left(1 - \frac{2}{5}\right)^3 = 16,200$$

or $16,200.

3. We have

$$S = R + R(1 + i) + R(1 + i)^2 + \cdots + R(1 + i)^{n-1}$$

Now, the sum on the right is easily seen to be the sum of the first n terms of a geometric progression with first term R and common ratio $(1 + i)$, so by virtue of formula (22),

$$S = \frac{R[1 - (1 + i)^n]}{1 - (1 + i)} = R\left[\frac{(1 + i)^n - 1}{i}\right]$$

U s i n g   T e c h n o l o g y

EXERCISE

A tractor purchased at a cost of $60,000 is to be depreciated over ten years, with a residual value of $0. With the double declining-balance method (see Example 9, page 176), its value at the end of n years is $V_1(n) = 60000(0.8)^n$ dollars. With straight-line depreciation, its value at the end of n years is $V_2(n) = 60000 - 6000n$. Use a graphing utility to sketch the graphs of V_1 and V_2 in the viewing rectangle $[0, 10] \times [0, 70,000]$. Comment on the relative merits of each method of depreciation.

CHAPTER FOUR SUMMARY OF PRINCIPAL FORMULAS AND TERMS

Formulas

1. Simple interest
$$A = P(1 + rt)$$

2. Compound interest:
   a. Accumulated amount $A = P(1 + i)^n$
   b. Present value $P = A(1 + i)^{-n}$
   c. Interest rate per compounding period $i = r/m$
   d. Number of conversion periods $n = mt$

3. Continuous compound interest:
   a. Accumulated amount $A = Pe^{rt}$
   b. Present value $P = Ae^{-rt}$

4. Effective rate of interest
$$r_{eff} = \left(1 + \frac{r}{m}\right)^m - 1$$

5. Annuities:
   a. Future value
$$S = R\left[\frac{(1 + i)^n - 1}{i}\right]$$
   b. Present value
$$P = R\left[\frac{1 - (1 + i)^{-n}}{i}\right]$$

6. Amortization payment
$$R = \frac{Pi}{1 - (1 + i)^{-n}}$$

7. Sinking fund payment
$$R = \frac{iS}{(1 + i)^n - 1}$$

Terms

Simple interest
Accumulated amount (future value)
Compound interest
Nominal rate (stated rate)
Effective rate
Present value
Future value
Annuity
Annuity certain

Ordinary annuity
Simple annuity
Future value S of an annuity
Present value P of an annuity
Sinking fund
Arithmetic progression
Common difference
Geometric progression
Common ratio

CHAPTER 4 REVIEW EXERCISES

1. Find the accumulated amount after four years if $5000 is invested at 10% per year compounded (a) annually, (b) semiannually, (c) quarterly, and (d) monthly.

2. Find the accumulated amount after eight years if $12,000 is invested at 6.5% per year compounded (a) annually, (b) semiannually, (c) quarterly, and (d) monthly.

3. Find the effective rate of interest corresponding to a nominal rate of 12% per year compounded (a) annually, (b) semiannually, (c) quarterly, and (d) monthly.

4. Find the effective rate of interest corresponding to a nominal rate of 11.5% per year compounded (a) annually, (b) semiannually, (c) quarterly, and (d) monthly.

5. Find the present value of $41,413 due in five years at an interest rate of 6.5% per year compounded quarterly.

6. Find the present value of $64,540 due in six years at an interest rate of 8% per year compounded monthly.

7. Find the amount (future value) of an ordinary annuity of $150 per quarter for seven years at 8% per year compounded quarterly.

8. Find the future value of an ordinary annuity of $120 per month for ten years at 9% per year compounded monthly.

9. Find the present value of an ordinary annuity of 36 payments of $250 each made monthly and earning interest at 9% per year compounded monthly.

10. Find the present value of an ordinary annuity of 60 payments of $5000 each made quarterly and earning interest at 8% per year compounded quarterly.

11. Find the payment R needed to amortize a loan of $22,000 at 8.5% per year compounded monthly with 36 monthly installments over a period of three years.

12. Find the payment R needed to amortize a loan of $10,000 at 9.2% per year compounded monthly with 36 monthly installments over a period of three years.

13. Find the payment R needed to accumulate $18,000 with 48 monthly installments over a period of four years at an interest rate of 6% per year compounded monthly.

14. Find the payment R needed to accumulate $15,000 with 60 monthly installments over a period of five

years at an interest rate of 7.2% per year compounded monthly.

15. Find the rate of interest per year compounded on a daily basis that is equivalent to 7.2% per year compounded monthly.

16. Find the rate of interest per year compounded on a daily basis that is equivalent to 9.6% per year compounded monthly.

17. The J.C.N. Media Corporation had sales of $1,750,000 in the first year of operation. If the sales increased by 14% per year thereafter, find the company's sales in the fourth year and the total sales over the first four years of operation.

18. The manager of a money-market fund has invested $4.2 million in certificates of deposit that pay interest at the rate of 5.4% per year compounded quarterly over a period of five years. How much will the investment be worth at the end of five years?

19. Ms. Kim invested a sum of money four years ago in a savings account that has since paid interest at the rate of 6.5% per year compounded monthly. Her investment is now worth $19,440.31. How much did she originally invest?

20. Mr. Sanchez invested $24,000 in a mutual fund five years ago. Today his investment is worth $34,616. Find the effective annual rate of return on his investment over the 5-year period.

21. The Blakes have decided to start a monthly savings program in order to provide for their son's college education. How much should they deposit at the end of each month in a savings account earning interest at the rate of 8% per year compounded monthly so that at the end of the tenth year the accumulated amount will be $40,000?

22. Mrs. Chang has contributed $200 at the end of each month into her company's employee retirement account for the past ten years. Her employer has matched her contribution each month. If the account has earned interest at the rate of 8% per year compounded monthly over the 10-year period, determine how much Mrs. Chang now has in her retirement account.

23. Ms. Lemsky has leased an auto for four years at $300 per month. If money is worth 5% per year compounded monthly, what is the equivalent cash payment (present value) of this annuity? (Assume that the payments are made at the end of each month.)

24. Ms. Cantwell made a down payment of $400 toward the purchase of new furniture. In order to pay the balance of the purchase price she has secured a loan from her bank at 12% per year compounded monthly. Under the terms of her finance agreement she is required to make payments of $75.32 at the end of each month for 24 months. What was the purchase price of the furniture?

25. The Turners have purchased a house for $150,000. They made an initial down payment of $30,000 and secured a mortgage with interest charged at the rate of 9% per year on the unpaid balance. (Interest computations are made at the end of each month.) Assume the loan is amortized over 30 years.
  a. What monthly payment will the Turners be required to make?
  b. What will their total interest payment be?
  c. What will their equity be (disregard depreciation) after ten years?

26. Refer to exercise 25. Assume the loan is amortized over 15 years.
  a. What monthly payment will the Turners be required to make?
  b. What will their total interest payment be?
  c. What will their equity be (disregard depreciation) after ten years?

27. The management of a corporation anticipates a capital expenditure of $500,000 in five years for the purpose of purchasing replacement machinery. In order to finance this purchase, a sinking fund that earns interest at the rate of 10% per year compounded quarterly will be set up. Determine the amount of each (equal) quarterly installment that should be deposited in the fund. (Assume that the payments are made at the end of each quarter.)

28. The management of a condominium association anticipates a capital expenditure of $120,000 in two years for the purpose of painting the exterior of the condominium. In order to pay for this maintenance, a sinking fund will be set up that will earn interest at the rate of 5.8% per year compounded monthly. Determine the amount of each (equal) monthly installment the association will be required to deposit into the fund at the end of each month for the next two years.

29. The outstanding balance on Mr. Baker's credit card account is $3200. The bank issuing the credit card is charging 18.6% per year compounded monthly. If Mr. Baker decides to pay off this balance in equal monthly installments at the end of each month for the next 18 months, how much will his monthly payment be?

30. Refer to exercise 29. What is the effective rate of interest the bank is charging Mr. Baker?

31. A hotel was purchased by a conglomerate for $4.5 million and sold five years later for $8.2 million. Find the annual rate of return (compounded continuously).

32. Find the present value of $119,346 due in four years at an interest rate of 10% per year compounded continuously.

CHAPTER 4   EXTENDED PROBLEMS AND PROJECTS

First, familiarize yourself with the greatest integer function by doing exercise 1. Then proceed to exercise 2.

1. The *greatest integer function* is defined by $f(x) = [x]$, where $[x]$ denotes the greatest integer that is less than or equal to x. (For example, $[7.4] = 7$, $[\pi] = 3$, $[\sqrt{3}] = 1$, and $[-1/2] = -1$.)
Sketch the graph of
  a. $f(x) = [x]$
  b. $g(x) = x - [x]$
  c. $h(x) = x[x]$
  d. $k(x) = [mx]$ (m, a positive integer)

2. Suppose you deposit a sum of $1000 in an account for a term of three years, earning interest at the rate of 8% per year compounded semiannually. Show that the

accumulated amount after t years is given by

$$A = 1000(1.04)^{[2t]}$$

Sketch the graph of A.

The next example requires the use of a graphics calculator or computer.

3. The formula

$$R = \frac{Pi}{1 - (1 + i)^{-n}} \qquad (25)$$

(formula (13), page 164) can be used to find the periodic payment R on a loan of P dollars to be amortized over n periods with interest charged at the rate of i per period.

Suppose you know the periodic payment, the loan amount P, and the number of periods in the term of the loan. Can you find the interest rate i, per period? Here's a suggestion (justify each step!) for solving this problem.

a. Show that formula (25) can be written in the form

$$i = \frac{R}{P}(1 - (1 + i)^{-n}) \qquad (26)$$

b. Show that equation (26) is equivalent to the following system of equations:

$$\begin{cases} y = i \\ y = \dfrac{R}{P}[1 - (1 + i)^{-n}] \end{cases} \qquad (27)$$

c. Use a graphing utility to solve system (27) for i by finding the point of intersection of the graphs of the two functions in (27). Here, the operation **ISECT** will prove useful. (See Section A.3, Finding the Points of Intersection of Two Graphs, in Appendix A.)

Use this procedure to solve the following exercise.

4. You are told that for a loan of $100,000 to be repaid in equal monthly installments over a period of 30 years, the size of each installment will be $733.76. If the bank charges interest at the rate of r percent per year on the unpaid balance, with interest computations made at the end of each month, find r. What is the effective rate of interest?

5

Systems of Linear Equations and Matrices

*T*he linear equations in two variables studied in Chapter 2 are readily extended to the case involving more than two variables. For example, a linear equation in three variables represents a plane in three-dimensional space. In this chapter we see how some real-world problems can be formulated in terms of systems of linear equations, and we develop two methods for solving these equations.

We also see how *matrices* (ordered rectangular arrays of numbers) can be used to write systems of linear equations in compact form. We then go on to consider some real-life applications of matrices.

5.1

Systems of Linear Equations—Introduction

Systems of Equations

Recall that in Section 2.4 we had to solve two simultaneous linear equations in order to find the *break-even point* and the *equilibrium point*. These are two examples of situations in which solving a real-world problem calls for solving a system of linear equations in two or more variables. In this chapter we take up a more systematic study of such systems.

We begin by considering a system of two linear equations in two variables. Recall that such a system may be written in the general form

$$ax + by = h$$

$$cx + dy = k \qquad \qquad (1)$$

where a, b, c, d, h, and k are real constants and neither a and b nor c and d are both zero.

Now let us study the nature of the solution of a system of equations in more detail. Recall that the graph of each equation in (1) is a straight line in the plane, so that geometrically the solution to the system is the point(s) of intersection of the two straight lines L_1 and L_2, represented by the first and second equations of the system.

Figure 1 depicts each of the three cases that may occur. The two lines L_1 and L_2 may

a. intersect at exactly one point

b. be parallel and coincident

c. be parallel and distinct

FIGURE 1

(a) A unique solution

(b) Infinitely many solutions

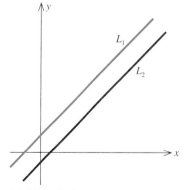

(c) No solution

Note that one and only one of these cases must occur. In the first case, the system has a unique solution corresponding to the single point of intersection of the two lines. In the second case, the system has infinitely many solutions corresponding to the points lying simultaneously on both lines. Finally, in the third case, the system has no solution, since the two lines do not intersect.

Let us illustrate each of these possibilities by considering some specific examples.

1. A system of equations with exactly one solution
Consider the system

$$2x - y = 1$$
$$3x + 2y = 12$$

Solving the first equation for y in terms of x, we obtain the equation

$$y = 2x - 1$$

Substituting this expression for y into the second equation yields

$$3x + 2(2x - 1) = 12$$
$$3x + 4x - 2 = 12$$
$$7x = 14$$
$$x = 2$$

Finally, substituting this value of x into the expression for y obtained earlier gives

$$y = 2(2) - 1 = 3$$

Therefore, the unique solution of the system is given by $x = 2$ and $y = 3$. Geometrically, the two lines represented by the two linear equations that make up the system intersect at the point $(2, 3)$ (Figure 2).

REMARK   We can check our result by substituting the values $x = 2$ and $y = 3$ into the equations. Thus

$$2(2) - (3) = 1 \qquad \checkmark$$
$$3(2) + 2(3) = 12 \qquad \checkmark$$

From the geometrical point of view, we have just verified that the point $(2, 3)$ lies on both lines.   □

FIGURE 2

A system of equations with one solution

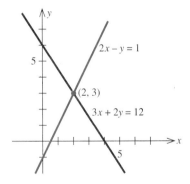

2. A system of equations with infinitely many solutions
Consider the system

$$2x - y = 1$$
$$6x - 3y = 3$$

Solving the first equation for y in terms of x, we obtain the equation

$$y = 2x - 1$$

Substituting this expression for y into the second equation gives

$$6x - 3(2x - 1) = 3$$
$$6x - 6x + 3 = 3$$

or
$$0 = 0$$

This result simply tells us that the second equation is equivalent to the first. (To see this, just multiply both sides of the first equation by

3.) Our computations have revealed that the system of two equations is equivalent to the single equation $2x - y = 1$. Thus, any ordered pair of numbers (x, y) satisfying the equation $2x - y = 1$ (or $y = 2x - 1$) constitutes a solution to the system.

In particular, by assigning the value t to x, where t is any real number, we find that $y = 2t - 1$, so the ordered pair $(t, 2t - 1)$ is a solution of the system. The variable t is called a **parameter.** For example, setting $t = 0$ gives the point $(0, -1)$ as a solution, and setting $t = 1$ gives the point $(1, 1)$ as another solution of the system. Since t represents any real number, there are infinitely many solutions to the system. Geometrically, the two equations in the system represent the same line, and all solutions of the system are points lying on the line (Figure 3). Such a system is said to be **dependent.**

3. ***A system of equations that has no solution***
Consider the system

$$2x - y = 1$$
$$6x - 3y = 12$$

The first equation is equivalent to $y = 2x - 1$. Substituting this expression for y into the second equation gives

$$6x - 3(2x - 1) = 12$$
$$6x - 6x + 3 = 12$$

or

$$0 = 9$$

which is clearly impossible. Thus, there is no solution to the system of equations. To interpret this situation geometrically, cast both equations in the slope-intercept form, obtaining

$$y = 2x - 1$$

and

$$y = 2x - 4$$

We see at once that the lines represented by these equations are parallel (each has slope 2) and distinct, since the first has intercept -1 and the second has intercept -4 (Figure 4). Systems with no solutions, such as this one, are said to be **inconsistent.**

REMARK   We have used the method of substitution in solving each of these systems. If you are familiar with the method of elimination, you might want to re-solve each of these systems using this method. We will study the method of elimination in detail in Section 5.2.   □

FIGURE 3

A system of equations with in-finitely many solutions; any point on the line is a solution

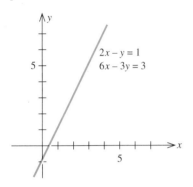

FIGURE 4

A system of equations with no solution

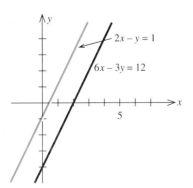

Application

In Section 2.4 we presented some real-world applications of systems involving two linear equations in two variables. Here is an example involving a system of three linear equations in three variables.

EXAMPLE 1   The Ace Novelty Company wishes to produce three types of souvenirs: types A, B, and C. To manufacture a type-A souvenir requires two minutes on machine I, one minute on machine II, and two minutes on machine III. A type-B souvenir requires one minute on machine I, three minutes on machine II, and one minute on machine III. A type-C souvenir requires one minute on machine I and two minutes each on machines II and III. There are three hours available on machine I, five hours available on machine II, and four hours available on machine III for processing the order. How many souvenirs of each type should Ace Novelty make in order to use all of the available time? Formulate, but do not solve, the problem.

Solution   The given information may be tabulated as follows:

	Type A	Type B	Type C	Time Available
Machine I	2	1	1	180
Machine II	1	3	2	300
Machine III	2	1	2	240

We have to determine the number of each of *three* types of souvenirs to be made. So, let x, y, and z denote the respective numbers of type-A, type-B, and type-C souvenirs to be made. The total amount of time that machine I is used is given by $2x + y + z$ minutes and must equal 180 minutes. This leads to the equation

$$2x + y + z = 180 \qquad \text{Time spent on machine I}$$

Similar considerations on the use of machines II and III lead to the following equations:

$$x + 3y + 2z = 300 \qquad \text{Time spent on machine II}$$

$$2x + y + 2z = 240 \qquad \text{Time spent on machine III}$$

Since the variables x, y, and z must satisfy simultaneously the three conditions represented by the three equations, the solution to the problem is found by solving the following system of linear equations:

$$2x + y + z = 180$$
$$x + 3y + 2z = 300$$
$$2x + y + 2z = 240$$

Solutions of Systems of Equations

We will complete the solution of the problem posed in Example 1 later on (page 203). For the moment let us look at the geometrical interpretation of a system of linear equations such as the system in Example 1 in order to gain some insight into the nature of the solution.

Now, a linear system comprising three linear equations in three variables x, y, and z has the general form

$$a_1x + b_1y + c_1z = d_1$$
$$a_2x + b_2y + c_2z = d_2$$
$$a_3x + b_3y + c_3z = d_3 \tag{2}$$

Just as a linear equation in two variables represents a straight line in the plane, it can be shown that a linear equation $ax + by + cz = d$ (a, b, and c not simultaneously equal to zero) in three variables represents a plane in three-dimensional space. Thus, each equation in system (2) represents a *plane* in three-dimensional space, and the *solution(s) of the system* is precisely the point(s) of intersection of the three planes defined by the three linear equations that make up the system. As before, the system has one and only one solution, infinitely many solutions, or no solution, depending on whether and how the planes intersect each other. Figure 5 illustrates each of these possibilities.

FIGURE 5

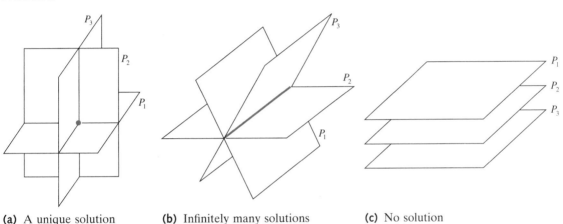

(a) A unique solution   **(b)** Infinitely many solutions   **(c)** No solution

In Figure 5a, the three planes intersect at a point corresponding to the situation in which system (2) has a unique solution. Figure 5b depicts the situation in which there are infinitely many solutions to the system. Here, the three planes intersect along a line, and the solutions are represented by the infinitely many points lying on this line. In Figure 5c, the three planes are parallel and distinct, so there is no point in common to all three planes, and system (2) has no solution in this case.

REMARK The situations depicted in Figure 5 are by no means exhaustive. You may consider various other orientations of the three planes that would illustrate the three possible outcomes in solving a system of linear equations involving three variables. □

LINEAR EQUATION IN n VARIABLES

A linear equation in n variables $x_1, x_2, \ldots, x_n$ is one of the form

$$a_1 x_1 + a_2 x_2 + \cdots + a_n x_n = c$$

where $a_1, a_2, \ldots, a_n$ (not all zero) and c are constants.

For example, the equation

$$3x_1 + 2x_2 - 4x_3 + 6x_4 = 8$$

is a linear equation in the four variables x_1, x_2, x_3, and x_4.

When the number of variables involved in a linear equation exceeds three, we no longer have the geometrical interpretation we had for the lower-dimensional spaces. Nevertheless, the algebraic concepts of the lower-dimensional spaces generalize to higher dimensions. For this reason, a linear equation in n variables $a_1 x_1 + a_2 x_2 + \cdots + a_n x_n = c$, where a_1, $a_2, \ldots, a_n$ are not all zero, is referred to as an *n-dimensional hyperplane*. We may interpret the solution(s) to a system comprising a finite number of such linear equations to be the *point(s) of intersection* of the hyperplanes defined by the equations that make up the system. As in the case of systems involving two or three variables, it can be shown that only three possibilities exist regarding the nature of the solution of such a system: (a) *a unique solution,* (b) *infinitely many solutions,* and (c) *no solution.*

SELF-CHECK EXERCISES 5.1

1. Determine whether the system of linear equations

$$2x - 3y = 12$$

$$x + 2y = 6$$

has (a) a unique solution, (b) infinitely many solutions, or (c) no solution. Find all solutions whenever they exist. Make a sketch of the set of lines described by the system.

2. A farmer has 200 acres of land suitable for cultivating crops A, B, and C. The cost per acre of cultivating crops A, B, and C is $40, $60, and $80, respectively. The farmer has $12,600 available for cultivation. Each acre of crop A requires 20 hours of labor, each acre of crop B requires 25 hours of labor, and each acre of crop C requires 40 hours of labor. The farmer has a maximum of 5,950 hours of labor available. If he wishes to use all of his cultivatable land, the entire budget, and all the labor available, how many acres of each crop should he plant? Formulate, but do not solve the problem.

Solutions to Self-Check Exercises 5.1 can be found on page 192.

5.1   EXERCISES

In exercises 1–12, determine whether each system of linear equations has (a) one and only one solution, (b) infinitely many solutions, or (c) no solution. Find all solutions whenever they exist.

1. $x - 3y = -1$
$4x + 3y = 11$

2. $2x - 4y = 5$
$3x + 2y = 6$

3. $x + 4y = 7$
$\frac{1}{2}x + 2y = 5$

4. $3x - 4y = 7$
$9x - 12y = 14$

5. $x + 2y = 7$
$2x - y = 4$

6. $\frac{3}{2}x - 2y = 4$
$x + \frac{1}{3}y = 2$

7. $2x - 5y = 10$
$6x - 15y = 30$

8. $5x - 6y = 8$
$10x - 12y = 16$

9. $4x - 5y = 14$
$2x + 3y = -4$

10. $\frac{5}{4}x - \frac{2}{3}y = 3$
$\frac{1}{4}x + \frac{5}{3}y = 6$

11. $2x - 3y = 6$
$6x - 9y = 12$

12. $\frac{2}{3}x + y = 5$
$\frac{1}{2}x + \frac{3}{4}y = \frac{15}{4}$

13. Determine the value of k for which the system of linear equations

$$2x - y = 3$$

$$4x + ky = 4$$

has no solution.

14. Determine the value of k for which the system of linear equations

$$3x + 4y = 12$$

$$x + ky = 4$$

has infinitely many solutions. Then find all the solutions corresponding to this value of k.

In exercises 15–25, formulate, but do not solve, the problem. You will be asked to solve these problems in the next section.

15. Agriculture The Johnson Farm has 500 acres of land allotted for cultivating corn and wheat. The cost of cultivating corn and wheat (including seeds and labor) is $42 and $30 per acre, respectively. Mr. Johnson has $18,600 available for cultivating these crops. If he wishes to use all the allotted land and his entire budget for cultivating these two crops, how many acres of each crop should he plant?

16. Investments Michael has a total of $2000 on deposit with two savings institutions. One pays simple interest at the rate of 6% per year, whereas the other pays simple interest at the rate of 8% per year. If Michael earned a total of $144 in interest during a single year, how much does he have on deposit in each institution?

17. Mixtures The Coffee Shoppe sells a coffee blend made from two coffees, one costing $2.50/lb and the other costing $3/lb. If the blended coffee sells for $2.80/lb, find how much of each coffee is used to obtain the desired blend. [*Hint:* Assume the weight of the blended coffee is 100 lb.]

18. Investments Kelly has a total of $30,000 invested in two municipal bonds that have yields of 8% and 10% simple interest per year, respectively. If the interest Kelly receives from the bonds in a year is $2,640, how much does she have invested in each bond?

19. Ridership The total number of passengers riding a certain city bus during the morning shift is 1000. If the child's fare is 25 cents, the adult fare is 75 cents, and the total revenue from the fares in the morning shift is $650, how many children and how many adults rode the bus during the morning shift?

20. Real Estate Cantwell Associates, a real estate developer, is planning to build a new apartment complex consisting of one-bedroom units and two- and three-bedroom townhouses. A total of 192 units is planned, and the number of family units (two- and three-

bedroom townhouses) will equal the number of one-bedroom units. If the number of one-bedroom units will be three times the number of three-bedroom units, find how many units of each type will be in the complex.

21. Investment Planning The annual interest on Mr. Carrington's three investments amounted to $21,600: 6% on a savings account, 8% on mutual funds, and 12% on money market certificates. If the amount of Carrington's investment in money market certificates was twice the amount of his investment in the savings account, and the interest earned from his investment in money market certificates was equal to the dividends he received from his investment in mutual funds, find how much money he placed in each type of investment.

22. Box-Office Receipts A theater has a seating capacity of 900 and charges $2 for children, $3 for students, and $4 for adults. At a certain screening with full attendance there were half as many adults as children and students combined. The receipts totaled $2800. How many children attended the show?

23. Management Decisions The management of Hartman Rent-A-Car has allocated $1 million to buy a fleet of new automobiles consisting of compact, intermediate, and full-size cars. Compacts cost $8,000 each, intermediate-size cars cost $12,000 each, and full-size cars cost $16,000 each. If Hartman purchases twice as many compacts as intermediate-size cars, and the total number of cars to be purchased is 100, determine how many cars of each type will be purchased. (Assume that the entire budget will be used.)

24. Investment Clubs The management of a private investment club has a fund of $200,000 earmarked for investment in stocks. To arrive at an acceptable overall level of risk, the stocks that management is considering have been classified into three categories: high-risk, medium-risk, and low-risk. Management estimates that high-risk stocks will have a rate of return of 15% per year; medium-risk stocks, 10% per year; and low-risk stocks, 6% per year. The investment in low-risk stocks is to be twice the sum of the investments in stocks of the other two categories. If the investment goal is to have an average rate of return of 9% per year on the total investment, determine how much the club should invest in each type of stock.

25. Diet Planning A dietician wishes to plan a meal around three foods. The percentage of the daily requirements of proteins, carbohydrates, and iron contained in each ounce of the three foods is summarized in the accompanying table.

	Food I	Food II	Food III
Percentage of Proteins	10	6	8
Percentage of Carbohydrates	10	12	6
Percentage of Iron	5	4	12

Determine how many ounces of each food the dietician should include in the meal to meet exactly the daily requirement of proteins, carbohydrates, and iron (100% of each).

SOLUTIONS TO SELF-CHECK EXERCISES 5.1

1. Solving the first equation for y in terms of x, we obtain

$$y = \frac{2}{3}x - 4$$

Next, substituting this result into the second equation of the system, we find

$$x + 2\left(\frac{2}{3}x - 4\right) = 6$$

$$x + \frac{4}{3}x - 8 = 6$$

$$\frac{7}{3}x = 14$$

or $x = 6$

Substituting this value of x into the expression for y obtained earlier, we have

$$y = \frac{2}{3}(6) - 4 = 0$$

Therefore, the system has the unique solution $x = 6$ and $y = 0$. Both lines are shown in the accompanying figure.

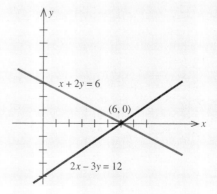

2. Let x, y, and z denote the number of acres of crop A, crop B, and crop C, respectively, to be cultivated. Then, the condition that all the cultivatable land be used translates into the equation

$$x + y + z = 200$$

Next, the total cost incurred in cultivating all three crops is $40x + 60y + 80z$ dollars, and since the entire budget is to be expended, we have

$$40x + 60y + 80z = 12,600$$

Finally, the amount of labor required to cultivate all three crops is $20x + 25y + 40z$ hours, and since all the available labor is to be used, we have

$$20x + 25y + 40z = 5,950$$

Thus, the solution is found by solving the following system of linear equations:

$$\begin{aligned} x + y + z &= 200 \\ 40x + 60y + 80z &= 12,600 \\ 20x + 25y + 40z &= 5,950 \end{aligned}$$

5.2 Solving Systems of Linear Equations I

The Gauss-Jordan Method

The method of substitution used in Section 5.1 is well suited to solving a system of linear equations when the number of linear equations and variables is small. But for large systems, the steps involved in the procedure become difficult to manage.

A suitable technique for solving systems of linear equations of any size is the **Gauss-Jordan elimination method.** One advantage of this technique is its adaptability to the computer. This method involves a sequence of operations on a system of linear equations to obtain at each stage an **equivalent system**—that is, a system having the same solution as the original system. The reduction is complete when the original system has been transformed so that it is in a certain standard form from which the solution can be easily read.

The operations of the Gauss-Jordan elimination method are:

1. Interchange any two equations.

2. Replace an equation by a nonzero constant multiple of itself.

3. Replace an equation by the sum of that equation and a constant multiple of any other equation.

To illustrate the Gauss-Jordan elimination method for solving systems of linear equations, let us apply it to the solution of the following system:

$$\begin{aligned} 2x + 4y &= 8 \\ 3x - 2y &= 4 \end{aligned}$$

We begin by working with the first, or x column. First, we transform the system into an equivalent system in which the coefficient of x in the first equation is 1:

$$2x + 4y = 8$$
$$3x - 2y = 4 \tag{3a}$$

$$x + 2y = 4 \qquad \text{Multiply the first equation in (3a)}$$
$$3x - 2y = 4 \qquad \text{by } \tfrac{1}{2} \text{ (operation 2).} \tag{3b}$$

Next, we eliminate x from the second equation.

$$x + 2y = 4 \qquad \text{Replace the second equation in (3b)}$$
$$\; - 8y = -8 \qquad \begin{array}{l} \text{by the sum of } -3 \text{ times the first} \\ \text{equation + the second equation} \\ \text{(operation 3).} \end{array} \tag{3c}$$

$$\begin{array}{r} -3x - 6y = -12 \\ 3x - 2y = 4 \\ \hline - 8y = -8 \end{array}$$

Then we obtain the following equivalent system, in which the coefficient of y in the second equation is 1.

$$x + 2y = 4 \qquad \text{Multiply the second equation in}$$
$$y = 1 \qquad \text{(3c) by } -\tfrac{1}{8} \text{ (operation 2).} \tag{3d}$$

Next, we eliminate y in the first equation.

$$x = 2 \qquad \text{Replace the first equation in (3d)}$$
$$\; y = 1 \qquad \begin{array}{l} \text{by the sum of } -2 \text{ times the second} \\ \text{equation + the first equation} \\ \text{(operation 3).} \end{array}$$

$$\begin{array}{r} x + 2y = 4 \\ - 2y = -2 \\ \hline x = 2 \end{array}$$

This system is now in standard form and we can read off the solution to (3a) as $x = 2$ and $y = 1$. We can also express this solution as $(2, 1)$ and interpret it geometrically as the point of intersection of the two lines represented by the two linear equations that make up the given system of equations.

Let us consider another example involving a system of three linear equations and three variables.

EXAMPLE 1   Solve the following system of equations:

$$2x + 4y + 6z = 22$$
$$3x + 8y + 5z = 27$$
$$-x + y + 2z = 2$$

Solution   First, we transform this system into an equivalent system in which the coefficient of x in the first equation is 1:

$$2x + 4y + 6z = 22$$
$$3x + 8y + 5z = 27$$
$$-x + y + 2z = 2 \tag{4a}$$

$$x + 2y + 3z = 11$$
$$3x + 8y + 5z = 27$$
$$-x + y + 2z = 2$$

Multiply the first equation in (4a) by $\frac{1}{2}$.

(4b)

Next, we eliminate the variable x from all equations except the first:

$$x + 2y + 3z = 11$$
$$2y - 4z = -6$$
$$-x + y + 2z = 2$$

Replace the second equation in (4b) by the sum of -3 times the first equation + the second equation.

$-3x - 6y - 9z = -33$
$\underline{3x + 8y + 5z = 27}$
$2y - 4z = -6$

(4c)

$$x + 2y + 3z = 11$$
$$2y - 4z = -6$$
$$3y + 5z = 13$$

Replace the third equation in (4c) by the sum of the first equation + the third equation.

$x + 2y + 3z = 11$
$\underline{-x + y + 2z = 2}$
$3y + 5z = 13$

(4d)

Then we transform the system (4d) into yet another equivalent system, in which the coefficient of y in the second equation is 1:

$$x + 2y + 3z = 11$$
$$y - 2z = -3$$
$$3y + 5z = 13$$

Multiply the second equation in (4d) by $\frac{1}{2}$.

(4e)

We now eliminate y from all equations except the second using operation 3 of the elimination method.

$$x + 7z = 17$$
$$y - 2z = -3$$
$$3y + 5z = 13$$

Replace the first equation in (4e) by the sum of the first equation + (-2) times the second equation.

$x + 2y + 3z = 11$
$\underline{- 2y + 4z = 6}$
$x + 7z = 17$

(4f)

$$x + 7z = 17$$
$$y - 2z = -3$$
$$11z = 22$$

Replace the third equation in (4f) by the sum of (-3) times the second equation + the third equation.

$-3y + 6z = 9$
$\underline{3y + 5z = 13}$
$11z = 22$

(4g)

Finally, multiplying the third equation by $1/11$ in (4g) leads to the system

$$x + 7z = 17$$
$$y - 2z = -3$$
$$z = 2$$

Eliminating z from all equations except the third (try it!) then leads to the system

$$
\begin{aligned}
x & = 3 \\
y & = 1 \\
z & = 2
\end{aligned}
$$

(4h)

In its final form, the solution to the given system of equations can be easily read off! We have $x = 3$, $y = 1$, and $z = 2$. Geometrically, the point $(3, 1, 2)$ lies in the intersection of the three planes described by the three equations comprising the given system. ■

Augmented Matrices

Observe from the preceding example that in each step of the reduction process, the variables x, y, and z play no significant role except as a reminder of the position of each coefficient in the system. With the aid of **matrices,** which are rectangular arrays of numbers, we can eliminate writing the variables at each step of the reduction and thus save ourselves a great deal of work. For example, the system

$$
\begin{aligned}
2x + 4y + 6z &= 22 \\
3x + 8y + 5z &= 27 \\
-x + y + 2z &= 2
\end{aligned}
$$

(5)

may be represented by the matrix

$$
\begin{bmatrix}
2 & 4 & 6 & 22 \\
3 & 8 & 5 & 27 \\
-1 & 1 & 2 & 2
\end{bmatrix}
$$

(6)

The augmented matrix representing system (5)

The submatrix, consisting of the first three columns of the matrix (6), is called the **coefficient matrix** of system (5). The matrix itself, (6), is referred to as the **augmented matrix** of system (5), since it is obtained by joining the matrix of coefficients to the column (matrix) of constants. The vertical line separates the column of constants from the matrix of coefficients.

The next example shows how much work you can save by using matrices instead of the standard representation of the systems of linear equations.

EXAMPLE 2 Write the augmented matrix corresponding to each equivalent system given in (4a) through (4h).

Solution The required sequence of augmented matrices follows.

Equivalent System	Augmented Matrix

a.
$$
\begin{aligned}
2x + 4y + 6z &= 22 \\
3x + 8y + 5z &= 27 \\
-x + y + 2z &= 2
\end{aligned}
\qquad
\begin{bmatrix}
2 & 4 & 6 & 22 \\
3 & 8 & 5 & 27 \\
-1 & 1 & 2 & 2
\end{bmatrix}
$$

(7a)

b. $\begin{aligned} x + 2y + 3z &= 11 \\ 3x + 8y + 5z &= 27 \\ -x + y + 2z &= 2 \end{aligned}$ $\left[\begin{array}{ccc|c} 1 & 2 & 3 & 11 \\ 3 & 8 & 5 & 27 \\ -1 & 1 & 2 & 2 \end{array}\right]$ **(7b)**

c. $\begin{aligned} x + 2y + 3z &= 11 \\ 2y - 4z &= -6 \\ -x + y + 2z &= 2 \end{aligned}$ $\left[\begin{array}{ccc|c} 1 & 2 & 3 & 11 \\ 0 & 2 & -4 & -6 \\ -1 & 1 & 2 & 2 \end{array}\right]$ **(7c)**

d. $\begin{aligned} x + 2y + 3z &= 11 \\ 2y - 4z &= -6 \\ 3y + 5z &= 13 \end{aligned}$ $\left[\begin{array}{ccc|c} 1 & 2 & 3 & 11 \\ 0 & 2 & -4 & -6 \\ 0 & 3 & 5 & 13 \end{array}\right]$ **(7d)**

e. $\begin{aligned} x + 2y + 3z &= 11 \\ y - 2z &= -3 \\ 3y + 5z &= 13 \end{aligned}$ $\left[\begin{array}{ccc|c} 1 & 2 & 3 & 11 \\ 0 & 1 & -2 & -3 \\ 0 & 3 & 5 & 13 \end{array}\right]$ **(7e)**

f. $\begin{aligned} x + 7z &= 17 \\ y - 2z &= -3 \\ 3y + 5z &= 13 \end{aligned}$ $\left[\begin{array}{ccc|c} 1 & 0 & 7 & 17 \\ 0 & 1 & -2 & -3 \\ 0 & 3 & 5 & 13 \end{array}\right]$ **(7f)**

g. $\begin{aligned} x + 7z &= 17 \\ y - 2z &= -3 \\ 11z &= 22 \end{aligned}$ $\left[\begin{array}{ccc|c} 1 & 0 & 7 & 17 \\ 0 & 1 & -2 & -3 \\ 0 & 0 & 11 & 22 \end{array}\right]$ **(7g)**

h. $\begin{aligned} x &= 3 \\ y &= 1 \\ z &= 2 \end{aligned}$ $\left[\begin{array}{ccc|c} 1 & 0 & 0 & 3 \\ 0 & 1 & 0 & 1 \\ 0 & 0 & 1 & 2 \end{array}\right]$ **(7h)**   ■

The coefficient matrix in (7h) is an example of a matrix in **row-reduced form.** In general, a coefficient matrix with m rows and n columns (called an $m \times n$ matrix) is in row-reduced form if it satisfies the following conditions.

ROW-REDUCED FORM

1. Each row consisting entirely of zeros lies below any other row having nonzero entries.
2. The first nonzero entry in each row is 1 (called a *leading 1*).
3. In any two successive (nonzero) rows, the leading 1 in the lower row lies to the right of the leading 1 in the upper row.
4. If a column contains a leading 1, then the other entries in that column are zeros.

EXAMPLE 3   Determine which of the following matrices are in row-reduced form. If a matrix is not in row-reduced form, state which condition is violated.

a. $\begin{bmatrix} 1 & 0 & 0 & | & 0 \\ 0 & 1 & 0 & | & 0 \\ 0 & 0 & 1 & | & 3 \end{bmatrix}$ **b.** $\begin{bmatrix} 1 & 0 & 0 & | & 4 \\ 0 & 1 & 0 & | & 3 \\ 0 & 0 & 0 & | & 0 \end{bmatrix}$ **c.** $\begin{bmatrix} 1 & 2 & 0 & | & 0 \\ 0 & 0 & 1 & | & 0 \\ 0 & 0 & 0 & | & 1 \end{bmatrix}$

d. $\begin{bmatrix} 0 & 1 & 2 & | & -2 \\ 1 & 0 & 0 & | & 3 \\ 0 & 0 & 1 & | & 2 \end{bmatrix}$ **e.** $\begin{bmatrix} 1 & 2 & 0 & | & 0 \\ 0 & 0 & 1 & | & 3 \\ 0 & 0 & 2 & | & 1 \end{bmatrix}$ **f.** $\begin{bmatrix} 1 & 0 & | & 4 \\ 0 & 3 & | & 0 \\ 0 & 0 & | & 0 \end{bmatrix}$

g. $\begin{bmatrix} 0 & 0 & 0 & | & 0 \\ 1 & 0 & 0 & | & 3 \\ 0 & 1 & 0 & | & 2 \end{bmatrix}$

Solution   The matrices in (a), (b), and (c) are in row-reduced form.

d. This matrix is not in row-reduced form. Conditions 3 and 4 are violated: The leading 1 in row 2 lies to the left of the leading 1 in row 1. Also, column 3 contains a leading 1 in row 3 and a nonzero element above it.

e. This matrix is not in row-reduced form. Conditions 2 and 4 are violated: The first nonzero entry in row 3 is a 2, not a 1. Also, column 3 contains a leading 1 and has a nonzero entry below it.

f. This matrix is not in row-reduced form. Condition 2 is violated: The first nonzero entry in row 2 is not a leading 1.

g. This matrix is not in row-reduced form. Condition 1 is violated: Row 1 consists of all zeros and does not lie below the other nonzero rows. ▪

The foregoing discussion suggests the following adaptation of the Gauss-Jordan elimination method in solving systems of linear equations using matrices. First, the three operations on the equations of a system (see page 193) translate into the following **row operations** on the corresponding augmented matrices.

ROW OPERATIONS

1. Interchange any two rows.

2. Replace any row by a nonzero constant multiple of itself.

3. Replace any row by the sum of that row and a constant multiple of any other row.

We obtained the augmented matrices in Example 2 by using the same operations that we used on the equivalent system of equations in Example 1.

In order to help us describe the Gauss-Jordan elimination method using matrices, let us introduce some terminology. We begin by defining what is meant by a **unit column.**

UNIT COLUMN

A column in the coefficient matrix is in unit form if one of the entries in the column is a 1 and the other entries are zeros.

For example, in the coefficient matrix of (7d), only the first column is in unit form; in the coefficient matrix of (7h), all three columns are in unit form. Now, the sequence of row operations that transforms the augmented matrix (7a) into the equivalent matrix (7d) in which the first column

$$2$$
$$3$$
$$-1$$

of (7a) is transformed into the unit column

$$1$$
$$0$$
$$0$$

is called **pivoting** the matrix about the element (number) 2. Similarly, we have pivoted about the element 2 in the second column of (7d), shown circled,

$$2$$
$$②$$
$$3$$

in order to obtain the augmented matrix (7g). Finally, pivoting about the element 11 in column 3 of (7g)

$$7$$
$$-2$$
$$⑪$$

leads to the augmented matrix (7h), in which all columns are in unit form. The element about which a matrix is pivoted is called the *pivot element*.

Before looking at the next example, let us introduce the following notation for the three types of row operations.

NOTATION FOR ROW OPERATIONS

Letting R_i denote the ith row of a matrix, we write:

Operation 1: $R_i \leftrightarrow R_j$ to mean: Interchange row i with row j
Operation 2: cR_i to mean: Replace row i with c times row i
Operation 3: $R_i + aR_j$ to mean: Replace row i with the sum of row i and a times row j

EXAMPLE 4   Pivot the matrix

$$\begin{bmatrix} ③ & 5 & | & 9 \\ 2 & 3 & | & 5 \end{bmatrix}$$

about the circled element.

Solution   Using the notation just introduced, we obtain

$$\begin{bmatrix} 3 & 5 & | & 9 \\ 2 & 3 & | & 5 \end{bmatrix} \xrightarrow{\frac{1}{3}R_1} \begin{bmatrix} 1 & \frac{5}{3} & | & 3 \\ 2 & 3 & | & 5 \end{bmatrix} \xrightarrow{R_2 - 2R_1} \begin{bmatrix} 1 & \frac{5}{3} & | & 3 \\ 0 & -\frac{1}{3} & | & -1 \end{bmatrix}$$

The first column, which originally contained the entry 3, is now in unit form, with a 1 where the pivot element used to be, and we are done.

Alternate Solution   In the first solution, we used operation 2 to obtain a 1 where the pivot element was originally. Alternatively, we can use operation 3 as follows:

$$\begin{bmatrix} 3 & 5 & | & 9 \\ 2 & 3 & | & 5 \end{bmatrix} \xrightarrow{R_1 - R_2} \begin{bmatrix} 1 & 2 & | & 4 \\ 2 & 3 & | & 5 \end{bmatrix} \xrightarrow{R_2 - 2R_1} \begin{bmatrix} 1 & 2 & | & 4 \\ 0 & -1 & | & -3 \end{bmatrix}$$

REMARK   In Example 4, the two matrices

$$\begin{bmatrix} 1 & \frac{5}{3} & | & 3 \\ 0 & -\frac{1}{3} & | & -1 \end{bmatrix} \quad \text{and} \quad \begin{bmatrix} 1 & 2 & | & 4 \\ 0 & -1 & | & -3 \end{bmatrix}$$

look quite different, but they are in fact equivalent. You can verify this by observing that they represent the systems of equations

$$x + \frac{5}{3}y = 3 \qquad\qquad x + 2y = 4$$
$$\text{and}$$
$$-\frac{1}{3}y = -1 \qquad\qquad -y = -3$$

respectively, and both have the same solution: $x = -2$ and $y = 3$. Example 4 also shows that we can sometimes avoid working with fractions by using the appropriate row operation.

A summary of the Gauss-Jordan method follows:

THE GAUSS-JORDAN ELIMINATION METHOD

1. Write the augmented matrix corresponding to the linear system.
2. Interchange rows (operation 1), if necessary, to obtain an augmented matrix in which the first entry in the first row is nonzero. Then pivot the matrix about this entry.
3. Interchange the second row with any row below it, if necessary, to obtain an augmented matrix in which the second entry in the second row is nonzero. Pivot the matrix about this entry.
4. Continue until the final matrix is in row-reduced form.

Before writing the augmented matrix, be sure to write all equations with the variables on the left and constant terms on the right of the equals sign. Also, make sure that the variables are in the same order in all equations.

EXAMPLE 5   Solve the system of linear equations given by

$$3x - 2y + 8z = 9$$
$$-2x + 2y + z = 3$$
$$x + 2y - 3z = 8 \qquad\qquad (8)$$

Solution   Using the Gauss-Jordan elimination method, we obtain the following sequence of equivalent augmented matrices:

$$\begin{bmatrix} ③ & -2 & 8 & | & 9 \\ -2 & 2 & 1 & | & 3 \\ 1 & 2 & -3 & | & 8 \end{bmatrix} \xrightarrow{R_1 + R_2} \begin{bmatrix} 1 & 0 & 9 & | & 12 \\ -2 & 2 & 1 & | & 3 \\ 1 & 2 & -3 & | & 8 \end{bmatrix}$$

$$\xrightarrow[R_3 - R_1]{R_2 + 2R_1} \begin{bmatrix} 1 & 0 & 9 & | & 12 \\ 0 & 2 & 19 & | & 27 \\ 0 & 2 & -12 & | & -4 \end{bmatrix}$$

$$\xrightarrow{R_2 \leftrightarrow R_3} \begin{bmatrix} 1 & 0 & 9 & | & 12 \\ 0 & ② & -12 & | & -4 \\ 0 & 2 & 19 & | & 27 \end{bmatrix}$$

$$\xrightarrow{\frac{1}{2}R_2} \begin{bmatrix} 1 & 0 & 9 & | & 12 \\ 0 & 1 & -6 & | & -2 \\ 0 & 2 & 19 & | & 27 \end{bmatrix}$$

$$\xrightarrow{R_3 - 2R_2} \begin{bmatrix} 1 & 0 & 9 & | & 12 \\ 0 & 1 & -6 & | & -2 \\ 0 & 0 & ㉛ & | & 31 \end{bmatrix}$$

$$\xrightarrow{\frac{1}{31}R_3} \begin{bmatrix} 1 & 0 & 9 & | & 12 \\ 0 & 1 & -6 & | & -2 \\ 0 & 0 & 1 & | & 1 \end{bmatrix}$$

$$\xrightarrow[R_2 + 6R_3]{R_1 - 9R_3} \begin{bmatrix} 1 & 0 & 0 & | & 3 \\ 0 & 1 & 0 & | & 4 \\ 0 & 0 & 1 & | & 1 \end{bmatrix}$$

The solution to system (8) is given by $x = 3$, $y = 4$, and $z = 1$ and may be verified by substitution into system (8) as follows:

$$3(3) - 2(4) + 8(1) = 9 \qquad \checkmark$$
$$-2(3) + 2(4) + 1 = 3 \qquad \checkmark$$
$$3 + 2(4) - 3(1) = 8 \qquad \checkmark$$

When searching for an element to serve as a pivot, it is important to keep in mind that you may work only with the row containing the potential pivot or any row *below* it. In order to see what can go wrong if this caution is not heeded, consider the following augmented matrix for some linear system:

$$\left[\begin{array}{ccc|c} 1 & 1 & 2 & 3 \\ 0 & 0 & 3 & 1 \\ 0 & 2 & 1 & -2 \end{array}\right]$$

Observe that column 1 is in unit form. The next step in the Gauss-Jordan elimination procedure calls for obtaining a nonzero element in the second position of the second row. If you use the first row (which is *above* the row under consideration) to help you obtain the pivot, you might proceed as follows:

$$\left[\begin{array}{ccc|c} 1 & 1 & 2 & 3 \\ 0 & 0 & 3 & 1 \\ 0 & 2 & 1 & -2 \end{array}\right] \xrightarrow{R_2 \leftrightarrow R_1} \left[\begin{array}{ccc|c} 0 & 0 & 3 & 1 \\ 1 & 1 & 2 & 3 \\ 0 & 2 & 1 & -2 \end{array}\right]$$

As you can see, not only have we obtained a nonzero element to serve as the next pivot, but it is already a 1, thus obviating the next step. This seems like a good move. But beware, we have undone some of our earlier work: The first column is no longer in the unit form where a 1 appears first. The correct move in this case is to interchange the second row with the third row.

The next example illustrates how to handle a situation in which the entry in the first row of the augmented matrix is zero.

EXAMPLE 6   Solve the system of linear equations given by

$$2y + 3z = 7$$
$$3x + 6y - 12z = -3$$
$$5x - 2y + 2z = -7$$

Solution   Using the Gauss-Jordan elimination method, we obtain the following sequence of equivalent augmented matrices:

$$\left[\begin{array}{ccc|c} 0 & 2 & 3 & 7 \\ 3 & 6 & -12 & -3 \\ 5 & -2 & 2 & -7 \end{array}\right] \xrightarrow{R_1 \leftrightarrow R_2} \left[\begin{array}{ccc|c} ③ & 6 & -12 & -3 \\ 0 & 2 & 3 & 7 \\ 5 & -2 & 2 & -7 \end{array}\right] \xrightarrow{\frac{1}{3}R_1} \left[\begin{array}{ccc|c} 1 & 2 & -4 & -1 \\ 0 & 2 & 3 & 7 \\ 5 & -2 & 2 & -7 \end{array}\right] \xrightarrow{R_3 - 5R_1}$$

$$\left[\begin{array}{ccc|c} 1 & 2 & -4 & -1 \\ 0 & ② & 3 & 7 \\ 0 & -12 & 22 & -2 \end{array}\right] \xrightarrow{\frac{1}{2}R_2} \left[\begin{array}{ccc|c} 1 & 2 & -4 & -1 \\ 0 & 1 & \frac{3}{2} & \frac{7}{2} \\ 0 & -12 & 22 & -2 \end{array}\right] \xrightarrow[R_3 + 12R_2]{R_1 - 2R_2} \left[\begin{array}{ccc|c} 1 & 0 & -7 & -8 \\ 0 & 1 & \frac{3}{2} & \frac{7}{2} \\ 0 & 0 & ㊵ & 40 \end{array}\right] \xrightarrow{\frac{1}{40}R_3}$$

$$\left[\begin{array}{ccc|c} 1 & 0 & -7 & -8 \\ 0 & 1 & \frac{3}{2} & \frac{7}{2} \\ 0 & 0 & 1 & 1 \end{array}\right] \xrightarrow[R_2 - \frac{3}{2}R_3]{R_1 + 7R_3} \left[\begin{array}{ccc|c} 1 & 0 & 0 & -1 \\ 0 & 1 & 0 & 2 \\ 0 & 0 & 1 & 1 \end{array}\right]$$

The solution to the system is given by $x = -1$, $y = 2$, and $z = 1$, and may be verified by substitution into the system.

Application

EXAMPLE 7   Complete the solution to Example 1, page 187.

Solution   To complete the solution of the problem posed in Example 1, recall that the mathematical formulation of the problem led to the following system of linear equations:

$$2x + y + z = 180$$
$$x + 3y + 2z = 300$$
$$2x + y + 2z = 240$$

where x, y, and z denote the respective numbers of type-A, type-B, and type-C souvenirs to be made.

Solving the foregoing system of linear equations by the Gauss-Jordan elimination method, we obtain the following sequence of equivalent augmented matrices:

$$
\begin{bmatrix}
\textcircled{2} & 1 & 1 & \bigm| & 180 \\
1 & 3 & 2 & \bigm| & 300 \\
2 & 1 & 2 & \bigm| & 240
\end{bmatrix}
\xrightarrow{R_1 \leftrightarrow R_2}
\begin{bmatrix}
1 & 3 & 2 & \bigm| & 300 \\
2 & 1 & 1 & \bigm| & 180 \\
2 & 1 & 2 & \bigm| & 240
\end{bmatrix}
$$

$$
\xrightarrow[R_3 - 2R_1]{R_2 - 2R_1}
\begin{bmatrix}
1 & 3 & 2 & \bigm| & 300 \\
0 & \textcircled{-5} & -3 & \bigm| & -420 \\
0 & -5 & -2 & \bigm| & -360
\end{bmatrix}
$$

$$
\xrightarrow{-\frac{1}{5}R_2}
\begin{bmatrix}
1 & 3 & 2 & \bigm| & 300 \\
0 & 1 & \frac{3}{5} & \bigm| & 84 \\
0 & -5 & -2 & \bigm| & -360
\end{bmatrix}
$$

$$
\xrightarrow[R_3 + 5R_2]{R_1 - 3R_2}
\begin{bmatrix}
1 & 0 & \frac{1}{5} & \bigm| & 48 \\
0 & 1 & \frac{3}{5} & \bigm| & 84 \\
0 & 0 & \textcircled{1} & \bigm| & 60
\end{bmatrix}
$$

$$
\xrightarrow[R_2 - \frac{3}{5}R_3]{R_1 - \frac{1}{5}R_3}
\begin{bmatrix}
1 & 0 & 0 & \bigm| & 36 \\
0 & 1 & 0 & \bigm| & 48 \\
0 & 0 & 1 & \bigm| & 60
\end{bmatrix}
$$

Thus, $x = 36$, $y = 48$, and $z = 60$; that is, Ace Novelty should make 36 type-A souvenirs, 48 type-B souvenirs, and 60 type-C souvenirs in order to use all available machine time.   ■

**SELF-CHECK
EXERCISES 5.2**

1. Solve the system of linear equations

$$2x + 3y + z = 6$$
$$x - 2y + 3z = -3$$
$$3x + 2y - 4z = 12$$

using the Gauss-Jordan elimination method.

2. A farmer has 200 acres of land suitable for cultivating crops A, B, and C. The cost per acre of cultivating crop A, crop B, and crop C is $40, $60, and $80, respectively. The farmer has $12,600 available for land cultivation. Each acre of crop A requires 20 hours of labor, each acre of crop B requires 25 hours of labor, and each acre of crop C requires 40 hours of labor. The farmer has a maximum of 5,950 hours of labor available. If he wishes to use all of his cultivatable land, the entire budget, and all the labor available, how many acres of each crop should he plant?

Solutions to Self-Check Exercises 5.2 can be found on page 207.

5.2   EXERCISES

In exercises 1–4, write the augmented matrix corresponding to the system of equations.

1. $2x - 3y = 7$
$3x + y = 4$

2. $3x + 7y - 8z = 5$
$x + 3z = -2$
$4x - 3y = 7$

3. $-y + 2z = 6$
$2x + 2y - 8z = 7$
$3y + 4z = 0$

4. $3x_1 + 2x_2 = 0$
$x_1 - x_2 + 2x_3 = 4$
$2x_2 - 3x_3 = 5$

In exercises 5–8, write the system of equations corresponding to the augmented matrix.

5. $\begin{bmatrix} 3 & 2 & | & -4 \\ 1 & -1 & | & 5 \end{bmatrix}$

6. $\begin{bmatrix} 0 & 3 & 2 & | & 4 \\ 1 & -1 & -2 & | & -3 \\ 4 & 0 & 3 & | & 2 \end{bmatrix}$

7. $\begin{bmatrix} 1 & 3 & 2 & | & 4 \\ 2 & 0 & 0 & | & 5 \\ 3 & -3 & 2 & | & 6 \end{bmatrix}$

8. $\begin{bmatrix} 2 & 3 & 1 & | & 6 \\ 4 & 3 & 2 & | & 5 \\ 0 & 0 & 0 & | & 0 \end{bmatrix}$

In exercises 9–18, indicate whether the matrix is in row-reduced form.

9. $\begin{bmatrix} 1 & 0 & | & 3 \\ 0 & 1 & | & -2 \end{bmatrix}$

10. $\begin{bmatrix} 1 & 1 & | & 3 \\ 0 & 0 & | & 0 \end{bmatrix}$

11. $\begin{bmatrix} 0 & 1 & | & 3 \\ 1 & 0 & | & 5 \end{bmatrix}$

12. $\begin{bmatrix} 0 & 1 & | & 3 \\ 0 & 0 & | & 5 \end{bmatrix}$

13. $\begin{bmatrix} 1 & 0 & 0 & | & 3 \\ 0 & 1 & 0 & | & 4 \\ 0 & 0 & 1 & | & 5 \end{bmatrix}$

14. $\begin{bmatrix} 1 & 0 & 0 & | & -1 \\ 0 & 1 & 0 & | & -2 \\ 0 & 0 & 2 & | & -3 \end{bmatrix}$

15. $\begin{bmatrix} 1 & 0 & 1 & | & 3 \\ 0 & 1 & 0 & | & 4 \\ 0 & 0 & -1 & | & 6 \end{bmatrix}$

16. $\begin{bmatrix} 1 & 0 & | & -10 \\ 0 & 1 & | & 2 \\ 0 & 0 & | & 0 \end{bmatrix}$

17. $\begin{bmatrix} 0 & 0 & 0 & | & 0 \\ 0 & 1 & 2 & | & 4 \\ 0 & 0 & 0 & | & 0 \end{bmatrix}$

18. $\begin{bmatrix} 1 & 0 & 0 & | & 3 \\ 0 & 1 & 0 & | & 6 \\ 0 & 0 & 0 & | & 4 \\ 0 & 0 & 1 & | & 5 \end{bmatrix}$

In exercises 19–26, pivot the system about the circled element.

19. $\begin{bmatrix} ②　4 & | & 8 \\ 3 & 1 & | & 2 \end{bmatrix}$

20. $\begin{bmatrix} 3 & 2 & | & 6 \\ ④　2 & | & 5 \end{bmatrix}$

21. $\begin{bmatrix} ⑴ & 2 & | & 3 \\ 6 & 4 & | & 2 \end{bmatrix}$

22. $\begin{bmatrix} ① & 3 & | & 4 \\ 2 & 4 & | & 6 \end{bmatrix}$

23. $\begin{bmatrix} ② & 4 & 6 & | & 12 \\ 2 & 3 & 1 & | & 5 \\ 3 & -1 & 2 & | & 4 \end{bmatrix}$

24. $\begin{bmatrix} 1 & 3 & 2 & | & 4 \\ ② & 4 & 8 & | & 6 \\ -1 & 2 & 3 & | & 4 \end{bmatrix}$

25. $\begin{bmatrix} 0 & 1 & 3 & | & 4 \\ 2 & 4 & ① & | & 3 \\ 5 & 6 & 2 & | & -4 \end{bmatrix}$

26. $\begin{bmatrix} 1 & 2 & 3 & | & 5 \\ 0 & ⑶ & 3 & | & 2 \\ 0 & 4 & -1 & | & 3 \end{bmatrix}$

In exercises 27–30, fill in the missing entries by performing the indicated row operations to obtain the row-reduced matrices.

27. $\begin{bmatrix} 3 & 9 & | & 6 \\ 2 & 1 & | & 4 \end{bmatrix} \xrightarrow{\frac{1}{3}R_1} \begin{bmatrix} \cdot & \cdot & | & \cdot \\ 2 & 1 & | & 4 \end{bmatrix} \xrightarrow{R_2 - 2R_1}$

$\begin{bmatrix} 1 & 3 & | & 2 \\ \cdot & \cdot & | & \cdot \end{bmatrix} \xrightarrow{-\frac{1}{5}R_2} \begin{bmatrix} 1 & 3 & | & 2 \\ \cdot & \cdot & | & \cdot \end{bmatrix} \xrightarrow{R_1 - 3R_2}$

$\begin{bmatrix} 1 & 0 & | & 2 \\ 0 & 1 & | & 0 \end{bmatrix}$

28. $\begin{bmatrix} 1 & 2 & | & 1 \\ 2 & 3 & | & -1 \end{bmatrix} \xrightarrow{R_2 - 2R_1} \begin{bmatrix} 1 & 2 & | & 1 \\ \cdot & \cdot & | & \cdot \end{bmatrix} \xrightarrow{-R_2}$

$\begin{bmatrix} 1 & 2 & | & 1 \\ \cdot & \cdot & | & \cdot \end{bmatrix} \xrightarrow{R_1 - 2R_2} \begin{bmatrix} 1 & 0 & | & -5 \\ 0 & 1 & | & 3 \end{bmatrix}$

29. $\begin{bmatrix} 1 & 3 & 1 & | & 3 \\ 3 & 8 & 3 & | & 7 \\ 2 & -3 & 1 & | & -10 \end{bmatrix} \xrightarrow[R_3 - 2R_1]{R_2 - 3R_1}$

$\begin{bmatrix} 1 & 3 & 1 & | & 3 \\ \cdot & \cdot & \cdot & | & \cdot \\ \cdot & \cdot & \cdot & | & \cdot \end{bmatrix} \xrightarrow{-R_2}$

$\begin{bmatrix} 1 & 3 & 1 & | & 3 \\ \cdot & \cdot & \cdot & | & \cdot \\ 0 & -9 & -1 & | & -16 \end{bmatrix} \xrightarrow[R_3 + 9R_2]{R_1 - 3R_2}$

$\begin{bmatrix} \cdot & \cdot & \cdot & | & \cdot \\ 0 & 1 & 0 & | & 2 \\ \cdot & \cdot & \cdot & | & \cdot \end{bmatrix} \xrightarrow[-R_3]{R_1 + R_3} \begin{bmatrix} 1 & 0 & 0 & | & -1 \\ 0 & 1 & 0 & | & 2 \\ 0 & 0 & 1 & | & -2 \end{bmatrix}$

30. $\begin{bmatrix} 0 & 1 & 3 & | & -4 \\ 1 & 2 & 1 & | & 7 \\ 1 & -2 & 0 & | & 1 \end{bmatrix} \xrightarrow{R_1 \leftrightarrow R_2}$

$\begin{bmatrix} \cdot & \cdot & \cdot & | & \cdot \\ \cdot & \cdot & \cdot & | & \cdot \\ 1 & -2 & 0 & | & 1 \end{bmatrix} \xrightarrow{R_3 - R_1} \begin{bmatrix} 1 & 2 & 1 & | & 7 \\ 0 & 1 & 3 & | & -4 \\ \cdot & \cdot & \cdot & | & \cdot \end{bmatrix}$

$\xrightarrow[R_3 + 4R_2]{R_1 + \frac{1}{2}R_2} \begin{bmatrix} \cdot & \cdot & \cdot & | & \cdot \\ 0 & 1 & 3 & | & -4 \\ \cdot & \cdot & \cdot & | & \cdot \end{bmatrix} \xrightarrow{\frac{1}{11}R_3}$

$\begin{bmatrix} 1 & 0 & \frac{1}{2} & | & 4 \\ 0 & 1 & 3 & | & -4 \\ \cdot & \cdot & \cdot & | & \cdot \end{bmatrix} \xrightarrow[R_2 - 3R_3]{R_1 - \frac{1}{2}R_3} \begin{bmatrix} 1 & 0 & 0 & | & 5 \\ 0 & 1 & 0 & | & 2 \\ 0 & 0 & 1 & | & -2 \end{bmatrix}$

31. Write a system of linear equations for the augmented matrix of exercise 27. Using the results of exercise 27, determine the solution of the system.

32. Repeat exercise 31 for the augmented matrix of exercise 28.

33. Repeat exercise 31 for the augmented matrix of exercise 29.

34. Repeat exercise 31 for the augmented matrix of exercise 30.

In exercises 35–50, solve the system of linear equations using the Gauss-Jordan elimination method.

35. $\begin{aligned} x - 2y &= 8 \\ 3x + 4y &= 4 \end{aligned}$ **36.** $\begin{aligned} 3x + y &= 1 \\ -7x - 2y &= -1 \end{aligned}$

37. $\begin{aligned} 2x - 3y &= -8 \\ 4x + y &= -2 \end{aligned}$ **38.** $\begin{aligned} 5x + 3y &= 9 \\ -2x + y &= -8 \end{aligned}$

39. $\begin{aligned} x + y + z &= 0 \\ 2x - y + z &= 1 \\ x + y - 2z &= 2 \end{aligned}$ **40.** $\begin{aligned} 2x + y - 2z &= 4 \\ x + 3y - z &= -3 \\ 3x + 4y - z &= 7 \end{aligned}$

41. $\begin{aligned} 2x + 2y + z &= 9 \\ x \quad\ + z &= 4 \\ 4y - 3z &= 17 \end{aligned}$ **42.** $\begin{aligned} 2x + 3y - 2z &= 10 \\ 3x - 2y + 2z &= 0 \\ 4x - y + 3z &= -1 \end{aligned}$

43. $\begin{aligned} - x_2 + x_3 &= 2 \\ 4x_1 - 3x_2 + 2x_3 &= 16 \\ 3x_1 + 2x_2 + x_3 &= 11 \end{aligned}$

44. $\begin{aligned} 2x + 4y - 6z &= 38 \\ x + 2y + 3z &= 7 \\ 3x - 4y + 4z &= -19 \end{aligned}$

45. $\begin{aligned} x_1 - 2x_2 + x_3 &= 6 \\ 2x_1 + x_2 - 3x_3 &= -3 \\ x_1 - 3x_2 + 3x_3 &= 10 \end{aligned}$

46. $\begin{aligned} 2x + 3y - 6z &= -11 \\ x - 2y + 3z &= 9 \\ 3x + y &= 7 \end{aligned}$ **47.** $\begin{aligned} 2x \quad\ + 3z &= -1 \\ 3x - 2y + z &= 9 \\ x + y + 4z &= 4 \end{aligned}$

48. $\begin{aligned} 2x_1 - x_2 + 3x_3 &= -4 \\ x_1 - 2x_2 + x_3 &= -1 \\ x_1 - 5x_2 + 2x_3 &= -3 \end{aligned}$

49. $\begin{aligned} x_1 - x_2 + 3x_3 &= 14 \\ x_1 + x_2 + x_3 &= 6 \\ -2x_1 - x_2 + x_3 &= -4 \end{aligned}$

50. $2x_1 - x_2 - x_3 = 0$
$3x_1 + 2x_2 + x_3 = 7$
$x_1 + 2x_2 + 2x_3 = 5$

The problems in exercises 51–59 correspond to those in exercises 15–23, Section 5.1. Use the results of your previous work to help you solve these problems.

51. Agriculture The Johnson Farm has 500 acres of land allotted for cultivating corn and wheat. The cost of cultivating corn and wheat (including seeds and labor) is $42 and $30 per acre, respectively. Mr. Johnson has $18,600 available for cultivating these crops. If he wishes to use all the allotted land and his entire budget for cultivating these two crops, how many acres of each crop should he plant?

52. Investments Michael has a total of $2000 on deposit with two savings institutions. One pays simple interest at the rate of 6% per year, whereas the other pays simple interest at the rate of 8% per year. If Michael earned a total of $144 in interest during a single year, how much does he have on deposit in each institution?

53. Mixtures The Coffee Shoppe sells a coffee blend made from two coffees, one costing $2.50/lb and the other costing $3/lb. If the blended coffee sells for $2.80/lb, find how much of each coffee is used to obtain the desired blend. [*Hint:* Assume the weight of the blended coffee is 100 lb.]

54. Investments Kelly has a total of $30,000 invested in two municipal bonds that have yields of 8% and 10% simple interest per year, respectively. If the interest Kelly receives from the bonds in a year is $2,640, how much does she have invested in each bond?

55. Ridership The total number of passengers riding a certain city bus during the morning shift is 1000. If the child's fare is 25 cents, the adult fare is 75 cents, and the total revenue from the fares in the morning shift is $650, how many children and how many adults rode the bus during the morning shift?

56. Real Estate Cantwell Associates, a real estate developer, is planning to build a new apartment complex consisting of one-bedroom units and two- and three-bedroom townhouses. A total of 192 units is planned, and the number of family units (two- and three-bedroom townhouses) will equal the number of one-bedroom units. If the number of one-bedroom units will be three times the number of three-bedroom units, find how many units of each type will be in the complex.

57. Investment Planning The annual interest on Mr. Carrington's three investments amounted to $21,600: 6% on a savings account, 8% on mutual funds, and 12% on money market certificates. If the amount of Carrington's investment in money market certificates was twice the amount of his investment in the savings account, and the interest earned from his investment in money market certificates was equal to the dividends he received from his investment in mutual funds, find how much money he placed in each type of investment.

58. Box-Office Receipts A theater has a seating capacity of 900 and charges $2 for children, $3 for students, and $4 for adults. At a certain screening with full attendance there were half as many adults as children and students combined. The receipts totaled $2800. How many children attended the show?

59. Management Decisions The management of Hartman Rent-A-Car has allocated $1 million to buy a fleet of new automobiles consisting of compact, intermediate, and full-size cars. Compacts cost $8,000 each, intermediate-size cars cost $12,000 each, and full-size cars cost $16,000 each. If Hartman purchases twice as many compacts as intermediate-size cars, and the total number of cars to be purchased is 100, determine how many cars of each type will be purchased. (Assume that the entire budget will be used.)

60. Investments Mr. and Mrs. Garcia have a total of $100,000 to be invested in stocks, bonds, and a money market account. The stocks have a rate of return of 12% per year, while the bonds and the money market account pay 8% and 4% per year, respectively. They have stipulated that the amount invested in the money market account should be equal to the sum of 20% of the amount invested in stocks and 10% of the amount invested in bonds. How should the Garcias allocate their resources if they require an annual income of $10,000 from their investments?

61. Investment Clubs The management of a private investment club has a fund of $200,000 earmarked for investment in stocks. To arrive at an acceptable overall level of risk, the stocks that management is considering have been classified into three categories: high-risk, medium-risk, and low-risk. Management estimates that high-risk stocks will have a rate of return of 15% per year; medium-risk stocks, 10% per year; and low-risk stocks, 6% per year. The investment in low-risk stocks is to be twice the sum of the investments in stocks of the other two categories. If the investment goal is to have an average rate of return of 9% per year on the total investment, determine how much the club should invest in each type of stock.

62. Diet Planning   A dietician wishes to plan a meal around three foods. The percentage of the daily requirements of proteins, carbohydrates, and iron contained in each ounce of the three foods is summarized in the accompanying table.

	Food I	Food II	Food III
Percentage of Proteins	10	6	8
Percentage of Carbohydrates	10	12	6
Percentage of Iron	5	4	12

Determine how many ounces of each food the dietician should include in the meal to meet exactly the daily requirement of proteins, carbohydrates, and iron (100% of each).

63. Box-Office Receipts   For the opening night at the Opera House, a total of 1000 tickets were sold. Front orchestra seats cost $80 apiece, rear orchestra seats cost $60 apiece, and front balcony seats cost $50 apiece. The combined number of tickets sold for the front orchestra and rear orchestra exceeded twice the number of front balcony tickets sold by 400. The total receipts for the performance were $62,800. Determine how many tickets of each type were sold.

64. Production Scheduling   A manufacturer of women's blouses makes three types of blouses: sleeveless, short-sleeve, and long-sleeve. The time required by each department to produce a dozen blouses of each type is shown in the accompanying table.

Department	Sleeveless	Short-sleeve	Long-sleeve
Cutting	9 min	12 min	15 min
Sewing	22 min	24 min	28 min
Packaging	6 min	8 min	8 min

The cutting, sewing, and packaging departments have available a maximum of 80, 160, and 48 labor hours, respectively, per day. How many dozens of each type of blouse can be produced each day if the plant is operated at full capacity?

SOLUTIONS TO SELF-CHECK EXERCISES 5.2

1. We obtain the following sequence of equivalent augmented matrices:

$$\begin{bmatrix} 2 & 3 & 1 & | & 6 \\ 1 & -2 & 3 & | & -3 \\ 3 & 2 & -4 & | & 12 \end{bmatrix} \xrightarrow{R_1 \leftrightarrow R_2} \begin{bmatrix} ① & -2 & 3 & | & -3 \\ 2 & 3 & 1 & | & 6 \\ 3 & 2 & -4 & | & 12 \end{bmatrix} \xrightarrow[R_3 - 3R_1]{R_2 - 2R_1}$$

$$\begin{bmatrix} 1 & -2 & 3 & | & -3 \\ 0 & 7 & -5 & | & 12 \\ 0 & 8 & -13 & | & 21 \end{bmatrix} \xrightarrow{R_2 \leftrightarrow R_3} \begin{bmatrix} 1 & -2 & 3 & | & -3 \\ 0 & ⑧ & -13 & | & 21 \\ 0 & 7 & -5 & | & 12 \end{bmatrix} \xrightarrow{R_2 - R_3}$$

$$\begin{bmatrix} 1 & -2 & 3 & | & -3 \\ 0 & 1 & -8 & | & 9 \\ 0 & 7 & -5 & | & 12 \end{bmatrix} \xrightarrow[R_3 - 7R_2]{R_1 + 2R_2} \begin{bmatrix} 1 & 0 & -13 & | & 15 \\ 0 & 1 & -8 & | & 9 \\ 0 & 0 & 51 & | & -51 \end{bmatrix} \xrightarrow{\frac{1}{51}R_3}$$

$$\begin{bmatrix} 1 & 0 & -13 & | & 15 \\ 0 & 1 & -8 & | & 9 \\ 0 & 0 & ① & | & -1 \end{bmatrix} \xrightarrow[R_2 + 8R_3]{R_1 + 13R_3} \begin{bmatrix} 1 & 0 & 0 & | & 2 \\ 0 & 1 & 0 & | & 1 \\ 0 & 0 & 1 & | & -1 \end{bmatrix}$$

The solution to the system is $x = 2$, $y = 1$, and $z = -1$.

2. Referring to the solution of exercise 2, Self-Check Exercises 5.1, we see that the problem reduces to solving the following system of linear equations:

$$\begin{aligned} x + y + z &= 200 \\ 40x + 60y + 80z &= 12600 \\ 20x + 25y + 40z &= 5950 \end{aligned}$$

Using the Gauss-Jordan elimination method, we have

$$\begin{bmatrix} 1 & 1 & 1 & | & 200 \\ 40 & 60 & 80 & | & 12600 \\ 20 & 25 & 40 & | & 5950 \end{bmatrix} \xrightarrow[R_3 - 20R_1]{R_2 - 40R_1}$$

$$\begin{bmatrix} 1 & 1 & 1 & | & 200 \\ 0 & 20 & 40 & | & 4600 \\ 0 & 5 & 20 & | & 1950 \end{bmatrix} \xrightarrow{\frac{1}{20}R_2} \begin{bmatrix} 1 & 1 & 1 & | & 200 \\ 0 & 1 & 2 & | & 230 \\ 0 & 5 & 20 & | & 1950 \end{bmatrix}$$

$$\xrightarrow[R_3 - 5R_2]{R_1 - R_2} \begin{bmatrix} 1 & 0 & -1 & | & -30 \\ 0 & 1 & 2 & | & 230 \\ 0 & 0 & 10 & | & 800 \end{bmatrix} \xrightarrow{\frac{1}{10}R_3}$$

$$\begin{bmatrix} 1 & 0 & -1 & | & -30 \\ 0 & 1 & 2 & | & 230 \\ 0 & 0 & 1 & | & 80 \end{bmatrix} \xrightarrow[R_2 - 2R_3]{R_1 + R_3} \begin{bmatrix} 1 & 0 & 0 & | & 50 \\ 0 & 1 & 0 & | & 70 \\ 0 & 0 & 1 & | & 80 \end{bmatrix}$$

From the last augmented matrix in reduced form we see that $x = 50$, $y = 70$, and $z = 80$. Therefore, the farmer should plant 50 acres of crop A, 70 acres of crop B, and 80 acres of crop C.

Using Technology

Solving a System of Linear Equations Using the Gauss-Jordan Method

The three matrix operations can be performed on a matrix using a graphing utility. The commands are as follows:

Operation	Calculator Function
$R_i \leftrightarrow R_j$	**rSwap**(A, i, j)
cR_i	**multR**(c, A, i)
$R_i + aR_j$	**mRAdd**(a, A, j, i)

(see Section A.5 in Appendix A).

When a row operation is performed on a matrix, the result is stored as **Ans** in the calculator. If another operation is performed on this matrix, then the matrix is erased. Should a mistake be made in the operation, then the previous matrix is lost. For this reason, the results of each operation should be stored. Do this by pressing **STO** (indicated by the symbol ▶ in Example 1), followed by the name of a matrix, and then **ENTER.** We will incorporate this process in the following example.

EXAMPLE 1   Use a graphing calculator to solve the following system of linear equations by the Gauss-Jordan method (see Example 5, Section 5.2).

$$3x - 2y + 8z = 9$$
$$-2x + 2y + z = 3$$
$$x + 2y - 3z = 8$$

Solution   Using the Gauss-Jordan method, we obtain the following sequence of equivalent matrices.

$$\begin{bmatrix} 3 & -2 & 8 & | & 9 \\ -2 & 2 & 1 & | & 3 \\ 1 & 2 & -3 & | & 8 \end{bmatrix} \xrightarrow{\textbf{mRAdd}(1, A, 2, 1) \blacktriangleright B} \begin{bmatrix} 1 & 0 & 9 & | & 12 \\ -2 & 2 & 1 & | & 3 \\ 1 & 2 & -3 & | & 8 \end{bmatrix} \xrightarrow{\textbf{mRAdd}(2, B, 1, 2) \blacktriangleright C}$$

$$\begin{bmatrix} 1 & 0 & 9 & | & 12 \\ 0 & 2 & 19 & | & 27 \\ 1 & 2 & -3 & | & 8 \end{bmatrix} \xrightarrow{\textbf{mRAdd}(-1, C, 1, 3) \blacktriangleright B} \begin{bmatrix} 1 & 0 & 9 & | & 12 \\ 0 & 2 & 19 & | & 27 \\ 0 & 2 & -12 & | & -4 \end{bmatrix} \xrightarrow{\textbf{multR}(\frac{1}{2}, B, 2) \blacktriangleright C}$$

$$\begin{bmatrix} 1 & 0 & 9 & | & 12 \\ 0 & 1 & 9.5 & | & 13.5 \\ 0 & 2 & -12 & | & -4 \end{bmatrix} \xrightarrow{\textbf{mRAdd}(-2, C, 2, 3) \blacktriangleright B} \begin{bmatrix} 1 & 0 & 9 & | & 12 \\ 0 & 1 & 9.5 & | & 13.5 \\ 0 & 0 & -31 & | & -31 \end{bmatrix} \xrightarrow{\textbf{multR}(-\frac{1}{31}, B, 3) \blacktriangleright C}$$

$$\begin{bmatrix} 1 & 0 & 9 & | & 12 \\ 0 & 1 & 9.5 & | & 13.5 \\ 0 & 0 & 1 & | & 1 \end{bmatrix} \xrightarrow{\textbf{mRAdd}(-9, C, 3, 1) \blacktriangleright B} \begin{bmatrix} 1 & 0 & 0 & | & 3 \\ 0 & 1 & 9.5 & | & 13.5 \\ 0 & 0 & 1 & | & 1 \end{bmatrix} \xrightarrow{\textbf{mRAdd}(-9.5, B, 3, 2) \blacktriangleright C}$$

$$\begin{bmatrix} 1 & 0 & 0 & | & 3 \\ 0 & 1 & 0 & | & 4 \\ 0 & 0 & 1 & | & 1 \end{bmatrix}$$

The last matrix is in row-reduced form, and we see that the solution of the system is $x = 3$, $y = 4$, and $z = 1$.   ■

Using rref (TI-85) to Solve a System of Linear Equations

The operation **rref** will transform an augmented matrix into one that is in row-reduced form. (See Section A.5 in Appendix A.) For example, using **rref,** we find

$$\begin{bmatrix} 3 & -2 & 8 & 9 \\ -2 & 2 & 1 & 3 \\ 1 & 2 & -3 & 8 \end{bmatrix} \xrightarrow{\text{rref } A} \begin{bmatrix} 1 & 0 & 0 & 3 \\ 0 & 1 & 0 & 4 \\ 0 & 0 & 1 & 1 \end{bmatrix}$$

as obtained earlier!

Using SIMULT (TI-85) to Solve a System of Equations

The **SIMULT** operation of a graphing utility can be used to solve a system of n linear equations in n variables, where n is an integer between 2 and 30. (See Section A.5 in Appendix A.)

EXERCISES

In exercises 1–6, use a graphing utility to solve the system of equations (a) by the Gauss-Jordan method, (b) using the **rref** *operation, and (c) using* **SIMULT.** *Express your answer accurate to four decimal places.*

1. $\begin{aligned} x_1 - 2x_2 + 2x_3 - 3x_4 &= -7 \\ 3x_1 + 2x_2 - x_3 + 5x_4 &= 22 \\ 2x_1 - 3x_2 + 4x_3 - x_4 &= -3 \\ 3x_1 - 2x_2 - x_3 + 2x_4 &= 12 \end{aligned}$

2. $\begin{aligned} 2x_1 - x_2 + 3x_3 - 2x_4 &= -2 \\ x_1 - 2x_2 + x_3 - 3x_4 &= 2 \\ x_1 - 5x_2 + 2x_3 + 3x_4 &= -6 \\ -3x_1 + 3x_2 - 4x_3 - 4x_4 &= 9 \end{aligned}$

3. $\begin{aligned} 2x_1 + x_2 + 3x_3 - x_4 &= 9 \\ -x_1 - 2x_2 \qquad - 3x_4 &= -1 \\ x_1 \qquad - 3x_3 + x_4 &= 10 \\ x_1 - x_2 - x_3 - x_4 &= 8 \end{aligned}$

4. $\begin{aligned} x_1 - 2x_2 - 2x_3 + x_4 &= 1 \\ 2x_1 - x_2 + 2x_3 + 3x_4 &= -2 \\ -x_1 - 5x_2 + 7x_3 - 2x_4 &= 3 \\ 3x_1 - 4x_2 + 3x_3 + 4x_4 &= -4 \end{aligned}$

5. $\begin{aligned} 2x_1 - 2x_2 + 3x_3 - x_4 + 2x_5 &= 16 \\ 3x_1 + x_2 - 2x_3 + x_4 - 3x_5 &= -11 \\ x_1 + 3x_2 - 4x_3 + 3x_4 - x_5 &= -13 \\ 2x_1 - x_2 + 3x_3 - 2x_4 + 2x_5 &= 15 \\ 3x_1 + 4x_2 - 3x_3 + 5x_4 - x_5 &= -10 \end{aligned}$

6. $\begin{aligned} 2.1x_1 - 3.2x_2 + 6.4x_3 + 7x_4 - 3.2x_5 &= 54.3 \\ 4.1x_1 + 2.2x_2 - 3.1x_3 - 4.2x_4 + 3.3x_5 &= -20.81 \\ 3.4x_1 - 6.2x_2 + 4.7x_3 + 2.1x_4 - 5.3x_5 &= 24.7 \\ 4.1x_1 + 7.3x_2 + 5.2x_3 + 6.1x_4 - 8.2x_5 &= 29.25 \\ 2.8x_1 + 5.2x_2 + 3.1x_3 + 5.4x_4 + 3.8x_5 &= 43.72 \end{aligned}$

5.3	**Solving Systems of Linear Equations II**

In this section we continue our study of systems of linear equations. More specifically, we look at systems that have infinitely many solutions and those that have no solution. We also study systems of linear equations in which the number of variables is not equal to the number of equations in the system.

Solution(s) of Linear Equations

Our first example illustrates the situation in which a system of linear equations has infinitely many solutions.

EXAMPLE 1   *A system of equations with an infinite number of solutions.*
Solve the system of linear equations given by

$$x + 2y - 3z = -2$$
$$3x - y - 2z = 1$$
$$2x + 3y - 5z = -3 \tag{9}$$

Solution   Using the Gauss-Jordan elimination method, we obtain the following sequence of equivalent augmented matrices:

$$\begin{bmatrix} ① & 2 & -3 & | & -2 \\ 3 & -1 & -2 & | & 1 \\ 2 & 3 & -5 & | & -3 \end{bmatrix} \xrightarrow[R_3 - 2R_1]{R_2 - 3R_1} \begin{bmatrix} 1 & 2 & -3 & | & -2 \\ 0 & ⑦ & 7 & | & 7 \\ 0 & -1 & 1 & | & 1 \end{bmatrix} \xrightarrow{-\frac{1}{7}R_2}$$

$$\begin{bmatrix} 1 & 2 & -3 & | & -2 \\ 0 & 1 & -1 & | & -1 \\ 0 & -1 & 1 & | & 1 \end{bmatrix} \xrightarrow[R_3 + R_2]{R_1 - 2R_2} \begin{bmatrix} 1 & 0 & -1 & | & 0 \\ 0 & 1 & -1 & | & -1 \\ 0 & 0 & 0 & | & 0 \end{bmatrix}$$

The last augmented matrix is in row-reduced form. Interpreting it as a system of linear equations gives

$$x - z = 0$$
$$y - z = -1$$

a system of two equations in the three variables x, y, and z.

Let us now single out one variable, say z, and solve for x and y in terms of it. We obtain

$$x = z$$
$$y = z - 1$$

If we assign a particular value to z, say $z = 0$, we obtain $x = 0$ and $y = -1$, giving the solution $(0, -1, 0)$ to system (9). By setting $z = 1$, we obtain the solution $(1, 0, 1)$. In general, if we set $z = t$, where t represents some real number, we obtain a solution given by $(t, t - 1, t)$. Since the parameter t may be any real number, we see that the system (9) has infinitely many solutions. Geometrically, the solutions of the system (9) lie on the straight line in three-dimensional space given by the intersection of the three planes determined by the three equations in the system. ■

REMARK   In Example 1 we chose the parameter to be z because it is more convenient to solve for x and y (both the x and y columns are in unit form) in terms of z. □

The next example shows what happens in the elimination procedure when the system does not have a solution.

EXAMPLE 2   *A system of equations that has no solution.* Solve the system of linear equations given by

$$x + y + z = 1$$
$$3x - y - z = 4$$
$$x + 5y + 5z = -1 \tag{10}$$

Solution Using the Gauss-Jordan elimination method, we obtain the following sequence of equivalent augmented matrices:

$$
\left[\begin{array}{ccc|c}
① & 1 & 1 & 1 \\
3 & -1 & -1 & 4 \\
1 & 5 & 5 & -1
\end{array}\right]
\xrightarrow[R_3 - R_1]{R_2 - 3R_1}
\left[\begin{array}{ccc|c}
1 & 1 & 1 & 1 \\
0 & -4 & -4 & 1 \\
0 & 4 & 4 & -2
\end{array}\right]
$$

$$
\xrightarrow{R_3 + R_2}
\left[\begin{array}{ccc|c}
1 & 1 & 1 & 1 \\
0 & -4 & -4 & 1 \\
0 & 0 & 0 & -1
\end{array}\right]
$$

Observe that the third row in the last matrix reads $0x + 0y + 0z = -1$—that is, $0 = -1$! We conclude therefore that system (10) is inconsistent and has no solution. Geometrically, we have a situation in which two of the planes intersect in a straight line but the third plane is parallel to this line of intersection of the two planes and does not intersect it. Consequently, there is no point of intersection of the three planes. ∎

Example 2 illustrates the following, more general, result of using the Gauss-Jordan elimination procedure.

SYSTEMS WITH NO SOLUTION

If there is a row in the augmented matrix containing all zeros to the left of the vertical line and a nonzero entry to the right of the line, then the system of equations has no solution.

It may have dawned on you that in all the previous examples we have dealt only with systems involving exactly the same number of linear equations as there are variables. However, systems in which the number of equations is different from the number of variables also occur in practice. Indeed, we will consider such systems in Examples 3 and 4.

The following theorem provides us with some preliminary information on a system of linear equations.

THEOREM 1

a. If the number of equations is greater than or equal to the number of variables in a linear system, then one of the following is true:
  i. The system has no solution.
  ii. The system has exactly one solution.
  iii. The system has infinitely many solutions.
b. If there are fewer equations than variables in a linear system, then the system either has no solution or it has infinitely many solutions.

REMARK Theorem 1 may be used to tell us—before we even begin to solve a problem—what the nature of the solution may be. □

Although we will not prove this theorem, you should recall that we illustrated part a geometrically for the case in which there are exactly as many equations (three) as there are variables. To show the validity of part b, let us once again consider the case in which a system has three variables. Now, if there is only one equation in the system, then it is clear that there are infinitely many solutions corresponding geometrically to all the points lying on the plane represented by the equation.

   Next, if there are two equations in the system, then *only* the following possibilities exist:

1. The two planes are parallel and distinct.

2. The two planes intersect in a straight line.

3. The two planes are coincident (the two equations define the same plane) (Figure 6).

FIGURE 6

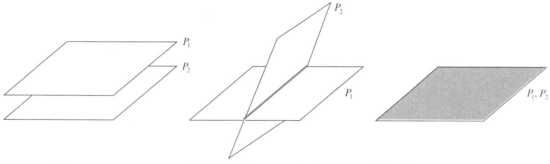

(a) No solution   **(b)** Infinitely many solutions   **(c)** Infinitely many solutions

Thus, either there is no solution or there are infinitely many solutions corresponding to the points lying on a line of intersection of the two planes or on a single plane determined by the two equations. In the case where two planes intersect in a straight line, the solutions will involve one parameter, and in the case where the two planes are coincident, the solutions will involve two parameters.

EXAMPLE 3   *A system with more equations than variables.* Solve the following system of linear equations:

$$x + 2y = \ 4$$
$$x - 2y = \ 0$$
$$4x + 3y = 12$$

Solution   We obtain the following sequence of equivalent augmented matrices:

$$
\begin{bmatrix}
① & 2 & | & 4 \\
1 & -2 & | & 0 \\
4 & 3 & | & 12
\end{bmatrix}
\xrightarrow[R_3 - 4R_1]{R_2 - R_1}
\begin{bmatrix}
1 & 2 & | & 4 \\
0 & ⊖4 & | & -4 \\
0 & -5 & | & -4
\end{bmatrix}
\xrightarrow{-\frac{1}{4}R_2}
$$

$$
\begin{bmatrix}
1 & 2 & | & 4 \\
0 & 1 & | & 1 \\
0 & -5 & | & -4
\end{bmatrix}
\xrightarrow[R_3 + 5R_2]{R_1 - 2R_2}
\begin{bmatrix}
1 & 0 & | & 2 \\
0 & 1 & | & 1 \\
0 & 0 & | & 1
\end{bmatrix}
$$

The last row of the row-reduced augmented matrix implies that $0 = 1$, which is impossible, so we conclude that the given system has no solution. Geometrically, the three lines defined by the three equations in the system do not intersect at a point. (To see this for yourself, draw the graphs of these equations.) ∎

EXAMPLE 4 *A system with more variables than equations.* Solve the following system of linear equations:

$$x + 2y - 3z + w = -2$$
$$3x - y - 2z - 4w = 1$$
$$2x + 3y - 5z + w = -3$$

Solution First, observe that the given system comprises three equations in four variables, and so by Theorem 1(b) the system has either no solution or it has infinitely many solutions. To solve it, we use the Gauss-Jordan method and obtain the following sequence of equivalent augmented matrices:

$$
\begin{bmatrix}
①& 2 & -3 & 1 & | & -2 \\
3 & -1 & -2 & -4 & | & 1 \\
2 & 3 & -5 & 1 & | & -3
\end{bmatrix}
\xrightarrow[R_3 - 2R_1]{R_2 - 3R_1}
\begin{bmatrix}
1 & 2 & -3 & 1 & | & -2 \\
0 & ⑦ & 7 & -7 & | & 7 \\
0 & -1 & 1 & -1 & | & 1
\end{bmatrix}
\xrightarrow{-\frac{1}{7}R_2}
$$

$$
\begin{bmatrix}
1 & 2 & -3 & 1 & | & -2 \\
0 & 1 & -1 & 1 & | & -1 \\
0 & -1 & 1 & -1 & | & 1
\end{bmatrix}
\xrightarrow[R_3 + R_2]{R_1 - 2R_2}
\begin{bmatrix}
1 & 0 & -1 & -1 & | & 0 \\
0 & 1 & -1 & 1 & | & -1 \\
0 & 0 & 0 & 0 & | & 0
\end{bmatrix}
$$

The last augmented matrix is in row-reduced form. Observe that the given system is equivalent to the system

$$x - z - w = 0$$
$$y - z + w = -1$$

of two equations in four variables. Thus, we may solve for two of the variables in terms of the other two. Letting $z = s$ and $w = t$ (s, t, parameters), we find that

$$x = s + t$$
$$y = s - t - 1$$
$$z = s$$
$$w = t$$

The solutions may be written in the form $(s + t, s - t - 1, s, t)$, where s and t are any real numbers. Geometrically, the three equations in the system represent three hyperplanes in four-dimensional space (since there are four variables) and their "points" of intersection lie in a two-dimensional subspace of fourspace (since there are two parameters). ∎

REMARK In Example 4 we assigned parameters to z and w rather than to x and y because x and y are readily solved in terms of z and w. □

Application

The following example illustrates a situation in which a system of linear equations has infinitely many solutions.

EXAMPLE 5   Figure 7 shows the flow of downtown traffic in a certain city during the rush hours on a typical weekday. The arrows indicate the direction of traffic flow on each one-way road, and the average number of vehicles entering and leaving each intersection per hour appears beside each road. Fifth and Sixth Avenues can handle up to 2000 vehicles per hour without causing congestion, whereas the maximum capacity of each of the two streets is 1000 vehicles per hour. The flow of traffic is controlled by traffic lights installed at each of the four intersections.

a. Write a general expression involving the rates of flow, x_1, x_2, x_3, x_4, and suggest two possible flow patterns that would ensure that there is no traffic congestion.

b. Suppose that the part of 4th Street between 5th Avenue and 6th Avenue is to be resurfaced and that traffic flow between the two junctions has to be slowed down to at most 300 vehicles per hour. Find two possible flow patterns that would result in a smooth flow of traffic.

FIGURE 7
Downtown traffic flow

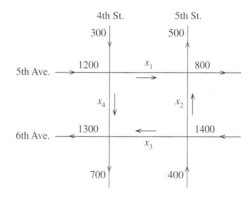

Solution

a. To avoid congestion, all traffic entering an intersection must also leave that intersection. Applying this condition to each of the four intersections in a clockwise direction beginning with the 5th Avenue and 4th Street intersection, we obtain the following equations:

$$1500 = x_1 + x_4$$
$$1300 = x_1 + x_2$$
$$1800 = x_2 + x_3$$
$$2000 = x_3 + x_4$$

This system of four linear equations in the four variables x_1, x_2, x_3, x_4 may be rewritten in the more standard form

$$\begin{aligned} x_1 \qquad\qquad\quad + x_4 &= 1500 \\ x_1 + x_2 \qquad\qquad\ &= 1300 \\ x_2 + x_3 \qquad &= 1800 \\ x_3 + x_4 &= 2000 \end{aligned}$$

Using the Gauss-Jordan elimination method to solve the system, we obtain

$$
\begin{bmatrix}
1 & 0 & 0 & 1 & | & 1500 \\
1 & 1 & 0 & 0 & | & 1300 \\
0 & 1 & 1 & 0 & | & 1800 \\
0 & 0 & 1 & 1 & | & 2000
\end{bmatrix}
\xrightarrow{R_2 - R_1}
\begin{bmatrix}
1 & 0 & 0 & 1 & | & 1500 \\
0 & 1 & 0 & -1 & | & -200 \\
0 & 1 & 1 & 0 & | & 1800 \\
0 & 0 & 1 & 1 & | & 2000
\end{bmatrix}
$$

$$
\xrightarrow{R_3 - R_2}
\begin{bmatrix}
1 & 0 & 0 & 1 & | & 1500 \\
0 & 1 & 0 & -1 & | & -200 \\
0 & 0 & 1 & 1 & | & 2000 \\
0 & 0 & 1 & 1 & | & 2000
\end{bmatrix}
$$

$$
\xrightarrow{R_4 - R_3}
\begin{bmatrix}
1 & 0 & 0 & 1 & | & 1500 \\
0 & 1 & 0 & -1 & | & -200 \\
0 & 0 & 1 & 1 & | & 2000 \\
0 & 0 & 0 & 0 & | & 0
\end{bmatrix}
$$

The last augmented matrix is in row-reduced form and is equivalent to a system of three linear equations in the four variables x_1, x_2, x_3, x_4. Thus, we may express three of the variables—say, x_1, x_2, x_3—in terms of x_4. Setting $x_4 = t$ (t, a parameter), we may write the infinitely many solutions of the system as

$$x_1 = 1500 - t$$

$$x_2 = -200 + t$$

$$x_3 = 2000 - t$$

$$x_4 = t$$

Observe that for a meaningful solution, $200 \leq t \leq 1000$, since x_1, x_2, x_3, and x_4 must all be nonnegative. For example, picking $t = 300$ gives the flow pattern

$$x_1 = 1200, \qquad x_2 = 100, \qquad x_3 = 1700, \qquad x_4 = 300$$

Selecting $t = 500$ gives the flow pattern

$$x_1 = 1000, \qquad x_2 = 300, \qquad x_3 = 1500, \qquad x_4 = 500$$

b. In this case, x_4 must not exceed 300. Again, using the results of (a), we find, upon setting $x_4 = t = 300$, the flow pattern

$$x_1 = 1200, \qquad x_2 = 100, \qquad x_3 = 1700, \qquad x_4 = 300$$

obtained earlier. Picking $t = 250$ gives the flow pattern

$$x_1 = 1250, \qquad x_2 = 50, \qquad x_3 = 1750, \qquad x_4 = 250$$ ■

SELF-CHECK
EXERCISES 5.3

1. The following augmented matrix in row-reduced form is equivalent to the augmented matrix of a certain system of linear equations. Use this result to solve the system of equations.

$$\begin{bmatrix} 1 & 0 & -1 & | & 3 \\ 0 & 1 & 5 & | & -2 \\ 0 & 0 & 0 & | & 0 \end{bmatrix}$$

2. Solve the system of linear equations

$$2x - 3y + z = 6$$
$$x + 2y + 4z = -4$$
$$x - 5y - 3z = 10$$

using the Gauss-Jordan elimination method.

3. Solve the system of linear equations

$$x - 2y + 3z = 9$$
$$2x + 3y - z = 4$$
$$x + 5y - 4z = 2$$

using the Gauss-Jordan elimination method.

Solutions to Self-Check Exercises 5.3 can be found on page 219.

5.3 EXERCISES

In exercises 1–12, given that the augmented matrix in row-reduced form is equivalent to the augmented matrix of a system of linear equations, (a) determine whether the system has a solution and (b) find the solution or solutions to the system, if they exist.

1. $\begin{bmatrix} 1 & 0 & 0 & | & 3 \\ 0 & 1 & 0 & | & -1 \\ 0 & 0 & 1 & | & 2 \end{bmatrix}$

2. $\begin{bmatrix} 1 & 0 & 0 & | & 3 \\ 0 & 1 & 0 & | & -2 \\ 0 & 0 & 1 & | & 1 \end{bmatrix}$

3. $\begin{bmatrix} 1 & 0 & | & 2 \\ 0 & 1 & | & 4 \\ 0 & 0 & | & 0 \end{bmatrix}$

4. $\begin{bmatrix} 1 & 0 & 0 & | & 3 \\ 0 & 1 & 0 & | & 1 \\ 0 & 0 & 0 & | & 0 \end{bmatrix}$

5. $\begin{bmatrix} 1 & 0 & 1 & | & 4 \\ 0 & 1 & 0 & | & -2 \end{bmatrix}$

6. $\begin{bmatrix} 1 & 0 & 0 & 0 & | & 3 \\ 0 & 1 & 1 & 0 & | & -1 \\ 0 & 0 & 0 & 1 & | & 2 \end{bmatrix}$

7. $\begin{bmatrix} 1 & 0 & 0 & 0 & | & 2 \\ 0 & 1 & 0 & 0 & | & 1 \\ 0 & 0 & 1 & 0 & | & 3 \\ 0 & 0 & 0 & 0 & | & 1 \end{bmatrix}$

8. $\begin{bmatrix} 1 & 0 & 0 & | & 4 \\ 0 & 1 & 0 & | & -1 \\ 0 & 0 & 1 & | & 3 \\ 0 & 0 & 0 & | & 1 \end{bmatrix}$

9.
$$\begin{bmatrix} 1 & 0 & 0 & 0 & | & 2 \\ 0 & 1 & 0 & 0 & | & -1 \\ 0 & 0 & 1 & 1 & | & 2 \\ 0 & 0 & 0 & 0 & | & 0 \end{bmatrix}$$

10.
$$\begin{bmatrix} 0 & 1 & 0 & 1 & | & 3 \\ 0 & 0 & 1 & -2 & | & 4 \\ 0 & 0 & 0 & 0 & | & 0 \\ 0 & 0 & 0 & 0 & | & 0 \end{bmatrix}$$

11.
$$\begin{bmatrix} 1 & 0 & 3 & 0 & | & 2 \\ 0 & 1 & -1 & 0 & | & 1 \\ 0 & 0 & 0 & 0 & | & 0 \\ 0 & 0 & 0 & 0 & | & 0 \end{bmatrix}$$

12.
$$\begin{bmatrix} 1 & 0 & 3 & -1 & | & 4 \\ 0 & 1 & -2 & 3 & | & 2 \\ 0 & 0 & 0 & 0 & | & 0 \\ 0 & 0 & 0 & 0 & | & 0 \end{bmatrix}$$

In exercises 13–32, solve the system of linear equations using the Gauss-Jordan elimination method.

13. $2x - y = 3$
$x + 2y = 4$
$2x + 3y = 7$

14. $x + 2y = 3$
$2x - 3y = -8$
$x - 4y = -9$

15. $3x - 2y = -3$
$2x + y = 3$
$x - 2y = -5$

16. $2x + 3y = 2$
$x + 3y = -2$
$x - y = 3$

17. $3x - 2y = 5$
$-x + 3y = -4$
$2x - 4y = 6$

18. $4x + 6y = 8$
$3x - 2y = -7$
$x + 3y = 5$

19. $x - 2y = 2$
$7x - 14y = 14$
$3x - 6y = 6$

20. $x + 2y + z = -2$
$-2x - 3y - z = 1$
$2x + 4y + 2z = -4$

21. $3x + 2y = 4$
$-\frac{3}{2}x - y = -2$
$6x + 4y = 8$

22. $3y + 2z = 4$
$2x - y - 3z = 3$
$2x + 2y - z = 7$

23. $2x_1 - x_2 + x_3 = -4$
$3x_1 - \frac{3}{2}x_2 + \frac{3}{2}x_3 = -6$
$-6x_1 + 3x_2 - 3x_3 = 12$

24. $x + y - 2z = -3$
$2x - y + 3z = 7$
$x - 2y + 5z = 0$

25. $x - 2y + 3z = 4$
$2x + 3y - z = 2$
$x + 2y - 3z = -6$

26. $x_1 - 2x_2 + x_3 = -3$
$2x_1 + x_2 - 2x_3 = 2$
$x_1 + 3x_2 - 3x_3 = 5$

27. $4x + y - z = 4$
$8x + 2y - 2z = 8$

28. $x_1 + 2x_2 + 4x_3 = 2$
$x_1 + x_2 + 2x_3 = 1$

29. $2x + y - 3z = 1$
$x - y + 2z = 1$
$5x - 2y + 3z = 6$

30. $3x - 9y + 6z = -12$
$x - 3y + 2z = -4$
$2x - 6y + 4z = 8$

31. $x + 2y - z = -4$
$2x + y + z = 7$
$x + 3y + 2z = 7$
$x - 3y + z = 9$

32. $3x - 2y + z = 4$
$x + 3y - 4z = -3$
$2x - 3y + 5z = 7$
$x - 8y + 9z = 10$

33. Management Decisions   The management of Hartman Rent-A-Car has allocated $840,000 to purchase 60 new automobiles to add to their existing fleet of rental cars. The buyer will choose from compact, midsized, and full-sized cars costing $10,000, $16,000, and $22,000 each, respectively. Find formulas giving the options available to the buyer. Give two specific options. (*Remark:* Your answers will not be unique.)

34. Nutrition   A dietician wishes to plan a meal around three foods. The meal is to include 8800 units of vitamin A, 3380 units of vitamin C, and 1020 units of calcium. The number of units of the vitamins and calcium in each ounce of the foods is summarized in the accompanying table.

	Food I	Food II	Food III
Vitamin A	400	1200	800
Vitamin C	110	570	340
Calcium	90	30	60

Determine the amount of each food the dietician should include in the meal in order to meet the vitamin and calcium requirements.

35. Nutrition   Refer to exercise 34. In planning for another meal, the dietician changes the requirement of vitamin C to 2160 units instead of 3380 units. All other requirements remain the same. Show that such a meal cannot be planned around the same foods.

36. Investments   Mr. and Mrs. Garcia have a total of $100,000 to be invested in stocks, bonds, and a money market account. The stocks have a rate of return of 12% per year, while the bonds and the money account pay 8% and 4% per year, respectively. They have stipulated that the amount invested in stocks should be equal to the sum of the amount invested in bonds and three times the amount invested in the money market account. How should the Garcias allocate their resources if they require an annual income of $10,000 from their investments?

37. Traffic Control   The accompanying figure shows the flow of traffic near a city's Civic Center during the rush hours on a typical weekday. Each road can handle a maximum of 1000 cars per hour without causing congestion. The flow of traffic is controlled by traffic lights at each of the five intersections.

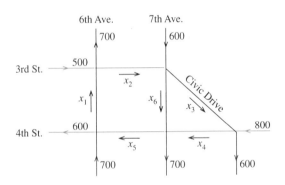

a. Set up a system of linear equations describing the traffic flow.
b. Solve the system devised in (a) and suggest two possible traffic-flow patterns that would ensure that there is no traffic congestion.
c. Suppose that 7th Avenue between 3rd and 4th Streets is soon to be closed for road repairs, and find one possible flow pattern that would result in a smooth flow of traffic.

38. Traffic Control   The accompanying figure shows the flow of downtown traffic during the rush hours on a

typical weekday. Each avenue can handle up to 1500 vehicles per hour without causing congestion, whereas the maximum capacity of each street is 1000 vehicles per hour. The flow of traffic is controlled by traffic lights at each of the six intersections.

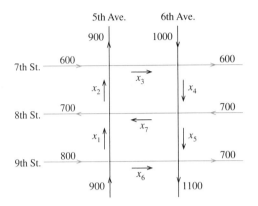

a. Set up a system of linear equations describing the traffic flow.
b. Solve the system devised in (a) and suggest two possible traffic-flow patterns that would ensure there is no traffic congestion.
c. Suppose that the traffic flow along 9th Street between 5th Avenue and 6th Avenue, x_6, is restricted due to sewer construction. What is the minimum permissible traffic flow along this road that will not result in traffic congestion?

39. Determine the value of k so that the following system of linear equations has a solution, and find the solution.

$$2x + 3y = 2$$
$$x + 4y = 6$$
$$5x + ky = 2$$

40. Determine the value of k so that the following system of linear equations has infinitely many solutions, and find the solutions.

$$3x - 2y + 4z = 12$$
$$-9x + 6y - 12z = k$$

SOLUTIONS TO SELF-CHECK EXERCISES 5.3

1. Let x, y, and z denote the variables. Then the given row-reduced augmented matrix tells us that the system of linear equations is equivalent to the two equations

$$x \quad - z = 3$$
$$y + 5z = -2$$

Letting $z = t$, where t is a parameter, we find the infinitely many solutions given by

$$x = \quad t + 3$$
$$y = -5t - 2$$
$$z = \quad t$$

2. We obtain the following sequence of equivalent augmented matrices:

$$\begin{bmatrix} 2 & -3 & 1 & | & 6 \\ 1 & 2 & 4 & | & -4 \\ 1 & -5 & -3 & | & 10 \end{bmatrix} \xrightarrow{R_1 \leftrightarrow R_2} \begin{bmatrix} 1 & 2 & 4 & | & -4 \\ 2 & -3 & 1 & | & 6 \\ 1 & -5 & -3 & | & 10 \end{bmatrix}$$

$$\xrightarrow[R_3 - R_1]{R_2 - 2R_1} \begin{bmatrix} 1 & 2 & 4 & | & -4 \\ 0 & -7 & -7 & | & 14 \\ 0 & -7 & -7 & | & 14 \end{bmatrix} \xrightarrow{-\frac{1}{7}R_2}$$

$$\begin{bmatrix} 1 & 2 & 4 & | & -4 \\ 0 & 1 & 1 & | & -2 \\ 0 & -7 & -7 & | & 14 \end{bmatrix} \xrightarrow[R_3 + 7R_2]{R_1 - 2R_2} \begin{bmatrix} 1 & 0 & 2 & | & 0 \\ 0 & 1 & 1 & | & -2 \\ 0 & 0 & 0 & | & 0 \end{bmatrix}$$

The last augmented matrix, which is in row-reduced form, tells us that the given system of linear equations is equivalent to the following system of two equations:

$$x \quad + 2z = \quad 0$$
$$y + \quad z = -2$$

Letting $z = t$, where t is a parameter, we see that the infinitely many solutions are given by

$$x = -2t$$
$$y = -\ t - 2$$
$$z = \quad t$$

3. We obtain the following sequence of equivalent augmented matrices.

$$\begin{bmatrix} \textcircled{1} & -2 & 3 & | & 9 \\ 2 & 3 & -1 & | & 4 \\ 1 & 5 & -4 & | & 2 \end{bmatrix} \xrightarrow[R_3 - R_1]{R_2 - 2R_1} \begin{bmatrix} 1 & -2 & 3 & | & 9 \\ 0 & 7 & -7 & | & -14 \\ 0 & 7 & -7 & | & -7 \end{bmatrix}$$

$$\xrightarrow{R_3 - R_2} \begin{bmatrix} 1 & -2 & 3 & | & 9 \\ 0 & 7 & -7 & | & -14 \\ 0 & 0 & 0 & | & 7 \end{bmatrix}$$

Since the last row of the final augmented matrix is equivalent to the equation $0 = 7$, a contradiction, we conclude that the given system has no solution.

U s i n g   T e c h n o l o g y

We can use the row operations of a graphing utility to solve a system of m linear equations in n unknowns by the Gauss-Jordan method, as we did in the previous Using Technology section. We can also use the **rref** operation to obtain the row-reduced form without going through all the steps of the Gauss-Jordan method. However, we cannot use the **SIMULT** function to solve a system where the number of equations and the number of variables are not the same. (See Section A.5 in Appendix A.)

EXERCISES

*In exercises 1–6, use a graphing utility to solve the system of equations (a) by the Gauss-Jordan method and (b) by using the **rref** operation.*

1.
$$2x_1 - x_2 - x_3 = 0$$
$$3x_1 - 2x_2 - x_3 = -1$$
$$-x_1 + 2x_2 - x_3 = 3$$
$$2x_2 - 2x_3 = 4$$

2.
$$3x_1 + x_2 - 4x_3 = 5$$
$$2x_1 - 3x_2 + 2x_3 = -4$$
$$-x_1 - 2x_2 + 4x_3 = 6$$
$$4x_1 + 3x_2 - 5x_3 = 9$$

3.
$$2x_1 + 3x_2 + 2x_3 + x_4 = -1$$
$$x_1 - x_2 + x_3 - 2x_4 = -8$$
$$5x_1 + 6x_2 - 2x_3 + 2x_4 = 11$$
$$x_1 + 3x_2 + 8x_3 + x_4 = -14$$

4.
$$x_1 - x_2 + 3x_3 - 6x_4 = 2$$
$$x_1 + x_2 + x_3 - 2x_4 = 2$$
$$-2x_1 - x_2 + x_3 + 2x_4 = 0$$

5.
$$x_1 + x_2 - x_3 - x_4 = -1$$
$$x_1 - x_2 + x_3 + 4x_4 = -6$$
$$3x_1 + x_2 - x_3 + 2x_4 = -4$$
$$5x_1 + x_2 - 3x_3 + x_4 = -9$$

6.
$$1.2x_1 - 2.3x_2 + 4.2x_3 + 5.4x_4 - 1.6x_5 = 4.2$$
$$2.3x_1 + 1.4x_2 - 3.1x_3 + 3.3x_4 - 2.4x_5 = 6.3$$
$$1.7x_1 + 2.6x_2 - 4.3x_3 + 7.2x_4 - 1.8x_5 = 7.8$$
$$2.6x_1 - 4.2x_2 + 8.3x_3 - 1.6x_4 + 2.5x_5 = 6.4$$

5.4   Matrices

Using Matrices to Represent Data

Many practical problems are solved by using arithmetic operations on the data associated with the problems. By properly organizing the data into *blocks* of numbers, we can then carry out these arithmetic operations in an orderly and efficient manner. In particular, this systematic approach enables us to use the computer to full advantage.

Let us begin by considering how the monthly output data of a manufacturer may be organized. The Acrosonic Company manufactures four different loudspeaker systems in three separate locations. The company's May output is summarized in Table 1.

TABLE 1

	Model A	Model B	Model C	Model D
Location I	320	280	460	280
Location II	480	360	580	0
Location III	540	420	200	880

Now, if we agree to preserve the relative location of each entry in Table 1, the set of data may be summarized further as follows:

$$\begin{bmatrix} 320 & 280 & 460 & 280 \\ 480 & 360 & 580 & 0 \\ 540 & 420 & 200 & 880 \end{bmatrix}$$

A matrix summarizing the data in Table 1.

The array of numbers displayed here is an example of a *matrix*. Observe that the numbers in the first row give the output of models A, B, C, and D of Acrosonic loudspeaker systems manufactured in location I; similarly, the numbers in the second and third rows give the respective outputs of these loudspeaker systems in locations II and III. The numbers in each column of the matrix give the outputs of a particular model of loudspeaker system manufactured in each of the company's three manufacturing locations.

More generally, recall that a matrix is an ordered rectangular array of real numbers. For example, each of the following arrays is a matrix:

$$A = \begin{bmatrix} 3 & 0 & -1 \\ 2 & 1 & 4 \end{bmatrix} \qquad B = \begin{bmatrix} 3 & 2 \\ 0 & 1 \\ -1 & 4 \end{bmatrix} \qquad C = \begin{bmatrix} 1 \\ 2 \\ 4 \\ 0 \end{bmatrix} \qquad D = \begin{bmatrix} 1 & 3 & 0 & 1 \end{bmatrix}$$

The real numbers that make up the array are called the *entries*, or *elements*, of the matrix. The entries in a row in the array are referred to as a *row* of the matrix, whereas the entries in a column in the array are referred to as a *column* of the matrix. Matrix A, for example, has two rows and three columns, which may be identified as follows:

$$\begin{array}{c} \\ \text{Row 1} \\ \text{Row 2} \end{array} \begin{array}{ccc} \text{Column 1} & \text{Column 2} & \text{Column 3} \\ \begin{bmatrix} 3 & 0 & -1 \\ 2 & 1 & 4 \end{bmatrix} \end{array}$$

A 2 × 3 matrix

The size, or *dimension*, of a matrix is described in terms of the number of rows and columns of the matrix. For example, matrix A has two rows and three columns and is said to have size 2 by 3, denoted 2×3. In general, a matrix having m rows and n columns is said to have size $m \times n$.

MATRIX

A matrix is an ordered rectangular array of numbers. A matrix with m rows and n columns has size $\boldsymbol{m \times n}$. The entry in the ith row and jth column is denoted by $\boldsymbol{a_{ij}}$.

A matrix of size $1 \times n$—a matrix having one row and n columns—is referred to as a row matrix, or *row vector*, of dimension n. For example, matrix D is a row vector of dimension 4. Similarly, a matrix having m rows and one column is referred to as a column matrix, or *column vector*, of dimension m. The matrix C is a column vector of dimension 4. Finally, an $n \times n$ matrix—that is, a matrix having the same number of rows as columns—is called a **square matrix.** For example, the matrix

$$\begin{bmatrix} -3 & 8 & 6 \\ 2 & \frac{1}{4} & 4 \\ 1 & 3 & 2 \end{bmatrix}$$

A 3 × 3 square matrix

is a square matrix of size 3×3, or simply of size 3.

EXAMPLE 1 Consider the matrix

$$P = \begin{bmatrix} 320 & 280 & 460 & 280 \\ 480 & 360 & 580 & 0 \\ 540 & 420 & 200 & 880 \end{bmatrix}$$

representing the output of loudspeaker systems of the Acrosonic Company discussed earlier (see Table 1).

a. What is the size of the matrix P?

b. Find a_{24} (the entry in the second row and fourth column of the matrix P) and give an interpretation of this number.

c. Find the sum of the entries that make up the first row of P and interpret the result.

d. Find the sum of the entries that make up the fourth column of P and interpret the result.

Solution

a. The matrix P has three rows and four columns and hence has size 3×4.

b. The required entry lies in the second row and fourth column and is the number zero. This means that no model D loudspeaker system was manufactured in location II in May.

c. The required sum is given by

$$320 + 280 + 460 + 280 = 1340$$

which gives the total number of loudspeaker systems manufactured in location I in May as 1340 units.

d. The required sum is given by

$$280 + 0 + 880 = 1160$$

giving the output of model D loudspeaker systems in all locations of the company in May as 1160 units. ◼

Equality of Matrices

Two matrices are said to be *equal* if they have the same size and their corresponding entries are equal. For example,

$$\begin{bmatrix} 2 & 3 & 1 \\ 4 & 6 & 2 \end{bmatrix} = \begin{bmatrix} (3-1) & 3 & 1 \\ 4 & (4+2) & 2 \end{bmatrix}$$

Also,
$$\begin{bmatrix} 1 & 3 & 5 \\ 2 & 4 & 3 \end{bmatrix} \neq \begin{bmatrix} 1 & 2 \\ 3 & 4 \\ 5 & 3 \end{bmatrix}$$

since the matrix on the left has size 2×3, whereas the matrix on the right has size 3×2, and

$$\begin{bmatrix} 2 & 3 \\ 4 & 6 \end{bmatrix} \neq \begin{bmatrix} 2 & 3 \\ 4 & 7 \end{bmatrix}$$

since the corresponding elements in the second row and second column of the two matrices are not equal.

EQUALITY OF MATRICES

Two matrices are equal if they have the same size and their corresponding entries are equal.

EXAMPLE 2   Solve the following matrix equation for x, y, and z:

$$\begin{bmatrix} 1 & x & 3 \\ 2 & y-1 & 2 \end{bmatrix} = \begin{bmatrix} 1 & 4 & z \\ 2 & 1 & 2 \end{bmatrix}$$

Solution   Since the corresponding elements of the two matrices must be equal, we find that $x = 4$, $z = 3$, and $y - 1 = 1$, or $y = 2$.   ∎

Addition and Subtraction

Two matrices A and B of the *same size* can be added or subtracted to produce a matrix of the same size. This is done by adding or subtracting the corresponding entries in the two matrices. For example,

$$\begin{bmatrix} 1 & 3 & 4 \\ -1 & 2 & 0 \end{bmatrix} + \begin{bmatrix} 1 & 4 & 3 \\ 6 & 1 & -2 \end{bmatrix} = \begin{bmatrix} 1+1 & 3+4 & 4+3 \\ -1+6 & 2+1 & 0+(-2) \end{bmatrix} = \begin{bmatrix} 2 & 7 & 7 \\ 5 & 3 & -2 \end{bmatrix}$$

Adding two matrices of the same size

$$\text{and} \begin{bmatrix} 1 & 2 \\ -1 & 3 \\ 4 & 0 \end{bmatrix} - \begin{bmatrix} 2 & -1 \\ 3 & 2 \\ -1 & 0 \end{bmatrix} = \begin{bmatrix} 1-2 & 2-(-1) \\ -1-3 & 3-2 \\ 4-(-1) & 0-0 \end{bmatrix} = \begin{bmatrix} -1 & 3 \\ -4 & 1 \\ 5 & 0 \end{bmatrix}$$

Subtracting two matrices of the same size

ADDITION AND SUBTRACTION OF MATRICES

Suppose A and B are two matrices of the same size.

1. The *sum* $A + B$ is the matrix obtained by adding the corresponding entries in the two matrices.

2. The *difference* $A - B$ is the matrix obtained by subtracting the corresponding entries in B from A.

EXAMPLE 3 The total output of the Acrosonic Company for June is summarized in Table 2.

TABLE 2

	Model A	Model B	Model C	Model D
Location I	210	180	330	180
Location II	400	300	450	40
Location III	420	280	180	740

The output for May was given earlier in Table 1. Find the total output of the company for May and June.

Solution As we saw earlier, the production matrix for the Acrosonic Company in May is given by

$$A = \begin{bmatrix} 320 & 280 & 460 & 280 \\ 480 & 360 & 580 & 0 \\ 540 & 420 & 200 & 880 \end{bmatrix}$$

Next, from Table 2, we see that the production matrix for June is given by

$$B = \begin{bmatrix} 210 & 180 & 330 & 180 \\ 400 & 300 & 450 & 40 \\ 420 & 280 & 180 & 740 \end{bmatrix}$$

Finally, the total output of the Acrosonic Company for May and June is given by the matrix

$$A + B = \begin{bmatrix} 320 & 280 & 460 & 280 \\ 480 & 360 & 580 & 0 \\ 540 & 420 & 200 & 880 \end{bmatrix} + \begin{bmatrix} 210 & 180 & 330 & 180 \\ 400 & 300 & 450 & 40 \\ 420 & 280 & 180 & 740 \end{bmatrix}$$

$$= \begin{bmatrix} 530 & 460 & 790 & 460 \\ 880 & 660 & 1030 & 40 \\ 960 & 700 & 380 & 1620 \end{bmatrix}$$

The following laws hold for matrix addition.

LAWS FOR MATRIX ADDITION

If A, B, and C are matrices of the same size, then

1. $A + B = B + A$ Commutative Law

2. $(A + B) + C = A + (B + C)$ Associative Law

The *commutative law* for matrix addition states that the order in which matrix addition is performed is immaterial. The *associative law* states that when adding three matrices together, we may first add A and B and then add the resulting sum to C. Equivalently, we can add A to the sum of B and C.

A *zero matrix* is one in which all entries are zero. The zero matrix O has the property that

$$A + O = O + A = A$$

for any matrix A having the same size as that of O. For example, the zero matrix of size 3×2 is

$$O = \begin{bmatrix} 0 & 0 \\ 0 & 0 \\ 0 & 0 \end{bmatrix}$$

If A is any 3×2 matrix, then

$$A + O = \begin{bmatrix} a_{11} & a_{12} \\ a_{21} & a_{22} \\ a_{31} & a_{32} \end{bmatrix} + \begin{bmatrix} 0 & 0 \\ 0 & 0 \\ 0 & 0 \end{bmatrix} = \begin{bmatrix} a_{11} & a_{12} \\ a_{21} & a_{22} \\ a_{31} & a_{32} \end{bmatrix} = A$$

where a_{ij} denotes the entry in the ith row and jth column of the matrix A.

The matrix obtained by interchanging the rows and columns of a given matrix A is called the *transpose* of A and is denoted A^T. For example, if

$$A = \begin{bmatrix} 1 & 2 & 3 \\ 4 & 5 & 6 \\ 7 & 8 & 9 \end{bmatrix}$$

then

$$A^T = \begin{bmatrix} 1 & 4 & 7 \\ 2 & 5 & 8 \\ 3 & 6 & 9 \end{bmatrix}$$

TRANSPOSE OF A MATRIX

If A is an $m \times n$ matrix with elements a_{ij}, then the **transpose** of A is the $n \times m$ matrix A^T with elements a_{ji}.

Scalar Multiplication

A matrix A may be multiplied by a real number, called a **scalar** in the context of matrix algebra. The *scalar product*, denoted by cA, is a matrix obtained by multiplying each entry of A by c. For example, the scalar product of the matrix

$$A = \begin{bmatrix} 3 & -1 & 2 \\ 0 & 1 & 4 \end{bmatrix}$$

by the scalar 3 is the matrix

$$3A = 3 \begin{bmatrix} 3 & -1 & 2 \\ 0 & 1 & 4 \end{bmatrix} = \begin{bmatrix} 9 & -3 & 6 \\ 0 & 3 & 12 \end{bmatrix}$$

SCALAR PRODUCT

If A is a matrix and c is a real number, then the **scalar product** cA is the matrix obtained by multiplying each entry of A by c.

EXAMPLE 4   Given

$$A = \begin{bmatrix} 3 & 4 \\ -1 & 2 \end{bmatrix} \quad \text{and} \quad B = \begin{bmatrix} 3 & 2 \\ -1 & 2 \end{bmatrix}$$

find the matrix X satisfying the *matrix equation* $2X + B = 3A$.

Solution   From the given equation $2X + B = 3A$, we find that

$$2X = 3A - B$$

$$= 3 \begin{bmatrix} 3 & 4 \\ -1 & 2 \end{bmatrix} - \begin{bmatrix} 3 & 2 \\ -1 & 2 \end{bmatrix}$$

$$= \begin{bmatrix} 9 & 12 \\ -3 & 6 \end{bmatrix} - \begin{bmatrix} 3 & 2 \\ -1 & 2 \end{bmatrix} = \begin{bmatrix} 6 & 10 \\ -2 & 4 \end{bmatrix}$$

so

$$X = \frac{1}{2} \begin{bmatrix} 6 & 10 \\ -2 & 4 \end{bmatrix} = \begin{bmatrix} 3 & 5 \\ -1 & 2 \end{bmatrix}$$

 EXAMPLE 5   The management of the Acrosonic Company has decided to increase its July production of loudspeaker systems by 10% (over its June output). Find a matrix giving the targeted production for July.

Solution   From the results of Example 3, we see that Acrosonic's total output for June may be represented by the matrix

$$B = \begin{bmatrix} 210 & 180 & 330 & 180 \\ 400 & 300 & 450 & 40 \\ 420 & 280 & 180 & 740 \end{bmatrix}$$

The required matrix is given by

$$(1.1)B = 1.1 \begin{bmatrix} 210 & 180 & 330 & 180 \\ 400 & 300 & 450 & 40 \\ 420 & 280 & 180 & 740 \end{bmatrix}$$

$$= \begin{bmatrix} 231 & 198 & 363 & 198 \\ 440 & 330 & 495 & 44 \\ 462 & 308 & 198 & 814 \end{bmatrix}$$

and is interpreted in the usual manner.

**SELF-CHECK
EXERCISES 5.4**

1. Perform the indicated operations:

$$\begin{bmatrix} 1 & 3 & 2 \\ -1 & 4 & 7 \end{bmatrix} - 3 \begin{bmatrix} 2 & 1 & 0 \\ 1 & 3 & 4 \end{bmatrix}$$

2. Solve the following matrix equation for x, y, and z:

$$\begin{bmatrix} x & 3 \\ z & 2 \end{bmatrix} + \begin{bmatrix} 2-y & z \\ 2-z & -x \end{bmatrix} = \begin{bmatrix} 3 & 7 \\ 2 & 0 \end{bmatrix}$$

3. Jack owns two gas stations, one downtown and the other in the Wilshire district. Over two consecutive days his gas stations recorded gasoline sales represented by the following matrices:

	Regular-unleaded	Unleaded-plus	Super-unleaded
Downtown	1200	750	650
Wilshire	1100	850	600

$$A =$$

and

	Regular-unleaded	Unleaded-plus	Super-unleaded
Downtown	1250	825	550
Wilshire	1150	750	750

$$B =$$

Find a matrix representing the total sales of the two gas stations over the 2-day period.

Solutions to Self-Check Exercises 5.4 can be found on page 230.

5.4   EXERCISES

In exercises 1–6, refer to the following matrices:

$$A = \begin{bmatrix} 2 & -3 & 9 & -4 \\ -11 & 2 & 6 & 7 \\ 6 & 0 & 2 & 9 \\ 5 & 1 & 5 & -8 \end{bmatrix}$$

$$B = \begin{bmatrix} 3 & -1 & 2 \\ 0 & 1 & 4 \\ 3 & 2 & 1 \\ -1 & 0 & 8 \end{bmatrix}$$

$$C = \begin{bmatrix} 1 & 0 & 3 & 4 & 5 \end{bmatrix}, \qquad D = \begin{bmatrix} 1 \\ 3 \\ -2 \\ 0 \end{bmatrix}$$

1. What is the size of A? of B? of C? of D?

2. Find a_{14}, a_{21}, a_{31}, and a_{43}.

3. Find b_{13}, b_{31}, and b_{43}.

4. Identify the row matrix. What is its transpose?

5. Identify the column matrix. What is its transpose?

6. Identify the square matrix. What is its transpose?

In exercises 7–12, refer to the following matrices:

$$A = \begin{bmatrix} -1 & 2 \\ 3 & -2 \\ 4 & 0 \end{bmatrix}, \quad B = \begin{bmatrix} 2 & 4 \\ 3 & 1 \\ -2 & 2 \end{bmatrix},$$

$$C = \begin{bmatrix} 3 & -1 & 0 \\ 2 & -2 & 3 \\ 4 & 6 & 2 \end{bmatrix}, \quad D = \begin{bmatrix} 2 & -2 & 4 \\ 3 & 6 & 2 \\ -2 & 3 & 1 \end{bmatrix}$$

7. What is the size of A? of B? of C? of D?

8. Explain why the matrix $A + C$ does not exist.

9. Compute $A + B$.

10. Compute $2A - 3B$.

11. Compute $C - D$.

12. Compute $4D - 2C$.

In exercises 13–20, perform the indicated operations.

13. $\begin{bmatrix} 6 & 3 & 8 \\ 4 & 5 & 6 \end{bmatrix} - \begin{bmatrix} 3 & -2 & -1 \\ 0 & -5 & -7 \end{bmatrix}$

14. $\begin{bmatrix} 2 & -3 & 4 & -1 \\ 3 & 1 & 0 & 0 \end{bmatrix} + \begin{bmatrix} 4 & 3 & -2 & -4 \\ 6 & 2 & 0 & -3 \end{bmatrix}$

15. $\begin{bmatrix} 1 & 4 & -5 \\ 3 & -8 & 6 \end{bmatrix} + \begin{bmatrix} 4 & 0 & -2 \\ 3 & 6 & 5 \end{bmatrix}$
$- \begin{bmatrix} 2 & 8 & 9 \\ -11 & 2 & -5 \end{bmatrix}$

16. $3 \begin{bmatrix} 1 & 1 & -3 \\ 3 & 2 & 3 \\ 7 & -1 & 6 \end{bmatrix} + 4 \begin{bmatrix} -2 & -1 & 8 \\ 4 & 2 & 2 \\ 3 & 6 & 3 \end{bmatrix}$

17. $\begin{bmatrix} 1.2 & 4.5 & -4.2 \\ 8.2 & 6.3 & -3.2 \end{bmatrix} - \begin{bmatrix} 3.1 & 1.5 & -3.6 \\ 2.2 & -3.3 & -4.4 \end{bmatrix}$

18. $\begin{bmatrix} 0.06 & 0.12 \\ 0.43 & 1.11 \\ 1.55 & -0.43 \end{bmatrix} - \begin{bmatrix} 0.77 & -0.75 \\ 0.22 & -0.65 \\ 1.09 & -0.57 \end{bmatrix}$

19. $\dfrac{1}{2} \begin{bmatrix} 1 & 0 & 0 & -4 \\ 3 & 0 & -1 & 6 \\ -2 & 1 & -4 & 2 \end{bmatrix}$
$+ \dfrac{4}{3} \begin{bmatrix} 3 & 0 & -1 & 4 \\ -2 & 1 & -6 & 2 \\ 8 & 2 & 0 & -2 \end{bmatrix}$
$- \dfrac{1}{3} \begin{bmatrix} 3 & -9 & -1 & 0 \\ 6 & 2 & 0 & -6 \\ 0 & 1 & -3 & 1 \end{bmatrix}$

20. $0.5 \begin{bmatrix} 1 & 3 & 5 \\ 5 & 2 & -1 \\ -2 & 0 & 1 \end{bmatrix} - 0.2 \begin{bmatrix} 2 & 3 & 4 \\ -1 & 1 & -4 \\ 3 & 5 & -5 \end{bmatrix}$
$+ 0.6 \begin{bmatrix} 3 & 4 & -1 \\ 4 & 5 & 1 \\ 1 & 0 & 0 \end{bmatrix}$

In exercises 21–24, solve for u, x, y, and z in the matrix equation.

21. $\begin{bmatrix} 2x - 2 & 3 & 2 \\ 2 & 4 & y - 2 \\ 2z & -3 & 2 \end{bmatrix} = \begin{bmatrix} 3 & u & 2 \\ 2 & 4 & 5 \\ 4 & -3 & 2 \end{bmatrix}$

22. $\begin{bmatrix} x & -2 \\ 3 & y \end{bmatrix} + \begin{bmatrix} -2 & z \\ -1 & 2 \end{bmatrix} = \begin{bmatrix} 4 & -2 \\ 2u & 4 \end{bmatrix}$

23. $\begin{bmatrix} 1 & x \\ 2y & -3 \end{bmatrix} - 4 \begin{bmatrix} 2 & -2 \\ 0 & 3 \end{bmatrix} = \begin{bmatrix} 3z & 10 \\ 4 & -u \end{bmatrix}$

24. $\begin{bmatrix} 1 & 2 \\ 3 & 4 \\ x & -1 \end{bmatrix} - 3 \begin{bmatrix} y - 1 & 2 \\ 1 & 2 \\ 4 & 2z + 1 \end{bmatrix} = 2 \begin{bmatrix} -4 & -u \\ 0 & -1 \\ 4 & 4 \end{bmatrix}$

In exercises 25 and 26, let

$$A = \begin{bmatrix} 2 & -4 & 3 \\ 4 & 2 & 1 \end{bmatrix}, \quad B = \begin{bmatrix} 4 & -3 & 2 \\ 1 & 0 & 4 \end{bmatrix},$$

and $\quad C = \begin{bmatrix} 1 & 0 & 2 \\ 3 & -2 & 1 \end{bmatrix}$

25. Verify by direct computation the validity of the commutative law for matrix addition.

26. Verify by direct computation the validity of the associative law for matrix addition.

In exercises 27–30, let

$$A = \begin{bmatrix} 3 & 1 \\ 2 & 4 \\ -4 & 0 \end{bmatrix} \quad and \quad B = \begin{bmatrix} 1 & 2 \\ -1 & 0 \\ 3 & 2 \end{bmatrix}$$

Verify each equation by direct computation.

27. $(3 + 5)A = 3A + 5A$

28. $2(4A) = (2 \cdot 4)A = 8A$

29. $4(A + B) = 4A + 4B$

30. $2(A - 3B) = 2A - 6B$

In exercises 31–34, find the transpose of the matrix.

31. $\begin{bmatrix} 3 & 2 & -1 & 5 \end{bmatrix}$

32. $\begin{bmatrix} 4 & 2 & 0 & -1 \\ 3 & 4 & -1 & 5 \end{bmatrix}$

33. $\begin{bmatrix} 1 & -1 & 2 \\ 3 & 4 & 2 \\ 0 & 1 & 0 \end{bmatrix}$

34. $\begin{bmatrix} 1 & 2 & 6 & 4 \\ 2 & 3 & 2 & 5 \\ 6 & 2 & 3 & 0 \\ 4 & 5 & 0 & 2 \end{bmatrix}$

35. Cholesterol Levels   Mr. Cross, Mr. Jones, and Mr. Smith each suffer from coronary heart disease. As part of their treatment, they were put on special low-cholesterol diets: Mr. Cross on diet I, Mr. Jones on diet II, and Mr. Smith on diet III. Progressive records of each patient's cholesterol level were kept. At the beginning of the first, second, third, and fourth months the cholesterol levels of the three patients were:

Cross: 220, 215, 210, and 205

Jones: 220, 210, 200, and 195

Smith: 215, 205, 195, and 190

Represent this information in a 3×4 matrix.

36. Bookstore Inventories   The Campus Bookstore's inventory of books is:

Hardcover: Textbooks, 5280; Fiction, 1680; Nonfiction, 2320; Reference, 1890

Paperback: Fiction, 2810; Nonfiction, 1490; Reference, 2070; Textbooks, 1940

The College Bookstore's inventory of books is:

Hardcover: Textbooks, 6340; Fiction, 2220; Nonfiction, 1790; Reference, 1980

Paperback: Fiction, 3100; Nonfiction, 1720; Reference, 2710; Textbooks, 2050

a. Represent Campus's inventory as a matrix A.

b. Represent College's inventory as a matrix B.

c. If the two companies decide to merge, write a matrix C that would represent the total inventory of the newly amalgamated company.

37. Banking   The numbers of three types of bank accounts on January 1 in the Central Bank and its branches are represented by matrix A.

	Checking accounts	Savings accounts	Fixed-deposit accounts
Main office	2820	1470	1120
A = West side branch	1030	520	480
East side branch	1170	540	460

The number and types of accounts opened during the first quarter are represented by matrix B, and the number and types of accounts closed during the same period are represented by matrix C. Thus,

$$B = \begin{bmatrix} 260 & 120 & 110 \\ 140 & 60 & 50 \\ 120 & 70 & 50 \end{bmatrix} \quad C = \begin{bmatrix} 120 & 80 & 80 \\ 70 & 30 & 40 \\ 60 & 20 & 40 \end{bmatrix}$$

a. Find matrix D, which represents the number of each type of account at the end of the first quarter at each location.

b. Because a new manufacturing plant is opening in the immediate area, it is anticipated that there will be a 10% increase in the number of accounts at each location during the second quarter. Write a matrix E to reflect this anticipated increase.

SOLUTIONS TO SELF-CHECK EXERCISES 5.4

1. $\begin{bmatrix} 1 & 3 & 2 \\ -1 & 4 & 7 \end{bmatrix} - 3 \begin{bmatrix} 2 & 1 & 0 \\ 1 & 3 & 4 \end{bmatrix} = \begin{bmatrix} 1 & 3 & 2 \\ -1 & 4 & 7 \end{bmatrix} - \begin{bmatrix} 6 & 3 & 0 \\ 3 & 9 & 12 \end{bmatrix}$

$$= \begin{bmatrix} -5 & 0 & 2 \\ -4 & -5 & -5 \end{bmatrix}$$

2. We are given

$$\begin{bmatrix} x & 3 \\ z & 2 \end{bmatrix} + \begin{bmatrix} 2-y & z \\ 2-z & -x \end{bmatrix} = \begin{bmatrix} 3 & 7 \\ 2 & 0 \end{bmatrix}$$

Performing the indicated operation on the left-hand side, we obtain

$$\begin{bmatrix} 2+x-y & 3+z \\ 2 & 2-x \end{bmatrix} = \begin{bmatrix} 3 & 7 \\ 2 & 0 \end{bmatrix}$$

By the equality of matrices, we have

$$2 + x - y = 3$$
$$3 + z = 7$$
$$2 - x = 0$$

from which we deduce that $x = 2$, $y = 1$, and $z = 4$.

3. The required matrix is

$$A + B = \begin{bmatrix} 1200 & 750 & 650 \\ 1100 & 850 & 600 \end{bmatrix} + \begin{bmatrix} 1250 & 825 & 550 \\ 1150 & 750 & 750 \end{bmatrix}$$

$$= \begin{bmatrix} 2450 & 1575 & 1200 \\ 2250 & 1600 & 1350 \end{bmatrix}$$

Using Technology

The graphing utility can be used to perform matrix addition, matrix subtraction, and scalar multiplication. It can also be used to find the transpose of a matrix. (See Sections A.6 and A.7 in Appendix A.)

EXERCISES

In exercises 1–8, refer to the following matrices, and use a graphic utility or computer software program to perform the indicated operations.

$$A = \begin{bmatrix} 1.2 & 3.1 & -5.4 & 2.7 \\ 4.1 & 3.2 & 4.2 & -3.1 \\ 1.7 & 2.8 & -5.2 & 8.4 \end{bmatrix}$$

$$B = \begin{bmatrix} 6.2 & -3.2 & 1.4 & -1.2 \\ 3.1 & 2.7 & -1.2 & 1.7 \\ 1.2 & -1.4 & -1.7 & 2.8 \end{bmatrix}$$

1. $12.5A$

2. $-8.4B$

3. $A - B$

4. $B - A$

5. $1.3A + 2.4B$

6. $2.1A - 1.7B$

7. $(A + B)^T$

8. $3A^T + 4B^T$

5.5   Multiplication of Matrices

Matrix Product

In Section 5.4 we saw how matrices of the same size may be added or subtracted and how a matrix may be multiplied by a scalar (real number), an operation referred to as scalar multiplication. In this section we will see how, with certain restrictions, one matrix may be multiplied by another matrix.

In order to define matrix multiplication, let us consider the following problem: On a certain day, Al's Service Station sold 1600 gallons of regular-unleaded, 1000 gallons of unleaded-plus, and 800 gallons of super-unleaded gasoline. If the price of gasoline on this day was \$1.04 for regular-unleaded, \$1.19 for unleaded-plus, and \$1.29 for super-unleaded gasoline, find the total revenue realized by Al's for that day.

The day's sale of gasoline may be represented by the matrix

$$A = \begin{bmatrix} 1600 & 1000 & 800 \end{bmatrix} \qquad \text{Row matrix (1} \times \text{3)}$$

Next, we let the unit selling price of regular-unleaded, unleaded-plus, and super-unleaded gasoline be the entries in the matrix

$$B = \begin{bmatrix} 1.04 \\ 1.19 \\ 1.29 \end{bmatrix} \qquad \text{Column matrix (3} \times \text{1)}$$

The first entry in matrix A gives the number of gallons of regular-unleaded gasoline sold, and the first entry in matrix B gives the selling price for each gallon of regular-unleaded gasoline, so their product $(1600)(1.04)$ gives the revenue realized from the sale of regular-unleaded gasoline for the day. A similar interpretation of the second and third entries in the two matrices suggests that we multiply the corresponding entries to obtain the respective revenues realized from the sale of regular-unleaded, unleaded-plus, and super-unleaded gasoline. Finally, the total revenue realized by Al's from the sale of gasoline is given by adding these products to obtain

$$(1600)(1.04) + (1000)(1.19) + (800)(1.29) = 3886$$

or \$3886.

This example suggests that if we have a row matrix of size $1 \times n$,

$$A = \begin{bmatrix} a_1 & a_2 & a_3 & \cdots & a_n \end{bmatrix}$$

and a column matrix of size $n \times 1$,

$$B = \begin{bmatrix} b_1 \\ b_2 \\ b_3 \\ \vdots \\ b_n \end{bmatrix}$$

we may define the *product* of A and B, written AB, by

$$AB = [a_1 \quad a_2 \quad a_3 \quad \cdots \quad a_n] \begin{bmatrix} b_1 \\ b_2 \\ b_3 \\ \vdots \\ b_n \end{bmatrix} = a_1 b_1 + a_2 b_2 + a_3 b_3 + \cdots + a_n b_n \qquad \textbf{\textit{(11)}}$$

EXAMPLE 1   Let

$$A = [1, \quad -2, \quad 3, \quad 5] \qquad \text{and} \qquad B = \begin{bmatrix} 2 \\ 3 \\ 0 \\ -1 \end{bmatrix}.$$

Then

$$AB = [1, \quad -2, \quad 3, \quad 5] \begin{bmatrix} 2 \\ 3 \\ 0 \\ -1 \end{bmatrix} = (1)(2) + (-2)(3) + (3)(0) + (5)(-1) = -9$$

EXAMPLE 2   Judy's stock holdings are given by the matrix

$$\begin{array}{ccc} \text{GM} & \text{IBM} & \text{BAC} \end{array}$$
$$A = [700 \quad 400 \quad 200]$$

At the close of trading on a certain day, the prices (in dollars per share) of these stocks are

$$B = \begin{bmatrix} 50 \\ 120 \\ 42 \end{bmatrix} \begin{array}{c} \text{GM} \\ \text{IBM} \\ \text{BAC} \end{array}$$

What is the total value of Judy's holdings as of that day?

Solution   Judy's holdings are worth

$$AB = [700 \quad 400 \quad 200] \begin{bmatrix} 50 \\ 120 \\ 42 \end{bmatrix} = (700)(50) + (400)(120) + (200)(42)$$

or $91,400.

Returning once again to the matrix product AB in equation (11), observe that the number of columns of the row matrix A is *equal* to the number of rows of the column matrix B. Observe further that the product matrix AB has size 1×1 (a real number may be thought of as a 1×1

matrix). Schematically,

Size of A Size of B

$(1 \times n)$ $(n \times 1)$

$\underline{\hspace{2cm}}$ (1×1) $\underline{\hspace{2cm}}$

Size of AB

More generally, if A is a matrix of size $m \times n$ and B is a matrix of size $n \times p$ (the number of columns of A equals the number of rows of B), then the **matrix product** of A and B, AB, is defined and is a matrix of size $m \times p$. Schematically,

Size of A Size of B

$(m \times n)$ $(n \times p)$

$\underline{\hspace{2cm}}$ $(m \times p)$ $\underline{\hspace{2cm}}$

Size of AB

Next, let us illustrate the mechanics of matrix multiplication by computing the product of a 2×3 matrix A and a 3×4 matrix B. Suppose

$$A = \begin{bmatrix} a_{11} & a_{12} & a_{13} \\ a_{21} & a_{22} & a_{23} \end{bmatrix}$$

and

$$B = \begin{bmatrix} b_{11} & b_{12} & b_{13} & b_{14} \\ b_{21} & b_{22} & b_{23} & b_{24} \\ b_{31} & b_{32} & b_{33} & b_{34} \end{bmatrix}$$

From the schematic

$\overline{\hspace{1cm}}$ same $\overline{\hspace{1cm}}$

Size of A (2×3) (3×4) Size of B

$\underline{\hspace{2cm}}$ (2×4) $\underline{\hspace{2cm}}$

Size of AB

we see that the matrix product $C = AB$ is defined (the number of columns of A equals the number of rows of B) and has size 2×4. Thus

$$C = \begin{bmatrix} c_{11} & c_{12} & c_{13} & c_{14} \\ c_{21} & c_{22} & c_{23} & c_{24} \end{bmatrix}$$

The entries of C are computed as follows: The entry c_{11} (the entry in the *first* row, *first* column of C) is the product of the row matrix comprising the entries from the *first* row of A and the column matrix comprising the *first* column of B. Thus,

$$c_{11} = \begin{bmatrix} a_{11} & a_{12} & a_{13} \end{bmatrix} \begin{bmatrix} b_{11} \\ b_{21} \\ b_{31} \end{bmatrix} = a_{11}b_{11} + a_{12}b_{21} + a_{13}b_{31}$$

The entry c_{12} (the entry in the *first* row, *second* column of C) is the product of the row matrix comprising the *first* row of A and the column matrix comprising the *second* column of B. Thus,

$$c_{12} = [a_{11} \quad a_{12} \quad a_{13}] \begin{bmatrix} b_{12} \\ b_{22} \\ b_{32} \end{bmatrix} = a_{11}b_{12} + a_{12}b_{22} + a_{13}b_{32}$$

The other entries in C are computed in a similar manner.

EXAMPLE 3   Let

$$A = \begin{bmatrix} 3 & 1 & 4 \\ -1 & 2 & 3 \end{bmatrix} \quad \text{and} \quad B = \begin{bmatrix} 1 & 3 & -3 \\ 4 & -1 & 2 \\ 2 & 4 & 1 \end{bmatrix}$$

Compute AB.

Solution   The size of matrix A is 2×3, and the size of matrix B is 3×3. Since the number of columns of matrix A is equal to the number of rows of matrix B, the matrix product $C = AB$ is defined. Furthermore, the size of matrix C is 2×3. Thus,

$$\begin{bmatrix} 3 & 1 & 4 \\ -1 & 2 & 3 \end{bmatrix} \begin{bmatrix} 1 & 3 & -3 \\ 4 & -1 & 2 \\ 2 & 4 & 1 \end{bmatrix} = \begin{bmatrix} c_{11} & c_{12} & c_{13} \\ c_{21} & c_{22} & c_{23} \end{bmatrix}$$

It remains now to determine the entries $c_{11}, c_{12}, c_{13}, c_{21}, c_{22},$ and c_{23}. We have

$$c_{11} = [3 \quad 1 \quad 4] \begin{bmatrix} 1 \\ 4 \\ 2 \end{bmatrix} = (3)(1) + (1)(4) + (4)(2) = 15$$

$$c_{12} = [3 \quad 1 \quad 4] \begin{bmatrix} 3 \\ -1 \\ 4 \end{bmatrix} = (3)(3) + (1)(-1) + (4)(4) = 24$$

$$c_{13} = [3 \quad 1 \quad 4] \begin{bmatrix} -3 \\ 2 \\ 1 \end{bmatrix} = (3)(-3) + (1)(2) + (4)(1) = -3$$

$$c_{21} = [-1 \quad 2 \quad 3] \begin{bmatrix} 1 \\ 4 \\ 2 \end{bmatrix} = (-1)(1) + (2)(4) + (3)(2) = 13$$

$$c_{22} = [-1 \quad 2 \quad 3] \begin{bmatrix} 3 \\ -1 \\ 4 \end{bmatrix} = (-1)(3) + (2)(-1) + (3)(4) = 7$$

and $c_{23} = \begin{bmatrix} -1 & 2 & 3 \end{bmatrix} \begin{bmatrix} -3 \\ 2 \\ 1 \end{bmatrix} = (-1)(-3) + (2)(2) + (3)(1) = 10$

so the required product AB is given by

$$AB = \begin{bmatrix} 15 & 24 & -3 \\ 13 & 7 & 10 \end{bmatrix}$$

EXAMPLE 4   If

$$A = \begin{bmatrix} 3 & 2 & 1 \\ -1 & 2 & 3 \\ 3 & 1 & 4 \end{bmatrix} \quad \text{and} \quad B = \begin{bmatrix} 1 & 3 & 4 \\ 2 & 4 & 1 \\ -1 & 2 & 3 \end{bmatrix}$$

then

$$AB = \begin{bmatrix} 3 \cdot 1 & + 2 \cdot 2 + 1 \cdot (-1) & 3 \cdot 3 & + 2 \cdot 4 + 1 \cdot 2 & 3 \cdot 4 & + 2 \cdot 1 + 1 \cdot 3 \\ (-1) \cdot 1 + 2 \cdot 2 + 3 \cdot (-1) & (-1) \cdot 3 + 2 \cdot 4 + 3 \cdot 2 & (-1) \cdot 4 + 2 \cdot 1 + 3 \cdot 3 \\ 3 \cdot 1 & + 1 \cdot 2 + 4 \cdot (-1) & 3 \cdot 3 & + 1 \cdot 4 + 4 \cdot 2 & 3 \cdot 4 & + 1 \cdot 1 + 4 \cdot 3 \end{bmatrix}$$

$$= \begin{bmatrix} 6 & 19 & 17 \\ 0 & 11 & 7 \\ 1 & 21 & 25 \end{bmatrix}$$

$$BA = \begin{bmatrix} 1 \cdot 3 & + 3 \cdot (-1) + 4 \cdot 3 & 1 \cdot 2 & + 3 \cdot 2 + 4 \cdot 1 & 1 \cdot 1 & + 3 \cdot 3 + 4 \cdot 4 \\ 2 \cdot 3 & + 4 \cdot (-1) + 1 \cdot 3 & 2 \cdot 2 & + 4 \cdot 2 + 1 \cdot 1 & 2 \cdot 1 & + 4 \cdot 3 + 1 \cdot 4 \\ (-1) \cdot 3 + 2 \cdot (-1) + 3 \cdot 3 & (-1) \cdot 2 + 2 \cdot 2 + 3 \cdot 1 & (-1) \cdot 1 + 2 \cdot 3 + 3 \cdot 4 \end{bmatrix}$$

$$= \begin{bmatrix} 12 & 12 & 26 \\ 5 & 13 & 18 \\ 4 & 5 & 17 \end{bmatrix}.$$

As the last example shows, in general, $AB \neq BA$ for any two square matrices A and B. However, the following laws are valid for matrix multiplication.

LAWS FOR MATRIX MULTIPLICATION

If the products and sums are defined for the matrices A, B, and C, then

1. $(AB)C = A(BC)$ Associative law
2. $A(B + C) = AB + AC$ Distributive law

The square matrix of size n having 1s along the main diagonal and zeros elsewhere is called the **identity matrix** of size n.

IDENTITY MATRIX

The identity matrix of size n is given by

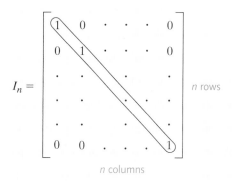

$$I_n = \begin{bmatrix} 1 & 0 & \cdot & \cdot & \cdot & 0 \\ 0 & 1 & \cdot & \cdot & \cdot & 0 \\ \cdot & \cdot & & & \cdot & \cdot \\ \cdot & \cdot & & & & \cdot \\ \cdot & \cdot & & & \cdot & \cdot \\ 0 & 0 & \cdot & \cdot & \cdot & 1 \end{bmatrix} \quad n \text{ rows}$$

n columns

The identity matrix has the property that $I_n A = A$ for any $n \times r$ matrix A, and $BI_n = B$ for any $s \times n$ matrix B. In particular, if A is a square matrix of size n, then

$$I_n A = AI_n = A$$

EXAMPLE 5   Let

$$A = \begin{bmatrix} 1 & 3 & 1 \\ -4 & 3 & 2 \\ 1 & 0 & 1 \end{bmatrix}$$

Then
$$I_3 A = \begin{bmatrix} 1 & 0 & 0 \\ 0 & 1 & 0 \\ 0 & 0 & 1 \end{bmatrix} \begin{bmatrix} 1 & 3 & 1 \\ -4 & 3 & 2 \\ 1 & 0 & 1 \end{bmatrix} = \begin{bmatrix} 1 & 3 & 1 \\ -4 & 3 & 2 \\ 1 & 0 & 1 \end{bmatrix} = A$$

$$AI_3 = \begin{bmatrix} 1 & 3 & 1 \\ -4 & 3 & 2 \\ 1 & 0 & 1 \end{bmatrix} \begin{bmatrix} 1 & 0 & 0 \\ 0 & 1 & 0 \\ 0 & 0 & 1 \end{bmatrix} = \begin{bmatrix} 1 & 3 & 1 \\ -4 & 3 & 2 \\ 1 & 0 & 1 \end{bmatrix} = A$$

so $I_3 A = AI_3$, confirming our result for this special case.   ■

Application

EXAMPLE 6   The Ace Novelty Company received an order from Magic World Amusement Park for 900 "Giant Pandas," 1200 "Saint Bernards," and 2000 "Big Birds." Ace's management decided that 500 Giant Pandas, 800 Saint Bernards, and 1300 Big Birds could be manufactured in their Los Angeles plant, and the balance of the order could be filled by their Seattle plant. Each Panda requires 1.5 square yards of plush, 30 cubic feet of stuffing, and 5 pieces of trim; each Saint Bernard requires 2 square yards of plush, 35 cubic feet of stuffing, and 8 pieces of trim; and each Big Bird requires 2.5 square yards of plush, 25 cubic feet of stuffing, and 15 pieces of trim. The plush costs $4.50 per square yard, the stuffing costs 10 cents per cubic foot, and the trim costs 25 cents per unit.

a. Find how much of each type of material must be purchased for each plant.

b. What is the total cost of materials incurred by each plant and the total cost of materials incurred by Ace Novelty in filling the order?

Solution The quantities of each type of stuffed animal to be produced at each plant location may be expressed as a 2×3 *production matrix P*. Thus,

$$P = \begin{array}{c} \text{L.A.} \\ \text{Seattle} \end{array} \begin{array}{ccc} \text{Pandas} & \text{St. Bernards} & \text{Birds} \\ \begin{bmatrix} 500 & 800 & 1300 \\ 400 & 400 & 700 \end{bmatrix} \end{array}$$

Similarly, we may represent the amount and type of material required to manufacture each type of animal by a 3×3 *activity matrix A*. Thus,

$$A = \begin{array}{c} \text{Pandas} \\ \text{St. Bernards} \\ \text{Birds} \end{array} \begin{array}{ccc} \text{Plush} & \text{Stuffing} & \text{Trim} \\ \begin{bmatrix} 1.5 & 30 & 5 \\ 2 & 35 & 8 \\ 2.5 & 25 & 15 \end{bmatrix} \end{array}$$

Finally, the unit cost for each type of material may be represented by the 3×1 *cost matrix C*:

$$C = \begin{array}{c} \text{Plush} \\ \text{Stuffing} \\ \text{Trim} \end{array} \begin{bmatrix} 4.50 \\ 0.10 \\ 0.25 \end{bmatrix}$$

a. The amount of each type of material required for each plant is given by the matrix PA. Thus,

$$PA = \begin{bmatrix} 500 & 800 & 1300 \\ 400 & 400 & 700 \end{bmatrix} \begin{bmatrix} 1.5 & 30 & 5 \\ 2 & 35 & 8 \\ 2.5 & 25 & 15 \end{bmatrix}$$

$$= \begin{array}{c} \text{L.A.} \\ \text{Seattle} \end{array} \begin{array}{ccc} \text{Plush} & \text{Stuffing} & \text{Trim} \\ \begin{bmatrix} 5,600 & 75,500 & 28,400 \\ 3,150 & 43,500 & 15,700 \end{bmatrix} \end{array}$$

b. The total cost of materials for each plant is given by the matrix PAC:

$$PAC = \begin{bmatrix} 5,600 & 75,500 & 28,400 \\ 3,150 & 43,500 & 15,700 \end{bmatrix} \begin{bmatrix} 4.50 \\ 0.10 \\ 0.25 \end{bmatrix}$$

$$= \begin{array}{c} \text{L.A.} \\ \text{Seattle} \end{array} \begin{bmatrix} 39,850 \\ 22,450 \end{bmatrix}$$

or \$39,850 for the L.A. plant and \$22,450 for the Seattle plant. Thus, the total cost of materials incurred by Ace Novelty is \$62,300. ∎

Matrix Representation

Example 7 shows how a system of linear equations may be written in a compact form with the help of matrices. (We will use this matrix equation representation in Section 5.6.)

EXAMPLE 7   Write the system of linear equations

$$2x - 4y + z = 6$$
$$-3x + 6y - 5z = -1$$
$$x - 3y + 7z = 0$$

in matrix form.

Solution   Let us write

$$A = \begin{bmatrix} 2 & -4 & 1 \\ -3 & 6 & -5 \\ 1 & -3 & 7 \end{bmatrix}, \quad X = \begin{bmatrix} x \\ y \\ z \end{bmatrix}, \quad B = \begin{bmatrix} 6 \\ -1 \\ 0 \end{bmatrix}$$

Note that A is just the 3×3 matrix of coefficients of the system, X is the 3×1 column matrix of unknowns (variables), and B is the 3×1 column matrix of constants. We now show that the required matrix representation of the system of linear equations is

$$AX = B$$

To see this, observe that

$$AX = \begin{bmatrix} 2 & -4 & 1 \\ -3 & 6 & -5 \\ 1 & -3 & 7 \end{bmatrix} \begin{bmatrix} x \\ y \\ z \end{bmatrix} = \begin{bmatrix} 2x - 4y + z \\ -3x + 6y - 5z \\ x - 3y + 7z \end{bmatrix}$$

Equating this 3×1 matrix with matrix B now gives

$$\begin{bmatrix} 2x - 4y + z \\ -3x + 6y - 5z \\ x - 3y + 7z \end{bmatrix} = \begin{bmatrix} 6 \\ -1 \\ 0 \end{bmatrix}$$

which, by matrix equality, is easily seen to be equivalent to the given system of linear equations. ■

**SELF-CHECK
EXERCISES 5.5**

1. Compute

$$\begin{bmatrix} 1 & 3 & 0 \\ 2 & 4 & -1 \end{bmatrix} \begin{bmatrix} 3 & 1 & 4 \\ 2 & 0 & 3 \\ 1 & 2 & -1 \end{bmatrix}$$

2. Write the following system of linear equations in matrix form:

$$y - 2z = 1$$
$$2x - y + 3z = 0$$
$$x \qquad + 4z = 7$$

3. On July 1, the stock holdings of Ash and Joan Robinson were given by the matrix

$$A = \begin{array}{c} \\ Ash \\ Joan \end{array} \begin{array}{cccc} \text{AT\&T} & \text{GTE} & \text{IBM} & \text{GM} \\ \begin{bmatrix} 2000 & 1000 & 500 & 5000 \\ 1000 & 500 & 2000 & 0 \end{bmatrix} \end{array}$$

and the closing prices of AT&T, GTE, IBM, and GM were \$24, \$47, \$150, and \$14 per share, respectively. Use matrix multiplication to determine the separate values of Ash's and Joan's stock holdings as of that date.

Solutions to Self-Check Exercises 5.5 can be found on page 243.

5.5 EXERCISES

In exercises 1–4, the sizes of matrices A and B are given. Find the size of AB and BA whenever they are defined.

1. A is of size 2×3 and B is of size 3×5

2. A is of size 3×4 and B is of size 4×3

3. A is of size 1×7 and B is of size 7×1

4. A is of size 4×4 and B is of size 4×4

5. Let A be a matrix of size $m \times n$ and B be a matrix of size $s \times t$. Find conditions on m, n, s, and t so that both matrix products AB and BA are defined.

6. Find condition(s) on the size of a matrix A so that A^2 (that is, AA) is defined.

In exercises 7–24, compute the indicated products.

7. $\begin{bmatrix} 1 & 2 \\ 3 & 0 \end{bmatrix} \begin{bmatrix} 1 \\ -1 \end{bmatrix}$

8. $\begin{bmatrix} -1 & 3 \\ 5 & 0 \end{bmatrix} \begin{bmatrix} 7 \\ 2 \end{bmatrix}$

9. $\begin{bmatrix} 3 & 1 & 2 \\ -1 & 2 & 4 \end{bmatrix} \begin{bmatrix} 4 \\ 1 \\ -2 \end{bmatrix}$

10. $\begin{bmatrix} 3 & 2 & -1 \\ 4 & -1 & 0 \\ -5 & 2 & 1 \end{bmatrix} \begin{bmatrix} 3 \\ -2 \\ 0 \end{bmatrix}$

11. $\begin{bmatrix} -1 & 2 \\ 3 & 1 \end{bmatrix} \begin{bmatrix} 2 & 4 \\ 3 & 1 \end{bmatrix}$

12. $\begin{bmatrix} 1 & 3 \\ -1 & 2 \end{bmatrix} \begin{bmatrix} 1 & 3 & 0 \\ 3 & 0 & 2 \end{bmatrix}$

13. $\begin{bmatrix} 2 & 1 & 2 \\ 3 & 2 & 4 \end{bmatrix} \begin{bmatrix} -1 & 2 \\ 4 & 3 \\ 0 & 1 \end{bmatrix}$

14. $\begin{bmatrix} -1 & 2 \\ 4 & 3 \\ 0 & 1 \end{bmatrix} \begin{bmatrix} 2 & 1 & 2 \\ 3 & 2 & 4 \end{bmatrix}$

15. $\begin{bmatrix} 0.1 & 0.9 \\ 0.2 & 0.8 \end{bmatrix} \begin{bmatrix} 1.2 & 0.4 \\ 0.5 & 2.1 \end{bmatrix}$

16. $\begin{bmatrix} 1.2 & 0.3 \\ 0.4 & 0.5 \end{bmatrix} \begin{bmatrix} 0.2 & 0.6 \\ 0.4 & -0.5 \end{bmatrix}$

17. $\begin{bmatrix} 6 & -3 & 0 \\ -2 & 1 & -8 \\ 4 & -4 & 9 \end{bmatrix} \begin{bmatrix} 1 & 0 & 0 \\ 0 & 1 & 0 \\ 0 & 0 & 1 \end{bmatrix}$

18. $\begin{bmatrix} 2 & 4 \\ -1 & -5 \\ 3 & -1 \end{bmatrix} \begin{bmatrix} 2 & -2 & 4 \\ 1 & 3 & -1 \end{bmatrix}$

19. $\begin{bmatrix} 3 & 0 & -2 & 1 \\ 1 & 2 & 0 & -1 \end{bmatrix} \begin{bmatrix} 2 & 1 & -1 \\ -1 & 2 & 0 \\ 0 & 0 & 1 \\ -1 & -2 & 2 \end{bmatrix}$

20. $\begin{bmatrix} 2 & 1 & -3 & 0 \\ 4 & -2 & -1 & 1 \\ -1 & 2 & 0 & 1 \end{bmatrix} \begin{bmatrix} 2 & -1 \\ 1 & 4 \\ 3 & -3 \\ 0 & -5 \end{bmatrix}$

21. $4 \begin{bmatrix} 1 & -2 & 0 \\ 2 & -1 & 1 \\ 3 & 0 & -1 \end{bmatrix} \begin{bmatrix} 1 & 3 & 1 \\ 1 & 4 & 0 \\ 0 & 1 & -2 \end{bmatrix}$

22. $3 \begin{bmatrix} 2 & -1 & 0 \\ 2 & 1 & 2 \\ 1 & 0 & -1 \end{bmatrix} \begin{bmatrix} 2 & 3 & 1 \\ 3 & -3 & 0 \\ 0 & 1 & -1 \end{bmatrix}$

23. $\begin{bmatrix} 1 & 0 \\ 0 & 1 \end{bmatrix} \begin{bmatrix} 4 & -3 & 2 \\ 7 & 1 & -5 \end{bmatrix} \begin{bmatrix} 1 & 0 & 0 \\ 0 & 1 & 0 \\ 0 & 0 & 1 \end{bmatrix}$

24. $2 \begin{bmatrix} 3 & 2 & -1 \\ 0 & 1 & 3 \\ 2 & 0 & 3 \end{bmatrix} \begin{bmatrix} 1 & 0 & 0 \\ 0 & 1 & 0 \\ 0 & 0 & 1 \end{bmatrix} \begin{bmatrix} 1 & 2 & 0 \\ 0 & -1 & -2 \\ 1 & 3 & 1 \end{bmatrix}$

In exercises 25 and 26, let

$$A = \begin{bmatrix} 1 & 0 & -2 \\ 1 & -3 & 2 \\ -2 & 1 & 1 \end{bmatrix}, \quad B = \begin{bmatrix} 3 & 1 & 0 \\ 2 & 2 & 0 \\ 1 & -3 & -1 \end{bmatrix},$$

and

$$C = \begin{bmatrix} 2 & -1 & 0 \\ 1 & -1 & 2 \\ 3 & -2 & 1 \end{bmatrix}$$

25. Verify the validity of the associative law for matrix multiplication.

26. Verify the validity of the distributive law for matrix multiplication.

27. Let

$$A = \begin{bmatrix} 1 & 2 \\ 3 & 4 \end{bmatrix} \quad \text{and} \quad B = \begin{bmatrix} 2 & 1 \\ 4 & 3 \end{bmatrix}$$

Compute AB and BA and hence deduce that matrix multiplication is, in general, not commutative.

28. Let

$$A = \begin{bmatrix} 0 & 3 & 0 \\ 1 & 0 & 1 \\ 0 & 2 & 0 \end{bmatrix}, \quad B = \begin{bmatrix} 2 & 4 & 5 \\ 3 & -1 & -6 \\ 4 & 3 & 4 \end{bmatrix},$$

and

$$C = \begin{bmatrix} 4 & 5 & 6 \\ 3 & -1 & -6 \\ 2 & 2 & 3 \end{bmatrix}$$

a. Compute AB.
b. Compute AC.
c. Using the results of (a) and (b), conclude that $AB = AC$ does not imply that $B = C$.

29. Let

$$A = \begin{bmatrix} 3 & 0 \\ 8 & 0 \end{bmatrix} \quad \text{and} \quad B = \begin{bmatrix} 0 & 0 \\ 4 & 5 \end{bmatrix}$$

Show that $AB = O$, thereby demonstrating that for matrix multiplication the equation $AB = O$ does not imply that one or both of the matrices A and B must be the zero matrix.

30. Let

$$A = \begin{bmatrix} 2 & 2 \\ -2 & -2 \end{bmatrix}$$

Show that $A^2 = O$. Compare this with the equation $a^2 = 0$, where a is a real number.

31. Find the matrix A such that

$$A \begin{bmatrix} 1 & 0 \\ -1 & 3 \end{bmatrix} = \begin{bmatrix} -1 & -3 \\ 3 & 6 \end{bmatrix}$$

Hint: Let $A = \begin{bmatrix} a & b \\ c & d \end{bmatrix}$.

32. Let

$$A = \begin{bmatrix} 3 & 1 \\ 0 & 2 \end{bmatrix} \quad \text{and} \quad B = \begin{bmatrix} 4 & -2 \\ 2 & 1 \end{bmatrix}$$

a. Compute $(A + B)^2$.
b. Compute $A^2 + 2AB + B^2$.
c. From the results of (a) and (b), show that, in general, $(A + B)^2 \neq A^2 + 2AB + B^2$.

33. Let

$$A = \begin{bmatrix} 2 & 4 \\ 5 & -6 \end{bmatrix} \quad \text{and} \quad B = \begin{bmatrix} 4 & 8 \\ -7 & 3 \end{bmatrix}$$

a. Find A^T and show that $(A^T)^T = A$.
b. Show that $(A + B)^T = A^T + B^T$.
c. Show that $(AB)^T = B^T A^T$.

34. Let

$$A = \begin{bmatrix} 1 & 3 \\ -2 & -1 \end{bmatrix} \quad \text{and} \quad B = \begin{bmatrix} 3 & -4 \\ 2 & -2 \end{bmatrix}$$

a. Find A^T and show that $(A^T)^T = A$.
b. Show that $(A + B)^T = A^T + B^T$.
c. Show that $(AB)^T = B^T A^T$.

In exercises 35–40, write the system of linear equations in matrix form.

35. $2x - 3y = 7$
$3x - 4y = 8$

36. $2x \quad\quad = 7$
$3x - 2y = 12$

37. $2x - 3y + 4z = 6$
$2y - 3z = 7$
$x - y + 2z = 4$

38. $x - 2y + 3z = -1$
$3x + 4y - 2z = 1$
$2x - 3y + 7z = 6$

39. $-x_1 + x_2 + x_3 = 0$
$2x_1 - x_2 - x_3 = 2$
$-3x_1 + 2x_2 + 4x_3 = 4$

40. $3x_1 - 5x_2 + 4x_3 = 10$
$4x_1 + 2x_2 - 3x_3 = -12$
$-x_1 \quad\quad + x_3 = -2$

41. Box-Office Receipts The Cinema Center comprises four theaters: Cinemas I, II, III, and IV. The admission price for one feature at the Center is $2 for children, $3 for students, and $4 for adults. The attendance for the Sunday matinee is given by the matrix

$$A = \begin{array}{c} \\ \text{Cinema I} \\ \text{Cinema II} \\ \text{Cinema III} \\ \text{Cinema IV} \end{array} \begin{array}{ccc} \text{Children} & \text{Students} & \text{Adults} \\ \left[\begin{array}{ccc} 225 & 110 & 50 \\ 75 & 180 & 225 \\ 280 & 85 & 110 \\ 0 & 250 & 225 \end{array}\right] \end{array}$$

Write a column vector B representing the admission prices. Then compute AB, the column vector showing the gross receipts for each theater. Finally, find the total revenue collected at the Cinema Center for admission that Sunday afternoon.

42. Real Estate Bond Brothers, Inc., a real estate developer, builds houses in three states. The projected number of units of each model to be built in each state is given by the matrix

$$A = \begin{array}{c} \\ \text{N.Y.} \\ \text{Conn.} \\ \text{Mass.} \end{array} \begin{array}{cccc} \multicolumn{4}{c}{\text{Model}} \\ \text{I} & \text{II} & \text{III} & \text{IV} \\ \left[\begin{array}{cccc} 60 & 80 & 120 & 40 \\ 20 & 30 & 60 & 10 \\ 10 & 15 & 30 & 5 \end{array}\right] \end{array}$$

The profits to be realized are $20,000, $22,000, $25,000, and $30,000, respectively, for each Model I, II, III, and IV house sold.
  a. Write a column matrix B representing the profit for each type of house.
  b. Find the total profit Bond Brothers expects to earn in each state if all the houses are sold.

43. Politics: Voter Affiliation Matrix A gives the percentage of eligible voters in the city of Newton, classified according to party affiliation and age group.

$$A = \begin{array}{c} \\ \text{Under 30} \\ \text{30 to 50} \\ \text{Over 50} \end{array} \begin{array}{ccc} \text{Dem.} & \text{Rep.} & \text{Ind.} \\ \left[\begin{array}{ccc} .50 & .30 & .20 \\ .45 & .40 & .15 \\ .40 & .50 & .10 \end{array}\right] \end{array}$$

The population of eligible voters in the city by age group is given by the matrix B:

$$\begin{array}{ccc} \text{Under 30} & \text{30 to 50} & \text{Over 50} \end{array}$$
$$B = [\,30{,}000 \quad 40{,}000 \quad 20{,}000\,]$$

Find a matrix giving the total number of eligible voters in the city who will vote Democrat, Republican, and Independent.

44. College Admissions A university admissions committee anticipates an enrollment of 8000 students in its freshman class next year. To satisfy admission quotas, incoming students have been categorized according to their sex and place of residence. The number of students in each category is given by the matrix

$$A = \begin{array}{c} \\ \text{In-state} \\ \text{Out-of-state} \\ \text{Foreign} \end{array} \begin{array}{cc} \text{Male} & \text{Female} \\ \left[\begin{array}{cc} 2700 & 3000 \\ 800 & 700 \\ 500 & 300 \end{array}\right] \end{array}$$

By using data accumulated in previous years, the admissions committee has determined that these students will elect to enter the College of Letters and Science, the College of Fine Arts, the School of Business Administration, and the School of Engineering according to the percentages that appear in the following matrix:

$$B = \begin{array}{c} \\ \text{Male} \\ \text{Female} \end{array} \begin{array}{cccc} \text{L. \& S.} & \text{Fine Arts} & \text{Bus. Ad.} & \text{Eng.} \\ \left[\begin{array}{cccc} .25 & .20 & .30 & .25 \\ .30 & .35 & .25 & .10 \end{array}\right] \end{array}$$

Find the matrix AB that shows the number of in-state, out-of-state, and foreign students expected to enter each discipline.

45. Production Planning Refer to Example 6. Suppose that Ace Novelty received an order from another amusement park for 1200 Pink Panthers, 1800 Giant Pandas, and 1400 Big Birds. The quantity of each type of stuffed animal to be produced at each plant is shown in the following production matrix:

$$P = \begin{array}{c} \\ \text{L.A.} \\ \text{Seattle} \end{array} \begin{array}{ccc} \text{Panthers} & \text{Pandas} & \text{Birds} \\ \begin{bmatrix} 700 & 1000 & 800 \\ 500 & 800 & 600 \end{bmatrix} \end{array}$$

Each Panther requires 1.3 square yards of plush, 20 cubic feet of stuffing, and 12 pieces of trim. Assume the materials required to produce the other two stuffed animals and the unit cost for each type of material are the same as those given in Example 6.

a. Find how much of each type of material must be purchased for each plant.

b. Find the total cost of materials that will be incurred at each plant.

c. Find the total cost of materials incurred by Ace Novelty in filling the order.

SOLUTIONS TO SELF-CHECK EXERCISES 5.5

1. We compute

$$\begin{bmatrix} 1 & 3 & 0 \\ 2 & 4 & -1 \end{bmatrix} \begin{bmatrix} 3 & 1 & 4 \\ 2 & 0 & 3 \\ 1 & 2 & -1 \end{bmatrix}$$

$$= \begin{bmatrix} 1(3) + 3(2) + 0(1) & 1(1) + 3(0) + 0(2) & 1(4) + 3(3) + 0(-1) \\ 2(3) + 4(2) - 1(1) & 2(1) + 4(0) - 1(2) & 2(4) + 4(3) - 1(-1) \end{bmatrix}$$

$$= \begin{bmatrix} 9 & 1 & 13 \\ 13 & 0 & 21 \end{bmatrix}$$

2. Let

$$A = \begin{bmatrix} 0 & 1 & -2 \\ 2 & -1 & 3 \\ 1 & 0 & 4 \end{bmatrix}, \qquad X = \begin{bmatrix} x \\ y \\ z \end{bmatrix}, \qquad B = \begin{bmatrix} 1 \\ 0 \\ 7 \end{bmatrix}$$

Then the given system may be written as the matrix equation

$$AX = B$$

3. Write

$$B = \begin{bmatrix} 24 \\ 47 \\ 150 \\ 14 \end{bmatrix} \begin{array}{l} \text{AT\&T} \\ \text{GTE} \\ \text{IBM} \\ \text{GM} \end{array}$$

and compute

$$AB = \begin{array}{c} \text{Ash} \\ \text{Joan} \end{array} \begin{bmatrix} 2000 & 1000 & 500 & 5000 \\ 1000 & 500 & 2000 & 0 \end{bmatrix} \begin{bmatrix} 24 \\ 47 \\ 150 \\ 14 \end{bmatrix}$$

$$= \begin{bmatrix} 240{,}000 \\ 347{,}500 \end{bmatrix} \begin{array}{l} \text{Ash} \\ \text{Joan} \end{array}$$

We conclude that Ash's stock holdings were worth $240,000 and Joan's stock holdings were worth $347,500 on July 1.

U s i n g   T e c h n o l o g y

The graphing utility can be used to perform matrix multiplication. In exercises 1–8, refer to the following matrices and use a graphic utility or computer software program to perform the indicated operations. Express your answers accurate to two decimal places.

$$A = \begin{bmatrix} 1.2 & 3.1 & -1.2 & 4.3 \\ 7.2 & 6.3 & 1.8 & -2.1 \\ 0.8 & 3.2 & -1.3 & 2.8 \end{bmatrix},$$

$$B = \begin{bmatrix} 0.7 & 0.3 & 1.2 & -0.8 \\ 1.2 & 1.7 & 3.5 & 4.2 \\ -3.3 & -1.2 & 4.2 & 3.2 \end{bmatrix},$$

$$C = \begin{bmatrix} 0.8 & 7.1 & 6.2 \\ 3.3 & -1.2 & 4.8 \\ 1.3 & 2.8 & -1.5 \\ 2.1 & 3.2 & -8.4 \end{bmatrix}$$

1. AC

2. CB

3. $(A + B)C$

4. $(2A + 3B)C$

5. $(2A - 3.1B)C$

6. $C(2.1A + 3.2B)$

7. $(AC)^T$

8. $(CA)^T$

In exercises 9–12, refer to the following matrices and use a graphic utility or computer software program to perform the indicated operations. Express your answers accurate to two decimal places.

$$A = \begin{bmatrix} 2 & 5 & -4 & 2 & 8 \\ 6 & 7 & 2 & 9 & 6 \\ 4 & 5 & 4 & 4 & 4 \\ 9 & 6 & 8 & 3 & 2 \end{bmatrix},$$

$$B = \begin{bmatrix} 2 & 6 & 7 & 5 \\ 3 & 4 & 6 & 2 \\ -5 & 8 & 4 & 3 \\ 8 & 6 & 9 & 5 \\ 4 & 7 & 8 & 8 \end{bmatrix},$$

$$C = \begin{bmatrix} 6.2 & 7.3 & -4.0 & 7.1 & 9.3 \\ 4.8 & 6.5 & 8.4 & -6.3 & 8.4 \\ 5.4 & 3.2 & 6.3 & 9.1 & -2.8 \\ 8.2 & 7.3 & 6.5 & 4.1 & 9.8 \\ 10.3 & 6.8 & 4.8 & -9.1 & 20.4 \end{bmatrix},$$

$$D = \begin{bmatrix} 4.6 & 3.9 & 8.4 & 6.1 & 9.8 \\ 2.4 & -6.8 & 7.9 & 11.4 & 2.9 \\ 7.1 & 9.4 & 6.3 & 5.7 & 4.2 \\ 3.4 & 6.1 & 5.3 & 8.4 & 6.3 \\ 7.1 & -4.2 & 3.9 & -6.4 & 7.1 \end{bmatrix},$$

9. Find AB and BA.

10. Find CD and DC. Is $CD = DC$?

11. Find $AC + AD$.

12. Find
   a. AC
   b. AD
   c. $A(C + D)$
   d. Is $A(C + D) = AC + AD$?

5.6 The Inverse of a Square Matrix

The Inverse of a Square Matrix

In this section, we discuss a procedure for finding the inverse of a matrix and show how the inverse can be used to help us solve a system of linear equations.

Recall that if a is a nonzero real number, then there exists a unique real number a^{-1} (that is, $1/a$) such that

$$a^{-1}a = \left(\frac{1}{a}\right)(a) = 1$$

The use of the inverse of a real number enables us to solve algebraic equations of the form

$$ax = b \tag{12}$$

For if $a \neq 0$, then $a^{-1} = 1/a$. Upon multiplying both sides of equation (12) by a^{-1}, we have

$$a^{-1}(ax) = a^{-1}b,$$

or

$$\left(\frac{1}{a}\right)(ax) = \frac{1}{a}(b)$$

and

$$x = \frac{b}{a}$$

For example, since the inverse of 2 is $2^{-1} = 1/2$, we can solve the equation

$$2x = 5$$

by multiplying both sides of the equation by $2^{-1} = 1/2$, giving

$$2^{-1}(2x) = 2^{-1} \cdot 5$$

or

$$x = \frac{5}{2}$$

We can use a similar procedure to solve the matrix equation

$$AX = B$$

where A, X, and B are matrices of the proper sizes. In order to do this, we need the matrix equivalent of the inverse of a real number. Such a matrix, whenever it exists, is called the *inverse of a matrix*.

INVERSE OF A MATRIX

Let A be a square matrix of size n. A square matrix A^{-1} of size n such that

$$A^{-1}A = AA^{-1} = I_n$$

is called the **inverse** of A.

Let us show that the matrix

$$A = \begin{bmatrix} 1 & 2 \\ 3 & 4 \end{bmatrix}$$

has as its inverse

$$A^{-1} = \begin{bmatrix} -2 & 1 \\ \frac{3}{2} & -\frac{1}{2} \end{bmatrix}$$

Since

$$AA^{-1} = \begin{bmatrix} 1 & 2 \\ 3 & 4 \end{bmatrix} \begin{bmatrix} -2 & 1 \\ \frac{3}{2} & -\frac{1}{2} \end{bmatrix} = \begin{bmatrix} 1 & 0 \\ 0 & 1 \end{bmatrix}$$

and

$$A^{-1}A = \begin{bmatrix} -2 & 1 \\ \frac{3}{2} & -\frac{1}{2} \end{bmatrix} \begin{bmatrix} 1 & 2 \\ 3 & 4 \end{bmatrix} = \begin{bmatrix} 1 & 0 \\ 0 & 1 \end{bmatrix} = I$$

we see that A^{-1} is the inverse of A, as asserted.

Not every square matrix has an inverse. A matrix that does not have an inverse is called **singular.** An example of a singular matrix is given by

$$B = \begin{bmatrix} 0 & 1 \\ 0 & 0 \end{bmatrix}$$

If B had an inverse given by

$$B^{-1} = \begin{bmatrix} a & b \\ c & d \end{bmatrix}$$

where $a, b, c,$ and d are some appropriate numbers, then, by the definition of an inverse, we would have $BB^{-1} = I$, that is,

$$\begin{bmatrix} 0 & 1 \\ 0 & 0 \end{bmatrix} \begin{bmatrix} a & b \\ c & d \end{bmatrix} = \begin{bmatrix} 1 & 0 \\ 0 & 1 \end{bmatrix}$$

or

$$\begin{bmatrix} c & d \\ 0 & 0 \end{bmatrix} = \begin{bmatrix} 1 & 0 \\ 0 & 1 \end{bmatrix}$$

which implies that $0 = 1$—an impossibility! This contradiction shows that B does not have an inverse.

A Method for Finding the Inverse of a Square Matrix

The methods of Section 5.5 can be used to find the inverse of a nonsingular matrix. In order to discover such an algorithm, let us find the inverse of the matrix A, given by

$$A = \begin{bmatrix} 1 & 2 \\ -1 & 3 \end{bmatrix}$$

Suppose A^{-1} exists and is given by

$$A^{-1} = \begin{bmatrix} a & b \\ c & d \end{bmatrix}$$

where $a, b, c,$ and d are to be determined. By the definition of an inverse, we have $AA^{-1} = I$—that is,

$$\begin{bmatrix} 1 & 2 \\ -1 & 3 \end{bmatrix} \begin{bmatrix} a & b \\ c & d \end{bmatrix} = \begin{bmatrix} 1 & 0 \\ 0 & 1 \end{bmatrix}$$

which simplifies to

$$\begin{bmatrix} a + 2c & b + 2d \\ -a + 3c & -b + 3d \end{bmatrix} = \begin{bmatrix} 1 & 0 \\ 0 & 1 \end{bmatrix}$$

But this matrix equation is equivalent to the two systems of linear equations

$$\left.\begin{array}{r} a + 2c = 1 \\ -a + 3c = 0 \end{array}\right\} \qquad \left.\begin{array}{r} b + 2d = 0 \\ -b + 3d = 1 \end{array}\right\}$$

with augmented matrices given by

$$\left[\begin{array}{cc|c} 1 & 2 & 1 \\ -1 & 3 & 0 \end{array}\right] \quad \text{and} \quad \left[\begin{array}{cc|c} 1 & 2 & 0 \\ -1 & 3 & 1 \end{array}\right]$$

Note that the matrices of coefficients of the two systems are identical. This suggests that we solve the two systems of simultaneous linear equations by writing the following augmented matrix, which we obtain by joining the coefficient matrix and the two columns of constants:

$$\left[\begin{array}{cc|cc} 1 & 2 & 1 & 0 \\ -1 & 3 & 0 & 1 \end{array}\right]$$

Using the Gauss-Jordan elimination method, we obtain the following sequence of equivalent matrices:

$$\left[\begin{array}{cc|cc} 1 & 2 & 1 & 0 \\ -1 & 3 & 0 & 1 \end{array}\right] \xrightarrow{R_2 + R_1} \left[\begin{array}{cc|cc} 1 & 2 & 1 & 0 \\ 0 & 5 & 1 & 1 \end{array}\right] \xrightarrow{\frac{1}{5}R_2}$$

$$\left[\begin{array}{cc|cc} 1 & 2 & 1 & 0 \\ 0 & 1 & \frac{1}{5} & \frac{1}{5} \end{array}\right] \xrightarrow{R_1 - 2R_2} \left[\begin{array}{cc|cc} 1 & 0 & \frac{3}{5} & -\frac{2}{5} \\ 0 & 1 & \frac{1}{5} & \frac{1}{5} \end{array}\right]$$

Thus, $a = 3/5$, $c = 1/5$, $b = -2/5$, and $d = 1/5$, giving

$$A^{-1} = \left[\begin{array}{cc} \frac{3}{5} & -\frac{2}{5} \\ \frac{1}{5} & \frac{1}{5} \end{array}\right]$$

The following computations verify that A^{-1} is indeed the inverse of A:

$$\left[\begin{array}{cc} 1 & 2 \\ -1 & 3 \end{array}\right]\left[\begin{array}{cc} \frac{3}{5} & -\frac{2}{5} \\ \frac{1}{5} & \frac{1}{5} \end{array}\right] = \left[\begin{array}{cc} 1 & 0 \\ 0 & 1 \end{array}\right] = \left[\begin{array}{cc} \frac{3}{5} & -\frac{2}{5} \\ \frac{1}{5} & \frac{1}{5} \end{array}\right]\left[\begin{array}{cc} 1 & 2 \\ -1 & 3 \end{array}\right]$$

This example suggests the following general algorithm for computing the inverse of a square matrix of size n when it exists.

FINDING THE INVERSE OF A MATRIX

Given the $n \times n$ matrix A:

1. Adjoin the $n \times n$ identity matrix I to obtain the augmented matrix

$$[A \,|\, I]$$

2. Use a sequence of row operations to reduce $[A \,|\, I]$ to the form

$$[I \,|\, B]$$

if possible.

The matrix B is the inverse of A.

REMARK Although matrix multiplication is not generally commutative, it is possible to prove that if $AB = I$, then $BA = I$ also. Hence, to verify that B is the inverse of A, it suffices to show that $AB = I$. □

EXAMPLE 1 Find the inverse of the matrix

$$A = \begin{bmatrix} 2 & 1 & 1 \\ 3 & 2 & 1 \\ 2 & 1 & 2 \end{bmatrix}$$

Solution We form the augmented matrix

$$\begin{bmatrix} 2 & 1 & 1 & \Big| & 1 & 0 & 0 \\ 3 & 2 & 1 & \Big| & 0 & 1 & 0 \\ 2 & 1 & 2 & \Big| & 0 & 0 & 1 \end{bmatrix}$$

and use the Gauss-Jordan elimination method to reduce it to the form $[I \,|\, B]$:

$$\begin{bmatrix} 2 & 1 & 1 & \Big| & 1 & 0 & 0 \\ 3 & 2 & 1 & \Big| & 0 & 1 & 0 \\ 2 & 1 & 2 & \Big| & 0 & 0 & 1 \end{bmatrix} \xrightarrow{R_1 - R_2} \begin{bmatrix} -1 & -1 & 0 & \Big| & 1 & -1 & 0 \\ 3 & 2 & 1 & \Big| & 0 & 1 & 0 \\ 2 & 1 & 2 & \Big| & 0 & 0 & 1 \end{bmatrix}$$

$$\xrightarrow[\substack{R_2 + 3R_1 \\ R_3 + 2R_1}]{-R_1} \begin{bmatrix} 1 & 1 & 0 & \Big| & -1 & 1 & 0 \\ 0 & -1 & 1 & \Big| & 3 & -2 & 0 \\ 0 & -1 & 2 & \Big| & 2 & -2 & 1 \end{bmatrix}$$

$$\xrightarrow[\substack{-R_2 \\ R_3 - R_2}]{R_1 + R_2} \begin{bmatrix} 1 & 0 & 1 & \Big| & 2 & -1 & 0 \\ 0 & 1 & -1 & \Big| & -3 & 2 & 0 \\ 0 & 0 & 1 & \Big| & -1 & 0 & 1 \end{bmatrix}$$

$$\xrightarrow[R_2 + R_3]{R_1 - R_3} \begin{bmatrix} 1 & 0 & 0 & \Big| & 3 & -1 & -1 \\ 0 & 1 & 0 & \Big| & -4 & 2 & 1 \\ 0 & 0 & 1 & \Big| & -1 & 0 & 1 \end{bmatrix}$$

The inverse of A is the matrix

$$A^{-1} = \begin{bmatrix} 3 & -1 & -1 \\ -4 & 2 & 1 \\ -1 & 0 & 1 \end{bmatrix}$$

We leave it to you to verify these results. ■

Example 2 illustrates what happens to the reduction process when a matrix A does not have an inverse.

EXAMPLE 2 Find the inverse of the matrix

$$A = \begin{bmatrix} 1 & 2 & 3 \\ 2 & 1 & 2 \\ 3 & 3 & 5 \end{bmatrix}$$

Solution We form the augmented matrix

$$\begin{bmatrix} 1 & 2 & 3 & | & 1 & 0 & 0 \\ 2 & 1 & 2 & | & 0 & 1 & 0 \\ 3 & 3 & 5 & | & 0 & 0 & 1 \end{bmatrix}$$

and use the Gauss-Jordan elimination method:

$$\begin{bmatrix} 1 & 2 & 3 & | & 1 & 0 & 0 \\ 2 & 1 & 2 & | & 0 & 1 & 0 \\ 3 & 3 & 5 & | & 0 & 0 & 1 \end{bmatrix} \xrightarrow[R_3 - 3R_1]{R_2 - 2R_1} \begin{bmatrix} 1 & 2 & 3 & | & 1 & 0 & 0 \\ 0 & -3 & -4 & | & -2 & 1 & 0 \\ 0 & -3 & -4 & | & -3 & 0 & 1 \end{bmatrix}$$

$$\xrightarrow[R_3 - R_2]{-R_2} \begin{bmatrix} 1 & 2 & 3 & | & 1 & 0 & 0 \\ 0 & 3 & 4 & | & 2 & -1 & 0 \\ 0 & 0 & 0 & | & -1 & -1 & 1 \end{bmatrix}$$

Since all entries in the last row of the 3×3 submatrix that comprises the left-hand side of the augmented matrix just obtained are all equal to zero, the latter cannot be reduced to the form $[I|B]$. Accordingly, we draw the conclusion that A is singular—that is, does not have an inverse. ■

More generally, we have the following criterion for determining when the inverse of a matrix does not exist.

MATRICES THAT HAVE NO INVERSES

If there is a row to the left of the vertical line in the augmented matrix containing all zeros, then the matrix does not have an inverse.

A Formula for the Inverse of a 2 × 2 Matrix

Before turning to some applications, we will show an alternative method that employs a formula for finding the inverse of a 2×2 matrix. This method will prove useful in many situations—we will see an application in Example 5. The derivation of this formula is left as an exercise (exercise 40).

FORMULA FOR THE INVERSE OF A 2 × 2 MATRIX

Let

$$A = \begin{bmatrix} a & b \\ c & d \end{bmatrix}$$

Suppose $D = ad - bc$ is not equal to zero. Then A^{-1} exists and is given by

$$A^{-1} = \frac{1}{D}\begin{bmatrix} d & -b \\ -c & a \end{bmatrix} \tag{13}$$

REMARK As an aid to memorizing the formula, note that D is the product of the elements along the main diagonal minus the product of the elements along the other diagonal:

$$\begin{bmatrix} a & b \\ c & d \end{bmatrix} \qquad\qquad D = ad - bc$$

Main diagonal

Next, the matrix

$$\begin{bmatrix} d & -b \\ -c & a \end{bmatrix}$$

is obtained by interchanging a and d and changing the signs of b and c. Finally, A^{-1} is obtained by dividing this matrix by D.   □

EXAMPLE 3   Find the inverse of

$$A = \begin{bmatrix} 1 & 2 \\ 3 & 4 \end{bmatrix}$$

Solution   We first compute $D = (1)(4) - (3)(2) = 4 - 6 = -2$. Next, we write the matrix

$$\begin{bmatrix} 4 & -2 \\ -3 & 1 \end{bmatrix}$$

Finally, dividing this matrix by D, we obtain

$$A^{-1} = \frac{1}{-2} \begin{bmatrix} 4 & -2 \\ -3 & 1 \end{bmatrix} = \begin{bmatrix} -2 & 1 \\ \frac{3}{2} & -\frac{1}{2} \end{bmatrix}$$ ■

Solving Systems of Equations with Inverses

We now show how the inverse of a matrix may be used to solve certain systems of linear equations in which the number of equations in the system is equal to the number of variables. For simplicity, let us illustrate the process for a system of three linear equations in three variables:

$$a_{11}x_1 + a_{12}x_2 + a_{13}x_3 = c_1$$
$$a_{21}x_1 + a_{22}x_2 + a_{23}x_3 = c_2$$
$$a_{31}x_1 + a_{32}x_2 + a_{33}x_3 = c_3 \qquad\qquad (14)$$

Let us write

$$A = \begin{bmatrix} a_{11} & a_{12} & a_{13} \\ a_{21} & a_{22} & a_{23} \\ a_{31} & a_{32} & a_{33} \end{bmatrix}, \qquad X = \begin{bmatrix} x_1 \\ x_2 \\ x_3 \end{bmatrix}, \qquad B = \begin{bmatrix} b_1 \\ b_2 \\ b_3 \end{bmatrix}$$

You should verify that the system (14) of linear equations may be written in the form of the matrix equation

$$AX = B \qquad\qquad (15)$$

If A is nonsingular, then the method of this section may be used to compute A^{-1}. Next, multiplying both sides of (15) by A^{-1} (on the left), we obtain

$$A^{-1}AX = A^{-1}B \qquad \text{or} \qquad IX = A^{-1}B \qquad \text{or} \qquad X = A^{-1}B$$

the desired solution to the problem.

In the case of a system of n equations with n unknowns, we have the following more general result.

USING INVERSES TO SOLVE SYSTEMS OF EQUATIONS

If $AX = B$ is a linear system of n equations in n unknowns, and if A^{-1} exists, then

$$X = A^{-1}B$$

is the unique solution of the system.

The use of inverses to solve systems of equations is particularly advantageous when we are required to solve more than one system of equations, $AX = B$, involving the same coefficient matrix, A, and different matrices of constants, B. As you will see in Examples 4 and 5, we need to compute A^{-1} just once in each case.

Applications

EXAMPLE 4   Solve the following systems of linear equations:

a. $\begin{aligned} 2x + y + z &= 1 \\ 3x + 2y + z &= 2 \\ 2x + y + 2z &= -1 \end{aligned}$ **b.** $\begin{aligned} 2x + y + z &= 2 \\ 3x + 2y + z &= -3 \\ 2x + y + 2z &= 1 \end{aligned}$

Solution   We may write the given systems of equations in the form

$$AX = B \qquad \text{and} \qquad AX = C$$

respectively, where

$$A = \begin{bmatrix} 2 & 1 & 1 \\ 3 & 2 & 1 \\ 2 & 1 & 2 \end{bmatrix}, \qquad X = \begin{bmatrix} x \\ y \\ z \end{bmatrix}, \qquad B = \begin{bmatrix} 1 \\ 2 \\ -1 \end{bmatrix}, \qquad C = \begin{bmatrix} 2 \\ -3 \\ 1 \end{bmatrix}$$

The inverse of the matrix A,

$$A^{-1} = \begin{bmatrix} 3 & -1 & -1 \\ -4 & 2 & 1 \\ -1 & 0 & 1 \end{bmatrix}$$

was found in Example 1. Using this result, we find that the solution of the first system is

$$X = A^{-1}B = \begin{bmatrix} 3 & -1 & -1 \\ -4 & 2 & 1 \\ -1 & 0 & 1 \end{bmatrix} \begin{bmatrix} 1 \\ 2 \\ -1 \end{bmatrix}$$

$$= \begin{bmatrix} (3)(1) + (-1)(2) + (-1)(-1) \\ (-4)(1) + (2)(2) + (1)(-1) \\ (-1)(1) + (0)(2) + (1)(-1) \end{bmatrix} = \begin{bmatrix} 2 \\ -1 \\ -2 \end{bmatrix}$$

or $x = 2$, $y = -1$, and $z = -2$.

The solution of the second system is

$$X = A^{-1}C = \begin{bmatrix} 3 & -1 & -1 \\ -4 & 2 & 1 \\ -1 & 0 & 1 \end{bmatrix} \begin{bmatrix} 2 \\ -3 \\ 1 \end{bmatrix} = \begin{bmatrix} 8 \\ -13 \\ -1 \end{bmatrix}$$

or $x = 8$, $y = -13$, and $z = -1$.   ◼

EXAMPLE 5   The management of Checkers Rent-A-Car plans to expand its fleet of rental cars for the next quarter by purchasing compact and full-size cars. The average cost of a compact is $7,000, and the average cost of a full-size car is $15,000.

a. If a total of 800 cars are to be purchased with a budget of $9 million, how many cars of each size will be acquired?

b. If the predicted demand calls for a total purchase of 1,000 cars with a budget of $10 million, how many cars of each type will be acquired?

Solution   Let x and y denote the number of compact and full-size cars to be purchased. Furthermore, let n denote the total number of cars to be acquired and b the amount of money budgeted for the purchase of these cars. Then

$$x + y = n$$

$$7,000x + 15,000y = b$$

This system of two equations in two variables may be written in the matrix form

$$AX = B$$

where

$$A = \begin{bmatrix} 1 & 1 \\ 7,000 & 15,000 \end{bmatrix}, \qquad X = \begin{bmatrix} x \\ y \end{bmatrix}, \qquad B = \begin{bmatrix} n \\ b \end{bmatrix}$$

Therefore,

$$X = A^{-1}B$$

Since A is a 2×2 matrix, its inverse may be found by using formula (13). We find $D = (1)(15,000) - (7,000)(1) = 8,000$, so

$$A^{-1} = \frac{1}{8000} \begin{bmatrix} 15,000 & -1 \\ -7,000 & 1 \end{bmatrix} = \begin{bmatrix} \frac{15000}{8000} & -\frac{1}{8000} \\ -\frac{7000}{8000} & \frac{1}{8000} \end{bmatrix}$$

Thus

$$X = \begin{bmatrix} \frac{15}{8} & -\frac{1}{8000} \\ -\frac{7}{8} & \frac{1}{8000} \end{bmatrix} \begin{bmatrix} n \\ b \end{bmatrix}$$

a. Here $n = 800$ and $b = 9,000,000$, so

$$X = A^{-1}B = \begin{bmatrix} \frac{15}{8} & -\frac{1}{8000} \\ -\frac{7}{8} & \frac{1}{8000} \end{bmatrix} \begin{bmatrix} 800 \\ 9,000,000 \end{bmatrix} = \begin{bmatrix} 375 \\ 425 \end{bmatrix}$$

Therefore, 375 compact cars and 425 full-size cars will be acquired in this case.

b. Here $n = 1,000$ and $b = 10,000,000$, so

$$X = A^{-1}B = \begin{bmatrix} \frac{15}{8} & -\frac{1}{8000} \\ -\frac{7}{8} & \frac{1}{8000} \end{bmatrix} \begin{bmatrix} 1,000 \\ 10,000,000 \end{bmatrix} = \begin{bmatrix} 625 \\ 375 \end{bmatrix}$$

Therefore, 625 compact cars and 375 full-size cars will be purchased in this case. ◼

**SELF-CHECK
EXERCISES 5.6**

1. Find the inverse of the matrix

$$A = \begin{bmatrix} 2 & 1 & -1 \\ 1 & 1 & -1 \\ -1 & -2 & 3 \end{bmatrix}$$

if it exists.

2. Solve the system of linear equations

$$2x + y - z = b_1$$
$$x + y - z = b_2$$
$$-x - 2y + 3z = b_3$$

where (a) $b_1 = 5$, $b_2 = 4$, $b_3 = -8$ and (b) $b_1 = 2$, $b_2 = 0$, $b_3 = 5$, by finding the inverse of the coefficient matrix.

Solutions to Self-Check Exercises 5.6 can be found on page 256.

5.6   EXERCISES

In exercises 1–4, show that the matrices are inverses of each other by showing that their product is the identity matrix I.

1. $\begin{bmatrix} 1 & -3 \\ 1 & -2 \end{bmatrix}$ and $\begin{bmatrix} -2 & 3 \\ -1 & 1 \end{bmatrix}$

2. $\begin{bmatrix} 4 & 5 \\ 2 & 3 \end{bmatrix}$ and $\begin{bmatrix} \frac{3}{2} & -\frac{5}{2} \\ -1 & 2 \end{bmatrix}$

3. $\begin{bmatrix} 3 & 2 & 3 \\ 2 & 2 & 1 \\ 2 & 1 & 1 \end{bmatrix}$ and $\begin{bmatrix} -\frac{1}{3} & -\frac{1}{3} & \frac{4}{3} \\ 0 & 1 & -1 \\ \frac{2}{3} & -\frac{1}{3} & -\frac{2}{3} \end{bmatrix}$

4. $\begin{bmatrix} 2 & 4 & -2 \\ -4 & -6 & 1 \\ 3 & 5 & -1 \end{bmatrix}$ and $\begin{bmatrix} \frac{1}{2} & -3 & -4 \\ -\frac{1}{2} & 2 & 3 \\ -1 & 1 & 2 \end{bmatrix}$

In exercises 5–16, find the inverse of the matrix, if it exists. Verify your answer.

5. $\begin{bmatrix} 2 & 5 \\ 1 & 3 \end{bmatrix}$

6. $\begin{bmatrix} 2 & 3 \\ 3 & 5 \end{bmatrix}$

7. $\begin{bmatrix} 3 & -3 \\ -2 & 2 \end{bmatrix}$

8. $\begin{bmatrix} 4 & 2 \\ 6 & 3 \end{bmatrix}$

9. $\begin{bmatrix} 2 & -3 & -4 \\ 0 & 0 & -1 \\ 1 & -2 & 1 \end{bmatrix}$ **10.** $\begin{bmatrix} 1 & -1 & 3 \\ 2 & 1 & 2 \\ -2 & -2 & 1 \end{bmatrix}$

11. $\begin{bmatrix} 4 & 2 & 2 \\ -1 & -3 & 4 \\ 3 & -1 & 6 \end{bmatrix}$ **12.** $\begin{bmatrix} 1 & 2 & 0 \\ -3 & 4 & -2 \\ -5 & 0 & -2 \end{bmatrix}$

13. $\begin{bmatrix} 1 & 4 & -1 \\ 2 & 3 & -2 \\ -1 & 2 & 3 \end{bmatrix}$ **14.** $\begin{bmatrix} 3 & -2 & 7 \\ -2 & 1 & 4 \\ 6 & -5 & 8 \end{bmatrix}$

15. $\begin{bmatrix} 1 & 1 & -1 & 1 \\ 2 & 1 & 1 & 0 \\ 2 & 1 & 0 & 1 \\ 2 & -1 & -1 & 3 \end{bmatrix}$ **16.** $\begin{bmatrix} 1 & 1 & 2 & 3 \\ 2 & 3 & 0 & -1 \\ 0 & 2 & -1 & 1 \\ 1 & 2 & 1 & 1 \end{bmatrix}$

In exercises 17–24, (a) write a matrix equation that is equivalent to the system of linear equations and (b) solve the system using the inverses found in exercises 5–16.

17. $2x + 5y = 3$
$x + 3y = 2$
(See exercise 5.)

18. $2x + 3y = 5$
$3x + 5y = 8$
(See exercise 6.)

19. $2x - 3y - 4z = 4$
$ -z = 3$
$x - 2y + z = -8$
(See exercise 9.)

20. $x_1 - x_2 + 3x_3 = 2$
$2x_1 + x_2 + 2x_3 = 2$
$-2x_1 - 2x_2 + x_3 = 3$
(See exercise 10.)

21. $x + 4y - z = 3$
$2x + 3y - 2z = 1$
$-x + 2y + 3z = 7$
(See exercise 13.)

22. $3x_1 - 2x_2 + 7x_3 = 6$
$-2x_1 + x_2 + 4x_3 = 4$
$6x_1 - 5x_2 + 8x_3 = 4$
(See exercise 14.)

23. $x_1 + x_2 - x_3 + x_4 = 6$
$2x_1 + x_2 + x_3 = 4$
$2x_1 + x_2 + x_4 = 7$
$2x_1 - x_2 - x_3 + 3x_4 = 9$
(See exercise 15.)

24. $x_1 + x_2 + 2x_3 + 3x_4 = 4$
$2x_1 + 3x_2 - x_4 = 11$
$ 2x_2 - x_3 + x_4 = 7$
$x_1 + 2x_2 + x_3 + x_4 = 6$
(See exercise 16.)

In exercises 25–32, (a) write each system of equations as a matrix equation and (b) solve the system of equations by using the inverse of the coefficient matrix.

25. $x + 2y = b_1$
$2x - y = b_2$
where   i. $b_1 = 14$, $b_2 = 5$
and   ii. $b_1 = 4$, $b_2 = -1$

26. $3x - 2y = b_1$
$4x + 3y = b_2$
where   i. $b_1 = -6$, $b_2 = 10$
and   ii. $b_1 = 3$, $b_2 = -2$

27. $x + 2y + z = b_1$
$x + y + z = b_2$
$3x + y + z = b_3$
where   i. $b_1 = 7$, $b_2 = 4$, $b_3 = 2$
and   ii. $b_1 = 5$, $b_2 = -3$, $b_3 = -1$

28. $x_1 + x_2 + x_3 = b_1$
$x_1 - x_2 + x_3 = b_2$
$x_1 - 2x_2 - x_3 = b_3$
where   i. $b_1 = 5$, $b_2 = -3$, $b_3 = -1$
and   ii. $b_1 = 1$, $b_2 = 4$, $b_3 = -2$

29. $3x + 2y - z = b_1$
$2x - 3y + z = b_2$
$x - y - z = b_3$
where   i. $b_1 = 2$, $b_2 = -2$, $b_3 = 4$
and   ii. $b_1 = 8$, $b_2 = -3$, $b_3 = 6$

30. $2x_1 + x_2 + x_3 = b_1$
$x_1 - 3x_2 + 4x_3 = b_2$
$-x_1 + x_3 = b_3$
where   i. $b_1 = 1$, $b_2 = 4$, $b_3 = -3$
and   ii. $b_1 = 2$, $b_2 = -5$, $b_3 = 0$

31. $x_1 + x_2 + x_3 + x_4 = b_1$
$x_1 - x_2 - x_3 + x_4 = b_2$
$ x_2 + 2x_3 + 2x_4 = b_3$
$x_1 + 2x_2 + x_3 - 2x_4 = b_4$
where   i. $b_1 = 1$, $b_2 = -1$, $b_3 = 4$, $b_4 = 0$
and   ii. $b_1 = 2$, $b_2 = 8$, $b_3 = 4$, $b_4 = -1$

32. $x_1 + x_2 + 2x_3 + x_4 = b_1$
$4x_1 + 5x_2 + 9x_3 + x_4 = b_2$
$3x_1 + 4x_2 + 7x_3 + x_4 = b_3$
$2x_1 + 3x_2 + 4x_3 + 2x_4 = b_4$
where   i. $b_1 = 3$, $b_2 = 6$, $b_3 = 5$, $b_4 = 7$
and   ii. $b_1 = 1$, $b_2 = -1$, $b_3 = 0$, $b_4 = -4$

33. Let

$$A = \begin{bmatrix} 2 & 3 \\ -4 & -5 \end{bmatrix}$$

a. Find A^{-1}.
b. Show that $(A^{-1})^{-1} = A$.

34. Let

$$A = \begin{bmatrix} 6 & -4 \\ -4 & 3 \end{bmatrix} \quad \text{and} \quad B = \begin{bmatrix} 3 & -5 \\ 4 & -7 \end{bmatrix}$$

a. Find AB, A^{-1}, B^{-1}.
b. Show that $(AB)^{-1} = B^{-1}A^{-1}$.

35. Let

$$A = \begin{bmatrix} 2 & -5 \\ 1 & -3 \end{bmatrix}, \quad B = \begin{bmatrix} 4 & 3 \\ 1 & 1 \end{bmatrix},$$

and

$$C = \begin{bmatrix} 2 & 3 \\ -2 & 1 \end{bmatrix}$$

a. Find ABC, A^{-1}, B^{-1}, C^{-1}.
b. Show that $(ABC)^{-1} = C^{-1}B^{-1}A^{-1}$.

36. Ticket Revenues Rainbow Harbor Cruises charges $8 per adult and $4 per child for a round-trip ticket. The records show that on a certain weekend, 1000 people took the cruise on Saturday and 800 people took the cruise on Sunday. The total receipts for Saturday were $6400, and the total receipts for Sunday were $4800. Determine how many adults and children took the cruise on Saturday and on Sunday.

37. Pricing Bel Air Publishing Company publishes a deluxe leather edition and a standard edition of its Daily Organizer. The company's marketing department estimates that x copies of the deluxe edition and y copies of the standard edition will be demanded per month when the unit prices are p dollars and q dollars, respectively, where x, y, p, and q are related by the following system of linear equations:

$$5x + \;\; y = 1000(70 - p)$$
$$x + 3y = 1000(40 - q)$$

Find the monthly demand for the deluxe edition and the standard edition when the unit prices are set according to the following schedules:
a. $p = 50$ and $q = 25$
b. $p = 45$ and $q = 25$
c. $p = 45$ and $q = 20$

38. Nutrition/Diet Planning Bob, a nutritionist with the University Medical Center, has been asked to prepare special diets for two patients, Susan and Tom. Bob has decided that Susan's meals should contain at least 400 mg of calcium, 20 mg of iron, and 50 mg of vitamin C, whereas Tom's meals should contain at least 350 mg of calcium, 15 mg of iron, and 40 mg

of vitamin C. Bob has also decided that the meals are to be prepared from three basic foods: food A, food B, and food C. The special nutritional contents of these foods are summarized in the accompanying table. Find how many ounces of each type of food should be used in a meal so that the minimum requirements of calcium, iron, and vitamin C are met for each patient's meals.

	Contents in mg/oz		
	Calcium	Iron	Vitamin C
Food A	30	1	2
Food B	25	1	5
Food C	20	2	4

39. Research Funding The Carver Foundation funds three nonprofit organizations engaged in alternate-energy research activities. From past data, the proportion of funds spent by each organization in research on solar energy, energy from harnessing the wind, and energy from the motion of ocean tides is given in the accompanying table.

	Proportion of Money Spent		
	Solar	Wind	Tides
Organization I	0.6	0.3	0.1
Organization II	0.4	0.3	0.3
Organization III	0.2	0.6	0.2

Find the amount awarded to each organization if the total amount spent by all three organizations on solar, wind, and tidal research is
a. $9.2 million, $9.6 million, and $5.2 million, respectively
b. $8.2 million, $7.2 million, and $3.6 million, respectively

40. Let

$$A = \begin{bmatrix} a & b \\ c & d \end{bmatrix}$$

a. Find A^{-1}.
b. Find the necessary condition for A to be nonsingular.
c. Verify that $AA^{-1} = A^{-1}A = I$.

**SOLUTIONS TO
SELF-CHECK
EXERCISES 5.6**

1. We form the augmented matrix

$$\left[\begin{array}{ccc|ccc} 2 & 1 & -1 & 1 & 0 & 0 \\ 1 & 1 & -1 & 0 & 1 & 0 \\ -1 & -2 & 3 & 0 & 0 & 1 \end{array}\right]$$

and row-reduce as follows:

$$\left[\begin{array}{ccc|ccc} 2 & 1 & -1 & 1 & 0 & 0 \\ 1 & 1 & -1 & 0 & 1 & 0 \\ -1 & -2 & 3 & 0 & 0 & 1 \end{array}\right] \xrightarrow{R_1 \leftrightarrow R_2}$$

$$\left[\begin{array}{ccc|ccc} 1 & 1 & -1 & 0 & 1 & 0 \\ 2 & 1 & -1 & 1 & 0 & 0 \\ -1 & -2 & 3 & 0 & 0 & 1 \end{array}\right] \xrightarrow[R_3 + R_1]{R_2 - 2R_1}$$

$$\left[\begin{array}{ccc|ccc} 1 & 1 & -1 & 0 & 1 & 0 \\ 0 & -1 & 1 & 1 & -2 & 0 \\ 0 & -1 & 2 & 0 & 1 & 1 \end{array}\right] \xrightarrow[\substack{-R_2 \\ R_3 - R_2}]{R_1 + R_2}$$

$$\left[\begin{array}{ccc|ccc} 1 & 0 & 0 & 1 & -1 & 0 \\ 0 & 1 & -1 & -1 & 2 & 0 \\ 0 & 0 & 1 & -1 & 3 & 1 \end{array}\right] \xrightarrow{R_2 + R_3}$$

$$\left[\begin{array}{ccc|ccc} 1 & 0 & 0 & 1 & -1 & 0 \\ 0 & 1 & 0 & -2 & 5 & 1 \\ 0 & 0 & 1 & -1 & 3 & 1 \end{array}\right]$$

From the preceding results, we see that

$$A^{-1} = \left[\begin{array}{ccc} 1 & -1 & 0 \\ -2 & 5 & 1 \\ -1 & 3 & 1 \end{array}\right]$$

2. a. We write the systems of linear equations in the matrix form

$$AX = B_1$$

where

$$A = \left[\begin{array}{ccc} 2 & 1 & -1 \\ 1 & 1 & -1 \\ -1 & -2 & 3 \end{array}\right], \qquad X = \left[\begin{array}{c} x \\ y \\ z \end{array}\right], \qquad B_1 = \left[\begin{array}{c} 5 \\ 4 \\ -8 \end{array}\right]$$

Now, using the results of exercise 1, we have

$$X = \left[\begin{array}{c} x \\ y \\ z \end{array}\right] = A^{-1}B_1 = \left[\begin{array}{ccc} 1 & -1 & 0 \\ -2 & 5 & 1 \\ -1 & 3 & 1 \end{array}\right]\left[\begin{array}{c} 5 \\ 4 \\ -8 \end{array}\right] = \left[\begin{array}{c} 1 \\ 2 \\ -1 \end{array}\right]$$

Therefore $x = 1$, $y = 2$, and $z = -1$.

b. Here A and X are as in (a), but

$$B_2 = \begin{bmatrix} 2 \\ 0 \\ 5 \end{bmatrix}$$

Therefore,

$$X = \begin{bmatrix} x \\ y \\ z \end{bmatrix} = A^{-1}B_2 = \begin{bmatrix} 1 & -1 & 0 \\ -2 & 5 & 1 \\ -1 & 3 & 1 \end{bmatrix}\begin{bmatrix} 2 \\ 0 \\ 5 \end{bmatrix} = \begin{bmatrix} 2 \\ 1 \\ 3 \end{bmatrix}$$

or $x = 2$, $y = 1$, and $z = 3$.

U s i n g   T e c h n o l o g y

The graphing utility can be used to find the inverse of a square matrix. (See Section A.7 in Appendix A.)

EXAMPLE 1 Use a graphing utility to find the inverse of

$$\begin{bmatrix} 1 & 3 & 5 \\ -2 & 2 & 4 \\ 5 & 1 & 3 \end{bmatrix}$$

Solution We first enter the given matrix as the matrix

$$A = \begin{bmatrix} 1 & 3 & 5 \\ -2 & 2 & 4 \\ 5 & 1 & 3 \end{bmatrix}$$

Then, recalling the matrix A and using the x^{-1} key, we find

$$A^{-1} = \begin{bmatrix} 0.1 & -0.2 & 0.1 \\ 1.3 & -1.1 & -0.7 \\ -0.6 & 0.7 & 0.4 \end{bmatrix}$$

EXERCISES

In exercises 1–4, refer to the following matrices and use a graphic utility or computer software program to find the inverse of the matrix. Express your answers accurate to two decimal places.

1. $\begin{bmatrix} 1.2 & 3.1 & -2.1 \\ 3.4 & 2.6 & 7.3 \\ -1.2 & 3.4 & -1.3 \end{bmatrix}$

2. $\begin{bmatrix} 4.2 & 3.7 & 4.6 \\ 2.1 & -1.3 & 2.4 \\ 1.8 & 7.6 & -2.3 \end{bmatrix}$

3. $\begin{bmatrix} 1.1 & 2.3 & 3.1 & 4.2 \\ 1.6 & 3.2 & 1.8 & 2.9 \\ 4.2 & 1.6 & 1.4 & 3.2 \\ 1.6 & 2.1 & 2.8 & 7.2 \end{bmatrix}$

4. $\begin{bmatrix} 2.1 & 3.2 & -1.4 & -3.2 \\ 6.2 & 7.3 & 8.4 & 1.6 \\ 2.3 & 7.1 & 2.4 & -1.3 \\ -2.1 & 3.1 & 4.6 & 3.7 \end{bmatrix}$

CHAPTER FIVE SUMMARY OF PRINCIPAL FORMULAS AND TERMS

Formulas

Laws for matrix addition:

Commutative law $\qquad\qquad A + B = B + A$

Associative law $\qquad\qquad (A + B) + C = A + (B + C)$

Laws for matrix multiplication:

Associative law $\qquad\qquad (AB)C = A(BC)$

Distributive law $\qquad\qquad A(B + C) = AB + AC$

Inverse of a 2×2 matrix

If $\qquad A = \begin{bmatrix} a & b \\ c & d \end{bmatrix}$

and $\qquad D = ad - bc \neq 0,$

then $\qquad A^{-1} = \dfrac{1}{D} \begin{bmatrix} d & -b \\ -c & a \end{bmatrix}$

Solution of system $AX = B$ $\qquad X = A^{-1}B$
(A, nonsingular)

Terms

Parameter	Matrix (*pl.:* matrices)
Dependent system	Row matrix
Inconsistent system	Column matrix
Equivalent system	Square matrix
Coefficient matrix	Transpose of a matrix
Augmented matrix	Scalar
Row-reduced form of a matrix	Scalar product
Row operations	Matrix product
Unit column	Identity matrix
Pivoting	Inverse of a matrix
Gauss-Jordan elimination method	Singular matrix
Size of a matrix	

CHAPTER 5 REVIEW EXERCISES

In exercises 1–4, perform the operation, if possible.

1. $\begin{bmatrix} 1 & 2 \\ -1 & 3 \\ 2 & 1 \end{bmatrix} + \begin{bmatrix} 1 & 0 \\ 0 & 1 \\ 1 & 2 \end{bmatrix}$

2. $\begin{bmatrix} -1 & 2 \\ 3 & 4 \end{bmatrix} - \begin{bmatrix} 1 & 2 \\ 5 & -2 \end{bmatrix}$

3. $\begin{bmatrix} -3 & 2 & 1 \end{bmatrix} \begin{bmatrix} 2 & 1 \\ -1 & 0 \\ 2 & 1 \end{bmatrix}$

4. $\begin{bmatrix} 1 & 3 & 2 \\ -1 & 2 & 3 \end{bmatrix} \begin{bmatrix} 1 \\ 4 \\ 2 \end{bmatrix}$

In exercises 5–8, find the values of the variables.

5. $\begin{bmatrix} 1 & x \\ y & 3 \end{bmatrix} = \begin{bmatrix} z & 2 \\ 3 & w \end{bmatrix}$

6. $\begin{bmatrix} 3 & x \\ y & 3 \end{bmatrix} \begin{bmatrix} 1 \\ 2 \end{bmatrix} = \begin{bmatrix} 7 \\ 4 \end{bmatrix}$

7. $\begin{bmatrix} 3 & a+3 \\ -1 & b \\ c+1 & d \end{bmatrix} = \begin{bmatrix} 3 & 6 \\ e+2 & 4 \\ -1 & 2 \end{bmatrix}$

8. $\begin{bmatrix} x & 3 & 1 \\ 0 & y & 2 \end{bmatrix} \begin{bmatrix} 1 & 1 \\ 3 & z \\ 4 & 2 \end{bmatrix} = \begin{bmatrix} 12 & 4 \\ 2 & 2 \end{bmatrix}$

In exercises 9–16, compute the expression, if possible, given that

$$A = \begin{bmatrix} 1 & 3 & 1 \\ -2 & 1 & 3 \\ 4 & 0 & 2 \end{bmatrix}, \qquad B = \begin{bmatrix} 2 & 1 & 3 \\ -2 & -1 & -1 \\ 1 & 4 & 2 \end{bmatrix},$$

and

$$C = \begin{bmatrix} 3 & -1 & 2 \\ 1 & 6 & 4 \\ 2 & 1 & 3 \end{bmatrix}$$

9. $2A + 3B$

10. $3A - 2B$

11. $2(3A)$

12. $2(3A - 4B)$

13. $A(B - C)$

14. $AB + AC$

15. $A(BC)$

16. $(\frac{1}{2})(CA - CB)$

In exercises 17–23, solve the system of linear equations using the Gauss-Jordan elimination method.

17. $2x - 3y = 5$
$3x + 4y = -1$

18. $3x + 2y = 3$
$2x - 4y = -14$

19. $x - y + 2z = 5$
$3x + 2y + z = 10$
$2x - 3y - 2z = -10$

20. $3x - 2y + 4z = 16$
$2x + y - 2z = -1$
$x + 4y - 8z = -18$

21. $3x - 2y + 4z = 11$
$2x - 4y + 5z = 4$
$x + 2y - z = 10$

22. $x - 2y + 3z + 4w = 17$
$2x + y - 2z - 3w = -9$
$3x - y + 2z - 4w = 0$
$4x + 2y - 3z + w = -2$

23. $3x - 2y + z = 4$
$x + 3y - 4z = -3$
$2x - 3y + 5z = 7$
$x - 8y + 9z = 10$

In exercises 24–31, find the inverse of the matrix (if it exists).

24. $A = \begin{bmatrix} 2 & 4 \\ 1 & 6 \end{bmatrix}$

25. $A = \begin{bmatrix} 3 & 1 \\ 1 & 2 \end{bmatrix}$

26. $A = \begin{bmatrix} 2 & 4 \\ 1 & -2 \end{bmatrix}$

27. $A = \begin{bmatrix} 3 & 4 \\ 2 & 2 \end{bmatrix}$

28. $A = \begin{bmatrix} 1 & 2 & 4 \\ 2 & 1 & 3 \\ -1 & 0 & 2 \end{bmatrix}$

29. $A = \begin{bmatrix} 2 & 3 & 1 \\ 1 & -1 & 2 \\ 1 & 2 & 1 \end{bmatrix}$

30. $A = \begin{bmatrix} 2 & 1 & -3 \\ 1 & 2 & -4 \\ 3 & 1 & -2 \end{bmatrix}$

31. $A = \begin{bmatrix} 1 & 2 & 4 \\ 3 & 1 & 2 \\ 1 & 0 & -6 \end{bmatrix}$

In exercises 32–35, compute the expression, if possible, given that

$$A = \begin{bmatrix} 1 & 2 \\ -1 & 2 \end{bmatrix}, \qquad B = \begin{bmatrix} 3 & 1 \\ 4 & 2 \end{bmatrix}, \qquad C = \begin{bmatrix} 1 & 1 \\ -1 & 2 \end{bmatrix}$$

32. $(ABC)^{-1}$

33. $(A^{-1}B)^{-1}$

34. $(A + B)^{-1}$

35. $(2A - C)^{-1}$

In exercises 36–39, write each system of linear equations in the form $AX = C$. Find A^{-1} and use the result to solve the system.

36. $x - 3y = -1$
$2x + 4y = 8$

37. $2x + 3y = -8$
$x - 2y = 3$

38. $2x - 3y + 4z = 17$
$x + 2y - 4z = -7$
$3x - y + 2z = 14$

39. $x - 2y + 4z = 13$
$2x + 3y - 2z = 0$
$x + 4y - 6z = -15$

40. Mr. Spaulding bought 10,000 shares of stock X, 20,000 shares of stock Y, and 30,000 shares of stock Z at a unit price of $20, $30, and $50 per share, respectively. Six months later, the closing prices of stocks X, Y, and Z were $22, $35, and $51 per share, respectively. Spaulding made no other stock transactions during the period in question. Compare the value of Mr. Spaulding's stock holdings at the time of purchase and six months later.

41. Ms. Newburg operates three self-service gasoline stations in different parts of town. On a certain day, Station A sold 600 gallons of premium, 1000 gallons of regular, 800 gallons of super-unleaded, and 1400 gallons of regular-unleaded gasoline; Station B sold 700 gallons of premium, 800 gallons of regular, 600 gallons of super-unleaded, and 1200 gallons of regular-unleaded gasoline; Station C sold 1200 gallons of premium, 800 gallons of regular, 1000 gallons of super-unleaded, and 900 gallons of regular-unleaded gasoline. If the price of gasoline on that day was $1.60 per gallon for premium, $1.20 per gallon for regular, $1.50 per gallon for super-unleaded, and $1.30 per gallon for regular-unleaded gasoline, use matrix algebra to find the total revenue at each station.

42. The Wildcat Oil Company has two refineries—one located in Houston and the other in Tulsa. The Houston refinery ships 60% of its petroleum to a Chicago distributor and 40% of its petroleum to a Los Angeles distributor. The Tulsa refinery ships 30% of its petro-leum to the Chicago distributor and 70% of its petroleum to the Los Angeles distributor. If, over the year, the Chicago distributor received 240,000 gallons of petroleum and the Los Angeles distributor received 460,000 gallons of petroleum, find the amount of petroleum produced at each of Wildcat's refineries.

43. Desmond Jewelry, Inc., wishes to produce three types of pendants: type A, type B, and type C. To manufacture a type-A pendant requires 2 minutes on machines I and II and 3 minutes on machine III. A type-B pendant requires 2 minutes on machine I, 3 minutes on machine II, and 4 minutes on machine III. A type-C pendant requires 3 minutes on machine I, 4 minutes on machine II, and 3 minutes on machine III. There are $3\frac{1}{2}$ hours available on machine I, $4\frac{1}{2}$ hours available on machine II, and 5 hours available on machine III. How many pendants of each type should Desmond make in order to use all the available time?

CHAPTER 5   EXTENDED PROBLEMS AND PROJECTS

1. Let

$$A = \begin{bmatrix} a & b \\ c & d \end{bmatrix}$$

   a. Find A^{-1}.
   b. Find the necessary condition for A to be nonsingular.
   c. Verify that $AA^{-1} = A^{-1}A = I$.

2. Figure 8 shows the routes of an international airline connecting five cities. A line joining two cities indicates that there is a direct flight between them. The same information can be represented by the matrix

	Singapore	Hong Kong	Taipei	Tokyo	L.A.
Singapore	0	1	1	0	0
Hong Kong	1	0	1	1	0
$A =$ Taipei	1	1	0	1	1
Tokyo	0	1	1	0	1
L.A.	0	0	1	1	0

where $a_{ji} = 1$ indicates there is a direct flight between the ith city and the jth city, and $a_{ij} = 0$ indicates there is no direct flight between the cities.
   a. Compute A^2.
   b. By referring to the figure, verify that the entry a_{ij} of A^2 represents the number of one-stop routes between the ith city and the jth city.

FIGURE 8

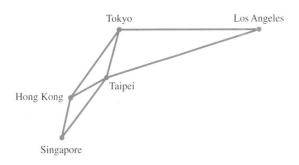

6

Linear Programming

*M*any practical problems involve maximizing or minimizing a function subject to certain constraints. For example, we may wish to maximize a profit function subject to certain limitations on the amount of material and labor available. Maximization or minimization problems that can be formulated in terms of a *linear* objective function and constraints in the form of linear inequalities are called *linear programming problems.* In this chapter we begin by looking at linear programming problems involving two variables. These problems are amenable to geometric analysis, and the method of solution introduced here will shed much light on the basic nature of a linear programming problem. Solving linear programming problems involving more than two variables requires algebraic techniques. One such technique, the *simplex method,* was developed by George Dantzig in the late 1940s and remains in wide use to this day.

Graphing Systems of Linear Inequalities in Two Variables

Graphing Linear Inequalities

In Chapter 2, we saw that a linear equation in two variables x and y

$$ax + by + c = 0 \qquad a, b \text{ not both equal to zero}$$

has a *solution set* that may be exhibited graphically as points on a straight line in the xy-plane. We now show that there is also a simple graphical representation for **linear inequalities** in two variables:

$$ax + by + c < 0 \qquad ax + by + c \leq 0$$
$$ax + by + c > 0 \qquad ax + by + c \geq 0$$

Before turning to a general procedure for graphing such inequalities, let us consider a specific example. Suppose we wish to graph

$$2x + 3y < 6 \tag{1}$$

We first graph the equation $2x + 3y = 6$, which is obtained from the given inequality by replacing the inequality "<" with an equality "=" (Figure 1).

FIGURE 1

A straight line divides the xy-plane into two half-planes.

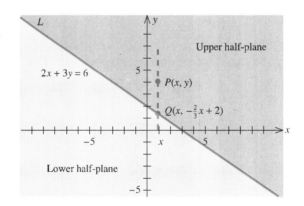

Observe that this line divides the xy-plane into two half-planes: an upper half-plane and a lower half-plane. Let us show that the upper half-plane is the graph of the linear inequality

$$2x + 3y > 6 \tag{2}$$

whereas the lower half-plane is the graph of the linear inequality

$$2x + 3y < 6 \tag{3}$$

To see this, let us write (2) and (3) in the equivalent forms

$$y > -\frac{2}{3}x + 2 \tag{4}$$

and

$$y < -\frac{2}{3}x + 2 \tag{5}$$

The equation of the line itself is

$$y = -\frac{2}{3}x + 2 \qquad (6)$$

Now pick any point $P(x, y)$ lying above the line L, and let Q be the point lying on L and directly below P (see Figure 1). Since Q lies on L, its coordinates must satisfy equation (6). In other words, Q has representation $Q(x, -\frac{2}{3}x + 2)$. Comparing the y-coordinates of P and Q and recalling that P lies above Q so that its y-coordinate must be larger than that of Q, we have

$$y > -\frac{2}{3}x + 2$$

But this inequality is just equation (4), or equivalently equation (2). Similarly, we can show that any point lying below L must satisfy equation (5) and therefore (3).

This analysis shows that the lower half-plane provides a solution to our problem (Figure 2). (The dotted line shows that the points on L do not belong to the solution set.) Observe that the two half-planes in question are mutually exclusive; that is, they do not have any points in common. Because of this, there is an alternative and easier method of determining the solution to the problem.

FIGURE 2

The set of points lying below the dotted line satisfies the given inequality.

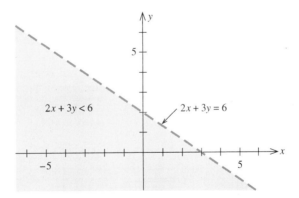

To determine the required half-plane, let us pick *any* point lying in one of the half-planes. For simplicity, let us pick the origin $(0, 0)$, which lies in the lower half-plane. Substituting $x = 0$ and $y = 0$ (the coordinates of this point) into the given inequality (1), we find

$$2(0) + 3(0) < 6$$

or $0 < 6$, which is certainly true. This tells us that the required half-plane is the half-plane containing the test point—namely, the lower half-plane.

Next, let us see what happens if we choose the point $(2, 3)$, which lies in the upper half-plane. Substituting $x = 2$ and $y = 3$ into the given inequality, we find

$$2(2) + 3(3) < 6$$

or $13 < 6$, which is false. This tells us that the upper half-plane is *not* the required half-plane, as expected. Note, too, that no point (x, y) lying on the line constitutes a solution to our problem, because of the *strict* inequality "$<$".

This discussion suggests the following procedure for graphing a linear inequality in two variables.

**PROCEDURE FOR
GRAPHING LINEAR
INEQUALITIES**

1. Draw the graph of the equation obtained for the given inequality by replacing the inequality sign with an equals sign. Use a dotted line if the problem involves a strict inequality, "$<$" or "$>$". Otherwise, use a solid line to indicate that the line itself constitutes part of the solution.

2. Pick a test point lying in one of the half-planes determined by the line sketched in step 1 and substitute the values of x and y into the given inequality. Use the origin whenever possible.

3. If the inequality is satisfied, the graph of the inequality includes the half-plane containing the test point. Otherwise, the solution includes the half-plane not containing the test point.

EXAMPLE 1   Determine the solution set for the inequality $2x + 3y \geq 6$.

Solution   Replacing the inequality "$\geq$" with an equality "$=$", we obtain the equation $2x + 3y = 6$, whose graph is the straight line shown in Figure 3. Instead of a dotted line as before, we use a solid line to show that all points on the line are also solutions to the problem. Picking the origin as our test point, we find $2(0) + 3(0) \geq 6$, or $0 \geq 6$, which is impossible. So we conclude that the solution set comprises the half-plane not containing the origin, including in this case the line given by $2x + 3y = 6$.

FIGURE 3
The set of points lying on the line and in the upper half-plane satisfies the given inequality.

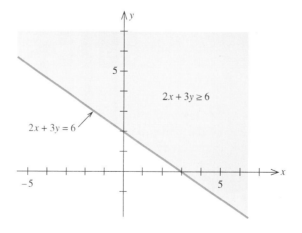

EXAMPLE 2   Graph $x \leq -1$.

Solution   The graph of $x = -1$ is the vertical line shown in Figure 4. Picking the origin $(0, 0)$ as a test point, we find $0 \leq -1$, which is false.

FIGURE 4

The set of points lying on the line x = −1 and in the left half-plane satisfies the given inequality.

FIGURE 5

The set of points in the lower half-plane satisfies x − 2y > 0.

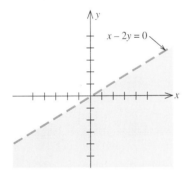

Therefore, the required solution is the *left* half-plane, which does not contain the origin.

EXAMPLE 3   Graph $x - 2y > 0$.

Solution   We first graph the equation $x - 2y = 0$, or $y = (\frac{1}{2})x$ (Figure 5). Since the origin lies on the line, we may not use it as a test point. (Why?) Let us pick (1, 2) as a test point. Substituting $x = 1$ and $y = 2$ into the given inequality, we find $1 - 2(2) > 0$, or $-3 > 0$, which is false. Therefore, the required solution is the half-plane that does not contain the test point—namely, the lower half-plane.

Graphing Systems of Linear Inequalities

By the *solution set of a system of linear inequalities* in the two variables x and y, we mean the set of all points (x, y) satisfying each inequality of the system. The graphical solution of such a system may be obtained by graphing the solution set for each inequality independently and then determining the region in common with each solution set.

EXAMPLE 4   Determine the solution set for the system

$$4x + 3y \geq 12$$
$$x - y \leq 0$$

Solution   Proceeding as in the previous examples, you should have no difficulty locating the half-planes determined by each of the linear inequalities that make up the system. These half-planes are shown in Figure 6. The intersection of the two half-planes is the shaded region. A point in this region is an element of the solution set for the given system. The point P, the intersection of the two straight lines determined by the equations, is found by solving the simultaneous equations

$$4x + 3y = 12$$
$$x - y = 0$$

FIGURE 6

The set of points in the shaded area satisfies the system

$$\begin{cases} 4x + 3y \geq 12 \\ x - y \leq 0 \end{cases}$$

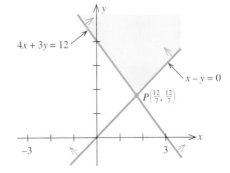

EXAMPLE 5   Sketch the solution set for the system

$$x + y - 6 \leq 0$$
$$2x + y - 8 \leq 0$$

$$x \geq 0$$

$$y \geq 0$$

Solution   The third inequality in the system defines the right half-plane—all points to the right of the y-axis plus all points lying on the y-axis itself. The fourth inequality in the system defines the upper half-plane, including the x-axis. The half-planes defined by the first and second inequalities are indicated by arrows in Figure 7. Thus, the required region, the intersection of the four half-planes defined by the four inequalities in the given system of linear inequalities, is the shaded region. The point P is found by solving the simultaneous equations $x + y - 6 = 0$ and $2x + y - 8 = 0$.

FIGURE 7
The set of points in the shaded region, including the x- and y-axes, satisfies the given inequalities.

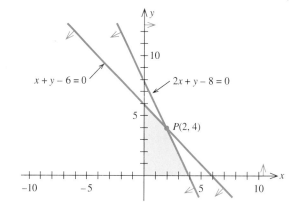

The solution set found in Example 5 is an example of a *bounded set*. Observe that the set can be enclosed by a circle. For example, if you draw a circle of radius 10 with center at the origin, you will see that the set lies entirely inside the circle. On the other hand, the solution set of Example 4 cannot be enclosed by a circle and is said to be *unbounded*.

BOUNDED AND UNBOUNDED SOLUTION SETS

A solution set of a system of linear inequalities is **bounded** if it can be enclosed by a circle. Otherwise, it is **unbounded.**

EXAMPLE 6   Determine the graphical solution set for the following system of linear inequalities:

$$2x + y \geq 50$$

$$x + 2y \geq 40$$

$$x \geq 0$$

$$y \geq 0$$

Solution   The required solution set is the unbounded region shown in Figure 8.

FIGURE 8
The solution set is an unbounded region.

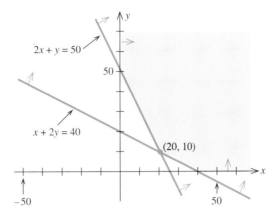

SELF-CHECK
EXERCISES 6.1

1. Determine graphically the solution set for the following system of inequalities:

$$x + 2y \leq 10$$

$$5x + 3y \leq 30$$

$$x \geq 0, y \geq 0$$

2. Determine graphically the solution set for the following system of inequalities:

$$5x + 3y \geq 30$$

$$x - 3y \leq \ 0$$

$$x \geq \ 2$$

Solutions to Self-Check Exercises 6.1 can be found on page 269.

  EXERCISES

In exercises 1–10, find the graphical solution of each inequality.

1. $4x - 8 < 0$

2. $3y + 2 > 0$

3. $x - y \leq 0$

4. $3x + 4y \leq -2$

5. $x \leq -3$

6. $y \geq -1$

7. $2x + y \leq 4$

8. $-3x + 6y \geq 12$

9. $4x - 3y \leq -24$

10. $5x - 3y \geq 15$

In exercises 11–18, write a system of linear inequalities that describes the shaded region.

11.

12.

13.

14.

15.

16.

17.

18.

In exercises 19–36, determine graphically the solution set for each system of inequalities and indicate whether the solution set is bounded or unbounded.

19. $2x + 4y > 16$
$-x + 3y \geq 7$

20. $3x - 2y > -13$
$-x + 2y > 5$

21. $x - y \leq 0$
$2x + 3y \geq 10$

22. $x + y \geq -2$
$3x - y \leq 6$

23. $x + 2y \geq 3$
$2x + 4y \leq -2$

24. $2x - y \geq 4$
$4x - 2y < -2$

25. $x + y \leq 6$
$0 \leq x \leq 3$
$y \geq 0$

26. $4x - 3y \leq 12$
$5x + 2y \leq 10$
$x \geq 0, y \geq 0$

27. $3x - 6y \leq 12$
$-x + 2y \leq 4$
$x \geq 0, y \geq 0$

28. $x + y \geq 20$
$x + 2y \geq 40$
$x \geq 0, y \geq 0$

29. $3x - 7y \geq -24$
$x + 3y \geq 8$
$x \geq 0, y \geq 0$

30. $3x + 4y \geq 12$
$2x - y \geq -2$
$0 \leq y \leq 3$
$x \geq 0$

31. $x + 2y \geq 3$
$5x - 4y \leq 16$
$0 \leq y \leq 2$
$x \geq 0$

32. $x + y \leq 4$
$2x + y \leq 6$
$2x - y \geq -1$
$x \geq 0, y \geq 0$

33. $6x + 5y \leq 30$
$3x + y \geq 6$
$x + y \geq 4$
$x \geq 0, y \geq 0$

34. $6x + 7y \leq 84$
$12x - 11y \leq 18$
$6x - 7y \leq 28$
$x \geq 0, y \geq 0$

35. $x - y \geq -6$
$x - 2y \leq -2$
$x + 2y \geq 6$
$x - 2y \geq -14$
$x \geq 0, y \geq 0$

36. $x - 3y \geq -18$
$3x - 2y \geq 2$
$x - 3y \leq -4$
$3x - 2y \leq 16$
$x \geq 0, y \geq 0$

SOLUTIONS TO SELF-CHECK EXERCISES 6.1

1. The required solution set is shown in the following figure:

The point P is found by solving the system of equations

$$x + 2y = 10$$

$$5x + 3y = 30$$

Solving the first equation for x in terms of y gives

$$x = 10 - 2y$$

Substituting this value of x into the second equation of the system gives

$$5(10 - 2y) + 3y = 30$$

$$50 - 10y + 3y = 30$$

$$-7y = -20$$

so $y = 20/7$. Substituting this value of y into the expression for x found earlier, we obtain

$$x = 10 - 2(\tfrac{20}{7}) = \tfrac{30}{7}$$

giving the point of intersection as $(30/7, 20/7)$.

2. The required solution set is shown in the following figure:

To find the coordinates of P, we solve the system

$$5x + 3y = 30$$

$$x - 3y = 0$$

Solving the second equation for x in terms of y and substituting this value of x in the first equation gives

$$5(3y) + 3y = 30$$

or $y = 5/3$. Substituting this value of y into the second equation gives $x = 5$. Next, the coordinates of Q are found by solving the system

$$5x + 3y = 30$$

$$x = 2$$

obtaining $x = 2$ and $y = 20/3$.

6.2   Linear Programming Problems

Many business and economic problems are concerned with optimizing (maximizing or minimizing) a function subject to a system of equalities or inequalities. The function to be optimized is called the **objective function.** Profit functions and cost functions are examples of objective functions. The system of equalities or inequalities to which the objective function is subjected reflects the constraints (for example, limitations on resources such as materials and labor) imposed on the solution(s) to the problem. Problems of this nature are called *mathematical programming problems.* In particular, problems in which both the objective function and the constraints are expressed as linear equations or inequalities are called *linear programming problems.*

**A LINEAR
PROGRAMMING
PROBLEM**

A **linear programming problem** consists of a linear objective function to be maximized or minimized subject to certain constraints in the form of linear equalities or inequalities.

Maximization Problems

As an example of a linear programming problem in which the objective function is to be maximized, let us consider the following simplified version of a production problem involving two variables.

EXAMPLE 1   The Ace Novelty Company wishes to produce two types of souvenirs: type A and type B. Each type-A souvenir will result in a profit of $1, and each type-B souvenir will result in a profit of $1.20. To manufacture a type-A souvenir requires 2 minutes on machine I and 1 minute on machine II. A type-B souvenir requires 1 minute on machine I and 3 minutes on machine II. There are 3 hours available on machine I and 5 hours available on machine II for processing the order. How many souvenirs of each type should Ace make in order to maximize profit?

Solution   As a first step toward the mathematical formulation of this problem, we tabulate the given information as shown in Table 1. Let x be the number of type-A souvenirs and y be the number of type-B souvenirs to be made. Then the total profit P, in dollars, is given by

$$P = x + 1.2y$$

which is the objective function to be maximized.

TABLE 1

	Type A	Type B	Time Available
Machine I	2 min	1 min	180 min
Machine II	1 min	3 min	300 min
Profit per Unit	$1	$1.20	

The total amount of time that machine I is used is given by $2x + y$ minutes and must not exceed 180 minutes. Thus, we have the inequality

$$2x + y \leq 180$$

Similarly, the total amount of time that machine II is used is $x + 3y$ minutes, which cannot exceed 300 minutes, so we are led to the inequality

$$x + 3y \leq 300$$

Finally, neither x nor y can be negative, so

$$x \geq 0$$

$$y \geq 0$$

To summarize, the problem at hand is one of maximizing the objective function $P = x + 1.2y$ subject to the system of inequalities

$$2x + \ y \le 180$$
$$x + 3y \le 300$$
$$x \ge 0$$
$$y \ge 0$$

The solution to this problem will be completed in Example 1, Section 6.3.
◼

Minimization Problems

In the following example of a linear programming problem, the objective function is to be minimized.

EXAMPLE 2   A nutritionist advises an individual who is suffering from iron and vitamin B deficiency to take at least 2400 mg of iron, 2100 mg of vitamin B-1 (Thiamine), and 1500 mg of vitamin B-2 (Riboflavin) over a period of time. Two vitamin pills are suitable, brand A and brand B. Each brand A pill contains 40 mg of iron, 10 mg of vitamin B-1, 5 mg of vitamin B-2, and costs 6 cents. Each brand B pill contains 10 mg of iron, 15 mg each of vitamins B-1 and B-2, and costs 8 cents (see Table 2). What combination of pills should the individual purchase in order to meet the minimum iron and vitamin requirements at the lowest cost?

TABLE 2

	Brand A	Brand B	Minimum Requirement
Iron	40 mg	10 mg	2400 mg
Vitamin B-1	10 mg	15 mg	2100 mg
Vitamin B-2	5 mg	15 mg	1500 mg
Cost per Pill	6¢	8¢	

Solution   Let x be the number of brand A pills and y be the number of brand B pills to be purchased. The cost C, measured in cents, is given by

$$C = 6x + 8y$$

and is the objective function to be minimized.

The amount of iron contained in x brand A pills and y brand B pills is given by $40x + 10y$ mg, and this must be greater than or equal to 2400 mg. This translates into the inequality

$$40x + 10y \ge 2400$$

Similar considerations involving the minimum requirements of vitamins B-1 and B-2 lead to the inequalities

$$10x + 15y \ge 2100$$

and
$$5x + 15y \ge 1500$$

respectively. Thus, the problem here is to minimize $C = 6x + 8y$ subject to

$$40x + 10y \geq 2400$$
$$10x + 15y \geq 2100$$
$$5x + 15y \geq 1500$$
$$x \geq 0, y \geq 0$$

The solution to this problem will be completed in Example 2, Section 6.3.

■

A Transportation Problem

EXAMPLE 3   Curtis-Roe Aviation Industries has two plants, I and II, that produce the Zephyr jet engines used in their light commercial airplanes. The maximum production capacities of these two plants are 100 units and 110 units per month, respectively. The engines are shipped to two of Curtis-Roe's main assembly plants, A and B. The shipping costs (in dollars) per engine from plants I and II to the main assembly plants A and B are as follows:

	To Assembly Plant	
From	**A**	**B**
Plant I	100	60
Plant II	120	70

In a certain month, assembly plant A needs 80 engines, whereas assembly plant B needs 70 engines. Find how many engines should be shipped from each plant to each main assembly plant if shipping costs are to be kept to a minimum.

Solution   Let x denote the number of engines shipped from plant I to assembly plant A, and let y denote the number of engines shipped from plant I to assembly plant B. Since the requirements of assembly plants A and B are 80 and 70 engines, respectively, the number of engines shipped from plant II to assembly plants A and B are $(80 - x)$ and $(70 - y)$, respectively. These numbers may be displayed in a schematic (Figure 9). With the aid of the schematic and the shipping cost schedule, we find that

FIGURE 9

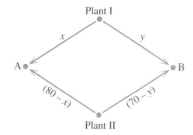

the total shipping costs incurred by Curtis-Roe are given by

$$C = 100x + 60y + 120(80 - x) + 70(70 - y)$$
$$= 14{,}500 - 20x - 10y$$

Next, the production constraints on plants I and II lead to the inequalities

$$x + y \leq 100$$

and $$(80 - x) + (70 - y) \leq 110$$

The last inequality simplifies to

$$x + y \geq 40$$

Also, the requirements of the two main assembly plants lead to the inequalities

$$x \geq 0, \qquad y \geq 0, \qquad 80 - x \geq 0, \qquad 70 - y \geq 0$$

The last two may be written as $x \leq 80$ and $y \leq 70$.

   Summarizing, we have the following linear programming problem: Minimize the objective (cost) function $C = 14{,}500 - 20x - 10y$ subject to the constraints

$$x + y \geq 40$$
$$x + y \leq 100$$
$$x \leq 80$$
$$y \leq 70$$

where $x \geq 0$ and $y \geq 0$.

   You will be asked to complete the solution to this problem in exercise 30, Section 6.3.   ■

A Warehouse Problem

EXAMPLE 4   The Acrosonic Company manufactures its model F loudspeaker systems in two separate locations, plant I and plant II. The output at plant I is at most 400 per month, whereas the output at plant II is at most 600 per month. These loudspeaker systems are shipped to three warehouses that serve as distribution centers for the company. In order for the warehouses to meet their orders, the minimum monthly requirements of warehouses A, B, and C are 200, 300, and 400, respectively. Shipping costs from plant I to warehouses A, B, and C are $20, $8, and $10 per loudspeaker system, respectively, and shipping costs from plant II to each of these warehouses are $12, $22, and $18, respectively. What should the shipping schedule be if Acrosonic wishes to meet the requirements of the distribution centers and at the same time keep its shipping costs to a minimum?

Solution   The respective shipping costs per loudspeaker system in dollars may be tabulated as in Table 3.

TABLE 3

Plant	Warehouse		
	A	B	C
I	20	8	10
II	12	22	18

Letting x_1 denote the number of loudspeaker systems shipped from plant I to warehouse A, x_2 the number shipped from plant I to warehouse B, and so on, leads to Table 4.

TABLE 4

Plant	Warehouse			Max. Prod.
	A	B	C	
I	x_1	x_2	x_3	400
II	x_4	x_5	x_6	600
Min. req.	200	300	400	

From Table 3 we see that the cost of shipping x_1 loudspeaker systems from plant I to warehouse A is $\$20x_1$, the cost of shipping x_2 loudspeaker systems from plant I to warehouse B is $\$8x_2$, and so on. Thus the total monthly shipping cost incurred by Acrosonic is given by

$$C = 20x_1 + 8x_2 + 10x_3 + 12x_4 + 22x_5 + 18x_6$$

Next, the production constraints on plants I and II lead to the inequalities

$$x_1 + x_2 + x_3 \leq 400$$

and

$$x_4 + x_5 + x_6 \leq 600$$

(see Table 4). Also, the minimum requirements of each of the three warehouses lead to the three inequalities

$$x_1 + x_4 \geq 200$$
$$x_2 + x_5 \geq 300$$

and

$$x_3 + x_6 \geq 400$$

Summarizing, we have the following linear programming problem: Minimize $C = 20x_1 + 8x_2 + 10x_3 + 12x_4 + 22x_5 + 18x_6$ subject to

$$x_1 + x_2 + x_3 \leq 400$$
$$x_4 + x_5 + x_6 \leq 600$$
$$x_1 + x_4 \geq 200$$
$$x_2 + x_5 \geq 300$$
$$x_3 + x_6 \geq 400$$
$$x_1 \geq 0, x_2 \geq 0, \ldots, x_6 \geq 0$$

The solution to this problem will be completed in Example 5, Section 6.5.

**SELF-CHECK
EXERCISE 6.2**

Mr. Balduzzi, proprietor of Luigi's Pizza Palace, allocates $9,000 a month for advertising in two newspapers, the *City Tribune* and the *Daily News*. The *City Tribune* charges $300 for a certain advertisement, whereas the *Daily News* charges $100 for the same ad. Mr. Balduzzi has stipulated that the ad is to appear in at least 15 but no more than 30 editions of the *Daily News* per month. The *City Tribune* has a daily circulation of 50,000, and the *Daily News* has a circulation of 20,000. Under these conditions, determine how many ads Balduzzi should place in each newspaper in order to reach the largest number of readers. Formulate but do not solve the problem. (This problem will be solved in exercise 32, Section 6.3).

Solutions to Self-Check Exercise 6.2 can be found on page 278.

6.2 EXERCISES

Formulate, but do not solve, each of the following exercises as a linear programming problem.

1. **Manufacturing—Production Scheduling** A company manufactures two products, A and B, on two machines I and II. It has been determined that the company will realize a profit of $3 on each unit of product A and a profit of $4 on each unit of product B. To manufacture a unit of product A requires 6 minutes on machine I and 5 minutes on machine II. To manufacture a unit of product B requires 9 minutes on machine I and 4 minutes on machine II. There are 5 hours of machine time available on machine I and 3 hours of machine time available on machine II in each work shift. How many units of each product should be produced in each shift to maximize the company's profit?

2. **Manufacturing—Production Scheduling** Kane Manufacturing has a division that produces two models of hibachis, model A and model B. To produce each model A hibachi requires 3 pounds of cast iron and 6 minutes of labor. To produce each model B hibachi requires 4 pounds of cast iron and 3 minutes of labor. The profit for each model A hibachi is $2, and the profit for each model B hibachi is $1.50. If 1000 pounds of cast iron and 20 hours of labor are available for the production of hibachis per day, how many hibachis of each model should the division produce in order to help maximize Kane's profits?

3. **Manufacturing—Production Scheduling** Refer to exercise 2. Because of a backlog of orders on model A hibachis, the manager of Kane Manufacturing has decided to produce at least 150 of these models a day. Operating under this additional constraint, how many hibachis of each model should Kane produce to maximize profit?

4. **Finance—Allocation of Funds** Madison Finance has a total of $20 million earmarked for homeowner and auto loans. On the average, homeowner loans will have a 10% annual rate of return, whereas auto loans will yield a 12% annual rate of return. Management has also stipulated that the total amount of homeowner loans should be greater than or equal to four times the total amount of automobile loans. Determine the total amount of loans of each type Madison should extend to each category in order to maximize its returns.

5. **Nutrition—Diet Planning** A nutritionist at the Medical Center has been asked to prepare a special diet for certain patients. She has decided that the meals should contain a minimum of 400 mg of calcium, 10 mg of iron, and 40 mg of vitamin C. She has further decided that the meals are to be prepared from foods A and B. Each ounce of food A contains 30 mg of calcium, 1 mg of iron, 2 mg of vitamin C, and 2 mg of cholesterol. Each ounce of food B contains 25 mg of calcium, 0.5 mg of iron, 5 mg of vitamin C, and 5 mg of cholesterol. Find how many ounces of each type of food should be used in a meal so that the cholesterol content is minimized and the minimum requirements of calcium, iron, and vitamin C are met.

6. **Agriculture—Crop Planning** A farmer has 150 acres of land suitable for cultivating crops A and B. The cost of cultivating crop A is $40 per acre, whereas that of crop B is $60 per acre. The farmer has a maximum of $7400 available for land cultivation. Each acre of crop A requires 20 hours of labor and each acre of crop B requires 25 hours of labor. The farmer has a maximum of 3300 hours of labor available. If he expects to make a profit of $150 per acre on crop A and $200 per acre on crop B, how many

acres of each crop should he plant in order to maximize his profit?

7. **Manufacturing—Shipping Costs** The TMA Company manufactures 19-inch color television picture tubes in two separate locations, location I and location II. The output at location I is at most 6000 tubes per month; the output at location II is at most 5000 per month. TMA is the main supplier of picture tubes to the Pulsar Corporation, its holding company, which has priority in having all its requirements met. In a certain month, Pulsar placed orders for 3000 and 4000 picture tubes to be shipped to two of its factories located in city A and city B, respectively. The shipping costs (in dollars) per picture tube from the two TMA plants to the two Pulsar factories are as follows:

Shipping Costs per Picture Tube

From	To Pulsar Factories	
	City A	City B
TMA (Loc. I)	$3	$2
TMA (Loc. II)	$4	$5

Find a shipping schedule that meets the requirements of both companies while keeping costs to a minimum.

8. **Social Programs Planning** AntiFam, a hunger-relief organization, has earmarked between $2 and $2.5 million, inclusive, for aid in two African countries, country A and country B. Country A is to receive between $1 and $1.5 million, inclusive, in aid, and country B is to receive at least $0.75 million in aid. It has been estimated that each dollar spent in country A will yield an effective return of 60 cents, whereas a dollar spent in country B will yield an effective return of 80 cents. How should the aid be allocated if the money is to be utilized most effectively according to these criteria? [*Hint:* If x and y denote the amount of money to be given to country A and country B, respectively, then the objective function to be maximized is $P = 0.6x + 0.8y$.]

9. **Manufacturing—Production Scheduling** A company manufactures products A, B, and C. Each product is processed in three departments: I, II, and III. The total available labor hours per week for departments I, II, and III are 900, 1080, and 840, respectively. The time requirements (in hours per unit) and profit per unit for each product are as follows:

	Product A	Product B	Product C
Dept. I	2	1	2
Dept. II	3	1	2
Dept. III	2	2	1
Profit	$18	$12	$15

How many units of each product should the company produce in order to maximize its profit?

10. **Advertising** As part of a campaign to promote its annual clearance sale, the Excelsior Company decided to buy television advertising time on Station KAOS. Excelsior's advertising budget was $102,000. Morning time costs $3,000 per minute, afternoon time costs $1,000 per minute, and evening (prime) time costs $12,000 per minute. Because of previous commitments, KAOS could not offer Excelsior more than 6 minutes of prime time or more than a total of 25 minutes of advertising time over the two weeks in which the commercials were to be run. KAOS estimated that morning commercials would be seen by 200,000 people, afternoon commercials would be seen by 100,000 people, and evening commercials would be seen by 600,000 people. How much morning, afternoon, and evening advertising time should Excelsior buy to maximize exposure of its commercials?

11. **Manufacturing—Production Scheduling** The Custom Office Furniture Company is introducing a new line of executive desks made from a specially selected grade of walnut. Initially, three different models—A, B, and C—are to be marketed. Each model A desk requires $1\frac{1}{4}$ hours for fabrication, 1 hour for assembly, and 1 hour for finishing; each model B desk requires $1\frac{1}{2}$ hours for fabrication, 1 hour for assembly, and 1 hour for finishing; each model C desk requires $1\frac{1}{2}$ hours, $\frac{3}{4}$ hour, and $\frac{1}{2}$ hour for fabrication, assembly, and finishing, respectively. The profit on each model A desk is $26, the profit on each model B desk is $28, and the profit on each model C desk is $24. The total time available in the fabrication department, the assembly department, and the finishing department in the first month of production is 310 hours, 205 hours, and 190 hours, respectively. To maximize Custom's profit, how many desks of each model should be made in the month?

12. **Manufacturing—Shipping Costs** The Acrosonic Company of Example 4 also manufactures a model G loudspeaker system in plants I and II. The output at plant I is at most 800 systems per month; the output at plant II is at most 600 per month. These

loudspeaker systems are also shipped to the three warehouses—A, B, and C—whose minimum monthly requirements are 500, 400, and 400, respectively. Shipping costs from plant I to warehouse A, warehouse B, and warehouse C are $16, $20, and $22 per loudspeaker system, respectively, and shipping costs from plant II to each of these warehouses are $18, $16, and $14, respectively. What shipping schedule will enable Acrosonic to meet the warehouses' requirements and at the same time keep its shipping costs to a minimum?

13. **Manufacturing—Shipping Costs** Steinwelt Piano manufactures uprights and consoles in two plants, plant I and plant II. The output of plant I is at most 300 per month; the output of plant II is at most 250 per month. These pianos are shipped to three warehouses that serve as distribution centers for the company. In order to fill current and projected future orders, warehouse A requires a minimum of 200 pianos per month, warehouse B requires at least 150 pianos per month, and warehouse C requires at least 200 pianos per month. The shipping cost of each piano from plant I to warehouse A, warehouse B, and warehouse C is $60, $60, and $80, respectively, and the shipping cost of each piano from plant II to warehouse A, warehouse B, and warehouse C is $80, $70, and $50, respectively. What shipping schedule will enable Steinwelt to meet the warehouses' requirements while keeping the shipping costs to a minimum?

14. **Manufacturing—Prefabricated Housing Production** Boise lumber has decided to enter the lucrative prefabricated housing business. Initially, it plans to offer three models: a standard model, a deluxe model, and a luxury model. Each house is prefabricated and partially assembled in the factory, and the final assembly is completed on site. The dollar amount of building material required, the amount of labor required in the factory for prefabrication and partial

assembly, the amount of on-site labor required, and the profit per unit are as follows:

	Standard Model	Deluxe Model	Luxury Model
Material ($)	6,000	8,000	10,000
Factory Labor (hr)	240	220	200
On-Site Labor (hr)	180	210	300
Profit ($)	3,400	4,000	5,000

For the first year's production, a sum of $8,200,000 is budgeted for the building material; the number of labor hours available for work in the factory (for prefabrication and partial assembly) is not to exceed 218,000 hours; and the amount of labor for on-site work is to be less than or equal to 237,000 labor hours. Determine how many houses of each type Boise should produce (market research has confirmed that there should be no problems with sales) to maximize its profit from this new venture.

15. **Manufacturing—Cold Formula Production** Bayer Pharmaceutical produces three kinds of cold formulas: formula I, formula II, and formula III. It takes 2.5 hours to produce 1,000 bottles of formula I, 3 hours to produce 1,000 bottles of formula II, and 4 hours to produce 1,000 bottles of formula III. The profits for each 1,000 bottles of formula I, formula II, and formula III are $180, $200, and $300, respectively. Suppose that for a certain production run there are enough ingredients on hand to make at most 9,000 bottles of formula I, 12,000 bottles of formula II, and 6,000 bottles of formula III. Furthermore, suppose that the time for the production run is limited to a maximum of 70 hours. Find how many bottles of each formula should be produced in this production run so that the profit is maximized.

SOLUTION TO SELF-CHECK EXERCISE 6.2

Let x denote the number of ads to be placed in the *City Tribune* and y the number to be placed in the *Daily News*. The total cost for placing x ads in the *City Tribune* and y ads in the *Daily News* is $300x + 100y$ dollars, and since the monthly budget is $9,000, we must have

$$300x + 100y \leq 9,000$$

Next, the condition that the ad must appear in at least 15 but no more than 30 editions of the *Daily News* translates into the inequalities

$$y \geq 15$$

and

$$y \leq 30$$

Finally, the objective function to be maximized is

$$P = 50,000x + 20,000y$$

To summarize, we have the following linear programming problem: Maximize $P = 50,000x + 20,000y$ subject to

$$300x + 100y \leq 9,000$$
$$y \geq 15$$
$$y \leq 30$$
$$x \geq 0, y \geq 0$$

6.3   Graphical Solution of Linear Programming Problems

The Graphical Method

Linear programming problems in two variables have relatively simple geometrical interpretations. For example, the system of linear constraints associated with a two-dimensional linear programming problem, unless it is inconsistent, defines a planar region whose boundary is composed of straight line segments and/or half-lines. Such problems are therefore amenable to graphical analysis.

Consider the following two-dimensional linear programming problem: Maximize $P = 3x + 2y$ subject to

$$2x + 3y \leq 12$$
$$2x + \ \ y \leq 8$$
$$x \geq 0, y \geq 0 \qquad\qquad\qquad (7)$$

The system of linear inequalities (7) defines the planar region S shown in Figure 10. Each point in S is a candidate for the solution of the problem at hand and is referred to as a **feasible solution.** The set S itself is referred

FIGURE 10

Each point in the feasible set S is a candidate for the optimal solution.

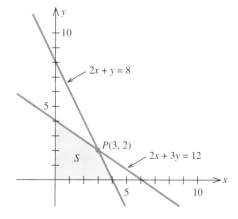

to as a **feasible set.** Our goal is to find from among all the points in the set S the point(s) that optimizes the objective function P. Such a feasible solution is called an **optimal solution** and constitutes the solution to the linear programming problem under consideration.

As noted earlier, each point $P(x, y)$ in S is a candidate for the optimal solution to the problem at hand. For example, the point $(1, 3)$ is easily seen to lie in S and is therefore in the running. The value of the objective function P at the point $(1, 3)$ is given by $P = 3(1) + 2(3) = 9$. Now, if we could compute the value of P corresponding to each point in S, then the point(s) in S that gave the largest value to P would constitute the solution set sought. Unfortunately, in most problems the number of candidates is either too large or, as in this problem, the number is infinite. Thus this method is at best unwieldy and at worst impractical.

Let us turn the question around. Instead of asking for the value of the objective function P at a feasible point, let us assign a value to the objective function P and ask whether there are feasible points that would correspond to the given value of P. To this end, suppose we assign a value of 6 to P. Then the objective function P becomes $3x + 2y = 6$, a linear equation in x and y, and, therefore, has a graph that is a straight line L_1 in the plane. In Figure 11 we have drawn the graph of this straight line superimposed on the feasible set S.

It is clear that each point on the straight line segment given by the intersection of the straight line L_1 and the feasible set S corresponds to the given value, 6, of P. Let us repeat the process, this time assigning a value of 10 to P. We obtain the equation $3x + 2y = 10$ and the line L_2 (see Figure 11), which suggests that there are feasible points that correspond to a larger value of P. Observe that the line L_2 is parallel to the line L_1, since both lines have slope equal to $-3/2$, which is easily seen by casting the corresponding equations in the slope-intercept form.

In general, by assigning different values to the objective function, we obtain a family of parallel lines, each with slope equal to $-3/2$. Furthermore, a line corresponding to a larger value of P lies further away from the origin than one with a smaller value of P. The implication is clear. To obtain the optimal solution(s) to the problem at hand, find the straight line, from this family of straight lines, that is farthest from the origin and still intersects the feasible set S. The required line is the one that passes through the point $P(3, 2)$ (see Figure 11), so the solution to the problem is given by $x = 3$, $y = 2$, resulting in a maximum value of $P = 3(3) + 2(2) = 13$.

That the optimal solution to this problem was found to occur at a vertex of the feasible set S is no accident. In fact, the result is a consequence of the following basic theorem on linear programming, which we state without proof.

FIGURE 11

A family of parallel lines that intersect the feasible set S

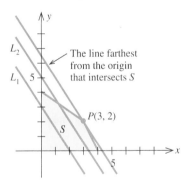

The line farthest from the origin that intersects S

THEOREM 1

If a linear programming problem has a solution, then it must occur at a vertex, or corner point, of the feasible set S associated with the problem. Furthermore, if the objective function P is optimized at two adjacent vertices of S, then it is optimized at every point on the line segment joining these vertices, in which case there are infinitely many solutions to the problem.

Theorem 1 tells us that our search for the solution(s) to a linear programming problem may be restricted to the examination of the set of vertices of the feasible set S associated with the problem. Since a feasible set S has finitely many vertices, the theorem suggests that the solution(s) to the linear programming problem may be found by inspecting the values of the objective function P at these vertices.

Although Theorem 1 sheds some light on the nature of the solution of a linear programming problem, it does not tell us when a linear programming problem has a solution. The following theorem states some conditions that will guarantee when a linear programming problem has a solution.

THEOREM 2: EXISTENCE OF A SOLUTION

Suppose we are given a linear programming problem with a feasible set S and an objective function $P = ax + by$.

1. If S is bounded, then P has both a maximum and a minimum value on S.

2. If S is unbounded and both a and b are nonnegative, then P has a minimum value on S provided that the constraints defining S include the inequalities $x \geq 0$ and $y \geq 0$.

3. If S is the empty set, then the linear programming problem has no solution; that is, P has neither a maximum nor a minimum value.

The **method of corners,** a simple procedure for solving linear programming problems based on Theorem 1, follows.

THE METHOD OF CORNERS

1. Graph the feasible set.

2. Find the coordinates of all corner points (vertices) of the feasible set.

3. Evaluate the objective function at each corner point.

4. Find the vertex that renders the objective function a maximum (minimum). If there is only one such vertex, then this vertex constitutes a unique solution to the problem. If the objective function is maximized (minimized) at two adjacent corner points of S, there are infinitely many optimal solutions given by the points on the line segment determined by these two vertices.

Applications

EXAMPLE 1 We are now in a position to complete the solution to the production problem posed in Example 1, Section 6.2. Recall that the mathematical formulation led to the following linear programming problem: Maximize $P = x + 1.2y$ subject to

$$2x + y \leq 180$$

$$x + 3y \leq 300$$

$$x \geq 0, y \geq 0$$

FIGURE 12
*The corner point that yields
the maximum profit is
C(48, 84).*

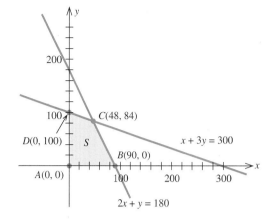

Solution   The feasible set S for the problem is shown in Figure 12. The vertices of the feasible set are $A(0, 0)$, $B(90, 0)$, $C(48, 84)$, and $D(0, 100)$. The values of P at these vertices may be tabulated as follows:

Vertex	$P = x + 1.2y$
$A(0, 0)$	0
$B(90, 0)$	90
$C(48, 84)$	148.8
$D(0, 100)$	120

From the table, we see that the maximum of $P = x + 1.2y$ occurs at the vertex $(48, 84)$ and has a value of 148.8. Recalling what the symbols x, y, and P represent, we conclude that Ace Novelty would maximize its profit (a figure of \$148.80) by producing 48 type-A souvenirs and 84 type-B souvenirs. ■

EXAMPLE 2   Complete the solution of the nutrition problem posed in Example 2, Section 6.2.

Solution   Recall that the mathematical formulation of the problem led to the following linear programming problem in two variables: Minimize $C = 6x + 8y$ subject to the constraints

$$40x + 10y \geq 2400$$

$$10x + 15y \geq 2100$$

$$5x + 15y \geq 1500$$

$$x \geq 0, y \geq 0$$

The feasible set S defined by the system of constraints is shown in Figure 13. The vertices of the feasible set S are $A(0, 240)$, $B(30, 120)$, $C(120, 60)$, and $D(300, 0)$. The values of the objective function C at these

FIGURE 13

The corner point that yields the minimum cost is B(30, 120).

vertices are given in the following table:

Vertex	$C = 6x + 8y$
$A(0, 240)$	1920
$B(30, 120)$	1140
$C(120, 60)$	1200
$D(300, 0)$	1800

From the table, we can see that the minimum for the objective function $C = 6x + 8y$ occurs at the vertex $B(30, 120)$ and has a value of 1140. Thus, the individual should purchase 30 brand A pills and 120 brand B pills at a minimum cost of \$11.40. ∎

EXAMPLE 3   Find the maximum and minimum of $P = 2x + 3y$ subject to the following system of linear inequalities:

$$2x + 3y \leq 30$$
$$y - x \leq 5$$
$$x + y \geq 5$$
$$x \leq 10$$
$$x \geq 0, y \geq 0$$

Solution   The feasible set S is shown in Figure 14. The vertices of the feasible set S are $A(5, 0)$, $B(10, 0)$, $C(10, 10/3)$, $D(3, 8)$, and $E(0, 5)$. The values of the objective function P at these vertices are given in the

FIGURE 14
Every point lying on the line segment joining C and D maximizes P.

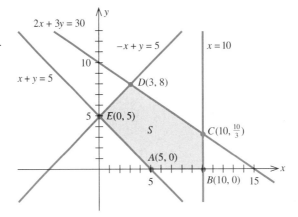

following table:

Vertex	$P = 2x + 3y$
$A(5,0)$	10
$B(10, 0)$	20
$C(10, 10/3)$	30
$D(3, 8)$	30
$E(0, 5)$	15

From the table, we see that the maximum for the objective function $P = 2x + 3y$ occurs at the vertices $C(10, 10/3)$ and $D(3, 8)$. This tells us that every point on the line segment joining the points $C(10, 10/3)$ and $D(3, 8)$ maximizes P, giving it a value of 30 at each of these points. From the table it is also clear that P is minimized at the point $(5, 0)$, where it attains a value of 10. ◼

We close this section by examining two situations in which a linear programming problem may have no solution.

EXAMPLE 4 Solve the following linear programming problem: Maximize $P = x + 2y$ subject to

$$-2x + \quad y \le 4$$

$$x - 3y \le 3$$

$$x \ge 0, y \ge 0$$

Solution The feasible set S for this problem is shown in Figure 15. Since the set S is unbounded (both x and y can take on arbitrarily large positive values), we see that we can make P as large as we please by making x and y large enough. Therefore, the problem has no solution. In this situation we say that the solution is unbounded. ◼

FIGURE 15
The maximization problem has no solution because the feasible set is unbounded.

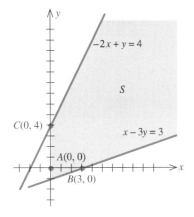

EXAMPLE 5   Solve the following linear programming problem: Maximize
$P = x + 2y$ subject to

$$x + 2y \leq 4$$

$$2x + 3y \geq 12$$

$$x \geq 0, y \geq 0$$

Solution   The half-planes described by the constraints (inequalities) have
no points in common (Figure 16). Therefore, there are no feasible points
and the problem has no solution. In this situation, we say that this problem
is *infeasible,* or *inconsistent.* (These situations are unlikely to occur in well-
posed problems arising from practical applications of linear programming.)

FIGURE 16

*The problem is inconsistent
because there is no point that
satisfies all the given inequalities.*

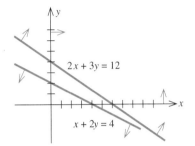

The method of corners is particularly effective in solving two-variable
linear programming problems with a small number of constraints, as the
preceding examples have amply demonstrated. Its effectiveness, however,
decreases rapidly as the number of variables and/or constraints increases.
For example, it may be shown that a linear programming problem in three
variables and five constraints may have up to ten feasible corner points.
The determination of the feasible corner points calls for the solution of
ten 3×3 systems of linear equations and then the verification, by the
substitution of each of these solutions into the system of constraints, that
each is in fact a feasible point. When the number of variables and constraints
goes up to five and ten, respectively (still a very small system from the
standpoint of applications in economics), the number of vertices to be
found and checked for feasible corner points increases dramatically to 252,
and each of these vertices is found by solving a 5×5 linear system! For this
reason, the method of corners is seldom used to solve linear programming
problems; its redeeming value lies in the fact that much insight is gained
into the nature of the solutions of linear programming problems through
its use in solving two-variable problems.

**SELF-CHECK
EXERCISES 6.3**

1. Use the method of corners to solve the following linear programming problem:
    Maximize $P = 4x + 5y$ subject to

    $$x + 2y \leq 10$$

    $$5x + 3y \leq 30$$

    $$x \geq 0, y \geq 0$$

2. Use the method of corners to solve the following linear programming problem:
Minimize $C = 5x + 3y$ subject to

$$5x + 3y \geq 30$$

$$x - 3y \leq 0$$

$$x \geq 2$$

Solutions to Self-Check Exercises 6.3 can be found on page 290.

6.3 EXERCISES

In exercises 1–6, find the optimal value(s) of the given objective function on the feasible set S.

1. $Z = 2x + 3y$

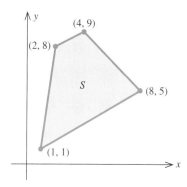

3. $Z = 3x + 4y$

2. $Z = 3x - y$

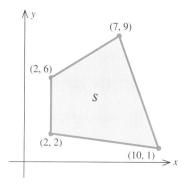

4. $Z = 7x + 9y$

5. $Z = x + 4y$

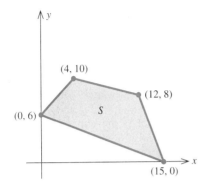

6. $Z = 3x + 2y$

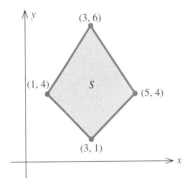

In exercises 7–21, solve each linear programming problem by the method of corners.

7. Maximize $P = 2x + 3y$ subject to

$$x + y \le 6$$
$$x \le 3$$
$$x \ge 0, y \ge 0$$

8. Maximize $P = 3x - 4y$ subject to

$$x + 3y \le 15$$
$$4x + y \le 16$$
$$x \ge 0, y \ge 0$$

9. Minimize $C = 2x + 10y$ subject to

$$5x + 2y \ge 40$$
$$x + 2y \ge 20$$
$$y \ge 3, x \ge 0$$

10. Minimize $C = 2x + 5y$ subject to

$$4x + y \ge 40$$
$$2x + y \ge 30$$
$$x + 3y \ge 30$$
$$x \ge 0, y \ge 0$$

11. Minimize $C = 6x + 3y$ subject to the constraints of exercise 10.

12. Maximize $P = 2x + 5y$ subject to

$$2x + y \le 16$$
$$2x + 3y \le 24$$
$$y \le 6$$
$$x \ge 0, y \ge 0$$

13. Minimize $C = 10x + 15y$ subject to

$$x + y \le 10$$
$$3x + y \ge 12$$
$$-2x + 3y \ge 3$$
$$x \ge 0, y \ge 0$$

14. Maximize $P = 2x + 5y$ subject to the constraints of exercise 13.

15. Maximize $P = 3x + 4y$ subject to

$$x + 2y \le 50$$
$$5x + 4y \le 145$$
$$2x + y \ge 25$$
$$y \ge 5, x \ge 0$$

16. Maximize $P = 4x - 3y$ subject to the constraints of exercise 15.

17. Maximize $P = 2x + 3y$ subject to

$$x + y \le 48$$
$$x + 3y \ge 60$$
$$9x + 5y \le 320$$
$$x \ge 10, y \ge 0$$

18. Minimize $C = 5x + 3y$ subject to the constraints of exercise 17.

19. Find the maximum and minimum of $P = 10x + 12y$ subject to

$$5x + 2y \ge 63$$
$$x + y \ge 18$$
$$3x + 2y \le 51$$
$$x \ge 0, y \ge 0$$

20. Find the maximum and minimum of $P = 4x + 3y$ subject to

$$3x + 5y \geq 20$$
$$3x + y \leq 16$$
$$-2x + y \leq 1$$
$$x \geq 0, y \geq 0$$

21. Find the maximum and minimum of $P = 2x + 4y$ subject to

$$x + y \leq 20$$
$$-x + y \leq 10$$
$$x \leq 10$$
$$x + y \geq 5$$
$$y \geq 5, x \geq 0$$

22. Manufacturing—Production Scheduling   A company manufactures two products, A and B, on two machines I and II. It has been determined that the company will realize a profit of $3 on each unit of product A and a profit of $4 on each unit of product B. To manufacture a unit of product A requires 6 minutes on machine I and 5 minutes on machine II. To manufacture a unit of product B requires 9 minutes on machine I and 4 minutes on machine II. There are 5 hours of machine time available on machine I and 3 hours of machine time available on machine II in each work shift. How many units of each product should be produced in each shift to maximize the company's profit?

23. Manufacturing—Production Scheduling   The National Business Machines Corporation manufactures two models of fax machines: A and B. Each model A costs $200 to make and each model B costs $300. The profits are $25 for each model A and $40 for each model B fax machine. If the total number of fax machines demanded per month does not exceed 2500 and the company has earmarked no more than $600,000 per month for manufacturing costs, find how many units of each model National should make each month in order to maximize its monthly profits.

24. Manufacturing—Production Scheduling   The Bata Aerobics Company manufactures two models of steppers used for aerobic exercises. To manufacture each luxury model requires 10 pounds of plastic and 10 minutes of labor. To manufacture each standard model requires 16 pounds of plastic and 8 minutes of labor. The profit for each luxury model is $40 and the profit for each standard model is $30. If 6000 pounds of plastic and 60 hours of labor are available for the production of the steppers per day, how many

steppers of each model should Bata produce in order to maximize its profits?

25. Manufacturing—Production Scheduling   Kane Manufacturing has a division that produces two models of hibachis, model A and model B. To produce each model A hibachi requires 3 pounds of cast iron and 6 minutes of labor. To produce each model B hibachi requires 4 pounds of cast iron and 3 minutes of labor. The profit for each model A hibachi is $2, and the profit for each model B hibachi is $1.50. If 1000 pounds of cast iron and 20 hours of labor are available for the production of hibachis per day, how many hibachis of each model should the division produce in order to help maximize Kane's profits?

26. Manufacturing—Production Scheduling   Refer to exercise 25. Because of a backlog of orders for model A hibachis, Kane's manager has decided to produce at least 150 of these models a day. Operating under this additional constraint, how many hibachis of each model should Kane produce to maximize profit?

27. Finance—Allocation of Funds   Madison Finance has a total of $20 million earmarked for homeowner and auto loans. On the average, homeowner loans will have a 10% annual rate of return, whereas auto loans will yield a 12% annual rate of return. Management has also stipulated that the total amount of homeowner loans should be greater than or equal to four times the total amount of automobile loans. Determine the total amount of loans of each type Madison should extend to each category in order to maximize its returns.

28. Nutrition—Diet Planning   A nutritionist at the Medical Center has been asked to prepare a special diet for certain patients. She has decided that the meals should contain a minimum of 400 mg of calcium, 10 mg of iron, and 40 mg of vitamin C. She has further decided that the meals are to be prepared from foods A and B. Each ounce of food A contains 30 mg of calcium, 1 mg of iron, 2 mg of vitamin C, and 2 mg of cholesterol. Each ounce of food B contains 25 mg of calcium, 0.5 mg of iron, 5 mg of vitamin C, and 5 mg of cholesterol. Find how many ounces of each type of food should be used in a meal so that the cholesterol content is minimized and the minimum requirements of calcium, iron, and vitamin C are met.

29. Agriculture—Crop Planning   A farmer has 150 acres of land suitable for cultivating crops A and B. The cost of cultivating crop A is $40 per acre, whereas that of crop B is $60 per acre. The farmer has a maximum of $7400 available for land cultivation. Each acre of crop A requires 20 hours of labor and each acre of crop B requires 25 hours of labor. The farmer has a maximum of 3300 hours of labor available. If he expects to make a profit of $150 per acre on crop A and $200 per acre on crop B, how many acres of each crop should he plant in order to maximize his profit?

30. Complete the solution to Example 3, Section 6.2.

31. **Manufacturing—Shipping Costs**   The TMA Company manufactures 19-inch color television picture tubes in two separate locations, locations I and II. The output at location I is at most 6000 tubes per month, whereas the output at location II is at most 5000 per month. TMA is the main supplier of picture tubes to the Pulsar Corporation, its holding company, which has priority in having all its requirements met. In a certain month, Pulsar placed orders for 3000 and 4000 picture tubes to be shipped to two of its factories located in city A and city B, respectively. The shipping costs (in dollars) per picture tube from the two TMA plants to the two Pulsar factories are as follows:

Shipping Costs per Picture Tube

| From | To Pulsar Factories | |
	City A	City B
TMA (Loc. I)	$3	$2
TMA (Loc. II)	$4	$5

Find a shipping schedule that meets the requirements of both companies while keeping costs to a minimum.

32. **Advertising—Maximizing Circulation**   Mr. Balduzzi, proprietor of Luigi's Pizza Palace, allocates $9,000 a month for advertising in two newspapers, the *City Tribune* and the *Daily News.* The *City Tribune* charges $300 for a certain advertisement, whereas the *Daily News* charges $100 for the same ad. Mr. Balduzzi has stipulated that the ad is to appear in at least 15 but no more than 30 editions of the *Daily News* per month. The *City Tribune* has a daily circulation of 50,000, and the *Daily News* has a circulation of 20,000. Under these conditions, determine how many ads Balduzzi should place in each newspaper in order to reach the largest number of readers.

33. **Social Programs Planning**   AntiFam, a hunger-relief organization, has earmarked between $2 and $2.5 million, inclusive, for aid in two African countries, country A and country B. Country A is to receive between $1 and $1.5 million, inclusive, in aid, and country B is to receive at least $0.75 million in aid. It has been estimated that each dollar spent in country A will yield an effective return of 60 cents, whereas a dollar spent in country B will yield an effective return of 80 cents. How should the aid be allocated if the money is to be utilized most effectively according to these criteria? [*Hint:* If x and y denote the amount of money to be given to country A and country B, respectively, then the objective function to be maximized is $P = 0.6x + 0.8y$.]

34. **Investment Planning**   Patricia has at most $30,000 to invest in securities in the form of corporate stocks. She has narrowed her choices to two groups of stocks: growth stocks that she assumes will yield a 15% return (dividends and capital appreciation) within a year, and speculative stocks that she assumes will yield a 25% return (mainly in capital appreciation) within a year. Determine how much she should invest in each group of stocks in order to maximize the return on her investments within a year if she has decided to invest at least three times as much in growth stocks as in speculative stocks.

36. **Veterinary Science**   A veterinarian has been asked to prepare a diet for a group of dogs to be used in a nutrition study at the School of Animal Science. It has been stipulated that each serving should be no larger than 8 ounces and must contain at least 29 units of nutrient I and 20 units of nutrient II. The vet has decided that the diet may be prepared from two brands of dog food: brand A and brand B. Each ounce of brand A contains 3 units of nutrient I and 4 units of nutrient II. Each ounce of brand B contains 5 units of nutrient I and 2 units of nutrient II. Brand A costs 3 cents per ounce and brand B costs 4 cents per ounce. Determine how many ounces of each brand of dog food should be used to meet the given requirements at a minimum cost.

36. **Market Research**   The Trendex Corporation, a telephone survey company, has been hired to conduct a television viewing poll among urban and suburban families in the Los Angeles area. The client has stipulated that a maximum of 1500 families are to be interviewed. At least 500 urban families must be interviewed, and at least half of the total number of families interviewed must be from the suburban area. For this service, Trendex will be paid $1500 plus $2 for each completed interview. From previous experience, Trendex has determined that it will incur an expense of $1.10 for each successful interview with an urban family and $1.25 for each successful interview with a suburban family. Find how many urban and suburban families Trendex should interview in order to maximize its profit.

**SOLUTIONS TO
SELF-CHECK
EXERCISES 6.3**

1. The feasible set S for the problem was graphed in the solution to exercise 1, Self-Check Exercises 6.1. It is reproduced in the accompanying figure.

The values of the objective function P at the vertices of S are summarized in the accompanying table.

Vertex	$P = 4x + 5y$
$A(0, 0)$	0
$B(6, 0)$	24
$C(\frac{30}{7}, \frac{20}{7})$	$\frac{220}{7} = 31\frac{3}{7}$
$D(0, 5)$	25

From the table, we see that the maximum for the objective function P is attained at the vertex $C(30/7, 20/7)$. Therefore, the solution to the problem is $x = 30/7$, $y = 20/7$, and $P = 31\frac{3}{7}$.

2. The feasible set S for the problem was graphed in the solution to exercise 2, Self-Check Exercises 6.1. It is reproduced in the accompanying figure.

Evaluating the objective function $C = 5x + 3y$ at each corner point, we obtain the table

Vertex	$C = 5x + 3y$
$A(5, \frac{5}{3})$	30
$B(2, \frac{20}{3})$	30

We conclude that the objective function is minimized at every point on the line segment joining the points (5, 5/3) and (2, 20/3) and the minimum value of C is 30.

The Simplex Method: Standard Maximization Problems

The Simplex Method

As mentioned earlier, the method of corners is not suitable for solving linear programming problems when the number of variables or constraints is large. Its major shortcoming is that a knowledge of all the corner points of the feasible set S associated with the problem is required. What we need is a method of solution that is based on a judicious selection of the corner points of the feasible set S, thereby reducing the number of points to be inspected. One such technique, called the simplex method, was developed in the late 1940s by George Dantzig and is based on the Gauss-Jordan elimination method. The simplex method is readily adaptable to the computer, which makes it ideally suitable for solving linear programming problems involving large numbers of variables and constraints.

Basically, the simplex method is an iterative procedure; that is, it is repeated over and over again. Beginning at some initial feasible solution (a corner point of the feasible set S, usually the origin) each iteration brings us to another corner point of S with an improved (but certainly no worse) value of the objective function. The iteration is terminated when the optimal solution is reached (if it exists).

In this section, we describe the simplex method for solving a large class of problems that are referred to as *standard maximization problems.*

Before stating a formal procedure for solving standard linear programming problems based on the simplex method, let us consider the following analysis of a two-variable problem. The ensuing discussion will clarify the general procedure and at the same time enhance our understanding of the simplex method by examining the motivation that led to the steps of the procedure.

A STANDARD LINEAR PROGRAMMING PROBLEM

A standard maximization problem is one in which

1. The objective function is to be maximized.
2. All the variables involved in the problem are nonnegative.
3. Each linear constraint may be written so that the expression involving the variables is less than or equal to a nonnegative constant.

Consider the linear programming problem presented at the beginning of Section 6.3:

Maximize

$$P = 3x + 2y \qquad\qquad (8)$$

subject to

$$2x + 3y \leq 12$$

$$2x + y \leq 8 \qquad\qquad (9)$$

$$x \geq 0, y \geq 0$$

FIGURE 17

The optimal solution occurs at C(3, 2).

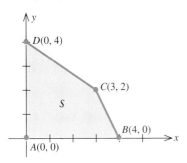

You can easily verify that this is a standard maximization problem. The feasible set S associated with this problem is reproduced in Figure 17, where we have labeled the four feasible corner points $A(0, 0)$, $B(4, 0)$, $C(3, 2)$, and $D(0, 4)$. Recall that the optimal solution to the problem occurs at the corner point $C(3, 2)$.

As a first step in the solution using the simplex method, we replace the system of inequality constraints (9) with a system of equality constraints. This may be accomplished by using nonnegative variables called **slack variables.** Let us begin by considering the inequality

$$2x + 3y \leq 12$$

Observe that the left-hand side of this equation is always less than or equal to the right-hand side. Therefore, by adding a nonnegative variable u to the left-hand side to compensate for this difference, we obtain the equality

$$2x + 3y + u = 12$$

For example, if $x = 1$ and $y = 1$ (you can see by referring to Figure 17 that the point $(1, 1)$ is a feasible point of S), then $u = 7$. Thus,

$$2(1) + 3(1) + 7 = 12$$

If $x = 2$ and $y = 1$ (the point $(2, 1)$ is also a feasible point of S), then $u = 5$. Thus,

$$2(2) + 3(1) + 5 = 12$$

The variable u is called a *slack variable.*

Similarly, the inequality $2x + y \leq 8$ is converted into the equation $2x + y + v = 8$ through the introduction of the slack variable v. The system of linear inequalities (9) may now be viewed as the system of linear equations

$$2x + 3y + u = 12$$

$$2x + y + v = 8$$

where x, y, u, and v are all nonnegative.

Finally, rewriting the objective function (8) in the form $-3x - 2y + P = 0$, where the coefficient of P is $+1$, we are led to the following system of linear equations:

$$2x + 3y + u = 12$$

$$2x + y + v = 8 \qquad\qquad (10)$$

$$-3x - 2y + P = 0$$

Since the system (10) comprises three linear equations in the five variables x, y, u, v, and P, we may solve for three of the variables in terms of the other two. Thus, there are infinitely many solutions to this system expressible in terms of two parameters. Our linear programming problem is now seen to be equivalent to the following: From among all the solutions of the system (10) for which x, y, u, and v are nonnegative (such solutions are called *feasible solutions*), determine the solution(s) that renders P a maximum.

The augmented matrix associated with system (10) is

$$
\begin{array}{ccccc}
x & y & u & v & P \\
\begin{bmatrix}
2 & 3 & 1 & 0 & 0 \\
2 & 1 & 0 & 1 & 0 \\
-3 & -2 & 0 & 0 & 1
\end{bmatrix}
&
\begin{array}{c}
12 \\
8 \\
0
\end{array}
\end{array}
\qquad \textbf{(11)}
$$

Observe that each of the u-, v-, and P-columns of the augmented matrix (11) is a unit column (see page 198). The variables associated with unit columns are called **basic variables;** all other variables are called **nonbasic variables.**

Now, the configuration of the augmented matrix (11) suggests that we solve for the basic variables u, v, and P in terms of the nonbasic variables x and y, obtaining

$$
\begin{aligned}
u &= 12 - 2x - 3y \\
v &= 8 - 2x - y \\
P &= 3x + 2y
\end{aligned}
\qquad \textbf{(12)}
$$

Of the infinitely many feasible solutions obtainable by assigning arbitrary nonnegative values to the parameters x and y, a particular solution is obtained by letting $x = 0$ and $y = 0$. In fact, this solution is given by

$$x = 0, \qquad y = 0, \qquad u = 12, \qquad v = 8, \qquad P = 0$$

Such a solution, obtained by setting the nonbasic variables equal to zero, is called a **basic solution** of the system. This particular solution corresponds to the corner point $A(0, 0)$ of the feasible set associated with the linear programming problem (Figure 17). Observe that $P = 0$ at this point.

Now, if the value of P could not be increased, we would have found the optimal solution to the problem at hand. To determine whether the value of P could, in fact, be improved, let us turn our attention to the objective function (8). Since both the coefficients of x and y are positive, the value of P could be improved by increasing x and/or y—that is, by moving away from the origin. Note that we arrive at the same conclusion by observing that the last row of the augmented matrix (11) contains entries that are *negative*. (Compare the original objective function, $P = 3x + 2y$, with the rewritten objective function, $-3x - 2y + P = 0$.)

Continuing our quest for an optimal solution, our next task is to determine whether it is more profitable to increase the value of x or that of y (increasing x and y simultaneously is more difficult). Since the coefficient of x is greater than that of y, a unit increase in the x-direction would result in a greater increase in the value of the objective function P than a unit increase in the y-direction. Thus, we should increase the value of x while holding y constant. How much can x be increased while holding $y = 0$? Upon setting $y = 0$ in the first two equations of (12), we see that

$$u = 12 - 2x$$
$$v = \;\;8 - 2x$$

(13)

Since u must be nonnegative, the first equation of (13) implies that x cannot exceed 12/2, or 6. The second equation of (13) and the nonnegativity of v implies that x cannot exceed 8/2, or 4. Thus, we conclude that x can be increased by at most 4.

Now, if we set $y = 0$ and $x = 4$ in system (12), we obtain the solution

$$x = 4, \qquad y = 0, \qquad u = 4, \qquad v = 0, \qquad P = 12$$

which is a basic solution to (10), this time with y and v as nonbasic variables. (Recall that the nonbasic variables are precisely the variables that are set equal to zero.)

Let us see how this basic solution may be found by working with the augmented matrix of the system. Since x is to replace v as a basic variable, our aim is to find an augmented matrix that is equivalent to the matrix (11) and has a configuration in which the x column is in the unit form

$$\begin{bmatrix} 0 \\ 1 \\ 0 \end{bmatrix}$$

replacing what is presently the form of the v column in (11). Now, this may be accomplished by pivoting about the circled number 2.

$$
\begin{array}{ccccc}
x & y & u & v & P
\end{array}
\left[\begin{array}{ccccc|c}
2 & 3 & 1 & 0 & 0 & 12 \\
\textcircled{2} & 1 & 0 & 1 & 0 & 8 \\
-3 & -2 & 0 & 0 & 1 & 0
\end{array}\right]
\xrightarrow{\frac{1}{2}R_2}
\begin{array}{ccccc}
x & y & u & v & P
\end{array}
\left[\begin{array}{ccccc|c}
2 & 3 & 1 & 0 & 0 & 12 \\
\textcircled{1} & \frac{1}{2} & 0 & \frac{1}{2} & 0 & 4 \\
-3 & -2 & 0 & 0 & 1 & 0
\end{array}\right]
\quad \textbf{(14)}
$$

$$
\xrightarrow[R_3 + 3R_2]{R_1 - 2R_2}
\begin{array}{ccccc}
x & y & u & v & P
\end{array}
\left[\begin{array}{ccccc|c}
0 & 2 & 1 & -1 & 0 & 4 \\
1 & \frac{1}{2} & 0 & \frac{1}{2} & 0 & 4 \\
0 & -\frac{1}{2} & 0 & \frac{3}{2} & 1 & 12
\end{array}\right]
\quad \textbf{(15)}
$$

Using (15), we now solve for the basic variables x, u, and P in terms of the nonbasic variables y and v, obtaining

$$x = \;\;4 - (1/2)y - (1/2)v$$
$$u = \;\;4 - \;\;\;\;\;2y + \;\;\;\;\;\;v$$
$$P = 12 + (1/2)y - (3/2)v$$

Setting the nonbasic variables y and v equal to zero gives

$$x = 4, \qquad y = 0, \qquad u = 4, \qquad v = 0, \qquad P = 12$$

as before.

Harley Lance Kaplan

Beta Industries is an investment firm, not a broad-based manufacturing conglomerate as its name implies. Its clients range from individuals with significant net worth to Fortune 500 corporations and nonprofit organizations with growing endowments. Kaplan and his partners help clients meet their long-term financial goals by recommending sound investment strategies.

Title: Registered Investment Advisor, Certified Financial Planner
Institution: Beta Industries

Beta Industries' clients "want to invest their money more efficiently." Kaplan helps them "identify their assets and define their financial objectives." He then assembles portfolios to meet their needs.

Kaplan must frequently educate clients on the potential risks and rewards investors face. He explains that, based on historical and statistical data and current risk factors, he can *project,* not predict, "possible future consequences." Even then, any investment strategy has to be tempered by an individual client's tolerance for risk.

Age often plays a role in a client's tolerance level. A 25-year-old investor with only $10,000 but many working years ahead might take greater risks, relying on a growth or an aggressive growth vehicle. (For a client with over $250,000 to invest, Kaplan recommends a personal money manager rather than a vehicle such as a mutual fund, where many investors contribute relatively small amounts.)

Data show that with a growth investment, a client risks gaining or losing as much as 10% to 25% of the principal. An aggressive growth investment can gain or lose up to 50%. Both types realize a solid rate of return over inflation, from a minimum of 10% to as much as 35%. The key to success is keeping the funds in place for at least five to seven years to ride out the cycles that typically affect the economy.

Kaplan's strategy for a 62-year-old, with only a few more working years, might rely heavily on the safety and liquidity offered by NOW accounts (combined checking and savings) and income vehicles such as bonds with moderate risk. With a $120,000 principal, Kaplan might recommend that 70% be placed into safety and income, with the balance invested as a hedge against inflation (20% into growth and 10% into aggressive growth). Inflation robs everyone of purchasing power, especially those on fixed incomes. For example, if inflation remains constant at 3% per year, $10,000 in interest earned by a retiree would buy only $7,374 in goods and services after ten years.

Investment vehicles vary widely. When choosing among investments, Kaplan uses a number of comparisons, including the net rate of return, how each investment affects a client's tax situation, and potential risk. For the client in the 40% tax bracket, the net return on a corporate bond paying 10% and a tax-free municipal paying 6% is the same. The only difference may be the level of risk each bond carries. All else being equal, the safer bond is the logical choice.

Through the years, Kaplan has learned to weigh not just tangible assets but intangible emotions—to tailor an investment strategy suited to each individual's needs.

We have now completed one iteration of the simplex procedure, and our search has brought us from the feasible corner point $A(0, 0)$, where $P = 0$, to the feasible corner point $B(4, 0)$, where P attained a value of 12, which is certainly an improvement! (See Figure 18.)

FIGURE 18
One iteration has taken us from $A(0, 0)$, where $P = 0$, to $B(4, 0)$, where $P = 12$.

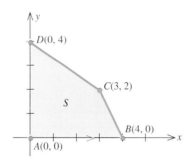

Before going on, let us introduce the following terminology. The circled element 2 in the first augmented matrix of (14), which was to be converted into a 1, is called a *pivot element.* The column containing the pivot element is called the *pivot column.* The pivot column is associated with a nonbasic variable, which is to be converted to a basic variable. Note that *the last entry in the pivot column is the negative number with the largest absolute value to the left of the vertical line in the last row*—precisely the criterion for choosing the direction of maximum increase in P.

The row containing the pivot element is called the *pivot row.* The pivot row can also be found by dividing each positive number in the pivot column into the corresponding number in the last column (the column of constants). *The pivot row is the one with the smallest ratio.* In the augmented matrix (14), the pivot row is the second row, since the ratio 8/2, or 4, is less than the ratio 12/2, or 6. (Compare this with the earlier analysis pertaining to the determination of the largest permissible increase in the value of x.)

The following is a summary of the procedure for selecting the pivot element:

SELECTING THE PIVOT ELEMENT

1. Select the pivot column: Locate the most negative entry to the left of the vertical line in the last row. The column containing this entry is the **pivot column.** (If there is more than one such column, choose any one.)

2. Select the pivot row: Divide each positive entry in the pivot column into its corresponding entry in the column of constants. The **pivot row** is the row corresponding to the smallest ratio thus obtained. (If there is more than one such entry, choose any one.)

3. The **pivot element** is the element common to both the pivot column and the pivot row.

Continuing with the solution to our problem, we observe that the last row of the augmented matrix (15) contains a negative number, namely, $-1/2$. This indicates that P is not maximized at the feasible corner point $B(4, 0)$ and another iteration is required. Without once again going into a detailed analysis, we proceed immediately to the selection of a pivot element. In accordance with the rules, we perform the necessary row operations as follows:

$$
\begin{array}{c}
\text{Pivot} \\
\text{row}
\end{array}
\rightarrow
\begin{array}{ccccc}
x & y & u & v & P
\end{array}
\left[
\begin{array}{ccccc|c}
0 & ② & 1 & -1 & 0 & 4 \\
1 & \frac{1}{2} & 0 & \frac{1}{2} & 0 & 4 \\
0 & -\frac{1}{2} & 0 & \frac{3}{2} & 1 & 12
\end{array}
\right]
\quad
\begin{array}{c}
\text{ratio} \\
\frac{4}{2} = 2 \\
\frac{4}{1/2} = 8
\end{array}
$$

$$\underset{\text{Pivot column}}{\uparrow}$$

$$
\xrightarrow{\frac{1}{2}R_1}
\begin{array}{ccccc}
x & y & u & v & P
\end{array}
\left[
\begin{array}{ccccc|c}
0 & ① & \frac{1}{2} & -\frac{1}{2} & 0 & 2 \\
1 & \frac{1}{2} & 0 & \frac{1}{2} & 0 & 4 \\
0 & -\frac{1}{2} & 0 & \frac{3}{2} & 1 & 12
\end{array}
\right]
$$

$$
\begin{array}{c}
R_2 - \frac{1}{2}R_1 \\
\xrightarrow{\hspace{1cm}} \\
R_3 + \frac{1}{2}R_1
\end{array}
\begin{array}{ccccc}
x & y & u & v & P
\end{array}
\left[
\begin{array}{ccccc|c}
0 & 1 & \frac{1}{2} & -\frac{1}{2} & 0 & 2 \\
1 & 0 & -\frac{1}{4} & \frac{3}{4} & 0 & 3 \\
0 & 0 & \frac{1}{4} & \frac{5}{4} & 1 & 13
\end{array}
\right]
$$

Interpreting the last augmented matrix in the usual fashion, we find the basic solution $x = 3$, $y = 2$, and $P = 13$. Since there are no negative entries in the last row, the solution is optimal and P cannot be increased further. The optimal solution is the feasible corner point $C(3, 2)$ [Figure 19]. Observe that this agrees with the solution we found using the method of corners in Section 6.3.

FIGURE 19

The next iteration has taken us from $B(4, 0)$, where $P = 12$, to $C(3, 2)$, where $P = 13$.

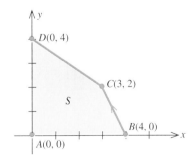

Having seen how the simplex method works, let us list the steps involved in the procedure. The first step is to set up the initial simplex tableau.

**SETTING UP THE
INITIAL SIMPLEX
TABLEAU**

1. Transform the system of linear inequalities into a system of linear equations by introducing slack variables.

2. Rewrite the objective function

$$P = c_1x_1 + c_2x_2 + \cdots + c_nx_n$$

in the form

$$-c_1x_1 - c_2x_2 - \cdots - c_nx_n + P = 0$$

where all the variables are on the left and the coefficient of P is $+1$. Write this equation below the equations of part 1.

3. Write the augmented matrix associated with this system of linear equations.

EXAMPLE 1   Set up the initial simplex tableau for the linear programming problem posed in Example 1, Section 6.2.

Solution   The problem at hand is to maximize

$$P = x + 1.2y$$

or, equivalently,

$$P = x + \frac{6}{5}y$$

subject to the constraints

$$2x + y \leq 180$$
$$x + 3y \leq 300 \qquad\qquad \textbf{\textit{(16)}}$$
$$x \geq 0, y \geq 0$$

This is a standard maximization problem and may be solved by the simplex method. Since the system (16) comprises two linear inequalities, we introduce the two slack variables u and v to convert it to the system of linear equations

$$2x + y + u \qquad\quad = 180$$
$$x + 3y \qquad + v \quad = 300$$

Next, by rewriting the objective function in the form

$$-x - \frac{6}{5}y + P = 0$$

where the coefficient of P is $+1$, and placing it below the system of equations, we obtain the system of linear equations

$$2x + y + u \qquad\qquad = 180$$
$$x + 3y \qquad + v \qquad = 300$$
$$-x - \frac{6}{5}y \qquad\qquad + P = \quad 0$$

The initial simplex tableau associated with this system is

x	y	u	v	P	Constant
2	1	1	0	0	180
1	3	0	1	0	300
-1	$-\frac{6}{5}$	0	0	1	0

Before completing the solution to the problem posed in Example 1, let us summarize the main steps of the simplex method.

THE SIMPLEX METHOD

1. *Set up the initial simplex tableau.*
2. *Determine whether the optimal solution has been reached by examining all of the entries in the last row to the left of the vertical line:*
   a. If all the entries are nonnegative, the optimal solution has been reached. Proceed to step 4.
   b. If there are one or more negative entries, the optimal solution has not been reached. Proceed to step 3.
3. *Perform the pivot operation:*
   Locate the pivot element and convert it to a 1 by dividing all the elements in the pivot row by the pivot element. Using row operations, convert the pivot column into a unit column by adding suitable multiples of the pivot row to each of the other rows as required. Return to step 2.
4. *Determine the optimal solution(s):*
   The value of the variable heading each unit column is given by the entry lying in the column of constants in the row containing the 1. The variables heading columns not in unit form are assigned the value zero.

EXAMPLE 2   Complete the solution to the problem discussed in Example 1.

Solution   The first step in our procedure, setting up the initial simplex tableau, was completed in Example 1. We continue with step 2.

Step 2   *Determine whether the optimal solution has been reached.*
First, refer to the initial simplex tableau:

x	y	u	v	P	Constant
2	1	1	0	0	180
1	3	0	1	0	300
-1	$-\frac{6}{5}$	0	0	1	0

(17)

Since there are negative entries in the last row of the initial simplex tableau, the initial solution is not optimal. We proceed to step 3.

Step 3 *Perform the following iterations.*

First, locate the pivot element:

a. Since the entry $-6/5$ is the most negative entry to the left of the vertical line in the last row of the initial simplex tableau, the second column in the tableau is the pivot column.

b. Divide each positive number of the pivot column into the corresponding entry in the column of constants and compare the ratios thus obtained. We see that the ratio 300/3 is less than the ratio 180/1, so the second row in the tableau is the pivot row.

c. The entry 3 lying in the pivot column and the pivot row is the pivot element.

Next, we convert this pivot element into a 1 by multiplying all the entries in the pivot row by 1/3. Then, using elementary row operations, we complete the conversion of the pivot column into a unit column.

The details of the iteration are recorded as follows:

	x	y	u	v	P	Constant	
	2	1	1	0	0	180	$\frac{180}{1} = 180$
Pivot row →	1	③	0	1	0	300	$\frac{300}{3} = 100$
	-1	$-\frac{6}{5}$	0	0	1	0	

↑
Pivot
column

$\xrightarrow{\frac{1}{3}R_2}$

	x	y	u	v	P	Constant
	2	1	1	0	0	180
	$\frac{1}{3}$	①	0	$\frac{1}{3}$	0	100
	-1	$-\frac{6}{5}$	0	0	1	0

$\xrightarrow[R_3 + \frac{6}{5}R_2]{R_1 - R_2}$

	x	y	u	v	P	Constant
	$\frac{5}{3}$	0	1	$-\frac{1}{3}$	0	80
	$\frac{1}{3}$	1	0	$\frac{1}{3}$	0	100
	$-\frac{3}{5}$	0	0	$\frac{2}{5}$	1	120

(18)

This completes one iteration. The last row of the simplex tableau contains a negative number, so an optimal solution has not been reached. Therefore, we repeat the iterative step once again as follows:

	x	y	u	v	P	Constant	
Pivot row →	⑤⁄₃	0	1	$-\frac{1}{3}$	0	80	$\frac{80}{5/3} = 48$
	$\frac{1}{3}$	1	0	$\frac{1}{3}$	0	100	$\frac{100}{1/3} = 300$
	$-\frac{3}{5}$	0	0	$\frac{2}{5}$	1	120	

↑
Pivot
column

$$\xrightarrow{\frac{3}{5}R_1}$$

	x	y	u	v	P	Constant
	①	0	$\frac{3}{5}$	$-\frac{1}{5}$	0	48
	$\frac{1}{3}$	1	0	$\frac{1}{3}$	0	100
	$-\frac{3}{5}$	0	0	$\frac{2}{5}$	1	120

$$\xrightarrow[R_3 + \frac{3}{5}R_1]{R_2 - \frac{1}{3}R_1}$$

x	y	u	v	P	Constant	
1	0	$\frac{3}{5}$	$-\frac{1}{5}$	0	48	
0	1	$-\frac{1}{5}$	$\frac{2}{5}$	0	84	
0	0	$\frac{9}{25}$	$\frac{7}{25}$	1	$148\frac{4}{5}$	**(19)**

The last row of the simplex tableau (19) contains no negative numbers, and we therefore conclude that the optimal solution has been reached.

Step 4   *Determine the optimal solution.*
Locate the basic variables in the final tableau. In this case, the basic variables (those heading unit columns) are x, y, and P. The value assigned to the basic variable x is the number 48, which is the entry lying in the column of constants and in the first row (the row that contains the 1).

x	y	u	v	P	Constant
①	0	$\frac{3}{5}$	$-\frac{1}{5}$	0	48
0	①	$-\frac{1}{5}$	$\frac{2}{5}$	0	84
0	0	$\frac{9}{25}$	$\frac{7}{25}$	①	$148\frac{4}{5}$

Similarly, we conclude that $y = 84$ and $P = 148.8$. Next, we note that the variables u and v are nonbasic and are accordingly assigned the values $u = 0$ and $v = 0$. These results agree with those obtained in Example 1, Section 6.3. ■

EXAMPLE 3   Maximize $P = 2x + 2y + z$ subject to

$$2x + \ \ y + 2z \le 14$$
$$2x + 4y + \ \ z \le 26$$
$$x + 2y + 3z \le 28$$
$$x \ge 0, y \ge 0, z \ge 0$$

Solution   Introducing the slack variables u, v, and w and rewriting the objective function in the standard form gives the system of linear equations

$$
\begin{aligned}
2x + \ \ y + 2z + u \ \ \ \ \ \ \ \ \ \ \ \ \ \ \ &= 14 \\
2x + 4y + \ \ z \ \ \ \ \ + v \ \ \ \ \ \ \ \ &= 26 \\
x + 2y + 3z \ \ \ \ \ \ \ \ \ + w \ \ \ \ &= 28 \\
-2x - 2y - \ \ z \ \ \ \ \ \ \ \ \ \ \ \ \ \ \ + P &= \ \ 0
\end{aligned}
$$

The initial simplex tableau is given by

x	y	z	u	v	w	P	Constant
2	1	2	1	0	0	0	14
2	4	1	0	1	0	0	26
1	2	3	0	0	1	0	28
−2	−2	−1	0	0	0	1	0

Since the most negative entry in the last row (-2) occurs twice, we may choose either the x or the y column as the pivot column. Choosing the x column as the pivot column and proceeding with the first iteration, we obtain the following sequence of tableaus:

Pivot row →

x	y	z	u	v	w	P	Constant		ratio
②	1	2	1	0	0	0	14		$\frac{14}{2} = 7$
2	4	1	0	1	0	0	26		$\frac{26}{2} = 13$
1	2	3	0	0	1	0	28		$\frac{28}{1} = 28$
−2	−2	−1	0	0	0	1	0		

↑ Pivot column

$\xrightarrow{\frac{1}{2}R_1}$

x	y	z	u	v	w	P	Constant
①	$\frac{1}{2}$	1	$\frac{1}{2}$	0	0	0	7
2	4	1	0	1	0	0	26
1	2	3	0	0	1	0	28
−2	−2	−1	0	0	0	1	0

$\xrightarrow[\substack{R_3 - R_1 \\ R_4 + 2R_1}]{R_2 - 2R_1}$

x	y	z	u	v	w	P	Constant
1	$\frac{1}{2}$	1	$\frac{1}{2}$	0	0	0	7
0	3	−1	−1	1	0	0	12
0	$\frac{3}{2}$	2	$-\frac{1}{2}$	0	1	0	21
0	−1	1	1	0	0	1	14

Since there is a negative number in the last row of the simplex tableau, we perform another iteration as follows:

	x	y	z	u	v	w	P	Constant
Pivot row →	1	$\frac{1}{2}$	1	$\frac{1}{2}$	0	0	0	7
	0	③	-1	-1	1	0	0	12
	0	$\frac{3}{2}$	2	$-\frac{1}{2}$	0	1	0	21
	0	-1	1	1	0	0	1	14

ratio
$\frac{7}{1/2} = 14$
$\frac{12}{3} = 4$
$\frac{21}{3/2} = 14$

↑
Pivot column

$\xrightarrow{\frac{1}{3}R_2}$

	x	y	z	u	v	w	P	Constant
	1	$\frac{1}{2}$	1	$\frac{1}{2}$	0	0	0	7
	0	①	$-\frac{1}{3}$	$-\frac{1}{3}$	$\frac{1}{3}$	0	0	4
	0	$\frac{3}{2}$	2	$-\frac{1}{2}$	0	1	0	21
	0	-1	1	1	0	0	1	14

$\xrightarrow[\substack{R_3 - \frac{3}{2}R_2 \\ R_4 + R_2}]{R_1 - \frac{1}{2}R_2}$

x	y	z	u	v	w	P	Constant
1	0	$\frac{7}{6}$	$\frac{2}{3}$	$-\frac{1}{6}$	0	0	5
0	1	$-\frac{1}{3}$	$-\frac{1}{3}$	$\frac{1}{3}$	0	0	4
0	0	$\frac{5}{2}$	0	$-\frac{1}{2}$	1	0	15
0	0	$\frac{2}{3}$	$\frac{2}{3}$	$\frac{1}{3}$	0	1	18

All entries in the last row are nonnegative, so we have reached the optimal solution. We conclude that $x = 5$, $y = 4$, $z = 0$, $u = 0$, $v = 0$, $w = 15$, and $P = 18$. ■

Application

EXAMPLE 4   The Ace Novelty Company has determined that the profits are $6, $5, and $4 for each type-A, type-B, and type-C souvenir that it plans to produce. To manufacture a type-A souvenir requires 2 minutes on machine I, 1 minute on machine II, and 2 minutes on machine III. A type-B souvenir requires 1 minute on machine I, 3 minutes on machine II, and 1 minute on machine III. A type-C souvenir requires 1 minute on machine I and 2 minutes on each of machines II and III. There are 3 hours available on machine I, 5 hours available on machine II, and 4 hours available on machine III for manufacturing these souvenirs each day. How many souvenirs of each type should Ace Novelty make per day in order to maximize its profit? (Compare with Example 1, Section 5.1.)

Solution   The given information may be tabulated as follows:

	Type A	Type B	Type C	Time Available (mins.)
Machine I	2	1	1	180
Machine II	1	3	2	300
Machine III	2	1	2	240
Profit per Unit ($)	6	5	4	

Let x, y, and z denote the respective numbers of type-A, type-B, and type-C souvenirs to be made. The total amount of time that machine I is used is given by $2x + y + z$ minutes and must not exceed 180 minutes. Thus, we have the inequality

$$2x + y + z \leq 180$$

Similar considerations on the use of machines II and III lead to the inequalities

$$x + 3y + 2z \leq 300$$
$$2x + y + 2z \leq 240$$

The profit resulting from the sale of the souvenirs produced is given by

$$P = 6x + 5y + 4z$$

The mathematical formulation of this problem has led to the following standard linear programming problem:

Maximize the objective (profit) function $P = 6x + 5y + 4z$ subject to the constraints

$$2x + y + z \leq 180$$
$$x + 3y + 2z \leq 300$$
$$2x + y + 2z \leq 240$$
$$x \geq 0, y \geq 0, z \geq 0$$

Introducing the slack variables u, v, and w gives the system of linear equations

$$2x + y + z + u \qquad\qquad = 180$$
$$x + 3y + 2z \quad + v \qquad = 300$$
$$2x + y + 2z \qquad + w \quad = 240$$
$$-6x - 5y - 4z \qquad\qquad + P = 0$$

The tableaus resulting from the use of the simplex algorithm are

	x	y	z	u	v	w	P	Constant		ratio
Pivot row →	②	1	1	1	0	0	0	180		$\frac{180}{2} = 90$
	1	3	2	0	1	0	0	300		$\frac{300}{1} = 300$
	2	1	2	0	0	1	0	240		$\frac{240}{2} = 120$
	−6	−5	−4	0	0	0	1	0		

Pivot column (↑ under x)

$\xrightarrow{\frac{1}{2}R_1}$

	x	y	z	u	v	w	P	Constant
	①	$\frac{1}{2}$	$\frac{1}{2}$	$\frac{1}{2}$	0	0	0	90
	1	3	2	0	1	0	0	300
	2	1	2	0	0	1	0	240
	−6	−5	−4	0	0	0	1	0

$\xrightarrow[\substack{R_2 - R_1 \\ R_3 - 2R_1 \\ R_4 + 6R_1}]{}$ Pivot row

	x	y	z	u	v	w	P	Constant		ratio
	1	$\frac{1}{2}$	$\frac{1}{2}$	$\frac{1}{2}$	0	0	0	90		$\frac{90}{1/2} = 180$
	0	⑤⁄₂	$\frac{3}{2}$	$-\frac{1}{2}$	1	0	0	210		$\frac{210}{5/2} = 84$
	0	0	1	−1	0	1	0	60		
	0	−2	−1	3	0	0	1	540		

Pivot column (↑ under y)

$\xrightarrow{\frac{2}{5}R_2}$

	x	y	z	u	v	w	P	Constant
	1	$\frac{1}{2}$	$\frac{1}{2}$	$\frac{1}{2}$	0	0	0	90
	0	①	$\frac{3}{5}$	$-\frac{1}{5}$	$\frac{2}{5}$	0	0	84
	0	0	1	−1	0	1	0	60
	0	−2	−1	3	0	0	1	540

$\xrightarrow[\substack{R_1 - \frac{1}{2}R_2 \\ R_4 + 2R_2}]{}$

	x	y	z	u	v	w	P	Constant
	1	0	$\frac{1}{5}$	$\frac{3}{5}$	$-\frac{1}{5}$	0	0	48
	0	1	$\frac{3}{5}$	$-\frac{1}{5}$	$\frac{2}{5}$	0	0	84
	0	0	1	−1	0	1	0	60
	0	0	$\frac{1}{5}$	$\frac{13}{5}$	$\frac{4}{5}$	0	1	708

From the final simplex tableau, we read off the solution

$$x = 48, \quad y = 84, \quad z = 0, \quad u = 0, \quad v = 0, \quad w = 60, \quad P = 708.$$

Thus, in order to maximize its profit, Ace Novelty should produce 48 type-A souvenirs, 84 type-B souvenirs, and no type-C souvenirs. The resulting profit is $708 per day. The value of the slack variable $w = 60$ tells us that 1 hour of the available time on machine III is left unused.   ∎

Interpreting Our Results   It is instructive to compare the results obtained here with those obtained in Example 7, Section 5.2. Recall that in order to use all of the machine time available on each of the three machines, Ace Novelty had to produce 36 type-A, 48 type-B, and 60 type-C souvenirs. This would have resulted in a profit of $696. Example 4 shows how, through the optimal use of equipment, a company may boost its profit and at the same time reduce machine wear!

Multiple Solutions and Unbounded Solutions

As we saw in Section 6.3, a linear programming problem may have infinitely many solutions. We also saw that a linear programming problem may have no solution or an unbounded solution. How do we spot each of these phenomena when using the simplex method to solve a problem?

A linear programming problem may have infinitely many solutions if and only if the last row to the left of the vertical line of the final simplex tableau has a zero in a column that is neither a unit column nor a column corresponding to a basic variable. Next, a linear programming problem will have no solution if the simplex method breaks down at some stage. For example, if at some stage there are no nonnegative ratios in our computation, then the linear programming problem has no solution.

SELF-CHECK EXERCISE 6.4

Solve the following linear programming problem by the simplex method: Maximize $P = 2x + 3y + 6z$ subject to

$$2x + 3y + z \leq 10$$
$$x + y + 2z \leq 8$$
$$2y + 3z \leq 6$$
$$x \geq 0, y \geq 0, z \geq 0$$

Solutions to Self-Check Exercise 6.4 can be found on page 310.

6.4   EXERCISES

In exercises 1–9, determine whether the simplex tableau is in final form. If so, find the solution to the associated regular linear programming problem. If not, find the pivot element to be used in the next iteration of the simplex method.

1.

x	y	u	v	P	Constant
0	1	$\frac{5}{7}$	$-\frac{1}{7}$	0	$\frac{20}{7}$
1	0	$-\frac{3}{7}$	$\frac{2}{7}$	0	$\frac{30}{7}$
0	0	$\frac{13}{7}$	$\frac{3}{7}$	1	$\frac{220}{7}$

2.

x	y	u	v	P	Constant
1	1	1	0	0	6
1	0	-1	1	0	2
3	0	5	0	1	30

3.

x	y	u	v	P	Constant
0	$\frac{1}{2}$	1	$-\frac{1}{2}$	0	2
1	$\frac{1}{2}$	0	$\frac{1}{2}$	0	4
0	$-\frac{1}{2}$	0	$\frac{3}{2}$	1	12

4.

x	y	z	u	v	w	P	Constant
3	0	5	1	1	0	0	28
2	1	3	0	1	0	0	16
2	0	8	0	3	0	1	48

5.

x	y	z	u	v	w	P	Constant
1	$-\frac{1}{3}$	0	$\frac{1}{3}$	0	$-\frac{2}{3}$	0	$\frac{1}{3}$
0	2	0	0	1	1	0	6
0	$\frac{2}{3}$	1	$\frac{1}{3}$	0	$\frac{1}{3}$	0	$\frac{13}{3}$
0	4	0	1	0	2	1	17

6.

x	y	z	u	v	w	P	Constant
$\frac{1}{2}$	0	$\frac{1}{4}$	1	$-\frac{1}{4}$	0	0	$\frac{19}{2}$
$\frac{1}{2}$	1	$\frac{3}{4}$	0	$\frac{1}{4}$	0	0	$\frac{21}{2}$
2	0	3	0	0	1	0	30
-1	0	$-\frac{1}{2}$	6	$\frac{3}{2}$	0	1	63

7.

x	y	z	s	t	u	v	P	Constant
$\frac{5}{2}$	3	0	1	0	0	-4	0	46
1	0	0	0	1	0	0	0	9
0	1	0	0	0	1	0	0	12
0	0	1	0	0	0	1	0	6
-180	-200	0	0	0	0	300	1	1800

8.

x	y	z	s	t	u	v	P	Constant
1	0	0	$\frac{2}{5}$	0	$-\frac{6}{5}$	$-\frac{8}{5}$	0	4
0	0	0	$-\frac{2}{5}$	1	$\frac{6}{5}$	$\frac{8}{5}$	0	5
0	1	0	0	0	1	0	0	12
0	0	1	0	0	0	1	0	6
0	0	0	72	0	-16	12	1	4920

9.

x	y	z	u	v	P	Constant
1	1	$\frac{3}{5}$	0	$\frac{1}{5}$	0	30
0	0	$-\frac{19}{5}$	1	$-\frac{3}{5}$	0	10
0	0	$\frac{26}{5}$	0	$\frac{2}{5}$	1	60

In exercises 10–23, solve each linear programming problem by the simplex method.

10. Maximize $P = 5x + 3y$ subject to

$$x + y \le 80$$
$$3x \le 90$$
$$x \ge 0, y \ge 0$$

11. Maximize $P = 10x + 12y$ subject to

$$x + 2y \le 12$$
$$3x + 2y \le 24$$
$$x \ge 0, y \ge 0$$

12. Maximize $P = 5x + 4y$ subject to

$$3x + 5y \le 78$$
$$4x + y \le 36$$
$$x \ge 0, y \ge 0$$

13. Maximize $P = 4x + 6y$ subject to

$$3x + y \le 24$$
$$2x + y \le 18$$
$$x + 3y \le 24$$
$$x \ge 0, y \ge 0$$

14. Maximize $P = 15x + 12y$ subject to

$$x + y \le 12$$
$$3x + y \le 30$$
$$10x + 7y \le 70$$
$$x \ge 0, y \ge 0$$

15. Maximize $P = 3x + 4y + 5z$ subject to

$$x + y + z \le 8$$
$$3x + 2y + 4z \le 24$$
$$x \ge 0, y \ge 0, z \ge 0$$

16. Maximize $P = 3x + 3y + 4z$ subject to

$$x + y + 3z \le 15$$
$$4x + 4y + 3z \le 65$$
$$x \ge 0, y \ge 0, z \ge 0$$

17. Maximize $P = 3x + 4y + z$ subject to

$$3x + 10y + 5z \le 120$$
$$5x + 2y + 8z \le 6$$
$$8x + 10y + 3z \le 105$$
$$x \ge 0, y \ge 0, z \ge 0$$

18. Maximize $P = x + 2y - z$ subject to

$$2x + y + z \le 14$$
$$4x + 2y + 3z \le 28$$
$$2x + 5y + 5z \le 30$$
$$x \ge 0, y \ge 0, z \ge 0$$

19. Maximize $P = 4x + 6y + 5z$ subject to

$$x + y + z \le 20$$
$$2x + 4y + 3z \le 42$$
$$2x \quad\quad + 3z \le 30$$
$$x \ge 0, y \ge 0, z \ge 0$$

20. Maximize $P = x + 4y - 2z$ subject to

$$3x + y - z \le 80$$
$$2x + y - z \le 40$$
$$-x + y + z \le 80$$
$$x \ge 0, y \ge 0, z \ge 0$$

21. Maximize $P = 12x + 10y + 5z$ subject to

$$2x + y + z \le 10$$
$$3x + 5y + z \le 45$$
$$2x + 5y + z \le 40$$
$$x \ge 0, y \ge 0, z \ge 0$$

22. Maximize $P = 2x + 6y + 6z$ subject to

$$2x + y + 3z \le 10$$
$$4x + y + 2z \le 56$$
$$6x + 4y + 3z \le 126$$
$$2x + y + z \le 32$$
$$x \ge 0, y \ge 0, z \ge 0$$

23. Maximize $P = 4x + y + 5z$ subject to

$$x + 2y + 5z \le 30$$
$$2x + y + z \le 10$$
$$3x + 2y + z \le 12$$
$$x + y + z \le 8$$
$$x \ge 0, y \ge 0, z \ge 0$$

24. Rework Example 3 using the y column as the pivot column in the first iteration of the simplex method.

25. Show that the following linear programming problem: Maximize $P = 2x + 2y - 4z$ subject to

$$3x + 3y - 2z \le 100$$
$$5x + 5y + 3z \le 150$$
$$x \ge 0, y \ge 0, z \ge 0$$

has optimal solutions $x = 30$, $y = 0$, $z = 0$, $P = 60$ and $x = 0$, $y = 30$, $z = 0$, and $P = 60$.

26. Manufacturing—Production Scheduling A company manufactures two products, A and B, on two machines, I and II. It has been determined that the company will realize a profit of $3 on each unit of product A and a profit of $4 on each unit of product B. To manufacture a unit of product A requires 6 minutes on machine I and 5 minutes on machine

II. To manufacture a unit of product B requires 9 minutes on machine I and 4 minutes on machine II. There are 5 hours of machine time available on machine I and 3 hours of machine time available on machine II in each work shift. How many units of each product should be produced in each shift to maximize the company's profit?

27. Manufacturing—Production Scheduling The National Business Machines Corporation manufactures two models of fax machines: A and B. Each model A costs $200 to make and each model B costs $300. The profits are $25 for each model A and $40 for each model B fax machine. If the total number of fax machines demanded per month does not exceed 2500 and the company has earmarked no more than $600,000 per month for manufacturing costs, find how many units of each model National should make each month in order to maximize its monthly profits.

28. Manufacturing—Production Scheduling Kane Manufacturing has a division that produces two models of hibachis, model A and model B. To produce each model A hibachi requires 3 pounds of cast iron and 6 minutes of labor. To produce each model B hibachi requires 4 pounds of cast iron and 3 minutes of labor. The profit for each model A hibachi is $2, and the profit for each model B hibachi is $1.50. If 1000 pounds of cast iron and 20 hours of labor are available for the production of hibachis per day, how many hibachis of each model should the division produce in order to help maximize Kane's profits?

29. Agriculture—Crop Planning A farmer has 150 acres of land suitable for cultivating crops A and B. The cost of cultivating crop A is $40 per acre, whereas that of crop B is $60 per acre. The farmer has a maximum of $7400 available for land cultivation. Each acre of crop A requires 20 hours of labor and each acre of crop B requires 25 hours of labor. The farmer has a maximum of 3300 hours of labor available. If he expects to make a profit of $150 per acre on crop A and $200 per acre on crop B, how many acres of each crop should he plant in order to maximize his profit?

30. Manufacturing—Production Scheduling The LaCrosse Iron Works makes two models of cast iron fireplace grates, model A and model B. Producing one model A grate requires 20 pounds of cast iron and 20 minutes of labor, whereas producing one model B grate requires 30 pounds of cast iron and 15 minutes of labor. The profit for a model A grate is $6 and the profit for a model B grate is $8. There are 7200 pounds of cast iron and 100 hours of labor available per week. Because of a surplus from the previous week, the proprietor has decided that he should make no more than 150 units of model A grates this week. Determine how many of each model he should make in order to maximize his profits.

31. **Manufacturing—Production Scheduling** A company manufactures products A, B, and C. Each product is processed in three departments: I, II, and III. The total available labor hours per week for departments I, II, and III are 900, 1080, and 840, respectively. The time requirements (in hours per unit) and profit per unit for each product are as follows:

	Product A	Product B	Product C
Dept. I	2	1	2
Dept. II	3	1	2
Dept. III	2	2	1
Profit	$18	$12	$15

How many units of each product should the company produce in order to maximize its profit?

32. **Advertising—Television Commercials** As part of a campaign to promote its annual clearance sale, the Excelsior Company decided to buy television advertising time on Station KAOS. Excelsior's television advertising budget was $102,000. Morning time costs $3,000 per minute, afternoon time costs $1,000 per minute, and evening (prime) time costs $12,000 per minute. Because of previous commitments, KAOS could not offer Excelsior more than 6 minutes of prime time or more than a total of 25 minutes of advertising time over the two weeks in which the commercials were to be run. KAOS estimated that morning commercials would be seen by 200,000 people, afternoon commercials would be seen by 100,000 people, and evening commercials would be seen by 600,000 people. How much morning, afternoon, and evening advertising time should Excelsior buy to maximize exposure of its commercials?

33. **Investments—Asset Allocation** Sharon has a total of $200,000 to invest in three types of mutual funds: growth, balanced, and income funds. Growth funds have a rate of return of 12% per year, balanced funds have a rate of return of 10% per year, and income funds have a return of 6% per year. The growth, balanced, and income mutual funds are assigned risk factors of 0.1, 0.06, and 0.02, respectively. Sharon has decided that at least half of her total portfolio is to be in income funds and at least a quarter of it in balanced funds. She has also decided that the average risk factor for her investment should not exceed 0.05.

Determine how much Sharon should invest in each type of fund in order to realize a maximum return on her investment. [*Hint:* The average risk factor for the investment is given by $0.1x + 0.06y + 0.02z \leq 0.05(x + y + z)$.]

34. **Manufacturing—Production Control** The Custom Office Furniture Company is introducing a new line of executive desks made from a specially selected grade of walnut. Initially, three models—A, B, and C—are to be marketed. Each model A desk requires $1\frac{1}{4}$ hours for fabrication, 1 hour for assembly, and 1 hour for finishing; each model B desk requires $1\frac{1}{2}$ hours for fabrication, 1 hour for assembly, and 1 hour for finishing; each model C desk requires $1\frac{1}{2}$ hours, $\frac{3}{4}$ hour, and $\frac{1}{2}$ hour for fabrication, assembly, and finishing, respectively. The profit on each model A desk is $26, the profit on each model B desk is $28, and the profit on each model C desk is $24. The total time available in the fabrication department, the assembly department, and the finishing department in the first month of production is 310 hours, 205 hours, and 190 hours, respectively. To maximize Custom's profit, how many desks of each model should be made in the month?

35. **Manufacturing—Prefabricated Housing Production** Boise Lumber has decided to enter the lucrative prefabricated housing business. Initially, it plans to offer three models: a standard model, a deluxe model, and a luxury model. Each house is prefabricated and partially assembled in the factory, and the final assembly is completed on site. The dollar amount of building material required, the amount of labor required in the factory for prefabrication and partial assembly, the amount of on-site labor required, and the profit per unit are as follows:

	Standard Model	Deluxe Model	Luxury Model
Material ($)	6,000	8,000	10,000
Factory Labor (hr)	240	220	200
On-site Labor (hr)	180	210	300
Profit ($)	3,400	4,000	5,000

For the first year's production, a sum of $8,200,000 is budgeted for the building material; the number of labor hours available for work in the factory (for prefabrication and partial assembly) is not to exceed 218,000 hours; and the amount of labor for on-site work is to be less than or equal to 237,000 labor hours. Determine how many houses of each type

Boise should produce (market research has confirmed that there should be no problems with sales) to maximize its profit from this new venture.

36. Manufacturing—Cold Formula Production   Bayer Pharmaceutical produces three kinds of cold formulas: formula I, formula II, and formula III. It takes 2.5 hours to produce 1,000 bottles of formula I, 3 hours to produce 1,000 bottles of formula II, and 4 hours to produce 1,000 bottles of formula III. The profits for each 1,000 bottles of formula I, formula II, and formula III are $180, $200, and $300, respectively. Suppose that for a certain production run there are enough ingredients on hand to make at most 9,000 bottles of formula I, 12,000 bottles of formula II, and 6,000 bottles of formula III. Furthermore, suppose that the time for the production run is limited to a maximum of 70 hours. Find how many bottles of each formula should be produced in this production run so that the profit is maximized.

SOLUTION TO SELF-CHECK EXERCISE 6.4

Introducing the slack variables u, v, and w, we obtain the system of linear equations

$$\begin{aligned}
2x + 3y + z + u &= 10 \\
x + y + 2z + v &= 8 \\
2y + 3z + w &= 6 \\
-2x - 3y - 6z + P &= 0
\end{aligned}$$

The initial simplex tableau and the successive tableaus resulting from the use of the simplex procedure follow:

	x	y	z	u	v	w	P	Constant	ratio	
	2	3	1	1	0	0	0	10	$\frac{10}{1} = 10$	$\xrightarrow{\frac{1}{3}R_3}$
	1	1	2	0	1	0	0	8	$\frac{8}{2} = 4$	
Pivot row →	0	2	③	0	0	1	0	6	$\frac{6}{3} = 2$	
	−2	−3	−6	0	0	0	1	0		

↑ Pivot column

	x	y	z	u	v	w	P	Constant	
	2	3	1	1	0	0	0	10	
	1	1	2	0	1	0	0	8	$\xrightarrow{\substack{R_1 - R_3 \\ R_2 - 2R_3 \\ R_4 + 6R_3}}$
	0	$\frac{2}{3}$	①	0	0	$\frac{1}{3}$	0	2	
	−2	−3	−6	0	0	0	1	0	

	x	y	z	u	v	w	P	Constant	ratio	
	2	$\frac{7}{3}$	0	1	0	$-\frac{1}{3}$	0	8	$\frac{8}{2} = 4$	$\xrightarrow{\substack{R_1 - 2R_2 \\ R_4 + 2R_2}}$
Pivot row →	①	$-\frac{1}{3}$	0	0	1	$-\frac{2}{3}$	0	4	$\frac{4}{1} = 4$	
	0	$\frac{2}{3}$	1	0	0	$\frac{1}{3}$	0	2		
	−2	1	0	0	0	2	1	12		

↑ Pivot column

x	y	z	u	v	w	P	Constant
0	3	0	1	-2	1	0	0
1	$-\frac{1}{3}$	0	0	1	$-\frac{2}{3}$	0	4
0	$\frac{2}{3}$	1	0	0	$\frac{1}{3}$	0	2
0	$\frac{1}{3}$	0	0	2	$\frac{2}{3}$	1	20

All entries in the last row are nonnegative and the tableau is final. We conclude that $x = 4$, $y = 0$, $z = 2$, and $P = 20$.

Using Technology

The graphing calculator can be used to solve a linear programming problem by the simplex method as illustrated in Example 1.

EXAMPLE 1 (Refer to Example 4, Section 6.4.) The problem reduces to the following linear programming problem:

Maximize $P = 6x + 5y + 4z$ subject to the constraints

$$2x + y + z \leq 180$$
$$x + 3y + 2z \leq 300$$
$$2x + y + 2z \leq 240$$
$$x \geq 0, y \geq 0, z \geq 0$$

Solution With u, v, and w as slack variables, we are led to the following sequence of simplex tableaus:

Pivot row →

	x	y	z	u	v	w	P	Constant	ratio	
Pivot row →	②	1	1	1	0	0	0	180	$\frac{180}{2} = 90$	multR($\frac{1}{2}$, A, 1) ▶ B
	1	3	2	0	1	0	0	300	$\frac{300}{1} = 300$	
	2	1	2	0	0	1	0	240	$\frac{240}{2} = 120$	
	-6	-5	-4	0	0	0	1	0		

↑
Pivot
column

x	y	z	u	v	w	P	Constant	
①	0.5	0.5	0.5	0	0	0	90	
1	3	2	0	1	0	0	300	mRAdd(-1, B, 1, 2) ▶ C
2	1	2	0	0	1	0	240	mRAdd(-2, C, 1, 3) ▶ B
-6	-5	-4	0	0	0	1	0	mRAdd(6, B, 1, 4) ▶ C

x	y	z	u	v	w	P	Constant
1	0.5	0.5	0.5	0	0	0	90
0	(2.5)	1.5	−0.5	1	0	0	210
0	0	1	−1	0	1	0	60
0	−2	−1	3	0	0	1	540

Pivot row → (points to second row)

$$\begin{array}{c} \text{ratio} \\ \frac{90}{0.5} = 180 \\ \frac{210}{2.5} = 84 \end{array}$$

$\xrightarrow{\textbf{mult}(\frac{1}{2.5}, C, 2) \blacktriangleright B}$

↑
Pivot
column

x	y	z	u	v	w	P	Constant
1	0.5	0.5	0.5	0	0	0	90
0	①	0.6	−0.2	0.4	0	0	84
0	0	1	−1	0	1	0	60
0	−2	−1	3	0	0	1	540

$\xrightarrow[\textbf{mRAdd}(2, C, 2, 4) \quad \blacktriangleright B]{\textbf{mRAdd}(-0.5, B, 2, 1) \blacktriangleright C}$

x	y	z	u	v	w	P	Constant
1	0	0.2	0.6	−0.2	0	0	48
0	1	0.6	−0.2	0.4	0	0	84
0	0	1	−1	0	1	0	60
0	0	0.2	2.6	0.8	0	1	708

The final simplex tableau is the same as the one obtained earlier. We see that $x = 48$, $y = 84$, $z = 0$, and $P = 708$. So Ace Novelty should produce 48 type-A souvenirs, 84 type-B souvenirs, and no type-C souvenirs—resulting in a profit of \$708 per day. ■

EXERCISES

In exercises 1–4, use a graphing calculator or a computer software program to solve the linear programming problem by the simplex method.

1. Maximize $P = 2x + 3y + 4z + 2w$ subject to

$$\begin{aligned} x + 2y + 3z + 2w &\leq 6 \\ 2x + 4y + z - w &\leq 4 \\ 3x + 2y - 2z + 3w &\leq 12 \\ x \geq 0, y \geq 0, z \geq 0, w &\geq 0 \end{aligned}$$

2. Maximize $P = 3x + 2y + 2z + w$ subject to

$$\begin{aligned} 2x + y - z + 2w &\leq 8 \\ 2x - y + 2z + 3w &\leq 20 \\ x + y + z + 2w &\leq 8 \\ 4x - 2y + z + 3w &\leq 24 \\ x \geq 0, y \geq 0, z \geq 0, w &\geq 0 \end{aligned}$$

3. Maximize $P = x + y + 2z + 3w$ subject to

$$\begin{aligned} 3x + 6y + 4z + 2w &\leq 12 \\ x + 4y + 8z + 4w &\leq 16 \\ 2x + y + 4z + w &\leq 10 \\ x \geq 0, y \geq 0, z \geq 0, w &\geq 0 \end{aligned}$$

4. Maximize $P = 2x + 4y + 3z + 5w$ subject to

$$\begin{aligned} x - 2y + 3z + 4w &\leq 8 \\ 2x + 2y + 4z + 6w &\leq 12 \\ 3x + 2y + z + 5w &\leq 10 \\ 2x + 8y - 2z + 6w &\leq 24 \\ x \geq 0, y \geq 0, z \geq 0, w &\geq 0 \end{aligned}$$

The Simplex Method: Standard Minimization Problems

Minimization with $\leq$ Constraints

In the last section we developed a procedure, called the simplex method, for solving standard linear programming problems. Recall that a standard maximization problem satisfies three conditions:

1. The objective function is to be maximized.
2. All the variables involved are nonnegative.
3. Each linear constraint may be written so that the expression involving the variables is less than or equal to a nonnegative constant.

   In this section we will see how the simplex method may be used to solve certain classes of problems that are not necessarily standard maximization problems. In particular, we will see how the modified procedure may be used to solve problems involving the minimization of objective functions.
   We begin by considering the class of linear programming problems that call for the minimization of objective functions but otherwise satisfy conditions (2) and (3) for standard maximization problems. The method used to solve these problems is illustrated in the following example.

EXAMPLE 1   Minimize $C = -2x - 3y$ subject to

$$5x + 4y \leq 32$$
$$x + 2y \leq 10$$
$$x \geq 0, y \geq 0$$

Solution   This problem involves the minimization of the objective function and is accordingly not a standard maximization problem. Note, however, that all other conditions for a standard maximization problem hold true. To solve a problem of this type we observe that minimizing the objective function C is equivalent to maximizing the objective function $P = -C$. Thus, the solution to this problem may be found by solving the following associated standard maximization problem: Maximize $P = 2x + 3y$ subject to the given constraints. Using the simplex method with u and v as slack variables, we obtain the following sequence of simplex tableaus:

	x	y	u	v	P	Constant	ratio
	5	4	1	0	0	32	$\frac{32}{4} = 8$
Pivot row →	1	②	0	1	0	10	$\frac{10}{2} = 5$
	-2	-3	0	0	1	0	

Pivot column ↑

	x	y	u	v	P	Constant
$\tfrac{1}{2}R_2 \longrightarrow$	5	4	1	0	0	32
	$\tfrac{1}{2}$	①	0	$\tfrac{1}{2}$	0	5
	-2	-3	0	0	1	0

	x	y	u	v	P	Constant	ratio
Pivot row $\longrightarrow$	③	0	1	-2	0	12	$\tfrac{12}{3} = 4$
$\dfrac{R_1 - 4R_2}{R_3 + 3R_2} \longrightarrow$	$\tfrac{1}{2}$	1	0	$\tfrac{1}{2}$	0	5	$\tfrac{5}{1/2} = 10$
	$-\tfrac{1}{2}$	0	0	$\tfrac{3}{2}$	1	15	

↑ Pivot column

	x	y	u	v	P	Constant
$\tfrac{1}{3}R_1 \longrightarrow$	①	0	$\tfrac{1}{3}$	$-\tfrac{2}{3}$	0	4
	$\tfrac{1}{2}$	1	0	$\tfrac{1}{2}$	0	5
	$-\tfrac{1}{2}$	0	0	$\tfrac{3}{2}$	1	15

	x	y	u	v	P	Constant
$\dfrac{R_2 - \tfrac{1}{2}R_1}{R_3 + \tfrac{1}{2}R_1} \longrightarrow$	1	0	$\tfrac{1}{3}$	$-\tfrac{2}{3}$	0	4
	0	1	$-\tfrac{1}{6}$	$\tfrac{5}{6}$	0	3
	0	0	$\tfrac{1}{6}$	$\tfrac{7}{6}$	1	17

The last tableau is in final form. The solution to the standard maximization problem associated with the given linear programming problem is $x = 4$, $y = 3$, and $P = 17$, so the required solution is given by $x = 4$, $y = 3$, and $C = -17$. You may verify that the solution is correct by using the method of corners. ■

The Dual Problem

Another special class of linear programming problems we encounter in practical applications is characterized by the following conditions:

1. The objective function is to be *minimized.*

2. All the variables involved are nonnegative.

3. Each linear constraint may be written so that the expression involving the variables is *greater than* or equal to a constant.

Such problems are called **standard minimization problems.**

A convenient method for solving this type of problem is based on the following observation. Each maximization linear programming problem is associated with a minimization problem, and vice versa. For the purpose of identification, the given problem is called the **primal problem,** whereas the problem related to it is called the **dual problem.** The following example illustrates the technique for constructing the dual of a given linear programming problem.

EXAMPLE 2 Write the dual problem associated with the following problem:

Minimize the objective function $C = 6x + 8y$ subject to the constraints

$$40x + 10y \geq 2400$$
$$10x + 15y \geq 2100$$
$$5x + 15y \geq 1500$$
$$x \geq 0, y \geq 0$$

Solution We first write down the following tableau for the given primal problem:

x	y	Constant
40	10	2400
10	15	2100
5	15	1500
6	8	

Next, we interchange the columns and rows of the foregoing tableau and head the three columns of the resulting array with the three variables u, v, and w, obtaining the tableau

u	v	w	Constant
40	10	5	6
10	15	15	8
2400	2100	1500	

Interpreting the last tableau as if it were part of the initial simplex tableau for a standard maximization problem, with the exception that the signs of the coefficients pertaining to the objective function are not reversed, we construct the required dual problem as follows:

Maximize the objective function $P = 2400u + 2100v + 1500w$ subject to the constraints

$$40u + 10v + 5w \leq 6$$
$$10u + 15v + 15w \leq 8$$

where $u \geq 0$, $v \geq 0$, and $w \geq 0$. ■

The connection between the solution of the primal problem and that of the dual problem is given by the following theorem. The theorem, attributed to John von Neumann (1903–1957), is stated without proof.

THE FUNDAMENTAL THEOREM OF DUALITY

A primal problem has a solution if and only if the corresponding dual problem has a solution. Furthermore, if a solution exists, then

1. the objective functions of both the primal and the dual problem attain the same optimal value

2. the optimal solution to the primal problem appears under the slack variables in the last row of the final simplex tableau associated with the dual problem

Armed with this theorem, we will solve the problem posed in Example 2.

EXAMPLE 3   Complete the solution to the problem posed in Example 2.

Solution   Observe that the dual problem associated with the given (primal) problem is a standard maximization problem. The solution may thus be found using the simplex algorithm. Introducing the slack variables x and y, we obtain the system of linear equations

$$
\begin{aligned}
40u + 10v + 5w + x \qquad\qquad &= 6 \\
10u + 15v + 15w \qquad + y \quad\ &= 8 \\
-2400u - 2100v - 1500w \qquad\qquad + P &= 0
\end{aligned}
$$

Continuing with the simplex algorithm, we obtain the sequence of simplex tableaus

	u	v	w	x	y	P	Constant	ratio
Pivot row →	④⓪	10	5	1	0	0	6	$\frac{6}{40} = \frac{3}{20}$
	10	15	15	0	1	0	8	$\frac{8}{10} = \frac{4}{5}$
	-2400	-2100	-1500	0	0	1	0	

↑
Pivot
column

	u	v	w	x	y	P	Constant
$\xrightarrow{\frac{1}{40}R_1}$	①	$\frac{1}{4}$	$\frac{1}{8}$	$\frac{1}{40}$	0	0	$\frac{3}{20}$
	10	15	15	0	1	0	8
	-2400	-2100	-1500	0	0	1	0

	u	v	w	x	y	P	Constant	ratio
$\xrightarrow[R_3 + 2400R_1]{R_2 - 10R_1}$	1	$\frac{1}{4}$	$\frac{1}{8}$	$\frac{1}{40}$	0	0	$\frac{3}{20}$	$\frac{3/20}{1/4} = \frac{3}{5}$
	0	㉕⁄₂	$\frac{55}{4}$	$-\frac{1}{4}$	1	0	$\frac{13}{2}$	$\frac{13/2}{25/2} = \frac{13}{25}$
	0	-1500	-1200	60	0	1	360	

	u	v	w	x	y	P	Constant
$\xrightarrow{\frac{2}{25}R_2}$	1	$\frac{1}{4}$	$\frac{1}{8}$	$\frac{1}{40}$	0	0	$\frac{3}{20}$
	0	①	$\frac{11}{10}$	$-\frac{1}{50}$	$\frac{2}{25}$	0	$\frac{13}{25}$
	0	-1500	-1200	60	0	1	360

	u	v	w	x	y	P	Constant
$R_1 - \frac{1}{4}R_2$	1	0	$-\frac{3}{20}$	$\frac{3}{100}$	$-\frac{1}{50}$	0	$\frac{1}{50}$
$R_3 + 1500R_2$	0	1	$\frac{11}{10}$	$-\frac{1}{50}$	$\frac{2}{25}$	0	$\frac{13}{25}$
	0	0	450	30	120	1	1140

Solution for the
primal problem

The last tableau is final. The fundamental theorem of duality tells us that the solution to the primal problem is $x = 30$ and $y = 120$ with a minimum value for C of 1140. Observe that the solution to the dual (maximization) problem may be read from the simplex tableau in the usual manner: $u = 1/50$, $v = 13/25$, $w = 0$, and $P = 1140$. Note that the maximum value of P is equal to the minimum value of C as guaranteed by the fundamental theorem of duality. The solution to the primal problem agrees with the solution of the same problem solved in Section 6.3, Example 2, using the method of corners. ◼

REMARK   We leave it to you to demonstrate that the dual of a standard minimization problem is always a standard maximization problem provided that the coefficients of the objective function in the primal problem are all nonnegative. Therefore, such problems can always be solved by applying the simplex method to solve the dual problem. In this book we assume that the nonnegativity condition on the coefficients of the objective function is satisfied.   ☐

EXAMPLE 4   Minimize the function $C = 3x + 2y$ subject to

$$
\begin{aligned}
8x + y &\geq 80 \\
3x + 2y &\geq 100 \\
x + 4y &\geq 80 \\
x \geq 0, y &\geq 0
\end{aligned}
$$

Solution   We begin by writing the dual problem associated with the given primal problem. First, we write down the following tableau for the primal problem:

x	y	Constant
8	1	80
3	2	100
1	4	80
3	2	

Next, interchanging the columns and rows of this tableau and heading the three columns of the resulting array with the three variables u, v, and w, we obtain the tableau

u	v	w	Constant
8	3	1	3
1	2	4	2
80	100	80	

Interpreting the last tableau as if it were part of the initial simplex tableau for a standard maximization problem, with the exception that the signs of the coefficients pertaining to the objective function are not reversed, we construct the dual problem as follows: Maximize the objective function $P = 80u + 100v + 80w$ subject to the constraints

$$8u + 3v + w \leq 3$$
$$u + 2v + 4w \leq 2$$

where $u \geq 0$, $v \geq 0$, and $w \geq 0$. Having constructed the dual problem, which is a standard maximization problem, we now solve it using the simplex method. Introducing the slack variables x and y, we obtain the system of linear equations

$$8u + 3v + w + x = 3$$
$$u + 2v + 4w + y = 2$$
$$-80u - 100v - 80w + P = 0$$

Continuing with the simplex algorithm, we obtain the sequence of simplex tableaus

	u	v	w	x	y	P	Constant	ratio
Pivot row →	8	3	1	1	0	0	3	$\frac{3}{3} = 1$
	1	②	4	0	1	0	2	$\frac{2}{2} = 1$
	-80	-100	-80	0	0	1	0	

Pivot column ↑

	u	v	w	x	y	P	Constant
	8	3	1	1	0	0	3
$\frac{1}{2}R_2$ →	$\frac{1}{2}$	①	2	0	$\frac{1}{2}$	0	1
	-80	-100	-80	0	0	1	0

	u	v	w	x	y	P	Constant	ratio
Pivot row ↳	⑬⁄₂	0	-5	1	$-\frac{3}{2}$	0	0	$\frac{0}{13/2} = 0$
$R_1 - 3R_2$ → $R_3 + 100R_2$	$\frac{1}{2}$	1	2	0	$\frac{1}{2}$	0	1	$\frac{1}{1/2} = 2$
	-30	0	120	0	50	1	100	

Pivot column ↑

$\xrightarrow{\frac{2}{13}R_1}$

u	v	w	x	y	P	Constant
①	0	$-\frac{10}{13}$	$\frac{2}{13}$	$-\frac{3}{13}$	0	0
$\frac{1}{2}$	1	2	0	$\frac{1}{2}$	0	1
-30	0	120	0	50	1	100

$\xrightarrow[R_3 + 30R_1]{R_2 - \frac{1}{2}R_1}$

u	v	w	x	y	P	Constant
1	0	$-\frac{10}{13}$	$\frac{2}{13}$	$-\frac{3}{13}$	0	0
0	1	$\frac{31}{13}$	$-\frac{1}{13}$	$\frac{8}{13}$	0	1
0	0	$\frac{1260}{13}$	$\frac{60}{13}$	$\frac{560}{13}$	1	100

Solution for the primal problem

The last tableau is final. The fundamental theorem of duality tells us that the solution to the primal problem is $x = 60/13$ and $y = 560/13$ with a minimum value for C of 100. ∎

Application

Our last example illustrates how the warehouse problem posed in Section 6.2 may be solved by duality.

EXAMPLE 5 Complete the solution to the warehouse problem given in Section 6.2, Example 4 (page 274).
 Minimize

$$C = 20x_1 + 8x_2 + 10x_3 + 12x_4 + 22x_5 + 18x_6 \qquad (20)$$

subject to

$$x_1 + x_2 + x_3 \qquad\qquad\qquad \le 400$$
$$x_4 + x_5 + x_6 \le 600$$
$$x_1 \qquad\quad + x_4 \qquad\qquad \ge 200$$
$$x_2 \qquad\quad + x_5 \qquad\quad \ge 300$$
$$x_3 \qquad\qquad + x_6 \ge 400$$
$$x_1 \ge 0,\ x_2 \ge 0, \ldots, x_6 \ge 0 \qquad (21)$$

Solution Upon multiplying each of the first two inequalities of (21) by -1, we obtain the following equivalent system of constraints in which each of the expressions involving the variables is greater than or equal to a constant:

$$-x_1 - x_2 - x_3 \qquad\qquad\qquad \ge -400$$
$$-x_4 - x_5 - x_6 \ge -600$$
$$x_1 \qquad\quad + x_4 \qquad\qquad \ge \quad 200$$
$$x_2 \qquad\quad + x_5 \qquad\quad \ge \quad 300$$
$$x_3 \qquad\qquad + x_6 \ge \quad 400$$

The problem may now be solved by duality. First we write the array of numbers

x_1	x_2	x_3	x_4	x_5	x_6	Constant
-1	-1	-1	0	0	0	-400
0	0	0	-1	-1	-1	-600
1	0	0	1	0	0	200
0	1	0	0	1	0	300
0	0	1	0	0	1	400
20	8	10	12	22	18	

Interchanging the rows and columns of this array of numbers and heading the five columns of the resulting array of numbers by the variables u_1, u_2, u_3, u_4, and u_5 leads to

u_1	u_2	u_3	u_4	u_5	Constant
-1	0	1	0	0	20
-1	0	0	1	0	8
-1	0	0	0	1	10
0	-1	1	0	0	12
0	-1	0	1	0	22
0	-1	0	0	1	18
-400	-600	200	300	400	

from which we construct the associated dual problem: Maximize $P = -400u_1 - 600u_2 + 200u_3 + 300u_4 + 400u_5$ subject to

$$
\begin{aligned}
-u_1 \quad\;\; + u_3 \qquad\qquad &\leq 20 \\
-u_1 \qquad\quad\; + u_4 \qquad &\leq\; 8 \\
-u_1 \qquad\qquad\quad + u_5 &\leq 10 \\
-u_2 + u_3 \qquad\qquad &\leq 12 \\
-u_2 \quad\;\; + u_4 \qquad &\leq 22 \\
-u_2 \qquad\quad\; + u_5 &\leq 18 \\
\end{aligned}
$$

$$u_1 \geq 0,\, u_2 \geq 0, \ldots,\, u_5 \geq 0$$

Solving the standard maximization problem by the simplex algorithm, we obtain the following sequence of tableaus ($x_1, x_2, \ldots, x_6$ are slack variables):

	u_1	u_2	u_3	u_4	u_5	x_1	x_2	x_3	x_4	x_5	x_6	P	Constant	ratio
	−1	0	1	0	0	1	0	0	0	0	0	0	20	—
	−1	0	0	1	0	0	1	0	0	0	0	0	8	—
Pivot row →	−1	0	0	0	①	0	0	1	0	0	0	0	10	10
	0	−1	1	0	0	0	0	0	1	0	0	0	12	—
	0	−1	0	1	0	0	0	0	0	1	0	0	22	—
	0	−1	0	0	1	0	0	0	0	0	1	0	18	18
	400	600	−200	−300	−400	0	0	0	0	0	0	1	0	

Pivot column (under u_5, −400)

	u_1	u_2	u_3	u_4	u_5	x_1	x_2	x_3	x_4	x_5	x_6	P	Constant	ratio
	−1	0	1	0	0	1	0	0	0	0	0	0	20	—
Pivot row →	−1	0	0	①	0	0	1	0	0	0	0	0	8	8
	−1	0	0	0	1	0	0	1	0	0	0	0	10	—
	0	−1	1	0	0	0	0	0	1	0	0	0	12	—
$\dfrac{R_6 - R_3}{R_7 + 400R_3}$	0	−1	0	1	0	0	0	0	0	1	0	0	22	22
	1	−1	0	0	0	0	0	−1	0	0	1	0	8	—
	0	600	−200	−300	0	0	0	400	0	0	0	1	4,000	

Pivot column (under u_4, −300)

	u_1	u_2	u_3	u_4	u_5	x_1	x_2	x_3	x_4	x_5	x_6	P	Constant	ratio
	−1	0	1	0	0	1	0	0	0	0	0	0	20	—
	−1	0	0	1	0	0	1	0	0	0	0	0	8	—
	−1	0	0	0	1	0	0	1	0	0	0	0	10	—
$\dfrac{R_5 - R_2}{R_7 + 300R_2}$	0	−1	1	0	0	0	0	0	1	0	0	0	12	—
	1	−1	0	0	0	0	−1	0	0	1	0	0	14	14
Pivot row →	①	−1	0	0	0	0	0	−1	0	0	1	0	8	8
	−300	600	−200	0	0	0	300	400	0	0	0	1	6,400	

Pivot column (under u_1, −300)

	u_1	u_2	u_3	u_4	u_5	x_1	x_2	x_3	x_4	x_5	x_6	P	Constant	ratio
	0	-1	1	0	0	1	0	-1	0	0	1	0	28	28
$R_1 + R_6$	0	-1	0	1	0	0	1	-1	0	0	1	0	16	—
$R_2 + R_6$	0	-1	0	0	1	0	0	0	0	0	1	0	18	—
$R_3 + R_6$														
$R_5 - R_6$	0	-1	①	0	0	0	0	0	1	0	0	0	12	12
$R_7 + 300R_6$	0	0	0	0	0	0	-1	1	0	1	-1	0	6	—
	1	-1	0	0	0	0	0	-1	0	0	1	0	8	—
	0	300	-200	0	0	0	300	100	0	0	300	1	8,800	

Pivot row (arrow pointing to row 4)

Pivot column (arrow pointing to u_3 column)

	u_1	u_2	u_3	u_4	u_5	x_1	x_2	x_3	x_4	x_5	x_6	P	Constant
	0	0	0	0	0	1	0	-1	-1	0	1	0	16
	0	-1	0	1	0	0	1	-1	0	0	1	0	16
	0	-1	0	0	1	0	0	0	0	0	1	0	18
$R_1 - R_4$	0	-1	1	0	0	0	0	0	1	0	0	0	12
$R_7 + 200R_4$	0	0	0	0	0	0	-1	1	0	1	-1	0	6
	1	-1	0	0	0	0	0	-1	0	0	1	0	8
	0	100	0	0	0	0	300	100	200	0	300	1	11,200

The last tableau is final, and we find that

$$x_1 = 0, \qquad x_2 = 300, \qquad x_3 = 100,$$
$$x_4 = 200, \qquad x_5 = 0, \qquad x_6 = 300, \qquad P = 11,200$$

Thus, to minimize shipping costs, Acrosonic should ship 300 loudspeaker systems from plant I to warehouse B, 100 systems from plant I to warehouse C, 200 systems from plant II to warehouse A, and 300 systems from plant II to warehouse C at a total cost of $11,200. ■

SELF-CHECK EXERCISES 6.5

1. Write the dual problem associated with the following problem: Minimize $C = 2x + 5y$ subject to

$$4x + y \geq 40$$
$$2x + y \geq 30$$
$$x + 3y \geq 30$$
$$x \geq 0, y \geq 0$$

2. Solve the primal problem posed in exercise 1.

Solutions to Self-Check Exercises 6.5 can be found on page 325.

6.5 EXERCISES

In exercises 1–6, use the technique developed in this section to solve the minimization problem.

1. Minimize $C = -2x + y$ subject to

$$x + 2y \le 6$$
$$3x + 2y \le 12$$
$$x \ge 0, y \ge 0$$

2. Minimize $C = -2x - 3y$ subject to

$$3x + 4y \le 24$$
$$7x - 4y \le 16$$
$$x \ge 0, y \ge 0$$

3. Minimize $C = -3x - 2y$ subject to the constraints of exercise 2.

4. Minimize $C = x - 2y + z$ subject to

$$x - 2y + 3z \le 10$$
$$2x + y - 2z \le 15$$
$$2x + y + 3z \le 20$$
$$x \ge 0, y \ge 0, z \ge 0$$

5. Minimize $C = 2x - 3y - 4z$ subject to

$$-x + 2y - z \le 8$$
$$x - 2y + 2z \le 10$$
$$2x + 4y - 3z \le 12$$
$$x \ge 0, y \ge 0, z \ge 0$$

6. Minimize $C = -3x - 2y - z$ subject to the constraints of exercise 5.

In exercises 7–10, you are given the final simplex tableau for the dual problem. Give the solution to the primal problem and to the associated dual problem.

7. Problem: Minimize $C = 8x + 12y$ subject to

$$x + 3y \ge 2$$
$$2x + 2y \ge 3$$
$$x \ge 0, y \ge 0$$

Final tableau:

u	v	x	y	P	Constant
0	1	$\frac{3}{4}$	$-\frac{1}{4}$	0	3
1	0	$-\frac{1}{2}$	$\frac{1}{2}$	0	2
0	0	$\frac{5}{4}$	$\frac{1}{4}$	1	13

8. Problem: Minimize $C = 3x + 2y$ subject to

$$5x + y \ge 10$$
$$2x + 2y \ge 12$$
$$x + 4y \ge 12$$
$$x \ge 0, y \ge 0$$

Final tableau:

u	v	w	x	y	P	Constant
1	0	$-\frac{3}{4}$	$\frac{1}{4}$	$-\frac{1}{4}$	0	$\frac{1}{4}$
0	1	$\frac{19}{8}$	$-\frac{1}{8}$	$\frac{5}{8}$	0	$\frac{7}{8}$
0	0	9	1	5	1	13

9. Problem: Minimize $C = 10x + 3y + 10z$ subject to

$$2x + y + 5z \ge 20$$
$$4x + y + z \ge 30$$
$$x \ge 0, y \ge 0, z \ge 0$$

Final tableau:

u	v	x	y	z	P	Constant
0	1	$\frac{1}{2}$	-1	0	0	2
1	0	$-\frac{1}{2}$	2	0	0	1
0	0	2	-9	1	0	3
0	0	5	10	0	1	80

10. Problem: Minimize $C = 2x + 3y$ subject to

$$x + 4y \ge 8$$
$$x + y \ge 5$$
$$2x + y \ge 7$$
$$x \ge 0, y \ge 0$$

Final tableau:

u	v	w	x	y	P	Constant
0	1	$\frac{7}{3}$	$\frac{4}{3}$	$-\frac{1}{3}$	0	$\frac{5}{3}$
1	0	$-\frac{1}{3}$	$-\frac{1}{3}$	$\frac{1}{3}$	0	$\frac{1}{3}$
0	0	2	4	1	1	11

In exercises 11–20, construct the dual problem associated with the primal problem. Solve the primal problem.

11. Minimize $C = 2x + 5y$ subject to

$$x + 2y \geq 4$$
$$3x + 2y \geq 6$$
$$x \geq 0, y \geq 0$$

12. Minimize $C = 3x + 2y$ subject to

$$2x + 3y \geq 90$$
$$3x + 2y \geq 120$$
$$x \geq 0, y \geq 0$$

13. Minimize $C = 6x + 4y$ subject to

$$6x + y \geq 60$$
$$2x + y \geq 40$$
$$x + y \geq 30$$
$$x \geq 0, y \geq 0$$

14. Minimize $C = 10x + y$ subject to

$$4x + y \geq 16$$
$$x + 2y \geq 12$$
$$x \geq 2$$
$$x \geq 0, y \geq 0$$

15. Minimize $C = 200x + 150y + 120z$ subject to

$$20x + 10y + z \geq 10$$
$$x + y + 2z \geq 20$$
$$x \geq 0, y \geq 0, z \geq 0$$

16. Minimize $C = 40x + 30y + 11z$ subject to

$$2x + y + z \geq 8$$
$$x + y - z \geq 6$$
$$x \geq 0, y \geq 0, z \geq 0$$

17. Minimize $C = 6x + 8y + 4z$ subject to

$$x + 2y + 2z \geq 10$$
$$2x + y + z \geq 24$$
$$x + y + z \geq 16$$
$$x \geq 0, y \geq 0, z \geq 0$$

18. Minimize $C = 12x + 4y + 8z$ subject to

$$2x + 4y + z \geq 6$$
$$3x + 2y + 2z \geq 2$$
$$4x + y + z \geq 2$$
$$x \geq 0, y \geq 0, z \geq 0$$

19. Minimize $C = 30x + 12y + 20z$ subject to

$$2x + 4y + 3z \geq 6$$
$$6x + z \geq 2$$
$$6y + 2z \geq 4$$
$$x \geq 0, y \geq 0, z \geq 0$$

20. Minimize $C = 8x + 6y + 4z$ subject to

$$2x + 3y + z \geq 6$$
$$x + 2y - 2z \geq 4$$
$$x + y + 2z \geq 2$$
$$x \geq 0, y \geq 0, z \geq 0$$

21. Shipping Costs   The Acrosonic Company also manufactures a model G loudspeaker system in plants I and II. The output at plant I is at most 800 per month, and the output at plant II is at most 600 per month. Model G loudspeaker systems are also shipped to the three warehouses—A, B, and C—whose minimum monthly requirements are 500, 400, and 400, respectively. Shipping costs from plant I to warehouse A, warehouse B, and warehouse C are $16, $20, and $22 per loudspeaker system, respectively, and shipping costs from plant II to each of these warehouses are $18, $16, and $14, respectively. What shipping schedule will enable Acrosonic to meet the requirements of the warehouses while keeping its shipping costs to a minimum?

22. Shipping Costs   The Steinwelt Piano Company manufactures uprights and consoles in two plants, plant I and plant II. The output of plant I is at most 300 per month, and the output of plant II is at most 250 per month. These pianos are shipped to three warehouses that serve as distribution centers for

Steinwelt. In order to fill current and projected future orders, warehouse A requires a minimum of 200 pianos per month, warehouse B requires at least 150 pianos per month, and warehouse C requires at least 200 pianos per month. The shipping cost of each piano from plant I to warehouse A, warehouse B, and warehouse C is $60, $60, and $80, respectively, and the shipping cost of each piano from plant II to warehouse A, warehouse B, and warehouse C is $80, $70, and $50, respectively. What shipping schedule will enable Steinwelt to meet the requirements of the warehouses while keeping the shipping costs to a minimum?

23. **Nutrition—Diet Planning**   The owner of the Health Juice-Bar wishes to prepare a low-calorie fruit juice with a high vitamin A and C content by blending orange juice and pink grapefruit juice. Each glass of the blended juice is to contain at least 1200 International units (I.U.) of vitamin A and 200 I.U. of vitamin C. One ounce of orange juice contains 60 I.U. of vitamin A, 16 I.U. of vitamin C, and 14 calories; each ounce of pink grapefruit juice contains 120 I.U. of vitamin A, 12 I.U. of vitamin C, and 11 calories. How many ounces of each juice should a glass of the blend contain if it is to meet the minimum vitamin requirements and at the same time contain a minimum number of calories?

24. **Product Control**   An oil company operates two refineries in a certain city. Refinery I has an output of 200, 100, and 100 barrels of low-, medium-, and high-grade oil per day, respectively. Refinery II has an output of 100, 200, and 600 barrels of low-, medium-, and high-grade oil per day, respectively. The company wishes to produce at least 1000, 1400, and 3000 barrels of low-, medium-, and high-grade oil to fill an order. If it costs $200 per day to operate refinery I and $300 per day to operate refinery II, determine how many days each refinery should be operated to meet the requirements of the order at minimum cost to the company.

SOLUTIONS TO SELF-CHECK EXERCISES 6.5

1. We first write down the following tableau for the (primal) problem:

x	y	Constant
4	1	40
2	1	30
1	3	30
2	5	0

Next, we interchange the columns and rows of the tableau and head the three columns of the resulting array with the three variables u, v, and w, obtaining the tableau

u	v	w	Constant
4	2	1	2
1	1	3	5
40	30	30	0

Interpreting the last tableau as if it were the initial tableau for a standard linear programming problem, with the exception that the signs of the coefficients pertaining to the objective function are not reversed, we construct the required dual problem as follows: Maximize $P = 40u + 30v + 30w$ subject to

$$4u + 2v + \ w \leq 2$$

$$u + \ v + 3w \leq 5$$

$$u \geq 0, v \geq 0, w \geq 0$$

2. We introduce slack variables x and y to obtain the system of linear equations

$$
\begin{aligned}
4u + 2v + w + x &= 2 \\
u + v + 3w + y &= 5 \\
-40u - 30v - 30w + P &= 0
\end{aligned}
$$

Using the simplex algorithm, we obtain the sequence of simplex tableaus

	u	v	w	x	y	P	Constant	ratio
Pivot row →	④	2	1	1	0	0	2	$\frac{2}{4} = \frac{1}{2}$
	1	1	3	0	1	0	5	$\frac{5}{1} = 5$
	−40	−30	−30	0	0	1	0	

$\xrightarrow{\frac{1}{4}R_1}$

Pivot column (under u)

	u	v	w	x	y	P	Constant
	①	$\frac{1}{2}$	$\frac{1}{4}$	$\frac{1}{4}$	0	0	$\frac{1}{2}$
	1	1	3	0	1	0	5
	−40	−30	−30	0	0	1	0

$\xrightarrow[R_3 + 40R_1]{R_2 - R_1}$

	u	v	w	x	y	P	Constant	ratio
	1	$\frac{1}{2}$	$\frac{1}{4}$	$\frac{1}{4}$	0	0	$\frac{1}{2}$	$\frac{1/2}{1/4} = 2$
Pivot row →	0	$\frac{1}{2}$	⑪⁄₄	$-\frac{1}{4}$	1	0	$\frac{9}{2}$	$\frac{9/2}{11/4} = \frac{18}{11}$
	0	−10	−20	10	0	1	20	

$\xrightarrow{\frac{4}{11}R_2}$

Pivot column (under w)

	u	v	w	x	y	P	Constant
	1	$\frac{1}{2}$	$\frac{1}{4}$	$\frac{1}{4}$	0	0	$\frac{1}{2}$
	0	$\frac{2}{11}$	①	$-\frac{1}{11}$	$\frac{4}{11}$	0	$\frac{18}{11}$
	0	−10	−20	10	0	1	20

$\xrightarrow[R_3 + 20R_2]{R_1 - \frac{1}{4}R_2}$

	u	v	w	x	y	P	Constant	ratio
Pivot row →	1	⑤⁄₁₁	0	$\frac{3}{11}$	$-\frac{1}{11}$	0	$\frac{1}{11}$	$\frac{1/11}{5/11} = \frac{1}{5}$
	0	$\frac{2}{11}$	1	$-\frac{1}{11}$	$\frac{4}{11}$	0	$\frac{18}{11}$	$\frac{18/11}{2/11} = 9$
	0	$-\frac{70}{11}$	0	$\frac{90}{11}$	$\frac{80}{11}$	1	$\frac{580}{11}$	

$\xrightarrow{\frac{11}{5}R_1}$

Pivot column (under v)

	u	v	w	x	y	P	Constant
	$\frac{11}{5}$	①	0	$\frac{3}{5}$	$-\frac{1}{5}$	0	$\frac{1}{5}$
	0	$\frac{2}{11}$	1	$-\frac{1}{11}$	$\frac{4}{11}$	0	$\frac{18}{11}$
	0	$-\frac{70}{11}$	0	$\frac{90}{11}$	$\frac{80}{11}$	1	$\frac{580}{11}$

$\xrightarrow[R_3 + \frac{70}{11}R_1]{R_2 - \frac{2}{11}R_1}$

u	v	w	x	y	P	Constant
$\frac{11}{5}$	1	0	$\frac{3}{5}$	$-\frac{1}{5}$	0	$\frac{1}{5}$
$-\frac{2}{5}$	0	1	$-\frac{1}{5}$	$\frac{2}{5}$	0	$\frac{8}{5}$
14	0	0	$\underbrace{12}$	6	1	54

Solution for the
primal problem

The last tableau is final, and the solution to the primal problem is $x = 12$ and $y = 6$ with a minimum value for C of 54.

U s i n g   T e c h n o l o g y

EXERCISES

In exercises 1–4, use a graphic calculator or a computer software program for the simplex method to solve the given linear programming problem using duality.

1. Minimize $C = x + y + 3z$ subject to

$$2x + y + 3z \geq 6$$
$$x + 2y + 4z \geq 8$$
$$3x + y - 2z \geq 4$$
$$x \geq 0, y \geq 0, z \geq 0$$

2. Minimize $C = 2x + 4y + z$ subject to

$$x + 2y + 4z \geq 7$$
$$3x + y - z \geq 6$$
$$x + 4y + 2z \geq 24$$
$$x \geq 0, y \geq 0, z \geq 0$$

3. Minimize $C = x + 1.2y + 3.5z$ subject to

$$2x + 3y + 5z \geq 12$$
$$3x + 1.2y - 2.2z \geq 8$$
$$1.2x + 3y + 1.8z \geq 14$$
$$x \geq 0, y \geq 0, z \geq 0$$

4. Minimize $C = 2.1x + 1.2y + z$ subject to

$$x + y - z \geq 5.2$$
$$x - 2.1y + 4.2z \geq 8.4$$
$$x \geq 0, y \geq 0, z \geq 0$$

**CHAPTER SIX
SUMMARY OF
PRINCIPAL TERMS**

Terms

Linear inequality
Bounded solution set
Unbounded solution set
Objective function
Linear programming problem
Feasible solution
Feasible set
Optimal solution
Method of corners
Simplex method
Standard maximization problem

Slack variable
Basic variable
Nonbasic variable
Basic solution
Pivot column
Pivot row
Pivot element
Simplex tableau
Standard minimization problem
Primal problem
Dual problem

CHAPTER 6 REVIEW EXERCISES

In exercises 1 and 2, find the optimal value(s) of the objective function on the feasible set S.

1. $Z = 2x + 3y$

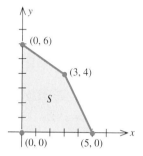

2. $Z = 4x + 3y$

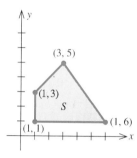

In exercises 3–12, use the method of corners to solve the linear programming problem.

3. Maximize $P = 3x + 5y$ subject to

$$2x + 3y \le 12$$
$$x + y \le 5$$
$$x \ge 0, y \ge 0$$

4. Maximize $P = 2x + 3y$ subject to

$$2x + y \le 12$$
$$x - 2y \le 1$$
$$x \ge 0, y \ge 0$$

5. Minimize $C = 2x + 5y$ subject to

$$x + 3y \le 15$$
$$4x + y \le 16$$
$$x \ge 0, y \ge 0$$

6. Minimize $C = 3x + 4y$ subject to

$$2x + y \ge 4$$
$$2x + 5y \ge 10$$
$$x \ge 0, y \ge 0$$

7. Maximize $P = 3x + 2y$ subject to

$$2x + y \le 16$$
$$2x + 3y \le 36$$
$$4x + 5y \ge 28$$
$$x \ge 0, y \ge 0$$

8. Maximize $P = 6x + 2y$ subject to

$$x + 2y \le 12$$
$$x + y \le 8$$
$$2x - 3y \ge 6$$
$$x \ge 0, y \ge 0$$

9. Minimize $C = 2x + 7y$ subject to

$$3x + 5y \ge 45$$
$$3x + 10y \ge 60$$
$$x \ge 0, y \ge 0$$

10. Minimize $C = 4x + y$ subject to

$$6x + y \ge 18$$
$$2x + y \ge 10$$
$$x + 4y \ge 12$$
$$x \ge 0, y \ge 0$$

11. Find the maximum and minimum of $Q = x + y$ subject to

$$5x + 2y \ge 20$$
$$x + 2y \ge 8$$
$$x + 4y \le 22$$
$$x \ge 0, y \ge 0$$

12. Find the maximum and minimum of $Q = 2x + 5y$ subject to

$$x + y \ge 4$$
$$-x + y \le 6$$
$$x + 3y \le 30$$
$$x \le 12$$
$$x \ge 0, y \ge 0$$

In exercises 13–24, use the simplex method to solve the linear programming problem.

13. Maximize $P = 3x + 4y$ subject to

$$x + 3y \le 15$$
$$4x + y \le 16$$
$$x \ge 0, y \ge 0$$

14. Maximize $P = 2x + 5y$ subject to

$$2x + y \le 16$$
$$2x + 3y \le 24$$
$$y \le 6$$
$$x \ge 0, y \ge 0$$

15. Maximize $P = 2x + 3y + 5z$ subject to

$$x + 2y + 3z \leq 12$$
$$x - 3y + 2z \leq 10$$
$$x \geq 0, y \geq 0, z \geq 0$$

16. Maximize $P = x + 2y + 3z$ subject to

$$2x + y + z \leq 14$$
$$3x + 2y + 4z \leq 24$$
$$2x + 5y - 2z \leq 10$$
$$x \geq 0, y \geq 0, z \geq 0$$

17. Minimize $C = 3x + 2y$ subject to

$$2x + 3y \geq 6$$
$$2x + y \geq 4$$
$$x \geq 0, y \geq 0$$

18. Minimize $C = x + 2y$ subject to

$$3x + y \geq 12$$
$$x + 4y \geq 16$$
$$x \geq 0, y \geq 0$$

19. Minimize $C = 24x + 18y + 24z$ subject to

$$3x + 2y + z \geq 4$$
$$x + y + 3z \geq 6$$
$$x \geq 0, y \geq 0, z \geq 0$$

20. Minimize $C = 4x + 2y + 6z$ subject to

$$x + 2y + z \geq 4$$
$$2x + y + 2z \geq 2$$
$$3x + 2y + z \geq 3$$
$$x \geq 0, y \geq 0, z \geq 0$$

21. Maximize $P = 3x - 4y$ subject to

$$x + y \leq 45$$
$$x - 2y \geq 10$$
$$x \geq 0, y \geq 0$$

22. Minimize $C = 2x + 3y$ subject to

$$x + y \leq 10$$
$$x + 2y \geq 12$$
$$2x + y \geq 12$$
$$x \geq 0, y \geq 0$$

23. Maximize $P = 2x + 3y$ subject to

$$2x + 5y \leq 20$$
$$-x + 5y \geq 5$$
$$x \geq 0, y \geq 0$$

24. Minimize $C = -3x - 4y$ subject to

$$x + y \leq 45$$
$$x + y \geq 15$$
$$x \leq 30$$
$$y \leq 25$$
$$x \geq 0, y \geq 0$$

In exercises 25–27, use the method of corners to solve the linear programming problem.

25. Financial Analysis   An investor has decided to commit no more than $80,000 to the purchase of the common stocks of two companies, Company A and Company B. He has also estimated that there is a chance of a 1% capital loss on his investment in Company A and a chance of a 4% loss on his investment in Company B, and he has decided that these losses should not exceed $2,000. On the other hand, he expects to make a 14% profit from his investment in Company A and a 20% profit from his investment in Company B. Determine how much he should invest in the stock of each company in order to maximize his investment returns.

26. Manufacturing—Production Scheduling   The Soundex Company produces two models of clock/radios. Model A requires 15 minutes of work on assembly line I and 10 minutes of work on assembly line II. Model B requires 10 minutes of work on assembly line I and 12 minutes of work on assembly line II. At most, 25 hours of assembly time on line I and 22 hours of assembly time on line II are available per day. It is anticipated that Soundex will realize a profit of $12 on model A and $10 on model B. How many clock/radios of each model should be produced per day in order to maximize Soundex's profit?

27. Manufacturing—Production Scheduling   Kane Manufacturing has a division that produces two models of hibachis, model A and model B. To produce each model A hibachi requires 3 pounds of cast iron and 6 minutes of labor. To produce each model B hibachi requires 4 pounds of cast iron and 3 minutes of labor. The profit for each model A hibachi is $2 and the profit for each model B hibachi is $1.50. One thousand pounds of cast iron and 20 hours of labor are available for the production of hibachis per day. Because of a backlog of orders for model B hibachis, Kane's manager has decided to produce at least 180 model B hibachis per day. How many hibachis of each model should Kane produce to maximize its profits?

In exercises 28–30, use the simplex method to solve the linear programming problem.

28. A company manufactures three products, A, B, and C, on two machines, I and II. It has been determined that the company will realize a profit of $4 on each unit of product A, $6 on each unit of product B, and $8 on each unit of product C. To manufacture a unit of product A requires 9 minutes on machine I and 6 minutes on machine II; to manufacture a unit of product B requires 12 minutes on machine I and 6 minutes on machine II; and to manufacture a unit of product C requires 18 minutes on machine I and 10 minutes on machine II. There are 6 hours of machine time available on machine I and 4 hours of machine time available on machine II in each work shift. How many units of each product should be produced in each shift in order to maximize the company's profit?

29. Mr. Moody has decided to invest at most $100,000 in securities in the form of corporate stocks. He has classified his options into three groups of stocks: blue-chip stocks that he assumes will yield a 10% return (dividends and capital appreciation) within a year, growth stocks that he assumes will yield a 15% return within a year, and speculative stocks that he assumes will yield a 20% return (mainly due to capital appreciation) within a year. Because of the relative risks involved in his investment, Mr. Moody has further decided that no more than 30% of his investment should be in growth and speculative stocks and at least 50% of his investment should be in blue-chip and speculative stocks. Determine how much Mr. Moody should invest in each group of stocks in the hope of maximizing the return on his investments.

30. Sandra has at most $200,000 to invest in stocks, bonds, and money-market funds. She expects annual yields of 15%, 10%, and 8%, respectively, on these investments. If Sandra wants at least $50,000 to be invested in money-market funds and requires that the amount invested in bonds be greater than or equal to the sum of her investments in stocks and money-market funds, determine how much she should invest in each vehicle in order to maximize the returns on her investments.

CHAPTER 6   EXTENDED PROBLEMS AND PROJECTS

1. Consider the regular linear programming problem:
   Maximize $Q = x + 2y$ subject to

   $$-2x + y \leq 4$$

   $$x - 3y \leq 3$$

   $$x \geq 0, y \geq 0$$

   a. Solve the problem graphically to show that it has an unbounded solution.
   b. Solve the problem using the simplex method. What part of the simplex method breaks down?

2. United Motors Corporation manufactures compact, mid-size, and full-size automobiles. For the next year the company plans to build at most 500,000 cars, of which a maximum of 80,000 will be full size. To comply with government fuel economy standards for passenger cars, the fleet must average at least 27 mpg (miles per gallon) to avoid extensive fines. United's engineering division has estimated that the compacts will average 35 mpg, mid-size models will average 24 mpg, and full-size models will average 20 mpg. Each compact can be expected to bring in a profit of $2,000, each mid-size model a profit of $3,000, and each full-size model a profit of $5,000.

   a. Suppose x, y, and z denote the number of compact, mid-size, and full-size cars to be built. Show that the condition that the fleet average be at least 27 mpg leads to the constraint $-8x + 3y + 7z \leq 0$.
   b. Use the simplex method to determine the number of models of each size United should build for the upcoming year if its profits are to be maximized.

Sets and Probability

7

*W*e often deal with well-defined collections of objects called *sets*. In this chapter we see how sets can be combined algebraically to yield other sets. We also look at some techniques for determining the number of elements in a set and for determining the number of ways the elements of a set can be arranged or combined.

Next we see how the notion of sets can be used to study probability. After giving the technical meaning of the term *probability*, we see how the rules of probability are applied to many real-life situations to compute the probability of the occurrence of certain events.

| 7.1 | ## Sets and Set Operations |

Set Terminology and Notation

We often deal with collections of different kinds of objects. For example, in conducting a study of the distribution of the weights of newborn infants, we might consider the collection of all infants born in the Massachusetts General Hospital during 1996. In a study of the fuel consumption of compact cars, we might be interested in the collection of compact cars manufactured by General Motors in the 1996 model year. Such collections are examples of *sets*. More specifically, a **set** is a well-defined collection of objects. Thus, a set is not just any collection of objects, but it must be well defined in the sense that if we are given an object, then we should be able to determine whether or not it belongs to the collection.

The objects of a set are called the **elements,** or *members,* of the set and are usually denoted by lowercase letters $a, b, c, \ldots$ The sets themselves are usually denoted by uppercase letters $A, B, C, \ldots$ The elements of a set may be displayed by listing each element between braces. For example, using **roster notation,** the set A consisting of the first three letters of the English alphabet is written

$$A = \{a, b, c\}$$

The set B of all letters of the alphabet may be written

$$B = \{a, b, c, \ldots, z\}$$

Another kind of set notation commonly used is **set-builder notation.** Here, a rule is given that describes the definite property or properties an object x must satisfy to qualify for membership in the set. Using this notation, the set B is written as

$$B = \{x \mid x \text{ is a letter of the English alphabet}\}$$

and is read "B is the set of all elements x such that x is a letter of the English alphabet."

If a is an element of a set A, we write $a \in A$, read "a belongs to A" or "a is an element of A." If, on the other hand, the element a does not belong to the set A, then we write $a \notin A$, read "a does not belong to A." For example, if $A = \{1, 2, 3, 4, 5\}$, then $3 \in A$ but $6 \notin A$.

SET EQUALITY

Two sets A and B are **equal,** written $A = B$, if and only if they have exactly the same elements.

EXAMPLE 1 Let A, B, and C be the sets

$$A = \{a, e, i, o, u\}$$
$$B = \{a, i, o, e, u\}$$
$$C = \{a, e, i, o\}$$

Then $A = B$, since they both contain exactly the same elements. Note that the order in which the elements are displayed is immaterial. Also, $A \neq C$, since $u \in A$ but $u \notin C$. Similarly, we conclude that $B \neq C$.

SUBSET

If every element of a set A is also an element of a set B, then we say that A is a **subset** of B and write $A \subseteq B$.

By this definition, two sets A and B are equal if and only if (1) $A \subseteq B$ and (2) $B \subseteq A$. You may verify this (see exercise 64).

EXAMPLE 2   Referring to Example 1, we find that $C \subseteq B$, since every element of C is also an element of B. Also, if D is the set

$$D = \{a, e, i, o, x\}$$

then D is not a subset of A, written $D \nsubseteq A$, since $x \in D$ but $x \notin A$. Observe that $A \nsubseteq D$ as well, since $u \in A$ but $u \notin D$.   ∎

If A and B are sets such that $A \subseteq B$ but $A \neq B$, then we say that A is a **proper subset** of B. In other words, a set A is a proper subset of a set B, written $A \subset B$, if (1) $A \subseteq B$ and (2) there exists at least one element in B that is not in A. The latter condition states that the set A is properly "smaller" than the set B.

EXAMPLE 3   Let $A = \{1, 2, 3, 4, 5, 6\}$ and $B = \{2, 4, 6\}$. Then B is a proper subset of A, since (1) $B \subseteq A$, which is easily verified, and (2) there exists at least one element in A that is not in B—for example, the element 1.   ∎

Notice that when we are referring to sets and subsets we use the symbols $\subset, \subseteq, \supset$, and $\supseteq$ to express the idea of "containment." However, when we wish to show that an element is contained in a set, we use the symbol $\in$. Thus, in Example 3, we would write $1 \in A$ and *not* $\{1\} \in A$.

EMPTY SET

The set that contains no elements is called the **empty set** and is denoted by $\varnothing$.

The empty set, $\varnothing$, is a subset of every set. To see this, observe that $\varnothing$ has no elements and, therefore, contains no element that is not also in A.

EXAMPLE 4   List all subsets of the set $A = \{a, b, c\}$.

Solution   There is one subset consisting of no elements—namely, the empty set $\varnothing$. Next, observe that there are three subsets consisting of one element,

$$\{a\}, \{b\}, \text{ and } \{c\}$$

three subsets consisting of two elements,

$$\{a, b\}, \{a, c\}, \text{ and } \{b, c\}$$

and one subset consisting of three elements, the set A itself. Therefore, the subsets of A are

$$\varnothing, \{a\}, \{b\}, \{c\}, \{a, b\}, \{a, c\}, \{b, c\}, \{a, b, c\}$$ ∎

In contrast with the empty set, we have, on the other extreme, the notion of a largest, or *universal,* set. A **universal set** is the set of all elements of interest in a particular discussion. It is the largest in the sense that all sets considered in the discussion of the problem are subsets of the universal set. Of course, different universal sets are associated with different problems, as shown in Example 5.

EXAMPLE 5

a. If the problem at hand is to determine the ratio of female to male students in a college, then a logical choice of a universal set is the set comprising the whole student body of the college.

b. If the problem is to determine the ratio of female to male students in the business department of the college in (a), then the set of all students in the business department may be chosen as the universal set.

A visual representation of sets is realized through the use of **Venn diagrams,** which are of considerable help in understanding the concepts introduced earlier as well as in solving problems involving sets. The universal set U is represented by a rectangle, and subsets of U are represented by regions lying inside the rectangle.

EXAMPLE 6   Use Venn diagrams to illustrate the following statements:

a. The sets A and B are equal.

b. The set A is a proper subset of the set B.

c. The sets A and B are not subsets of each other.

Solution   The respective Venn diagrams are shown in Figures 1a, b, and c.

FIGURE 1

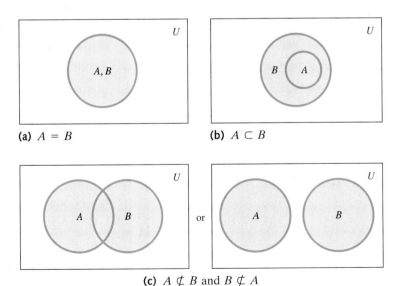

(a) $A = B$ **(b)** $A \subset B$

(c) $A \not\subset B$ and $B \not\subset A$

Set Operations

Having introduced the concept of a set, our next task is to consider operations on sets—that is, to consider ways in which sets may be combined to yield other sets. These operations enable us to combine sets in much the same way the operations of addition and multiplication enable us to combine numbers to obtain other numbers. In what follows, all sets are assumed to be subsets of a given universal set U.

SET UNION

Let A and B be sets. The **union** of A and B, written $A \cup B$, is the set of all elements that belong to either A or B or both.

$$A \cup B = \{x \mid x \in A \text{ or } x \in B \text{ or both}\}$$

The shaded portion of the Venn diagram (Figure 2) depicts the set $A \cup B$.

FIGURE 2
Set union

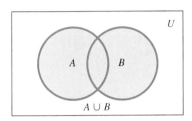

EXAMPLE 7 If $A = \{a, b, c\}$ and $B = \{a, c, d\}$, then $A \cup B = \{a, b, c, d\}$. ∎

SET INTERSECTION

Let A and B be sets. The set of elements in common with the sets A and B, written $A \cap B$, is called the **intersection** of A and B.

$$A \cap B = \{x \mid x \in A \text{ and } x \in B\}$$

The shaded portion of the Venn diagram (Figure 3) depicts the set $A \cap B$.

FIGURE 3
Set intersection

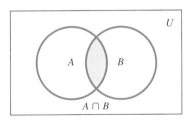

EXAMPLE 8 Let $A = \{a, b, c\}$ and $B = \{a, c, d\}$. Then $A \cap B = \{a, c\}$. (Compare this result with Example 7.) ∎

EXAMPLE 9 Let $A = \{1, 3, 5, 7, 9\}$ and $B = \{2, 4, 6, 8, 10\}$. Then $A \cap B = \emptyset$. ∎

The two sets of Example 9 have null intersection. In general, the sets A and B are said to be **disjoint** if they have no elements in common—that is, if $A \cap B = \emptyset$.

EXAMPLE 10   If U is the set of all students in the classroom and $M = \{x \in U \mid x \text{ is male}\}$ and $F = \{x \in U \mid x \text{ is female}\}$, then $F \cap M = \emptyset$, and F and M are disjoint.   ■

COMPLEMENT OF A SET

If U is a universal set and A is a subset of U, then the set of all elements in U that are not in A is called the **complement** of A and is denoted A^c.

$$A^c = \{x \mid x \in U, x \notin A\}$$

The shaded portion of the Venn diagram (Figure 4) shows the set A^c.

FIGURE 4
Set complementation

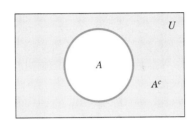

EXAMPLE 11   Let $U = \{1, 2, 3, 4, 5, 6, 7, 8, 9, 10\}$ and $A = \{2, 4, 6, 8, 10\}$. Then $A^c = \{1, 3, 5, 7, 9\}$.   ■

The following rules hold for the operation of complementation. See whether you can verify them.

SET COMPLEMENTATION

If U is a universal set and A is a subset of U, then

a. $U^c = \emptyset$ **b.** $\emptyset^c = U$ **c.** $(A^c)^c = A$

d. $A \cup A^c = U$ **e.** $A \cap A^c = \emptyset$

The following rules govern the operations on sets.

SET OPERATIONS

Let U be a universal set. If A, B, and C are arbitrary subsets of U, then

$A \cup B = B \cup A$	*Commutative law for union*
$A \cap B = B \cap A$	*Commutative law for intersection*
$A \cup (B \cup C) = (A \cup B) \cup C$	*Associative law for union*
$A \cap (B \cap C) = (A \cap B) \cap C$	*Associative law for intersection*
$A \cup (B \cap C)$	
$\quad = (A \cup B) \cap (A \cup C)$	*Distributive law for union*
$A \cap (B \cup C)$	
$\quad = (A \cap B) \cup (A \cap C)$	*Distributive law for intersection*

There are two additional rules, referred to as De Morgan's Laws, that govern the operations on sets.

DE MORGAN'S LAWS

Let A and B be sets. Then

$$(A \cup B)^c = A^c \cap B^c \qquad (1)$$
$$(A \cap B)^c = A^c \cup B^c \qquad (2)$$

Equation (1) states that the complement of the union of two sets is equal to the intersection of their complements. Equation (2) states that the complement of the intersection of two sets is equal to the union of their complements.

We will not prove De Morgan's Laws here, but the plausibility of (2) is illustrated in the following example.

FIGURE 5
$(A \cap B)^c$

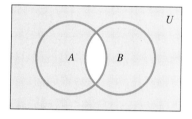

EXAMPLE 12   Using Venn diagrams, show that

$$(A \cap B)^c = A^c \cup B^c$$

Solution $(A \cap B)^c$ is the set of elements in U but not in $A \cap B$ and is thus the shaded region shown in Figure 5. Next, A^c and B^c are shown in Figures 6a and b. By referring once again to Figure 5, we can easily see that their union, $A^c \cup B^c$, is equivalent to $(A \cap B)^c$.

FIGURE 6
$A^c \cup B^c$ *is the set obtained by joining (a) and (b).*

(a)

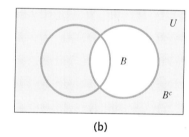

(b)

EXAMPLE 13   Let $U = \{1, 2, 3, 4, 5, 6, 7, 8, 9, 10\}$, $A = \{1, 2, 4, 8, 9\}$, and $B = \{3, 4, 5, 6, 8\}$. Verify by direct computation that $(A \cup B)^c = A^c \cap B^c$.

Solution $A \cup B = \{1, 2, 3, 4, 5, 6, 8, 9\}$, so $(A \cup B)^c = \{7, 10\}$. However, $A^c = \{3, 5, 6, 7, 10\}$ and $B^c = \{1, 2, 7, 9, 10\}$, so $A^c \cap B^c = \{7, 10\}$. The required result follows.

Application

EXAMPLE 14   Let U denote the set of all cars in a dealer's lot and

$$A = \{x \in U \mid x \text{ is equipped with automatic transmission}\}$$
$$B = \{x \in U \mid x \text{ is equipped with air conditioning}\}$$
$$C = \{x \in U \mid x \text{ is equipped with antilock brakes}\}$$

Find an expression in terms of A, B, and C for each of the following sets:

a. The set of cars with at least one of the given options.

b. The set of cars with exactly one of the given options.

c. The set of cars with automatic transmission and antilock brakes but no air conditioning.

Solution

a. The set of cars with at least one of the given options is $A \cup B \cup C$ (Figure 7a).

b. The set of cars with automatic transmission only is given by $A \cap B^c \cap C^c$. Similarly, we find that the set of cars with air conditioning only is given by $B \cap C^c \cap A^c$, and the set of cars with antilock brakes only is given by $C \cap A^c \cap B^c$. Thus, the set of cars with exactly one of the given options is $(A \cap B^c \cap C^c) \cup (B \cap C^c \cap A^c) \cup (C \cap A^c \cap B^c)$ (Figure 7b).

c. The set of cars with automatic transmission and antilock brakes but no air conditioning is given by $A \cap C \cap B^c$ (Figure 7c).

FIGURE 7

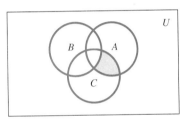

(a) The set of cars with at least one option

(b) The set of cars with exactly one option

(c) The set of cars with automatic transmission and antilock brakes but no air conditioning

SELF-CHECK EXERCISES 7.1

1. Let $U = \{1, 2, 3, 4, 5, 6, 7\}$, $A = \{1, 2, 3\}$, $B = \{3, 4, 5, 6\}$, and $C = \{2, 3, 4\}$. Find the following sets:

    a. A^c
    b. $A \cup B$
    c. $B \cap C$

    d. $(A \cup B) \cap C$
    e. $(A \cap B) \cup C$
    f. $A^c \cap (B \cup C)^c$

2. Let U denote the set of all members of the House of Representatives. Let

$$D = \{x \in U \mid x \text{ is a Democrat}\}$$

$$R = \{x \in U \mid x \text{ is a Republican}\}$$

$$F = \{x \in U \mid x \text{ is a female}\}$$

$$L = \{x \in U \mid x \text{ is a lawyer by training}\}$$

Describe each of the following sets in words.

    a. $D \cap F$
    b. $F^c \cap R$
    c. $D \cap F \cap L^c$

Solutions to Self-Check Exercises 7.1 can be found on page 341.

7.1 EXERCISES

In exercises 1–4, write the set in set-builder notation.

1. The set of gold medalists in the 1992 Summer Olympic Games

2. The set of football teams in the NFL

3. $\{3, 4, 5, 6, 7\}$

4. $\{1, 3, 5, 7, 9, 11, \ldots, 39\}$

In exercises 5–8, list the elements of the set in roster notation.

5. $\{x \mid x$ is a digit in the number 352,646$\}$

6. $\{x \mid x$ is a letter in the word *HIPPOPOTAMUS*$\}$

7. $\{x \mid 2 - x = 4; x,$ an integer$\}$

8. $\{x \mid 2 - x = 4; x,$ a fraction$\}$

In exercises 9–14, state whether the statement is true or false.

9. a. $\{a, b, c\} = \{c, a, b\}$ **b.** $A \in A$

10. a. $\varnothing \in A$ **b.** $A \subset A$

11. a. $0 \in \varnothing$ **b.** $0 = \varnothing$

12. a. $\{\varnothing\} = \varnothing$ **b.** $\{a, b\} \in \{a, b, c\}$

13. $\{$Chevrolet, Pontiac, Buick$\} \subset \{$General Motors$\}$

14. $\{x \mid x$ is a silver medalist in the 1992 Summer Olympic Games$\} = \varnothing$

In exercises 15 and 16, let $A = \{1, 2, 3, 4, 5\}$. Determine whether the statement is true or false.

15. a. $2 \in A$ **b.** $A \subseteq \{2, 4, 6\}$

16. a. $0 \in A$ **b.** $\{1, 3, 5\} \in A$

17. Let $A = \{1, 2, 3\}$. Which of the following sets are equal to A?
  a. $\{2, 1, 3\}$ **b.** $\{3, 2, 1\}$ **c.** $\{0, 1, 2, 3\}$

18. Let $A = \{a, e, l, t, r\}$. Which of the following sets are equal to A?
  a. $\{x \mid x$ is a letter of the word *later*$\}$
  b. $\{x \mid x$ is a letter of the word *latter*$\}$
  c. $\{x \mid x$ is a letter of the word *relate*$\}$

19. List all subsets of the following sets:
  a. $\{1, 2\}$ **b.** $\{1, 2, 3\}$ **c.** $\{1, 2, 3, 4\}$

20. List all subsets of the set $A = \{$IBM, U.S. Steel, Union Carbide, Boeing$\}$. Which of these are proper subsets of A?

In exercises 21–24, find the smallest possible set (that is, the set with the least number of elements) that contains the given sets as subsets.

21. $\{1, 2\}, \{1, 3, 4\}, \{4, 6, 8, 10\}$

22. $\{1, 2, 4\}, \{a, b\}$

23. $\{$Jill, John, Jack$\}, \{$Susan, Sharon$\}$

24. $\{$GM, Ford, Chrysler$\}, \{$Daimler-Benz, Volkswagen$\}, \{$Toyota, Datsun$\}$

25. Use Venn diagrams to represent the following relationships:
  a. $A \subset B$ and $B \subset C$
  b. $A \subset U$ and $B \subset U$, where A and B have no elements in common
  c. The sets A, B, and C are equal

26. Let U denote the set of all students who applied for admission to the freshman class at Faber College for the upcoming academic year and let

$A = \{x \in U \mid x$ is a successful applicant$\}$

$B = \{x \in U \mid x$ is a female student who enrolled in the freshman class$\}$

$C = \{x \in U \mid x$ is a male student who enrolled in the freshman class$\}$

  a. Use Venn diagrams to represent the sets U, A, B, and C.
  b. Determine whether the following statements are true or false.
    i. $A \subseteq B$ **ii.** $B \subset A$ **iii.** $C \subset B$

In exercises 27 and 28, shade the portion of the accompanying figure that represents each set.

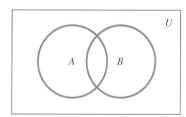

27. a. $A \cap B^c$ **b.** $A^c \cap B$
28. a. $A^c \cap B^c$ **b.** $(A \cup B)^c$

In exercises 29–32, shade the portion of the accompanying figure that represents each set.

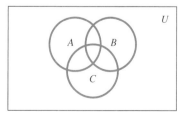

29. a. $A \cup B \cup C$ **b.** $A \cap B \cap C$
30. a. $A \cap B \cap C^c$ **b.** $A^c \cap B \cap C$
31. a. $A^c \cap B^c \cap C^c$ **b.** $(A \cup B)^c \cap C$
32. a. $A \cup (B \cap C)^c$ **b.** $(A \cup B \cup C)^c$

In exercises 33–36, let $U = \{1, 2, 3, 4, 5, 6, 7, 8, 9, 10\}$, $A = \{1, 3, 5, 7, 9\}$, $B = \{2, 4, 6, 8, 10\}$, *and* $C = \{1, 2, 4, 5, 8, 9\}$. *Find each set.*

33. a. A^c **b.** $B \cup C$ **c.** $C \cup C^c$

34. a. $C \cap C^c$ **b.** $(A \cap C)^c$ **c.** $A \cup (B \cap C)$

35. a. $(A \cap B) \cup C$ **b.** $(A \cup B \cup C)^c$
     c. $(A \cap B \cap C)^c$

36. a. $A^c \cap (B \cap C^c)$ **b.** $(A \cup B^c) \cup (B \cap C^c)$
     c. $(A \cup B)^c \cap C^c$

In exercises 37 and 38, determine whether the given pairs of sets are disjoint.

37. a. $\{1, 2, 3, 4\}, \{4, 5, 6, 7\}$
     b. $\{a, c, e, g\}, \{b, d, f\}$

38. a. $\varnothing, \{1, 3, 5\}$ **b.** $\{0, 1, 3, 4\}, \{0, 2, 5, 7\}$

In exercises 39–42, let U *denote the set of all employees at the Universal Life Insurance Company. Let*

$$T = \{x \in U \mid x \text{ drinks tea}\}$$
$$C = \{x \in U \mid x \text{ drinks coffee}\}$$

Describe each set in words.

39. a. T^c **b.** C^c

40. a. $T \cup C$ **b.** $T \cap C$

41. a. $T \cap C^c$ **b.** $T^c \cap C$

42. a. $T^c \cap C^c$ **b.** $(T \cup C)^c$

In exercises 43–46, let U *denote the set of all employees in a hospital. Let*

$$N = \{x \in U \mid x \text{ is a nurse}\}$$
$$D = \{x \in U \mid x \text{ is a doctor}\}$$
$$A = \{x \in U \mid x \text{ is an administrator}\}$$
$$M = \{x \in U \mid x \text{ is a male}\}$$
$$F = \{x \in U \mid x \text{ is a female}\}$$

Describe each set in words.

43. a. D^c **b.** N^c

44. a. $N \cup D$ **b.** $N \cap M$

45. a. $D \cap M^c$ **b.** $D \cap A$

46. a. $N \cap F$ **b.** $(D \cup N)^c$

In exercises 47 and 48, let U *denote the set of all senators in Congress. Let*

$$D = \{x \in U \mid x \text{ is a Democrat}\}$$
$$R = \{x \in U \mid x \text{ is a Republican}\}$$
$$F = \{x \in U \mid x \text{ is a female}\}$$
$$L = \{x \in U \mid x \text{ is a lawyer}\}$$

Write the set that represents each statement.

47. a. The set of all Democrats who are female
     b. The set of all Republicans who are male and are not lawyers

48. a. The set of all Democrats who are female or are lawyers
     b. The set of all senators who are not Democrats or are lawyers

In exercises 49 and 50, let U *denote the set of all students in the Business College of a certain university. Let*

$$A = \{x \in U \mid x \text{ had taken a course in accounting}\}$$
$$B = \{x \in U \mid x \text{ had taken a course in economics}\}$$
$$C = \{x \in U \mid x \text{ had taken a course in marketing}\}$$

Write the set that represents each statement.

49. a. The set of students who have not had a course in economics
     b. The set of students who have had courses in accounting and economics
     c. The set of students who have had courses in accounting and economics but not marketing

50. a. The set of students who have had courses in economics but not courses in accounting or marketing
     b. The set of students who have had at least one of the three courses
     c. The set of students who have had all three courses

In exercises 51–56, use Venn diagrams to illustrate each statement.

51. $A \subset A \cup B; B \subset A \cup B$

52. $A \cap B \subset A; A \cap B \subset B$

53. $A \cup (B \cup C) = (A \cup B) \cup C$

54. $A \cap (B \cap C) = (A \cap B) \cap C$

55. $A \cap (B \cup C) = (A \cap B) \cup (A \cap C)$

56. $(A \cup B)^c = A^c \cap B^c$

In exercises 57 and 58, let $U = \{1, 2, 3, 4, 5, 6, 7, 8, 9, 10\}$, $A = \{1, 3, 5, 7, 9\}$, $B = \{1, 2, 4, 7, 8\}$, *and* $C = \{2, 4, 6, 8\}$. *Verify each equation by direct computation.*

57. a. $A \cup (B \cup C) = (A \cup B) \cup C$
  b. $A \cap (B \cap C) = (A \cap B) \cap C$

58. a. $A \cap (B \cup C) = (A \cap B) \cup (A \cap C)$
  b. $(A \cup B)^c = A^c \cap B^c$

In exercises 59–62, refer to the accompanying figure and find the points that belong to each set.

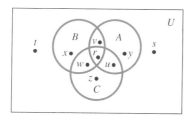

59. a. $A \cup B$ **b.** $A \cap B$

60. a. $A \cap (B \cup C)$ **b.** $(B \cap C)^c$

61. a. $(B \cup C)^c$ **b.** A^c

62. a. $(A \cap B) \cap C^c$ **b.** $(A \cup B \cup C)^c$

63. Suppose that $A \subset B$ and $B \subset C$, where A and B are any two sets. What conclusion can be drawn regarding the sets A and C?

64. Verify the assertion that two sets A and B are equal if and only if (a) $A \subseteq B$ and (b) $B \subseteq A$.

SOLUTIONS TO SELF-CHECK EXERCISES 7.1

1. a. A^c is the set of elements in U but not in A. Therefore,

$$A^c = \{4, 5, 6, 7\}$$

b. $A \cup B$ consists of all elements in A and/or B. So,

$$A \cup B = \{1, 2, 3, 4, 5, 6\}$$

c. $B \cap C$ is the set of all elements in both B and C. Therefore,

$$B \cap C = \{3, 4\}$$

d. Using the result from (b), we find

$$(A \cup B) \cap C = \{1, 2, 3, 4, 5, 6\} \cap \{2, 3, 4\}$$

$$= \{2, 3, 4\}$$

e. First we compute

$$A \cap B = \{3\}$$

Next, since $(A \cap B) \cup C$ is the set of all elements in $(A \cap B)$ and/or C, we conclude that

$$(A \cap B) \cup C = \{3\} \cup \{2, 3, 4\}$$

$$= \{2, 3, 4\}$$

f. From (a), we have $A^c = \{4, 5, 6, 7\}$. Next, we compute

$$B \cup C = \{3, 4, 5, 6\} \cup \{2, 3, 4\}$$

$$= \{2, 3, 4, 5, 6\}$$

from which we deduce that

$$(B \cup C)^c = \{1, 7\} \qquad \text{The set of elements in } U \text{ but not in } B \cup C$$

Finally, using these results, we obtain

$$A^c \cap (B \cup C)^c = \{4, 5, 6, 7\} \cap \{1, 7\} = \{7\}$$

2. **a.** $D \cap F$ denotes the set of all elements in both D and F. Since an element in D is a Democrat and an element in F is a female representative, we see that $D \cap F$ is the set of all female Democrats in the House of Representatives.

b. Since F^c is the set of male representatives and R is the set of Republicans, we see that $F^c \cap R$ is the set of male Republicans in the House of Representatives.

c. L^c is the set of representatives who are not lawyers by training. Therefore, $D \cap F \cap L^c$ is the set of female Democratic representatives who are not lawyers by training.

7.2 The Number of Elements in a Finite Set

Counting the Elements in a Set

The solution to some problems in mathematics calls for finding the number of elements in a set. Such problems are called *counting problems* and constitute a field of study known as **combinatorics.** Our study of combinatorics, however, is restricted to the results that will be required for our work in probability later on.

The number of elements in a finite set is determined by simply counting the elements in the set. If A is a set, then $n(A)$ denotes the number of elements in A. For example, if

$$A = \{1, 2, 3, \ldots, 20\}, \qquad B = \{a, b\}, \qquad C = \{8\}$$

then $n(A) = 20$, $n(B) = 2$, and $n(C) = 1$.

The empty set has no elements in it, so $n(\emptyset) = 0$. Another result that is easily seen to be true is the following: If A and B are disjoint sets, then

$$n(A \cup B) = n(A) + n(B) \tag{3}$$

EXAMPLE 1 If $A = \{a, c, d\}$ and $B = \{b, e, f, g\}$, then $n(A) = 3$ and $n(B) = 4$, so $n(A) + n(B) = 7$. However, $A \cup B = \{a, b, c, d, e, f, g\}$ and $n(A \cup B) = 7$. Thus, equation (3) holds true in this case. Note that $A \cap B = \emptyset$. ∎

In the general case, A and B need not be disjoint, which leads us to the formula

$$n(A \cup B) = n(A) + n(B) - n(A \cap B) \tag{4}$$

To see this, we observe that the set $A \cup B$ may be viewed as the union of three mutually disjoint sets with x, y, and z elements, respectively (Figure 8). This figure shows that

$$n(A \cup B) = x + y + z$$

FIGURE 8
$n(A \cup B) = x + y + z$

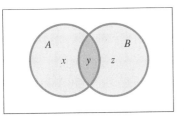

Also, $n(A) = x + y$

and $n(B) = y + z$

so $n(A) + n(B) = (x + y) + (y + z)$

$$= (x + y + z) + y$$

$$= n(A \cup B) + n(A \cap B) \qquad \small{n(A \cap B) = y}$$

Thus, solving for $n(A \cup B)$, we obtain

$$n(A \cup B) = n(A) + n(B) - n(A \cap B)$$

the desired result.

EXAMPLE 2 Let $A = \{a, b, c, d, e\}$ and $B = \{b, d, f, h\}$. Verify equation (4) directly.

Solution $A \cup B = \{a, b, c, d, e, f, h\}$ so $n(A \cup B) = 7$

$A \cap B = \{b, d\}$ so $n(A \cap B) = 2$

Furthermore, $n(A) = 5$ and $n(B) = 4$

so $n(A) + n(B) - n(A \cap B) = 5 + 4 - 2 = 7 = n(A \cup B)$

■

Applications

EXAMPLE 3 In a survey of 100 coffee drinkers it was found that 70 take sugar, 60 take cream, and 50 take both sugar and cream with their coffee. How many coffee drinkers take sugar or cream with their coffee?

Solution Let U denote the set of 100 coffee drinkers surveyed, and let

$$A = \{x \in U \mid x \text{ takes sugar}\}$$

$$B = \{x \in U \mid x \text{ takes cream}\}$$

Then $n(A) = 70, n(B) = 60$, and $n(A \cap B) = 50$. The set of coffee drinkers who take sugar or cream with their coffee is given by $A \cup B$. Using equation (4), we find

$$n(A \cup B) = n(A) + n(B) - n(A \cap B)$$

$$= 70 + 60 - 50 = 80$$

so 80 out of the 100 coffee drinkers surveyed take cream or sugar with their coffee.

■

An equation similar to (4) may be derived for the case that involves any finite number of finite sets. For example, a relationship involving the number of elements in the sets A, B, and C is given by

$$n(A \cup B \cup C) = n(A) + n(B) + n(C) - n(A \cap B)$$

$$- n(A \cap C) - n(B \cap C) + n(A \cap B \cap C) \qquad \textbf{(5)}$$

For a proof of this result, see exercise 32.

As useful as equations such as (5) are, in practice it is often easier to attack a problem directly with the aid of Venn diagrams, as shown by the following example.

EXAMPLE 4   A leading cosmetics manufacturer advertises its products in three magazines: *Cosmopolitan, McCalls,* and the *Ladies Home Journal.* A survey of 500 customers by the manufacturer reveals the following information:

180 learned of its products from *Cosmopolitan*

200 learned of its products from *McCalls*

192 learned of its products from the *Ladies Home Journal*

84 learned of its products from *Cosmopolitan* and *McCalls*

52 learned of its products from *Cosmopolitan* and the *Ladies Home Journal*

64 learned of its products from *McCalls* and the *Ladies Home Journal*

38 learned of its products from all three magazines

How many of the customers saw the manufacturer's advertisement in

a. at least one magazine?

b. exactly one magazine?

Solution   Let U denote the set of all customers surveyed, and let

$$C = \{x \in U \mid x \text{ learned of the products from } Cosmopolitan\}$$

$$M = \{x \in U \mid x \text{ learned of the products from } McCalls\}$$

$$L = \{x \in U \mid x \text{ learned of the products from the } Ladies\ Home\ Journal\}$$

The result that 38 customers learned of the products from all three magazines translates into $n(C \cap M \cap L) = 38$ (Figure 9a). Next, the result that 64 learned of the products from *McCalls* and the *Ladies Home Journal* translates into $n(M \cap L) = 64$. This leaves

$$64 - 38 = 26$$

who learned of the products from *only McCalls* and the *Ladies Home Journal* (Figure 9b). Similarly, $n(C \cap L) = 52$, so

$$52 - 38 = 14$$

learned of the products from *only Cosmopolitan* and the *Ladies Home Journal,* and $n(C \cap M) = 84$, so

$$84 - 38 = 46$$

FIGURE 9

(a) All three magazines

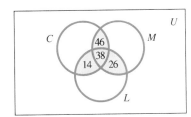

(b) Two or more magazines

learned of the products from *only Cosmopolitan* and *McCalls*. These numbers appear in the appropriate regions in Figure 9b.

Continuing, we have $n(L) = 192$, so the number who learned of the products from the *Ladies Home Journal* only is given by

$$192 - 14 - 38 - 26 = 114$$

(See Figure 10.) Similarly, $n(M) = 200$, so

$$200 - 46 - 38 - 26 = 90$$

learned of the products from *only McCalls,* and $n(C) = 180$, so

$$180 - 14 - 38 - 46 = 82$$

learned of the products from *only Cosmopolitan.* Finally,

$$500 - (90 + 26 + 114 + 14 + 82 + 46 + 38) = 90$$

learned of the products from other sources.

We are now in a position to answer questions (a) and (b).

a. Referring to Figure 10, we see that the number of customers who learned of the products from at least one magazine is given by

$$n(C \cup M \cup L) = 90 + 26 + 114 + 14 + 82 + 46 + 38 = 410$$

b. The number of customers who learned of the products from exactly one magazine (Figure 11) is given by

$$n(L \cap C^c \cap M^c) + n(M \cap C^c \cap L^c) + n(C \cap L^c \cap M^c)$$
$$= 90 + 114 + 82 = 286 \qquad \blacksquare$$

FIGURE 10
At least one magazine

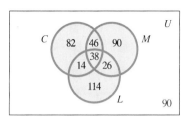

FIGURE 11
Exactly one magazine

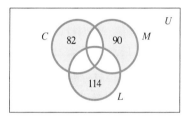

SELF-CHECK EXERCISES 7.2

1. Let A and B be subsets of a universal U and suppose $n(U) = 100$, $n(A) = 60$, $n(B) = 40$, and $n(A \cap B) = 20$. Compute

  a. $n(A \cup B)$ **b.** $n(A \cap B^c)$ **c.** $n(A^c \cap B)$

2. In a recent survey of 1000 readers of *Video Magazine* it was found that 900 own at least one videocassette recorder (VCR) in the VHS format, 240 own at least one VCR in the Beta format, and 160 own VCRs in both formats. How many of the readers surveyed own VCRs in the VHS format only? How many of the readers surveyed do not own a VCR in either format?

Solutions to Self-Check Exercises 7.2 can be found on page 348.

7.2   EXERCISES

In exercises 1 and 2, verify the equation

$$n(A \cup B) = n(A) + n(B)$$

for the given disjoint sets.

1. $A = \{a, e, i, o, u\}$ and $B = \{g, h, k, l, m\}$

2. $A = \{x \mid x \text{ is a whole number between 0 and 4}\}$
$B = \{x \mid x \text{ is a negative integer greater than } -4\}$

3. Let $A = \{2, 4, 6, 8\}$ and $B = \{6, 7, 8, 9, 10\}$. Compute
  a. $n(A)$ **b.** $n(B)$
  c. $n(A \cup B)$ **d.** $n(A \cap B)$

4. Verify directly that $n(A \cup B) = n(A) + n(B) - n(A \cap B)$ for the sets in exercise 3.

5. Let $A = \{a, e, i, o, u\}$ and $B = \{b, d, e, o, u\}$. Verify by direct computation that $n(A \cup B) = n(A) + n(B) - n(A \cap B)$.

6. If $n(A) = 15$, $n(A \cap B) = 5$, and $n(A \cup B) = 30$, what is $n(B)$?

7. If $n(A) = 10$, $n(A \cup B) = 15$, and $n(B) = 8$, what is $n(A \cap B)$?

In exercises 8 and 9, let A and B be subsets of a universal set U and suppose $n(U) = 200$, $n(A) = 100$, $n(B) = 80$, and $n(A \cap B) = 40$. Compute:

8. a. $n(A \cup B)$ **b.** $n(A^c)$
    c. $n(A \cap B^c)$

9. a. $n(A^c \cap B)$ **b.** $n(B^c)$
    c. $n(A^c \cap B^c)$

10. Find $n(A \cup B)$ given that $n(A) = 6$, $n(B) = 10$, and $n(A \cap B) = 3$.

11. If $n(B) = 6$, $n(A \cup B) = 14$, and $n(A \cap B) = 3$, find $n(A)$.

12. If $n(A) = 4$, $n(B) = 5$, and $n(A \cup B) = 9$, find $n(A \cap B)$.

13. If $n(A) = 16$, $n(B) = 16$, $n(C) = 14$, $n(A \cap B) = 6$, $n(A \cap C) = 5$, $n(B \cap C) = 6$, and $n(A \cup B \cup C) = 31$, find $n(A \cap B \cap C)$.

14. If $n(A) = 12$, $n(B) = 12$, $n(A \cap B) = 5$, $n(A \cap C) = 5$, $n(B \cap C) = 4$, $n(A \cap B \cap C) = 2$, and $n(A \cup B \cup C) = 25$, find $n(C)$.

15. A survey of 1000 subscribers to the *Los Angeles Times* revealed that 900 people subscribe to the daily morning edition and 500 subscribe to both the daily morning and the Sunday editions. How many subscribe to the Sunday edition? How many subscribe to the Sunday edition only?

16. Of 100 clock/radios sold recently in a department store, 70 had FM circuitry and 90 had AM circuitry. How many radios had both FM and AM circuitry? How many could receive FM transmission only? How many could receive AM transmission only?

17. On a certain day the Wilton County Jail had 190 prisoners. Of these, 130 were accused of felonies and 121 were accused of misdemeanors. How many prisoners were accused of both a felony and a misdemeanor?

18. Consumer Surveys   In a recent survey of 200 members of a local sports club, 100 members indicated

that they plan to attend the next Summer Olympic Games, 60 indicated that they plan to attend the next Winter Olympic Games, and 40 indicated that they plan to attend both games. How many members of the club plan to attend
a. at least one of the two games?
b. exactly one of the games?
c. the Summer Olympic Games only?
d. none of the games?

19. Consumer Surveys   In a survey of 120 consumers conducted in a shopping mall, 80 consumers indicated that they buy brand A of a certain product, 68 buy brand B, and 42 buy both brands. Determine the number of consumers participating in the survey who buy
a. at least one of these brands
b. exactly one of these brands
c. only brand A
d. none of these brands

20. Commuter Trends   Of 50 employees of a store located in downtown Boston, 18 people take the subway to work, 12 take the bus, and 7 take both the subway and the bus. Determine the number of employees who
a. take the subway or the bus to work
b. take only the bus to work
c. take either the bus or the subway to work
d. get to work by some other means

21. Investing   In a poll conducted among 200 active investors it was found that 120 use discount brokers, 126 use full-service brokers, and 64 use both discount and full-service brokers. Determine the number of investors who
a. use at least one kind of broker
b. use exactly one kind of broker
c. use only discount brokers
d. don't use a broker

In exercises 22–25, let A, B, and C be subsets of a universal set U and suppose $n(U) = 100$, $n(A) = 28$, $n(B) = 30$, $n(C) = 34$, $n(A \cap B) = 8$, $n(A \cap C) = 10$, $n(B \cap C) = 15$, and $n(A \cap B \cap C) = 5$. Compute:

22. a. $n(A \cup B \cup C)$ **b.** $n(A^c \cap B \cap C)$

23. a. $n[A \cap (B \cup C)]$ **b.** $n[A \cap (B \cup C)^c]$

24. a. $n(A^c \cap B^c \cap C^c)$ **b.** $n[A^c \cap (B \cup C)]$

25. a. $n[A \cup (B \cap C)]$ **b.** $n(A^c \cap B^c \cap C^c)^c$

26. Student Dropout Rate   Data released by the Department of Education regarding the rate (percentage) of ninth-grade students who don't graduate showed that out of 50 states,

12 states had an increase in the dropout rate during the past two years

15 states had a dropout rate of at least 30% during the past two years

21 states had an increase in the dropout rate and/or a dropout rate of at least 30% during the past two years

How many states
a. had both a dropout rate of at least 30% and an increase in the dropout rate over the 2-year period?
b. had a dropout rate that was less than 30% but that had increased over the 2-year period?

27. Economic Surveys   A survey of the opinions of 10 leading economists in a certain country showed that because oil prices were expected to drop,

7 had lowered their estimate of the consumer inflation rate

8 had raised their estimate of the gross national product growth rate

2 had lowered their estimate of the consumer inflation rate but had not raised their estimate of the gross national product growth rate

in that country over the next 12 months. How many economists had both lowered their estimate of the consumer inflation rate and raised their estimate of the gross national product growth rate for that period?

28. SAT Scores   Results of a Department of Education survey of SAT test scores in 22 states showed that

10 states had an average composite test score of at least 900

15 states had an increase of at least 10 points in the average composite score

8 states had both an average composite SAT score of at least 900 and an increase in the average composite score of at least 10 points

How many of the 22 states
a. had composite scores less than 900 and showed an increase of at least 10 points over the 3-year period?
b. had composite scores of at least 900 and did not show an increase of at least 10 points over the 3-year period?

29. Student Reading Habits   A survey of 100 college students who frequent the reading lounge of a university revealed the following results:

40 read *Time* magazine

30 read *Newsweek*

25 read *U.S. News and World Report*

15 read *Time magazine and Newsweek*

12 read *Time* magazine and *U.S. News and World Report*

10 read *Newsweek* and *U.S. News and World Report*

4 read all three magazines

How many of the students surveyed read
a. at least one magazine?
b. exactly one magazine?
c. exactly two magazines?
d. none of these magazines?

30. Student Surveys   To help plan the number of meals to be prepared in a college cafeteria, a survey was conducted and the following data were obtained:

130 students ate breakfast

180 students ate lunch

275 students ate dinner

68 students ate breakfast and lunch

112 students ate breakfast and dinner

90 students ate lunch and dinner

58 students ate all three meals

How many of the students
a. ate at least one meal in the cafeteria?
b. ate exactly one meal in the cafeteria?
c. ate only dinner in the cafeteria?
d. ate exactly two meals in the cafeteria?

31. Consumer Surveys   The 120 consumers of exercise 19 were also asked about their buying preferences concerning another product that is sold in the market under three labels. The results were:

12 buy only those sold under label A

25 buy only those sold under label B

26 buy only those sold under label C

15 buy only those sold under labels A and B

10 buy only those sold under labels A and C

12 buy only those sold under labels B and C

8 buy the product sold under all three labels

How many of the consumers surveyed buy the product sold under
a. at least one of the three labels?
b. labels A and B but not C?
c. label A?
d. none of these labels?

32. Derive equation (5). [*Hint:* Equation (4) may be written as $n(D \cup E) = n(D) + n(E) - n(D \cap E)$. Now, put $D = A \cup B$ and $E = C$. Use (4) again if necessary.]

**SOLUTIONS TO
SELF-CHECK
EXERCISES 7.2**

1. Refer to the following Venn diagram.

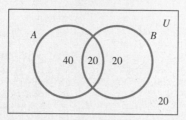

Using this result, we see that

a. $n(A \cup B) = 40 + 20 + 20 = 80$
b. $n(A \cap B^c) = 40$
c. $n(A^c \cap B) = 20$

2. Let U denote the set of all readers surveyed, and let

$$A = \{x \in U \mid x \text{ owns at least one VCR in the VHS format}\}$$

$$B = \{x \in U \mid x \text{ owns at least one VCR in the Beta format}\}$$

Then the result that 160 of the readers own VCRs in both formats gives $n(A \cap B) = 160$. Also, $n(A) = 900$ and $n(B) = 240$. Using this information, we obtain the following Venn diagram:

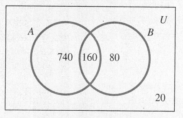

From the Venn diagram we see that the number of readers who own VCRs in the VHS format only is given by

$$n(A \cap B^c) = 740$$

The number of readers who do not own a VCR in either format is given by

$$n(A^c \cap B^c) = 20$$

7.3   The Multiplication Principle

The Fundamental Principle of Counting

The solution of certain problems requires more sophisticated counting techniques than those developed in the previous section. We will look at some such techniques in this and the following section. We begin by stating a fundamental principle of counting called the **multiplication principle.**

THE MULTIPLICATION PRINCIPLE

Suppose that there are *m* ways of performing a task T_1 and *n* ways of performing a task T_2. Then there are *mn* ways of performing the task T_1 followed by the task T_2.

EXAMPLE 1 There are three trunk roads connecting town A and town B and two trunk roads connecting town B and town C.

a. Use the multiplication principle to find the number of ways a journey from town A to town C via town B may be completed.

b. Verify (a) directly by exhibiting all possible routes.

Solution

a. Since there are three ways of performing the first task (going from town A to town B) followed by two ways of performing the second task (going from town B to town C), the multiplication principle says that there are $3 \cdot 2$, or 6, ways to complete a journey from town A to town C via town B.

b. Label the trunk roads connecting town A and town B with the roman numerals I, II, and III, and the trunk roads connecting town B and town C with the lowercase letters *a* and *b*. A schematic of this is shown in Figure 12.

FIGURE 12
Roads from A to C

Then the routes from town A to town C via town B may be exhibited with the aid of a **tree diagram** (Figure 13). If we follow all of the branches from the initial point A to the right-hand edge of the tree, we obtain

FIGURE 13

Tree diagram displaying the possible routes from town A to town C

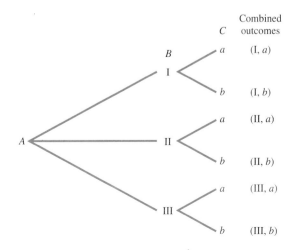

the six routes represented by the six ordered pairs

$$(\text{I}, a), \quad (\text{I}, b), \quad (\text{II}, a), \quad (\text{II}, b), \quad (\text{III}, a), \quad (\text{III}, b)$$

where (I, a) means that the journey from town A to town B is made on trunk road I with the rest of the journey from town B to town C to be completed on trunk road a, and so forth.   ■

EXAMPLE 2   Diners at Angelo's Spaghetti Bar may select their entree from 6 varieties of pasta and 28 choices of sauce. How many such combinations are there that consist of one variety of pasta and one kind of sauce?

Solution   There are 6 ways of choosing a pasta followed by 28 ways of choosing a sauce, so by the multiplication principle, there are $6 \cdot 28$, or 168, combinations of this pasta dish.   ■

The multiplication principle may be easily extended, which leads to the **generalized multiplication principle:**

GENERALIZED MULTIPLICATION PRINCIPLE

Suppose a task T_1 can be performed in N_1 ways, a task T_2 can be performed in N_2 ways, ..., and finally a task T_n can be performed in N_n ways. Then the number of ways of performing the tasks T_1, T_2, ..., T_n in succession is given by the product

$$N_1 N_2 \cdots N_n$$

We now illustrate the application of the generalized multiplication principle to several diverse situations.

EXAMPLE 3   A coin is tossed three times and the sequence of heads and tails is recorded.

a. Use the generalized multiplication principle to determine the number of outcomes of this activity.

b. Exhibit all the sequences by means of a tree diagram.

Solution

a. The coin may land in two ways. Therefore, in three tosses the number of outcomes (sequences) is given by $2 \cdot 2 \cdot 2$, or 8.

b. Let H and T denote the outcomes "a head" and "a tail," respectively. Then the required sequences may be obtained as shown in Figure 14,

FIGURE 14
Tree diagram displaying possible outcomes of three consecutive coin tosses

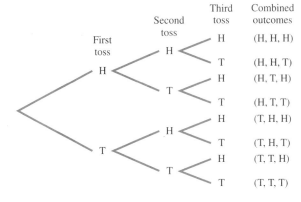

giving the sequence as HHH, HHT, HTH, HTT, THH, THT, TTH, and TTT.

Applications

EXAMPLE 4 A combination lock is unlocked by dialing a sequence of numbers, first to the left, then to the right, and to the left again. If there are ten digits on the dial, determine the number of possible combinations.

Solution There are ten choices for the first number, followed by ten for the second and ten for the third, so by the generalized multiplication principle, there are $10 \cdot 10 \cdot 10$, or 1000, possible combinations.

EXAMPLE 5 An investor has decided to purchase shares in the stocks of three companies: one engaged in aerospace activities, one involved in energy development, and an electronics company. After some research, the account executive of a brokerage firm has recommended that the investor consider stocks from five aerospace companies, three companies engaged in energy development, and four electronics companies. In how many ways may the investor select the group of three companies from the executive's list?

Solution The investor has five choices for selecting an aerospace company, three choices for selecting a company engaged in energy development, and four choices for selecting an electronics company. Therefore, by the generalized multiplication principle, there are $5 \cdot 3 \cdot 4$, or 60, ways in which she can select a group of three companies, one from each industry group.

EXAMPLE 6 Tom is planning to leave for New York City from Washington, D.C., on Monday morning and has decided that he will either fly or take the train. There are five flights and two trains departing for New York City from Washington on that morning. When he returns on Sunday afternoon, Tom plans to either fly or hitch a ride with a friend. There are two flights departing from New York City to Washington that afternoon. In how many ways can Tom complete this round trip?

Solution There are seven ways Tom can go from Washington, D.C., to New York City (five by plane and two by train). On the return trip, Tom can travel in three ways (two by plane and one by car). Therefore, by the multiplication principle, Tom can complete the round trip in (7)(3), or 21, ways.

SELF-CHECK EXERCISES 7.3

1. Encore Travel, Inc., offers a "Theater Week in London" package originating from New York City. There is a choice of eight flights departing from New York City per week, a choice of five hotel accommodations, and a choice of one complimentary ticket to one of eight shows. How many such travel packages can one choose from?

2. The Cafe Napolean offers a dinner special on Wednesdays consisting of a choice of two entrees (Beef Bourguignon and Chicken Basquaise); one dinner

salad; one French roll; a choice of three vegetables; a choice of a carafe of Burgundy, Rosé, or Chablis wine; a choice of coffee or tea; and a choice of six French pastries for dessert. How many combinations of dinner specials are there?

Solutions to Self-Check Exercises 7.3 can be found on page 353.

7.3 EXERCISES

1. **Rental Rates** Lynbrook West, an apartment complex financed by the State Housing Finance Agency, consists of one-, two-, three-, and four-bedroom units. The rental rate for each type of unit—low, moderate, or market—is determined by the income of the tenant. How many different rates are there?

2. **Commuter Passes** Five different types of monthly commuter passes are offered by a city's local transit authority for three different groups of passengers: youths, adults, and senior citizens. How many different kinds of passes must be printed each month?

3. **Blackjack** In the game of Blackjack, a two-card hand consisting of an ace and either a face card or a 10 is called a "blackjack." If a standard 52-card deck is used, determine how many blackjack hands can be dealt.

4. **Coin Tosses** A coin is tossed four times and the sequence of heads and tails is recorded.
   a. Use the generalized multiplication principle to determine the number of outcomes of this activity.
   b. Exhibit all the sequences by means of a tree diagram.

5. **Wardrobe Selection** A female executive selecting her wardrobe purchased two blazers, four blouses, and three skirts in coordinating colors. How many ensembles consisting of a blazer, a blouse, and a skirt can she create from this collection?

6. **Commuter Options** There are four commuter trains and three express buses departing from city A to city B in the morning and three commuter trains and three express buses operating on the return trip in the evening. In how many ways can a commuter from city A to city B complete a daily round trip via bus and/or train?

7. **Psychology Experiments** A psychologist has constructed the following maze for use in an experiment. The maze is constructed so that a rat must pass through a series of one-way doors. How many different paths are there from start to finish?

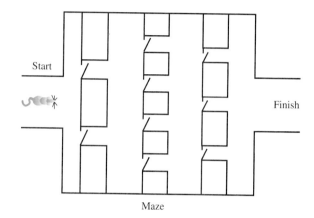

Maze

8. **Union Bargaining Issues** In a survey conducted by a union, members were asked to rate the importance of the following issues:

   Reduced working hours

   Increased fringe benefits

   Improved working conditions

   Five different responses were allowed for each issue. Among completed surveys, how many different responses to this survey were possible?

9. **Health Plan Options** A new state employee is offered a choice of ten basic health plans, three dental plans, and two vision care plans. How many different health care plans are there to choose from if one plan is selected from each category?

10. **Code Words** How many three-letter code words can be constructed from the first ten letters of the Greek alphabet if no repetitions are allowed?

11. **Computer Dating** A computer dating service uses the results of its compatibility survey for arranging dates. The survey consists of 50 questions, each having 5 possible answers. How many different responses are possible if every question is answered?

12. Serial Numbers   Computers manufactured by a certain company have a serial number consisting of a letter of the alphabet followed by a four-digit number. If all the serial numbers of this type have been used, how many sets have already been manufactured?

13. Menu Selections   Two soups, five entrees, and three desserts are listed on the "Special" menu at the Neptune Restaurant. How many different selections consisting of one soup, one entree, and one dessert can a customer choose from this menu?

14. Automobile Selection   An automobile manufacturer has three different subcompact cars in the line. Customers selecting one of these cars have a choice of three engine sizes, four body styles, and three color schemes. How many different selections can a customer make?

15. ATM Cards   To gain access to his account, a customer using an automatic teller machine (ATM) must enter a four-digit code. If repetition of the same four digits is not allowed (for example, 5555), how many possible combinations are there?

16. Television Viewing Polls   An opinion poll is to be conducted among cable TV viewers. Six multiple-choice questions, each with four possible answers, will be asked. In how many different ways can a viewer complete the poll if exactly one response is given to each question?

17. Political Polls   An opinion poll was conducted by the Morris Polling Group. Respondents were classified according to their sex (M or F), political affiliation (D, I, R), and the region of the country in which they reside (NW, W, C, S, E, NE).
  a. Use the generalized multiplication principle to determine the number of possible classifications.
  b. Construct a tree diagram to exhibit all possible classifications of females.

18. License Plate Numbers   A certain state uses license plates consisting of three letters of the alphabet followed by three digits. How many different license plate numbers can be formed?

19. Exams   An exam consists of ten true-or-false questions. Assuming that every question is answered, in how many different ways can a student complete the exam? In how many ways may the exam be completed if a penalty is imposed for each incorrect answer, so that a student may leave some questions unanswered?

20. Warranty Numbers   A warranty identification number for a certain product consists of a letter of the alphabet followed by a five-digit number. How many possible identification numbers are there if the first digit of the five-digit number must be nonzero?

21. Lotteries   In a state lottery there are 15 finalists eligible for the Big Money Draw. In how many ways can the first, second, and third prizes be awarded if no ticket holder may win more than one prize?

22. Telephone Numbers
  a. How many seven-digit telephone numbers are possible if the first digit must be nonzero?
  b. How many international direct-dialing numbers are possible if each number consists of a three-digit area code (the first digit of which must be nonzero) and a number of the type described in (a)?

23. Slot Machines   A "lucky dollar" is one of the nine symbols printed on each reel of a slot machine with three reels. A player receives one of various payouts whenever one or more "lucky dollars" appear in the window of the machine. Find the number of winning combinations for which the machine gives a payoff. [*Hint:* (a) Compute the number of ways in which the nine symbols on the first, second, and third wheels can appear in the window slot and (b) compute the number of ways in which the eight symbols other than the "lucky dollar" can appear in the window slot. The difference $(a - b)$ is the number of ways in which the "lucky dollar" can appear in the window slot. Why?]

**SOLUTIONS TO
SELF-CHECK
EXERCISES 7.3**

1. A tourist has a choice of eight flights, five hotel accommodations, and eight tickets. By the generalized multiplication principle, there are $8 \cdot 5 \cdot 8$, or 320, travel packages.

2. There is a choice of two entrees, one dinner salad, one French roll, a choice of three vegetables, a choice of three wines, a choice of two nonalcoholic beverages, and a choice of six pastries. Therefore, by the generalized multiplication principle, there are $2 \cdot 1 \cdot 1 \cdot 3 \cdot 3 \cdot 2 \cdot 6$, or 216, combinations of dinner specials.

7.4   Permutations and Combinations

Permutations

In this section, we apply the generalized multiplication principle to the solution of two types of counting problems. Both types involve the determination of the number of ways of arranging the elements of a set and both play an important role in the solution of problems in probability.

We begin by considering the permutations of a set. Specifically, given a set of distinct objects, a **permutation** of the set is an arrangement of these objects in a *definite order.* To see why the order in which objects are arranged is important in certain practical situations, suppose that the winning number for the first prize in a raffle is 9237. Then the number 2973, although it contains the same digits as the winning number, cannot be a first prize winner (Figure 15). Here the four objects—the numbers 9, 2, 3, and 7—are arranged in a different order; one arrangement is associated with the winning number for the first prize and the other is not.

FIGURE 15
The same digits appear on each ticket, but the order of the digits is different.

EXAMPLE 1   Let $A = \{a, b, c\}$.

a. Find the number of permutations of A.

b. List all the permutations of A with the aid of a tree diagram.

Solution

a. Each permutation of A consists of a sequence of the three letters a, b, c. Therefore we may think of such a sequence as being constructed by filling in each of the three blanks

$$\underline{\hspace{1.5cm}} \quad \underline{\hspace{1.5cm}} \quad \underline{\hspace{1.5cm}}$$

with one of the three letters. Now, there are three ways in which we may fill the first blank—we may choose a, b, or c. Having selected a letter for the first blank, there are two letters left for the second blank. Finally, there is but one way left to fill the third blank. Schematically, we have

$$\underline{\hspace{0.8cm}3\hspace{0.8cm}} \quad \underline{\hspace{0.8cm}2\hspace{0.8cm}} \quad \underline{\hspace{0.8cm}1\hspace{0.8cm}}$$

Invoking the generalized multiplication principle, we conclude that there are $3 \cdot 2 \cdot 1$, or 6, permutations of the set A.

b. The tree diagram associated with this problem appears in Figure 16, and the six permutations of A are abc, acb, bac, bca, cab, and cba.

FIGURE 16
Permutations of three objects

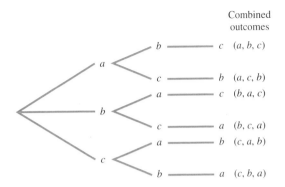

REMARK   Notice that order is taken into account when the possible outcomes are listed in the tree diagram in Example 1. Thus, (a, b, c) and (a, c, b) are two different arrangements.   □

EXAMPLE 2   Find the number of ways a baseball team consisting of nine people can arrange themselves in a line for a group picture.

Solution   We want to determine the number of permutations of the nine members of the baseball team. Each permutation in this situation consists of an arrangement of the nine team members in a line. The nine positions can be represented by nine blanks. Thus

Position   1   2   3   4   5   6   7   8   9

There are nine ways to choose from among the nine players to fill the first position. Next there are eight players left, which gives us eight ways to fill the second position. Proceeding in a similar manner, we find that there are seven ways to fill the third position, and so on. Schematically, we have

Number of ways to   9   8   7   6   5   4   3   2   1
fill each position

Invoking the generalized multiplication principle, we conclude that there are $9 \cdot 8 \cdot 7 \cdot 6 \cdot 5 \cdot 4 \cdot 3 \cdot 2 \cdot 1$, or 362,880, ways the baseball team can be arranged for the picture.   ■

  Whenever we are asked to determine the number of ways the objects of a set can be arranged in a line, order is important. For example, if we take a picture of two baseball players, A and B, then the two players can line up for the picture in two ways, AB or BA, and the two pictures will be different.

Pursuing the same line of argument used in solving the problems in the last two examples, we can derive an expression for the number of ways of permuting a set A of n distinct objects taken n at a time. In fact, each permutation may be viewed as being obtained by filling each of n blanks with one and only one element from the set. There are n ways of filling the first blank, followed by $(n - 1)$ ways of filling the second blank, and so on, so by the generalized multiplication principle there are

$$n(n - 1)(n - 2) \cdots 3 \cdot 2 \cdot 1$$

ways of permuting the elements of the set A.

Before stating this result formally, let us introduce a notation that will enable us to write in a compact form many of the expressions that follow. We use the symbol $n!$, read "*n*-factorial," to denote the product of the first n natural numbers.

***n*-FACTORIAL**

For any natural number n

$$n! = n(n-1)(n-2)\cdots \cdot 3 \cdot 2 \cdot 1$$

and $0! = 1$

For example,

$$1! = 1$$
$$2! = 2 \cdot 1 = 2$$
$$3! = 3 \cdot 2 \cdot 1 = 6$$
$$4! = 4 \cdot 3 \cdot 2 \cdot 1 = 24$$
$$5! = 5 \cdot 4 \cdot 3 \cdot 2 \cdot 1 = 120$$
$$\vdots$$

and $10! = 10 \cdot 9 \cdot 8 \cdot 7 \cdot 6 \cdot 5 \cdot 4 \cdot 3 \cdot 2 \cdot 1 = 3,628,800$

Using this notation, we may express *the number of permutations of n distinct objects taken n at a time, P(n, n), as*

$$P(n, n) = n!$$

In many situations we are interested in determining the number of ways of permuting n distinct objects taken r at a time, where $r \leq n$. To derive a formula for computing the number of ways of permuting a set consisting of n distinct objects taken r at a time, we observe that each such permutation may be viewed as being obtained by filling each of r blanks with precisely one element from the set. Now there are n ways of filling the first blank, followed by $(n-1)$ ways of filling the second blank, and so on. Finally, there are $(n - r + 1)$ ways of filling the rth blank. We may represent this argument schematically:

Number of ways $\underset{1\text{st}}{n}$ $\underset{2\text{nd}}{n-1}$ $\underset{3\text{rd}}{n-2}$ $\cdots$ $\underset{r\text{th}}{n-r+1}$
Position

Using the generalized multiplication principle, we conclude that *the number of ways of permuting n distinct objects taken r at a time, P(n, r), is given by*

$$P(n, r) = \underbrace{n(n-1)(n-2)\cdots(n-r+1)}_{r \text{ factors}}$$

Since

$$n(n-1)(n-2) \cdots (n-r+1)$$

$$= [n(n-1)(n-2) \cdots (n-r+1)] \cdot \frac{[(n-r)(n-r-1) \cdots \cdot 3 \cdot 2 \cdot 1]}{[(n-r)(n-r-1) \cdots \cdot 3 \cdot 2 \cdot 1]}$$

↑
Here we are multiplying by 1.

$$= \frac{[n(n-1)(n-2) \cdots (n-r+1)][(n-r)(n-r-1) \cdots \cdot 3 \cdot 2 \cdot 1]}{(n-r)(n-r-1) \cdots \cdot 3 \cdot 2 \cdot 1}$$

$$= \frac{n!}{(n-r)!}$$

we have the following formula:

**PERMUTATIONS OF
n DISTINCT OBJECTS**

The number of *permutations* of n distinct objects taken r at a time is

$$P(n,r) = \frac{n!}{(n-r)!} \tag{6}$$

REMARK   When $r = n$, (6) reduces to

$$P(n,n) = \frac{n!}{0!} = \frac{n!}{1} = n! \qquad \text{Note that } 0! = 1.$$

In other words, the number of permutations of a set of n distinct objects, taken all together, is $n!$. □

EXAMPLE 3   Compute (a) $P(4,4)$ and (b) $P(4,2)$ and interpret your results.

Solution

a. $P(4,4) = \dfrac{4!}{(4-4)!} = \dfrac{4!}{0!} = \dfrac{4!}{1} = \dfrac{4 \cdot 3 \cdot 2 \cdot 1}{1} = 24$ Recall that $0! = 1$.

This gives the number of permutations of 4 objects taken 4 at a time.

b. $P(4,2) = \dfrac{4!}{(4-2)!} = \dfrac{4!}{2!} = \dfrac{4 \cdot 3 \cdot 2 \cdot 1}{2 \cdot 1} = 12$

and this is the number of permutations of 4 objects taken 2 at a time.

EXAMPLE 4   Let $A = \{a, b, c, d\}$.

a. Use equation (6) to compute the number of permutations of the set A taken two at a time.

b. Display the permutations of (a) with the aid of a tree diagram.

Solution

a. Here $n = 4$ and $r = 2$, so the required number of permutations is given by

$$P(4, 2) = \frac{4!}{(4 - 2)!} = \frac{4!}{2!} = \frac{4 \cdot 3 \cdot 2 \cdot 1}{2 \cdot 1} = 4 \cdot 3$$

or 12.

b. The tree diagram associated with the problem is shown in Figure 17, and the permutations of A taken two at a time are

$$ab, \quad ac, \quad ad, \quad ba, \quad bc, \quad bd, \quad ca, \quad cb, \quad cd, \quad da, \quad db, \quad dc$$

FIGURE 17
Permutations of four objects taken two at a time

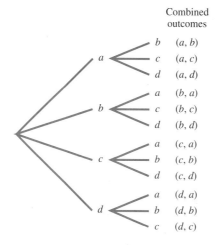

Combined outcomes

a — b (a, b)
a — c (a, c)
a — d (a, d)
b — a (b, a)
b — c (b, c)
b — d (b, d)
c — a (c, a)
c — b (c, b)
c — d (c, d)
d — a (d, a)
d — b (d, b)
d — c (d, c)

EXAMPLE 5 Find the number of ways a chairman, a vice-chairman, a secretary, and a treasurer can be chosen from a committee of eight members.

Solution The problem is equivalent to finding the number of permutations of eight distinct objects taken four at a time. Therefore, the number of ways of choosing the four officials from the committee of eight members is given by

$$P(8, 4) = \frac{8!}{(8 - 4)!} = \frac{8!}{4!} = 8 \cdot 7 \cdot 6 \cdot 5 = 1680$$

or 1680 ways.

The permutations considered thus far have been those involving sets of *distinct* objects. In many situations we are interested in finding the number of permutations of a set of objects in which not all of the objects are distinct.

**PERMUTATIONS OF
n OBJECTS, NOT ALL
DISTINCT**

Given a set of n objects in which n_1 objects are alike and of one kind, n_2 objects are alike and of another kind, . . . , and, finally, n_r objects are alike and of yet another kind so that

$$n_1 + n_2 + \cdots + n_r = n$$

then the number of permutations of these n objects taken n at a time is given by

$$\frac{n!}{n_1!n_2!\cdots n_r!} \tag{7}$$

To establish (7), let us denote the number of such permutations by x. Now, if we *think* of the n_1 objects as being distinct, then they may be permuted in $n_1!$ ways. Similarly, if we *think* of the n_2 objects as being distinct, then they may be permuted in $n_2!$ ways, and so on. Therefore, if we *think* of the n objects as being distinct, then, by the generalized multiplication principle, there are $x \cdot n_1! \cdot n_2! \cdot \cdots \cdot n_r!$ permutations of these objects. But, the number of permutations of a set of n distinct objects taken n at a time is just equal to $n!$. Therefore, we have

$$x(n_1! \cdot n_2! \cdot \cdots \cdot n_r!) = n!$$

from which we deduce that

$$x = \frac{n!}{n_1!n_2!\cdots n_r!}$$

EXAMPLE 6   Find the number of permutations that can be formed from all the letters in the word *ATLANTA*.

Solution   There are seven objects (letters) involved, so $n = 7$. However, three of them are alike and of one kind (the three A's), and two of them are alike and of another kind (the two T's), so that in this case, $n_1 = 3$, $n_2 = 2$, $n_3 = 1$, and $n_4 = 1$. Therefore, using equation (7), the number of required permutations is given by

$$\frac{7!}{3!2!1!1!} = \frac{7 \cdot 6 \cdot 5 \cdot 4 \cdot 3 \cdot 2 \cdot 1}{3 \cdot 2 \cdot 1 \cdot 2 \cdot 1}$$

or 420.

EXAMPLE 7   Weaver and Kline, a stock brokerage firm, has received nine inquiries regarding new accounts. In how many ways can these inquiries be directed to three of the firm's account executives if each account executive is to handle three inquiries?

Solution   If we think of the nine inquiries as being slots arranged in a row with inquiry 1 on the left and inquiry 9 on the right, then the problem can be thought of as one of filling each slot with a business card from an account executive. Then nine business cards would be used, of which three are alike and of one kind, three are alike and of another kind, and three are alike and of yet another kind. Thus, using (7) with $n = 9$, $n_1 = n_2 = n_3 = 3$, we see that the number of ways of assigning the inquiries is given by

$$\frac{9!}{3!3!3!} = \frac{9 \cdot 8 \cdot 7 \cdot 6 \cdot 5 \cdot 4 \cdot 3 \cdot 2 \cdot 1}{3 \cdot 2 \cdot 1 \cdot 3 \cdot 2 \cdot 1 \cdot 3 \cdot 2 \cdot 1}$$

or 1680 ways.

Combinations

Up to now, we have dealt with permutations of a set; that is, we have dealt with arrangements of the objects of a set in which the *order* of the elements is taken into consideration. In many situations we are interested in determining the number of ways of selecting r objects from a set of n objects without any regard to the order in which the objects are selected. Such a subset is called a **combination.**

For example, if we are interested in knowing the number of 5-card poker hands that can be dealt from a standard deck of 52 cards, then the order in which the poker hand is dealt is unimportant (Figure 18). In this situation, we are interested in determining the number of combinations of 5 cards (objects) selected from a deck (set) of 52 cards (objects). (We will solve this problem later.)

FIGURE 18

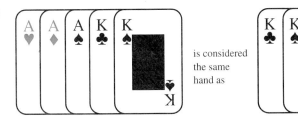

is considered the same hand as

To derive a formula for determining the number of combinations of n objects taken r at a time, written

$$C(n, r) \quad \text{or} \quad \binom{n}{r}$$

we observe that each of the $C(n, r)$ combinations of r objects can be permuted in $r!$ ways (Figure 19).

FIGURE 19

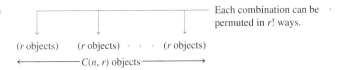

Each combination can be permuted in $r!$ ways.

(r objects)   (r objects) $\cdot$ $\cdot$ $\cdot$ (r objects)

$\longleftarrow$———$C(n, r)$ objects———$\longrightarrow$

Thus, by the multiplication principle, the product $r!\,C(n, r)$ gives the number of permutations of n objects taken r at a time; that is,

$$r!\,C(n, r) = P(n, r)$$

from which we find

$$C(n, r) = \frac{P(n, r)}{r!}$$

or, using (6),

$$C(n, r) = \frac{n!}{r!(n - r)!}$$

COMBINATIONS OF n OBJECTS

The number of *combinations* of n distinct objects taken r at a time is given by

$$C(n, r) = \frac{n!}{r!(n - r)!} \qquad \text{where } r \le n \qquad \textbf{(8)}$$

EXAMPLE 8   Compute (a) $C(4, 4)$ and (b) $C(4, 2)$ and interpret the results.

Solution

a. $C(4, 4) = \dfrac{4!}{4!(4 - 4)!} = \dfrac{4!}{4!0!} = 1$ Recall that 0! = 1.

This gives 1 as the number of combinations of 4 distinct objects taken 4 at a time.

b. $C(4, 2) = \dfrac{4!}{2!(4 - 2)!} = \dfrac{4!}{2!2!} = \dfrac{4 \cdot 3}{2} = 6$

This gives 6 as the number of combinations of 4 distinct objects taken 2 at a time.

Applications

EXAMPLE 9   A Senate investigation subcommittee of four members is to be selected from a Senate committee of ten members. Determine the number of ways this can be done.

Solution   Since the order in which the members of the subcommittee are selected is unimportant, the number of ways of choosing the subcommittee is given by $C(10, 4)$, the number of combinations of ten objects taken four at a time. But

$$C(10, 4) = \frac{10!}{4!(10 - 4)!} = \frac{10!}{4!6!} = \frac{10 \cdot 9 \cdot 8 \cdot 7}{4 \cdot 3 \cdot 2 \cdot 1} = 210$$

so there are 210 ways of choosing such a subcommittee.

REMARK   Remember, a combination is a selection of objects *without* regard to order. Thus, in Example 9, we used a combination formula rather than a permutation formula to solve the problem, because the order of selection was not important; that is, it did not matter whether a member of the subcommittee was selected first, second, third, or fourth.   ☐

EXAMPLE 10   How many poker hands of 5 cards can be dealt from a standard deck of 52 cards?

Solution   The order in which the 5 cards are dealt is not important. The number of ways of dealing a poker hand of 5 cards from a standard deck of 52 cards is given by $C(52, 5)$, the number of combinations of 52 objects taken 5 at a time. Now,

$$C(52, 5) = \frac{52!}{5!(52 - 5)!} = \frac{52!}{5!47!}$$

$$= \frac{52 \cdot 51 \cdot 50 \cdot 49 \cdot 48}{5 \cdot 4 \cdot 3 \cdot 2 \cdot 1}$$

$$= 2{,}598{,}960$$

so there are 2,598,960 ways of dealing such a poker hand. ■

The next several examples show that solving a counting problem often involves the repeated application of equation (6) and/or (8), possibly in conjunction with the multiplication principle.

EXAMPLE 11   The members of a string quartet comprising two violinists, a violist, and a cellist are to be selected from a group of six violinists, three violists, and two cellists, respectively.

a. In how many ways can the string quartet be formed?

b. In how many ways can the string quartet be formed if one of the violinists is to be designated as the first violinist and the other is to be designated as the second violinist?

Solution

a. Since the order in which each musician is selected is not important, we use combinations. The violinists may be selected in $C(6, 2)$, or 15, ways; the violist may be selected in $C(3, 1)$, or 3, ways; and the cellist may be selected in $C(2, 1)$, or 2, ways. By the multiplication principle, there are $15 \cdot 3 \cdot 2$, or 90, ways of forming the string quartet.

b. The order in which the violinists are selected is important here. Consequently, the number of ways of selecting the violinists is given by $P(6, 2)$, or 30, ways. The number of ways of selecting the violist and the cellist are, of course, 3 and 2, respectively. Therefore, the number of ways in which the string quartet may be formed is given by $30 \cdot 3 \cdot 2$, or 180, ways. ■

REMARK   The solution of Example 11 involves both a permutation and a combination. When we select two violinists from six violinists, order is not important, and we use a combination formula to solve the problem. However, when one of the violinists is designated as a first violinist, order is important, and we use a permutation formula to solve the problem. □

EXAMPLE 12   Refer to Example 5, Section 7.3. Suppose that the investor has decided to purchase shares in the stocks of two aerospace companies, two companies involved in energy development, and two electronics companies. In how many ways may the investor select the group of six companies for the investment from the recommended list of five aerospace companies, three companies engaged in energy development, and four electronics companies?

Solution   There are $C(5, 2)$ ways in which the investor may select the aerospace companies, $C(3, 2)$ ways in which she may select the companies involved in energy development, and $C(4, 2)$ ways in which she may select the electronics companies as investments. By the generalized multiplication principle, there are

$$C(5, 2)C(3, 2)C(4, 2) = \frac{5!}{2!3!} \cdot \frac{3!}{2!1!} \cdot \frac{4!}{2!2!}$$

$$= \frac{5 \cdot 4}{2} \cdot 3 \cdot \frac{4 \cdot 3}{2} = 180$$

ways of selecting the group of six companies for her investment. ■

EXAMPLE 13 The Futurists, a rock group, are planning a concert tour with performances to be given in five cities: San Francisco, Los Angeles, San Diego, Denver, and Las Vegas. In how many ways can they arrange their itinerary

a. if there are no restrictions?

b. if the three performances in California must be given consecutively?

Solution

a. The order is important here, and we see that there are

$$P(5, 5) = 5! = 120$$

ways of arranging their itinerary.

b. First, note that there are $P(3, 3)$ ways of choosing between performing in California and in the two cities outside that state. Next, there are $P(3, 3)$ ways of arranging their itinerary in the three cities in California. Therefore, by the multiplication principle, there are

$$P(3, 3) \cdot P(3, 3) = \frac{3!}{(3 - 3)!} \cdot \frac{3!}{(3 - 3)!} = (6)(6)$$

or 36, ways of arranging their itinerary. ■

EXAMPLE 14 The U.N. Security Council consists of 5 permanent members and 10 nonpermanent members. Decisions made by the council require 9 votes for passage. However, any permanent member may veto a measure and thus block its passage. In how many ways can a measure be passed if all 15 members of the Council vote (no abstentions)?

Solution If a measure is to be passed, then all 5 permanent members must vote for passage of that measure. This can be done in $C(5, 5)$, or 1, way.

Next, observe that since 9 votes are required for passage of a measure, *at least* 4 of the 10 nonpermanent members must also vote for its passage. To determine the number of ways this can be done, notice that there are $C(10, 4)$ ways in which exactly 4 of the nonpermanent members could vote for passage of a measure, $C(10, 5)$ ways in which exactly 5 of them could vote for passage of a measure, and so on. Finally, there are $C(10, 10)$ ways in which all 10 nonpermanent members could vote for passage of a measure. Therefore, there are

$$C(10, 4) + C(10, 5) + \cdots + C(10, 10)$$

ways in which at least 4 of the 10 nonpermanent members could vote for a measure. So, by the multiplication principle, the number of ways a measure can be passed is

$$C(5, 5) \cdot [C(10, 4) + C(10, 5) + \cdots + C(10, 10)]$$

$$= (1) \left[\frac{10!}{4!6!} + \frac{10!}{5!5!} + \cdots + \frac{10!}{10!0!} \right]$$

$$= (1)(210 + 252 + 210 + 120 + 45 + 10 + 1) = 848$$

or 848 ways.

SELF-CHECK EXERCISES 7.4

1. Evaluate **a.** 5!   **b.** $C(7, 4)$ **c.** $P(6, 2)$

2. A space shuttle crew consists of a shuttle commander, a pilot, 3 engineers, a scientist, and a civilian. The shuttle commander and pilot are to be chosen from 8 candidates, the 3 engineers from 12 candidates, the scientist from 5 candidates, and the civilian from 2 candidates. How many such space shuttle crews can be formed?

Solutions to Self-Check Exercises 7.4 can be found on page 368.

7.4   EXERCISES

In exercises 1–22, evaluate the given expression.

1. 3(5!)

2. 2(7!)

3. $\dfrac{5!}{2!3!}$

4. $\dfrac{6!}{4!2!}$

5. $P(5, 5)$

6. $P(6, 6)$

7. $P(5, 2)$

8. $P(5, 3)$

9. $P(n, 1)$

10. $P(k, 2)$

11. $C(6, 6)$

12. $C(8, 8)$

13. $C(7, 4)$

14. $C(9, 3)$

15. $C(5, 0)$

16. $C(6, 5)$

17. $C(9, 6)$

18. $C(10, 3)$

19. $C(n, 2)$

20. $C(7, r)$

21. $P(n, n - 2)$

22. $C(n, n - 2)$

In exercises 23–30, classify each problem according to whether it involves a permutation or a combination.

23. In how many ways can the letters of the word *glacier* be arranged?

24. In the eighth-grade dance class, there are 10 girls and 14 boys. In how many ways can these students be paired off to form dance couples consisting of one boy and one girl?

25. As part of a quality-control program, 3 record-o-phones are selected at random for testing from each 100 phones produced by the manufacturer. In how many ways can this test batch be chosen?

26. How many three-digit numbers can be formed using the numerals in the set {3, 2, 7, 9} if repetition is not allowed?

27. In how many ways can nine different books be arranged on a shelf?

28. A member of a book club wishes to purchase two books from a selection of eight books recommended for a certain month. In how many ways can she choose them?

29. How many five-card poker hands can be dealt consisting of three queens and a pair?

30. Four couples are to be seated at an oval dining table. If each couple must be seated together, how many possible seating arrangements are there?

31. How many four-letter permutations can be formed from the first four letters of the alphabet?

32. How many three-letter permutations can be formed from the first five letters of the alphabet?

33. In how many ways can four students be seated in a row of four seats?

34. In how many ways can five people line up at a checkout counter in a supermarket?

35. How many different batting orders can be formed for a nine-member baseball team?

36. In how many ways can the names of six candidates for political office be listed on a ballot?

37. In how many ways can a member of a hiring committee select 3 of 12 job applicants for further consideration?

38. In how many ways can an investor select four mutual funds for his investment portfolio from a recommended list of eight mutual funds?

39. Find the number of distinguishable permutations that can be formed from the letters of the word *ANTARCTICA*.

40. Find the number of distinguishable permutations that can be formed from the letters of the word *PHILIPPINES*.

41. Management Decisions In how many ways can a supermarket chain select 3 out of 12 possible sites for the construction of new supermarkets?

42. Book Selections A student is given a reading list of ten books from which he must select two for an outside reading requirement. In how many ways can he make his selections?

43. Quality Control In how many ways can a quality-control engineer select a sample of 3 transistors for testing from a batch of 100 transistors?

44. Study Groups A group of five students studying for a bar exam have formed a study group. Each member of the group will be responsible for preparing a study outline for one of five courses. In how many different ways can the five courses be assigned to the members of the group?

45. Television Programming In how many ways can a television programming director schedule six different commercials in the six time slots allocated to commercials during a 1-hour program?

46. Waiting Lines Seven people arrive at the ticket counter of a cinema at the same time. In how many ways can they line up to purchase their tickets?

47. Management Decisions Weaver and Kline, a stock brokerage firm, has received 6 inquiries regarding new accounts. In how many ways can these inquiries

be directed to its 12 account executives if each executive handles no more than 1 inquiry?

48. Car Pools A company car that has a seating capacity of six is to be used by six employees who have formed a car pool. If only four of these employees can drive, how many possible seating arrangements are there for the group?

49. Book Displays At a college library exhibition of faculty publications, three mathematics books, four social science books, and three biology books will be displayed on a shelf. (Assume that none of the books are alike.)
 a. In how many ways can the ten books be arranged on the shelf?
 b. In how many ways can the ten books be arranged on the shelf if books on the same subject matter are placed together?

50. Seating In how many ways can four married couples attending a concert be seated in a row of eight seats
 a. if there are no restrictions?
 b. if each married couple is seated together?
 c. if the members of each sex are seated together?

Advertisement

1	2	3	4
5	6	7	8
9	10	11	12
13	14	15	16
17	18	19	20

51. Newspaper Advertisements Four items from 5 different departments of the Metro Department Store will be featured in a one-page newspaper advertisement as shown in the following diagram:
 a. In how many different ways can the 20 featured items be arranged on the page?
 b. If items from the same department must be in the same row, how many arrangements are possible?

52. Management Decisions The C & J Realty Company has received 12 inquiries from prospective home buyers. In how many ways can the inquiries be directed to 4 of the firm's real estate agents if each agent handles 3 inquiries?

53. Sports   In the women's tennis tournament at Wimbledon, two finalists, A and B, are competing for the title, which will be awarded to the first player to win two sets. In how many different ways can the match be completed?

54. Sports   In the men's tennis tournament at Wimbledon, two finalists, A and B, are competing for the title, which will be awarded to the first player to win three sets. In how many different ways can the match be completed?

55. U.N. Voting   Refer to Example 14. In how many ways can a measure be passed if two particular permanent and two particular nonpermanent members of the Council abstain from voting?

56. Jury Selection   In how many different ways can a panel of 12 jurors and 2 alternate jurors be chosen from a group of 30 prospective jurors?

57. Teaching Assistantships   Twelve graduate students have applied for three available teaching assistantships. In how many ways can the assistantships be awarded among these applicants
   a. if no preference is given to any student?
   b. if one particular student must be awarded an assistantship?
   c. if the group of applicants includes seven men and five women and it is stipulated that at least one woman must be awarded an assistantship?

58. Exams   A student taking an examination is required to answer 10 out of 15 questions.
   a. In how many ways can the 10 questions be selected?
   b. In how many ways can the 10 questions be selected if exactly 2 of the first 3 questions must be answered?

59. Contract Bidding   The U.B.S. Television Company is considering bids submitted by seven different firms for three different contracts. In how many ways can the contracts be awarded among these firms if no firm is to receive more than two contracts?

60. Senate Committees   In how many ways can a subcommittee of four be chosen from a Senate committee of five Democrats and four Republicans
   a. if all members are eligible?
   b. if the subcommittee must consist of two Republicans and two Democrats?

61. Course Selection   A student planning her curriculum for the upcoming year must select one of five business courses, one of three mathematics courses, two of six elective courses, and either one of four history courses or one of three social science courses. How many different curricula are available for her consideration?

C **62.** Personnel Selection   The J.C.L. Computer Company has five vacancies in its executive trainee program. In how many ways can the company select five trainees from a group of ten female and ten male applicants
   a. if the vacancies may be filled by any combination of men and women?
   b. if the vacancies must be filled by two men and three women?

63. Drivers' Tests   The State Motor Vehicle Department requires learners to pass a written test on the motor vehicle laws of the state. The exam consists of ten true-or-false questions, of which eight must be answered correctly to qualify for a permit. In how many different ways can a learner who answers all the questions on the exam qualify for a permit?

C **64.** Quality Control   The Goodman Tire Company has 32 tires of a particular size and grade in stock, 2 of which are defective. If a set of 4 tires is to be selected,
   a. how many different selections can be made?
   b. how many different selections can be made that do not include any defective tires?

A list of poker hands ranked in order from the highest to the lowest is shown in the following table, along with a description and example of each hand. Use the table to answer exercises 65–70.

Hand	Description	Example
Straight flush	5 cards in sequence in the same suit	A ♥ 2 ♥ 3 ♥ 4 ♥ 5 ♥
Four of a kind	4 cards of the same rank and any other card	K ♥ K ♦ K ♠ K ♣ 2 ♥
Full house	3 of a kind and a pair	3 ♥ 3 ♦ 3 ♣ 7 ♥ 7 ♦
Flush	5 cards of the same suit that are not all in sequence	5 ♥ 6 ♥ 9 ♥ J ♥ K ♥
Straight	5 cards in sequence but not all of the same suit	10 ♥ J ♦ Q ♣ K ♠ A ♥

Hand	Description	Example
Three of a kind	3 cards of the same rank and 2 unmatched cards	K ♥ K ♦ K ♠ 2 ♥ 4 ♦
Two pair	2 cards of the same rank and 2 cards of any other rank with an unmatched card	K ♥ K ♦ 2 ♥ 2 ♠ 4 ♣
One pair	2 cards of the same rank and 3 unmatched cards	K ♥ K ♦ 5 ♥ 2 ♠ 4 ♥

If a 5-card poker hand is dealt from a well-shuffled deck of 52 cards, how many different hands consist of the following.

65. A straight flush? (Note that an ace may be played as either a high or a low card in a straight sequence—that is, A, 2, 3, 4, 5, or 10, J, Q, K, A. Hence, there are 10 possible sequences for a straight in one suit.)

66. A straight (but not a straight flush)?

c **67.** A flush (but not a straight flush)?

68. Four of a kind?

69. A full house?

c **70.** Two pairs?

71. **Bus Routing** The following is a schematic diagram of a city's street system between the points A and B. The City Transit Authority is in the process of selecting a route from A to B along which to provide bus service. If the company's intention is to keep the route as short as possible, how many routes must be considered?

Street system

72. **Sports** In the World Series, one National League team and one American League team compete for the coveted title that is awarded to the first team to win four games. In how many different ways can the series of games be completed?

73. **Voting Quorums** A quorum (minimum) of 6 voting members is required at all meetings of the Curtis Townhomes Owners Association. If there are a total of 12 voting members in the group, find the number of ways this quorum can be formed.

74. **Circular Permutations** Suppose n distinct objects are arranged in a circle. Show that the number of (different) circular arrangements of the n objects is $(n - 1)!$ [*Hint:* Consider the arrangements of the 5 letters A, B, C, D, and E as shown in the accompanying figure. The permutations $ABCDE$, $BCDEA$, $CDEAB$, $DEABC$, and $EABCD$ are not distinguishable. Generalize this observation to the case of n objects.]

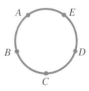

75. Refer to exercise 74. In how many ways can five TV commentators be seated at a round table for a discussion?

76. Refer to exercise 74. In how many ways can four men and four women be seated at a round table at a dinner party if each guest is seated next to members of the opposite sex?

77. At the end of Section 6.3, we mentioned that a linear programming problem in 3 variables and 8 constraints may have up to 56 feasible corner points and that the determination of these feasible corner points calls for the solution of 56 3×3 systems of linear equations. Prove this assertion.

78. Refer to exercise 77. Show that a linear programming problem in 5 variables and 15 constraints may have up to 3003 feasible corner points. The assertion was also made at the end of Section 6.3.

SOLUTIONS TO SELF-CHECK EXERCISES 7.4

1. **a.** $5! = 5 \cdot 4 \cdot 3 \cdot 2 \cdot 1 = 120$

   b. $C(7, 4) = \dfrac{7!}{4!3!} = \dfrac{7 \cdot 6 \cdot 5}{3 \cdot 2 \cdot 1} = 35$

   c. $P(6, 2) = \dfrac{6!}{4!} = 6 \cdot 5 = 30$

2. There are $P(8, 2)$ ways of picking the shuttle commander and pilot (the order *is* important here), $C(12, 3)$ ways of picking the engineers (the order is not important here), $C(5, 1)$ ways of picking the scientist, and $C(2, 1)$ ways of picking the civilian. By the multiplication principle, there are

$$P(8, 2) \cdot C(12, 3) \cdot C(5, 1) \cdot C(2, 1)$$

$$= \frac{8!}{6!} \cdot \frac{12!}{9!3!} \cdot \frac{5!}{4!1!} \cdot \frac{2!}{1!1!}$$

$$= \frac{(8)(7)(12)(11)(10)(5)(2)}{(3)(2)}$$

$$= 123{,}200$$

or 123,200 ways a crew could be selected.

Using Technology

The graphing utility can be used to calculate factorials, permutations, and combinations with relative ease. The graphing utility is, therefore, an indispensable tool in solving counting problems involving large numbers of objects. (See Section A.8 in Appendix A.)

EXAMPLE 1   Use a graphing utility to find (a) 12!, (b) $P(52, 5)$, (c) $C(38, 10)$.

Solution

a. Using the factorial function, we find that $12! = 479{,}001{,}600$.

b. Using the nPr function, we have

$$P(52, 5) = 52 \text{ nPr } 5 = 311{,}875{,}200$$

c. Using the nCr function, we obtain

$$C(38, 10) = 38 \text{ nCr } 10 = 472{,}733{,}756$$

EXERCISES

In exercises 1–10, evaluate the expression.

1. $15!$

2. $20!$

3. $4(18!)$

4. $\dfrac{30!}{18!}$

5. $P(52, 7)$

6. $P(24, 8)$

7. $C(52, 7)$

8. $C(26, 8)$

9. $P(10, 4) \cdot C(12, 6)$

10. $P(20, 5) \cdot C(9, 3) \cdot C(8, 4)$

11. A mathematics professor uses a computerized test bank to prepare her final exam. If 25 different problems are available for the first 3 exam questions, 40 different problems available for the next 5 questions, and 30 different problems available for the last 2 questions, how many different 10-question exams can she set? (Assume that the order of the questions within each group is not important.)

12. The S & S Brokerage Company has received 100 inquiries from prospective clients. In how many ways can the inquiries be directed to 5 of the firm's brokers if each broker handles 20 inquiries?

7.5   Experiments, Sample Spaces, and Events

Terminology

A number of specialized terms are used in the study of probability. We begin by defining the term *experiment*.

EXPERIMENT

An **experiment** is an activity with observable results.

The results of the experiment are called the **outcomes** of the experiment. Three examples of experiments are:

■ Tossing a coin and observing whether it falls "heads" or "tails"

■ Casting a die and observing which of the numbers 1, 2, 3, 4, 5, or 6 shows up

■ Testing a spark plug from a batch of 100 spark plugs and observing whether or not it is defective

In our discussion of experiments, we will use the following terms:

SAMPLE POINT, SAMPLE SPACE, AND EVENT

Sample point: an outcome of an experiment

Sample space: the set consisting of all possible sample points of an experiment

Event: a subset of a sample space of an experiment

The sample space of an experiment is a universal set whose elements are precisely the outcomes of the experiment, and the events of the experiment are the subsets of the universal set. A sample space associated with an experiment that has a finite number of possible outcomes (sample points) is called a **finite sample space.**

Since the events of an experiment are subsets of a universal set (the sample space of the experiment), we may use the results for set theory to help us study probability. We begin by explaining the roles played by the empty set and a universal set when viewed as events associated with an experiment. The empty set, $\varnothing$, is called the *impossible event;* it cannot occur since the $\varnothing$ has no elements (outcomes). Next, the universal set S is referred to as the *certain event;* it must occur since S contains all the outcomes of the experiment.

This terminology is illustrated in the next several examples.

EXAMPLE 1   Describe the sample space associated with the experiment of tossing a coin and observing whether it falls "heads" or "tails." What are the events of this experiment?

Solution   The two outcomes are "heads" and "tails," and the required sample space is given by $S = \{H, T\}$, where H denotes the outcome "heads" and T denotes the outcome "tails." The events of the experiment, the subsets of S, are:

$$\varnothing, \quad \{H\}, \quad \{T\}, \quad S$$

Note that we have included the impossible event, $\varnothing$, and the certain event, S.

Since the events of an experiment are subsets of the sample space of the experiment, we may talk about the *union* and *intersection* of any two events; we can also consider the *complement* of an event with respect to the sample space.

**THE UNION OF
TWO EVENTS**

The **union** of the two events E and F is the event $E \cup F$.

Thus, the event $E \cup F$ comprises the set of outcomes of E and/or F.

**THE INTERSECTION
OF TWO EVENTS**

The **intersection** of the two events E and F is the event $E \cap F$.

Thus, the event $E \cap F$ comprises the set of outcomes of E and F.

**THE COMPLEMENT
OF AN EVENT**

The **complement** of an event E is the event E^c.

Thus, the event E^c is the set comprising all the outcomes in the sample space S that are not in E.

Venn diagrams depicting the union, intersection, and complementation of events are shown in Figure 20.

These concepts are illustrated in the following example.

FIGURE 20

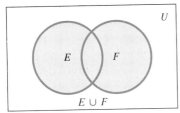

$E \cup F$

(a) The union of two events

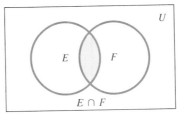

$E \cap F$

(b) The intersection of two events

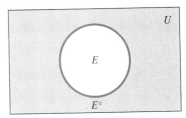

E^c

(c) The complement of the event E

EXAMPLE 2   Consider the experiment of casting a die and observing the number that falls uppermost. Let $S = \{1, 2, 3, 4, 5, 6\}$ denote the sample space of the experiment and $E = \{2, 4, 6\}$ and $F = \{1, 3\}$ be events of this experiment. Compute (a) $E \cup F$, (b) $E \cap F$, and (c) F^c. Interpret your results.

Solution

a. $E \cup F = \{1, 2, 3, 4, 6\}$ and is the event that the outcome of the experiment is a 1, a 2, a 3, a 4, or a 6.

b. $E \cap F = \varnothing$ is the impossible event; the number appearing uppermost when a die is cast cannot be both even and odd at the same time.

c. $F^c = \{2, 4, 5, 6\}$ is precisely the event that the event F does not occur.

If two events cannot occur at the same time, they are said to be mutually exclusive. Using set notation, we have the following definition.

MUTUALLY EXCLUSIVE EVENTS

E and F are **mutually exclusive** if $E \cap F = \varnothing$.

As before, we may use Venn diagrams to illustrate these events. In this case, the two mutually exclusive events are depicted as two nonintersecting circles (Figure 21).

FIGURE 21

Mutually exclusive events

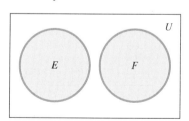

EXAMPLE 3   An experiment consists of tossing a coin three times and observing the resulting sequence of "heads" and "tails."

a. Describe the sample space S of the experiment.

b. Determine the event E that exactly two heads appear.

c. Determine the event F that at least one head appears.

Solution

a. The sample points may be obtained with the aid of a tree diagram (Figure 22). The required sample space S is given by

$$S = \{\text{HHH, HHT, HTH, HTT, THH, THT, TTH, TTT}\}$$

FIGURE 22

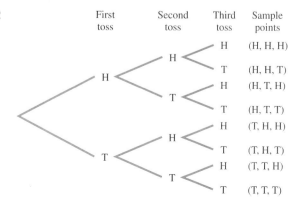

b. By scanning the sample space S obtained in (a), we see that the outcomes in which exactly two heads appear are given by the event

$$E = \{HHT, HTH, THH\}$$

c. Proceeding as in (b), we find

$$F = \{HHH, HHT, HTH, HTT, THH, THT, TTH\}$$

EXAMPLE 4   An experiment consists of casting a pair of dice and observing the number that falls uppermost on each die.

a. Describe an appropriate sample space S for this experiment.

b. Determine the events $E_2, E_3, E_4, \ldots, E_{12}$ that the sum of the numbers falling uppermost is 2, 3, 4, $\ldots$, 12, respectively.

Solution

a. We may represent each outcome of the experiment by an ordered pair of numbers, the first representing the number that appears uppermost on the first die and the second representing the number that appears uppermost on the second die. In order to distinguish between the two dice, think of the first die as being red and the second as being green. Since there are six possible outcomes for each die, the multiplication principle implies that there are $6 \cdot 6$, or 36, elements in the sample space:

$$S = \{(1, 1), (1, 2), (1, 3), (1, 4), (1, 5), (1, 6),$$
$$(2, 1), (2, 2), (2, 3), (2, 4), (2, 5), (2, 6),$$
$$(3, 1), (3, 2), (3, 3), (3, 4), (3, 5), (3, 6),$$
$$(4, 1), (4, 2), (4, 3), (4, 4), (4, 5), (4, 6),$$
$$(5, 1), (5, 2), (5, 3), (5, 4), (5, 5), (5, 6),$$
$$(6, 1), (6, 2), (6, 3), (6, 4), (6, 5), (6, 6)\}$$

b. With the aid of the results of (a), we obtain the required list of events in Table 1.

TABLE 1

Sum of Uppermost Numbers	Event
2	$E_2 = \{(1, 1)\}$
3	$E_3 = \{(1, 2), (2, 1)\}$
4	$E_4 = \{(1, 3), (2, 2), (3, 1)\}$
5	$E_5 = \{(1, 4), (2, 3), (3, 2), (4, 1)\}$
6	$E_6 = \{(1, 5), (2, 4), (3, 3), (4, 2), (5, 1)\}$
7	$E_7 = \{(1, 6), (2, 5), (3, 4), (4, 3), (5, 2), (6, 1)\}$
8	$E_8 = \{(2, 6), (3, 5), (4, 4), (5, 3), (6, 2)\}$
9	$E_9 = \{(3, 6), (4, 5), (5, 4), (6, 3)\}$
10	$E_{10} = \{(4, 6), (5, 5), (6, 4)\}$
11	$E_{11} = \{(5, 6), (6, 5)\}$
12	$E_{12} = \{(6, 6)\}$

Applications

EXAMPLE 5   The manager of a local cinema records the number of patrons attending a first-run movie at the 1 P.M. screening. The theater has a seating capacity of 500.

a. What is an appropriate sample space for this experiment?
b. Describe the event E that fewer than 50 people attend the screening.
c. Describe the event F that the theater is more than half full at the screening.

Solution

a. The number of patrons at the screening (the outcome) could run from 0 to 500. Therefore, a sample space for this experiment is

$$S = \{0, 1, 2, 3, \ldots, 500\}$$

b. $E = \{0, 1, 2, 3, \ldots, 49\}$
c. $F = \{251, 252, 253, \ldots, 500\}$

EXAMPLE 6   An experiment consists of studying the composition of a three-child family in which the children were born at different times.

a. Describe an appropriate sample space S for this experiment.
b. Describe the event E that there are two girls and a boy in the family.
c. Describe the event F that the oldest child is a girl.
d. Describe the event G that the oldest child is a girl and the youngest child is a boy.

Solution

a. The sample points of the experiment may be obtained with the aid of the tree diagram shown in Figure 23, where b denotes a boy and g denotes a girl. We see from the tree diagram that the required sample space is given by

$$S = \{(b, b, b), (b, b, g), (b, g, b), (b, g, g), (g, b, b),$$

$$(g, b, g), (g, g, b), (g, g, g)\}$$

FIGURE 23

Tree diagram for three-child families

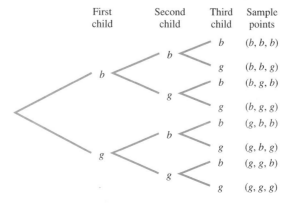

First child	Second child	Third child	Sample points
		b	(b, b, b)
	b	g	(b, b, g)
b		b	(b, g, b)
	g	g	(b, g, g)
		b	(g, b, b)
	b	g	(g, b, g)
g		b	(g, g, b)
	g	g	(g, g, g)

Using the tree diagram, we find that

b. $E = \{(b, g, g), (g, b, g), (g, g, b)\}$

c. $F = \{(g, b, b), (g, b, g), (g, g, b), (g, g, g)\}$

d. $G = \{(g, b, b), (g, g, b)\}$

The next example shows that sample spaces may be infinite.

EXAMPLE 7   The Ever-Brite Battery Company is developing a high-amperage, high-capacity battery as a source for powering electric cars. The battery is tested by installing it in a prototype electric car and running the car with a fully charged battery on a test track at a constant speed of 55 mph until the car runs out of power. The distance covered by the car is then observed.

a. What is the sample space for this experiment?

b. Describe the event E that the driving range under test conditions is less than 150 miles.

c. Describe the event F that the driving range is between 200 and 250 miles, inclusive.

Solution

a. Since the distance d covered by the car in any run may be given by any nonnegative number, the sample space S is given by

$$S = \{d \mid d \geq 0\}$$

b. The event E is given by

$$E = \{d \mid d < 150\}$$

c. The event F is given by

$$F = \{d \mid 200 \leq d \leq 250\}$$

**SELF-CHECK
EXERCISES 7.5**

1. A sample of three apples taken from Cavallero's Fruit Stand are examined to determine whether they are good or rotten.
   a. What is an appropriate sample space for this experiment?
   b. Describe the event E that exactly one of the apples picked is rotten.
   c. Describe the event F that the first apple picked is rotten.

2. Refer to exercise 1.
   a. Find $E \cup F$.
   b. Find $E \cap F$.
   c. Find F^c.
   d. Are the events E and F mutually exclusive?

Solutions to Self-Check Exercises 7.5 can be found on page 377.

7.5 EXERCISES

In exercises 1–6, let $S = \{a, b, c, d, e, f\}$ be a sample space of an experiment, and let $E = \{a, b\}$, $F = \{a, d, f\}$, and $G = \{b, c, e\}$ be events of this experiment.

1. Find the events $E \cup F$ and $E \cap F$.

2. Find the events $F \cup G$ and $F \cap G$.

3. Find the events F^c and $E \cap G^c$.

4. Find the events E^c and $F^c \cap G$.

5. Are the events E and F mutually exclusive?

6. Are the events $E \cup F$ and $E \cap F^c$ mutually exclusive?

In exercises 7–12, let $S = \{1, 2, 3, 4, 5, 6\}$, $E = \{2, 4, 6\}$, $F = \{1, 3, 5\}$, and $G = \{5, 6\}$.

7. Find the event $E \cup F \cup G$.

8. Find the event $E \cap F \cap G$.

9. Find the event $(E \cup F \cup G)^c$.

10. Find the event $(E \cap F \cap G)^c$.

11. Are the events E and F mutually exclusive?

12. Are the events F and G mutually exclusive?

In exercises 13–18, let S be any sample space and E, F, and G be any three events associated with the experiment. Describe the given events using the symbols $\cup$, $\cap$, and ^{c}.

13. The event that E and/or F occurs.

14. The event that both E and F occur.

15. The event that G does not occur.

16. The event that E but not F occurs.

17. The event that none of the events E, F, and G occurs.

18. The event that E occurs but neither of the events F or G occurs.

19. Let $S = \{a, b, c\}$ be a sample space of an experiment with outcomes a, b, and c. List all the events of this experiment.

20. Let $S = \{1, 2, 3\}$ be a sample space associated with an experiment.
   a. List all events of this experiment.
   b. How many subsets of S contain the number 3?
   c. How many subsets of S contain either the number 2 or the number 3?

21. An experiment consists of selecting a card from a standard deck of playing cards and noting whether it is black (B) or red (R).
   a. Describe an appropriate sample space for this experiment.
   b. What are the events of this experiment?

22. An experiment consists of selecting a letter at random from the letters in the word *MASSACHU-SETTS* and observing the outcomes.
   a. What is an appropriate sample space for this experiment?
   b. Describe the event "the letter selected is a vowel."

23. An experiment consists of tossing a coin and casting a die and observing the outcomes.
   a. Describe an appropriate sample space for this experiment.
   b. Describe the event "a head is tossed and an even number is cast."

24. An experiment consists of spinning the hand of the numbered disc shown in the following figure and observing the region in which the pointer stops. (If the needle stops on a line, the result is discounted and the needle is spun again.)

   a. What is an appropriate sample space for this experiment?
   b. Describe the event "the spinner points to the number 2."
   c. Describe the event "the spinner points to an odd number."

25. **Quality Control** A sample of three transistors taken from a local electronics store were examined to determine whether they were defective (d) or nondefective (n). What is an appropriate sample space for this experiment?

26. **Blood Typing** Human blood is classified by the presence or absence of three main antigens (A, B, and Rh). When a blood specimen is typed, the presence of the A and/or B antigen is indicated by listing the letter A and/or the letter B. If neither the A nor B antigen is present, the letter O is used to represent this case. The presence or absence of the Rh antigen is indicated by the symbols $+$ or $-$, respectively. Thus, if a blood specimen is classified as AB^+, it contains the A and the B antigens as well as the Rh antigen. Similarly, O^- blood contains none of the three antigens. Using this information, determine the sample space corresponding to the different blood groups.

27. **Game Shows** In a television game show, the winner is asked to select three prizes from five different prizes, A, B, C, D, and E.
   a. Describe a sample space of possible outcomes (order is not important).
   b. How many points are there in the sample space corresponding to a selection that includes A?

c. How many points are there in the sample space corresponding to a selection that includes A and B?

d. How many points are there in the sample space corresponding to a selection that includes either A or B?

28. Automatic Tellers   The manager of a local bank observes how long it takes a customer to complete his transactions at the automatic bank teller.

a. Describe an appropriate sample space for this experiment.

b. Describe the event that it takes a customer between 2 and 3 minutes to complete his transactions at the automatic bank teller.

29. Common Stocks   Mr. Robbins purchased shares of a machine tool company and shares of an airline company. Let E be the event that the shares of the machine tool company increase in value over the next six months, and let F be the event that the shares of the airline company increase in value over the next six months. Using the symbols $\cup$, $\cap$, and ^{c}, describe the following events.

a. The shares in the machine tool company do not increase in value.

b. The shares in both the machine tool company and the airline company do not increase in value.

c. The shares of at least one of the two companies increase in value.

d. The shares of only one of the two companies increase in value.

30. Customer Service Surveys   The customer service department of the Universal Instruments Company, manufacturer of the Galaxy home computer, conducted a survey among customers who had returned their purchase registration cards. Purchasers of its deluxe model home computer were asked to report the length of time (t) in days before service was required.

a. Describe a sample space corresponding to this survey.

b. Describe the event E that a home computer required service before a period of 90 days had elapsed.

c. Describe the event F that a home computer did not require service before a period of 1 year had elapsed.

31. Assembly Time Studies   A time study was conducted by the production manager of the Vista Vision Corporation to determine the length of time in minutes required by an assembly worker to complete a certain task during the assembly of its Pulsar color television sets.

a. Describe a sample space corresponding to this time study.

b. Describe the event E that an assembly worker took 2 minutes or less to complete the task.

c. Describe the event F that an assembly worker took more than 2 minutes to complete the task.

32. Political Polls   An opinion poll is conducted among a state's electorate to determine the relationship between their income levels and their stands on a proposition aimed at reducing state income taxes. Voters are classified as belonging to either the low-, middle-, or upper-income group. They are asked whether they favor, oppose, or are undecided about the proposition. Let the letters L, M, and U represent the low-, middle-, and upper-income groups, respectively, and let the letters f, o, and u represent the responses—favor, oppose, and undecided, respectively.

a. Describe a sample space corresponding to this poll.

b. Describe the event E_1 that a respondent favors the proposition.

c. Describe the event E_2 that a respondent opposes the proposition and does not belong to the low-income group.

d. Describe the event E_3 that a respondent does not favor the proposition and does not belong to the upper-income group.

33. Quality Control   As part of a quality-control procedure, an inspector at Bristol Farms randomly selects ten eggs from each consignment of eggs he receives and records the number of broken eggs.

a. What is an appropriate sample space for this experiment?

b. Describe the event E that at most three eggs are broken.

c. Describe the event F that at least five eggs are broken.

34. Political Polls   In the opinion poll of exercise 32, the voters were also asked to indicate their political affiliations—Democrat, Republican, or Independent. As before, let the letters L, M, and U represent the low-, middle-, and upper-income groups, respectively, and let the letters D, R, and I represent Democrat, Republican, and Independent, respectively.

a. Describe a sample space corresponding to this poll.

b. Describe the event E_1 that a respondent is a Democrat.

c. Describe the event E_2 that a respondent belongs to the upper-income group and is a Republican.

d. Describe the event E_3 that a respondent belongs to the middle-income group and is not a Democrat.

35. **Shuttle Bus Usage**   A certain airport hotel operates a shuttle bus service between the hotel and the airport. The maximum capacity of a bus is 20 passengers. On alternate trips of the shuttle bus over a period of one week, the hotel manager kept a record of the number of passengers arriving at the hotel in each bus.
    a. What is an appropriate sample space for this experiment?
    b. Describe the event E that a shuttle bus carried fewer than ten passengers.
    c. Describe the event F that a shuttle bus arrived with a full capacity.

36. **Sports**   Eight players, A, B, C, D, E, F, G, and H, are competing in a series of elimination matches of a tennis tournament in which the winner of each preliminary match will advance to the semifinals and the winner of the semifinals will advance to the finals. An outline of the scheduled matches follows. Describe a sample space listing the possible participants in the finals.

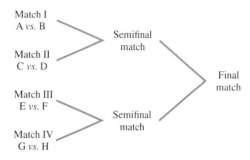

37. Let S be a sample space for an experiment. Show that if E is any event of an experiment, then E and E^c are mutually exclusive.

38. Let S be a sample space for an experiment, and let E and F be events of this experiment. Show that the events $E \cup F$ and $E^c \cap F^c$ are mutually exclusive. [*Hint:* Use De Morgan's Law.]

39. Let S be a sample space of an experiment with n outcomes. Determine the number of events of this experiment.

SOLUTIONS TO SELF-CHECK EXERCISES 7.5

1. a. Let g denote a good apple and r a rotten apple. Thus, the required sample points may be obtained with the aid of a tree diagram (compare with Example 3). The required sample space is given by

$$S = \{ggg, ggr, grg, grr, rgg, rgr, rrg, rrr\}$$

   b. By scanning the sample space S obtained in (a), we identify the outcomes in which exactly one apple is rotten. We find

$$E = \{ggr, grg, rgg\}$$

   c. Proceeding as in (b), we find

$$F = \{rgg, rgr, rrg, rrr\}$$

2. Using the results of exercise 1, we find
   a. $E \cup F = \{ggr, grg, rgg, rgr, rrg, rrr\}$
   b. $E \cap F = \{rgg\}$
   c. F^c is the set of outcomes in S but not in F. Thus,

$$F^c = \{ggg, ggr, grg, grr\}$$

   d. Since $E \cap F \neq \varnothing$, we conclude that E and F are not mutually exclusive.

7.6   Probability

Finding the Probability of an Event

Let us return to the coin-tossing experiment. The sample space of this experiment is given by $S = \{H, T\}$, where the sample points H and T correspond to the two possible outcomes, a *head* and a *tail*. If the coin is *unbiased*, then there is *one chance out of two* of obtaining a head (or a tail) and we say that the *probability* of tossing a head (tail) is 1/2, abbreviated

$$P(H) = \frac{1}{2}, \qquad P(T) = \frac{1}{2}$$

An alternative method of obtaining the values of $P(H)$ and $P(T)$ is based on continued experimentation and does not depend on the assumption that the two outcomes are equally likely. Table 2 summarizes the results of such an exercise.

TABLE 2

As the number of trials increases, the relative frequency approaches 0.5.

Number of Tosses (n)	Number of Heads (m)	Relative Frequency of Heads (m/n)
10	4	0.4000
100	58	0.5800
1,000	492	0.4920
10,000	5,034	0.5034
20,000	10,024	0.5012
40,000	20,032	0.5008

Observe that the relative frequencies (column 3) differ considerably when the number of trials is small, but as the number of trials becomes very large, the relative frequency approaches the number 0.5. This result suggests that we assign to $P(H)$ the value 1/2, as before.

More generally, consider an experiment that may be repeated over and over again under independent and similar conditions. Suppose that in n trials, an event E occurs m times. We call the ratio m/n the **relative frequency** of the event E after n repetitions. If this relative frequency approaches some value $P(E)$ as n becomes larger and larger, then $P(E)$ is called the **empirical probability** of E. Thus, the probability $P(E)$ of an event occurring is a measure of the proportion of the time that the event E will occur in the long run. Observe that this method of computing the probability of a head occurring is effective even in the case when a biased coin is used in the experiment.

The **probability of an event** is a number that lies between 0 and 1. In general, the larger the probability of an event, the more likely the event will occur. Thus, an event with a probability of 0.8 is more likely to occur than an event with a probability of 0.6. An event with a probability of 1/2, or 0.5, has a "fifty-fifty," or equal, chance of occurring.

Now suppose we are given an experiment and wish to determine the probabilities associated with certain events of the experiment. This problem could be solved by computing $P(E)$ directly for each event E of interest. However, in practice, the number of events that we may be interested in is usually quite large, so this approach is not satisfactory.

The following approach is particularly suitable when the sample space of an experiment is finite.* Let S be a finite sample space with n outcomes; that is,

$$S = \{s_1, s_2, s_3, \ldots, s_n\}$$

Then the events

$$\{s_1\}, \{s_2\}, \{s_3\}, \ldots, \{s_n\}$$

which consist of exactly one point, are called **simple,** or **elementary,** events of the experiment. They are elementary in the sense that any (nonempty) event of the experiment may be obtained by taking a finite union of suitable elementary events. The simple events of an experiment are also *mutually exclusive;* that is, given any two simple events of the experiment, only one can occur.

By assigning probabilities to each of the simple events, we obtain the results shown in Table 3. This table is called a **probability distribution** for the experiment. The function P, which assigns a probability to each of the simple events, is called a **probability function.**

TABLE 3

A probability distribution

Simple Event	Probability**
$\{s_1\}$	$P(s_1)$
$\{s_2\}$	$P(s_2)$
$\{s_3\}$	$P(s_3)$
$\vdots$	$\vdots$
$\{s_n\}$	$P(s_n)$

The numbers $P(s_1), P(s_2), \ldots, P(s_n)$ have the following properties:

1. $0 \le P(s_i) \le 1 \qquad i = 1, 2, \ldots, n$

2. $P(s_1) + P(s_2) + \cdots + P(s_n) = 1$

3. $P(\{s_i\} \cup \{s_j\}) = P(s_i) + P(s_j) \qquad i \ne j \ (i = 1, 2, \ldots, n; j = 1, 2, \ldots, n)$

The first property simply states that the probability of a simple event must be between 0 and 1 inclusive. The second property states that the sum of the probabilities of all simple events of the sample space is 1. This follows from the fact that the event S is certain to occur. The third property states that the probability of the union of two mutually exclusive events is given by the sum of their probabilities.

As we saw earlier, there is no unique method for assigning probabilities to the simple events of an experiment. In practice, the methods used to determine these probabilities may range from theoretical considerations of

*For the remainder of the chapter we will assume that all sample spaces are finite.

**For simplicity, we will use the notation $P(s_i)$ instead of the technically more correct $P(\{s_i\})$.

the problem on the one extreme to the reliance on "educated guesses" on the other.

Sample spaces in which the outcomes are equally likely are called **uniform sample spaces.** Assigning probabilities to the simple events in these spaces is relatively easy.

PROBABILITY OF AN EVENT IN A UNIFORM SAMPLE SPACE

If

$$S = \{s_1, s_2, \ldots, s_n\}$$

is the sample space for an experiment in which the outcomes are equally likely, then we assign the probabilities

$$P(s_1) = P(s_2) = \cdots = P(s_n) = \frac{1}{n}$$

to each of the simple events $s_1, s_2, \ldots, s_n$.

EXAMPLE 1   A fair die is cast and the number that falls uppermost is observed. Determine the probability distribution for the experiment.

Solution   The sample space for the experiment is $S = \{1, 2, 3, 4, 5, 6\}$ and the simple events are accordingly given by the sets {1}, {2}, {3}, {4}, {5}, and {6}. Since the die is assumed to be fair, the six outcomes are equally likely. We therefore assign a probability of 1/6 to each of the simple events and obtain the following probability distribution shown in Table 4.

TABLE 4
A probability distribution

Simple Event	Probability
{1}	$\frac{1}{6}$
{2}	$\frac{1}{6}$
{3}	$\frac{1}{6}$
{4}	$\frac{1}{6}$
{5}	$\frac{1}{6}$
{6}	$\frac{1}{6}$

The next example shows how the *relative frequency* interpretation of probability lends itself to the computation of probabilities.

EXAMPLE 2   Refer to Example 7, Section 7.5. The data shown in Table 5 were obtained in tests involving 200 test runs. Each run was made with a fully charged battery.

a. Describe an appropriate sample space for this experiment.

b. Find the probability distribution for this experiment.

TABLE 5

Data obtained during 200 test runs of an electric car

Distance Covered in Miles (x)	Frequency of Occurrence
$0 \leq x \leq 50$	4
$50 < x \leq 100$	10
$100 < x \leq 150$	30
$150 < x \leq 200$	100
$200 < x \leq 250$	40
$250 < x$	16

Solution

a. Let s_1 denote the outcome that the distance covered by the car does not exceed 50 miles; let s_2 denote the outcome that the distance covered by the car is greater than 50 miles but does not exceed 100 miles; and so on. Finally, let s_6 denote the outcome that the distance covered by the car is greater than 250 miles. Then the required sample space is given by

$$S = \{s_1, s_2, s_3, s_4, s_5, s_6\}$$

b. To compute the probability distribution for the experiment, we turn to the relative frequency interpretation of probability. Accepting the inaccuracies inherent in a relatively small number of trials (200 runs), we will take the probability of s_1 occurring as

$$P(s_1) = \frac{\text{Number of trials in which } s_1 \text{ occurs}}{\text{Total number of trials}}$$

$$= \frac{4}{200} = 0.02$$

In a similar manner, we assign probabilities to the other simple events, obtaining the probability distribution shown in Table 6.

TABLE 6

A probability distribution

Simple Event	Probability
$\{s_1\}$	0.02
$\{s_2\}$	0.05
$\{s_3\}$	0.15
$\{s_4\}$	0.50
$\{s_5\}$	0.20
$\{s_6\}$	0.08

We are now in a position to give a procedure for computing the probability $P(E)$ of an arbitrary event E of an experiment.

FINDING THE PROBABILITY OF AN EVENT *E*

1. Determine a sample space S associated with the experiment.
2. Assign probabilities to the simple events of S.
3. If $E = \{s_1, s_2, s_3, \ldots, s_n\}$ where $\{s_1\}, \{s_2\}, \{s_3\}, \ldots, \{s_n\}$ are simple events, then

$$P(E) = P(s_1) + P(s_2) + P(s_3) + \cdots + P(s_n)$$

If E is the empty set, $\varnothing$, then $P(E) = 0$.

The principle stated in step 3 is called the **addition principle** and is a consequence of property 3 of the probability function (page 379). This principle allows us to find the probabilities of all the other events once the probabilities of the simple events are known.

 The addition rule applies *only* to the addition of probabilities of simple events.

FIGURE 24
The event that the two dice show the same number

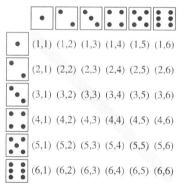

FIGURE 25
The event that the sum of the numbers on the two dice is 6

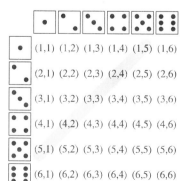

Applications

EXAMPLE 3   A pair of fair dice is cast.

a. Calculate the probability that the two dice show the same number.
b. Calculate the probability that the sum of the numbers of the two dice is six.

Solution   From the results of Example 4, Section 7.5, we see that the sample space S of the experiment consists of 36 outcomes:

$$S = \{(1, 1), (1, 2), \ldots, (6, 5), (6, 6)\}$$

Since both dice are fair, each of the 36 outcomes is equally likely. Accordingly, we assign the probability of 1/36 to each simple event. We are now in a position to answer the questions posed.

a. The event that the two dice show the same number is given by

$$E = \{(1, 1), (2, 2), (3, 3)\,(4, 4), (5, 5), (6, 6)\} \qquad \text{See Figure 24.}$$

Therefore, by the addition principle, the probability that the two dice show the same number is given by

$$\begin{aligned}
P(E) &= P[(1, 1)] + P[(2, 2)] + \cdots + P[(6, 6)] \\
&= \frac{1}{36} + \frac{1}{36} + \cdots + \frac{1}{36} \qquad \text{Six terms} \\
&= \frac{1}{6}
\end{aligned}$$

b. The event that the sum of the numbers of the two dice is six is given by

$$E_6 = \{(1, 5), (2, 4), (3, 3), (4, 2), (5, 1)\} \qquad \text{See Figure 25.}$$

Therefore, the probability that the sum of the numbers on the two dice is six is given by

$$P(E_6) = P[(1, 5)] + P[(2, 4)] + P[(3, 3)] + P[(4, 2)] + P[(5, 1)]$$

$$= \frac{1}{36} + \frac{1}{36} + \cdots + \frac{1}{36} \qquad \text{Five terms}$$

$$= \frac{5}{36}$$

■

EXAMPLE 4   Consider the experiment by the Ever-Brite Company in Example 2. What is the probability that the prototype car will travel more than 150 miles on a fully charged battery?

Solution   Using the results of Example 2, we see that the event that the car will travel more than 150 miles on a fully charged battery is given by $E = \{s_4, s_5, s_6\}$. Therefore, the probability that the car will travel more than 150 miles on one charge is given by

$$P(E) = P(s_4) + P(s_5) + P(s_6)$$

or, using the probability distribution for the experiment obtained in Example 2,

$$P(E) = 0.50 + 0.20 + 0.08 = 0.78$$

■

SELF-CHECK EXERCISES 7.6

1. A biased die was cast repeatedly, and the results of the experiment are summarized in the following table.

Outcome	1	2	3	4	5	6
Frequency of Occurrence	142	173	158	175	162	190

Using the relative frequency interpretation of probability, find the probability distribution for this experiment.

2. In an experiment conducted to study the effectiveness of an eye-level third brake light in the prevention of rear-end collisions, 250 of the 500 highway patrol cars of a certain state were equipped with such lights. At the end of the 1-year trial period, the records revealed that for those equipped with a third brake light, there were 14 incidents of rear-end collision. There were 22 such incidents involving the cars not equipped with the accessory. Based on these data, what is the probability that a highway patrol car equipped with a third brake light will be rear-ended within a 1-year period? What is the probability that a car not so equipped will be rear-ended within a 1-year period?

Solutions to Self-Check Exercises 7.6 can be found on page 387.

7.6 EXERCISES

In exercises 1–8, list the simple events associated with each experiment.

1. A nickel and a dime are tossed and the result of heads or tails is recorded for each coin.

2. A card is selected at random from a standard 52-card deck and its suit—hearts (h), diamonds (d), spades (s), or clubs (c)—is recorded.

3. Opinion Polls   An opinion poll is conducted among a group of registered voters. Their political affiliation—Democrat (D), Republican (R), or Independent (I)—and their sex—male (m) or female (f)—are recorded.

4. Quality Control   As part of a quality-control procedure, eight circuit boards are checked and the number of defectives is recorded.

5. Movie Attendance   In a survey conducted to determine whether movie attendance is increasing (i), decreasing (d), or holding steady (s) among various sectors of the population, participants are classified as follows:

   Group 1: those aged 10–19

   Group 2: those aged 20–29

   Group 3: those aged 30–39

   Group 4: those aged 40–49

   Group 5: those 50 and over

   The response of each participant and his or her age group are recorded.

6. Durable Goods Orders   Data concerning durable goods orders are obtained each month by an economist. A record is kept for a 1-year period of any increase (i), decrease (d), or unchanged movement (u) in the number of durable goods orders for each month as compared to the number of such orders in the same month in the previous year.

7. Blood Types   Blood tests are given as a part of the admission procedure at the Monterey Garden Community Hospital. The blood type of each patient (A, B, AB, or O) and the presence or absence of the Rh factor in each patient's blood (Rh$^+$ or Rh$^-$) are recorded.

8. Meteorology   A meteorologist preparing a weather map classifies the expected average temperature in each of five neighboring states for the upcoming week as follows:

   a. More than 10° below average
   b. Normal to 10° below average
   c. Higher than normal to 10° above average
   d. More than 10° above average

   Using each state's abbreviation and the categories—(a), (b), (c), and (d)—the meteorologist records these data.

9. Grade Distributions   The grade distribution for a certain class is shown in the following table. Find the probability distribution associated with these data.

Grade	A	B	C	D	F
Frequency of Occurrence	4	10	18	6	2

10. Blood Types   The percentage of the general population that has each blood type is shown in the following table.

Blood Type	A	B	AB	O
Percentage of Population	41%	12%	3%	44%

Determine the probability distribution associated with these data.

11. Traffic Surveys   The number of cars entering a tunnel leading to an airport in a major city over a period of 200 peak hours was observed and the following data were obtained:

Number of Cars (x)	Frequency of Occurrence
$0 < x \leq 200$	15
$200 < x \leq 400$	20
$400 < x \leq 600$	35
$600 < x \leq 800$	70
$800 < x \leq 1000$	45
$x > 1000$	15

a. Describe an appropriate sample space for this experiment.
b. Find the probability distribution for this experiment.

12. **Product Surveys** The accompanying data were obtained from a survey of 1500 Americans who were asked: How safe are American-made consumer products?

Rating	A (Very safe)	B (Somewhat safe)
Number of Respondents	285	915

Rating	C (Not too safe)	D (Not safe at all)
Number of Respondents	225	30

Rating	E (Don't know)
Number of Respondents	45

Determine the probability distribution associated with these data.

13. **Political Views** In a poll conducted among 2000 college freshmen to ascertain the political views of college students, the accompanying data were obtained.

Political Views	A (Far left)	B (Liberal)
Number of Respondents	52	398

Political Views	C (Middle-of-the-road)
Number of Respondents	1140

Political Views	D (Conservative)	E (Far right)
Number of Respondents	386	24

Determine the probability distribution associated with these data.

14. **Service-Utilization Studies** The Metro Telephone Company of Belmont compiled the accompanying information during a service-utilization study pertaining to the number of customers using their Dial-the-Time service from 7 A.M. to 9 A.M. on a certain weekday morning. Using these data, find the probability distribution associated with the experiment.

Number of Calls Received/Minute	Frequency of Occurrence
10	6
11	15
12	12
13	3
14	12
15	36
16	24
17	0
18	6
19	6

15. **Assembly Time Studies** The results of a time study conducted by the production manager of the Ace Novelty Company are shown in the accompanying table, where the number of space action-figures produced each quarter hour during an 8-hour workday has been tabulated. Find the probability distribution associated with this experiment.

Number of Figures Produced (in dozens)	Frequency of Occurrence
30	4
31	0
32	6
33	8
34	6
35	4
36	4

16. Electronic Mail Services   The number of subscribers to five leading electronic mail services is shown in the accompanying table.

Company	A	B	C
Subscribers	300,000	200,000	120,000

Company	D	E
Subscribers	80,000	60,000

Find the probability distribution associated with these data.

c 17. Corrective Lens Use   According to Mediamark Research, Inc., 84 million out of 179 million adults in the United States correct their vision by using prescription eyeglasses, bifocals, or contact lenses. (Some respondents use more than one type.) What is the probability that an adult selected at random from the adult population uses corrective lenses?

c 18. Correctional Supervision   A study conducted by the Corrections Department of a certain state revealed that 163,605 people out of a total adult population of 1,778,314 were under correctional supervision (on probation, parole, or in jail). What is the probability that a person selected at random from the adult population in that state is under correctional supervision?

19. Lightning Deaths   According to data obtained from the National Weather Service, 376 of the 439 people killed by lightning in the United States between 1985 and 1992 were men. (Job and recreational habits of men make them more vulnerable to lightning.) Assuming that this trend holds in the future, find the probability that
a. a person killed by lightning is a male
b. a person killed by lightning is a female

20. Quality Control   One light bulb is selected at random from a lot of 120 light bulbs of which 5% are defective. What is the probability that the light bulb selected is defective?

21. Efforts to Stop Shoplifting   According to a survey of 176 retailers, 46% of them use electronic tags as protection against shoplifting and employee theft. If one of these retailers is selected at random, what is the probability that he or she uses electronic tags as antitheft devices?

22. If a ball is selected at random from an urn containing three red balls, two white balls, and five blue balls, what is the probability that it will be a white ball?

23. If one card is drawn at random from a standard 52-card deck, what is the probability that the card drawn is
a. a diamond?   **b.** a black card?
c. an ace?

24. A pair of fair dice is cast. What is the probability that
a. the sum of the numbers shown uppermost is less than five?
b. at least one 6 is cast?

25. Traffic Lights   What is the probability of arriving at a traffic light when it is red if the red signal flashes for 30 seconds, the yellow signal for 5 seconds, and the green signal for 45 seconds?

26. Roulette   What is the probability that a roulette ball will come to rest on an even number other than 0 or 00? (Assume that there are 38 equally likely outcomes consisting of the numbers 1–36, 0, and 00.)

27. Refer to exercise 9. What is the probability that a student selected at random from this class received a passing grade (D or better)?

28. Refer to exercise 11. What is the probability that more than 600 cars will enter the airport tunnel during a peak hour?

c 29. Disposition of Criminal Cases   Of the 98 first-degree murder cases from 1990 through the first half of 1992 in the Suffolk superior court, 9 cases were thrown out of the system, 62 cases were plea bargained, and 27 cases went to trial. What is the probability that
a. a case selected at random was settled through plea bargaining?
b. a case selected at random went to trial?

c 30. Sweepstakes   One hundred thousand entries have been received in a sweepstakes sponsored by the Gemini Paper Products Company. If 1 grand prize, 5 first prizes, 25 second prizes, and 500 third prizes are to be awarded, what is the probability that
a. a person who has submitted one entry will win the grand prize?
b. a person who has submitted one entry will win a prize?

In exercises 31–34, determine whether the given experiment has a sample space with equally likely outcomes.

31. A loaded die is cast and the number appearing uppermost on the die is recorded.

32. Two fair dice are cast and the sum of the numbers appearing uppermost is recorded.

33. A ball is selected at random from an urn containing six black balls and six red balls and the color of the ball is recorded.

34. A weighted coin is thrown and the outcome of heads or tails is recorded.

35. Let $S = \{s_1, s_2, s_3, s_4, s_5\}$ be the sample space associated with an experiment having the following probability distribution.

Outcome	$\{s_1\}$	$\{s_2\}$	$\{s_3\}$	$\{s_4\}$	$\{s_5\}$
Probability	$\frac{1}{14}$	$\frac{3}{14}$	$\frac{6}{14}$	$\frac{2}{14}$	$\frac{2}{14}$

Find the probability of the event
a. $A = \{s_1, s_2, s_4\}$ **b.** $B = \{s_1, s_5\}$
c. $C = S$

36. Let $S = \{s_1, s_2, s_3, s_4, s_5, s_6\}$ be the sample space associated with an experiment having the following probability distribution.

Outcome	$\{s_1\}$	$\{s_2\}$	$\{s_3\}$	$\{s_4\}$	$\{s_5\}$	$\{s_6\}$
Probability	$\frac{1}{12}$	$\frac{1}{4}$	$\frac{1}{12}$	$\frac{1}{6}$	$\frac{1}{3}$	$\frac{1}{12}$

Find the probability of the event
a. $A = \{s_1, s_3\}$ **b.** $B = \{s_2, s_4, s_5, s_6\}$
c. $C = S$

37. Political Polls An opinion poll was conducted among a group of registered voters in a certain state concerning a proposition aimed at limiting state and local taxes. Results of the poll indicated that 35% of the voters favored the proposition, 32% were against it, and the remaining group were undecided. If the results of the poll are assumed to be representative of the opinions of the state's electorate, what is the probability that

a. a registered voter selected at random from the electorate favors the proposition?
b. a registered voter selected at random from the electorate is undecided about the proposition?

38. Consider the composition of a three-child family in which the children were born at different times. Assume that a girl is as likely as a boy at each birth.
a. What is the probability that there are two girls and a boy in the family?
b. What is the probability that the oldest child is a girl?
c. What is the probability that the oldest child is a girl and the youngest child is a boy?

39. Airfone Usage The number of planes in the fleets of five leading airlines that contain Airfones is shown in the accompanying table.

Airline	Number of Planes with Airfones	Size of Fleet
A	50	295
B	40	325
C	31	167
D	29	50
E	25	248

a. If a plane is selected at random from airline A, what is the probability that it contains an Airfone?
b. If a plane is selected at random from the entire fleet of the five airlines, what is the probability that it contains an Airfone?

SOLUTIONS TO SELF-CHECK EXERCISES 7.6

1. $P(1) = \dfrac{\text{Number of trials in which a 1 appears uppermost}}{\text{Total number of trials}}$

$= \dfrac{142}{1000}$

$= 0.142$

Similarly, we compute $P(2), \ldots, P(6)$, obtaining the following probability distribution.

Outcome	1	2	3	4	5	6
Probability	0.142	0.173	0.158	0.175	0.162	0.190

2. The probability that a highway patrol car equipped with a third brake light will be rear-ended within a 1-year period is given by

$$\frac{\text{Number of rear-end collisions involving cars}}{\text{Total number of such cars}} = \frac{14}{250} = 0.056$$

The probability that a highway patrol car not equipped with a third brake light will be rear-ended within a 1-year period is given by

$$\frac{\text{Number of rear-end collisions involving cars}}{\text{Total number of such cars}} = \frac{22}{250} = 0.088$$

| 7.7 | **Rules of Probability** |

Properties of the Probability Function and Their Applications

In this section we examine some of the properties of the probability function and look at the role they play in solving certain problems. We begin by looking at the generalization of the three properties of the probability function, which were stated for simple events in the last section. Let S be a sample space of an experiment and suppose E and F are events of the experiment. Then we have

PROPERTY 1

$$P(E) \geq 0 \text{ for any } E$$

PROPERTY 2

$$P(S) = 1$$

PROPERTY 3

If E and F are mutually exclusive (that is, only one of them can occur, or equivalently, $E \cap F = \varnothing$), then

$$P(E \cup F) = P(E) + P(F) \qquad \text{See Figure 26.}$$

FIGURE 26
If E and F are mutually exclusive events, then $P(E \cup F) = P(E) + P(F)$.

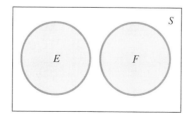

Property 3 may be easily extended to the case involving any finite number of mutually exclusive events. Thus, if $E_1, E_2, \ldots, E_n$ are mutually exclusive events, then

$$P(E_1 \cup E_2 \cup \cdots \cup E_n) = P(E_1) + P(E_2) + \cdots + P(E_n)$$

TABLE 7
Probability distribution

Score (x)	Probability
$x > 700$	0.01
$600 < x \leq 700$	0.07
$500 < x \leq 600$	0.19
$400 < x \leq 500$	0.23
$300 < x \leq 400$	0.31
$x \leq 300$	0.19

EXAMPLE 1 The superintendent of a metropolitan school district has estimated the probabilities associated with the SAT verbal scores of students from that district. The results are shown in Table 7. If a student is selected at random, what is the probability that his or her SAT verbal score will be

a. more than 400?

b. less than or equal to 500?

c. greater than 400 but less than or equal to 600?

Solution Let A, B, C, D, E, and F denote, respectively, the event that the score is greater than 700, greater than 600 but less than or equal to 700, greater than 500 but less than or equal to 600, and so forth. Then, these events are mutually exclusive. Therefore

a. The probability that the student's score will be more than 400 is given by

$$P(D \cup C \cup B \cup A) = P(D) + P(C) + P(B) + P(A)$$
$$= 0.23 + 0.19 + 0.07 + 0.01$$
$$= 0.5$$

b. The probability that the student's score will be less than or equal to 500 is given by

$$P(D \cup E \cup F) = P(D) + P(E) + P(F)$$
$$= 0.23 + 0.31 + 0.19$$
$$= 0.73$$

c. The probability that the student's score will be greater than 400 but less than or equal to 600 is given by

$$P(C \cup D) = P(C) + P(D)$$
$$= 0.19 + 0.23$$
$$= 0.42$$

Property 3 holds if and only if E and F are mutually exclusive. In the general case, we have

PROPERTY 4 (ADDITION RULE)

If E and F are any two events of an experiment, then

$$P(E \cup F) = P(E) + P(F) - P(E \cap F). \qquad \text{See Figure 27.}$$

FIGURE 27
$P(E \cup F) = P(E) + P(F) - P(E \cap F)$

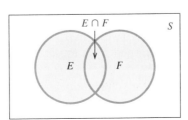

To derive the property, we use equation (4) to see that

$$n(E \cup F) = n(E) + n(F) - n(E \cap F)$$

where E and F are events of an experiment with sample space S. Dividing both sides of this equation by $n(S)$, we obtain

$$\frac{n(E \cup F)}{n(S)} = \frac{n(E)}{n(S)} + \frac{n(F)}{n(S)} - \frac{n(E \cap F)}{n(S)}$$

Recalling the definition of the probability of an event then leads to

$$P(E \cup F) = P(E) + P(F) - P(E \cap F)$$

as we wish to show.

REMARK Observe that when E and F are mutually exclusive—that is, when $E \cap F = \varnothing$—then the equation of property 4 reduces to that of property 3. In other words, if E and F are mutually exclusive events, then

$$P(E \cup F) = P(E) + P(F)$$

If E and F are not mutually exclusive events, then

$$P(E \cup F) = P(E) + P(F) - P(E \cap F) \qquad \square$$

EXAMPLE 2 A card is drawn from a well-shuffled deck of 52 playing cards. What is the probability that it is an ace or a spade?

Solution Let E denote the event that the card drawn is an ace, and let F denote the event that the card drawn is a spade. Then

$$P(E) = \frac{4}{52} \qquad \text{and} \qquad P(F) = \frac{13}{52}$$

Furthermore, E and F are not mutually exclusive events. In fact, $E \cap F$ is the event that the card drawn is an ace of spades. Consequently,

$$P(E \cap F) = \frac{1}{52}$$

The event that a card drawn is an ace or a spade is $E \cup F$, with probability given by

$$P(E \cup F) = P(E) + P(F) - P(E \cap F)$$

$$= \frac{4}{52} + \frac{13}{52} - \frac{1}{52} = \frac{4}{13} \qquad \text{See Figure 28.}$$

This result, of course, can be obtained by arguing that 16 of the 52 cards are either spades or aces of other suits.   ∎

FIGURE 28
$P(E \cup F) = P(E) + P(F) - P(E \cap F)$

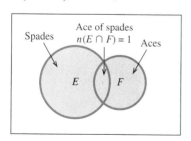

EXAMPLE 3 The quality-control department of the Vista Vision Corporation, manufacturer of the Pulsar 19-inch color television set, has determined from records obtained from the company's service centers that 3% of the sets sold experience video problems, 1% experience audio problems, and 0.1% experience both video as well as audio problems before the expiration of the 90-day warranty. Find the probability that a set purchased by a consumer will experience video or audio problems before the warranty expires.

FIGURE 29
$P(E \cup F) = P(E) + P(F) -$
$P(E \cap F)$

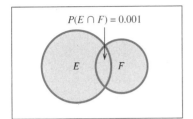

$P(E \cap F) = 0.001$

Solution Let E denote the event that a set purchased will experience video problems within 90 days, and let F denote the event that a set purchased will experience audio problems within 90 days. Then

$$P(E) = 0.03, \qquad P(F) = 0.01, \qquad P(E \cap F) = 0.001$$

The event that a set purchased will experience video problems or audio problems before the warranty expires is $E \cup F$, and the probability of this event is given by

$$P(E \cup F) = P(E) + P(F) - P(E \cap F) \qquad \text{See Figure 29.}$$
$$= 0.03 + 0.01 - 0.001$$
$$= 0.039 \qquad \blacksquare$$

Another property of the probability function that is of considerable aid in computing probability follows.

**PROPERTY 5
(RULE OF
COMPLEMENTS)**

If E is an event of an experiment, and E^c denotes the complement of E, then

$$P(E^c) = 1 - P(E)$$

Property 5 is an immediate consequence of properties 2 and 3. Indeed, we have $E \cup E^c = S$ and $E \cap E^c = \varnothing$, so

$$1 = P(S) = P(E \cup E^c) = P(E) + P(E^c)$$

and, therefore, $P(E^c) = 1 - P(E)$

EXAMPLE 4   Refer to Example 3. What is the probability that a Pulsar 19-inch color television set bought by a consumer will not experience video or audio difficulties before the warranty expires?

Solution Let E denote the event that a set bought by a consumer will experience video or audio difficulties before the warranty expires. Then the event that the set will not experience either problem before the warranty expires is given by E^c, with probability

$$P(E^c) = 1 - P(E)$$
$$= 1 - 0.039$$
$$= 0.961 \qquad \blacksquare$$

Computations Involving the Rules of Probability

We close this section by looking at two additional samples that illustrate the rules of probability.

EXAMPLE 5   Let E and F be two mutually exclusive events and suppose $P(E) = 0.1$ and $P(F) = 0.6$. Compute

a. $P(E \cap F)$ **b.** $P(E \cup F)$ **c.** $P(E^c)$ **d.** $P(E^c \cap F^c)$
e. $P(E^c \cup F^c)$

Solution

a. Since the events E and F are mutually exclusive—that is, $E \cap F = \varnothing$—we have $P(E \cap F) = 0$.

b. $P(E \cup F) = P(E) + P(F)$ Since E and F are mutually exclusive

$= 0.1 + 0.6$

$= 0.7$.

c. $P(E^c) = 1 - P(E)$ Property 5

$= 1 - 0.1$

$= 0.9$

d. Observe that, by De Morgan's Law, $E^c \cap F^c = (E \cup F)^c$, so

$$P(E^c \cap F^c) = P[(E \cup F)^c]$$ See Figure 30.

$$= 1 - P(E \cup F)$$ Property 5

$$= 1 - 0.7$$ Using the result of (b)

$$= 0.3$$

e. Again, using De Morgan's Law, we find

$$P(E^c \cup F^c) = P[(E \cap F)^c]$$

$$= 1 - P(E \cap F)$$

$$= 1 - 0$$ Using the result of (a)

$$= 1$$

FIGURE 30

$P(E^c \cap F^c) = P[(E \cup F)^c]$

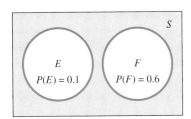

EXAMPLE 6   Let E and F be two events of an experiment with sample space S. Suppose $P(E) = 0.2$, $P(F) = 0.1$, and $P(E \cap F) = 0.05$. Compute

a. $P(E \cup F)$ **b.** $P(E^c \cap F^c)$

c. $P(E^c \cap F)$ [*Hint:* Draw a Venn diagram.]

Solution

a. $P(E \cup F) = P(E) + P(F) - P(E \cap F)$ Property 4

$= 0.2 + 0.1 - 0.05$

$= 0.25$

b. Using De Morgan's Law, we have

$$P(E^c \cap F^c) = P[(E \cup F)^c]$$

$$= 1 - P(E \cup F)$$ Property 5

$$= 1 - 0.25$$ Using the result of (a)

$$= 0.75$$

c. From the Venn diagram describing the relationship between E, F, and S (Figure 31), we have

$$P(E^c \cap F) = 0.05$$ The shaded subset is the event $E^c \cap F$.

FIGURE 31

$P(E^c \cap F)$: *the probability that the event F, but not the event E, will occur*

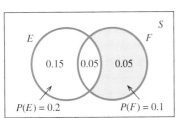

This result may also be obtained by using the relationship

$$P(E^c \cap F) = P(F) - P(E \cap F)$$
$$= 0.1 - 0.05$$
$$= 0.05$$

as before.   ■

**SELF-CHECK
EXERCISES 7.7**

1. Let E and F be events of an experiment with sample space S. Suppose $P(E) = 0.4$, $P(F) = 0.5$, and $P(E \cap F) = 0.1$. Compute

   a. $P(E \cup F)$ **b.** $P(E \cap F^c)$

2. Ms. Garcia wishes to sell or lease a condominium through a realty company. The realtor estimates that the probability of finding a buyer within a month of the date the property is listed for sale or lease is 0.3, the probability of finding a lessee is 0.8, and the probability of finding both a buyer and a lessee is 0.1. Determine the probability that the property will be sold or leased within one month from the date the property is listed for sale or lease.

Solutions to Self-Check Exercises 7.7 can be found on page 396.

7.7 EXERCISES

A pair of dice is cast and the number that appears uppermost on each die is observed. In exercises 1–6, refer to this experiment and find the probability of the given event.

1. The sum of the numbers is an even number.

2. The sum of the numbers is either 7 or 11.

3. A pair of 1s is thrown.

4. A double is thrown.

5. One die shows a 6 and the other a number less than 3.

6. The sum of the numbers is at least 4.

An experiment consists of selecting a card at random from a 52-card deck. In exercises 7–12, refer to this experiment and find the probability of the given event.

7. A king of diamonds is drawn.

8. A diamond or a king is drawn.

9. A face card is drawn.

10. A red face card is drawn.

11. An ace is not drawn.

12. A black face card is not drawn.

13. Five hundred people have purchased raffle tickets. What is the probability that a person holding one ticket will win the first prize? What is the probability that he or she will not win the first prize?

14. TV Households   The results of a recent television survey of American TV households revealed that 87 out of every 100 TV households have at least one remote control. What is the probability that a randomly selected TV household does not have at least one remote control?

In exercises 15–24, explain why the statement is incorrect.

15. The sample space associated with an experiment is given by $S = \{a, b, c\}$, where $P(a) = 0.3$, $P(b) = 0.4$, and $P(c) = 0.4$.

16. The probability that a bus will arrive late at the Civic Center is 0.35, and the probability that it will be on time or early is 0.60.

17. A person participates in a weekly office pool in which he has one chance in ten of winning the purse. If he participates for five weeks in succession, the probability of winning at least one purse is 5/10.

18. The probability that a certain stock will increase in value over a period of one week is 0.6. Therefore, the probability that the stock will decrease in value is 0.4.

19. A red die and a green die are tossed. The probability that a 6 will appear uppermost on the red die is 1/6 and the probability that a 1 will appear uppermost on the green die is 1/6. Hence, the probability that the red die will show a 6 or the green die will show a 1 is 1/6 + 1/6.

20. Joanne, a high school senior, has applied for admission to four colleges, A, B, C, and D. She has estimated that the probability that she will be accepted for admission by A, B, C, and D is 0.5, 0.3, 0.1, and 0.08, respectively. Thus, the probability that she will be accepted for admission by at least one college is $P(A) + P(B) + P(C) + P(D) = 0.5 + 0.3 + 0.1 + 0.08 = 0.98$.

21. The sample space associated with an experiment is given by $S = \{a, b, c, d, e\}$. The events $E = \{a, b\}$ and $F = \{c, d\}$ are mutually exclusive. Hence, the events E^c and F^c are mutually exclusive.

22. A 5-card poker hand is dealt from a 52-card deck. Let A denote the event that a flush is dealt and let B be the event that a straight is dealt. Then the events A and B are mutually exclusive.

23. Mr. Owens, an optician, estimates that the probability that a customer coming into his store will purchase one or more pairs of glasses but not contact lenses is 0.40, and the probability that he will purchase one or more pairs of contact lenses but not glasses is 0.25. Hence, Owens concludes that the probability that a customer coming into his store will purchase neither a pair of glasses nor a pair of contact lenses is 0.35.

24. There are eight grades in the Garfield Elementary School. If a student is selected at random from the school, then the probability that the student is in the first grade is 1/8.

25. Let E and F be two events that are mutually exclusive and suppose $P(E) = 0.2$ and $P(F) = 0.5$. Compute
  a. $P(E \cap F)$ b. $P(E \cup F)$
  c. $P(E^c)$ d. $P(E^c \cap F^c)$

26. Let E and F be two events of an experiment with sample space S. Suppose $P(E) = 0.6$, $P(F) = 0.4$, and $P(E \cap F) = 0.2$. Compute
  a. $P(E \cup F)$ b. $P(E^c)$
  c. $P(F^c)$ d. $P(E^c \cap F)$

27. Let $S = \{s_1, s_2, s_3, s_4\}$ be the sample space associated with an experiment having the probability distribu-

tion shown in the accompanying table. If $A = \{s_1, s_2\}$ and $B = \{s_1, s_3\}$, find
  a. $P(A)$, $P(B)$ b. $P(A^c)$, $P(B^c)$
  c. $P(A \cap B)$ d. $P(A \cup B)$

Outcome	Probability
$\{s_1\}$	$\frac{1}{8}$
$\{s_2\}$	$\frac{3}{8}$
$\{s_3\}$	$\frac{1}{4}$
$\{s_4\}$	$\frac{1}{4}$

28. Let $S = \{s_1, s_2, s_3, s_4, s_5, s_6\}$ be the sample space associated with an experiment having the probability distribution shown in the accompanying table. If $A = \{s_1, s_2\}$ and $B = \{s_1, s_5, s_6\}$, find
  a. $P(A)$, $P(B)$ b. $P(A^c)$, $P(B^c)$
  c. $P(A \cap B)$ d. $P(A \cup B)$
  e. $P(A^c \cap B^c)$ f. $P(A^c \cup B^c)$

Outcome	Probability
$\{s_1\}$	$\frac{1}{3}$
$\{s_2\}$	$\frac{1}{8}$
$\{s_3\}$	$\frac{1}{6}$
$\{s_4\}$	$\frac{1}{6}$
$\{s_5\}$	$\frac{1}{12}$
$\{s_6\}$	$\frac{1}{8}$

29. **Teacher Attitudes**   In a survey of 2140 teachers in a certain metropolitan area conducted by a nonprofit organization regarding teacher attitudes, the following data were obtained:

900 said that lack of parental support is a problem

890 said that abused or neglected children are problems

680 said that malnutrition or students in poor health is a problem

120 said that lack of parental support and abused or neglected children are problems

110 said that lack of parental support and malnutrition or poor health are problems

140 said that abused or neglected children and malnutrition or poor health are problems

40 said that lack of parental support, abuse or neglect, and malnutrition or poor health are problems

What is the probability that a teacher selected at random from this group said that lack of parental support is the only problem hampering a student's schooling? [*Hint:* Draw a Venn diagram.]

30. Course Enrollments Among 500 freshmen pursuing a business degree at a university, 320 are enrolled in an economics course, 225 are enrolled in a mathematics course, and 140 are enrolled in both an economics and a mathematics course. What is the probability that a freshman selected at random from this group is enrolled in

a. an economics and/or a mathematics course?
b. exactly one of these two courses?
c. neither an economics course nor a mathematics course?

31. Consumer Surveys A leading manufacturer of kitchen appliances advertised its products in two magazines: *Good Housekeeping* and the *Ladies Home Journal.* A survey of 500 customers revealed that 140 learned of its products from *Good Housekeeping,* 130 learned of its products from the *Ladies Home Journal,* and 80 learned of its products from both magazines. What is the probability that a person selected at random from this group saw the manufacturer's advertisement in

a. both magazines?
b. at least one of the two magazines?
c. exactly one magazine?

32. Study Habits Students at a certain university were asked to state how many hours per week they spent studying in the library. Results of the survey revealed the following information.

Time Spent (in hours) (x)	Percentage of Students
$0 \le x \le 1$	32.3
$1 < x \le 4$	40.7
$4 < x \le 10$	16.5
$x > 10$	10.5

Find the probability distribution associated with these data. What is the probability that a student selected at random at the university studied in the library

a. more than 4 hours per week?
b. no more than 10 hours per week?

33. Assembly Time Studies A time study was conducted by the production manager of the Universal Instruments Company to determine how much time it took an assembly worker to complete a certain task during the assembly of its Galaxy home computers. Results of the study indicated that 20% of the workers were able to complete the task in less than 3 minutes,

60% of the workers were able to complete the task in 4 minutes or less, and 10% of the workers required more than 5 minutes to complete the task. If an assembly line worker is selected at random from this group, what is the probability that

a. he or she will be able to complete the task in 5 minutes or less?
b. he or she will not be able to complete the task within 4 minutes?
c. the time taken for the worker to complete the task will be between 3 and 4 minutes (inclusive)?

34. Plans to Keep Cars In a survey conducted to see how long Americans keep their cars, 2000 automobile owners were asked how long they plan to keep their present cars. The results of the survey follow.

Number of Years Car Is Kept (x)	Number of Respondents
$0 \le x < 1$	60
$1 \le x < 3$	440
$3 \le x < 5$	360
$5 \le x < 7$	340
$7 \le x < 10$	240
$x \ge 10$	560

Find the probability distribution associated with these data. What is the probability that an automobile owner selected at random from those surveyed plans to keep his or her present car

a. less than five years?
b. more than three years?

35. Gun-Control Laws A poll was conducted among 250 residents of a certain city regarding tougher gun-control laws. The results of the poll are shown in the accompanying table.

	Own Only a Handgun	Own Only a Rifle	Own a Handgun and a Rifle	Own Neither	Total
Favor Tougher Laws	0	12	0	138	150
Oppose Tougher Laws	58	5	25	0	88
No Opinion	0	0	0	12	12
Total	58	17	25	150	250

If one of the participants in this poll is selected at random, what is the probability that he or she

a. favors tougher gun-control laws?

b. owns a handgun?

c. owns a handgun but not a rifle?

d. favors tougher gun-control laws and does not own a handgun?

SOLUTIONS TO SELF-CHECK EXERCISES 7.7

1. a. Using property 4, we find

$$P(E \cup F) = P(E) + P(F) - P(E \cap F)$$

$$= 0.4 + 0.5 - 0.1$$

$$= 0.8$$

b. From the accompanying Venn diagram, in which the subset $E \cap F^c$ is shaded, we see that

$$P(E \cap F^c) = 0.3$$

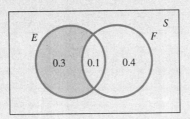

The result may also be obtained by using the relationship

$$P(E \cap F^c) = P(E) - P(E \cap F)$$

$$= 0.4 - 0.1 = 0.3$$

2. Let E denote the event that the property will be sold within one month of the date it is listed for sale or lease, and let F denote the event that the property will be leased within the same time period. Then

$$P(E) = 0.3, \qquad P(F) = 0.8, \qquad P(E \cap F) = 0.1$$

The probability of the event that the property will be sold or leased within one month of the date it is listed for sale or lease is given by

$$P(E \cup F) = P(E) + P(F) - P(E \cap F)$$

$$= 0.3 + 0.8 - 0.1 = 1$$

that is, a certainty.

CHAPTER SEVEN SUMMARY OF PRINCIPAL FORMULAS AND TERMS

Formulas

1. Commutative laws

$$A \cup B = B \cup A$$
$$A \cap B = B \cap A$$

2. Associative laws

$$A \cup (B \cup C) = (A \cup B) \cup C$$
$$A \cap (B \cap C) = (A \cap B) \cap C$$

3. Distributive laws

$$A \cup (B \cap C)$$
$$= (A \cup B) \cap (A \cup C)$$
$$A \cap (B \cup C)$$
$$= (A \cap B) \cup (A \cap C)$$

4. De Morgan's laws

$$(A \cup B)^c = A^c \cap B^c$$
$$(A \cap B)^c = A^c \cup B^c$$

5. Number of elements in the union of two finite sets

$$n(A \cup B) = n(A) + n(B) - n(A \cap B)$$

6. Permutation of n distinct objects

$$P(n, r) = \frac{n!}{(n - r)!}$$

7. Permutation of n objects, not all distinct, taken r at a time

$$\frac{n!}{n_1! n_2! \cdots n_r!}$$

8. Combination of n distinct objects

$$C(n, r) = \frac{n!}{r!(n - r)!}$$

9. Probability of an event

$$P(E) = \frac{n(E)}{n(S)}$$

10. Probability of the union of two mutually exclusive events

$$P(E \cup F) = P(E) + P(F)$$

11. Addition Rule

$$P(E \cup F) = P(E) + P(F) - P(E \cap F)$$

12. Rule of Complements

$$P(E^c) = 1 - P(E)$$

Terms

Set
Element of a set
Roster notation
Set-builder notation
Set equality
Subset
Proper subset
Empty set
Universal set
Venn diagram
Set union
Set intersection
Disjoint set
Set complementation
Combinatorics
Multiplication principle
Tree diagram
Generalized multiplication
  principle
Permutation

n-Factorial
Combination
Experiment
Outcome
Sample point
Sample space
Event of an experiment
Finite sample space
Union of two events
Intersection of two events
Complement of an event
Mutually exclusive events
Relative frequency
Empirical probability
Probability of an event
Elementary (simple) events
Probability distribution
Probability function
Uniform sample space
Addition principle

CHAPTER 7 REVIEW EXERCISES

In exercises 1–4, list the elements of the given set in roster notation.

1. $\{x \mid 3x - 2 = 7; x, \text{ an integer}\}$

2. $\{x \mid x \text{ is a letter of the word } TALLAHASSEE\}$

3. The set whose elements are the even numbers between 3 and 11

4. $\{x \mid (x - 3)(x + 4) = 0; x, \text{ a negative integer}\}$

Let $A = \{a, c, e, r\}$. In exercises 5–8, determine whether the given set is equal to A.

5. $\{r, e, c, a\}$

6. $\{x \mid x \text{ is a letter of the word } career\}$

7. $\{x \mid x \text{ is a letter of the word } racer\}$

8. $\{x \mid x \text{ is a letter of the word } cares\}$

In exercises 9–12, shade the portion of the accompanying figure that represents the set.

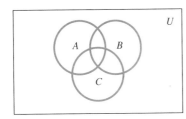

9. $A \cup (B \cap C)$

10. $(A \cap B \cap C)^c$

11. $A^c \cap B^c \cap C^c$

12. $A^c \cap (B^c \cup C^c)$

Let $U = \{a, b, c, d, e\}$, $A = \{a, b\}$, $B = \{b, c, d\}$, and $C = \{a, d, e\}$. In exercises 13–16, verify the equation by direct computation.

13. $A \cup (B \cup C) = (A \cup B) \cup C$

14. $A \cap (B \cap C) = (A \cap B) \cap C$

15. $A \cap (B \cup C) = (A \cap B) \cup (A \cap C)$

16. $A \cup (B \cap C) = (A \cup B) \cap (A \cup C)$

Let $U = \{$all participants in a consumer behavior survey conducted by a national polling group$\}$

$A = \{$consumers who avoided buying a product because it is not recyclable$\}$

$B = \{$consumers who used cloth rather than disposable diapers$\}$

$C = \{$consumers who boycotted a company's products because of their record on the environment$\}$

and $D = \{$consumers who voluntarily recycled their garbage$\}$

In exercises 17–20, describe each set in words.

17. $A \cap C$

18. $A \cup D$

19. $B^c \cap D$

20. $C^c \cup D^c$

Let A and B be subsets of a universal set U and suppose $n(U) = 350$, $n(A) = 120$, $n(B) = 80$, and $n(A \cap B) = 50$. In exercises 21–26, find the number of elements in each set.

21. $n(A \cup B)$

22. $n(A^c)$

23. $n(B^c)$

24. $n(A^c \cap B)$

25. $n(A \cap B^c)$

26. $n(A^c \cap B^c)$

In exercises 27–30, evaluate each expression.

27. $C(20, 18)$

28. $P(9, 7)$

29. $C(5, 3) \cdot P(4, 2)$

30. $4 \cdot P(5, 3) \cdot C(7, 4)$

31. Let E and F be two mutually exclusive events and suppose $P(E) = 0.4$ and $P(F) = 0.2$. Compute
    a. $P(E \cap F)$ **b.** $P(E \cup F)$
    c. $P(E^c)$ **d.** $P(E^c \cap F^c)$
    e. $P(E^c \cup F^c)$

32. Let E and F be two events of an experiment with sample space S. Suppose $P(E) = 0.3$, $P(F) = 0.2$, and $P(E \cap F) = 0.15$. Compute
    a. $P(E \cup F)$ **b.** $P(E^c \cap F^c)$
    c. $P(E^c \cap F)$

33. Suppose a die is loaded and it was determined that the probability distribution associated with the experiment of casting the die and observing which number falls uppermost is given by

Simple Event	Probability
{1}	0.20
{2}	0.12
{3}	0.16
{4}	0.18
{5}	0.15
{6}	0.19

a. What is the probability of the number being even?

b. What is the probability of the number being either a 1 or a 6?

a. What is the probability of the number being less than 4?

34. A comparison of 5 major credit cards showed that

3 offered cash advances

3 offered extended payments for *all* goods and services purchased

2 required an annual fee of less than $35

2 offered both cash advances and extended payments

1 offered extended payments and had an annual fee less than $35

no card had an annual fee less than $35 and offered both cash advances and extended payments

How many cards had an annual fee less than $35 and offered cash advances? (Assume that every card had at least 1 of the 3 mentioned features.)

35. The Department of Foreign Languages of a liberal arts college conducted a survey of its recent graduates to determine the foreign language courses they had taken while undergraduates at the college. Of the 480 graduates

200 had at least one year of Spanish

178 had at least one year of French

140 had at least one year of German

33 had at least one year of Spanish and French

24 had at least one year of Spanish and German

18 had at least one year of French and German

3 had at least one year of all three languages

How many of the graduates had

a. at least one year of at least one of the three languages?

b. at least one year of exactly one of the three languages?

c. less than a year of any of the three languages?

36. In how many ways can six different compact discs be arranged on a shelf?

37. In how many ways can three pictures be selected from a group of six different pictures?

38. In an election being held by the Associated Students Organization there are six candidates for president, four for vice-president, five for secretary, and six for treasurer. How many different possible outcomes are there for this election?

39. How many three-digit numbers can be formed from the numerals in the set {1, 2, 3, 4 ,5}
   a. if repetition of digits is not allowed?
   b. if repetition of digits is allowed?

40. Find the number of distinguishable permutations that can be formed from the letters of each word.
   a. *CINCINNATI*   b. *HONOLULU*

41. From a standard 52-card deck, how many 5-card poker hands can be dealt consisting of
   a. 5 clubs?   b. 3 kings and a pair?

42. In how many ways can seven students be assigned seats in a row containing seven desks
   a. if there are no restrictions?
   b. if two of the students must not be seated next to each other?

43. There are eight seniors and six juniors in the Math Club at Jefferson High School. In how many ways can a math team consisting of four seniors and two juniors be selected from the members of the Math Club?

44. A sample of 4 balls is to be selected at random from an urn containing 15 balls numbered 1 to 15. If 6 balls are green, 5 are white, and 4 are black,
   a. how many different samples can be selected?
   b. how many samples can be selected that contain at least 1 white ball?

45. From a shipment of 60 transistors, 5 of which are defective, a sample of 4 transistors is selected at random.
   a. In how many different ways can the sample be selected?
   b. How many samples contain 3 defective transistors?
   c. How many samples do not contain any defective transistors?

46. The quality-control department of the Starr Communications Company, the manufacturer of video-game cartridges, has determined from records that 1.5% of the cartridges sold have video defects, 0.8% have audio defects, and 0.4% have both audio and video defects. What is the probability that a cartridge purchased by a customer

a. will have a video or audio defect?

b. will not have a video or audio defect?

CHAPTER 7   EXTENDED PROBLEMS AND PROJECTS

This project requires the use of a graphing utility.

a. You can use the **rand** (random number) function of a graphing utility to simulate a coin-tossing experiment as follows: Generate random numbers between 0 and 1 and let each number generated represent the outcome of the experiment, with a number less than 0.5 representing tails and a number greater than or equal to 0.5 representing heads. Record the number of tosses, the number of heads, and the relative frequency of heads (see Table 2, page 378). Does the relative frequency approach 0.5 as the number of trials gets larger and larger?

b. Can you devise a similar project for simulating an experiment involving casting a fair die?

Additional Topics in Probability

8

*I*n this chapter we develop additional techniques for computing the probabilities of certain events and we take a look at descriptive statistics. We begin by looking at problems involving large sample spaces. In Section 8.2, we consider the effect that the occurrence of prior events has on the probability of an event occurring, and in Section 8.3 we learn how to compute the probabilities of certain events after the occurrence of an event. Throughout our discussion, we will see how these techniques are applied to many practical problems in fields as diverse as quality control and medical research. In the rest of the chapter we take a glimpse at statistics, the branch of mathematics concerned with the collection, analysis, and interpretation of data.

8.1   Use of Counting Techniques in Probability

Further Applications of Counting Techniques

As we have seen many times before, a problem in which the underlying sample space has a small number of elements may be solved by first determining all such sample points. For problems involving sample spaces with a large number of sample points, however, this approach is neither practical nor desirable.

In this section we see how the counting techniques studied in Chapter 7 may be employed to help us solve problems in which the associated sample spaces contain large numbers of sample points. In particular, we restrict our attention to the study of **uniform sample spaces**—that is, sample spaces in which the outcomes are equally likely. For such spaces we have the following result:

COMPUTING THE PROBABILITY OF AN EVENT IN A UNIFORM SAMPLE SPACE

Let S be a uniform sample space and let E be any event. Then,

$$P(E) = \frac{\text{Number of favorable outcomes in } E}{\text{Number of possible outcomes in } S} = \frac{n(E)}{n(S)} \qquad (1)$$

EXAMPLE 1   An unbiased coin is tossed six times.

a. What is the probability that the coin will land heads exactly three times?

b. What is the probability that the coin will land heads at most three times?

c. What is the probability that the coin will land heads on the first and the last toss?

Solution

a. Each outcome of the experiment may be represented as a sequence of heads and tails. Using the generalized multiplication principle, we see that the number of outcomes of this experiment is given by 2^6, or 64. Let E denote the event that the coin lands heads exactly three times. Since there are $C(6, 3)$ ways this can occur, we see that the required probability is

$$P(E) = \frac{n(E)}{n(S)} = \frac{C(6, 3)}{64} = \frac{\dfrac{6!}{3!3!}}{64} \qquad \text{\textit{S, sample space of the experiment}}$$

$$= \frac{\dfrac{6 \cdot 5 \cdot 4}{3 \cdot 2}}{64} = \frac{20}{64} = \frac{5}{16} = 0.3125$$

b. Let F denote the event that the coin lands heads at most three times. Then $n(F)$ is given by the sum of the number of ways the coin lands heads zero times (no heads!), the number of ways it lands heads exactly once, the number of ways it lands heads exactly twice, and the number of ways it lands heads exactly three times. That is,

$$n(F) = C(6, 0) + C(6, 1) + C(6, 2) + C(6, 3)$$

$$= \frac{6!}{0!6!} + \frac{6!}{1!5!} + \frac{6!}{2!4!} + \frac{6!}{3!3!}$$

$$= 1 + 6 + \frac{(6)(5)}{2} + \frac{(6)(5)(4)}{(3)(2)} = 42$$

Therefore, the required probability is

$$P(F) = \frac{n(F)}{n(S)} = \frac{42}{64} = \frac{21}{32} \approx 0.66$$

c. Let F denote the event that the coin land heads on the first and the last toss. Then $n(F) = 1 \cdot 2 \cdot 2 \cdot 2 \cdot 2 \cdot 1 = 2^4$, so the probability that this event occurs is

$$P(F) = \frac{2^4}{2^6}$$

$$= \frac{1}{2^2}$$

$$= \frac{1}{4}$$

∎

EXAMPLE 2   Two cards are selected at random from a well-shuffled pack of 52 playing cards.

a. What is the probability that they are both aces?

b. What is the probability that neither of them is an ace?

Solution

a. The experiment consists of selecting 2 cards from a pack of 52 playing cards. Since the order in which the cards are selected is immaterial, the sample points are combinations of 52 cards taken 2 at a time. Now, there are $C(52, 2)$ ways of selecting 52 cards taken 2 at a time, so the number of elements in the sample space S is given by $C(52, 2)$. Next, we observe that there are $C(4, 2)$ ways of selecting 2 aces from the 4 in the deck. Therefore, if E denotes the event that the cards selected are both aces, then

$$P(E) = \frac{n(E)}{n(S)}$$

$$= \frac{C(4, 2)}{C(52, 2)} = \frac{\dfrac{4!}{2!2!}}{\dfrac{52!}{2!50!}}$$

$$= \frac{1}{221}$$

b. Let F denote the event that neither of the two cards selected is an ace. Since there are $C(48, 2)$ ways of selecting two cards, neither of which is an ace, we find that

$$P(F) = \frac{n(F)}{n(S)} = \frac{C(48, 2)}{C(52, 2)} = \frac{\dfrac{48!}{2!46!}}{\dfrac{52!}{2!50!}} = \frac{48 \cdot 47}{2} \cdot \frac{2}{52 \cdot 51}$$

$$= \frac{188}{221}$$

EXAMPLE 3   A bin in the hi-fi department of Building 20, a bargain outlet, contains 100 blank cassette tapes of which 10 are known to be defective. If a customer selects 6 of these cassette tapes, determine the probability that

a. 2 of them are defective   **b.** at least 1 of them is defective

Solution

a. There are $C(100, 6)$ ways of selecting a set of 6 cassette tapes from the 100, and this gives $n(S)$, the number of outcomes in the sample space associated with the experiment. Next, we observe that there are $C(10, 2)$ ways of selecting a set of 2 defective cassette tapes from the 10 defective cassette tapes and $C(90, 4)$ ways of selecting a set of 4 nondefective cassette tapes from the 90 nondefective cassette tapes (Figure 1). Thus, by the multiplication principle, there are $C(10, 2) \cdot C(90, 4)$ ways of selecting 2 defective and 4 nondefective cassette tapes. Therefore, the probability of selecting 6 cassette tapes of which 2 are defective is given by

FIGURE 1

A sample of 6 tapes selected from 90 nondefective tapes and 10 defective tapes

$$\frac{C(10, 2) \cdot C(90, 4)}{C(100, 6)} = \frac{\dfrac{10!}{2!8!} \dfrac{90!}{4!86!}}{\dfrac{100!}{6!94!}}$$

$$= \frac{10 \cdot 9}{2} \cdot \frac{90 \cdot 89 \cdot 88 \cdot 87}{4 \cdot 3 \cdot 2 \cdot 1} \cdot \frac{6 \cdot 5 \cdot 4 \cdot 3 \cdot 2 \cdot 1}{100 \cdot 99 \cdot 98 \cdot 97 \cdot 96 \cdot 95}$$

$$\approx 0.096$$

b. Let E denote the event that none of the cassette tapes selected is defective. Then E^c gives the event that at least 1 of the cassette tapes is defective. But, by the Rule of Complements,

$$P(E^c) = 1 - P(E)$$

To compute $P(E)$, we observe that there are $C(90, 6)$ ways of selecting a set of 6 cassette tapes that are nondefective. Therefore,

$$P(E) = \frac{C(90, 6)}{C(100, 6)}$$

and

$$P(E^c) = 1 - \frac{C(90, 6)}{C(100, 6)}$$

$$= 1 - \frac{\dfrac{90!}{6!84}}{\dfrac{100!}{6!94!}}$$

$$= 1 - \frac{90 \cdot 89 \cdot 88 \cdot 87 \cdot 86 \cdot 85}{6 \cdot 5 \cdot 4 \cdot 3 \cdot 2 \cdot 1} \cdot \frac{6 \cdot 5 \cdot 4 \cdot 3 \cdot 2 \cdot 1}{100 \cdot 99 \cdot 98 \cdot 97 \cdot 96 \cdot 95}$$

$$\approx 0.48 \qquad\blacksquare$$

The Birthday Problem

EXAMPLE 4 A group of 5 people is selected at random. What is the probability that at least 2 of them have the same birthday?

Solution For simplicity we will assume that none of the 5 people was born on February 29 of a leap year. Since the 5 people were selected at random, we may also assume that each of them is equally likely to have any of the 365 days of a year as his or her birthday. If we let A, B, C, D, and E represent the 5 people, then an outcome of the experiment may be represented by (a, b, c, d, e), where the numbers a, b, c, d, and e give the birthdays of A, B, C, D, and E, respectively.

We first observe that since there are 365 possibilities for each of the dates a, b, c, d, and e, the multiplication principle implies that there are

$$\boxed{365} \cdot \boxed{365} \cdot \boxed{365} \cdot \boxed{365} \cdot \boxed{365}$$
$$\quad a \qquad b \qquad c \qquad d \qquad e$$

or 365^5, outcomes of the experiment. Therefore,

$$n(S) = 365^5$$

where S denotes the sample space of the experiment.

Next, let E denote the event that 2 or more of the 5 people have the same birthday. It is now necessary to compute $P(E)$. However, a direct computation of $P(E)$ is relatively difficult. It is much easier to compute $P(E^c)$, where E^c is the event that no 2 of the 5 people have the same birthday, and then use the relation

$$P(E) = 1 - P(E^c)$$

To compute $P(E^c)$, observe that there are 365 ways (corresponding to the 365 dates) on which A's birthday can occur followed by 364 ways on which B's birthday could occur if B were not to have the same birthday as A, and so on. Therefore, by the generalized multiplication principle,

$$n(E^c) = \underset{\substack{\text{A's}\\ \text{birthday}}}{365} \cdot \underset{\substack{\text{B's}\\ \text{birthday}}}{364} \cdot \underset{\substack{\text{C's}\\ \text{birthday}}}{363} \cdot \underset{\substack{\text{D's}\\ \text{birthday}}}{362} \cdot \underset{\substack{\text{E's}\\ \text{birthday}}}{361}$$

Thus,
$$P(E^c) = \frac{n(E^c)}{n(S)}$$

$$= \frac{365 \cdot 364 \cdot 363 \cdot 362 \cdot 361}{365^5}$$

and
$$P(E) = 1 - P(E^c)$$

$$= 1 - \frac{365 \cdot 364 \cdot 363 \cdot 362 \cdot 361}{365^5}$$

$$\approx 0.027$$ ■

We can extend the result obtained in Example 4 to the general case involving r people. In fact, if E denotes the event that at least 2 of the r people have the same birthday, an argument similar to that used in Example 4 leads to the result

$$P(E) = 1 - \frac{365 \cdot 364 \cdot 363 \cdot \cdots \cdot (365 - r + 1)}{365^r}$$

By letting r take on the values 5, 10, 15, 20, ... , 50, in turn, we obtain the probabilities that at least 2 of 5, 10, 15, 20, ... , 50 people, respectively, have the same birthday. These results are summarized in Table 1.

TABLE 1
Probability that at least two people in a randomly selected group of r people have the same birthday

r	5	10	15	20	22	23	25	30	40	50
$P(E)$	0.027	0.117	0.253	0.411	0.476	0.507	0.569	0.706	0.891	0.970

The results show that in a group of 23 randomly selected people, the chances are greater than 50% that at least 2 of them will have the same birthday. In a group of 50 people, it is an excellent bet that at least 2 people in the group will have the same birthday.

**SELF-CHECK
EXERCISES 8.1**

1. Four balls are selected at random without replacement from an urn containing 10 white balls and 8 red balls. What is the probability that all the chosen balls are white?

2. A box contains 20 microchips of which 4 are substandard. If 2 of the chips are taken from the box, what is the probability that they are both substandard?

Solutions to Self-Check Exercises 8.1 can be found on page 409.

8.1 EXERCISES

c *A calculator is recommended for this exercise set. An un-biased coin is tossed five times. In exercises 1–4, find the probability of the given event.*

1. The coin lands heads all five times.

2. The coin lands heads exactly once.

3. The coin lands heads at least once.

4. The coin lands heads more than once.

Two cards are selected at random without replacement from a well-shuffled deck of 52 playing cards. In exercises 5–8, find the probability of the given event.

5. A pair is drawn.

6. A pair is not drawn.

7. Two black cards are drawn.

8. Two cards of the same suit are drawn.

Four balls are selected at random without replacement from an urn containing three white balls and five blue balls. In exercises 9–12, find the probability of the given event.

9. Two of the balls are white and two are blue.

10. All of the balls are blue.

11. Exactly three of the balls are blue.

12. Two or three of the balls are white.

Assume that the probability of a boy being born is the same as the probability of a girl being born. In exercises 13–16, find the probability that a family with three children will have the given composition.

13. Two boys and one girl

14. At least one girl

15. No girls

16. The two oldest children are girls.

17. An exam consists of ten true-or-false questions. If a student guesses at every answer, what is the probability that he or she will answer exactly six questions correctly?

18. **Personnel Selection** Jacobs & Johnson, Inc., an accounting firm, employs 14 accountants, of whom 8 are CPAs. If a delegation of 3 accountants is randomly selected from the firm to attend a conference, what is the probability that 3 CPAs will be selected?

19. **Quality Control** Two light bulbs are selected at random from a lot of 24, of which 4 are defective. What is the probability that
   a. both of the light bulbs are defective?
   b. at least 1 of the light bulbs is defective?

20. A customer at Cavallaro's Fruit Stand picks a sample of 3 oranges at random from a crate containing 60 oranges, of which 4 are rotten. What is the probability that the sample contains 1 or more rotten oranges?

21. **Quality Control** A shelf in the Metro Department Store contains 80 typewriter cartridge ribbons for a popular portable typewriter. Six of the cartridges are defective. If a customer selects 2 of these cartridges at random from the shelf, determine the probability that
   a. both are defective
   b. at least 1 is defective

22. **Quality Control** Electronic baseball games manufactured by Tempco Electronics are shipped in lots of 24. Before shipping, a quality-control inspector randomly selects a sample of 8 from each lot for testing. If the sample contains any defective games, the entire lot is rejected. What is the probability that a lot containing exactly 2 defective games will still be shipped?

23. **Personnel Selection** The City Transit Authority plans to hire 12 new bus drivers. From a group of 100 qualified applicants, of which 60 are men and 40 are women, 12 names are to be selected by lot. If Mary and John Lewis are among the 100 qualified applicants,
   a. what is the probability that Mary's name will be selected? That both Mary's and John's names will be selected?
   b. and it is stipulated that an equal number of men and women are to be selected (6 men from the group of 60 men and 6 women from the group of 40 women), what is the probability that Mary's name will be selected? That Mary's and John's names will be selected?

24. **Public Housing** The City Housing Authority has received 50 applications from qualified applicants for 8 low-income apartments. Three of the apartments are on the north side of town and 5 are on the south side. If the apartments are to be assigned by means of a lottery, what is the probability that
   a. a specific qualified applicant will be selected for one of these apartments?
   b. two specific qualified applicants will be selected for apartments on the same side of town?

25. A student studying for a vocabulary test knows the meanings of 12 words from a list of 20 words. If the test contains 10 words from the study list, what is the probability that at least 8 of the words on the test are words that the student knows?

26. **Drivers' Tests** Four different written driving tests are administered by the City Motor Vehicle Department. One of these four tests is selected at random for each applicant for a driver's license. If a group consisting of two women and three men apply for a license, what is the probability that
   a. exactly two of the five will take the same test?
   b. the two women will take the same test?

27. Brand Selection A druggist wishes to select three brands of aspirin to sell in his store. He has five major brands to choose from: brands A, B, C, D, and E. If he selects the three brands at random, what is the probability that he will select

a. brand B?   b. brands B and C?

c. at least one of the two brands B and C?

28. Blackjack In the game of blackjack, a 2-card hand consisting of an ace and a face card or a 10 is called a blackjack.

a. If a player is dealt 2 cards from a standard deck of 52 well-shuffled cards, what is the probability that the player will receive a blackjack?

b. If a player is dealt 2 cards from 2 well-shuffled standard decks, what is the probability that the player will receive a blackjack?

29. Slot Machines Refer to exercise 23, Section 7.3, where the "Lucky Dollar" slot machine was described. What is the probability that the three "Lucky Dollar" symbols will appear in the window of the slot machine?

30. Roulette In 1959 a world record was set for the longest run on an ungaffed (fair) roulette wheel at the El San Juan Hotel in Puerto Rico. The number 10 appeared 6 times in a row. What is the probability of the occurrence of this event? (Assume that there are 38 equally likely outcomes consisting of the numbers 1–36, 0, and 00.)

In "The Numbers Game," a state lottery, four numbers are drawn with replacement from an urn containing the digits 0–9, inclusive. In exercises 31–34, find the probability of a ticket holder having the indicated winning ticket.

31. All four digits in exact order (the grand prize)

32. Two specified, consecutive digits in exact order (the first two digits, the middle two digits, or the last two digits)

33. One specified digit in exact order (the first, second, third, or fourth digit)

34. All four digits in any order

A list of poker hands ranked in order from the highest to the lowest is shown in the accompanying table, along with a description and example of each hand. Use the table to answer exercises 35–40.

Hand	Description	Example
Straight flush	5 cards in sequence in the same suit	A ♥ 2 ♥ 3 ♥ 4 ♥ 5 ♥

Hand	Description	Example
Four of a kind	4 cards of the same rank and any other card	K ♥ K ♦ K ♠ K ♣ 2 ♥
Full house	3 of a kind and a pair	3 ♥ 3 ♦ 3 ♠ 7 ♣ 7 ♥
Flush	5 cards of the same suit that are not all in sequence	5 ♥ 6 ♥ 9 ♥ J ♥ K ♥
Straight	5 cards in sequence but not all of the same suit	10 ♥ J ♦ Q ♣ K ♠ A ♥
Three of a kind	3 cards of the same rank and 2 unmatched cards	K ♥ K ♦ K ♠ 2 ♥ 4 ♦
Two pair	2 cards of the same rank and 2 cards of any other rank with an unmatched card	K ♥ K ♦ 2 ♥ 2 ♠ 4 ♣
One pair	2 cards of the same rank and 3 unmatched cards	K ♥ K ♦ 5 ♥ 2 ♠ 4 ♥

If a 5-card poker hand is dealt from a well-shuffled deck of 52 cards, what is the probability of being dealt the given hand?

35. A straight flush (Note that an ace may be played as either a high or low card in a straight sequence—that is, A, 2, 3, 4, 5 or 10, J, Q, K, A. Hence, there are 10 possible sequences for a straight in one suit.)

36. A straight (but not a straight flush)

37. A flush (but not a straight flush)

38. Four of a kind

39. A full house

40. Two pairs

41. Zodiac Signs   There are 12 signs of the Zodiac: Aries, Taurus, Gemini, Cancer, Leo, Virgo, Libra, Scorpio, Sagittarius, Capricorn, Aquarius, and Pisces. Each sign corresponds to a different calendar period of approximately one month. Assuming that a person is just as likely to be born under one sign as another, what is the probability that in a group of 5 people
   a. at least 2 of them have the same sign?
   b. at least 2 of them were born under the sign of Aries?

42. Birthday Problem   What is the probability that at least two justices of the U.S. Supreme Court have the same birthday?

43. Birthday Problem   Fifty people are selected at random. What is the probability that none of the people in this group have the same birthday?

44. Birthday Problem   During an episode of the ''Tonight Show,'' Johnny Carson related ''The Birthday Problem'' to the audience—noting that, in a group of 50 or more people, probabilists have calculated that the probability of at least 2 people having the same birthday is very high. To illustrate this point, he proceeded to conduct his own experiment. A person selected at random from the audience was asked to state his birthday. Mr. Carson then asked if anyone in the audience had the same birthday. The response was negative. He repeated the experiment. Once again, the response was negative. These results, observed Mr. Carson, were contrary to expectations. In a later episode of the show Mr. Carson explained why this experiment had been improperly conducted. Explain why Mr. Carson failed to illustrate the point he was trying to make in the earlier episode.

SOLUTIONS TO SELF-CHECK EXERCISES 8.1

1. The probability that all 4 balls selected are white is given by

$$\frac{\text{The number of ways of selecting 4 white balls from the 10 in the urn}}{\text{The number of ways of selecting any 4 balls from the 18 balls in the urn}} = \frac{C(10,4)}{C(18,4)}$$

$$= \frac{\dfrac{10!}{4!6!}}{\dfrac{18!}{4!14!}}$$

$$= \frac{10 \cdot 9 \cdot 8 \cdot 7}{4 \cdot 3 \cdot 2} \cdot \frac{4 \cdot 3 \cdot 2}{18 \cdot 17 \cdot 16 \cdot 15}$$

$$= 0.069$$

2. The probability that both chips are substandard is given by

$$\frac{\text{The number of ways of choosing any 2 of the 4 substandard chips}}{\text{The number of ways of choosing any 2 of the 20 chips}} = \frac{C(4,2)}{C(20,2)}$$

$$= \frac{\dfrac{4!}{2!2!}}{\dfrac{20!}{2!18!}}$$

$$= \frac{4 \cdot 2}{2} \cdot \frac{2}{20 \cdot 19}$$

$$= 0.032$$

Lilli Meiselman

Buying women's clothes for a chain of 50 discount clothing stores is a demanding job, requiring Meiselman to be part fashion arbiter and part accountant. She must balance her fashion choices against a bottom line that has to show a profit. To make her job even more difficult, Meiselman's selections are judged by women from northern New England to the midwestern states. What sells in one area may not sell in another. An incorrect choice can be a costly mistake.

Title: Buyer

Meiselman's job is challenging. She visits New York at least twice a month to buy clothes from a number of different manufacturers. She works with the store's in-house advertising agency, planning advertising as well as approving copy. Meiselman notes that the vast majority of the ads appear in local newspapers, the store's principal advertising medium.

For all the demands on her time, Meiselman enjoys her work. "It's gratifying to see things I bought for the stores being sold," she says. Although Meiselman decides how much to spend on particular items and which styles, sizes, and colors to select, her decisions are the end result of detailed plans that guide her buying decisions.

Based on the previous year's sales, Meiselman will "work with a departmental planner" to develop a seasonal merchandise plan (in retail there are only two seasons, fall and spring, each covering a six-month time span).

As part of that merchandising plan, Meiselman must determine how much inventory is required per month. If she needs $600,000 worth of inventory at the beginning of June, she subtracts projected sales of $400,000 from the May inventory of $800,000, leaving a balance of $400,000. To meet her June inventory goal of $600,000, she has to increase her inventory by an additional $200,000.

The finances have to be coordinated with actual quantities of suits, dresses, blouses, outerwear, and so on. Using a sales-to-stock ratio, Meiselman estimates that perhaps half of a particular item will sell in any given month. To meet her financial goals, she needs sufficient quantity on hand to sell. For example, Meiselman would need 4,000 raincoats in April retailing at $99 each to reach a $200,000 sales goal if the stores' sales were expected to reach a 50% sales-to-stock ratio.

Sales volume is the key to success in any discount business, whether clothing or home appliances. Meiselman's merchandise is marked up or down depending on that volume. Her goal is a 45% markup on all merchandise. If a line of suits doesn't sell, she may mark it down, achieving only a 35% markup over the wholesale price. To balance out the loss, she marks up another line of items so that her *average* markup hits 45%.

With its high overhead, the chain has to generate sufficient sales volume and profits to stay in business. Meiselman's fashion choices have helped make that goal a continuing reality.

Conditional Probability and Independent Events

Conditional Probability

Three cities, A, B, and C, are vying to play host to the Summer Olympic Games in the year 2000. If each city has the same chance of winning the right to host the Games, then the probability of city A hosting the Games is 1/3. Suppose that city B then decides to pull out of contention because of fiscal problems. Then it would seem that city A's chances of playing host will increase. In fact, if each of the two remaining cities have equal chances of winning, then the probability of city A playing host to the Games is 1/2.

In general, the probability of an event is affected by the occurrence of other events and/or by the knowledge of information relevant to the event. Basically, the injection of conditions into a problem modifies the underlying sample space of the original problem. This in turn leads to a change in the probability of the event.

EXAMPLE 1   Two cards are drawn without replacement from a well-shuffled deck of 52 playing cards.

a. What is the probability that the first card drawn is an ace?

b. What is the probability that the second card drawn is an ace given that the first card drawn was not an ace?

c. What is the probability that the second card drawn is an ace given that the first card drawn was an ace?

Solution

a. The sample space here consists of 52 equally likely outcomes, 4 of which are aces. Therefore, the probability that the first card drawn is an ace is 4/52.

b. Having drawn the first card, there are 51 cards left in the deck. In other words, for the second phase of the experiment, we are working in a *reduced* sample space. If the first card drawn was not an ace, then this modified sample space of 51 points contains 4 "favorable" outcomes (the 4 aces), so the probability that the second card drawn is an ace is given by 4/51.

c. If the first card drawn was an ace, then there are 3 aces left in the deck of 51 playing cards, so the probability that the second card drawn is an ace is given by 3/51.   ■

Observe that in Example 1 the occurrence of the first event reduces the size of the original sample space. The information concerning the first card drawn also leads us to the consideration of modified sample spaces: in (b) the deck contained 4 aces, and in (c) the deck contained 3 aces.

The probability found in (b) or (c) of Example 1 is known as a *conditional probability*, since it is the probability of an event occurring given that another event has already occurred. For example, in (b) we computed the probability of the event that the second card drawn is an ace, given the event that the first card drawn was not an ace. In general, given two events A and B of an experiment, one may, under certain circumstances, compute the probability of the event B given that the event A has already occurred. This probability, denoted by $P(B \mid A)$, is called the *conditional probability of B given A*.

A formula for computing the conditional probability of B given A may be discovered with the aid of a Venn diagram. Consider an experiment with a uniform sample space S and suppose A and B are two events of the experiment (Figure 2).

FIGURE 2

(a) Original sample space

(b) Reduced sample space A.
The shaded area is $A \cap B$.

The condition that the event A has occurred tells us that the possible outcomes of the experiment in the second phase are restricted to those outcomes (elements) in the set A. In other words, we may work with the reduced sample space A instead of the original sample space S in the experiment. Next, we observe that, with respect to the reduced sample space A, the outcomes favorable to the event B are precisely those elements in the set $A \cap B$. Consequently, the conditional probability of B given A is

$$P(B \mid A) = \frac{\text{Number of elements in } A \cap B}{\text{Number of elements in } A}$$

$$= \frac{n(A \cap B)}{n(A)} \qquad n(A) \neq 0$$

Dividing the numerator and the denominator by $n(S)$, the number of elements in S, we have

$$P(B \mid A) = \frac{\dfrac{n(A \cap B)}{n(S)}}{\dfrac{n(A)}{n(S)}}$$

which is equivalent to the following formula:

CONDITIONAL PROBABILITY OF AN EVENT

If A and B are events in an experiment and $P(A) \neq 0$, then the **conditional probability** that the event B will occur given that the event A has already occurred is

$$P(B \mid A) = \frac{P(A \cap B)}{P(A)} \tag{2}$$

EXAMPLE 2   A pair of fair dice is cast. What is the probability that the sum of the numbers falling uppermost is 7 if it is known that one of the numbers is a 5?

Solution   Let A denote the event that the sum of the numbers falling uppermost is 7 and let B denote the event that one of the numbers is a 5. From the results of Example 4, Section 7.5, we find that

$$A = \{(6, 1), (5, 2), (4, 3), (3, 4), (2, 5), (1, 6)\}$$

and

$$B = \{(5, 1), (5, 2), (5, 3), (5, 4), (5, 5), (5, 6),$$
$$(1, 5), (2, 5), (3, 5), (4, 5), (6, 5)\}$$

so that $A \cap B = \{(5, 2), (2, 5)\}$ See Figure 3.

FIGURE 3
$A \cap B = \{(5, 2), (2, 5)\}$

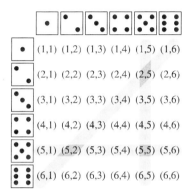

(1,1)	(1,2)	(1,3)	(1,4)	(1,5)	(1,6)
(2,1)	(2,2)	(2,3)	(2,4)	(2,5)	(2,6)
(3,1)	(3,2)	(3,3)	(3,4)	(3,5)	(3,6)
(4,1)	(4,2)	(4,3)	(4,4)	(4,5)	(4,6)
(5,1)	(5,2)	(5,3)	(5,4)	(5,5)	(5,6)
(6,1)	(6,2)	(6,3)	(6,4)	(6,5)	(6,6)

Since the dice are fair, each outcome of the experiment is equally likely; therefore

$$P(A \cap B) = \frac{2}{36} \quad \text{and} \quad P(B) = \frac{11}{36} \quad \text{Recall that } n(S) = 36.$$

Thus, the probability that the sum of the numbers falling uppermost is 7 given that one of the numbers is a 5 is, by virtue of (2),

$$P(A \mid B) = \frac{\dfrac{2}{36}}{\dfrac{11}{36}} = \frac{2}{11}$$

EXAMPLE 3   In a recently conducted test by the U.S. Army, it was found that of 1000 new recruits, 600 men and 400 women, 50 of the men and 4 of the women were red-green colorblind. Given that a recruit selected at random from this group is red-green colorblind, what is the probability that the recruit is a male?

Solution   Let C denote the event that a randomly selected subject is red-green colorblind and let M denote the event that the subject is a male recruit. Since 54 of the 1000 subjects are colorblind, we may take

$$P(C) = \frac{54}{1000} = 0.054$$

Therefore, by equation (2), the probability that a subject is male given that the subject is red-green colorblind is

$$P(M \mid C) = \frac{P(M \cap C)}{P(C)} \qquad P(M \cap C) = \frac{50}{1000} = 0.05$$

$$= \frac{0.05}{0.054} = \frac{25}{27}$$

In certain problems, the probability of an event B occurring given that A has occurred, written $P(B \mid A)$, is known and we wish to find the probability of A *and* B occurring. The solution to such a problem is facilitated by the use of the following formula:

THE PRODUCT RULE $P(A \cap B) = P(A) \cdot P(B \mid A)$ if $P(A) \neq 0$ **(3)**

This formula is obtained from (2) by multiplying both sides of the equation by $P(A)$. We illustrate the use of the Product Rule in the next several examples.

EXAMPLE 4   There are 300 seniors in Jefferson High School, of which 140 are males. It is known that 80% of the males and 60% of the females have their driver's license. If a student is selected at random from this senior class,

a. what is the probability that the student is a male who has a driver's license?

b. what is the probability that the student is a female who does not have a driver's license?

Solution

a. Let M denote the event that the student is a male and let D denote the event that the student has a driver's license. Then

$$P(M) = \frac{140}{300} \quad \text{and} \quad P(D \mid M) = 0.8$$

Now, the event that the student selected at random is a male and has a driver's license is $M \cap D$, and, by the Product Rule, the probability of this event occurring is given by

$$P(M \cap D) = P(M) \cdot P(D \mid M)$$
$$= \left(\frac{140}{300}\right)(0.8) = \frac{28}{75}$$

b. Let F denote the event that the student is a female and let D be as before. Then D^c is the event that the student does not have a driver's license. We have

$$P(F) = \frac{160}{300} \quad \text{and} \quad P(D^c \mid F) = 1 - 0.6 = 0.4$$

Note that we have used the complement rule in the computation of $P(D^c \mid F)$. Now, the event that the student selected at random is a female and does not have a driver's license is $F \cap D^c$, so by the Product Rule, the probability of this event occurring is given by

$$P(F \cap D^c) = P(F) \cdot P(D^c \mid F)$$
$$= \left(\frac{160}{300}\right)(0.4) = \frac{16}{75}$$

EXAMPLE 5   Two cards are drawn without replacement from a well-shuffled deck of 52 playing cards. What is the probability that the first card drawn is an ace and the second card drawn is a face card?

Solution   Let A denote the event that the first card drawn is an ace and let F denote the event that the second card drawn is a face card. Then $P(A) = 4/52$. After drawing the first card, there are 51 cards left in the deck, of which 12 are face cards. Therefore, the probability of drawing a face card given that the first card drawn was an ace is given by

$$P(F \mid A) = \frac{12}{51}$$

By the Product Rule, the probability that the first card drawn is an ace and the second card drawn is a face card is given by

$$P(A \cap F) = P(A) \cdot P(F \mid A)$$
$$= \left(\frac{4}{52}\right)\left(\frac{12}{51}\right) = \frac{4}{221}$$

The Product Rule may be generalized to the case involving any finite number of events. For example, in the case involving the three events E, F, and G, it may be shown that

$$P(E \cap F \cap G) = P(E) \cdot P(F \mid E) \cdot P(G \mid E \cap F) \qquad \textbf{(4)}$$

More on Tree Diagrams

Formula (4) and its generalizations may be used to help us solve problems that involve *finite stochastic processes*. More specifically, a **finite stochastic process** is an experiment consisting of a finite number of stages in which the outcomes and associated probabilities of each stage depend on the outcomes and associated probabilities of the preceding stages.

    We can use tree diagrams to help us solve problems involving finite stochastic processes. Consider, for example, the experiment consisting of drawing 2 cards without replacement from a well-shuffled deck of 52 playing cards. What is the probability that the second card drawn is a face card?

    We may think of this experiment as a stochastic process with two stages. The events associated with the first stage are F, that the card drawn is a face card, and F^c, that the card drawn is not a face card. Since there are 12 face cards, we have

$$P(F) = \frac{12}{52} \qquad \text{and} \qquad P(F^c) = 1 - \frac{12}{52} = \frac{40}{52}$$

The outcomes of this trial, together with the associated probabilities, may be represented along two "branches" of a tree diagram, as shown in Figure 4.

    In the second trial, we again have two events: G, that the card drawn is a face card, and G^c, that the card drawn is not a face card. But the outcome of the second trial depends on the outcome of the first trial. For example, if the first card drawn was a face card, then the event G that the second card drawn is a face card has probability given by the *conditional probability* $P(G \mid F)$. Since the occurrence of a face card in the first draw leaves 11 face cards in a deck of 51 cards for the second draw, we see that

FIGURE 4

F is the probability that a face card is drawn.

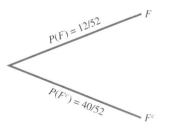

$$P(G \mid F) = \frac{11}{51} \qquad \begin{array}{l}\text{The probability of drawing a} \\ \text{face card given that a face card} \\ \text{has already been drawn}\end{array}$$

Similarly, the occurrence of a face card in the first draw leaves 40 that are other than face cards in a deck of 51 cards for the second draw. Therefore, the probability of drawing other than a face card in the second draw given that the first card drawn is a face card is

$$P(G^c \mid F) = \frac{40}{51}$$

Using these results, we extend the tree diagram of Figure 4 by displaying another two branches of the tree growing from its upper branch (Figure 5).

FIGURE 5

G is the probability that the second card drawn is a face card.

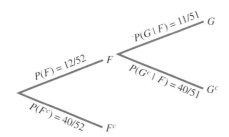

To complete the tree diagram, we compute $P(G \mid F^c)$ and $P(G^c \mid F^c)$, the conditional probabilities that the second card drawn is a face card and other than a face card, respectively, given that the first card drawn is not a face card. We find that

$$P(G \mid F^c) = \frac{12}{51} \qquad \text{and} \qquad P(G^c \mid F^c) = \frac{39}{51}$$

This leads to the completion of the tree diagram, shown in Figure 6, where the branches of the tree that lead to the two outcomes of interest have been highlighted.

FIGURE 6

Tree diagram showing the two trials of the experiment

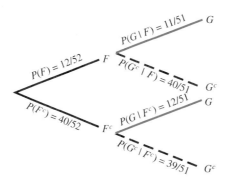

Having constructed the tree diagram associated with the problem, we are now in a position to answer the question posed earlier—namely, "What is the probability of the second card being a face card?" Observe that Figure 6 shows the two ways in which a face card may result in the second draw—namely, the two Gs on the extreme right of the diagram.

Now, by the Product Rule, the probability that the second card drawn is a face card and the first card drawn is a face card (this is represented by the upper branch) is

$$P(G \cap F) = P(F) \cdot P(G \mid F)$$

Similarly, the probability that the second card drawn is a face card and the first card drawn is other than a face card (this corresponds to the other branch) is

$$P(G \cap F^c) = P(F^c) \cdot P(G \mid F^c)$$

Observe that each of these probabilities is obtained by taking the *product of the probabilities appearing on the respective branch.* Since $G \cap F$ and $G \cap F^c$ are mutually exclusive events (why?), the probability that the second card drawn is a face card is given by

$$P(G \cap F) + P(G \cap F^c) = P(F) \cdot P(G \mid F) + P(F^c) \cdot P(G \mid F^c)$$

or, upon replacing the probabilities on the right of the expression by their numerical values,

$$P(G \cap F) + P(G \cap F^c) = \left(\frac{12}{52}\right)\left(\frac{11}{51}\right) + \left(\frac{40}{52}\right)\left(\frac{12}{51}\right)$$

$$= \frac{3}{13}$$

EXAMPLE 6   The picture tubes for the Pulsar 19-inch color television sets are manufactured in three locations and then shipped to the main plant of the Vista Vision Corporation for final assembly. Plants A, B, and C supply 50%, 30%, and 20%, respectively, of the picture tubes used by the company. The quality-control department of the company has determined that 1% of the picture tubes produced by plant A are defective, whereas 2% of the picture tubes produced by plants B and C are defective. What is the probability that a randomly selected Pulsar 19-inch color television set will have a defective picture tube?

Solution   Let A, B, and C denote the events that the set chosen has a picture tube manufactured in plant A, plant B, and plant C, respectively. Also, let D denote the event that a set has a defective picture tube. Using the given information, we draw the tree diagram shown in Figure 7.

FIGURE 7

Tree diagram showing the probabilities of producing defective picture tubes at each plant

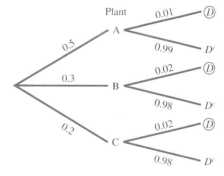

(The events that result in a set with a defective picture tube being selected are circled.) Taking the product of the probabilities along each branch leading to such an event and adding them yields the probability that a set chosen at random has a defective picture tube. Thus, the required probability is given by

$$(0.5)(0.01) + (0.3)(0.02) + (0.2)(0.02) = 0.005 + 0.006 + 0.004$$

$$= 0.015 \qquad ∎$$

EXAMPLE 7   A box contains eight 9-volt transistor batteries of which two are known to be defective. The batteries are selected one at a time without replacement and tested until a nondefective one is found. What is the probability that the number of batteries tested is (a) one, (b) two, (c) three?

Solution   We may view this experiment as a multistage process with up to three stages. In the first stage, a battery is selected with a probability of 6/8 of it being nondefective and a probability of 2/8 of it being defective. If the battery selected is good, the experiment is terminated. Otherwise, a second battery is selected with probabilities of 6/7 and 1/7, respectively, of being nondefective and defective. If the second battery selected is good, the experiment is terminated. Otherwise, a third battery is selected with probabilities of 1 and 0, respectively, of being nondefective and defective. The tree diagram associated with this experiment is shown in Figure 8, where N denotes the event that the battery selected is nondefective, and D denotes the event that the battery selected is defective.

FIGURE 8

In this experiment, batteries are selected until a nondefective one is found.

With the aid of the tree diagram we see that (a) the probability that only one battery is selected is 6/8, (b) the probability that two batteries are selected is (2/8)(6/7), or 3/14, and (c) the probability that three batteries are selected is (2/8)(1/7)(1) = 1/28. ∎

Independent Events

Let us return to the experiment of drawing 2 cards in succession without replacement from a well-shuffled deck of 52 playing cards considered in Example 5. Let E denote the event that the first card drawn is not a face card and let F denote the event that the second card drawn is a face card. It is intuitively clear that the events E and F are *not* independent of each other, since whether or not the first card drawn is a face card affects the likelihood that the second card drawn is a face card.

Next, let us consider the experiment of tossing a coin twice and observing the outcomes: If H denotes the event that the first toss produces "heads" and T denotes the event that the second toss produces "tails," then it is intuitively clear that H and T *are* independent of each other, since the outcome of the first toss does not affect the outcome of the second.

In general, two events A and B are *independent* if the outcome of one does not affect the outcome of the other. Thus, if A and B are independent

events, then

$$P(A \mid B) = P(A) \qquad \text{or} \qquad P(B \mid A) = P(B)$$

Using the Product Rule, we can find a simple test to determine the independence of two events. Suppose that A and B are independent and $P(A) \neq 0$ and $P(B) \neq 0$. Then,

$$P(B \mid A) = P(B)$$

so by the Product Rule, we have

$$P(A \cap B) = P(A) \cdot P(B \mid A) = P(A) \cdot P(B)$$

Conversely, if this equation holds, then it can be seen that $P(B \mid A) = P(B)$; that is, A and B are independent. Accordingly, we have the following *test for the independence of two events.*

TEST FOR THE INDEPENDENCE OF TWO EVENTS

Two events A and B are **independent** if and only if

$$P(A \cap B) = P(A) \cdot P(B) \tag{5}$$

Do not confuse *independent* events with *mutually exclusive* events. The former pertains to how the occurrence of one event affects the occurrence of another event, whereas the latter pertains to the question of whether the events can occur at the same time.

EXAMPLE 8   Consider the experiment consisting of tossing a fair coin twice and observing the outcomes. Show that the event of "heads" in the first toss and "tails" in the second toss are independent events.

Solution   Let A denote the event that the outcome of the first toss is a *head* and let B denote the event that the outcome of the second toss is a *tail*. The sample space of the experiment is

$$S = \{(H, H), (H, T), (T, H), (T, T)\}$$

and $A = \{(H, H), (H, T)\}$ and $B = \{(H, T), (T, T)\}$

so that $A \cap B = \{(H, T)\}$

Next, we compute

$$P(A \cap B) = \tfrac{1}{4}, \qquad P(A) = \tfrac{1}{2}, \qquad P(B) = \tfrac{1}{2}$$

and observe that equation (5) is satisfied in this case so that A and B are independent events, as we set out to show. ∎

EXAMPLE 9   A survey conducted by an independent agency for the National Lung Society found that of 2000 women, 680 were heavy smokers and 50 had emphysema. Of those who had emphysema, 42 were also heavy smokers. Using the data in this survey, determine whether the events "being a heavy smoker" and "having emphysema" are independent events.

Solution   Let A denote the event that a woman is a heavy smoker and let B denote the event that a woman has emphysema. Then the probability that a woman is a heavy smoker and has emphysema is given by

$$P(A \cap B) = \frac{42}{2000} = 0.021$$

Next $P(A) = \frac{680}{2000} = 0.34$ and $P(B) = \frac{50}{2000} = 0.025$

so that $P(A) \cdot P(B) = (0.34)(0.025) = 0.0085$

Since $P(A \cap B) \neq P(A) \cdot P(B)$, we conclude that A and B are not independent events.

The solution of many practical problems involves more than two independent events. In such cases we use the following result.

If $E_1, E_2, \ldots, E_n$ are independent events, then
$$P(E_1 \cap E_2 \cap \cdots \cap E_n) = P(E_1) \cdot P(E_2) \cdot \cdots \cdot P(E_n) \quad \textbf{(6)}$$

Formula (6) states that the probability of the simultaneous occurrence of n independent events is equal to the product of the probabilities of the n events.

It is important to note that the mere requirement that the n events $E_1, E_2, \ldots, E_n$ satisfy (6) is not sufficient to guarantee that the n events are indeed independent. However, a criterion does exist for determining the independence of n events and may be found in more advanced texts on probability.

EXAMPLE 10   It is known that the three events A, B, and C are independent and $P(A) = 0.2$, $P(B) = 0.4$, and $P(C) = 0.5$.

a. Compute $P(A \cap B)$.   **b.** Compute $P(A \cap B \cap C)$.

Solution   Using formulas (5) and (6), we find

a. $P(A \cap B) = P(A) \cdot P(B)$

$\qquad\qquad\quad = (0.2)(0.4) = 0.08$

b. $P(A \cap B \cap C) = P(A) \cdot P(B) \cdot P(C)$

$\qquad\qquad\qquad = (0.2)(0.4)(0.5) = 0.04$

EXAMPLE 11   The Acrosonic model F loudspeaker system has four loudspeaker components: a woofer, a midrange, a tweeter, and an electrical crossover. The quality-control manager of Acrosonic has determined that on the average 1% of the woofers, 0.8% of the midranges, and 0.5% of the tweeters are defective, while 1.5% of the electrical crossovers are defective. Determine the probability that an Acrosonic model F loudspeaker system selected at random coming off the assembly line and before final inspection is not defective. Assume that the defects in the manufacturing of the components are unrelated.

Solution   Let A, B, C, and D denote, respectively, the events that the woofer, the midrange, the tweeter, and the electrical crossover are defective. Then,

$$P(A) = 0.01, \qquad P(B) = 0.008, \qquad P(C) = 0.005, \qquad P(D) = 0.015$$

and the probabilities of the corresponding complementary events are

$$P(A^c) = 0.99, \qquad P(B^c) = 0.992, \qquad P(C^c) = 0.995, \qquad P(D^c) = 0.985$$

The event that a loudspeaker system selected at random is not defective is given by $A^c \cap B^c \cap C^c \cap D^c$, and because the events A, B, C, and D (and therefore also A^c, B^c, C^c, and D^c) are assumed to be independent, we find that the required probability is given by

$$P(A^c \cap B^c \cap C^c \cap D^c) = P(A^c) \cdot P(B^c) \cdot P(C^c) \cdot P(D^c)$$

$$= (0.99)(0.992)(0.995)(0.985)$$

$$\approx 0.96$$

SELF-CHECK EXERCISES 8.2

1. Let A and B be events in a sample space S such that $P(A) = 0.4$, $P(B) = 0.8$, and $P(A \cap B) = 0.3$. Find
   a. $P(A \mid B)$ **b.** $P(B \mid A)$

2. Three friends—Alice, Betty, and Cathy—are unmarried and are 30, 35, and 40 years old, respectively. While they were having tea one afternoon, Alice recalled reading an article by Yale sociologist Neil Bennett in which he concluded that women who are still unmarried at age 30 have only a 20% chance of marrying, those who are still unmarried at 35 have a 5.4% chance, and those who are still unmarried at 40 have a 1.3% chance. Betty wondered what the probability was that all three of them would eventually "tie the knot." Cathy, a statistician, pulled out her pocket calculator and answered Betty's question. What was her answer?

Solutions to Self-Check Exercises 8.2 can be found on page 425.

---

1. Let A and B be events in a sample space S such that $P(A) = 0.6$, $P(B) = 0.5$, and $P(A \cap B) = 0.2$. Find
   a. $P(A \mid B)$ **b.** $P(B \mid A)$

2. Let A and B be two events in a sample space S such that $P(A) = 0.4$, $P(B) = 0.6$, and $P(A \cap B) = 0.3$. Find
   a. $P(A \mid B)$ **b.** $P(B \mid A)$

3. Let A and B be two events in a sample space S such that $P(A) = 0.6$ and $P(B \mid A) = 0.5$. Find $P(A \cap B)$.

4. Let A and B be the events described in exercise 1. Find
   a. $P(A \mid B^c)$ [*Hint:* $(A \cap B^c) \cup (A \cap B) = A$]
   b. $P(B \mid A^c)$

In exercises 5–8, determine whether the given events A and B are independent.

5. $P(A) = 0.3$, $P(B) = 0.6$, $P(A \cap B) = 0.18$

6. $P(A) = 0.6$, $P(B) = 0.8$, $P(A \cap B) = 0.2$

7. $P(A) = 0.5$, $P(B) = 0.7$, $P(A \cup B) = 0.85$

8. $P(A^c) = 0.3$, $P(B^c) = 0.4$, $P(A \cap B) = 0.42$

9. If A and B are independent events and $P(A) = 0.4$ and $P(B) = 0.6$, find
   a. $P(A \cap B)$ **b.** $P(A \cup B)$

10. If A and B are independent events and $P(A) = 0.35$ and $P(B) = 0.45$, find
    a. $P(A \cap B)$ **b.** $P(A \cup B)$

11. The accompanying tree diagram represents an experiment consisting of two trials.

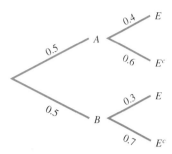

Use the diagram to find
a. $P(A)$
b. $P(E \mid A)$
c. $P(A \cap E)$
d. $P(E)$
e. Does $P(A \cap E) = P(A) \cdot P(E)$?
f. Are A and E independent events?

12. The accompanying tree diagram represents an experiment consisting of two trials. Use the diagram to find
a. $P(A)$
b. $P(E \mid A)$
c. $P(A \cap E)$
d. $P(E)$
e. Does $P(A \cap E) = P(A) \cdot P(E)$?
f. Are A and E independent events?

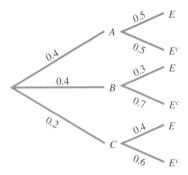

13. An experiment consists of two trials. The outcomes of the first trial are A and B, with probabilities of occurring of 0.4 and 0.6. There are also two outcomes C and D in the second trial, with probabilities of 0.3 and 0.7. Draw a tree diagram representing this experiment. Use this diagram to find
a. $P(A)$
b. $P(C \mid A)$
c. $P(A \cap C)$
d. $P(C)$
e. Does $P(A \cap C) = P(A) \cdot P(C)$?
f. Are A and C independent events?

14. An experiment consists of two trials. The outcomes of the first trial are A, B, and C, with probabilities of occurring of 0.2, 0.5, and 0.3, respectively. The outcomes of the second trial are E and F, with probabilities of occurring of 0.6 and 0.4. Draw a tree dia-

gram representing this experiment. Use this diagram to find
a. $P(B)$
b. $P(F \mid B)$
c. $P(B \cap F)$
d. $P(F)$
e. Does $P(B \cap F) = P(B) \cdot P(F)$?
f. Are B and F independent events?

15. A pair of fair dice is cast. What is the probability that the sum of the numbers falling uppermost is less than nine, if it is known that one of the numbers is a 6?

16. Product Reliability   The probability that a battery will last 10 hours or more is 0.80 and the probability that it will last 15 hours or more is 0.15. Given that a battery has lasted 10 hours, find the probability that it will last 15 hours or more.

17. Two cards are drawn without replacement from a well-shuffled deck of 52 playing cards.
a. What is the probability that the first card drawn is a heart?
b. What is the probability that the second card drawn is a heart given that the first card drawn was not a heart?
c. What is the probability that the second card drawn is a heart given that the first card drawn was a heart?

18. Five black balls and four white balls are placed in an urn. Two balls are then drawn in succession. What is the probability that the second ball drawn is a white ball
a. if the second ball is drawn without replacing the first?
b. if the first ball is replaced before the second is drawn?

19. Auditing Tax Returns   A tax specialist has estimated the probability that a tax return selected at random will be audited is 0.02. Furthermore, he estimates the probability that an audited return will result in additional assessments being levied on the taxpayer is 0.60. What is the probability that a tax return selected at random will result in additional assessments being levied on the taxpayer?

20. Student Enrollment   At a certain medical school, 1/7 of the students are from a minority group. Of those students who belong to a minority group, 1/3 are black.
a. What is the probability that a student selected at random from this medical school is black?
b. What is the probability that a student selected at random from this medical school is black if it is known that the student is a member of a minority group?

21. Educational Level of Voters In a survey of 1000 eligible voters selected at random, it was found that 80 had a college degree. Additionally, it was found that 80% of those who had a college degree voted in the last presidential election, whereas 55% of the people who did not have a college degree voted in the last presidential election. Assuming that the poll is representative of all eligible voters, find the probability that an eligible voter selected at random

a. had a college degree and voted in the last presidential election

b. did not have a college degree and did not vote in the last presidential election

c. voted in the last presidential election

d. did not vote in the last presidential election

22. Three cards are drawn without replacement from a well-shuffled deck of 52 playing cards. What is the probability that the third card drawn is a diamond?

23. A coin is tossed three times. What is the probability that the coin will land

a. heads at least twice?

b. heads on the second toss given that heads were thrown on the first toss?

c. heads on the third toss given that tails were thrown on the first toss?

24. In a three-child family, what is the probability that all three children are girls given that one of the children is a girl? (Assume that the probability of a boy being born is the same as the probability of a girl being born.)

25. Quality Control An automobile manufacturer obtains the microprocessors used to regulate fuel consumption in its automobiles from three microelectronic firms, A, B, and C. The quality-control department of the company has determined that 1% of the microprocessors produced by firm A are defective, 2% of those produced by firm B are defective, and 1.5% of those produced by firm C are defective. Firms A, B, and C supply 45%, 25%, and 30%, respectively, of the microprocessors used by the company. What is the probability that a randomly selected automobile manufactured by the company will have a defective microprocessor?

26. Car Theft Figures obtained from a city's police department seem to indicate that, of all motor vehicles reported as stolen, 64% were stolen by professionals, whereas 36% were stolen by amateurs (primarily for joy rides). Of those vehicles presumed stolen by professionals, 24% were recovered within 48 hours, 16% were recovered after 48 hours, and 60% were never recovered. Of those vehicles presumed stolen by am-

ateurs, 38% were recovered within 48 hours, 58% were recovered after 48 hours, and 4% were never recovered.

a. Draw a tree diagram representing these data.

b. What is the probability that a vehicle stolen by a professional in this city will be recovered within 48 hours?

c. What is the probability that a vehicle stolen in this city will never be recovered?

27. Housing Loans The chief loan officer of the La Crosse Home Mortgage Company summarized the housing loans extended by the company in 1993 according to type and term of the loan. Her list shows that 70% of the loans were fixed-rate mortgages (F), 25% were adjustable-rate mortgages (A), and 5% belong to some other category (O) (mostly second trust-deed loans and loans extended under the graduated payment plan). Of the fixed-rate mortgages, 80% were 30-year loans and 20% were 15-year loans; of the adjustable-rate mortgages, 40% were 30-year loans and 60% were 15-year loans; finally, of the other loans extended, 30% were 20-year loans, 60% were 10-year loans, and 10% were for a term of 5 years or less.

a. Draw a tree diagram representing this experiment.

b. What is the probability that a home loan extended by La Crosse has an adjustable rate and is for a term of 15 years?

c. What is the probability that a home loan extended by La Crosse is for a term of 15 years?

28. College Admissions The admissions office of a private university released the following admission data for the preceding academic year: From a pool of 3900 male applicants, 40% were accepted by the university and of these, 40% subsequently enrolled. Additionally, from a pool of 3600 female applicants 45% were accepted by the university and of these, 40% subsequently enrolled. Find the probability that

a. a male applicant will be accepted by and subsequently will enroll in the university

b. a student who applies for admission will be accepted by the university

c. a student who applies for admission will be accepted by the university and subsequently will enroll

29. Quality Control Suppose that a box contains two defective Christmas tree lights that have been inadvertently mixed with eight nondefective lights. If the lights are selected one at a time without replacement and tested until both defective lights are found, what is the probability that both defective lights will be found after three trials?

30. Quality Control   It is estimated that 0.80% of a large consignment of eggs in a certain supermarket is broken.
  a. What is the probability that a customer who randomly selects a dozen of these eggs receives at least one broken egg?
  b. What is the probability that a customer who selects these eggs at random will have to check three cartons before finding a carton without any broken eggs? (Each carton contains a dozen eggs.)

31. Student Financial Aid   The accompanying data were obtained from the financial aid office of a certain university.

	Receiving Financial Aid	Not Receiving Financial Aid	Total
Undergraduates	4,222	3,898	8,120
Graduates	1,879	731	2,610
Total	6,101	4,629	10,730

Let A be the event that a student selected at random from this university is an undergraduate student and let B be the event that a student selected at random is receiving financial aid.
  a. Find each of the following probabilities:

  $P(A)$, $P(B)$, $P(A \cap B)$, $P(B \mid A)$, $P(B \mid A^c)$

  b. Are the events A and B independent events?

32. Employee Education and Income   The personnel department of the Franklin National Life Insurance Company compiled the accompanying data regarding the income and education of its employees.

	Income $30,000 or Below	Income Above $30,000
Noncollege Graduate	2040	840
College Graduate	400	720

Let A be the event that an employee has a college degree and B the event that an employee's income is more than $30,000.
  a. Find each of the following probabilities:

  $P(A)$, $P(B)$, $P(A \cap B)$, $P(B \mid A)$, $P(B \mid A^c)$

  b. Are the events A and B independent events?

33. Two cards are drawn without replacement from a well-shuffled deck of 52 cards. Let A be the event that the first card drawn is a heart and let B be the event that the second card drawn is a red card. Show that the events A and B are dependent events.

34. Medical Research   A nationwide survey conducted by the National Cancer Society revealed the following information: Of 10,000 people surveyed, 3,200 were "heavy coffee drinkers" and 160 had cancer of the pancreas. Of those who had cancer of the pancreas, 132 were heavy coffee drinkers. Using the data in this survey, determine whether the events "being a heavy coffee drinker" and "having cancer of the pancreas" are independent events.

35. Mail Delivery   Suppose that the probability that your mail will be delivered before 2 P.M. on a delivery day is 0.90. What is the probability that your mail will be delivered before 2 P.M.
  a. for two consecutive delivery days?
  b. for three consecutive delivery days?

36. Reliability of Security Systems   Before being allowed to enter a maximum-security area at a military installation, a person must pass three identification tests: a voice-pattern test, a fingerprint test, and a handwriting test. If the reliability of the first test is 97%, the reliability of the second test is 98.5%, and that of the third is 98.5%, what is the probability that this security system will allow an improperly identified person to enter the maximum-security area?

37. Quality Control   Copykwik, Inc., has four photocopy machines, A, B, C, and D. The probability that a given machine will break down on a particular day is

$$P(A) = \frac{1}{50}, \quad P(B) = \frac{1}{60}, \quad P(C) = \frac{1}{75}, \quad P(D) = \frac{1}{40}$$

Assuming independence, what is the probability on a particular day that
  a. all four machines will break down?
  b. none of the machines will break down?
  c. exactly one machine will break down?

38. Product Reliability   The proprietor of Cunningham's Hardware Store has decided to install floodlights on the premises as a measure against vandalism and theft. If the probability is 0.01 that a certain brand of floodlight will burn out within a year, find the minimum number of floodlights that must be installed to ensure that the probability that at least one of them will remain functional within the year is at least 0.99999. (Assume that the floodlights operate independently.)

SOLUTIONS TO SELF-CHECK EXERCISES 8.2

1. a. $P(A \mid B) = \dfrac{P(A \cap B)}{P(B)}$

$= \dfrac{0.3}{0.8} = \dfrac{3}{8}$

b. $P(B \mid A) = \dfrac{P(A \cap B)}{P(A)}$

$= \dfrac{0.3}{0.4} = \dfrac{3}{4}$

2. Let A, B, and C denote the events that three women who are still unmarried at ages 30, 35, and 40 *will* marry eventually. Then it seems reasonable to assume that these events are independent, with $P(A) = 0.20$, $P(B) = 0.054$, and $P(C) = 0.013$. Based on these figures, the probability that all three women will be married eventually is given by

$$P(A)P(B)P(C) = (0.2)(0.054)(0.013)$$

$$= 0.00014$$

8.3 Bayes' Theorem

A Posteriori Probabilities

Suppose three machines, A, B, and C, produce similar engine components, Machine A produces 45% of the total components, machine B produces 30%, and machine C, 25%. For the usual production schedule, 6% of the components produced by machine A do not meet established specifications; for machine B and machine C, the corresponding figures are 4% and 3%. One component is selected at random from the total output and is found to be defective. What is the probability that the component selected was produced by machine A?

The answer to this question is found by calculating the probability *after* the outcomes of the experiment have been observed. Such probabilities are called *a posteriori* **probabilities** as opposed to *a priori* **probabilities**—probabilities that give the likelihood that an event *will* occur, which was the subject of the last several sections.

Returning to the example under consideration, we need to determine the *a posteriori* probability for the event that the component selected was produced by machine A. To this end, let A, B, and C denote the event that a component is produced by machine A, machine B, and machine C, respectively. We may represent this experiment with a Venn diagram (Figure 9).

The three mutually exclusive events A, B, and C form a *partition* of the sample space S. That is, aside from being mutually exclusive, their union is precisely S. The event D that a component is defective is the shaded area. Again referring to Figure 9, we see that

1. The event D may be expressed as

$$D = (A \cap D) \cup (B \cap D) \cup (C \cap D)$$

2. The event that a component is defective and is produced by machine A is given by $A \cap D$.

FIGURE 9

D is the event that a defective component is produced by machine A, machine B, or machine C.

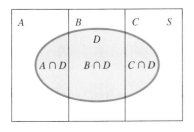

Thus, the *a posteriori* probability that a defective component selected was produced by machine A is given by

$$P(A \mid D) = \frac{n(A \cap D)}{n(D)}$$

Upon dividing both the numerator and the denominator by $n(S)$ and observing that the events $A \cap D$, $B \cap D$, and $C \cap D$ are mutually exclusive, we obtain

$$P(A \mid D) = \frac{P(A \cap D)}{P(D)}$$

$$= \frac{P(A \cap D)}{P(A \cap D) + P(B \cap D) + P(C \cap D)} \qquad (7)$$

Next, using the Product Rule, we may express

$$P(A \cap D) = P(A) \cdot P(D \mid A)$$

$$P(B \cap D) = P(B) \cdot P(D \mid B)$$

and $$P(C \cap D) = P(C) \cdot P(D \mid C)$$

so that equation (7) may be expressed in the form

$$P(A \mid D) = \frac{P(A) \cdot P(D \mid A)}{P(A) \cdot P(D \mid A) + P(B) \cdot P(D \mid B) + P(C) \cdot P(D \mid C)} \qquad (8)$$

which is a special case of a result known as **Bayes' Theorem.**

Observe that the expression on the right of (8) involves the probabilities $P(A)$, $P(B)$, $P(C)$ and the conditional probabilities $P(D \mid A)$, $P(D \mid B)$, and $P(D \mid C)$, all of which may be calculated in the usual fashion. In fact, by displaying these quantities on a tree diagram, we obtain Figure 10. We may compute the required probability by substituting the relevant quantities into equation (8), or we may use the following device:

$$P(A \mid D) = \frac{\text{Product of probabilities along the limb through } A}{\text{Sum of products of the probabilities along each limb terminating at } D}$$

FIGURE 10

A tree diagram displaying the probabilities that a defective component is produced by machine A, machine B, or machine C

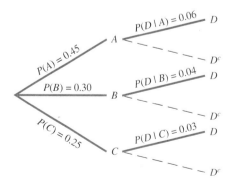

In either case, we obtain

$$P(A \mid D) = \frac{(0.45)(0.06)}{(0.45)(0.06) + (0.3)(0.04) + (0.25)(0.03)}$$

$$= 0.58$$

Before looking at any further examples, let us state the general form of Bayes' Theorem.

BAYES' THEOREM

Let $A_1, A_2, \ldots, A_n$ be a partition of a sample space S and let E be an event of the experiment such that $P(E) \neq 0$. Then the *a posteriori* probability $P(A_i \mid E)$ $(1 \leq i \leq n)$ is given by

$$P(A_i \mid E) = \frac{P(A_i) \cdot P(E \mid A_i)}{P(A_1) \cdot P(E \mid A_1) + P(A_2) \cdot P(E \mid A_2) + \cdots + P(A_n) \cdot P(E \mid A_n)} \qquad \textbf{(9)}$$

Applications

 EXAMPLE 1   The picture tubes for the Pulsar 19-inch color television sets are manufactured in three locations and then shipped to the main plant of the Vista Vision Corporation for final assembly. Plants A, B, and C supply 50%, 30%, and 20%, respectively, of the picture tubes used by Vista Vision. The quality control department of the company has determined that 1% of the picture tubes produced by plant A are defective, whereas 2% of the picture tubes produced by plants B and C are defective. If a Pulsar 19-inch color television set is selected at random and the picture tube is found to be defective, what is the probability that the picture tube was manufactured in plant C? Compare with Example 6, Section 8.2, page 417.

Solution   Let A, B, and C denote the event that the set chosen has a picture tube manufactured in plant A, plant B, and plant C, respectively. Also, let D denote the event that a set has a defective picture tube. Using the given information, we may draw the tree diagram shown in Figure 11. Next, using equation (9), we find that the required *a posteriori* probability is given by

$$P(C \mid D) = \frac{P(C) \cdot P(D \mid C)}{P(A) \cdot P(D \mid A) + P(B) \cdot P(D \mid B) + P(C) \cdot P(D \mid C)}$$

$$= \frac{(0.2)(0.02)}{(0.5)(0.01) + (0.3)(0.02) + (0.2)(0.02)}$$

$$\approx 0.27$$

FIGURE 11
$P(C \mid D) =$

$$\frac{\textit{Product of probabilities of branches to D through C}}{\textit{Sum of product of probabilites of branches leading to D}}$$

EXAMPLE 2　A study was conducted in a large metropolitan area to determine the annual incomes of married couples in which the husbands were the sole providers and of those in which the husbands and wives were both employed. Table 2 gives the results of this study.

TABLE 2

Annual Family Income ($)	% of Married Couples	% of Income Group with Both Spouses Working
75,000 and over	4	65
50,000–74,999	10	73
35,000–49,999	21	68
25,000–34,999	24	63
15,000–24,999	30	43
Under 15,000	11	28

a. What is the probability that a couple selected at random from this area has two incomes?

b. If a randomly chosen couple has two incomes, what is the probability that the annual income of this couple is over $75,000?

c. If a randomly chosen couple has two incomes, what is the probability that the annual income of this couple is greater than $24,999?

Solution　Let A denote the event that the annual income of the couple is $75,000 and over; let B denote the event that the annual income is between $50,000 and $74,999; let C denote the event that the annual income is between $35,000 and $49,999, and so on; finally, let F denote the event that the annual income is less than $15,000 and let T denote the event that both spouses work. The probabilities of the occurrence of these events are displayed in the following tree diagram (Figure 12).

FIGURE 12

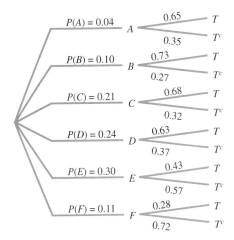

a. The probability that a couple selected at random from this group has two incomes is given by

$$P(T) = P(A) \cdot P(T \mid A) + P(B) \cdot P(T \mid B) + P(C) \cdot P(T \mid C)$$

$$+ P(D) \cdot P(T \mid D) + P(E) \cdot P(T \mid E) + P(F) \cdot P(T \mid F)$$

$$= (0.04)(0.65) + (0.10)(0.73) + (0.21)(0.68) + (0.24)(0.63)$$

$$+ (0.30)(0.43) + (0.11)(0.28)$$

$$= 0.5528$$

b. Using the results of (a) and Bayes' Theorem, we find that the probability that a randomly chosen couple has an annual income over $75,000 given that both spouses are working is

$$P(A \mid T) = \frac{P(A) \cdot P(T \mid A)}{P(T)} = \frac{(0.04)(0.65)}{0.5528}$$

$$= 0.047$$

c. The probability that a randomly chosen couple has an annual income greater than $24,999 given that both spouses are working is

$$P(A \mid T) + P(B \mid T) + P(C \mid T) + P(D \mid T)$$

$$= \frac{P(A) \cdot P(T \mid A) + P(B) \cdot P(T \mid B) + P(C) \cdot P(T \mid C) + P(D) \cdot P(T \mid D)}{P(T)}$$

$$= \frac{(0.04)(0.65) + (0.1)(0.73) + (0.21)(0.68) + (0.24)(0.63)}{0.5528}$$

$$= 0.711$$

■

**SELF-CHECK
EXERCISES 8.3**

1. The accompanying tree diagram represents a two-stage experiment. Use the diagram to find $P(B \mid D)$.

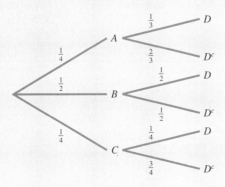

2. In a recent presidential election, it was estimated that the probability that the Republican candidate would be elected was 3/5 and therefore the probability that the Democratic candidate would be elected was 2/5 (the two Independent candidates were given little chance of being elected). It was also estimated that if the Republican candidate were elected, then the probability that research for a new manned bomber would continue was 4/5. But if the Democratic candidate were successful, then the probability that the research would continue was 3/10. Research was terminated shortly after the successful presidential candidate took office. What is the probability that the Republican candidate won that election?

Solutions to Self-Check Exercises 8.3 can be found on page 434.

8.3 EXERCISES

In exercises 1–3, refer to the accompanying Venn diagram. An experiment in which the three mutually exclusive events A, B, and C form a partition of the uniform sample space S is depicted in the diagram.

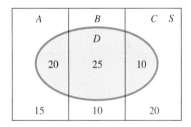

1. Draw a tree diagram using the information given in the Venn diagram illustrating the probabilities of the events A, B, C, and D.

2. Find: **a.** $P(D)$ **b.** $P(A \mid D)$

3. Find: **a.** $P(D^c)$ **b.** $P(B \mid D^c)$

In exercises 4–6, refer to the accompanying Venn diagram. An experiment in which the three mutually exclusive events A, B, and C form a partition of the uniform sample space S is depicted in the diagram.

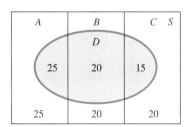

4. Draw a tree diagram using the information given in the Venn diagram illustrating the probabilities of the events A, B, C, and D.

5. Find: **a.** $P(D)$ **b.** $P(B \mid D)$

6. Find: **a.** $P(D^c)$ **b.** $P(B \mid D^c)$

7. The accompanying tree diagram represents a two-stage experiment. Use the diagram to find
     a. $P(A) \cdot P(D \mid A)$ **b.** $P(B) \cdot P(D \mid B)$
     c. $P(A \mid D)$

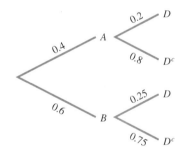

8. The accompanying tree diagram represents a two-stage experiment. Use the diagram to find
     a. $P(A) \cdot P(D \mid A)$ **b.** $P(B) \cdot P(D \mid B)$
     c. $P(A \mid D)$

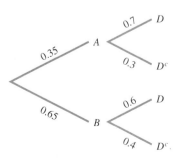

9. The accompanying tree diagram represents a two-stage experiment. Use the diagram to find
     a. $P(A) \cdot P(D \mid A)$ **b.** $P(B) \cdot P(D \mid B)$
     c. $P(C) \cdot P(D \mid C)$ **d.** $P(A \mid D)$

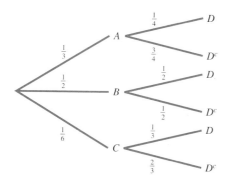

10. The accompanying tree diagram represents a two-stage experiment. Use this diagram to find
   a. $P(A \cap D)$ **b.** $P(B \cap D)$
   c. $P(C \cap D)$ **d.** $P(D)$
   e. Verify:

$P(A \mid D)$

$= \dfrac{P(A \cap D)}{P(D)}$

$= \dfrac{P(A) \cdot P(D \mid A)}{P(A) \cdot P(D \mid A) + P(B) \cdot P(D \mid B) + P(C) \cdot P(D \mid C)}$

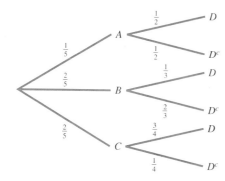

In exercises 11–14, refer to the following experiment: Two cards are drawn in succession without replacement from a standard deck of 52 cards.

11. What is the probability that the first card is a heart given that the second card is a heart?

12. What is the probability that the first card is a heart given that the second card is a diamond?

13. What is the probability that the first card is a jack given that the second card is an ace?

14. What is the probability that the first card is a face card given that the second card is an ace?

In exercises 15–18, refer to the following experiment: Urn A contains four white and six black balls. Urn B contains

three white and five black balls. A ball is drawn from urn A and then transferred to urn B. A ball is then drawn from urn B.

15. Represent the probabilities associated with this two-stage experiment in the form of a tree diagram.

16. What is the probability that the transferred ball was white given that the second ball drawn was white?

17. What is the probability that the transferred ball was black given that the second ball drawn was white?

18. What is the probability that the transferred ball was black given that the second ball drawn was black?

19. **Politics**   The 1992 U.S. Senate is composed of 57 Democrats and 43 Republicans. Sixty-six percent of the Democrats have served in the military, whereas 64% of the Republicans have seen military service. If a senator selected at random has served in the military, what is the probability that he is a Republican? (*Note:* No congresswoman has served in the military.)

20. **Quality Control**   Jansen Electronics has four machines that produce an identical component for use in its videocassette players. The proportion of the components produced by each machine and the probability of that component being defective are shown in the accompanying table. What is the probability that
   a. a component selected at random is defective?
   b. a defective component was produced by machine I?
   c. a defective component was produced by machine II?

Machine	Proportion of Components Produced	Probability of Defective Component
I	0.15	0.04
II	0.30	0.02
III	0.35	0.02
IV	0.20	0.03

21. An experiment consists of randomly selecting one of three coins, tossing it, and observing the outcome—heads or tails. The first coin is a two-headed coin, the second is a biased coin such that $P(H) = 0.75$, and the third is a fair coin.
   a. What is the probability that the coin that is tossed will show heads?

b. If the coin selected shows heads, what is the probability that this coin is the fair coin?

[c] *A calculator is recommended for the remainder of this exercise set.*

22. Reliability of Medical Tests A medical test has been designed to detect the presence of a certain disease. Among those who have the disease, the probability that the disease will be detected by the test is 0.95. However, the probability that the test will erroneously indicate the presence of the disease in those who do not actually have it is 0.04. It is estimated that 4% of the population who take this test have the disease.
   a. If the test administered to an individual is positive, what is the probability that the person actually has the disease?
   b. If an individual takes the test twice and both times the test is positive, what is the probability that the person actually has the disease?

23. Reliability of Medical Tests Refer to exercise 22. Suppose that 20% of the people who were referred to a clinic for the test did in fact have the disease. If the test administered to an individual from this group is positive, what is the probability that the person actually has the disease?

24. Quality Control A desk lamp produced by the Luminar Company was found to be defective. The company has three factories where the lamps are manufactured. The percentage of the total number of desk lamps produced by each factory and the probability that a lamp manufactured by that factory is defective are shown in the accompanying table. What is the probability that the defective lamp was manufactured in factory III?

Factory	Percentage of Total Production	Probability of Defective Component
I	0.35	0.015
II	0.35	0.01
III	0.30	0.02

25. Auto-Accident Rates An insurance company has compiled the accompanying data relating the age of drivers and the accident rate (the probability of being involved in an accident during a 1-year period) for drivers within that group.

Age Group	Percentage of Insured Drivers	Accident Rate
Under 25	0.16	0.055
25–44	0.40	0.025
45–64	0.30	0.02
65 and over	0.14	0.04

   a. What is the probability that an insured driver will be involved in an accident?
   b. What is the probability that an insured driver who is involved in an accident is under 25?

26. Seat-Belt Compliance Data compiled by the Highway Patrol Department regarding the use of seat belts by drivers in a certain area after the passage of a compulsory seat-belt law are shown in the accompanying table.

Drivers	Percentage of Drivers in Group	Percentage of Group Stopped for Moving Violation
Group I (using seat belts)	0.64	0.002
Group II (not using seat belts)	0.36	0.005

If a driver in that area is stopped for a moving violation, what is the probability that
   a. he or she will have a seat belt on?
   b. he or she will not have a seat belt on?

27. Medical Research Based on data obtained from the National Institute of Dental Research, it has been determined that 42% of twelve-year-olds have never had a cavity, 34% of thirteen-year-olds have never had a cavity, and 28% of fourteen-year-olds have never had a cavity. If a child is selected at random from a group of 24 junior high school students comprising 6 twelve-year-olds, 8 thirteen-year-olds, and 10 fourteen-year-olds and this child does not have a cavity, what is the probability that this child is fourteen years old?

28. Voting Patterns In a recent senatorial election, 50% of the voters in a certain district were registered as Democrats, 35% were registered as Republicans, and 15% were registered as Independents. The incumbent Democratic senator was reelected over her Republican and Independent opponents. Exit polls indicated that she gained 75% of the Democratic vote, 25% of the Republican vote, and 30% of the Independent vote. Assuming that the exit poll is accurate, what is the probability that a vote for the incumbent was cast by a registered Republican?

29. **Crime Rates** Data compiled by the Department of Justice on the number of people arrested for serious crimes (murder, forcible rape, robbery, and so on) in 1988 revealed that 89% were male and 11% were female. Of the males, 30% were under 18, whereas 27% of the females arrested were under 18.
   a. What is the probability that a person arrested for a serious crime in 1988 was under 18?
   b. If a person arrested for a serious crime in 1988 was known to be under 18, what is the probability that the person is female?

30. **Opinion Polls** A poll was conducted among 500 registered voters in a certain area regarding their position on a national lottery to raise revenue for the government. The results of the poll are shown in the accompanying table.

Sex	Percentage of Voters Polled	Percentage Favoring Lottery	Percentage Not Favoring Lottery	Percentage Expressing No Opinion
Male	0.51	0.62	0.32	0.06
Female	0.49	0.68	0.28	0.04

What is the probability that a registered voter who
a. favored a national lottery was a woman?
b. expressed no opinion regarding the lottery was a woman?

31. **Customer Surveys** The sales department of the Thompson Drug Company released the accompanying data concerning the sales of a certain pain reliever manufactured by the company.

Pain Reliever	Percentage of Drug Sold	Percentage of Group Sold in Extra-Strength Dosage
Group I (capsule form)	0.57	0.38
Group II (tablet form)	0.43	0.31

If a customer purchased the extra-strength dosage of this drug, what is the probability that it was in capsule form?

32. **Selection of Supreme Court Judges** In a past presidential election, it was estimated that the probability that the Republican candidate would be elected was 3/5 and therefore the probability that the Democratic candidate would be elected was 2/5 (the two Independent candidates were given little chance of being elected). It was also estimated that if the Republican candidate were elected, the probabilities that a conservative, moderate, or liberal judge would be appointed to the Supreme Court (one retirement was expected during the presidential term) were 1/2, 1/3, and 1/6, respectively. If the Democratic candidate were elected, the probabilities that a conservative, moderate, or liberal judge would be appointed to the Supreme Court would be 1/8, 3/8, and 1/2, respectively. A conservative judge *was* appointed to the Supreme Court during the presidential term. What is the probability that the Democratic candidate was elected?

33. **Personnel Selection** Applicants for temporary office work at the Carter Temporary Help Agency who have successfully completed a typing test are then placed in suitable positions by Ms. Dwyer and Ms. Newberg. Employers who hire temporary help through the agency return a card indicating satisfaction or dissatisfaction with the work performance of those hired. From past experience it is known that 80% of the employees placed by Ms. Dwyer are rated as satisfactory, whereas 70% of those placed by Ms. Newberg are rated as satisfactory. Ms. Newberg places 55% of the temporary office help at the agency and Ms. Dwyer the remaining 45%. If a Carter office worker is rated unsatisfactory, what is the probability that he or she was placed by Ms. Newberg?

34. **College Majors** The Office of Admissions and Records of a large western university released the accompanying information concerning the contemplated majors of its freshman class.

Major	Percentage of Freshmen Choosing This Major	Percentage of Females	Percentage of Males
Business	0.24	0.38	0.62
Humanities	0.08	0.60	0.40
Education	0.08	0.66	0.34
Social science	0.07	0.58	0.42
Natural sciences	0.09	0.52	0.48
Other	0.44	0.48	0.52

a. What is the probability that a student selected at random from the freshman class is a female?

b. What is the probability that a business student selected at random from the freshman class is a male?

c. What is the probability that a female student selected at random from the freshman class is majoring in business?

35. Medical Diagnoses A study was conducted among a certain group of union members whose health insurance policies required second opinions prior to surgery. Of those members whose doctors advised them to have surgery, 20% were informed by a second doctor that no surgery was needed. Of these, 70% took the second doctor's opinion and did not go through with the surgery. Of the members who were advised to have surgery by both doctors, 95% went through with the surgery. What is the probability that a union member who had surgery was advised to do so by a second doctor?

36. Age Distribution of Renters A study conducted by the Metro Housing Agency in a midwestern city re-

vealed the accompanying information concerning the age distribution of renters within the city.

Age	Percentage of Adult Population	Percentage of Group Who Are Renters
21–44	0.51	0.58
45–64	0.31	0.45
65 and over	0.18	0.60

a. What is the probability that an adult selected at random from this population is a renter?

b. If a renter is selected at random, what is the probability that he or she is in the 21–44 age bracket?

c. If a renter is selected at random, what is the probability that he or she is 45 years of age or older?

SOLUTIONS TO SELF-CHECK EXERCISES 8.3

1. By Bayes' Theorem, we have, using the probabilities given in the tree diagram,

$$P(B \mid D) = \frac{P(B)P(D \mid B)}{P(A)P(D \mid A) + P(B)P(D \mid B) + P(C)P(D \mid C)}$$

$$= \frac{\left(\frac{1}{2}\right)\left(\frac{1}{2}\right)}{\left(\frac{1}{4}\right)\left(\frac{1}{3}\right) + \left(\frac{1}{2}\right)\left(\frac{1}{2}\right) + \left(\frac{1}{4}\right)\left(\frac{1}{4}\right)}$$

$$= \frac{12}{19}$$

2. Let R and D, respectively, denote the event that the Republican and the Democratic candidate won the presidential election. Then $P(R) = 3/5$ and $P(D) = 2/5$. Also, let C denote the event that research for the new manned bomber would continue. These data may be exhibited as in the accompanying tree diagram.

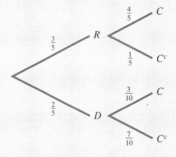

Using Bayes' Theorem, we find that the probability that the Republican candidate had won the election is given by

$$P(R \mid C^c) = \frac{P(R)P(C^c \mid R)}{P(R)P(C^c \mid R) + P(D)P(C^c \mid D)}$$

$$= \frac{\left(\frac{3}{5}\right)\left(\frac{1}{5}\right)}{\left(\frac{3}{5}\right)\left(\frac{1}{5}\right) + \left(\frac{2}{5}\right)\left(\frac{7}{10}\right)} = \frac{3}{10}$$

8.4 Distributions of Random Variables

Random Variables

In many situations it is desirable to assign numerical values to the outcomes of an experiment. For example, if an experiment consists of casting a die and observing the face that lands uppermost, then it is natural to assign the numbers 1, 2, 3, 4, 5, and 6, respectively, to the outcomes *one, two, three, four, five,* and *six* of the experiment. If we let X denote the outcome of the experiment, then X assumes one of the numbers. Because the values assumed by X depend on the outcomes of a chance experiment, the outcome X is referred to as a *random variable.*

RANDOM VARIABLE

A **random variable** is a rule that assigns a number to each outcome of a chance experiment.

More precisely, a random variable is a function with domain given by the set of outcomes of a chance experiment and range contained in the set of real numbers.

EXAMPLE 1 A coin is tossed three times. Let the random variable X denote the number of heads that occur in the three tosses.

a. List the outcomes of the experiment; that is, find the domain of the function X.

b. Find the value assigned to each outcome of the experiment by the random variable X.

c. Find the event comprising the outcomes to which a value of 2 has been assigned by X. This event is written $(X = 2)$ and is the event comprising the outcomes in which two heads occur.

Solution

a. From the results of Example 3, Section 7.5 (page 371), we see that the set of outcomes of the experiment is given by the sample space

$$S = \{\text{HHH, HHT, HTH, THH, HTT, THT, TTH, TTT}\}$$

b. The outcomes of the experiment are displayed in the first column of Table 3. The corresponding value assigned to each such outcome by the random variable X (the number of heads) appears in the second column.

c. With the aid of Table 3, we see that the event $(X = 2)$ is given by the set

$$\{\text{HHT, HTH, THH}\}$$

■

TABLE 3
Number of heads in three coin tosses

Outcome	Value of X
HHH	3
HHT	2
HTH	2
THH	2
HTT	1
THT	1
TTH	1
TTT	0

EXAMPLE 2 A coin is tossed repeatedly until a head occurs. Let the random variable Y denote the number of coin tosses in the experiment. What are the values of Y?

TABLE 4
Number of coin tosses before heads appears

Outcome	Value of Y
H	1
TH	2
TTH	3
TTTH	4
TTTTH	5
⋮	⋮

Solution   The outcomes of the experiment make up the infinite set

$$S = \{\text{H, TH, TTH, TTTH, TTTTH, } \ldots\}$$

These outcomes of the experiment are displayed in the first column of Table 4. The corresponding values assumed by the random variable Y (the number of tosses) appear in the second column.   ◼

EXAMPLE 3   A disposable flashlight is turned on until its battery runs out. Let the random variable Z denote the length (in hours) of the life of the battery. What values may Z assume?

Solution   The values assumed by Z may be any nonnegative real numbers; that is, the possible values of Z comprise the interval $0 \le Z < \infty$.   ◼

One advantage of working with random variables rather than working directly with the outcomes of an experiment is that random variables are functions that may be added, subtracted, and multiplied. Because of this, results developed in the field of algebra and other areas of mathematics may be used freely to help us solve problems in the fields of probability and statistics.

A random variable is classified into three categories depending on the set of values it assumes. A random variable is called **finite discrete** if it assumes only finitely many values. For example, the random variable X of Example 1 is finite discrete since it may assume values only from the finite set $\{0, 1, 2, 3\}$ of numbers. Next, a random variable is said to be **infinite discrete** if it takes on infinitely many values, which may be arranged in a sequence. For example, the random variable Y of Example 2 is infinite discrete since it assumes values from the set $\{1, 2, 3, 4, 5, \ldots\}$, which has been arranged in the form of an infinite sequence. Finally, a random variable is called **continuous** if the values it may assume comprise an interval of real numbers. For example, the random variable Z of Example 3 is continuous since the values it may assume comprise the interval of nonnegative real numbers. For the remainder of this section *all random variables will be assumed to be finite discrete.*

Probability Distributions of Random Variables

In Section 7.6 we learned how to construct the probability distribution for an experiment. There, the probability distribution took the form of a table that gave the probabilities associated with the outcomes of an experiment. Since the random variable associated with an experiment is related to the outcomes of the experiment, it is clear that we should be able to construct a probability distribution associated with the *random variable* rather than one associated with the outcomes of the experiment. Such a distribution is called the **probability distribution of a random variable** and may be given in the form of a formula or displayed in a table that gives the distinct (numerical) values of the random variable X and the probabilities associated with these values. Thus, if $x_1, x_2, \ldots, x_n$ are the values assumed by the random variable X, with associated probabilities $P(X = x_1)$, $P(X = x_2)$, $\ldots, P(X = x_n)$, respectively, then the required probability distribution of the random variable X, where $p_i = P(X = x_i)$, $i = 1, 2, \ldots, n$, may be expressed in the form of a table:

*Probability distribution
for the random variable X*

x	$P(X = x)$
x_1	p_1
x_2	p_2
x_3	p_3
$\vdots$	$\vdots$
x_n	p_n

The next several examples illustrate the construction of probability distributions.

EXAMPLE 4   Find the probability distribution of the random variable associated with the experiment of Example 1.

Solution   From the results of Example 1, we see that the values assumed by the random variable X are 0, 1, 2, and 3, corresponding to the events of 0, 1, 2, and 3 heads occurring, respectively. Referring to Table 3 once again, we see that the outcome associated with the event $(X = 0)$ is given by the set {TTT}. Consequently, the probability associated with the random variable X when it assumes the value 0 is given by

$$P(X = 0) = \tfrac{1}{8} \qquad \text{Note that } n(S) = 8.$$

Next, observe that the event $(X = 1)$ is given by the set {HTT, THT, TTH}, so

$$P(X = 1) = \tfrac{3}{8}$$

In a similar manner we may compute $P(X = 2)$ and $P(X = 3)$, which gives the probability distribution shown in Table 5.

TABLE 5
*Probability
distribution*

x	$P(X = x)$
0	$\tfrac{1}{8}$
1	$\tfrac{3}{8}$
2	$\tfrac{3}{8}$
3	$\tfrac{1}{8}$

EXAMPLE 5   Let X denote the random variable that gives the sum of the faces that fall uppermost when two fair dice are cast. Find the probability distribution of X.

Solution   The values assumed by the random variable X are 2, 3, 4, . . . , 12, corresponding to the events $E_2, E_3, E_4, \ldots, E_{12}$ (see Example 4, Section 7.5). Next, the probabilities associated with the random variable X when X assumes the values 2, 3, 4, . . . , 12 are precisely the probabilities $P(E_2)$, $P(E_3), \ldots, P(E_{12})$, respectively, and may be computed in much the same way as the solution to Example 3, Section 7.6. Thus,

$$P(X = 2) = P(E_2) = \frac{1}{36}, \qquad P(X = 3) = P(E_3) = \frac{2}{36}, \qquad \text{and so on}$$

The required probability distribution of X is given in Table 6.

TABLE 6
Probability distribution for the random variable that gives the sum of the faces of two dice

x	2	3	4	5	6	7	8	9	10	11	12
$P(X = x)$	$\tfrac{1}{36}$	$\tfrac{2}{36}$	$\tfrac{3}{36}$	$\tfrac{4}{36}$	$\tfrac{5}{36}$	$\tfrac{6}{36}$	$\tfrac{5}{36}$	$\tfrac{4}{36}$	$\tfrac{3}{36}$	$\tfrac{2}{36}$	$\tfrac{1}{36}$

EXAMPLE 6   The following data give the number of cars observed waiting in line at the beginning of 2-minute intervals between 3 and 5 P.M. on a certain Friday at the drive-in teller of the Westwood Savings Bank and the corresponding frequency of occurrence. Find the probability distribution of the random variable X, where X denotes the number of cars observed waiting in line.

Number of Cars	Frequency of Occurrence
0	2
1	9
2	16
3	12
4	8
5	6
6	4
7	2
8	1

TABLE 7
Probability distribution

x	$P(X = x)$
0	0.03
1	0.15
2	0.27
3	0.20
4	0.13
5	0.10
6	0.07
7	0.03
8	0.02

Solution   Dividing each number on the right of the given table by 60 (the sum of these numbers) gives the respective probabilities (here, we use the relative frequency interpretation of probability) associated with the random variable X when X assumes the values 0, 1, 2, . . . , 8. For example,

$$P(X = 0) = \frac{2}{60} \approx 0.03$$

$$P(X = 1) = \frac{9}{60} = 0.15$$

$$\vdots$$

and so on

The resulting probability distribution is shown in Table 7.   ■

Histograms

A probability distribution of a random variable may be exhibited graphically by means of a **histogram.** To construct a histogram of a particular probability distribution, first locate the values of the random variable on the number line. Then, above each such number, erect a rectangle with width 1 and height equal to the probability associated with that value of the random variable. For example, the histogram of the probability distribution appearing in Table 5 is shown in Figure 13. The histograms of the probability distributions of Examples 5 and 6 are constructed in a similar manner and are displayed in Figures 14 and 15, respectively.

Observe that in each histogram, the area of a rectangle associated with a value of a random variable X gives precisely the probability associated with that value of X. This follows because each such rectangle, by construc-

FIGURE 13

Histogram showing the probability distribution for the number of heads occurring in three coin tosses

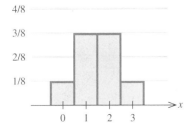

FIGURE 14

Histogram showing the probability distribution for the sum of the faces of two dice

FIGURE 15

Histogram showing the probability distribution for the number of cars waiting in line

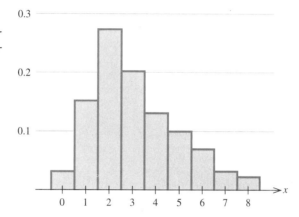

tion, has width 1 and height corresponding to the probability associated with the value of the random variable. Another consequence arising from the method of construction of a histogram is that *the probability associated with more than one value of the random variable X is given by the sum of the areas of the rectangles associated with those values of X.* For example, in the coin-tossing experiment of Example 1, the event of obtaining at least two heads, which corresponds to the event $(X = 2)$ or $(X = 3)$, is given by

$$P(X = 2) + P(X = 3)$$

and may be obtained from the histogram depicted in Figure 13 by adding the areas associated with the values 2 and 3, respectively, of the random variable X. We obtain

$$P(X = 2) + P(X = 3) = (1)(\tfrac{3}{8}) + (1)(\tfrac{1}{8}) = \tfrac{1}{2}$$

This result provides us with a method of computing the probabilities of events directly from the knowledge of a histogram of the probability distribution of the random variable associated with the experiment.

EXAMPLE 7   Suppose that the probability distribution of a random variable X is represented by the histogram shown in Figure 16. Identify that part of the histogram whose area gives the probability $P(10 \leq X \leq 20)$. Do not evaluate the result.

Solution   The event $(10 \leq X \leq 20)$ is the event comprising outcomes related to the values 10, 11, 12, ... , 20 of the random variable X. The

FIGURE 16

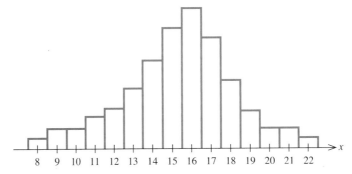

FIGURE 17
$P(10 \leq X \leq 20)$

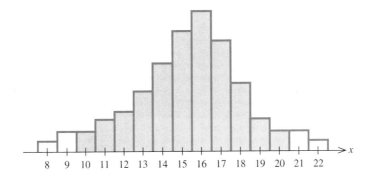

probability of this event $P(10 \leq X \leq 20)$ is therefore given by the shaded area of the histogram in Figure 17. ■

SELF-CHECK EXERCISES 8.4

1. Three balls are selected at random without replacement from an urn containing four black balls and five white balls. Let the random variable X denote the number of black balls drawn.
   a. List the outcomes of the experiment.
   b. Find the value assigned to each outcome of the experiment by the random variable X.
   c. Find the event comprising the outcomes to which a value of 2 has been assigned by X.

2. The following data, extracted from the records of the Dover Public Library, give the number of books borrowed by the library's members over a 1-month period.

Number of Books	0	1	2	3	4	5	6	7	8
Frequency of Occurrence	780	300	412	205	98	54	57	30	6

   a. Find the probability distribution of the random variable X, where X denotes the number of books checked out over a 1-month period by a randomly chosen member.
   b. Draw the histogram representing this probability distribution.

Solutions to Self-Check Exercises 8.4 can be found on page 443.

8.4 EXERCISES

1. Three balls are selected at random without replacement from an urn containing four green balls and six red balls. Let the random variable X denote the number of green balls drawn.
   a. List the outcomes of the experiment.
   b. Find the value assigned to each outcome of the experiment by the random variable X.
   c. Find the event comprising the outcomes to which a value of 3 has been assigned by X.

2. A coin is tossed four times. Let the random variable X denote the number of tails that occur.
   a. List the outcomes of the experiment.
   b. Find the value assigned to each outcome of the experiment by the random variable X.
   c. Find the event comprising the outcomes to which a value of 2 has been assigned by X.

3. A die is cast repeatedly until a 6 falls uppermost. Let the random variable X denote the number of times the die is cast. What are the values that X may assume?

4. Cards are selected one at a time without replacement from a well-shuffled deck of 52 cards until an ace is drawn. Let X denote the random variable that gives the number of cards drawn. What values may X assume?

5. Let X denote the random variable that gives the sum of the faces that fall uppermost when two fair dice are cast. Find $P(X = 7)$.

6. Two cards are drawn from a well-shuffled deck of 52 playing cards. Let X denote the number of aces drawn. Find $P(X = 2)$.

In exercises 7–12, give the range of values that the random variable X may assume and classify the random variable as finite discrete, infinite discrete, or continuous.

7. X = The number of times a die is thrown until a 2 appears

8. X = The number of defective watches in a sample of eight watches

9. X = The distance a commuter travels to work

10. X = The number of hours a child watches television on a given day

11. X = The number of times an accountant takes the CPA examination before passing

12. X = The number of boys in a four-child family.

13. The probability distribution of the random variable X is shown in the accompanying table.

x		-10	-5	0	5
$P(X = x)$		0.20	0.15	0.05	0.1

x		10	15	20
$P(X = x)$		0.25	0.1	0.15

Find:
   a. $P(X = -10)$ b. $P(X \geq 5)$
   c. $P(-5 \leq X \leq 5)$ d. $P(X \leq 20)$

14. The probability distribution of the random variable X is shown in the accompanying table.

x		-5	-3	-2
$P(X = x)$		0.17	0.13	0.33

x		0	2	3
$P(X = x)$		0.16	0.11	0.10

Find:
   a. $P(X \leq 0)$ b. $P(X \leq -3)$
   c. $P(-2 \leq X \leq 2)$

15. Suppose that a probability distribution of a random variable X is represented by the accompanying histogram. Shade that part of the histogram whose area gives the probability $P(17 \leq X \leq 20)$.

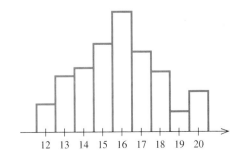

16. Exams An examination consisting of 10 true-or-false questions was taken by a class of 100 students. The probability distribution of the random variable X, where X denotes the number of questions answered correctly by a randomly chosen student, is represented by the accompanying histogram. The rectangle with base centered on the number 8 is missing. What should the height of this rectangle be?

17. Two dice are cast. Let the random variable X denote the number that falls uppermost on the first die and let Y denote the number that falls uppermost on the second die.
  a. Find the probability distributions of X and Y.
  b. Find the probability distribution of $X + Y$.

18. Distribution of Families by Size A survey was conducted by the Public Housing Authority in a certain community among 1000 families to determine the distribution of families by size. The results follow.

Family Size	2	3	4	5
Frequency of Occurrence	350	200	245	125

Family Size	6	7	8
Frequency of Occurrence	66	10	4

  a. Find the probability distribution of the random variable X where X denotes the number of persons in a randomly chosen family.
  b. Draw the histogram corresponding to the probability distribution found in (a).

19. Waiting Lines The accompanying data were obtained in a study conducted by the manager of the Sav-More Supermarket. In this study, the number of customers waiting in line at the express checkout at the beginning of each 3-minute interval between 9 A.M. and 12 noon on Saturday was observed.

Number of Customers	0	1	2	3	4
Frequency of Occurrence	1	4	2	7	14

Number of Customers	5	6	7	8	9	10
Frequency of Occurrence	8	10	6	3	4	1

  a. Find the probability distribution of the random variable X where X denotes the number of customers observed waiting in line.
  b. Draw the histogram representing this probability distribution.

20. Money-Market Rates The rates paid by 30 financial institutions on a certain day for money-market deposit accounts are shown in the accompanying table.

Rate (%)	6	6.25	6.55	6.56
Number of Institutions	1	7	7	1

Rate (%)	6.58	6.60	6.65	6.85
Number of Institutions	1	8	3	2

Let the random variable X denote the interest paid by a randomly chosen financial institution on its money-market deposit accounts and find the probability distribution associated with these data.

21. Television Pilots After the private screening of a new television pilot, audience members were asked to rate the new show on a scale of 1 to 10 (10 being the highest rating). From a group of 140 people, the accompanying responses were obtained.

Rating	1	2	3	4	5
Frequency of Occurrence	1	4	3	11	23

Rating	6	7	8	9	10
Frequency of Occurrence	21	28	29	16	4

Let the random variable X denote the rating given to the show by a randomly chosen audience member. Find the probability distribution associated with these data.

1. a. Using the accompanying tree diagram, we see that the outcomes of the experiment are

$$S = \{(B, B, B), (B, B, W), (B, W, B), (B, W, W),$$

$$(W, B, B), (W, B, W), (W, W, B), (W, W, W)\}$$

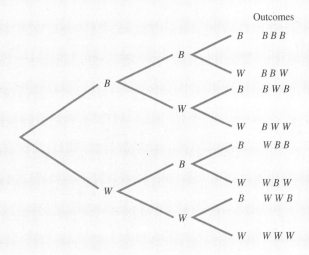

b. Using the results of (a), we obtain the values assigned to the outcomes of the experiment as follows:

Outcome	BBB	BBW	BWB	BWW
Value	3	2	2	1

Outcome	WBB	WBW	WWB	WWW
Value	2	1	1	0

c. The required event is {BBW, BWB, WBB}.

2. a. We divide each number in the bottom row of the given table by 1942 (the sum of these numbers) to obtain the probabilities associated with the random variable X when X takes on the values 0, 1, 2, 3, 4, 5, 6, 7, and 8.

For example,

$$P(X = 0) = \frac{780}{1942} \simeq 0.402$$

$$P(X = 1) = \frac{300}{1942} \simeq 0.154$$

The required probability distribution and histogram follow.

x	0	1	2	3	4
P(X = x)	0.402	0.154	0.212	0.106	0.050

x	5	6	7	8
P(X = x)	0.028	0.029	0.015	0.003

b.

8.5   Expected Value

Mean

The average value of a set of numbers is a familiar notion to most people. For example, to compute the average of the four numbers

$$12, 16, 23, 37$$

we simply add these numbers and divide the resulting sum by 4, giving the required average as

$$\frac{12 + 16 + 23 + 37}{4} = \frac{88}{4}$$

or 22. In general, we have the following definition:

AVERAGE OR MEAN   The **average**, or **mean**, of the n numbers

$$x_1, x_2, \ldots, x_n$$

is $\bar{x}$ (read "x bar"), where

$$\bar{x} = \frac{x_1 + x_2 + \cdots + x_n}{n}$$

EXAMPLE 1   Refer to Example 6, Section 8.4. Find the average number of cars waiting in line at the bank's drive-in teller observed at the beginning of each 2-minute interval during the period in question.

Solution   The number of cars, together with its corresponding frequency of occurrence, are reproduced in Table 8.

TABLE 8

Number of Cars	0	1	2	3	4	5	6	7	8
Frequency of Occurrence	2	9	16	12	8	6	4	2	1

Observe that the number 0 (of cars) occurs twice, the number 1 occurs 9 times, and so on. There are altogether

$$2 + 9 + 16 + 12 + 8 + 6 + 4 + 2 + 1 = 60$$

numbers to be averaged. Therefore, the required average is given by

$$\frac{0 \cdot 2 + 1 \cdot 9 + 2 \cdot 16 + 3 \cdot 12 + 4 \cdot 8 + 5 \cdot 6 + 6 \cdot 4 + 7 \cdot 2 + 8 \cdot 1}{60} \approx 3.1 \qquad (10)$$

or approximately 3.1 cars.

Expected Value

Let us reconsider the expression in (10) that gives the average of the frequency distribution shown in Table 8. Dividing each term by the denominator, the expression may be rewritten in the form

$$0 \cdot \left(\frac{2}{60}\right) + 1 \cdot \left(\frac{9}{60}\right) + 2 \cdot \left(\frac{16}{60}\right) + 3 \cdot \left(\frac{12}{60}\right) + 4 \cdot \left(\frac{8}{60}\right) + 5 \cdot \left(\frac{6}{60}\right)$$
$$+ 6 \cdot \left(\frac{4}{60}\right) + 7 \cdot \left(\frac{2}{60}\right) + 8 \cdot \left(\frac{1}{60}\right)$$

Observe that each term in the sum is a product of two factors; the first factor is the value assumed by the random variable X, where X denotes the number of cars observed waiting in line, and the second factor is just the probability associated with that value of the random variable. This observation suggests the following general method for calculating the *expected value* (that is, the average, or mean) of a random variable X that assumes a finite number of values from the knowledge of its probability distribution.

EXPECTED VALUE OF A RANDOM VARIABLE X

Let X denote a random variable that assumes the values x_1, x_2, $\ldots$, x_n with associated probabilities p_1, p_2, $\ldots$, p_n, respectively. Then the **expected value** of X, $E(X)$, is given by

$$E(X) = x_1 p_1 + x_2 p_2 + \cdots + x_n p_n \qquad (11)$$

REMARK   The numbers x_1, x_2, $\ldots$, x_n may be positive, zero, or negative. For example, such a number will be positive if it represents a profit and negative if it represents a loss. $\square$

TABLE 9
Probability distribution

x	$P(X = x)$
0	0.03
1	0.15
2	0.27
3	0.20
4	0.13
5	0.10
6	0.07
7	0.03
8	0.02

EXAMPLE 2   Re-solve Example 1 by using the probability distribution associated with the experiment, reproduced in Table 9.

Solution   Let X denote the number of cars observed waiting in line. Then the average number of cars observed waiting in line is given by the expected value of X—that is, by

$$E(X) = 0 \cdot (0.03) + 1 \cdot (0.15) + 2 \cdot (0.27) + 3 \cdot (0.20) + 4 \cdot (0.13)$$
$$+ 5 \cdot (0.10) + 6 \cdot (0.07) + 7 \cdot (0.03) + 8 \cdot (0.02)$$
$$= 3.1$$

or 3.1 cars, which agrees with the earlier result.   ■

The expected value of a random variable X is a measure of central tendency of the probability distribution associated with X. In repeated trials of an experiment with random variable X, the average of the observed values of X gets closer and closer to the expected value of X as the number of trials gets larger and larger. Geometrically, the expected value of a random variable X has the following simple interpretation: If a laminate is made of the histogram of a probability distribution associated with a random variable X, then the expected value of X corresponds to the point on the base of the laminate at which the latter will balance perfectly when the point is directly over a fulcrum (Figure 18).

FIGURE 18
Expected value of a random variable X

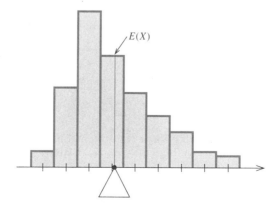

EXAMPLE 3   Let X denote the random variable that gives the sum of the faces that fall uppermost when two fair dice are cast. Find the expected value, $E(X)$, of X.

TABLE 10
Probability distribution

x	P(X = x)
2	$\frac{1}{36}$
3	$\frac{2}{36}$
4	$\frac{3}{36}$
5	$\frac{4}{36}$
6	$\frac{5}{36}$
7	$\frac{6}{36}$
8	$\frac{5}{36}$
9	$\frac{4}{36}$
10	$\frac{3}{36}$
11	$\frac{2}{36}$
12	$\frac{1}{36}$

Solution   The probability distribution of X, reproduced in Table 10, was found in Example 5, Section 8.4. Using this result, we find

$$E(X) = 2(\tfrac{1}{36}) + 3(\tfrac{2}{36}) + 4(\tfrac{3}{36}) + 5(\tfrac{4}{36}) + 6(\tfrac{5}{36}) + 7(\tfrac{6}{36})$$
$$+ 8(\tfrac{5}{36}) + 9(\tfrac{4}{36}) + 10(\tfrac{3}{36}) + 11(\tfrac{2}{36}) + 12(\tfrac{1}{36})$$
$$= 7$$

Note that, because of the symmetry of the histogram of the probability distribution with respect to the vertical line $x = 7$, the result could have been obtained by merely inspecting Figure 19.

FIGURE 19
Histogram showing the probability distribution for the sum of the faces of two dice

Applications

The next example shows how we can use the concept of expected value to help us make the best investment decision.

EXAMPLE 4   A group of private investors intends to purchase one of two motels currently being offered for sale in a certain city. The terms of sale of the two motels are similar, although the Regina Inn has 52 rooms and is in a slightly better location than the Merlin Motor Lodge, which has 60 rooms. Records obtained for each motel reveal that the occupancy rates, with corresponding probabilities, during the May–September tourist season are as shown in the accompanying tables.

Regina Inn

Occupancy Rate	0.80	0.85	0.90	0.95	1.00
Probability	0.19	0.22	0.31	0.23	0.05

Merlin Motor Lodge

Occupancy Rate	0.75	0.80	0.85	0.90	0.95	1.00
Probability	0.35	0.21	0.18	0.15	0.09	0.02

The average profit per day for each occupied room at the Regina Inn is $10, whereas the average profit per day for each occupied room at the Merlin Motor Lodge is $9.

a. Find the average number of rooms occupied per day at each motel.

b. If the investors' objective is to purchase the motel that generates the higher daily profit, which motel should they purchase? (Compare the expected daily profit of the two motels.)

Solution

a. Let X denote the occupancy rate at the Regina Inn. Then the average daily occupancy rate at the Regina Inn is given by the expected value of X, that is, by

$$E(X) = (0.80)(0.19) + (0.85)(0.22) + (0.90)(0.31)$$
$$+ (0.95)(0.23) + (1.00)(0.05)$$
$$= 0.8865$$

The average number of rooms occupied per day at the Regina is

$$(0.8865)(52) \approx 46.1$$

or 46.1 rooms. Similarly, letting Y denote the occupancy rate at the Merlin Motor Lodge, we have

$$E(Y) = (0.75)(0.35) + (0.80)(0.21) + (0.85)(0.18) + (0.90)(0.15)$$
$$+ (0.95)(0.09) + (1.00)(0.02)$$
$$= 0.8240$$

The average number of rooms occupied per day at the Merlin is

$$(0.8240)(60) \approx 49.4$$

or 49.4 rooms.

b. The expected daily profit at the Regina is given by

$$46.1(10) = 461$$

or \$461. The expected daily profit at the Merlin is given by

$$(49.4)(9) \approx 445$$

or \$445. From these results we conclude that the investors should purchase the Regina Inn, which is expected to yield a higher daily profit. ◼

EXAMPLE 5   The Island Club is holding a fund-raising raffle. Ten thousand tickets have been sold for \$2 each. There will be a first prize of \$3000, 3 second prizes of \$1000 each, 5 third prizes of \$500 each, and 20 consolation prizes of \$100 each. Letting X denote the net winnings (that is, winnings less the cost of the ticket) associated with the tickets, find $E(X)$. Interpret your results.

Solution   The values assumed by X are $(0 - 2)$, $(100 - 2)$, $(500 - 2)$, $(1000 - 2)$, and $(3000 - 2)$—that is, -2, 98, 498, 998, and 2998, which correspond, respectively, to the value of a losing ticket, a consolation prize, a third prize, and so on. The probability distribution of X may be calculated in the usual manner and appears in Table 11.

TABLE 11
Probability distribution for a raffle

x	P(X = x)
−2	0.9971
98	0.0020
498	0.0005
998	0.0003
2998	0.0001

Using the table, we find

$$E(X) = (-2)(0.9971) + 98(0.0020) + 498(0.0005)$$
$$+ \ 998(0.0003) + 2998(0.0001)$$
$$= -0.95$$

This expected value gives the long-run average loss (negative gain) of a holder of one ticket; that is, if one participated in such a raffle by purchasing one ticket each time, in the long run, one may expect to lose, on the average, 95 cents per raffle. ∎

EXAMPLE 6   In the game of roulette as played in Las Vegas casinos, the wheel is divided into 38 compartments numbered 1 through 36, 0, and 00. One-half of the numbers 1 through 36 are red, the other half black, and 0 and 00 are green (Figure 20). Of the many types of bets that may be placed, one type involves betting on the outcome of the color of the winning number. For example, one may place a certain sum of money on *red*. If the winning number is red, one wins an amount equal to the bet placed and loses the bet otherwise. Find the expected value of the winnings on a $1 bet placed on *red*.

FIGURE 20
Roulette wheel

Solution   Let X be a random variable whose values are 1 and −1, which correspond to a win and a loss. The probabilities associated with the values 1 and −1 are 18/38 and 20/38, respectively. Therefore, the expected value is given by

$$E(X) = 1(\tfrac{18}{38}) + (-1)(\tfrac{20}{38}) = -\tfrac{2}{38}$$
$$\approx -0.053$$

Thus, if one places a $1 bet on *red* over and over again, one may expect to lose, on the average, approximately 5 cents per bet in the long run. ∎

Examples 5 and 6 illustrate games that are not "fair." Of course, most participants in such games are aware of this fact and participate in them for other reasons. In a fair game, neither party has an advantage, a condition that translates into the condition that $E(X) = 0$, where X takes on the values of a player's winnings.

EXAMPLE 7   Mike and Bill play a card game with a standard deck of 52 cards. Mike selects a card from a well-shuffled deck and receives A dollars from Bill if the card selected is a diamond; otherwise, Mike pays Bill a dollar. Determine the value of A if the game is to be fair.

Solution   Let X denote a random variable whose values are associated with Mike's winnings. Then X takes on the value A with probability $P(X = A) = 1/4$ (since there are 13 diamonds in the deck) if Mike wins and takes on the value -1 with probability $P(X = -1) = 3/4$ if Mike loses. Since the game is to be a fair one, the expected value $E(X)$ of Mike's winnings must be equal to zero, that is,

$$E(X) = A\left(\frac{1}{4}\right) + (-1)\left(\frac{3}{4}\right) = 0$$

Solving this equation for A gives $A = 3$. Thus, the card game will be fair if Bill makes a $3 payoff for a winning bet of $1 placed by Mike.   ∎

Odds

In everyday parlance, the probability of the occurrence of an event is often stated in terms of the *odds in favor of* (or *odds against*) the occurrence of the event. For example, one often hears statements such as "the odds that the Dodgers will win the World Series this season are 7 to 5" and "the odds that it will not rain tomorrow are 3 to 2." We will return to these examples later on. But first, let us look at a definition that ties these two concepts together.

ODDS IN FAVOR OF AND ODDS AGAINST

If $P(E)$ is the probability of an event E occurring, then

1. the **odds in favor of** E occurring are

$$\frac{P(E)}{1 - P(E)} = \frac{P(E)}{P(E^c)} \qquad P(E) \neq 1 \qquad \text{(12a)}$$

2. the **odds against** E occurring are

$$\frac{1 - P(E)}{P(E)} = \frac{P(E^c)}{P(E)} \qquad P(E) \neq 0 \qquad \text{(12b)}$$

REMARKS

1. Notice that the odds in favor of the occurrence of an event are given by the ratio of the probability of the event occurring to the probability of the event not occurring. The odds against the occurrence of an event are given by the reciprocal of the odds in favor of the occurrence of the event.

2. Whenever possible, odds are expressed as ratios of whole numbers. If the odds in favor of E are a/b, we say the odds in favor of E are a to b. If the odds against E occurring are b/a, we say the odds against E are b to a.   □

EXAMPLE 8   Find the odds in favor of winning a bet on *red* in American roulette. What are the odds against winning a bet on *red*?

Solution   The probability of winning a bet here—the probability that the ball lands in a red compartment—is given by $P = 18/38$. Therefore, using equation (12a), we see that the odds in favor of winning a bet on *red* are

$$\frac{P(E)}{1 - P(E)} = \frac{\dfrac{18}{38}}{1 - \dfrac{18}{38}} \qquad \text{\small E, event of winning a bet on red}$$

$$= \frac{\dfrac{18}{38}}{\dfrac{38 - 18}{38}}$$

$$= \frac{18}{38} \cdot \frac{38}{20}$$

$$= \frac{18}{20} = \frac{9}{10}$$

or 9 to 10. Next, using (12b), we see that the odds against winning a bet on *red* are 10/9, or 10 to 9.   ■

Now, suppose the odds in favor of the occurrence of an event are a to b. Then (12a) gives

$$\frac{a}{b} = \frac{P(E)}{1 - P(E)}$$

$$a(1 - P(E)) = bP(E) \qquad \text{\small Cross-multiplying}$$

$$a - aP(E) = bP(E)$$

$$a = (a + b)P(E)$$

and
$$P(E) = \frac{a}{a + b},$$

which leads us to the following result.

PROBABILITY OF AN EVENT (GIVEN THE ODDS)

If the odds in favor of an event E occurring are a to b, then the probability of E occurring is

$$P(E) = \frac{a}{a + b} \qquad\qquad \textbf{(13)}$$

Formula (13) is often used to determine subjective probabilities, as the next example shows.

EXAMPLE 9   Consider each of the following statements.

a. "The odds that the Dodgers will win the World Series this season are 7 to 5."

b. "The odds that it will not rain tomorrow are 3 to 2."

Express each of these odds as a probability of the event occurring.

Solution

a. Using (13) with $a = 7$ and $b = 5$ gives the required probability as

$$\frac{7}{7 + 5} = \frac{7}{12} \approx 0.5833$$

b. Here the event is that it will not rain tomorrow. Using formula (13) with $a = 3$ and $b = 2$, we conclude that the probability that it will not rain tomorrow is

$$\frac{3}{3 + 2} = \frac{3}{5} = 0.6$$

In concluding this section, we should mention that in addition to the mean there are two other measures of central tendency of a set of numerical data, the median and the mode of a set of numbers. The *median* is the middle value in a set of data arranged in increasing or decreasing order (when there are an odd number of entries). If there are an even number of entries, the median is the mean of the two middle numbers. The *mode* is the value that occurs most frequently in a set of data. (See exercises 34 and 35.)

SELF-CHECK EXERCISES 8.5

1. Find the expected value of a random variable X having the following probability distribution.

x	-4	-3	-1	0	1	2
$P(X = x)$	0.10	0.20	0.25	0.10	0.25	0.10

2. The developer of Shoreline Condominiums has provided the following estimate of the probability that 20, 25, 30, 35, 40, 45, or 50 of the townhouses will be sold within the first month they are offered for sale.

Number of Units	20	25	30	35	40	45	50
Probability	0.05	0.10	0.30	0.25	0.15	0.10	0.05

How many townhouses can the developer expect to sell within the first month they are put on the market?

Solutions to Self-Check Exercises 8.5 can be found on page 456.

8.5 EXERCISES

C *A calculator is recommended for this exercise set.*

1. During the first year at a university that uses a 4-point grading system, a freshman took ten 3-credit

courses and received two A's, three B's, four C's, and one D.
a. Compute this student's grade-point average.
b. Let the random variable X denote the number of

points corresponding to a given letter grade. Find the probability distribution of the random variable X and compute $E(X)$, the expected value of X.

2. Records kept by the chief dietitian at the university cafeteria over a 30-week period show the following weekly consumption of milk (in gallons).

Milk (in gallons)	200	205	210	215	220
Number of Weeks	3	4	6	5	4

Milk (in gallons)	225	230	235	240
Number of Weeks	3	2	2	1

a. Find the average number of gallons of milk consumed per week in the cafeteria.
b. Let the random variable X denote the number of gallons of milk consumed in a week at the cafeteria. Find the probability distribution of the random variable X and compute $E(X)$, the expected value of X.

3. Find the expected value of a random variable X having the following probability distribution.

x	-5	-1	0
$P(X = x)$	0.12	0.16	0.28

x	1	5	8
$P(X = x)$	0.22	0.12	0.1

4. Find the expected value of a random variable X having the following probability distribution.

x	0	1	2
$P(X = x)$	1/8	1/4	3/16

x	3	4	5
$P(X = x)$	1/4	1/16	1/8

5. The daily earnings X of an employee who works on a commission basis are given by the following

probability distribution. Find the employee's expected earnings.

x (in $)	0	25	50	75
$P(X = x)$	0.07	0.12	0.17	0.14

x (in $)	100	125	150
$P(X = x)$	0.28	0.18	0.04

6. In a four-child family, what is the expected number of boys? (Assume that the probability of a boy being born is the same as the probability of a girl being born.)

7. Based on past experience, the manager of the Video-Rama Store has compiled the following table, which gives the probabilities that a customer who enters the Video-Rama Store will buy 0, 1, 2, 3, or 4 video-cassettes. How many videocassettes can a customer entering this store be expected to buy?

Number of Videocassettes	0	1	2	3	4
Probability	0.42	0.36	0.14	0.05	0.03

8. If a sample of three batteries is selected from a lot of ten, of which two are defective, what is the expected number of defective batteries?

9. **Auto Accidents**   The number of accidents that occur at a certain intersection known as "Five Corners" on a Friday afternoon between the hours of 3 P.M. and 6 P.M., along with the corresponding probabilities, are shown in the following table. Find the expected number of accidents during the period in question.

Number of Accidents	0	1	2	3	4
Probability	0.935	0.03	0.02	0.01	0.005

10. **Expected Demand**   The owner of a newsstand in a college community estimates the weekly demand for a certain magazine as follows:

Quantity Demanded	10	11	12
Probability	0.05	0.15	0.25

Quantity Demanded	13	14	15
Probability	0.30	0.20	0.05

Find the number of issues of the magazine that the newsstand owner can expect to sell per week.

11. **Expected Product Reliability** A bank has two automatic tellers at its main office and two at each of its three branches. The number of machines that break down on a given day, along with the corresponding probabilities, are shown in the following table.

Number of Machines that Break Down	0	1	2	3	4
Probability	0.43	0.19	0.12	0.09	0.04

Number of Machines that Break Down	5	6	7	8
Probability	0.03	0.03	0.02	0.05

Find the expected number of machines that will break down on a given day.

12. **Lotteries** In a lottery, 5000 tickets are sold for $1 each. One first prize of $2000, 1 second prize of $500, 3 third prizes of $100, and 10 consolation prizes of $25 are to be awarded. What are the expected net earnings of a person who buys one ticket?

13. **Life Insurance Premiums** A man wishes to purchase a 5-year term life insurance policy that will pay the beneficiary $20,000 in the event that the man's death occurs during the next five years. Using life insurance tables, he determines that the probability that he will live another five years is 0.96. What is the minimum amount that he can expect to pay for his premium? [*Hint:* The minimum premium occurs when the insurance company's expected profit is zero.]

14. **Expected Gain** A woman purchased a $10,000, 1-year term life insurance policy for $130. Assuming that the probability that she will live another year is 0.992, find the company's expected gain.

15. **Life Insurance Policies** As a fringe benefit, Mr. Taylor receives a $25,000 life insurance policy from his employer. The probability that Mr. Taylor will live another year is 0.9935. If he purchases the same coverage for himself, find the minimum amount that he could expect to pay for the policy.

16. **Expected Profit** A buyer for Discount Fashions, an outlet for women's apparel, is considering buying a batch of clothing for $64,000. She estimates that the company will be able to sell it for $80,000, $75,000, or $70,000 with probabilities of 0.30, 0.60, and 0.10, respectively. Based on these estimates, what will the company's expected gross profit be?

17. **Investment Analysis** The proprietor of Midland Construction Company has to decide between two projects. He estimates that the first project will yield a profit of $180,000 with a probability of 0.7 or a profit of $150,000 with a probability of 0.3; the second project will yield a profit of $220,000 with a probability of 0.6 or a profit of $80,000 with a probability of 0.4. Which project should the proprietor choose if he wants to maximize his expected profit?

18. **Cable Television** The management of Multi-Vision, Inc., a cable TV company, intends to submit a bid for the cable television rights in one of two cities, A or B. If the company obtains the rights to city A, the probability of which is 0.2, the estimated profit over the next ten years is $10 million; if the company obtains the rights to city B, the probability of which is 0.3, the estimated profit over the next ten years is $7 million. The cost of submitting a bid for rights in city A is $250,000 and that of city B is $200,000. By comparing the expected profits for each venture, determine whether the company should bid for the rights in city A or city B.

19. **Expected Auto Sales** Mr. Hunt intends to purchase one of two car dealerships currently for sale in a certain city. Records obtained from each of the two dealers reveal that their weekly volume of sales, with corresponding probabilities, are as follows:

Dahl Motors

Number of Cars Sold per Week	5	6	7	8
Probability	0.05	0.09	0.14	0.24

Number of Cars Sold per Week	9	10	11	12
Probability	0.18	0.14	0.11	0.05

Farthington Auto Sales

Number of Cars Sold per Week	5	6	7
Probability	0.08	0.21	0.31

Number of Cars Sold per Week	8	9	10
Probability	0.24	0.10	0.06

The average profit per car at Dahl Motors is $362, and the average profit per car at Farthington Auto Sales is $436.

a. Find the average number of cars sold per week at each dealership.

b. If Mr. Hunt's objective is to purchase the dealership that generates the higher weekly profit, which dealership should he purchase? (Compare the expected weekly profit for each dealership.)

20. Expected Home Sales   Ms. Leonard, a real estate broker, is relocating in a large metropolitan area where she has received job offers from realty company A and realty company B. The number of houses she expects to sell in a year at each firm and the associated probabilities are shown in the following tables.

Company A

Number of Houses Sold	12	13	14	15	16
Probability	0.02	0.03	0.05	0.07	0.07

Number of Houses Sold	17	18	19	20
Probability	0.16	0.17	0.13	0.11

Number of Houses Sold	21	22	23	24
Probability	0.09	0.06	0.03	0.01

Company B

Number of Houses Sold	6	7	8	9	10
Probability	0.01	0.04	0.07	0.06	0.11

Number of Houses Sold	11	12	13	14
Probability	0.12	0.19	0.17	0.13

Number of Houses Sold	15	16	17	18
Probability	0.04	0.03	0.02	0.01

The average price of a house in the locale of company A is $104,000, whereas the average price of a house in the locale of company B is $177,000. If Ms. Leonard will receive a 3% commission on sales at both companies, which job offer should she accept to maximize her expected yearly commission?

21. Weather Predictions   If the probability that it will rain tomorrow is 0.3,

a. what are the odds that it will rain tomorrow?

b. what are the odds that it will not rain tomorrow?

22. Roulette   In American roulette as described in Example 6, a player may bet on a split (two adjacent numbers). In this case, if the player bets $1 and either number comes up, the player wins $17 and gets his $1 back. If neither comes up, he loses his $1 bet. Find the expected value of the winnings on a $1 bet placed on a split.

23. Roulette   If a player placed a $1 bet on red and a $1 bet on black in a single play in American roulette, what would the expected value of his winnings be?

24. Roulette   In European roulette the wheel is divided into 37 compartments numbered 1 through 36 and 0. (In American roulette there are 38 compartments numbered 1 through 36, 0, and 00.) Find the expected value of the winnings on a $1 bet placed on red in European roulette.

25. The probability of an event E occurring is 0.8. What are the odds in favor of E occurring? What are the odds against E occurring?

26. The probability of an event E not occurring is 0.6. What are the odds in favor of E occurring? What are the odds against E occurring?

27. The odds in favor of an event E occurring are 9 to 7. What is the probability of E occurring?

28. The odds against an event E occurring are 2 to 3. What is the probability of E not occurring?

29. Odds   Joan, a computer sales representative, feels that the odds are 8 to 5 that she will clinch the sale of a minicomputer to a certain company. What is the (subjective) probability that Joan will make the sale?

30. Sports   Steffi feels that the odds in favor of her winning her tennis match tomorrow are 7 to 5. What is the (subjective) probability that she will win her match tomorrow?

31. Sports   If a sports forecaster states that the odds of a certain boxer winning a match are 4 to 3, what is the probability that the boxer will win the match?

32. Odds   Bob, the proprietor of Midland Lumber, feels that the odds in favor of a business deal going through are 9 to 5. What is the (subjective) probability that this deal will not materialize?

33. Roulette
a. Show that for any number c,

$$E(cX) = cE(X)$$

b. Use this result to find the expected loss if a gambler bets $300 on red in a single play in American roulette. [*Hint:* Use the results of Example 6.]

34. Exam Scores   In an examination given to a class of 20 students, the following test scores were obtained:

$$40, 45, 50, 50, 55, 60, 60, 75, 75, 80,$$

$$80, 85, 85, 85, 85, 90, 90, 95, 95, 100$$

a. Find the mean, or average, score, the mode, and the median score.
b. Which of these three measures of central tendency do you think is the least representative of the set of scores?

35. Wage Rates   The frequency distribution of the hourly wage rates among blue-collar workers in a certain factory is given in the following table. Find the mean, or average, wage rate, the mode, and the median wage rate of these workers.

Wage Rate (in $)	5.70	5.80	5.90
Frequency	60	90	75

Wage Rate (in $)	6.00	6.10	6.20
Frequency	120	60	45

SOLUTIONS TO SELF-CHECK EXERCISES 8.5

1. $E(X) = (-4)(0.10) + (-3)(0.20) + (-1)(0.25)$

$$+ (0)(0.10) + (1)(0.25) + (2)(0.10)$$

$$= -0.8$$

2. Let X denote the number of townhouses that will be sold within one month of being put on the market. Then, the number of townhouses the developer expects to sell within one month is given by the expected value of X—that is, by

$$E(X) = 20(0.05) + 25(0.10) + 30(0.30) + 35(0.25)$$

$$+ 40(0.15) + 45(0.10) + 50(0.05)$$

$$= 34.25$$

or 34 townhouses.

8.6

Variance and Standard Deviation

Variance

The mean, or expected value, of a random variable enables us to express an important property of the probability distribution associated with the random variable in terms of a single number. But the knowledge of the location, or central tendency, of a probability distribution alone is usually not enough to give a reasonably accurate picture of the probability distribution. Consider, for example, the two probability distributions whose histograms appear in Figure 21. Both distributions have the same expected value, or mean, $\mu = 4$ (the Greek letter μ is read "mu"). Note that the probability distribution with the histogram shown in Figure 21a is closely concentrated about its mean μ, whereas the one with the histogram shown in Figure 21b is widely dispersed or spread about its mean.

FIGURE 21

The histograms of two probability distributions

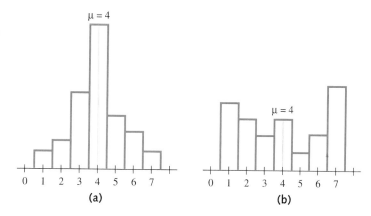

(a)   (b)

As another example, suppose Mr. David Horowitz, host of the popular television show "The Consumer Advocate," decides to demonstrate the accuracy of the weights of two popular brands of potato chips. Ten packages of potato chips of each brand are selected at random and weighed carefully. The results are as follows:

Weight in ounces

| Brand A | 16.1 | 16 | 15.8 | 16 | 15.9 | 16.1 | 15.9 | 16 | 16 | 16.2 |
| Brand B | 16.3 | 15.7 | 15.8 | 16.2 | 15.9 | 16.1 | 15.7 | 16.2 | 16 | 16.1 |

In Example 3 we will verify that the mean weights for each of the two brands is 16 ounces. However, a cursory examination of the data now shows that the weights of the brand B packages exhibit much greater dispersion about the mean than those of brand A.

One measure of the degree of dispersion, or spread, of a probability distribution about its mean is given by the variance of the random variable associated with the probability distribution. A probability distribution with a small spread about its mean will have a small variance, whereas one with a larger spread will have a larger variance. Thus, the variance of the random variable associated with the probability distribution whose histogram appears in Figure 21a is smaller than the variance of the random variable associated with the probability distribution whose histogram is shown in Figure 21b (see Example 1). Also, as we will see in Example 3, the variance of the random variable associated with the weights of the brand A potato chips is smaller than that of the random variable associated with the weights of the brand B potato chips. (This observation was made earlier.)

We now define the variance of a random variable.

**VARIANCE
OF A RANDOM
VARIABLE X**

Suppose a random variable has the probability distribution

x	x_1	x_2	x_3	$\cdots$	x_n
P(X = x)	p_1	p_2	p_3	$\cdots$	p_n

and expected value

$$E(X) = \mu$$

Then the **variance** of the random variable X is

$$\text{Var}(X) = p_1(x_1 - \mu)^2 + p_2(x_2 - \mu)^2 + \cdots + p_n(x_n - \mu)^2 \qquad \textbf{(14)}$$

Let's look a little closer at (14). First, note that the numbers

$$x_1 - \mu, x_2 - \mu, \ldots, x_n - \mu \qquad \textbf{(15)}$$

measure the *deviations* of x_1, x_2, ..., x_n from μ, respectively. Thus the numbers

$$(x_1 - \mu)^2, (x_2 - \mu)^2, \ldots, (x_n - \mu)^2 \qquad \textbf{(16)}$$

measure the squares of the deviations of $x_1, x_2, \ldots, x_n$ from μ, respectively. Next, by multiplying each of the numbers in equation (14) by the probability associated with each value of the random variable X, the numbers are weighted accordingly so that their sum is a measure of the variance of X about its mean. An attempt to define the variance of a random variable about its mean in a similar manner using the deviations in (14) rather than their squares would not be fruitful, since some of the deviations may be positive whereas others may be negative, and (because of cancellations) the sum will not give a satisfactory measure of the variance of the random variable.

EXAMPLE 1   Find the variance of the random variable X and of the random variable Y whose probability distributions are shown in the following table. These are the probability distributions associated with the histograms shown in Figures 21a and 21b.

x	P(X = x)	y	P(Y = y)
1	0.05	1	0.2
2	0.075	2	0.15
3	0.2	3	0.1
4	0.375	4	0.15
5	0.15	5	0.05
6	0.1	6	0.1
7	0.05	7	0.25

Solution   The mean of the random variable X is given by

$$\mu_X = (1)(0.05) + (2)(0.075) + (3)(0.2) + (4)(0.375) + (5)(0.15)$$
$$+ (6)(0.1) + (7)(0.05)$$
$$= 4$$

Therefore, using (14) and the data from the probability distribution of X, we find that the variance of X is given by

$$\text{Var}\,(X) = (0.05)(1 - 4)^2 + (0.075)(2 - 4)^2 + (0.2)(3 - 4)^2$$
$$+ (0.375)(4 - 4)^2 + (0.15)(5 - 4)^2$$
$$+ (0.1)(6 - 4)^2 + (0.05)(7 - 4)^2$$
$$= 1.95$$

Next, we find that the mean of the random variable Y is given by

$$\mu_Y = (1)(0.2) + (2)(0.15) + (3)(0.1) + (4)(0.15) + (5)(0.05)$$
$$+ (6)(0.1) + (7)(0.25)$$
$$= 4$$

and so the variance of Y is given by

$$\text{Var}\,(Y) = (0.2)(1 - 4)^2 + (0.15)(2 - 4)^2 + (0.1)(3 - 4)^2$$
$$+ (0.15)(4 - 4)^2 + (0.05)(5 - 4)^2$$
$$+ (0.1)(6 - 4)^2 + (0.25)(7 - 4)^2$$
$$= 5.2$$

Note that Var (X) is smaller than Var (Y), which confirms the earlier observations about the spread, or dispersion, of the probability distribution of X and Y, respectively.   ◾

Standard Deviation

Because (14), which gives the variance of the random variable X, involves the squares of the deviations, the unit of measurement of Var (X) is the square of the unit of measurement of the values of X. For example, if the values assumed by the random variable X are measured in units of a gram, then Var (X) will be measured in units involving the *square* of a gram. To remedy this situation, one normally works with the square root of Var (X) rather than Var (X) itself. The former is called the *standard deviation of X*.

STANDARD DEVIATION OF A RANDOM VARIABLE X

The **standard deviation** of a random variable X, σ (pronounced "sigma"), is defined by

$$\sigma = \sqrt{\text{Var}\,(X)}$$
$$= \sqrt{p_1(x_1 - \mu)^2 + p_2(x_2 - \mu)^2 + \cdots + p_n(x_n - \mu)^2} \qquad \textbf{\textit{(17)}}$$

where $x_1, x_2, \ldots, x_n$ denote the values assumed by the random variable X and $p_1 = P(X = x_1)$, $p_2 = P(X = x_2)$, $\ldots$, $p_n = P(X = x_n)$.

EXAMPLE 2   Find the standard deviations of the random variables X and Y of Example 1.

Solution   From the results of Example 1, we have Var $(X) = 1.95$ and Var $(Y) = 5.2$. Taking their respective square roots, we have

$$\sigma_X = \sqrt{1.95}$$

$$\approx 1.40$$

and
$$\sigma_Y = \sqrt{5.2}$$

$$\approx 2.28$$

   EXAMPLE 3   Let X and Y denote the random variables whose values are the weights of the brand A and brand B potato chips, respectively (see page 457). Compute the means and standard deviations of X and Y and interpret your results.

Solution   The probability distributions of X and Y may be computed from the given data as follows:

Brand A

x	Relative Frequency of Occurrence	$P(X = x)$
15.8	1	0.1
15.9	2	0.2
16.0	4	0.4
16.1	2	0.2
16.2	1	0.1

Brand B

y	Relative Frequency of Occurrence	$P(Y = y)$
15.7	2	0.2
15.8	1	0.1
15.9	1	0.1
16.0	1	0.1
16.1	2	0.2
16.2	2	0.2
16.3	1	0.1

The means of X and Y are given by

$$\mu_X = (0.1)(15.8) + (0.2)(15.9) + (0.4)(16.0) + (0.2)(16.1)$$
$$+ (0.1)(16.2)$$
$$= 16$$
$$\mu_Y = (0.2)(15.7) + (0.1)(15.8) + (0.1)(15.9) + (0.1)(16.0)$$
$$+ (0.2)(16.1) + (0.2)(16.2) + (0.1)(16.3)$$
$$= 16$$

Therefore,

$$\text{Var}(X) = (0.1)(15.8 - 16)^2 + (0.2)(15.9 - 16)^2 + (0.4)(16 - 16)^2$$
$$+ (0.2)(16.1 - 16)^2 + (0.1)(16.2 - 16)^2$$
$$= 0.012$$

and

$$\begin{aligned}
\text{Var}\,(Y) = {} & (0.2)(15.7 - 16)^2 + (0.1)(15.8 - 16)^2 + (0.1)(15.9 - 16)^2 \\
& + (0.1)(16 - 16)^2 + (0.2)(16.1 - 16)^2 + (0.2)(16.2 - 16)^2 \\
& + (0.1)(16.3 - 16)^2 \\
= {} & 0.042
\end{aligned}$$

so that the required standard deviations are

$$\sigma_X = \sqrt{\text{Var}\,(X)}$$
$$= \sqrt{0.012}$$
$$\approx 0.11$$

and

$$\sigma_Y = \sqrt{\text{Var}\,(Y)}$$
$$= \sqrt{0.042}$$
$$\approx 0.20$$

The mean of X and that of Y are both equal to 16. Therefore, the average weight of a package of potato chips of either brand is 16 ounces. However, the standard deviation of Y is greater than that of X. This tells us that the weights of the packages of brand B potato chips are more widely dispersed about the common mean of 16 than are those of brand A. ■

The standard deviation of a random variable X may be used in statistical estimations. For example, the following result, derived by the Russian mathematician P. L. Chebychev (1821–1894), gives the proportion of the values of X lying within k standard deviations of the expected value of X.

Chebychev's Inequality

CHEBYCHEV'S INEQUALITY

Let X be a random variable with expected value μ and standard deviation σ. Then the probability that a randomly chosen outcome of the experiment lies between $\mu - k\sigma$ and $\mu + k\sigma$ is at least $1 - (1/k^2)$, that is,

$$P(\mu - k\sigma \leq X \leq \mu + k\sigma) \geq 1 - \frac{1}{k^2} \qquad (18)$$

In order to shed some light on this result, let us take $k = 2$ in inequality (18) and compute

$$P(\mu - 2\sigma \leq X \leq \mu + 2\sigma) = 1 - \frac{1}{2^2} = 1 - \frac{1}{4} = 0.75$$

This tells us that at least 75% of the outcomes of the experiment lie within 2 standard deviations of the mean (Figure 22). Taking $k = 3$ in (18), we have

$$P(\mu - 3\sigma \leq X \leq \mu + 3\sigma) = 1 - \frac{1}{3^2} = 1 - \frac{1}{9} = \frac{8}{9} \approx 0.89$$

FIGURE 22

At least 75% of the outcomes fall within this interval.

This tells us that at least 89% of the outcomes of the experiment lie within 3 standard deviations of the mean (Figure 23).

FIGURE 23

At least 89% of the outcomes fall within this interval.

EXAMPLE 4   A probability distribution has a mean of 10 and a standard deviation of 1.5. Use Chebychev's inequality to estimate the probability that an outcome of the experiment lies between 7 and 13.

Solution   Here $\mu = 10$ and $\sigma = 1.5$. Next, to determine the value of k, note that $\mu - k\sigma = 7$ and $\mu + k\sigma = 13$. Substituting the appropriate values for μ and σ, we find $k = 2$. Using Chebychev's inequality (18), we see that the probability that an outcome of the experiment lies between 7 and 13 is given by

$$P(7 \le X \le 13) \ge 1 - \left(\frac{1}{2^2}\right)$$

$$= \frac{3}{4}$$

that is, at least 75%.

REMARK   The results of Example 4 tell us that at least 75% of the outcomes of the experiment lie between $10 - 2\sigma$ and $10 + 2\sigma$—that is, between 7 and 13.

EXAMPLE 5   The Great Northwest Lumber Company employs 400 workers in its mills. It has been estimated that X, the random variable measuring the number of mill workers who have industrial accidents during a 1-year period, is normally distributed with a mean of 40 and a standard deviation of 6. Using Chebychev's inequality, estimate the probability that the number of workers who will have an industrial accident over a 1-year period is between 30 and 50, inclusive.

Solution   Here $\mu = 40$ and $\sigma = 6$. We wish to estimate $P(30 \le X \le 50)$. In order to use Chebychev's inequality (18), we first determine the value of k from the equation

$$\mu - k\sigma = 30 \qquad \text{or} \qquad \mu + k\sigma = 50$$

Since $\mu = 40$ and $\sigma = 6$ in this case, we see that k satisfies

$$40 - 6k = 30 \qquad \text{and} \qquad 40 + 6k = 50$$

from which we deduce that $k = 5/3$. Thus, the probability that the number of mill workers who will have an industrial accident during a 1-year period is between 30 and 50 is given by

$$P(30 \le X \le 50) \ge 1 - \frac{1}{(5/3)^2}$$

$$= \frac{16}{25}$$

that is, at least 64%.

**SELF-CHECK
EXERCISES 8.6**

1. Compute the mean, variance, and standard deviation of the random variable X with probability distribution as follows.

x	-4	-3	-1	0	2	5
$P(X = x)$	0.1	0.1	0.2	0.3	0.1	0.2

2. James recorded the following travel times (the length of time in minutes it took him to drive to work) on ten consecutive days:

55 50 52 48 50 52 46 48 50 51

Calculate the mean and standard deviation of the random variable X associated with these data.

Solutions to Self-Check Exercises 8.6 can be found on page 466.

8.6 EXERCISES

c *A calculator is recommended for this exercise set.*

In exercises 1–6, the probability distribution of a random variable X is given. Compute the mean, variance, and standard deviation of X.

1.

x	1	2	3	4
$P(X = x)$	0.4	0.3	0.2	0.1

2.

x	-4	-2	0	2	4
$P(X = x)$	0.1	0.2	0.3	0.1	0.3

3.

x	-2	-1	0	1	2
$P(X = x)$	1/16	4/16	6/16	4/16	1/16

4.

x	10	11	12	13	14	15
$P(X = x)$	1/8	2/8	1/8	2/8	1/8	1/8

5.

x	430	480	520	565	580
$P(X = x)$	0.1	0.2	0.4	0.2	0.1

6.

x	-198	-195	-193	-188	-185
$P(X = x)$	0.15	0.30	0.10	0.25	0.20

7. The following histograms represent the probability distributions of the random variables X and Y. Determine by inspection which probability distribution has the larger variance.

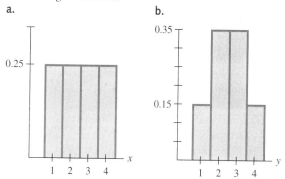

8. The following histograms represent the probability distributions of the random variables X and Y. Determine by inspection which probability distribution has the larger variance.

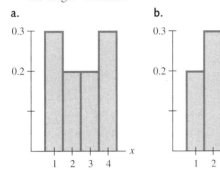

a.   b.

In exercises 9 and 10, find the variance of the probability distribution for the given histogram.

9.

10.

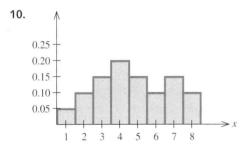

11. An experiment consists of casting an eight-sided die (numbered 1 through 8) and observing the number that appears uppermost. Find the mean and variance of this experiment.

12. **Driving Age Requirements**   The minimum age requirement for a regular driver's license differs from state to state. The frequency distribution for this age requirement in the 50 states is given in the following table.

Minimum Age	15	16	17	18	19	21
Frequency of Occurrence	1	15	4	28	1	1

a. Describe a random variable X that is associated with these data.

b. Find the probability distribution for the random variable X.

c. Compute the mean, variance, and standard deviation of X.

13. **Birth Rates**   The birth rates in the United States for the years 1981–1990 are given in the following table. (The birth rate is the number of live births per 1000 population.)

Year	1981	1982	1983	1984
Birth Rate (number/1000)	15.9	15.5	15.5	15.7

Year	1985	1986	1987
Birth Rate (number/1000)	15.7	15.6	15.7

Year	1988	1989	1990
Birth Rate (number/1000)	15.9	16.2	16.7

a. Describe a random variable X that is associated with these data.

b. Find the probability distribution for the random variable X.

c. Compute the mean, variance, and standard deviation of X.

14. **Investment Analysis**   Mr. Hunt is considering two business ventures. The anticipated returns (in thousands of dollars) of each venture are described by the following probability distributions.

Venture A

Earnings	Probability
−20	0.3
40	0.4
50	0.3

Venture B

Earnings	Probability
−15	0.2
30	0.5
40	0.3

a. Compute the mean and variance for each venture.
b. Which investment would provide Mr. Hunt with the higher expected return (the greater mean)?
c. In which investment would the element of risk be less (that is, which probability distribution has the smaller variance)?

15. Investment Analysis Ms. Walters is considering investing $10,000 in two mutual funds. The anticipated returns from price appreciation and dividends (in hundreds of dollars) are described by the following probability distributions.

Mutual Fund A

Returns	Probability
−4	0.2
8	0.5
10	0.3

Mutual Fund B

Returns	Probability
−2	0.2
6	0.4
8	0.4

a. Compute the mean and variance associated with the returns for each mutual fund.
b. Which investment would provide Ms. Walters with the higher expected return (the greater mean)?
c. In which investment would the element of risk be less (that is, which probability distribution has the smaller variance)?

16. The distribution of the number of chocolate chips in a cookie is shown in the following table. Find the mean and the variance of the number of chocolate chips in a cookie.

Number of Chocolate Chips (x)	0	1	2
P(X = x)	0.01	0.03	0.05

Number of Chocolate Chips (x)	3	4	5
P(X = x)	0.11	0.13	0.24

Number of Chocolate Chips (x)	6	7	8
P(X = x)	0.22	0.16	0.05

17. Formula (14) can also be expressed in the form

$$\text{Var}(X) = (p_1x_1^2 + p_2x_2^2 + \cdots + p_nx_n^2) - \mu^2$$

Find the variance of the distribution of exercise 1 using this formula.

18. Find the variance of the distribution of exercise 16 using the formula

$$\text{Var}(X) = (p_1x_1^2 + p_2x_2^2 + \cdots + p_nx_n^2) - \mu^2$$

19. Housing Prices A survey was conducted by the market research department of the National Real Estate Company among 500 prospective buyers in a large metropolitan area to determine the maximum price a prospective buyer would be willing to pay for a house. From the data collected, the following distribution was obtained. Compute the mean, variance, and standard deviation of the maximum price that these buyers were willing to pay for a house.

Maximum Price Considered, x (in thousands of dollars)	P(X = x)
80	$\frac{10}{500}$
90	$\frac{20}{500}$
100	$\frac{75}{500}$
110	$\frac{85}{500}$
120	$\frac{70}{500}$
150	$\frac{90}{500}$
180	$\frac{90}{500}$
200	$\frac{55}{500}$
250	$\frac{5}{500}$

20. A probability distribution has a mean of 20 and a standard deviation of 3. Use Chebychev's inequality to estimate the probability that an outcome of the experiment lies between
a. 15 and 25   b. 10 and 30

21. A probability distribution has a mean of 42 and a standard deviation of 2. Use Chebychev's inequality to estimate the probability that an outcome of the experiment lies between
a. 38 and 46   b. 32 and 52

22. A probability distribution has a mean of 50 and a standard deviation of 1.4. Use Chebychev's inequality to find the value of c, which guarantees that the probability is at least 96% that an outcome of the experiment lies between $50 - c$ and $50 + c$.

23. Suppose X is a random variable with mean μ and standard deviation σ. If a large number of trials are observed, at least what percentage of these values is expected to lie between $\mu - 2\sigma$ and $\mu + 2\sigma$?

24. Product Reliability   A Christmas tree light has an expected life of 200 hours and a standard deviation of 2 hours.

   a. Estimate the probability that one of these Christmas tree lights will last between 190 and 210 hours.
   b. Suppose 150,000 of these Christmas tree lights are used by a large city as part of its Christmas decorations. Estimate the number of lights that will require replacement between 180 and 220 hours of use.

25. Product Reliability   The expected lifetime of the deluxe model hair dryer produced by the Roland Electric Company has a mean life of 24 months and a standard deviation of 3 months. Find the probability that one of these hair dryers will last between 20 and 28 months.

26. Starting Salaries   The mean annual starting salary of a new graduate in a certain profession is $30,000 with a standard deviation of $500. What is the probability that the starting salary of a new graduate in this profession will be between $28,000 and $32,000?

27. Quality Control   Sugar packaged by a certain machine has a mean weight of 5 pounds and a standard deviation of 0.02 pound. For what values of c can the manufacturer of the machinery claim that the sugar packaged by this machine has a weight between $5 - c$ and $5 + c$ pounds with probability at least 96%?

SOLUTIONS TO SELF-CHECK EXERCISES 8.6

1. The mean of the random variable X is

$$\mu = (-4)(0.1) + (-3)(0.1) + (-1)(0.2)$$
$$+ (0)(0.3) + (2)(0.1) + (5)(0.2)$$
$$= 0.3$$

The variance of X is

$$\text{Var}\,(X) = (0.1)(-4 - 0.3)^2 + (0.1)(-3 - 0.3)^2$$
$$+ (0.2)(-1 - 0.3)^2 + (0.3)(0 - 0.3)^2$$
$$+ (0.1)(2 - 0.3)^2 + (0.2)(5 - 0.3)^2$$
$$= 8.01$$

The standard deviation of X is

$$\sigma = \sqrt{\text{Var}\,(X)} = \sqrt{8.01} \approx 2.83$$

2. We first compute the probability distribution of X from the given data as follows:

x	Relative Frequency of Occurrence	$P(X = x)$
46	1	0.1
48	2	0.2
50	3	0.3
51	1	0.1
52	2	0.2
55	1	0.1

The mean of X is

$$\mu = (0.1)(46) + (0.2)(48) + (0.3)(50)$$
$$+ (0.1)(51) + (0.2)(52) + (0.1)(55)$$
$$= 50.2$$

The variance of X is

$$\text{Var}(X) = (0.1)(46 - 50.2)^2 + (0.2)(48 - 50.2)^2$$
$$+ (0.3)(50 - 50.2)^2 + (0.1)(51 - 50.2)^2$$
$$+ (0.2)(52 - 50.2)^2 + (0.1)(55 - 50.2)^2$$
$$= 5.76$$

from which we deduce the standard deviation

$$\sigma = \sqrt{5.76}$$
$$= 2.4$$

CHAPTER EIGHT SUMMARY OF PRINCIPAL FORMULAS AND TERMS

Formulas

1. Conditional probability $\qquad P(B \mid A) = \dfrac{P(A \cap B)}{P(A)}$

2. Product Rule $\qquad P(A \cap B) = P(A) \cdot P(B \mid A)$

3. Test for independence $\qquad P(A \cap B) = P(A) \cdot P(B)$

4. Mean of n numbers $\qquad \bar{x} = \dfrac{x_1 + x_2 + \cdots + x_n}{n}$

5. Expected value $\qquad E(X) = x_1 p_1 + x_2 p_2 + \cdots + x_n p_n$

6. Odds in favor of E occurring $\qquad \dfrac{P(E)}{P(E^c)}$

7. Odds against E occurring $\qquad \dfrac{P(E^c)}{P(E)}$

8. Probability of an event occurring given the odds $\qquad \dfrac{a}{a + b}$

9. Variance of a random variable

$$\text{Var}(X) = p_1(x_1 - \mu)^2$$
$$+ p_2(x_2 - \mu)^2 + \cdots$$
$$+ p_n(x_n - \mu)^2$$

10. Standard deviation of a random variable $\qquad \sigma = \sqrt{\text{Var}(X)}$

11. Chebychev's inequality

$$P(\mu - k\sigma \le X \le \mu + k\sigma)$$
$$\ge 1 - \dfrac{1}{k^2}$$

Terms

Conditional probability

Finite stochastic process

Independent events

A posteriori probability

A priori probability	Histogram
Bayes' Theorem	Average (mean)
Random variable	Expected value
Finite discrete random variable	Odds in favor of
Infinite discrete random variable	Odds against
Continuous random variable	Variance
Probability distribution of a random variable	Standard deviation

CHAPTER 8 REVIEW EXERCISES

The accompanying tree diagram represents an experiment consisting of two trials. In exercises 1–5, use the diagram to find the given probability.

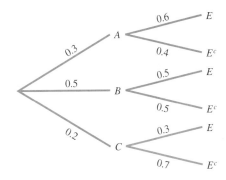

1. $P(A \cap E)$

2. $P(B \cap E)$

3. $P(C \cap E)$

4. $P(A \mid E)$

5. $P(E)$

6. Let E and F be two events and suppose $P(E) = 0.35$, $P(F) = 0.55$, and $P(E \cup F) = 0.70$. Find $P(E \mid F)$.

7. An experiment consists of tossing a fair coin three times and observing the outcomes. Let A be the event that at least one head is thrown and B the event that at most two tails are thrown.
 a. Find $P(A)$.
 b. Find $P(B)$.
 c. Are A and B independent events?

8. In a group of 20 ballpoint pens on a shelf in the stationery department of the Metro Department Store, 2 are known to be defective. If a customer selects 3 of these pens, what is the probability that
 a. at least 1 is defective?
 b. no more than 1 is defective?

9. Five people are selected at random. What is the probability that none of the people in this group were born on the same day of the week?

10. A pair of fair dice is cast. What is the probability that the sum of the numbers falling uppermost is 8 if it is known that the two numbers are different?

Three cards are drawn at random without replacement from a standard deck of 52 playing cards. In exercises 11–15, find the probability of each event.

11. All three cards are aces.

12. All three cards are face cards.

13. The second and third cards are red.

14. The second card is black, given that the first card was red.

15. The second card is a club, given that the first card was black.

16. Three balls are selected at random without replacement from an urn containing three white balls and four blue balls. Let the random variable X denote the number of blue balls drawn.
 a. List the outcomes of this experiment.
 b. Find the value assigned to each outcome of this experiment by the random variable X.
 c. Find the probability distribution of the random variable associated with this experiment.
 d. Draw the histogram representing this distribution.

17. A man purchased a $25,000, 1-year term life insurance policy for $375. Assuming that the probability that he will live for another year is 0.989, find the company's expected gain.

18. The probability distribution of a random variable X is shown in the accompanying table.

x	$P(X = x)$
0	0.1
1	0.1
2	0.2
3	0.3
4	0.2
5	0.1

a. Compute $P(1 \le X \le 4)$.
b. Compute the mean and standard deviation of X.

19. The heights of 4000 women who participated in a recent survey were found to have a mean of 64.5 inches and a standard deviation of 2.5 inches. Use Chebychev's inequality to estimate the probability that the height of a woman who participated in the survey will fall within 2 standard deviations of the mean; that is, that her height will be between 59.5 and 69.5 inches.

20. Of 320 male and 280 female employees at the home office of the Gibraltar Insurance Company, 160 of the men and 190 of the women are on flextime (flexible working hours). Given that an employee selected at random from this group is on flextime, what is the probability that the employee is a male?

21. In a manufacturing plant, three machines, A, B, and C, produce 40%, 35%, and 25%, respectively, of the total production. The company's quality control department has determined that 1% of the items produced by machine A, 1.5% of the items produced by machine B, and 2% of the items produced by machine C are defective. If an item is selected at random and found to be defective, what is the probability that it was produced by machine B?

22. Applicants who wish to be admitted to a certain professional school in a large university are required to take a screening test that was devised by an educational testing service. From past results, the testing service has estimated that 70% of all applicants are eligible for admission and that 92% of those who are eligible for admission pass the exam, whereas 12% of those who are ineligible for admission pass the exam. Using these results, find the probability that
a. an applicant for admission passed the exam
b. an applicant for admission who passed the exam was actually ineligible

CHAPTER 8 EXTENDED PROBLEMS AND PROJECTS

1. Refer to the Birthday Problem of Example 4, Section 8.1. Compute the probabilities that at least 2 of 5, 10, 15, 20, . . . , 50 people, respectively, have the same birthday. These probabilities are shown in Table 1 (page 406).

2. If you have opportunity, conduct the experiment described in Example 4, Section 8.1, in a group of at least 30 people.

The Derivative

9

*I*n this chapter we begin the study of differential calculus. Historically, differential calculus was developed in response to the problem of finding the tangent line to an arbitrary curve. But it quickly became apparent that solving this problem provided mathematicians with a method of solving many practical problems involving the rate of change of one quantity with respect to another. The basic tool used in differential calculus is the *derivative* of a function. The concept of the derivative is based, in turn, on a more fundamental notion—that of the *limit* of a function.

9.1 Limits

Introduction to Calculus

Historically, the development of calculus by Isaac Newton (1642–1727) and Gottfried Wilhelm Leibniz (1646–1716) resulted from the investigation of the following problems:

1. Finding the tangent line to a curve at a given point on the curve (Figure 1a).

2. Finding the area of a planar region bounded by an arbitrary curve (Figure 1b).

FIGURE 1

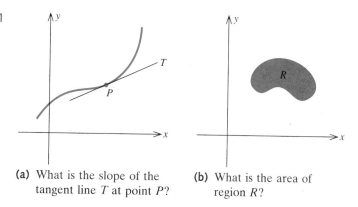

(a) What is the slope of the tangent line T at point P?

(b) What is the area of region R?

The tangent-line problem might appear to be unrelated to any practical applications of mathematics, but as you will see later, the problem of finding the *rate of change* of one quantity with respect to another is mathematically equivalent to the geometric problem of finding the slope of the *tangent line* to a curve at a given point on the curve. It is precisely the discovery of the relationship between these two problems that spurred the development of calculus in the seventeenth century and made it such an indispensable tool for solving practical problems. The following are a few examples of such problems.

- Finding the velocity of an object
- Finding the rate of change of a bacteria population with respect to time
- Finding the rate of change of a company's profit with respect to time
- Finding the rate of change of a travel agency's revenue with respect to the agency's expenditure for advertising

The study of the tangent-line problem led to the creation of *differential calculus,* which relies on the concept of the *derivative* of a function. The study of the area problem led to the creation of *integral calculus,* which relies on the concept of the *antiderivative,* or *integral,* of a function. (The derivative of a function and the integral of a function are intimately related, as you will see in Section 11.4.) Both the derivative of a function and the integral of a function are defined in terms of a more fundamental concept: the limit, our next topic.

A Real-Life Example

Suppose that a car traveling along a straight road covers a distance of s feet in t seconds, where s and t are related by the function

$$s = f(t) = 2t^2 \qquad 0 \le t \le 10 \tag{1}$$

Then the position of the car at time $t = 0, 1, 2, 3, \ldots, 10$, measured from its initial position, is

$$f(0) = 0, \quad f(1) = 2, \quad f(2) = 8, \quad f(3) = 18, \ldots, \quad f(10) = 200$$

feet (Figure 2).

FIGURE 2

Position of a car at time t

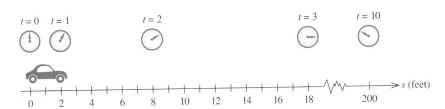

Suppose we want to know the velocity of the car when $t = 2$. This is just the speed of the car that would be indicated on the speedometer at that precise instant. Offhand, calculating this quantity using only equation (1) appears to be an impossible task, but consider what quantities we *can* compute using this relationship. Obviously, we can compute the position of the car at any time t, as we did earlier for some selected values of t. Using these values we can then compute the *average velocity* of the car over an interval of time. For example, the average velocity of the car over the interval [2, 4] is given by

$$\frac{\text{Distance covered}}{\text{Time elapsed}} = \frac{f(4) - f(2)}{4 - 2}$$

$$= \frac{2(4^2) - 2(2^2)}{2}$$

$$= \frac{32 - 8}{2} = 12$$

or 12 ft/sec. This is not quite the velocity of the car at $t = 2$, but it is an approximation. Can we do better? Intuitively, the smaller the time interval we pick (with endpoint $t = 2$),* the better the average velocity over that time interval will approximate the actual velocity of the car at $t = 2$.

To formulate this process mathematically, let $t > 2$. Then the distance covered by the car over the time interval [2, t] is $(f(t) - f(2))$ feet, and the time elapsed is $(t - 2)$ seconds. Therefore, the average velocity of the car over the time interval [2, t] is given by

$$\frac{f(t) - f(2)}{t - 2} = \frac{2t^2 - 2(2)^2}{t - 2}$$

$$= \frac{2(t^2 - 4)}{t - 2} \tag{2}$$

---

*Actually, any interval containing $t = 2$ will suffice.

Equation (2) may be used to generate a sequence of average velocities of the car over some appropriate intervals. By choosing the values of t closer and closer to $t = 2$, we obtain a sequence that approximates with increasing accuracy the *instantaneous velocity* of the car at $t = 2$.

Before going on, let us consider a sample calculation. Taking the sequence $t = 3, 2.5, 2.1, 2.01, 2.001$, and 2.0001, which evidently approaches $t = 2$, we obtain (with the help of a calculator)

$$\text{Average velocity over } [2, 3] \text{ is } \frac{2[(3)^2 - 4]}{3 - 2} = 10, \text{ or } 10 \text{ ft/sec}$$

$$\text{Average velocity over } [2, 2.5] \text{ is } \frac{2[(2.5)^2 - 4]}{2.5 - 2} = 9, \text{ or } 9 \text{ ft/sec}$$

and so forth. These results are summarized in Table 1.

TABLE 1

	t approaches 2 from the right					
t	3	2.5	2.1	2.01	2.001	2.0001
Average velocity over [2, *t*]	10	9	8.2	8.02	8.002	8.0002

average velocity approaches 8 from the right

From Table 1, we see that the average velocities of the car over the time intervals $[2, 3], [2, 2.5], \ldots, [2, 2.0001]$ seem to approach the number 8. These computations suggest that the instantaneous velocity of the car at $t = 2$ is 8 ft/sec.

This example illustrates the mechanics of the limit process. Let us summarize what we have done. Given the function

$$g(t) = \frac{2(t^2 - 4)}{t - 2}$$

which is just the function in equation (2), we were required to determine the value that the function g approached as t approached the fixed number 2. We took the sequence of values of t approaching $t = 2$ from the right-hand side and saw that $g(t)$ approached the number 8 as t approached $t = 2$. Similarly, if we take the sequence $t = 1, 1.5, 1.9, 1.99, 1.999$, and 1.9999, which approaches $t = 2$ from the left, we obtain the results given in Table 2.

TABLE 2

	t approaches 2 from the left					
t	1	1.5	1.9	1.99	1.999	1.9999
g(*t*)	6	7	7.8	7.98	7.998	7.9998

g(*t*) approaches 8 from the left

Observe that $g(t)$ approaches the number 8 as t approaches $t = 2$, this time from the left-hand side. In this situation, we say that the *limit of the function*

$$g(t) = \frac{2(t^2 - 4)}{t - 2}$$

as t approaches 2 is 8, written

$$\lim_{t \to 2} \frac{2(t^2 - 4)}{t - 2} = 8$$

The graph of the function g shown in Figure 3 confirms this observation.

FIGURE 3
As t approaches t = 2 from either direction, g(t) approaches y = 8.

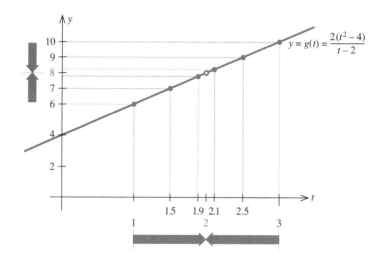

$$y = g(t) = \frac{2(t^2 - 4)}{t - 2}$$

Observe that the point $t = 2$ is not in the domain of the function g (for this reason, the point $(2, 8)$ is missing from the graph of g). This, however, is inconsequential because the value, if any, of $g(t)$ at $t = 2$ plays no role in computing the limit.

This example leads to the following informal definition of the **limit of a function.**

LIMIT OF A FUNCTION

The function f has the limit L as x approaches a, written

$$\lim_{x \to a} f(x) = L$$

if the values $f(x)$ can be made as close to the number L as we please by taking x sufficiently close to (but not equal to) a.

Evaluating the Limit of a Function

Let us now consider some examples involving the evaluation of limits.

EXAMPLE 1   Let $f(x) = x^3$ and evaluate $\lim_{x \to 2} f(x)$.

Solution   The graph of f is shown in Figure 4. You can see that $f(x)$ can be made as close to the number 8 as we please by taking x sufficiently close to 2. Therefore,

$$\lim_{x \to 2} x^3 = 8$$

FIGURE 4

*f(x) is close to 8 whenever x
is close to 2.*

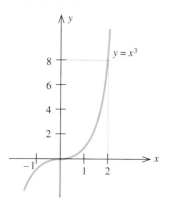

EXAMPLE 2 Let

$$g(x) = \begin{cases} x + 2 & \text{if } x \neq 1 \\ 1 & \text{if } x = 1 \end{cases}$$

Evaluate $\lim_{x \to 1} g(x)$.

Solution The domain of g is the set of all real numbers. From the graph of g shown in Figure 5, we see that $g(x)$ can be made as close to 3 as we please by taking x sufficiently close to 1. Therefore

$$\lim_{x \to 1} g(x) = 3$$

Observe that $g(1) = 1$, which is not equal to the limit of the function g as x approaches 1. (Once again, the value of $g(x)$ at $x = 1$ has no bearing on the existence or value of the limit of g as x approaches 1.) ■

FIGURE 5

$\lim_{x \to 1} g(x) = 3$

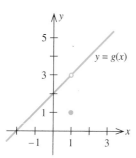

EXAMPLE 3 Evaluate the limit of the following functions as x approaches the indicated point.

a. $f(x) = \begin{cases} -1 & \text{if } x < 0 \\ 1 & \text{if } x \geq 0 \end{cases};\quad x = 0$ **b.** $g(x) = \dfrac{1}{x^2};\quad x = 0$

Solution The graphs of the functions f and g are shown in Figure 6.

a. Referring to Figure 6a, we see that no matter how close x is to $x = 0$, $f(x)$ takes on the values 1 or -1 depending on whether x is positive or negative. Thus, there is no *single* real number L that $f(x)$ approaches as x approaches zero. We conclude that the limit of $f(x)$ does *not* exist as x approaches zero.

b. Referring to Figure 6b, we see that as x approaches $x = 0$ (from either side), $g(x)$ increases without bound and thus does not approach any specific real number. We conclude, accordingly, that the limit of $g(x)$ does not exist as x approaches zero.

FIGURE 6

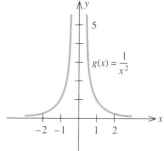

(a) $\lim_{x \to 0} f(x)$ does not exist   **(b)** $\lim_{x \to 0} g(x)$ does not exist   ■

Until now, we have relied on knowing the actual values of a function or the graph of a function near $x = a$ to help us evaluate the limit of the function $f(x)$ as x approaches a. The following properties of limits, which we list without proof, enable us to evaluate limits of functions algebraically.

**THEOREM 1
PROPERTIES OF
LIMITS**

Suppose that

$$\lim_{x \to a} f(x) = L \quad \text{and} \quad \lim_{x \to a} g(x) = M$$

Then

1. $\lim_{x \to a} [f(x)]^r = [\lim_{x \to a} f(x)]^r = L^r \qquad$ *r, a real number*

2. $\lim_{x \to a} cf(x) = c \lim_{x \to a} f(x) = cL \qquad$ *c, a real number*

3. $\lim_{x \to a} [f(x) \pm g(x)] = \lim_{x \to a} f(x) \pm \lim_{x \to a} g(x) = L \pm M$

4. $\lim_{x \to a} [f(x)g(x)] = [\lim_{x \to a} f(x)][\lim_{x \to a} g(x)] = LM$

5. $\lim_{x \to a} \dfrac{f(x)}{g(x)} = \dfrac{\lim_{x \to a} f(x)}{\lim_{x \to a} g(x)} = \dfrac{L}{M} \qquad$ *Provided $M \neq 0$*

EXAMPLE 4   Use Theorem 1 to evaluate the following limits.

a. $\lim_{x \to 2} x^3$

b. $\lim_{x \to 4} 5x^{3/2}$

c. $\lim_{x \to 1} (5x^4 - 2)$

d. $\lim_{x \to 3} 2x^3 \sqrt{x^2 + 7}$

e. $\lim_{x \to 2} \dfrac{2x^2 + 1}{x + 1}$

Solution

a. $\lim_{x \to 2} x^3 = [\lim_{x \to 2} x]^3 \qquad$ Property 1

$\qquad \qquad = 2^3 = 8 \qquad$ $\lim_{x \to 2} x = 2$

b. $\lim_{x \to 4} 5x^{3/2} = 5 [\lim_{x \to 4} x^{3/2}] \qquad$ Property 2

$\qquad \qquad = 5(4)^{3/2} = 40$

c. $\lim_{x \to 1} (5x^4 - 2) = \lim_{x \to 1} 5x^4 - \lim_{x \to 1} 2 \qquad$ Property 3

To evaluate $\lim_{x \to 1} 2$, observe that the constant function $g(x) = 2$ has value
2 for all values of x. Therefore, $g(x)$ must approach the limit 2 as x
approaches $x = 1$ (or any other point for that matter!). Therefore,

$$\lim_{x \to 1} (5x^4 - 2) = 5(1)^4 - 2$$

$$= 3$$

d. $\lim_{x \to 3} 2x^3 \sqrt{x^2 + 7} = 2 \lim_{x \to 3} x^3 \sqrt{x^2 + 7} \qquad$ Property 2

$\qquad \qquad = 2 \lim_{x \to 3} x^3 \lim_{x \to 3} \sqrt{x^2 + 7} \qquad$ Property 4

$\qquad \qquad = 2(3)^3 \sqrt{3^2 + 7} \qquad$ Property 1

$\qquad \qquad = 2(27)\sqrt{16} = 216$

e. $\lim_{x \to 2} \dfrac{2x^2 + 1}{x + 1} = \dfrac{\lim_{x \to 2} (2x^2 + 1)}{\lim_{x \to 2} (x + 1)} \qquad$ Property 5

$\qquad \qquad = \dfrac{2(2)^2 + 1}{2 + 1} = \dfrac{9}{3} = 3$

Indeterminate Forms

Let us emphasize once again that property 5 of limits is valid only when the limit of the function that appears in the denominator is not equal to zero at the point in question.

If the numerator has a limit different from zero and the denominator has a limit equal to zero, then the limit of the quotient does not exist at the point in question. This is the case with the function $g(x) = 1/x^2$ in Example 3b. Here, as x approaches zero, the numerator approaches 1 but the denominator approaches zero, so that the quotient becomes arbitrarily large. Thus, as observed earlier, the limit does not exist.

Next, consider

$$\lim_{x \to 2} \frac{2(x^2 - 4)}{x - 2}$$

which we evaluated earlier by looking at the values of the function for x near $x = 2$. If we attempt to evaluate this expression by applying property 5 of limits, we see that both the numerator and denominator of the function

$$\frac{2(x^2 - 4)}{x - 2}$$

approach zero as x approaches 2; that is, we obtain an expression of the form 0/0. In this event, we say that the limit of the quotient $f(x)/g(x)$ as x approaches 2 has the **indeterminate form 0/0.**

We will need to evaluate limits of this type when we discuss the derivative of a function, a fundamental concept in the study of calculus. As the name suggests, the meaningless expression 0/0 does not provide us with a solution to our problem. One strategy that can be used to solve this type of problem follows.

STRATEGY FOR EVALUATING INDETERMINATE FORMS

1. Replace the given function with an appropriate one that takes on the same values as the original function everywhere except at $x = a$.

2. Evaluate the limit of this function as x approaches a.

Examples 5 and 6 illustrate this strategy.

EXAMPLE 5   Evaluate

$$\lim_{x \to 2} \frac{2(x^2 - 4)}{x - 2}$$

Solution   Since both the numerator and the denominator of this expression approach zero as x approaches 2, we have the indeterminate form 0/0. We rewrite

$$\frac{2(x^2 - 4)}{x - 2} = \frac{2(x - 2)(x + 2)}{(x - 2)}$$

which, upon canceling the common factors, is equivalent to $2(x + 2)$. Next, we replace $2(x^2 - 4)/(x - 2)$ with $2(x + 2)$ and take the limit as x approaches

2, obtaining

$$\lim_{x\to2}\frac{2(x^2-4)}{x-2}=\lim_{x\to2}2(x+2)=8$$

The graphs of the functions

$$f(x)=\frac{2(x^2-4)}{x-2}\qquad\text{and}\qquad g(x)=2(x+2)$$

are shown in Figure 7a and b. Observe that the graphs are identical except when $x=2$. The function g is defined for all values of x and, in particular, its value at $x=2$ is $g(2)=2(2+2)=8$. Thus, the point $(2,8)$ is on the graph of g. However, the function f is not defined at $x=2$. Since $f(x)=g(x)$ for all values of x except $x=2$, it follows that the graph of f must look exactly like the graph of g, with the exception that the point $(2,8)$ is missing from the graph of f.

FIGURE 7
The graphs of $f(x)$ and $g(x)$ are identical except at the point $(2,8)$.

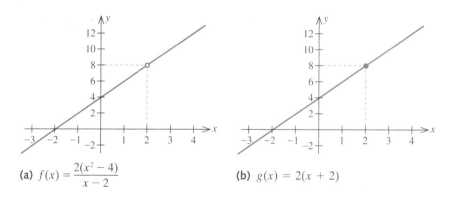

(a) $f(x)=\dfrac{2(x^2-4)}{x-2}$ (b) $g(x)=2(x+2)$

This illustrates graphically why we can evaluate the limit of f by evaluating the limit of the "equivalent" function g. ∎

REMARK   Notice that this is the same limit that we evaluated earlier when we discussed the instantaneous velocity of a car at a specified time.   □

EXAMPLE 6   Evaluate

$$\lim_{x\to-1}\frac{x^2-x-2}{x+1}$$

Solution   Since both the numerator and the denominator of this expression approach zero as x approaches -1, we have the indeterminate form $0/0$. We proceed as follows:

$$\lim_{x\to-1}\frac{x^2-x-2}{x+1}=\lim_{x\to-1}\frac{(x+1)(x-2)}{x+1}\qquad\text{Factoring the numerator}$$

$$=\lim_{x\to-1}(x-2)\qquad\text{Canceling the common factors}$$

$$=-3\qquad\qquad\blacksquare$$

Limits at Infinity

Up to now we have studied the limit of a function as x approaches a (finite) number a. There are occasions, however, when we want to know whether

$f(x)$ approaches a unique number as x increases without bound. Consider, for example, the function P, giving the number of fruit flies (*Drosophila*) in a container under controlled laboratory conditions, as a function of time t. The graph of P is shown in Figure 8. You can see from the graph of P that, as t increases without bound (gets larger and larger), $P(t)$ approaches the number 400. This number, called the *carrying capacity* of the environment, is determined by the amount of living space and food available, as well as other environmental factors.

FIGURE 8

The graph of P(t) gives the population of fruit flies in a laboratory experiment.

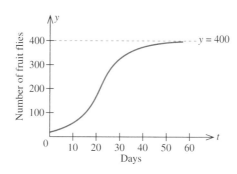

As another example, suppose that we are given the function

$$f(x) = \frac{2x^2}{1 + x^2}$$

and we want to determine what happens to $f(x)$ as x gets larger and larger. Picking the sequence of numbers 1, 2, 5, 10, 100, and 1000 and computing the corresponding values of $f(x)$, we obtain the following table of values:

x	1	2	5	10	100	1000
$f(x)$	1	1.6	1.92	1.98	1.9998	1.999998

FIGURE 9

The graph of $y = \dfrac{2x^2}{1 + x^2}$ *has a horizontal asymptote at* $y = 2$.

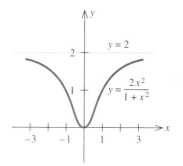

From the table, we see that as x gets larger and larger, $f(x)$ gets closer and closer to 2. The graph of the function f shown in Figure 9 confirms this observation. We call the line $y = 2$ a *horizontal asymptote.**

In this situation we say that the limit of the function

$$f(x) = \frac{2x^2}{1 + x^2}$$

as x increases without bound is 2, written

$$\lim_{x \to \infty} \frac{2x^2}{1 + x^2} = 2$$

In general, we have the following definition of a **limit of a function at infinity.**

---

*We discuss asymptotes in greater detail in Section 10.3.

**LIMIT OF
A FUNCTION
AT INFINITY**

The function f has the limit L as x increases without bound (or, as x approaches infinity), written

$$\lim_{x \to \infty} f(x) = L$$

if $f(x)$ can be made arbitrarily close to L by taking x large enough.
    Similarly, the function f has the limit M as x decreases without bound (or, as x approaches negative infinity), written

$$\lim_{x \to -\infty} f(x) = M$$

if $f(x)$ can be made arbitrarily close to M by taking x to be negative and sufficiently large in absolute value.

EXAMPLE 7   Let f and g be the functions

$$f(x) = \begin{cases} -1 & \text{if } x < 0 \\ 1 & \text{if } x \geq 0 \end{cases} \quad \text{and} \quad g(x) = \frac{1}{x^2}$$

Evaluate

a. $\lim_{x \to \infty} f(x)$ and $\lim_{x \to -\infty} f(x)$ **b.** $\lim_{x \to \infty} g(x)$ and $\lim_{x \to -\infty} g(x)$

Solution   The graphs of $f(x)$ and $g(x)$ are shown in Figure 10. Referring to the graphs of the respective functions, we see that

a. $\lim_{x \to \infty} f(x) = 1$ and $\lim_{x \to -\infty} f(x) = -1$

b. $\lim_{x \to \infty} \frac{1}{x^2} = 0$ and $\lim_{x \to -\infty} \frac{1}{x^2} = 0$

FIGURE 10

(a) $\lim_{x \to \infty} f(x) = 1$ and
$\lim_{x \to -\infty} f(x) = -1$

(b) $\lim_{x \to \infty} g(x) = 0$ and
$\lim_{x \to -\infty} g(x) = 0$

All of the properties of limits listed in Theorem 1 are valid when a is replaced by ∞ or $-\infty$. In addition, we have the following property for the limit at infinity.

THEOREM 2

For all $n > 0$

$$\lim_{x \to \infty} \frac{1}{x^n} = 0 \quad \text{and} \quad \lim_{x \to -\infty} \frac{1}{x^n} = 0$$

provided that $\frac{1}{x^n}$ is defined.

In evaluating the limit at infinity of a rational function, the following technique is often used: *Divide the numerator and denominator of the expression by x^n, where n is the highest power present in the denominator of the expression.*

EXAMPLE 8 Evaluate

$$\lim_{x \to \infty} \frac{x^2 - x + 3}{2x^3 + 1}$$

Solution Since the limits of both the numerator and the denominator do not exist as x approaches infinity, the property pertaining to the limit of a quotient (property 5) is not applicable. Using the technique just described, we divide the numerator and denominator of the rational expression by x^3, the highest power of x present in the denominator. Thus,

$$\lim_{x \to \infty} \frac{x^2 - x + 3}{2x^3 + 1} = \lim_{x \to \infty} \frac{\dfrac{1}{x} - \dfrac{1}{x^2} + \dfrac{3}{x^3}}{2 + \dfrac{1}{x^3}}$$

$$= \frac{0}{2} \qquad \text{Using Theorem 2}$$

$$= 0 \qquad \blacksquare$$

EXAMPLE 9 Let

$$f(x) = \frac{3x^2 + 8x - 4}{2x^2 + 4x - 5}$$

Compute $\lim_{x \to \infty} f(x)$ if it exists.

Solution Again, we see that property 5 is not applicable. Dividing the numerator and the denominator by x^2, the highest power of x present in the denominator, we obtain

$$\lim_{x \to \infty} \frac{3x^2 + 8x - 4}{2x^2 + 4x - 5} = \lim_{x \to \infty} \frac{3 + \dfrac{8}{x} - \dfrac{4}{x^2}}{2 + \dfrac{4}{x} - \dfrac{5}{x^2}}$$

$$= \frac{\lim\limits_{x \to \infty} 3 + 8 \lim\limits_{x \to \infty} \dfrac{1}{x} - 4 \lim\limits_{x \to \infty} \dfrac{1}{x^2}}{\lim\limits_{x \to \infty} 2 + 4 \lim\limits_{x \to \infty} \dfrac{1}{x} - 5 \lim\limits_{x \to \infty} \dfrac{1}{x^2}}$$

$$= \frac{3 + 0 - 0}{2 + 0 - 0}$$

$$= \frac{3}{2} \qquad \text{Using Theorem 2} \qquad \blacksquare$$

EXAMPLE 10 Let $f(x) = \dfrac{2x^3 - 3x^2 + 1}{x^2 + 2x + 4}$ and evaluate

a. $\lim\limits_{x \to \infty} f(x)$ **b.** $\lim\limits_{x \to -\infty} f(x)$

Solution

a. Dividing the numerator and the denominator of the rational expression by x^2, the highest power of x present in the denominator, we obtain

$$\lim_{x \to \infty} \frac{2x^3 - 3x^2 + 1}{x^2 + 2x + 4} = \lim_{x \to \infty} \frac{2x - 3 + \dfrac{1}{x^2}}{1 + \dfrac{2}{x} + \dfrac{4}{x^2}}$$

Since the numerator becomes arbitrarily large while the denominator approaches 1 as x approaches infinity, we see that the quotient $f(x)$ gets larger and larger as x approaches infinity. In other words, the limit does not exist. We indicate this by writing

$$\lim_{x \to \infty} \frac{2x^3 - 3x^2 + 1}{x^2 + 2x + 4} = \infty$$

b. Once again, dividing both the numerator and the denominator by x^2, we obtain

$$\lim_{x \to -\infty} \frac{2x^3 - 3x^2 + 1}{x^2 + 2x + 4} = \lim_{x \to -\infty} \frac{2x - 3 + \dfrac{1}{x^2}}{1 + \dfrac{2}{x} + \dfrac{4}{x^2}}$$

In this case, the numerator becomes arbitrarily large in magnitude but negative in sign, while the denominator approaches 1 as x approaches negative infinity. Therefore, the quotient $f(x)$ decreases without bound and the limit does not exist. We indicate this by writing

$$\lim_{x \to -\infty} \frac{2x^3 - 3x^2 + 1}{2x^2 + 2x + 4} = -\infty$$

∎

Example 11 gives an application of the concept of the limit of a function at infinity.

EXAMPLE 11 The Custom Office Company makes a line of executive desks. It is estimated that the total cost of making x Senior Executive Model desks is $C(x) = 100x + 200{,}000$ dollars per year, so that the average cost of making x desks is given by

$$\overline{C}(x) = \frac{C(x)}{x} = \frac{100x + 200{,}000}{x} = 100 + \frac{200{,}000}{x}.$$

dollars per desk. Evaluate $\lim_{x \to \infty} \overline{C}(x)$ and interpret your results.

Solution

$$\lim_{x \to \infty} \overline{C}(x) = \lim_{x \to \infty} \left(100 + \frac{200{,}000}{x} \right)$$

$$= \lim_{x \to \infty} 100 + \lim_{x \to \infty} \frac{200{,}000}{x} = 100$$

A sketch of the graph of the function $\overline{C}(x)$ appears in Figure 11. The result we obtained is fully expected if we consider its economic implications. Note

FIGURE 11
As the level of production increases, the average cost approaches $100 per desk.

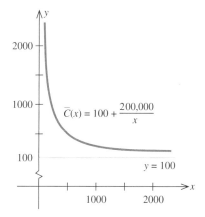

$$\overline{C}(x) = 100 + \frac{200,000}{x}$$

$y = 100$

that as the level of production increases, the fixed cost per desk produced, represented by the term $(200,000/x)$, drops steadily. The average cost should approach a constant unit cost of production—$100 in this case. ■

SELF-CHECK EXERCISES 9.1

1. Find the indicated limit if it exists.

a. $\displaystyle\lim_{x\to 3} \frac{\sqrt{x^2+7} + \sqrt{3x-5}}{x+2}$

b. $\displaystyle\lim_{x\to -1} \frac{x^2-x-2}{2x^2-x-3}$

2. The average cost per disc (in dollars) incurred by the Herald Record Company in pressing x compact audio discs is given by the average cost function

$$\overline{C}(x) = 1.8 + \frac{3000}{x}$$

Evaluate $\displaystyle\lim_{x\to\infty} \overline{C}(x)$ and interpret your result.

Solutions to Self-Check Exercises 9.1 can be found on page 488.

9.1 EXERCISES

In exercises 1–8, use the graph of the function f to determine $\displaystyle\lim_{x\to a} f(x)$ at the indicated value of a, if it exists.

1.

$a = -2$

2.

$a = 1$

3.

4.

5.

6.

7.

8.

In exercises 9–16, complete the table by computing $f(x)$ at the given values of x. Use these results to estimate the indicated limit, if it exists.

9. $f(x) = x^2 + 1$; $\lim\limits_{x \to 2} f(x)$

x	1.9	1.99	1.999	2.001	2.01	2.1
f(x)						

10. $f(x) = 2x^2 - 1$; $\lim\limits_{x \to 1} f(x)$

x	0.9	0.99	0.999	1.001	1.01	1.1
f(x)						

11. $f(x) = \dfrac{|x|}{x}$; $\lim\limits_{x \to 0} f(x)$

x	−0.1	−0.01	−0.001	0.001	0.01	0.1
f(x)						

12. $f(x) = \dfrac{|x - 1|}{x - 1}$; $\lim\limits_{x \to 1} f(x)$

x	0.9	0.99	0.999	1.001	1.01	1.1
f(x)						

13. $f(x) = \dfrac{1}{(x - 1)^2}$; $\lim\limits_{x \to 1} f(x)$

x	0.9	0.99	0.999	1.001	1.01	1.1
f(x)						

14. $f(x) = \dfrac{1}{x - 2}$; $\lim\limits_{x \to 2} f(x)$

x	1.9	1.99	1.999	2.001	2.01	2.1
f(x)						

15. $f(x) = \dfrac{x^2 + x - 2}{x - 1}$; $\lim\limits_{x \to 1} f(x)$

x	0.9	0.99	0.999	1.001	1.01	1.1
f(x)						

16. $f(x) = \dfrac{x - 1}{x - 1}$; $\lim\limits_{x \to 1} f(x)$

x	0.9	0.99	0.999	1.001	1.01	1.1
f(x)						

In exercises 17–34, find the indicated limit.

17. $\lim\limits_{x \to 2} 3$

18. $\lim\limits_{x \to -2} -3$

19. $\lim\limits_{x \to 3} x$

20. $\lim\limits_{x \to -2} -3x$

21. $\lim\limits_{x \to 1} (1 - 2x^2)$

22. $\lim\limits_{t \to 3} (4t^2 - 2t + 1)$

23. $\lim\limits_{x \to 1} (2x^3 - 3x^2 + x + 2)$

24. $\lim\limits_{x \to 0} (4x^5 - 20x^2 + 2x + 1)$

25. $\lim\limits_{s \to 0} (2s^2 - 1)(2s + 4)$

26. $\lim\limits_{x \to 2} (x^2 + 1)(x^2 - 4)$

27. $\lim\limits_{x \to 2} \dfrac{2x + 1}{x + 2}$

28. $\lim\limits_{x \to 1} \dfrac{x^3 + 1}{2x^3 + 2}$

29. $\lim\limits_{x \to 2} \sqrt{x + 2}$

30. $\lim\limits_{x \to -2} \sqrt[3]{5x + 2}$

31. $\lim\limits_{x \to -3} \sqrt{2x^4 + x^2}$

32. $\lim\limits_{x \to 2} \sqrt{\dfrac{2x^3 + 4}{x^2 + 1}}$

33. $\lim\limits_{x \to -1} \dfrac{\sqrt{x^2 + 8}}{2x + 4}$

34. $\lim\limits_{x \to 3} \dfrac{x\sqrt{x^2 + 7}}{2x - \sqrt{2x + 3}}$

In exercises 35–42, find the indicated limit given that $\lim\limits_{x \to a} f(x) = 3$ *and* $\lim\limits_{x \to a} g(x) = 4$.

35. $\lim\limits_{x \to a} [f(x) - g(x)]$

36. $\lim\limits_{x \to a} 2f(x)$

37. $\lim\limits_{x \to a} [2f(x) - 3g(x)]$

38. $\lim\limits_{x \to a} [f(x)g(x)]$

39. $\lim\limits_{x \to a} \sqrt{g(x)}$

40. $\lim\limits_{x \to a} \sqrt[3]{5f(x) + 3g(x)}$

41. $\lim\limits_{x \to a} \dfrac{2f(x) - g(x)}{f(x)g(x)}$

42. $\lim\limits_{x \to a} \dfrac{g(x) - f(x)}{f(x) + \sqrt{g(x)}}$

In exercises 43–56, find the indicated limit, if it exists.

43. $\lim\limits_{x \to 1} \dfrac{x^2 - 1}{x - 1}$

44. $\lim\limits_{x \to -2} \dfrac{x^2 - 4}{x + 2}$

45. $\lim\limits_{x \to 0} \dfrac{x^2 - x}{x}$

46. $\lim\limits_{x \to 0} \dfrac{2x^2 - 3x}{x}$

47. $\lim\limits_{x \to -5} \dfrac{x^2 - 25}{x + 5}$

48. $\lim\limits_{b \to -3} \dfrac{b + 1}{b + 3}$

49. $\lim\limits_{x \to 1} \dfrac{x}{x - 1}$

50. $\lim\limits_{x \to 2} \dfrac{x + 2}{x - 2}$

51. $\lim\limits_{x \to -2} \dfrac{x^2 - x - 6}{x^2 + x - 2}$

52. $\lim\limits_{z \to 2} \dfrac{z^3 - 8}{z - 2}$

53. $\lim\limits_{x \to 1} \dfrac{\sqrt{x} - 1}{x - 1}$ $\left[\text{Hint: Multiply by } \dfrac{\sqrt{x} + 1}{\sqrt{x} + 1}.\right]$

54. $\lim\limits_{x \to 4} \dfrac{x - 4}{\sqrt{x} - 2}$ [Hint: See exercise 53.]

55. $\lim\limits_{x \to 1} \dfrac{x - 1}{x^3 + x^2 - 2x}$

56. $\lim\limits_{x \to -2} \dfrac{4 - x^2}{2x^2 + x^3}$

In exercises 57–62, use the graph of the function f to determine $\lim\limits_{x \to \infty} f(x)$ *and* $\lim\limits_{x \to -\infty} f(x)$, *if they exist.*

57.

58.

59.

60.

$f(x) = \dfrac{x^2}{x^2 + 1}$

x	−5	−10	−100	−1000
f(x)				

C **65.** $f(x) = 3x^3 - x^2 + 10;$ $\displaystyle\lim_{x\to\infty} f(x)$ and $\displaystyle\lim_{x\to-\infty} f(x)$

x	1	5	10	100	1000
f(x)					

61.

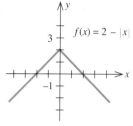

$f(x) = 2 - |x|$

x	−1	−5	−10	−100	−1000
f(x)					

66. $f(x) = \dfrac{|x|}{x};$ $\displaystyle\lim_{x\to\infty} f(x)$ and $\displaystyle\lim_{x\to-\infty} f(x)$

x	1	10	100	−1	−10	−100
f(x)						

62.

$f(x) = \begin{cases} \sqrt{-x} & \text{if } x \le 0 \\ \dfrac{x}{x+1} & \text{if } x > 0 \end{cases}$

In exercises 67–74, find the indicated limits, if they exist.

67. $\displaystyle\lim_{x\to\infty} \dfrac{3x + 2}{x - 5}$

68. $\displaystyle\lim_{x\to-\infty} \dfrac{4x^2 - 1}{x + 2}$

In exercises 63–66, complete the table by computing f(x) at the given values of x. Use the results to guess at the indicated limits, if they exist.

69. $\displaystyle\lim_{x\to-\infty} \dfrac{3x^3 + x^2 + 1}{x^3 + 1}$

70. $\displaystyle\lim_{x\to\infty} \dfrac{2x^2 + 3x + 1}{x^4 - x^2}$

C **63.** $f(x) = \dfrac{1}{x^2 + 1};$ $\displaystyle\lim_{x\to\infty} f(x)$ and $\displaystyle\lim_{x\to-\infty} f(x)$

71. $\displaystyle\lim_{x\to-\infty} \dfrac{x^4 + 1}{x^3 - 1}$

72. $\displaystyle\lim_{x\to\infty} \dfrac{4x^4 - 3x^2 + 1}{2x^4 + x^3 + x^2 + x + 1}$

x	1	10	100	1000
f(x)				

x	−1	−10	−100	−1000
f(x)				

73. $\displaystyle\lim_{x\to\infty} \dfrac{x^5 - x^3 + x - 1}{x^6 + 2x^2 + 1}$

74. $\displaystyle\lim_{x\to\infty} \dfrac{2x^2 - 1}{x^3 + x^2 + 1}$

75. Toxic Waste   A city's main well was recently found to be contaminated with trichloroethylene, a cancer-causing chemical, as a result of an abandoned chemical dump leaching chemicals into the water. A proposal submitted to the selectmen of the city indicates that the cost, measured in millions of dollars, of removing $x\%$ of the toxic pollutant is given by

C **64.** $f(x) = \dfrac{2x}{x + 1};$ $\displaystyle\lim_{x\to\infty} f(x)$ and $\displaystyle\lim_{x\to-\infty} f(x)$

$$C(x) = \dfrac{0.5x}{100 - x} \qquad 0 < x < 100$$

x	1	10	100	1000
f(x)				

a. Find the cost of removing 50% of the pollutant; 60%; 70%; 80%; 90%; 95%.
b. Evaluate

$$\lim_{x \to 100} \frac{0.5x}{100 - x}$$

and interpret your result.

76. Average Cost   The average cost per disc in dollars incurred by the Herald Record Company in pressing x video discs is given by the average cost function

$$\overline{C}(x) = 2.2 + \frac{2500}{x}$$

Evaluate $\lim_{x \to \infty} \overline{C}(x)$ and interpret your result.

77. Concentration of a Drug in the Bloodstream   The concentration of a certain drug in a patient's bloodstream t hours after injection is given by

$$C(t) = \frac{0.2t}{t^2 + 1}$$

milligrams per cubic centimeter. Evaluate $\lim_{t \to \infty} C(t)$ and interpret your result.

78. Box Office Receipts   The total worldwide box office receipts for a long-running blockbuster movie are approximated by the function

$$T(x) = \frac{120x^2}{x^2 + 4}$$

where $T(x)$ is measured in millions of dollars and x is the number of months since the movie's release.
a. What are the total box office receipts after the first month? The second month? The third month?
b. What will the movie gross in the long run?

79. Population Growth   A major corporation is building a 4325-acre complex of homes, offices, stores, schools, and churches in the rural community of Glen Cove. As a result of this development, the planners have estimated that Glen Cove's population t years from now (in thousands) will be given by

$$P(t) = \frac{25t^2 + 125t + 200}{t^2 + 5t + 40}$$

a. What is the current population of Glen Cove?
b. What will the population be in the long run?

[c] 80. Driving Costs   A study of driving costs of 1992 model subcompact (four-cylinder) cars found that the average cost (car payments, gas, insurance, up-keep, and depreciation), measured in cents per mile, is approximated by the function

$$C(x) = \frac{2010}{x^{2.2}} + 17.80$$

where x denotes the number of miles (in thousands) the car is driven in a year.
a. What is the average cost of driving a subcompact car 5,000 miles a year? 10,000 miles a year? 15,000 miles a year? 20,000 miles a year? 25,000 miles a year?
b. Use (a) to help sketch the graph of the function C.
c. What happens to the average cost as the number of miles driven increases without bound?
Source: Runzheimer International Study

81. Speed of a Chemical Reaction   Certain proteins, known as enzymes, serve as catalysts for chemical reactions in living things. In 1913, Leonor Michaelis and L. M. Menten discovered the following formula, giving the initial speed V (in moles per liter per second) at which the reaction begins in terms of the amount of substrate x (the substance being acted upon, measured in moles per liter):

$$V = \frac{ax}{x + b}$$

where a and b are positive constants. Evaluate

$$\lim_{x \to \infty} \frac{ax}{x + b}$$

and interpret your result.

82. Show that the effective rate of interest $\bar{r}_{\text{eff}}$ that corresponds to a nominal interest rate r per year compounded continuously is given by

$$\bar{r}_{\text{eff}} = e^r - 1$$

[*Hint:* From equation (6), Section 4.1, we see that the effective rate r_{eff} corresponding to a nominal interest rate r per year compounded m times a year is given by

$$r_{\text{eff}} = \left(1 + \frac{r}{m}\right)^m - 1$$

Let m tend to infinity in this expression.]

83. Investment Analysis   Refer to exercise 82. Bank A pays interest on deposits at a 7% annual rate compounded quarterly, and Bank B pays interest on deposits at a $7\frac{1}{8}$% annual rate compounded continuously. Which bank has the higher effective rate of interest?

**SOLUTIONS TO
SELF-CHECK
EXERCISES 9.1**

1. **a.** $\displaystyle\lim_{x\to 3}\frac{\sqrt{x^2+7}+\sqrt{3x-5}}{x+2}=\frac{\sqrt{3^2+7}+\sqrt{3(3)-5}}{3+2}$

$$=\frac{\sqrt{16}+\sqrt{4}}{5}$$

$$=\frac{6}{5}$$

b. Letting x approach -1 leads to the indeterminate form 0/0. Thus we proceed as follows:

$$\lim_{x\to -1}\frac{x^2-x-2}{2x^2-x-3}=\lim_{x\to -1}\frac{(x+1)(x-2)}{(x+1)(2x-3)} \qquad \text{Factoring}$$

$$=\lim_{x\to -1}\frac{x-2}{2x-3} \qquad \begin{array}{l}\text{Canceling the}\\ \text{common factors}\end{array}$$

$$=\frac{-1-2}{2(-1)-3} \qquad \text{Substituting } x=-1$$

$$=\frac{3}{5}$$

2. $\displaystyle\lim_{x\to\infty}\overline{C}(x)=\lim_{x\to\infty}\left(1.8+\frac{3000}{x}\right)$

$$=\lim_{x\to\infty}1.8+\lim_{x\to\infty}\frac{3000}{x}$$

$$=1.8$$

Our computation reveals that as the production of audio discs increases "without bound," the average cost drops and approaches a unit cost of $1.80 per disc.

Using Technology

A graphing utility can be used to help us find the limit of a function, if it exists, as illustrated in the following examples.

EXAMPLE 1   Let $f(x)=\dfrac{x^3-1}{x-1}$.

a. Plot the graph of f in the viewing rectangle $[-2, 2] \times [0, 4]$.

b. Use **ZOOM** to find $\displaystyle\lim_{x\to 1}\frac{x^3-1}{x-1}$.

c. Verify your result by evaluating the limit algebraically.

Solution

a. The graph of f in the viewing rectangle $[-2, 2] \times [0, 4]$ is shown in Figure 1.

FIGURE 1

The graph of $f(x) = \dfrac{x^3 - 1}{x - 1}$
in the viewing rectangle
$[-2, 2] \times [0, 4]$

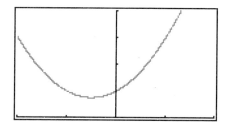

b. Using **ZIN** repeatedly, we see that the y-value approaches 3 as the x-value approaches 1. We conclude, accordingly, that

$$\lim_{x \to 1} \frac{x^3 - 1}{x - 1} = 3$$

c. We compute

$$\lim_{x \to 1} \frac{x^3 - 1}{x - 1} = \lim_{x \to 1} \frac{(x - 1)(x^2 + x + 1)}{x - 1}$$

$$= \lim_{x \to 1} (x^2 + x + 1) = 3 \qquad \blacksquare$$

REMARK   If you attempt to find the limit in Example 1 by using **EVAL** to find the value of $f(x)$ when $x = 1$, you will see that the calculator does not display the y-value. This happens because the point $x = 1$ is not in the domain of f.   □

EXAMPLE 2   Use **ZOOM** to find $\lim_{x \to 0} (1 + x)^{1/x}$.

Solution   We first plot the graph of $f(x) = (1 + x)^{1/x}$ in a suitable viewing rectangle. Figure 2 shows a plot of f in the rectangle $[-1, 1] \times [0, 4]$. Then using **ZIN** repeatedly, we see that $\lim_{x \to 0} (1 + x)^{1/x} \approx 2.71828$.

FIGURE 2

The graph of $f(x) = (1 + x)^{1/x}$
in the viewing rectangle
$[-1, 1] \times [0, 4]$

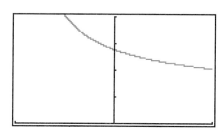

The limit of $f(x) = (1 + x)^{1/x}$ as x approaches zero, denoted by the letter e, plays a very important role in the study of mathematics and its applications (see Section 9.7). Thus

$$\lim_{x \to 0} (1 + x)^{1/x} = e$$

where, as we have just seen, $e \approx 2.71828$.   ■

EXERCISES

*In exercises 1–10, use a graphing utility to find the indicated limit by first plotting the graph of the function in a suitable viewing rectangle and then using the **ZOOM** feature of the calculator.*

1. $\lim_{x \to 1} \dfrac{2x^3 - 2x^2 + 3x - 3}{x - 1}$

2. $\lim_{x \to -2} \dfrac{2x^3 + 3x^2 - x + 2}{x + 2}$

3. $\lim_{x \to -1} \dfrac{x^3 + 1}{x + 1}$

4. $\lim_{x \to -1} \dfrac{x^4 - 1}{x + 1}$

5. $\lim_{x \to 1} \dfrac{x^3 - x^2 - x + 1}{x^3 - 3x + 2}$

6. $\lim_{x \to 2} \dfrac{x^3 + 2x^2 - 16}{2x^3 - x^2 + 2x - 16}$

7. $\lim_{x \to 0} \dfrac{\sqrt{x + 1} - 1}{x}$

8. $\lim_{x \to 0} \dfrac{(x + 4)^{3/2} - 8}{x}$

9. $\lim_{x \to 0} (1 + 2x)^{1/x}$

10. $\lim_{x \to 0} \dfrac{2^x - 1}{x}$

11. Use a graphing utility to show that $\lim_{x \to 3} \dfrac{2}{x - 3}$ does not exist.

12. Use a graphing utility to show that $\lim_{x \to 2} \dfrac{x^3 - 2x + 1}{x - 2}$ does not exist.

9.2   Continuity

One-Sided Limits

Consider the function f defined by

$$f(x) = \begin{cases} x - 1 & \text{if } x < 0 \\ x + 1 & \text{if } x \geq 0 \end{cases}$$

From the graph of f shown in Figure 12, we see that the function f does not have a limit as x approaches zero, because no matter how close x is to zero, $f(x)$ takes on values that are close to 1 if x is positive and values that are close to -1 if x is negative. Therefore, $f(x)$ cannot be close to a single number L—no matter how close x is to zero. Now if we restrict x to be greater than zero (to the right of zero), then we see that $f(x)$ can be made as close to the number 1 as we please by taking x sufficiently close to zero. In this situation, we say that the right-hand limit of f as x approaches zero (from the right) is 1, written

$$\lim_{x \to 0^+} f(x) = 1$$

FIGURE 12
The function f does not have a limit as x approaches zero.

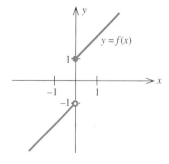

James H. Chesebro, M.D.

In 1992 Chesebro was appointed professor of medicine at Harvard Medical School, after having conducted cardiovascular research at the Mayo Clinic for a number of years. In addition, he was made Associate Director for Research in the Cardiac Unit at Massachusetts General Hospital, one of Harvard's teaching affiliates.

Institution: Harvard Medical School, Massachusetts General Hospital

For over 20 years, James Chesebro has worked as an investigative cardiologist, diagnosing and treating patients suffering from heart and blood vessel diseases. He specializes in researching ways to prevent blood clots from narrowing the heart's arteries. Even when patients have coronary bypass operations to correct problems, stresses Chesebro, "the piece of vein used to detour around blocked arteries is prone to plug up with clots."

Throughout his career, Chesebro has continued to investigate the body's clotting mechanisms as well as substances that may prevent, or even dissolve, clots. A study conducted in 1982 showed that bypass patients taking dipyridamole and aspirin were nearly three times less likely to develop clots in vein grafts than patients given placebos.

Recently, Chesebro has been working with a substance called hirudin derived from leech saliva. Hirudin has proved quite beneficial in blocking arterial blood clot formations.

Determining the correct medication and dosage involves intense quantitative research. Colleagues from other branches of science such as nuclear medicine, molecular biology, and biochemistry play key roles in the research process. Chesebro and his colleagues have to understand how the heart's physiology and pharmacological variables such as distribution rates, concentration levels, elimination rates, and biological effects of particular substances dictate the choice and dosage of medication.

In considering whether a particular medication will bind effectively with specific body cells, researchers must determine bond tightness, speed, and length of time to reach optimal concentration. Researchers rely on complicated equations such as "integrating the area under a curve to find the amount of medication in the body at a given time," notes Chesebro. Without calculus, these equations can't be solved.

Whether physicians rely on medication or surgical procedures such as balloon angioplasty to open clogged passages, other factors play a major role in the outcome. Cholesterol and blood sugar levels, patient age, blood pressure, the geometry of the blockage, and the velocity and turbulence of blood flow all warrant consideration. According to Chesebro "linear modeling can be used to predict the contribution of patient variables and local blood vessel variables in the outcome of opening blocked arteries."

The bottom line? Calculus has been a key contributor to Chesebro's success in preventing disabling arterial blood clots.

Similarly, we see that $f(x)$ can be made as close to the number -1 as we please by taking x sufficiently close to, but to the left of, zero. In this situation, we say that the left-hand limit of f as x approaches zero (from the left) is -1, written

$$\lim_{x \to 0^-} f(x) = -1$$

These limits are called one-sided limits. More generally, we have the following definitions.

ONE-SIDED LIMITS

The function f has the right-hand limit L as x approaches a from the right, written

$$\lim_{x \to a^+} f(x) = L$$

if the values $f(x)$ can be made as close to L as we please by taking x sufficiently close to (but not equal to) a and to the right of a.

Similarly, the function f has the left-hand limit M as x approaches a from the left, written

$$\lim_{x \to a^-} f(x) = M$$

if the values $f(x)$ can be made as close to M as we please by taking x sufficiently close to (but not equal to) a and to the left of a.

The connection between one-sided limits and the two-sided limit defined earlier is given by the following theorem.

THEOREM 3

Let f be a function that is defined for all values of x close to $x = a$ with the possible exception of a itself. Then

$$\lim_{x \to a} f(x) = L \quad \text{if and only if} \quad \lim_{x \to a^+} f(x) = \lim_{x \to a^-} f(x) = L$$

Thus, the two-sided limit exists if and only if the one-sided limits exist and are equal.

EXAMPLE 1   Let

$$f(x) = \begin{cases} \sqrt{x} & \text{if } x > 0 \\ -x & \text{if } x \le 0 \end{cases} \qquad \text{and} \qquad g(x) = \begin{cases} -1 & \text{if } x < 0 \\ 1 & \text{if } x \ge 0 \end{cases}$$

a. Show that $\lim_{x \to 0} f(x)$ exists by studying the one-sided limits of f as x approaches $x = 0$.

b. Show that $\lim_{x \to 0} g(x)$ does not exist.

Solution

a. We find

$$\lim_{x \to 0^+} f(x) = \lim_{x \to 0^+} \sqrt{x} = 0$$

and

$$\lim_{x \to 0^-} f(x) = \lim_{x \to 0^-} (-x) = 0$$

so

$$\lim_{x \to 0} f(x) = 0 \qquad \text{Figure 13a}$$

b. We have

$$\lim_{x \to 0^-} g(x) = -1 \qquad \text{and} \qquad \lim_{x \to 0^+} g(x) = 1$$

and since these one-sided limits are not equal, we conclude that $\lim_{x \to 0} g(x)$ does not exist (Figure 13b).

FIGURE 13

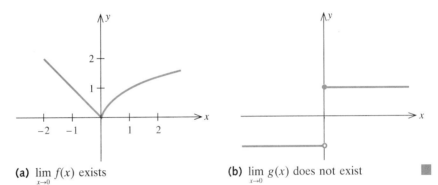

(a) $\lim_{x \to 0} f(x)$ exists   **(b)** $\lim_{x \to 0} g(x)$ does not exist

Continuous Functions

Continuous functions will play an important role throughout most of our study of calculus. Loosely speaking, a function is continuous at a point if the graph of the function at that point is devoid of holes, gaps, jumps, or breaks. Consider, for example, the graph of the function f depicted in Figure 14.

FIGURE 14
The graph of this function is not continuous at $x = a$, $x = b$, $x = c$, and $x = d$.

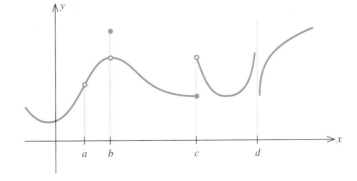

Let us take a closer look at the behavior of f at or near each of the points $x = a$, $x = b$, $x = c$, and $x = d$. First, note that f is not defined at

$x = a$; that is, the point $x = a$ is not in the domain of f, thereby resulting in a "hole" in the graph of f. Next, observe that the value of f at b, $f(b)$, is not equal to the limit of $f(x)$ as x approaches b, resulting in a "jump" in the graph of f at that point. The function f does not have a limit at $x = c$ since the left-hand and right-hand limits of $f(x)$ are not equal, also resulting in a jump in the graph of f at that point. Finally, the limit of f does not exist at $x = d$, resulting in a break in the graph of f. The function f is *discontinuous* at each of these points. It is *continuous* at all other points.

CONTINUITY AT A POINT

A function f is **continuous at the point** $x = a$ if the following conditions are satisfied:

1. $f(a)$ is defined,   **2.** $\lim_{x \to a} f(x)$ exists,   **3.** $\lim_{x \to a} f(x) = f(a)$.

Thus, a function f is continuous at the point $x = a$ if the limit of f at the point $x = a$ exists and has the value $f(a)$. Geometrically, f is continuous at the point $x = a$ if proximity of x to a implies the proximity of $f(x)$ to $f(a)$.

If f is not continuous at $x = a$, then f is said to be **discontinuous** at $x = a$. Also, f is **continuous on an interval** if f is continuous at every point in the interval.

Figure 15 depicts the graph of a continuous function on the interval (a, b). Notice that the graph of the function over the stated interval can be sketched without lifting one's pencil from the paper.

FIGURE 15

The graph of f is continuous on the interval (a, b).

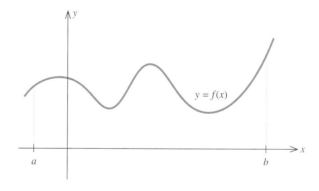

EXAMPLE 2   Find the values of x for which each of the following functions is continuous.

a. $f(x) = x + 2$ **b.** $g(x) = \dfrac{x^2 - 4}{x - 2}$ **c.** $h(x) = \begin{cases} x + 2 & \text{if } x \neq 2 \\ 1 & \text{if } x = 2 \end{cases}$

d. $F(x) = \begin{cases} -1 & \text{if } x < 0 \\ 1 & \text{if } x \geq 0 \end{cases}$ **e.** $G(x) = \begin{cases} \dfrac{1}{x} & \text{if } x > 0 \\ -1 & \text{if } x \leq 0 \end{cases}$

The graph of each function is shown in Figure 16.

FIGURE 16

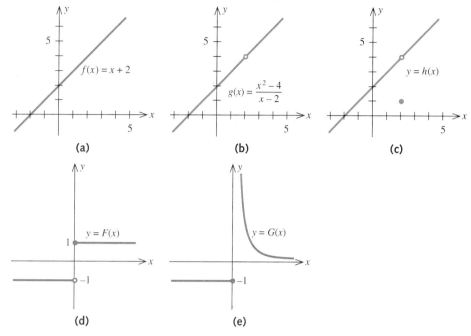

(a)   (b)   (c)

(d)   (e)

Solution

a. The function f is continuous everywhere, because the three conditions for continuity are satisfied for all values of x.

b. The function g is discontinuous at the point $x = 2$, because g is not defined at that point. It is continuous everywhere else.

c. The function h is discontinuous at $x = 2$ because the third condition for continuity is violated; the limit of $h(x)$ as x approaches 2 exists and has the value 4, but this limit is not equal to $h(2) = 1$. It is continuous for all other values of x.

d. The function F is continuous everywhere except at the point $x = 0$, where the limit of $F(x)$ fails to exist as x approaches zero (see Example 3a, Section 9.1).

e. Since the limit of $G(x)$ does not exist as x approaches zero, we conclude that G fails to be continuous at $x = 0$. The function G is continuous at all other points. ◼

Continuity of Polynomial and Rational Functions

Among the functions that occur frequently in practice are polynomial and rational functions. The following result tells us under what conditions these functions are continuous. (A proof is sketched in exercise 59, page 501.)

CONTINUITY OF POLYNOMIAL AND RATIONAL FUNCTIONS

1. A polynomial function $y = P(x)$ is continuous at every point x.

2. A rational function $R(x) = p(x)/q(x)$ is continuous at every point x where $q(x) \neq 0$.

EXAMPLE 3   Find the values of x for which each of the following functions is continuous.

a. $f(x) = 3x^3 + 2x^2 - x + 10$ **b.** $g(x) = \dfrac{8x^{10} - 4x + 1}{x^2 + 1}$

c. $h(x) = \dfrac{4x^3 - 3x^2 + 1}{x^2 - 3x + 2}$

Solution

a. The function f is a polynomial function of degree 3, so $f(x)$ is continuous for all values of x.

b. The function g is a rational function. Observe that the denominator of g, namely $x^2 + 1$, is never equal to zero. Therefore, we conclude that g is continuous for all values of x.

c. The function h is a rational function. In this case, however, the denominator of h is equal to zero at $x = 1$ and $x = 2$, which can be seen by factoring it. Thus,

$$x^2 - 3x + 2 = (x - 2)(x - 1)$$

We conclude, therefore, that h is continuous everywhere except at $x = 1$ and $x = 2$, where it is discontinuous. ∎

Applications

Up to this point, most of the applications we have discussed involved functions that are continuous everywhere. In Example 4, we will consider an application from the field of educational psychology that involves a discontinuous function.

EXAMPLE 4   Figure 17 depicts the learning curve associated with a certain individual. Beginning with no knowledge of the subject being taught, the individual makes steady progress toward understanding it over the time interval $0 \le t < t_1$. In this instance, the individual's progress slows down as we approach time t_1 because he fails to grasp a particularly difficult concept. All of a sudden, a breakthrough occurs at time t_1, propelling his knowledge of the subject to a higher level. The curve is discontinuous at t_1.

FIGURE 17
A learning curve that is discontinuous at $t = t_1$

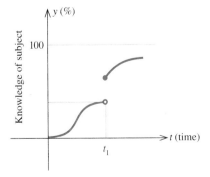

We close this section by pointing out a property of continuous functions that will play a very important role in calculus (Theorem 4).

THEOREM 4

Suppose that a continuous function f assumes the values $f(a)$ and $f(b)$ at two points $x = a$ and $x = b$ with $a < b$. If $f(a)$ and $f(b)$ have opposite signs, then there must be at least one point $x = c$, with $a < c < b$, where $f(c) = 0$ (see Figure 18).

FIGURE 18
f(a) and f(b) have opposite signs.

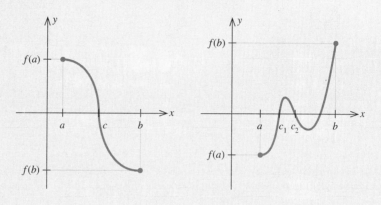

FIGURE 19
A discontinuous function

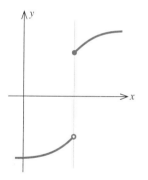

Geometrically, this property states that if the graph of a continuous function goes from above the x-axis to below the x-axis or vice versa, it must *cross* the x-axis. This, of course, is not necessarily true if the function is not continuous (Figure 19).

EXAMPLE 5   Let $f(x) = x^3 + x + 1$.

a. Show that f is continuous for all values of x.

b. Compute $f(-1)$ and $f(1)$ and use the results to deduce that there must be at least one point $x = c$ where c lies in the interval $(-1, 1)$ and $f(c) = 0$.

Solution

a. The function f is a polynomial function of degree 3 and is therefore continuous everywhere.

b. $f(-1) = (-1)^3 + (-1) + 1 = -1$
    $f(1) = 1^3 + 1 + 1 = 3$

Since $f(-1)$ and $f(1)$ have opposite signs, Theorem 4 tells us that there must be at least one point $x = c$ with $-1 < c < 1$ such that $f(c) = 0$. ■

SELF-CHECK EXERCISES 9.2

1. Evaluate $\lim_{x \to -1^-} f(x)$ and $\lim_{x \to -1^+} f(x)$, where

$$f(x) = \begin{cases} 1 & \text{if } x < -1 \\ 1 + \sqrt{x + 1} & \text{if } x \geq -1 \end{cases}$$

Does $\lim_{x \to -1} f(x)$ exist?

2. Determine the values of x for which the given function is discontinuous. At each point of discontinuity, indicate which condition(s) for continuity are violated. Sketch the graph of the function.

a. $f(x) = \begin{cases} -x^2 + 1 & \text{if } x \le 1 \\ x - 1 & \text{if } x > 1 \end{cases}$

b. $g(x) = \begin{cases} -x + 1 & \text{if } x < -1 \\ 2 & \text{if } -1 < x \le 1 \\ -x + 3 & \text{if } x > 1 \end{cases}$

Solutions to Self-Check Exercises 9.2 can be found on page 501.

9.2 EXERCISES

In exercises 1–8, use the graph of the function f to find $\lim\limits_{x \to a^-} f(x)$, $\lim\limits_{x \to a^+} f(x)$ and $\lim\limits_{x \to a} f(x)$ at the indicated value of a, if the limit exists.

1.

$a = 2$

2.

$a = 3$

3.

$a = -1$

4.

$a = 1$

5.

$a = 1$

6.

$a = 0$

7.

$a = 0$

8.

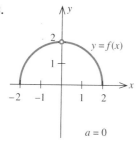

$a = 0$

In exercises 9–26, find the indicated one-sided limit, if it exists.

9. $\lim\limits_{x \to 1^+} (2x + 4)$

10. $\lim\limits_{x \to 1^-} (3x - 4)$

11. $\lim\limits_{x \to 2^-} \dfrac{x - 3}{x + 2}$

12. $\lim\limits_{x \to 1^+} \dfrac{x + 2}{x + 1}$

13. $\lim\limits_{x \to 0^+} \dfrac{1}{x}$

14. $\lim\limits_{x \to 0^-} \dfrac{1}{x}$

15. $\lim\limits_{x \to 0^+} \dfrac{x - 1}{x^2 + 1}$

16. $\lim\limits_{x \to 2^+} \dfrac{x + 1}{x^2 - 2x + 3}$

17. $\lim\limits_{x \to 0^+} \sqrt{x}$

18. $\lim\limits_{x \to 2^+} 2\sqrt{x - 2}$

19. $\lim\limits_{x \to -2^+} (2x + \sqrt{2 + x})$

20. $\lim\limits_{x \to -5^+} x(1 + \sqrt{5 + x})$

21. $\lim\limits_{x \to 1^-} \dfrac{1 + x}{1 - x}$

22. $\lim\limits_{x \to 1^+} \dfrac{1 + x}{1 - x}$

23. $\lim\limits_{x \to 2^-} \dfrac{x^2 - 4}{x - 2}$

24. $\lim\limits_{x \to -3^+} \dfrac{\sqrt{x + 3}}{x^2 + 1}$

25. $\lim\limits_{x \to 3^+} \dfrac{x^2 - 9}{x + 3}$

26. $\lim\limits_{x \to -2^+} \dfrac{\sqrt[3]{x + 10}}{2x^2 + 1}$

In exercises 27–34, determine the values of x, if any, at which each function is discontinuous. At each point of discontinuity, state the condition(s) for continuity that are violated.

27.

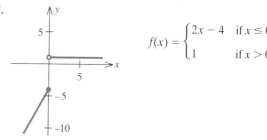

$f(x) = \begin{cases} 2x - 4 & \text{if } x \le 0 \\ 1 & \text{if } x > 0 \end{cases}$

28.

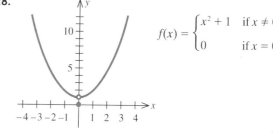

$f(x) = \begin{cases} x^2 + 1 & \text{if } x \ne 0 \\ 0 & \text{if } x = 0 \end{cases}$

29.

$f(x) = \begin{cases} x + 5 & \text{if } x \le 0 \\ -x^2 + 5 & \text{if } x > 0 \end{cases}$

30.

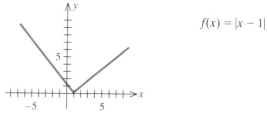

$f(x) = |x - 1|$

31.

$f(x) = \begin{cases} x + 5 & \text{if } x < 0 \\ 2 & \text{if } x = 0 \\ -x^2 + 5 & \text{if } x > 0 \end{cases}$

32.

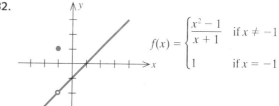

$f(x) = \begin{cases} \dfrac{x^2 - 1}{x + 1} & \text{if } x \ne -1 \\ 1 & \text{if } x = -1 \end{cases}$

33.

$$f(x) = \begin{cases} -|x| + 1 & \text{if } x \neq 0 \\ 0 & \text{if } x = 0 \end{cases}$$

34.

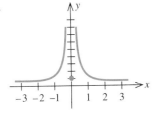

$$f(x) = \begin{cases} \dfrac{1}{x^2} & \text{if } x \neq 0 \\ 1 & \text{if } x = 0 \end{cases}$$

In exercises 35–44, find the values of x for which the function is continuous.

35. $f(x) = 2x^2 + x - 1$

36. $f(x) = x^3 - 2x^2 + x - 1$

37. $f(x) = \dfrac{2}{x^2 + 1}$

38. $f(x) = \dfrac{x}{2x^2 + 1}$

39. $f(x) = \dfrac{2}{2x - 1}$

40. $f(x) = \dfrac{x + 1}{x - 1}$

41. $f(x) = \dfrac{2x + 1}{x^2 + x - 2}$

42. $f(x) = \dfrac{x - 1}{x^2 + 2x - 3}$

43. $f(x) = \begin{cases} x & \text{if } x \leq 1 \\ 2x - 1 & \text{if } x > 1 \end{cases}$

44. $f(x) = \begin{cases} -x + 1 & \text{if } x \leq -1 \\ x + 1 & \text{if } x > -1 \end{cases}$

In exercises 45–48, determine all values of x at which the function is discontinuous.

45. $f(x) = \dfrac{2x}{x^2 - 1}$

46. $f(x) = \dfrac{1}{(x - 1)(x - 2)}$

47. $f(x) = \dfrac{x^2 - 2x}{x^2 - 3x + 2}$

48. $f(x) = \dfrac{x^2 - 3x + 2}{x^2 - 2x}$

49. Inventory Control   As part of an optimal inventory policy, the manager of an office supply company orders 500 reams of xerographic copy paper every 20 days. The accompanying graph shows the *actual* inventory level of copy paper in an office supply store during the first 60 business days of 1992. Determine the values of t for which the "inventory function" is discontinuous and give an interpretation of the graph.

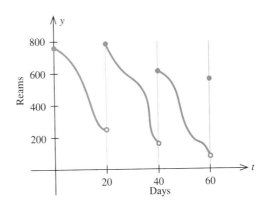

50. Learning Curves   The following graph describes the progress Michael made in solving a problem correctly during a mathematics quiz. Here y denotes the percentage of work completed and x is measured in minutes. Give an interpretation of the graph.

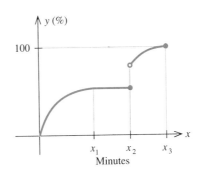

51. Ailing Financial Institutions   The Franklin Savings and Loan Company acquired two ailing financial institutions in 1992. One of them was acquired at time $t = T_1$, and the other was acquired at time $t = T_2$ ($t = 0$ corresponds to the beginning of 1992). The following graph shows the total amount of money on deposit with Franklin:

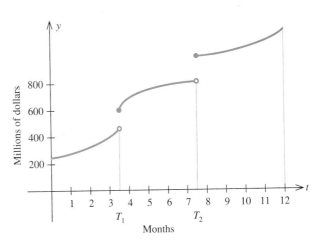

Explain the significance of the discontinuities of the function at T_1 and T_2.

52. Energy Consumption The following graph shows the amount of home heating oil remaining in a 200-gallon tank over a 120-day period ($t = 0$ corresponds to October 1):

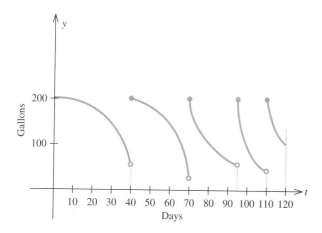

Explain why the function is discontinuous at $t = 40$, $t = 70$, $t = 95$, and $t = 110$.

53. Administration of an Intravenous Solution A dextrose solution is being administered to a patient intravenously. The 1-liter bottle holding the solution is removed and replaced by another as soon as the contents drop to approximately 5% of the initial (1-liter) amount. The rate of discharge is constant, and it takes 6 hours to discharge 95% of the contents of a full bottle. Draw a graph showing the amount of dextrose solution in a bottle in the IV system over a 24-hour period, assuming that we started with a full bottle.

54. Commissions The base salary of a salesman working on commission is $12,000. For each $50,000 of sales beyond $100,000, he is paid a $1,000 commission. Sketch a graph showing his earnings as a function of the level of his sales x. Determine the values of x for which the function f is discontinuous.

In exercises 55–58 (a) show that the function f is continuous for all values of x in the interval [a, b] and (b) prove that f must have at least one zero in the interval (a, b) by showing that f(a) and f(b) have opposite signs.

55. $f(x) = x^2 - 6x + 8$; $a = 1$, $b = 3$

56. $f(x) = x^3 - 2x^2 + 3x + 2$; $a = -1$, $b = 1$

57. $f(x) = 2x^3 - 3x^2 - 36x + 14$; $a = 0$, $b = 1$

58. $f(x) = 2x^{5/3} - 5x^{4/3}$; $a = 14$, $b = 16$

59. a. Prove that a polynomial function $y = P(x)$ is continuous at every point x. Follow these steps:
  1. Use properties 2 and 3 of continuous functions to establish that the function $g(x) = x^n$, where n is a positive integer, is continuous everywhere.
  2. Use properties 1 and 5 to show that $f(x) = cx^n$, where c is a constant and n is a positive integer, is continuous everywhere.
  3. Use property 4 to complete the proof of the result.

b. Prove that a rational function $R(x) = p(x)/q(x)$ is continuous at every point x where $q(x) \neq 0$. [*Hint:* Use the result of (a) and property 6.]

SOLUTIONS TO SELF-CHECK EXERCISES 9.2

1. For $x < -1$, $f(x) = 1$, and so

$$\lim_{x \to -1^-} f(x) = \lim_{x \to -1^-} 1 = 1$$

For $x \geq -1$, $f(x) = 1 + \sqrt{x + 1}$, and so

$$\lim_{x \to -1^+} f(x) = \lim_{x \to -1^+} (1 + \sqrt{x + 1}) = 1$$

Since the left-hand and right-hand limits of f exist as x approaches $x = -1$ and both are equal to 1, we conclude that

$$\lim_{x \to -1} f(x) = 1$$

2. a. The graph of *f* is as follows:

We see that *f* is continuous everywhere.

b. The graph of *g* is as follows:

Since *g* is not defined at $x = -1$, it is discontinuous there. It is continuous everywhere else.

Using Technology

You can very often recognize the points of discontinuity of a function *f* by examining its graph. For example, Figure 1 shows the graph of $f(x) = \dfrac{x}{x^2 - 1}$ obtained using a graphing utility. It is evident that *f* is discontinuous at $x = -1$ and $x = 1$. This observation is also borne out by the fact that both these points are not in the domain of *f*.

FIGURE 1

The graph of $f(x) = \dfrac{x}{x^2 - 1}$

in the viewing rectangle
$[-4, 4] \times [-10, 10]$

Consider the function

$$g(x) = \frac{2x^3 + x^2 - 7x - 6}{x^2 - x - 2}$$

Using a graphing utility we obtain the graph of g shown in Figure 2. An examination of this graph does not reveal any points of discontinuity. But if we factor both the numerator and the denominator of the rational expression, we see that

$$g(x) = \frac{(x + 1)(x - 2)(2x + 3)}{(x + 1)(x - 2)}$$

$$= 2x + 3$$

provided $x \neq -1$ and $x \neq 2$, so that its graph, in fact, looks like that shown in Figure 3.

FIGURE 2

The graph of $g(x) =$
$\dfrac{2x^3 + x^2 - 7x - 6}{x^2 - x - 2}$ *in the standard viewing rectangle*

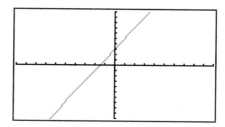

FIGURE 3

The graph of g has holes at $(-1, 1)$ and $(2, 7)$.

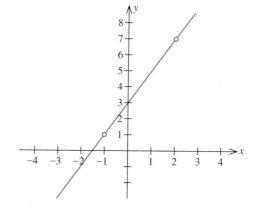

This example shows the limitation of the graphing utility and reminds us of the importance of studying functions analytically!

EXERCISES

In exercises 1–10, use a graphing utility to plot the graph of f and to spot the points of discontinuity of f. Then use analytical means to verify your observation and find all points of discontinuity.

1. $f(x) = \dfrac{2}{x^2 - x}$

2. $f(x) = \dfrac{2x + 1}{x^2 + x - 2}$

3. $f(x) = \dfrac{\sqrt{x}}{x^2 - x - 2}$

4. $f(x) = \dfrac{3}{\sqrt{x}(x + 1)}$

5. $f(x) = \dfrac{6x^3 + x^2 - 2x}{2x^2 - x}$

6. $f(x) = \dfrac{2x^3 - x^2 - 13x - 6}{2x^2 - 5x - 3}$

7. $f(x) = \dfrac{2x^4 - 3x^3 - 2x^2}{2x^2 - 3x - 2}$

8. $f(x) = \dfrac{6x^4 - x^3 + 5x^2 - 1}{6x^2 - x - 1}$

9. $f(x) = \dfrac{x^3 + x^2 - 2x}{x^4 + 2x^3 - x - 2}$

[*Hint*: $x^4 + 2x^3 - x - 2 = (x^3 - 1)(x + 2)$]

10. $f(x) = \dfrac{x^3 - x}{x^{4/3} - x + x^{1/3} - 1}$

[*Hint*: $x^{4/3} - x + x^{1/3} - 1 = (x^{1/3} - 1)(x + 1)$]

Can you explain why part of the graph is missing?

| 9.3 | ## The Derivative |

An Intuitive Example

We mentioned in Section 9.1 that the problem of finding the *rate of change* of one quantity with respect to another is mathematically equivalent to the problem of finding the *slope of the tangent line* to a curve at a given point on the curve. Before going on to establish this relationship, let us show its plausibility by looking at it from an intuitive point of view.

Consider the motion of the car discussed in Section 9.1. Recall that the position of the car at any time t is given by

$$s = f(t) = 2t^2 \qquad 0 \le t \le 10$$

where s is measured in feet and t in seconds. The graph of the function f is sketched in Figure 20.

Observe that the graph of f rises slowly at first but more rapidly as t increases, reflecting the fact that the speed of the car is increasing with time. This observation suggests a relationship between the speed of the car at any time t and the *steepness* of the curve at the point corresponding to this value of t. Thus it would appear that we can solve the problem of finding the speed of the car at any time if we can find a way to measure the steepness of the curve at any point on the curve.

To discover a yardstick that will measure the steepness of a curve, consider the graph of a function f such as the one shown in Figure 21a. Think of the curve as representing a stretch of roller coaster track (Figure 21b). When the car is at the point P on the curve, a passenger sitting erect in the car and looking straight ahead will have a line of sight that is parallel to the line T, the tangent to the curve at P.

FIGURE 20

Graph showing the position s of a car at time t

FIGURE 21

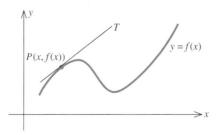

(a) T is the tangent line to the curve at P.

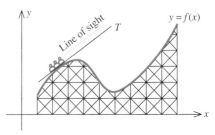

(b) T is parallel to the line of sight.

As Figure 21a suggests, the steepness of the curve—that is, the rate at which y is increasing or decreasing with respect to x—is given by the slope of the tangent line to the graph of f at the point $P(x, f(x))$. But for now we will show how this relationship can be used to estimate the rate of change of a function from its graph.

EXAMPLE 1   The graph of the function $y = N(t)$, shown in Figure 22, gives the number of social security beneficiaries from the beginning of 1990 ($t = 0$) through the year 2045 ($t = 55$). Use the graph of $y = N(t)$ to estimate the rate at which the number of social security beneficiaries will be growing at the beginning of the year 2000 ($t = 10$) and at the beginning of 2025 ($t = 35$). (Assume that the rate of change of the function N at any value of t is given by the slope of the tangent line at the point $P(t, N(t))$.)

FIGURE 22

The number of social security beneficiaries from 1990 through 2045. We can use the slope of the tangent line at the indicated points to estimate the rate at which the number of social security beneficiaries will be changing.

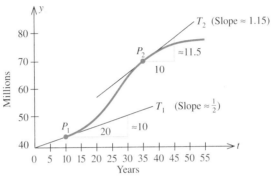

Source: Social Security Administration

Solution   From Figure 22, we see that the slope of the tangent line T_1 to the graph of $y = N(t)$ at $P_1(10, 44.7)$ is approximately 0.5. This tells us that the quantity y is increasing at the rate of 1/2 unit per unit increase in t, when $t = 10$. In other words, at the beginning of the year 2000, the number of social security beneficiaries will be increasing at the rate of approximately 0.5 million, or 500,000, per year.

     The slope of the tangent line T_2 at $P_2(35, 71.9)$ is approximately 1.15. This tells us that at the beginning of 2025, the number of social security beneficiaries will be growing at the rate of approximately 1.15 million, or 1,150,000, per year.   ■

Slope of a Tangent Line

In Example 1, we answered the questions raised by drawing the graph of the function N and estimating the position of the tangent lines. Ideally, however, we would like to solve a problem analytically whenever possible. In order to do this we need a precise definition of the slope of a tangent line to a curve.

     To define the tangent line to a curve C at a point P on the curve, fix P and let Q be any point on C distinct from P (Figure 23). The straight line passing through P and Q is called a **secant line**. Now, as the point Q is allowed to move toward P along the curve, the secant line through P and Q rotates about the fixed point P and approaches a fixed line through P. This fixed line, which is the limiting position of the secant lines through

FIGURE 23

As Q approaches P along the curve C, the secant lines approach the tangent line T.

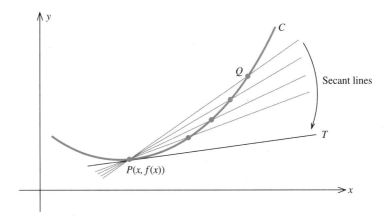

P and Q as Q approaches P, is the **tangent line** to the graph of f at the point P.

We can describe the process more precisely as follows. Suppose the curve C is the graph of a function f defined by $y = f(x)$. Then the point P is described by $P(x, f(x))$ and the point Q by $Q(x + h, f(x + h))$, where h is some appropriate nonzero number (Figure 24a). Observe that we can make Q approach P along the curve C by letting h approach zero (Figure 24b).

FIGURE 24

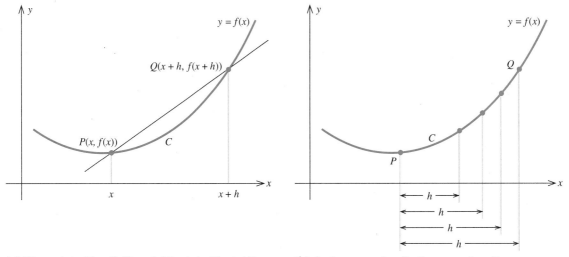

(a) The points $P(x, f(x))$ and $Q(x + h, f(x + h))$ **(b)** As h approaches 0, Q approaches P.

Next, using the formula for the slope of a line, we can write the slope of the secant line passing through $P(x, f(x))$ and $Q(x + h, f(x + h))$ as

$$\frac{f(x + h) - f(x)}{(x + h) - x} = \frac{f(x + h) - f(x)}{h} \tag{3}$$

As observed earlier, Q approaches P and, therefore, the secant line through P and Q approaches the tangent line T as h approaches zero. Consequently, we might expect that the slope of the secant line would approach the slope of the tangent line T as h approaches zero. This leads to the following definition.

SLOPE OF A TANGENT LINE

The slope of the tangent line to the graph f at the point $P(x, f(x))$ is given by

$$\lim_{h \to 0} \frac{f(x+h) - f(x)}{h} \qquad (4)$$

if it exists.

Rates of Change

We will now show that the problem of finding the slope of the tangent line to the graph of a function f at the point $P(x, f(x))$ is mathematically equivalent to the problem of finding the rate of change of f at x. To see this, suppose that we are given a function f that describes the relationship between the two quantities x and y:

$$y = f(x)$$

The number $f(x + h) - f(x)$ measures the change in y that corresponds to a change h in x (Figure 25).

FIGURE 25
$f(x + h) - f(x)$ is the change in y that corresponds to a change h in x.

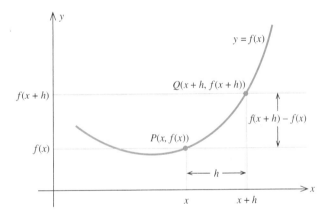

Then, the **difference quotient**

$$\frac{f(x+h) - f(x)}{h} \qquad (5)$$

measures the *average rate of change of y with respect to x* over the interval $[x, x + h]$. For example, if y measures the position of a car at time x, then (5) gives the average velocity of the car over the time interval $[x, x + h]$.

Observe that the difference quotient (5) is the same as equation (3). We conclude that the difference quotient (5) also measures the slope of the secant line that passes through the two points $P(x, f(x))$ and $Q(x + h, f(x + h))$ lying on the graph of $y = f(x)$. Next, by taking the limit of the difference quotient (5) as h goes to zero—that is, by evaluating

$$\lim_{h \to 0} \frac{f(x + h) - f(x)}{h} \qquad (6)$$

we obtain the rate of change of f at x. For example, if y measures the position of a car at time x, then (6) gives the velocity of the car at time x. For emphasis, the rate of change of a function f at x is often called the *instantaneous rate of change of f at x*. This distinguishes it from the average rate of change of f, which is computed over an *interval* $[x, x + h]$ rather than at a *point x*.

Observe that the limit (6) is the same as limit (4). Therefore, the limit of the difference quotient also measures the slope of the tangent line to the graph of $y = f(x)$ at the point $(x, f(x))$.

The following summarizes this discussion.

AVERAGE AND INSTANTANEOUS RATES OF CHANGE

Average rate of change of f over the interval $[x, x + h]$ or slope of the secant line to the graph of f through the points $(x, f(x))$ and $(x + h, f(x + h))$, is

$$\frac{f(x + h) - f(x)}{h}. \qquad (7)$$

Instantaneous rate of change of f at x, or *slope of the tangent line* to the graph of f at $(x, f(x))$, is

$$\lim_{h \to 0} \frac{f(x + h) - f(x)}{h} \qquad (8)$$

The Derivative

The limit (4), or (8), which measures both the slope of the tangent line to the graph of $y = f(x)$ at the point $P(x, f(x))$ and the (instantaneous) rate of change of f at x is given a special name: the *derivative of f at x*.

THE DERIVATIVE OF A FUNCTION

The derivative of a function f with respect to x is the function f' (read "f prime of x"), defined by

$$f'(x) = \lim_{h \to 0} \frac{f(x + h) - f(x)}{h}. \qquad (9)$$

The domain of f' is the set of all x where the limit exists.

Thus, the derivative of a function f is a function f' that gives the slope of the tangent line to the graph of f at *any* point $(x, f(x))$, and also the rate of change of f at x (Figure 26).

Other notations for the derivative of f include:

$$D_x f(x) \qquad \text{Read "dee sub x of f of x"}$$

$$\frac{dy}{dx} \qquad \text{Read "dee y dee x"}$$

FIGURE 26
The slope of the tangent line at $P(x, f(x))$ is $f'(x)$; f changes at the rate of $f'(x)$ units per unit change in x at x.

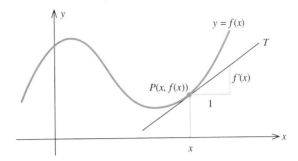

and y' Read "y prime"

The last two are used when the rule for f is written in the form $y = f(x)$.

The calculation of the derivative of f is facilitated using the following four-step process.

THE FOUR-STEP PROCESS FOR FINDING $f'(x)$

1. Compute $f(x + h)$.
2. Form the difference $f(x + h) - f(x)$.
3. Form the quotient $\dfrac{f(x + h) - f(x)}{h}$.
4. Compute $f'(x) = \lim\limits_{h \to 0} \dfrac{f(x + h) - f(x)}{h}$.

EXAMPLE 2   Find the slope of the tangent line to the graph of $f(x) = 3x + 5$ at any point $(x, f(x))$.

Solution   The slope of the tangent line at any point on the graph of f is given by the derivative of f at x. To find the derivative, we use the four-step process:

Step 1 $f(x + h) = 3(x + h) + 5 = 3x + 3h + 5$
Step 2 $f(x + h) - f(x) = (3x + 3h + 5) - (3x + 5) = 3h$
Step 3 $\dfrac{f(x + h) - f(x)}{h} = \dfrac{3h}{h} = 3$
Step 4 $f'(x) = \lim\limits_{h \to 0} \dfrac{f(x + h) - f(x)}{h} = \lim\limits_{h \to 0} 3 = 3$

We expect this result, since the tangent line to any point on a straight line must coincide with the line itself and, therefore, must have the same slope as the line. In this case, the graph of f is a straight line with slope 3.

EXAMPLE 3   Let $f(x) = x^2 - 4x$.

a. Compute $f'(x)$.
b. Find the point on the graph of f where the tangent line to the curve is horizontal.

FIGURE 27

The tangent line to the graph of $y = x^2 - 4x$ at $(2, -4)$ is $y = -4$.

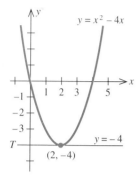

c. Sketch the graph of f and the tangent line to the curve at the point found in (b).

d. What is the rate of change of f at this point?

Solution

a. To find $f'(x)$, we use the four-step process.

Step 1 $f(x + h) = (x + h)^2 - 4(x + h) = x^2 + 2xh + h^2 - 4x - 4h$

Step 2 $f(x + h) - f(x) = x^2 + 2xh + h^2 - 4x - 4h - (x^2 - 4x)$

$$= 2xh + h^2 - 4h = h(2x + h - 4)$$

Step 3 $\dfrac{f(x + h) - f(x)}{h} = \dfrac{h(2x + h - 4)}{h} = 2x + h - 4$

Step 4 $f'(x) = \lim_{h \to 0} \dfrac{f(x + h) - f(x)}{h} = \lim_{h \to 0} (2x + h - 4) = 2x - 4$

b. At a point on the graph of f where the tangent line to the curve is horizontal and hence has slope zero, the derivative f' of f is zero. Accordingly, to find such point(s) we set $f'(x) = 0$, which gives $2x - 4 = 0$, or $x = 2$. The corresponding value of y is given by $y = f(2) = -4$, and the required point is $(2, -4)$.

c. The graph of f and the tangent line are shown in Figure 27.

d. The rate of change of f at $x = 2$ is zero.

EXAMPLE 4   Let $f(x) = \dfrac{1}{x}$.

a. Compute $f'(x)$.

b. Find the slope of the tangent line T to the graph of f at the point where $x = 1$.

c. Find an equation of the tangent line T in (b).

Solution

a. To find $f'(x)$, we use the four-step process.

Step 1 $f(x + h) = \dfrac{1}{x + h}$

Step 2 $f(x + h) - f(x) = \dfrac{1}{x + h} - \dfrac{1}{x} = \dfrac{x - (x + h)}{x(x + h)} = -\dfrac{h}{x(x + h)}$

Step 3 $\dfrac{f(x + h) - f(x)}{h} = -\dfrac{h}{x(x + h)} \cdot \dfrac{1}{h} = -\dfrac{1}{x(x + h)}$

Step 4 $f'(x) = \lim_{h \to 0} \dfrac{f(x + h) - f(x)}{h} = \lim_{h \to 0} -\dfrac{1}{x(x + h)} = -\dfrac{1}{x^2}$

FIGURE 28

The tangent line to the graph of $f(x) = \dfrac{1}{x}$ at $(1, 1)$

b. The slope of the tangent line T to the graph of f where $x = 1$ is given by $f'(1) = -1$.

c. When $x = 1$, $y = f(1) = 1$ and T is tangent to the graph of f at the point $(1, 1)$. From (b), we know that the slope of T is -1. Thus, an equation of T is

$$y - 1 = -1(x - 1)$$

or $y = -x + 2$ Figure 28

Applications

EXAMPLE 5   Suppose that the distance (in feet) covered by a car moving along a straight road t seconds after starting from rest is given by the function $f(t) = 2t^2$ $(0 \le t \le 30)$.

a. Calculate the average velocity of the car over the time intervals [22, 23], [22, 22.1], and [22, 22.01].

b. Calculate the (instantaneous) velocity of the car when $t = 22$.

c. Compare the results obtained in (a) with that obtained in (b).

Solution

a. We first compute the average velocity (average rate of change of f) over the interval $[t, t + h]$ using (7). We find

$$\frac{f(t + h) - f(t)}{h} = \frac{2(t + h)^2 - 2t^2}{h}$$

$$= \frac{2t^2 + 4th + 2h^2 - 2t^2}{h}$$

$$= 4t + 2h$$

Next, using $t = 22$ and $h = 1$, we find that the average velocity of the car over the time interval [22, 23] is

$$4(22) + 2(1) = 90$$

or 90 feet per second. Similarly, using $t = 22$, $h = 0.1$, and $h = 0.01$, we find that its average velocities over the time intervals [22, 22.1] and [22, 22.01] are 88.2 and 88.02 feet per second, respectively.

b. Using the limit (8), we see that the instantaneous velocity of the car at any time t is given by

$$\lim_{h \to 0} \frac{f(t + h) - f(t)}{h} = \lim_{h \to 0} (4t + 2h) \qquad \text{Using the results from (a) or using the four-step process}$$

$$= 4t$$

In particular, the velocity of the car 22 seconds from rest ($t = 22$) is given by

$$v = 4(22)$$

or 88 feet per second.

c. The computations in (a) show that as the time intervals over which the average velocity of the car are computed become smaller and smaller, the average velocities over these intervals do approach 88 ft/sec, the instantaneous velocity of the car at $t = 22$.   ∎

EXAMPLE 6   The management of the Titan Tire Company has determined that the weekly demand function for their Super Titan tires is given by

$$p = f(x) = 144 - x^2$$

where p is measured in dollars and x is measured in units of a thousand (Figure 29).

FIGURE 29

The graph of the demand function $p = 144 - x^2$

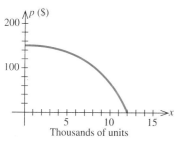

a. Find the average rate of change in the unit price of a tire if the quantity demanded is between 5000 and 6000 tires. Between 5000 and 5100 tires. Between 5000 tires and 5010 tires.

b. What is the instantaneous rate of change of the unit price when the quantity demanded is 5000 units? Give an interpretation of the result.

Solution

a. The average rate of change of the unit price of a tire if the quantity demanded is between x and $x + h$ is

$$\frac{f(x + h) - f(x)}{h} = \frac{[144 - (x + h)^2] - (144 - x^2)}{h}$$

$$= \frac{144 - x^2 - 2x(h) - h^2 - 144 + x^2}{h}$$

$$= -2x - h$$

To find the average rate of change of the unit price of a tire when the quantity demanded is between 5000 and 6000 tires (that is, over the interval [5, 6]), we take $x = 5$ and $h = 1$, obtaining

$$-2(5) - 1 = -11$$

or $-\$11$ per 1000 tires. (Remember, x is measured in units of a thousand.) Similarly, taking $h = 0.1$, and $h = 0.01$ with $x = 5$, we find that the average rates of change of the unit price when the quantities demanded are between 5000 and 5100 and between 5000 and 5010 are $-\$10.10$ and $-\$10.01$ per 1000 tires, respectively.

b. The instantaneous rate of change of the unit price of a tire when the quantity demanded is x units is given by

$$\lim_{h \to 0} \frac{f(x + h) - f(x)}{h} = \lim_{h \to 0} (-2x - h) \qquad \text{Using the results from (a)}$$

$$= -2x$$

In particular, the instantaneous rate of change of the unit price per tire when the quantity demanded is 5000 is given by $-2(5)$, or $-\$10$ per 1000 tires. Our result tells us that at the level when the quantity demanded is 5000 tires, the unit price of a tire is dropping at the rate of $\$10$ per thousand tires.   ◼

The derivative of a function provides us with a tool for measuring the rate of change of one quantity with respect to another. Table 3, at the top of page 513, lists several other applications involving this limit.

Differentiability and Continuity

In practical applications, one encounters functions that fail to be **differentiable**—that is, do not have a derivative at certain values in the domain of the function f. It can be shown that a continuous function f fails to be differentiable at a point $x = a$ when the graph of f makes an abrupt change of direction at that point. We call such a point a "corner." A function also

TABLE 3

x stands for	y stands for	$\dfrac{f(a + h) - f(a)}{h}$ measures the	$\displaystyle\lim_{h \to 0} \dfrac{f(a + h) - f(a)}{h}$ measures the
time	*concentration of a drug* in the bloodstream at time x	average rate of change in the concentration of the drug over the time interval $[a, a + h]$	instantaneous rate of change in the concentration of the drug in the bloodstream at time $x = a$
number of items sold	*revenue* at a sales level of x units	average rate of change in the revenue when the sales level is between $x = a$ and $x = a + h$	instantaneous rate of change in the revenue when the sales level is a units
time	*volume of sales* at time x	average rate of change in the volume of sales over the time interval $[a, a + h]$	instantaneous rate of change in the volume of sales at time $x = a$
time	*population* of *Drosophila* (fruit flies) at time x	average rate of growth of the fruit fly population over the time interval $[a, a + h]$	instantaneous rate of change of the fruit fly population at time $x = a$
temperature in a chemical reaction	*amount of product formed in the chemical reaction* when the temperature is x degrees	average rate of formation of chemical product over the temperature range $[a, a + h]$	instantaneous rate of formation of chemical product when the temperature is a degrees

fails to be differentiable at a point where the tangent line is vertical, since the slope of a vertical line is undefined. These cases are illustrated in Figure 30.

FIGURE 30

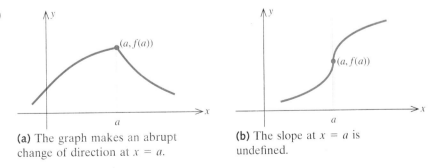

(a) The graph makes an abrupt change of direction at $x = a$.

(b) The slope at $x = a$ is undefined.

The next example illustrates a function that is not differentiable at a point.

EXAMPLE 7   Mary works at the B&O department store, where, on a weekday, she is paid $6 per hour for the first 8 hours and $9 per hour for

FIGURE 31
The function f is not differentiable at (8, 48).

overtime. The function

$$f(x) = \begin{cases} 6x & 0 \le x \le 8 \\ 9x - 24 & 8 < x \end{cases}$$

gives Mary's earnings on a weekday in which she worked x hours. Sketch the graph of the function f and explain why it is not differentiable at $x = 8$.

Solution The graph of f is shown in Figure 31. Observe that the graph of f has a corner at $x = 8$ and consequently is not differentiable at $x = 8$. ∎

We close this section by mentioning the connection between the continuity and the differentiability of a function at a given value $x = a$ in the domain of f. By reexamining the function of Example 7, it becomes clear that f is continuous everywhere and, in particular, when $x = 8$. This shows that, in general, the continuity of a function at a point $x = a$ does not necessarily imply the differentiability of the function at that point. The converse, however, is true: If a function f is differentiable at a point $x = a$, then it is continuous there.

DIFFERENTIABILITY AND CONTINUITY

If a function is differentiable at $x = a$, then it is continuous at $x = a$.

For a proof of this result, see exercise 43, page 518.

EXAMPLE 8 Figure 32 depicts a portion of the graph of a function. Explain why the function fails to be differentiable at each of the points $x = a, b, c, d, e, f$, and g.

FIGURE 32
The graph of this function is not differentiable at the points a–g.

Solution The function fails to be differentiable at the points $x = a, b$, and c because it is discontinuous at each of these points. The derivative of the function does not exist at $x = d, e$, and f because it has a kink at each of these points. Finally, the function is not differentiable at $x = g$ because the tangent line is vertical at that point. ∎

SELF-CHECK
EXERCISES 9.3

1. Let $f(x) = -x^2 - 2x + 3$.
   a. Find the derivative f' of f using the definition of the derivative.
   b. Find the slope of the tangent line to the graph of f at the point $(0, 3)$.
   c. Find the rate of change of f when $x = 0$.
   d. Find an equation of the tangent line to the graph of f at the point $(0, 3)$.
   e. Sketch the graph of f and the tangent line to the curve at the point $(0, 3)$.

2. The losses (in millions of dollars) due to bad loans extended chiefly in agriculture, real estate, shipping, and energy by the Franklin Bank are estimated to be

$$A = f(t) = -t^2 + 10t + 30 \qquad 0 \le t \le 10$$

where t is the time in years ($t = 0$ corresponds to the beginning of 1988). How fast were the losses mounting at the beginning of 1991? At the beginning of 1993? How fast will the losses be mounting at the beginning of 1995? Interpret your results.

Solutions to Self-Check Exercises 9.3 can be found on page 518.

9.3 EXERCISES

1. **Average Weight of an Infant**   The following graph shows the weight measurements of the average infant from the time of birth ($t = 0$) through age two ($t = 24$). By computing the slopes of the respective tangent lines, estimate the rate of change of the average infant's weight when $t = 3$ and when $t = 18$. What is the average rate of change in the average infant's weight over its first year of life?

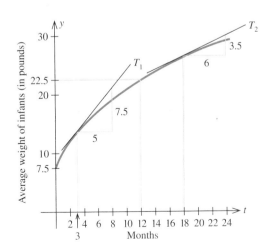

2. **Forestry**   The following graph shows the volume of wood produced in a single-species forest. Here $f(t)$

is measured in cubic meters per hectare and t is measured in years. By computing the slopes of the respective tangent lines, estimate the rate at which the wood grown is changing at the beginning of the 10th year and at the beginning of the 30th year.

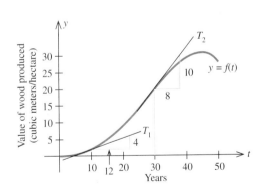

3. **TV Viewing Patterns**   The following graph, based on data supplied by the A. C. Nielsen Company, shows the percentage of U.S. households watching television during a 24-hr period on a weekday ($t = 0$ corresponds to 6 A.M.). By computing the slopes of the respective tangent lines, estimate the rate of change of the percentage of households watching television at 4 P.M. and 11 P.M.

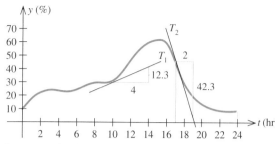

Source: A. C. Nielsen Company

4. Crop Yield Productivity and yield of cultivated crops are often reduced by insect pests. The following graph shows the relationship between the yield of a certain crop, $f(x)$, as a function of the density of aphids x. (Aphids are small insects that suck plant juices.) Here $f(x)$ is measured in kg/4000 sq meters, and x is measured in hundreds of aphids per bean stem. By computing the slopes of the respective tangent lines, estimate the rate of change of the crop yield with respect to the density of aphids when that density is 200 aphids per bean stem. When it is 800 aphids per bean stem.

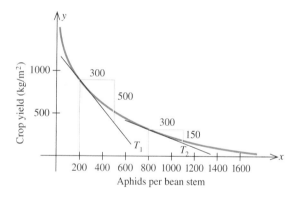

Aphids per bean stem

In exercises 5–12, use the four-step process to find the slope of the tangent line to the graph of the given function at any point.

5. $f(x) = 13$

6. $f(x) = -6$

7. $f(x) = 2x + 7$

8. $f(x) = 8 - 4x$

9. $f(x) = 3x^2$

10. $f(x) = -\frac{1}{2}x^2$

11. $f(x) = -x^2 + 3x$

12. $f(x) = 2x^2 + 5x$

In exercises 13–18, find the slope of the tangent line to the graph of each function at the given point and determine an equation of the tangent line.

13. $f(x) = 2x + 7$ at $(2, 11)$

14. $f(x) = -3x + 4$ at $(-1, 7)$

15. $f(x) = 3x^2$ at $(1, 3)$

16. $f(x) = 3x - x^2$ at $(-2, -10)$

17. $f(x) = -\frac{1}{x}$ at $\left(3, -\frac{1}{3}\right)$

18. $f(x) = \frac{3}{2x}$ at $\left(1, \frac{3}{2}\right)$

19. Let $f(x) = 2x^2 + 1$.
   a. Find the derivative f' of f.
   b. Find an equation of the tangent line to the curve at the point $(1, 3)$.
   c. Sketch the graph of f.

20. Let $f(x) = x^2 + 6x$.
   a. Find the derivative f' of f.
   b. Find the point on the graph of f where the tangent line to the curve is horizontal. [*Hint:* Find the value of x for which $f'(x) = 0$.]
   c. Sketch the graph of f and the tangent line to the curve at the point found in (b).

21. Let $f(x) = x^2 - 2x + 1$.
   a. Find the derivative f' of f.
   b. Find the point on the graph of f where the tangent line to the curve is horizontal.
   c. Sketch the graph of f and the tangent line to the curve at the point found in (b).
   d. What is the rate of change of f at this point?

22. Let $f(x) = \dfrac{1}{x - 1}$.
   a. Find the derivative f' of f.
   b. Find an equation of the tangent line to the curve at the point $(-1, -\frac{1}{2})$.
   c. Sketch the graph of f.

23. Let $y = f(x) = x^2 + x$.
   a. Find the average rate of change of y with respect to x in the interval from $x = 2$ to $x = 3$. In the interval from $x = 2$ to $x = 2.5$. In the interval from $x = 2$ to $x = 2.1$.
   b. Find the (instantaneous) rate of change of y at $x = 2$.
   c. Compare the results obtained in (a) with that of (b).

24. Let $y = f(x) = x^2 - 4x$.
   a. Find the average rate of change of y with respect to x in the interval from $x = 3$ to $x = 4$. In the interval from $x = 3$ to $x = 3.5$. In the interval from $x = 3$ to $x = 3.1$.
   b. Find the (instantaneous) rate of change of y at $x = 3$.
   c. Compare the results obtained in (a) with that of (b).

25. Velocity of a Car   Suppose that the distance s (in feet) covered by a car moving along a straight road t seconds after starting from rest is given by the function $f(t) = 2t^2 + 48t$.

   a. Calculate the average velocity of the car over the time intervals [20, 21], [20, 20.1], and [20, 20.01].

   b. Calculate the (instantaneous) velocity of the car when $t = 20$.

   c. Compare the results of (a) with that of (b).

26. Velocity of a Ball Thrown into the Air   A ball is thrown straight up with an initial velocity of 128 ft/sec, so that its height (in feet) after t seconds is given by $s(t) = 128t - 16t^2$.

   a. What is the average velocity of the ball over the time intervals [2, 3], [2, 2.5], and [2, 2.1]?

   b. What is the instantaneous velocity at time $t = 2$?

   c. What is the instantaneous velocity at time $t = 5$? Is the ball rising or falling at this time?

   d. When will the ball hit the ground?

27. Cost of Producing Surfboards   The total cost $C(x)$ (in dollars) incurred by the Aloha Company in manufacturing x surfboards a day is given by

$$C(x) = -10x^2 + 300x + 130 \qquad 0 \le x \le 15$$

   a. Find $C'(x)$.

   b. What is the rate of change of the total cost when the level of production is ten surfboards a day?

   c. What is the average cost Aloha incurs in manufacturing ten surfboards a day?

28. Effect of Advertising on Profit   The quarterly profit of Cunningham Realty (in thousands of dollars) is given by

$$P(x) = -\frac{1}{3}x^2 + 7x + 30 \quad 0 \le x \le 50$$

where x (in thousands of dollars) is the amount of money Cunningham spends on advertising per quarter.

   a. Find $P'(x)$.

   b. What is the rate of change of Cunningham's quarterly profit if the amount it spends on advertising is $10,000 per quarter ($x = 10$)? $30,000 per quarter ($x = 30$)?

29. A Country's GDP   The gross domestic product (GDP) of a certain country is projected to be

$$N(t) = t^2 + 2t + 50 \qquad 0 \le t \le 5$$

billion dollars t years from now. What will the rate of change of the country's GDP be two years from now? Four years from now?

30. Growth of Bacteria   Under a set of controlled laboratory conditions, the size of the population of a certain bacteria culture at time t (in minutes) is described by the function

$$P = f(t) = 3t^2 + 2t + 1$$

Find the rate of population growth at $t = 10$ minutes.

In exercises 31–34, let x and f(x) represent the given quantities. Fix x = a and let h be a small positive number. Give an interpretation of the quantities

$$\frac{f(a + h) - f(a)}{h} \qquad and \qquad \lim_{h \to 0} \frac{f(a + h) - f(a)}{h}$$

31. x denotes time and $f(x)$ denotes the population of seals at time x.

32. x denotes time and $f(x)$ denotes the prime interest rate at time x.

33. x denotes time and $f(x)$ denotes a country's industrial production.

34. x denotes the level of production of a certain commodity and $f(x)$ denotes the total cost incurred in producing x units of the commodity.

In each of exercises 35–40, the graph of a function is shown. For each function, state whether or not (a) $f(x)$ has a limit at $x = a$, (b) $f(x)$ is continuous at $x = a$, and (c) $f(x)$ is differentiable at $x = a$. Justify your answers.

35.   **36.**

37.   **38.**

39.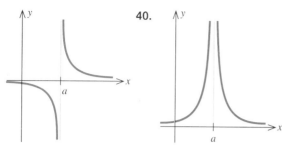

40.

41. Sketch the graph of the function $f(x) = |x + 1|$ and show that the function does not have a derivative at $x = -1$.

42. Sketch the graph of the function $f(x) = 1/(x - 1)$ and show that the function does not have a derivative at $x = 1$.

43. Show that if a function f is differentiable at a point $x = a$, then f must be continuous at that point. *Hint:* Write

$$f(x) - f(a) = \left[\frac{f(x) - f(a)}{x - a} \right] (x - a)$$

Use the product rule for limits and the definition of the derivative to show that

$$\lim_{x \to a} [f(x) - f(a)] = 0$$

SOLUTIONS TO SELF-CHECK EXERCISES 9.3

1. a. $f'(x) = \lim_{h \to 0} \dfrac{f(x + h) - f(x)}{h}$

$$= \lim_{h \to 0} \frac{[-(x + h)^2 - 2(x + h) + 3] - (-x^2 - 2x + 3)}{h}$$

$$= \lim_{h \to 0} \frac{-x^2 - 2xh - h^2 - 2x - 2h + 3 + x^2 + 2x - 3}{h}$$

$$= \lim_{h \to 0} \frac{h(-2x - h - 2)}{h}$$

$$= \lim_{h \to 0} (-2x - h - 2) = -2x - 2$$

b. From the result of (a), we see that the slope of the tangent line to the graph of f at any point $(x, f(x))$ is given by

$$f'(x) = -2x - 2$$

In particular, the slope of the tangent line to the graph of f at (0, 3) is

$$f'(0) = -2$$

c. The rate of change of f when $x = 0$ is given by $f'(0) = -2$, or -2 units per unit change in x.

d. Using the result from (b), we see that an equation of the required tangent line is

$$y - 3 = -2(x - 0)$$

or

$$y = -2x + 3$$

e.

2. The rate of change of the losses at any time t is given by

$$f'(t) = \lim_{h \to 0} \frac{f(t+h) - f(t)}{h}$$

$$= \lim_{h \to 0} \frac{[-(t+h)^2 + 10(t+h) + 30] - (-t^2 + 10t + 30)}{h}$$

$$= \lim_{h \to 0} \frac{-t^2 - 2th - h^2 + 10t + 10h + 30 + t^2 - 10t - 30}{h}$$

$$= \lim_{h \to 0} \frac{h(-2t - h + 10)}{h}$$

$$= \lim_{h \to 0} (-2t - h + 10)$$

$$= -2t + 10.$$

Therefore, the rate of change of the losses suffered by the bank at the beginning of 1991 ($t = 3$) was

$$f'(3) = -2(3) + 10 = 4$$

that is, the losses were increasing at the rate of $4 million per year. At the beginning of 1993 ($t = 5$),

$$f'(5) = -2(5) + 10 = 0$$

and we see that the growth in losses due to bad loans was zero at this point in time. At the beginning of 1995 ($t = 7$),

$$f'(7) = -2(7) + 10 = -4$$

and we conclude that the losses will be decreasing at the rate of $4 million per year.

Using Technology

Graphing a Function and Its Tangent Lines

We can use a graphing utility to plot the graph of a function f and the tangent line at any point on the graph.

EXAMPLE 1   Let $f(x) = x^2 - 4x$.

a. Find an equation of the tangent line to the graph of f at the point $(3, -3)$.

b. Plot both the graph of f and the tangent line found in (a) on the same set of axes.

Solution

a. The slope of the tangent line at any point on the graph of f is given by $f'(x)$. But from Example 3 (page 509), we find $f'(x) = 2x - 4$. Using this result, we see that the slope of the required tangent line is

$$f'(3) = 2(3) - 4 = 2$$

Finally, using the point-slope form of the equation of a line, we find that an equation of the tangent line is

$$y - (-3) = 2(x - 3)$$
$$y + 3 = 2x - 6$$

or $y = 2x - 9$.

b. The graph of f in the standard viewing rectangle and the tangent line of interest are shown in Figure 1.

FIGURE 1
The graph of $f(x) = x^2 - 4x$ and the tangent line $y = 2x - 9$ in the standard viewing rectangle

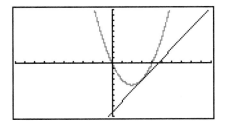

Finding the Derivative of a Function at a Given Point

The **dy/dx** operation in a graphing utility can be used to give an approximate value of the derivative of a function for a given value of x. (See Section A.11 in Appendix A.)

EXAMPLE 2 Let $f(x) = \sqrt{x}$.

a. Use the **dy/dx** operation to find the derivative of f at $(4, 2)$.

b. Find an equation of the tangent line to the graph of f at $(4, 2)$.

c. Plot the graph of f and the tangent line on the same set of axes.

Solution

a. If we plot the graph of f in the standard viewing rectangle, then the cursor can be moved to the x-value 3.9682539683 or 4.126984127. Since we want the derivative of f when $x = 4$, we use **ZIN** repeatedly to enable us to move the cursor sufficiently close to the point $(4, 2)$ on the graph of f. For example, zooming in three times and with the cursor at the point $(3.9998759921, 1.9997479839)$ (you might not obtain the same point!), we find, upon using the **dy/dx** operation, that $f'(4) \approx 0.24998450045$. Therefore, we may take $f'(4) = \frac{1}{4}$.

b. An equation of the required tangent line is

$$y - 2 = \frac{1}{4}(x - 4)$$

or $y = \frac{1}{4}x + 1$.

c. The graph of f and the tangent line in the viewing rectangle $[0, 15] \times [0, 4]$ is shown in Figure 2.

FIGURE 2
The graph of $f(x) = \sqrt{x}$ and the tangent line $y = \dfrac{1}{4}x + 1$
in the viewing rectangle
$[0, 15] \times [0, 4]$

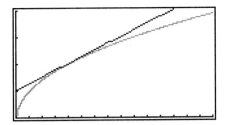

We can also use the **der1** function of a graphing utility to calculate the first derivative of a function at a given point (with answer exact to 14 digits) without graphing the function. (See Section A.11 in Appendix A.)

EXAMPLE 3   Use the **der1** function to find the derivative of $f(x) = \sqrt{x}$ at the point $(4, 2)$.

Solution   We find the value of f' at $x = 4$ to be

$$f'(x) = \text{der1}\,(x\hat{\ }.5, x, 4) = 0.25$$

as obtained in Example 2.

EXERCISES

In exercises 1–10, (a) find an equation of the tangent line to the graph of f at the indicated point, and (b) use a graphing utility to plot the graph of f and the tangent line on the same set of axes. Use a suitable viewing rectangle.

1. $f(x) = 4x - 3$; $(2, 5)$

2. $f(x) = -2x + 5$; $(1, 3)$

3. $f(x) = 2x^2 + x$; $(-2, 6)$

4. $f(x) = -x^2 + 2x$; $(1, 1)$

5. $f(x) = 2x^2 + x - 3$; $(2, 7)$

6. $f(x) = -3x^2 + 2x - 1$; $(1, -2)$

7. $f(x) = x + \dfrac{1}{x}$; $(1, 2)$

8. $f(x) = x - \dfrac{1}{x}$; $(1, 0)$

9. $f(x) = \sqrt{x}$; $(4, 2)$

10. $f(x) = \dfrac{1}{\sqrt{x}}$; $\left(4, \dfrac{1}{2}\right)$

*In exercises 11–20, (a) use the **dy/dx** operation of a graphing utility to find the derivative of f for the given value of x (to two desired places of accuracy), (b) find an equation of the tangent line to the graph of f at the indicated point, and (c) plot the graph of f and the tangent line on the same set of axes. Use a suitable viewing rectangle.*

11. $f(x) = x^3 + x + 1$; $x = 1$; $(1, 3)$

12. $f(x) = -2x^3 + 3x^2 + 2$; $x = -1$; $(-1, 7)$

13. $f(x) = x^4 - 3x^2 + 1$; $x = 2$; $(2, 5)$

14. $f(x) = -x^4 + 3x + 1$; $x = 1$; $(1, 3)$

15. $f(x) = x - \sqrt{x}$; $x = 4$; $(4, 2)$

16. $f(x) = x^{3/2} - x$; $x = 4$; $(4, 4)$

17. $f(x) = \dfrac{1}{x + 1}$; $x = 1$; $\left(1, \dfrac{1}{2}\right)$

18. $f(x) = \dfrac{x}{x + 1}$; $x = 3$; $\left(3, \dfrac{3}{4}\right)$

19. $f(x) = x\sqrt{x^2 + 1}$; $x = 2$; $(2, 2\sqrt{5})$

20. $f(x) = \dfrac{x}{\sqrt{x^2 + 1}}$; $x = 1$; $\left(1, \dfrac{\sqrt{2}}{2}\right)$

21. Use the **der1** operation to find the derivative of the function of exercise 11 when $x = 1$.

22. Use the **der1** operation to find the derivative of the function of exercise 12 when $x = -1$.

23. Use the **der1** operation to find the derivative of the function of exercise 13 when $x = 2$.

24. Use the **der1** operation to find the derivative of the function of exercise 14 when $x = 1$.

25. Use the **der1** operation to find the derivative of the function of exercise 15 when $x = 4$.

26. Use the **der1** operation to find the derivative of the function of exercise 16 when $x = 4$.

27. Use the **der1** operation to find the derivative of the function of exercise 17 when $x = 1$.

28. Use the **der1** operation to find the derivative of the function of exercise 18 when $x = 3$.

29. Use the **der1** operation to find the derivative of the function of exercise 19 when $x = 2$.

30. Use the **der1** operation to find the derivative of the function of exercise 20 when $x = 1$.

9.4   Basic Rules of Differentiation

Four Basic Rules

The method used in Section 9.3 for computing the derivative of a function is based on a faithful interpretation of the definition of the derivative as the limit of a quotient. Thus, to find the rule for the derivative f' of a function f, we first computed the difference quotient

$$\frac{f(x + h) - f(x)}{h}$$

and then evaluated its limit as h approached zero. As you have probably observed, this method is tedious even for relatively simple functions.

The main purpose of this chapter is to derive certain rules that will simplify the process of finding the derivative of a function. Throughout this book we will use the notation

$$\frac{d}{dx}[f(x)]$$

[read "dee, dee x of f of x"] to mean "the derivative of f with respect to x at x." In stating the rules of differentiation, we assume that the functions f and g are differentiable.

**RULE 1:
DERIVATIVE
OF A CONSTANT**

$$\frac{d}{dx}(c) = 0 \qquad c, \text{a constant}$$

The derivative of a constant function is equal to zero.

We can see this from a geometric viewpoint by recalling that the graph of a constant function is a straight line parallel to the x-axis (Figure 33).

FIGURE 33

The slope of the tangent line to the graph of $f(x) = c$, where c is a constant, is zero.

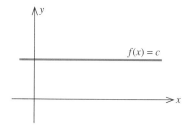

Since the tangent line to a straight line at any point on the line coincides with the straight line itself, its slope (as given by the derivative of $f(x) = c$) must be zero. We can also use the definition of the derivative to prove this result by computing

$$f'(x) = \lim_{h \to 0} \frac{f(x + h) - f(x)}{h}$$

$$= \lim_{h \to 0} \frac{c - c}{h}$$

$$= \lim_{h \to 0} 0 = 0$$

EXAMPLE 1

a. If $f(x) = 28$, then

$$f'(x) = \frac{d}{dx}(28) = 0$$

b. If $f(x) = -2$, then

$$f'(x) = \frac{d}{dx}(-2) = 0$$

RULE 2: THE POWER RULE

If n is any real number, then $\dfrac{d}{dx}(x^n) = nx^{n-1}$.

Let us verify the Power Rule for the special case $n = 2$. If $f(x) = x^2$, then

$$f'(x) = \frac{d}{dx}(x^2) = \lim_{h \to 0} \frac{f(x + h) - f(x)}{h}$$

$$= \lim_{h \to 0} \frac{(x + h)^2 - x^2}{h}$$

$$= \lim_{h \to 0} \frac{x^2 + 2xh + h^2 - x^2}{h}$$

$$= \lim_{h \to 0} \frac{2xh + h^2}{h} = \lim_{h \to 0} (2x + h) = 2x$$

as we set out to show.

The proof of the Power Rule for the general case is not easy to prove and will be omitted. However, you will be asked to prove the rule for the special case $n = 3$ in exercise 59, page 531.

EXAMPLE 2

a. If $f(x) = x$, then

$$f'(x) = \frac{d}{dx}(x) = 1 \cdot x^{1-1} = x^0 = 1$$

b. If $f(x) = x^8$, then

$$f'(x) = \frac{d}{dx}(x^8) = 8x^7$$

c. If $f(x) = x^{5/2}$, then

$$f'(x) = \frac{d}{dx}(x^{5/2}) = \frac{5}{2}x^{3/2}$$

In order to differentiate a function whose rule involves a radical, we first rewrite the rule using fractional powers. The resulting expression can then be differentiated using the Power Rule.

EXAMPLE 3   Find the derivative of the given function.

a. $f(x) = \sqrt{x}$ **b.** $g(x) = \dfrac{1}{\sqrt[3]{x}}$

Solution

a. Rewriting $\sqrt{x}$ in the form $x^{1/2}$, we obtain

$$f'(x) = \frac{d}{dx}(x^{1/2})$$

$$= \frac{1}{2}x^{-1/2} = \frac{1}{2x^{1/2}} = \frac{1}{2\sqrt{x}}$$

b. Rewriting $\dfrac{1}{\sqrt[3]{x}}$ in the form $x^{-1/3}$, we obtain

$$g'(x) = \frac{d}{dx}(x^{-1/3})$$

$$= -\frac{1}{3}x^{-4/3} = -\frac{1}{3x^{4/3}}$$

**RULE 3:
DERIVATIVE
OF A CONSTANT
MULTIPLE OF A
FUNCTION**

$$\frac{d}{dx}[cf(x)] = c\frac{d}{dx}[f(x)] \qquad c, \text{ a constant}$$

The derivative of a constant times a differentiable function is equal to the constant times the derivative of the function.

This result follows from the following computations.
     If $g(x) = cf(x)$, then

$$g'(x) = \lim_{h \to 0}\frac{g(x+h) - g(x)}{h} = \lim_{h \to 0}\frac{cf(x+h) - cf(x)}{h}$$

$$= c\lim_{h \to 0}\frac{f(x+h) - f(x)}{h}$$

$$= cf'(x)$$

EXAMPLE 4

a. If $f(x) = 5x^3$, then

$$f'(x) = \frac{d}{dx}(5x^3) = 5\frac{d}{dx}(x^3)$$

$$= 5(3x^2) = 15x^2$$

b. If $f(x) = \dfrac{3}{\sqrt{x}}$, then

$$f'(x) = \frac{d}{dx}(3x^{-1/2}) \qquad \text{Rewriting } \frac{3}{\sqrt{x}} = \frac{3}{x^{1/2}} = 3x^{-1/2}$$

$$= 3\left(-\frac{1}{2}x^{-3/2}\right) = -\frac{3}{2x^{3/2}}$$

RULE 4:
THE SUM RULE

$$\frac{d}{dx}[f(x) \pm g(x)] = \frac{d}{dx}[f(x)] \pm \frac{d}{dx}[g(x)]$$

The derivative of the sum or difference of two differentiable functions is equal to the sum or difference of their derivatives.

This result may be extended to the sum and difference of any finite number of differentiable functions. Let us verify the rule for a sum of two functions.
    If $s(x) = f(x) + g(x)$, then

$$s'(x) = \lim_{h \to 0} \frac{s(x + h) - s(x)}{h}$$

$$= \lim_{h \to 0} \frac{[f(x + h) + g(x + h)] - [f(x) + g(x)]}{h}$$

$$= \lim_{h \to 0} \frac{[f(x + h) - f(x)] + [g(x + h) - g(x)]}{h}$$

$$= \lim_{h \to 0} \frac{f(x + h) - f(x)}{h} + \lim_{h \to 0} \frac{g(x + h) - g(x)}{h}$$

$$= f'(x) + g'(x)$$

EXAMPLE 5   Find the derivatives of the following functions:

a. $f(x) = 4x^5 + 3x^4 - 8x^2 + x + 3$ **b.** $g(t) = \dfrac{t^2}{5} + \dfrac{5}{t^3}$

Solution

a. $f'(x) = \dfrac{d}{dx}(4x^5 + 3x^4 - 8x^2 + x + 3)$

$$= \frac{d}{dx}(4x^5) + \frac{d}{dx}(3x^4) - \frac{d}{dx}(8x^2) + \frac{d}{dx}(x) + \frac{d}{dx}(3)$$

$$= 20x^4 + 12x^3 - 16x + 1$$

b. Here the independent variable is t instead of x, so we differentiate with respect to t. Thus,

$$g'(t) = \frac{d}{dt}\left(\frac{1}{5}t^2 + 5t^{-3}\right) \qquad \text{Rewriting } \frac{1}{t^3} \text{ as } t^{-3}$$

$$= \frac{2}{5}t - 15t^{-4}$$

$$= \frac{2t^5 - 75}{5t^4} \qquad \text{Rewriting } t^{-4} \text{ as } \frac{1}{t^4} \text{ and simplifying} \qquad \blacksquare$$

EXAMPLE 6   Find the slope and an equation of the tangent line to the graph of $f(x) = 2x + \dfrac{1}{\sqrt{x}}$ at the point $(1, 3)$.

Solution   The slope of the tangent line at any point on the graph of f is given by

$$f'(x) = \frac{d}{dx}\left(2x + \frac{1}{\sqrt{x}}\right)$$

$$= \frac{d}{dx}(2x + x^{-1/2}) \qquad \text{Rewriting } \frac{1}{\sqrt{x}} = \frac{1}{x^{1/2}} = x^{-1/2}$$

$$= 2 - \frac{1}{2}x^{-3/2} \qquad \text{Using the Sum Rule}$$

$$= 2 - \frac{1}{2x^{3/2}}$$

In particular, the slope of the tangent line to the graph of f at $(1, 3)$ [where $x = 1$] is

$$f'(1) = 2 - \frac{1}{2(1^{3/2})} = 2 - \frac{1}{2} = \frac{3}{2}$$

Using the point-slope form of the equation of a line with slope 3/2 and the point $(1, 3)$, we see that an equation of the tangent line is

$$y - 3 = \frac{3}{2}(x - 1) \qquad (y - y_1) = m(x - x_1)$$

or, upon simplification,

$$y = \frac{3}{2}x + \frac{3}{2} \qquad\qquad \blacksquare$$

Applications

EXAMPLE 7   A group of marine biologists at the Neptune Institute of Oceanography recommended that a series of conservation measures be carried out over the next decade to save a certain species of whale from extinction. After implementing the conservation measures, the population of this species is expected to be

$$N(t) = 3t^3 + 2t^2 - 10t + 600 \qquad 0 \le t \le 10$$

where $N(t)$ denotes the population at the end of year t. Find the rate of growth of the whale population when $t = 2$ and $t = 6$. How large will the whale population be eight years after implementing the conservation measures?

Solution The rate of growth of the whale population at any time t is given by

$$N'(t) = 9t^2 + 4t - 10$$

In particular, when $t = 2$ and $t = 6$, we have

$$N'(2) = 9(2)^2 + 4(2) - 10$$
$$= 34$$
$$N'(6) = 9(6)^2 + 4(6) - 10$$
$$= 338$$

so the whale population's rate of growth will be 34 whales per year after two years and 338 per year after six years.

The whale population at the end of the eighth year will be

$$N(8) = 3(8)^3 + 2(8)^2 - 10(8) + 600$$
$$= 2184$$

or 2184 whales. The graph of the function N appears in Figure 34. Note the rapid growth of the population in the later years, as the conservation measures begin to pay off, compared to the growth in the early years. ∎

FIGURE 34
The whale population after year t is given by N(t).

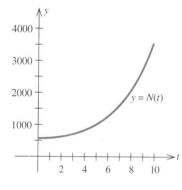

EXAMPLE 8 The altitude of a rocket (in feet) t seconds into flight is given by

$$s = f(t) = -t^3 + 96t^2 + 195t + 5 \qquad t \geq 0$$

a. Find an expression v for the rocket's velocity at any time t.

b. Compute the rocket's velocity when $t = 0, 30, 50, 65,$ and 70. Interpret your results.

c. Using the results from the solution to (b) and the observation that at the highest point in its trajectory the rocket's velocity is zero, find the maximum altitude attained by the rocket.

Solution

a. The rocket's velocity at any time t is given by

$$v = f'(t) = -3t^2 + 192t + 195$$

b. The rocket's velocity when $t = 0, 30, 50, 65,$ and 70 is given by

$$f'(0) = -3(0)^2 + 192(0) + 195 = 195$$
$$f'(30) = -3(30)^2 + 192(30) + 195 = 3255$$
$$f'(50) = -3(50)^2 + 192(50) + 195 = 2295$$
$$f'(65) = -3(65)^2 + 192(65) + 195 = 0$$
$$f'(70) = -3(70)^2 + 192(70) + 195 = -1065$$

or 195, 3255, 2295, 0, and −1065 feet per second.

FIGURE 35

The rocket's altitude t seconds into flight is given by f(t).

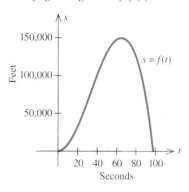

Thus, the rocket has an initial velocity of 195 ft/sec at $t = 0$ and accelerates to a velocity of 3255 ft/sec at $t = 30$. Fifty seconds into the flight, the rocket's velocity is 2295 ft/sec, which is less than the velocity at $t = 30$. This means that the rocket begins to decelerate after an initial period of acceleration.

The deceleration continues: the velocity is 0 ft/sec at $t = 65$ and -1065 ft/sec when $t = 70$. This figure tells us that 70 seconds into flight the rocket is heading back to Earth with a speed of 1065 ft/sec.

c. The results of (b) show that the rocket's velocity is zero when $t = 65$. At this instant, the rocket's maximum altitude is

$$s = f(65) = -(65)^3 + 96(65)^2 + 195(65) + 5$$

$$= 143,655$$

or 143,655 feet. A sketch of the graph of f appears in Figure 35. ∎

SELF-CHECK EXERCISES 9.4

1. Find the derivative of each of the following functions using the rules of differentiation.

a. $f(x) = 1.5x^2 + 2x^{1.5}$

b. $g(x) = 2\sqrt{x} + \dfrac{3}{\sqrt{x}}$

2. Let $f(x) = 2x^3 - 3x^2 + 2x - 1$.

a. Compute $f'(x)$.

b. What is the slope of the tangent line to the graph of f when $x = 2$?

c. What is the rate of change of the function f at $x = 2$?

3. A certain country's gross domestic product (GDP) (in millions of dollars) is described by the function

$$G(t) = -2t^3 + 45t^2 + 20t + 6000 \qquad 0 \le t \le 11$$

where $t = 0$ corresponds to the beginning of 1986.

a. At what rate was the GDP changing at the beginning of 1991? At the beginning of 1993? At the beginning of 1996?

b. What was the average rate of growth of the GDP over the period 1991 to 1996?

Solutions to Self-Check Exercises 9.4 can be found on page 531.

9.4 EXERCISES

In exercises 1–34, find the derivative of the function f by using the rules of differentiation.

1. $f(x) = -3$

2. $f(x) = 365$

3. $f(x) = x^5$

4. $f(x) = x^7$

5. $f(x) = x^{2.1}$

6. $f(x) = x^{0.8}$

7. $f(x) = 3x^2$

8. $f(x) = -2x^3$

9. $f(r) = \pi r^2$

10. $f(r) = \frac{4}{3}\pi r^3$

11. $f(x) = 9x^{1/3}$

12. $f(x) = \frac{5}{4}x^{4/5}$

13. $f(x) = 3\sqrt{x}$

14. $f(u) = \dfrac{2}{\sqrt{u}}$

15. $f(x) = 7x^{-12}$

16. $f(x) = 0.3x^{-1.2}$

17. $f(x) = 5x^2 - 3x + 7$

18. $f(x) = x^3 - 3x^2 + 1$

19. $f(x) = -x^3 + 2x^2 - 6$

20. $f(x) = x^4 - 2x^2 + 5$

21. $f(x) = 0.03x^2 - 0.4x + 10$

22. $f(x) = 0.002x^3 - 0.05x^2 + 0.1x - 20$

23. $f(x) = \dfrac{x^3 - 4x^2 + 3}{x}$

24. $f(x) = \dfrac{x^3 + 2x^2 + x - 1}{x}$

25. $f(x) = 4x^4 - 3x^{5/2} + 2$

26. $f(x) = 5x^{4/3} - \frac{2}{3}x^{3/2} + x^2 - 3x + 1$

27. $f(x) = 3x^{-1} + 4x^{-2}$

28. $f(x) = -\frac{1}{3}(x^{-3} - x^6)$

29. $f(t) = \dfrac{4}{t^4} - \dfrac{3}{t^3} + \dfrac{2}{t}$

30. $f(x) = \dfrac{5}{x^3} - \dfrac{2}{x^2} - \dfrac{1}{x} + 200$

31. $f(x) = 2x - 5\sqrt{x}$

32. $f(t) = 2t^2 + \sqrt{t^3}$

33. $f(x) = \dfrac{2}{x^2} - \dfrac{3}{x^{1/3}}$

34. $f(x) = \dfrac{3}{x^3} + \dfrac{4}{\sqrt{x}} + 1$

35. Let $f(x) = 2x^3 - 4x$. Find (**a**) $f'(-2)$, (**b**) $f'(0)$, and (**c**) $f'(2)$.

36. Let $f(x) = 4x^{5/4} + 2x^{3/2} + x$. Find (**a**) $f'(0)$ and (**b**) $f'(16)$.

In exercises 37–40, find the slope and an equation of the tangent line to the graph of the function f at the specified point.

37. $f(x) = 2x^2 - 3x + 4$; $(2, 6)$

38. $f(x) = -\frac{5}{3}x^2 + 2x + 2$; $(-1, -\frac{5}{3})$

39. $f(x) = x^4 - 3x^3 + 2x^2 - x + 1$; $(1, 0)$

40. $f(x) = \sqrt{x} + \dfrac{1}{\sqrt{x}}$; $\left(4, \dfrac{5}{2}\right)$

41. Let $f(x) = x^3$.
  a. Find the point on the graph of f where the tangent line is horizontal.
  b. Sketch the graph of f and draw the horizontal tangent line.

42. Let $f(x) = x^3 - 4x^2$. Find the point(s) on the graph of f where the tangent line is horizontal.

43. Let $f(x) = x^3 + 1$.
  a. Find the point(s) on the graph of f where the slope of the tangent line is equal to 12.
  b. Find the equation(s) of the tangent line(s) of (a).
  c. Sketch the graph of f showing the tangent line(s).

44. Let $f(x) = \frac{2}{3}x^3 + x^2 - 12x + 6$. Find the values of x for which (**a**) $f'(x) = -12$, (**b**) $f'(x) = 0$, and (**c**) $f'(x) = 12$.

45. Let $f(x) = \frac{1}{4}x^4 - \frac{1}{3}x^3 - x^2$. Find the point(s) on the graph of f where the slope of the tangent line is equal to (**a**) $-2x$, (**b**) 0, and (**c**) $10x$.

46. **Growth of a Cancerous Tumor** The volume of a spherical cancer tumor is given by the function

$$V(r) = \frac{4}{3}\pi r^3$$

where r is the radius of the tumor in centimeters. Find the rate of change in the volume of the tumor when
  a. $r = \frac{2}{3}$ cm   **b.** $r = \frac{5}{4}$ cm

47. **Velocity of Blood in an Artery** The velocity (in centimeters per second) of blood r centimeters from the central axis of an artery is given by

$$v(r) = k(R^2 - r^2)$$

where k is a constant and R is the radius of the artery (see the accompanying figure). Suppose that $k = 1000$ and $R = 0.2$ cm. Find $v(0.1)$ and $v'(0.1)$ and interpret your results.

Blood vessel

48. **Worker Efficiency** An efficiency study conducted for the Elektra Electronics Company showed that the number of "Space Commander" walkie-talkies assembled by the average worker t hours after starting work at 8 A.M. is given by

$$N(t) = -t^3 + 6t^2 + 15t$$

  a. Find the rate at which the average worker will be assembling walkie-talkies t hours after starting work.

b. At what rate will the average worker be assembling walkie-talkies at 10 A.M.? At 11 A.M.?

c. How many walkie-talkies will the average worker assemble between 10 A.M. and 11 A.M.?

49. Consumer Price Index An economy's consumer price index (CPI) is described by the function

$$I(t) = -0.2t^3 + 3t^2 + 100 \qquad 0 \le t \le 10$$

where $t = 0$ corresponds to 1983.

a. At what rate was the CPI changing in 1988? In 1990? In 1993?

b. What was the average rate of increase in the CPI over the period from 1988 to 1993?

50. Effect of Advertising on Sales The relationship between the amount of money x that the Cannon Precision Instruments Corporation spends on advertising and the company's total sales $S(x)$ is given by the function

$$S(x) = -0.002x^3 + 0.6x^2 + x + 500 \qquad 0 \le x \le 200$$

where x is measured in thousands of dollars. Find the rate of change of the sales with respect to the amount of money spent on advertising. Are Cannon's total sales increasing at a faster rate when the amount of money spent on advertising is (a) \$100,000 or (b) \$150,000?

51. Population Growth A study prepared for a Sunbelt town's chamber of commerce projected that the town's population in the next three years will grow according to the rule

$$P(t) = 50{,}000 + 30t^{3/2} + 20t$$

where $P(t)$ denotes the population t months from now. How fast will the population be increasing nine months from now? Sixteen months from now?

52. Curbing Population Growth Five years ago, the government of a Pacific island state launched an extensive propaganda campaign toward curbing the country's population growth. According to the Census Department, the population (measured in thousands of people) for the following four years was

$$P(t) = -\frac{1}{3}t^3 + 64t + 3000$$

where t is measured in years and $t = 0$ at the start of the campaign. Find the rate of change of the population at the end of years one, two, three, and four. Was the plan working?

53. Conservation of Species A certain species of turtle faces extinction because dealers collect truckloads of turtle eggs to be sold as aphrodisiacs. After severe conservation measures are implemented, it is hoped that the turtle population will grow according to the rule

$$N(t) = 2t^3 + 3t^2 - 4t + 1000 \qquad 0 \le t \le 10$$

where $N(t)$ denotes the population at the end of year t. Find the rate of growth of the turtle population when $t = 2$ and $t = 8$. What will the population be ten years after the conservation measures are implemented?

54. Flight of a Rocket The altitude (in feet) of a rocket t seconds into flight is given by

$$s = f(t) = -2t^3 + 114t^2 + 480t + 1 \qquad t \ge 0$$

a. Find an expression v for the rocket's velocity at any time t.

b. Compute the rocket's velocity when $t = 0, 20, 40,$ and 60. Interpret your results.

c. Using the results from the solution to (b), find the maximum altitude attained by the rocket. [*Hint:* At its highest point, the velocity of the rocket is zero.]

55. Stopping Distance of a Racing Car During a test by the editors of an auto magazine, the stopping distance s (in feet) of the MacPherson X-2 racing car conformed to the rule

$$s = f(t) = 120t - 15t^2 \qquad t \ge 0$$

where t was the time (in seconds) after the brakes were applied.

a. Find an expression for the car's velocity v at any time t.

b. What was the car's velocity when the brakes were first applied?

c. What was the car's stopping distance for that particular test? [*Hint:* The stopping time is found by setting $v = 0$.]

C **56. Demand Functions** The demand function for the Luminar desk lamp is given by

$$p = f(x) = -0.1x^2 - 0.4x + 35$$

where x is the quantity demanded (measured in thousands) and p is the unit price in dollars.

a. Find $f'(x)$.

b. What is the rate of change of the unit price when the quantity demanded is 10,000 units ($x = 10$)? What is the unit price at that level of demand?

C **57. Average Speed of a Vehicle on a Highway** The average speed of a vehicle on a stretch of Route 134 between 6 A.M. and 10 A.M. on a typical weekday is approximated by the function

$$f(t) = 20t - 40\sqrt{t} + 50 \qquad 0 \le t \le 4$$

where $f(t)$ is measured in miles per hour and t is measured in hours, $t = 0$ corresponding to 6 A.M.

a. Compute $f'(t)$.

b. Compute $f(0)$, $f(1)$, and $f(2)$ and interpret your results.

c. Compute $f'(\frac{1}{2})$, $f'(1)$, and $f'(2)$ and interpret your results.

58. Health-Care Spending Despite efforts at cost containment, the cost of the medicare program is increasing at a high rate. Two major reasons for this increase are an aging population and the constant development and extensive use by physicians of new technologies. Based on data from the Health Care Financing Administration and the U.S. Census Bureau, health-care spending through the year 2000 may be approximated by the function

$$S(t) = 0.02836t^3 - 0.05167t^2 + 9.60881t + 41.9$$

$$0 \leq t \leq 35$$

where $S(t)$ is the spending in billions of dollars and t is measured in years, with $t = 0$ corresponding to the beginning of 1965.

a. Find an expression for the rate of change of health-care spending at any time t.

b. How fast was health-care spending changing at the beginning of 1980? How fast will health-care spending be changing at the beginning of 2000?

c. What was the amount of health-care spending at the beginning of 1980? What will the amount of health-care spending be at the beginning of 2000?

Source: Health Care Financing Administration and the U.S. Census Bureau

59. Prove the Power Rule (Rule 2) for the special case $n = 3$. *Hint:* Compute

$$\lim_{h \to 0} \left[\frac{(x + h)^3 - x^3}{h} \right]$$

SOLUTIONS TO SELF-CHECK EXERCISES 9.4

1. a. $f'(x) = \dfrac{d}{dx}(1.5x^2) + \dfrac{d}{dx}(2x^{1.5})$

$\qquad = (1.5)(2x) + (2)(1.5x^{0.5})$

$\qquad = 3x + 3\sqrt{x} = 3(x + \sqrt{x})$

b. $g'(x) = \dfrac{d}{dx}(2x^{1/2}) + \dfrac{d}{dx}(3x^{-1/2})$

$\qquad = (2)\left(\dfrac{1}{2}x^{-1/2}\right) + (3)\left(-\dfrac{1}{2}x^{-3/2}\right)$

$\qquad = x^{-1/2} - \dfrac{3}{2}x^{-3/2}$

$\qquad = \dfrac{1}{2}x^{-3/2}(2x - 3) = \dfrac{2x - 3}{2x^{3/2}}$

2. a. $f'(x) = \dfrac{d}{dx}(2x^3) - \dfrac{d}{dx}(3x^2) + \dfrac{d}{dx}(2x) - \dfrac{d}{dx}(1)$

$\qquad = (2)(3x^2) - (3)(2x) + 2$

$\qquad = 6x^2 - 6x + 2$

b. The slope of the tangent line to the graph of f when $x = 2$ is given by

$$f'(2) = 6(2)^2 - 6(2) + 2 = 14$$

c. The rate of change of f at $x = 2$ is given by $f'(2)$. Using the results of (b), we see that $f'(2)$ is 14 units per unit change in x.

3. a. The rate at which the GDP was changing at any time t ($0 < t < 11$) is given by

$$G'(t) = -6t^2 + 90t + 20$$

In particular, the rates of change of the GDP at the beginning of the years 1991 ($t = 5$), 1993 ($t = 7$), and 1996 ($t = 10$) are given by

$$G'(5) = 320, \qquad G'(7) = 356, \qquad G'(10) = 320$$

respectively; that is, by $320 million per year, $356 million per year, and $320 million per year, respectively.

b. The average rate of growth of the GDP over the period from the beginning of 1991 ($t = 5$) to the beginning of 1996 ($t = 10$) is given by

$$\frac{G(10) - G(5)}{10 - 5} = \frac{[-2(10)^3 + 45(10)^2 + 20(10) + 6000]}{5}$$

$$- \frac{[-2(5)^3 + 45(5)^2 + 20(5) + 6000]}{5}$$

$$= \frac{8700 - 6975}{5}$$

or $345 million per year.

Using Technology

Finding the Rate of Change of a Function

Recall that in Using Technology, Section 9.3 (pages 520–521) we used the **dy/dx** operation of a graphing utility to help us obtain an approximate value of the derivative of a function at a given value of x. Since the derivative of a function $f(x)$ measures the rate of change of the function with respect to x, the **dy/dx** operation can be used to answer questions pertaining to the rate of change of one quantity y with respect to another quantity x, where $y = f(x)$, for a specific value of x.

EXAMPLE 1   Let $y = 3t^3 + 2\sqrt{t}$.

a. Use the **dy/dx** operation of a graphing utility to find how fast y is changing with respect to t when $t = 1$.

b. Verify the result of (a) using the rules of differentiation of this section.

Solution

a. Using **ZIN** repeatedly so that the cursor is sufficiently close to $x = 1$ (we use x for t in the graphing utility), we find $dy/dx \approx 10$. Thus, y is changing at the rate of 10 units per unit change in t.

b. Here $f(t) = 3t^3 + 2t^{1/2}$ and

$$f'(t) = 9t^2 + 2\left(\frac{1}{2}t^{-1/2}\right) = 9t^2 + \frac{1}{\sqrt{t}}$$

Using this result, we see that when $t = 1$, y is changing at the rate of

$$f'(1) = 9(1^2) + \frac{1}{\sqrt{1}} = 10$$

or 10 units per unit change in t, as obtained earlier.

EXAMPLE 2   According to the American Heart Association, the treatment benefit for heart attacks depends on the time to treatment and is described by the function

$$f(t) = \frac{-16.94t + 203.28}{t + 2.0328} \qquad 0 \le t \le 12$$

where t is measured in hours and $f(t)$ is a percentage.

a. Use a graphing utility to graph the function f using the viewing rectangle $[0, 13] \times [0, 120]$.

b. Use a graphing utility to approximate the derivative of f when $t = 0$. When $t = 2$.

c. Interpret the results obtained in (b).

Solution

a. The graph of f is shown in Figure 1.

FIGURE 1

The graph of f in the viewing rectangle $[0, 13] \times [0, 120]$

Source: American Heart Association

b. Using the **dy/dx** operation with the cursor set at $x = 0$, we see that $dy/dx \approx -57.5266$; that is, f is dropping at the rate of approximately 58% per hour. Next, using **ZIN** repeatedly so that the cursor is sufficiently close to $x = 2$, we find $dy/dx \approx -14.6165$; that is, f is dropping at the rate of approximately 15% per hour.

c. The results of (b) show that the treatment benefit drops off at the rate of 58% per hour at the time when the heart attack first occurs and falls off at the rate of 15% per hour when the time to treatment is two hours. Thus, it is extremely urgent that a patient suffering a heart attack receive medical attention as soon as possible. ∎

EXERCISES

In exercises 1–6, use the **dy/dx** *operation of a graphing utility to find the rate of change of $f(x)$ at the given value of x. Give your answer accurate to four decimal places.*

1. $f(x) = 4x^5 - 3x^3 + 2x^2 + 1; x = 0.5$

2. $f(x) = -x^5 + 4x^2 + 3; x = 0.4$

3. $f(x) = x - 2\sqrt{x}; x = 3$

4. $f(x) = \dfrac{\sqrt{x} - 1}{x}; x = 2$

5. $f(x) = x^{1/2} - x^{1/3}; x = 1.2$

6. $f(x) = 2x^{5/4} + x; x = 2$

7. Carbon Monoxide in the Atmosphere   The projected average global atmospheric concentration of carbon monoxide is approximated by the function

$$f(t) = 0.881443t^4 - 1.45533t^3 + 0.695876t^2$$
$$+ 2.87801t + 293$$
$$0 \le t \le 4$$

where t is measured in 40-year intervals with $t = 0$ corresponding to the beginning of 1860 and $f(t)$ is measured in parts per million by volume.

a. Use a graphing utility to plot the graph of f in the viewing rectangle $[0, 4] \times [280, 400]$.

b. Use a graphing utility to estimate how fast the projected average global atmospheric concentration of carbon monoxide was changing at the beginning of the year 1900 ($t = 1$). At the beginning of 1990 ($t = 3.5$).

Source: "Beyond the Limits," Meadows et al.

8. **Growth of HMOs**   Based on data compiled by the Group Health Association of America, the number of people receiving their care in an HMO (Health Maintenance Organization) from the beginning of 1984 through 1994 is approximated by the function

$$f(t) = 0.0514t^3 - 0.853t^2 + 6.8147t + 15.6524$$

$$0 \le t \le 11$$

where $f(t)$ gives the number of people in millions and t is measured in years, with $t = 0$ corresponding to the beginning of 1984.

a. Use a graphing utility to plot the graph of f in the viewing window $[0, 12] \times [0, 80]$.

b. How fast was the number of people receiving their care in an HMO growing at the beginning of 1992?

Source: Group Health Association of America

9. **Home Sales**   According to the Greater Boston Real Estate Board-Multiple Listing Service, the average number of days a single-family home remains for sale from listing to accepted offer is approximated by the

function

$$f(t) = 0.0171911t^4 - 0.662121t^3 + 6.18083t^2$$

$$- 8.97086t + 53.3357$$

$$0 \le t \le 10$$

where t is measured in years, with $t = 0$ corresponding to the beginning of 1984.

a. Use a graphing utility to plot the graph of f in the viewing rectangle $[0, 12] \times [0, 120]$.

b. How fast is the average number of days a single-family home remains for sale from listing to accepted offer changing at the beginning of 1984 ($t = 0$)? At the beginning of 1988 ($t = 4$)?

Source: Greater Boston Real Estate Board-Multiple Listing Service

10. **Manufacturing Capacity**   Data obtained from the Federal Reserve shows that the annual change in manufacturing capacity between 1988 and 1994 is given by

$$f(t) = 0.0388889t^3 - 0.283333t^2$$

$$+ 0.477778t + 2.04286$$

$$0 \le t \le 6$$

where $f(t)$ is a percentage and t is measured in years, with $t = 0$ corresponding to the beginning of 1988.

a. Use a graphing utility to plot the graph of f in the viewing rectangle $[0, 8] \times [0, 4]$.

b. How fast was $f(t)$ changing at the beginning of 1990 ($t = 2$)? At the beginning of 1992 ($t = 4$)?

Source: Federal Reserve

9.5   The Product and Quotient Rules; Higher-Order Derivatives

In this section, we will study two more rules of differentiation, known as the *Product Rule* and the *Quotient Rule*.

The Product Rule

The derivative of the product of two differentiable functions is given by the following rule:

RULE 5: THE PRODUCT RULE
$$\frac{d}{dx}[f(x)g(x)] = f(x)g'(x) + g(x)f'(x)$$

The derivative of the product of two functions is the first function times the derivative of the second plus the second function times the derivative of the first.

The Product Rule may be extended to the case involving the product of any finite number of functions (see exercise 69, p. 546). The verification of the Product Rule is given at the end of this section.

 The derivative of the product of two functions is not given by the product of the derivatives of the functions; that is, in general

$$\frac{d}{dx}[f(x)g(x)] \neq f'(x)g'(x)$$

EXAMPLE 1   Find the derivative of the function

$$f(x) = (2x^2 - 1)(x^3 + 3)$$

Solution   By the Product Rule,

$$f'(x) = (2x^2 - 1)\frac{d}{dx}(x^3 + 3) + (x^3 + 3)\frac{d}{dx}(2x^2 - 1)$$

$$= (2x^2 - 1)(3x^2) + (x^3 + 3)(4x)$$

$$= 6x^4 - 3x^2 + 4x^4 + 12x$$

$$= 10x^4 - 3x^2 + 12x$$

$$= x(10x^3 - 3x + 12)$$

EXAMPLE 2   Differentiate (that is, find the derivative of) the function

$$f(x) = x^3(\sqrt{x} + 1)$$

Solution   First, we express the function in exponential form, obtaining

$$f(x) = x^3(x^{1/2} + 1)$$

By the Product Rule,

$$f'(x) = x^3\frac{d}{dx}(x^{1/2} + 1) + (x^{1/2} + 1)\frac{d}{dx}x^3$$

$$= x^3\left(\frac{1}{2}x^{-1/2}\right) + (x^{1/2} + 1)(3x^2)$$

$$= \frac{1}{2}x^{5/2} + 3x^{5/2} + 3x^2$$

$$= \frac{7}{2}x^{5/2} + 3x^2$$

REMARK   We can also solve the problem by first expanding the product before differentiating f. Examples for which this is not possible will be considered in Section 9.6, where the true value of the Product Rule will be appreciated.

The Quotient Rule

The derivative of the quotient of two differentiable functions is given by the following rule:

RULE 6: THE QUOTIENT RULE

$$\frac{d}{dx}\left[\frac{f(x)}{g(x)}\right] = \frac{g(x)f'(x) - f(x)g'(x)}{[g(x)]^2} \qquad g(x) \neq 0$$

As an aid to remembering this expression, observe that it has the following form:

$$\frac{d}{dx}\left[\frac{f(x)}{g(x)}\right] = \frac{(\text{denominator})\left(\begin{array}{c}\text{derivative of}\\\text{numerator}\end{array}\right) - (\text{numerator})\left(\begin{array}{c}\text{derivative of}\\\text{denominator}\end{array}\right)}{\text{square of denominator}}$$

For a proof of the Quotient Rule, see exercise 70, page 546.

The derivative of a quotient is not equal to the quotient of the derivatives; that is,

$$\frac{d}{dx}\left[\frac{f(x)}{g(x)}\right] \neq \frac{f'(x)}{g'(x)}$$

For example, if $f(x) = x^3$ and $g(x) = x^2$, then

$$\frac{d}{dx}\left[\frac{f(x)}{g(x)}\right] = \frac{d}{dx}\left(\frac{x^3}{x^2}\right) = \frac{d}{dx}(x) = 1$$

which is not equal to

$$\frac{f'(x)}{g'(x)} = \frac{\dfrac{d}{dx}(x^3)}{\dfrac{d}{dx}(x^2)} = \frac{3x^2}{2x} = \frac{3}{2}x$$

EXAMPLE 3   Find $f'(x)$ if $f(x) = \dfrac{x}{2x - 4}$.

Solution   Using the Quotient Rule, we obtain

$$f'(x) = \frac{(2x - 4)\dfrac{d}{dx}(x) - x\dfrac{d}{dx}(2x - 4)}{(2x - 4)^2}$$

$$= \frac{(2x - 4)(1) - x(2)}{(2x - 4)^2}$$

$$= \frac{2x - 4 - 2x}{(2x - 4)^2} = -\frac{4}{(2x - 4)^2} = -\frac{1}{(x - 2)^2}$$

EXAMPLE 4   Find $f'(x)$ if $f(x) = \dfrac{x^2 + 1}{x^2 - 1}$.

Solution   By the Quotient Rule,

$$f'(x) = \frac{(x^2 - 1)\dfrac{d}{dx}(x^2 + 1) - (x^2 + 1)\dfrac{d}{dx}(x^2 - 1)}{(x^2 - 1)^2}$$

$$= \frac{(x^2 - 1)(2x) - (x^2 + 1)(2x)}{(x^2 - 1)^2}$$

$$= \frac{2x^3 - 2x - 2x^3 - 2x}{(x^2 - 1)^2}$$

$$= -\frac{4x}{(x^2 - 1)^2}$$

EXAMPLE 5   Find $h'(x)$ if

$$h(x) = \frac{\sqrt{x}}{x^2 + 1}$$

Solution   Rewrite $h(x)$ in the form $h(x) = \dfrac{x^{1/2}}{x^2 + 1}$. By the Quotient Rule, we find

$$h'(x) = \frac{(x^2 + 1)\dfrac{d}{dx}(x^{1/2}) - x^{1/2}\dfrac{d}{dx}(x^2 + 1)}{(x^2 + 1)^2}$$

$$= \frac{(x^2 + 1)\left(\dfrac{1}{2}x^{-1/2}\right) - x^{1/2}(2x)}{(x^2 + 1)^2}$$

$$= \frac{\dfrac{1}{2}x^{-1/2}(x^2 + 1 - 4x^2)}{(x^2 + 1)^2} \qquad \text{Factoring out } \dfrac{1}{2}x^{-1/2} \text{ from the numerator}$$

$$= \frac{1 - 3x^2}{2\sqrt{x}(x^2 + 1)^2} \qquad \text{Rewriting } \dfrac{1}{2}x^{-1/2} \text{ as } \dfrac{1}{2x^{1/2}} \text{ and simplifying}$$

Applications

EXAMPLE 6   The sales (in millions of dollars) of a laser disc recording of a hit movie t years from the date of release is given by

$$S(t) = \frac{5t}{t^2 + 1}$$

a. Find the rate at which the sales are changing at time t.

b. How fast are the sales changing at the time the laser discs are released ($t = 0$)? Two years from the date of release?

Solution

a. The rate at which the sales are changing at time t is given by $S'(t)$. Using the Quotient Rule, we obtain

$$S'(t) = \frac{d}{dt}\left[\frac{5t}{t^2 + 1}\right] = 5\frac{d}{dt}\left[\frac{t}{t^2 + 1}\right]$$

$$= 5\left[\frac{(t^2 + 1)(1) - t(2t)}{(t^2 + 1)^2}\right]$$

$$= 5\left[\frac{t^2 + 1 - 2t^2}{(t^2 + 1)^2}\right] = \frac{5(1 - t^2)}{(t^2 + 1)^2}$$

b. The rate at which the sales are changing at the time the laser discs are released is given by

$$S'(0) = \frac{5(1 - 0)}{(0 + 1)^2} = 5$$

that is, they are increasing at the rate of $5 million per year.

Two years from the date of release, the sales are changing at the rate of

$$S'(2) = \frac{5(1 - 4)}{(4 + 1)^2} = -\frac{3}{5} = -0.6$$

that is, decreasing at the rate of $600,000 per year.

The graph of the function S is shown in Figure 36.

FIGURE 36

After a spectacular rise, the sales begin to taper off.

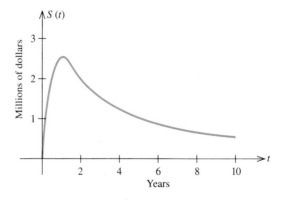

EXAMPLE 7   When organic waste is dumped into a pond, the oxidation process that takes place reduces the pond's oxygen content. However, given time, nature will restore the oxygen content to its natural level. Suppose that the oxygen content t days after organic waste has been dumped into the pond is given by

$$f(t) = 100\left[\frac{t^2 - 4t + 4}{t^2 + 4}\right] \qquad 0 \leq t < \infty$$

percent of its normal level.

a. Derive a general expression that gives the rate of change of the pond's oxygen level at any time t.

b. How fast is the pond's oxygen content changing one day after the organic waste has been dumped? Two days after? Three days after?

Solution

a. The rate of change of the pond's oxygen level at any time t is given by the derivative of the function f. Thus, the required expression is

$$f'(t) = 100 \frac{d}{dt} \left[\frac{t^2 - 4t + 4}{t^2 + 4} \right]$$

$$= 100 \left[\frac{(t^2 + 4)(2t - 4) - (t^2 - 4t + 4)(2t)}{(t^2 + 4)^2} \right]$$

$$= 100 \left[\frac{2t^3 - 4t^2 + 8t - 16 - 2t^3 + 8t^2 - 8t}{(t^2 + 4)^2} \right]$$

$$= \frac{100(4t^2 - 16)}{(t^2 + 4)^2} = \frac{400(t^2 - 4)}{(t^2 + 4)^2}$$

b. The rate at which the pond's oxygen content is changing one day after the organic waste has been dumped is given by

$$f'(1) = \frac{400(1 - 4)}{(1 + 4)^2} = -48$$

that is, it is dropping at the rate of 48% per day. Two days after, the rate is

$$f'(2) = \frac{400(4 - 4)}{(4 + 4)^2} = 0$$

that is, it is neither increasing nor decreasing. Three days after, the rate is

$$f'(3) = \frac{400(3^2 - 4)}{(3^2 + 4)^2} = 11.83$$

that is, the oxygen content is increasing at the rate of 11.83% per day, and the restoration process has indeed begun. ∎

Higher-Order Derivatives

The derivative f' of a function f is also a function. As such, the differentiability of f' may be considered. Thus, the function f' has a derivative f'' at a point x in the domain of f' if the limit of the quotient

$$\frac{f'(x + h) - f'(x)}{h}$$

exists as h approaches zero. In other words, it is the derivative of the first derivative.

The function f'' obtained in this manner is called the **second derivative** of the function f, just as the derivative f' of f is often called the first derivative of f. Continuing in this fashion, we are led to considering the third, fourth, and higher-order derivatives of f whenever they exist. Notations for the first, second, third, and, in general, nth derivatives of a function f at a point x are

$$f'(x), f''(x), f'''(x), \ldots, f^{(n)}(x)$$

or

$$D^1 f(x), D^2 f(x), D^3 f(x), \ldots, D^n f(x)$$

If f is written in the form $y = f(x)$, then the notations for its derivatives are

$$y', y'', y''', \ldots, y^{(n)}$$

$$\frac{dy}{dx}, \frac{d^2 y}{dx^2}, \frac{d^3 y}{dx^3}, \ldots, \frac{d^n y}{dx^n}$$

or

$$D^1 y, D^2 y, D^3 y, \ldots, D^n y$$

respectively.

EXAMPLE 8   Find the derivatives of all orders of the polynomial function $f(x) = x^5 - 3x^4 + 4x^3 - 2x^2 + x - 8$.

Solution   We have

$$f'(x) = 5x^4 - 12x^3 + 12x^2 - 4x + 1$$

$$f''(x) = \frac{d}{dx} f'(x) = 20x^3 - 36x^2 + 24x - 4$$

$$f'''(x) = \frac{d}{dx} f''(x) = 60x^2 - 72x + 24$$

$$f^{(4)}(x) = \frac{d}{dx} f'''(x) = 120x - 72$$

$$f^{(5)}(x) = \frac{d}{dx} f^{(4)}(x) = 120$$

and, in general,

$$f^{(n)}(x) = 0 \quad \text{for } n > 5$$

EXAMPLE 9   Find the third derivative of the function f defined by $y = x^{2/3}$. What is its domain?

Solution   We have

$$y' = \frac{2}{3} x^{-1/3}$$

$$y'' = \left(\frac{2}{3}\right)\left(-\frac{1}{3}\right) x^{-4/3} = -\frac{2}{9} x^{-4/3}$$

so the required derivative is

$$y''' = \left(-\frac{2}{9}\right)\left(-\frac{4}{3}\right) x^{-7/3} = \frac{8}{27} x^{-7/3} = \frac{8}{27 x^{7/3}}$$

The common domain of the functions f', f'', and f''' is the set of all real numbers except $x = 0$. The domain of $y = x^{2/3}$ is the set of all real numbers. The graph of the function $y = x^{2/3}$ appears in Figure 37.

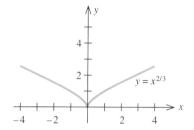

FIGURE 37
The graph of the function
$y = x^{2/3}$

REMARK   Always simplify an expression before differentiating it to obtain the next order derivative. ☐

Applications

Just as the derivative of a function f at a point x measures the rate of change of the function f at that point, the second derivative of f (the derivative of f') measures the rate of change of the derivative f' of the function f. The third derivative of the function f, f''', measures the rate of change of f'', and so on.

In Chapter 10 we will discuss applications involving the geometric interpretation of the second derivative of a function. The following example gives an interpretation of the second derivative in a familiar role.

EXAMPLE 10   (Refer to Example 5, page 511.) The distance s (in feet) covered by a car moving along a straight road t seconds after starting from rest is given by the function $s = 2t^2 \, (0 \le t \le 30)$. What is the car's acceleration at the end of 30 seconds?

Solution   The velocity of the car t seconds from rest is given by

$$v = \frac{ds}{dt} = \frac{d}{dt}(2t^2) = 4t$$

The acceleration of the car t seconds from rest is given by the rate of change of the velocity of t; that is,

$$a = \frac{d}{dt}v = \frac{d}{dt}\left(\frac{ds}{dt}\right) = \frac{d^2s}{dt^2} = \frac{d}{dt}(4t) = 4$$

or 4 feet per second per second, normally abbreviated, 4 ft/sec². ▨

EXAMPLE 11   A ball is thrown straight up into the air from the roof of a building. The height of the ball as measured from the ground is given by

$$s = -16t^2 + 24t + 120$$

where s is measured in feet and t in seconds. Find the velocity and acceleration of the ball 3 seconds after it is thrown into the air.

Solution   The velocity v and acceleration a of the ball at any time t are given by

$$v = \frac{ds}{dt} = \frac{d}{dt}(-16t^2 + 24t + 120) = -32t + 24$$

and

$$a = \frac{d^2t}{dt^2} = \frac{d}{dt}\left(\frac{ds}{dt}\right) = \frac{d}{dt}(-32t + 24) = -32$$

Therefore, the velocity of the ball 3 seconds after it is thrown into the air is

$$v = -32(3) + 24 = -72$$

That is, the ball is falling downward at a speed of 72 ft/sec. The acceleration of the ball is 32 ft/sec² downward at any time during the motion.   ■

Another interpretation of the second derivative of a function—this time from the field of economics—follows. Suppose that the consumer price index (CPI) of an economy between the years a and b is described by the function $I(t)$ ($a \le t \le b$) (Figure 38). Then the first derivative of I, $I'(t)$, gives the rate of inflation of the economy at any time t. The second derivative of I, $I''(t)$, gives the *rate of change of the inflation rate* at any time t. Thus, when the economist or politician claims that "inflation is slowing," what he or she is saying is that the rate of inflation is decreasing. Mathematically, this is equivalent to noting that the second derivative $I''(t)$ is negative at the time t under consideration. Observe that $I'(t)$ could be positive at a point in time when $I''(t)$ is negative (see Example 12). Thus, one may not draw the conclusion from the aforementioned quote that prices of goods and services are about to drop!

EXAMPLE 12   An economy's consumer price index (CPI) is described by the function

$$I(t) = -0.2t^3 + 3t^2 + 100 \qquad 0 \le t \le 9$$

where $t = 0$ corresponds to the year 1987. Compute $I'(6)$ and $I''(6)$ and use these results to show that even though the CPI was rising at the beginning of 1993, "inflation was moderating" at that point in time.

Solution   We find

$$I'(t) = -0.6t^2 + 6t$$

and

$$I''(t) = -1.2t + 6$$

so

$$I'(6) = -0.6(6)^2 + 6(6) = 14.4$$

and

$$I''(6) = -1.2(6) + 6 = -1.2$$

Our computations reveal that at the beginning of 1993 ($t = 6$), the CPI was increasing at the rate of 14.4 points per year, whereas the rate of the inflation rate was decreasing by 1.2 points per year. Thus, inflation was moderating at that point in time (Figure 39). In Section 10.2, we will see that relief actually began in early 1992.   ■

Verification of the Product Rule

We will now verify the Product Rule. If $p(x) = f(x)g(x)$, then

$$p'(x) = \lim_{h \to 0} \frac{p(x + h) - p(x)}{h}$$

$$= \lim_{h \to 0} \frac{f(x + h)g(x + h) - f(x)g(x)}{h}$$

FIGURE 38

The CPI of a certain economy from year a to year b is given by I(t).

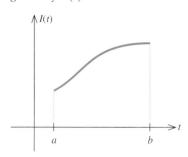

FIGURE 39

The CPI of an economy is given by I(t).

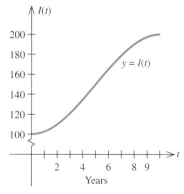

Years

By adding $-f(x + h)g(x) + f(x + h)g(x)$ (which is zero!) to the numerator and factoring, we have

$$p'(x) = \lim_{h \to 0} \frac{f(x + h)[g(x + h) - g(x)] + g(x)[f(x + h) - f(x)]}{h}$$

$$= \lim_{h \to 0} \left\{ f(x + h) \left[\frac{g(x + h) - g(x)}{h} \right] + g(x) \left[\frac{f(x + h) - f(x)}{h} \right] \right\}$$

$$= \lim_{h \to 0} f(x + h) \left[\frac{g(x + h) - g(x)}{h} \right]$$

$$+ \lim_{h \to 0} g(x) \left[\frac{f(x + h) - f(x)}{h} \right] \qquad \text{By property 3 of limits}$$

$$= \lim_{h \to 0} f(x + h) \cdot \lim_{h \to 0} \frac{g(x + h) - g(x)}{h}$$

$$+ \lim_{h \to 0} g(x) \cdot \lim_{h \to 0} \frac{f(x + h) - f(x)}{h} \qquad \text{By property 4 of limits}$$

$$= f(x)g'(x) + g(x)f'(x)$$

Observe that in the second from the last link in the chain of equalities, we have used the fact that $\lim_{h \to 0} f(x + h) = f(x)$ because f is continuous at x.

SELF-CHECK EXERCISES 9.5

1. Find the derivative of $f(x) = \dfrac{2x + 1}{x^2 - 1}$.

2. What is the slope of the tangent line to the graph of

$$f(x) = (x^2 + 1)(2x^3 - 3x^2 + 1)$$

at the point $(2, 25)$? How fast is the function f changing when $x = 2$?

c 3. The total sales of the Security Products Corporation in its first two years of operation are given by

$$S = f(t) = \frac{0.3t^3}{1 + 0.4t^2} \qquad 0 \le t \le 2$$

where S is measured in millions of dollars and $t = 0$ corresponds to the date Security Products began operations. How fast were the sales increasing at the beginning of the company's second year of operation?

Solutions to Self-Check Exercises 9.5 can be found on page 547.

9.5   EXERCISES

In exercises 1–30, find the derivative of the function.

1. $f(x) = 2x(x^2 + 1)$

2. $f(x) = 3x^2(x - 1)$

3. $f(t) = (t - 1)(2t + 1)$

4. $f(x) = (2x + 3)(3x - 4)$

5. $f(x) = (3x + 1)(x^2 - 2)$

6. $f(x) = (x + 1)(2x^2 - 3x + 1)$

7. $f(x) = (x^3 - 1)(x + 1)$

8. $f(x) = (x^3 - 12x)(3x^2 + 2x)$

9. $f(w) = (w^3 - w^2 + w - 1)(w^2 + 2)$

10. $f(x) = \frac{1}{5}x^5 + (x^2 + 1)(x^2 - x - 1) + 28$

11. $f(x) = (5x^2 + 1)(2\sqrt{x} - 1)$

12. $f(t) = (1 + \sqrt{t})(2t^2 - 3)$

13. $f(x) = (x^2 - 5x + 2)\left(x - \frac{2}{x}\right)$

14. $f(x) = (x^3 + 2x + 1)\left(2 + \frac{1}{x^2}\right)$

15. $f(x) = \frac{1}{x - 2}$ **16.** $g(x) = \frac{3}{2x + 4}$

17. $f(x) = \frac{x - 1}{2x + 1}$ **18.** $f(t) = \frac{1 - 2t}{1 + 3t}$

19. $f(x) = \frac{1}{x^2 + 1}$ **20.** $f(u) = \frac{u}{u^2 + 1}$

21. $f(s) = \frac{s^2 - 4}{s + 1}$ **22.** $f(x) = \frac{x^3 - 2}{x^2 + 1}$

23. $f(x) = \frac{\sqrt{x}}{x^2 + 1}$ **24.** $f(x) = \frac{x^2 + 1}{\sqrt{x}}$

25. $f(x) = \frac{x^2 + 2}{x^2 + x + 1}$ **26.** $f(x) = \frac{x + 1}{2x^2 + 2x + 3}$

27. $f(x) = \frac{(x + 1)(x^2 + 1)}{x - 2}$

28. $f(x) = (3x^2 - 1)\left(x^2 - \frac{1}{x}\right)$

29. $f(x) = \frac{x}{x^2 - 4} - \frac{x - 1}{x^2 + 4}$

30. $f(x) = \frac{x + \sqrt{3x}}{3x - 1}$

In exercises 31–34, find the derivative of the function and evaluate $f'(x)$ at the given value of x.

31. $f(x) = (2x - 1)(x^2 + 3); \ x = 1$

32. $f(x) = \frac{2x + 1}{2x - 1}; x = 2$

33. $f(x) = \frac{x}{x^4 - 2x^2 - 1}; x = -1$

34. $f(x) = (\sqrt{x} + 2x)(x^{3/2} - x); \ x = 4$

In exercises 35–38, find the slope and an equation of the tangent line to the graph of the function f at the specified point.

35. $f(x) = (x^3 + 1)(x^2 - 2); \ (2, 18)$

36. $f(x) = \frac{x^2}{x + 1}; \left(2, \frac{4}{3}\right)$

37. $f(x) = \frac{x + 1}{x^2 + 1}; (1, 1)$

38. $f(x) = \frac{1 + 2x^{1/2}}{1 + x^{3/2}}; \left(4, \frac{5}{9}\right)$

In exercises 39–44, find the first and second derivatives of the function.

39. $f(x) = 4x^2 - 2x + 1$

40. $f(x) = -0.2x^2 + 0.3x + 4$

41. $f(x) = 2x^3 - 3x^2 + 1$

42. $g(x) = -3x^3 + 24x^2 + 6x - 64$

43. $h(t) = t^4 - 2t^3 + 6t^2 - 3t + 10$

44. $f(x) = x^5 - x^4 + x^3 - x^2 + x - 1$

In exercises 45–48, find the third derivative of the function.

45. $f(x) = 3x^4 - 4x^3$

46. $f(x) = 3x^5 - 6x^4 + 2x^2 - 8x + 12$

47. $f(x) = \frac{1}{x}$ **48.** $f(x) = \frac{2}{x^2}$

49. Find an equation of the tangent line to the graph of the function $f(x) = (x^3 + 1)(3x^2 - 4x + 2)$ at the point $(1, 2)$.

50. Find an equation of the tangent line to the graph of the function $f(x) = \frac{3x}{x^2 - 2}$ at the point $(2, 3)$.

51. Let $f(x) = (x^2 + 1)(2 - x)$. Find the point(s) on the graph of f where the tangent line is horizontal.

52. Let $f(x) = \frac{x}{x^2 + 1}$. Find the point(s) on the graph of f where the tangent line is horizontal.

53. Find the point(s) on the graph of the function $f(x) = (x^2 + 6)(x - 5)$ where the slope of the tangent line is equal to -2.

54. Find the point(s) on the graph of the function $f(x) = \dfrac{x+1}{x-1}$ where the slope of the tangent line is equal to $-\frac{1}{2}$.

55. Concentration of a Drug in the Bloodstream   The concentration of a certain drug in a patient's bloodstream t hours after injection is given by

$$C(t) = \frac{0.2t}{t^2 + 1}$$

  a. Find the rate at which the concentration of the drug is changing with respect to time.
  b. How fast is the concentration changing $\frac{1}{2}$ hour after the injection? 1 hour after the injection? 2 hours after the injection?

56. Cost of Removing Toxic Waste   A city's main well was recently found to be contaminated with trichloroethylene, a cancer-causing chemical, as a result of an abandoned chemical dump leaching chemicals into the water. A proposal submitted to the city's selectmen indicates that the cost, measured in millions of dollars, of removing $x\%$ of the toxic pollutant is given by

$$C(x) = \frac{0.5x}{100 - x}$$

Find $C'(80)$, $C'(90)$, $C'(95)$, and $C'(99)$ and interpret your results.

57. Drug Dosages   Thomas Young has suggested the following rule for calculating the dosage of medicine for children 1 to 12 years old. If a denotes the adult dosage (in mg), and if t is the child's age (in years), then the child's dosage is given by

$$D(t) = \frac{at}{t + 12}$$

Suppose that the adult dosage of a substance is 500 mg. Find an expression that gives the rate of change of a child's dosage with respect to the child's age. What is the rate of change of a child's dosage with respect to his or her age for a 6-year-old child? For a 10-year-old child?

58. Effect of Bactericide   The number of bacteria $N(t)$ in a certain culture t minutes after an experimental bactericide is introduced obeys the rule

$$N(t) = \frac{10{,}000}{1 + t^2} + 2{,}000$$

Find the rate of change of the number of bacteria in the culture one minute after and two minutes after

the bactericide is introduced. What is the population of the bacteria in the culture one minute after the bactericide is introduced? Two minutes after?

59. Demand Functions   The demand function for the Sicard wristwatch is given by

$$d(x) = \frac{50}{0.01x^2 + 1} \qquad 0 \le x \le 20$$

where x (measured in units of a thousand) is the quantity demanded per week and $d(x)$ is the unit price in dollars.
  a. Find $d'(x)$.
  b. Find $d'(5)$, $d'(10)$, and $d'(15)$ and interpret your results.

60. Learning Curves   From experience, the Emory Secretarial School knows that the average student taking Advanced Typing will progress according to the rule

$$N(t) = \frac{60t + 180}{t + 6} \qquad t \ge 0$$

where $N(t)$ measures the number of words per minute the student can type after t weeks in the course.
  a. Find an expression for $N'(t)$.
  b. Compute $N'(t)$ for $t = 1, 3, 4$, and 7 and interpret your results.
  c. Sketch the graph of the function N. Does it confirm the results obtained in (b)?
  d. What will the average student's typing speed be at the end of the 12-week course?

61. Box Office Receipts   The total worldwide box office receipts for a long-running movie are approximated by the function

$$T(x) = \frac{120x^2}{x^2 + 4}$$

where $T(x)$ is measured in millions of dollars and x is the number of years since the movie's release. How fast are the total receipts changing one year after the movie's release? Three years after its release? Five years after its release?

62. A Falling Object   During the construction of an office building, a hammer is accidentally dropped from a height of 256 ft. The distance the hammer falls in t seconds is $s = 16t^2$. What is the hammer's velocity when it strikes the ground? What is its acceleration?

63. Acceleration of a Car   The distance s (in feet) covered by a car t seconds after starting from is given by

$$s = 20t + 8t^2 - t^3 \qquad 0 \le t \le 6$$

Find a general expression for the car's acceleration at any time t ($0 \le t \le 6$). Show that the car is decelerating $2\frac{2}{3}$ seconds after starting from rest.

64. Crime Rates   The number of major crimes committed in Bronxville between 1985 and 1992 is approximated by the function

$$N(t) = -0.1t^3 + 1.5t^2 + 100 \qquad 0 \le t \le 7$$

where $N(t)$ denotes the number of crimes committed in year t and $t = 0$ corresponds to the year 1985. Enraged by the dramatic increase in the crime rate, Bronxville's citizens, with the help of the local police, organized "Neighborhood Crime Watch" groups in early 1989 to combat this menace.
  a. Verify that the crime rate was increasing from 1985 through 1992. [*Hint:* Compute $N'(0)$, $N'(1)$, ..., $N'(7)$.]
  b. Show that the Neighborhood Crime Watch program was working by computing $N''(4)$, $N''(5)$, $N''(6)$, and $N''(7)$.

65. GDP of a Developing Country   A developing country's gross domestic product (GDP) from 1986 to 1994 is approximated by the function

$$G(t) = -0.2t^3 + 2.4t^2 + 60 \qquad 0 \le t \le 8$$

where $G(t)$ is measured in billions of dollars and $t = 0$ corresponds to the year 1986.
  a. Compute $G'(0)$, $G'(1)$, ..., $G'(8)$.
  b. Compute $G''(0)$, $G''(1)$, ..., $G''(8)$.
  c. Using the results obtained in (a) and (b), show that after a spectacular growth rate in the early years, the growth of the GDP cooled off.

66. Test Flight of a VTOL   In a test flight of the McCord Terrier, McCord Aviation's experimental VTOL (vertical take-off and landing) aircraft, it was determined that t seconds after lift-off, when the craft was operated in the vertical take-off mode, its altitude (in feet) was

$$h(t) = \frac{1}{16}t^4 - t^3 + 4t^2 \qquad 0 \le t \le 8$$

  a. Find an expression for the craft's velocity at time t.
  b. Find the craft's velocity when $t = 0$ (the initial velocity), $t = 4$, and $t = 8$.
  Find an expression for the craft's acceleration at time t.
  ind the craft's acceleration when $t = 0$, $t = 4$, e $t = 8$.
  ' the craft's height when $t = 0$, $t = 4$, and $t = 8$.

67. U.S. Census   According to the U.S. Census Bureau, the number of Americans aged 45 to 54 will be approximately

$$N(t) = -0.00233t^4 + 0.00633t^3 - 0.05417t^2$$
$$+ 1.3467t + 25$$

million people in year t, where $t = 0$ corresponds to the beginning of 1990. Compute $N'(10)$ and $N''(10)$ and interpret your results.
Source: U.S. Census Bureau

68. Air Purification   During testing of a certain brand of air purifier, it was determined that the amount of smoke remaining t minutes after the start of the test was

$$A(t) = 100 - 17.63t + 1.915t^2 - 0.1316t^3$$
$$+ 0.00468t^4 - 0.00006t^5$$

percent of the original amount. Compute $A'(10)$ and $A''(10)$ and interpret your results.
Source: Consumers Report

69. Extend the Product Rule for differentiation to the following case involving the product of three differentiable functions: Let $h(x) = u(x)v(x)w(x)$ and show that $h'(x) = u(x)v(x)w'(x) + u(x)v'(x)w(x) + u'(x)v(x)w(x)$. [*Hint:* Let $f(x) = u(x)v(x)$, $g(x) = w(x)$, and $h(x) = f(x)g(x)$, and apply the Product Rule to the function h.]

70. Prove the Quotient Rule for differentiation (Rule 6). [*Hint:* Verify the following steps.]

  a. $\dfrac{k(x+h) - k(x)}{h} = \dfrac{f(x+h)g(x) - f(x)g(x+h)}{hg(x+h)g(x)}$

  b. By adding $(-f(x)g(x) + f(x)g(x))$ to the numerator and simplifying,

$$\frac{k(x+h) - k(x)}{h} = \frac{1}{g(x+h)g(x)} \times$$

$$\left\{ \left[\frac{f(x+h) - f(x)}{h} \right] \cdot g(x) - \left[\frac{g(x+h) - g(x)}{h} \right] \cdot f(x) \right\}$$

  c. $k'(x) = \lim\limits_{h \to 0} \dfrac{k(x+h) - k(x)}{h}$

$$= \frac{g(x)f'(x) - f(x)g'(x)}{[g(x)]^2}$$

1. We use the Quotient Rule to obtain

$$h'(x) = \frac{(x^2 - 1)\dfrac{d}{dx}(2x + 1) - (2x + 1)\dfrac{d}{dx}(x^2 - 1)}{(x^2 - 1)^2}$$

$$= \frac{(x^2 - 1)(2) - (2x + 1)(2x)}{(x^2 - 1)^2}$$

$$= \frac{2x^2 - 2 - 4x^2 - 2x}{(x^2 - 1)^2}$$

$$= \frac{-2x^2 - 2x - 2}{(x^2 - 1)^2}$$

$$= -\frac{2(x^2 + x + 1)}{(x^2 - 1)^2}$$

2. The slope of the tangent line to the graph of f at any point is given by

$$f'(x) = (x^2 + 1)\frac{d}{dx}(2x^3 - 3x^2 + 1)$$

$$+ (2x^3 - 3x^2 + 1)\frac{d}{dx}(x^2 + 1)$$

$$= (x^2 + 1)(6x^2 - 6x) + (2x^3 - 3x^2 + 1)(2x)$$

In particular, the slope of the tangent line to the graph of f when $x = 2$ is

$$f'(2) = (2^2 + 1)[6(2^2) - 6(2)]$$

$$+ [2(2^3) - 3(2^2) + 1][2(2)]$$

$$= 60 + 20 = 80$$

Note that it is not necessary to simplify the expression for $f'(x)$, since we are required only to evaluate the expression at $x = 2$. We also conclude, from this result, that the function f is changing at the rate of 80 units per unit change in x when $x = 2$.

3. The rate at which the company's total sales are changing at any time t is given by

$$S'(t) = \frac{(1 + 0.4t^2)\dfrac{d}{dt}(0.3t^3) - (0.3t^3)\dfrac{d}{dt}(1 + 0.4t^2)}{(1 + 0.4t^2)^2}$$

$$= \frac{(1 + 0.4t^2)(0.9t^2) - (0.3t^3)(0.8t)}{(1 + 0.4t^2)^2}$$

Therefore, at the beginning of the second year of operation, Security Products' sales were increasing at the rate of

$$S'(1) = \frac{(1 + 0.4)(0.9) - (0.3)(0.8)}{(1 + 0.4)^2} = 0.520408$$

or $520,408 per year.

Using Technology

The **dy/dx** operation of a graphing utility can be used to find the derivative of any differentiable algebraic function.

Finding the Second Derivative of a Function at a Given Point

The **der2** function of a graphing utility can be used to find the second derivative of a function at a given point. (See Section A.12 in Appendix A.)

EXAMPLE 1 Use the **der2** function to find the second derivative of $f(x) = \sqrt{x}$ when $x = 4$.

Solution We find

$$f''(4) = \text{der2 } (x^{\wedge}0.5, x, 4) = -0.03125$$

EXERCISES

*In exercises 1–6, use the **dy/dx** operation of a graphing utility to find the rate of change of $f(x)$ at the given value of x. Give your answer accurate to four decimal places.*

1. $f(x) = (2x^2 + 1)(x^3 + 3x + 4);\quad x = -0.5$

2. $f(x) = (\sqrt{x} + 1)(2x^2 + x - 3);\quad x = 1.5$

3. $f(x) = \dfrac{\sqrt{x} - 1}{\sqrt{x} + 1};\quad x = 3$

4. $f(x) = \dfrac{\sqrt{x}(x^2 + 4)}{x^3 + 1};\quad x = 4$

5. $f(x) = \dfrac{\sqrt{x}(1 + x^{-1})}{x + 1};\quad x = 1$

6. $f(x) = \dfrac{x^2(2 + \sqrt{x})}{1 + \sqrt{x}};\quad x = 1$

7. New Construction Jobs   The president of a major housing construction company claims that the number of construction jobs created in the next t months is given by

$$f(t) = 1.42 \left(\frac{7t^2 + 140t + 700}{3t^2 + 80t + 550} \right)$$

where $f(t)$ is measured in millions of jobs per year. At what rate will construction jobs be created one year from now, assuming her projection is correct?

In exercises 8–15, find the value of the second derivative of f at the given value of x.

8. $f(x) = 2x^3 - 3x^2 + 1;\quad x = -1$

9. $f(x) = 2.5x^5 - 3x^3 + 1.5x + 4;\quad x = 2.1$

10. $f(x) = 2.1x^{3.1} - 4.2x^{1.7} + 4.2;\quad x = 1.4$

11. $f(x) = 1.7x^{4.2} - 3.2x^{1.3} + 4.2x - 3.2;\quad x = 2.2$

12. $f(x) = \dfrac{x^2 + 2x - 5}{x^3 + 1};\quad x = 2.1$

13. $f(x) = \dfrac{x^3 + x + 2}{2x^2 - 5x + 4};\quad x = 1.2$

14. $f(x) = \dfrac{x^{1/2} + 2x^{3/2} + 1}{2x^{1/2} + 3};\quad x = 0.5$

15. $f(x) = \dfrac{\sqrt{x} - 1}{2x + \sqrt{x} + 4};\quad x = 2.3$

9.6   The Chain Rule

This section introduces another rule of differentiation called the *Chain Rule*. When used in conjunction with the rules of differentiation developed in the last two sections, the Chain Rule enables us to greatly enlarge the class of functions we are able to differentiate.

Composition of Functions

Many functions are built up from other, generally simpler, functions through a process known as the *composition of functions.* Consider for example, the function h whose rule is given by $h(x) = \sqrt{x^2 - 1}$. Let f and g be functions defined by the rules $f(x) = x^2 - 1$ and $g(x) = \sqrt{x}$. Evaluating the function g at the point $f(x)$ (remember that for each real number x in the domain of f, $f(x)$ is simply a real number), we find that

$$g(f(x)) = \sqrt{f(x)} = \sqrt{x^2 - 1}$$

which is just the rule defining the function h!

In general, the *composition of a function g with a function f* is defined as the function h whose value is found by evaluating the function g at the point $f(x)$; that is,

$$h(x) = g(f(x))$$

The function h is called a *composite function* and is commonly denoted by $h = g \circ f$ (read "g circle f" or "g composed with f"). The interpretation of the function h as a machine is illustrated in Figure 40, and its interpretation as a mapping is shown in Figure 41.

FIGURE 40
The composite function $h = g \circ f$ is found by evaluating f at x and then g at $f(x)$, in that order.

FIGURE 41
The function $h = g \circ f$ viewed as a mapping

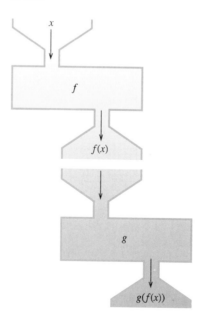

EXAMPLE 1   Let $f(x) = x^2 - 1$ and $g(x) = \sqrt{x} + 1$. Compute

a. the rule for the composite function $g \circ f$

b. the rule for the composite function $f \circ g$

Solution

a. To find the rule for the composite function $g \circ f$, evaluate the function g at $f(x)$. Therefore,

$$(g \circ f)(x) = g(f(x)) = \sqrt{f(x)} + 1 = \sqrt{x^2 - 1} + 1$$

b. To find the rule for the composite function $f \circ g$, evaluate the function f at $g(x)$. Thus,

$$(f \circ g)(x) = f(g(x)) = (g(x))^2 - 1 = (\sqrt{x} + 1)^2 - 1$$
$$= x + 2\sqrt{x} + 1 - 1 = x + 2\sqrt{x}$$

Example 1 reminds us that, in general, $g \circ f$ is different from $f \circ g$, and care must be taken regarding the order when computing a composite function.

 EXAMPLE 2   An environmental impact study conducted for Oxnard's City Environmental Management Department indicates that, under existing environmental protection laws, the level of carbon monoxide present in the air due to pollution from automobile exhaust will be $0.01x^{2/3}$ parts per million when the number of motor vehicles is x thousand. A separate study conducted by a state government agency estimates that t years from now the number of motor vehicles in Oxnard will be $0.2t^2 + 4t + 64$ thousand.

a. Find an expression for the concentration of carbon monoxide in the air due to automobile exhaust t years from now.

b. What will the level of concentration be five years from now?

Solution

a. The level of carbon monoxide present in the air due to pollution from automobile exhaust is described by the function $g(x) = 0.01x^{2/3}$, where x is the number (in thousands) of motor vehicles. But the number of motor vehicles x (in thousands) t years from now may be estimated by the rule $f(t) = 0.2t^2 + 4t + 64$. Therefore, the concentration of carbon monoxide due to automobile exhaust t years from now is given by

$$C(t) = g(f(t)) = 0.01(0.2t^2 + 4t + 64)^{2/3}$$

parts per million.

b. The level of concentration five years from now will be

$$C(5) = 0.01[0.2(5)^2 + 4(5) + 64]^{2/3}$$
$$= (0.01)89^{2/3} \approx 0.20$$

or 0.20 parts per million.

The Chain Rule

Consider the function $h(x) = (x^2 + x + 1)^2$. If we were to compute $h'(x)$ using only the rules of differentiation from the previous sections, then our approach might be to expand $h(x)$. Thus,

$$h(x) = (x^2 + x + 1)^2 = (x^2 + x + 1)(x^2 + x + 1)$$
$$= x^4 + 2x^3 + 3x^2 + 2x + 1$$

from which we find

$$h'(x) = 4x^3 + 6x^2 + 6x + 2$$

But what about the function $H(x) = (x^2 + x + 1)^{100}$? The same technique may be used to find the derivative of the function H, but the amount of work involved in this case would be prodigious! Consider, also, the function $G(x) = \sqrt{x^2 + 1}$. For each of the two functions H and G, the rules of differentiation of the previous sections cannot be applied directly to compute the derivatives H' and G'.

Observe that both H and G are composite functions; that is, each is composed of, or built up from, simpler functions. For example, the function H is composed of the two simpler functions $f(x) = x^2 + x + 1$ and $g(x) = x^{100}$ as follows:

$$H(x) = g[f(x)] = [f(x)]^{100}$$
$$= (x^2 + x + 1)^{100}$$

In a similar manner, we see that the function G is composed of the two simpler functions $f(x) = x^2 + 1$ and $g(x) = \sqrt{x}$. Thus,

$$G(x) = g[f(x)] = \sqrt{f(x)}$$
$$= \sqrt{x^2 + 1}$$

As a first step toward finding the derivative h' of a composite function $h = g \circ f$ defined by $h(x) = g[f(x)]$, we write

$$u = f(x) \qquad \text{and} \qquad y = g[f(x)] = g(u)$$

The dependency of h on g and f is illustrated in Figure 42. Since u is a function of x, we may compute the derivative of u with respect to x, if f is a differentiable function, obtaining $du/dx = f'(x)$. Next, if g is a differentiable function of u, we may compute the derivative of g with respect to u, obtaining $dy/du = g'(u)$. Now, since the function h is composed of the function g and the function f, we might suspect that the rule $h'(x)$ for the derivative h' of h will be given by an expression that involves the rules for the derivatives of f and g. But how do we combine these derivatives to yield h'?

FIGURE 42

The composite function
$h(x) = g[f(x)]$

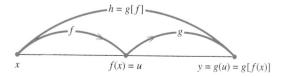

This question can be answered by interpreting the derivative of each function as giving the rate of change of that function. For example, suppose that $u = f(x)$ changes three times as fast as x; that is,

$$f'(x) = \frac{du}{dx} = 3$$

and $y = g(u)$ changes twice as fast as u; that is,

$$g'(u) = \frac{dy}{du} = 2$$

Then we would expect $y = h(x)$ to change six times as fast as x; that is,

$$h'(x) = g'(u)f'(x) = (2)(3) = 6$$

or equivalently,

$$\frac{dy}{dx} = \frac{dy}{du} \cdot \frac{du}{dx} = (2)(3) = 6$$

This observation suggests the following result, which we state without proof.

RULE 7:
THE CHAIN RULE

If $h(x) = g[f(x)]$, then

$$h'(x) = \frac{d}{dx}g(f(x)) = g'(f(x))f'(x) \tag{10}$$

Equivalently, if we write $y = h(x) = g(u)$, where $u = f(x)$, then

$$\frac{dy}{dx} = \frac{dy}{du} \cdot \frac{du}{dx} \tag{11}$$

REMARKS

1. If we label the composite function h in the following manner

Inside function
↓
$$h(x) = g[f(x)]$$
↑
Outside function

then $h'(x)$ is just the *derivative* of the "outside function" *evaluated at* the "inside function" times the *derivative* of the "inside function."

2. Equation (11) can be remembered by observing that if we "cancel" the du's, then

$$\frac{dy}{dx} = \frac{dy}{du} \cdot \frac{du}{dx} = \frac{dy}{dx}$$

□

The Chain Rule for Powers of Functions

Many composite functions have the special form $h(x) = g(f(x))$, where g is defined by the rule $g(x) = x^n$ (n, a real number); that is,

$$h(x) = [f(x)]^n$$

In other words, the function h is given by the power of a function f. The functions

$$h(x) = (x^2 + x + 1)^2, \qquad H = (x^2 + x + 1)^{100}, \qquad G = \sqrt{x^2 + 1}$$

discussed earlier are examples of this type of composite function. By using the following corollary of the Chain Rule, the General Power Rule, we are able to find the derivative of this type of function much more easily than by using the Chain Rule directly.

THE GENERAL POWER RULE

If the function f is differentiable and $h(x) = [f(x)]^n$ (n, a real number), then

$$h'(x) = \frac{d}{dx}[f(x)]^n = n[f(x)]^{n-1}f'(x) \qquad (12)$$

To see this, we observe that $h(x) = g(f(x))$, where $g(x) = x^n$, so that, by virtue of the Chain Rule, we have

$$h'(x) = g'(f(x))f'(x)$$
$$= n[f(x)]^{n-1}f'(x)$$

since $g'(x) = nx^{n-1}$.

EXAMPLE 3   Find $H'(x)$, if

$$H(x) = (x^2 + x + 1)^{100}$$

Solution   Using the General Power Rule, we obtain

$$H'(x) = 100(x^2 + x + 1)^{99} \frac{d}{dx}(x^2 + x + 1)$$

$$= 100(x^2 + x + 1)^{99}(2x + 1)$$

Observe the bonus that comes from using the General Power Rule: The answer is completely factored.   ■

EXAMPLE 4   Differentiate the function $G(x) = \sqrt{x^2 + 1}$.

Solution   We rewrite the function $G(x)$ as

$$G(x) = (x^2 + 1)^{1/2}$$

and apply the General Power Rule, obtaining

$$G'(x) = \frac{1}{2}(x^2 + 1)^{-1/2} \frac{d}{dx}(x^2 + 1)$$

$$= \frac{1}{2}(x^2 + 1)^{-1/2} \cdot 2x = \frac{x}{\sqrt{x^2 + 1}}$$ ■

EXAMPLE 5   Differentiate the function $f(x) = x^2(2x + 3)^5$.

Solution   Applying the Product Rule followed by the General Power Rule, we obtain

$$f'(x) = x^2 \frac{d}{dx}(2x + 3)^5 + (2x + 3)^5 \frac{d}{dx}(x^2)$$

$$= (x^2)5(2x + 3)^4 \cdot \frac{d}{dx}(2x + 3) + (2x + 3)^5(2x)$$

$$= 5x^2(2x + 3)^4(2) + 2x(2x + 3)^5$$

$$= 2x(2x + 3)^4(5x + 2x + 3) = 2x(7x + 3)(2x + 3)^4 \qquad \blacksquare$$

EXAMPLE 6   Find $f'(x)$ if $f(x) = \dfrac{1}{(4x^2 - 7)^2}$.

Solution   Rewriting $f(x)$ and then applying the General Power Rule, we obtain

$$f'(x) = \frac{d}{dx}\left[\frac{1}{(4x^2 - 7)^2}\right] = \frac{d}{dx}(4x^2 - 7)^{-2}$$

$$= -2(4x^2 - 7)^{-3}\frac{d}{dx}(4x^2 - 7)$$

$$= -2(4x^2 - 7)^{-3}(8x) = -\frac{16x}{(4x^2 - 7)^3} \qquad \blacksquare$$

EXAMPLE 7   Find the slope of the tangent line to the graph of the function

$$f(x) = \left(\frac{2x + 1}{3x + 2}\right)^3$$

at the point $(0, \frac{1}{8})$.

Solution   The slope of the tangent line to the graph of f at any point is given by $f'(x)$. To compute $f'(x)$, we use the General Power Rule followed by the Quotient Rule, obtaining

$$f'(x) = 3\left(\frac{2x + 1}{3x + 2}\right)^2 \frac{d}{dx}\left(\frac{2x + 1}{3x + 2}\right)$$

$$= 3\left(\frac{2x + 1}{3x + 2}\right)^2 \left[\frac{(3x + 2)(2) - (2x + 1)(3)}{(3x + 2)^2}\right]$$

$$= 3\left(\frac{2x + 1}{3x + 2}\right)^2 \left[\frac{6x + 4 - 6x - 3}{(3x + 2)^2}\right]$$

$$= \frac{3(2x + 1)^2}{(3x + 2)^4}$$

In particular, the slope of the tangent line to the graph of f at $(0, \frac{1}{8})$ is given by

$$f'(0) = \frac{3(0 + 1)^2}{(0 + 2)^4} = \frac{3}{16} \qquad \blacksquare$$

Applications

EXAMPLE 8 The membership of The Fitness Center, which opened a few years ago, is approximated by the function

$$N(t) = 100(64 + 4t)^{2/3} \qquad 0 \le t \le 52$$

where $N(t)$ gives the number of members at the beginning of week t.

a. Find $N'(t)$.

b. How fast was the center's membership increasing initially ($t = 0$)?

c. How fast was the membership increasing at the beginning of the 40th week?

d. What was the membership when the center first opened? At the beginning of the 40th week?

Solution

a. Using the General Power Rule, we obtain

$$N'(t) = \frac{d}{dt}[100(64 + 4t)^{2/3}]$$

$$= 100\frac{d}{dt}(64 + 4t)^{2/3}$$

$$= 100\left(\frac{2}{3}\right)(64 + 4t)^{-1/3}\frac{d}{dt}(64 + 4t)$$

$$= \frac{200}{3}(64 + 4t)^{-1/3}(4)$$

$$= \frac{800}{3(64 + 4t)^{1/3}}$$

b. The rate at which the membership was increasing when the center first opened is given by

$$N'(0) = \frac{800}{3(64)^{1/3}} \approx 66.7$$

or approximately 67 people per week.

c. The rate at which the membership was increasing at the beginning of the 40th week is given by

$$N'(40) = \frac{800}{3(64 + 160)^{1/3}} \approx 43.9$$

or approximately 44 people per week.

d. The membership when the center first opened is given by

$$N(0) = 100(64)^{2/3} = 100(16)$$

or approximately 1600 people. The membership at the beginning of the 40th week is given by

$$N(40) = 100(64 + 160)^{2/3} \approx 3688.4$$

or approximately 3688 people.

FIGURE 43

Cross section of the aorta

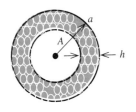

EXAMPLE 9   Arteriosclerosis begins during childhood when plaque (soft masses of fatty material) forms in the arterial walls, blocking the flow of blood through the arteries and leading to heart attacks, strokes, and gangrene. Suppose the idealized cross section of the aorta is circular with radius a cm, and that by year t the thickness of the plaque (assume that it is uniform) is $h = f(t)$ cm (Figure 43). Then the area of the opening is given by $A = \pi(a - h)^2$ square centimeters.

Suppose that the radius of an individual's artery is 1 cm ($a = 1$) and the thickness of the plaque in year t is given by

$$h = g(t) = 1 - 0.01(10{,}000 - t^2)^{1/2} \text{ cm}$$

Since the area of the arterial opening is given by

$$A = f(h) = \pi(1 - h)^2$$

the rate at which A is changing with respect to time is given by

$$\frac{dA}{dt} = \frac{dA}{dh} \cdot \frac{dh}{dt} = f'(h) \cdot g'(t)$$ By the Chain Rule

$$= 2\pi(1 - h)(-1)[-0.01(\tfrac{1}{2})(10{,}000 - t^2)^{-1/2}(-2t)]$$ Using the Chain Rule twice

$$= -2\pi(1 - h)\left[\frac{0.01t}{(10{,}000 - t^2)^{1/2}}\right]$$

$$= -\frac{0.02\pi(1 - h)t}{\sqrt{10{,}000 - t^2}}$$

For example, when $t = 50$,

$$h = g(50) = 1 - 0.01(10{,}000 - 2{,}500)^{1/2} \approx 0.134$$

so that

$$\frac{dA}{dt} = -\frac{0.02\pi(1 - 0.134)50}{\sqrt{10{,}000 - 2{,}500}} \approx -0.03$$

that is, the area of the arterial opening is decreasing at the rate of 0.03 cm^2/year.

1. Find the derivative of

$$f(x) = -\frac{1}{\sqrt{2x^2 - 1}}$$

2. Suppose that the life expectancy at birth (in years) of a female in a certain country is described by the function

$$g(t) = 50.02(1 + 1.09t)^{0.1} \qquad 0 \le t \le 150$$

where t is measured in years and $t = 0$ corresponds to the beginning of 1900.
   a. What is the life expectancy at birth of a female born at the beginning of 1980? At the beginning of the year 2000?
   b. How fast is the life expectancy at birth of a female born at any time t changing?

Solutions to Self-Check Exercises 9.6 can be found on page 559.

9.6 EXERCISES

In exercises 1–6, find the rules for the composite functions f ∘ g and g ∘ f.

1. $f(x) = x^2 + x + 1$; $g(x) = x^2$

2. $f(x) = 3x^2 + 2x + 1$; $g(x) = x + 3$

3. $f(x) = \sqrt{x} + 1$; $g(x) = x^2 - 1$

4. $f(x) = 2\sqrt{x} + 3$; $g(x) = x^2 + 1$

5. $f(x) = \dfrac{x}{x^2 + 1}$; $g(x) = \dfrac{1}{x}$

6. $f(x) = \sqrt{x + 1}$; $g(x) = \dfrac{1}{x - 1}$

In exercises 7–10, evaluate h(2) where h = g ∘ f.

7. $f(x) = x^2 + x + 1$; $g(x) = x^2$

8. $f(x) = \sqrt[3]{x^2 - 1}$; $g(x) = 3x^3 + 1$

9. $f(x) = \dfrac{1}{2x + 1}$; $g(x) = \sqrt{x}$

10. $f(x) = \dfrac{1}{x - 1}$; $g(x) = x^2 + 1$

In exercises 11–16, find functions f and g such that h = g ∘ f. (Note: The answer is not unique.)

11. $h(x) = (2x^3 + x^2 + 1)^5$

12. $h(x) = (3x^2 - 4)^{-3}$

13. $h(x) = \sqrt{x^2 - 1}$

14. $h(x) = (2x - 3)^{3/2}$

15. $h(x) = \dfrac{1}{x^2 - 1}$

16. $h(x) = \dfrac{1}{\sqrt{x^2 - 4}}$

In exercises 17–58, find the derivative of the function.

17. $f(x) = (2x - 1)^4$

18. $f(x) = (1 - x)^3$

19. $f(x) = (x^2 + 2)^5$

20. $f(t) = 2(t^3 - 1)^5$

21. $f(x) = (2x - x^2)^3$

22. $f(x) = 3(x^3 - x)^4$

23. $f(x) = (2x + 1)^{-2}$

24. $f(t) = \tfrac{1}{2}(2t^2 + t)^{-3}$

25. $f(x) = (x^2 - 4)^{3/2}$

26. $f(t) = (3t^2 - 2t + 1)^{3/2}$

27. $f(x) = \sqrt{3x - 2}$

28. $f(t) = \sqrt{3t^2 - t}$

29. $f(x) = \sqrt[3]{1 - x^2}$

30. $f(x) = \sqrt{2x^2 - 2x + 3}$

31. $f(x) = \dfrac{1}{(2x + 3)^3}$

32. $f(x) = \dfrac{2}{(x^2 - 1)^4}$

33. $f(t) = \dfrac{1}{\sqrt{2t - 3}}$

34. $f(x) = \dfrac{1}{\sqrt{2x^2 - 1}}$

35. $y = \dfrac{1}{(4x^4 + x)^{3/2}}$

36. $f(t) = \dfrac{4}{\sqrt[3]{2t^2 + t}}$

37. $f(x) = (3x^2 + 2x + 1)^{-2}$

38. $f(t) = (5t^3 + 2t^2 - t + 4)^{-3}$

39. $f(x) = (x^2 + 1)^3 - (x^3 + 1)^2$

40. $f(t) = (2t - 1)^4 + (2t + 1)^4$

41. $f(t) = (t^{-1} - t^{-2})^3$

42. $f(v) = (v^{-3} + 4v^{-2})^3$

43. $f(x) = \sqrt{x + 1} + \sqrt{x - 1}$

44. $f(u) = (2u + 1)^{3/2} + (u^2 - 1)^{-3/2}$

45. $f(x) = 2x^2(3 - 4x)^4$

46. $h(t) = t^2(3t + 4)^3$

47. $f(x) = (x - 1)^2(2x + 1)^4$

48. $g(u) = (1 + u^2)^5(1 - 2u^2)^8$

49. $f(x) = \left(\dfrac{x + 3}{x - 2}\right)^3$

50. $f(x) = \left(\dfrac{x + 1}{x - 1}\right)^5$

51. $s(t) = \left(\dfrac{t}{2t + 1}\right)^{3/2}$

52. $g(s) = \left(s^2 + \dfrac{1}{s}\right)^{3/2}$

53. $g(u) = \sqrt{\dfrac{u + 1}{3u + 2}}$

54. $g(x) = \sqrt{\dfrac{2x + 1}{2x - 1}}$

55. $f(x) = \dfrac{x^2}{(x^2 - 1)^4}$

56. $g(u) = \dfrac{2u^2}{(u^2 + u)^3}$

57. $h(x) = \dfrac{(3x^2 + 1)^3}{(x^2 - 1)^4}$

58. $g(t) = \dfrac{(2t - 1)^2}{(3t + 2)^4}$

In exercises 59–62, find dy/du, du/dx, and dy/dx.

59. $y = u^{4/3}$ and $u = 3x^2 - 1$

60. $y = \sqrt{u}$ and $u = 7x - 2x^2$

61. $y = u^{-2/3}$ and $u = 2x^3 - x + 1$

62. $y = 2u^2 + 1$ and $u = x^2 + 1$

In exercises 63–68, find the first and second derivatives of the function.

63. $f(x) = (x^2 + 2)^5$

64. $g(t) = t^2(3t + 1)^4$

65. $f(x) = (2x^2 + 2)^{7/2}$

66. $h(w) = (w^2 + 2w + 4)^{5/2}$

67. $f(x) = \dfrac{x}{2x + 1}$ **68.** $g(t) = \dfrac{t^2}{t - 1}$

In exercises 69 and 70, find an equation of the tangent line to the graph of the function at the given point.

69. $f(x) = (1 - x)(x^2 - 1)^2; (2, -9)$

70. $f(x) = \left(\dfrac{x + 1}{x - 1}\right)^2; (3, 4)$

71. Television Viewing The number of viewers of a television series introduced several years ago is approximated by the function

$$N(x) = (60 + 2x)^{2/3} \qquad 1 \le x \le 26$$

where $N(x)$ (measured in millions) denotes the number of weekly viewers of the series in the xth week. Find the rate of increase of the weekly audience at the end of the 2nd week and at the end of the 12th week. How many viewers were there in the 2nd week? In the 24th week?

72. Male Life Expectancy Suppose that the life expectancy of a male at birth in a certain country is described by the function

$$f(t) = 46.9(1 + 1.09t)^{0.1} \qquad 0 \le t \le 150$$

where t is measured in years and $t = 0$ corresponds to the beginning of 1900. How long can a male born at the beginning of the year 2000 in that country expect to live? What is the rate of change of the life expectancy of a male born in that country at the beginning of the year 2000?

73. Concentration of Carbon Monoxide in the Air According to a joint study conducted by Oxnard's Environmental Management Department and a state government agency, the concentration of carbon monoxide in the air due to automobile exhaust t years from now is given by

$$C(t) = 0.01(0.2t^2 + 4t + 64)^{2/3}$$

parts per million.
a. Find the rate at which the level of carbon monoxide is changing with respect to time.
b. Find the rate at which the level of carbon monoxide will be changing five years from now.

74. Continuing Education Enrollment The registrar of Kellogg University estimates that the total student enrollment in the Continuing Education division will be given by

$$N(t) = -\dfrac{20,000}{\sqrt{1 + 0.2t}} + 21,000$$

where $N(t)$ denotes the number of students enrolled in the division t years from now. Find an expression for $N'(t)$. How fast is the student enrollment increasing currently? How fast will it be increasing five years from now?

75. Air Pollution According to the South Coast Air Quality Management District, the level of nitrogen dioxide, a brown gas that impairs breathing, present in the atmosphere on a certain May day in downtown Los Angeles is approximated by

$$A(t) = 0.03t^3(t - 7)^4 + 60.2 \qquad 0 \le t \le 7$$

where $A(t)$ is measured in pollutant standard index and t is measured in hours, with $t = 0$ corresponding to 7 A.M.
a. Find $A'(t)$.
b. Find $A'(1)$, $A'(3)$, and $A'(4)$ and interpret your results.

76. Effect of Luxury Tax on Consumption Government economists of a developing country determined that the purchase of imported perfume is related to a proposed "luxury tax" by the formula

$$N(x) = \sqrt{10,000 - 40x - 0.02x^2} \qquad 0 \le x \le 200$$

where $N(x)$ measures the percentage of normal consumption of perfume when a "luxury tax" of x percent is imposed on it. Find the rate of change of $N(x)$ for taxes of 10%, 100%, and 150%.

77. Pulse Rate of an Athlete The pulse rate (the number of heartbeats per minute) of a long-distance runner t seconds after leaving the starting line is given by

$$P(t) = \dfrac{300\sqrt{\frac{1}{2}t^2 + 2t + 25}}{t + 25} \qquad t \ge 0$$

Compute $P'(t)$. How fast is the athlete's pulse rate increasing 10 seconds into the run? 60 seconds into the run? 2 minutes into the run? What is her pulse rate 2 minutes into the run?

78. Thurstone Learning Model Psychologist L. L. Thurstone suggested the following relationship between the learning time T and the length of a list n:

$$T = f(n) = An\sqrt{n - b}$$

where A and b are constants that depend on the person and the task.
a. Compute dT/dn and interpret your result.
b. Suppose that for a certain person and a certain task, $A = 4$ and $b = 4$. Compute $f'(13)$ and $f'(29)$ and interpret your results.

79. Oil Spills In calm waters, the oil spilling from the ruptured hull of a grounded tanker spreads in all directions. Assuming that the area polluted is a circle and that its radius is increasing at a rate of 2 ft/sec, determine how fast the area is increasing when the radius of the circle is 40 feet.

80. Arteriosclerosis Refer to Example 9, page 556. Suppose that the radius of an individual's artery is 1 cm and the thickness of the plaque (in cm) t years from now is given by

$$h = g(t) = \frac{0.5t^2}{t^2 + 10} \qquad 0 \le t \le 10$$

How fast will the arterial opening be decreasing five years from now?

81. Demand for PCs The quantity demanded per month, x, of a certain make of personal computer (PC) is related to the average unit price, p, (in dollars) of personal computers by the equation

$$x = f(p) = \frac{100}{9} \sqrt{810,000 - p^2}$$

It is estimated that t months from now, the average price of a PC will be given by

$$p(t) = \frac{400}{1 + \frac{1}{8}\sqrt{t}} + 200 \qquad 0 \le t \le 60$$

dollars. Find the rate at which the quantity demanded per month of the PCs will be changing 16 months from now.

SOLUTIONS TO SELF-CHECK EXERCISES 9.6

1. Rewriting, we have

$$f(x) = -(2x^2 - 1)^{-1/2}$$

Using the General Power Rule, we find

$$f'(x) = -\frac{d}{dx}(2x^2 - 1)^{-1/2}$$

$$= -\left(-\frac{1}{2}\right)(2x^2 - 1)^{-3/2}\frac{d}{dx}(2x^2 - 1)$$

$$= \frac{1}{2}(2x^2 - 1)^{-3/2}(4x)$$

$$= \frac{2x}{(2x^2 - 1)^{3/2}}$$

2. a. The life expectancy at birth of a female born at the beginning of 1980 is given by

$$g(80) = 50.02[1 + 1.09(80)]^{0.1} \approx 78.29$$

or approximately 78 years. Similarly, the life expectancy at birth of a female born at the beginning of the year 2000 is given by

$$g(100) = 50.02[1 + 1.09(100)]^{0.1} \approx 80.04$$

or approximately 80 years.

b. The rate of change of life expectancy at birth of a female born at any time t is given by $g'(t)$. Using the General Power Rule, we have

$$g'(t) = 50.02\frac{d}{dt}(1 + 1.09t)^{0.1}$$

$$= (50.02)(0.1)(1 + 1.09t)^{-0.9}\frac{d}{dt}(1 + 1.09t)$$

$$= (50.02)(0.1)(1.09)(1 + 1.09t)^{-0.9}$$

$$= 5.45218(1 + 1.09t)^{-0.9}$$

$$= \frac{5.45218}{(1 + 1.09t)^{0.9}}$$

U s i n g   T e c h n o l o g y

The **dy/dx** operation of a graphing utility will yield the derivative of differentiable composite functions. Use this fact in solving the following exercises.

EXERCISES

*In exercises 1–6, use the **dy/dx** operation of a graphing utility to find the rate of change of f(x) at the given value of x. Give your answer accurate to four decimal places.*

1. $f(x) = \sqrt{x^2 - x^4}$; $x = 0.5$

2. $f(x) = x - \sqrt{1 - x^2}$; $x = 0.4$

3. $f(x) = x\sqrt{1 - x^2}$; $x = 0.2$

4. $f(x) = (x + \sqrt{x^2 + 4})^{3/2}$; $x = 1$

5. $f(x) = \dfrac{\sqrt{1 + x^2}}{x^3 + 2}$; $x = -1$

6. $f(x) = \dfrac{x^3}{1 + (1 + x^2)^{3/2}}$; $x = 3$

7. Population Growth   A major corporation is building a 4325-acre complex of homes, offices, stores, schools, and churches in the rural community of Glen Cove. As a result of this development, the planners have estimated that Glen Cove's population (in thousands)

t years from now will be given by

$$P(t) = \frac{25t^2 + 125t + 200}{t^2 + 5t + 40}$$

a. What will the population be ten years from now?
b. At what rate will the population be increasing ten years from now?

8. Worldwide Production of Vehicles   According to *Automotive News*, the worldwide production of vehicles between 1960 and 1990 is given by the function

$$f(t) = 16.5\sqrt{1 + 2.2t} \qquad 0 \le t \le 3$$

where $f(t)$ is measured in units of a million and *t* is measured in decades with $t = 0$ corresponding to the beginning of 1960. What was the rate of change of the worldwide production of vehicles at the beginning of 1970? At the beginning of 1980?

Source: Automotive News

9.7

Differentiation of Exponential and Logarithmic Functions

The Derivative of the Exponential Function

In order to study the effects of budget deficit–reduction plans at different income levels, it is important to know the income distribution of American families. According to the House Budget Committee and the U.S. Census Bureau, the graph of *f* shown in Figure 44 gives the number of American

FIGURE 44
The graph of f shows the number of families versus their annual income.

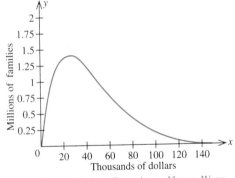

Source: House Budget Committee, House Ways and Means Committee, and U.S. Census Bureau

families y (in millions) as a function of their annual income x (in thousands of dollars) in 1990.

Observe that the graph of f rises very quickly and then tapers off. From the graph of f, you can see that the bulk of American families earned less than \$100,000 per year. In fact, 95% of U.S. families earned less than \$102,358 per year in 1990. (We will refer to this model again in Using Technology at the end of this section.)

To analyze mathematical models involving exponential and logarithmic functions in greater detail, we need to develop rules for computing the derivative of these functions. We begin by looking at the rule for computing the derivative of the exponential function.

RULE 8: DERIVATIVE OF THE EXPONENTIAL FUNCTION

$$\frac{d}{dx} e^x = e^x$$

Thus, the derivative of the exponential function with base e is equal to the function itself. To demonstrate the validity of this rule, we compute

$$f'(x) = \lim_{h \to 0} \frac{f(x + h) - f(x)}{h}$$

$$= \lim_{h \to 0} \frac{e^{x+h} - e^x}{h}$$

$$= \lim_{h \to 0} \frac{e^x(e^h - 1)}{h} \qquad \text{Writing } e^{x+h} = e^x e^h \text{ and factoring}$$

$$= e^x \lim_{h \to 0} \frac{e^h - 1}{h} \qquad \text{Why?}$$

To evaluate

$$\lim_{h \to 0} \frac{e^h - 1}{h}$$

let us refer to Table 4, which is constructed with the aid of a calculator. From the table, we see that

$$\lim_{h \to 0} \frac{e^h - 1}{h} = 1$$

TABLE 4

h	0.1	0.01	0.001	-0.1	-0.01	-0.001
$\dfrac{e^h - 1}{h}$	1.0517	1.0050	1.0005	0.9516	0.9950	0.9995

For a visual confirmation of this result, see the example in Using Technology at the end of this section. (Although a rigorous proof of this fact is possible, it is beyond the scope of this book.) Using this result, we conclude that

$$f'(x) = e^x \cdot 1 = e^x$$

as we set out to show.

EXAMPLE 1 Compute the derivative of each of the following functions:

a. $f(x) = x^2 e^x$ **b.** $g(t) = (e^t + 2)^{3/2}$

Solution

a. The Product Rule gives

$$f'(x) = \frac{d}{dx}(x^2 e^x)$$

$$= x^2 \frac{d}{dx}(e^x) + e^x \frac{d}{dx}(x^2)$$

$$= x^2 e^x + e^x(2x)$$

$$= xe^x(x + 2)$$

b. Using the General Power Rule, we find

$$g'(t) = \frac{3}{2}(e^t + 2)^{1/2} \frac{d}{dt}(e^t + 2)$$

$$= \frac{3}{2}(e^t + 2)^{1/2} e^t$$

$$= \frac{3}{2}e^t(e^t + 2)^{1/2}$$ ■

Applying the Chain Rule to Exponential Functions

To enlarge the class of exponential functions to be differentiated, we appeal to the Chain Rule to obtain the following rule for differentiating composite functions of the form $h(x) = e^{f(x)}$. An example of such a function is $h(x) = e^{x^2 - 2x}$. Here $f(x) = x^2 - 2x$.

RULE 9:   THE CHAIN RULE FOR EXPONENTIAL FUNCTIONS

If $f(x)$ is a differentiable function, then

$$\frac{d}{dx}(e^{f(x)}) = e^{f(x)} f'(x)$$

To see this, observe that if $h(x) = g[f(x)]$, where $g(x) = e^x$, then by virtue of the Chain Rule,

$$h'(x) = g'(f(x))f'(x) = e^{f(x)} f'(x)$$

since $g'(x) = e^x$.

As an aid to remembering the Chain Rule for exponential functions, observe that it has the following form:

$$\frac{d}{dx}(e^{f(x)}) = e^{f(x)} \cdot \text{Derivative of exponent}$$

$$\underset{\text{Same}}{\underline{}}$$

EXAMPLE 2   Find the derivative of each of the following functions:

a. $f(x) = e^{2x}$ **b.** $y = e^{-3x}$ **c.** $g(t) = e^{2t^2+t}$

Solution

a. $f'(x) = e^{2x} \dfrac{d}{dx}(2x) = e^{2x} \cdot 2 = 2e^{2x}$

b. $\dfrac{dy}{dx} = e^{-3x} \dfrac{d}{dx}(-3x) = -3e^{-3x}$

c. $g'(t) = e^{2t^2+t} \cdot \dfrac{d}{dt}(2t^2 + t) = (4t + 1)e^{2t^2+t}$ ∎

EXAMPLE 3   Differentiate the function $y = xe^{-2x}$.

Solution   Using the Product Rule, followed by the Chain Rule, we find

$$\frac{dy}{dx} = x \frac{d}{dx}e^{-2x} + e^{-2x}\frac{d}{dx}(x)$$

$$= xe^{-2x}\frac{d}{dx}(-2x) + e^{-2x} \qquad \text{Using the Chain Rule on the first term}$$

$$= -2xe^{-2x} + e^{-2x}$$

$$= e^{-2x}(1 - 2x) \qquad \qquad ∎$$

EXAMPLE 4   Differentiate the function $g(t) = \dfrac{e^t}{e^t + e^{-t}}$.

Solution   Using the Quotient Rule, followed by the Chain Rule, we find

$$g'(t) = \frac{(e^t + e^{-t})\dfrac{d}{dt}(e^t) - e^t \dfrac{d}{dt}(e^t + e^{-t})}{(e^t + e^{-t})^2}$$

$$= \frac{(e^t + e^{-t})e^t - e^t(e^t - e^{-t})}{(e^t + e^{-t})^2}$$

$$= \frac{e^{2t} + 1 - e^{2t} + 1}{(e^t + e^{-t})^2} \qquad e^0 = 1$$

$$= \frac{2}{(e^t + e^{-t})^2} \qquad \qquad ∎$$

EXAMPLE 5   In Section 3.3, we discussed some practical applications of the exponential function

$$Q(t) = Q_0 e^{kt}$$

where Q_0 and k are positive constants and $t \in [0, \infty)$. A quantity $Q(t)$ growing according to this law experiences *exponential growth*. Show that for a quantity $Q(t)$ experiencing exponential growth, the rate of growth of the quantity $Q'(t)$ at any time t is directly proportional to the amount of the quantity present.

Solution Using the Chain Rule for exponential functions, we compute the derivative Q' of the function Q. Thus,

$$Q'(t) = Q_0 e^{kt} \frac{d}{dt}(kt)$$

$$= Q_0 e^{kt}(k)$$

$$= k Q_0 e^{kt}$$

$$= k Q(t) \qquad Q(t) = Q_0 e^{kt}$$

which is the desired conclusion.   ∎

The Derivative of ln x

Let us now turn our attention to the differentiation of logarithmic functions.

RULE 10:
DERIVATIVE
OF ln x

$$\frac{d}{dx} \ln |x| = \frac{1}{x} \qquad x \neq 0$$

To derive Rule 10, suppose that $x > 0$ and write $f(x) = \ln x$ in the equivalent form

$$x = e^{f(x)}$$

Differentiating both sides of the equation with respect to x, we find, using the Chain Rule,

$$1 = e^{f(x)} \cdot f'(x)$$

from which we see that

$$f'(x) = \frac{1}{e^{f(x)}}$$

or, since $e^{f(x)} = x$,

$$f'(x) = \frac{1}{x}$$

as we set out to show. You are asked to prove the rule for the case $x < 0$ in exercise 76, page 569.

EXAMPLE 6   Compute the derivative of each of the following functions:

a. $f(x) = x \ln x$ **b.** $g(x) = \dfrac{\ln x}{x}$

Solution

a. Using the Product Rule, we obtain

$$f'(x) = \frac{d}{dx}(x \ln x) = x \frac{d}{dx}(\ln x) + (\ln x) \frac{d}{dx}(x)$$

$$= x \left(\frac{1}{x} \right) + \ln x$$

$$= 1 + \ln x$$

b. Using the Quotient Rule, we obtain

$$g'(x) = \frac{x \dfrac{d}{dx}(\ln x) - (\ln x)\dfrac{d}{dx}(x)}{x^2}$$

$$= \frac{x\left(\dfrac{1}{x}\right) - \ln x}{x^2} = \frac{1 - \ln x}{x^2}$$

The Chain Rule for Logarithmic Functions

To enlarge the class of logarithmic functions to be differentiated, we appeal once more to the Chain Rule to obtain the following rule for differentiating composite functions of the form $h(x) = \ln f(x)$, where $f(x)$ is assumed to be a positive differentiable function.

RULE 11: THE CHAIN RULE FOR LOGARITHMIC FUNCTIONS

If $f(x)$ is a differentiable function, then

$$\frac{d}{dx}[\ln f(x)] = \frac{f'(x)}{f(x)} \qquad f(x) > 0$$

To see this, observe that $h(x) = g[f(x)]$, where $g(x) = \ln x \ (x > 0)$. Since $g'(x) = 1/x$, we have, using the Chain Rule,

$$h'(x) = g'(f(x))f'(x)$$

$$= \frac{1}{f(x)}f'(x)$$

$$= \frac{f'(x)}{f(x)}$$

Observe that in the special case $f(x) = x$, $h(x) = \ln x$, so the derivative of h is, by Rule 10, given by $h'(x) = 1/x$.

EXAMPLE 7 Find the derivative of the function $f(x) = \ln(x^2 + 1)$.

Solution Using Rule 11, we see immediately that

$$f'(x) = \frac{\dfrac{d}{dx}(x^2 + 1)}{x^2 + 1}$$

$$= \frac{2x}{x^2 + 1}$$

When differentiating functions involving logarithms, the rules of logarithms may be used to advantage, as shown in Examples 8 and 9.

EXAMPLE 8   Differentiate the function $y = \ln[(x^2 + 1)(x^3 + 2)^6]$.

Solution   We first rewrite the given function using the properties of logarithms:

$$y = \ln[(x^2 + 1)(x^3 + 2)^6]$$

$$= \ln(x^2 + 1) + \ln(x^3 + 2)^6 \qquad \ln mn = \ln m + \ln n$$

$$= \ln(x^2 + 1) + 6\ln(x^3 + 2) \qquad \ln m^n = n\ln m$$

Differentiating and using Rule 11, we obtain

$$y' = \frac{\dfrac{d}{dx}(x^2 + 1)}{x^2 + 1} + \frac{6\dfrac{d}{dx}(x^3 + 2)}{x^3 + 2}$$

$$= \frac{2x}{x^2 + 1} + \frac{6(3x^2)}{x^3 + 2}$$

$$= \frac{2x}{x^2 + 1} + \frac{18x^2}{x^3 + 2}$$

EXAMPLE 9   Find the derivative of the function $g(t) = \ln(t^2 e^{-t^2})$.

Solution   Here again, to save a lot of work, we first simplify the given expression using the properties of logarithms. We have

$$g(t) = \ln(t^2 e^{-t^2})$$

$$= \ln t^2 + \ln e^{-t^2} \qquad \ln mn = \ln m + \ln n$$

$$= 2\ln t - t^2 \qquad \ln m^n = n\ln m \text{ and } \ln e = 1$$

Therefore, $$g'(t) = \frac{2}{t} - 2t = \frac{2(1 - t^2)}{t}$$

 Applications

Examples 10 and 11 involve finding the rate of change of an exponential function.

EXAMPLE 10   An industrial asset is being depreciated at a rate so that its book value t years from now will be

$$V(t) = 50{,}000e^{-0.4t}$$

dollars. How fast will the book value of the asset be changing three years from now?

Solution   The rate of change of the book value of the asset t years from now is

$$V'(t) = 50{,}000 \frac{d}{dt}e^{-0.4t}$$

$$= 50{,}000(-0.4)e^{-0.4t} = -20{,}000e^{-0.4t}$$

Therefore, three years from now the book value of the asset will be changing at the rate of

$$V'(3) = -20{,}000e^{-0.4(3)} = -20{,}000e^{-1.2} \approx -6{,}023.88$$

that is, decreasing at the rate of approximately $6,024 per year.

EXAMPLE 11   According to the Internet Society, Internet connections are proliferating at an ever-increasing rate. The number of host computers (in millions) is estimated to be

$$N(t) = 3.45e^{0.64t} \qquad 0 \le t \le 6$$

in years t ($t = 0$ corresponds to the beginning of 1994). How fast is the number of host computers growing at the beginning of 1996? At the beginning of 1999? *Source: Internet Society*

Solution   The rate of growth of host computers at time t is given by

$$N'(t) = (3.45)(0.64)e^{0.64t}$$
$$= 2.208e^{0.64t}$$

In particular, the rate of growth of host computers at the beginning of 1996 is given by

$$N'(2) = 2.208e^{0.64(2)} = 7.94138$$

or approximately 7.94 million computers per year. The rate of growth of host computers at the beginning of 1999 is given by

$$N'(5) = 2.208e^{0.64(5)} = 54.16783$$

or approximately 54.17 million computers per year.   ∎

SELF-CHECK EXERCISES 9.7

1. Find the first and second derivatives of $f(x) = xe^{-x}$.

2. Find an equation of the tangent line to the graph of $f(x) = x \ln(2x + 3)$ at the point $(-1, 0)$.

Solutions to Self-Check Exercises 9.7 can be found on page 569.

9.7 EXERCISES

In exercises 1–28, find the derivative of the given function.

1. $f(x) = e^{3x}$

2. $f(x) = 3e^x$

3. $g(t) = e^{-t}$

4. $f(x) = e^{-2x}$

5. $f(x) = e^x + x$

6. $f(x) = 2e^x - x^2$

7. $f(x) = x^3 e^x$

8. $f(u) = u^2 e^{-u}$

9. $f(x) = \dfrac{2e^x}{x}$

10. $f(x) = \dfrac{x}{e^x}$

11. $f(x) = 3(e^x + e^{-x})$

12. $f(x) = \dfrac{e^x + e^{-x}}{2}$

13. $f(w) = \dfrac{e^w + 1}{e^w}$

14. $f(x) = \dfrac{e^x}{e^x + 1}$

15. $f(x) = 2e^{3x-1}$

16. $f(t) = 4e^{3t+2}$

17. $h(x) = e^{-x^2}$

18. $f(x) = e^{x^2-1}$

19. $f(x) = 3e^{-1/x}$

20. $f(x) = e^{1/(2x)}$

21. $f(x) = (e^x + 1)^{25}$

22. $f(x) = (4 - e^{-3x})^3$

23. $f(x) = e^{\sqrt{x}}$

24. $f(t) = -e^{-\sqrt{2t}}$

25. $f(x) = (x - 1)e^{3x+2}$

26. $f(s) = (s^2 + 1)e^{-s^2}$

27. $f(x) = \dfrac{e^x - 1}{e^x + 1}$

28. $g(t) = \dfrac{e^{-t}}{1 + t^2}$

In exercises 29–32, find the second derivative of the given function.

29. $f(x) = e^{-4x} + 2e^{3x}$

30. $f(t) = 3e^{-2t} - 5e^{-t}$

31. $f(x) = 2xe^{3x}$

32. $f(t) = t^2 e^{-2t}$

33. Find an equation of the tangent line to the graph of $y = e^{2x-3}$ at the point $(\frac{3}{2}, 1)$.

34. Find an equation of the tangent line to the graph of $y = e^{-x^2}$ at the point $(1, 1/e)$.

In exercises 35–62, find the derivative of the function.

35. $f(x) = 5 \ln x$

36. $f(x) = \ln 5x$

37. $f(x) = \ln(x + 1)$

38. $g(x) = \ln(2x + 1)$

39. $f(x) = \ln x^8$

40. $h(t) = 2 \ln t^5$

41. $f(x) = \ln \sqrt{x}$

42. $f(x) = \ln(\sqrt{x} + 1)$

43. $f(x) = \ln \dfrac{1}{x^2}$

44. $f(x) = \ln \dfrac{1}{2x^3}$

45. $f(x) = \ln(4x^2 - 6x + 3)$

46. $f(x) = \ln(3x^2 - 2x + 1)$

47. $f(x) = \ln \dfrac{2x}{x + 1}$

48. $f(x) = \ln \dfrac{x + 1}{x - 1}$

49. $f(x) = x^2 \ln x$

50. $f(x) = 3x^2 \ln 2x$

51. $f(x) = \dfrac{2 \ln x}{x}$

52. $f(x) = \dfrac{3 \ln x}{x^2}$

53. $f(u) = \ln(u - 2)^3$

54. $f(x) = \ln(x^3 - 3)^4$

55. $f(x) = \sqrt{\ln x}$

56. $f(x) = \sqrt{\ln x} + x$

57. $f(x) = (\ln x)^3$

58. $f(x) = 2(\ln x)^{3/2}$

59. $f(x) = \ln(x^3 + 1)$

60. $f(x) = \ln \sqrt{x^2 - 4}$

61. $f(x) = e^x \ln x$

62. $f(x) = e^x \ln \sqrt{x + 3}$

In exercises 63–66, find the second derivative of the function.

63. $f(x) = \ln 2x$

64. $f(x) = \ln(x + 5)$

65. $f(x) = \ln(x^2 + 2)$

66. $f(x) = (\ln x)^2$

67. Find an equation of the tangent line to the graph of $y = x \ln x$ at the point $(1, 0)$.

68. Find an equation of the tangent line to the graph of $y = \ln x^2$ at the point $(2, \ln 4)$.

c *A calculator is recommended for the remainder of these exercises.*

69. Sales Promotion The Lady Bug, a women's clothing chain store, found that t days after the end of a sales promotion the volume of sales was given by

$$S(t) = 20{,}000(1 + e^{-0.5t}) \qquad 0 \le t \le 5$$

dollars. Find the rate of change of The Lady Bug's sales volume when $t = 1$, $t = 2$, $t = 3$, and $t = 4$.

70. Energy Consumption of Appliances The average energy consumption of the typical refrigerator/freezer manufactured by York Industries is approximately

$$C(t) = 1486e^{-0.073t} + 500 \qquad 0 \le t \le 20$$

kilowatt hours per year, where t is measured in years, with $t = 0$ corresponding to 1972.

a. What was the average energy consumption of the York refrigerator/freezer at the beginning of 1972?

b. Prove that the average energy consumption of the York refrigerator/freezer is decreasing over the years in question.

c. All refrigerator/freezers manufactured as of January 1, 1990, must meet the 950-kilowatt-hours-per-year maximum energy-consumption standard set by the National Appliance Conservation Act. Does the York refrigerator/freezer meet this requirement?

71. Polio Immunization Polio, a once-feared killer, declined markedly in the United States in the 1950s after Jonas Salk developed the inactivated polio vaccine and mass immunization of children took place. The number of polio cases in the United States from the beginning of 1959 to the beginning of 1963 is approximated by the function

$$N(t) = 5.3e^{0.095t^2 - 0.85t} \qquad 0 \le t \le 4$$

where $N(t)$ gives the number of polio cases (in thousands) and t is measured in years, with $t = 0$ corresponding to the beginning of 1959. How fast was the number of polio cases changing at the beginning of 1959? At the beginning of 1962? [Comment: Following the introduction of the oral vaccine developed by Dr. Albert B. Sabin in 1963, polio in the United States has, for all practical purposes, been eliminated.]

Source: World Health Organization

72. Price of Perfume The monthly demand for a certain brand of perfume is given by the demand equation

$$p = 100e^{-0.0002x} + 150$$

where p denotes the retail unit price in dollars and x denotes the quantity (in 1-ounce bottles) demanded.

a. Find the rate of change of the price per bottle when $x = 1000$. When $x = 2000$.

b. What is the price per bottle when $x = 1000$? When $x = 2000$?

73. Price of Wine The monthly demand for a certain brand of table wine is given by the demand equation

$$p = 240 \left(1 - \dfrac{3}{3 + e^{-0.0005x}}\right)$$

where p denotes the wholesale price per case (in dollars) and x denotes the number of cases demanded.

a. Find the rate of change of the price per case when $x = 1000$.
b. What is the price per case when $x = 1000$?

74. **Oil Used to Fuel Productivity** A study on worldwide oil use was prepared for a major oil company. The study predicted that the amount of oil used to fuel productivity in a certain country is given by

$$f(t) = 1.5 + 1.8te^{-1.2t} \qquad 0 \le t \le 4$$

where $f(t)$ denotes the number of barrels per \$1000 of economic output and t is measured in decades

($t = 0$ corresponds to 1965). Compute $f'(0)$, $f'(1)$, $f'(2)$, and $f'(3)$ and interpret your results.

75. **Percentage of Population Relocating** Based on data obtained from the Census Bureau, the manager of Plymouth Van Lines estimates that the percentage of the total population relocating in year t ($t = 0$ corresponds to the year 1960) may be approximated by the formula

$$P(t) = 20.6e^{-0.009t} \qquad 0 \le t \le 35$$

Compute $P'(10)$, $P'(20)$, and $P'(30)$ and interpret your results.
Source: U.S. Census Bureau

76. Prove thuat $d/dx \ln|x| = 1/x$ ($x \ne 0$) for the case $x < 0$.

SOLUTIONS TO SELF-CHECK EXERCISES 9.7

1. Using the Product Rule, we obtain

$$f'(x) = x\frac{d}{dx}e^{-x} + e^{-x}\frac{d}{dx}x$$

$$= -xe^{-x} + e^{-x} = (1 - x)e^{-x}$$

Using the Product Rule once again, we obtain

$$f''(x) = (1 - x)\frac{d}{dx}e^{-x} + e^{-x}\frac{d}{dx}(1 - x)$$

$$= (1 - x)(-e^{-x}) + e^{-x}(-1)$$

$$= -e^{-x} + xe^{-x} - e^{-x} = (x - 2)e^{-x}$$

2. The slope of the tangent line to the graph of f at any point $(x, f(x))$ lying on the graph of f is given by $f'(x)$. Using the Product Rule, we find

$$f'(x) = \frac{d}{dx}[x\ln(2x + 3)]$$

$$= x\frac{d}{dx}\ln(2x + 3) + \ln(2x + 3) \cdot \frac{d}{dx}(x)$$

$$= x\left(\frac{2}{2x + 3}\right) + \ln(2x + 3) \cdot 1$$

$$= \frac{2x}{2x + 3} + \ln(2x + 3)$$

In particular, the slope of the tangent line to the graph of f at the point $(-1, 0)$ is

$$f'(-1) = \frac{-2}{-2 + 3} + \ln 1 = -2$$

Therefore, using the point-slope form of the equation of a line, we see that a required equation is

$$y - 0 = -2(x + 1)$$

or

$$y = -2x - 2$$

U s i n g   T e c h n o l o g y

EXAMPLE   At the beginning of Section 9.7, we demonstrated via a table of values of $(e^h - 1)/h$ for selected values of h the plausibility of the result

$$\lim_{h \to 0} \frac{e^h - 1}{h} = 1$$

In order to obtain a visual confirmation of this result, we plot the graph of

$$f(x) = \frac{e^x - 1}{x}$$

in the viewing rectangle $[-1, 1] \times [0, 2]$ (see Figure 1). From the graph of f, we see that $f(x)$ appears to approach 1 as x approaches 0.

FIGURE 1
The graph of f in the viewing rectangle $[-1, 1] \times [0, 2]$

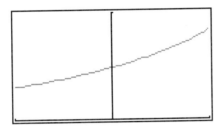

The **dx/dx** operation of a graphing utility will yield the derivative of an exponential or logarithmic function for any value of x, just as it did for algebraic functions.

EXERCISES

In exercises 1–6, use the **dy/dx** *operation of a graphing utility to find the rate of change of f(x) at the given value of x. Give your answer accurate to four decimal places.*

1. $f(x) = x^3 e^{-1/x}; \quad x = -1$

2. $f(x) = (\sqrt{x} + 1)^{3/2} e^{-x}; \quad x = 0.5$

3. $f(x) = x^3 \sqrt{\ln x}; \quad x = 2$

4. $f(x) = \dfrac{\sqrt{x} \ln x}{x + 1}; \quad x = 3.2$

5. $f(x) = e^{-x} \ln(2x + 1); \quad x = 0.5$

6. $f(x) = \dfrac{e^{-\sqrt{x}}}{\ln(x^2 + 1)}; \quad x = 1$

7. **Income of American Families**   Using data compiled by the House Budget Committee, the House Ways and Means Committee, and the U.S. Census Bureau, it is estimated that the number of American families y (in millions) who earned x thousand dollars in 1990 is related by the equation

$$y = 0.1584x e^{-0.0000016x^3 + 0.00011x^2 - 0.04491x} \qquad x > 0$$

a. Use a graphing utility to plot the graph of the equation in the viewing rectangle $[0, 150] \times [0, 2]$.

b. How fast is y changing with respect to x when $x = 10$? When $x = 50$? Interpret your results.

Source: House Budget Committee, House Ways and Means Committee, and U.S. Census Bureau

8. **World Population Growth**   According to a study conducted by the United Nations Population Division, the world population in billions is approximated by the function

$$f(t) = \frac{12}{1 + 3.74914 e^{-1.42804t}} \qquad 0 \le t \le 4$$

where t is measured in half-centuries, with $t = 0$ corresponding to the beginning of 1950.

a. Use a graphing utility to plot the graph of f in the viewing rectangle $[0, 5] \times [0, 14]$.

b. How fast is the world population expected to increase at the beginning of the year 2000?

Source: United Nations Population Division

9. Percentage of Females in the Labor Force Based on data from the United States Census Bureau, the chief economist of Manpower, Inc., constructed the following formula, giving the percentage of the total female population in the civilian labor force, $P(t)$, at the beginning of the tth decade ($t = 0$ corresponds to the year 1900):

$$P(t) = \frac{74}{1 + 2.6e^{-0.166t + 0.04536t^2 - 0.0066t^3}} \qquad 0 \le t \le 11$$

Assume this trend continues for the rest of the twentieth century.

a. What will the percentage of the total female population in the civilian labor force be at the beginning of the year 2000?

b. What will the growth rate of the percentage of the total female population in the civilian labor force be at the beginning of the year 2000?

Source: U.S. Census Bureau

9.8 Marginal Functions in Economics

Marginal analysis is the study of the rate of change of economic quantities. For example, an economist is not merely concerned with the value of an economy's gross domestic product (GDP) at a given time, but is equally concerned with the rate at which it is growing or declining. In the same vein, a manufacturer is not only interested in the total cost corresponding to a certain level of production of a commodity, but is also interested in the rate of change of the total cost with respect to the level of production, and so on. Let us begin with an example to explain the meaning of the adjective *marginal,* as used by economists.

Cost Functions

 EXAMPLE 1 Suppose that the total cost in dollars incurred per week by the Polaraire Company for manufacturing x refrigerators is given by the total cost function

$$C(x) = 8000 + 200x - 0.2x^2 \qquad 0 \le x \le 400$$

a. What is the actual cost incurred for manufacturing the 251st refrigerator?
b. Find the rate of change of the total cost with respect to x when $x = 250$.
c. Compare the results obtained in (a) and (b).

Solution

a. The actual cost incurred in producing the 251st refrigerator is the difference between the total cost incurred in producing the first 251 refrigerators and the total cost of producing the first 250 refrigerators:

$$C(251) - C(250) = [8000 + 200(251) - 0.2(251)^2]$$
$$- [8000 + 200(250) - 0.2(250)^2]$$
$$= 45{,}599.8 - 45{,}500$$
$$= 99.80 \quad \text{or} \quad \$99.80$$

b. The rate of change of the total cost function C with respect to x is given by the derivative of C; that is, $C'(x) = 200 - 0.4x$. Thus, when the level of production is 250 refrigerators, the rate of change of the total cost

with respect to x is given by

$$C'(250) = 200 - 0.4(250)$$
$$= 100 \quad \text{or} \quad \$100$$

c. From the solution to (a), we know that the actual cost for producing the 251st refrigerator is \$99.80. This answer is very closely approximated by the answer to (b), \$100. To see why this is so, observe that the difference $C(251) - C(250)$ may be written in the form

$$\frac{C(251) - C(250)}{1} = \frac{C(250 + 1) - C(250)}{1}$$
$$= \frac{C(250 + h) - C(250)}{h}$$

where $h = 1$. In other words, the difference $C(251) - C(250)$ is precisely the average rate of change of the total cost function C over the interval [250, 251], or, equivalently, the slope of the secant line through the points (250, 45,500) and (251, 45,599.8). On the other hand, the number $C'(250) = 100$ is the instantaneous rate of change of the total cost function C at $x = 250$, or, equivalently, the slope of the tangent line to the graph of C at $x = 250$.

Now when h is small, the average rate of change of the function C is a good approximation to the instantaneous rate of change of the function C, or, equivalently, the slope of the secant line through the points in question is a good approximation to the slope of the tangent line through the point in question. Thus, we may expect

$$C(251) - C(250) = \frac{C(251) - C(250)}{1} = \frac{C(250 + h) - C(250)}{h}$$
$$\approx \lim_{h \to 0} \frac{C(250 + h) - C(250)}{h} = C'(250)$$

which is precisely the case in this example.   ■

The actual cost incurred in producing an additional unit of a certain commodity given that a plant is already at a certain level of operation is called the *marginal cost*. Knowing this cost is very important to management in their decision-making processes. As we saw in Example 1, the marginal cost is approximated by the rate of change of the total cost function evaluated at the appropriate point. For this reason, economists have defined the **marginal cost function** to be the derivative of the corresponding total cost function. In other words, if C is a total cost function, then the marginal cost function is defined to be its derivative C'. Thus, the adjective *marginal* is synonymous with *derivative of*.

 EXAMPLE 2   A subsidiary of the Elektra Electronics Company manufactures a programmable pocket calculator. Management determined that the daily total cost of producing these calculators (in dollars) is given by

$$C(x) = 0.0001x^3 - 0.08x^2 + 40x + 5000$$

where x stands for the number of calculators produced.

a. Find the marginal cost function.

b. What is the marginal cost when $x = 200, 300, 400$, and 600?

c. Interpret your results.

Solution

a. The marginal cost function C' is given by the derivative of the total cost function C. Thus,

$$C'(x) = 0.0003x^2 - 0.16x + 40$$

b. The marginal cost when $x = 200, 300, 400$, and 600 is given by

$$C'(200) = 0.0003(200)^2 - 0.16(200) + 40 = 20$$

$$C'(300) = 0.0003(300)^2 - 0.16(300) + 40 = 19$$

$$C'(400) = 0.0003(400)^2 - 0.16(400) + 40 = 24$$

$$C'(600) = 0.0003(600)^2 - 0.16(600) + 40 = 52$$

or $20, $19, $24, and $52, respectively.

c. From the results of (b), we see that Elektra's actual cost for producing the 201st calculator is approximately $20. The actual cost incurred for producing one additional calculator when the level of production is already 300 calculators is approximately $19, and so on. Observe that when the level of production is already 600 units, the actual cost of producing one additional unit is approximately $52. The higher cost for producing this additional unit when the level of production is 600 units may be the result of several factors, among them excessive costs incurred because of overtime or higher maintenance, production breakdown caused by greater stress and strain on the equipment, and so on. The graph of the total cost function appears in Figure 45. ■

FIGURE 45

The cost of producing x calculators is given by C(x).

Average Cost Functions

Let us now introduce another marginal concept closely related to the marginal cost. Let $C(x)$ denote the total cost incurred in producing x units of a certain commodity. Then the *average cost* of producing x units of the commodity is obtained by dividing the total production cost by the number of units produced. This leads to the following definition.

AVERAGE COST FUNCTION

Suppose $C(x)$ is a total cost function. Then the **average cost function**, denoted by $\overline{C}(x)$, is

$$\frac{C(x)}{x}$$

The derivative $\overline{C}'(x)$ of the average cost function, called the **marginal average cost function,** measures the rate of change of the average cost function with respect to the number of units produced.

EXAMPLE 3 The total cost of producing x units of a certain commodity is given by

$$C(x) = 400 + 20x \text{ dollars}$$

a. Find the average cost function $\overline{C}$.

b. Find the marginal average cost function $\overline{C}'$.

c. Interpret the results obtained in (a) and (b).

Solution

a. The average cost function is given by

$$\overline{C}(x) = \frac{C(x)}{x} = \frac{400 + 20x}{x}$$

$$= 20 + \frac{400}{x}$$

b. The marginal average cost function is

$$\overline{C}'(x) = -\frac{400}{x^2}$$

FIGURE 46

As the level of production increases, the average cost approaches $20.

$y = \overline{C}(x) = 20 + \dfrac{400}{x}$

$y = 20$

c. Since the marginal average cost function is negative for all admissible values of x, the rate of change of the average cost function is negative for all $x > 0$; that is, $\overline{C}(x)$ decreases as x increases. However, the graph of $\overline{C}$ always lies above the horizontal line $y = 20$, but it approaches the line, since

$$\lim_{x \to \infty} \overline{C}(x) = \lim_{x \to \infty} \left(20 + \frac{400}{x} \right) = 20$$

A sketch of the graph of the function $\overline{C}(x)$ appears in Figure 46. This result is fully expected if we consider the economic implications. Note that as the level of production increases, the fixed cost per unit of production, represented by the term $(400/x)$, drops steadily. The average cost approaches the constant unit of production, which is $20 in this case. ∎

EXAMPLE 4   Once again consider the subsidiary of the Elektra Electronics Company. The daily total cost for producing its programmable calculators is given by

$$C(x) = 0.0001x^3 - 0.08x^2 + 40x + 5000$$

dollars, where x stands for the number of calculators produced (see Example 2).

a. Find the average cost function $\overline{C}$.

b. Find the marginal average cost function $\overline{C}'$. Compute $\overline{C}'(500)$.

c. Sketch the graph of the function $\overline{C}$ and interpret the results obtained in (a) and (b).

Solution

a. The average cost function is given by

$$\overline{C}(x) = \frac{C(x)}{x} = 0.0001x^2 - 0.08x + 40 + \frac{5000}{x}$$

b. The marginal average cost function is given by

$$\overline{C}'(x) = 0.0002x - 0.08 - \frac{5000}{x^2}$$

Also, $\overline{C}'(500) = 0.0002(500) - 0.08 - \frac{5000}{(500)^2} = 0$

c. To sketch the graph of the function $\overline{C}$, observe that if x is a small positive number, then $\overline{C}(x) > 0$. Furthermore, $\overline{C}(x)$ becomes arbitrarily large as x approaches zero from the right, since the term $(5000/x)$ becomes arbitrarily large as x approaches zero. Next, the result $\overline{C}'(500) = 0$ obtained in (b) tells us that the tangent line to the graph of the function $\overline{C}$ is horizontal at the point $(500, 35)$ on the graph. Finally, plotting the points on the graph corresponding to, say, $x = 100, 200, 300, \ldots, 900$, we obtain the sketch in Figure 47. As expected, the average cost drops as the level of production increases. But in this case, as opposed to the case in Example 3, the average cost reaches a minimum value of $35, corresponding to a production level of 500, and *increases* thereafter.

This phenomenon is typical in situations where the marginal cost increases from some point on as production increases, as in Example 2. This situation is in contrast to that of Example 3, in which the marginal cost remains constant at any level of production. ∎

FIGURE 47

The average cost reaches a minimum of $35 when 500 calculators are produced.

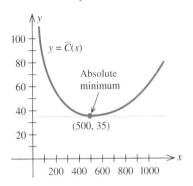

Revenue Functions

Another marginal concept, the *marginal revenue function,* is associated with the revenue function R given by

$$R(x) = px \qquad (13)$$

where x is the number of units sold of a certain commodity and p is the unit selling price. In general, however, the unit selling price p of a commodity is related to the quantity x of the commodity demanded. This relationship, $p = f(x)$, is called a *demand equation* (see Section 2.5). Solving the demand equation for p in terms of x, we obtain the unit price function f, given by

$$p = f(x)$$

Thus, the revenue function R is given by

$$R(x) = px = xf(x)$$

where f is the unit price function. The derivative R' of the function R, called the **marginal revenue function,** measures the rate of change of the revenue function.

EXAMPLE 5   Suppose that the relationship between the unit price p in dollars and the quantity demanded x of the Acrosonic model F loudspeaker system is given by the equation

$$p = -0.02x + 400 \qquad 0 \le x \le 20{,}000$$

a. Find the revenue function R.

b. Find the marginal revenue function R'.

c. Compute $R'(2000)$ and interpret your result.

Solution

a. The revenue function R is given by

$$R(x) = px$$

$$= x(-0.02x + 400)$$

$$= -0.02x^2 + 400x \qquad 0 \le x \le 20{,}000$$

b. The marginal revenue function R' is given by

$$R'(x) = -0.04x + 400$$

c.

$$R'(2000) = -0.04(2000) + 400 = 320$$

Thus, the actual revenue to be realized from the sale of the 2001st loudspeaker system is approximately $320. ■

Profit Functions

Our final example of a marginal function involves the profit function. The profit function P is given by

$$P(x) = R(x) - C(x)$$

where R and C are the revenue and cost functions and x is the number of units of a commodity produced and sold. The **marginal profit function $P'(x)$** measures the rate of change of the profit function P and provides us with a good approximation of the actual profit or loss realized from the sale of the $(x + 1)$st unit of the commodity (assuming the xth unit has been sold).

EXAMPLE 6   Refer to Example 5. Suppose that the cost of producing x units of the Acrosonic model F loudspeaker is

$$C(x) = 100x + 200{,}000$$

dollars.

a. Find the profit function P.
b. Find the marginal profit function P'.
c. Compute $P'(2000)$ and interpret your result.
d. Sketch the graph of the profit function P.

Solution

a. From the solution to Example 5a, we have

$$R(x) = -0.02x^2 + 400x$$

Thus, the required profit function P is given by

$$P(x) = R(x) - C(x)$$

$$= (-0.02x^2 + 400x) - (100x + 200{,}000)$$

$$= -0.02x^2 + 300x - 200{,}000$$

b. The marginal profit function P' is given by

$$P'(x) = -0.04x + 300$$

c. $$P'(2000) = -0.04(2000) + 300 = 220$$

Thus, the actual profit realized from the sale of the 2001st loudspeaker system is approximately $220.

d. The graph of the profit function P appears in Figure 48.

FIGURE 48

The total profit made when x loudspeakers are produced is given by P(x).

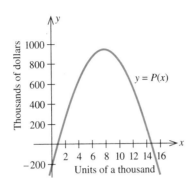

SELF-CHECK
EXERCISE 9.8

[c] The weekly demand for Pulsar VCRs (videocassette recorders) is given by the demand equation

$$p = -0.02x + 300 \qquad 0 \le x \le 15,000$$

where p denotes the wholesale unit price in dollars and x denotes the quantity demanded. The weekly total cost function associated with manufacturing these VCRs is

$$C(x) = 0.000003x^3 - 0.04x^2 + 200x + 70,000$$

dollars.

a. Find the revenue function R and the profit function P.
b. Find the marginal cost function C', the marginal revenue function R', and the marginal profit function P'.
c. Find the marginal average cost function $\overline{C}'$.
d. Compute $C'(3000)$, $R'(3000)$, and $P'(3000)$ and interpret your results.

Solutions to Self-Check Exercise 9.8 can be found on page 579.

9.8   **EXERCISES**

[c] *A calculator is recommended for exercises 1–20.*

1. Marginal Cost   The total weekly cost in dollars incurred by the Lincoln Record Company in pressing x long-playing records is

$$C(x) = 2000 + 2x - 0.0001x^2 \qquad 0 \le x \le 6000$$

a. What is the actual cost incurred in producing the 1001st record? The 2001st record?
b. What is the marginal cost when $x = 1000$? When $x = 2000$?

2. Marginal Cost   A division of Ditton Industries manufactures the Futura model microwave oven. The daily cost (in dollars) of producing these microwave ovens is

$$C(x) = 0.0002x^3 - 0.06x^2 + 120x + 5000$$

where x stands for the number of units produced.
   a. What is the actual cost incurred in manufacturing the 101st oven? The 201st oven? The 301st oven?
   b. What is the marginal cost when $x = 100$? When $x = 200$? When $x = 300$?

3. Marginal Average Cost   The Custom Office Company makes a line of executive desks. It is estimated that the total cost for making x units of their Senior Executive Model is

$$C(x) = 100x + 200,000$$

dollars per year.
   a. Find the average cost function $\overline{C}$.
   b. Find the marginal average cost function $\overline{C}\,'$.
   c. What happens to $\overline{C}(x)$ when x is very large? Interpret your results.

4. Marginal Average Cost   The management of the Thermo-Master Company, whose Mexico subsidiary manufactures an indoor-outdoor thermometer, has estimated that the total weekly cost in dollars for producing x thermometers is

$$C(x) = 5000 + 2x$$

   a. Find the average cost function $\overline{C}$.
   b. Find the marginal average cost function $\overline{C}\,'$.
   c. Interpret your results.

5. Find the average cost function $\overline{C}$ and the marginal average cost function $\overline{C}\,'$ associated with the total cost function C of exercise 1.

6. Find the average cost function $\overline{C}$ and the marginal average cost function $\overline{C}\,'$ associated with the total cost function C of exercise 2.

7. Marginal Revenue   The Williams Commuter Air Service realizes a monthly revenue of

$$R(x) = 8000x - 100x^2$$

dollars when the price charged per passenger is x dollars.
   a. Find the marginal revenue R'.
   b. Compute $R'(39)$, $R'(40)$, and $R'(41)$. What do your results imply?

8. Marginal Revenue   The management of the Acrosonic Company plans to market the Electro-Stat, an electrostatic speaker system. The marketing department has determined that the demand for these speakers is

$$p = -0.04x + 800 \qquad 0 \le x \le 20,000$$

where p denotes the speaker's unit price (in dollars) and x denotes the quantity demanded.
   a. Find the revenue function R.
   b. Find the marginal revenue function R'.
   c. Compute $R'(5000)$ and interpret your result.

9. Marginal Profit   Lynbrook West, an apartment complex, has 100 two-bedroom units. The monthly profit (in dollars) realized from renting x apartments is

$$P(x) = -10x^2 + 1760x - 50,000$$

   a. What is the actual profit realized from renting the 51st unit, assuming that 50 units have already been rented?
   b. Compute the marginal profit when $x = 50$ and compare your results with that obtained in (a).

10. Marginal Profit   Refer to exercise 8. Acrosonic's production department estimates that the total cost (in dollars) incurred in manufacturing x Electro-Stat speaker systems in the first year of production will be

$$C(x) = 200x + 300,000$$

   a. Find the profit function P.
   b. Find the marginal profit function P'.
   c. Compute $P'(5000)$ and $P'(8000)$.
   d. Sketch the graph of the profit function and interpret your results.

11. Marginal Cost, Revenue, and Profit   The weekly demand for the Pulsar 25 color console television is

$$p = 600 - 0.05x \qquad 0 \le x \le 12,000$$

where p denotes the wholesale unit price in dollars and x denotes the quantity demanded. The weekly total cost function associated with manufacturing the Pulsar 25 is given by

$$C(x) = 0.000002x^3 - 0.03x^2 + 400x + 80,000$$

where $C(x)$ denotes the total cost incurred in producing x sets.
   a. Find the revenue function R and the profit function P.
   b. Find the marginal cost function C', the marginal revenue function R', and the marginal profit function P'.
   c. Compute $C'(2000)$, $R'(2000)$, and $P'(2000)$ and interpret your results.
   d. Sketch the graphs of the functions C, R, and P and interpret (b) and (c) using the graphs obtained.

12. Marginal Cost, Revenue, and Profit   The Pulsar Corporation also manufactures a series of 19-inch color television sets. The quantity x of these sets demanded each week is related to the wholesale unit price p by the equation

$$p = -0.006x + 180$$

The weekly total cost incurred by Pulsar for producing x sets is

$$C(x) = 0.000002x^3 - 0.02x^2 + 120x + 60,000$$

dollars. Answer the questions in exercise 11 for these data.

13. Marginal Average Cost   Find the average cost function $\overline{C}$ associated with the total cost function C of exercise 11.
  a. What is the marginal average cost function $\overline{C}\,'$?
  b. Compute $\overline{C}\,'(5,000)$ and $\overline{C}\,'(10,000)$ and interpret your results.
  c. Sketch the graph of $\overline{C}$.

14. Marginal Average Cost   Find the average cost function $\overline{C}$ associated with the total cost function C of exercise 12.
  a. What is the marginal average cost function $\overline{C}\,'$?
  b. Compute $\overline{C}\,'(5,000)$ and $\overline{C}\,'(10,000)$ and interpret your results.

15. Marginal Revenue   The quantity of Sicard wristwatches demanded per month is related to the unit price by the equation

$$p = \frac{50}{0.01x^2 + 1} \qquad 0 \le x \le 20$$

where p is measured in dollars and x in units of a thousand.
  a. Find the revenue function R.
  b. Find the marginal revenue function R'.
  c. Compute $R'(2)$ and interpret your result.

16. Marginal Propensity to Consume   The consumption function of the U.S. economy for 1929 to 1941 is

$$C(x) = 0.712x + 95.05$$

where $C(x)$ is the personal consumption expenditure and x is the personal income, both measured in billions of dollars. Find the rate of change of consumption with respect to income, dC/dx. This quantity is called the *marginal propensity to consume*.

17. Marginal Propensity to Consume   Refer to exercise 16. Suppose that a certain economy's consumption function is

$$C(x) = 0.873x^{1.1} + 20.34$$

where $C(x)$ and x are measured in billions of dollars. Find the marginal propensity to consume when $x = 10$.

18. Marginal Propensity to Save   Suppose that $C(x)$ measures an economy's personal consumption expenditure and x the personal income, both in billions of dollars. Then

$$S(x) = x - C(x) \qquad \text{Income minus consumption}$$

measures the economy's savings corresponding to an income of x billion dollars. Show that

$$\frac{dS}{dx} = 1 - \frac{dC}{dx}$$

The quantity dS/dx is called the *marginal propensity to save*.

19. Refer to exercise 18. For the consumption function of exercise 16, find the marginal propensity to save.

20. Refer to exercise 18. For the consumption function of exercise 17, find the marginal propensity to save when $x = 10$.

SOLUTION TO SELF-CHECK EXERCISE 9.8

a.

$$R(x) = px$$
$$= x(-0.02x + 300)$$
$$= -0.02x^2 + 300x \qquad 0 \le x \le 15,000$$
$$P(x) = R(x) - C(x)$$
$$= -0.02x^2 + 300x$$
$$\quad - (0.000003x^3 - 0.04x^2 + 200x + 70,000)$$
$$= -0.000003x^3 + 0.02x^2 + 100x - 70,000$$

b.
$$C'(x) = 0.000009x^2 - 0.08x + 200$$

$$R'(x) = -0.04x + 300$$

and

$$P'(x) = -0.000009x^2 + 0.04x + 100$$

c. The average cost function is

$$\overline{C}(x) = \frac{C(x)}{x}$$

$$= \frac{0.000003x^3 - 0.04x^2 + 200x + 70{,}000}{x}$$

$$= 0.000003x^2 - 0.04x + 200 + \frac{70{,}000}{x}$$

Therefore, the marginal average cost function is

$$\overline{C}'(x) = 0.000006x - 0.04 - \frac{70{,}000}{x^2}$$

d. Using the results from (b), we find

$$C'(3000) = 0.000009(3000)^2 - 0.08(3000) + 200$$

$$= 41$$

that is, when the level of production is already 3000 VCRs, the actual cost of producing one additional VCR is approximately $41. Next,

$$R'(3000) = -0.04(3000) + 300 = 180$$

that is, the actual revenue to be realized from selling the 3001st VCR is approximately $180. Finally,

$$P'(3000) = -0.000009(3000)^2 + 0.04(3000) + 100$$

$$= 139$$

that is, the actual profit realized from selling the 3001st VCR is approximately $139.

CHAPTER NINE SUMMARY OF PRINCIPAL FORMULAS AND TERMS

Formulas

1. Average rate of change of f over $[x, x + h]$
or
Slope of the secant line to the graph of f through $(x, f(x))$ and $(x + h, f(x + h))$
or
Difference quotient

$$\frac{f(x + h) - f(x)}{h}$$

2. Instantaneous rate of change of f at $(x, f(x))$

or

Slope of tangent line to the graph of f at $(x, f(x))$

or

Derivative of f at x

$$\lim_{h \to 0} \frac{f(x+h) - f(x)}{h}$$

3. Derivative of a constant

$$\frac{d}{dx}(c) = 0 \quad (c \text{ a constant})$$

4. Power Rule

$$\frac{d}{dx}(x^n) = nx^{n-1}$$

5. Constant Multiple Rule

$$\frac{d}{dx}(cu) = c\frac{du}{dx} \quad (c \text{ a constant})$$

6. Sum Rule

$$\frac{d}{dx}(u \pm v) = \frac{du}{dx} \pm \frac{dv}{dx}$$

7. Product Rule

$$\frac{d}{dx}(uv) = u\frac{dv}{dx} + v\frac{du}{dx}$$

8. Quotient Rule

$$\frac{d}{dx}\left(\frac{u}{v}\right) = \frac{v\dfrac{du}{dx} - u\dfrac{dv}{dx}}{v^2}$$

9. Chain Rule

$$\frac{dy}{dx} = \frac{dy}{du} \cdot \frac{du}{dx}$$

10. General Power Rule

$$\frac{d}{dx}(u^n) = nu^{n-1}\frac{du}{dx}$$

11. Derivative of the exponential function

$$\frac{d}{dx}(e^x) = e^x$$

12. Chain rule for exponential functions

$$\frac{d}{dx}(e^u) = e^u\frac{du}{dx}$$

13. Derivative of the logarithmic function

$$\frac{d}{dx}\ln|x| = \frac{1}{x}$$

14. Chain rule for logarithmic functions

$$\frac{d}{dx}\ln|u| = \frac{1}{u}\frac{du}{dx}$$

15. Average cost function

$$\overline{C}(x) = \frac{C(x)}{x}$$

16. Revenue function

$$R(x) = px$$

17. Profit function

$$P(x) = R(x) - C(x)$$

Terms

Limit of a function

Indeterminate form 0/0

Limit of a function at infinity

One-sided limits

Right-hand limit of a function

Left-hand limit of a function

Continuity of a function at a point

Continuity of a function on an interval

Discontinuous function

Secant line

Tangent line to the graph of f

Differentiable function

Second derivative

Composition of functions

Marginal cost function

Marginal average cost function

Marginal revenue function

Marginal profit function

CHAPTER 9 REVIEW EXERCISES

In exercises 1–14, find the indicated limits, if they exist.

1. $\lim_{x \to 0} (5x - 3)$

2. $\lim_{x \to 1} (x^2 + 1)$

3. $\lim_{x \to -1} (3x^2 + 4)(2x - 1)$

4. $\lim_{x \to 3} \dfrac{x - 3}{x + 4}$

5. $\lim_{x \to 2} \dfrac{x + 3}{x^2 - 9}$

6. $\lim_{x \to -2} \dfrac{x^2 - 2x - 3}{x^2 + 5x + 6}$

7. $\lim_{x \to 3} \sqrt{2x^3 - 5}$

8. $\lim_{x \to 3} \dfrac{4x - 3}{\sqrt{x + 1}}$

9. $\lim_{x \to 1^+} \dfrac{x - 1}{x(x - 1)}$

10. $\lim_{x \to 1^-} \dfrac{\sqrt{x} - 1}{x - 1}$

11. $\lim_{x \to \infty} \dfrac{x^2}{x^2 - 1}$

12. $\lim_{x \to -\infty} \dfrac{x + 1}{x}$

13. $\lim_{x \to \infty} \dfrac{3x^2 + 2x + 4}{2x^2 - 3x + 1}$

14. $\lim_{x \to -\infty} \dfrac{x^2}{x + 1}$

15. Sketch the graph of the function

$$f(x) = \begin{cases} 2x - 3 & \text{if } x \le 2 \\ -x + 3 & \text{if } x > 2 \end{cases}$$

and evaluate $\lim_{x \to a^+} f(x)$, $\lim_{x \to a^-} f(x)$, and $\lim_{x \to a} f(x)$ at the point $a = 2$, if the limits exist.

16. Sketch the graph of the function

$$f(x) = \begin{cases} 4 - x & \text{if } x \le 2 \\ x + 2 & \text{if } x > 2 \end{cases}$$

and evaluate $\lim_{x \to a^+} f(x)$, $\lim_{x \to a^-} f(x)$, and $\lim_{x \to a} f(x)$ at the point $a = 2$, if the limits exist.

In exercises 17–20, determine all values of x for which the function is discontinuous.

17. $g(x) = \begin{cases} x + 3 & \text{if } x \ne 2 \\ 0 & \text{if } x = 2 \end{cases}$

18. $f(x) = \dfrac{3x + 4}{4x^2 - 2x - 2}$

19. $f(x) = \begin{cases} \dfrac{1}{(x + 1)^2} & \text{if } x \ne -1 \\ 2 & \text{if } x = -1 \end{cases}$

20. $f(x) = \dfrac{|2x|}{x}$

21. Let $y = x^2 + 2$.
   a. Find the average rate of change of y with respect to x in the intervals [1, 2], [1, 1.5], and [1, 1.1].
   b. Find the (instantaneous) rate of change of y at $x = 1$.

22. Use the definition of the derivative to find the slope of the tangent line to the graph of the function $f(x) = 3x + 5$ at any point $P(x, f(x))$ on the graph.

23. Use the definition of the derivative to find the slope of the tangent line to the graph of the function $f(x) = -x^2$ at the point $(2, -4)$ and determine an equation of the tangent line.

24. The graph of the function f is shown in the accompanying figure.

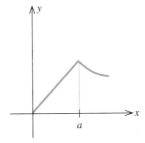

   a. Is f continuous at $x = a$? Why?
   b. Is f differentiable at $x = a$? Justify your answer.

In exercises 25–72, find the derivative of the function.

25. $f(x) = 3x^5 - 2x^4 + 3x^2 - 2x + 1$

26. $f(x) = 4x^6 + 2x^4 + 3x^2 - 2$

27. $g(x) = -2x^{-3} + 3x^{-1} + 2$

28. $f(t) = 2t^2 - 3t^3 - t^{-1/2}$

29. $g(t) = 2t^{-1/2} + 4t^{-3/2} + 2$

30. $h(x) = x^2 + \dfrac{2}{x}$

31. $f(t) = t + \dfrac{2}{t} + \dfrac{3}{t^2}$

32. $g(s) = 2s^2 - \dfrac{4}{s} + \dfrac{2}{\sqrt{s}}$

33. $h(x) = x^2 - \dfrac{2}{x^{3/2}}$

34. $f(x) = \dfrac{x+1}{2x-1}$

35. $g(t) = \dfrac{t^2}{2t^2+1}$

36. $h(t) = \dfrac{\sqrt{t}}{\sqrt{t}+1}$

37. $f(x) = \dfrac{\sqrt{x}-1}{\sqrt{x}+1}$

38. $f(t) = \dfrac{t}{2t^2+1}$

39. $f(x) = \dfrac{x^2(x^2+1)}{x^2-1}$

40. $f(x) = (2x^2 + x)^3$

41. $f(x) = (3x^3 - 2)^8$

42. $h(x) = (\sqrt{x} + 2)^5$

43. $f(t) = \sqrt{2t^2 + 1}$

44. $g(t) = \sqrt[3]{1 - 2t^3}$

45. $s(t) = (3t^2 - 2t + 5)^{-2}$

46. $f(x) = (2x^3 - 3x^2 + 1)^{-3/2}$

47. $f(x) = xe^{2x}$

48. $f(t) = \sqrt{t}e^t + t$

49. $g(t) = \sqrt{t}e^{-2t}$

50. $g(x) = e^x\sqrt{1 + x^2}$

51. $y = \dfrac{e^{2x}}{1 + e^{-2x}}$

52. $f(x) = e^{2x^2 - 1}$

53. $f(x) = xe^{-x^2}$

54. $g(x) = (1 + e^{2x})^{3/2}$

55. $f(x) = x^2e^x + e^x$

56. $g(t) = t\ln t$

57. $f(x) = \ln(e^{x^2} + 1)$

58. $f(x) = \dfrac{x}{\ln x}$

59. $f(x) = \dfrac{\ln x}{x+1}$

60. $y = (x + 1)e^x$

61. $y = \ln(e^{4x} + 3)$

62. $f(r) = \dfrac{re^r}{1 + r^2}$

63. $f(x) = \dfrac{\ln x}{1 + e^x}$

64. $g(x) = \dfrac{e^{x^2}}{1 + \ln x}$

65. $h(x) = \left(x + \dfrac{1}{x}\right)^2$

66. $h(x) = \dfrac{1+x}{(2x^2 + 1)^2}$

67. $h(t) = (t^2 + t)^4(2t^2)$

68. $f(x) = (2x + 1)^3(x^2 + x)^2$

69. $g(x) = \sqrt{x}(x^2 - 1)^3$

70. $f(x) = \dfrac{x}{\sqrt{x^3 + 2}}$

71. $h(x) = \dfrac{\sqrt{3x + 2}}{4x - 3}$

72. $f(t) = \dfrac{\sqrt{2t + 1}}{(t + 1)^3}$

In exercises 73–82, find the second derivative of the function.

73. $f(x) = 2x^4 - 3x^3 + 2x^2 + x + 4$

74. $g(x) = \sqrt{x} + \dfrac{1}{\sqrt{x}}$

75. $h(t) = \dfrac{t}{t^2 + 4}$

76. $f(x) = xe^{-2x}$

77. $h(x) = \dfrac{e^x}{1 + e^x}$

78. $y = x\ln x$

79. $y = \ln(3x + 1)$

80. $f(x) = (x^3 + x + 1)^2$

81. $f(x) = \sqrt{2x^2 + 1}$

82. $f(t) = t(t^2 + 1)^3$

83. Find $h'(0)$ if $h(x) = g(f(x))$, $g(x) = x + \dfrac{1}{x}$, and $f(x) = e^x$.

84. Find $h'(1)$ if $h(x) = g(f(x))$, $g(x) = \dfrac{x+1}{x-1}$, and $f(x) = \ln x$.

85. Let $f(x) = 2x^3 - 3x^2 - 16x + 3$.
 a. Find the points on the graph of f at which the slope of the tangent line is equal to -4.
 b. Find the equation(s) of the tangent line(s) of (a).

86. Let $f(x) = \frac{1}{3}x^3 + \frac{1}{2}x^2 - 4x + 1$.
 a. Find the points on the graph of f at which the slope of the tangent line is equal to -2.
 b. Find the equation(s) of the tangent line(s) of (a).

87. Find an equation of the tangent line to the graph of $y = \sqrt{4 - x^2}$ at the point $(1, \sqrt{3})$.

88. Find an equation of the tangent line to the graph of $y = x(x + 1)^5$ at the point $(1, 32)$.

89. Find an equation of the tangent line to the graph of $y = e^{-2x}$ at the point $(1, e^{-2})$.

90. Find an equation of the tangent line to the graph of $y = xe^{-x}$ at the point $(1, e^{-1})$.

91. Find the third derivative of the function

$$f(x) = \dfrac{1}{2x - 1}$$

What is its domain?

92. The average cost (in dollars) of producing x units of a certain commodity is given by

$$\overline{C}(x) = 20 + \dfrac{400}{x}$$

Evaluate $\lim\limits_{x \to \infty} \overline{C}(x)$ and interpret your results.

93. The number of subscribers to CNC Cable Television in the town of Randolph is approximated by the

function

$$N(x) = 1000(1 + 2x)^{1/2} \qquad 1 \le x \le 30$$

where $N(x)$ denotes the number of subscribers to the service in the xth week. Find the rate of increase in the number of subscribers at the end of the twelfth week.

94. The total weekly cost in dollars incurred by the Herald Record Company in pressing x video discs is given by the total cost function

$$C(x) = 2500 + 2.2x \qquad 0 \le x \le 8000$$

   a. What is the marginal cost when $x = 1000$? When $x = 2000$?
   b. Find the average cost function $\overline{C}$ and the marginal average cost function $\overline{C}\,'$.
   c. Using the results from (b), show that the average cost incurred by Herald in pressing a video disc approaches \$2.20 per disc when the level of production is high enough.

95. The marketing department of Telecon Corporation has determined that the demand for their cordless phones obeys the relationship

$$p = -0.02x + 600 \qquad 0 \le x \le 30,000$$

where p denotes the phone's unit price (in dollars) and x denotes the quantity demanded.
   a. Find the revenue function R.
   b. Find the marginal revenue function R'.
   c. Compute $R'(10,000)$ and interpret your result.

96. The weekly demand for the Lectro-Copy photocopying machine is given by the demand equation

$$p = 2000 - 0.04x \qquad 0 \le x \le 50,000$$

where p denotes the wholesale unit price in dollars and x denotes the quantity demanded. The weekly total cost function for manufacturing these copiers is given by

$$C(x) = 0.000002x^3 - 0.02x^2 + 1,000x + 120,000$$

where $C(x)$ denotes the total cost incurred in producing x units.
   a. Find the revenue function R, the profit function P, and the average cost function $\overline{C}$.
   b. Find the marginal cost function C', the marginal revenue function R', the marginal profit function P', and the marginal average cost function $\overline{C}\,'$.
   c. Compute $C'(3000)$, $R'(3000)$, and $P'(3000)$.
   d. Compute $\overline{C}\,'(5000)$ and $\overline{C}\,'(8000)$ and interpret your results.

CHAPTER 9   EXTENDED PROBLEMS AND PROJECTS

1. Let $f(x) = x^2$.
   a. Verify by direct substitution that the points $Q_1(3, 9)$, $Q_2(2, 4)$, $Q_3(1.5, 2.25)$, $Q_4(1.1, 1.21)$, and $Q_5(1.01, 1.0201)$ lie on the graph of f.
   b. i. Compute the slopes of the secant lines S_1, S_2, S_3, S_4, and S_5 through $P(1, 1)$ and each of the points Q_1, Q_2, Q_3, Q_4, and Q_5, respectively.
   ii. Use these values to guess at the slope of the tangent line T to the graph of f at the point $P(1, 1)$.
   c. Let $P(x, x^2)$ be a point that lies on the graph of f and let Q be a neighboring point on the graph with x-coordinate $x + h$.
   i. Find the y-coordinate of the point Q by evaluating $f(x + h)$.
   ii. Use the equation

$$m = \frac{f(x + h) - f(x)}{h}$$

   to show that the slope of the secant line through P and Q is $2x + h$.

   iii. Find the slope of the tangent line T at P by evaluating $\lim_{h \to 0} (2x + h)$.
   iv. Use the result of part c(iii) to find the slope of the tangent line at the point $P(1, 1)$. How does this compare with the results obtained in part b(ii)?

2. The occupancy rate of the all-suite Wonderland Hotel, located near an amusement park, is given by the function

$$r(t) = \frac{10}{81}t^3 - \frac{10}{3}t^2 + \frac{200}{9}t + 55 \qquad 0 \le r \le 100$$

where t is measured in months and $t = 0$ corresponds to the beginning of January. Management has estimated that the monthly revenue (in thousands of dollars) is approximated by the function

$$R(r) = -\frac{3}{5000}r^3 + \frac{9}{50}r^2 \qquad 0 \le r \le 100$$

where r is the occupancy rate.

a. What is the hotel's occupancy rate at the beginning of January? At the beginning of June?

b. What is the hotel's monthly revenue at the beginning of January? At the beginning of June?

c. Find an expression that gives the rate of change of Wonderland's occupancy rate with respect to time.

d. Find an expression that gives the rate of change of Wonderland's monthly revenue with respect to time.

e. What is the rate of change of Wonderland's monthly revenue with respect to time at the beginning of January? At the beginning of June? [*Hint:* Use the Chain Rule to find $R'(r(0))r'(0)$ and $R'(r(6))r'(6)$.]

3. Suppose a quantity $Q(t)$ does not exceed some number C, called the *carrying capacity of the environment.*

Furthermore, suppose the rate of growth $Q(t)$ is *jointly proportional* to its current size and the difference between its upper bound and its current size; that is,

$$\frac{dQ}{dt} = kQ(C - Q) \qquad (1)$$

where k is a constant of proportionality.

a. Study equation (1) and explain why $Q(t)$ exhibits exponential growth initially.

b. Explain why the growth of $Q(t)$ slows down as $Q(t)$ approaches C.

c. Use your qualitative analysis of (1) to make a sketch of the graph of the function Q. Assume that $Q(0) = Q_0$, where Q_0 is smaller than C.

REMARK The solution of equation (1) is the logistic function described on page 133. ☐

Applications of the Derivative

*T*his chapter further explores the power of the derivative, which we use to help analyze the properties of functions. The information obtained can then be used to accurately sketch graphs of functions. We will also see how the derivative is used in solving a large class of optimization problems, including finding what level of production will yield a maximum profit for a company; finding what level of production will result in minimal cost to a company; finding the maximum height attained by a rocket; finding the maximum velocity at which air is expelled when a person coughs; and a host of other problems.

10.1 Applications of the First Derivative

Determining the Intervals Where a Function Is Increasing or Decreasing

According to a study by the U.S. Department of Energy and the Shell Development Company, a typical car's fuel economy as a function of its speed is described by the graph shown in Figure 1. Observe that the fuel economy $f(x)$ in miles per gallon (mpg) improves as x, the vehicle's speed in miles per hour (mph), increases from 0 to 42, and then drops as the speed increases beyond 42 mph. We use the terms *increasing* and *decreasing* to describe the behavior of a function as we move from left to right along its graph.

FIGURE 1

A typical car's fuel economy improves as the speed at which it is driven increases from 0 mph to 42 mph and drops at speeds greater than 42 mph.

Source: U.S. Department of Energy and Shell Development Co.

More precisely, we have the following definitions.

INCREASING AND DECREASING FUNCTIONS

A function f is **increasing** on an interval (a, b) if for any two numbers x_1 and x_2 in (a, b), $f(x_1) < f(x_2)$ whenever $x_1 < x_2$ (Figure 2a).

A function f is **decreasing** on an interval (a, b) if for any two numbers x_1 and x_2 in (a, b), $f(x_1) > f(x_2)$ whenever $x_1 < x_2$ (Figure 2b).

FIGURE 2

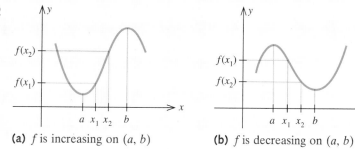

(a) f is increasing on (a, b)

(b) f is decreasing on (a, b)

We say that f is *increasing at a point c* if there exists an interval (a, b) containing c such that f is increasing on (a, b). Similarly, we say that f is *decreasing at a point c* if there exists an interval (a, b) containing c such that f is decreasing on (a, b).

Since the rate of change of a function at a point $x = c$ is given by the derivative of the function at that point, the derivative lends itself naturally to being a tool for determining the intervals where a differentiable function is increasing or decreasing. Indeed, as we saw in Chapter 9, the derivative of a function at a point measures both the slope of the tangent line to the graph of the function at that point and the rate of change of the function at the same point. In fact, at a point where the derivative is positive, the slope of the tangent line to the graph is positive and the function is increasing. At a point where the derivative is negative, the slope of the tangent line to the graph is negative and the function is decreasing (Figure 3).

FIGURE 3

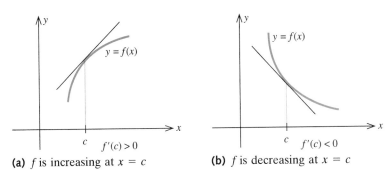

(a) f is increasing at $x = c$ (b) f is decreasing at $x = c$

These observations lead to the following important theorem, which we state without proof.

THEOREM 1

a. If $f'(x) > 0$ for each value of x in an interval (a, b), then f is increasing on (a, b).

b. If $f'(x) < 0$ for each value of x in an interval (a, b), then f is decreasing on (a, b).

c. If $f'(x) = 0$ for each value of x in an interval (a, b), then f is constant on (a, b).

FIGURE 4

The graph of f falls on $(-\infty, 0)$ where $f'(x) < 0$ and rises on $(0, \infty)$ where $f'(x) > 0$.

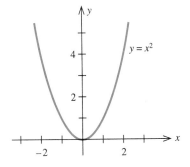

EXAMPLE 1   Find the interval where the function $f(x) = x^2$ is increasing and the interval where it is decreasing.

Solution   The derivative of $f(x) = x^2$ is $f'(x) = 2x$. Since

$$f'(x) = 2x > 0 \quad \text{if } x > 0$$

and

$$f'(x) = 2x < 0 \quad \text{if } x < 0$$

f is increasing on the interval $(0, \infty)$ and decreasing on the interval $(-\infty, 0)$ (Figure 4). ∎

Recall that the graph of a continuous function cannot have any breaks. As a consequence, a continuous function cannot change sign unless it equals zero for some value of x. This observation suggests the following procedure for determining the sign of the derivative f' of a function f, and hence the intervals where the function f is increasing and where it is decreasing.

DETERMINING THE INTERVALS WHERE A FUNCTION IS INCREASING OR DECREASING

1. Find all values of x for which $f'(x) = 0$ or f' is discontinuous, and identify the open intervals determined by these points.
2. Select a test point c in each interval found in step 1 and determine the sign of $f'(c)$ in that interval.
   a. If $f'(c) > 0$, f is increasing on that interval.
   b. If $f'(c) < 0$, f is decreasing on that interval.

EXAMPLE 2 Determine the intervals where the function $f(x) = x^3 - 3x^2 - 24x + 32$ is increasing and where it is decreasing.

Solution

1. The derivative of f is

$$f'(x) = 3x^2 - 6x - 24 = 3(x + 2)(x - 4)$$

and it is continuous everywhere. The zeros of $f'(x)$ are $x = -2$ and $x = 4$, and these points divide the real line into the intervals $(-\infty, -2)$, $(-2, 4)$, and $(4, \infty)$.
2. To determine the sign of $f'(x)$ in the intervals $(-\infty, -2)$, $(-2, 4)$, and $(4, \infty)$, compute $f'(x)$ at a convenient test point in each interval. The results are shown in the following table.

Interval	Test point c	$f'(c)$	Sign of $f'(x)$
$(-\infty, -2)$	-3	21	$+$
$(-2, 4)$	0	-24	$-$
$(4, \infty)$	5	21	$+$

FIGURE 5
Sign diagram for f'

Using these results, we obtain the sign diagram shown in Figure 5. We conclude that f is increasing on the intervals $(-\infty, -2)$ and $(4, \infty)$ and is decreasing on the interval $(-2, 4)$. Figure 6 shows the graph of f.

FIGURE 6
The graph of f rises on $(-\infty, -2)$, falls on $(-2, 4)$, and rises again on $(4, \infty)$.

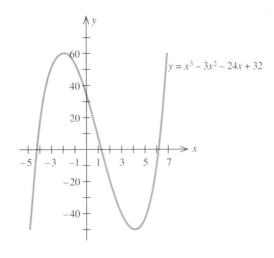

REMARK Do not be concerned with how the graphs in this section are obtained. We will learn how to sketch these graphs later. However, if you are familiar with the use of a graphing utility you may go ahead and verify each graph. □

FIGURE 7
Sign diagram for f'

FIGURE 8
f decreases on $(-\infty, 0)$ *and increases on* $(0, \infty)$.

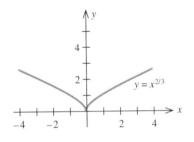

FIGURE 9
f' does not change sign as we move across x = 0.

FIGURE 10
The graph of f rises on $(-\infty, -1)$, *falls on* $(-1, 0)$ *and* $(0, 1)$, *and rises again on* $(1, \infty)$.

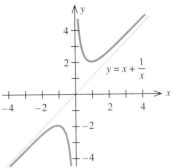

EXAMPLE 3   Find the interval where the function $f(x) = x^{2/3}$ is increasing and the interval where it is decreasing.

Solution

1. The derivative of f is

$$f'(x) = \frac{2}{3}x^{-1/3} = \frac{2}{3x^{1/3}}$$

The function f' is not defined at $x = 0$, so f' is discontinuous there. It is continuous everywhere else. Furthermore, f' is not equal to zero anywhere. The point $x = 0$ divides the real line (the domain of f) into the intervals $(-\infty, 0)$ and $(0, \infty)$.

2. Pick a test point, say $x = -1$, in the interval $(-\infty, 0)$ and compute

$$f'(-1) = -\frac{2}{3}$$

Since $f'(-1) < 0$, we see that $f'(x) < 0$ on $(-\infty, 0)$. Next, we pick a test point, say $x = 1$, in the interval $(0, \infty)$ and compute

$$f'(1) = \frac{2}{3}$$

Since $f'(1) > 0$, we see that $f'(x) > 0$ on $(0, \infty)$. Figure 7 shows these results in the form of a sign diagram.

We conclude that f is decreasing on the interval $(-\infty, 0)$ and increasing on the interval $(0, \infty)$. The graph of f, shown in Figure 8, confirms these results.   ■

EXAMPLE 4   Find the intervals where the function $f(x) = x + 1/x$ is increasing and where it is decreasing.

Solution

1. The derivative of f is

$$f'(x) = 1 - \frac{1}{x^2} = \frac{x^2 - 1}{x^2}$$

Since f' is not defined at $x = 0$, it is discontinuous there. Furthermore, $f'(x)$ is equal to zero when $x^2 - 1 = 0$ or $x = \pm 1$. These values of x partition the domain of f' into the open intervals $(-\infty, -1)$, $(-1, 0)$, $(0, 1)$, and $(1, \infty)$, where the sign of f' is different from zero.

2. To determine the sign of f' in each of these intervals, we compute $f'(x)$ at the test points $x = -2$, $-\frac{1}{2}$, $\frac{1}{2}$, and 2, respectively, obtaining $f'(-2) = \frac{3}{4}$, $f'(-\frac{1}{2}) = -3$, $f'(\frac{1}{2}) = -3$, and $f'(2) = \frac{3}{4}$. From the sign diagram for f' (Figure 9), we conclude that f is increasing on $(-\infty, -1)$ and $(1, \infty)$ and decreasing on $(-1, 0)$ and $(0, 1)$.

The graph of f appears in Figure 10. Note that f' does not change sign as we move across the point of discontinuity, $x = 0$. (Compare this with Example 3.)   ■

Example 4 reminds us that we must not automatically conclude that the derivative f' must change sign when we move across a point of discontinuity or a zero of f'.

Relative Extrema

In addition to helping us determine where the graph of a function is increasing and decreasing, the first derivative may be used to help us locate certain "high points" and "low points" on the graph of f. Knowing these points is invaluable in sketching the graphs of functions and solving optimization problems. These "high points" and "low points" correspond to the *relative (local) maxima* and *relative minima* of a function. They are so called because they are the highest or the lowest points when compared to points nearby.

The graph shown in Figure 11 gives the U.S. budget deficit from 1980 ($t = 0$) through 1991. The relative maxima and the relative minima of the function f are indicated on the graph.

FIGURE 11
U.S. budget deficit from 1980 to 1991

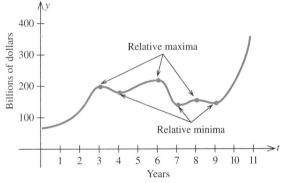

Source: Office of Management and Budget

More generally, we have the following definition:

RELATIVE MAXIMUM   A function f has a **relative maximum** at $x = c$ if there exists an open interval (a, b) containing c such that $f(x) \leq f(c)$ for all x in (a, b).

Geometrically, this means that there is *some* interval containing $x = c$ such that no point on the graph of f with its x-coordinate in that interval can lie above the point $(c, f(c))$; that is, $f(c)$ is the largest value of $f(x)$ in some interval around $x = c$. Figure 12 depicts the graph of a function f that has a relative maximum at $x = x_1$ and another at $x = x_3$.

Observe that all the points on the graph of f with x-coordinates in the interval I_1 containing x_1 (shown in blue) lie on or below the point $(x_1, f(x_1))$. This is also true for the point $(x_3, f(x_3))$ and the interval I_3. Thus, even though there are points on the graph of f that are "higher" than the points $(x_1, f(x_1))$ and $(x_3, f(x_3))$, the latter points are "highest" relative to points in their respective neighborhoods (intervals). Points on the graph of a

FIGURE 12
f has a relative maximum at
$x = x_1$ *and* $x = x_3$.

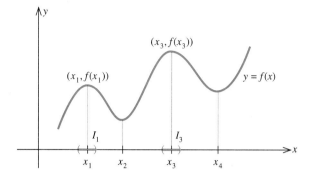

function f that are "highest" and "lowest" with respect to *all* points in the domain of f will be studied in Section 10.4.

The definition of the relative minimum of a function parallels that of the relative maximum of a function.

RELATIVE MINIMUM

A function f has a **relative minimum** at $x = c$ if there exists an open interval (a, b) containing c such that $f(x) \geq f(c)$ for all x in (a, b).

The graph of the function f, depicted in Figure 12, has a relative minimum at $x = x_2$ and another at $x = x_4$.

Finding the Relative Extrema

We refer to the relative maximum and relative minimum of a function as the **relative extrema** of that function. As a first step in our quest for the relative extrema of a function, we will consider functions that have derivatives at such points. Suppose that f is a function that is differentiable in some interval (a, b) that contains a point $x = c$, and that f has a relative maximum at $x = c$ (Figure 13a).

Observe that the slope of the tangent line to the graph of f must change from positive to negative as we move across the point $x = c$ from left to right. Therefore, the tangent line to the graph of f at the point $(c, f(c))$ must be horizontal; that is, $f'(c) = 0$ (Figure 13a).

FIGURE 13

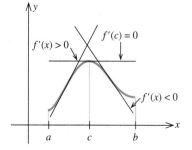

(a) f has a relative maximum at $x = c$

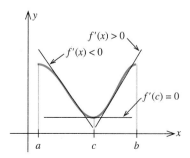

(b) f has a relative minimum at $x = c$

Using a similar argument, it may be shown that the derivative f' of a differentiable function f must also be equal to zero at a point $x = c$, where f has a relative minimum (Figure 13b).

This analysis reveals an important characteristic of the relative extrema of a differentiable function f: *At any point c where f has a relative extremum,* $f'(c) = 0$.

Before we develop a procedure for finding such points, a few words of caution are in order.

First, this result tells us that if a differentiable function f has a relative extremum at a point $x = c$, then $f'(c) = 0$. The converse of this statement—if $f'(c) = 0$ at some point $x = c$, then f must have a relative extremum at that point—is not true. Consider, for example, the function $f(x) = x^3$. Here $f'(x) = 3x^2$, so $f'(0) = 0$. Yet f has neither a relative maximum nor a relative minimum at $x = 0$ (Figure 14).

Second, our result assumes that the function is differentiable and thus has a derivative at a point that gives rise to a relative extremum. The functions $f(x) = |x|$ and $g(x) = x^{2/3}$ demonstrate that a relative extremum of a function may exist at a point at which the derivative does not exist. Both these functions fail to be differentiable at $x = 0$, but each has a relative minimum there. Figure 15 shows the graphs of these functions. Note that the slopes of the tangent lines change from negative to positive as we move across $x = 0$, just as in the case of a function that is differentiable at a value of x that gives rise to a relative minimum.

FIGURE 14
$f'(0) = 0$, *but f does not have a relative extremum at* (0, 0).

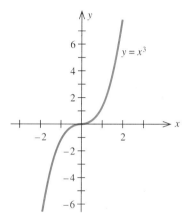

FIGURE 15
Each of these functions has a relative extremum at (0, 0), *but the derivative does not exist there.*

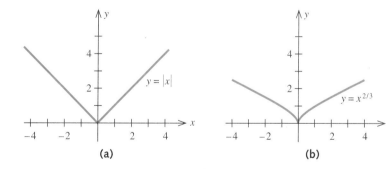

(a)   (b)

We refer to a point in the domain of f that *may* give rise to a relative extremum as a *critical point*.

CRITICAL POINT OF *f*

A **critical point** of a function f is any point x in the domain of f such that $f'(x) = 0$ or $f'(x)$ does not exist.

Figure 16 depicts the graph of a function that has critical points at $x = a$, b, c, d, and e. Observe that $f'(x) = 0$ at $x = a$, b, and c. Next, since there is a corner at $x = d$, $f'(x)$ does not exist there. Finally, $f'(x)$ does not exist at $x = e$, because the tangent line there is vertical. Also, observe that the

FIGURE 16
Critical points of f

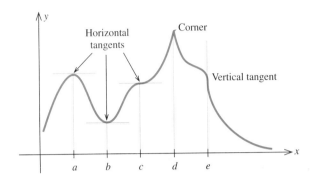

critical points $x = a$, b, and d give rise to relative extrema of f, whereas the critical points $x = c$ and $x = e$ do not.

Having defined what a critical point is, we can now state a formal procedure for finding the relative extrema of a continuous function that is differentiable everywhere except at isolated values of x. Incorporated into the procedure is the so-called **First Derivative Test,** which enables us to determine whether or not a point gives rise to a relative maximum or a relative minimum of the function f.

PROCEDURE FOR FINDING RELATIVE EXTREMA (THE FIRST DERIVATIVE TEST)

1. Determine the critical points of f.
2. Determine the sign of $f'(x)$ to the left and right of each critical point.
   a. If $f'(x)$ changes sign from *positive* to *negative* as we move across a critical point $x = c$, then $f(c)$ is a relative maximum.
   b. If $f'(x)$ changes sign from *negative* to *positive* as we move across a critical point $x = c$, then $f(c)$ is a relative minimum.
   c. If $f'(x)$ does not change sign as we move across a critical point $x = c$, then $f(c)$ is not a relative extremum.

FIGURE 17
f has a relative minimum at $x = 0$.

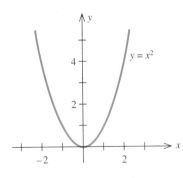

EXAMPLE 5 Find the relative maxima and relative minima of the function $f(x) = x^2$.

Solution The derivative of $f(x) = x^2$ is given by $f'(x) = 2x$. Setting $f'(x) = 0$ yields $x = 0$ as the only critical point of f. Since

$$f'(x) < 0 \quad \text{if } x < 0$$

and

$$f'(x) > 0 \quad \text{if } x > 0$$

we see that $f'(x)$ changes sign from negative to positive as we move across the critical point $x = 0$. Thus we conclude that $f(0) = 0$ is a relative minimum of f (Figure 17).

EXAMPLE 6 Find the relative maxima and relative minima of the function $f(x) = x^{2/3}$ (see Example 3).

Solution The derivative of f is $f'(x) = (2/3)x^{-1/3}$. As noted in Example 3, f' is not defined at $x = 0$, is continuous everywhere else, and is not equal to zero in its domain. Thus, $x = 0$ is the only critical point of the function f.

FIGURE 18
Sign diagram for f′

f′ not defined at x = 0

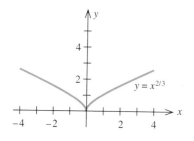

The sign diagram obtained in Example 3 is reproduced in Figure 18. We can see that the sign of $f'(x)$ changes from negative to positive as we move across $x = 0$ from left to right. Thus, an application of the First Derivative Test tells us that $f(0) = 0$ is a relative minimum of f (Figure 19).   ■

FIGURE 19

f has a relative minimum at x = 0.

EXAMPLE 7   Find the relative maxima and relative minima of the function

$$f(x) = x^3 - 3x^2 - 24x + 32$$

Solution   The derivative of f is

$$f'(x) = 3x^2 - 6x - 24 = 3(x + 2)(x - 4)$$

and it is continuous everywhere. The zeros of $f'(x)$, $x = -2$ and $x = 4$, are the only critical points of the function f. The sign diagram for f' is shown in Figure 20. Examine the two critical points $x = -2$ and $x = 4$ for a relative extremum using the First Derivative Test and the sign diagram for f':

1. *The critical point $x = -2$*: Since the function $f'(x)$ changes sign from positive to negative as we move across $x = -2$ from left to right, we conclude that a relative maximum of f occurs at $x = -2$. The value of $f(x)$ when $x = -2$ is

$$f(-2) = (-2)^3 - 3(-2)^2 - 24(-2) + 32 = 60$$

2. *The critical point $x = 4$*: $f'(x)$ changes sign from negative to positive as we move across $x = 4$ from left to right, so $f(4) = -48$ is a relative minimum of f. The graph of f appears in Figure 21.

FIGURE 20

Sign diagram for f′

FIGURE 21

f has a relative maximum at x = −2 and a relative minimum at x = 4.

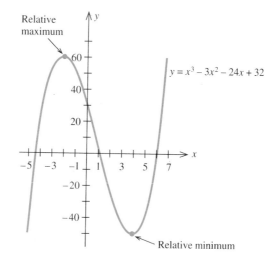

EXAMPLE 8   Find the relative maxima and the relative minima of the function

$$f(x) = x + \frac{1}{x}$$

FIGURE 22

x = 0 is not a critical point because f is not defined at x = 0.

f' is not defined at x = 0

$+ + + + 0 - - - - - 0 + + + +$

−1   0   1

FIGURE 23

$f(x) = x + \dfrac{1}{x}$

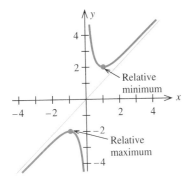

Relative minimum

Relative maximum

FIGURE 24

Sign diagram for f'

$+ + + + + + + + 0 - - - - -$

0   1

FIGURE 25

The graph of $f(x) = xe^{-x}$ is increasing on $(-\infty, 1)$ and decreasing on $(1, \infty)$.

Solution   The derivative of f is

$$f'(x) = 1 - \frac{1}{x^2} = \frac{x^2 - 1}{x^2} = \frac{(x + 1)(x - 1)}{x^2}$$

Since f' is equal to zero at $x = -1$ and $x = 1$, these are critical points for the function f. Next, observe that f' is discontinuous at $x = 0$. However, because f is *not defined at that point*, the point $x = 0$ does not qualify as a critical point of f. Figure 22 shows the sign diagram for f'.

Since $f'(x)$ changes sign from positive to negative as we move across $x = -1$ from left to right, the First Derivative Test implies that $f(-1) = -2$ is a relative maximum of the function f. Next, $f'(x)$ changes sign from negative to positive as we move across $x = 1$ from left to right, so $f(1) = 2$ is a relative minimum of the function f. The graph of f appears in Figure 23. Note that this function has a relative maximum that lies below its relative minimum.

EXAMPLE 9   Let $f(x) = xe^{-x}$ and find (a) the interval where f is increasing or decreasing and (b) the relative extrema of f.

Solution   The derivative of f is

$$f'(x) = \frac{d}{dx}(xe^{-x}) = x\frac{d}{dx}(e^{-x}) + e^{-x}\frac{d}{dx}(x) \qquad \text{Using the Product Rule}$$

$$= xe^{-x}(-1) + e^{-x} = (1 - x)e^{-x}$$

Observe that f' is continuous everywhere. Next, setting $f'(x) = 0$ gives $x = 1$ (remember that $e^{-x} \neq 0$ for all values of x). Therefore, $x = 0$ is the sole critical point of f. The sign diagram of f' is shown in Figure 24.

From the sign diagram, we see that f is increasing on $(-\infty, 1)$ and decreasing on $(1, \infty)$. Also, using this diagram and the First Derivative Test, we see that the critical point $x = 1$ gives rise to a relative maximum of f with value $f(1) = e^{-1} = 1/e$. The graph of f shown in Figure 25 confirms our results.

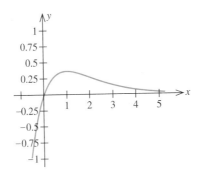

Applications

EXAMPLE 10   The profit function of the Acrosonic Company is given by

$$P(x) = -0.02x^2 + 300x - 200{,}000$$

dollars, where x is the number of Acrosonic Model F loudspeaker systems produced. Find where the function P is increasing and where it is decreasing.

FIGURE 26

The profit function is increasing on (0, 7500) *and decreasing on* (7500, ∞).

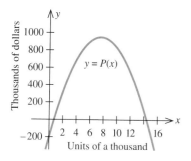

Solution   The derivative P' of the function P is

$$P'(x) = -0.04x + 300 = -0.04(x - 7500)$$

Thus, $P'(x) = 0$ when $x = 7500$. Furthermore, $P'(x) > 0$ for x in the interval (0, 7500), and $P'(x) < 0$ for x in the interval (7500, ∞). This means that the profit function P is increasing on (0, 7500) and decreasing on (7500, ∞) [Figure 26]. ∎

EXAMPLE 11   The number of major crimes committed in the city of Bronxville from 1987 to 1994 is approximated by the function

$$N(t) = -0.1t^3 + 1.5t^2 + 100 \qquad 0 \le t \le 7$$

where $N(t)$ denotes the number of crimes committed in year t and $t = 0$ corresponds to the beginning of 1987. Find where the function N is increasing and where it is decreasing.

Solution   The derivative N' of the function N is

$$N'(t) = -0.3t^2 + 3t = -0.3t(t - 10)$$

Since $N'(t) > 0$ for t in the interval (0, 7), the function N is increasing throughout that interval (Figure 27).

FIGURE 27

The number of crimes, $N(t)$, *was increasing over the 7-year interval.*

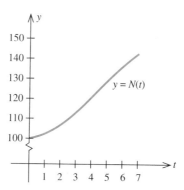

SELF-CHECK EXERCISES 10.1

1. Find the intervals where the function $f(x) = \frac{2}{3}x^3 - x^2 - 12x + 3$ is increasing and the intervals where it is decreasing.

2. Find the relative extrema of $f(x) = \frac{x^2}{1 - x^2}$.

Solutions to Self-Check Exercises 10.1 can be found on page 600.

10.1   EXERCISES

In exercises 1–8, you are given the graph of a function f. Determine the intervals where f is increasing, constant, or decreasing.

1.

2.

3.

4.

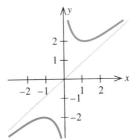

33. $f(x) = \ln x^2$

34. $f(x) = \dfrac{\ln x}{x}$

35. $f(x) = \dfrac{x^2 - 1}{x}$

36. $h(x) = \dfrac{x^2}{x - 1}$

37. $f(x) = \dfrac{1}{(x - 1)^2}$

38. $g(x) = \dfrac{x}{(x + 1)^2}$

*In exercises 39–46, you are given the graph of a function f.
Determine the relative maxima and relative minima, if any.*

5.

6.

39.

40.

7.

8.

41.

42.

*In exercises 9–38, find the interval(s) where each function
is increasing and the interval(s) where it is decreasing.*

43.

44.

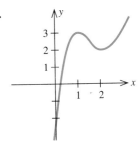

9. $f(x) = 3x + 5$

10. $f(x) = 4 - 5x$

11. $f(x) = x^2 - 3x$

12. $f(x) = 2x^2 + x + 1$

13. $g(x) = x - x^3$

14. $f(x) = x^3 - 3x^2$

15. $g(x) = x^3 + 3x^2 + 1$

16. $f(x) = x^3 - 3x + 4$

17. $f(x) = \frac{1}{3}x^3 - 3x^2 + 9x + 20$

18. $f(x) = \frac{2}{3}x^3 - 2x^2 - 6x - 2$

19. $h(x) = x^4 - 4x^3 + 10$

20. $g(x) = x^4 - 2x^2 + 4$

45.

21. $f(x) = \dfrac{1}{x - 2}$

22. $h(x) = \dfrac{1}{2x + 3}$

23. $h(t) = \dfrac{t}{t - 1}$

24. $g(t) = \dfrac{2t}{t^2 + 1}$

25. $f(x) = x^{3/5}$

26. $f(x) = x^{2/3} + 5$

27. $f(x) = \sqrt{x + 1}$

28. $f(x) = (x - 5)^{2/3}$

29. $f(x) = \sqrt{16 - x^2}$

30. $g(x) = x\sqrt{x + 1}$

31. $f(x) = e^{-x^2/2}$

32. $f(x) = x^2 e^{-x}$

46.

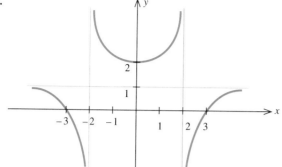

In exercises 47–70, find the relative maxima and relative minima, if any, of each function.

47. $f(x) = x^2 - 4x$

48. $g(x) = x^2 + 3x + 8$

49. $f(x) = \frac{1}{2}x^2 - 2x + 4$

50. $h(t) = -t^2 + 6t + 6$

51. $f(x) = x^{2/3} + 2$

52. $f(x) = x^{5/3}$

53. $g(x) = x^3 - 3x^2 + 4$

54. $f(x) = x^3 - 3x + 6$

55. $F(x) = \frac{1}{3}x^3 - x^2 - 3x + 4$

56. $f(x) = \frac{1}{2}x^4 - x^2$

57. $g(x) = x^4 - 4x^3 + 8$

58. $h(x) = \frac{1}{2}x^4 - 3x^2 + 4x - 8$

59. $f(x) = 3x^4 - 2x^3 + 4$

60. $F(t) = 3t^5 - 20t^3 + 20$

61. $g(x) = \dfrac{x+1}{x}$

62. $h(x) = \dfrac{x}{x+1}$

63. $f(x) = x + \dfrac{9}{x} + 2$

64. $g(x) = 2x^2 + \dfrac{4000}{x} + 10$

65. $f(x) = \dfrac{x}{1+x^2}$

66. $g(x) = \dfrac{x}{x^2-1}$

67. $f(x) = \dfrac{x^2}{x^2-4}$

68. $g(t) = \dfrac{t^2}{1+t^2}$

69. $f(x) = (x-1)^{2/3}$

70. $g(x) = x\sqrt{x-4}$

71. A stone is thrown straight up from the roof of an 80-foot building. The distance of the stone from the ground at any time t (in seconds) is given by

$$h(t) = -16t^2 + 64t + 80$$

feet. When is the stone rising and when is it falling? If the stone were to miss the building, when would

it hit the ground? [*Hint:* The stone is on the ground when $h(t) = 0$.] Sketch the graph of h.

72. Profit Functions   The Mexico subsidiary of the Thermo-Master Company manufactures an indoor-outdoor thermometer. Management estimates that the profit realizable by the company for the manufacture and sale of x units of thermometers per week is

$$P(x) = -0.001x^2 + 8x - 5000$$

dollars. Find the intervals where the profit function P is increasing and the intervals where P is decreasing.

73. Flight of a Rocket   The height (in feet) attained by a rocket t seconds into flight is given by the function

$$h(t) = -\frac{1}{3}t^3 + 16t^2 + 33t + 10$$

When is the rocket rising and when is it descending?

74. Environment of Forests   Following the lead of the National Wildlife Federation, the Department of the Interior of a South American country began to record an index of environmental quality that measured progress and decline in the environmental quality of its forests. The index for the years 1984 through 1994 is approximated by the function

$$I(t) = \frac{1}{3}t^3 - \frac{5}{2}t^2 + 80 \qquad 0 \le t \le 10$$

where $t = 0$ corresponds to the year 1984. Find the intervals where the function I is increasing and the intervals where it is decreasing. Interpret your results.
Source: World Almanac

75. Average Speed of a Highway Vehicle   The average speed of a vehicle on a stretch of Route 134 between 6 A.M. and 10 A.M. on a typical weekday is approximated by the function

$$f(t) = 20t - 40\sqrt{t} + 50 \qquad 0 \le t \le 4$$

where $f(t)$ is measured in miles per hour and t is measured in hours, with $t = 0$ corresponding to 6 A.M. Find the interval where f is increasing and the interval where f is decreasing and interpret your results.

76. Average Cost   The average cost in dollars incurred by the Lincoln Record Company per week in pressing x long-playing records is given by

$$\overline{C}(x) = -0.0001x + 2 + \frac{2000}{x} \qquad 0 < x \le 6000$$

Show that $\overline{C}(x)$ is always decreasing over the interval $(0, 6000)$.

77. Air Pollution According to the South Coast Air Quality Management District, the level of nitrogen dioxide, a brown gas that impairs breathing, present in the atmosphere on a certain May day in downtown Los Angeles is approximated by

$$A(t) = 0.03t^3(t - 7)^4 + 60.2 \qquad 0 \le t \le 7$$

where $A(t)$ is measured in pollutant standard index (PSI) and t is measured in hours with $t = 0$ corresponding to 7 A.M. At what time of day is the air pollution increasing and at what time is it decreasing?

78. Projected Social Security Funds Based on data from the Social Security Administration, the estimated cash in the Social Security retirement and disability trust funds in the five decades beginning with the year 1990 is given by

$$A(t) = -96.6t^4 + 403.6t^3 + 660.9t^2$$
$$+ 250 \qquad 0 \le t \le 5$$

where $A(t)$ is measured in billions of dollars and t is measured in decades, with $t = 0$ corresponding to the year 1990. Find the interval where A is increasing and the interval where A is decreasing and interpret your results. [*Hint:* Use the quadratic formula.]
Source: Social Security Administration

79. Learning Curves The Emory Secretarial School finds from experience that the average student taking Advanced Typing will progress according to the rule

$$N(t) = \frac{60t + 180}{t + 6} \qquad t \ge 0$$

where $N(t)$ measures the number of words per minute the student can type after t weeks in the course. Compute $N'(t)$ and use this result to show that the function N is increasing on the interval $(0, \infty)$.

80. Drug Concentration in the Blood The concentration of a certain drug in a patient's body t hours after injection is given by

$$C(t) = \frac{t^2}{2t^3 + 1} \qquad 0 \le t \le 4$$

milligrams per cubic centimeter. When is the concentration of the drug increasing and when is it decreasing?

81. Air Pollution The amount of nitrogen dioxide, a brown gas that impairs breathing, present in the atmosphere on a certain May day in the city of Long Beach is approximated by

$$A(t) = \frac{136}{1 + 0.25(t - 4.5)^2} + 28 \qquad 0 \le t \le 11$$

where $A(t)$ is measured in pollutant standard index (PSI) and t is measured in hours, with $t = 0$ corresponding to 7 A.M. Find the intervals where A is increasing and where A is decreasing and interpret your results.

82. Prison Overcrowding The 1980s saw a trend toward old-fashioned punitive deterrence as opposed to the more liberal penal policies and community-based corrections popular in the 1960s and early 1970s. As a result, prisons became more crowded and the gap between the number of people in prison and the prison capacity widened. Based on figures from the U.S. Department of Justice, the number of prisoners (in thousands) in federal and state prisons is approximated by the function

$$N(t) = 3.5t^2 + 26.7t + 436.2 \qquad 0 \le t \le 10$$

where t is measured in years and $t = 0$ corresponds to 1984. The number of inmates for which prisons were designed is given by

$$C(t) = 24.3t + 365 \qquad 0 \le t \le 10$$

where $C(t)$ is measured in thousands and t has the same meaning as before. Show that the gap between the number of prisoners and the number for which the prisons were designed has been widening at any time t. [*Hint:* First write a function G that gives the gap between the number of prisoners and the number for which the prisons were designed at any time t. Then show that $G'(t) > 0$ for all values of t in the interval $(0, 10)$.]
Source: U.S. Dept. of Justice

SOLUTIONS TO SELF-CHECK EXERCISES 10.1

1. The derivative of f is

$$f'(x) = 2x^2 - 2x - 12 = 2(x + 2)(x - 3)$$

and it is continuous everywhere. The zeros of $f'(x)$ are $x = -2$ and $x = 3$. The sign diagram of f' is shown in the accompanying figure. We conclude that f is increasing on the intervals $(-\infty, -2)$ and $(3, \infty)$ and decreasing on the interval $(-2, 3)$.

2. The derivative of f is

$$f'(x) = \frac{(1 - x^2)\dfrac{d}{dx}(x^2) - x^2\dfrac{d}{dx}(1 - x^2)}{(1 - x^2)^2}$$

$$= \frac{(1 - x^2)(2x) - x^2(-2x)}{(1 - x^2)^2} = \frac{2x}{(1 - x^2)^2}$$

and it is continuous everywhere except at $x = \pm 1$. Since $f'(x)$ is equal to zero at $x = 0$, $x = 0$ is a critical point of f. Next, observe that $f'(x)$ is discontinuous at $x = \pm 1$, but since these points are not in the domain of f, they do not qualify as critical points of f. Finally, from the sign diagram of f' shown in the accompanying figure, we conclude that $f(0) = 0$ is a relative minimum of f.

Using Technology

A graphing utility is an effective tool for analyzing the properties of functions. This is especially true when we also bring into play the power of calculus, as the following examples show.

EXAMPLE 1   Let $f(x) = 2.4x^4 - 8.2x^3 + 2.7x^2 + 4x + 1$.

a. Use a graphing utility to plot the graph of f.

b. Find the intervals where f is increasing and the intervals where f is decreasing.

c. Find the relative extrema of f.

Solution

a. The graph of f in the viewing rectangle $[-2, 4] \times [-10, 10]$ is shown in Figure 1.

b. We compute

$$f'(x) = 9.6x^3 - 24.6x^2 + 5.4x + 4$$

and observe that f' is continuous everywhere. So the critical points of f occur at values of x where $f'(x) = 0$. In order to solve this last equation, observe that $f'(x)$ is a *polynomial function* of degree 3. The easiest way

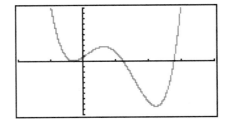

to solve the polynomial equation

$$9.6x^3 - 24.6x^2 + 5.4x + 4 = 0$$

is to use **POLY.** We find

$$x_1 \approx 2.22564943249, \qquad x_2 \approx 0.63272944121, \qquad x_3 \approx -0.295878873696$$

Referring to Figure 1, we conclude that f is decreasing on $(-\infty, -0.2959)$ and $(0.6327, 2.2256)$ (correct to four decimal places), and f is increasing on $(-0.2959, 0.6327)$ and $(2.2256, \infty)$.

c. Using **EVAL,** we find the value of f at each of the critical points found in (b). Upon referring to Figure 1 once again, we see that $f(x_3) \approx 0.2836$ and $f(x_1) \approx -8.2366$ are relative minimum values of f, and $f(x_2) \approx 2.9194$ is a relative maximum value of f. ■

REMARK   The equation $f'(x) = 0$ in Example 1 is a polynomial equation, and so it is easily solved using **POLY.** We could also solve the equation using **ROOT,** but it would require much more work. For equations that are *not* polynomial equations, however, our only choice is to use **ROOT.** ☐

If the derivative of a function is difficult to compute or simplify and we do not require great precision in the solution, we can find the relative extrema of the function using a combination of **ZOOM** and **TRACE.** This technique, which does not require the use of the derivative of f, is illustrated in the following example. (See Sections A.10 and A.13 in Appendix A.)

EXAMPLE 2   Let $f(x) = x^{1/3}(x^2 + 1)^{-3/2}e^{-x}$.

a. Use a graphing utility to plot the graph of f.

b. Find the relative extrema of f.

Solution

a. The graph of f in the viewing rectangle $[-4, 2] \times [-2, 1]$ is shown in Figure 2.

b. From the graph of f in Figure 2, we see that f has relative maxima when $x \approx -2$ and $x \approx 0.25$, and a relative minimum when $x \approx -0.75$. To obtain a better approximation of the first relative maximum, we use **ZIN** with the cursor at approximately the point on the graph corresponding to $x \approx -2$. Then, using **TRACE,** we see that a relative maximum occurs when $x \approx -2.12$ with value $y \approx -0.83$. Similarly, we find the other relative maximum where $x \approx 0.21$ with value $y \approx 0.45$. Repeating the procedure, we find the relative minimum at the point where $x \approx -0.75$ and $y \approx -0.98$. ∎

Finally, we comment that if you have access to a computer and software such as Derive, Maple, or Mathematica, then symbolic differentiation will yield the derivative $f'(x)$ of any differentiable function. This software will also solve the equation $f'(x) = 0$ with ease. Thus, the use of a computer will simplify even more greatly the analysis of functions.

EXERCISES

In exercises 1–4, use a graphing utility to find (a) the intervals where f is increasing and the intervals where f is decreasing, and (b) the relative extrema of f. Express your answers accurate to four decimal places.

1. $f(x) = 3.4x^4 - 6.2x^3 + 1.8x^2 + 3x - 2$

2. $f(x) = 1.8x^4 - 9.1x^3 + 5x - 4$

3. $f(x) = 2x^5 - 5x^3 + 8x^2 - 3x + 2$

4. $f(x) = 3x^5 - 4x^2 + 3x - 1$

*In exercises 5–8, use the **ZOOM** and **TRACE** features of a graphing utility to find (a) the intervals where f is increasing and the intervals where f is decreasing, and (b) the relative extrema of f. Express your answers accurate to two decimal places.*

5. $f(x) = (2x + 1)^{1/3}(x^2 + 1)^{-2/3}$

6. $f(x) = [x^2(x^3 - 1)]^{1/3} + \dfrac{1}{x}$

7. $f(x) = e^{-x}\sqrt{x^2 + 1} + x^3$

8. $f(x) = \dfrac{xe^{-x^2} + x^{3/2}}{x^2 + 1}$

9. Rate of Bank Failures The Federal Deposit Insurance Company (FDIC) estimates that the rate at which banks were failing between 1982 and 1994 is given by

$$f(t) = 0.063447t^4 - 1.953283t^3 + 14.632576t^2$$
$$- 6.684704t + 47.458874 \qquad 0 \le t \le 12$$

where $f(t)$ is measured in the number of banks per year and t is measured in years with $t = 0$ corresponding to the beginning of 1982.
a. Use a graphing utility to plot the graph of f in the viewing rectangle $[0, 13] \times [0, 220]$.
b. Determine the intervals where f is increasing and where f is decreasing and interpret your result.
c. Find the relative maximum of f and interpret your result.
Source: Federal Deposit Insurance Corporation

10. Manufacturing Capacity Data obtained from the Federal Reserve show that the annual increase in manufacturing capacity between 1988 and 1994 is given by

$$f(t) = 0.0388889t^3 - 0.283333t^2 + 0.477778t$$
$$+ 2.04286 \qquad 0 \le t \le 6$$

where $f(t)$ is a percentage and t is measured in years with $t = 0$ corresponding to the beginning of 1988.
a. Use a graphing utility to plot the graph of f in the viewing rectangle $[0, 8] \times [0, 4]$.
b. Determine the intervals where f is increasing and where f is decreasing and interpret your result.
Source: Federal Reserve

11. Home Sales According to the Greater Boston Real Estate Board-Multiple Listing Service, the average number of days a single-family home remains for sale from listing to accepted offer is approximated by the function

$$f(t) = 0.0171911t^4 - 0.662121t^3 + 6.18083t^2$$
$$- 8.97086t + 53.3357 \qquad 0 \le t \le 10$$

where t is measured in years, with $t = 0$ corresponding to the beginning of 1984.

a. Use a graphing utility to plot the graph of f in the viewing rectangle $[0, 12] \times [0, 120]$.

b. Use the graph of f to find, approximately, the intervals where f is increasing and the inter-

vals where f is decreasing. What does this result tell us about the sales of single-family homes in the greater Boston area from 1984 through 1994?

Source: Greater Boston Real Estate Board-Multiple Listing Service

---

10.2 Applications of the Second Derivative

Determining the Intervals of Concavity

Consider the graphs shown in Figure 28, which give the estimated population of the world and of the United States through the year 2000. Both graphs are rising, indicating that both the U.S. population and the world population will continue to increase through the year 2000. But observe that the graph in Figure 28a opens upward, whereas the graph in Figure 28b opens downward. What is the significance of this? To answer this question, let us look at the slopes of the tangent lines to various points on each graph (Figure 29).

FIGURE 28

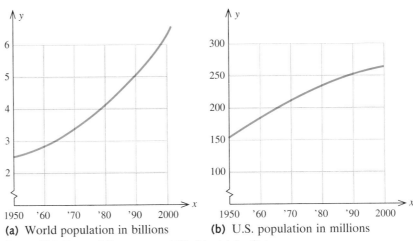

(a) World population in billions   (b) U.S. population in millions

Source: U.S. Dept. of Commerce and Worldwatch Institute

FIGURE 29

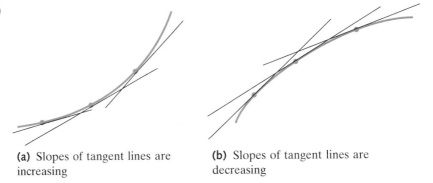

(a) Slopes of tangent lines are increasing

(b) Slopes of tangent lines are decreasing

In Figure 29a we see that the slopes of the tangent lines to the graph are increasing as we move from left to right. Since the slope of the tangent line to the graph at a point on the graph measures the rate of change of the function at that point, we conclude that the world population is not only increasing through the year 2000 but is increasing at an *increasing* pace. A similar analysis of Figure 29b reveals that the U.S. population is increasing, but at a *decreasing* pace.

The shape of a curve can be described using the notion of *concavity*:

CONCAVITY OF A FUNCTION f

Let the function f be differentiable on an interval (a, b). Then

1. f is **concave upward** on (a, b) if f' is increasing on (a, b).

2. f is **concave downward** on (a, b) if f' is decreasing on (a, b).

Geometrically, a curve is concave upward if it lies above its tangent lines (Figure 30a). Similarly, a curve is concave downward if it lies below its tangent lines (Figure 30b).

FIGURE 30

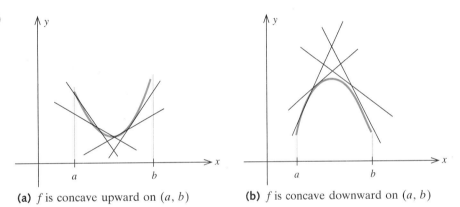

(a) f is concave upward on (a, b) **(b)** f is concave downward on (a, b)

We also say that f is *concave upward at a point $x = c$* if there exists an interval (a, b) containing c in which f is concave upward. Similarly, we say that f is *concave downward at a point $x = c$* if there exists an interval (a, b) containing c in which f is concave downward.

If a function f has a second derivative f'', we can use f'' to determine the intervals of concavity of the function. Recall that $f''(x)$ measures the rate of change of the slope $f'(x)$ of the tangent line to the graph of f at the point $(x, f(x))$. Thus, if $f''(x) > 0$ on an interval (a, b), then the slopes of the tangent lines to the graph of f are increasing on (a, b) and so f is concave upward on (a, b). Similarly, if $f''(x) < 0$ on (a, b), then f is concave downward on (a, b). These observations suggest the following theorem.

THEOREM 2

a. If $f''(x) > 0$ for each value of x in (a, b), then f is concave upward on (a, b).

b. If $f''(x) < 0$ for each value of x in (a, b), then f is concave downward on (a, b).

The following procedure, based on the conclusions of Theorem 2, may be used to determine the intervals of concavity of a function.

DETERMINING THE INTERVALS OF CONCAVITY OF f

1. Determine the values of x for which f'' is zero or where f'' is not defined, and identify the open intervals determined by these points.
2. Determine the sign of f'' in each interval found in step 1. To do this, compute $f''(c)$, where c is any conveniently chosen test point in the interval.
   a. If $f''(c) > 0$, f is concave upward on that interval.
   b. If $f''(c) < 0$, f is concave downward on that interval.

FIGURE 31
Sign diagram for f"

EXAMPLE 1   Determine where the function $f(x) = x^3 - 3x^2 - 24x + 32$ is concave upward and where it is concave downward.

Solution   Here, $f'(x) = 3x^2 - 6x - 24$

so $f''(x) = 6x - 6 = 6(x - 1)$

and f'' is defined everywhere. Setting $f''(x) = 0$ gives $x = 1$. The sign diagram of f'' appears in Figure 31. We conclude that f is concave downward on the interval $(-\infty, 1)$ and is concave upward on the interval $(1, \infty)$. Figure 32 shows the graph of f.

FIGURE 32
f is concave downward on $(-\infty, 1)$ and concave upward on $(1, \infty)$.

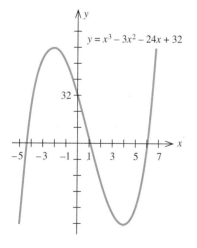

EXAMPLE 2   Determine the intervals where the function $f(x) = x + 1/x$ is concave upward and where it is concave downward.

Solution   We have $f'(x) = 1 - \dfrac{1}{x^2}$

and $f''(x) = \dfrac{2}{x^3}$

We deduce from the sign diagram for f'' (Figure 33) that the function f is concave downward on the interval $(-\infty, 0)$ and concave upward on the interval $(0, \infty)$. The graph of f is sketched in Figure 34.

FIGURE 34
f is concave down on $(-\infty, 0)$ and concave up on $(0, \infty)$.

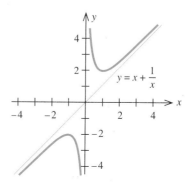

FIGURE 33
The sign diagram for f"

FIGURE 35
The graph of S has a point of inflection at (50, 2700).

Inflection Points

Figure 35 shows the total sales S of a manufacturer of automobile air conditioners versus the amount of money x that the company spends on advertising its product. Notice that the graph of the continuous function $y = S(x)$ changes concavity—from upward to downward—at the point (50, 2700). This point is called an *inflection point* of S. To understand the significance of this inflection point, observe that the total sales increase rather slowly at first, but as more money is spent on advertising, the total sales increase rapidly. This rapid increase reflects the effectiveness of the company's ads. However, a point is soon reached after which any additional advertising expenditure results in increased sales but at a slower rate of increase. This point, commonly known as the *point of diminishing returns*, is the point of inflection of the function S. We will return to this example later.

Let us now state formally the definition of an inflection point.

INFLECTION POINT

A point on the graph of a differentiable function f at which the concavity changes is called an **inflection point.**

Observe that the graph of a function crosses its tangent line at a point of inflection (Figure 36).

FIGURE 36
At each point of inflection, the graph of a function crosses its tangent line.

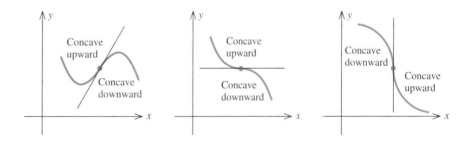

The following procedure may be used to find inflection points.

FINDING INFLECTION POINTS

1. Compute $f''(x)$.
2. Determine the points in the domain of f for which $f''(x) = 0$ or $f''(x)$ does not exist.
3. Determine the sign of $f''(x)$ to the left and right of each point $x = c$ found in step 2. If there is a change in the sign of $f''(x)$ as we move across the point $x = c$, then $(c, f(c))$ is an inflection point of f.

FIGURE 37
Sign diagram for f"

FIGURE 38
f has an inflection point at (0, 0).

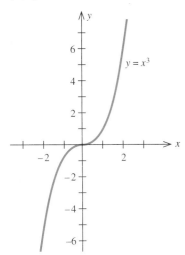

FIGURE 39
Sign diagram for f"

FIGURE 40
f has an inflection point at (1, 0).

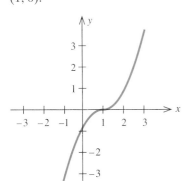

The points determined in step 2 are only *candidates* for the inflection points of f. For example, you can easily verify that $f''(0) = 0$ if $f(x) = x^4$, but a sketch of the graph of f will show that $(0, 0)$ is not an inflection point of f.

EXAMPLE 3   Find the points of inflection of the function $f(x) = x^3$.

Solution

$$f'(x) = 3x^2$$

so

$$f''(x) = 6x$$

Observe that f'' is continuous everywhere and is zero if $x = 0$. The sign diagram of f'' is shown in Figure 37. From this diagram, we see that $f''(x)$ changes sign as we move across $x = 0$. Thus the point $(0, 0)$ is an inflection point of the function f (Figure 38).   ■

EXAMPLE 4   Determine the intervals where the function $f(x) = (x - 1)^{5/3}$ is concave upward and where it is concave downward and find the inflection points of f.

Solution   The first derivative of f is

$$f'(x) = \frac{5}{3}(x - 1)^{2/3}$$

and the second derivative of f is

$$f''(x) = \frac{10}{9}(x - 1)^{-1/3} = \frac{10}{9(x - 1)^{1/3}}$$

We see that f'' is not defined at $x = 1$. Furthermore, $f''(x)$ is not equal to zero anywhere. The sign diagram of f'' is shown in Figure 39. From the sign diagram, we see that f is concave downward on $(-\infty, 1)$ and concave upward on $(1, \infty)$. Next, since $x = 1$ does lie in the domain of f, our computations also reveal that the point $(1, 0)$ is an inflection point of f (Figure 40).   ■

EXAMPLE 5   Determine the intervals where the function

$$f(x) = \frac{1}{x^2 + 1}$$

is concave upward and where it is concave downward and find the inflection points of f.

Solution   The first derivative of f is

$$f'(x) = \frac{d}{dx}(x^2 + 1)^{-1} = -2x(x^2 + 1)^{-2} \qquad \text{Using the General Power Rule}$$

$$= -\frac{2x}{(x^2 + 1)^2}$$

Next, using the Quotient Rule, we find

$$f''(x) = \frac{(x^2 + 1)^2(-2) + (2x)2(x^2 + 1)(2x)}{(x^2 + 1)^4}$$

$$= \frac{(x^2 + 1)[-2(x^2 + 1) + 8x^2]}{(x^2 + 1)^4} = \frac{(x^2 + 1)(6x^2 - 2)}{(x^2 + 1)^4}$$

$$= \frac{2(3x^2 - 1)}{(x^2 + 1)^3} \qquad \text{Canceling the common factors}$$

Observe that f'' is continuous everywhere and is zero if

$$3x^2 - 1 = 0$$

$$x^2 = \frac{1}{3}$$

FIGURE 41

Sign diagram for f''

or $x = \pm\sqrt{3}/3$. The sign diagram for f'' is shown in Figure 41. From the sign diagram for f'', we see that f is concave upward on $(-\infty, -\sqrt{3}/3) \cup (\sqrt{3}/3, \infty)$ and concave downward on $(-\sqrt{3}/3, \sqrt{3}/3)$. Also, observe that $f''(x)$ changes sign as we move across the points $x = -\sqrt{3}/3$ and $x = \sqrt{3}/3$. Since

$$f\left(-\frac{\sqrt{3}}{3}\right) = \frac{1}{\frac{1}{3} + 1} = \frac{3}{4} \qquad \text{and} \qquad f\left(\frac{\sqrt{3}}{3}\right) = \frac{3}{4}$$

FIGURE 42

The graph of $f(x) = \dfrac{1}{x^2 + 1}$

is concave upward on

$$\left(-\infty, -\frac{\sqrt{3}}{3}\right) \cup \left(\frac{\sqrt{3}}{3}, \infty\right)$$

and concave downward on

$$\left(-\frac{\sqrt{3}}{3}, \frac{\sqrt{3}}{3}\right).$$

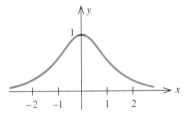

we see that the points $(-\sqrt{3}/3, 3/4)$ and $(\sqrt{3}/3, 3/4)$ are inflection points of f. The graph of f is shown in Figure 42. ■

Applications

Examples 6 and 7 illustrate familiar interpretations of the significance of the inflection point of a function.

EXAMPLE 6   The total sales S, in thousands of dollars, of the Arctic Air Corporation, a manufacturer of automobile air conditioners, is related to the amount of money x the company spends on advertising its products by the formula

$$S = -0.01x^3 + 1.5x^2 + 200 \qquad 0 \le x \le 100$$

where x is measured in thousands of dollars. Find the inflection point of the function S.

Solution   The first two derivatives of S are given by

$$S' = -0.03x^2 + 3x$$

and

$$S'' = -0.06x + 3$$

Setting $S'' = 0$ gives $x = 50$ as the only candidate for an inflection point of S. Moreover, since

$$S'' > 0 \quad \text{for} \quad x < 50$$

and

$$S'' < 0 \quad \text{for} \quad x > 50$$

FIGURE 43

The graph of $S(x)$ has a point of inflection at (50, 2700).

Thousands of dollars

the point (50, 2700) is an inflection point of the function S. The graph of S appears in Figure 43. Notice that this is the graph of the function we discussed earlier. ■

EXAMPLE 7   An economy's consumer price index (CPI) is described by the function

$$I(t) = -0.2t^3 + 3t^2 + 100 \qquad 0 \le t \le 9$$

where $t = 0$ corresponds to the year 1987. Find the point of inflection of the function I and discuss its significance.

Solution   The first two derivatives of I are given by

$$I'(t) = -0.6t^2 + 6t$$

and $$I''(t) = -1.2t + 6 = -1.2(t - 5)$$

Setting $I''(t) = 0$ gives $t = 5$ as the only candidate for an inflection point of I. Next, we observe that

$$I'' > 0 \quad \text{for} \quad t < 5$$

and $$I'' < 0 \quad \text{for} \quad t > 5,$$

so the point $(5, 150)$ is an inflection point of I. The graph of I is sketched in Figure 44.

FIGURE 44
The graph of $I(t)$ has a point of inflection at $(5, 150)$.

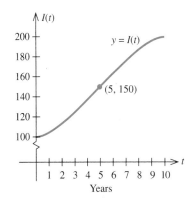

Since the second derivative of I measures the rate of change of the inflation rate, our computations reveal that the rate of inflation had, in fact, peaked at $t = 5$. Thus, relief actually began at the beginning of 1992.   ■

The Second Derivative Test

We will now show how the second derivative f'' of a function f can be used to help us determine whether a critical point of f is a relative extremum of f. Figure 45a shows the graph of a function that has a relative maximum

FIGURE 45

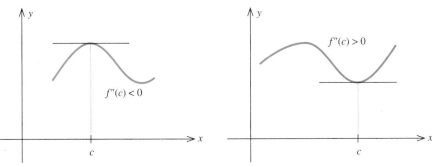

(a) f has a relative maximum at $x = c$ **(b)** f has a relative minimum at $x = c$

at $x = c$. Observe that f is concave downward at that point. Similarly, Figure 45b shows that at a relative minimum of f the graph is concave upward. But from our previous work we know that f is concave downward at $x = c$ if $f''(c) < 0$ and f is concave upward at $x = c$ if $f''(c) > 0$. These observations suggest the following alternative procedure for determining whether a critical point of f gives rise to a relative extremum of f. This result is called the Second Derivative Test and is applicable when f'' exists.

THE SECOND DERIVATIVE TEST

1. Compute $f'(x)$ and $f''(x)$.

2. Find all the critical points of f at which $f'(x) = 0$.

3. Compute $f''(c)$ for each such critical point c.
    a. If $f''(c) < 0$, then f has a relative maximum at c.
    b. If $f''(c) > 0$, then f has a relative minimum at c.
    c. If $f''(c) = 0$, the test fails; that is, it is inconclusive.

REMARK   As stated in step 3(c), the Second Derivative Test does not yield a conclusion if $f''(c) = 0$ or if $f''(c)$ does not exist. In other words, $x = c$ may give rise to a relative extremum or an inflection point (see exercise 83, page 616). In such cases, you should revert to the First Derivative Test. □

EXAMPLE 8   Determine the relative extrema of the function

$$f(x) = x^3 - 3x^2 - 24x + 32$$

using the Second Derivative Test. (See Example 7, Section 10.1.)

Solution   We have

$$f'(x) = 3x^2 - 6x - 24 = 3(x + 2)(x - 4)$$

so $f'(x) = 0$ gives $x = -2$ and $x = 4$, the critical points of f, as in Example 7. Next, we compute

$$f''(x) = 6x - 6 = 6(x - 1)$$

Since

$$f''(-2) = 6(-2 - 1) = -18 < 0$$

the Second Derivative Test implies that $f(-2) = 60$ is a relative maximum of f. Also,

$$f''(4) = 6(4 - 1) = 18 > 0$$

and the Second Derivative Test implies that $f(4) = -48$ is a relative minimum of f, which confirms the results obtained earlier.   ■

Comparing the First and Second Derivative Tests

Notice that both the First Derivative Test and the Second Derivative Test are used to classify the critical points of f. What are the pros and cons of the two tests? Since the Second Derivative Test is applicable only when f''

exists, it is less versatile than the First Derivative Test. For example, it cannot be used to locate the relative minimum $f(0) = 0$ of the function $f(x) = x^{2/3}$.

Furthermore, the Second Derivative Test is inconclusive when f'' is equal to zero at a critical point of f, whereas the First Derivative Test always yields positive conclusions. The Second Derivative Test is also inconvenient to use when f'' is difficult to compute. On the plus side, if it is easy to compute f'', then we use the Second Derivative Test, since it involves just the evaluation of f'' at the critical point(s) of f. Also, the conclusions of the Second Derivative Test are important in theoretical work.

We close this section by summarizing the different roles played by the first derivative f' and the second derivative f'' of a function f in determining the properties of the graph of f. The first derivative f' tells us where f is increasing and where f is decreasing, whereas the second derivative f'' tells us where f is concave upward and where f is concave downward. These different properties of f are reflected by the signs of f' and f'' in the interval of interest. The following table shows the general characteristics of the function f for various possible combinations of the signs of f' and f'' in the interval (a, b).

Signs of f' and f''	Properties of the Graph of f	General Shape of the Graph of f
$f'(x) > 0$	f increasing	
$f''(x) > 0$	f concave upward	
$f'(x) > 0$	f increasing	
$f''(x) < 0$	f concave downward	
$f'(x) < 0$	f decreasing	
$f''(x) > 0$	f concave upward	
$f'(x) < 0$	f decreasing	
$f''(x) < 0$	f concave downward	

SELF-CHECK EXERCISES 10.2

1. Determine where the function $f(x) = 4x^3 - 3x^2 + 6$ is concave upward and where it is concave downward.

2. Using the Second Derivative Test, if applicable, find the relative extrema of the function $f(x) = 2x^3 - \frac{1}{2}x^2 - 12x - 10$.

3. A certain country's gross domestic product (GDP) [in millions of dollars] in year t is described by the function

$$G(t) = -2t^3 + 45t^2 + 20t + 6000 \qquad 0 \le t \le 11$$

where $t = 0$ corresponds to the beginning of the year 1985. Find the inflection point of the function G and discuss its significance.

Solutions to Self-Check Exercises 10.2 can be found on page 617.

10.2   EXERCISES

In exercises 1–8, you are given the graph of a function f. Determine the intervals where f is concave upward and where it is concave downward. Also, find all inflection points of f, if any.

1.

2.

3.

4.

5.

6.

7.

8.

In exercises 9–12, determine which graph—a, b, or c—is the graph of the function f with the specified properties.

9. $f(2) = 1$, $f'(2) > 0$, and $f''(2) < 0$

a.

b.

c.

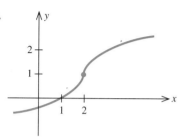

10. $f(1) = 2$, $f'(x) > 0$ on $(-\infty, 1) \cup (1, \infty)$, and $f''(1) = 0$

a.

b.

c.

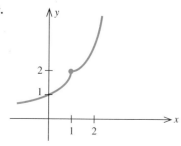

11. $f'(0)$ is undefined, f is decreasing on $(-\infty, 0)$, f is concave downward on $(0, 3)$, and f has an inflection point at $x = 3$

a.

b.

c.

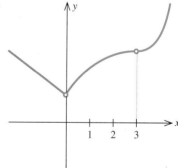

12. f is decreasing on $(-\infty, 2)$, increasing on $(2, \infty)$, f is concave upward on $(1, \infty)$, and has inflection points at $x = 0$ and $x = 1$

a.

b.

c.

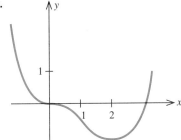

In exercises 13–18, show that the function is concave upward wherever it is defined.

13. $f(x) = 4x^2 - 12x + 7$

14. $g(x) = x^4 + \frac{1}{2}x^2 + 6x + 10$

15. $h(x) = \dfrac{1}{x^2}$ **16.** $f(x) = \dfrac{1}{x^4}$

17. $h(x) = \sqrt{x^2 + 4}$ **18.** $g(x) = -\sqrt{4 - x^2}$

In exercises 19–42, determine where the function is concave upward and where it is concave downward.

19. $f(x) = 2x^2 - 3x + 4$

20. $g(x) = -x^2 + 3x + 4$

21. $f(x) = x^3 - 1$ **22.** $g(x) = x^3 - x$

23. $f(x) = x^4 - 6x^3 + 2x + 8$

24. $f(x) = 3x^4 - 6x^3 + x - 8$

25. $f(x) = x^{4/7}$ **26.** $f(x) = \sqrt[3]{x}$

27. $f(x) = \sqrt{4 - x}$ **28.** $g(x) = \sqrt{x - 2}$

29. $f(x) = \dfrac{1}{x - 2}$ **30.** $g(x) = \dfrac{x}{x + 1}$

31. $f(x) = \dfrac{1}{2 + x^2}$ **32.** $g(x) = \dfrac{x}{1 + x^2}$

33. $h(t) = \dfrac{t^2}{t - 1}$ **34.** $f(x) = \dfrac{x + 1}{x - 1}$

35. $g(x) = x + \dfrac{1}{x^2}$ **36.** $h(r) = -\dfrac{1}{(r - 2)^2}$

37. $g(t) = (2t - 4)^{1/3}$ **38.** $f(x) = (x - 2)^{2/3}$

39. $f(x) = xe^x$ **40.** $f(x) = \dfrac{e^x - e^{-x}}{2}$

41. $f(x) = \dfrac{\ln x}{x}$ **42.** $f(x) = x^2 + \ln x^2$

In exercises 43–56, find the inflection points, if any, of each function.

43. $f(x) = x^3 - 2$ **44.** $g(x) = x^3 - 6x$

45. $f(x) = 6x^3 - 18x^2 + 12x - 15$

46. $g(x) = 2x^3 - 3x^2 + 18x - 8$

47. $f(x) = 3x^4 - 4x^3 + 1$

48. $f(x) = x^4 - 2x^3 + 6$

49. $g(t) = \sqrt[3]{t}$ **50.** $f(x) = \sqrt[5]{x}$

51. $f(x) = (x - 1)^3 + 2$ **52.** $f(x) = (x - 2)^{4/3}$

53. $f(x) = xe^{-2x}$ **54.** $f(x) = 2e^{-x^2}$

55. $f(x) = \ln(x^2 + 1)$ **56.** $f(x) = x^2 \ln x$

In exercises 57–74, find the relative extrema, if any, of each function. Use the Second Derivative Test, if applicable.

57. $f(x) = -x^2 + 2x + 4$ **58.** $g(x) = 2x^2 + 3x + 7$

59. $f(x) = 2x^3 + 1$ **60.** $g(x) = x^3 - 6x$

61. $f(x) = \frac{1}{3}x^3 - 2x^2 - 5x - 10$

62. $f(x) = 2x^3 + 3x^2 - 12x - 4$

63. $g(t) = t + \dfrac{9}{t}$ **64.** $f(t) = 2t + \dfrac{3}{t}$

65. $f(x) = \dfrac{x}{1 - x}$ **66.** $f(x) = \dfrac{2x}{x^2 + 1}$

67. $f(t) = t^2 - \dfrac{16}{t}$ **68.** $g(x) = x^2 + \dfrac{2}{x}$

69. $g(s) = \dfrac{s}{1 + s^2}$ **70.** $g(x) = \dfrac{1}{1 + x^2}$

71. $f(x) = x^2 e^x$ **72.** $g(t) = e^{t^2 - 2t}$

73. $g(x) = x - \ln x$ **74.** $f(x) = \ln(x^2 + 1)$

75. **Effect of Budget Cuts on Drug-Related Crimes**   The following graphs were used by a police commissioner to illustrate what effect a budget cut would have on crime in the city. The number $N_1(t)$ gives the projected number of drug-related crimes in the next 12 months. The number $N_2(t)$ gives the projected number of drug-related crimes in the same time frame if next year's budget is cut.

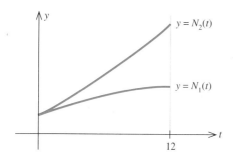

a. Explain why $N_1'(t)$ and $N_2'(t)$ are both positive on the interval $(0, 12)$.

b. What are the signs of $N_1''(t)$ and $N_2''(t)$ on the interval $(0, 12)$?

c. Interpret the results of (b).

76. Demand for RNs   The following graph gives the total number of "help wanted" ads for RNs (registered nurses) in 22 cities over the last 12 months as a function of time t (t measured in months).

a. Explain why $N'(t)$ is positive on the interval $(0, 12)$.

b. Determine the signs of $N''(t)$ on the interval $(0, 6)$ and the interval $(6, 12)$.

c. Interpret the results of (b).

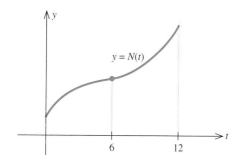

77. Effect of Advertising on Sales   The total sales S, in thousands of dollars, of the Cannon Precision Instruments Corporation is related to the amount of money x that Cannon spends on advertising its products by the function

$$S(x) = -0.002x^3 + 0.6x^2 + x + 500 \qquad 0 \le x \le 200$$

where x is measured in thousands of dollars. Find the inflection point of the function S and discuss its significance.

78. Forecasting Profits   As a result of increasing energy costs, the growth rate of the profit of the 4-year-old Venice Glassblowing Company has begun to decline. Venice's management, after consulting with energy experts, decides to implement certain energy-conservation measures aimed at cutting energy bills. The general manager reports that, according to his calculations, the growth rate of Venice's profit should once again be on the increase within four years. If Venice's

profit (in hundreds of dollars) x years from now is given by the function

$$P(x) = x^3 - 9x^2 + 40x + 50 \qquad 0 \le x \le 8$$

determine whether the general manager's forecast will be accurate. [*Hint:* Find the inflection point of the function P and study the concavity of P.]

79. Worker Efficiency   An efficiency study conducted for the Elektra Electronics Company showed that the number of Space Commander walkie-talkies assembled by the average worker t hours after starting work at 8 A.M. is given by

$$N(t) = -t^3 + 6t^2 + 15t \qquad 0 \le t \le 4$$

At what time during the morning shift is the average worker performing at peak efficiency?

C   80. Cost of Producing Calculators   A subsidiary of Elektra Electronics manufactures a programmable calculator. Management determines that the daily cost $C(x)$ of producing these calculators (in dollars) is

$$C(x) = 0.0001x^3 - 0.08x^2 + 40x + 5000$$

where x stands for the number of calculators produced. Find the inflection point of the function C and interpret your result.

C   81. Flight of a Rocket   The altitude (in feet) of a rocket t seconds into flight is given by

$$s = f(t) = -t^3 + 54t^2 + 480t + 6$$

Find the point of inflection of the function f and interpret your result. What is the maximum velocity attained by the rocket?

C   82. Air Pollution   The level of ozone, an invisible gas that irritates and impairs breathing, present in the atmosphere on a certain May day in the city of Riverside was approximated by

$$A(t) = 1.0974t^3 - 0.0915t^4 \qquad 0 \le t \le 11$$

where $A(t)$ is measured in pollutant standard index (PSI) and t is measured in hours, with $t = 0$ corresponding to 7 A.M. Use the Second Derivative Test to show that the function A has a relative maximum at approximately $t = 9$. Interpret your results.

83. Consider the functions $f(x) = x^3$, $g(x) = x^4$, and $h(x) = -x^4$.

a. Show that $x = 0$ is a critical point of each of the functions f, g, and h.

b. Show that the second derivative of each of the functions f, g, and h equals zero at $x = 0$.

c. Show that f has neither a relative maximum nor a relative minimum at $x = 0$, that g has a relative minimum at $x = 0$, and that h has a relative maximum at $x = 0$.

SOLUTIONS TO SELF-CHECK EXERCISES 10.2

1. We first compute

$$f'(x) = 12x^2 - 6x$$

and

$$f''(x) = 24x - 6 = 6(4x - 1)$$

Observe that f'' is continuous everywhere and has a zero at $x = \frac{1}{4}$. The sign diagram of f'' is shown in the accompanying figure.

From the sign diagram for f'' we see that f is concave upward on $(\frac{1}{4}, \infty)$ and concave downward on $(-\infty, \frac{1}{4})$.

2. First, we find the critical points of f by solving the equation

$$f'(x) = 6x^2 - x - 12 = 0$$

that is,

$$(3x + 4)(2x - 3) = 0$$

giving $x = -\frac{4}{3}$ and $x = \frac{3}{2}$. Next, we compute

$$f''(x) = 12x - 1$$

Since

$$f''\left(-\frac{4}{3}\right) = 12\left(-\frac{4}{3}\right) - 1 = -17 < 0$$

the Second Derivative Test implies that $f\left(-\frac{4}{3}\right) = \frac{10}{27}$ is a relative maximum of f. Also

$$f''\left(\frac{3}{2}\right) = 12\left(\frac{3}{2}\right) - 1 = 17 > 0$$

and we see that $f(\frac{3}{2}) = -\frac{179}{8}$ is a relative minimum.

3. We compute the second derivative of G. Thus,

$$G'(t) = -6t^2 + 90t + 20$$

$$G''(t) = -12t + 90$$

Now G'' is continuous everywhere and $G''(t) = 0$ when $t = \frac{15}{2}$, giving $t = \frac{15}{2}$ as the only candidate for an inflection point of G. Since $G''(t) > 0$ for $t < \frac{15}{2}$ and $G''(t) < 0$ for $t > \frac{15}{2}$, we see that the point $(\frac{15}{2}, \frac{15675}{2})$ is an inflection point of G. The results of our computations tell us that the country's GDP was increasing most rapidly at the beginning of July 1992.

Using Technology

The **INFLC** operation of a graphing utility can be used to find the inflection points of a function and hence the intervals where the graph of the function is concave upward and the intervals where it is concave downward. (See Section A.16 in Appendix A.)

EXAMPLE 1   Let $f(x) = 2.5x^5 - 12.4x^3 + 4.2x^2 - 5.2x + 4$.

a. Use a graphing utility to plot the graph of f.
b. Use **INFLC** to find the inflection points of f.
c. Find the intervals where f is concave upward and where it is concave downward.

Solution

a. The graph of f using the viewing rectangle $[-3, 3] \times [-25, 60]$ is shown in Figure 1.

FIGURE 1
The graph of f in the viewing rectangle $[-3, 3] \times [-25, 60]$

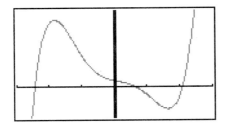

b. From Figure 1, we see that f has three inflection points—one occurring at the point where the x-coordinate is approximately -1, another at the point where $x \approx 0$, and the third at the point where $x \approx 1$. To find the first inflection point, we use the **INFLC** operation, moving the cursor to the point on the graph of f where $x \approx -1$. We obtain the point $(-1.2728, 34.6395)$ (accurate to four decimal places). Next, setting the cursor near the point $x = 0$ yields the inflection point $(0.1139, 3.4440)$. Finally, with the cursor set at $x = 1$, we obtain the third inflection point $(1.1589, -10.4594)$.

c. From the results of (b), we see that f is concave upward on the intervals $(-1.2728, 0.1139)$ and $(1.1589, \infty)$, and concave downward on $(-\infty, -1.2728)$ and $(0.1139, 1.1589)$. ■

EXERCISES

In exercises 1–8, use a graphing utility to find (a) the intervals where f is concave upward and the intervals where f is concave downward, and (b) the inflection points of f. Express your answers accurate to four decimal places.

1. $f(x) = 1.8x^4 - 4.2x^3 + 2.1x + 2$

2. $f(x) = -2.1x^4 + 3.1x^3 + 2x^2 - x + 1.2$

3. $f(x) = 1.2x^5 - 2x^4 + 3.2x^3 - 4x + 2$

4. $f(x) = -2.1x^5 + 3.2x^3 - 2.2x^2 + 4.2x - 4$

5. $f(x) = x^3(x^2 + 1)^{-1/3}$

6. $f(x) = x^2(x^3 - 1)^3$

7. $f(x) = \dfrac{e^x}{x} + x^3$

8. $f(x) = xe^{-x} + \dfrac{1}{x}$

9. Growth of HMOs   Based on data compiled by the Group Health Association of America, the number of people receiving their care in an HMO (Health Maintenance Organization) from the beginning of 1984 through 1994 is approximated by the function

$$f(t) = 0.0514t^3 - 0.853t^2 + 6.8147t + 15.6524$$

$$0 < t \le 11$$

where $f(t)$ gives the number of of people in millions and t is measured in years, with $t = 0$ corresponding to the beginning of 1984.

a. Use a graphing utility to plot the graph of f in the viewing rectangle $[0, 12] \times [0, 120]$.
b. Find the points of inflection of f.
c. At what time in the given time interval was the number of people receiving their care at an HMO increasing fastest?

Source: Group Health Association of America

10. Manufacturing Capacity Data obtained from the Federal Reserve show that the annual increase in manufacturing capacity between 1988 and 1994 is given by

$$f(t) = 0.0388889t^3 - 0.283333t^2 + 0.477778t$$

$$+ 2.04286 \qquad 0 \le t \le 6$$

where $f(t)$ is a percentage and t is measured in years, with $t = 0$ corresponding to the beginning of 1988.
   a. Use a graphing utility to plot the graph of f in the viewing rectangle $[0, 8] \times [0, 4]$.
   b. Find the point of inflection and interpret your result.
   Source: Federal Reserve

11. Time on the Market According to the Greater Boston Real Estate Board-Multiple Listing Service, the average number of days a single-family home remains for sale from listing to accepted offer is approximated by the function

$$f(t) = 0.0171911t^4 - 0.662121t^3 + 6.18083t^2$$

$$- 8.97086t + 53.3357 \qquad 0 \le t \le 10$$

where t is measured in years, with $t = 0$ corresponding to the beginning of 1984.
   a. Use a graphing utility to plot the graph of f in the viewing rectangle $[0, 12] \times [0, 120]$.
   b. Find the points of inflection and interpret your result.
   Source: Greater Boston Real Estate Board-Multiple Listing Service

12. Our Growing Numbers According to a study conducted by the United Nations Population Division, the world population in billions is approximated by the function

$$f(t) = \frac{12}{1 + 3.74914e^{-1.42804t}} \qquad 0 \le t \le 4$$

where t is measured in half-centuries, with $t = 0$ corresponding to the beginning of 1950.
   a. Use a graphing utility to plot the graph of f in the viewing rectangle $[0, 5] \times [0, 14]$.
   b. Estimate the intervals where f is concave upward and where it is concave downward.
   c. Find the point of inflection and interpret your result.

10.3 Curve Sketching

A Real-Life Example

As we have seen on numerous occasions, the graph of a function is a useful aid for visualizing the function's properties. From a practical point of view, the graph of a function also gives, at one glance, a complete summary of all the information captured by the function.

Consider, for example, the graph of the function giving the Dow Jones Industrial Average (DJIA) on Black Monday, October 19, 1987 (Figure 46). Here $t = 0$ corresponds to 8:30 A.M., when the market was open for business, and $t = 7.5$ corresponds to 4 P.M., the closing time. The following information may be gleaned from studying the graph.

The graph is *decreasing* rapidly from $t = 0$ to $t = 1$, reflecting the sharp drop in the index in the first hour of trading. The point $(1, 2047)$ is a *relative minimum* point of the function, and this turning point coincides with the start of an aborted recovery. The short-lived rally, represented by the portion of the graph that is *increasing* on the interval $(1, 2)$, quickly fizzled out at $t = 2$ (10:30 A.M.). The *relative maximum* point $(2, 2150)$ marks the highest point of the recovery. The function is decreasing in the rest of the interval. The point $(4, 2006)$ is an *inflection point* of the function; it shows that there was a temporary respite at $t = 4$ (12:30 P.M.). However, selling pressure continued unabated, and the Dow Jones Industrial Index continued to fall until the closing bell. Finally, the graph also shows that the index opened at the high of the day ($f(0) = 2247$ is the *absolute maximum* of

FIGURE 46
The Dow Jones Industrial Average on Black Monday

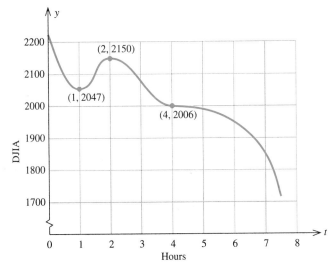

Source: Wall Street Journal

the function) and closed at the low of the day ($f(\frac{15}{2}) = 1739$ is the *absolute minimum* of the function), a drop of 508 points!*

Before we turn our attention to the actual task of sketching the graph of a function, let us look at some properties of graphs that will be helpful in this connection.

Vertical Asymptotes

Before going on, you might want to review the material on one-sided limits and the limit at infinity of a function (Sections 9.1 and 9.2).

Consider the graph of the function

$$f(x) = \frac{x+1}{x-1}$$

shown in Figure 47. Observe that $f(x)$ increases without bound (tends to infinity) as x approaches $x = 1$ from the right; that is,

$$\lim_{x \to 1^+} \frac{x+1}{x-1} = \infty$$

FIGURE 47
The graph of f has a vertical asymptote at x = 1.

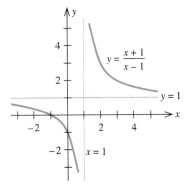

---

*Absolute maxima and absolute minima of functions are covered in Section 10.4.

You can verify this by taking a sequence of values of x approaching $x = 1$ from the right and looking at the corresponding values of $f(x)$.

Here is another way of looking at the situation: Observe that if x is a number that is a little larger than 1, then both $(x + 1)$ and $(x - 1)$ are positive, so that $(x + 1)/(x - 1)$ is also positive. As x approaches $x = 1$, the numerator $(x + 1)$ approaches the number 2, but the denominator $(x - 1)$ approaches zero, so the quotient $(x + 1)/(x - 1)$ approaches infinity, as observed earlier. The line $x = 1$ is called a *vertical asymptote* of the graph of f.

For the function $f(x) = (x + 1)/(x - 1)$, you can show that

$$\lim_{x \to 1^-} \frac{x + 1}{x - 1} = -\infty$$

and this tells us how $f(x)$ approaches the asymptote $x = 1$ from the left.

More generally, we have the following definition:

VERTICAL ASYMPTOTES

The line $x = a$ is a **vertical asymptote** of the graph of a function f if either

$$\lim_{x \to a^+} f(x) = \infty \qquad \text{or} -\infty$$

or

$$\lim_{x \to a^-} f(x) = \infty \qquad \text{or} -\infty$$

REMARK   Although a vertical asymptote of a graph is not part of the graph, it serves as a useful aid for sketching the graph.   □

For rational functions

$$f(x) = \frac{P(x)}{Q(x)}$$

there is a simple criterion for determining whether the graph of f has any vertical asymptotes.

FINDING VERTICAL ASYMPTOTES OF RATIONAL FUNCTIONS

Suppose that f is a rational function

$$f(x) = \frac{P(x)}{Q(x)}$$

where P and Q are polynomial functions. Then the line $x = a$ is a vertical asymptote of the graph of f if $Q(a) = 0$ but $P(a) \neq 0$.

For the function

$$f(x) = \frac{x + 1}{x - 1}$$

considered earlier, $P(x) = x + 1$ and $Q(x) = x - 1$. Observe that $Q(1) = 0$ but $P(1) = 2 \neq 0$, so $x = 1$ is a vertical asymptote of the graph of f.

EXAMPLE 1   Find the vertical asymptotes of the graph of the function

$$f(x) = \frac{x^2}{4 - x^2}$$

Solution   The function f is a rational function with $P(x) = x^2$ and $Q(x) = 4 - x^2$. The zeros of Q are found by solving

$$4 - x^2 = 0$$

that is,

$$(2 - x)(2 + x) = 0$$

giving $x = -2$ and $x = 2$. These are candidates for the vertical asymptotes of the graph of f. Examining $x = -2$, we compute $P(-2) = (-2)^2 = 4 \neq 0$, and we see that $x = -2$ is indeed a vertical asymptote of the graph of f. Similarly, we find $P(2) = 2^2 = 4 \neq 0$, and so $x = 2$ is also a vertical asymptote of the graph of f. The graph of f sketched in Figure 48 confirms these results.

FIGURE 48
$x = -2$ and $x = 2$ are vertical asymptotes of the graph of f.

 Recall that in order for the line $x = a$ to be a vertical asymptote of the graph of a rational function f, *only* the denominator of $f(x)$ must be equal to zero at $x = 0$. If *both* $P(a)$ and $Q(a)$ are equal to zero, then $x = a$ need *not* be a vertical asymptote. For example, look at the function $f(x) = \frac{2(x^2 - 4)}{x - 2}$, whose graph appears in Figure 7a, page 478.

Horizontal Asymptotes

Let us return to the function f defined by

$$f(x) = \frac{x + 1}{x - 1}$$

(Figure 49).

Observe that $f(x)$ approaches the horizontal line $y = 1$ as x approaches infinity, and, in this case, $f(x)$ approaches $y = 1$ as x approaches minus infinity as well. The line $y = 1$ is called a *horizontal asymptote* of the graph of f. More generally, we have the following definition.

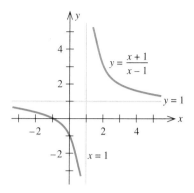

HORIZONTAL ASYMPTOTES

The line $y = b$ is a **horizontal asymptote** of the graph of a function f if either

$$\lim_{x \to \infty} f(x) = b \quad \text{or} \quad \lim_{x \to -\infty} f(x) = b$$

For the function

$$f(x) = \frac{x + 1}{x - 1}$$

we see that

$$\lim_{x \to \infty} \frac{x + 1}{x - 1} = \lim_{x \to \infty} \frac{1 + \dfrac{1}{x}}{1 - \dfrac{1}{x}} \qquad \text{Divide numerator and denominator by } x.$$

$$= 1$$

Also,

$$\lim_{x \to -\infty} \frac{x + 1}{x - 1} = \lim_{x \to -\infty} \frac{1 + \dfrac{1}{x}}{1 - \dfrac{1}{x}}$$

$$= 1$$

In either case, we conclude that $y = 1$ is a horizontal asymptote of the graph of f, as observed earlier.

EXAMPLE 2   Find the horizontal asymptotes of the graph of the function

$$f(x) = \frac{x^2}{4 - x^2}$$

Solution   We compute

$$\lim_{x \to \infty} \frac{x^2}{4 - x^2} = \lim_{x \to \infty} \frac{1}{\dfrac{4}{x^2} - 1} \qquad \text{Dividing numerator and denominator by } x^2$$

$$= -1$$

and so $y = -1$ is a horizontal asymptote, as before. The graph of f sketched in Figure 50 confirms this result.

FIGURE 50
The graph of f has a horizontal asymptote at $y = -1$.

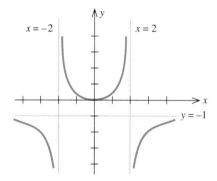

We next state an important property of polynomial functions.

> A polynomial function has no vertical or horizontal asymptotes.

To see this, note that a polynomial function, $P(x)$, can be written as a rational function with denominator equal to 1. Thus,

$$P(x) = \frac{P(x)}{1}$$

Since the denominator is never equal to zero, P has no vertical asymptotes. Next, if P is a polynomial of degree greater than or equal to 1, then

$$\lim_{x \to \infty} P(x) \qquad \text{and} \qquad \lim_{x \to -\infty} P(x)$$

are either infinity or minus infinity; that is, they do not exist. Therefore, P has no horizontal asymptotes.

In the last several sections we saw how the first and second derivatives of a function are used to reveal various properties of the graph of a function f. We will now show how this information can be used to help us sketch the graph of f. We begin by giving a general procedure for curve sketching.

A GUIDE TO CURVE SKETCHING

1. Determine the domain of f.
2. Find the x- and y-intercepts of f.*
3. Determine the behavior of f for large absolute values of f.
4. Find all horizontal and vertical asymptotes of f.
5. Determine the intervals where f is increasing and where f is decreasing.

---

*The equation $f(x) = 0$ may be difficult to solve, in which case one may decide against finding the x-intercepts.

6. Find the relative extrema of f.
7. Determine the concavity of f.
8. Find the inflection points of f.
9. Plot a few additional points to help further identify the shape of the graph of f, and sketch the graph.

We will now illustrate the techniques of curve sketching with several examples.

Two Step-by-Step Examples

EXAMPLE 3 Sketch the graph of the function
$$y = f(x) = x^3 - 6x^2 + 9x + 2$$

Solution Obtain the following information on the graph of f.

1. The domain of f is the interval $(-\infty, \infty)$.
2. By setting $x = 0$, we find that the y-intercept is 2. The x-intercept is found by setting $y = 0$, which, in this case, leads to a cubic equation. Since the solution is not readily found, we will not use this information.
3. Since
$$\lim_{x \to -\infty} f(x) = \lim_{x \to -\infty} (x^3 - 6x^2 + 9x + 2) = -\infty$$
and
$$\lim_{x \to \infty} f(x) = \lim_{x \to \infty} (x^3 - 6x^2 + 9x + 2) = \infty$$

we see that f decreases without bound as x decreases without bound, and f increases without bound as x increases without bound.
4. Since f is a polynomial function, there are no asymptotes.
$$f'(x) = 3x^2 - 12x + 9 = 3(x^2 - 4x + 3)$$
$$= 3(x - 3)(x - 1)$$

FIGURE 51

Sign diagram for f'

5. Setting $f'(x) = 0$ gives $x = 1$ or $x = 3$. The sign diagram for f' shows that f is increasing on the intervals $(-\infty, 1)$ and $(3, \infty)$ and decreasing on the interval $(1, 3)$ (Figure 51).
6. From the results of step 5, we see that $x = 1$ and $x = 3$ are critical points of f. Furthermore, f' changes sign from positive to negative as we move across $x = 1$, so a relative maximum of f occurs at $x = 1$. Similarly, we see that a relative minimum of f occurs at $x = 3$. Now,
$$f(1) = 1 - 6 + 9 + 2 = 6$$
and
$$f(3) = 3^3 - 6(3)^2 + 9(3) + 2 = 2$$

so $f(1) = 6$ is a relative maximum of f and $f(3) = 2$ is a relative minimum of f.

FIGURE 52

Sign diagram for f''

7.
$$f''(x) = 6x - 12 = 6(x - 2)$$

which is equal to zero when $x = 2$. The sign diagram of f'' shows that f is concave downward on the interval $(-\infty, 2)$ and concave upward on the interval $(2, \infty)$ (Figure 52).

FIGURE 53
We first plot the intercept, the relative extrema, and the inflection point.

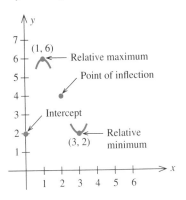

8. From the results of step 7, we see that f'' changes sign as we move across the point $x = 2$. Next,

$$f(2) = 2^3 - 6(2)^2 + 9(2) + 2 = 4$$

and so the required inflection point of f is $(2, 4)$.

9. Summarizing, we have

Domain	$(-\infty, \infty)$
Intercept	$(0, 2)$
$\lim\limits_{x \to -\infty} f(x); \lim\limits_{x \to \infty} f(x)$	$-\infty; \infty$
Asymptotes	None
Intervals where f is ↗ or ↘	↗ on $(-\infty, 1) \cup (3, \infty)$; ↘ on $(1, 3)$
Relative extrema	Rel. max. at $(1, 6)$; rel. min. at $(3, 2)$
Concavity	Downward on $(-\infty, 2)$; upward on $(2, \infty)$
Point of inflection	$(2, 4)$

In general, it is a good idea to start graphing by plotting the intercept, relative extrema, and inflection point (Figure 53). Then, using the rest of the information, we complete the graph of f as sketched in Figure 54. ■

FIGURE 54
The graph of
$y = x^3 - 6x^2 + 9x + 2$

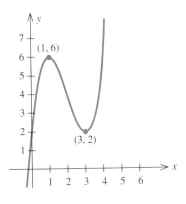

EXAMPLE 4 Sketch the graph of the function

$$y = f(x) = \frac{x + 1}{x - 1}$$

Solution Obtain the following information:

1. f is undefined when $x = 1$, so the domain of f is the set of all real numbers other than $x = 1$.

2. Setting $y = 0$ gives -1, the x-intercept of f. Next, setting $x = 0$ gives -1 as the y-intercept of f.

3. Earlier we found that

$$\lim_{x \to \infty} \frac{x + 1}{x - 1} = 1 \qquad \text{and} \qquad \lim_{x \to -\infty} \frac{x + 1}{x - 1} = 1$$

(see page 623). Consequently, we see that $f(x)$ approaches the line $y = 1$ as $|x|$ becomes arbitrarily large. For $x > 1$, $f(x) > 1$ and $f(x)$ approaches the line $y = 1$ from above. For $x < 1$, $f(x) < 1$, so $f(x)$ approaches the line $y = 1$ from below.

4. The straight line $x = 1$ is a vertical asymptote of the graph of f. Also, from the results of step 3, we conclude that $y = 1$ is a horizontal asymptote of the graph of f.

5.
$$f'(x) = \frac{(x - 1)(1) - (x + 1)(1)}{(x - 1)^2} = -\frac{2}{(x - 1)^2}$$

and is discontinuous at $x = 1$.

FIGURE 55

The sign diagram for f'

f' is not defined here

The sign diagram of f' shows that $f'(x) < 0$ whenever it is defined. Thus, f is decreasing on the intervals $(-\infty, 1)$ and $(1, \infty)$ (Figure 55).

6. From the results of step 5, we see that there are no critical points of f, since $f'(x)$ is never equal to zero for any value of x in the domain of f.

7.
$$f''(x) = \frac{d}{dx}[-2(x-1)^{-2}] = 4(x-1)^{-3} = \frac{4}{(x-1)^3}$$

The sign diagram of f'' shows immediately that f is concave downward on the interval $(-\infty, 1)$ and concave upward on the interval $(1, \infty)$ (Figure 56).

8. From the results of step 7, we see that there are no candidates for inflection points of f, since $f''(x)$ is never equal to zero for any value of x in the domain of f. Hence f has no inflection points.

9. Summarizing, we have

FIGURE 56

The sign diagram for f''

f'' is not defined here

Domain	$(-\infty, 1) \cup (1, \infty)$
Intercepts	$(-1, 0)$; $(0, -1)$
$\lim\limits_{x \to -\infty} f(x)$; $\lim\limits_{x \to \infty} f(x)$	1; 1
Asymptotes	$x = 1$ is a vertical asymptote; $y = 1$ is a horizontal asymptote
Intervals where f is ↗ or ↘	↘ on $(-\infty, 1) \cup (1, \infty)$
Relative extrema	None
Concavity	Downward on $(-\infty, 1)$; upward on $(1, \infty)$
Points of inflection	None

The graph of f is sketched in Figure 57.

FIGURE 57

The graph of f has a horizontal asymptote at y = 1 and a vertical asymptote at x = 1.

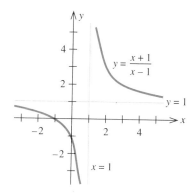

$$y = \frac{x+1}{x-1}$$

$y = 1$

$x = 1$

SELF-CHECK EXERCISES 10.3

1. Find the horizontal and vertical asymptotes of the graph of the function

$$f(x) = \frac{2x^2}{x^2 - 1}$$

2. Sketch the graph of the function
$$f(x) = \tfrac{2}{3}x^3 - 2x^2 - 6x + 4$$

Solutions to Self-Check Exercises 10.3 can be found on page 631.

Solutions to Self-Check Exercises 10.3 can be found on page 631.

10.3 EXERCISES

In exercises 1–10, find the horizontal and vertical asymptotes of the graph.

1.

$y = 2 - x^{2/3}$

2.

$y = \dfrac{1}{(x+1)^2}$

3.

$y = \dfrac{1}{x^3}$

4.

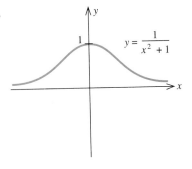

$y = \dfrac{1}{x^2 + 1}$

5.

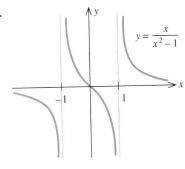

$y = \dfrac{x}{x^2 - 1}$

6.

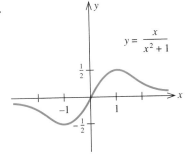

$y = \dfrac{x}{x^2 + 1}$

7.

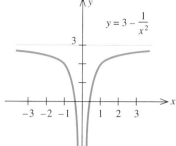

$y = 3 - \dfrac{1}{x^2}$

8.

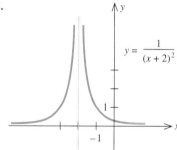

$y = \dfrac{1}{(x + 2)^2}$

9.

$y = e^{-x^2}$

10.

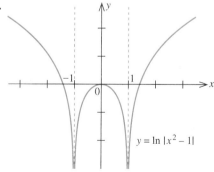

$y = \ln |x^2 - 1|$

In exercises 11–28, find the horizontal and vertical asymptotes of the graph of the function. (You need not sketch the graph.)

11. $f(x) = \dfrac{1}{x}$ **12.** $f(x) = \dfrac{1}{x + 2}$

13. $f(x) = -\dfrac{2}{x^2}$ **14.** $g(x) = \dfrac{1}{1 + 2x^2}$

15. $f(x) = \dfrac{x - 1}{x + 1}$ **16.** $g(t) = \dfrac{t + 1}{2t - 1}$

17. $h(x) = x^3 - 3x^2 + x + 1$

18. $g(x) = 2x^3 + x^2 + 1$

19. $f(t) = \dfrac{t^2}{t^2 - 9}$ **20.** $g(x) = \dfrac{x^3}{x^2 - 4}$

21. $f(x) = \dfrac{3x}{x^2 - x - 6}$ **22.** $g(x) = \dfrac{2x}{x^2 + x - 2}$

23. $g(t) = 2 + \dfrac{5}{(t - 2)^2}$ **24.** $f(x) = 1 + \dfrac{2}{x - 3}$

25. $f(x) = \dfrac{x^2 - 2}{x^2 - 4}$ **26.** $h(x) = \dfrac{2 - x^2}{x^2 + x}$

27. $g(x) = \dfrac{x^3 - x}{x(x + 1)}$

28. $f(x) = \dfrac{x^4 - x^2}{x(x - 1)(x + 2)}$

In exercises 29–56, sketch the graph of the function using the curve-sketching guide of this section.

29. $f(x) = x^2 - 2x + 3$ **30.** $g(x) = 4 - 3x - 2x^3$

31. $f(x) = 2x^3 + 1$ **32.** $h(x) = x^3 - 3x + 1$

33. $f(t) = 2t^3 - 15t^2 + 36t - 20$

34. $f(x) = -2x^3 + 3x^2 + 12x + 2$

35. $f(t) = 3t^4 + 4t^3$

36. $h(x) = \frac{3}{2}x^4 - 2x^3 - 6x^2 + 8$

37. $f(x) = \sqrt{x^2 + 5}$ **38.** $f(t) = \sqrt{t^2 - 4}$

39. $f(x) = \sqrt[3]{x^2}$ **40.** $g(x) = \frac{1}{2}x - \sqrt{x}$

41. $f(x) = \dfrac{1}{x + 1}$ **42.** $g(x) = \dfrac{2}{x - 1}$

43. $g(x) = \dfrac{x}{x - 1}$ **44.** $h(x) = \dfrac{x + 2}{x - 2}$

45. $g(x) = \dfrac{x}{x^2 - 4}$ **46.** $f(t) = \dfrac{t^2}{1 + t^2}$

47. $f(t) = e^t - t$ **48.** $h(x) = \dfrac{e^x + e^{-x}}{2}$

49. $f(x) = 2 - e^{-x}$ **50.** $f(x) = \dfrac{3}{1 + e^{-x}}$

51. $f(x) = \ln(x - 1)$ **52.** $f(x) = 2x - \ln x$

53. $f(x) = \dfrac{x^2 - 9}{x^2 - 4}$ **54.** $g(t) = -\dfrac{t^2 - 2}{t - 1}$

55. $g(x) = (x + 2)^{3/2} + 1$ **56.** $h(x) = (x - 1)^{2/3} + 1$

57. Cost of Removing Toxic Pollutants   A city's main well was recently found to be contaminated with trichloroethylene (a cancer-causing chemical) as a result of an abandoned chemical dump leaching chemicals into the water. A proposal submitted to the city's selectmen indicated that the cost, measured in millions of dollars, of removing $x\%$ of the toxic pollutants is given by

$$C(x) = \frac{0.5x}{100 - x}$$

a. Find the vertical asymptote of $C(x)$.
b. Is it possible to remove 100% of the toxic pollutant from the water?

58. Average Cost of Producing Video Discs   The average cost per disc (in dollars) incurred by the Herald Record Company in pressing x video discs is given by the average cost function

$$\overline{C}(x) = 2.2 + \frac{2500}{x}$$

a. Find the horizontal asymptote of $\overline{C}(x)$.
b. What is the limiting value of the average cost?

59. Concentration of a Drug in the Bloodstream   The concentration of a certain drug in a patient's bloodstream t hours after injection is given by

$$C(t) = \frac{0.2t}{t^2 + 1}$$

milligrams per cubic centimeter.
a. Find the horizontal asymptote of $C(t)$.
b. Interpret your result.

60. Effect of Enzymes on Chemical Reactions   Certain proteins, known as enzymes, serve as catalysts for chemical reactions in living things. In 1913, Leonor Michaelis and L. M. Menten discovered the following formula giving the initial speed V (in moles per liter per second) at which the reaction begins in terms of the amount of substrate x (the substance that is being acted upon, measured in moles per liter):

$$V = \frac{ax}{x + b}$$

where a and b are positive constants.
a. Find the horizontal asymptote of V.
b. Interpret your result.

61. Worker Efficiency   An efficiency study showed that the total number of cordless telephones assembled by an average worker at Delphi Electronics t hours after starting work at 8 A.M. is given by

$$N(t) = -\frac{1}{2}t^3 + 3t^2 + 10t \qquad 0 \le t \le 4$$

Sketch the graph of the function N and interpret your results.

62. GDP of a Developing Country   A developing country's gross domestic product (GDP) from 1988 to 1996 is approximated by the function

$$G(t) = -0.2t^3 + 2.4t^2 + 60 \qquad 0 \le t \le 8$$

where $G(t)$ is measured in billions of dollars and $t = 0$ corresponds to the year 1988. Sketch the graph of the function G and interpret your results.

63. Concentration of a Drug in the Bloodstream   The concentration of a certain drug in a patient's bloodstream t hours after injection is given by

$$C(t) = \frac{0.2t}{t^2 + 1}$$

milligrams per cubic centimeter. Sketch the graph of the function C and interpret your results.

64. Oxygen Content of a Pond   When organic waste is dumped into a pond, the oxidation process that takes place reduces the pond's oxygen content. However, given time, nature will restore the oxygen content to its natural level. Suppose that the oxygen content t days after organic waste has been dumped into the pond is given by

$$f(t) = 100 \left(\frac{t^2 - 4t + 4}{t^2 + 4} \right) \qquad 0 \le t < \infty$$

percent of its normal level. Sketch the graph of the function f and interpret your results.

65. Box Office Receipts   The total worldwide box office receipts for a long-running movie are approximated by the function

$$T(x) = \frac{120x^2}{x^2 + 4}$$

where $T(x)$ is measured in millions of dollars and x is the number of years since the movie's release. Sketch the graph of the function T and interpret your results.

66. Cost of Removing Toxic Pollutants   A city's main well was recently found to be contaminated with trichloroethylene, a cancer-causing chemical, as a result of an abandoned chemical dump leaching chemicals into the water. A proposal submitted to the city's selectmen indicates that the cost, measured in millions of dollars, of removing $x\%$ of the toxic pollutant is given by

$$C(x) = \frac{0.5x}{100 - x}$$

Sketch the graph of the function C and interpret your results.

SOLUTIONS TO SELF-CHECK EXERCISES 10.3

1. Since

$$\lim_{x \to \infty} \frac{2x^2}{x^2 - 1} = \lim_{x \to \infty} \frac{2}{1 - \dfrac{1}{x^2}}$$

Divide the numerator and denominator by x^2.

$$= 2$$

we see that $y = 2$ is a horizontal asymptote. Next, since

$$x^2 - 1 = (x + 1)(x - 1) = 0$$

implies $x = -1$ or $x = 1$, these are candidates for the vertical asymptotes of f. Since the numerator of f is not equal to zero for $x = -1$ or $x = 1$, we conclude that $x = -1$ and $x = 1$ are vertical asymptotes of the graph of f.

2. We obtain the following information on the graph of f.
   (1) The domain of f is the interval $(-\infty, \infty)$.
   (2) By setting $x = 0$, we find the y-intercept is 4.
   (3) Since

   $$\lim_{x \to -\infty} f(x) = \lim_{x \to -\infty} \left(\frac{2}{3}x^3 - 2x^2 - 6x + 4 \right) = -\infty$$

   and

   $$\lim_{x \to \infty} f(x) = \lim_{x \to \infty} \left(\frac{2}{3}x^3 - 2x^2 - 6x + 4 \right) = \infty$$

   we see that $f(x)$ decreases without bound as x decreases without bound, and $f(x)$ increases without bound as x increases without bound.
   (4) Since f is a polynomial function, there are no asymptotes.
   (5)

   $$f'(x) = 2x^2 - 4x - 6 = 2(x^2 - 2x - 3)$$
   $$= 2(x + 1)(x - 3)$$

   Setting $f'(x) = 0$ gives $x = -1$ or $x = 3$. The accompanying sign diagram for f' shows that f is increasing on the intervals $(-\infty, -1)$ and $(3, \infty)$ and decreasing on $(-1, 3)$.

(6) From the results of step 5, we see that $x = -1$ and $x = 3$ are critical points of f. Furthermore, the sign diagram of f' tells us that $x = -1$ gives rise to a relative maximum of f and $x = 3$ gives rise to a relative minimum of f. Now,

$$f(-1) = \frac{2}{3}(-1)^3 - 2(-1)^2 - 6(-1) + 4 = \frac{22}{3}$$

and

$$f(3) = \frac{2}{3}(3)^3 - 2(3)^2 - 6(3) + 4 = -14$$

so $f(-1) = \frac{22}{3}$ is a relative maximum of f and $f(3) = -14$ is a relative minimum of f.

(7) $$f''(x) = 4x - 4 = 4(x - 1)$$

which is equal to zero when $x = 1$. The accompanying sign diagram of f'' shows that f is concave downward on the interval $(-\infty, 1)$ and concave upward on the interval $(1, \infty)$.

(8) From the results of step 7, we see that $x = 1$ is the only candidate for an inflection point of f. Since $f''(x)$ changes sign as we move across the point $x = 1$ and

$$f(1) = \frac{2}{3}(1)^3 - 2(1)^2 - 6(1) + 4 = -\frac{10}{3}$$

we see that the required inflection point is $(1, -\frac{10}{3})$.

(9) Summarizing this information, we have

Domain	$(-\infty, \infty)$
Intercept	$(0, 4)$
Intervals where f is $\nearrow$ or $\searrow$	$\nearrow$ on $(-\infty, -1) \cup (3, \infty)$; $\searrow$ on $(-1, 3)$
Relative extrema	Rel. max. at $(-1, \frac{22}{3})$; rel. min. at $(3, -14)$
Concavity	Downward on $(-\infty, 1)$; upward on $(1, \infty)$
Point of inflection	$(1, -\frac{10}{3})$
$\lim\limits_{x \to -\infty} f(x); \lim\limits_{x \to \infty} f(x)$	$-\infty$; ∞
Asymptotes	None

The graph of f is sketched in the accompanying figure.

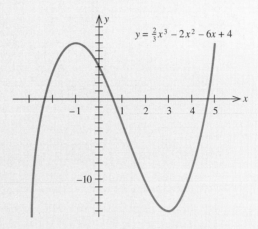

$$y = \tfrac{2}{3}x^3 - 2x^2 - 6x + 4$$

U s i n g   T e c h n o l o g y

One of the main purposes of studying Section 10.3 is to see how the many concepts of calculus come together to paint a picture of a function. The techniques of graphing also play a very practical role. For example, using the techniques of graphing developed in Section 10.3, you can tell if the graph of a function generated by a graphing utility is reasonably complete. Furthermore, these techniques can often reveal details that are missing from a graph.

EXAMPLE 1   Consider the function $f(x) = 2x^3 - 3.5x^2 + x - 10$. A plot of the graph of f in the standard viewing rectangle is shown in Figure 1. Since the domain of f is the interval $(-\infty, \infty)$, we see that Figure 1 does not reveal the part of the graph to the left of the y-axis. This suggests that we enlarge the viewing rectangle accordingly. Figure 2 shows the graph of f in the viewing rectangle $[-10, 10] \times [-20, 10]$.

FIGURE 1

The graph of f in the standard viewing rectangle

The behavior of f for large values of f ($\lim\limits_{x \to -\infty} f(x) = -\infty$ and $\lim\limits_{x \to \infty} f(x) = \infty$) suggests that this viewing rectangle has captured a sufficiently complete picture of f. Next, an analysis of the first derivative of f,

$$f'(x) = 6x^2 - 7x + 1 = (6x - 1)(x - 1)$$

FIGURE 2
The graph of f in the viewing rectangle
$[-10, 10] \times [-20, 10]$

reveals that f has critical values at $x = 1/6$ and $x = 1$. In fact, a sign diagram of f' shows that f has a relative maximum at $x = 1/6$ and a relative minimum at $x = 1$, details that are not revealed in the graph of f shown in Figure 2. In order to examine this portion of the graph of f, we use, say, the viewing rectangle $[-1, 2] \times [-11, -9]$. The resulting graph of f is shown in Figure 3, and it certainly reveals the hitherto missing details! Thus, through an interaction of calculus and a graphing utility, we are able to obtain a good picture of the properties of f.

FIGURE 3
The graph of f in the viewing rectangle
$[-1, 2] \times [-11, -9]$

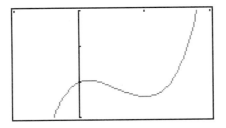

Finding x-Intercepts

As noted in Section 10.3, it is not always easy to find the x-intercepts of the graph of a function. But this information is very important in applications. The use of the **POLY** and **ROOT** operations of a graphing utility will solve the equation $f(x) = 0$ quite easily and hence yield the x-intercepts of the graph of a function. (See Sections A.14 and A.15 in Appendix A.)

EXAMPLE 2   Let $f(x) = x^3 - 3x^2 + x + 1.5$.

a. Find the x-intercepts of the graph of f using **POLY**.

b. Find the x-intercepts of the graph of f using **ROOT**.

Solution

a. Observe that f is a polynomial function of degree 3, and so we may use **POLY** to solve the equation $x^3 - 3x^2 + x + 1.5 = 0$ ($f(x) = 0$). We find that the solutions (x-intercepts) are

$$x_1 \approx -0.525687120865, \quad x_2 \approx 1.2586520225, \quad x_3 \approx 2.26703509836$$

b. Using the graph of f (Figure 4), we see that $x_1 \approx -0.5$, $x_2 \approx 1$, and $x_3 \approx 2$. Using **ROOT** and these values of x as initial guesses, we find

$$x_1 \approx -0.5256871209, \quad x_2 \approx 1.2586520225, \quad x_3 \approx 2.2670350984$$

FIGURE 4
The graph of
$f(x) = x^3 - 3x^2 + x + 1.5$

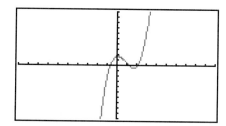

REMARK   POLY will solve a polynomial equation $f(x) = 0$, where f is a polynomial function. **ROOT** however, will solve equations $f(x) = 0$ even if f is not a polynomial.   □

EXERCISES

In exercises 1–4, use the method of Example 1 to analyze the function. (Note: Your answers will not be unique.)

1. $f(x) = 4x^3 - 4x^2 + x + 10$

2. $f(x) = x^3 + 2x^2 + x - 12$

3. $f(x) = \frac{1}{2}x^4 + x^3 + \frac{1}{2}x^2 - 10$

4. $f(x) = 2.25x^4 - 4x^3 + 2x^2 + 2$

In exercises 5–12, find the x-intercepts of the graph of f. Give your answer accurate to four decimal places.

5. $f(x) = 0.2x^3 - 1.2x^2 + 0.8x + 2.1$

6. $f(x) = -0.5x^3 + 1.7x^2 - 1.2$

7. $f(x) = 0.3x^4 - 1.2x^3 + 0.8x^2 + 1.1x - 2$

8. $f(x) = -0.2x^4 + 0.8x^3 - 2.1x + 1.2$

9. $f(x) = 2x^2 - \sqrt{x + 1} - 3$

10. $f(x) = x - \sqrt{1 - x^2}$

11. $f(x) = e^x - 2x - 2$

12. $f(x) = \ln(1 + x^2) + 2x - 3$

10.4   Optimization I

Absolute Extrema

The graph of the function f in Figure 58 shows the average age of cars in use in the United States from the beginning of 1946 ($t = 0$) to the beginning

FIGURE 58
$f(t)$ gives the average age of cars in use in year t, t in [0, 44].

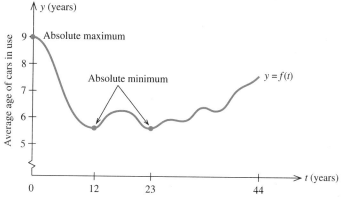

Source: American Automobile Association

of 1990 ($t = 44$). Observe that the highest average age of cars in use during this period is 9 years, whereas the lowest average age of cars in use during the same period is $5\frac{1}{2}$ years. The number 9, the largest value of $f(t)$ for all values of t in the interval $[0, 44]$ (the domain of f) is called the *absolute maximum value of f* on that interval. The number $5\frac{1}{2}$, the smallest value of $f(t)$ for all values of t in $[0, 44]$, is called the *absolute minimum value of f* on that interval. Notice, too, that the absolute maximum value of f is attained at the endpoint $t = 0$ of the interval, whereas the absolute minimum value of f is attained at the two interior points $t = 12$ (corresponding to 1958) and $t = 23$ (corresponding to 1969).

A precise definition of the **absolute extrema** (absolute maximum or absolute minimum) of a function follows.

THE ABSOLUTE EXTREMA OF A FUNCTION f

If $f(x) \leq f(c)$ for all x in the domain of f, then $f(c)$ is called the **absolute maximum value** of f.

If $f(x) \geq f(c)$ for all x in the domain of f, then $f(c)$ is called the **absolute minimum value** of f.

Figure 59 shows the graphs of several functions and gives the absolute maximum and absolute minimum of each function, if they exist.

FIGURE 59

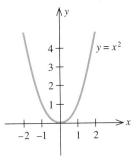

(a) $f(0) = 0$ is the absolute minimum of f; f has no absolute maximum

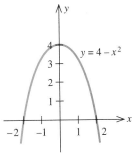

(b) $f(0) = 4$ is the absolute maximum of f; f has no absolute minimum

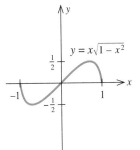

(c) $f\left(\dfrac{\sqrt{2}}{2}\right) = \dfrac{1}{2}$ is the absolute maximum of f.

$f\left(-\dfrac{\sqrt{2}}{2}\right) = -\dfrac{1}{2}$ is the absolute minimum of f.

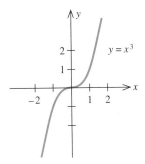

(d) f has no absolute extrema.

Absolute Extrema on a Closed Interval

As the preceding examples show, a continuous function defined on an arbitrary interval does not always have an absolute maximum or an absolute minimum. But an important case arises often in practical applications in which both the absolute maximum and the absolute minimum of a function are guaranteed to exist. This occurs when a continuous function is defined on a *closed* interval. Let us state this important result in the form of a theorem, whose proof we will omit.

THEOREM 3

If a function f is continuous on a closed interval $[a, b]$, then f has both an absolute maximum value and an absolute minimum value on $[a, b]$.

Observe that if an absolute extremum of a continuous function f occurs at a point in an open interval (a, b), then it must be a relative extremum of f and hence its x-coordinate must be a critical point of f. Otherwise, the absolute extremum of f must occur at one or both of the endpoints of the interval $[a, b]$. A typical situation is illustrated in Figure 60.

FIGURE 60

The relative minimum of f at x_3 is the absolute minimum of f. The right endpoint b gives rise to the absolute maximum value $f(b)$ of f.

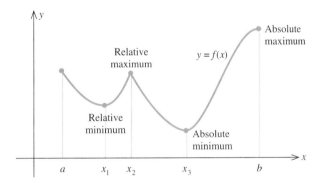

Here x_1, x_2, and x_3 are critical points of f. The absolute minimum of f occurs at x_3, which lies in the open interval (a, b) and is a critical point of f. The absolute maximum of f occurs at b, an endpoint. This observation suggests the following procedure for finding the absolute extrema of a continuous function on a closed interval.

FINDING THE ABSOLUTE EXTREMA OF f ON A CLOSED INTERVAL

1. Find the critical points of f that lie in (a, b).
2. Compute the value of f at each of these critical points of f and compute $f(a)$ and $f(b)$.
3. The absolute maximum value and absolute minimum value of f will correspond to the largest and smallest numbers, respectively, found in step 2.

FIGURE 61

F has an absolute minimum value of 0 and an absolute maximum value of 4.

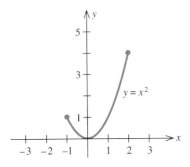

FIGURE 61

F has an absolute minimum value of 0 and an absolute maximum value of 4.

EXAMPLE 1 Find the absolute extrema of the function $F(x) = x^2$ defined on the interval $[-1, 2]$.

Solution The function F is continuous on the closed interval $[-1, 2]$ and differentiable on the open interval $(-1, 2)$. The derivative of F is

$$F'(x) = 2x$$

so $x = 0$ is the only critical point of F. Next, evaluate $F(x)$ at $x = -1$, $x = 0$, and $x = 2$. Thus,

$$F(-1) = 1, \qquad F(0) = 0, \qquad F(2) = 4$$

It follows that 0 is the absolute minimum value of F and 4 is the absolute maximum value of F. The graph of F, which appears in Figure 61, confirms our results.   ■

EXAMPLE 2 Find the absolute extrema of the function

$$f(x) = x^3 - 2x^2 - 4x + 4$$

defined on the interval $[0, 3]$.

Solution The function f is continuous on the closed interval $[0, 3]$ and differentiable on the open interval $(0, 3)$. The derivative of f is

$$f'(x) = 3x^2 - 4x - 4 = (3x + 2)(x - 2)$$

and it is equal to zero when $x = -\frac{2}{3}$ and $x = 2$. Since the point $x = -\frac{2}{3}$ lies outside the interval $[0, 3]$, it is dropped from further consideration and $x = 2$ is seen to be the sole critical point of f. Next, we evaluate $f(x)$ at the critical point of f as well as at the endpoints of f, obtaining

$$f(0) = 4, \qquad f(2) = -4, \qquad f(3) = 1$$

From these results, we conclude that -4 is the absolute minimum value of f and 4 is the absolute maximum value of f. The graph of f, which appears in Figure 62, confirms our results. Observe that the absolute maximum of f occurs at the endpoint $x = 0$ of the interval $[0, 3]$, while the absolute minimum of f occurs at $x = 2$, which is a point in the interval $(0, 3)$.

FIGURE 62

f has an absolute maximum value of 4 and an absolute minimum value of −4.

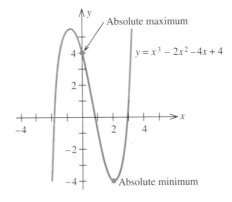

■

FIGURE 63

f has an absolute minimum value of $f(0) = 0$ and an absolute maximum value of $f(8) = 4$.

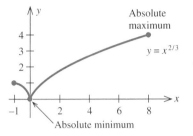

EXAMPLE 3 Find the absolute maximum and absolute minimum values of the function $f(x) = x^{2/3}$ on the interval $[-1, 8]$.

Solution The derivative of f is

$$f'(x) = \frac{2}{3}x^{-1/3} = \frac{2}{3x^{1/3}}$$

Note that f' is not defined at $x = 0$, is continuous everywhere else, and does not equal zero for all x. Therefore, $x = 0$ is the only critical point of f. Evaluating $f(x)$ at $x = -1$, 0, and 8, we obtain

$$f(-1) = 1, \qquad f(0) = 0, \qquad f(8) = 4$$

We conclude that the absolute minimum value of f is 0, attained at $x = 0$, and the absolute maximum value of f is 4, attained at $x = 8$ (Figure 63).

Applications

Many real-world applications call for finding the absolute maximum value or the absolute minimum value of a given function. For example, management is interested in finding what level of production will yield the maximum profit for a company; a farmer is interested in finding the right amount of fertilizer to maximize crop yield; a doctor is interested in finding the maximum concentration of a drug in a patient's body and the time at which it occurs; and an engineer is interested in finding the dimensions of a container with a specified shape and volume that can be constructed at a minimum cost.

EXAMPLE 4 The Acrosonic Company's total profit from manufacturing and selling x units of their Model F loudspeaker systems is given by

$$P(x) = -0.02x^2 + 300x - 200{,}000 \qquad 0 \le x \le 20{,}000$$

dollars. How many units of the loudspeaker system must Acrosonic produce to maximize its profits?

Solution To find the absolute maximum of P on $[0, 20{,}000]$, first find the critical points of P on the interval $(0, 20{,}000)$. To do this, compute

$$P'(x) = -0.04x + 300$$

Solving the equation $P'(x) = 0$ gives $x = 7{,}500$. Next, evaluate $P(x)$ at $x = 7{,}500$ as well as at the endpoints $x = 0$ and $x = 20{,}000$ of the interval $[0, 20{,}000]$, obtaining

$$P(0) = -200{,}000$$

$$P(7{,}500) = 925{,}000$$

and $$P(20{,}000) = -2{,}200{,}000$$

From these computations we see that the absolute maximum value of the function P is 925,000. Thus, by producing 7,500 units, Acrosonic will realize a maximum profit of $925,000. The graph of P is sketched in Figure 64.

FIGURE 64

P has an absolute maximum at (7,500, 925,000).

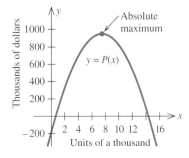

EXAMPLE 5   When a person coughs, the trachea, or windpipe, contracts, allowing air to be expelled at a maximum velocity. It can be shown that, during a cough, the velocity v of airflow is given by the function

$$v = f(r) = kr^2(R - r)$$

where r is the trachea's radius (in centimeters) during a cough, R is the trachea's normal radius (in centimeters), and k is a positive constant that depends on the length of the trachea. Find the radius r for which the velocity of airflow is greatest.

Solution   To find the absolute maximum of f on $[0, R]$, first find the critical points of f on the interval $(0, R)$. We compute

$$f'(r) = 2kr(R - r) - kr^2 \qquad \text{Using the Product Rule}$$

$$= -3kr^2 + 2kRr = kr(-3r + 2R)$$

Setting $f'(r) = 0$ gives $r = 0$ or $r = \frac{2}{3}R$, and so $r = \frac{2}{3}R$ is the sole critical point of f ($r = 0$ is an endpoint). Evaluating $f(r)$ at $r = \frac{2}{3}R$, as well as at the endpoints $r = 0$ and $r = R$, we obtain

$$f(0) = 0$$

$$f\left(\frac{2}{3}R\right) = \frac{4k}{27}R^3$$

and $\qquad\qquad\qquad f(R) = 0$

from which we deduce that the velocity of airflow is greatest when the radius of the contracted trachea is $\frac{2}{3}R$—that is, when the radius is contracted by approximately 33%. The graph of the function f is shown in Figure 65. ■

FIGURE 65
The velocity of airflow is greatest when the radius of the contracted trachea is $\frac{2}{3}R$.

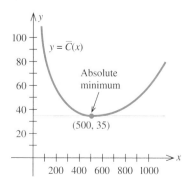

EXAMPLE 6   The daily average cost function (in dollars per unit) of the Elektra Electronics Company is given by

$$\overline{C}(x) = 0.0001x^2 - 0.08x + 40 + \frac{5{,}000}{x} \qquad x > 0$$

where x stands for the number of programmable calculators Elektra produces. Show that a production level of 500 units per day results in a minimum average cost for the company.

FIGURE 66
The minimum average cost is $35/unit.

Solution   The domain of the function $\overline{C}$ is the interval $(0, \infty)$, which is not closed. In order to solve the problem, we resort to the graphic method. Using the techniques of graphing from the last section, we sketch the graph of $\overline{C}$ (Figure 66).
   Now,

$$\overline{C}'(x) = 0.0002x - 0.08 - \frac{5{,}000}{x^2}$$

Substituting the given value of x, 500, into $\overline{C}'(x)$ gives $\overline{C}'(500) = 0$, so $x = 500$ is a critical point of $\overline{C}$. Next,

$$\overline{C}''(x) = 0.0002 + \frac{10{,}000}{x^3}$$

Thus, $$\overline{C}''(500) = 0.0002 + \frac{10,000}{(500)^3} > 0$$

and by the Second Derivative Test, a relative minimum of the function $\overline{C}$ occurs at the point $x = 500$. Furthermore, $\overline{C}''(x) > 0$ for $x > 0$, which implies that the graph of $\overline{C}$ is concave upward everywhere, so the relative minimum of $\overline{C}$ must be the absolute minimum of $\overline{C}$. The minimum average cost is given by

$$\overline{C}(500) = 0.0001(500)^2 - 0.08(500) + 40 + \frac{5,000}{500}$$

$$= 35$$

or $35 per unit. ∎

Our final example involves finding the absolute maximum of an exponential function.

EXAMPLE 7   The present value of the market price of the Blakely Office Building is given by

$$P(t) = 300,000e^{-0.09t + \sqrt{t}/2} \qquad 0 \le t \le 10$$

Find the optimal present value of the building's market price.

Solution   To find the maximum value of P over $[0, 10]$, we compute

$$P'(t) = 300,000e^{-0.09t + \sqrt{t}/2}\frac{d}{dt}\left(-0.09t + \frac{1}{2}t^{1/2}\right)$$

$$= 300,000e^{-0.09t + \sqrt{t}/2}\left(-0.09 + \frac{1}{4}t^{-1/2}\right)$$

Setting $P'(t) = 0$ gives

$$-0.09 + \frac{1}{4t^{1/2}} = 0$$

since $e^{-0.09t + \sqrt{t}/2}$ is never zero for any value of t. Solving this equation, we find

$$\frac{1}{4t^{1/2}} = 0.09$$

$$t^{1/2} = \frac{1}{4(0.09)}$$

$$= \frac{1}{0.36}$$

or $$t \approx 7.72$$

the sole critical point of the function P. Finally, evaluating $P(t)$ at the critical point as well as at the endpoints of $[0, 10]$, we have

t	$P(t)$
0	300,000
7.72	600,779
10	592,838

We conclude, accordingly, that the optimal present value of the property's market price is $600,779 and that this will occur 7.72 years from now.

**SELF-CHECK
EXERCISES 10.4**

1. Find the absolute extrema of $f(x) = 3x^4 + 4x^3 + 1$ on $[-2, 1]$.

c 2. The operating rate (expressed as a percentage) of factories, mines, and utilities in a certain region of the country on the tth day of the year 1996 is given by the function

$$f(t) = 80 + \frac{1,200t}{t^2 + 40,000} \qquad 0 \le t \le 250$$

On which day of the first 250 days of 1996 was the manufacturing capacity operating rate highest?

Solutions to Self-Check Exercises 10.4 can be found on page 646.

10.4 EXERCISES

In exercises 1–8, you are given the graph of a function f defined on the indicated interval. Find the absolute maximum and the absolute minimum of f, if they exist.

1.

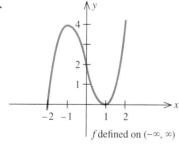

f defined on $(-\infty, \infty)$

3.

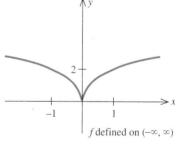

f defined on $(-\infty, \infty)$

2.

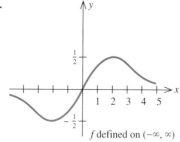

f defined on $(-\infty, \infty)$

4.

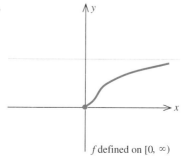

f defined on $[0, \infty)$

5.

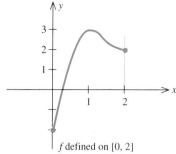

f defined on [0, 2]

6.

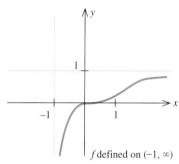

f defined on (−1, ∞)

7.

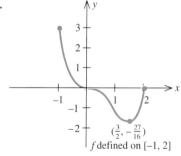

$\left(\frac{3}{2}, -\frac{27}{16}\right)$

f defined on [−1, 2]

8.

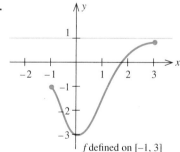

f defined on [−1, 3]

In exercises 9–42, find the absolute maximum value and the absolute minimum value, if any, of the function.

9. $f(x) = 2x^2 + 3x - 4$ **10.** $g(x) = -x^2 + 4x + 3$

11. $h(x) = x^{1/3}$ **12.** $f(x) = x^{2/3}$

13. $f(x) = \dfrac{1}{1 + x^2}$ **14.** $f(x) = \dfrac{x}{1 + x^2}$

15. $f(x) = x^2 - 2x - 3$ on $[-2, 3]$

16. $g(x) = x^2 - 2x - 3$ on $[0, 4]$

17. $f(x) = -x^2 + 4x + 6$ on $[0, 5]$

18. $f(x) = -x^2 + 4x + 6$ on $[3, 6]$

19. $f(x) = x^3 + 3x^2 - 1$ on $[-3, 2]$

20. $g(x) = x^3 + 3x^2 - 1$ on $[-3, 1]$

21. $g(x) = 3x^4 + 4x^3$ on $[-2, 1]$

22. $f(x) = \frac{1}{2}x^4 - \frac{2}{3}x^3 - 2x^2 + 3$ on $[-2, 3]$

23. $f(x) = \dfrac{x + 1}{x - 1}$ on $[0, 3]$ **24.** $g(t) = \dfrac{t}{t - 1}$ on $[2, 4]$

25. $f(x) = 4x + \dfrac{1}{x}$ on $[1, 3]$ **26.** $f(x) = 9x - \dfrac{1}{x}$ on $[1, 3]$

27. $f(x) = \frac{1}{2}x^2 - 2\sqrt{x}$ on $[0, 3]$

28. $g(x) = \frac{1}{8}x^2 - 4\sqrt{x}$ on $[0, 9]$

29. $f(x) = 3x^{2/3} - 2x$ on $[0, 3]$

30. $g(x) = x^2 + 2x^{2/3}$ on $[-2, 2]$

31. $f(x) = x^{2/3}(x^2 - 4)$ on $[-1, 2]$

32. $f(x) = x^{2/3}(x^2 - 4)$ on $[-1, 3]$

33. $f(x) = \dfrac{x}{x^2 + 2}$ on $[-1, 2]$

34. $f(x) = e^{-x^2}$ on $[-1, 1]$

35. $h(x) = e^{x^2 - 4}$ on $[-2, 2]$

36. $g(x) = (2x - 1)e^{-x}$ on $[0, 4]$

37. $f(x) = xe^{-x^2}$ on $[0, 2]$

38. $f(x) = x - \ln x$ on $[\frac{1}{2}, 3]$

39. $g(x) = x/\ln x$ on $[2, 5]$

40. $f(x) = \dfrac{1}{x^2 + 2x + 5}$ on $[-2, 1]$

41. $f(x) = \dfrac{x}{\sqrt{x^2 + 1}}$ on $[-1, 1]$

42. $g(x) = x\sqrt{4 - x^2}$ on $[0, 2]$

43. A stone is thrown straight up from the roof of an 80-foot building. The height of the stone at any time t (in seconds), measured from the ground, is given by

$$h(t) = -16t^2 + 64t + 80$$

feet. What is the maximum height the stone reaches?

44. Maximizing Profits   Lynbrook West, an apartment complex, has 100 two-bedroom units. The monthly profit realized from renting out x apartments is given by

$$P(x) = -10x^2 + 1{,}760x - 50{,}000$$

dollars. Find how many units should be rented out in order to maximize the monthly rental profit. What is the maximum monthly profit realizable?

45. Maximizing Profits   The estimated monthly profit realizable by the Cannon Precision Instruments Corporation for manufacturing and selling x units of its Model M1 camera is

$$P(x) = -0.04x^2 + 240x - 10{,}000$$

dollars. Determine how many cameras Cannon should produce per month in order to maximize its profits.

46. Flight of a Rocket   The altitude (in feet) attained by a model rocket t seconds into flight is given by the function

$$h(t) = -\frac{1}{3}t^3 + 4t^2 + 20t + 2$$

Find the maximum altitude attained by the rocket.

47. GDP of a Developing Country   A developing country's gross domestic product (GDP) from 1988 to 1996 is approximated by the function

$$G(t) = -0.2t^3 + 2.4t^2 + 60 \qquad 0 \le t \le 8$$

where $G(t)$ is measured in billions of dollars and $t = 0$ corresponds to the year 1988. Show that the growth rate of the country's GDP was maximal in 1992.

48. Crime Rates   The number of major crimes committed in the city of Bronxville between 1987 and 1994 is approximated by the function

$$N(t) = -0.1t^3 + 1.5t^2 + 100 \qquad 0 \le t \le 7$$

where $N(t)$ denotes the number of crimes committed in year t ($t = 0$ corresponds to the year 1987). Enraged by the dramatic increase in the crime rate, the citizens of Bronxville, with the help of the local police, organized "Neighborhood Crime Watch" groups in early 1991 to combat this menace. Show that the growth in the crime rate was maximal in 1992, giving credence to the claim that the Neighborhood Crime Watch program was working.

49. Maximizing Profits   The management of Trappee and Sons, Inc., producers of the famous Texa-Pep hot sauce, estimate that their profit from the daily production and sale of x cases (each case consisting of 24 bottles) of the hot sauce is given by

$$P(x) = -0.000002x^3 + 6x - 400$$

dollars. What is the largest possible profit Trappee can make in one day?

50. Maximizing Profits   The quantity demanded per month of the Walter Serkin recording of Beethoven's *Moonlight Sonata,* manufactured by Phonola Record Industries, is related to the price per record. The equation

$$p = -0.00042x + 6 \qquad 0 \le x \le 12{,}000$$

where p denotes the unit price in dollars and x is the number of records demanded, relates the demand to the price. The total monthly cost for pressing and packaging x copies of this classical recording is given by

$$C(x) = 600 + 2x - 0.00002x^2 \qquad 0 \le x \le 20{,}000$$

dollars. Determine how many copies Phonola should produce per month in order to maximize its profits. [*Hint:* The revenue is $R(x) = px$ and the profit is $P(x) = R(x) - C(x)$.]

51. Maximizing Profit   A manufacturer of tennis rackets finds that the total cost $C(x)$ [in dollars] of manufacturing x rackets per day is given by $C(x) = 400 + 4x + 0.0001x^2$. Each racket can be sold at a price of p dollars, where p is related to x by the demand equation $p = 10 - 0.0004x$. If all the rackets that are manufactured can be sold, find the daily level of production that will yield a maximum profit for the manufacturer.

52. Maximizing Profit   The weekly demand for the Pulsar 25-inch color console television is given by the demand equation

$$p = -0.05x + 600 \qquad 0 \le x \le 12{,}000$$

where p denotes the wholesale unit price in dollars and x denotes the quantity demanded. The weekly total cost function associated with manufacturing these sets is given by

$$C(x) = 0.000002x^3 - 0.03x^2 + 400x + 80{,}000$$

where $C(x)$ denotes the total cost incurred in producing x sets. Find the level of production that will yield a maximum profit for the manufacturer. [*Hint:* Use the quadratic formula.]

53. Minimizing Average Costs   Suppose that the total cost function for manufacturing a certain product is $C(x) = 0.2(0.01x^2 + 120)$ dollars, where x represents the number of units produced. Find the level of production that will minimize the average cost.

C **54.** Minimizing Production Costs   The total monthly cost in dollars incurred by Cannon Precision Instruments Corporation for manufacturing x units of the Model M1 camera is given by the function

$$C(x) = 0.0025x^2 + 80x + 10,000$$

a. Find the average cost function $\overline{C}$.
b. Find the level of production that results in the smallest average production cost.
c. Find the level of production for which the average cost is equal to the marginal cost.
d. Compare the result of (c) with that of (b).

C **55.** Minimizing Production Costs   The daily total cost in dollars incurred by Trappee and Sons, Inc., for producing x cases of Texa-Pep hot sauce is given by the function

$$C(x) = 0.000002x^3 + 5x + 400$$

Using this function, answer the questions posed in exercise 54.

C **56.** Maximum Oil Production   It has been estimated that the total production of oil from a certain oil well is given by

$$T(t) = -1,000(t + 10)e^{-0.1t} + 10,000$$

thousand barrels t years after production has begun. Determine the year when the oil well will be producing at maximum capacity.

C **57.** Optimal Selling Time   The present value of a piece of waterfront property purchased by an investor is given by the function

$$P(t) = 80,000e^{\sqrt{t}/2 - 0.09t} \qquad 0 \le t \le 8$$

where $P(t)$ is measured in dollars and t is the time in years from the present. Determine the optimal time (based on present value) for the investor to sell the property. What is the property's optimal present value?

58. Oxygen Content of a Pond   When organic waste is dumped into a pond, the oxidation process that takes place reduces the pond's oxygen content. However, given time, nature will restore the oxygen content to its natural level. Suppose that the oxygen content t days after organic waste has been dumped into the pond is given by

$$f(t) = 100 \left[\frac{t^2 - 4t + 4}{t^2 + 4} \right] \qquad 0 \le t < \infty$$

percent of its normal level.
a. When is the level of oxygen content lowest?
b. When is the rate of oxygen regeneration greatest?

59. Air Pollution   The amount of nitrogen dioxide, a brown gas that impairs breathing, present in the at-

mosphere on a certain May day in the city of Long Beach is approximated by

$$A(t) = \frac{136}{1 + 0.25(t - 4.5)^2} + 28 \qquad 0 \le t \le 11$$

where $A(t)$ is measured in pollutant standard index (PSI) and t is measured in hours, with $t = 0$ corresponding to 7 A.M. Determine the time of day when the pollution is at its highest level.

60. Maximizing Revenue   Suppose that the quantity demanded per week of a certain dress is related to the unit price p by the demand equation $p = \sqrt{800 - x}$, where p is in dollars and x is the number of dresses made. How many dresses should be made and sold per week in order to maximize the revenue? [Hint: $R(x) = px$.]

61. Maximizing Revenue   The quantity demanded per month of the Sicard wristwatch is related to the unit price by the equation

$$p = \frac{50}{0.01x^2 + 1} \qquad 0 \le x \le 20$$

where p is measured in dollars and x is measured in units of a thousand. How many watches must be sold to yield a maximum revenue?

62. Social Security Surplus   Based on data from the Social Security Administration, the estimated cash in the social security retirement and disability trust funds in the five decades beginning with the year 1990 is given by

$$A(t) = -96.6t^4 + 403.6t^3 + 660.9t^2 + 250 \quad 0 \le t \le 5$$

where $A(t)$ is measured in billions of dollars and t is measured in decades, with $t = 0$ corresponding to the year 1990. Show that the social security surplus will be at its highest level at approximately the beginning of the year 2030. [Hint: Use the quadratic formula.]
Source: Social Security Administration

63. Velocity of Blood   According to a law discovered by the nineteenth-century physician Jean Louis Marie Poiseuille, the velocity (in centimeters per second) of blood r centimeters from the central axis of an artery is given by

$$v(r) = k(R^2 - r^2)$$

where k is a constant and R is the radius of the artery. Show that the velocity of blood is greatest along the central axis.

64. Energy Expended by a Fish   It has been conjectured that a fish swimming a distance of L ft at a speed of v ft/sec relative to the water and against a current flowing at the rate of u ft/sec ($u < v$) expends a total energy given by

$$E(v) = \frac{aLv^3}{v - u}$$

where E is measured in foot-pounds (ft-lbs), and a is a constant. Find the speed v at which the fish must swim in order to minimize the total energy expended. (*Note:* This result has been verified by biologists.)

SOLUTIONS TO SELF-CHECK EXERCISES 10.4

1. The function f is continuous on the interval $[-2, 1]$. It is also differentiable on the open interval $(-2, 1)$. The derivative of f is

$$f'(x) = 12x^3 + 12x^2 = 12x^2(x + 1)$$

and it is continuous on $(-2, 1)$. Setting $f'(x) = 0$ gives $x = -1$ and $x = 0$ as critical points of f. Evaluating $f(x)$ at these critical points of f as well as at the endpoints of the interval $[-2, 1]$, we obtain

$$f(-2) = 17, \quad f(-1) = 0, \quad f(0) = 1, \quad f(1) = 8$$

From these results, we see that 0 is the absolute minimum value of f and 17 is the absolute maximum value of f.

2. The problem is solved by finding the absolute maximum of the function f on $[0, 250]$. Differentiating $f(t)$, we obtain

$$f'(t) = \frac{(t^2 + 40,000)(1,200) - 1,200t(2t)}{(t^2 + 40,000)^2}$$

$$= \frac{-1,200\,(t^2 - 40,000)}{(t^2 + 40,000)^2}$$

Upon setting $f'(t) = 0$ and solving the resulting equation, we obtain $t = -200$ or 200. Since -200 lies outside the interval $[0, 250]$, we are interested only in the critical point $t = 200$ of f. Evaluating $f(t)$ at $t = 0$, $t = 200$, and $t = 250$, we find

$$f(0) = 80, \quad f(200) = 83, \quad f(250) = 82.93$$

We conclude that the manufacturing capacity operating rate was the highest on the 200th day of 1996—that is, a little past the middle of July of 1996.

Using Technology

The **FMIN** and **FMAX** operations of a graphing utility can be used to find the absolute minimum and the absolute maximum values of a continuous function f on the closed interval [xMin, xMax]. (See Section A.17 in Appendix A.)

EXAMPLE 1   Let $f(x) = \dfrac{2x + 4}{(x^2 + 1)^{3/2}}$.

a. Use a graphing utility to plot the graph of f in the viewing rectangle $[-3, 3] \times [-1, 5]$.

b. Find the absolute maximum and absolute minimum values of f on the interval $[-3, 3]$. Express your answers accurate to four decimal places.

Solution

a. The graph of *f* is shown in Figure 1.

FIGURE 1
The graph of f in the viewing
rectangle $[-3, 3] \times [-1, 5]$

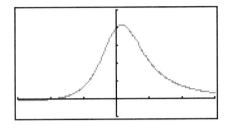

b. Using **FMIN,** we find the absolute minimum value of *f* to be -0.0632. Next, using **FMAX,** we find the absolute value to be 4.1593 occurring at $x \approx 0.1583$. ◼

We can also use the **fMin** and **fMax** functions of a graphing utility to find the absolute maximum and the absolute minimum values of a function on a closed interval without graphing the function.

EXAMPLE 2   Use the **fMin** and **fMax** functions to find the absolute maximum and the absolute minimum values of

$$f(x) = \frac{2x + 4}{(x^2 + 1)^{3/2}}$$

on the interval $[-3, 3]$.

Solution   We find that the absolute minimum value of *f* occurs at

$$\text{fMin }(2x + 4)/(x{\char94}2 + 1){\char94}1.5, x, -3, 3) = -2.99999519032$$

and the absolute maximum value of *f* occurs at

$$\text{fMax }(2x + 4)/(x{\char94}2 + 1){\char94}1.5, x, -3, 3) = 0.158311741276$$

Then using the **evalF** function, we find the absolute minimum value of *f* to be

$$\text{evalF}((2x + 4)/(x{\char94}2 + 1){\char94}1.5, x, -2.99999519032) = -0.063245522784$$

and the absolute maximum value of *f* to be

$$\text{evalF}((2x + 4)/(x{\char94}2 + 1){\char94}1.5, x, 0.158311741276) = 4.15928406072$$

as obtained in Example 1. ◼

EXERCISES

In exercises 1–6, use a graphing utility to find the absolute maximum and the absolute minimum values of f in the given interval (a) using the method of Example 1 and (b) using the method of Example 2. Express your answers accurate to four decimal places.

1. $f(x) = 3x^4 - 4.2x^3 + 6.1x - 2;\ [-2, 3]$

2. $f(x) = 2.1x^4 - 3.2x^3 + 4.1x^2 + 3x - 4;\ [-1, 2]$

3. $f(x) = \dfrac{2x^3 - 3x^2 + 1}{x^2 + 2x - 8};\ [-3, 1]$

4. $f(x) = \sqrt{x}(x^3 - 4)^2;\ [0.5, 1]$

5. $f(x) = e^{-x} \ln(x^2 + 1);\ [-2, 2]$

6. $f(x) = x^2 e^{-2x};\ [-1, 1]$

7. Rate of Bank Failures   The Federal Deposit Insurance Company (FDIC) estimates that the rate at which banks were failing between 1982 and 1994 is given by

$$f(t) = 0.063447t^4 - 1.953283t^3 + 14.632576t^2$$
$$- 6.684704t + 47.458874 \qquad 0 \le t \le 12$$

where $f(t)$ is the number of banks per year and t is measured in years, with $t = 0$ corresponding to the beginning of 1982.

a. Use a graphing utility to plot the graph of f in the viewing rectangle $[0, 12] \times [0, 220]$.
b. What is the highest rate of bank failures during the period in question?

Source: Federal Deposit Insurance Corporation

8. **Boston's Daily Temperature for 1993**   The average daily temperature in Boston for 1993 is given by

$$f(t) = 0.0434841t^4 - 1.18523t^3 + 9.38548t^2$$
$$- 17.7553t + 38.9272 \qquad 0 \le t \le 12$$

where $f(t)$ is measured in degrees Fahrenheit and t is in months, with $t = 0$ corresponding to the beginning of 1993. Find the absolute extrema for f and interpret your results.

Source: Robert Lautzenheiser, Climatologist

9. **Time on the Market**   According to the Greater Boston Real Estate Board-Multiple Listing Service, the average number of days a single-family home remains for sale from listing to accepted offer is approximated by the function

$$f(t) = 0.0171911t^4 - 0.662121t^3 + 6.18083t^2$$
$$- 8.97086t + 53.3357 \qquad 0 \le t \le 10$$

where t is measured in years, with $t = 0$ corresponding to the beginning of 1984.

a. Use a graphing utility to plot the graph of f in the viewing rectangle $[0, 12] \times [0, 120]$.
b. Find the absolute maximum value and the absolute minimum value of f in the interval $[0, 12]$. Interpret your results.

Source: Greater Boston Real Estate Board-Multiple Listing Service

---

| 10.5 | Optimization II |

Section 10.4 outlined how to find the solution to certain optimization problems in which the objective function is given. In this section, we consider problems in which we are required to first find the appropriate function to be optimized. The following guidelines will be useful for solving these problems.

GUIDELINES FOR SOLVING OPTIMIZATION PROBLEMS

1. Assign a letter to each variable mentioned in the problem. If appropriate, draw and label a figure.

2. Find an expression for the quantity to be optimized.

3. Use the conditions given in the problem to write the quantity to be optimized as a function f of *one* variable. Note any restrictions to be placed on the domain of f from physical considerations of the problem.

4. Optimize the function f over its domain using the methods of Section 10.4.

REMARK   In carrying out step 4, remember that if the function f to be optimized is continuous on a closed interval, then the absolute maximum and absolute minimum of f are, respectively, the largest and smallest values of $f(x)$ on the set comprising the critical points of f and the endpoints of the interval. If the domain of f is not a closed interval, then we resort to the graphical method.   □

Optimization Problems

EXAMPLE 1   A man wishes to have a rectangular-shaped garden in his backyard. He has 50 feet of fencing material with which to enclose his garden. Find the dimensions for the largest garden he can have if he uses all of the fencing material.

Solution

FIGURE 67

What is the maximum rectangular area that can be enclosed with 50 feet of fencing?

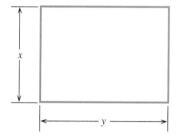

Step 1   Let x and y denote the dimensions (in feet) of two adjacent sides of the garden (Figure 67) and let A denote its area.

Step 2   The area of the garden,

$$A = xy \qquad (1)$$

is the quantity to be maximized.

Step 3   The perimeter of the rectangle, $(2x + 2y)$ feet, must equal 50 feet. Therefore, we have the equation

$$2x + 2y = 50$$

Next, solving this equation for y in terms of x yields

$$y = 25 - x \qquad (2)$$

which, when substituted into equation (1), gives

$$A = x(25 - x)$$
$$= -x^2 + 25x$$

(Remember, the function to be optimized must involve just one variable.) Since the sides of the rectangle must be nonnegative, we must have $x \geq 0$ and $y = 25 - x \geq 0$; that is, we must have $0 \leq x \leq 25$. Thus the problem is reduced to that of finding the absolute maximum of $A = f(x) = -x^2 + 25x$ on the closed interval $[0, 25]$.

Step 4   Observe that f is continuous on $[0, 25]$. So the absolute maximum value of f must occur at the endpoint(s) or at the critical point(s) of f. The derivative of the function A is given by

$$A' = f'(x) = -2x + 25$$

Setting $A' = 0$ gives

$$-2x + 25 = 0$$

or $x = 12.5$, as the critical point of A. Next, we evaluate the function $A = f(x)$ at $x = 12.5$ and at the endpoints $x = 0$ and $x = 25$ of the interval $[0, 25]$, obtaining

$$f(0) = 0, \quad f(12.5) = 156.25, \quad f(25) = 0$$

We see that the absolute maximum value of the function f is 156.25. From equation (2) we see that when $x = 12.5$, the value of y is given by $y = 12.5$. Thus, the garden would be of maximum area (156.25 square feet) if it were in the form of a square with sides 12.5 feet long.   ∎

EXAMPLE 2   By cutting away identical squares from each corner of a rectangular piece of cardboard and folding up the resulting flaps, the cardboard may be turned into an open box. If the cardboard is 16 inches long and 10 inches wide, find the dimensions of the box that will yield the maximum volume.

Solution

Step 1   Let x denote the length (in inches) of one side of each of the identical squares to be cut out of the cardboard (Figure 68), and let V denote the volume of the resulting box.

FIGURE 68

The dimensions of the open box are $(16 - 2x)$ by $(10 - 2x)$ by x inches.

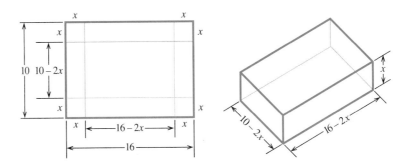

Step 2   The dimensions of the box are $(16 - 2x)$ inches long, $(10 - 2x)$ inches wide, and x inches high. Therefore, its volume,

$$V = (16 - 2x)(10 - 2x)x$$

$$= 4(x^3 - 13x^2 + 40x)   \text{Expanding the expression}$$

cubic inches, is the quantity to be maximized.

Step 3   Since each side of the box must be nonnegative, x must satisfy the inequalities $x \geq 0$, $16 - 2x \geq 0$, and $10 - 2x \geq 0$. This set of inequalities is satisfied if $0 \leq x \leq 5$. Thus the problem at hand is equivalent to that of finding the absolute maximum of

$$V = f(x) = 4(x^3 - 13x^2 + 40x)$$

on the closed interval $[0, 5]$.

Step 4   Observe that f is continuous on $[0, 5]$, so the absolute maximum value of f must be attained at the endpoint(s) or at the critical point(s) of f.

Differentiating $f(x)$, we obtain

$$f'(x) = 4(3x^2 - 26x + 40)$$

$$= 4(3x - 20)(x - 2)$$

Upon setting $f'(x) = 0$ and solving the resulting equation for x, we obtain $x = \frac{20}{3}$ or $x = 2$. Since $\frac{20}{3}$ lies outside the interval $[0, 5]$, it is no longer considered and we are interested only in the critical point $x = 2$ of f. Next, evaluating $f(x)$ at $x = 0$, $x = 5$ (the endpoints of the interval $[0, 5]$), and $x = 2$, we obtain

$$f(0) = 0, \qquad f(2) = 144, \qquad f(5) = 0$$

Thus the volume of the box is maximized by taking $x = 2$. The dimensions of the box are $12'' \times 6'' \times 2''$, and the volume is 144 cubic inches. ■

EXAMPLE 3   A city's Metropolitan Transit Authority (MTA) operates a subway line for commuters from a certain suburb to the downtown metropolitan area. Currently, an average of 6000 passengers a day take the trains, paying a fare of $1.50 per ride. The Board of the MTA, contemplating raising the fare to $1.75 per ride in order to generate a larger revenue, engages the services of a consulting firm. The firm's study reveals that for each 25-cent increase in fare, the ridership will be reduced by an average of 1000 passengers a day. Thus the consulting firm recommends that MTA stick to the current fare of $1.50 per ride, which already yields a maximum revenue. Show that the consultants are correct.

Solution

Step 1   Let x denote the number of passengers per day, p denote the fare per ride, and R be MTA's revenue.

Step 2   In order to find a relationship between x and p, observe that the given data imply that when $x = 6000$, $p = 1.5$, and when $x = 5000$, $p = 1.75$. Therefore, the points $(6000, 1.5)$ and $(5000, 1.75)$ lie on a straight line. (Why?) To find the linear relationship between p and x, use the point-slope form of the equation of a straight line. Now, the slope of the line is

$$m = \frac{1.75 - 1.50}{5000 - 6000} = -0.00025$$

Therefore, the required equation is

$$p - 1.5 = -0.00025(x - 6000)$$
$$= -0.00025x + 1.5$$
$$p = -0.00025x + 3$$

Therefore, the revenue,

$$R = f(x) = xp = -0.00025x^2 + 3x \qquad \text{Number of riders times unit fare}$$

is the quantity to be maximized.

Step 3   Since both p and x must be nonnegative, we see that $0 \le x \le 12{,}000$, and the problem is that of finding the absolute maximum of the function f on the closed interval $[0, 12{,}000]$.

Step 4   Observe that f is continuous on $[0, 12{,}000]$. To find the critical point of R, we compute

$$f'(x) = -0.0005x + 3$$

and set it equal to zero, giving $x = 6000$. Evaluating the function f at $x = 6000$, as well as at the endpoints $x = 0$ and $x = 12{,}000$, yields

$$f(0) = 0$$
$$f(6000) = 9000$$
$$f(12{,}000) = 0$$

f has an absolute maximum of 9000 when x = 6000.

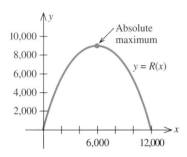

We conclude that a maximum revenue of $9000 per day is realized when the ridership is 6000 per day. The optimum price of the fare per ride is therefore $1.50, as recommended by the consultants. The graph of the revenue function R is shown in Figure 69.   ∎

An Inventory Problem

One problem faced by many companies is that of controlling the inventory of goods carried. Ideally, the manager must ensure that the company has sufficient stock to meet customer demand at all times. At the same time she must make sure that this is accomplished without overstocking (incurring unnecessary storage costs) and also without having to place orders too frequently (incurring reordering costs).

EXAMPLE 4   The Dixie Import-Export Company is the sole agent for the Excalibur 250-cc motorcycle. Management estimates that the demand for these motorcycles is 10,000 per year and that they will sell at a uniform rate throughout the year. The cost incurred in ordering each shipment of motorcycles is $10,000 and the cost per year of storing each motorcycle is $200.

Dixie's management faces the following problem: Ordering too many motorcycles at one time ties up valuable storage space and increases the storage cost. On the other hand, placing orders too frequently increases the ordering costs. How large should each order be, and how often should orders be placed, to minimize ordering and storage costs?

Solution   Let x denote the number of motorcycles in each order (the lot size). Then, assuming that each shipment arrives just as the previous shipment has been sold, the average number of motorcycles in storage during the year is $x/2$. You can see that this is the case by examining Figure 70. Thus, Dixie's storage cost for the year is given by $200(x/2)$, or $100x$ dollars.

As each lot is depleted, the new lot arrives. The average inventory level is x/2 if x is the lot size.

Next, since the company requires 10,000 motorcycles for the year and since each order is for x motorcycles, the number of orders required is

$$\frac{10{,}000}{x}$$

This gives an ordering cost of

$$10{,}000\left(\frac{10{,}000}{x}\right) = \frac{100{,}000{,}000}{x}$$

dollars for the year. Thus, the total yearly cost incurred by Dixie, which includes the ordering and storage costs attributed to the sale of these motorcycles, is given by

$$C(x) = 100x + \frac{100{,}000{,}000}{x}$$

The problem is reduced to finding the absolute minimum of the function C in the interval $(0, 10{,}000]$. To accomplish this, we compute

$$C'(x) = 100 - \frac{100{,}000{,}000}{x^2}$$

Setting $C'(x) = 0$ and solving the resulting equation, we obtain $x = \pm 1{,}000$. Since the number $-1{,}000$ is outside the domain of the function C, it is rejected, leaving $x = 1{,}000$ as the only critical point of C. Next, we find

$$C''(x) = \frac{200{,}000{,}000}{x^3}$$

Since $C''(1{,}000) > 0$, the Second Derivative Test implies that the critical point $x = 1{,}000$ is a relative minimum of the function C (Figure 71). Also, since $C''(x) > 0$ for all x in $(0, 10{,}000]$, the function C is concave upward everywhere so that the point $x = 1{,}000$ also gives the absolute minimum of C. Thus, in order to minimize the ordering and storage costs, Dixie should place $10{,}000/1{,}000$, or 10, orders a year, each for a shipment of 1,000 motorcycles.

FIGURE 71
C has an absolute minimum at (1,000, 200,000).

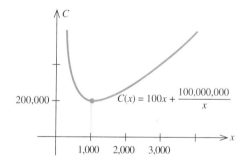

Solutions to Self-Check Exercises 10.5 can be found on page 656.

SELF-CHECK EXERCISES 10.5

1. A man wishes to have an enclosed vegetable garden in his backyard. If the garden is to be a rectangular area of 300 square feet, find the dimensions of the garden that will minimize the amount of fencing material needed.

2. The demand for the Super Titan tires is 1,000,000 per year. The set-up cost for each production run is $4000, and the manufacturing cost is $20 per tire. The cost of storing each tire over the year is $2. Assuming uniformity of demand throughout the year and instantaneous production, determine how many tires should be manufactured per production run in order to keep the production cost to a minimum.

10.5 EXERCISES

1. **Enclosing the Largest Area** The owner of the Rancho Los Feliz has 3000 yards of fencing material with which to enclose a rectangular piece of grazing land along the straight portion of a river. If fencing is not required along the river, what are the dimensions of the largest area he can enclose? What is this area?

2. **Enclosing the Largest Area** Refer to exercise 1. As an alternative plan, the owner of the Rancho Los Feliz might use the 3000 yards of fencing material to enclose the rectangular piece of grazing land along the straight portion of the river and then subdivide it by means of a fence running parallel to the sides. Again, no fencing is required along the river. What are the dimensions of the largest area that can be enclosed? What is this area? (See the accompanying figure.)

River

3. **Minimizing Construction Costs** The management of the UNICO department store has decided to enclose an 800-square-foot area outside the building for displaying potted plants and flowers. One side will be formed by the external wall of the store, two sides will be constructed of pine boards, and the fourth side will be made of galvanized steel fencing material (see the accompanying figure). If the pine board fencing costs $6 per running foot and the steel fencing costs $3 per running foot, determine the dimensions of the enclosure that can be erected at minimum cost.

Wood

Store   Steel

Wood

4. **Packaging** By cutting away identical squares from each corner of a rectangular piece of cardboard and folding up the resulting flaps, an open box may be made. If the cardboard is 15 inches long and 8 inches wide, find the dimensions of the box that will yield the maximum volume.

5. **Metal Fabrication** If an open box is made from a tin sheet 8 inches square by cutting out identical squares from each corner and bending up the resulting flaps, determine the dimensions of the largest box that can be made.

6. **Minimizing Construction Costs** If an open box has a square base and a volume of 108 cubic inches, and is constructed from a tin sheet, find the dimensions of such a box, assuming a minimum amount of material is used in its construction.

C 7. **Minimizing Construction Costs** What are the dimensions of a closed rectangular box that has a square cross section, a capacity of 128 cubic inches, and is constructed using the least amount of material?

8. **Minimizing Construction Costs** A rectangular box is to have a square base and a volume of 20 cubic feet (see the accompanying figure). If the material for the base costs 30 cents per square foot, the material for the sides costs 10 cents per square foot, and the material for the top costs 20 cents per square foot, determine the dimensions of the box that can be constructed at minimum cost.

C 9. **Parcel Post Regulations** Postal regulations specify that a parcel sent by parcel post may have a combined length and girth of no more than 108 inches. Find the dimensions of a rectangular package that has a square cross section and largest volume that may be sent through the mail. What is the volume of such a package? [*Hint:* The length plus the girth is $4x + h$ (see the accompanying figure).]

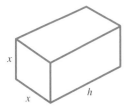

10. Book Design   A production editor at Saunders-Roe Publishing decided that the pages of a book should have one-inch margins at the top and bottom and half-inch margins on the sides. She further stipulated that each page should have an area of 50 square inches (see the accompanying figure). Determine the page dimensions that will result in the maximum printed area on the page.

c 11. Parcel Post Regulations   Postal regulations specify that a parcel sent by parcel post may have a combined length and girth of no more than 108 inches. Find the dimensions of the cylindrical package of greatest volume that may be sent through the mail (see the accompanying figure). What is the volume of such a package? Compare with exercise 9. [*Hint:* The length plus the girth is $2\pi r + l$.]

c 12. Product Design   The cabinet that will enclose the Acrosonic Model D loudspeaker system will be rectangular and will have an internal volume of 2.4 cubic feet. For aesthetic reasons, it has been decided that the height of the cabinet is to be 1.5 times its width. If the top, bottom, and sides of the cabinet are constructed of veneer costing 40 cents per square foot, and the front (ignore the cutouts in the baffle) and rear are constructed of particle board costing 20 cents

per square foot, what are the dimensions of the enclosure that can be constructed at a minimum cost?

13. Designing a Norman Window   A norman window has the shape of a rectangle surmounted by a semicircle (see the accompanying figure). If a norman window is to have a perimeter of 28 ft, what should its dimensions be in order to allow the maximum amount of light through the window?

14. Optimal Charter Flight Fare   If exactly 200 people sign up for a charter flight, the Leisure World Travel Agency charges $300 per person. However, if more than 200 people sign up for the flight (assume this is the case), then each fare is reduced by $1 for each additional person. Determine how many passengers will result in a maximum revenue for the travel agency. What is the maximum revenue? What would the fare per passenger be in this case? [*Hint:* Let x denote the number of passengers beyond 200. Show that the revenue function R is given by $R(x) = (200 + x)(300 - x)$.]

15. Maximizing Yield   An apple orchard has an average yield of 36 bushels of apples per tree if tree density is 22 trees per acre. For each unit increase in tree density, the yield decreases by 2 bushels. Find how many trees should be planted in order to maximize the yield.

16. Strength of a Beam   A wooden beam has a rectangular cross section of height h inches and width w inches (see the accompanying figure). The strength S of the beam is directly proportional to its width and the square of its height. What are the dimensions of the cross section of the strongest beam that can be cut from a round log of diameter 24 inches? [*Hint:* $S = kh^2w$, where k is a constant of proportionality.]

17. Minimizing Construction Costs   In the following diagram, S represents the position of a power relay station located on a straight coast and E shows the location of a marine biology experimental station on an island. A cable is to be laid connecting the relay station with the experimental station. If the cost of running the cable on land is $1 per running foot, and the cost of running the cable under water is $3 per running foot, locate the point P that will result in a minimum cost (solve for x).

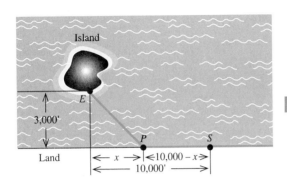

18. Flights of Birds   During daylight hours, some birds fly more slowly over water than over land because some of their energy is expended in overcoming the downdrafts of air over open bodies of water. Suppose a bird that flies at a constant speed of 4 mph over water and 6 mph over land starts its journey at the point E on an island and ends at its nest N on the shore of the mainland, as shown in the accompanying figure. Find the location of the point P that allows the bird to complete its journey in the minimum time (solve for x).

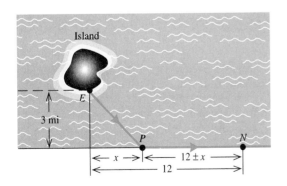

19. Optimal Speed of a Truck   A truck gets $400/x$ miles per gallon when driven at a constant speed of x miles per hour (between 50 and 70 miles per hour). If the price of fuel is $1 per gallon and the driver is paid $8 an hour, at what speed between 50 and 70 miles per hour is it most economical to drive?

20. Inventory Control and Planning   The demand for motorcycle tires imported by the Dixie Import-Export Company is 40,000 per year and may be assumed to be uniform throughout the year. The cost of ordering a shipment of tires is $400, and the cost of storing each tire for a year is $2. Determine how many tires should be in each shipment if the ordering and storage costs are to be minimized. (Assume that each shipment arrives just as the previous one has been sold.)

21. Inventory Control and Planning   The McDuff Preserves Company expects to bottle and sell 2,000,000 32-ounce jars of jam. The company orders its containers from the Consolidated Bottle Company. The cost of ordering a shipment of bottles is $200, and the cost of storing each empty bottle for a year is 40 cents. Determine how many orders McDuff should place per year and how many bottles should be in each shipment if the ordering and storage costs are to be minimized. (Assume that each shipment of bottles is used up before the next shipment arrives.)

22. Inventory Control and Planning   The Neilsen Cookie Company sells its assorted butter cookies in containers that have a net content of 1 pound. The estimated demand for the cookies is 1,000,000 units. The set-up cost for each production run is $500, and the manufacturing cost is 50 cents for each container of cookies. The cost of storing each container of cookies over the year is 40 cents. Assuming uniformity of demand throughout the year and instantaneous production, determine how many containers of cookies Neilsen should produce per production run in order to minimize the production cost. [*Hint:* Following the method of Example 4, show that the total production cost is given by the function

$$C(x) = \frac{500,000,000}{x} + 0.2x + 500,000$$

Then minimize the function C on the interval $(0, 1,000,000)$.]

SOLUTIONS TO SELF-CHECK EXERCISES 10.5

1. Let x and y (measured in feet) denote the length and width of the rectangular garden (see the accompanying figure). Since the area is to be 300 square feet, we have

$$xy = 300$$

Next, the amount of fencing to be used is given by the perimeter, and this quantity is to be minimized. Thus, we want to minimize

$$2x + 2y$$

or, since $y = \dfrac{300}{x}$ (obtained by solving for y in the first equation), we see that the expression to be minimized is

$$f(x) = 2x + 2\left(\frac{300}{x}\right)$$

$$= 2x + \frac{600}{x}$$

for positive values of x. Now,

$$f'(x) = 2 - \frac{600}{x^2}$$

Setting $f'(x) = 0$ yields $x = -\sqrt{300}$ or $x = \sqrt{300}$. We consider only the critical point $x = \sqrt{300}$, since $-\sqrt{300}$ lies outside the interval $(0, \infty)$. We then compute

$$f''(x) = \frac{1200}{x^3}$$

Since

$$f''(\sqrt{300}) > 0$$

the Second Derivative Test implies that a relative minimum of f occurs at $x = \sqrt{300}$. In fact, since $f''(x) > 0$ for all x in $(0, \infty)$, we conclude that $x = \sqrt{300}$ gives rise to the absolute minimum of f. The corresponding value of y, obtained by substituting this value of x into the equation $xy = 300$, is $y = \sqrt{300}$. Therefore, the required dimensions of the vegetable garden are approximately 17.3 ft $\times$ 17.3 ft.

2. Let x denote the number of tires in each production run. Then the average number of tires in storage is $x/2$, so the storage cost incurred by the company is $2(x/2)$, or x dollars. Next, since the company needs to manufacture 1,000,000 tires for the year in order to meet the demand, the number of production runs is $1,000,000/x$. This gives set-up costs amounting to

$$4,000\left(\frac{1,000,000}{x}\right) = \frac{4,000,000,000}{x}$$

dollars for the year. Thus, the total yearly cost incurred by the company is given by

$$C(x) = x + \frac{4,000,000,000}{x}$$

Differentiating $C(x)$, we find

$$C'(x) = 1 - \frac{4,000,000,000}{x^2}$$

Setting $C'(x) = 0$ gives $x = 63,246$ as the critical point in the interval $(0, 1,000,000)$. Next, we find

$$C''(x) = \frac{8,000,000,000}{x^3}$$

Since $C''(x) > 0$ for all $x > 0$, we see that C is concave upward for all $x > 0$. Furthermore, $C''(63,246) > 0$ implies that $x = 63,246$ gives rise to a relative minimum of C (by the Second Derivative Test). Since C is always concave upward for $x > 0$, $x = 63,246$ gives the absolute minimum of C. Therefore, the company should manufacture 63,246 tires in each production run.

CHAPTER TEN SUMMARY OF PRINCIPAL TERMS

Terms

Increasing function

Decreasing function

Relative maximum

Relative minimum

Relative extrema

Critical point

First Derivative Test

Concave upward

Concave downward

Inflection point

Second Derivative Test

Vertical asymptote

Horizontal asymptote

Absolute extrema

Absolute maximum value

Absolute minimum value

CHAPTER 10 REVIEW EXERCISES

In exercises 1–12, (a) find the intervals where the given function f is increasing and where it is decreasing, (b) find the relative extrema of f, (c) find the intervals where f is concave upward and where it is concave downward, and (d) find the inflection points, if any, of f.

1. $f(x) = \frac{1}{3}x^3 - x^2 + x - 6$

2. $f(x) = (x - 2)^3$

3. $f(x) = x^4 - 2x^2$

4. $f(x) = x + \frac{4}{x}$

5. $f(x) = \frac{x^2}{x - 1}$

6. $f(x) = \sqrt{x - 1}$

7. $f(x) = (1 - x)^{1/3}$

8. $f(x) = x\sqrt{x - 1}$

9. $f(x) = \frac{2x}{x + 1}$

10. $f(x) = \frac{-1}{1 + x^2}$

11. $f(x) = (4 - x)e^x$

12. $f(x) = x^2 \ln x$

In exercises 13–22, sketch the graph of the function using the curve-sketching guide of Section 10.3.

13. $f(x) = x^2 - 5x + 5$ **14.** $f(x) = -2x^2 - x + 1$

15. $g(x) = 2x^3 - 6x^2 + 6x + 1$

16. $g(x) = \frac{1}{3}x^3 - x^2 + x - 3$

17. $h(x) = x\sqrt{x - 2}$ **18.** $h(x) = \frac{2x}{1 + x^2}$

19. $f(x) = \frac{x - 2}{x + 2}$ **20.** $f(x) = x - \frac{1}{x}$

21. $f(x) = xe^{-2x}$ **22.** $f(x) = x^2 - \ln x$

In exercises 23–26, find the horizontal and vertical asymptotes of the graphs of the given functions. Do not sketch the graphs.

23. $f(x) = \frac{1}{2x + 3}$ **24.** $f(x) = \frac{2x}{x + 1}$

25. $f(x) = \frac{5x}{x^2 - 2x - 8}$ **26.** $f(x) = \frac{x^2 + x}{x(x - 1)}$

In exercises 27–37, find the absolute maximum value and the absolute minimum value, if any, of the given function.

27. $f(x) = 2x^2 + 3x - 2$ **28.** $g(t) = \sqrt{25 - t^2}$

29. $f(x) = \frac{1}{3}x^3 - x^2 + x + 1$ on $[0, 2]$

30. $h(t) = t^3 - 6t^2$ on $[2, 5]$

31. $g(x) = \frac{x}{x^2 + 1}$ on $[0, 5]$

32. $f(x) = x - \frac{1}{x}$ on $[1, 3]$

33. $h(t) = 8t - \frac{1}{t^2}$ on $[1, 3]$

34. $f(t) = te^{-t}$ on $[-2, 2]$

35. $g(t) = \frac{\ln t}{t}$ on $[1, 2]$

36. $f(s) = s\sqrt{1 - s^2}$ on $[-1, 1]$

37. $f(x) = \frac{x^2}{x - 1}$ on $[-1, 3]$

38. Odyssey Travel Agency's monthly profit (measured in thousands of dollars) depends on the amount of money x spent on advertising per month according to the rule

$$P(x) = -x^2 + 8x + 20$$

where x is also measured in thousands of dollars. What should Odyssey's monthly advertising budget be in order to maximize its monthly profits?

39. The Department of the Interior of an African country began to record an index of environmental quality to measure progress or decline in the environmental quality of its wildlife. The index for the years 1984 through 1994 is approximated by the function

$$I(t) = \frac{50t^2 + 600}{t^2 + 10} \qquad 0 \le t \le 10$$

a. Compute $I'(t)$ and show that $I(t)$ is decreasing on the interval $(0, 10)$.
b. Compute $I''(t)$. Study the concavity of the graph of I.
c. Sketch the graph of I.
d. Interpret your results.

40. The weekly demand for video discs manufactured by the Herald Record Company is given by

$$p = -0.0005x^2 + 60$$

where p denotes the unit price in dollars and x denotes the quantity demanded. The weekly total cost function associated with producing these discs is given by

$$C(x) = -0.001x^2 + 18x + 4000$$

where $C(x)$ denotes the total cost incurred in pressing x discs. Find the production level that will yield a maximum profit for the manufacturer. [*Hint:* Use the quadratic formula.]

41. The total monthly cost (in dollars) incurred by the Carlota Music Company in manufacturing x units of its Professional Series guitars is given by the function

$$C(x) = 0.001x^2 + 100x + 4000$$

a. Find the average cost function $\overline{C}$.
b. Determine the production level that will result in the smallest average production cost.

42. The average worker at Wakefield Avionics, Inc., can assemble

$$N(t) = -2t^3 + 12t^2 + 2t \qquad 0 \le t \le 4$$

ready-to-fly radio-controlled model airplanes t hours into the 8 A.M. to 12 noon morning shift. At what time during this shift is the average worker performing at peak efficiency?

43. You wish to construct a closed rectangular box that has a volume of 4 cubic feet. The length of the base of the box will be twice as long as its width. The material for the top and bottom of the box costs 30 cents per square foot. The material for the sides of the box costs 20 cents per square foot. Find the dimensions of the least expensive box that can be constructed.

44. The Lehen Vinters Company imports a certain brand of beer. The demand, which may be assumed to be uniform, is 800,000 cases per year. The cost of ordering a shipment of beer is $500, and the cost of storing each case of beer for a year is $2. Determine how many cases of beer should be in each shipment if the ordering and storage costs are to be kept to a minimum. (Assume that each shipment of beer arrives just as the previous one has been sold.)

CHAPTER 10   EXTENDED PROBLEMS AND PROJECTS

1. One condition that must be satisfied before Theorem 3 is applicable is that the function f must be continuous on the closed interval $[a, b]$. Define a function f on the closed interval $[-1, 1]$ by

$$f(x) = \begin{cases} \dfrac{1}{x} & \text{if } x \in [-1, 1] \quad x \neq 0 \\ 0 & \text{if } x = 0 \end{cases}$$

a. Show that f is not continuous at $x = 0$.
b. Show that $f(x)$ does not attain an absolute maximum or absolute minimum on the interval $[-1, 1]$.
c. Confirm your results by sketching the function f.

2. An investor purchased a piece of waterfront property. Because of the development of a marina in the vicinity, the market value of the property is expected to increase according to the rule

$$V(t) = 80,000 e^{\sqrt{t}/2}$$

where $V(t)$ is measured in dollars and t is the time in years from the present.

a. If the rate of inflation is expected to be 9% compounded continuously for the next eight years, find an expression for the present value $P(t)$ of the market price of the property valid for the next eight years.
b. What is the market value of the property expected to be in four years?
c. What is the optimal time (based on present value) for the investor to sell the property?
d. What is the optimal present value of the property?

Integration

*D*ifferential calculus is concerned with the problem of finding the rate of change of one quantity with respect to another. In this chapter we begin the study of the other branch of calculus known as integral calculus. Here we are interested in precisely the opposite problem: If we know the rate of change of one quantity with respect to another, can we find the relationship between these two quantities? The principal tool used in the study of integral calculus is the *antiderivative* of a function, and we develop rules for antidifferentiation, or *integration,* as the process of finding the antiderivative is called. We will also show that a link is established between differential and integral calculus—via the Fundamental Theorem of Calculus.

Antiderivatives and the Rules of Integration

Antiderivatives

In Chapter 9 we considered this problem:

Given the position of a car at any time t, can we find its velocity at time t?

As we saw, the velocity of the car is given by $f'(t)$, where f is the function giving the position of the car at any time t (Figure 1). The function f' is just the derivative of f.

FIGURE 1
The position of a car moving along a straight road at any time t is given by $s = f(t)$.

$f(0)$ $f(t)$ s

In Chapter 11, we consider precisely the opposite problem:

Given the velocity of a car at any time t, can we find its position at time t?

Put another way, if we know the car's velocity function f', can we find its position function f?

In order to solve this problem, we need the concept of an *antiderivative* of a function.

ANTIDERIVATIVE   A function F is an **antiderivative** of f on an interval I if $F'(x) = f(x)$ for all x in I.

Thus, an antiderivative of a function f is a function F whose derivative is f. For example, $F(x) = x^2$ is an antiderivative of $f(x) = 2x$ because

$$F'(x) = \frac{d}{dx}(x^2) = 2x = f(x)$$

and $F(x) = x^3 + 2x + 1$ is an antiderivative of $f(x) = 3x^2 + 2$ because

$$F'(x) = \frac{d}{dx}(x^3 + 2x + 1) = 3x^2 + 2 = f(x)$$

EXAMPLE 1   Let $F(x) = \frac{1}{3}x^3 - 2x^2 + x - 1$. Show that F is an antiderivative of $f(x) = x^2 - 4x + 1$.

Solution   Differentiating the function F, we obtain

$$F'(x) = x^2 - 4x + 1 = f(x)$$

and the desired result follows.

EXAMPLE 2   Let $F(x) = x$, $G(x) = x + 2$, and $H(x) = x + C$, where C is a constant. Show that F, G, and H are all antiderivatives of the function f defined by $f(x) = 1$.

Solution Since

$$F'(x) = \frac{d}{dx}(x) = 1 = f(x)$$

$$G'(x) = \frac{d}{dx}(x + 2) = 1 = f(x)$$

and $$H'(x) = \frac{d}{dx}(x + C) = 1 = f(x)$$

we see that F, G, and H are indeed antiderivatives of f.

Example 2 shows that once an antiderivative G of a function f is known, then another antiderivative of f may be found by adding an arbitrary constant to the function G. The following theorem states that no function other than one obtained in this manner can be an antiderivative of f. (We omit the proof.)

THEOREM 1 Let G be an antiderivative of a function f. Then every antiderivative F of f must be of the form $F(x) = G(x) + C$, where C is a constant.

Returning to Example 2, we see that there are infinitely many antiderivatives of the function $f(x) = 1$. We obtain each one by specifying the constant C in the function $F(x) = x + C$. Figure 2 shows the graphs of some of these antiderivatives for selected values of C. These graphs constitute part of a family of infinitely many parallel straight lines, each having a slope equal to 1. This result is expected, since there are infinitely many curves (straight lines) with a given slope equal to 1. The antiderivatives $F(x) = x + C$ (C, a constant) are precisely the functions representing this family of straight lines.

FIGURE 2
The graphs of some antiderivatives of $f(x) = 1$

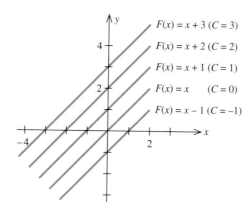

$F(x) = x + 3 \ (C = 3)$
$F(x) = x + 2 \ (C = 2)$
$F(x) = x + 1 \ (C = 1)$
$F(x) = x \quad (C = 0)$
$F(x) = x - 1 \ (C = -1)$

EXAMPLE 3 Prove that the function $G(x) = x^2$ is an antiderivative of the function $f(x) = 2x$. Write a general expression for the antiderivatives of f.

Solution Since $G'(x) = 2x = f(x)$, we have shown that $G(x) = x^2$ is an antiderivative of $f(x) = 2x$. By Theorem 1, every antiderivative of the

FIGURE 3

The graphs of some anti-derivatives of $f(x) = 2x$

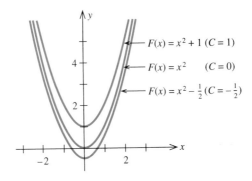

function $f(x) = 2x$ has the form $F(x) = x^2 + C$, where C is some constant. The graphs of a few of the antiderivatives of f are shown in Figure 3.

Application

The next example involves a familiar interpretation of the antiderivative.

EXAMPLE 4   The velocity of a car (in ft/sec) t seconds after starting from rest and moving in a straight line is given by

$$f(t) = 2t \qquad 0 \le t \le 20$$

Find the position of the car at any time t.

Solution   Let $s = F(t)$ be the function that describes the position of the car at any time t. Then the velocity of the car at any time t is given by

$$v = \frac{ds}{dt} = F'(t)$$

But the car's velocity at time t is given by $f(t) = 2t$, so we have

$$F'(t) = f(t)$$

and this problem is equivalent to finding the antiderivative F of the function $f(t) = 2t$. Using the result of Example 3, we have

$$s = F(t) = t^2 + C \qquad C, \text{ a constant}$$

Next, observe that when $t = 0$, the position $s = 0$, so that we have

$$0 = 0^2 + C$$

or $C = 0$. Thus the required expression for the position of the car at any time t is given by

$$s = F(t) = t^2 \qquad \blacksquare$$

Notice that the "initial condition" helped us select the required function from the infinitely many antiderivatives of the function f. This situation is typical of problems involving antidifferentiation.

The Indefinite Integral

The process of finding all antiderivatives of a function is called **antidifferentiation,** or **integration.** We use the symbol $\int$, called an *integral sign,* to indicate that the operation of integration is to be performed on some function f. Thus,

$$\int f(x)\,dx = F(x) + C$$

(read "the indefinite integral of $f(x)$ with respect to x equals $F(x)$ plus C") tells us that the **indefinite integral** of f is the family of functions given by $F(x) + C$, where $F'(x) = f(x)$. The function f to be integrated is called the **integrand** and the constant C is called a **constant of integration.** The expression dx following the integrand $f(x)$ reminds us that the operation is performed with respect to x. If the independent variable is t, we write $\int f(t)\,dt$ instead. In this sense both t and x are "dummy variables."

Using this notation, the results of Examples 2 and 3 may be written as

$$\int 1\,dx = x + C \qquad \text{and} \qquad \int 2x\,dx = x^2 + K$$

where C and K are arbitrary constants.

Basic Integration Rules

Our next task is to develop some rules for finding the indefinite integral of a given function f. Because integration and differentiation are reverse operations, we discover many of the rules of integration by first making an "educated guess" at the antiderivative F of the function f to be integrated. Then this result is verified by demonstrating that $F' = f$.

RULE 1 INDEFINITE INTEGRAL OF A CONSTANT	$\int k\,dx = kx + C \qquad k, \text{a constant}$

To prove this result, observe that

$$F'(x) = \frac{d}{dx}(kx + C) = k$$

EXAMPLE 5 Evaluate each of the following indefinite integrals.

a. $\displaystyle\int 2\,dx$ **b.** $\displaystyle\int \pi^2\,dx$

Solution Each of the integrands has the form $f(x) = k$, where k is a constant. Applying Rule 1 in each case yields

a. $\displaystyle\int 2\,dx = 2x + C$ **b.** $\displaystyle\int \pi^2\,dx = \pi^2 x + C$ ■

Next, from the rule of differentiation,

$$\frac{d}{dx}x^n = nx^{n-1}$$

we obtain the following rule of integration.

RULE 2   THE POWER RULE

$$\int x^n \, dx = \frac{1}{n+1} x^{n+1} + C \qquad n \neq -1$$

An antiderivative of a power function is another power function obtained from the integrand by increasing its power by 1 and dividing the resulting expression by the new power.

To prove this result, observe that

$$F'(x) = \frac{d}{dx}\left[\frac{1}{n+1} x^{n+1} + C \right]$$

$$= \frac{n+1}{n+1} x^n$$

$$= x^n$$

$$= f(x)$$

EXAMPLE 6   Evaluate each of the following indefinite integrals:

a. $\displaystyle\int x^3 \, dx$ **b.** $\displaystyle\int x^{3/2} \, dx$ **c.** $\displaystyle\int \frac{1}{x^{3/2}} \, dx$

Solution   Each integrand is a power function with exponent $n \neq -1$. Applying Rule 2 in each case yields the following results:

a. $\displaystyle\int x^3 \, dx = \frac{1}{4} x^4 + C$

b. $\displaystyle\int x^{3/2} \, dx = \frac{1}{5/2} x^{5/2} + C = \frac{2}{5} x^{5/2} + C$

c. $\displaystyle\int \frac{1}{x^{3/2}} \, dx = \int x^{-3/2} \, dx = \frac{1}{-1/2} x^{-1/2} + C = -2x^{-1/2} + C$

These results may be verified by differentiating each of the antiderivatives and showing that the result is equal to the corresponding integrand.   ■

The next rule tells us that a constant factor may be moved through an integral sign.

RULE 3   INDEFINITE INTEGRAL OF A CONSTANT MULTIPLE OF A FUNCTION

$$\int cf(x) \, dx = c \int f(x) \, dx \qquad c, \text{a constant}$$

The indefinite integral of a constant multiple of a function is equal to the constant multiple of the indefinite integral of the function.

This result follows from the corresponding rule of differentiation (see Rule 3, Section 9.4).

Only a constant can be "moved out" of an integral sign. For example, it would be incorrect to write

$$\int x^2 \, dx = x^2 \int 1 \, dx$$

In fact, $\int x^2 \, dx = \frac{1}{3}x^3 + C$, whereas $x^2 \int 1 \, dx = x^2(x + C) = x^3 + Cx^2$.

EXAMPLE 7

a. $\displaystyle\int 2t^3 \, dt$ **b.** $\displaystyle\int -3x^{-2} \, dx$

Solution Each integrand has the form $cf(x)$, where c is a constant. Applying Rule 3, we obtain

a. $\displaystyle\int 2t^3 \, dt = 2\int t^3 \, dt = 2\left[\frac{1}{4}t^4 + K\right] = \frac{1}{2}t^4 + 2K = \frac{1}{2}t^4 + C$

where $C = 2K$. From now on we will write the constant of integration as C, since any nonzero multiple of an arbitrary constant is an arbitrary constant.

b. $\displaystyle\int -3x^{-2} \, dx = -3\int x^{-2} \, dx = (-3)(-1)x^{-1} + C = \frac{3}{x} + C$ ∎

RULE 4
THE SUM RULE

$$\int [f(x) + g(x)] \, dx = \int f(x) \, dx + \int g(x) \, dx$$

and

$$\int [f(x) - g(x)] \, dx = \int f(x) \, dx - \int g(x) \, dx$$

The indefinite integral of a sum (difference) of two integrable functions is equal to the sum (difference) of their indefinite integrals.

This result is easily extended to the case involving the sum and difference of any finite number of functions. As in Rule 3, the proof of Rule 4 follows from the corresponding rule of differentiation (see Rule 4, Section 9.4).

EXAMPLE 8 Evaluate the indefinite integral

$$\int (3x^5 + 4x^{3/2} - 2x^{-1/2}) \, dx$$

Solution Applying the extended version of Rule 4, we find that

$$\int (3x^5 + 4x^{3/2} - 2x^{-1/2}) \, dx$$

$$= \int 3x^5 \, dx + \int 4x^{3/2} \, dx - \int 2x^{-1/2} \, dx$$

$$= 3\int x^5 \, dx + 4\int x^{3/2} \, dx - 2\int x^{-1/2} \, dx \qquad \text{Rule 3}$$

$$= (3)\left(\frac{1}{6}\right)x^6 + (4)\left(\frac{2}{5}\right)x^{5/2} - (2)(2)x^{1/2} + C \qquad \text{Rule 2}$$

$$= \frac{1}{2}x^6 + \frac{8}{5}x^{5/2} - 4x^{1/2} + C$$

∎

Observe that we have combined the three constants of integration, which arise from evaluating the three indefinite integrals, to obtain one constant C. After all, the sum of three arbitrary constants is also an arbitrary constant.

**RULE 5
THE INDEFINITE
INTEGRAL OF THE
EXPONENTIAL
FUNCTION**

$$\int e^x \, dx = e^x + C$$

The indefinite integral of the exponential function with base e is equal to the function itself (except, of course, for the constant of integration).

EXAMPLE 9   Evaluate the indefinite integral

$$\int (2e^x - x^3) \, dx$$

Solution   We have

$$\int (2e^x - x^3) \, dx = \int 2e^x \, dx - \int x^3 \, dx$$

$$= 2 \int e^x \, dx - \int x^3 \, dx$$

$$= 2e^x - \frac{1}{4}x^4 + C$$

The last rule of integration in this section covers the integration of the function $f(x) = x^{-1}$. Remember that this function constituted the only exceptional case in the integration of the power function $f(x) = x^n$ (see Rule 2).

**RULE 6
THE INDEFINITE
INTEGRAL OF
THE FUNCTION
$f(x) = x^{-1}$**

$$\int x^{-1} \, dx = \int \frac{1}{x} \, dx = \ln |x| + C \qquad x \neq 0$$

To prove Rule 6, observe that

$$\frac{d}{dx} \ln |x| = \frac{1}{x} \qquad \text{See Rule 10, Section 9.7.}$$

EXAMPLE 10   Evaluate the indefinite integral

$$\int \left(2x + \frac{3}{x} + \frac{4}{x^2} \right) dx$$

Solution

$$\int \left(2x + \frac{3}{x} + \frac{4}{x^2}\right) dx = \int 2x \, dx + \int \frac{3}{x} \, dx + \int \frac{4}{x^2} \, dx$$

$$= 2\int x \, dx + 3\int \frac{1}{x} \, dx + 4\int x^{-2} \, dx$$

$$= 2\left(\frac{1}{2}\right)x^2 + 3\ln|x| + 4(-1)x^{-1} + C$$

$$= x^2 + 3\ln|x| - \frac{4}{x} + C$$

In many applications, we are given the derivative f' of a function f and we are required to find f. In such instances, a condition on f must be specified. This is illustrated in Examples 11–13.

EXAMPLE 11 Find the function f if it is known that

$$f'(x) = 3x^2 - 4x + 8 \quad \text{and} \quad f(1) = 9$$

Solution Integrating the function f', we find

$$f(x) = \int f'(x) \, dx$$

$$= \int (3x^2 - 4x + 8) \, dx$$

$$= x^3 - 2x^2 + 8x + C$$

Using the condition $f(1) = 9$, we have

$$9 = f(1) = 1^3 - 2(1)^2 + 8(1) + C = 7 + C$$

or $\quad\quad C = 2$

Therefore, the required function f is given by $f(x) = x^3 - 2x^2 + 8x + 2$.

Additional Applications

EXAMPLE 12 The management of the Staedtler Office Equipment Company determined that the daily marginal revenue function associated with producing and selling their battery-operated pencil sharpeners is given by

$$R'(x) = -0.0006x + 6$$

where x denotes the number of units produced and sold and $R'(x)$ is measured in dollars per unit.

a. Determine the revenue function $R(x)$ associated with producing and selling these pencil sharpeners.

b. What is the demand equation that relates the wholesale unit price of the pencil sharpeners to the quantity demanded?

Solution

a. The revenue function R is found by integrating the marginal revenue function $R'(x)$. Thus,

$$R(x) = \int R'(x)\, dx$$

$$= \int (-0.0006x + 6)\, dx$$

$$= -0.0003x^2 + 6x + C$$

where C is a constant. To determine the value of C, observe that the total revenue realized by Staedtler is nil when the production and sales level is zero; that is, $R(0) = 0$. This condition implies that

$$R(0) = -0.0003(0)^2 + 6(0) + C = 0$$

or $C = 0$. Thus, the required revenue function is given by

$$R(x) = -0.0003x^2 + 6x$$

b. Let p denote the unit wholesale price of the pencil sharpeners. Then

$$R(x) = px$$

Thus, $$p = \frac{R(x)}{x} = -0.0003x + 6$$

so the required demand equation is

$$p = -0.0003x + 6$$

EXAMPLE 13   The current circulation of *Investor's Digest* is 2000 copies per week. Circulation is expected to grow at the rate of

$$5 + 2t^{2/3}$$

copies per week, t weeks from now, for the next three years. Determine what the digest's circulation will be 125 weeks from now.

Solution   Let $S(t)$ denote the digest's circulation t weeks from now. Then $S'(t)$, the rate of change in the circulation per week, is given by

$$S'(t) = 5 + 2t^{2/3}$$

Therefore, $$S(t) = \int (5 + 2t^{2/3})\, dt$$

$$= 5t + (2)\left(\frac{3}{5}\right)t^{5/3} + C$$

$$= 5t + \frac{6}{5}t^{5/3} + C$$

To determine the value of C, observe that the circulation at $t = 0$ is 2000. Using this condition, we have

$$S(0) = 2000 = 5(0) + \frac{6}{5}(0) + C$$

and $C = 2000$, so

$$S(t) = 5t + \frac{6}{5}t^{5/3} + 2000$$

Therefore, 125 weeks from now the circulation will be

$$S(125) = 5(125) + \frac{6}{5}(125)^{5/3} + 2000 = 6375$$

or 6375 copies per week.   ■

**SELF-CHECK
EXERCISES 11.1**

1. Evaluate $\int \left(\frac{1}{\sqrt{x}} - \frac{2}{x} + 3e^x \right) dx$.

2. Find the rule for the function f given that (1) the slope of the tangent line to the graph of f at any point $P(x, f(x))$ is given by the expression $3x^2 - 6x + 3$ and (2) the graph of f passes through the point $(2, 9)$.

c **3.** Suppose that United Motors' share of the new cars sold in a certain country is changing at the rate of

$$f(t) = -0.01875t^2 + 0.15t - 1.2 \qquad 0 \le t \le 12$$

percent at year t ($t = 0$ corresponds to the beginning of 1984). The company's market share at the beginning of 1984 was 48.4%. What was United Motors' market share at the beginning of 1996?

Solutions to Self-Check Exercises 11.1 can be found on page 674.

11.1 EXERCISES

In exercises 1–4, verify directly that F is an antiderivative of f.

1. $F(x) = \frac{1}{3}x^3 + 2x^2 - x + 2; \ f(x) = x^2 + 4x - 1$

2. $F(x) = xe^x + \pi; \ f(x) = e^x(1 + x)$

3. $F(x) = \sqrt{2x^2 - 1}; \ f(x) = \dfrac{2x}{\sqrt{2x^2 - 1}}$

4. $F(x) = x \ln x - x; \ f(x) = \ln x$

In exercises 5–8, (a) verify that G is an antiderivative of f, (b) find all antiderivatives of f, and (c) sketch the graphs of a few of the family of antiderivatives found in (b).

5. $G(x) = 2x; \ f(x) = 2$

6. $G(x) = 2x^2; \ f(x) = 4x$

7. $G(x) = \frac{1}{3}x^3; \ f(x) = x^2$

8. $G(x) = e^x; \ f(x) = e^x$

In exercises 9–50, find the indefinite integral.

9. $\int 6\,dx$

10. $\int \sqrt{2}\,dx$

11. $\int x^3\,dx$

12. $\int 2x^5\,dx$

13. $\int x^{-4}\,dx$

14. $\int 3t^{-7}\,dt$

15. $\int x^{2/3}\,dx$

16. $\int 2u^{3/4}\,du$

17. $\int x^{-5/4}\,dx$

18. $\int 3x^{-2/3}\,dx$

19. $\int \frac{2}{x^2}\,dx$

20. $\int \frac{1}{3x^5}\,dx$

21. $\int \pi \sqrt{t}\, dt$

22. $\int \dfrac{3}{\sqrt{t}}\, dt$

23. $\int (3 - 2x)\, dx$

24. $\int (1 + u + u^2)\, du$

25. $\int (x^2 + x + x^{-3})\, dx$

26. $\int (0.3t^2 + 0.02t + 2)\, dt$

27. $\int 4e^x\, dx$

28. $\int (1 + e^x)\, dx$

29. $\int (1 + x + e^x)\, dx$

30. $\int (2 + x + 2x^2 + e^x)\, dx$

31. $\int \left(4x^3 - \dfrac{2}{x^2} - 1\right) dx$

32. $\int \left(6x^3 + \dfrac{3}{x^2} - x\right) dx$

33. $\int (x^{5/2} + 2x^{3/2} - x)\, dx$

34. $\int (t^{3/2} + 2t^{1/2} - 4t^{-1/2})\, dt$

35. $\int \left(\sqrt{x} + \dfrac{3}{\sqrt{x}}\right) dx$

36. $\int \left(\sqrt[3]{x^2} - \dfrac{1}{x^2}\right) dx$

37. $\int \left(\dfrac{u^3 + 2u^2 - u}{3u}\right) du$

$$\left[\text{Hint:}\, \dfrac{u^3 + 2u^2 - u}{3u} = \dfrac{1}{3}u^2 + \dfrac{2}{3}u - \dfrac{1}{3}\right]$$

38. $\int \dfrac{x^4 - 1}{x^2}\, dx$ $\left[\text{Hint:}\, \dfrac{x^4 - 1}{x^2} = x^2 - x^{-2}\right]$

39. $\int (2t + 1)(t - 2)\, dt$

40. $\int u^{-2}(1 - u^2 + u^4)\, du$

41. $\int \dfrac{1}{x^2}(x^4 - 2x^2 + 1)\, dx$

42. $\int \sqrt{t}(t^2 + t - 1)\, dt$

43. $\int \dfrac{ds}{(s + 1)^{-2}}$

44. $\int \left(\sqrt{x} + \dfrac{3}{x} - 2e^x\right) dx$

45. $\int (e^t + t^e)\, dt$

46. $\int \left(\dfrac{1}{x^2} - \dfrac{1}{\sqrt[3]{x^2}} + \dfrac{1}{\sqrt{x}}\right) dx$

47. $\int \left(\dfrac{x^3 + x^2 - x + 1}{x^2}\right) dx$ [*Hint:* Simplify the integrand first.]

48. $\int \dfrac{t^3 + \sqrt[3]{t}}{t^2}\, dt$ [*Hint:* Simplify the integrand first.]

49. $\int \dfrac{(\sqrt{x} - 1)^2}{x^2}\, dx$ [*Hint:* Simplify the integrand first.]

50. $\int (x + 1)^2 \left(1 - \dfrac{1}{x}\right) dx$ [*Hint:* Simplify the integrand first.]

From the given information in exercises 51–54, find the function f.

51. $f'(x) = 3x^2 + 4x - 1;\ f(2) = 9$

52. $f'(x) = e^x - 2x;\ f(0) = 2$

53. $f'(x) = \dfrac{x + 1}{x};\ f(1) = 1$

54. $f'(x) = 1 + e^x + \dfrac{1}{x};\ f(1) = 3 + e$

In exercises 55–58, find the function f given that the slope of the tangent line to the graph of f at any point $(x, f(x))$ is $f'(x)$ and that the graph of f passes through the given point.

55. $f'(x) = \tfrac{1}{2}x^{-1/2};\ (2, \sqrt{2})$

56. $f'(t) = t^2 - 2t + 3;\ (1, 2)$

57. $f'(x) = e^x + x;\ (0, 3)$

58. $f'(x) = \dfrac{2}{x} + 1;\ (1, 2)$

59. Velocity of a Car The velocity of a car (in ft/sec) t seconds after starting from rest is given by the function

$$f(t) = 2\sqrt{t} \qquad 0 \le t \le 30$$

Find the car's position at any time t.

60. Cost of Producing Clocks The Lorimar Watch Company manufactures travel clocks. The daily marginal cost function associated with producing these clocks is

$$C'(x) = 0.000009x^2 - 0.009x + 8$$

where $C'(x)$ is measured in dollars per unit and x denotes the number of units produced. Management has determined that the daily fixed cost incurred in producing these clocks is $120. Find the total cost incurred by Lorimar in producing the first 500 travel clocks per day.

61. Revenue Functions The management of the Lorimar Watch Company has determined that the daily marginal revenue function associated with producing and selling their travel clocks is given by

$$R'(x) = -0.009x + 12$$

where x denotes the number of units produced and sold and $R'(x)$ is measured in dollars per unit.
a. Determine the revenue function $R(x)$ associated with producing and selling these clocks.
b. What is the demand equation that relates the wholesale unit price with the quantity of travel clocks demanded?

62. Profit Functions The Cannon Precision Instruments Corporation makes an automatic electronic flash with Thyrister circuitry. The estimated marginal profit associated with producing and selling these electronic flashes is

$$(-0.004x + 20)$$

dollars per unit per month when the production level is x units per month. Cannon's fixed cost for producing and selling these electronic flashes is $16,000 per month. At what level of production does Cannon realize a maximum profit? What is the maximum monthly profit?

63. Cost of Producing Guitars The Carlota Music Company estimates that the marginal cost of manufacturing its Professional Series guitars is

$$C'(x) = 0.002x + 100$$

dollars per month when the level of production is x guitars per month. The fixed costs incurred by Carlota are $4000 per month. Find the total monthly cost incurred by Carlota in manufacturing x guitars per month.

64. Quality Control As part of a quality-control program, the chess sets manufactured by the Jones Brothers Company are subjected to a final inspection before packing. The rate of increase in the number of sets checked per hour by an inspector t hours into the 8 A.M. to 12 noon morning shift is approximately

$$N'(t) = -3t^2 + 12t + 45 \qquad 0 \le t \le 4$$

Find an expression $N(t)$ that approximates the number of sets inspected at the end of t hours. [*Hint:* $N(0) = 0$.] How many sets does the average inspector check during a morning shift?

65. Ballast Dropped from a Balloon Ballast is dropped from a stationary hot-air balloon that is hovering at an altitude of 400 ft. Its velocity after t seconds is $-32t$ ft/sec.
a. Find the height $h(t)$ of the ballast from the ground at time t. [*Hint:* $h'(t) = -32t$ and $h(0) = 400$.]
b. When will the ballast strike the ground?

c. Find the velocity of the ballast when it hits the ground.

Ballast

66. Cable TV Subscribers A study conducted by Tele-Cable, Inc., estimates that the number of cable television subscribers will grow at the rate of

$$100 + 210t^{3/4}$$

new subscribers per month t months from the start date of the service. If 5000 subscribers signed up for the service before the starting date, how many subscribers will there be 16 months from that date?

67. Air Pollution On an average summer day, the level of carbon monoxide in a city's air is 2 parts per million. An environmental protection agency's study predicts that, unless more stringent measures are taken to protect the city's atmosphere, the concentration of carbon monoxide present in the air will increase at the rate of

$$0.003t^2 + 0.06t + 0.1$$

parts per million per year t years from now. If no further pollution-control efforts are made, what will the concentration of carbon monoxide be on an average summer day five years from now?

68. Population Growth The development of Astro World ("The Amusement Park of the Future") on the outskirts of a city will increase the city's population at the rate of

$$4500\sqrt{t} + 1000$$

people per year t years from the start of construction. The population before construction is 30,000. Determine the projected population nine years after construction of the park has begun.

69. Ozone Pollution The rate of change of the level of ozone, an invisible gas that is an irritant and impairs breathing, present in the atmosphere on a certain

May day in the city of Riverside is given by

$$R(t) = 3.2922t^2 - 0.366t^3 \qquad 0 \le t < 11$$

(measured in pollutant standard index per hour). Here t is measured in hours, with $t = 0$ corresponding to 7 A.M. Find the ozone level $A(t)$ at any time t assuming that at 7 A.M. it is zero. [*Hint:* $A'(t) = R(t)$ and $A(0) = 0$.]

c **70. Surface Area of a Human**   Empirical data suggest that the surface area of a 180-cm-tall human body changes at the rate of

$$S'(W) = 0.131773W^{-0.575}$$

square meters per kilogram, where W is the weight of the body in kilograms. If the surface area of a 180-cm-tall human body weighing 70 kg is 1.886277 square meters, what is the surface area of a human body of the same height weighing 75 kg?

c **71. Flight of a Rocket**   The velocity, in feet per second, of a rocket t seconds into vertical flight is given by

$$v(t) = -3t^2 + 192t + 120$$

Find an expression $h(t)$ that gives the rocket's altitude, in feet, t seconds after lift-off. What is the alti-

tude of the rocket 30 seconds after lift-off? [*Hint:* $h'(t) = v(t); h(0) = 0$.]

72. Flow of Blood in an Artery   Nineteenth-century physician Jean Louis Marie Poiseuille discovered that the rate of change of the velocity of blood r cm from the central axis of an artery (in cm/sec/cm) is given by

$$a(r) = -kr$$

where k is a constant. If the radius of an artery is R cm (see the accompanying diagram), find an expression for the velocity of blood as a function of r. [*Hint:* $v'(r) = a(r)$ and $v(R) = 0$. (Why?)]

Blood vessel

SOLUTIONS TO SELF-CHECK EXERCISES 11.1

1. $\displaystyle \int \left(\frac{1}{\sqrt{x}} - \frac{2}{x} + 3e^x \right) dx = \int \left(x^{-1/2} - \frac{2}{x} + 3e^x \right) dx$

$$= \int x^{-1/2}\, dx - 2 \int \frac{1}{x}\, dx + 3 \int e^x\, dx$$

$$= 2x^{1/2} - 2 \ln |x| + 3e^x + C$$

$$= 2\sqrt{x} - 2 \ln |x| + 3e^x + C$$

2. The slope of the tangent line to the graph of the function f at any point $P(x, f(x))$ is given by the derivative f' of f. Thus, the first condition implies that

$$f'(x) = 3x^2 - 6x + 3$$

which, upon integration, yields

$$f(x) = \int (3x^2 - 6x + 3)\, dx$$

$$= x^3 - 3x^2 + 3x + k$$

where k is the constant of integration.

To evaluate k, we use condition (2), which implies that $f(2) = 9$, or

$$9 = f(2) = 2^3 - 3(2)^2 + 3(2) + k$$

or $k = 7$. Hence, the required rule of definition of the function f is

$$f(x) = x^3 - 3x^2 + 3x + 7$$

3. Let $M(t)$ denote United Motors' market share at year t. Then

$$M(t) = \int f(t)\, dt$$

$$= \int (-0.01875t^2 + 0.15t - 1.2)\, dt$$

$$= -0.00625t^3 + 0.075t^2 - 1.2t + C$$

To determine the value of C, we use the condition $M(0) = 48.4$, obtaining $C = 48.4$. Therefore,

$$M(t) = -0.00625t^3 + 0.075t^2 - 1.2t + 48.4$$

In particular, United Motors' market share of new cars at the beginning of 1996 is given by

$$M(12) = -0.00625(12)^3 + 0.075(12)^2$$

$$-1.2(12) + 48.4 = 34$$

or 34%.

11.2 Integration by Substitution

In Section 11.1 we developed certain rules of integration that are closely related to the corresponding rules of differentiation in Chapter 9. In this section, we introduce a method of integration called the *method of substitution,* which is related to the Chain Rule for differentiating functions. When used in conjunction with the rules of integration developed earlier, the method of substitution is a powerful tool for integrating a large class of functions.

How the Method of Substitution Works

Consider the indefinite integral

$$\int 2(2x + 4)^5\, dx \qquad\qquad (1)$$

One way of evaluating this integral is to expand the expression $(2x + 4)^5$ and then integrate the resulting integrand term by term. As an alternative approach, let us see if we can simplify the integral by making a change of variable. Write

$$u = 2x + 4$$

with differential*

$$du = 2\, dx$$

If we substitute these quantities into equation (1), we obtain

$$\int 2(2x + 4)^5\, dx = \int (2x + 4)^5 (2\, dx) = \int u^5\, du$$

$$\uparrow \text{Rewriting} \qquad\qquad \begin{cases} u = 2x + 4 \\ du = 2\, dx \end{cases}$$

---

*If $u = f(x)$, the *differential* of u, written du, is $du = f'(x)\, dx$.

Now, the last integral involves a power function and is easily evaluated using Rule 2 of Section 11.1. Thus,

$$\int u^5 \, du = \frac{1}{6} u^6 + C$$

Therefore, using this result and replacing u by $2x + 4$, we obtain

$$\int 2(2x + 4)^5 \, dx = \frac{1}{6}(2x + 4)^6 + C$$

We can verify that the foregoing result is indeed correct by computing

$$\frac{d}{dx}\left[\frac{1}{6}\left(2x + 4\right)^6\right] = \frac{1}{6} \cdot 6(2x + 4)^5(2) \qquad \text{Using the Chain Rule}$$

$$= 2(2x + 4)^5$$

and observing that the last expression is just the integrand of (1).

The Method of Integration by Substitution

In order to see why the approach used in evaluating the integral in equation (1) is successful, write

$$f(x) = x^5 \qquad \text{and} \qquad g(x) = 2x + 4$$

Then $g'(x) = 2 \, dx$. Furthermore, the integrand of (1) is just the composition of f and g. Thus,

$$(f \circ g)(x) = f(g(x))$$

$$= [g(x)]^5 = (2x + 4)^5$$

Therefore, (1) can be written as

$$\int f(g(x))g'(x) \, dx \qquad\qquad (2)$$

Next, let us show that an integral having the form (2) can always be written as

$$\int f(u) \, du \qquad\qquad (3)$$

Suppose F is an antiderivative of f. By the Chain Rule, we have

$$\frac{d}{dx}[F(g(x))] = F'(g(x))g'(x)$$

and therefore,

$$\int F'(g(x))g'(x) \, dx = F(g(x)) + C$$

Letting $F' = f$ and making the substitution $u = g(x)$, we have

$$\int f(g(x))g'(x) \, dx = F(u) + C = \int F'(u) \, du = \int f(u) \, du$$

as we wished to show. Thus, if the transformed integral is readily evaluated, as is the case with the integral (1), then the method of substitution will prove successful.

John L. Higgins

As the concierge for a small, ten-story office building in Boston, Higgins supervises daily operations. From his post in the lobby, he resolves tenant problems, oversees building maintenance and security, and coordinates the budget. Part building manager, lobby security guard, and friend, Higgins brings an upbeat quality to his job, joking with tenants as they enter and leave.

Title: Concierge

Institution: RVM&G Property Managers

Higgins's responsibility is to provide building services that "assist the tenants in their day-to-day work." Although not the official building manager, he oversees outside contractors, requesting bids and outlining the scope of the work to be done. He ensures that all such work is within budget and performed correctly. Since the building doesn't have a security guard, Higgins also monitors lobby traffic.

If a major problem arises, tenants feel comfortable approaching Higgins, knowing he'll do his best to resolve it. He stresses that "positive tenant relations" are critical to the building's success. In Higgins's book, "the tenant is always right."

Each month Higgins chronicles building operations in a report to the owner, listing the status of capital improvements, ongoing maintenance, significant tenant complaints, and any variances in the building's accounts. To determine a variance, he has to balance each account monthly. If an account's funds are too high or too low, Higgins has to explain the fluctuation. Math at this level requires simple arithmetic skills, but the calculations become more complicated when it comes to computing a tenant's base rent and future increases.

First, the building's operating costs—utility bills, staff salaries, and vendor fees (cleaning, exterminating, window washing, and so on)—have to be calculated. For example, if the costs add up to $420,000 per year, Higgins divides the total by the building's 70,000 square feet to calculate a general operating expense of $6 per square foot. Dividing the annual real estate tax of $280,000 by the square footage yields another $4 per square foot.

If the owner agrees to modify a new tenant's offices, construction costs—which can run into tens of thousands of dollars—have to be factored in. The construction costs are then amortized like a mortgage over the life of a typical five-year lease. The principal and interest are calculated so that the tenant pays a fixed amount each month until construction costs are repaid.

Miscellaneous costs (broker, legal, and architectural fees) and the owner's expected profit margin round out the tenant's base for the first year. The base is multiplied by the number of square feet a tenant occupies to determine monthly rent. As operating costs and real estate taxes increase in subsequent years, Higgins has to refigure each tenant's monthly rent using the original base as his starting point.

Before we look at more examples, let us summarize the steps involved in integration by substitution.

INTEGRATION BY SUBSTITUTION

Step 1   Let $u = g(x)$, where $g(x)$ is part of the integrand, usually the "inside function" of the composite function $f(g(x))$.

Step 2   Compute $du = g'(x)\, dx$.

Step 3   Use the substitution $u = g(x)$ and $du = g'(x)\, dx$ to convert the *entire* integral into one involving *only u.*

Step 4   Evaluate the resulting integral.

Step 5   Replace u by $g(x)$ to obtain the final solution as a function of x.

REMARK   Sometimes we need to consider different choices of g for the substitution $u = g(x)$ in order to carry out step 3 and/or step 4.   ☐

EXAMPLE 1   Evaluate $\int 2x(x^2 + 3)^4\, dx$.

Solution

Step 1   Observe that the integrand involves the composite function $(x^2 + 3)^4$ with "inside function" $g(x) = x^2 + 3$. So, we choose $u = x^2 + 3$.

Step 2   Compute $du = 2x\, dx$.

Step 3   Making the substitution $u = x^2 + 3$ and $du = 2x\, dx$, we obtain

$$\int 2x(x^2 + 3)^4\, dx = \int (x^2 + 3)^4 (2x\, dx) = \int u^4\, du$$

$$\uparrow$$
Rewriting

an integral involving only the variable u.

Step 4   Evaluate

$$\int u^4\, du = \frac{1}{5}u^5 + C$$

Step 5   Replacing u by $x^2 + 3$, we obtain

$$\int 2x(x^2 + 3)^4\, dx = \frac{1}{5}(x^2 + 3)^5 + C$$

EXAMPLE 2   Evaluate $\int 3\sqrt{3x + 1}\, dx$.

Solution

Step 1   The integrand involves the composite function $\sqrt{3x + 1}$ with "inside function" $g(x) = 3x + 1$. So, let $u = 3x + 1$.

Step 2   Compute $du = 3\, dx$.

Step 3.   Making the substitution $u = 3x + 1$ and $du = 3\, dx$, we obtain

$$\int 3\sqrt{3x + 1}\, dx = \int \sqrt{3x + 1}(3\, dx) = \int \sqrt{u}\, du$$

an integral involving only the variable u.

Step 4   Evaluate

$$\int \sqrt{u}\ du = \int u^{1/2}\ du = \frac{2}{3} u^{3/2} + C$$

Step 5   Replacing u by $3x + 1$, we obtain

$$\int 3\sqrt{3x + 1}\ dx = \frac{2}{3}(3x + 1)^{3/2} + C$$

EXAMPLE 3   Evaluate $\int x^2(x^3 + 1)^{3/2}\ dx$.

Solution

Step 1   The integrand contains the composite function $(x^3 + 1)^{3/2}$ with "inside function" $g(x) = x^3 + 1$. So, let $u = x^3 + 1$.

Step 2   Compute $du = 3x^2\ dx$.

Step 3   Making the substitution $u = x^3 + 1$ and $du = 3x^2\ dx$, or $x^2\ dx = \frac{1}{3}\ du$, we obtain

$$\int x^2(x^3 + 1)^{3/2}\ dx = \int (x^3 + 1)^{3/2}(x^2\ dx)$$

$$= \int u^{3/2}\left(\frac{1}{3}\ du\right) = \frac{1}{3}\int u^{3/2}\ du$$

an integral involving only the variable u.

Step 4   We evaluate

$$\frac{1}{3}\int u^{3/2}\ du = \frac{1}{3} \cdot \frac{2}{5} u^{5/2} + C = \frac{2}{15} u^{5/2} + C$$

Step 5   Replacing u by $x^3 + 1$, we obtain

$$\int x^2(x^3 + 1)^{3/2}\ dx = \frac{2}{15}(x^3 + 1)^{5/2} + C$$

In the remaining examples, we will drop the practice of labeling the steps involved in evaluating each integral.

EXAMPLE 4   Evaluate $\int e^{-3x}\ dx$.

Solution   Let $u = -3x$ so that $du = -3\ dx$, or $dx = -\frac{1}{3}\ du$. Then

$$\int e^{-3x}\ dx = \int e^u\left(-\frac{1}{3}\ du\right) = -\frac{1}{3}\int e^u\ du$$

$$= -\frac{1}{3} e^u + C = -\frac{1}{3} e^{-3x} + C$$

EXAMPLE 5   Evaluate $\displaystyle\int \frac{x}{3x^2 + 1}\, dx$.

Solution   Let $u = 3x^2 + 1$. Then $du = 6x\, dx$, or $x\, dx = \frac{1}{6}\, du$. Making the appropriate substitutions, we have

$$\int \frac{x}{3x^2 + 1}\, dx = \int \frac{1/6}{u}\, du$$

$$= \frac{1}{6}\int \frac{1}{u}\, du$$

$$= \frac{1}{6}\ln |u| + C$$

$$= \frac{1}{6}\ln (3x^2 + 1) + C \qquad \text{since } 3x^2 + 1 > 0 \qquad ∎$$

EXAMPLE 6   Evaluate $\displaystyle\int \frac{(\ln x)^2}{2x}\, dx$.

Solution   Let $u = \ln x$.

Then $$du = \frac{d}{dx}(\ln x)\, dx = \frac{1}{x}\, dx$$

Thus, $$\int \frac{(\ln x)^2}{2x}\, dx = \frac{1}{2}\int \frac{(\ln x)^2}{x}\, dx$$

$$= \frac{1}{2}\int u^2\, du$$

$$= \frac{1}{6}u^3 + C$$

$$= \frac{1}{6}(\ln x)^3 + C \qquad ∎$$

Applications

Examples 7 and 8 show how the method of substitution can be used in practical situations.

EXAMPLE 7   In 1988, the head of the research and development department of the Soloron Corporation claimed that the cost of producing solar cell panels would drop at the rate of

$$\frac{58}{(3t + 2)^2} \qquad 0 \le t \le 10$$

dollars per peak watt for the next t years, with $t = 0$ corresponding to the beginning of the year 1988. (A peak watt is the power produced at noon on a sunny day.) In 1988, the panels, which are used for photovoltaic power systems, cost $10 per peak watt. Find an expression giving the cost per peak watt of producing solar cell panels at the beginning of year t. What will the cost be at the beginning of 1998?

Solution Let $C(t)$ denote the cost per peak watt for producing solar cell panels at the beginning of year t. Then

$$C'(t) = -\frac{58}{(3t+2)^2}$$

Integrating, we find that

$$C(t) = \int \frac{-58}{(3t+2)^2}\, dt$$

$$= -58 \int (3t+2)^{-2}\, dt$$

Let $u = 3t + 2$ so that

$$du = 3\, dt \qquad \text{or} \qquad dt = \frac{1}{3}\, du$$

Then
$$C(t) = -58 \left(\frac{1}{3}\right) \int u^{-2}\, du$$

$$= -\frac{58}{3}(-1)u^{-1} + k$$

$$= \frac{58}{3(3t+2)} + k$$

where k is an arbitrary constant. To determine the value of k, note that the cost per peak watt of producing solar cell panels at the beginning of 1988 ($t = 0$) was 10, or $C(0) = 10$. This gives

$$C(0) = \frac{58}{3(2)} + k = 10$$

or $k = \frac{1}{3}$. Therefore, the required expression is given by

$$C(t) = \frac{58}{3(3t+2)} + \frac{1}{3}$$

$$= \frac{58 + (3t+2)}{3(3t+2)}$$

$$= \frac{t+20}{3t+2}$$

The cost per peak watt for producing solar cell panels at the beginning of 1998 is given by

$$C(10) = \frac{10+20}{3(10)+2} \approx 0.94$$

or approximately $.94 per peak watt. ■

 EXAMPLE 8   A study prepared by the marketing department of the Universal Instruments Company forecasts that, after its new line of Galaxy Home Computers is introduced into the market, sales will grow at the rate of

$$2000 - 1500e^{-0.05t} \qquad 0 \le t \le 60$$

units per month. Find an expression that gives the total number of computers that will sell t months after they become available on the market. How many computers will Universal sell in the first year they are on the market?

Solution   Let $N(t)$ denote the total number of computers that may be expected to be sold t months after their introduction in the market. Then, the rate of growth of sales is given by $N'(t)$ units per month. Thus,

$$N'(t) = 2000 - 1500e^{-0.05t}$$

so that

$$N(t) = \int (2000 - 1500e^{-0.05t})\, dt$$

$$= \int 2000\, dt - 1500 \int e^{-0.05t}\, dt$$

Upon integrating the second integral by the method of substitution, we obtain

$$N(t) = 2{,}000t + \frac{1{,}500}{0.05} e^{-0.05t} + C \qquad \text{Let } u = -0.05t;\ \text{then } du = -0.05\, dt.$$

$$= 2{,}000t + 30{,}000e^{-0.05t} + C$$

To determine the value of C, note that the number of computers sold at the end of month 0 is nil, so $N(0) = 0$. This gives

$$N(0) = 30{,}000 + C = 0 \qquad \text{Since } e^0 = 1$$

or $C = -30{,}000$. Therefore, the required expression is given by

$$N(t) = 2{,}000t + 30{,}000e^{-0.05t} - 30{,}000$$

$$= 2{,}000t + 30{,}000(e^{-0.05t} - 1)$$

The number of computers that Universal can expect to sell in the first year is given by

$$N(12) = 2{,}000(12) + 30{,}000(e^{-0.05(12)} - 1)$$

$$\approx 10{,}464$$

or 10,464 units.

SELF-CHECK EXERCISES 11.2

1. Evaluate $\int \sqrt{2x + 5}\, dx$.

2. Evaluate $\int \dfrac{x^2}{(2x^3 + 1)^{3/2}}\, dx$.

3. Evaluate $\int xe^{2x^2-1}\, dx$.

 4. According to a joint study conducted by Oxnard's Environmental Management Department and a state government agency, the concentration of carbon monoxide in the air due to automobile exhaust is increasing at the rate given by

$$f(t) = \frac{8(0.1t + 1)}{300(0.2t^2 + 4t + 64)^{1/3}}$$

parts per million per year t. Currently, the concentration of carbon monoxide due to automobile exhaust is 0.16 part per million. Find an expression giving the concentration of carbon monoxide t years from now.

Solutions to Self-Check Exercises 11.2 can be found on page 685.

11.2 EXERCISES

In exercises 1–50, evaluate the indefinite integral.

1. $\int 4(4x + 3)^4 \, dx$

2. $\int 4x(2x^2 + 1)^7 \, dx$

3. $\int (x^3 - 2x)^2(3x^2 - 2) \, dx$

4. $\int (3x^2 - 2x + 1)(x^3 - x^2 + x)^4 \, dx$

5. $\int \dfrac{4x}{(2x^2 + 3)^3} \, dx$

6. $\int \dfrac{3x^2 + 2}{(x^3 + 2x)^2} \, dx$

7. $\int 3t^2\sqrt{t^3 + 2} \, dt$

8. $\int 3t^2(t^3 + 2)^{3/2} \, dt$

9. $\int (x^2 - 1)^9 x \, dx$

10. $\int x^2(2x^3 + 3)^4 \, dx$

11. $\int \dfrac{x^4}{1 - x^5} \, dx$

12. $\int \dfrac{x^2}{\sqrt{x^3 - 1}} \, dx$

13. $\int \dfrac{2}{x - 2} \, dx$

14. $\int \dfrac{x^2}{x^3 - 3} \, dx$

15. $\int \dfrac{0.3x - 0.2}{0.3x^2 - 0.4x + 2} \, dx$

16. $\int \dfrac{2x^2 + 1}{0.2x^3 + 0.3x} \, dx$

17. $\int \dfrac{x}{3x^2 - 1} \, dx$

18. $\int \dfrac{x^2 - 1}{x^3 - 3x + 1} \, dx$

19. $\int e^{-2x} \, dx$

20. $\int e^{-0.02x} \, dx$

21. $\int e^{2-x} \, dx$

22. $\int e^{2t+3} \, dt$

23. $\int xe^{-x^2} \, dx$

24. $\int x^2 e^{x^3-1} \, dx$

25. $\int (e^x - e^{-x}) \, dx$

26. $\int (e^{2x} + e^{-3x}) \, dx$

27. $\int \dfrac{e^x}{1 + e^x} \, dx$

28. $\int \dfrac{e^{2x}}{1 + e^{2x}} \, dx$

29. $\int \dfrac{e^{\sqrt{x}}}{\sqrt{x}} \, dx$

30. $\int \dfrac{e^{-1/x}}{x^2} \, dx$

31. $\int \dfrac{e^{3x} + x^2}{(e^{3x} + x^3)^3} \, dx$

32. $\int \dfrac{e^x - e^{-x}}{(e^x + e^{-x})^{3/2}} \, dx$

33. $\int e^{2x}(e^{2x} + 1)^3 \, dx$

34. $\int e^{-x}(1 + e^{-x}) \, dx$

35. $\int \dfrac{\ln 5x}{x} \, dx$

36. $\int \dfrac{(\ln u)^3}{u} \, du$

37. $\int \dfrac{1}{x \ln x} \, dx$

38. $\int \dfrac{1}{x(\ln x)^2} \, dx$

39. $\int \dfrac{\sqrt{\ln x}}{x} \, dx$

40. $\int \dfrac{(\ln x)^{7/2}}{x} \, dx$

41. $\int \left(xe^{x^2} - \dfrac{x}{x^2 + 2} \right) dx$

42. $\int \left(xe^{-x^2} + \dfrac{e^x}{e^x + 3} \right) dx$

43. $\int \dfrac{x + 1}{\sqrt{x} - 1} \, dx$ [*Hint:* Let $u = \sqrt{x} - 1$.]

44. $\int \dfrac{e^{-u} - 1}{e^{-u} + u} \, du$ [*Hint:* Let $v = e^{-u} + u$.]

45. $\int x(x - 1)^5 \, dx$ [*Hint:* $u = x - 1$ implies $x = u + 1$.]

46. $\int \dfrac{t}{t + 1} \, dt$ $\left[\textit{Hint:} \dfrac{t}{t + 1} = 1 - \dfrac{1}{t + 1}. \right]$

47. $\int \dfrac{1 - \sqrt{x}}{1 + \sqrt{x}} \, dx$ [*Hint:* Let $u = 1 + \sqrt{x}$.]

48. $\int \dfrac{1 + \sqrt{x}}{1 - \sqrt{x}} \, dx$ [*Hint:* Let $u = 1 - \sqrt{x}$.]

49. $\int v^2(1 - v)^6 \, dv$ [*Hint:* Let $u = 1 - v$.]

50. $\int x^3(x^2 + 1)^{3/2} \, dx$ [*Hint:* Let $u = x^2 + 1$.]

In exercises 51–54, find the function f given that the slope of the tangent line to the graph of f at any point $(x, f(x))$ is $f'(x)$ and that the graph of f passes through the given point.

51. $f'(x) = 5(2x - 1)^4$; $(1, 3)$

52. $f'(x) = \dfrac{3x^2}{2\sqrt{x^3 - 1}}$; $(1, 1)$

53. $f'(x) = -2xe^{-x^2+1}$; $(1, 0)$

54. $f'(x) = 1 - \dfrac{2x}{x^2 + 1}$; $(0, 2)$

C **55. Student Enrollment** The registrar of Kellogg University estimates that the total student enrollment in the Continuing Education division will grow at the rate of

$$N'(t) = 2000(1 + 0.2t)^{-3/2}$$

students per year t years from now. If the current student enrollment is 1000, find an expression giving the total student enrollment t years from now. What will the student enrollment be five years from now?

C **56. TV Viewers: Newsmagazine Shows** The number of viewers of a weekly TV newsmagazine show, introduced in the 1993 season, has been increasing at the rate of

$$3\left(2 + \frac{1}{2}x\right)^{-1/3} \qquad 1 \le x \le 6$$

million viewers per year in its xth year on the air. The number of viewers of the program during its first year on the air is given by $9(5/2)^{2/3}$ million. Find how many viewers are expected in the 1998 season.

57. Demand: Ladies' Boots   The rate of change of the unit price p (in dollars) of Apex ladies' boots is given by

$$p'(x) = \frac{-250x}{(16 + x^2)^{3/2}}$$

where x is the quantity demanded daily in units of a hundred. Find the demand function for these boots if the quantity demanded daily is 300 pairs ($x = 3$) when the unit price is $50 per pair.

C 58. Supply: Ladies' Boots   The rate of change of the unit price p (in dollars) of Apex ladies' boots is given by

$$p'(x) = \frac{240x}{(5 - x)^2}$$

where x is the number of pairs that the supplier will make available in the market daily when the unit price is $p per pair. Find the supply equation for these boots if the quantity the supplier is willing to make available is 200 pairs daily ($x = 2$) when the unit price is $50 per pair.

C 59. Oil Spills   In calm waters, the oil spilling from the ruptured hull of a grounded tanker forms an oil slick that is circular in shape. If the radius r of the circle is increasing at the rate of

$$r'(t) = \frac{30}{\sqrt{2t + 4}}$$

feet per minute t minutes after the rupture occurs, find an expression for the radius at any time t. How large is the polluted area 16 minutes after the rupture occurred? [*Hint:* $r(0) = 0$.]

C 60. Life Expectancy of a Female   Suppose that in a certain country the life expectancy at birth of a female is changing at the rate of

$$g'(t) = \frac{5.45218}{(1 + 1.09t)^{0.9}}$$

years per year. Here t is measured in years and $t = 0$ corresponds to the beginning of 1900. Find an expression $g(t)$ giving the life expectancy at birth (in

years) of a female in that country if the life expectancy at the beginning of 1900 is 50.02 years. What is the life expectancy at birth of a female born in the year 2000 in that country?

C 61. Average Birth Height of Boys   Using data collected at Kaiser Hospital, pediatricians estimate that the average height of male children changes at the rate of

$$h'(t) = \frac{52.8706e^{-0.3277t}}{(1 + 2.449e^{-0.3277t})^2}$$

inches per year, where the child's height $h(t)$ is measured in inches and t, the child's age, is measured in years, with $t = 0$ corresponding to the age at birth. Find an expression $h(t)$ for the average height of a boy at age t if the height at birth of an average child is 19.4 inches. What is the height of an average eight-year-old boy?

C 62. Learning Curves   The average student enrolled in the 20-week Advanced Shorthand course at the American Institute of Stenography progresses according to the rule

$$N'(t) = 6e^{-0.05t} \qquad 0 \le t \le 20$$

where $N'(t)$ measures the rate of change in the number of words per minute of dictation the student takes in shorthand after t weeks in the course. Assuming that the average student enrolled in this course begins with a dictation speed of 60 words per minute, find an expression $N(t)$ that gives the dictation speed of the student after t weeks in the course.

C 63. Sales: Loudspeakers   Two thousand pairs of Acrosonic model F loudspeaker systems were sold in the first year they appeared in the market. Since then, sales of these loudspeaker systems have been growing at the rate of

$$f'(t) = 2000(3 - 2e^{-t})$$

units per year, where t denotes the number of years these systems have been on the market. Determine the number of systems that were sold in the first five years after their introduction.

C 64. Amount of Glucose in the Bloodstream   Suppose that a patient is given a continuous intravenous infusion of glucose at a constant rate of r milligrams per minute. Then, the rate at which the amount of glucose in the bloodstream is changing at time t due to this infusion is given by

$$A'(t) = re^{-at}$$

milligrams per minute, where a is a positive constant associated with the rate at which excess glucose is eliminated from the bloodstream and is dependent on the patient's metabolism rate. Derive an expression for the amount of glucose in the bloodstream at time t. [*Hint:* $A(0) = 0$.]

65. Concentration of a Drug in an Organ A drug is carried into an organ of volume V cm^3 by a liquid that enters the organ at the rate of a cm^3/sec and leaves it at the rate of b cm^3/sec. The concentration of the drug in the liquid entering the organ is c g/cm^3. If the concentration of the drug in the organ

at time t is increasing at the rate of

$$x'(t) = \frac{1}{V}(ac - bx_0)e^{-bt/V}$$

g/cm^3 per second, and the concentration of the drug in the organ initially is x_0 g/cm^3, show that the concentration of the drug in the organ at time t is given by

$$x(t) = \frac{ac}{b} + \left(x_0 - \frac{ac}{b}\right)e^{-bt/V}$$

SOLUTIONS TO SELF-CHECK EXERCISES 11.2

1. Let $u = 2x + 5$. Then $du = 2\,dx$, or $dx = \frac{1}{2}\,du$. Making the appropriate substitutions, we have

$$\int \sqrt{2x+5}\,dx = \int \sqrt{u}\left(\frac{1}{2}\,du\right) = \frac{1}{2}\int u^{1/2}\,du$$

$$= \frac{1}{2}\left(\frac{2}{3}\right)u^{3/2} + C$$

$$= \frac{1}{3}(2x+5)^{3/2} + C$$

2. Let $u = 2x^3 + 1$, so that $du = 6x^2\,dx$, or $x^2\,dx = \frac{1}{6}\,du$. Making the appropriate substitutions, we have

$$\int \frac{x^2}{(2x^3+1)^{3/2}}\,dx = \int \frac{\left(\frac{1}{6}\right)du}{u^{3/2}} = \frac{1}{6}\int u^{-3/2}\,du$$

$$= \left(\frac{1}{6}\right)(-2)u^{-1/2} + C$$

$$= -\frac{1}{3}(2x^3+1)^{-1/2} + C$$

$$= -\frac{1}{3\sqrt{2x^3+1}} + C$$

3. Let $u = 2x^2 - 1$, so that $du = 4x\,dx$, or $x\,dx = \frac{1}{4}\,du$. Then

$$\int xe^{2x^2-1}\,dx = \frac{1}{4}\int e^u\,du$$

$$= \frac{1}{4}e^u + C$$

$$= \frac{1}{4}e^{2x^2-1} + C$$

4. Let $C(t)$ denote the concentration of carbon monoxide in the air due to automobile exhaust t years from now. Then

$$C'(t) = f(t) = \frac{8(0.1t+1)}{300(0.2t^2+4t+64)^{1/3}}$$

$$= \frac{8}{300}(0.1t+1)(0.2t^2+4t+64)^{-1/3}$$

Integrating, we find

$$C(t) = \int \frac{8}{300}(0.1t + 1)(0.2t^2 + 4t + 64)^{-1/3}\, dt$$

$$= \frac{8}{300}\int (0.1t + 1)(0.2t^2 + 4t + 64)^{-1/3}\, dt$$

Let $u = 0.2t^2 + 4t + 64$, so that $du = (0.4t + 4)\, dt = 4(0.1t + 1)\, dt$, or

$$(0.1t + 1)\, dt = \frac{1}{4}\, du$$

Then
$$C(t) = \frac{8}{300}\left(\frac{1}{4}\right)\int u^{-1/3}\, du$$

$$= \frac{1}{150}\left(\frac{3}{2}u^{2/3}\right) + k$$

$$= 0.01(0.2t^2 + 4t + 64)^{2/3} + k$$

where k is an arbitrary constant. To determine the value of k, we use the condition $C(0) = 0.16$, obtaining

$$C(0) = 0.16 = 0.01(64)^{2/3} + k$$

or
$$0.16 = 0.16 + k$$

and $k = 0$. Therefore,

$$C(t) = 0.01(0.2t^2 + 4t + 64)^{2/3}$$

11.3   Area and the Definite Integral

An Intuitive Look

Suppose that a certain state's annual rate of petroleum consumption over a four-year period is constant and is given by the function

$$f(t) = 1.2 \qquad 0 \le t \le 4$$

where t is measured in years and $f(t)$ in millions of barrels per year. Then the state's total petroleum consumption over the period of time in question is

$$(1.2)(4 - 0) \qquad \text{Rate of consumption} \cdot \text{time elapsed}$$

or 4.8 million barrels. If you examine the graph of f shown in Figure 4, you will see that this total is just the area of the rectangular region bounded above by the graph of f, below by the t-axis, and to the left and right by the vertical lines $t = 0$ (the y-axis) and $t = 4$, respectively.

Figure 5 shows the actual petroleum consumption of a certain New England state over a four-year period from 1990 ($t = 0$) to 1994 ($t = 4$). Observe that the rate of consumption is not constant; that is, the function f is not a constant function. What is the state's total petroleum consumption over this four-year period? It seems reasonable to conjecture that it is given by the "area" of the region bounded above by the graph of f, below by

FIGURE 4
The total petroleum consump-
tion is given by the area of
the rectangular region.

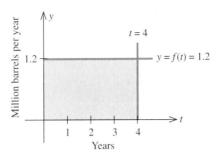

FIGURE 5
The daily petroleum consump-
tion is given by the "area" of
the shaded region.

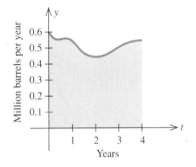

the *t*-axis, and to the left and right by the vertical lines $t = 0$ and $t = 4$, respectively.

This example raises two questions:

1. What is the "area" of the region shown in Figure 5?
2. How do we compute this area?

The Area Problem

The preceding example touches on the second fundamental problem in calculus: Calculate the area of the region bounded by the graph of a nonnegative function *f*, the *x*-axis, and the vertical lines $x = a$ and $x = b$ (Figure 6). This area is called the *area under the graph of f* on the interval $[a, b]$, or from *a* to *b*.

Defining Area—Two Examples

FIGURE 6
The area under the graph of
f on [a, b]

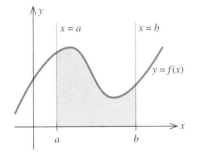

Just as we used the slopes of secant lines (quantities that we could compute) to help us define the slope of the tangent line to a point on the graph of a function, we will now adopt a parallel approach and use the areas of rectangles (quantities that we can compute) to help us define the area under the graph of a function. We begin by looking at a specific example.

EXAMPLE 1   Let $f(x) = x^2$ and consider the region *R* under the graph of *f* on the interval $[0, 1]$ (Figure 7a). In order to obtain an approximation of the area of *R*, let us construct four nonoverlapping rectangles as follows: Divide the interval $[0, 1]$ into four subintervals

$$\left[0, \frac{1}{4}\right], \qquad \left[\frac{1}{4}, \frac{1}{2}\right], \qquad \left[\frac{1}{2}, \frac{3}{4}\right], \qquad \left[\frac{3}{4}, 1\right]$$

of equal length 1/4. Next, construct four rectangles with these subintervals as bases and with heights given by the values of the function at the midpoints

$$\frac{1}{8}, \qquad \frac{3}{8}, \qquad \frac{5}{8}, \qquad \frac{7}{8}$$

of each subinterval. Then each of these rectangles has width 1/4 and height

$$f\left(\frac{1}{8}\right), \qquad f\left(\frac{3}{8}\right), \qquad f\left(\frac{5}{8}\right), \qquad f\left(\frac{7}{8}\right)$$

respectively (Figure 7b).

FIGURE 7

The area of the region under the graph of f on [a, b] in (a) is approximated by the sum of the areas of the four rectangles in (b).

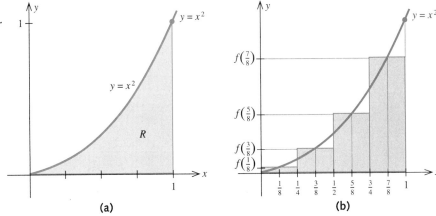

FIGURE 7

The area of the region under the graph of f on [a, b] in (a) is approximated by the sum of the areas of the four rectangles in (b).

If we approximate the area A of S by the sum of the areas of the four rectangles, we obtain

$$A \approx \frac{1}{4}f\left(\frac{1}{8}\right) + \frac{1}{4}f\left(\frac{3}{8}\right) + \frac{1}{4}f\left(\frac{5}{8}\right) + \frac{1}{4}f\left(\frac{7}{8}\right)$$

$$= \frac{1}{4}\left[f\left(\frac{1}{8}\right) + f\left(\frac{3}{8}\right) + f\left(\frac{5}{8}\right) + f\left(\frac{7}{8}\right)\right]$$

$$= \frac{1}{4}\left[\left(\frac{1}{8}\right)^2 + \left(\frac{3}{8}\right)^2 + \left(\frac{5}{8}\right)^2 + \left(\frac{7}{8}\right)^2\right] \qquad \text{Recall that } f(x) = x^2.$$

$$= \frac{1}{4}\left(\frac{1}{64} + \frac{9}{64} + \frac{25}{64} + \frac{49}{64}\right) = \frac{21}{64}$$

or approximately 0.328125 square units. ◼

Following the procedure of Example 1, we can obtain approximations of the area of the region R using any number n of rectangles ($n = 4$ in Example 1). Figure 8a shows the approximation of the area A of R using 8 rectangles ($n = 8$), and Figure 8b shows the approximation of the area A of R using 16 rectangles.

FIGURE 8

As n increases, the number of rectangles increases and the approximation improves.

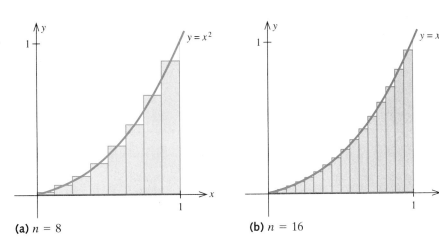

(a) $n = 8$ (b) $n = 16$

These figures suggest that the approximations seem to get better as n increases. This is borne out by the results given in Table 1, which were obtained using a computer.

TABLE 1

Number of rectangles n	4	8	16	32	64	100	200
Approximation of A	0.328125	0.332031	0.333008	0.333252	0.333313	0.333325	0.333331

Our computations seem to suggest that the approximations approach the number 1/3 as n gets larger and larger. This result suggests that we *define* the area of the region under the graph of $f(x) = x^2$ on the interval $[0, 1]$ to be 1/3 square units.

In Example 1 we chose the *midpoint* of each subinterval as the point at which to evaluate $f(x)$ to obtain the height of the approximating rectangle. Let us consider another example, this time choosing the *left endpoint* of each subinterval.

EXAMPLE 2   Let R be the region under the graph of $f(x) = 16 - x^2$ on the interval $[1, 3]$. Find an approximation of the area A of R using four subintervals of $[1, 3]$ of equal length and picking the left endpoint of each subinterval to evaluate $f(x)$ to obtain the height of the approximating rectangle.

Solution   The graph of f is sketched in Figure 9a. Since the length of $[1, 3]$ is 2, we see that the length of each subinterval is 2/4, or 1/2. Therefore, the four subintervals are

$$\left[1, \frac{3}{2}\right], \quad \left[\frac{3}{2}, 2\right], \quad \left[2, \frac{5}{2}\right], \quad \left[\frac{5}{2}, 3\right]$$

The left endpoints of these subintervals are $1, \frac{3}{2}, 2,$ and $\frac{5}{2}$, respectively, so the heights of the approximating rectangles are $f(1), f(\frac{3}{2}), f(2),$ and $f(\frac{5}{2})$,

FIGURE 9
The area of R in (a) is approximated by the sum of the areas of the four rectangles in (b).

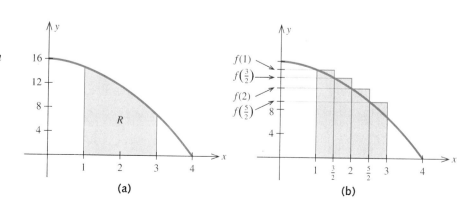

(a)   (b)

respectively (Figure 9b). Therefore, the required approximation is

$$A \approx \frac{1}{2}f(1) + \frac{1}{2}f\left(\frac{3}{2}\right) + \frac{1}{2}f(2) + \frac{1}{2}f\left(\frac{5}{2}\right)$$

$$= \frac{1}{2}\left[f(1) + f\left(\frac{3}{2}\right) + f(2) + f\left(\frac{5}{2}\right) \right]$$

$$= \frac{1}{2}\left\{ [16 - (1)^2] + \left[16 - \left(\frac{3}{2}\right)^2 \right] \right.$$

$$\left. + [16 - (2)^2] + \left[16 - \left(\frac{5}{2}\right)^2 \right] \right\} \qquad \text{Recall that } f(x) = 16 - x^2.$$

$$= \frac{1}{2}\left(15 + \frac{55}{4} + 12 + \frac{39}{4} \right) = \frac{101}{4},$$

or approximately 25.25 square units.   ◼

Table 2 shows the approximations of the area A of the region R of Example 2 when n rectangles are used for the approximation and the heights of the approximating rectangles are found by evaluating $f(x)$ at the left endpoints.

TABLE 2

Number of rectangles n	4	10	100	1,000	10,000	50,000	100,000
Approximation of A	25.2500	24.1200	23.4132	23.3413	23.3341	23.3335	23.3334

Once again we see that the approximations seem to approach a unique number as n gets larger and larger—this time the number is $23\frac{1}{3}$. This result suggests that we *define* the area of the region under the graph of $f(x) = 16 - x^2$ on the interval $[1, 3]$ to be $23\frac{1}{3}$ square units.

Defining Area—The General Case

Examples 1 and 2 point the way to defining the area A under the graph of an arbitrary but continuous and nonnegative function f on an interval $[a, b]$ (Figure 10a). Divide the interval $[a, b]$ into n subintervals of equal length $\Delta x = (b - a)/n$. Next, pick n arbitrary points $x_1, x_2, \ldots, x_n$, called *representative points,* from the first, second, $\ldots$, and nth subintervals, respectively (Figure 10b). Then, approximating the area A of the region R by the n rectangles of width Δx and heights $f(x_1), f(x_2), \ldots, f(x_n)$, so that the areas of the rectangles are $f(x_1)\Delta x, f(x_2)\Delta x, \ldots, f(x_n)\Delta x$, we have

$$A \approx f(x_1)\Delta x + f(x_2)\Delta x + \cdots + f(x_n)\Delta x$$

The sum on the right-hand side of this expression is called a **Riemann sum** in honor of the German mathematician Bernhard Riemann (1826–1866). Now, as the earlier examples seem to suggest, the Riemann sum will

The area of the region under the graph of f on [a, b] in (a) is approximated by the sum of the areas of the n rectangles shown in (b).

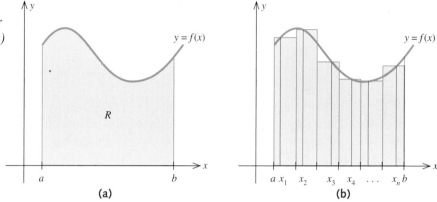

(a)   (b)

approach a unique number as n becomes arbitrarily large.* We define this number to be the area A of the region R.

THE AREA UNDER THE GRAPH OF A FUNCTION

Let f be a nonnegative continuous function on $[a, b]$. Then the area of the region under the graph of f is

$$A = \lim_{n \to \infty} [f(x_1) + f(x_2) + \cdots + f(x_n)]\Delta x \qquad (4)$$

where $x_1, x_2, \ldots, x_n$ are arbitrary points in the n subintervals of $[a, b]$ of equal width $\Delta x = (b - a)/n$.

The Definite Integral

As we have just seen, the area under the graph of a continuous *nonnegative* function f on an interval $[a, b]$ is defined by the limit of the Riemann sum

$$\lim_{n \to \infty} [f(x_1)\Delta x + f(x_2)\Delta x + \cdots + f(x_n)\Delta x]$$

We now turn our attention to the study of limits of Riemann sums involving functions that are not necessarily nonnegative. Such limits arise in many applications of calculus.

For example, the calculation of the distance covered by a body traveling along a straight line involves evaluating a limit of this form. The computation of the total revenue realized by a company over a certain time period, the calculation of the total amount of electricity consumed in a typical home over a 24-hour period, the average concentration of a drug in a body over a certain interval of time, and the volume of a solid—all involve limits of this type.

---

*Even though we chose the representative points to be the midpoints of the subintervals in Example 1 and the left endpoints in Example 2, it can be shown that each of the respective sums will always approach a unique number as n approaches infinity.

We begin with the following definition.

DEFINITE INTEGRAL

Let f be defined on $[a, b]$. If

$$\lim_{n \to \infty} [f(x_1)\Delta x + f(x_2)\Delta x + \cdots + f(x_n)\Delta x]$$

exists for all choices of representative points $x_1, x_2, \ldots, x_n$ in the n subintervals of $[a, b]$ of equal width $\Delta x = (b - a)/n$, then this limit is called the **definite integral** of *f from a to b* and is denoted by $\int_a^b f(x)\, dx$. Thus,

$$\int_a^b f(x)\, dx = \lim_{n \to \infty} [f(x_1)\Delta x + f(x_2)\Delta x + \cdots + f(x_n)\Delta x] \qquad \textbf{(5)}$$

The number a is the **lower limit of integration** and the number b is the **upper limit of integration.**

REMARKS

1. If f is nonnegative, then the limit in equation (5) is the same as the limit in (4) and, therefore, the definite integral gives the area under the graph of f on $[a, b]$.
2. The limit in (5) is denoted by the integral sign $\int$ because, as we will see later, the definite integral and the antiderivative of a function f are related.
3. It is important to realize that the definite integral $\int_a^b f(x)\, dx$ is a number, whereas the indefinite integral $\int f(x)\, dx$ represents a family of functions (the antiderivatives of f).
4. If the limit in (5) exists, we say that f is *integrable* on the interval $[a, b]$.

When Is a Function Integrable?

The following theorem, which we state without proof, guarantees that a continuous function is integrable.

INTEGRABILITY OF A FUNCTION

Let f be continuous on $[a, b]$. Then f is integrable on $[a, b]$; that is, the definite integral $\int_a^b f(x)\, dx$ exists.

Geometric Interpretation of the Definite Integral

If f is nonnegative and integrable on $[a, b]$, then we have the following geometric interpretation of the definite integral $\int_a^b f(x)\, dx$.

GEOMETRIC INTERPRETATION OF $\int_a^b f(x)\, dx$ FOR $f(x) \geq 0$ ON $[a, b]$

If f is nonnegative and continuous on $[a, b]$, then

$$\int_a^b f(x)\, dx \qquad \textbf{(6)}$$

is equal to the area of the region under the graph of f on $[a, b]$ (see Figure 11).

FIGURE 11

If $f(x) \geq 0$ on $[a, b]$, then $\int_a^b f(x)\, dx$ = area under the graph of f on $[a, b]$.

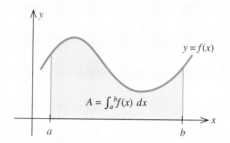

Next, let's extend our geometric interpretation of the definite integral to include the case where f assumes both positive as well as negative values on $[a, b]$. Consider a typical Riemann sum of the function f,

$$f(x_1)\Delta x + f(x_2)\Delta x + \cdots + f(x_n)\Delta x$$

corresponding to a partition of $[a, b]$ into n subintervals of equal width $(b - a)/n$, where $x_1, x_2, \ldots, x_n$ are representative points in the subintervals. The sum consists of n terms in which a positive term corresponds to the area of a rectangle of height $f(x_k)$ [for some positive integer k] lying above the x-axis, and a negative term corresponds to the area of a rectangle of height $-f(x_k)$ lying below the x-axis. (See Figure 12, which depicts a situation with $n = 6$.)

FIGURE 12

The positive terms in the Riemann sum are associated with the areas of the rectangles that lie above the x-axis, and the negative terms are associated with the areas of those that lie below the x-axis.

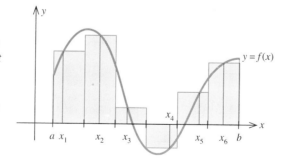

As n gets larger and larger, the sums of the areas of the rectangles lying above the x-axis seem to give a better and better approximation of the area of the region lying above the x-axis (see Figure 13). Similarly, the sums of the areas of those rectangles lying below the x-axis seem to give a better and better approximation of the area of the region lying below the x-axis. These observations suggest the following geometric interpretation of the definite integral for an arbitrary continuous function on an interval $[a, b]$.

FIGURE 13
As n gets larger, the approximations get better. Here n = 12, and we are approximating with twice as many rectangles as in Figure 12.

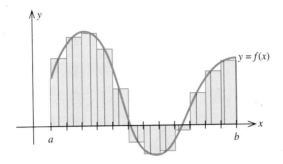

GEOMETRIC INTERPRETATION OF $\int_a^b f(x)\, dx$ ON $[a, b]$

If f is continuous on $[a, b]$ then

$$\int_a^b f(x)\, dx$$

is equal to the area of the region above $[a, b]$ minus the area of the region below $[a, b]$ (see Figure 14).

FIGURE 14
$\int_a^b f(x)\, dx$ = area of R_1 − area of R_2 + area of R_3.

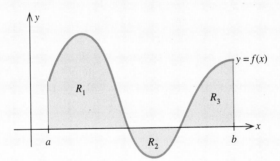

SELF-CHECK EXERCISE 11.3

Find an approximation of the area of the region R under the graph of $f(x) = 2x^2 + 1$ on the interval $[0, 3]$ using four subintervals of $[0, 3]$ of equal length and picking the midpoint of each subinterval as a representative point.

The Solution to Self-Check Exercise 11.3 can be found on page 696.

11.3 EXERCISES

In exercises 1 and 2, find an approximation of the area of the region R under the graph of f by computing the Riemann sum of f corresponding to the partition of the interval into the subintervals shown in the accompanying figures. In each case, use the midpoints of the subintervals as the representative points.

1.

2.

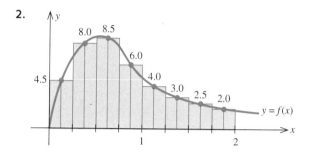

3. Let $f(x) = 3x$.
   a. Sketch the region R under the graph of f on the interval $[0, 2]$ and find its exact area using geometry.
   b. Use a Riemann sum with four subintervals of equal length ($n = 4$) to approximate the area of R. Choose the representative points to be the left endpoints of the subintervals.
   c. Repeat (b) with eight subintervals of equal length ($n = 8$).
   d. Compare the approximations obtained in (b) and (c) with the exact area found in (a). Do the approximations improve with larger n?

4. Repeat exercise 3 choosing the representative points to be the right endpoints of the subintervals.

5. Let $f(x) = 4 - 2x$.
   a. Sketch the region R under the graph of f on the interval $[0, 2]$ and find its exact area using geometry.
   b. Use a Riemann sum with five subintervals of equal length ($n = 5$) to approximate the area of R. Choose the representative points to be the left endpoints of the subintervals.
   c. Repeat (b) with ten subintervals of equal length ($n = 10$).
   d. Compare the approximations obtained in (b) and (c) with the exact area found in (a). Do the approximations improve with larger n?

6. Repeat exercise 5 choosing the representative points to be the right endpoints of the subintervals.

7. Let $f(x) = x^2$ and compute the Riemann sum of f over the interval $[2, 4]$ using
   a. two subintervals of equal length ($n = 2$)
   b. five subintervals of equal length ($n = 5$)
   c. ten subintervals of equal length ($n = 10$)
   d. In each case, choose the representative points to be the midpoints of the subintervals. Can you guess at the area of the region under the graph of f on the interval $[2, 4]$?

8. Repeat exercise 7 choosing the representative points to be the left endpoints of the subintervals.

9. Repeat exercise 7 choosing the representative points to be the right endpoints of the subintervals.

10. Let $f(x) = x^3$ and compute the Riemann sum of f over the interval $[0, 1]$ using
   a. two subintervals of equal length ($n = 2$)
   b. five subintervals of equal length ($n = 5$)
   c. ten subintervals of equal length ($n = 10$)
   d. In each case, choose the representative points to be the midpoints of the subintervals. Can you guess at the area of the region under the graph of f on the interval $[0, 1]$?

11. Repeat exercise 10 choosing the representative points to be the left endpoints of the subintervals.

12. Repeat exercise 10 choosing the representative points to be the right endpoints of the subintervals.

In exercises 13–16, find an approximation of the area of the region R under the graph of the given function f on the interval [a, b]. In each case, use n subintervals and choose the representative points as indicated.

13. $f(x) = x^2 + 1$; $[0, 2]$; $n = 5$; midpoints

14. $f(x) = 4 - x^2$; $[-1, 2]$; $n = 6$; left endpoints

15. $f(x) = \dfrac{1}{x}$; $[1, 3]$; $n = 4$; right endpoints

16. $f(x) = e^x$; $[0, 3]$; $n = 5$; midpoints

17. Real Estate The accompanying figure (a) shows a vacant lot with a 100-foot frontage in a development. To estimate its area, we introduce a coordinate system so that the x-axis coincides with the edge of the straight road forming the lower boundary of the property as shown in figure (b). Then thinking of the upper boundary of the property as the graph of a continuous function f over the interval $[0, 100]$, we see that the problem is mathematically equivalent to that of finding the area under the graph of f on $[0, 100]$. To estimate the area of the lot using a Riemann sum, we divide the interval $[0, 100]$ into five equal subintervals of length 20 feet. Then, using surveyor's equipment, we measure the distance from the midpoint of each of these subintervals to the upper boundary of the property. These measurements give the values of $f(x)$ at $x = 10, 30, 50, 70,$ and 90. What is the approximate area of the lot?

Road

(a)

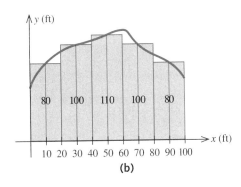

(b)

18. Real Estate Use the technique of exercise 17 to obtain an estimate of the area of the vacant lot shown in the accompanying figure.

Road

(a)

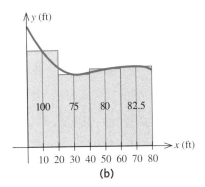

(b)

**SOLUTION TO
SELF-CHECK
EXERCISE 11.3**

The length of each subinterval is 3/4. Therefore, the four subintervals are

$$\left[0, \frac{3}{4}\right], \quad \left[\frac{3}{4}, \frac{3}{2}\right], \quad \left[\frac{3}{2}, \frac{9}{4}\right], \quad \left[\frac{9}{4}, 3\right]$$

The representative points are $\frac{3}{8}, \frac{9}{8}, \frac{15}{8},$ and $\frac{21}{8}$, respectively. Therefore, the required approximation is

$$A = \frac{3}{4}f\left(\frac{3}{8}\right) + \frac{3}{4}f\left(\frac{9}{8}\right) + \frac{3}{4}f\left(\frac{15}{8}\right) + \frac{3}{4}f\left(\frac{21}{8}\right)$$

$$= \frac{3}{4}\left[f\left(\frac{3}{8}\right) + f\left(\frac{9}{8}\right) + f\left(\frac{15}{8}\right) + f\left(\frac{21}{8}\right)\right]$$

$$= \frac{3}{4}\left\{\left[2\left(\frac{3}{8}\right)^2 + 1\right] + \left[2\left(\frac{9}{8}\right)^2 + 1\right] + \left[2\left(\frac{15}{8}\right)^2 + 1\right] + \left[2\left(\frac{21}{8}\right)^2 + 1\right]\right\}$$

$$= \frac{3}{4}\left(\frac{41}{32} + \frac{113}{32} + \frac{257}{32} + \frac{473}{32}\right) = \frac{663}{32}$$

or approximately 20.72 square units.

11.4   The Fundamental Theorem of Calculus

The Fundamental Theorem of Calculus

In Section 11.3 we defined the definite integral of an arbitrary continuous function on an interval $[a, b]$ as a limit of Riemann sums. Calculating the value of a definite integral by actually taking the limit of such sums is tedious and in most cases impractical. It is important to realize that the numerical results we obtained in Examples 1 and 2 of Section 11.3 were *approximations* of the respective areas of the regions in question, even though these results enabled us to *conjecture* what the actual areas might be. Fortunately, there is a much better way of finding the exact value of a definite integral.

The following theorem shows how to evaluate the definite integral of a continuous function provided we can find an antiderivative of that function. Because of its importance in establishing the relationship between differentiation and integration, this theorem—discovered independently by Sir Isaac Newton (1642–1727) in England and Gottfried Wilhelm Leibniz (1646–1716) in Germany—is called the *Fundamental Theorem of Calculus*.

**THE FUNDAMENTAL
THEOREM OF
CALCULUS**

Let f be continuous on $[a, b]$. Then

$$\int_a^b f(x)\, dx = F(b) - F(a) \qquad (7)$$

where F is any antiderivative of f; that is $F'(x) = f(x)$.

We will explain why this theorem is true at the end of this section.

When applying the Fundamental Theorem of Calculus, it is convenient to use the notation

$$F(x) \bigg|_a^b = F(b) - F(a)$$

For example, using this notation, equation (7) is written

$$\int_a^b f(x)\, dx = F(x) \bigg|_a^b = F(b) - F(a)$$

EXAMPLE 1   Let R be the region under the graph of $f(x) = x$ on the interval $[1, 3]$. Use the Fundamental Theorem of Calculus to find the area A of R and verify your result by elementary means.

Solution   The region R is shown in Figure 15a. Since f is nonnegative on $[1, 3]$, the area of R is given by the definite integral of f from 1 to 3; that is

$$A = \int_1^3 x\, dx$$

In order to evaluate the definite integral, observe that an antiderivative of $f(x) = x$ is $F(x) = \frac{1}{2}x^2 + C$, where C is an arbitrary constant. Therefore,

FIGURE 15
The area of R can be computed in two different ways.

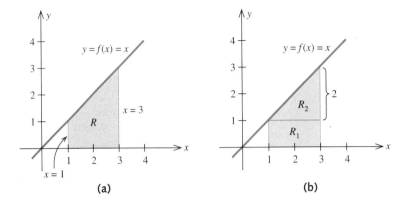

(a)   (b)

by the Fundamental Theorem of Calculus, we have

$$A = \int_1^3 x \, dx = \frac{1}{2}x^2 + C \Big|_1^3$$

$$= \left(\frac{9}{2} + C\right) - \left(\frac{1}{2} + C\right) = 4 \text{ square units}$$

To verify this result by elementary means, observe that the area A is the area of the rectangle R_1 (width $\times$ height) plus the area of the triangle R_2 ($\frac{1}{2}$ base $\times$ height) [see Figure 15b]; that is

$$2(1) + \frac{1}{2}(2)(2) = 2 + 2 = 4$$

which agrees with the result obtained earlier.   ■

Observe that in evaluating the definite integral in Example 1, the constant of integration "dropped out." This is true in general, for if $F(x) + C$ denotes an antiderivative of some function f, then

$$F(x) + C \Big|_a^b = [F(b) + C] - [F(a) + C]$$

$$= F(b) + C - F(a) - C$$

$$= F(b) - F(a)$$

With this fact in mind, we may, in all future computations involving the evaluations of a definite integral, drop the constant of integration from our calculations.

Finding the Area Under a Curve

Having seen how effective the Fundamental Theorem of Calculus is in helping us find the area of simple regions, we will now use it to find the area of more complicated regions.

EXAMPLE 2   In Section 11.3, we conjectured that the area of the region R under the graph of $f(x) = x^2$ on the interval $[0, 1]$ was 1/3 square units. Use the Fundamental Theorem of Calculus to verify this conjecture.

Solution   The region R is reproduced in Figure 16. Observe that f is non-negative on $[0, 1]$, so the area of R is given by $A = \int_0^1 x^2\, dx$. Since an antiderivative of $f(x) = x^2$ is $F(x) = \frac{1}{3}x^3$, we see, using the Fundamental Theorem of Calculus, that

$$A = \int_0^1 x^2\, dx = \frac{1}{3}x^3\,\bigg|_0^1 = \frac{1}{3}(1) - \frac{1}{3}(0) = \frac{1}{3}\ \text{square units}$$

as we wished to show.

FIGURE 16
The area of R is $\int_0^1 x^2\, dx = \frac{1}{3}$.

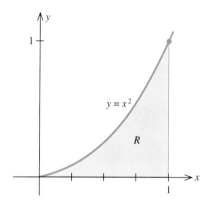

EXAMPLE 3   Find the area of the region R under the graph of $y = x^2 + 1$ from $x = -1$ to $x = 2$.

FIGURE 17
Area of $R = \int_{-1}^2 (x^2 + 1)\, dx$

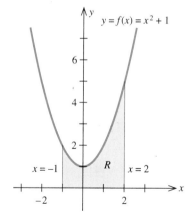

Solution   The region R under consideration is shown in Figure 17. Using the Fundamental Theorem of Calculus, we find that the required area is

$$\int_{-1}^2 (x^2 + 1)\, dx = \left(\frac{1}{3}x^3 + x\right)\bigg|_{-1}^2$$

$$= \left[\frac{1}{3}(8) + 2\right] - \left[\frac{1}{3}(-1)^3 + (-1)\right] = 6$$

or 6 square units.

Evaluating Definite Integrals

In Examples 4 and 5, we use the rules of integration of Section 11.1 to help us evaluate the definite integrals.

EXAMPLE 4   Evaluate $\int_1^3 (3x^2 + e^x)\, dx$.

Solution

$$\int_1^3 (3x^2 + e^x)\, dx = x^3 + e^x\,\bigg|_1^3$$

$$= (27 + e^3) - (1 + e) = 26 + e^3 - e$$

EXAMPLE 5   Evaluate $\int_1^2 \left(\dfrac{1}{x} - \dfrac{1}{x^2} \right) dx$.

Solution

$$\int_1^2 \left(\frac{1}{x} - \frac{1}{x^2} \right) dx = \int_1^2 \left(\frac{1}{x} - x^{-2} \right) dx$$

$$= \ln |x| + \frac{1}{x} \Big|_1^2$$

$$= \left(\ln 2 + \frac{1}{2} \right) - (\ln 1 + 1)$$

$$= \ln 2 - \frac{1}{2} \qquad \text{Recall } \ln 1 = 0.$$

Applications

EXAMPLE 6   The management of Staedtler Office Equipment has determined that the daily marginal cost function associated with producing battery-operated pencil sharpeners is given by

$$C'(x) = 0.000006x^2 - 0.006x + 4$$

where $C'(x)$ is measured in dollars per unit and x denotes the number of units produced. Management has also determined that the daily fixed cost incurred in producing these pencil sharpeners is $100. Find Staedtler's daily total cost for producing (a) the first 500 units and (b) the 201st through 400th units.

Solution

a. Since $C'(x)$ is the marginal cost function, its antiderivative $C(x)$ is the total cost function. The daily fixed cost incurred in producing the pencil sharpeners is $C(0)$ dollars. Since the daily fixed cost is given as $100, we have $C(0) = 100$. We are required to find $C(500)$. Let us compute $C(500) - C(0)$, the net change in the total cost function $C(x)$ over the interval $[0, 500]$. Using the Fundamental Theorem of Calculus, we find

$$C(500) - C(0) = \int_0^{500} C'(x)\, dx$$

$$= \int_0^{500} (0.000006x^2 - 0.006x + 4)\, dx$$

$$= 0.000002x^3 - 0.003x^2 + 4x \Big|_0^{500}$$

$$= [0.000002(500)^3 - 0.003(500)^2 + 4(500)]$$

$$\quad - [0.000002(0)^3 - 0.003(0)^2 + 4(0)]$$

$$= 1500$$

Therefore $C(500) = 1500 + C(0) = 1500 + 100 = 1600$, so the total cost incurred daily by Staedtler in producing 500 pencil sharpeners is $1600.

b. The daily total cost incurred by Staedtler in producing the 201st through 400th units of battery-operated pencil sharpeners is given by

$$C(400) - C(200) = \int_{200}^{400} C'(x)\, dx$$

$$= \int_{200}^{400} (0.000006x^2 - 0.006x + 4)\, dx$$

$$= 0.000002x^3 - 0.003x^2 + 4x \,\Big|_{200}^{400}$$

$$= 552, \quad \text{or} \quad \$552$$

Since $C'(x)$ is nonnegative for x in the interval $[0, \infty)$, we have the following geometrical interpretation of the two definite integrals in Example 6: $\int_0^{500} C'(x)\, dx$ is the area of the region under the graph of the function C' from $x = 0$ to $x = 500$, shown in Figure 18a, and $\int_{200}^{400} C'(x)\, dx$ is the area of the region shown in Figure 18b from $x = 200$ to $x = 400$.

FIGURE 18

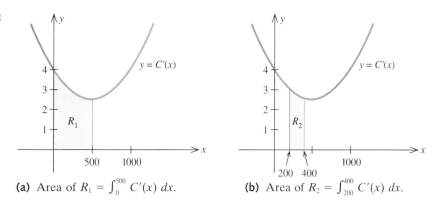

(a) Area of $R_1 = \int_0^{500} C'(x)\, dx$.   **(b)** Area of $R_2 = \int_{200}^{400} C'(x)\, dx$.

EXAMPLE 7   An efficiency study conducted for the Elektra Electronics Company showed that the rate at which Space Commander walkie-talkies are assembled by the average worker t hours after starting work at 8 A.M. is given by the function

$$f(t) = -3t^2 + 12t + 15 \qquad 0 \le t \le 4$$

Determine how many walkie-talkies can be assembled by the average worker in the first hour of the morning shift. How many units can the average worker be expected to assemble in the second hour of the morning shift?

Solution   Let $N(t)$ denote the number of walkie-talkies assembled by the average worker t hours after starting work in the morning shift. Then we have

$$N'(t) = f(t) = -3t^2 + 12t + 15$$

Therefore, the number of units assembled by the average worker in the first hour of the morning shift is

$$N(1) - N(0) = \int_0^1 N'(t)\, dt = \int_0^1 (-3t^2 + 12t + 15)\, dt$$

$$= -t^3 + 6t^2 + 15t \Big|_0^1 = -1 + 6 + 15$$

$$= 20, \quad \text{or} \quad 20 \text{ units}$$

EXAMPLE 8   A certain city's rate of electricity consumption is expected to grow exponentially with a growth constant of $k = 0.04$. If the present rate of consumption is 40 million kilowatt-hours (kwh) per year, what should the total production of electricity be over the next three years in order to meet the projected demand?

Solution   If $R(t)$ denotes the expected rate of consumption of electricity t years from now, then

$$R(t) = 40e^{0.04t}$$

million kwh per year. Next, if $C(t)$ denotes the expected total consumption of electricity over a period of t years, then

$$C'(t) = R(t)$$

Therefore, the total consumption of electricity expected over the next three years is given by

$$\int_0^3 C'(t)\, dt = \int_0^3 40e^{0.04t}\, dt$$

$$= \frac{40}{0.04} e^{0.04t} \Big|_0^3$$

$$= 1000(e^{0.12} - 1)$$

$$= 127.5 \quad \text{or} \quad 127.5 \text{ million kwh}$$

the amount that must be produced over the next three years in order to meet the demand.

Validity of the Fundamental Theorem of Calculus

In order to demonstrate the plausibility of the Fundamental Theorem of Calculus for the case where f is nonnegative on an interval $[a, b]$, let us define an "area function" A as follows. Let $A(t)$ denote the area of the region R under the graph of $y = f(x)$ from $x = a$ to $x = t$, where $a \leq t \leq b$ (Figure 19).

If h is a small positive number, then $A(t + h)$ is the area of the region under the graph of $y = f(x)$ from $x = a$ to $x = t + h$. Therefore, the difference

$$A(t + h) - A(t)$$

is the area under the graph of $y = f(x)$ from $x = t$ to $x = t + h$ (Figure 20). Now, the area of this last region can be approximated by the area of the rectangle of width h and height $f(t)$—that is, by the expression $h \cdot f(t)$

FIGURE 19

$A(t) = $ The area under the graph of f from $x = a$ to $x = t$

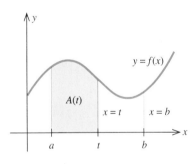

FIGURE 20

$A(t + h) - A(t) = $ The area under the graph of f from $x = t$ to $x = t + h$

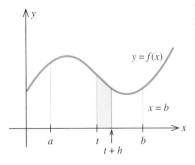

FIGURE 21

Area of rectangle = h · f(t)

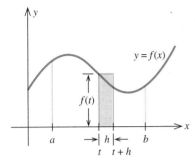

FIGURE 21

Area of rectangle = h · f(t)

(Figure 21). Thus,

$$A(t + h) - A(t) \approx h \cdot f(t)$$

where the approximations improve as h is taken to be smaller and smaller. Dividing both sides of the foregoing relationship by h, we obtain

$$\frac{A(t + h) - A(t)}{h} \approx f(t)$$

Taking the limit as h approaches zero, we find, by the definition of the derivative, that the left-hand side is

$$\lim_{h \to 0} \frac{A(t + h) - A(t)}{h} = A'(t)$$

The right-hand side, which is independent of h, remains constant throughout the limiting process. Because the approximation becomes exact as h approaches zero, we find that

$$A'(t) = f(t)$$

Since the foregoing equation holds for all values of t in the interval $[a, b]$, we have shown that the *area function A* is an antiderivative of the function $f(x)$. By Theorem 1 of Section 11.1, we conclude that $A(x)$ must have the form

$$A(x) = F(x) + C$$

where F is any antiderivative of f and C is an arbitrary constant. To determine the value of C, observe that $A(a) = 0$. This condition implies that

$$A(a) = F(a) + C = 0$$

FIGURE 22

The area of R is given by A(b).

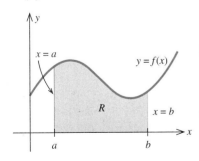

or $C = -F(a)$. Next, since the area of the region R is $A(b)$ (Figure 22), we see that the required area is

$$A(b) = F(b) + C$$
$$= F(b) - F(a)$$

Since the area of the region R is

$$\int_a^b f(x)\, dx$$

we have

$$\int_a^b f(x)\, dx = F(b) - F(a)$$

as we set out to show.

SELF-CHECK EXERCISES 11.4

1. Evaluate

$$\int_0^2 (x + e^x)\, dx$$

c **2.** The daily marginal profit function associated with producing and selling Texa-Pep hot sauce is

$$P'(x) = -0.000006x^2 + 6$$

where x denotes the number of cases (each case contains 24 bottles) produced and sold daily and $P'(x)$ is measured in dollars per unit. The fixed cost is $400.

a. What is the total profit realizable from producing and selling 1000 cases of Texa-Pep per day?

b. What is the additional profit realizable if the production and sale of Texa-Pep is increased from 1000 to 1200 cases per day?

Solutions to Self-Check Exercises 11.4 can be found on page 705.

11.4   EXERCISES

In exercises 1–4, find the area of the region under the graph of the function f on the interval [a, b] using the Fundamental Theorem of Calculus. Then verify your result using geometry.

1. $f(x) = 2$; $[1, 4]$ **2.** $f(x) = 4$; $[-1, 2]$

3. $f(x) = 2x$; $[1, 3]$ **4.** $f(x) = -\frac{1}{4}x + 1$; $[1, 4]$

In exercises 5–16, find the area of the region under the graph of the function f on the interval [a, b].

5. $f(x) = 2x + 3$; $[-1, 2]$ **6.** $f(x) = 4x - 1$; $[2, 4]$

7. $f(x) = -x^2 + 4$; $[-1, 2]$

8. $f(x) = 4x - x^2$; $[0, 4]$

9. $f(x) = \frac{1}{x}$; $[1, 2]$ **10.** $f(x) = \frac{1}{x^2}$; $[2, 4]$

11. $f(x) = \sqrt{x}$; $[1, 9]$ **12.** $f(x) = x^3$; $[1, 3]$

13. $f(x) = 1 - \sqrt[3]{x}$; $[-8, -1]$

14. $f(x) = \frac{1}{\sqrt{x}}$; $[1, 9]$

15. $f(x) = e^x$; $[0, 2]$ **16.** $f(x) = e^x - x$; $[1, 2]$

In exercises 17–40, evaluate the definite integral.

17. $\int_2^4 3\, dx$ **18.** $\int_{-1}^2 -2\, dx$

19. $\int_1^3 (2x + 3)\, dx$ **20.** $\int_{-1}^0 (4 - x)\, dx$

21. $\int_{-1}^3 2x^2\, dx$ **22.** $\int_0^2 8x^3\, dx$

23. $\int_{-2}^2 (x^2 - 1)\, dx$ **24.** $\int_1^4 \sqrt{u}\, du$

25. $\int_1^8 4x^{1/3}\, dx$ **26.** $\int_1^4 2x^{-3/2}\, dx$

27. $\int_0^1 (x^3 - 2x^2 + 1)\, dx$ **28.** $\int_1^2 (t^5 - t^3 + 1)\, dt$

29. $\int_2^4 \frac{1}{x}\, dx$ **30.** $\int_1^3 \frac{2}{x}\, dx$

31. $\int_0^4 x(x^2 - 1)\, dx$ **32.** $\int_0^2 (x - 4)(x - 1)\, dx$

33. $\int_1^3 (t^2 - t)^2\, dt$ **34.** $\int_{-1}^1 (x^2 - 1)^2\, dx$

35. $\int_{-3}^{-1} \frac{1}{x^2}\, dx$ **36.** $\int_1^2 \frac{2}{x^3}\, dx$

37. $\int_1^4 \left(\sqrt{x} - \frac{1}{\sqrt{x}}\right) dx$ **38.** $\int_0^1 \sqrt{2x}(\sqrt{x} + \sqrt{2})\, dx$

39. $\int_1^4 \frac{3x^3 - 2x^2 + 4}{x^2}\, dx$ **40.** $\int_1^2 \left(1 + \frac{1}{u} + \frac{1}{u^2}\right) du$

c **41.** **Marginal Cost** A division of Ditton Industries manufactures a deluxe toaster oven. Management has determined that the daily marginal cost function associated with producing these toaster ovens is given by

$$C'(x) = 0.0003x^2 - 0.12x + 20$$

where $C'(x)$ is measured in dollars per unit and x denotes the number of units produced. Management has also determined that the daily fixed cost incurred in the production is $800.

a. Find the total cost incurred by Ditton in producing the first 300 units of these toaster ovens per day.

b. What is the total cost incurred by Ditton in producing the 201st through 300th units per day?

c **42. Marginal Revenue**   The management of Ditton Industries has determined that the daily marginal revenue function associated with selling x units of their deluxe toaster ovens is given by

$$R'(x) = -0.1x + 40$$

where $R'(x)$ is measured in dollars per unit.
a. Find the daily total revenue realized from the sale of 200 units of the toaster ovens.
b. Find the additional revenue realized when the production (and sales) level is increased from 200 to 300 units.

c **43. Marginal Profit**   Refer to exercise 41. The daily marginal profit function associated with the production and sale of the deluxe toaster ovens is known to be

$$P'(x) = -0.0003x^2 + 0.02x + 20$$

where x denotes the number of units manufactured and sold daily and $P'(x)$ is measured in dollars per unit.
a. Find the total profit realizable from the manufacture and sale of 200 units of the toaster ovens per day. [*Hint:* $P(200) - P(0) = \int_0^{200} P'(x)\,dx$, $P(0) = -800$.]
b. What is the additional daily profit realizable if the production and sale of the toaster ovens is increased from 200 to 220 units per day?

c **44. Efficiency Studies**   Tempco Electronics, a division of Tempco Toys, Inc., manufactures an electronic football game. An efficiency study showed that the rate at which the games are assembled by the average worker t hours after starting work at 8 A.M. is

$$-\frac{3}{2}t^2 + 6t + 20 \qquad 0 \le t \le 4$$

units per hour.
a. Find the total number of games the average worker can be expected to assemble in the 4-hour morning shift.

b. How many units can the average worker be expected to assemble in the first hour of the morning shift? In the second hour of the morning shift?

c **45. Speedboat Racing**   In a recent pretrial run for the world water speed record, the velocity of the *Sea Falcon II* t seconds after firing the booster rocket was given by

$$v(t) = -t^2 + 20t + 440 \qquad 0 \le t \le 20$$

feet per second. Find the distance covered by the boat over the 20-second period after the booster rocket was activated. [*Hint:* The distance is given by $\int_0^{20} v(t)\,dt$.]

c **46. Aging Population**   According to the U.S. Census Bureau, the number of Americans aged 45 to 54 (which stood at 25 million at the beginning of 1990) will grow at the rate of

$$R(t) = 0.00933t^3 + 0.019t^2 - 0.10833t + 1.3467$$

million people per year, t years from the beginning of 1990. How many Americans aged 45 to 54 will be added to the population between 1990 and the year 2000?
Source: U.S. Census Bureau

c **47. Air Purification**   To test air purifiers, engineers run a purifier in a smoke-filled 10×20-foot room. While conducting a test for a certain brand of air purifier, it was determined that the amount of smoke in the room was decreasing at the rate of

$$R(t) = 0.00032t^4 - 0.01872t^3 + 0.3948t^2$$

$$- 3.83t + 17.63 \qquad 0 \le t \le 20$$

percent of the (original) amount of the smoke per minute, t minutes after the start of the test. How much smoke was left in the room five minutes after the start of the test? Ten minutes after the start of the test?
Source: Consumers Reports

SOLUTIONS TO SELF-CHECK EXERCISES 11.4

1.

$$\int_0^2 (x + e^x)\,dx = \frac{1}{2}x^2 + e^x \Big|_0^2$$

$$= \left[\frac{1}{2}(2)^2 + e^2\right] - \left[\frac{1}{2}(0) + e^0\right]$$

$$= 2 + e^2 - 1$$

$$= e^2 + 1$$

2. a. We want $P(1000)$. But

$$P(1000) - P(0) = \int_0^{1000} P'(x)\, dx = \int_0^{1000} (-0.000006x^2 + 6)\, dx$$

$$= -0.000002x^3 + 6x \bigg|_0^{1000}$$

$$= -0.000002(1000)^3 + 6(1000)$$

$$= 4000$$

So, $P(1000) = 4000 + P(0) = 4000 - 400$, or \$3600 per day $[P(0) = -C(0)]$.

b. The additional profit realizable is given by

$$\int_{1000}^{1200} P'(x)\, dx = -0.000002x^3 + 6x \bigg|_{1000}^{1200}$$

$$= [-0.000002(1200)^3 + 6(1200)]$$

$$- [-0.000002(1000)^3 + 6(1000)]$$

$$= 3744 - 4000$$

$$= 256$$

that is, the company sustains a loss of \$256 per day if production is increased to 1200 cases a day.

Using Technology

The definite integral $\int f(x)\, dx$ can be found using the **fnInt** (function integral) operation or the $\int f(x)$ (numerical integral) operation. (See Section A.18 in Appendix A.)

EXAMPLE 1   Use (a) the **fnInt** operation and (b) the $\int f(x)$ operation of a graphing utility to evaluate

$$\int_{-1}^2 \frac{2x + 4}{(x^2 + 1)^{3/2}}\, dx$$

Solution

a. Using the **fnInt** operation by entering

$$\text{fnInt}((2x + 4)/(x^2 + 1)^1.5, x, -1, 2)$$

we find

$$\int_{-1}^2 \frac{2x + 4}{(x^2 + 1)^{3/2}}\, dx \approx 6.92592225992$$

b. Using the $\int f(x)$ operation, we find

$$\int_{-1}^2 \frac{2x + 4}{(x^2 + 1)^{3/2}}\, dx \approx 6.9259222599$$

EXERCISES

In exercises 1–4, use a graphing utility to find the area of the region under the graph of f on the interval $[a, b]$. Express your answer to four decimal places.

1. $f(x) = 0.002x^5 + 0.032x^4 - 0.2x^2 + 2$; $[-1.1, 2.2]$

2. $f(x) = x\sqrt{x^3 + 1}$; $[1, 2]$

3. $f(x) = \sqrt{x}e^{-x}$; $[0, 3]$

4. $f(x) = \dfrac{\ln x}{\sqrt{1 + x^2}}$; $[1, 2]$

In exercises 5–10, use a graphing utility to evaluate the definite integral.

5. $\displaystyle\int_{-1.2}^{2.3} (0.2x^4 - 0.32x^3 + 1.2x - 1)\, dx$

6. $\displaystyle\int_1^3 x(x^4 - 1)^{3.2}\, dx$

7. $\displaystyle\int_0^2 \dfrac{3x^3 + 2x^2 + 1}{2x^2 + 3}\, dx$

8. $\displaystyle\int_1^2 \dfrac{\sqrt{x} + 1}{2x^2 + 1}\, dx$

9. $\displaystyle\int_0^2 \dfrac{e^x}{\sqrt{x^2 + 1}}\, dx$

10. $\displaystyle\int_1^3 e^{-x} \ln(x^2 + 1)\, dx$

11. Rework exercise 46, Exercises 11.4.

12. Rework exercise 47, Exercises 11.4.

11.5 Evaluating Definite Integrals

This section continues our discussion of the applications of the Fundamental Theorem of Calculus.

Properties of the Definite Integral

Before going on, we list the following useful properties of the definite integral, some of which parallel the rules of integration of Section 11.1.

PROPERTIES OF THE DEFINITE INTEGRAL

Let f and g be integrable functions; then

1. $\displaystyle\int_a^a f(x)\, dx = 0$

2. $\displaystyle\int_a^b f(x)\, dx = -\int_b^a f(x)\, dx$

3. $\displaystyle\int_a^b cf(x)\, dx = c\int_a^b f(x)\, dx \qquad c,\ \text{a constant}$

4. $\displaystyle\int_a^b [f(x) \pm g(x)]\, dx = \int_a^b f(x)\, dx \pm \int_a^b g(x)\, dx$

5. $\displaystyle\int_a^b f(x)\, dx = \int_a^c f(x)\, dx + \int_c^b f(x)\, dx \qquad a < c < b$

Property 5 states that if c is a number lying between a and b so that the interval $[a, b]$ is divided into the intervals $[a, c]$ and $[c, b]$, then the integral of f over the interval $[a, b]$ may be expressed as the sum of the integral of f over the interval $[a, c]$ and the integral of f over the interval $[c, b]$.

Property 5 has the following geometrical interpretation when f is nonnegative. By definition

$$\int_a^b f(x)\, dx$$

is the area of the region under the graph of $y = f(x)$ from $x = a$ to $x = b$ (Figure 23). Similarly, we interpret the definite integrals

$$\int_a^c f(x)\, dx \qquad \text{and} \qquad \int_c^b f(x)\, dx$$

as the areas of the regions under the graph of $y = f(x)$ from $x = a$ to $x = c$ and from $x = c$ to $x = b$, respectively. Since the two regions do not overlap, we see that

$$\int_a^b f(x)\, dx = \int_a^c f(x)\, dx + \int_c^b f(x)\, dx$$

FIGURE 23
$$\int_a^b f(x)\,dx =$$
$$\int_a^c f(x)\,dx + \int_c^b f(x)\,dx$$

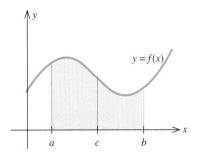

The Method of Substitution for Definite Integrals

Our first example shows two approaches generally used when evaluating a definite integral using the method of substitution.

EXAMPLE 1 Evaluate

$$\int_0^4 x\sqrt{9 + x^2}\,dx$$

Solution

Method 1 We first find the corresponding indefinite integral

$$I = \int x\sqrt{9 + x^2}\,dx$$

Make the substitution

$$u = 9 + x^2$$

so that
$$du = \frac{d}{dx}(9 + x^2)\,dx$$

$$= 2x\,dx$$

or
$$x\,dx = \frac{1}{2}\,du \qquad \text{Dividing both sides by 2}$$

Then

$$I = \int \frac{1}{2}\sqrt{u}\,du = \frac{1}{2}\int u^{1/2}\,du$$

$$= \frac{1}{3}u^{3/2} + C = \frac{1}{3}(9 + x^2)^{3/2} + C \qquad \begin{array}{l}\text{Substituting}\\ 9 + x^2 \text{ for } u\end{array}$$

Using this result, we now evaluate the given definite integral.

$$\int_0^4 x\sqrt{9 + x^2}\,dx = \frac{1}{3}(9 + x^2)^{3/2}\,\Big|_0^4$$

$$= \frac{1}{3}[(9 + 16)^{3/2} - 9^{3/2}]$$

$$= \frac{1}{3}(125 - 27) = \frac{98}{3} = 32\frac{2}{3}$$

Method 2   *Changing the Limits of Integration:* As before, we make the substitution

$$u = 9 + x^2 \qquad (8)$$

so that $$du = 2x \, dx$$

or $$x \, dx = \frac{1}{2} du$$

Next, observe that the given definite integral is evaluated *with respect to x* with the range of integration given by the interval [0, 4]. If we perform the integration *with respect to u* via the substitution (8), then we must adjust the range of integration to reflect the fact that the integration is being performed with respect to the new variable u. To determine the proper range of integration, note that when $x = 0$, (8) implies that

$$u = 9 + 0^2 = 9$$

which gives the required lower limit of integration with respect to u. Similarly, when $x = 4$,

$$u = 9 + 16 = 25$$

is the required upper limit of integration with respect to u. Thus, the range of integration when the integration is performed with respect to u is given by the interval [9, 25]. Therefore, we have

$$\int_0^4 x\sqrt{1 + x^2} \, dx = \int_9^{25} \frac{1}{2} \sqrt{u} \, du = \frac{1}{2} \int_9^{25} u^{1/2} \, du$$

$$= \frac{1}{3} u^{3/2} \Big|_9^{25} = \frac{1}{3}(25^{3/2} - 9^{3/2})$$

$$= \frac{1}{3}(125 - 27) = \frac{98}{3} = 32\frac{2}{3}$$

which agrees with the result obtained using Method 1.   ∎

EXAMPLE 2   Evaluate

$$\int_0^2 xe^{2x^2} \, dx$$

Solution   Let $u = 2x^2$ so that $du = 4x \, dx$, or $x \, dx = \frac{1}{4} du$. When $x = 0$, $u = 0$, and when $x = 2$, $u = 8$. This gives the lower and upper limits of integration with respect to u. Making the indicated substitutions, we find

$$\int_0^2 xe^{2x^2} \, dx = \int_0^8 \frac{1}{4} e^u \, du = \frac{1}{4} e^u \Big|_0^8 = \frac{1}{4}(e^8 - 1)$$ ∎

EXAMPLE 3   Evaluate

$$\int_0^1 \frac{x^2}{x^3 + 1} \, dx$$

Solution   Let $u = x^3 + 1$ so that $du = 3x^2\, dx$, or $x^2\, dx = \frac{1}{3}\, du$. When $x = 0$, $u = 1$, and when $x = 1$, $u = 2$. This gives the lower and upper limits of integration with respect to u. Making the indicated substitutions, we find

$$\int_0^1 \frac{x^2}{x^3 + 1}\, dx = \frac{1}{3}\int_1^2 \frac{du}{u} = \frac{1}{3}\ln |u|\,\Big|_1^2$$

$$= \frac{1}{3}(\ln 2 - \ln 1) = \frac{1}{3}\ln 2 \qquad ■$$

Finding the Area Under a Curve

EXAMPLE 4   Find the area of the region R under the graph of $f(x) = e^{(1/2)x}$ from $x = -1$ to $x = 1$.

Solution   The region R is shown in Figure 24. Its area is given by

$$A = \int_{-1}^1 e^{(1/2)x}\, dx$$

In order to evaluate this integral, we make the substitution

$$u = \frac{1}{2}x$$

so that $$du = \frac{1}{2}\, dx \qquad \text{or} \qquad dx = 2\, du$$

When $x = -1$, $u = -\frac{1}{2}$, and when $x = 1$, $u = \frac{1}{2}$. Making the indicated substitutions, we obtain

$$A = \int_{-1}^1 e^{(1/2)x}\, dx = 2\int_{-1/2}^{1/2} e^u\, du$$

$$= 2e^u\,\Big|_{-1/2}^{1/2}$$

$$= 2(e^{1/2} - e^{-1/2})$$

or approximately 2.08 square units.   ■

Average Value of a Function

The *average value* of a function over an interval provides us with an application of the definite integral. Recall that the average value of a set of n numbers is the number

$$\frac{y_1 + y_2 + \cdots + y_n}{n}$$

Now, suppose that f is a continuous function defined on $[a, b]$. Let us divide the interval $[a, b]$ into n subintervals of equal length $(b - a)/n$. Choose points $x_1, x_2, \ldots, x_n$ in the first, second, $\ldots$, and nth subintervals, respectively. Then the average value of the numbers $f(x_1)$, $f(x_2)$, $\ldots$, $f(x_n)$, given by

$$\frac{f(x_1) + f(x_2) + \cdots + f(x_n)}{n}$$

FIGURE 24
Area of $R = \int_{-1}^1 e^{(1/2)x}\, dx$

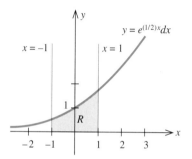

is an approximation of the average of all the values of $f(x)$ on the interval $[a, b]$. This expression can be written in the form

$$\frac{(b-a)}{(b-a)}\left[f(x_1) \cdot \frac{1}{n} + f(x_2) \cdot \frac{1}{n} + \cdots + f(x_n) \cdot \frac{1}{n}\right]$$

$$= \frac{1}{b-a}\left[f(x_1) \cdot \frac{b-a}{n} + f(x_2) \cdot \frac{b-a}{n} + \cdots + f(x_n) \cdot \frac{b-a}{n}\right]$$

$$= \frac{1}{b-a}[f(x_1)\Delta x + f(x_2)\Delta x + \cdots + f(x_n)\Delta x] \tag{9}$$

As n gets larger and larger, the expression (9) approximates the average value of $f(x)$ over $[a, b]$ with increasing accuracy. But the sum inside the brackets in (9) is a Riemann sum of the function f over $[a, b]$. In view of this, we have

$$\lim_{n \to \infty}\left[\frac{f(x_1) + f(x_2) + \cdots + f(x_n)}{n}\right]$$

$$= \frac{1}{b-a}\lim_{n \to \infty}[f(x_1)\Delta x + f(x_2)\Delta x + \cdots + f(x_n)\Delta x]$$

$$= \frac{1}{b-a}\int_a^b f(x)\,dx$$

This discussion motivates the following definition.

THE AVERAGE VALUE OF A FUNCTION

Suppose that f is integrable on $[a, b]$. Then the **average value of f** over $[a, b]$ is

$$\frac{1}{b-a}\int_a^b f(x)\,dx$$

EXAMPLE 5 Find the average value of the function $f(x) = \sqrt{x}$ over the interval $[0, 4]$.

Solution The required average value is given by

$$\frac{1}{4-0}\int_0^4 \sqrt{x}\,dx = \frac{1}{4}\int_0^4 x^{1/2}\,dx$$

$$= \frac{1}{6}x^{3/2}\,\Big|_0^4$$

$$= \frac{4}{3}$$

Applications

EXAMPLE 6 The interest rates charged by Madison Finance on auto loans for used cars over a certain 6-month period in 1993 are approximated by

the function

$$r(t) = -\frac{1}{12}t^3 + \frac{7}{8}t^2 - 3t + 12 \qquad 0 \le t \le 6$$

where t is measured in months and $r(t)$ is the annual percentage rate. What is the average rate on auto loans extended by Madison over the 6-month period?

Solution   The average rate over the 6-month period in question is given by

$$\frac{1}{6-0}\int_0^6 \left(-\frac{1}{12}t^3 + \frac{7}{8}t^2 - 3t + 12\right)dt$$

$$= \frac{1}{6}\left(-\frac{1}{48}t^4 + \frac{7}{24}t^3 - \frac{3}{2}t^2 + 12t\right)\Big|_0^6$$

$$= \frac{1}{6}\left[-\frac{1}{48}(6^4) + \frac{7}{24}(6^3) - \frac{3}{2}(6^2) + 12(6)\right]$$

$$= 9$$

or 9% per year.   ◼

EXAMPLE 7   The amount of a certain drug in a patient's body t days after it has been administered is

$$C(t) = 5e^{-0.2t}$$

units. Determine the average amount of the drug present in the patient's body for the first four days after the drug has been administered.

Solution   The average amount of the drug present in the patient's body for the first four days after it has been administered is given by

$$\frac{1}{4-0}\int_0^4 5e^{-0.2t}\,dt = \frac{5}{4}\int_0^4 e^{-0.2t}\,dt$$

$$= \frac{5}{4}\left[\left(-\frac{1}{0.2}\right)e^{-0.2t}\Big|_0^4\right]$$

$$= \frac{5}{4}(-5e^{-0.8} + 5)$$

$$\approx 3.44$$

or approximately 3.44 units.   ◼

We now give a geometrical interpretation of the average value of a function f over an interval $[a, b]$. Suppose that $f(x)$ is nonnegative so that the definite integral

$$\int_a^b f(x)\,dx$$

gives the area under the graph of f from $x = a$ to $x = b$ (Figure 25). Observe that, in general, the "height" $f(x)$ varies from point to point. Can we replace $f(x)$ by a constant function $g(x) = k$ (which has constant height) such that the areas under each of the two functions f and g are the same?

FIGURE 25
The average value of f over
[a, b] is k.

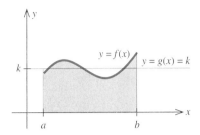

If so, since the area under the graph of g from $x = a$ to $x = b$ is $k(b - a)$, we have

$$k(b - a) = \int_a^b f(x)\,dx$$

or $\qquad\qquad\qquad k = \dfrac{1}{b - a} \int_a^b f(x)\,dx$

so that k is the average value of f over $[a, b]$. Thus, the average value of a function f over an interval $[a, b]$ is the height of a rectangle with base of length $(b - a)$ that has the same area under the graph of f from $x = a$ to $x = b$.

SELF-CHECK
EXERCISES 11.5

1. Evaluate

$$\int_0^2 \sqrt{2x + 5}\,dx$$

2. Find the average value of the function $f(x) = 1 - x^2$ over the interval $[-1, 2]$.

3. The median price of a house in a southwestern state between January 1, 1991, and January 1, 1996, is approximated by the function

$$f(t) = t^3 - 7t^2 + 17t + 130 \qquad 0 \le t \le 5$$

where $f(t)$ is measured in thousands of dollars and t is expressed in years ($t = 0$ corresponds to the beginning of 1991). Determine the average median price of a house over that time interval.

Solutions to Self-Check Exercises 11.5 can be found on page 716.

11.5 EXERCISES

In exercises 1–28, evaluate the definite integral.

1. $\displaystyle\int_0^2 x(x^2 - 1)^3\,dx$

2. $\displaystyle\int_0^1 x^2(2x^3 - 1)^4\,dx$

3. $\displaystyle\int_0^1 x\sqrt{5x^2 + 4}\,dx$

4. $\displaystyle\int_1^3 x\sqrt{3x^2 - 2}\,dx$

5. $\displaystyle\int_0^2 x^2(x^3 + 1)^{3/2}\,dx$

6. $\displaystyle\int_1^5 (2x - 1)^{5/2}\,dx$

7. $\displaystyle\int_0^1 \frac{1}{\sqrt{2x + 1}}\,dx$

8. $\displaystyle\int_0^2 \frac{x}{\sqrt{x^2 + 5}}\,dx$

9. $\displaystyle\int_1^2 (2x - 1)^4\,dx$

10. $\displaystyle\int_1^2 (2x + 4)(x^2 + 4x - 8)^3\,dx$

11. $\int_{-1}^{1} x^2(x^3 + 1)^4 \, dx$

12. $\int_{-1}^{1} (x^3 + \frac{3}{4})(x^4 + 3x)^{-2} \, dx$

13. $\int_{1}^{5} x\sqrt{x - 1} \, dx$

14. $\int_{1}^{4} x\sqrt{x + 1} \, dx$ [*Hint:* Let $u = x + 1$.]

15. $\int_{0}^{2} xe^{x^2} \, dx$ **16.** $\int_{0}^{1} e^{-x} \, dx$

17. $\int_{0}^{1} (e^{2x} + x^2 + 1) \, dx$ **18.** $\int_{0}^{2} (e^t - e^{-t}) \, dt$

19. $\int_{-1}^{1} xe^{x^2+1} \, dx$ **20.** $\int_{0}^{4} \frac{e^{\sqrt{x}}}{\sqrt{x}} \, dx$

21. $\int_{3}^{6} \frac{2}{x - 2} \, dx$ **22.** $\int_{0}^{1} \frac{x}{1 + 2x^2} \, dx$

23. $\int_{1}^{2} \frac{x^2 + 2x}{x^3 + 3x^2 - 1} \, dx$ **24.** $\int_{0}^{1} \frac{e^x}{1 + e^x} \, dx$

25. $\int_{1}^{2} \left(4e^{2u} - \frac{1}{u}\right) du$ **26.** $\int_{1}^{2} \left(1 + \frac{1}{x} + e^x\right) dx$

27. $\int_{1}^{2} \left(2e^{-4x} - \frac{1}{x^2}\right) dx$ **28.** $\int_{1}^{2} \frac{\ln x}{x} \, dx$

In exercises 29–38, find the average value of the function f over the indicated interval [a, b].

29. $f(x) = 2x + 3$; $[0, 2]$

30. $f(x) = 8 - x$; $[1, 4]$

31. $f(x) = 2x^2 - 3$; $[1, 3]$

32. $f(x) = 4 - x^2$; $[-2, 3]$

33. $f(x) = x^2 + 2x - 3$; $[-1, 2]$

34. $f(x) = x^3$; $[-1, 1]$

35. $f(x) = \sqrt{2x + 1}$; $[0, 4]$

36. $f(x) = e^{-x}$; $[0, 4]$ **37.** $f(x) = xe^{x^2}$; $[0, 2]$

38. $f(x) = \frac{1}{x + 1}$; $[0, 2]$

C **39. World Production of Coal** A study proposed in 1980 by researchers from the major producers and consumers of the world's coal concluded that coal could and must play an important role in fueling global economic growth over the next 20 years. The world production of coal in 1980 was 3.5 billion metric tons. If output were to increase at the rate of $3.5e^{0.05t}$ billion metric tons per year in year t ($t = 0$ corresponding to 1980), determine how much coal will be produced worldwide between 1980 and the end of the century.

C **40. Newton's Law of Cooling** A bottle of white wine at room temperature (68°F) is placed in a refrigerator

at 4 P.M. Its temperature after t hours is changing at the rate of

$$-18e^{-0.6t}$$

degrees Fahrenheit per hour. By how many degrees will the temperature of the wine have dropped by 7 P.M.? What will the temperature of the wine be at 7 P.M.?

C **41. Net Investment Flow** The net investment flow (rate of capital formation) of the giant conglomerate LTF Incorporated is projected to be

$$t\sqrt{\frac{1}{2}t^2 + 1}$$

million dollars per year in year t. Find the accruement on the company's capital stock in the second year. *Hint:* The amount is given by

$$\int_{1}^{2} t\sqrt{\frac{1}{2}t^2 + 1} \, dt$$

C **42. Oil Production** Based on a preliminary report by a geological survey team, it is estimated that a newly discovered oil field can be expected to produce oil at the rate of

$$R(t) = \frac{600t^2}{t^3 + 32} + 5 \qquad 0 \leq t \leq 20$$

thousand barrels per year, t years after production begins. Find the amount of oil that the field can be expected to yield during the first five years of production, assuming that the projection holds true.

C **43. Depreciation: Double Declining-Balance Method** Suppose that a tractor purchased at a price of $60,000 is to be depreciated by the *double declining-balance method* over a period of ten years. It can be shown that the rate at which the book value will be decreasing is given by

$$R(t) = 13388.61e^{-0.22314t} \qquad 0 \leq t \leq 10$$

dollars per year at year t. Find the amount by which the book value of the tractor will depreciate over the first five years of its life.

44. Velocity of a Car A car moves along a straight road in such a way that its velocity (in ft/sec) at any time t (in sec) is given by

$$v(t) = 3t\sqrt{16 - t^2} \qquad 0 \leq t \leq 4$$

Find the distance traveled by the car in the four seconds from $t = 0$ to $t = 4$.

C **45. Average Temperature** The temperature (in °F) in Boston over a 12-hour period on a certain December

day was given by

$$T = -0.05t^3 + 0.4t^2 + 3.8t + 5.6 \qquad 0 \le t \le 12$$

where t is measured in hours, with $t = 0$ corresponding to 6 A.M. Determine the average temperature on that day over the 12-hour period from 6 A.M. to 6 P.M.

C **46.** **Whale Population**   A group of marine biologists estimates that if certain conservation measures are implemented the population of an endangered species of whale will be

$$N(t) = 3t^3 + 2t^2 - 10t + 600 \qquad 0 \le t \le 10$$

where $N(t)$ denotes the population at the end of year t. Find the average population of the whales over the next ten years.

C **47.** **Cable TV Subscribers**   The manager of the Tele-Star Cable Television Service estimates that the total number of subscribers to the service in a certain city t years from now will be

$$N(t) = -\frac{40,000}{\sqrt{1 + 0.2t}} + 50,000$$

Find the average number of cable television subscribers over the next five years if this prediction holds true.

48. **Average Yearly Sales**   The sales of the Universal Instruments Company in the first t years of its operation are approximated by the function

$$S(t) = t\sqrt{0.2t^2 + 4}$$

where $S(t)$ is measured in millions of dollars. What were Universal's average yearly sales over its first five years of operation?

49. **Concentration of a Drug in the Bloodstream**   The concentration of a certain drug in a patient's bloodstream t hours after injection is

$$C(t) = \frac{0.2t}{t^2 + 1}$$

milligrams per cubic centimeter. Determine the average concentration of the drug in the patient's bloodstream over the first four hours after the drug is injected.

50. Refer to exercise 44. Find the average velocity of the car over the time interval $[0, 4]$.

51. **Flow of Blood in an Artery**   According to a law discovered by nineteenth-century physician Jean Louis Marie Poiseuille, the velocity of blood (in cm/sec) r centimeters from the central axis of an artery is given

by

$$v(r) = k(R^2 - r^2)$$

where k is a constant and R is the radius of the artery. Find the average velocity of blood along a radius of the artery (see the accompanying figure). [*Hint:* Evaluate $\dfrac{1}{R}\displaystyle\int_0^R v(r)\,dr$.]

Blood vessel

52. Prove property 1 of the definite integral. [*Hint:* Let F be an antiderivative of f, and use the definition of the definite integral.]

53. Prove property 2 of the definite integral. [*Hint:* See exercise 52.]

54. Verify by direct computation that

$$\int_1^3 x^2\,dx = -\int_3^1 x^2\,dx$$

55. Prove property 3 of the definite integral. [*Hint:* See exercise 52.]

56. Verify by direct computation that

$$\int_1^9 2\sqrt{x}\,dx = 2\int_1^9 \sqrt{x}\,dx$$

57. Verify by direct computation that

$$\int_0^1 (1 + x - e^x)\,dx = \int_0^1 dx + \int_0^1 x\,dx - \int_0^1 e^x\,dx$$

What properties of the definite integral are demonstrated in this exercise?

58. Verify by direct computation that

$$\int_0^3 (1 + x^3)\,dx = \int_0^1 (1 + x^3)\,dx + \int_1^3 (1 + x^3)\,dx$$

What property of the definite integral is demonstrated here?

59. Verify by direct computation that

$$\int_0^3 (1 + x^3)\,dx = \int_0^1 (1 + x^3)\,dx$$
$$+ \int_1^2 (1 + x^3)\,dx + \int_2^3 (1 + x^3)\,dx$$

hence showing that property 5 may be extended.

SOLUTIONS TO
SELF-CHECK
EXERCISES 11.5

1. Let $u = 2x + 5$. Then $du = 2\,dx$, or $dx = \frac{1}{2}du$. Also, when $x = 0$, $u = 5$ and when $x = 2$, $u = 9$. Therefore,

$$\int_0^2 \sqrt{2x + 5}\,dx = \int_0^2 (2x + 5)^{1/2}\,dx$$

$$= \frac{1}{2}\int_5^9 u^{1/2}\,du$$

$$= \left(\frac{1}{2}\right)\left(\frac{2}{3}u^{3/2}\right)\Big|_5^9$$

$$= \frac{1}{3}[9^{3/2} - 5^{3/2}]$$

$$= \frac{1}{3}(27 - 5\sqrt{5})$$

2. The required average value is given by

$$\frac{1}{2 - (-1)}\int_{-1}^2 (1 - x^2)\,dx = \frac{1}{3}\int_{-1}^2 (1 - x^2)\,dx$$

$$= \frac{1}{3}\left(x - \frac{1}{3}x^3\right)\Big|_{-1}^2$$

$$= \frac{1}{3}\left[\left(2 - \frac{8}{3}\right) - \left(-1 + \frac{1}{3}\right)\right]$$

$$= 0$$

3. The average median price of a house over the stated time interval is given by

$$\frac{1}{5 - 0}\int_0^5 (t^3 - 7t^2 + 17t + 130)\,dt = \frac{1}{5}\left(\frac{1}{4}t^4 - \frac{7}{3}t^3 + \frac{17}{2}t^2 + 130t\right)\Big|_0^5$$

$$= \frac{1}{5}\left[\frac{1}{4}(5)^4 - \frac{7}{3}(5)^3 + \frac{17}{2}(5)^2 + 130(5)\right]$$

$$= 145.417$$

or $145,417.

11.6 Area Between Two Curves

Suppose that a certain country's petroleum consumption is expected to grow at the rate of $f(t)$ million barrels per year, t years from now, for the next five years. Then the country's total petroleum consumption over the period of time in question is given by the area under the graph of f on the interval $[0, 5]$ (Figure 26).

Next, suppose that because of the implementation of certain energy-conservation measures, the rate of growth of petroleum consumption is expected to be $g(t)$ million barrels per year instead. Then the country's projected total petroleum consumption over the 5-year period is given by the area under the graph of g on the interval $[0, 5]$ (Figure 27).

FIGURE 26

At a rate of consumption f(t) million barrels per year, the total petroleum consumption is given by the area of the region under the graph of f.

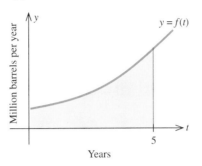

FIGURE 27

At a rate of consumption of g(t) million barrels per year, the total petroleum consumption is given by the area of the region under the graph of g.

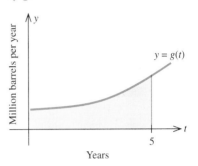

Therefore, the area of the shaded region S lying between the graphs of f and g on the interval $[0, 5]$ (Figure 28) gives the amount of petroleum that would be saved over the 5-year period because of the conservation measures. But the area of S is given by

Area under the graph of f on $[a, b]$ − Area under the graph of g on $[a, b]$

$$= \int_0^5 f(t)\, dt - \int_0^5 g(t)\, dt$$

$$= \int_0^5 [f(t) - g(t)]\, dt \qquad \text{By property 4, Section 11.5}$$

FIGURE 28

The area of S gives the amount of petroleum that would be saved over the 5-year period.

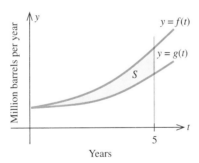

This example shows that some practical problems can be solved by finding the area of a region between two curves, which, in turn, can be found by evaluating an appropriate definite integral.

Finding the Area Between Two Curves

We now turn our attention to the general problem of finding the area of a plane region bounded both above and below by the graphs of functions. First, consider the situation in which the graph of one function lies above that of another. More specifically, let R be the region in the xy-plane (Figure 29) that is bounded above by the graph of a continuous function f; below by a continuous function g, where $f(x) \geq g(x)$ on $[a, b]$; and to the left and

FIGURE 29
Area of
$R = \int_a^b [f(x) - g(x)]\, dx$

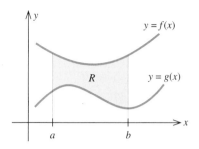

right by the vertical lines $x = a$ and $x = b$, respectively. From the figure, we see that

$$\text{Area of } R = \text{Area under } f(x) - \text{area under } g(x)$$

$$= \int_a^b f(x)\, dx - \int_a^b g(x)\, dx$$

$$= \int_a^b [f(x) - g(x)]\, dx$$

upon using property 4 of the definite integral.

THE AREA BETWEEN TWO CURVES

Let f and g be continuous functions such that $f(x) \geq g(x)$ on the interval $[a, b]$. Then the area of the region bounded above by $y = f(x)$ and below by $y = g(x)$ on $[a, b]$ is given by

$$\int_a^b [f(x) - g(x)]\, dx \qquad \qquad \textbf{(10)}$$

Even though we assumed that both f and g were nonnegative in the derivation of (10), it may be shown that this equation is valid if f and g are not nonnegative (see exercise 48, page 726). Also, observe that if $g(x)$ is 0 for all x—that is, when the lower boundary of the region R is the x-axis—equation (10) gives the area of the region under the curve $y = f(x)$ from $x = a$ to $x = b$, as we would expect.

EXAMPLE 1   Find the area of the region bounded by the x-axis, the graph of $y = -x^2 + 4x - 8$, and the lines $x = -1$ and $x = 4$.

Solution   The region R under consideration is shown in Figure 30. We can view R as the region bounded above by the graph of $f(x) = 0$ (the x-axis) and below by the graph of $g(x) = -x^2 + 4x - 8$ on $[-1, 4]$. Therefore, the area of R is given by

FIGURE 30
Area of $R = -\int_{-1}^4 g(x)\, dx$

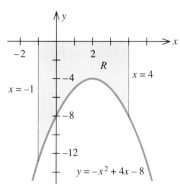

$$\int_a^b [f(x) - g(x)]\, dx = \int_{-1}^4 [0 - (-x^2 + 4x - 8)]\, dx$$

$$= \int_{-1}^4 (x^2 - 4x + 8)\, dx$$

$$= \frac{1}{3}x^3 - 2x^2 + 8x \,\Big|_{-1}^4$$

$$= \left[\frac{1}{3}(64) - 2(16) + 8(4)\right] - \left[\frac{1}{3}(-1) - 2(1) + 8(-1)\right]$$

$$= 31\frac{2}{3} \quad \text{or} \quad 31\frac{2}{3} \text{ square units}$$

EXAMPLE 2   Find the area of the region R bounded by the graphs of

$$f(x) = 2x - 1 \qquad \text{and} \qquad g(x) = x^2 - 4$$

and the vertical lines $x = 1$ and $x = 2$.

Solution   We first sketch the graphs of the functions $f(x) = 2x - 1$ and $g(x) = x^2 - 4$, and the vertical lines $x = 1$ and $x = 2$, and identify the region R whose area is to be calculated (Figure 31).

Since the graph of f always lies above that of g for x in the interval $[1, 2]$, we see by (10) that the required area is given by

$$\int_1^2 [f(x) - g(x)] \, dx = \int_1^2 [(2x - 1) - (x^2 - 4)] \, dx$$

$$= \int_1^2 (-x^2 + 2x + 3) \, dx$$

$$= -\frac{1}{3}x^3 + x^2 + 3x \, \Big|_1^2$$

$$= \left(-\frac{8}{3} + 4 + 6\right) - \left(-\frac{1}{3} + 1 + 3\right) = \frac{11}{3}$$

or $3\frac{2}{3}$ square units.

FIGURE 31
Area of
$R = \int_1^2 [f(x) - g(x)] \, dx$

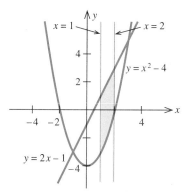

EXAMPLE 3   Find the area of the region R that is completely enclosed by the graphs of the functions

$$f(x) = 2x - 1 \qquad \text{and} \qquad g(x) = x^2 - 4$$

Solution   The region R is shown in Figure 32. First we find the points of intersection of the two curves. To do this, we solve the system that comprises the two equations $y = 2x - 1$ and $y = x^2 - 4$. Equating the two values of y gives

$$x^2 - 4 = 2x - 1$$

$$x^2 - 2x - 3 = 0$$

$$(x + 1)(x - 3) = 0$$

so $x = -1$ or $x = 3$. That is, the two curves intersect when $x = -1$ and $x = 3$.

Observe that we could also view the region R as the region bounded above by the graph of the function $f(x) = 2x - 1$, below by the graph of the function $g(x) = x^2 - 4$, and to the left and right by the vertical lines $x = -1$ and $x = 3$, respectively.

Next, since the graph of the function f always lies above that of the function g on $[-1, 3]$, we can use (10) to compute the desired area:

$$\int_a^b [f(x) - g(x)] \, dx = \int_{-1}^3 [(2x - 1) - (x^2 - 4)] \, dx$$

$$= \int_{-1}^3 (-x^2 + 2x + 3) \, dx$$

$$= -\frac{1}{3}x^3 + x^2 + 3x \, \Big|_{-1}^3$$

$$= (-9 + 9 + 9) - \left(\frac{1}{3} + 1 - 3\right) = \frac{32}{3}$$

FIGURE 32
Area of
$R = \int_{-1}^3 [f(x) - g(x)] \, dx$

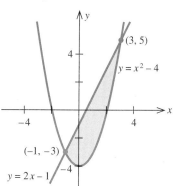

or $10\frac{2}{3}$ square units.

EXAMPLE 4   Find the area of the region R bounded by the graphs of the functions

$$f(x) = x^2 - 2x - 1 \qquad \text{and} \qquad g(x) = -e^x - 1$$

and the vertical lines $x = -1$ and $x = 1$.

FIGURE 33
Area of
$R = \int_{-1}^{1} [f(x) - g(x)]\, dx$

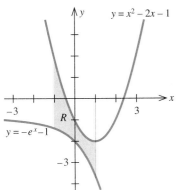

Solution   The region R is shown in Figure 33. Since the graph of the function f always lies above that of the function g, the area of the region R is given by

$$\int_{a}^{b} [f(x) - g(x)]\, dx = \int_{-1}^{1} [(x^2 - 2x - 1) - (-e^x - 1)]\, dx$$

$$= \int_{-1}^{1} (x^2 - 2x + e^x)\, dx$$

$$= \frac{1}{3}x^3 - x^2 + e^x \Big|_{-1}^{1}$$

$$= \left(\frac{1}{3} - 1 + e\right) - \left(-\frac{1}{3} - 1 + e^{-1}\right)$$

$$= \frac{2}{3} + e - \frac{1}{e} \text{ square units} \qquad \blacksquare$$

Equation (10), which gives the area of the region between the curves $y = f(x)$ and $y = g(x)$ for $a \le x \le b$, is valid when the graph of the function f lies above that of the function g over the interval $[a, b]$. Example 5 shows how to use (10) to find the area of a region when the latter condition does not hold.

EXAMPLE 5   Find the area of the region bounded by the graph of the function $f(x) = x^3$, the x-axis, and the lines $x = -1$ and $x = 1$.

Solution   The region R under consideration can be thought of as comprising the two subregions R_1 and R_2, as shown in Figure 34.

FIGURE 34
Area of R_1 = Area of R_2

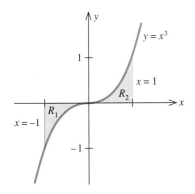

Recall that the x-axis is represented by the function $g(x) = 0$. Since $g(x) \ge f(x)$ on $[-1, 0]$, we see that the area of R_1 is given by

$$\int_{a}^{b} [g(x) - f(x)]\, dx = \int_{-1}^{0} (0 - x^3)\, dx = -\int_{-1}^{0} x^3\, dx$$

$$= -\frac{1}{4}x^4 \Big|_{-1}^{0} = 0 - \left(-\frac{1}{4}\right) = \frac{1}{4}$$

To find the area of R_2, we observe that $f(x) \ge g(x)$ on $[0, 1]$, so it is given by

$$\int_{a}^{b} [f(x) - g(x)]\, dx = \int_{0}^{1} (x^3 - 0)\, dx = \int_{0}^{1} x^3\, dx$$

$$= \frac{1}{4}x^4 \Big|_{0}^{1} = \left(\frac{1}{4}\right) - 0 = \frac{1}{4}$$

Therefore, the area of R is $\frac{1}{4} + \frac{1}{4}$, or $\frac{1}{2}$, square units.

By making use of symmetry, we could have obtained the same result by computing

$$-2 \int_{-1}^{0} x^3\, dx \qquad \text{or} \qquad 2 \int_{0}^{1} x^3\, dx$$

as you may verify. $\qquad \blacksquare$

EXAMPLE 6   Find the area of the region completely enclosed by the graphs of the functions

$$f(x) = x^3 - 3x + 3 \qquad \text{and} \qquad g(x) = x + 3$$

Solution   First sketch the graphs of $y = x^3 - 3x + 3$ and $y = x + 3$, and then identify the required region R. We can view the region R as being composed of the two subregions R_1 and R_2, as shown in Figure 35. By solving the equations $y = x + 3$ and $y = x^3 - 3x + 3$ simultaneously, we find the points of intersection of the two curves. Equating the two values of y, we have

$$x^3 - 3x + 3 = x + 3$$

$$x^3 - 4x = 0$$

or

$$x(x^2 - 4) = 0$$

$$x(x + 2)(x - 2) = 0 \qquad \text{or} \qquad x = 0, -2, \text{ or } 2$$

Hence, the points of intersection of the two curves are $(-2, 1)$, $(0, 3)$, and $(2, 5)$.

For $-2 \le x \le 0$, we see that the graph of the function f lies above that of the function g, so the area of the region R_1 is, by virtue of (10),

$$\int_{-2}^{0} [(x^3 - 3x + 3) - (x + 3)] \, dx = \int_{-2}^{0} (x^3 - 4x) \, dx$$

$$= \frac{1}{4}x^4 - 2x^2 \Big|_{-2}^{0}$$

$$= -(4 - 8)$$

$$= 4 \qquad \text{or} \qquad 4 \text{ square units}$$

For $0 \le x \le 2$, the graph of the function g lies above that of the function f, and the area of R_2 is given by

$$\int_{0}^{2} [(x + 3) - (x^3 - 3x + 3)] \, dx = \int_{0}^{2} (-x^3 + 4x) \, dx$$

$$= -\frac{1}{4}x^4 + 2x^2 \Big|_{0}^{2}$$

$$= -4 + 8$$

$$= 4 \qquad \text{or} \qquad 4 \text{ square units}$$

Therefore, the required area is the sum of the area of the two regions $R_1 + R_2$—that is, $4 + 4$, or 8 square units.

FIGURE 35
Area of R_1 + area of R_2 = $\int_{-2}^{0} [f(x) - g(x)] \, dx$ + $\int_{0}^{2} [g(x) - f(x)] \, dx$

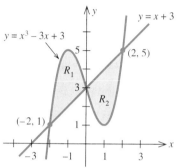

Application

EXAMPLE 7   In a 1989 study for a developing country's Economic Development Board, government economists and energy experts concluded that if the Energy Conservation Bill were implemented in 1990, the country's oil consumption for the next five years would be expected to grow in accordance with the model

$$R(t) = 20e^{0.05t}$$

where t is measured in years ($t = 0$ corresponding to the year 1990) and $R(t)$ in millions of barrels per year. Without the government-imposed conservation measures, however, the expected rate of growth of oil consumption would be given by

$$R_1(t) = 20e^{0.08t}$$

millions of barrels per year. Using these models, determine how much oil would have been saved from 1990 through 1995 if the bill had been implemented.

Solution   Under the Energy Conservation Bill, the total amount of oil that would have been consumed between 1990 and 1995 is given by

$$\int_0^5 R(t)\, dt = \int_0^5 20e^{0.05t}\, dt \qquad (11)$$

Without the bill, the total amount of oil that would have been consumed between 1990 and 1995 is given by

$$\int_0^5 R_1(t)\, dt = \int_0^5 20e^{0.08t}\, dt \qquad (12)$$

Equation (11) may be interpreted as the area of the region under the curve $y = R(t)$ from $t = 0$ to $t = 5$. Similarly, we interpret (12) as the area of the region under the curve $y = R_1(t)$ from $t = 0$ to $t = 5$. Furthermore, note that the graph of $y = R_1(t) = 20e^{0.08t}$ always lies on or above the graph of $y = R(t) = 20e^{0.05t}$ ($t \geq 0$). Thus, the area of the shaded region S in Figure 36 shows the amount of oil that would have been saved from 1990 to 1995 if the Energy Conservation Bill had been implemented. But the area of the region S is given by

$$\int_0^5 [R_1(t) - R(t)]\, dt = \int_0^5 [20e^{0.08t} - 20^{0.05t}]\, dt$$

$$= 20 \int_0^5 (e^{0.08t} - e^{0.05t})\, dt$$

$$= 20 \left(\frac{e^{0.08t}}{0.08} - \frac{e^{0.05t}}{0.05} \right) \Bigg|_0^5$$

$$= 20 \left[\left(\frac{e^{0.4}}{0.08} - \frac{e^{0.25}}{0.05} \right) - \left(\frac{1}{0.08} - \frac{1}{0.05} \right) \right]$$

$$\approx 9.3 \qquad \text{or} \qquad \text{approximately 9.3 square units}$$

Thus, the amount of oil that would be saved is 9.3 million barrels.   ■

FIGURE 36
Area of
$S = \int_0^5 [R_1(t) - R(t)]\, dt$

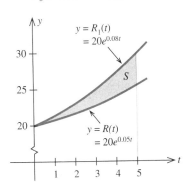

**SELF-CHECK
EXERCISES 11.6**

1. Find the area of the region bounded by the graphs of $f(x) = x^2 + 2$ and $g(x) = 1 - x$ and the vertical lines $x = 0$ and $x = 1$.

2. Find the area of the region completely enclosed by the graphs of $f(x) = -x^2 + 6x + 5$ and $g(x) = x^2 + 5$.

c 3. The management of the Kane Corporation, which operates a chain of hotels, expects its profits to grow at the rate of $1 + t^{2/3}$ million dollars per year t

years from now. However, with renovations and improvements of existing hotels and proposed acquisitions of new hotels, Kane's profits are expected to grow at the rate of $t - 2\sqrt{t} + 4$ million dollars per year in the next decade. What additional profits are expected over the next ten years if the group implements the proposed plans?

Solutions to Self-Check Exercises 11.6 can be found on page 726.

11.6 EXERCISES

In exercises 1–8, find the area of the shaded region.

1.

$y = x^3 - 6x^2$

$y = 0$

6

2.

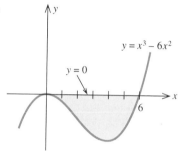

$y = x^4 - 2x^3$

$y = 0$

3.

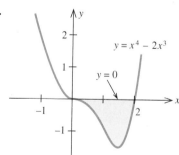

$y = x\sqrt{1 - x^2}$

$y = 0$

$y = 0$

4.

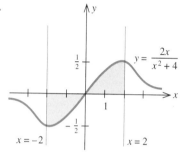

$y = \dfrac{2x}{x^2 + 4}$

$x = -2$ $x = 2$

5.

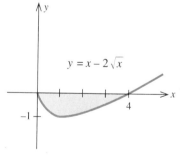

$y = x - 2\sqrt{x}$

4

6.

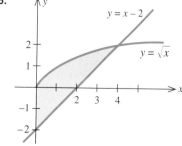

$y = x - 2$

$y = \sqrt{x}$

7.

8.

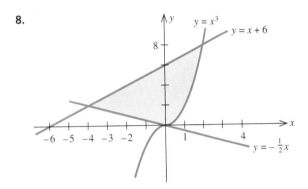

In exercises 9–16, sketch the graph and find the area of the region bounded below by the graph of each function and above by the x-axis from $x = a$ to $x = b$.

9. $f(x) = -x^2$; $a = -1$, $b = 2$

10. $f(x) = x^2 - 4$; $a = -2$, $b = 2$

11. $f(x) = x^2 - 5x + 4$; $a = 1$, $b = 3$

12. $f(x) = x^3$; $a = -1$, $b = 0$

13. $f(x) = -1 - \sqrt{x}$; $a = 0$, $b = 9$

14. $f(x) = \frac{1}{2}x - \sqrt{x}$; $a = 0$, $b = 4$

15. $f(x) = -e^{(1/2)x}$; $a = -2$, $b = 4$

16. $f(x) = -xe^{-x^2}$; $a = 0$, $b = 1$

In exercises 17–26, sketch the graphs of the functions f and g and find the area of the region enclosed by these graphs and the vertical lines $x = a$ and $x = b$.

17. $f(x) = x^2 + 3$, $g(x) = 1$; $a = 1$, $b = 3$

18. $f(x) = x + 2$, $g(x) = x^2 - 4$; $a = -1$, $b = 2$

19. $f(x) = -x^2 + 2x + 3$, $g(x) = -x + 3$; $a = 0$, $b = 2$

20. $f(x) = 9 - x^2$, $g(x) = 2x + 3$; $a = -1$, $b = 1$

21. $f(x) = x^2 + 1$, $g(x) = \frac{1}{3}x^3$; $a = -1$, $b = 2$

22. $f(x) = \sqrt{x}$, $g(x) = -\frac{1}{2}x - 1$; $a = 1$, $b = 4$

23. $f(x) = \frac{1}{x}$, $g(x) = 2x - 1$; $a = 1$, $b = 4$

24. $f(x) = x^2$, $g(x) = \frac{1}{x^2}$; $a = 1$, $b = 3$

25. $f(x) = e^x$, $g(x) = \frac{1}{x}$; $a = 1$, $b = 2$

26. $f(x) = x$, $g(x) = e^{2x}$; $a = 1$, $b = 3$

In exercises 27–34, sketch the graph and find the area of the region bounded by the graph of the function f and the lines $y = 0$, $x = a$, and $x = b$.

27. $f(x) = x$; $a = -1$, $b = 2$

28. $f(x) = x^2 - 2x$; $a = -1$, $b = 1$

29. $f(x) = -x^2 + 4x - 3$; $a = -1$, $b = 2$

30. $f(x) = x^3 - x^2$; $a = -1$, $b = 1$

31. $f(x) = x^3 - 4x^2 + 3x$; $a = 0$, $b = 2$

32. $f(x) = 4x^{1/3} + x^{4/3}$; $a = -1$, $b = 8$

33. $f(x) = e^x - 1$; $a = -1$, $b = 3$

34. $f(x) = xe^{x^2}$; $a = 0$, $b = 2$

In exercises 35–40, sketch the graph and find the area of the region completely enclosed by the graphs of the functions f and g.

35. $f(x) = x + 2$ and $g(x) = x^2 - 4$

36. $f(x) = -x^2 + 4x$ and $g(x) = 2x - 3$

37. $f(x) = x^2$ and $g(x) = x^3$

38. $f(x) = x^3 - 6x^2 + 9x$ and $g(x) = x^2 - 3x$

39. $f(x) = \sqrt{x}$ and $g(x) = x^2$

40. $f(x) = 2x$ and $g(x) = x\sqrt{x + 1}$

41. Effect of Advertising on Revenue In the accompanying figure, the function f gives the rate of change of Odyssey Travel's revenue with respect to the amount x it spends on advertising with their current

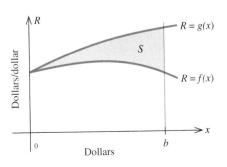

advertising agency. By engaging the services of a different advertising agency, it is expected that Odyssey's revenue will grow at the rate given by the function g. Give an interpretation of the area A of the region S, and find an expression for A in terms of a definite integral involving f and g.

42. Pulse Rate During Exercise   In the accompanying figure, the function f gives the rate of increase of an individual's pulse rate when he walked a prescribed course on a treadmill six months ago. The function g gives the rate of increase of his pulse rate when he recently walked the same prescribed course. Give an interpretation of the area A of the region S and find an expression for A in terms of a definite integral involving f and g.

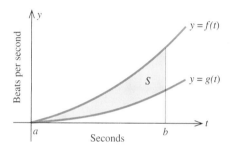

43. Air Purification   In order to study the effectiveness of air purifiers in removing smoke, engineers run each purifier in a smoke-filled 10×20-foot room. In the accompanying figure, the function f gives the rate of change of the smoke level per minute, t minutes after the start of the test, when a Brand A purifier is used. The function g gives the rate of change of the smoke level per minute when a Brand B purifier is used. Give an interpretation of the area of the region S, and find an expression for the area of S in terms of a definite integral involving f and g.

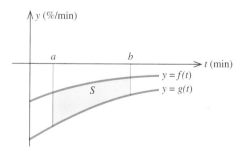

44. Turbo-Charged Engine vs. Standard Engine   In tests conducted by Auto Test Magazine on two identical models of the Phoenix Elite—one equipped with a standard engine and the other with a turbo-charger—it was found that the acceleration of the former is given by

$$a = f(t) = 4 + 0.8t \qquad 0 \le t \le 12$$

feet per second per second, t seconds after starting from rest at full throttle, while the acceleration of the latter is given by

$$a = g(t) = 4 + 1.2t + 0.03t^2 \qquad 0 \le t \le 12$$

feet per second per second. How much faster is the turbo-charged model moving than the model with the standard engine, at the end of a 10-second test run at full throttle?

C 45. Alternative Energy Sources   Because of the increasingly important role played by coal as a viable alternative energy source, the production of coal has been growing at the rate of

$$3.5e^{0.05t}$$

billion metric tons per year t years from 1980 (which corresponds to $t = 0$). Had it not been for the energy crisis, the rate of production of coal since 1980 might have been only

$$3.5e^{0.01t}$$

billion metric tons per year t years from 1980. Determine how much additional coal will be produced between 1980 and the end of the century as an alternate energy source.

C 46. Effect of TV Advertising on Car Sales   Carl Williams, the new proprietor of Carl Williams Auto Sales, estimates that with extensive television advertising, car sales over the next several years could be increasing at the rate of

$$5e^{0.3t}$$

thousand cars per year t years from now, instead of at the current rate of

$$(5 + 0.5t^{3/2})$$

thousand cars per year t years from now. Find how many more cars Mr. Williams expects to sell over the next five years by implementing his advertising plans.

C 47. Population Growth   In an endeavor to curb population growth in a Southeast Asian island state, the government has decided to launch an extensive propaganda campaign. Without curbs, the government expects the rate of population growth to have been

$$60e^{0.02t}$$

thousand people per year t years from now, over the next five years. However, successful implementation of the proposed campaign is expected to result in a population growth rate of

$$-t^2 + 60$$

thousand people per year t years from now, over the next five years. Assuming that the campaign is mounted, how many fewer people will there be in that country five years from now than there would have been if no curbs were imposed?

48. Show that the area of a region R bounded above by the graph of a function f and below by the graph of a function g from $x = a$ to $x = b$ is given by

$$\int_a^b [f(x) - g(x)]\, dx$$

[*Hint:* The validity of the formula was verified earlier for the case when both f and g were nonnegative. Now let f and g be two functions such that $f(x) \geq g(x)$ for $a \leq x \leq b$. Then there exists some nonnegative constant c such that the curves $y = f(x) + c$ and $y = g(x) + c$ are translated in the y-direction in such

a way that the region R' has the same area as the region R (see the accompanying figures). Show that the area of R' is given by

$$\int_a^b \{[f(x) + c] - [g(x) + c]\}\, dx$$

$$= \int_a^b [f(x) - g(x)]\, dx$$

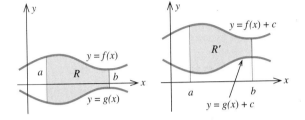

SOLUTIONS TO
SELF-CHECK
EXERCISES 11.6

1. The region in question is shown in the accompanying figure. Since the graph of the function f lies above that of the function g for $0 \leq x \leq 1$, we see that

the required area is given by

$$\int_0^1 [(x^2 + 2) - (1 - x)]\, dx = \int_0^1 (x^2 + x + 1)\, dx$$

$$= \frac{1}{3}x^3 + \frac{1}{2}x^2 + x \,\Big|_0^1$$

$$= \frac{1}{3} + \frac{1}{2} + 1$$

$$= \frac{11}{6}$$

or $\frac{11}{6}$ square units.

2. The region in question is shown in the accompanying figure. To find the points of intersection of the two curves, we solve the equation

$$-x^2 + 6x + 5 = x^2 + 5$$

$$2x^2 - 6x = 0$$

$$2x(x - 3) = 0$$

giving $x = 0$ or $x = 3$. Therefore, the points of intersection are $(0, 5)$ and $(3, 14)$. Since the graph of f always lies above that of g for $0 \leq x \leq 3$, we see

that the required area is given by

$$\int_0^3 [(-x^2 + 6x + 5) - (x^2 + 5)] \, dx = \int_0^3 (-2x^2 + 6x) \, dx$$

$$= -\frac{2}{3}x^3 + 3x^2 \Big|_0^3$$

$$= -18 + 27$$

$$= 9$$

or 9 square units.

3. The additional profits realizable over the next ten years are given by

$$\int_0^{10} [(t - 2\sqrt{t} + 4) - (1 + t^{2/3})] \, dt$$

$$= \int_0^{10} (t - 2t^{1/2} + 3 - t^{2/3}) \, dt$$

$$= \frac{1}{2}t^2 - \frac{4}{3}t^{3/2} + 3t - \frac{3}{5}t^{5/3} \Big|_0^{10}$$

$$= \frac{1}{2}(10)^2 - \frac{4}{3}(10)^{3/2} + 3(10) - \frac{3}{5}(10)^{5/3}$$

$$\approx 9.99$$

or approximately $10 million.

U s i n g   T e c h n o l o g y

The **fnInt** (function integral) operation can be used to find the area between two curves. We do this by using the **fnInt** operation to evaluate an appropriate definite integral or the sum (difference) of appropriate definite integrals. In the following example, the **ISECT** (intersection) operation is also used to advantage to help us find the limits of integration.

EXAMPLE 1 Use a graphing utility to find the area of the region R that is completely enclosed by the graphs of the functions

$$f(x) = 2x^3 - 8x^2 + 4x - 3 \quad \text{and} \quad g(x) = 3x^2 + 10x - 11$$

Solution The graphs of f and g in the viewing rectangle $[-3, 4] \times [-20, 5]$ are shown in Figure 1.

FIGURE 1
The region R is completely enclosed by the graphs of f and g.

Using **ISECT,** we find the x-coordinates of the points of intersection of the two graphs to be approximately -1.04 and 0.65, respectively. Since the graph of f lies above that of g on the interval $[-1.04, 0.65]$, we see that the area of R is given by

$$A = \int_{-1.04}^{0.65} [(2x^3 - 8x^2 + 4x - 3) - (3x^2 + 10x - 11)]\, dx$$

$$= \int_{-1.04}^{0.65} (2x^3 - 11x^2 - 6x + 8)\, dx$$

Using the **fnInt** operation by entering

$$\text{fnInt} (2x\,\hat{}\,3 - 11x\,\hat{}\,2 - 6x + 8, x, -1.04, 0.65)$$

we find $A \approx 9.87$ and so the area of R is approximately 9.87 square units. ∎

EXERCISES

In exercises 1–6, use a graphing utility to (a) plot the graphs of the functions f and g and (b) find the area of the region enclosed by these graphs and the vertical lines $x = a$ and $x = b$. Express your answers accurate to four decimal places.

1. $f(x) = x^3(x - 5)^4$, $g(x) = 0$, $a = 1$, $b = 3$

2. $f(x) = x - \sqrt{1 - x^2}$, $g(x) = 0$, $a = -\dfrac{1}{2}$, $b = \dfrac{1}{2}$

3. $f(x) = x^{1/3}(x + 1)^{1/2}$, $g(x) = x^{-1}$, $a = 1.2$, $b = 2$

4. $f(x) = 2$, $g(x) = \ln(1 + x^2)$, $a = -1$, $b = 1$

5. $f(x) = \sqrt{x}$, $g(x) = \dfrac{x^2 - 3}{x^2 + 1}$, $a = 0$, $b = 3$

6. $f(x) = \dfrac{4}{x^2 + 1}$, $g(x) = x^4$, $a = -1$, $b = 1$

In exercises 7–12, use a graphing utility to (a) plot the graphs of the functions f and g and (b) find the area of the region totally enclosed by the graphs of these functions.

7. $f(x) = 2x^3 - 8x^2 + 4x - 3$ and
$g(x) = -3x^2 + 10x - 10$

8. $f(x) = x^4 - 2x^2 + 2$ and $g(x) = 4 - 2x^2$

9. $f(x) = 2x^3 - 3x^2 + x + 5$ and $g(x) = e^{2x} - 3$

10. $f(x) = \frac{1}{2}x^2 - 3$ and $g(x) = \ln x$

11. $f(x) = xe^{-x}$ and $g(x) = x - 2\sqrt{x}$

12. $f(x) = e^{-x^2}$ and $g(x) = x^4$

11.7 Applications of the Definite Integral to Business and Economics

In this section, we consider several applications of the definite integral in the fields of business and economics.

Consumers' and Producers' Surplus

We begin by deriving a formula for computing the consumers' surplus. Suppose $p = D(x)$ is the demand function that relates the unit price p of a commodity to the quantity x demanded of it. Furthermore, suppose that a fixed unit market price has been established for the commodity and that corresponding to this unit price, the quantity demanded is $\bar{x}$ units (Figure 37). Then those consumers who would be willing to pay a unit price higher than $\bar{p}$ for the commodity would in effect experience a savings. This difference between what the consumers *would* be willing to pay for $\bar{x}$ units of the commodity and what they *actually* pay for them is called the *consumers' surplus.*

FIGURE 37
$D(x)$ is a demand function.

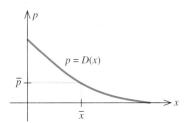

To derive a formula for computing the consumers' surplus, divide the interval $[0, \bar{x}]$ into n subintervals, each of length $\Delta x = \bar{x}/n$, and denote the right endpoints of these subintervals by $x_1, x_2, \ldots, x_n = \bar{x}$ (Figure 38).

Referring once again to Figure 38, we observe that there are consumers who would pay a unit price of at least $D(x_1)$ dollars for the first Δx units of the commodity instead of the market price of $\bar{p}$ dollars per unit. The savings to these consumers is approximated by

$$D(x_1)\Delta x - \bar{p}\Delta x = [D(x_1) - \bar{p}]\Delta x$$

which is the area of the rectangle r_1. Pursuing the same line of reasoning, we find that the savings to the consumers who would be willing to pay a unit price of at least $D(x_2)$ dollars for the next Δx units (from x_1 through x_2) of the commodity, instead of the market price of $\bar{p}$ dollars per unit, is

FIGURE 38
*Approximating consumers'
surplus by the sum of the
rectangles $r_1, r_2, \ldots, r_n$*

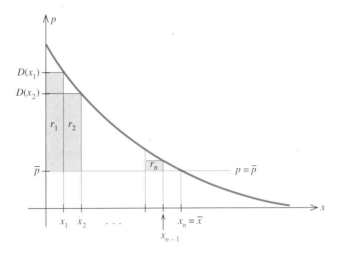

approximated by

$$D(x_2)\Delta x - \overline{p}\Delta x = [D(x_2) - \overline{p}]\Delta x$$

Continuing, we approximate the total savings to the consumers in purchasing $\overline{x}$ units of the commodity by the sum

$$[D(x_1) - \overline{p}]\Delta x + [D(x_2) - \overline{p}]\Delta x + \cdots + [D(x_n) - \overline{p}]\Delta x$$

$$= [D(x_1) + D(x_2) + \cdots + D(x_n)]\Delta x - \underbrace{[\overline{p}\Delta x + \overline{p}\Delta x + \cdots + \overline{p}\Delta x]}_{n \text{ terms}}$$

$$= [D(x_1) + D(x_2) + \cdots + D(x_n)]\,\Delta x - n\overline{p}\Delta x$$

$$= [D(x_1) + D(x_2) + \cdots + D(x_n)]\,\Delta x - \overline{p}\,\overline{x}$$

Now, the first term in the last expression is the Riemann sum of the demand function $p = D(x)$ over the interval $[0, \overline{x}]$ with representative points x_1, $x_2, \ldots, x_n$. Letting n approach infinity, we obtain the following formula for the consumers' surplus CS.

**CONSUMERS'
SURPLUS**

The **consumers' surplus** is given by

$$CS = \int_0^{\overline{x}} D(x)\,dx - \overline{p}\,\overline{x} \tag{13}$$

where D is the demand function, $\overline{p}$ is the unit market price, and $\overline{x}$ is the quantity sold.

It is the area of the region bounded above by the demand curve $p = D(x)$ and below by the straight line $p = \overline{p}$ from $x = 0$ to $x = \overline{x}$ (Figure 39). We can also see this if we rewrite (13) in the form

$$\int_0^{\overline{x}} [D(x) - \overline{p}]\,dx$$

and interpret the result geometrically.

FIGURE 39
Consumers' surplus

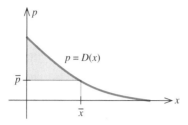

Analogously, we can derive a formula for computing the producers' surplus. Suppose that $p = S(x)$ is the supply equation that relates the unit price p of a certain commodity to the quantity x that the supplier will make available in the market at that price.

FIGURE 40
S(x) is a supply function.

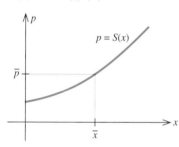

Again, suppose that a fixed market price $\bar{p}$ has been established for the commodity and that, corresponding to this unit price, a quantity of $\bar{x}$ units will be made available in the market by the supplier (Figure 40). Then the suppliers who would be willing to make the commodity available at a lower price stand to gain from the fact that the market price is set as such. The difference between what the suppliers actually receive and what they would be willing to receive is called the *producers' surplus*. Proceeding in a manner similar to the derivation of the equation for computing the consumers' surplus, we find that the producers' surplus PS is defined as follows:

THE PRODUCERS' SURPLUS

The **producers' surplus** is given by

$$PS = \bar{p}\bar{x} - \int_0^{\bar{x}} S(x)\, dx \qquad (14)$$

where $S(x)$ is the supply function, $\bar{p}$ is the unit market price, and $\bar{x}$ is the quantity supplied.

Geometrically, the producers' surplus is given by the area of the region bounded above by the straight line $p = \bar{p}$ and below by the supply curve $p = S(x)$ (Figure 41).

FIGURE 41
Producers' surplus

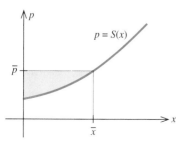

We can also show that the last statement is true by converting (14) to the form

$$\int_0^{\bar{x}} [\bar{p} - S(x)]\, dx$$

and interpreting the definite integral geometrically.

EXAMPLE 1   The demand function for a certain make of ten-speed bicycle is given by

$$p = D(x) = -0.001x^2 + 250$$

where p is the unit price in dollars and x is the quantity demanded in units of a thousand. The supply function for these bicycles is given by

$$p = S(x) = 0.0006x^2 + 0.02x + 100$$

where p stands for the unit price in dollars and x stands for the number of bicycles that the supplier will put on the market, in units of a thousand. Determine the consumers' surplus and the producers' surplus if the market price of a bicycle is set at the equilibrium price.

Solution   Recall that the equilibrium price is the unit price of the commodity when market equilibrium occurs. We determine the equilibrium price by solving for the point of intersection of the demand curve and the supply curve (Figure 42). To solve the system of equations

$$p = -0.001x^2 + 250$$

and

$$p = 0.0006x^2 + 0.02x + 100$$

we simply substitute the first equation into the second, obtaining

$$0.0006x^2 + 0.02x + 100 = -0.001x^2 + 250$$

$$0.0016x^2 + 0.02x - 150 = 0$$

$$16x^2 + 200x - 1,500,000 = 0$$

$$2x^2 + 25x - 187,500 = 0$$

Factoring this last equation, we obtain

$$(2x + 625)(x - 300) = 0$$

Thus, $x = -\frac{625}{2}$ or $x = 300$. The first number lies outside the interval of interest, so we are left with the solution $x = 300$, with a corresponding value of

$$p = -0.001(300)^2 + 250 = 160$$

Thus, the equilibrium point is (300, 160); that is, the equilibrium quantity is 300,000 and the equilibrium price is \$160. Setting the market price at \$160 per unit and using (13) with $\bar{p} = 160$ and $\bar{x} = 300$, we find that the consumers' surplus is given by

$$CS = \int_0^{300} (-0.001x^2 + 250)\, dx - (160)(300)$$

$$= \left(-\frac{1}{3,000}x^3 + 250x \right) \Bigg|_0^{300} - 48,000$$

$$= -\frac{300^3}{3,000} + (250)(300) - 48,000$$

$$= 18,000 \qquad \text{or} \qquad \$18,000,000$$

FIGURE 42

Consumers' surplus and producers' surplus when market price = equilibrium price

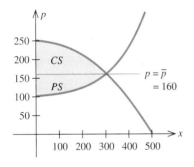

(Recall that x is measured in units of a thousand.) Next, using (14), we find that the producer's surplus is given by

$$PS = (160)(300) - \int_0^{300} (0.0006x^2 + 0.02x + 100)\, dx$$

$$= 48{,}000 - (0.0002x^3 + 0.01x^2 + 100x)\Big|_0^{300}$$

$$= 48{,}000 - [(0.0002)(300)^3 + (0.01)(300)^2 + 100(300)]$$

$$= 11{,}700 \quad \text{or} \quad \$11{,}700{,}000 \qquad \blacksquare$$

The Future and Present Value of an Income Stream

To introduce the notion of the future and the present value of an income stream, suppose that a firm generates a stream of income over a period of time—the revenue generated by a large chain of retail stores over a 5-year period, for example. As the income is realized, it is reinvested and earns interest at a fixed rate. The *accumulated future income stream* over the 5-year period is the amount of money the firm ends up with at the end of that period.

The definite integral can be used to determine this *accumulated,* or *total, future income stream* over a period of time. The total future value of an income stream gives us a way to measure the value of such a stream. To find the total future value of an income stream, suppose that

$$R(t) = \text{Rate of income generation at any time } t \qquad \text{Dollars per year}$$

$$r = \text{Interest rate compounded continuously}$$

and $T = \text{Term}$ In years

Let us divide the time interval $[0, T]$ into n subintervals of equal length $\Delta t = T/n$ and denote the right endpoints of these intervals by $t_1, t_2, \ldots, t_n = T$, as shown in Figure 43.

If R is a continuous function on $[0, T]$, then $R(t)$ will not differ by much from $R(t_1)$ in the subinterval $[0, t_1]$ provided that the subinterval is small (which is true if n is large). Therefore, the income generated over the time interval $[0, t_1]$ is approximately

$$R(t_1)\Delta t \qquad \text{Constant rate of income} \cdot \text{length of time}$$

dollars. The future value of this amount, T years from now, calculated as if it were earned at time t_1, is

$$[R(t_1)\Delta t]e^{r(T-t_1)} \qquad \text{Equation (3), Section 11.1}$$

dollars. Similarly, the income generated over the time interval $[t_1, t_2]$ is approximately $P(t_2)\Delta t$ dollars and has a future value, T years from now, of approximately

$$[R(t_2)\Delta t]e^{r(T-t_2)}$$

dollars. Therefore, the sum of the future values of the income stream generated over the time interval $[0, T]$ is approximately

$$R(t_1)e^{r(T-t_1)}\Delta t + R(t_2)e^{r(T-t_2)}\Delta t + \cdots + R(t_n)e^{r(T-t_n)}\Delta t$$

$$= e^{rT}[R(t_1)e^{-rt_1}\Delta t + R(t_2)e^{-rt_2}\Delta t + \cdots + R(t_n)e^{-rt_n}\Delta t]$$

FIGURE 43

The time interval $[0, T]$ is partitioned into n subintervals.

$$0 \quad t_1 \quad t_2 \quad t_3 \cdots t_{n-1} \quad t_n = T$$

dollars. But this sum is just the Riemann sum of the function $e^{rT}R(t)e^{-rt}$ over the interval $[0, T]$ with representative points $t_1, t_2, \ldots, t_n$. Letting n approach infinity, we obtain the following result.

ACCUMULATED OR TOTAL FUTURE VALUE OF AN INCOME STREAM

The **accumulated,** or **total, future value** after T years of an income stream of $R(t)$ dollars per year, earning interest at the rate of r per year compounded continuously, is given by

$$A = e^{rT} \int_0^T R(t)e^{-rt}\, dt \qquad (15)$$

EXAMPLE 2   Crystal Car Wash recently bought an automatic car-washing machine that is expected to generate \$40,000 in revenue per year, t years from now, for the next five years. If the income is reinvested in a business earning interest at the rate of 12% per year compounded continuously, find the total accumulated value of this income stream at the end of five years.

Solution   We are required to find the total future value of the given income stream after five years. Using (15) with $R(t) = 40,000$, $r = 0.12$, and $T = 5$, we see that the required value is given by

$$e^{0.12(5)} \int_0^5 40,000e^{-0.12t}\, dt$$

$$= e^{0.6}\left[-\frac{40,000}{0.12}e^{-0.12t}\right]\Bigg|_0^5 \qquad \text{Integrate using the substitution } u = -0.12t.$$

$$= -\frac{40,000e^{0.6}}{0.12}(e^{-0.6} - 1) \approx 274,039.60,$$

or approximately \$274,040.

Another way of measuring the value of an income stream is by considering its present value. The *present value* of an income stream of $R(t)$ dollars per year over a term of T years, earning interest at the rate of r per year compounded continuously, is the principal P that will yield the same accumulated value as the income stream itself when P is invested today for a period of T years at the same rate of interest. In other words,

$$Pe^{rT} = e^{rT} \int_0^T R(t)e^{-rt}\, dt$$

Dividing both sides of the equation by e^{rT} gives the following result.

PRESENT VALUE OF AN INCOME STREAM

The **present value of an income stream** of $R(t)$ dollars per year, earning interest at the rate of r per year compounded continuously, is given by

$$PV = \int_0^T R(t)e^{-rt}\, dt \qquad (16)$$

EXAMPLE 3   The owner of a local cinema is considering two alternative plans for renovating and improving the theater. Plan A calls for an immediate cash outlay of $250,000, whereas plan B requires an immediate cash outlay of $180,000. It has been estimated that adopting plan A would result in a net income stream generated at the rate of

$$f(t) = 630,000$$

dollars per year, whereas adopting plan B would result in a net income stream generated at the rate of

$$g(t) = 580,000$$

dollars per year for the next three years. If the prevailing interest rate for the next five years were 10% per year, which plan would generate a higher net income by the end of three years?

Solution   Since the initial outlay is $250,000, we find, using equation (16), with $R(t) = 630,000$, $r = 0.1$, and $T = 3$, that the present value of the net income under plan A is given by

$$\int_0^3 630,000 e^{-0.1t}\, dt - 250,000$$

$$= \frac{630,000}{-0.1} e^{-0.1t}\Big|_0^3 - 250,000 \qquad \text{Integrate using the substitution } u = -0.1t.$$

$$= -6,300,000 e^{-0.3} + 6,300,000 - 250,000$$

$$\approx 1,382,845$$

or approximately $1,382,845.

To find the present value of the net income under plan B, we use (16) with $R(t) = 580,000$, $r = 0.1$, and $T = 3$, obtaining

$$\int_0^3 580,000 e^{-0.1t}\, dt - 180,000$$

dollars. Proceeding as in the previous computation, we see that the required value is $1,323,254 (exercise 8, page 740).

Comparing the present value of each plan, we conclude that plan A would generate a higher net income by the end of three years.   ■

REMARK   The function R in Example 3 is a constant function. If R is not a constant function, then we may need more sophisticated techniques of integration to evaluate the integral in (16).   □

The Amount and Present Value of an Annuity

An annuity is a sequence of payments made at regular time intervals. The time period in which these payments are made is called the *term* of the annuity. Although the payments need not be equal in size, they are equal in many important applications, and we will assume that they are equal in our discussion. Examples of annuities are regular deposits to a savings account, monthly home mortgage payments, and monthly insurance payments.

The *amount of an annuity* is the sum of the payments plus the interest earned. A formula for computing the amount of an annuity A can be derived with the help of (15). Let

$$P = \text{Size of each payment in the annuity}$$

$$r = \text{Interest rate compounded continuously}$$

$$T = \text{Term of the annuity (in years)}$$

and $$m = \text{Number of payments per year}$$

The payments into the annuity constitute a constant income stream of $R(t) = mP$ dollars per year. With this value of $R(t)$, (15) yields

$$A = e^{rT} \int_0^T R(t) e^{-rt}\, dt = e^{rT} \int_0^T mP e^{-rt}\, dt$$

$$= mP e^{rT} \left[-\frac{e^{-rt}}{r} \right]\Bigg|_0^T$$

$$= mP e^{rT} \left[-\frac{e^{-rT}}{r} + \frac{1}{r} \right] = \frac{mP}{r}(e^{rT} - 1)$$

This leads us to the following formula.

AMOUNT OF AN ANNUITY

The **amount of an annuity** is

$$A = \frac{mP}{r}(e^{rT} - 1) \tag{17}$$

where P, r, T, and m are as defined earlier.

 EXAMPLE 4   On January 1, 1985, Mr. Chapman deposited $2000 into an Individual Retirement Account (IRA) paying interest at the rate of 10% per year compounded continuously. Assuming that he deposits $2000 annually into the account, how much will he have in his IRA at the beginning of the year 2001?

Solution   We use (17), with $P = 2{,}000$, $r = 0.1$, $T = 16$, and $m = 1$, obtaining

$$A = \frac{2{,}000}{0.1}\left(e^{1.6} - 1\right)$$

$$\approx 79{,}060.65$$

Thus, Mr. Chapman will have approximately $79,061 in his account at the beginning of the year 2001. ■

Using (16), we can derive the following formula for the *present value* of an annuity.

PRESENT VALUE OF AN ANNUITY

The **present value of an annuity** is given by

$$PV = \frac{mP}{r}(1 - e^{-rT}) \tag{18}$$

where P, r, T, and m are as defined earlier.

EXAMPLE 5   Mr. Carson, the proprietor of a hardware store, wants to establish a fund from which he will withdraw $1000 per month for the next ten years. If the fund earns interest at the rate of 9% per year compounded continuously, how much money does he need to establish the fund?

Solution   We want to find the present value of an annuity with $P = 1,000$, $r = 0.09$, $T = 10$, and $m = 12$. Using (18), we find

$$PV = \frac{12,000}{0.09}(1 - e^{-(0.09)(10)})$$

$$\approx 79,124.04$$

Thus, Mr. Carson needs approximately $79,124 to establish the fund.

Lorentz Curves and Income Distributions

One method used by economists to study the distribution of income in a society is based on the **Lorentz curve,** named after American statistician M. D. Lorentz. To describe the Lorentz curve, let $f(x)$ denote the proportion of the total income received by the poorest $100x\%$ of the population for $0 \leq x \leq 1$. Using this terminology, $f(0.3) = 0.1$ simply states that the lowest 30% of the income recipients receive 10% of the total income.

The function f has the following properties:

1. The domain of f is $[0, 1]$.
2. The range of f is $[0, 1]$.
3. $f(0) = 0$ and $f(1) = 1$.
4. $f(x) \leq x$ for every x in $[0, 1]$.
5. f is increasing on $[0, 1]$.

The first two properties follow from the fact that both x and $f(x)$ are fractions of a whole. Property 3 is a statement that 0% of the income recipients receive 0% of the total income and 100% of the income recipients receive 100% of the total income. Property 4 follows from the fact that the lowest $100x\%$ of the income recipients cannot receive more than $100x\%$ of the total income. A typical Lorentz curve is shown in Figure 44.

FIGURE 44

A Lorentz curve

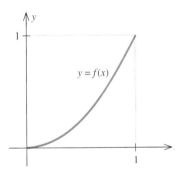

FIGURE 45

The Lorentz curve
$f(x) = \frac{19}{20}x^2 + \frac{1}{20}x$

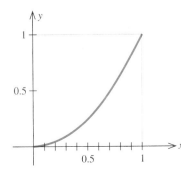

EXAMPLE 6   A developing country's income distribution is described by the function

$$f(x) = \frac{19}{20}x^2 + \frac{1}{20}x$$

a. Sketch the Lorentz curve for the given function.
b. Compute $f(0.2)$ and $f(0.8)$ and interpret your results.

Solution

a. The Lorentz curve is shown in Figure 45.

b.
$$f(0.2) = \frac{19}{20}(0.2)^2 + \frac{1}{20}(0.2) = 0.048$$

so that the lowest 20% of the people receive 4.8% of the total income.

$$f(0.8) = \frac{19}{20}(0.8)^2 + \frac{1}{20}(0.8) = 0.648$$

so that the lowest 80% of the people receive 64.8% of the total income.

■

Next, let us consider the Lorentz curve described by the function $y = f(x) = x$. Since exactly $100x\%$ of the total income is received by the lowest $100x\%$ of income recipients, the line $y = x$ is called the *line of complete equality*. For example, 10% of the total income is received by the lowest 10% of income recipients, 20% of the total income is received by the lowest 20% of income recipients, and so on. Now, it is evident that the closer a Lorentz curve is to this line, the more equitable the income distribution is among the income recipients. But the proximity of a Lorentz curve to the line of complete equality is reflected by the area between the Lorentz curve and the line $y = x$ (Figure 46). The closer the curve is to the line, the smaller the enclosed area.

This observation suggests that we may define a number, called the *coefficient of inequality*, or *Gini Index*, of a Lorentz curve, as the ratio of the area between the line of complete equality and the Lorentz curve to the area under the line of complete equality. Since the area under the line of complete equality is $\frac{1}{2}$, we see that the coefficient of inequality is given by the following formula.

FIGURE 46

The closer the Lorentz curve is to the line, the more equitable the income distribution.

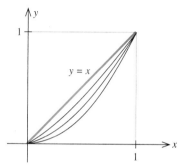

COEFFICIENT OF INEQUALITY OF A LORENTZ CURVE

The **coefficient of inequality** of a Lorentz curve is

$$L = 2\int_0^1 [x - f(x)]\, dx \qquad (19)$$

The coefficient of inequality is a number between 0 and 1. For example, a coefficient of zero implies that the income distribution is perfectly uniform.

EXAMPLE 7   In a study conducted by a certain country's Economic Development Board with regard to the income distribution of certain segments of the country's work force, it was found that the Lorentz curves for the distribution of income of medical doctors and of movie actors and actresses are described by the functions

$$f(x) = \frac{14}{15}x^2 + \frac{1}{15}x \qquad \text{and} \qquad g(x) = \frac{5}{8}x^4 + \frac{3}{8}x$$

respectively. Compute the coefficient of inequality for each Lorentz curve. Which profession has a more equitable income distribution?

Solution   The required coefficients of inequality are, respectively,

$$L_1 = 2\int_0^1 \left[x - \left(\frac{14}{15}x^2 + \frac{1}{15}x \right) \right] dx = 2\int_0^1 \left(\frac{14}{15}x - \frac{14}{15}x^2 \right) dx$$

$$= \frac{28}{15}\int_0^1 (x - x^2)\, dx = \frac{28}{15}\left(\frac{1}{2}x^2 - \frac{1}{3}x^3 \right)\Big|_0^1$$

$$= \frac{14}{45} \approx 0.311$$

and $$L_2 = 2 \int_0^1 \left[x - \left(\frac{5}{8} x^4 + \frac{3}{8} x \right) \right] dx = 2 \int_0^1 \left(\frac{5}{8} x - \frac{5}{8} x^4 \right) dx$$

$$= \frac{5}{4} \int_0^1 (x - x^4) \, dx = \frac{5}{4} \left(\frac{1}{2} x^2 - \frac{1}{5} x^5 \right) \Big|_0^1$$

$$= \frac{15}{40} \approx 0.375$$

We conclude that in this country the incomes of medical doctors are more evenly distributed than the incomes of movie actors and actresses.   ■

**SELF-CHECK
EXERCISE 11.7**

c The demand function for a certain make of exercise bicycle that is sold exclusively through cable television is

$$p = d(x) = \sqrt{9 - 0.02x}$$

where p is the unit price in hundreds of dollars and x is the quantity demanded per week. The corresponding supply function is given by

$$p = s(x) = \sqrt{1 + 0.02x}$$

where p has the same meaning as before and x is the number of exercise bicycles the supplier will make available at price p. Determine the consumers' surplus and the producers' surplus if the unit price is set at the equilibrium price.

The Solution to Self-Check Exercise 11.7 can be found on page 741.

11.7 EXERCISES

c *A calculator is recommended for this exercise set.*

1. Consumers' Surplus   The demand function for a certain make of cartridge typewriter ribbon is given by

$$p = -0.01x^2 - 0.1x + 6$$

where p is the unit price in dollars and x is the quantity demanded each week, measured in units of a thousand. Determine the consumers' surplus if the market price is set at $4 per cartridge.

2. Consumers' Surplus   The demand function for a certain brand of compact discs is given by

$$p = -0.01x^2 - 0.2x + 8$$

where p is the wholesale unit price in dollars and x is the quantity demanded each week, measured in units of a thousand. Determine the consumers' surplus if the wholesale market price is set at $5 per disc.

3. Consumers' Surplus   It is known that the quantity demanded of a certain make of portable hair dryer is x hundred units per week, and the corresponding wholesale unit price is

$$p = \sqrt{225 - 5x}$$

dollars. Determine the consumers' surplus if the wholesale market price is set at $10 per unit.

4. Producers' Surplus   The supplier of the portable hair dryers in exercise 3 will make x hundred units of hair dryers available in the market when the wholesale unit price is

$$p = \sqrt{36 + 1.8x}$$

dollars. Determine the producers' surplus if the wholesale market price is set at $9 per unit.

5. **Producers' Surplus**   The supply function for the compact discs of exercise 2 is given by

$$p = 0.01x^2 + 0.1x + 3$$

where p is the unit wholesale price in dollars and x stands for the quantity that will be made available in the market by the supplier, measured in units of a thousand. Determine the producers' surplus if the wholesale market price is set at the equilibrium price.

6. **Consumers' and Producers' Surplus**   The management of the Titan Tire Company has determined that the quantity demanded x of their Super Titan tires per week is related to the unit price p by the relation

$$p = 144 - x^2$$

where p is measured in dollars and x is measured in units of a thousand. Titan will make x units of the tires available in the market if the unit price is

$$p = 48 + \frac{1}{2}x^2$$

dollars. Determine the consumers' surplus and the producers' surplus when the market unit price is set at the equilibrium price.

7. **Consumers' and Producers' Surplus**   The quantity demanded x (in units of a hundred) of the Mikado miniature cameras per week is related to the unit price p (in dollars) by

$$p = -0.2x^2 + 80$$

and the quantity x (in units of a hundred) that the supplier is willing to make available in the market is related to the unit price p (in dollars) by

$$p = 0.1x^2 + x + 40$$

If the market price is set at the equilibrium price, find the consumers' surplus and the producers' surplus.

8. Refer to Example 3, page 735. Verify that

$$\int_0^3 580,000e^{-0.1t}\, dt - 180,000 \approx 1,323,254$$

9. **Present Value of an Investment**   Suppose that an investment is expected to generate income at the rate of

$$R(t) = 200,000$$

dollars per year for the next five years. Find the present value of this investment if the prevailing interest rate is 8% per year compounded continuously.

10. **Franchises**   Ms. Gordon purchased a 15-year franchise for a computer outlet store that is expected to generate income at the rate of

$$R(t) = 400,000$$

dollars per year. If the prevailing interest rate is 10% per year compounded continuously, find the present value of the franchise.

11. **The Amount of an Annuity**   Find the amount of an annuity if $250 a month is paid into it for a period of 20 years earning interest at the rate of 8% per year compounded continuously.

12. **The Amount of an Annuity**   Find the amount of an annuity if $400 a month is paid into it for a period of 20 years earning interest at the rate of 8% per year compounded continuously.

13. **The Amount of an Annuity**   Mr. Jorgenson deposits $150 per month in a savings account paying 8% per year compounded continuously. Estimate the amount that will be in his account after 15 years.

14. **Custodial Accounts**   The Armstrongs wish to establish a custodial account to finance their children's education. If they deposit $200 monthly for ten years in a savings account paying 9% per year compounded continuously, find how much their savings account will be worth at the end of this period.

15. **IRA Accounts**   Refer to Example 4, page 736. Suppose that Mr. Chapman makes his IRA payment on April 1, 1985, and annually thereafter. If interest is paid at the same initial rate, approximately how much will Mr. Chapman have in his account at the beginning of the year 2001?

16. **Present Value of an Annuity**   Estimate the present value of an annuity if payments are $800 monthly for 12 years and the account earns interest at the rate of 10% per year compounded continuously.

17. **Present Value of an Annuity**   Estimate the present value of an annuity if payments are $1200 monthly for 15 years and the account earns interest at the rate of 10% per year compounded continuously.

18. **Lottery Payments**   A state lottery commission pays the winner of the "Million Dollar" lottery 20 annual installments of $50,000 each. If the prevailing interest rate is 8% per year compounded continuously, find the present value of the winning ticket.

19. **Reverse Annuity Mortgages**   Mr. Sinclair wishes to supplement his retirement income by $300 per month for the next ten years. He plans to obtain a reverse annuity mortgage (RAM) on his home to meet this need. Estimate the amount of the mortgage he will

require if the prevailing interest rate is 12% per year compounded continuously.

20. Reverse Annuity Mortgage Refer to exercise 19. Miss Santos wishes to supplement her retirement income by $400 per month for the next 15 years by obtaining a RAM. Estimate the amount of the mortgage she will require if the prevailing interest rate is 9% per year compounded continuously.

21. Lorentz Curves A certain country's income distribution is described by the function

$$f(x) = \frac{15}{16}x^2 + \frac{1}{16}x$$

a. Sketch the Lorentz curve for this function.
b. Compute $f(0.4)$ and $f(0.9)$ and interpret your results.

22. Lorentz Curves In a study conducted by a certain country's Economic Development Board, it was found that the Lorentz curve for the distribution of income of college teachers was described by the function

$$f(x) = \frac{13}{14}x^2 + \frac{1}{14}x$$

and that of lawyers by the function

$$g(x) = \frac{9}{11}x^4 + \frac{2}{11}x$$

a. Compute the coefficient of inequality for each Lorentz curve.
b. Which profession has a more equitable income distribution?

23. Lorentz Curves A certain country's income distribution is described by the function

$$f(x) = \frac{14}{15}x^2 + \frac{1}{15}x$$

a. Sketch the Lorentz curve for this function.
b. Compute $f(0.3)$ and $f(0.7)$.

24. Lorentz Curves In a study conducted by a certain country's Economic Development Board, it was found that the Lorentz curve for the distribution of income of stockbrokers was described by the function

$$f(x) = \frac{11}{12}x^2 + \frac{1}{12}x$$

and that of high school teachers by the function

$$g(x) = \frac{5}{6}x^2 + \frac{1}{6}x$$

a. Compute the coefficient of inequality for each Lorentz curve.
b. Which profession has a more equitable income distribution?

SOLUTION TO SELF-CHECK EXERCISE 11.7

We find the equilibrium price and equilibrium quantity by solving the system of equations

$$p = \sqrt{9 - 0.02x}$$

and

$$p = \sqrt{1 + 0.02x}$$

simultaneously. Substituting the first equation into the second, we have

$$\sqrt{9 - 0.02x} = \sqrt{1 + 0.02x}$$

Squaring both sides of the equation then leads to

$$9 - 0.02x = 1 + 0.02x$$

giving

$$x = 200$$

Therefore,

$$p = \sqrt{9 - 0.02(200)}$$
$$= \sqrt{5} \approx 2.24$$

and the equilibrium price is $224 and the equilibrium quantity is 200. The consumers' surplus is given by

$$CS = \int_0^{200} \sqrt{9 - 0.02x}\, dx - (2.24)(200)$$

$$= \int_0^{200} (9 - 0.02x)^{1/2}\, dx - 448 \qquad \text{Integrating by substitution}$$

$$= -\frac{1}{0.02}\left(\frac{2}{3}\right)(9 - 0.02x)^{3/2}\Big|_0^{200} - 448$$

$$= -\frac{1}{0.03}(5^{3/2} - 9^{3/2}) - 448$$

$$\approx 79.32$$

or approximately \$7,932.

Next, the producers' surplus is given by

$$PS = (2.24)(200) - \int_0^{200} \sqrt{1 + 0.02x}\, dx$$

$$= 448 - \int_0^{200} (1 + 0.02x)^{1/2}\, dx$$

$$= 448 - \frac{1}{0.02}\left(\frac{2}{3}\right)(1 + 0.02x)^{3/2}\Big|_0^{200}$$

$$= 448 - \frac{1}{0.03}(5^{3/2} - 1)$$

$$\approx 108.66$$

or approximately \$10,866.

Using Technology

1. Resolve Example 1, Section 11.7 using a graphing utility. [*Hint:* Use the **ISECT** operation to find the equilibrium quantity and the equilibrium price. Use the **fnInt** operation to evaluate the definite integral.]

2. Resolve exercise 7, Exercises 11.7, using a graphing utility. [*Hint:* See exercise 1.]

3. The demand function for a certain brand of travel alarm clocks is given by

$$p = -0.01x^2 - 0.3x + 10$$

where p is the wholesale unit price in dollars and x is the quantity demanded each month, measured in units of a thousand. The supply function for this brand of clocks is given by

$$p = -0.01x^2 + 0.2x + 4$$

where p has the same meaning as before and x is the quantity, in thousands, the supplier will make available in the marketplace per month. Determine the consumers' surplus and the producers' surplus when the market unit price is set at the equilibrium price.

4. The quantity demanded of a certain make of compact disc organizer is x thousand units per week, and the corresponding wholesale unit price is

$$p = \sqrt{400 - 8x}$$

dollars. The supplier of the organizers will make x thousand units available in the market when the unit wholesale price is

$$p = 0.02x^2 + 0.04x + 5$$

dollars. Determine the consumers' surplus and the producers' surplus when the market unit price is set at the equilibrium price.

CHAPTER ELEVEN SUMMARY OF PRINCIPAL FORMULAS AND TERMS

Formulas

1. Indefinite integral of a constant

$$\int k\, du = ku + C$$

2. Power Rule

$$\int u^n\, du = \frac{u^{n+1}}{n+1} + C$$

3. Constant multiple rule

$$\int kf(u)\, du = k \int f(u)\, du$$

$(k$ a constant$)$

4. Sum Rule

$$\int [f(u) \pm g(u)]\, du$$
$$= \int f(u)\, du \pm \int g(u)\, du$$

5. Indefinite integral of the exponential function

$$\int e^u\, du = e^u + C$$

6. Indefinite integral of $f(u) = \dfrac{1}{u}$

$$\int \frac{du}{u} = \ln |u| + C$$

7. Method of substitution

$$\int f(g(x))g'(x)\, dx = \int f(u)\, du$$

8. Fundamental Theorem of Calculus

$$\int_a^b f(x)\, dx$$
$$= F(b) - F(a),\ F'(x) = f(x)$$

9. Area between two curves

$$\int_a^b [f(x) - g(x)]\, dx, f(x) \geq g(x)$$

10. Definite integral as the limit of a sum

$$\int_a^b f(x)\, dx = \lim_{n\to\infty} S_n,$$

where S_n is a Riemann sum

11. Average value of f over $[a, b]$

$$\frac{1}{b-a} \int_a^b f(x)\, dx$$

12. Consumers' surplus

$$CS = \int_0^{\bar{x}} D(x)\, dx - \bar{p}\bar{x}$$

13. Producers' surplus

$$PS = \bar{p}\bar{x} - \int_0^{\bar{x}} S(x)\, dx$$

14. Accumulated (future) value of an income stream

$$A = e^{rT} \int_0^T R(t)e^{-rt}\, dt$$

15. Present value of an income stream

$$PV = \int_0^T R(t)e^{-rt}\, dt$$

16. Present value of an annuity

$$PV = \frac{mP}{r}(1 - e^{-rT})$$

17. Amount of an annuity

$$A = \frac{mP}{r}(e^{rT} - 1)$$

18. Coefficient of inequality of a Lorentz curve

$$L = 2\int_0^1 [x - f(x)]\, dx$$

Terms

Antiderivative	Riemann sum
Antidifferentiation	Definite integral
Integration	Lower limit of integration
Indefinite integral	Upper limit of integration
Integrand	Lorentz curve
Constant of integration	

CHAPTER 11 REVIEW EXERCISES

In exercises 1–20, evaluate each indefinite integral.

1. $\int (x^3 + 2x^2 - x)\, dx$

2. $\int (\frac{1}{3}x^3 - 2x^2 + 8)\, dx$

3. $\int \left(x^4 - 2x^3 + \frac{1}{x^2}\right) dx$

4. $\int (x^{1/3} - \sqrt{x} + 4)\, dx$

5. $\int x(2x^2 + x^{1/2})\, dx$

6. $\int (x^2 + 1)(\sqrt{x} - 1)\, dx$

7. $\int \left(x^2 - x + \frac{2}{x} + 5\right) dx$

8. $\int \sqrt{2x + 1}\, dx$

9. $\int (3x - 1)(3x^2 - 2x + 1)^{1/3}\, dx$

10. $\int x^2(x^3 + 2)^{10}\, dx$

11. $\int \frac{x - 1}{x^2 - 2x + 5}\, dx$

12. $\int 2e^{-2x}\, dx$

13. $\int (x + \frac{1}{2})e^{x^2 + x + 1}\, dx$

14. $\int \frac{e^{-x} - 1}{(e^{-x} + x)^2}\, dx$

15. $\int \frac{(\ln x)^5}{x}\, dx$

16. $\int \frac{\ln x^2}{x}\, dx$

17. $\int x^3(x^2 + 1)^{10}\, dx$

18. $\int x\sqrt{x + 1}\, dx$

19. $\int \frac{x}{\sqrt{x - 2}}\, dx$

20. $\int \frac{3x}{\sqrt{x + 1}}\, dx$

In exercises 21–32, evaluate each definite integral.

21. $\int_0^1 (2x^3 - 3x^2 + 1)\, dx$

22. $\int_0^2 (4x^3 - 9x^2 + 2x - 1)\, dx$

23. $\int_1^4 (\sqrt{x} + x^{-3/2})\, dx$

24. $\int_0^1 20x(2x^2 + 1)^4\, dx$

25. $\int_{-1}^0 12(x^2 - 2x)(x^3 - 3x^2 + 1)^3\, dx$

26. $\int_4^7 x\sqrt{x - 3}\, dx$

27. $\int_0^2 \frac{x}{x^2 + 1}\, dx$

28. $\int_0^1 \frac{dx}{(5 - 2x)^2}$

29. $\int_0^2 \frac{4x}{\sqrt{1 + 2x^2}}\, dx$

30. $\int_0^2 xe^{(-1/2)x^2}\, dx$

31. $\int_{-1}^0 \frac{e^{-x}}{(1 + e^{-x})^2}\, dx$

32. $\int_1^e \frac{\ln x}{x}\, dx$

In exercises 33–36, find the function f given that the slope of the tangent line to the graph at any point $(x, f(x))$ is $f'(x)$ and that the graph of f passes through the given point.

33. $f'(x) = 3x^2 - 4x + 1$; $(1, 1)$

34. $f'(x) = \frac{x}{\sqrt{x^2 + 1}}$; $(0, 1)$

35. $f'(x) = 1 - e^{-x}$; $(0, 2)$

36. $f'(x) = \dfrac{\ln x}{x}$; $(1, -2)$

37. Let $f(x) = -2x^2 + 1$, and compute the Riemann sum of f over the interval $[1, 2]$ by partitioning the interval into five subintervals of the same length $(n = 5)$, where the points p_i $(1 \le i \le 5)$ are taken to be the *right* endpoints of the respective subintervals.

38. The management of the National Electric Corporation has determined that the daily marginal cost function associated with producing their automatic drip coffeemakers is given by

$$C'(x) = 0.00003x^2 - 0.03x + 20$$

where $C'(x)$ is measured in dollars per unit and x denotes the number of units produced. Management has also determined that the daily fixed cost incurred in producing these coffeemakers is $500. What is the total cost incurred by National in producing the first 400 coffeemakers per day?

39. Refer to exercise 38. Management has also determined that the daily marginal revenue function associated with producing and selling their coffeemakers is given by

$$R'(x) = -0.03x + 60$$

where x denotes the number of units produced and sold and $R'(x)$ is measured in dollars per unit.
  a. Determine the revenue function $R(x)$ associated with producing and selling these coffeemakers.
  b. What is the demand equation relating the wholesale unit price to the quantity of coffeemakers demanded?

40. The Franklin National Life Insurance Company purchased a new computer for $200,000. If the rate at which the computer's resale value changes is given by the function

$$V'(t) = 3800(t - 10)$$

where t is the length of time since the purchase date and $V'(t)$ is measured in dollars per year, find an expression $V(t)$ that gives the resale value of the computer after t years. How much would the computer cost after six years?

41. The marketing department of the Vista Vision Corporation forecasts that sales of their new line of projection television systems will grow at the rate of

$$3000 - 2000e^{-0.04t} \qquad 0 \le t \le 24$$

units per month once they are introduced into the market. Find an expression giving the total number of units of the projection television systems that Vista may expect to sell t months from the time they are put on the market. How many units of the television systems can Vista expect to sell during the first year?

42. Due to the increasing cost of fuel, the manager of the City Transit Authority estimates that the number of commuters using the city subway system will increase at the rate of

$$3000(1 + 0.4t)^{-1/2} \qquad 0 \le t \le 36$$

per month t months from now. If 100,000 commuters are currently using the system, find an expression giving the total number of commuters who will be using the subway t months from now. How many commuters will be using the subway six months from now?

43. The management of a division of Ditton Industries has determined that the daily marginal cost function associated with producing their hot-air corn poppers is given by

$$C'(x) = 0.00003x^2 - 0.03x + 10$$

where $C'(x)$ is measured in dollars per unit and x denotes the number of units manufactured. Management has also determined that the daily fixed cost incurred in producing these corn poppers is $600. Find the total cost incurred by Ditton in producing the first 500 corn poppers.

44. In 1980 the world produced 3.5 billion metric tons of coal. If output increased at the rate of

$$3.5e^{0.04t}$$

billion metric tons per year in year t ($t = 0$ corresponds to 1980), determine how much coal was produced worldwide between 1980 and the end of 1985.

45. Find the area of the region under the curve $y = 3x^2 + 2x + 1$ from $x = -1$ to $x = 2$.

46. Find the area of the region under the curve $y = e^{2x}$ from $x = 0$ to $x = 2$.

47. Find the area of the region bounded by the graph of the function $y = 1/x^2$, the x-axis, and the lines $x = 1$ and $x = 3$.

48. Find the area of the region bounded by the curve $y = -x^2 - x + 2$ and the x-axis.

49. Find the area of the region bounded by the graphs of the functions $f(x) = e^x$ and $g(x) = x$ and the vertical lines $x = 0$ and $x = 2$.

50. Find the area of the region that is completely enclosed by the graphs of $f(x) = x^4$ and $g(x) = x$.

51. Find the area of the region between the curve $y = x(x - 1)(x - 2)$ and the x-axis.

52. Using current production techniques, the rate of oil production from a certain oil well t years from now is estimated to be

$$R_1(t) = 100e^{0.05t}$$

thousand barrels per year. Using a new production technique, however, it is estimated that the rate of oil production from that oil well t years from now will be

$$R_2(t) = 100e^{0.08t}$$

thousand barrels per year. Determine how much additional oil will be produced over the next ten years if the new technique is adopted.

53. Find the average value of the function

$$f(x) = \frac{x}{\sqrt{x^2 + 16}}$$

over the interval $[0, 3]$.

54. The demand function for a brand of blank videocassettes is given by

$$p = -0.01x^2 - 0.2x + 23$$

where p is the wholesale unit price in dollars and x is the quantity demanded each week, measured in units of a thousand. Determine the consumers' surplus if the wholesale unit price is $8 per cassette.

55. The quantity demanded x (in units of a hundred) of the Sportsman 5×7 tents, per week, is related to the unit price p (in dollars) by the relation

$$p = -0.1x^2 - x + 40$$

The quantity x (in units of a hundred) that the supplier is willing to make available in the market is related to the unit price by the relation

$$p = 0.1x^2 + 2x + 20$$

If the market price is set at the equilibrium price, find the consumers' surplus and the producers' surplus.

56. Mr. Cunningham plans to deposit $4000 per year in his Keogh Retirement Account. If interest is compounded continuously at the rate of 8% per year, how much will he have in his retirement account after 20 years?

57. Miss Parker sold her house under an installment contract whereby the buyer gave her a down payment of $9000 and agreed to make monthly payments of $925 per month for 30 years. If the prevailing interest rate is 12% per year compounded continuously, find the present value of the purchase price of the house.

58. Ms. Carlson purchased a 10-year franchise for a health spa that is expected to generate income at the rate of

$$P(t) = 80,000$$

dollars per year. If the prevailing interest rate is 10% per year compounded continuously, find the present value of the franchise.

59. A certain country's income distribution is described by the function

$$f(x) = \frac{17}{18}x^2 + \frac{1}{18}x$$

a. Sketch the Lorentz curve for this function.
b. Compute $f(0.3)$ and $f(0.6)$ and interpret your results.
c. Compute the coefficient of inequality for this Lorentz curve.

60. The population of a certain Sun Belt city, currently 80,000, is expected to grow exponentially in the next five years with a growth constant of 0.05. If the prediction comes true, what will the average population of the city be over the next five years?

CHAPTER 11   EXTENDED PROBLEMS AND PROJECTS

The graph in the accompanying figure shows the consumption of petroleum in the United States in quadrillion BTU, from 1976 to 1990 (Figure 1).

In order to estimate the average consumption of petroleum during the 14-year period, we need to evaluate the integral

$$\frac{1}{14}\int_0^{14} f(t)\, dt \tag{1}$$

where f is the function whose graph appears in Figure 1 and $t = 0$ corresponds to 1976.

Since we do not have an algebraic rule defining the

FIGURE 1

FIGURE 2

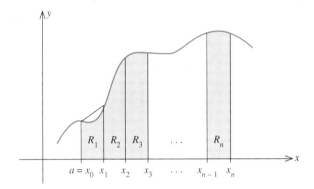

integrand f for all values of t in $[0, 14]$ (in fact, we are given its values only at a set of discrete points in that interval), we cannot find an antiderivative of f. Therefore, the Fundamental Theorem of Calculus is of no help here. One way to obtain an approximation of (1) is to use an appropriate Riemann sum. Another way is to use the Trapezoidal Rule, which you are asked to derive in Problem 1.

1. **The Trapezoidal Rule**   Suppose f is defined on $[a, b]$, but we know the values of f only at the points

$$x_0 = a, \; x_1 = a + h, \; x_2 = a + 2h, \ldots, x_n = a + nh$$

where $h = \dfrac{b - a}{n}$, n a positive integer. Then

$$\int_0^b f(x) \, dx \approx \frac{h}{2} \left[f(x_0) + 2f(x_1) + 2f(x_2) \right.$$

$$\left. + \cdots + 2f(x_{n-1}) + f(x_n) \right]$$

To derive the Trapezoidal Rule, refer to Figure 2.

Observe that

$$\int_a^b f(x) \, dx = \text{Area of } R_1 + \text{area of } R_2$$

$$+ \cdots + \text{area of } R_n$$

Referring to Figure 2, you can see that R_1 is approximated by the area of the trapezoid shown. Thus

$$\text{Area of } R_1 \approx \left[\frac{f(x_1) + f(x_2)}{2} \right] h$$

Approximate the areas of the regions $R_2, \ldots, R_n$ in a similar manner to derive the Trapezoidal Rule.

2. Apply the Trapezoidal Rule to solve the problem given at the beginning of these projects.

Calculus of Several Variables

12.1	Functions of Several Variables
12.2	Partial Derivatives
12.3	Maxima and Minima of Functions of Several Variables

*U*p to now we have dealt with functions involving one variable. In many real-life situations, however, we encounter quantities that depend on two or more quantities. For example, the Consumer Price Index (CPI) compiled every month by the Bureau of Labor Statistics depends on the price of more than 95,000 consumer items from gas to groceries. In order to study such relationships, we need the notion of a function of several variables. Next, generalizing the concept of the derivative of a function of one variable, we are led to the idea of the *partial derivatives* of a function of two or more variables. Partial derivatives enable us to study the rate of change of a function with respect to one variable while holding all the other variables constant. We will also learn how to find the maximum and/or minimum values of a function of several variables. For example, we will learn how a manufacturer can maximize her profits by producing the right amount of each of her products.

12.1 Functions of Several Variables

Up to now, our study of calculus has been restricted to functions of one variable. In many practical situations, however, the formulation of a problem results in a mathematical model that involves a function of two or more variables. For example, suppose that the Ace Novelty Company determines that the profits are $6, $5, and $4 for three types of souvenirs it produces. Let x, y, and z denote the number of type A, type B, and type C souvenirs to be made; then the company's profit is given by

$$P = 6x + 5y + 4z$$

and P is a function of the three variables x, y, and z.

Functions of Two Variables

Although this chapter deals with real-valued functions of several variables, most of our definitions and results are stated in terms of a function of two variables. One reason for adopting this approach, as you will soon see, is that there is a geometrical interpretation for this special case, which serves as an important visual aid. We can then draw upon the experience gained from studying the two-variable case to help us understand the concepts and results connected with the more general case, which, by and large, is just a simple extension of the lower-dimensional case.

A FUNCTION OF TWO VARIABLES

A real-valued **function of two variables**, f, consists of

1. a set A of ordered pairs of real numbers (x, y) called the **domain** of the function and
2. a rule that associates with each ordered pair in the domain of f one and only one real number, denoted by $z = f(x, y)$.

The variables x and y are called *independent variables*, and the variable z, which is dependent on the values of x and y, is referred to as a *dependent variable*.

As in the case of a real-valued function of one real variable, the number $z = f(x, y)$ is called the *value of f* at the point (x, y). And, unless specified, the domain of the function f will be taken to be the largest possible set for which the rule defining f is meaningful.

EXAMPLE 1 Let f be the function defined by

$$f(x, y) = x + xy + y^2 + 2$$

Compute $f(0, 0)$, $f(1, 2)$, and $f(2, 1)$.

Solution We have

$$f(0, 0) = 0 + (0)(0) + 0^2 + 2 = 2$$
$$f(1, 2) = 1 + (1)(2) + 2^2 + 2 = 9$$

and
$$f(2, 1) = 2 + (2)(1) + 1^2 + 2 = 7$$

The domain of a function of two variables $f(x, y)$ is a set of ordered pairs of real numbers and may therefore be viewed as a subset of the xy-plane.

EXAMPLE 2   Find the domain of each of the following functions.

a. $f(x, y) = x^2 + y^2$ **b.** $g(x, y) = \dfrac{2}{x - y}$

c. $h(x, y) = \sqrt{1 - x^2 - y^2}$

Solution

a. $f(x, y)$ is defined for all real values of x and y, so the domain of the function f is the set of all points (x, y) in the xy-plane.

b. $g(x, y)$ is defined for all $x \neq y$, so the domain of the function g is the set of all points in the xy-plane except those lying on the line $y = x$ (Figure 1a).

FIGURE 1

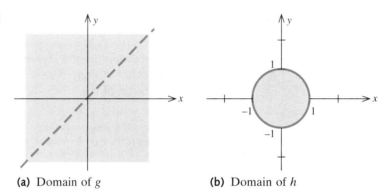

(a) Domain of g **(b)** Domain of h

c. We require that $1 - x^2 - y^2 \geq 0$ or $x^2 + y^2 \leq 1$, which is just the set of all points (x, y) lying on and inside the circle of radius 1 with center at the origin (Figure 1b).   ■

Applications

EXAMPLE 3   The Acrosonic Company manufactures a bookshelf loudspeaker system that may be bought fully assembled or in a kit. The demand equations that relate the unit prices, p and q, to the quantities demanded weekly, x and y, of the assembled and kit versions of the loudspeaker systems are given by

$$p = 300 - \frac{1}{4}x - \frac{1}{8}y \qquad \text{and} \qquad q = 240 - \frac{1}{8}x - \frac{3}{8}y$$

a. What is the weekly total revenue function $R(x, y)$?

b. What is the domain of the function R?

Solution

a. The weekly revenue realizable from the sale of x units of the assembled speaker systems at p dollars per unit is given by xp dollars. Similarly, the weekly revenue realizable from the sale of y units of the kits at q

dollars per unit is given by yq dollars. Therefore, the weekly total revenue function R is given by

$$R(x, y) = xp + yq$$

$$= x\left(300 - \frac{1}{4}x - \frac{1}{8}y\right) + y\left(240 - \frac{1}{8}x - \frac{3}{8}y\right)$$

$$= -\frac{1}{4}x^2 - \frac{3}{8}y^2 - \frac{1}{4}xy + 300x + 240y$$

b. To find the domain of the function R, let us observe that the quantities x, y, p, and q must be nonnegative. This observation leads to the following system of linear inequalities:

$$300 - \frac{1}{4}x - \frac{1}{8}y \geq 0$$

$$240 - \frac{1}{8}x - \frac{3}{8}y \geq 0$$

$$x \geq 0$$

$$y \geq 0$$

The domain of the function R is sketched in Figure 2.

FIGURE 2
The domain of $R(x, y)$

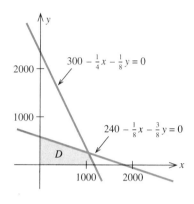

EXAMPLE 4   The monthly payment that amortizes a loan of A dollars in t years when the interest rate is r per year is given by

$$P = f(A, r, t) = \frac{Ar}{12\left[1 - \left(1 + \dfrac{r}{12}\right)^{-12t}\right]}$$

Find the monthly payment for a home mortgage of \$90,000 to be amortized over 30 years when the interest rate is 10% per year.

Solution   Letting $A = 90{,}000$, $r = 0.1$, and $t = 30$, we find the required monthly payment to be

$$P = f(90{,}000, 0.1, 30) = \frac{90{,}000(0.1)}{12\left[1 - \left(1 + \dfrac{0.1}{12}\right)^{-360}\right]}$$

$$\approx 789.81$$

or approximately \$789.81.

Graphs of Functions of Two Variables

In order to graph a function of two variables, we need a three-dimensional coordinate system. This is readily constructed by adding a third axis to the plane Cartesian coordinate system in such a way that the three resulting axes are mutually perpendicular and intersect at O. Observe that, by construction, the zeros of the three number scales coincide at the origin of the **three-dimensional Cartesian coordinate system** (Figure 3).

   A point in three-dimensional space can now be represented uniquely in this coordinate system by an **ordered triple** of numbers (x, y, z), and,

FIGURE 3
The three-dimensional Cartesian coordinate system

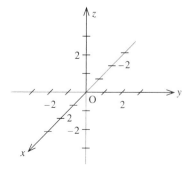

conversely, every ordered triple of real numbers (x, y, z) represents a point in three-dimensional space (Figure 4a). For example, the points $A(2, 3, 4)$, $B(1, -2, -2)$, $C(2, 4, 0)$, and $D(0, 0, 4)$ are shown in Figure 4b.

FIGURE 4

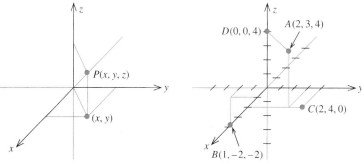

(a) A point in three-dimensional space

(b) Some sample points in three-dimensional space

FIGURE 5

The graph of a function in three-dimensional space

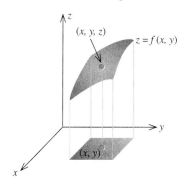

Now, if $f(x, y)$ is a function of two variables x and y, the domain of f is a subset of the xy-plane. Let $z = f(x, y)$ so that there is one and only one point $(x, y, z) \equiv (x, y, f(x, y))$ associated with each point (x, y) in the domain of f. The totality of all such points makes up the *graph* of the function f and is, except for certain degenerate cases, a surface in three-dimensional space (Figure 5).

In interpreting the graph of a function $f(x, y)$, one often thinks of the value $z = f(x, y)$ of the function at the point (x, y) as the "height" of the point (x, y, z) on the graph of f. If $f(x, y) > 0$, then the point (x, y, z) is $f(x, y)$ units above the xy-plane; if $f(x, y) < 0$, then the point (x, y, z) is $|f(x, y)|$ units below the xy-plane.

In general, it is quite difficult to draw the graph of a function of two variables. But techniques have been developed that enable us to generate such graphs with minimum effort using a computer. Figure 6 shows the computer-generated graphs of two functions.

FIGURE 6

Two computer-generated graphs of functions of two variables

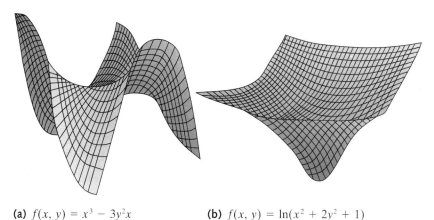

(a) $f(x, y) = x^3 - 3y^2x$

(b) $f(x, y) = \ln(x^2 + 2y^2 + 1)$

Level Curves

As mentioned earlier, the graph of a function of two variables is often difficult to sketch, and we will not develop a systematic procedure for sketching it. Instead, we will describe a method that is used in constructing topographical maps. This method is relatively easy to apply and conveys sufficient information to enable one to obtain a feel for the graph of the function.

Suppose that $f(x, y)$ is a function of two variables x and y, with a graph as shown in Figure 7. If c is some value of the function f, then the equation $f(x, y) = c$ describes a curve lying on the plane $z = c$ called the *trace* of the graph of f in the plane $z = c$. If this trace is projected onto the xy-plane, the resulting curve in the xy-plane is called a **level curve.** By drawing the level curves corresponding to several admissible values of c, we obtain a **contour map.** Observe that, by construction, every point on a particular

FIGURE 7

The graph of the function $z = f(x, y)$ and its intersection with the plane $z = c$

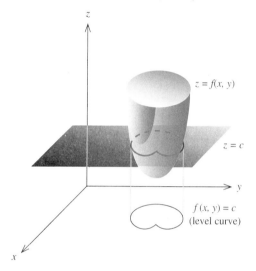

level curve corresponds to a point in the surface $z = f(x, y)$ that is a certain fixed distance from the xy-plane. Thus, by elevating or depressing the level curves that make up the contour map in one's mind, it is possible to obtain a feel for the general shape of the surface represented by the function f. Figure 8a shows a part of a mountain range with one peak, and Figure 8b is the associated contour map.

FIGURE 8

(a) A peak on a mountain range

(b) A contour map for the mountain peak

EXAMPLE 5   Sketch a contour map for the function $f(x, y) = x^2 + y^2$.

Solution   The level curves are the graphs of the equation $x^2 + y^2 = c$ for nonnegative numbers c. Taking $c = 0, 1, 4, 9$, and 16, for example, we obtain

$$c = 0 \ : x^2 + y^2 = 0$$

$$c = 1 \ : x^2 + y^2 = 1$$

$$c = 4 \ : x^2 + y^2 = 4 = 2^2$$

$$c = 9 \ : x^2 + y^2 = 9 = 3^2$$

$$c = 16: x^2 + y^2 = 16 = 4^2$$

The five level curves are concentric circles with center at the origin and radius given by $r = 0, 1, 2, 3,$ and 4, respectively (Figure 9a). A sketch of the graph of $f(x, y) = x^2 + y^2$ is included for your reference in Figure 9b.

FIGURE 9

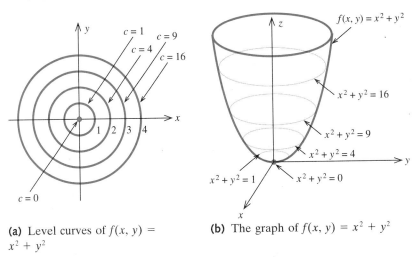

(a) Level curves of $f(x, y) = x^2 + y^2$

(b) The graph of $f(x, y) = x^2 + y^2$

EXAMPLE 6   Sketch the level curves for the function $f(x, y) = 2x^2 - y$ corresponding to $z = -2, -1, 0, 1,$ and 2.

FIGURE 10

Level curves for $f(x, y) = 2x^2 - y$

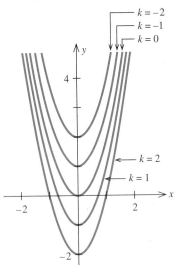

Solution   The level curves are the graphs of the equation $2x^2 - y = k$ or $y = 2x^2 - k$ for $k = -2, -1, 0, 1,$ and 2. The required level curves are shown in Figure 10.

Level curves of functions of two variables are found in many practical applications. For example, if $f(x, y)$ denotes the temperature at a location within the continental United States with longitude x and latitude y at a certain time of day, then the temperature at the point (x, y) is given by the "height" of the surface, represented by $z = f(x, y)$. In this situation, the level curve $f(x, y) = k$ is a curve superimposed on a map of the United States connecting points having the same temperature at a given time (Figure 11). These level curves are called *isotherms*.

Similarly, if $f(x, y)$ gives the barometric pressure at the location (x, y), then the level curves of the function f are called *isobars*, lines connecting points having the same barometric pressure at a given time.

As a final example, suppose that $P(x, y, z)$ is a function of three variables $x, y,$ and z giving the profit realized when $x, y,$ and z units of three products A, B, and C, respectively, are produced and sold. Then the equation $P(x, y, z) = k$, where k is a constant, represents a surface in three-

FIGURE 11
Isotherms: curves connecting points that have the same temperature

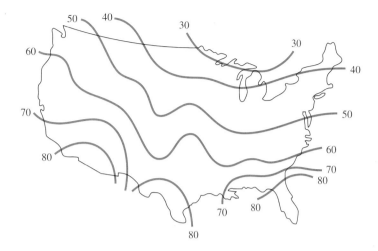

dimensional space called a **level surface** of *P*. In this situation, the level surface represented by $P(x, y, z) = k$ represents the product mix that results in a profit of exactly *k* dollars. Such a level surface is called an **isoprofit surface.**

SELF-CHECK EXERCISES 12.1

1. Let $f(x, y) = x^2 - 3xy + \sqrt{x + y}$. Compute $f(1, 3)$ and $f(-1, 1)$. Is the point $(-1, 0)$ in the domain of *f*?

2. Find the domain of $f(x, y) = \dfrac{1}{x} + \dfrac{1}{x - y} - e^{x+y}$.

c 3. The Odyssey Travel Agency has a monthly advertising budget of $20,000. Odyssey's management estimates that if they spend *x* dollars on newspaper advertising and *y* dollars on television advertising, then the monthly revenue will be

$$f(x, y) = 30x^{1/4}y^{3/4}$$

dollars. What will the monthly revenue be if Odyssey spends $5,000 per month on newspaper ads and $15,000 per month on television ads? If Odyssey spends $4,000 per month on newspaper ads and $16,000 per month on television ads?

Solutions to Self-Check Exercises 12.1 can be found on page 758.

12.1   EXERCISES

1. Let $f(x, y) = 2x + 3y - 4$. Compute $f(0, 0)$, $f(1, 0)$, $f(0, 1)$, $f(1, 2)$, and $f(2, -1)$.

2. Let $g(x, y) = 2x^2 - y^2$. Compute $g(1, 2)$, $g(2, 1)$, $g(1, 1)$, $g(-1, 1)$, and $g(2, -1)$.

3. Let $f(x, y) = x^2 + 2xy - x + 3$. Compute $f(1, 2)$, $f(2, 1)$, $f(-1, 2)$, and $f(2, -1)$.

4. Let $h(x, y) = (x + y)/(x - y)$. Compute $h(0, 1)$, $h(-1, 1)$, $h(2, 1)$, and $h(\pi, -\pi)$.

5. Let $g(s, t) = 3s\sqrt{t} + t\sqrt{s} + 2$. Compute $g(1, 2)$, $g(2, 1)$, $g(0, 4)$, and $g(4, 9)$.

6. Let $f(x, y) = xye^{x^2+y^2}$. Compute $f(0, 0)$, $f(0, 1)$, $f(1, 1)$, and $f(-1, -1)$.

7. Let $h(s, t) = s \ln t - t \ln s$. Compute $h(1, e)$, $h(e, 1)$, and $h(e, e)$.

8. Let $f(u, v) = (u^2 + v^2)e^{uv^2}$. Compute $f(0, 1)$, $f(-1, -1)$, $f(a, b)$, and $f(b, a)$.

9. Let $g(r, s, t) = re^{s/t}$. Compute $g(1, 1, 1)$, $g(1, 0, 1)$, and $g(-1, -1, -1)$.

10. Let $g(u, v, w) = (ue^{vw} + ve^{uw} + we^{uv})/(u^2 + v^2 + w^2)$. Compute $g(1, 2, 3)$ and $g(3, 2, 1)$.

In exercises 11–18, find the domain of the function.

11. $f(x, y) = 2x + 3y$

12. $g(x, y, z) = x^2 + y^2 + z^2$

13. $h(u, v) = \dfrac{uv}{u - v}$ **14.** $f(s, t) = \sqrt{s^2 + t^2}$

15. $g(r, s) = \sqrt{rs}$ **16.** $f(x, y) = e^{-xy}$

17. $h(x, y) = \ln(x + y - 5)$

18. $h(u, v) = \sqrt{4 - u^2 - v^2}$

In exercises 19–24, sketch the level curves of the function corresponding to the given values of z.

19. $f(x, y) = 2x + 3y;\ z = -2, -1, 0, 1, 2$

20. $f(x, y) = -x^2 + y;\ z = -2, -1, 0, 1, 2$

21. $f(x, y) = 2x^2 + y;\ z = -2, -1, 0, 1, 2$

22. $f(x, y) = xy;\ z = -4, -2, 2, 4$

23. $f(x, y) = \sqrt{16 - x^2 - y^2};\ z = 0, 1, 2, 3, 4$

24. $f(x, y) = e^x - y;\ z = -2, -1, 0, 1, 2$

25. The volume of a cylindrical tank of radius r and height h is given by

$$V = f(r, h) = \pi r^2 h$$

Find the volume of a cylindrical tank of radius 1.5 ft and height 4 ft.

26. **IQs** The IQ (intelligence quotient) of a person whose mental age is m years and whose chronological age is c years is defined as

$$f(m, c) = \frac{100m}{c}$$

What is the IQ of a nine-year-old child who has a mental age of 13.5 years?

27. **Poiseuille's Law** Poiseuille's law states that the resistance R, measured in dynes, of blood flowing in a blood vessel of length l and radius r (both in centime-

ters) is given by

$$R = f(l, r) = \frac{kl}{r^4}$$

where k is the viscosity of blood (in dyne-sec/cm^2). What is the resistance, in terms of k, of blood flowing through an arteriole 4 cm long and of radius 0.1 cm?

28. **Revenue Functions** The Country Workshop manufactures both finished and unfinished furniture for the home. The estimated quantities demanded each week of its roll-top desks in the finished and unfinished versions are x and y units when the corresponding unit prices are

$$p = 200 - \frac{1}{5}x - \frac{1}{10}y$$

$$q = 160 - \frac{1}{10}x - \frac{1}{4}y$$

dollars, respectively.
a. What is the weekly total revenue function $R(x, y)$?
b. Find the domain of the function R.

29. For the total revenue function $R(x, y)$ of exercise 28, compute $R(100, 60)$ and $R(60, 100)$. Interpret your results.

30. **Revenue Functions** The Weston Publishing Company publishes a deluxe edition and a standard edition of its English language dictionary. Weston's management estimates that the number of deluxe editions demanded is x copies per day, and the number of standard editions demanded is y copies per day when the unit prices are

$$p = 20 - 0.005x - 0.001y$$

$$q = 15 - 0.001x - 0.003y$$

dollars, respectively.
a. Find the daily total revenue function $R(x, y)$.
b. Find the domain of the function R.

31. For the total revenue function $R(x, y)$ of exercise 30, compute $R(300, 200)$ and $R(200, 300)$. Interpret your results.

32. **Volume of a Gas** The volume of a certain mass of gas is related to its pressure and temperature by the formula

$$V = \frac{30.9T}{P}$$

where the volume V is measured in liters, the temperature T is measured in degrees Kelvin (obtained by adding $273°$ to the Celsius temperature), and the pressure P is measured in millimeters of mercury pressure.

a. Find the domain of the function V.
b. Calculate the volume of the gas at standard temperature and pressure—that is, when $T = 273°$K and $P = 760$ mm of mercury.

c 33. **Surface Area of a Human Body** An empirical formula by E. F. Dubois relates the surface area S of a human body (in square meters) to its weight W in kilograms and its height H in centimeters. The formula, given by

$$S = 0.007184 W^{0.425} H^{0.725}$$

is used by physiologists in metabolism studies.
a. Find the domain of the function S.
b. What is the surface area of a human body that weighs 70 kilograms and has a height of 178 centimeters?

c 34. **Arson for Profit** A study of arson for profit was conducted by a team of paid civilian experts and police detectives appointed by the mayor of a large city. It was found that the number of suspicious fires in that city in 1992 was very closely related to the concentration of tenants in the city's public housing and to the level of reinvestment in the area in conventional mortgages by the ten largest banks. In fact, the number of fires was closely approximated by the formula

$$N(x, y) = \frac{100(1000 + 0.03 x^2 y)^{1/2}}{(5 + 0.2y)^2}$$

$$0 \le x \le 150; 5 \le y \le 35$$

where x denotes the number of persons per census tract and y denotes the level of reinvestment in the area in cents per dollar deposited. Using this formula, estimate the total number of suspicious fires in the districts of the city where the concentration of public housing tenants was 100 per census tract and the level of reinvestment was 20 cents per dollar deposited.

c 35. **Continuously Compounded Interest** If a principal of P dollars is deposited in an account earning interest at the rate of r per year compounded continuously, then the accumulated amount at the end of t years is given by

$$A = f(P, r, t) = Pe^{rt}$$

dollars. Find the accumulated amount at the end of three years if a sum of $10,000 is deposited in an account earning interest at the rate of 10% per year.

c 36. **Home Mortgages** The monthly payment that amortizes a loan of A dollars in t years when the interest

rate is r per year is given by

$$P = f(A, r, t) = \frac{Ar}{12\left[1 - \left(1 + \dfrac{r}{12}\right)^{-12t}\right]}$$

a. Find the monthly payment for a home mortgage of $100,000 that will be amortized over 30 years with an interest rate of 8% per year. With an interest rate of 10% per year.
b. Find the monthly payment for a home mortgage of $100,000 that will be amortized over 20 years with an interest rate of 8% per year.

c 37. **Home Mortgages** Suppose that a home buyer secures a bank loan of A dollars to purchase a house. If the interest rate charged is r per year and the loan is to be amortized in t years, then the principal repayment at the end of i months is given by

$$B = f(A, r, t, i)$$

$$= A\left[\frac{\left(1 + \dfrac{r}{12}\right)^{i} - 1}{\left(1 + \dfrac{r}{12}\right)^{12t} - 1}\right] \quad 0 \le i \le 12t$$

Suppose that the Blakelys borrow a sum of $80,000 from a bank to help finance the purchase of a house and the bank charges interest at a rate of 9% per year. If the Blakelys agree to repay the loan in equal installments over 30 years, how much will they owe the bank after the 60th payment (5 years)? After the 240th payment (20 years)?

c 38. **Force Generated by a Centrifuge** A centrifuge is a machine designed for the specific purpose of subjecting materials to a sustained centrifugal force. The actual amount of centrifugal force, F, expressed in dynes (1 gram of force = 980 dynes) is given by

$$F = f(M, S, R) = \frac{\pi^2 S^2 MR}{900}$$

where S is in revolutions per minute (rpm), M is in grams, and R is in centimeters. Show that an object revolving at the rate of 600 rpm in a circle with radius of 10 cm generates a centrifugal force that is approximately 40 times gravity.

39. **Wilson Lot Size Formula** The Wilson lot size formula in economics states that the optimal quantity Q of goods for a store to order is given by

$$Q = f(C, N, h) = \sqrt{\frac{2CN}{h}}$$

where C is the cost of placing an order, N is the number of items the store sells per week, and h is the weekly holding cost for each item. Find the most economical quantity of ten-speed bicycles to order

if it costs the store $20 to place an order, $5 to hold a bicycle for a week, and the store expects to sell 40 bicycles a week.

SOLUTIONS TO SELF-CHECK EXERCISES 12.1

1. $f(1, 3) = 1^2 - 3(1)(3) + \sqrt{1 + 3} = -6$
$f(-1, 1) = (-1)^2 - 3(-1)(1) + \sqrt{-1 + 1} = 4$

The point $(-1, 0)$ is not in the domain of f because the term $\sqrt{x + y}$ is not defined when $x = -1$ and $y = 0$. In fact, the domain of f consists of all real values of x and y that satisfy the inequality $x + y \geq 0$, the shaded half-plane shown in the accompanying figure.

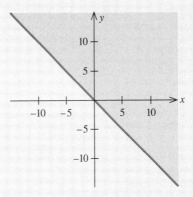

2. Since division by zero is not permitted, we see that $x \neq 0$ and $x - y \neq 0$. Therefore, the domain of f is the set of all points in the xy-plane not containing the y-axis ($x = 0$) and the straight line $x = y$.

3. If Odyssey spends $5,000 per month on newspaper ads ($x = 5,000$) and $15,000 per month on television ads ($y = 15,000$), then its monthly revenue will be given by

$$f(5,000, 15,000) = 30(5,000)^{1/4}(15,000)^{3/4}$$

$$\approx 341,926.06$$

or approximately $341,925. If the agency spends $4,000 per month on newspaper ads and $16,000 per month on television ads, then its monthly revenue will be given by

$$f(4,000, 16,000) = 30(4,000)^{1/4}(16,000)^{3/4}$$

$$\approx 339,411.25$$

or approximately $339,411.

12.2   Partial Derivatives

Partial Derivatives

For a function $f(x)$ of one variable x, there is no ambiguity when we speak about the rate of change of $f(x)$ with respect to x, since x must be constrained

FIGURE 12

We can approach a point in the plane from infinitely many directions.

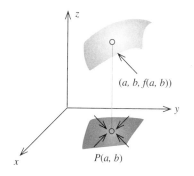

to move along the x-axis. The situation becomes more complicated, however, when we study the rate of change of a function of two or more variables. For example, the domain D of a function of two variables $f(x, y)$ is a subset of the plane (Figure 12), so if $P(a, b)$ is any point in the domain of f, there are infinitely many directions from which one can approach the point P. We may therefore ask for the rate of change of f at P along any of these directions.

However, we will not deal with this general problem. Instead, we will restrict ourselves to studying the rate of change of the function $f(x, y)$ at a point $P(a, b)$ in each of two *preferred directions*—namely, the direction parallel to the x-axis and the direction parallel to the y-axis. Let $y = b$, where b is a constant, so that $f(x, b)$ is a function of the one variable x. Since the equation $z = f(x, y)$ is the equation of a surface, the equation $z = f(x, b)$ is the equation of the curve C on the surface formed by the intersection of the surface and the plane $y = b$ (Figure 13).

FIGURE 13

The curve C is formed by the intersection of the plane $y = b$ with the surface $z = f(x, y)$.

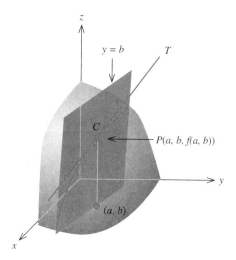

Because $f(x, b)$ is a function of one variable x, we may compute the derivative of f with respect to x at $x = a$. This derivative, obtained by keeping the variable y fixed and differentiating the resulting function $f(x, y)$ with respect to x, is called the **first partial derivative** of f with respect to x at (a, b), written

$$\frac{\partial z}{\partial x}(a, b) \qquad \text{or} \qquad \frac{\partial f}{\partial x}(a, b) \qquad \text{or} \qquad f_x(a, b)$$

Thus, $$\frac{\partial z}{\partial x}(a, b) = \frac{\partial f}{\partial x}(a, b) = f_x(a, b) = \lim_{h \to 0} \frac{f(a + h, b) - f(a, b)}{h}$$

provided that the limit exists. The first partial derivative of f with respect to x at (a, b) measures both the slope of the tangent line T to the curve C and the rate of change of the function f in the x-direction when $x = a$ and $y = b$. We also write

$$\left. \frac{\partial f}{\partial x} \right|_{(a,b)} \equiv f_x(a, b)$$

Similarly, we define the first partial derivative of f with respect to y at (a, b), written

$$\frac{\partial z}{\partial y}(a, b) \quad \text{or} \quad \frac{\partial f}{\partial y}(a, b) \quad \text{or} \quad f_y(a, b)$$

as the derivative obtained by keeping the variable x fixed and differentiating the resulting function $f(x, y)$ with respect to y. That is,

$$\frac{\partial z}{\partial y}(a, b) = \frac{\partial f}{\partial y}(a, b) = f_y(a, b)$$

$$= \lim_{k \to 0} \frac{f(a, b + k) - f(a, b)}{k}$$

if the limit exists. The first partial derivative of f with respect to y at (a, b) measures both the slope of the tangent line T to the curve C, obtained by holding x constant (Figure 14), and the rate of change of the function f in the y-direction when $x = a$ and $y = b$. We write

$$\left. \frac{\partial f}{\partial y} \right|_{(a,b)} \equiv f_y(a, b)$$

FIGURE 14
The first partial derivative of f with respect to y at (a, b) measures the slope of the tangent line T to the curve C with x held constant.

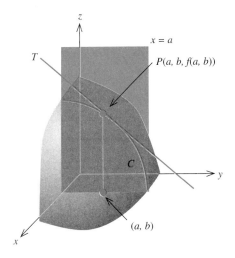

Before looking at some examples, let us summarize these definitions.

FIRST PARTIAL DERIVATIVES OF $f(x, y)$

Suppose that $f(x, y)$ is a function of the two variables x and y. Then, the first partial derivative of f with respect to x at the point (x, y) is

$$\frac{\partial f}{\partial x} = \lim_{h \to 0} \frac{f(x + h, y) - f(x, y)}{h}$$

provided the limit exists. The first partial derivative of f with respect to y at the point (x, y) is

$$\frac{\partial f}{\partial y} = \lim_{k \to 0} \frac{f(x, y + k) - f(x, y)}{k}$$

provided the limit exists.

EXAMPLE 1   Find the partial derivatives $\partial f/\partial x$ and $\partial f/\partial y$ of the function

$$f(x, y) = x^2 - xy^2 + y^3$$

What is the rate of change of the function f in the x-direction at the point
$(1, 2)$? What is the rate of change of the function f in the y-direction at
the point $(1, 2)$?

Solution   To compute $\partial f/\partial x$, think of the variable y as a constant and
differentiate the resulting function of x with respect to x. Let us write

$$f(x, y) = x^2 - xy^2 + y^3$$

where the variable y to be treated as a constant is shown in color. Then,

$$\frac{\partial f}{\partial x} = 2x - y^2$$

To compute $\partial f/\partial y$, think of the variable x as being fixed—that is, as a
constant—and differentiate the resulting function of y with respect to y.
In this case,

$$f(x, y) = x^2 - xy^2 + y^3$$

so that $$\frac{\partial f}{\partial y} = -2xy + 3y^2$$

The rate of change of the function f in the x-direction at the point $(1, 2)$
is given by

$$f_x(1, 2) = \frac{\partial f}{\partial x}\bigg|_{(1,2)} = 2(1) - 2^2 = -2$$

that is, f decreases 2 units for each unit increase in the x-direction, y being
kept constant ($y = 2$). The rate of change of the function f in the y-direction
at the point $(1, 2)$ is given by

$$f_y(1, 2) = \frac{\partial f}{\partial y}\bigg|_{(1,2)} = -2(1)(2) + 3(2)^2 = 8$$

that is, f increases 8 units for each unit increase in the y-direction, x being
kept constant ($x = 1$).   ■

EXAMPLE 2   Compute the first partial derivatives of each of the following functions.

a. $f(x, y) = \dfrac{xy}{x^2 + y^2}$ **b.** $g(s, t) = (s^2 - st + t^2)^5$

c. $h(u, v) = e^{u^2 - v^2}$ **d.** $f(x, y) = \ln(x^2 + 2y^2)$

Solution

a. To compute $\partial f/\partial x$, think of the variable y as a constant. Thus,

$$f(x, y) = \frac{xy}{x^2 + y^2}$$

so that, upon using the Quotient Rule, we have

$$\frac{\partial f}{\partial x} = \frac{(x^2 + y^2)y - xy(2x)}{(x^2 + y^2)^2}$$

$$= \frac{y(y^2 - x^2)}{(x^2 + y^2)^2}$$

upon simplification and factorization. To compute $\partial f/\partial y$, think of the variable x as a constant. Thus,

$$f(x, y) = \frac{xy}{x^2 + y^2}$$

so that, upon using the Quotient Rule once again, we obtain

$$\frac{\partial f}{\partial y} = \frac{(x^2 + y^2)x - xy(2y)}{(x^2 + y^2)^2}$$

$$= \frac{x(x^2 - y^2)}{(x^2 + y^2)^2}$$

b. To compute $\partial g/\partial s$, we treat the variable t as if it were a constant. Thus,

$$g(s, t) = (s^2 - st + t^2)^5$$

Using the General Power Rule, we find

$$\frac{\partial g}{\partial s} = 5(s^2 - st + t^2)^4 \cdot (2s - t)$$

$$= 5(2s - t)(s^2 - st + t^2)^4$$

To compute $\partial g/\partial t$, we treat the variable s as if it were a constant. Thus

$$g(s, t) = (s^2 - st + t^2)^5$$

$$\frac{\partial g}{\partial t} = 5(s^2 - st + t^2)^4(-s + 2t)$$

$$= 5(2t - s)(s^2 - st + t^2)^4$$

c. To compute $\partial h/\partial u$, think of the variable v as a constant. Thus,

$$h(u, v) = e^{u^2 - v^2}$$

Using the Chain Rule for exponential functions, we have

$$\frac{\partial h}{\partial u} = e^{u^2 - v^2} \cdot 2u$$

$$= 2ue^{u^2 - v^2}$$

Next, we treat the variable u as if it were a constant,

$$h(u, v) = e^{u^2 - v^2}$$

and we obtain

$$\frac{\partial h}{\partial v} = e^{u^2 - v^2} \cdot (-2v)$$

$$= -2v e^{u^2 - v^2}$$

d. To compute $\partial f / \partial x$, think of the variable y as a constant. Thus,

$$f(x, y) = \ln(x^2 + 2y^2)$$

so that the Chain Rule for logarithmic functions gives

$$\frac{\partial f}{\partial x} = \frac{2x}{x^2 + 2y^2}$$

Next, treating the variable x as if it were a constant, we find

$$f(x, y) = \ln(x^2 + 2y^2)$$

$$\frac{\partial f}{\partial y} = \frac{4y}{x^2 + 2y^2}$$ ◼

To compute the partial derivative of a function of several variables with respect to one variable, say x, we think of the other variables as if they were constants and differentiate the resulting function with respect to x.

EXAMPLE 3   Compute the first partial derivatives of the function

$$w = f(x, y, z) = xyz - xe^{yz} + x \ln y$$

Solution   Here we have a function of three variables x, y, and z, and we are required to compute

$$\frac{\partial f}{\partial x}, \quad \frac{\partial f}{\partial y}, \quad \frac{\partial f}{\partial z}$$

To compute f_x, we think of the other two variables, y and z, as fixed, and we differentiate the resulting function of x with respect to x, thereby obtaining

$$f_x = yz - e^{yz} + \ln y$$

To compute f_y, we think of the other two variables, x and z, as constants, and we differentiate the resulting function of y with respect to y. We then obtain

$$f_y = xz - xze^{yz} + \frac{x}{y}$$

Finally, to compute f_z, we treat the variables x and y as constants and differentiate the function f with respect to z, obtaining

$$f_z = xy - xye^{yz}$$ ◼

The Cobb-Douglas Production Function

For an economic interpretation of the first partial derivatives of a function of two variables, let us turn our attention to the function

$$f(x, y) = ax^b y^{1-b} \qquad (1)$$

where a and b are positive constants with $0 < b < 1$. This function is called the **Cobb-Douglas production function.** Here x stands for the amount of money expended for labor, y stands for the cost of capital equipment (buildings, machinery, and other tools of production), and the function f measures the output of the finished product (in suitable units) and is called, accordingly, the *production function.*

The partial derivative f_x is called the **marginal productivity of labor.** It measures the rate of change of production with respect to the amount of money expended for labor, with the level of capital expenditure held constant. Similarly, the partial derivative f_y, called the **marginal productivity of capital,** measures the rate of change of production with respect to the amount expended on capital, with the level of labor expenditure held fixed.

EXAMPLE 4   A certain country's production in the early years following World War II is described by the function

$$f(x, y) = 30x^{2/3}y^{1/3}$$

units, when x units of labor and y units of capital were used.

a. Compute f_x and f_y.

b. What is the marginal productivity of labor and the marginal productivity of capital when the amounts expended on labor and capital are 125 units and 27 units, respectively?

c. Should the government have encouraged capital investment rather than increasing expenditure on labor to increase the country's productivity?

Solution

a.
$$f_x = 30 \cdot \frac{2}{3} x^{-1/3} y^{1/3} = 20 \left(\frac{y}{x} \right)^{1/3}$$

$$f_y = 30x^{2/3} \cdot \frac{1}{3} y^{-2/3} = 10 \left(\frac{x}{y} \right)^{2/3}$$

b. The required marginal productivity of labor is given by

$$f_x(125, 27) = 20 \left(\frac{27}{125} \right)^{1/3} = 20 \left(\frac{3}{5} \right)$$

or 12 units per unit increase in labor expenditure (capital expenditure is held constant at 27 units). The required marginal productivity of capital is given by

$$f_y(125, 27) = 10 \left(\frac{125}{27} \right)^{2/3} = 10 \left(\frac{25}{9} \right),$$

or $27\frac{7}{9}$ units per unit increase in capital expenditure (labor outlay is held constant at 125 units).

c. From the results of (b), we see that a unit increase in capital expenditure resulted in a much faster increase in productivity than a unit increase in labor expenditure would have. Therefore, the government should have encouraged increased spending on capital rather than on labor during the early years of reconstruction.

Substitute and Complementary Commodities

For another application of the first partial derivatives of a function of two variables in the field of economics, let us consider the relative demands of two commodities. We say that the two commodities are *substitute* (competitive) *commodities* if a decrease in the demand for one results in an increase in the demand for the other. Examples of competitive commodities are coffee and tea. However, two commodities are referred to as *complementary commodities* if a decrease in the demand for one results in a decrease in the demand for the other as well. Examples of complementary commodities are automobiles and tires.

We will now derive a criterion for determining whether two commodities A and B are substitute or complementary. Suppose that the demand equations that relate the quantities demanded, x and y, to the unit prices, p and q, of the two commodities are given by

$$x = f(p, q) \quad \text{and} \quad y = g(p, q)$$

Let us consider the partial derivative $\partial f / \partial p$. Since f is the demand function for commodity A, we see that, for fixed q, f is typically a decreasing function of p; that is, $\partial f / \partial p < 0$. Now, if the two commodities were substitute commodities, then the quantity demanded of commodity B would increase with respect to p, that is $\partial g / \partial p > 0$. A similar argument with p fixed shows that if A and B are substitute commodities, then $\partial f / \partial q > 0$. Thus, the two commodities A and B are substitute commodities if

$$\frac{\partial f}{\partial q} > 0 \quad \text{and} \quad \frac{\partial g}{\partial p} > 0$$

Similarly, A and B are complementary commodities if

$$\frac{\partial f}{\partial q} < 0 \quad \text{and} \quad \frac{\partial g}{\partial p} < 0$$

SUBSTITUTE AND COMPLEMENTARY COMMODITIES

Two commodities A and B are **substitute commodities** if

$$\frac{\partial f}{\partial q} > 0 \quad \text{and} \quad \frac{\partial g}{\partial p} > 0 \tag{2}$$

Two commodities A and B are **complementary commodities** if

$$\frac{\partial f}{\partial q} < 0 \quad \text{and} \quad \frac{\partial g}{\partial p} < 0 \tag{3}$$

EXAMPLE 5   Suppose that the daily demand for butter is given by

$$x = f(p, q) = \frac{3q}{1 + p^2}$$

and the daily demand for margarine is given by

$$y = g(p, q) = \frac{2p}{1 + \sqrt{q}} \qquad p > 0, q > 0$$

where p and q denote the prices per pound (in dollars) of butter and margarine, respectively, and x and y are measured in millions of pounds.

Determine whether these two commodities are substitute, complementary, or neither.

Solution We compute

$$\frac{\partial f}{\partial q} = \frac{3}{1 + p^2}$$

and

$$\frac{\partial g}{\partial p} = \frac{2}{1 + \sqrt{q}}$$

Since

$$\frac{\partial f}{\partial q} > 0 \quad \text{and} \quad \frac{\partial g}{\partial p} > 0$$

for all values of $p > 0$ and $q > 0$, we conclude that butter and margarine are substitute commodities.   ◼

Second-Order Partial Derivatives

The first partial derivatives $f_x(x, y)$ and $f_y(x, y)$ of a function $f(x, y)$ of the two variables x and y are also functions of x and y. As such, we may differentiate each of the functions f_x and f_y to obtain the **second-order partial derivatives** of f. Thus, differentiating the function f_x with respect to x leads to the second partial derivative

$$f_{xx} \equiv \frac{\partial^2 f}{\partial x^2} = \frac{\partial}{\partial x}(f_x)$$

However, differentiation of f_x with respect to y leads to the second partial derivative

$$f_{xy} \equiv \frac{\partial^2 f}{\partial y \partial x} = \frac{\partial}{\partial y}(f_x)$$

Similarly, differentiation of the function f_y with respect to x and with respect to y leads to

$$f_{yx} \equiv \frac{\partial^2 f}{\partial x \partial y} = \frac{\partial}{\partial x}(f_y)$$

and

$$f_{yy} \equiv \frac{\partial^2 f}{\partial y^2} = \frac{\partial}{\partial y}(f_y)$$

respectively. Note that, in general, it is not true that $f_{xy} = f_{yx}$. However, in most practical applications f_{xy} and f_{yx} are equal.

EXAMPLE 6 Find the second-order partial derivatives of the function

$$f(x, y) = x^3 - 3x^2 y + 3xy^2 + y^2$$

Solution The first partial derivatives of f are

$$f_x = \frac{\partial}{\partial x}(x^3 - 3x^2 y + 3xy^2 + y^2)$$

$$= 3x^2 - 6xy + 3y^2$$

$$f_y = \frac{\partial}{\partial y}(x^3 - 3x^2 y + 3xy^2 + y^2)$$

$$= -3x^2 + 6xy + 2y$$

Therefore,
$$f_{xx} = \frac{\partial}{\partial x}(f_x) = \frac{\partial}{\partial x}(3x^2 - 6xy + 3y^2)$$
$$= 6x - 6y = 6(x - y)$$
$$f_{xy} = \frac{\partial}{\partial y}(f_x) = \frac{\partial}{\partial y}(3x^2 - 6xy + 3y^2)$$
$$= -6x + 6y = 6(y - x)$$
$$f_{yx} = \frac{\partial}{\partial x}(f_y) = \frac{\partial}{\partial x}(-3x^2 + 6xy + 2y)$$
$$= -6x + 6y = 6(y - x)$$
$$f_{yy} = \frac{\partial}{\partial y}(f_y) = \frac{\partial}{\partial y}(-3x^2 + 6xy + 2y)$$
$$= 6x + 2$$

EXAMPLE 7 Find the second-order partial derivatives of the function
$$f(x, y) = e^{xy^2}$$

Solution We have
$$f_x = \frac{\partial}{\partial x}(e^{xy^2})$$
$$= y^2 e^{xy^2}$$
and
$$f_y = \frac{\partial}{\partial y}(e^{xy^2})$$
$$= 2xy e^{xy^2}$$

so the required second-order partial derivatives of f are
$$f_{xx} = \frac{\partial}{\partial x}(f_x) = \frac{\partial}{\partial x}(y^2 e^{xy^2})$$
$$= y^4 e^{xy^2}$$
$$f_{xy} = \frac{\partial}{\partial y}(f_x) = \frac{\partial}{\partial y}(y^2 e^{xy^2})$$
$$= 2y e^{xy^2} + 2xy^3 e^{xy^2}$$
$$= 2y e^{xy^2}(1 + xy^2)$$
$$f_{yx} = \frac{\partial}{\partial x}(f_y) = \frac{\partial}{\partial x}(2xy e^{xy^2})$$
$$= 2y e^{xy^2} + 2xy^3 e^{xy^2}$$
$$= 2y e^{xy^2}(1 + xy^2)$$
$$f_{yy} = \frac{\partial}{\partial y}(f_y) = \frac{\partial}{\partial y}(2xy e^{xy^2})$$
$$= 2x e^{xy^2} + (2xy)(2xy)e^{xy^2}$$
$$= 2x e^{xy^2}(1 + 2xy^2)$$

SELF-CHECK EXERCISES 12.2

1. Compute the first partial derivatives of $f(x, y) = x^3 - 2xy^2 + y^2 - 8$.

2. Find the first partial derivatives of $f(x, y) = x \ln y + ye^x - x^2$ at $(0, 1)$ and interpret your results.

3. Find the second-order partial derivatives of the function of exercise 1 above.

4. A certain country's production is described by the function

$$f(x, y) = 60x^{1/3}y^{2/3}$$

when x units of labor and y units of capital are used.
   a. What is the marginal productivity of labor and the marginal productivity of capital when the amounts expended on labor and capital are 125 units and 8 units, respectively?
   b. Should the government encourage capital investment rather than increased expenditure on labor at this time in order to increase the country's productivity?

Solutions to Self-Check Exercises 12.2 can be found on page 770.

12.2 EXERCISES

In exercises 1–22, find the first partial derivatives of the function.

1. $f(x, y) = 2x + 3y + 5$

2. $f(x, y) = 2xy$

3. $g(x, y) = 2x^2 + 4y + 1$

4. $f(x, y) = 1 + x^2 + y^2$

5. $f(x, y) = \dfrac{2y}{x^2}$

6. $f(x, y) = \dfrac{x}{1 + y}$

7. $g(u, v) = \dfrac{u - v}{u + v}$

8. $f(x, y) = \dfrac{x^2 - y^2}{x^2 + y^2}$

9. $f(s, t) = (s^2 - st + t^2)^3$

10. $g(s, t) = s^2t + st^{-3}$

11. $f(x, y) = (x^2 + y^2)^{2/3}$

12. $f(x, y) = x\sqrt{1 + y^2}$

13. $f(x, y) = e^{xy+1}$

14. $f(x, y) = (e^x + e^y)^5$

15. $f(x, y) = x \ln y + y \ln x$

16. $f(x, y) = x^2e^{y^2}$

17. $g(u, v) = e^u \ln v$

18. $f(x, y) = \dfrac{e^{xy}}{x + y}$

19. $f(x, y, z) = xyz + xy^2 + yz^2 + zx^2$

20. $g(u, v, w) = \dfrac{2uvw}{u^2 + v^2 + w^2}$

21. $h(r, s, t) = e^{rst}$

22. $f(x, y, z) = xe^{y/z}$

In exercises 23–32, evaluate the first partial derivatives of the function at the given point.

23. $f(x, y) = x^2y + xy^2$; $(1, 2)$

24. $f(x, y) = x^2 + xy + y^2 + 2x - y$; $(-1, 2)$

25. $f(x, y) = x\sqrt{y} + y^2$; $(2, 1)$

26. $g(x, y) = \sqrt{x^2 + y^2}$; $(3, 4)$

27. $f(x, y) = \dfrac{x}{y}$; $(1, 2)$

28. $f(x, y) = \dfrac{x + y}{x - y}$; $(1, -2)$

29. $f(x, y) = e^{xy}$; $(1, 1)$

30. $f(x, y) = e^x \ln y$; $(0, e)$

31. $f(x, y, z) = x^2yz^3$; $(1, 0, 2)$

32. $f(x, y, z) = x^2y^2 + z^2$; $(1, 1, 2)$

In exercises 33–40, find the second-order partial derivatives of the function. In each case, show that the mixed partial derivatives f_{xy} and f_{yx} are equal.

33. $f(x, y) = x^2y + xy^3$

34. $f(x, y) = x^3 + x^2y + x + 4$

35. $f(x, y) = x^2 - 2xy + 2y^2 + x - 2y$

36. $f(x, y) = x^3 + x^2y^2 + y^3 + x + y$

37. $f(x, y) = \sqrt{x^2 + y^2}$

38. $f(x, y) = x\sqrt{y} + y\sqrt{x}$

39. $f(x, y) = e^{-x/y}$

40. $f(x, y) = \ln(1 + x^2y^2)$

41. Productivity of a Country The productivity of a South American country is given by the function

$$f(x, y) = 20x^{3/4}y^{1/4}$$

when x units of labor and y units of capital are used.
  a. What is the marginal productivity of labor and the marginal productivity of capital when the amounts expended on labor and capital are 256 units and 16 units, respectively?
  b. Should the government encourage capital investment rather than increased expenditure on labor at this time in order to increase the country's productivity?

42. Productivity of a Country The productivity of a country in Western Europe is given by the function

$$f(x, y) = 40x^{4/5}y^{1/5}$$

when x units of labor and y units of capital are used.
  a. What is the marginal productivity of labor and the marginal productivity of capital when the amounts expended on labor and capital are 32 units and 243 units, respectively?
  b. Should the government encourage capital investment rather than increased expenditure on labor at this time in order to increase the country's productivity?

43. Land Prices The rectangular region R shown in the accompanying figure represents a city's financial district. The price of land within the district is approximated by the function

$$p(x, y) = 200 - 10\left(x - \frac{1}{2}\right)^2 - 15(y - 1)^2$$

where $p(x, y)$ is the price of land at the point (x, y) in dollars per square foot and x and y are measured in miles. Compute $\dfrac{\partial p}{\partial x}(0, 1)$ and $\dfrac{\partial p}{\partial y}(0, 1)$ and interpret your results.

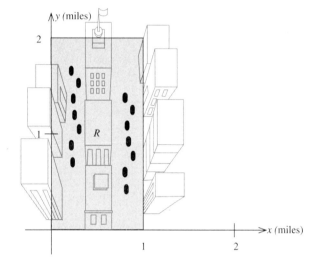

44. Complementary and Substitute Commodities In a survey conducted by *Home Entertainment* magazine it was determined that the demand equation for videocassette recorders (VCRs) is given by

$$x = f(p, q) = 10{,}000 - 10p + 0.2q^2$$

and the demand equation for videodisc players is given by

$$y = g(p, q) = 5000 + 0.8p^2 - 20q$$

where p and q denote the unit prices (in dollars) for the cassette recorders and disc players, respectively, and x and y denote the number of cassette recorders and disc players demanded per week. Determine whether these two products are substitute, complementary, or neither.

45. Complementary and Substitute Commodities In the survey mentioned in exercise 44, it was determined that the demand equation for VCRs is given by

$$x = f(p, q) = 10{,}000 - 10p - e^{0.5q}$$

and the demand equation for blank tapes for the VCRs is given by

$$y = g(p, q) = 50{,}000 - 4{,}000q - 10p$$

where p and q denote the unit prices, respectively, and x and y denote the number of VCRs and the number of blank VCR tapes demanded per week.

Determine whether these two products are substitute, complementary, or neither.

46. Complementary and Substitute Commodities Refer to exercise 28, Exercises 12.1. Show that the finished and unfinished home furniture manufactured by the Country Workshop are substitute commodities. [*Hint:* Solve the system of equations for x and y in terms of p and q.]

C **47. Revenue Functions** The total weekly revenue (in dollars) of the Country Workshop associated with manufacturing and selling their roll-top desks is given by the function

$$R(x, y) = -0.2x^2 - 0.25y^2 - 0.2xy + 200x + 160y$$

where x denotes the number of finished units and y denotes the number of unfinished units manufactured and sold per week. Compute $\partial R/\partial x$ and $\partial R/\partial y$ when $x = 300$ and $y = 250$. Interpret your results.

C **48. Profit Functions** The monthly profit (in dollars) of the Bond and Barker Department Store depends on the level of inventory x (in thousands of dollars) and the floor space y (in thousands of square feet) available for display of the merchandise, as given by

the equation

$$P(x, y) = -0.02x^2 - 15y^2 + xy$$
$$+ 39x + 25y - 20{,}000$$

Compute $\partial P/\partial x$ and $\partial P/\partial y$ when $x = 4000$ and $y = 150$. Interpret your results. Repeat with $x = 5000$ and $y = 150$.

C **49. Volume of a Gas** The volume V (in liters) of a certain mass of gas is related to its pressure P (in millimeters of mercury) and its temperature T (in degrees Kelvin) by the law

$$V = \frac{30.9T}{P}$$

Compute $\partial V/\partial T$ and $\partial V/\partial P$ when $T = 300$ and $P = 800$. Interpret your results.

50. Surface Area of a Human Body The formula

$$S = 0.007184W^{0.425}H^{0.725}$$

gives the surface area S of a human body (in square meters) in terms of its weight W in kilograms and its height H in centimeters. Compute $\partial S/\partial W$ and $\partial S/\partial H$ when $W = 70$ kg and $H = 180$ cm. Interpret your results.

SOLUTIONS TO SELF-CHECK EXERCISES 12.2

1.
$$f_x = \frac{\partial f}{\partial x} = 3x^2 - 2y^2$$

$$f_y = \frac{\partial f}{\partial y} = -2x(2y) + 2y$$

$$= 2y(1 - 2x)$$

2.
$$f_x = \ln y + ye^x - 2x; \quad f_y = \frac{x}{y} + e^x$$

In particular,
$$f_x(0, 1) = \ln 1 + 1e^0 - 2(0) = 1$$

$$f_y(0, 1) = \frac{0}{1} + e^0 = 1$$

The results tell us that at the point $(0, 1)$, $f(x, y)$ increases 1 unit for each unit increase in the x-direction, y being kept constant; $f(x, y)$ also increases 1 unit for each unit increase in the y-direction, x being kept constant.

3. From the results of exercise 1,

$$f_x = 3x^2 - 2y^2$$

Therefore,

$$f_{xx} = \frac{\partial}{\partial x}(3x^2 - 2y^2) = 6x$$

and
$$f_{xy} = \frac{\partial}{\partial y}(3x^2 - 2y^2) = -4y$$

Also, from the results of exercise 1,

$$f_y = 2y(1 - 2x)$$

and so

$$f_{yx} = \frac{\partial}{\partial x}[2y(1 - 2x)] = -4y$$

and
$$f_{yy} = \frac{\partial}{\partial y}[2y(1 - 2x)] = 2(1 - 2x)$$

4. **a.** The marginal productivity of labor when the amount expended on labor and capital are x and y units, respectively, is given by

$$f_x(x, y) = 60\left(\frac{1}{3}x^{-2/3}\right)y^{2/3} = 20\left(\frac{y}{x}\right)^{2/3}$$

In particular, the required marginal productivity of labor is given by

$$f_x(125, 8) = 20\left(\frac{8}{125}\right)^{2/3} = 20\left(\frac{4}{25}\right)$$

or 3.2 units per unit increase in labor expenditure, capital expenditure being held constant at 8 units. Next, we compute

$$f_y(x, y) = 60x^{1/3}\left(\frac{2}{3}y^{-1/3}\right) = 40\left(\frac{x}{y}\right)^{1/3}$$

and deduce that the required marginal productivity of capital is given by

$$f_y(125, 8) = 40\left(\frac{125}{8}\right)^{1/3} = 40\left(\frac{5}{2}\right)$$

or 100 units per unit increase in capital expenditure, labor expenditure being held constant at 125 units.

b. The results of (a) tell us that the government should encourage increased spending on capital rather than on labor.

Using Technology

Finding Partial Derivatives at a Given Point

Suppose $f(x, y)$ is a function of two variables and we wish to compute

$$f_x(a, b) = \left.\frac{\partial f}{\partial x}\right|_{(a,b)}$$

Recall that in computing $\partial f/\partial x$, we think of y as being fixed. But in this situation we are evaluating $\partial f/\partial x$ at (a, b). Therefore, we set y equal to b. Doing this leads to the function g of one variable, x, defined by

$$g(x) = f(x, b)$$

It follows from the definition of the partial derivative that

$$f_x(a, b) = g'(a)$$

Thus, the value of the partial derivative $\partial f/\partial x$ at a given point (a, b) can be found by evaluating the derivative of a function of one variable. In particular, the latter can be found by using the **dy/dx** operation of a graphing utility. We find $f_y(a, b)$ in a similar manner.

EXAMPLE 1   Let $f(x, y) = (1 + xy^2)^{3/2}e^{x^2y}$. Find (a) $f_x(1, 2)$ and (b) $f_y(1, 2)$.

Solution

a. Define $g(x) = f(x, 2) = (1 + 4x)^{3/2}e^{2x^2}$. Using the **dy/dx** operation to find $g'(1)$, we obtain

$$f_x(1, 2) = g'(1) \approx 429.5832249$$

b. Define $h(y) = f(1, y) = (1 + y^2)^{3/2}e^y$. Using the **dy/dx** operation to find $h'(2)$, we obtain

$$f_y(1, 2) = h'(2) \approx 181.74674899 \qquad \blacksquare$$

EXERCISES

In exercises 1–6, use a graphing utility to compute $\dfrac{\partial f}{\partial x}$ and $\dfrac{\partial f}{\partial y}$ at the given point.

1. $f(x, y) = \sqrt{x}(2 + xy^2)^{1/3}; (1, 2)$

2. $f(x, y) = \sqrt{xy}(1 + 2xy)^{2/3}; (1, 4)$

3. $f(x, y) = \dfrac{x + y^2}{1 + x^2y}; (1, 2)$

4. $f(x, y) = \dfrac{xy^2}{(\sqrt{x} + \sqrt{y})^2}; (4, 1)$

5. $f(x, y) = e^{-xy^2}(x + y)^{1/3}; (1, 1)$

6. $f(x, y) = \dfrac{\ln(\sqrt{x} + y^2)}{x^2 + y^2}; (4, 1)$

| 12.3 | # Maxima and Minima of Functions of Several Variables |

Maxima and Minima

In Chapter 10 we saw how solving a practical problem formulated in terms of a function f of one variable often centers on determining an extreme value of f with respect to that variable. For example, we can solve the problem of finding a firm's production level x that will yield a maximum profit by finding the absolute maximum value of the profit function $P(x)$ with respect to x. On the other hand, to find a manufacturer's production level x that will result in a minimal cost of operation, we find the absolute minimum value of the cost function $C(x)$ with respect to x.

The notion of an extreme value of a function plays an equally important role in the case of a function of several variables. As in the case of a function of one variable, it is important to distinguish between the concept of a relative maximum (relative minimum) of f and that of an absolute maximum (absolute minimum) of f. More specifically, a function $f(x, y)$ of two variables has a **relative maximum** at a point (a, b) in the domain of f if $f(x, y) \leq f(a, b)$ for all points (x, y) that are sufficiently close to the point

(a, b). Similarly, f has a **relative minimum** at (a, b) if $f(x, y) \geq f(a, b)$ for all points (x, y) that are sufficiently close to the point (a, b). Geometrically, this means that no point on the graph of f corresponding to points (x, y) lying sufficiently close to the point (a, b) can be above (or below) the point $(a, b, f(a, b))$. Figure 15 depicts the graph of a function f of two variables that has a relative maximum at the point (a, b) and a relative minimum at the point (c, d).

FIGURE 15

f has a relative maximum at
(a, b) and a relative minimum
at (c, d).

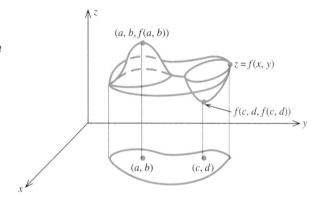

A function $f(x, y)$ of two variables has an **absolute maximum** at a point (a, b) if

$$f(x, y) \leq f(a, b)$$

for *all* points (x, y) in the domain of f. Similarly, a function $f(x, y)$ has an **absolute minimum** at a point (a, b) if

$$f(x, y) \geq f(a, b)$$

for *all* points (x, y) in the domain of f. Geometrically, the absolute maximum and the absolute minimum of a function f are, respectively, the highest and the lowest points on the graph of f.

Just as in the case of a function of one variable, a relative extremum (relative maximum or relative minimum) may or may not be an absolute extremum. However, in order to simplify matters, we will assume that whenever an absolute extremum exists, it will occur at a point where f has a relative extremum.

Just as the first and second derivatives play an important role in determining the relative extrema of a function of one variable, the first and second partial derivatives are powerful tools for locating and classifying the relative extrema of functions of several variables.

Suppose now that a differentiable function $f(x, y)$ of two variables has a relative maximum (relative minimum) at a point (a, b) in the domain of f. From Figure 16, it is clear that at the point (a, b) the slope of the "tangent lines" to the surface in any direction must be zero. In particular, this implies that both

$$\frac{\partial f}{\partial x}(a, b) \qquad \text{and} \qquad \frac{\partial f}{\partial y}(a, b)$$

must be zero.

FIGURE 16

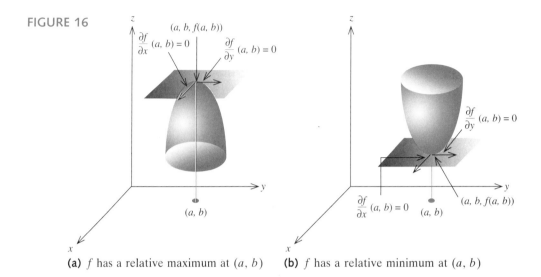

(a) f has a relative maximum at (a, b) **(b)** f has a relative minimum at (a, b)

The point (a, b) is called a **critical point** of the function f. In case we are tempted to conclude that a critical point of a function f must automatically be a relative extremum of f, let us consider the graph of the function f in Figure 17. Here we have both

$$\frac{\partial f}{\partial x}(a, b) = 0 \qquad \text{and} \qquad \frac{\partial f}{\partial y}(a, b) = 0$$

FIGURE 17
The point $(a, b, f(a, b))$ is called a saddle point.

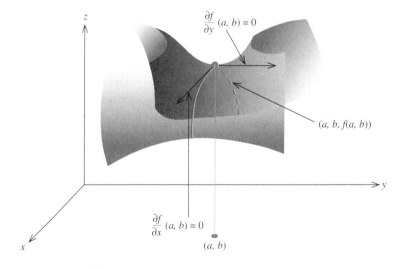

The point $(a, b, f(a, b))$ is neither a relative maximum nor a relative minimum of the function f, since there are points nearby that are higher and others that are lower than it. Such a point is called a **saddle point.** Thus, as in the case of a function of one variable, we may conclude that a critical point of a function of two (or more) variables is only a candidate for a relative extremum of f.

To determine the nature of a critical point of a function $f(x, y)$ of two variables, we make use of the second partial derivatives of f. The resulting test, which aids us in classifying these points, is called the **Second Derivative Test** and is incorporated in the following procedure for finding and classifying the relative extrema of f.

DETERMINING RELATIVE EXTREMA

1. Find the critical points of $f(x, y)$ by solving the system of simultaneous equations

$$f_x = 0$$
$$f_y = 0$$

2. The Second Derivative Test: Let

$$D(x, y) = f_{xx}f_{yy} - f_{xy}^2$$

Then

a. $D(a, b) > 0$ and $f_{xx}(a, b) < 0$ implies that $f(x, y)$ has a **relative maximum** at the point (a, b).

b. $D(a, b) > 0$ and $f_{xx}(a, b) > 0$ implies that $f(x, y)$ has a **relative minimum** at the point (a, b).

c. $D(a, b) < 0$ implies that $f(x, y)$ has neither a relative maximum nor a relative minimum at the point (a, b).

d. $D(a, b) = 0$ implies that the test is inconclusive, so some other technique must be used to solve the problem.

EXAMPLE 1 Find the relative extrema of the function

$$f(x, y) = x^2 + y^2$$

Solution We have

$$f_x = 2x$$

and

$$f_y = 2y$$

To find the critical point(s) of f, we set $f_x = 0$ and $f_y = 0$ and solve the resulting system of simultaneous equations

$$2x = 0$$
$$2y = 0$$

obtaining $x = 0$, $y = 0$, or $(0, 0)$, as the sole critical point of f. Next, we apply the Second Derivative Test to determine the nature of the critical point $(0, 0)$. We compute

$$f_{xx} = 2, \qquad f_{xy} = 0, \qquad f_{yy} = 2$$

and

$$D(x, y) = f_{xx}f_{yy} - f_{xy}^2 = (2)(2) - 0 = 4$$

In particular, $D(0, 0) = 4$. Since $D(0, 0) > 0$ and $f_{xx}(0, 0) = 2 > 0$, we conclude that $f(x, y)$ has a relative minimum at the point $(0, 0)$. The relative minimum value, 0, also happens to be the absolute minimum of f. The graph of the function f, shown in Figure 18, confirms these results.

FIGURE 18
The graph of
$f(x, y) = x^2 + y^2$

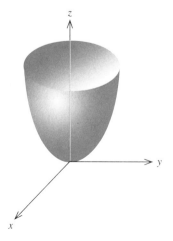

EXAMPLE 2   Find the relative extrema of the function

$$f(x, y) = 3x^2 - 4xy + 4y^2 - 4x + 8y + 4$$

Solution   We have

$$f_x = 6x - 4y - 4$$

and

$$f_y = -4x + 8y + 8$$

To find the critical points of f, we set $f_x = 0$ and $f_y = 0$ and solve the resulting system of simultaneous equations

$$6x - 4y = 4$$

$$-4x + 8y = -8$$

Multiplying the first equation by 2 and the second equation by 3, we obtain the equivalent system

$$12x - 8y = 8$$

$$-12x + 24y = -24$$

Adding the two equations gives $16y = -16$, or $y = -1$. We substitute this value for y into either equation in the system to get $x = 0$. Thus, the only critical point of f is the point $(0, -1)$. Next, we apply the Second Derivative Test to determine whether the point $(0, -1)$ gives rise to a relative extremum of f. We compute

$$f_{xx} = 6, \qquad f_{xy} = -4, \qquad f_{yy} = 8$$

and

$$D(x, y) = f_{xx}f_{yy} - f_{xy}^2 = (6)(8) - (-4)^2 = 32$$

Since $D(0, -1) = 32 > 0$ and $f_{xx}(0, -1) = 6 > 0$, we conclude that $f(x, y)$ has a relative minimum at the point $(0, -1)$. The value of $f(x, y)$ at the point $(0, -1)$ is given by

$$f(0, -1) = 3(0)^2 - 4(0)(-1) + 4(-1)^2 - 4(0) + 8(-1) + 4 = 0 \qquad \blacksquare$$

EXAMPLE 3   Find the relative extrema of the function

$$f(x, y) = 4y^3 + x^2 - 12y^2 - 36y + 2$$

Solution   To find the critical points of f, we set $f_x = 0$ and $f_y = 0$ simultaneously, obtaining

$$f_x = 2x = 0$$

$$f_y = 12y^2 - 24y - 36 = 0$$

The first equation implies that $x = 0$. The second equation implies that

$$y^2 - 2y - 3 = 0$$

$$(y + 1)(y - 3) = 0$$

that is, $y = -1$ or 3. Therefore, there are two critical points of the function f, namely, $(0, -1)$ and $(0, 3)$.

Next, we apply the Second Derivative Test to determine the nature of each of the two critical points. We compute

$$f_{xx} = 2, \qquad f_{xy} = 0, \qquad f_{yy} = 24y - 24 = 24(y - 1)$$

Therefore,

$$D(x, y) = f_{xx}f_{yy} - f_{xy}^2 = 48(y - 1)$$

For the point $(0, -1)$,

$$D(0, -1) = 48(-1 - 1) = -96 < 0$$

Since $D(0, -1) < 0$, we conclude that the point $(0, -1)$ gives a saddle point of f. For the point $(0, 3)$,

$$D(0, 3) = 48(3 - 1) = 96 > 0$$

Since $D(0, 3) > 0$ and $f_{xx}(0, 3) > 0$, we conclude that the function f has a relative minimum at the point $(0, 3)$. Furthermore, since

$$f(0, 3) = 4(3)^3 + (0)^2 - 12(3)^2 - 36(3) + 2$$
$$= -106$$

we see that the relative minimum value of f is -106. ■

Applications

As in the case of a practical optimization problem involving a function of one variable, the solution to an optimization problem involving a function of several variables calls for finding the *absolute* extremum of the function. Determining the absolute extremum of a function of several variables is more difficult than merely finding the relative extrema of the function. However, in many situations, the absolute extremum of a function actually coincides with the largest relative extremum of the function that occurs in the interior of its domain. We will assume that the problems considered here belong to this category. Furthermore, the existence of the absolute extremum (solution) of a practical problem is often deduced from the geometric or physical nature of the problem.

EXAMPLE 4 The total weekly revenue (in dollars) that the Acrosonic Company realizes in producing and selling its bookshelf loudspeaker systems is given by

$$R(x, y) = -\frac{1}{4}x^2 - \frac{3}{8}y^2 - \frac{1}{4}xy + 300x + 240y$$

where x denotes the number of fully assembled units and y denotes the number of kits produced and sold per week. The total weekly cost attributable to the production of these loudspeakers is

$$C(x, y) = 180x + 140y + 5000$$

dollars, where x and y have the same meaning as before. Determine how many assembled units and how many kits Acrosonic should produce per week to maximize its profit.

Solution The contribution to Acrosonic's weekly profit stemming from the production and sale of the bookshelf loudspeaker systems is given by

$$P(x, y) = R(x, y) - C(x, y)$$

$$= \left(-\frac{1}{4}x^2 - \frac{3}{8}y^2 - \frac{1}{4}xy + 300x + 240y \right) - (180x + 140y + 5000)$$

$$= -\frac{1}{4}x^2 - \frac{3}{8}y^2 - \frac{1}{4}xy + 120x + 100y - 5000$$

To find the relative maximum of the profit function $P(x, y)$, we first locate the critical point(s) of P. Setting $P_x(x, y)$ and $P_y(x, y)$ equal to zero, we obtain

$$P_x = -\frac{1}{2}x - \frac{1}{4}y + 120 = 0$$

$$P_y = -\frac{3}{4}y - \frac{1}{4}x + 100 = 0$$

Solving the first of these equations for y yields

$$y = -2x + 480$$

which, upon substitution into the second equation, yields

$$-\frac{3}{4}(-2x + 480) - \frac{1}{4}x + 100 = 0$$

$$6x - 1440 - x + 400 = 0$$

or $$x = 208$$

We substitute this value of x into the equation $y = -2x + 480$ to get

$$y = 64$$

Therefore, the function P has the sole critical point $(208, 64)$. To show that the point $(208, 64)$ is a solution to our problem, we use the Second Derivative Test. We compute

$$P_{xx} = -\frac{1}{2}, \qquad P_{xy} = -\frac{1}{4}, \qquad P_{yy} = -\frac{3}{4}$$

So

$$D(x, y) = \left(-\frac{1}{2}\right)\left(-\frac{3}{4}\right) - \left(-\frac{1}{4}\right)^2 = \frac{3}{8} - \frac{1}{16} = \frac{5}{16}$$

In particular, $D(208, 64) = \frac{5}{16} > 0$.

Since $D(208, 64) > 0$ and $P_{xx}(208, 64) < 0$, the point $(208, 64)$ yields a relative maximum of P. This relative maximum is also the absolute maximum of P. We conclude that Acrosonic can maximize its weekly profit by manufacturing 208 assembled units and 64 kits of their bookshelf loudspeaker systems. The maximum weekly profit realizable from the production and sale of these loudspeaker systems is given by

$$P(208, 64) = -\frac{1}{4}(208)^2 - \frac{3}{8}(64)^2 - \frac{1}{4}(208)(64)$$

$$+ 120(208) + 100(64) - 5{,}000$$

$$= 10{,}680$$

or $10,680.

FIGURE 19

Locating a site for a television relay station

EXAMPLE 5   A television relay station will serve towns A, B, and C, whose relative locations are shown in Figure 19. Determine a site for the location of the station if the sum of the squares of the distances from each town to the site is minimized.

Solution Suppose the required site is located at the point $P(x, y)$. With the aid of the distance formula, we find that the square of the distance from town A to the site is

$$(x - 30)^2 + (y - 20)^2$$

The respective distances from towns B and C to the site are found in a similar manner, so that the sum of the squares of the distances from each town to the site is given by

$$f(x, y) = (x - 30)^2 + (y - 20)^2 + (x + 20)^2$$
$$+ (y - 10)^2 + (x - 10)^2 + (y + 10)^2$$

To find the relative minimum of $f(x, y)$, we first find the critical point(s) of f. Using the Chain Rule to find $f_x(x, y)$ and $f_y(x, y)$ and setting each equal to zero, we obtain

$$f_x = 2(x - 30) + 2(x + 20) + 2(x - 10) = 6x - 40 = 0$$

and $\qquad f_y = 2(y - 20) + 2(y - 10) + 2(y + 10) = 6y - 40 = 0$

from which we deduce that $(\frac{20}{3}, \frac{20}{3})$ is the sole critical point of f. Since

$$f_{xx} = 6, \qquad f_{xy} = 0, \qquad f_{yy} = 6$$

we have

$$D(x, y) = f_{xx}f_{yy} - f_{xy}^2 = (6)(6) - 0 = 36$$

Since $D(\frac{20}{3}, \frac{20}{3}) > 0$ and $f_{xx}(\frac{20}{3}, \frac{20}{3}) > 0$, we conclude that the point $(\frac{20}{3}, \frac{20}{3})$ yields a relative minimum of f. Thus, the required site has coordinates $x = \frac{20}{3}$ and $y = \frac{20}{3}$. ■

SELF-CHECK
EXERCISES 12.3

1. Let $f(x, y) = 2x^2 + 3y^2 - 4xy + 4x - 2y + 3$.
   a. Find the critical point of f.
   b. Use the Second Derivative Test to classify the nature of the critical point.
   c. Find the relative extremum of f if it exists.

2. The Robertson Controls Company manufactures two basic models of setback thermostats: a standard mechanical thermostat and a deluxe electronic thermostat. Robertson's monthly revenue (in hundreds of dollars) is

$$R(x, y) = -\frac{1}{8}x^2 - \frac{1}{2}y^2 - \frac{1}{4}xy + 20x + 60y$$

where x (in units of a hundred) denotes the number of mechanical thermostats manufactured and y (in units of a hundred) denotes the number of electronic thermostats manufactured per month. The total monthly cost incurred in producing these thermostats is

$$C(x, y) = 7x + 20y + 280$$

hundred dollars. Find how many thermostats of each model Robertson should manufacture per month in order to maximize its profits. What is the maximum profit?

Solutions to Self-Check Exercises 12.3 can be found on page 782.

12.3   EXERCISES

In exercises 1–20, find the critical point(s) of the given functions. Then use the Second Derivative Test to classify the nature of each of these point(s), if possible. Finally, determine the relative extrema of each function.

1. $f(x, y) = 1 - 2x^2 - 3y^2$

2. $f(x, y) = x^2 - xy + y^2 + 1$

3. $f(x, y) = x^2 - y^2 - 2x + 4y + 1$

4. $f(x, y) = 2x^2 + y^2 - 4x + 6y + 3$

5. $f(x, y) = x^2 + 2xy + 2y^2 - 4x + 8y - 1$

6. $f(x, y) = x^2 - 4xy + 2y^2 + 4x + 8y - 1$

7. $f(x, y) = 2x^3 + y^2 - 9x^2 - 4y + 12x - 2$

8. $f(x, y) = 2x^3 + y^2 - 6x^2 - 4y + 12x - 2$

9. $f(x, y) = x^3 + y^2 - 2xy + 7x - 8y + 4$

10. $f(x, y) = 2y^3 - 3y^2 - 12y + 2x^2 - 6x + 2$

11. $f(x, y) = x^3 - 3xy + y^3 - 2$

12. $f(x, y) = x^3 - 2xy + y^2 + 5$

13. $f(x, y) = xy + \dfrac{4}{x} + \dfrac{2}{y}$

14. $f(x, y) = \dfrac{x}{y^2} + xy$ **15.** $f(x, y) = x^2 - e^{y^2}$

16. $f(x, y) = e^{x^2 - y^2}$ **17.** $f(x, y) = e^{x^2 + y^2}$

18. $f(x, y) = e^{xy}$

19. $f(x, y) = \ln(1 + x^2 + y^2)$

20. $f(x, y) = xy + \ln x + 2y^2$

21. Maximizing Profit   The total weekly revenue (in dollars) of the Country Workshop realized in manufacturing and selling its roll-top desks is given by

$$R(x, y) = -0.2x^2 - 0.25y^2 - 0.2xy + 200x + 160y$$

where x denotes the number of finished units and y denotes the number of unfinished units manufactured and sold per week. The total weekly cost attributable to the manufacture of these desks is given by

$$C(x, y) = 100x + 70y + 4000$$

dollars. Determine how many finished units and how many unfinished units the company should manufacture per week in order to maximize its profit. What is the maximum profit realizable?

22. Maximizing Profit   The total daily revenue (in dollars) that the Weston Publishing Company realizes in publishing and selling its English language dictionaries is given by

$$R(x, y) = -0.005x^2 - 0.003y^2 - 0.002xy$$
$$+ 20x + 15y$$

where x denotes the number of deluxe copies and y denotes the number of standard copies published and sold daily. The total daily cost of publishing these dictionaries is given by

$$C(x, y) = 6x + 3y + 200$$

dollars. Determine how many deluxe copies and how many standard copies Weston should publish per day to maximize its profits. What is the maximum profit realizable?

23. Maximum Price   The rectangular region R shown in the accompanying figure represents the financial district of a city. The price of land within the district is approximated by the function

$$p(x, y) = 200 - 10\left(x - \frac{1}{2}\right)^2 - 15(y - 1)^2$$

where $p(x, y)$ is the price of land at the point (x, y) in dollars per square foot and x and y are measured in miles. At what point within the financial district is the price of land highest?

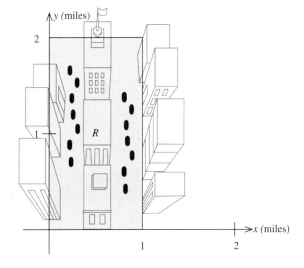

24. Maximizing Profit   C & G Imports, Inc., imports two brands of white wine, one from Germany and the other from Italy. The German wine costs $4 a bottle, and the Italian wine can be obtained for $3 a bottle. It has been estimated that if the German wine retails at p dollars per bottle and the Italian wine is sold for q dollars per bottle, then

$$2000 - 150p + 100q$$

bottles of the German wine and

$$1000 + 80p - 120q$$

bottles of the Italian wine will be sold per week. Determine the unit price for each brand that will allow C & G to realize the largest possible weekly profit.

25. **Determining the Optimal Site**   An auxiliary electric power station will serve three communities, A, B, and C, whose relative locations are shown in the accompanying figure. Determine where the power station should be located if the sum of the squares of the distances from each community to the site is minimized.

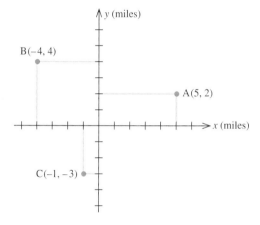

26. **Packaging**   An open rectangular box having a volume of 108 cubic inches is to be constructed from a tin sheet. Find the dimensions of such a box if the amount of material used in its construction is to be minimal. [*Hint:* Let the dimensions of the box be x'' by y'' by z''. Then

$$xyz = 108$$

and the amount of material used is given by

$$S = xy + 2yz + 2xz$$

Show that

$$S = f(x, y) = xy + \frac{216}{x} + \frac{216}{y}$$

Minimize $f(x, y)$.]

27. **Packaging**   Postal regulations specify that the combined length and girth of a parcel sent by parcel post may not exceed 108 inches. Find the dimensions of the rectangular package that would have the greatest possible volume under these regulations. [*Hint:* Let the dimensions of the box be x'' by y'' by z'' (see the figure). Then

$$2x + 2z + y = 108$$

and the volume

$$V = xyz$$

Show that

$$V = f(x, z) = 108xz - 2x^2z - 2xz^2$$

Maximize $f(x, z)$.]

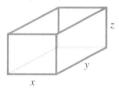

28. **Minimizing Heating and Cooling Costs**   A building in the shape of a rectangular box is to have a volume of 12,000 cubic feet (see figure). It is estimated that the annual heating and cooling costs will be $2 per square foot for the top, $4 per square foot for the front and back, and $3 per square foot for the sides. Find the dimensions of the building that will result in a minimal annual heating and cooling cost. What is the minimal annual heating and cooling cost?

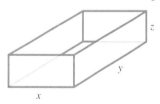

29. **Packaging**   An open box having a volume of 48 cubic inches is to be constructed. If the box is to include a partition that is parallel to a side of the box, as shown in the figure, and the amount of material used is to be minimal, what should the dimensions of the box be?

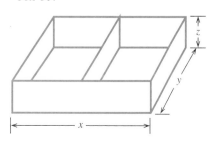

1. a. To find the critical point(s) of f, we solve the system of equations

$$f_x = 4x - 4y + 4 = 0$$

$$f_y = -4x + 6y - 2 = 0$$

obtaining $x = -2$ and $y = -1$. Thus, the only critical point of f is the point $(-2, -1)$.

b. We have $f_{xx} = 4$, $f_{xy} = -4$, and $f_{yy} = 6$, so

$$D(x, y) = f_{xx}f_{yy} - f_{xy}^2$$

$$= (4)(6) - (-4)^2 = 8$$

Since $D(-2, -1) > 0$ and $f_{xx}(-2, -1) > 0$, we conclude that f has a relative minimum at the point $(-2, -1)$.

c. The relative minimum value of $f(x, y)$ at the point $(-2, -1)$ is

$$f(-2, -1) = 2(-2)^2 + 3(-1)^2 - 4(-2)(-1) + 4(-2) - 2(-1) + 3$$

$$= 0$$

2. Robertson's monthly profit is

$$P(x, y) = R(x, y) - C(x, y)$$

$$= \left(-\frac{1}{8}x^2 - \frac{1}{2}y^2 - \frac{1}{4}xy + 20x + 60y\right) - (7x + 20y + 280)$$

$$= -\frac{1}{8}x^2 - \frac{1}{2}y^2 - \frac{1}{4}xy + 13x + 40y - 280$$

The critical point of P is found by solving the system

$$P_x = -\frac{1}{4}x - \frac{1}{4}y + 13 = 0$$

$$P_y = -\frac{1}{4}x - y + 40 = 0$$

giving $x = 16$ and $y = 36$. Thus $(16, 36)$ is the critical point of P. Next,

$$P_{xx} = -\frac{1}{4}, \qquad P_{xy} = -\frac{1}{4}, \qquad P_{yy} = -1$$

and

$$D(x, y) = f_{xx}f_{yy} - f_{xy}^2$$

$$= \left(-\frac{1}{4}\right)(-1) - \left(-\frac{1}{4}\right)^2 = \frac{3}{16}$$

Since $D(16, 36) > 0$ and $P_{xx}(16, 36) < 0$, the point $(16, 36)$ yields a relative maximum of P. We conclude that the monthly profit is maximized by manufacturing 1600 mechanical and 3600 electronic setback thermostats per month. The maximum monthly profit realizable is

$$P(16, 36) = -\frac{1}{8}(16)^2 - \frac{1}{2}(36)^2 - \frac{1}{4}(16)(36) + 13(16) + 40(36) - 280$$

$$= 544$$

or $54,400.

CHAPTER TWELVE SUMMARY OF PRINCIPAL TERMS

Terms

Three-dimensional Cartesian coordinate system
Ordered triple
Level curve
Contour map
Level surface
Function of two variables:
  Domain
  First partial derivative
  Second-order partial derivative
  Relative maximum
  Relative minimum
  Absolute maximum
  Absolute minimum
  Critical point
  Saddle point
  Second Derivative Test

Cobb-Douglas production function
Marginal productivity of labor
Marginal productivity of capital
Substitute commodities
Complementary commodities

CHAPTER 12 REVIEW EXERCISES

1. Let $f(x, y) = xy/(x^2 + y^2)$. Compute $f(0, 1)$, $f(1, 0)$, and $f(1, 1)$. Does $f(0, 0)$ exist?

2. Let $f(x, y) = \dfrac{xe^y}{1 + \ln xy}$. Compute $f(1, 1)$, $f(1, 2)$, and $f(2, 1)$. Does $f(1, 0)$ exist?

3. Let $h(x, y, z) = xye^z + (x/y)$. Compute $h(1, 1, 0)$, $h(-1, 1, 1)$, and $h(1, -1, 1)$.

4. Find the domain of the function $f(u, v) = \dfrac{u}{u - v}$.

5. Find the domain of the function $f(x, y) = \dfrac{x - y}{x + y}$.

6. Find the domain of the function $f(x, y) = x\sqrt{y} + y\sqrt{1 - x}$.

7. Find the domain of the function $f(x, y, z) = \dfrac{xy\sqrt{z}}{(1 - x)(1 - y)(1 - z)}$.

In exercises 8–11, sketch the level curves of the function corresponding to the given values of z.

8. $z = f(x, y) = 2x + 3y$; $z = -2, -1, 0, 1, 2$

9. $z = f(x, y) = y - x^2$; $z = -2, -1, 0, 1, 2$

10. $z = f(x, y) = \sqrt{x^2 + y^2}$; $z = 0, 1, 2, 3, 4$

11. $z = f(x, y) = e^{xy}$; $z = 1, 2, 3$

In exercises 12–21, compute the first partial derivatives of the function.

12. $f(x, y) = x^2y^3 + 3xy^2 + \dfrac{x}{y}$

13. $f(x, y) = x\sqrt{y} + y\sqrt{x}$

14. $f(u, v) = \sqrt{uv^2 - 2u}$

15. $f(x, y) = \dfrac{x - y}{y + 2x}$ **16.** $g(x, y) = \dfrac{xy}{x^2 + y^2}$

17. $h(x, y) = (2xy + 3y^2)^5$

18. $f(x, y) = (xe^y + 1)^{1/2}$

19. $f(x, y) = (x^2 + y^2)e^{x^2 + y^2}$

20. $f(x, y) = \ln(1 + 2x^2 + 4y^4)$

21. $f(x, y) = \ln\left(1 + \dfrac{x^2}{y^2}\right)$

In exercises 22–27, compute the second-order partial derivatives of the function.

22. $f(x, y) = x^3 - 2x^2y + y^2 + x - 2y$

23. $f(x, y) = x^4 + 2x^2y^2 - y^4$

24. $f(x, y) = (2x^2 + 3y^2)^3$ **25.** $g(x, y) = \dfrac{x}{x + y^2}$

26. $g(x, y) = e^{x^2 + y^2}$ **27.** $h(s, t) = \ln\left(\dfrac{s}{t}\right)$

28. Let $f(x, y, z) = x^3 y^2 z + xy^2 z + 3xy - 4z$. Compute $f_x(1, 1, 0)$, $f_y(1, 1, 0)$, and $f_z(1, 1, 0)$ and interpret your results.

In exercises 29–34, find the critical point(s) of the function. Then use the Second Derivative Test to classify the nature of the points, if possible. Finally, determine the relative extrema of each function.

29. $f(x, y) = 2x^2 + y^2 - 8x - 6y + 4$

30. $f(x, y) = x^2 + 3xy + y^2 - 10x - 20y + 12$

31. $f(x, y) = x^3 - 3xy + y^2$

32. $f(x, y) = x^3 + y^2 - 4xy + 17x - 10y + 8$

33. $f(x, y) = e^{2x^2 + y^2}$

34. $f(x, y) = \ln(x^2 + y^2 - 2x - 2y + 4)$

35. A division of Ditton Industries makes a 16-speed and a 10-speed electric blender. The company's management estimates that x units of the 16-speed model and y units of the 10-speed model are demanded daily when the unit prices are

$$p = 80 - 0.02x - 0.1y$$

and $$q = 60 - 0.1x - 0.05y$$

dollars, respectively.
   a. Find the daily total revenue function $R(x, y)$.
   b. Find the domain of the function R.
   c. Compute $R(100, 300)$ and interpret your result.

36. In a survey conducted by *Home Entertainment* magazine it was determined that the demand equation for

compact disc (CD) players is given by

$$x = f(p, q) = 900 - 9p - e^{0.4q}$$

whereas the demand equation for compact audio discs is given by

$$y = g(p, q) = 20{,}000 - 3{,}000q - 4p$$

where p and q denote the unit prices (in dollars) for the CD players and audio discs, respectively, and x and y denote the number of CD players and audio discs demanded per week. Determine whether these two products are substitute, complementary, or neither.

37. The Odyssey Travel Agency's monthly revenue depends on the amount of money x (in thousands of dollars) spent on advertising per month and the number of agents y in its employ in accordance with the rule

$$R(x, y) = -x^2 - 0.5y^2 + xy + 8x + 3y + 20$$

Determine the amount of money the agency should spend per month and the number of agents it should employ in order to maximize its monthly revenue.

38. The owner of the Rancho Grande wants to enclose a rectangular piece of grazing land along the straight portion of a river and then subdivide it using a fence running parallel to the sides. No fencing is required along the river. If the material for the sides costs $3 per running yard and the material for the divider costs $2 per running yard, what will the dimensions of a 303,750-sq yd pasture be if the cost of fencing material is kept to a minimum?

CHAPTER 12   EXTENDED PROBLEMS AND PROJECTS

1. According to the definitions of the first partial derivatives of $f(x, y)$ given on pages 760–761, we have

$$\frac{\partial f}{\partial x} = \lim_{h \to 0} \frac{f(x + h, y) - f(x, y)}{h}$$

and

$$\frac{\partial f}{\partial y} = \lim_{k \to 0} \frac{f(x, y + k) - f(x, y)}{k}$$

provided the limits exist. Using these definitions, compute $\dfrac{\partial f}{\partial x}$ and $\dfrac{\partial f}{\partial y}$ for

$$f(x, y) = 2xy^2$$

and verify your results using the method of Section 12.2.

APPENDIX A

Procedures for the TI-85 Graphics Calculator

In this appendix, we follow the notations used in the *TI-85 Graphics Calculator Guidebook*. Consult your manual for details to realize the full potential of your calculator.

A.1 Graphing a Single Function

1. Enter the function into the function table as follows:

   a. Press $\boxed{\text{GRAPH}}$ and select $\langle \mathbf{y(x)} \rangle$. The function table

   $$y1 =$$

   will be displayed. (If a function has been entered previously, pressing $\boxed{\text{CLEAR}}$ will clear the display.)

   b. Enter the function. For example, to graph the function $f(x) = x^3(x - 3)^4$ in Example 2 on page 89, we enter x^3(x − 3)^4 next to y1 = .

2. Select the viewing rectangle as follows:

   a. Select $\langle \mathbf{RANGE} \rangle$. The range table will be displayed.

   b. Enter the values for the viewing rectangle. For example, the viewing rectangle $[-1, 5] \times [-40, 40]$ of Example 2, page 89, has the following values:

   $$xMin = -1$$
   $$xMax = 5$$
   $$xScl = 1$$
   $$yMin = -40$$
   $$yMax = 40$$
   $$yScl = 1$$

   Press $\boxed{\text{ENTER}}$ after entering each value.

3. Display the desired graph by selecting $\langle \mathbf{GRAPH} \rangle$.

A.2   Evaluating a Function

Method A

1. Display the graph of the function. (See the procedure for graphing a single function, Section A.1.)
2. Press $\boxed{\text{MORE}}$ twice to display the function **EVAL**. Then select ⟨**EVAL**⟩. The display

$$\text{Eval } x =$$

appears on the bottom of the screen.
3. Enter the value of x and press $\boxed{\text{ENTER}}$. The desired value of $f(x)$ will be displayed on the bottom of the screen.

Method B (Without displaying the graph of the function)

1. Press $\boxed{\text{2nd}}$ $\boxed{\text{CALC}}$.
2. Select ⟨**evalF**⟩. The screen displays

$$\text{evalF(}$$

Enter the function, the variable, and the value of x at which the function is to be evaluated. For example, to evaluate $f(2)$, where $f(x) = x^2 + 1$, we enter

$$\text{evalF } (x^2 + 1, x, 2).$$

Then press $\boxed{\text{ENTER}}$. The value of f at $x = 2$, 5, will be displayed on the screen.

A.3   Finding the Points of Intersection of Two Graphs

1. Plot the graphs of the two functions. To do this, follow step 1 of the procedure for graphing a single function (A.1) to enter the first function. The press $\boxed{\text{ENTER}}$ to display

$$y2 =$$

in the function table. Enter the second function. Then proceed to step 2 of the procedure for graphing a function (A.1).
2. Select the **ISECT** operation as follows:
   a. Press $\boxed{\text{MORE}}$ once to display the function **MATH** in the menu. Then select ⟨**MATH**⟩. Press $\boxed{\text{MORE}}$ to obtain the **ISECT** function display.
   b. Select ⟨**ISECT**⟩ to use this operation.
3. To find coordinates of a point of intersection, move the cursor close to the point of intersection on one of the graphs and then press $\boxed{\text{ENTER}}$. The cursor will automatically jump to a point on the other graph. Press $\boxed{\text{ENTER}}$ again and the display ISECT will appear at the bottom of the screen together with the coordinates of the point of intersection of the two graphs.

A.4   Entering a Matrix

1. Press $\boxed{\text{2nd}}$ $\boxed{\text{MATRX}}$ to call the matrix menu. Then select ⟨**EDIT**⟩ to use the **EDIT** function.

2. Enter the name of the matrix. For example, pressing A will give the matrix the label "A." Press $\boxed{\text{ENTER}}$.

3. Enter the size of the matrix. For example, if A has order 3×4, press $\boxed{3}$, $\boxed{\text{ENTER}}$, $\boxed{4}$, $\boxed{\text{ENTER}}$.

4. Enter the elements of the matrix row by row, starting wtih the first row.

5. Return to calculator mode by pressing $\boxed{\text{2nd}}$ $\boxed{\text{QUIT}}$. To enter more than one matrix, repeat the above procedure as necessary.

REMARK   To view a matrix, press $\boxed{\text{2nd}}$ $\boxed{\text{MATRX}}$. Then select the name of the matrix to be viewed. Press $\boxed{\text{ENTER}}$. If necessary, use the right or left cursor keys to move any hidden part of the matrix into view.   ☐

A.5   Using Row Operations on Matrices, rref, and SIMULT

1. Using Row Operations

Enter the matrix A and go back to the calculator mode by pressing $\boxed{\text{2nd}}$ $\boxed{\text{QUIT}}$. Press $\boxed{\text{2nd}}$ $\boxed{\text{MATRX}}$ and then select ⟨**OPS**⟩. Press $\boxed{\text{MORE}}$ to display the menu for the row operations on the screen. The calculator is now ready to perform the following operations:

   a. Operation 1 ($R_i \leftrightarrow R_j$): Select ⟨**rSwap**⟩ to use the **rSwap** (row swap) operation. The command is rSwap(A, i, j). Press $\boxed{\text{ENTER}}$.

   b. Operation 2 (cR_i): Select ⟨**multR**⟩ to use the **multR** (row multiplication) operation. The command is multR(c, A, i). Press $\boxed{\text{ENTER}}$.

   c. Operation 3 ($R_i + aR_j$): Select ⟨**mRAdd**⟩ to use the **mRAdd** (multiply and add row) operation. The command is mRAdd(a, A, j, i). Press $\boxed{\text{ENTER}}$.

2. Using rref

After the matrix A is entered, go back to the calculator mode by pressing $\boxed{\text{2nd}}$ $\boxed{\text{QUIT}}$. Press $\boxed{\text{2nd}}$ $\boxed{\text{MATRX}}$ and then select ⟨**OPS**⟩ to display the menu. Select ⟨**rref**⟩ to use the **rref** (reduced row echelon form) operation. Enter A and then press $\boxed{\text{ENTER}}$. The row-reduced (equivalent) matrix is displayed.

3. Using SIMULT

Press $\boxed{\text{2nd}}$ $\boxed{\text{SIMULT}}$. Enter the number of variables (and equations) in the system. Press $\boxed{\text{ENTER}}$. Then enter the coefficients of the system (written in standard form), row by row, starting with the first. Select ⟨**SOLVE**⟩ to solve the system.

A.6

Operations on Matrices—Addition, Subtraction, Multiplication by a Scalar, and Matrix Multiplication

To perform any of the above operations on matrices, first enter the matrices A, B, C, and so on, and go back to the calculator mode. Then press $\boxed{2\text{nd}}$ $\boxed{\text{MATRX}}$. The calculator is now ready to perform the following operations:

1. **Addition of two matrices:** To find the sum $A + B$ of two matrices A and B, press, in succession, A, $\boxed{+}$, B, and $\boxed{\text{ENTER}}$.

2. **Subtraction of two matrices:** To find the difference $A - B$ of two matrices A and B press, in succession, A, $\boxed{-}$, B, and $\boxed{\text{ENTER}}$.

3. **Multiplication of a matrix by a scalar:** To find the matrix cA, enter the scalar c. Then press A and $\boxed{\text{ENTER}}$.

4. **Multiplication of two matrices:** To find the product AB of two matrices A and B, press, in succession, A $\boxed{\times}$, B, $\boxed{\text{ENTER}}$.

REMARK   More than two of the above operations may be combined into a single statement. For example, we can compute $(cA + dB)C$ by entering $(cA + dB) \times C$ and pressing $\boxed{\text{ENTER}}$.   □

A.7

Finding the Transpose of a Matrix; Finding the Inverse of an $n \times n$ Matrix

1. Finding the transpose of a matrix: Enter the matrix as the matrix A and go back to the calculator mode. Press $\boxed{2\text{nd}}$ $\boxed{\text{MATRX}}$ and then recall the matrix A. Then select ⟨**MATH**⟩ followed by ⟨^{T}⟩ and press $\boxed{\text{ENTER}}$.

2. Finding the inverse of a matrix: Enter the matrix as the matrix A and go back to the calculator mode. Press $\boxed{2\text{nd}}$ $\boxed{\text{MATRX}}$. Recall the matrix A, press $\boxed{2\text{nd}}$ $\boxed{x^{-1}}$ and then press $\boxed{\text{ENTER}}$.

A.8

Evaluating $n!$, $P(n, r)$, and $C(n, r)$

To use any of the operations $n!$, $P(n, r)$, or $C(n, r)$, press $\boxed{2\text{nd}}$ $\boxed{\text{MATH}}$ to display the Math menu. Then select ⟨**PROB**⟩ to use the **PROB** functions.

1. **To evaluate $n!$:** Enter the number n, then select ⟨!⟩ to use the factorial function. Press $\boxed{\text{ENTER}}$ and the answer will appear on the right-hand side of the screen.

2. **To evaluate $P(n, r)$:** Call for the **PROB** functions as before. Then enter the number n. Select ⟨**nPr**⟩ to use the **P(n, r)** function. Next, enter the number r. Finally, press $\boxed{\text{ENTER}}$ to obtain the answer, which will be displayed on the right-hand side of the screen.

3. **To evaluate $C(n, r)$:** Proceed as in part 2, but select ⟨**nCr**⟩ instead of ⟨**nPr**⟩.

A.9

Generating a Random Number

Press $\boxed{2\text{nd}}$ $\boxed{\text{MATH}}$ to call the Math menu. Then select ⟨**PROB**⟩ to use the **PROB** menu. Observe that the **rand** function is displayed at the bottom of the screen. To generate a random number between a and c, enter $a +$ b, where $b = c - a$. Select ⟨**rand**⟩ to obtain the display

$$a + b \text{ rand}$$

Then press ENTER to obtain the random number. For example, to generate a random number between 2 and 6, enter $2 + 4$.

A.10 Using the ZOOM Function

1. Display the graph of the function. (See the procedure for graphing a single function, Section A.1.)
2. Select ⟨**ZOOM**⟩. To zoom-in select ⟨**ZIN**⟩.
3. Use the left, right, up, or down arrow keys to move the cursor near the point of interest. Press ENTER.
4. To zoom-in again, move the cursor close to the desired point and press ENTER once more.

A.11 Finding the Derivative of a Function at a Given Point

Method A

1. Display the graph of the function. (See the procedure for graphing a single function, Section A.1.)
2. Move the cursor as close as possible to the point at which the derivative is to be evaluated by using **ZIN** repeatedly, if necessary.
3. Find dy/dx as follows: Press MORE to display the function **MATH** in the menu. Then select ⟨**MATH**⟩ followed by ⟨**dy/dx**⟩. Press ENTER and the value of dy/dx will be displayed at the bottom of the screen.

Method B (Without displaying the graph of the function)

1. Press 2nd CALC.
2. Select ⟨**der1**⟩. The screen displays

$$\text{der1(}$$

Enter the function, the variable, and the value of x at which the derivative is to be evaluated. For example, to evaluate $f'(4)$, where $f(x) = \sqrt{x}$, we enter

$$\text{der1 (x\textasciicircum.5, x, 4)}$$

Then press ENTER. The value of the derivative, 0.25, will appear on the screen.

A.12 Finding the Second Derivative of a Function at a Given Point

1. Press 2nd CALC.
2. Select ⟨**der2**⟩. The screen displays

$$\text{der2 (}$$

Enter the function, the variable, and the value of x at which the second derivative is to be evaluated. For example, to find $f''(4)$, where $f(x) = \sqrt{x}$, enter

$$\mathrm{der2(x^{.}5, x, 4)}$$

Then press $\boxed{\text{ENTER}}$. The answer -0.03125 will appear on the screen.

A.13   Using the TRACE Function

1. Display the graph of the function. (See the procedure for graphing a single function, Section A.1.)

2. Select ⟨**TRACE**⟩. The cursor will appear on the graph.

3. By pressing the left or right arrow keys repeatedly, the cursor can be made to move along the graph. The coordinates of the point indicated by the cursor appear at the bottom of the screen.

REMARK   If more than one graph is displayed on the screen, then the cursor can be made to jump from one graph to the other by pressing the up or down arrow keys.

A.14   Finding the Roots of a Polynomial

1. Press $\boxed{\text{2nd}}$ $\boxed{\text{POLY}}$. The screen displays

POLY

order =

Enter the degree of the polynomial. Then press $\boxed{\text{ENTER}}$. For example, if $n = 3$, then the display

$$\mathrm{a3x^{\hat{}}3 + \cdots + a1x + a0 = 0}$$

$$\mathrm{a3 =}$$

$$\mathrm{a2 =}$$

$$\mathrm{a1 =}$$

$$\mathrm{a0 =}$$

appears.

2. Enter the values of the coefficients of the polynomial. Then select ⟨**SOLVE**⟩. The roots x_1, x_2, x_3 of the polynomial equation will appear on the screen.

A.15   Finding the Roots of an Equation

1. Display the graph of the function.

2. Select the **ROOT** function as follows:
   a. Press $\boxed{\text{MORE}}$ once to display the function **MATH** in the menu. Then select ⟨**MATH**⟩ to obtain the **ROOT** function display.
   b. Select ⟨**ROOT**⟩.

3. To find a root, move the cursor close to the point at which the graph of *f* crosses the *x*-axis. Then press ENTER. The desired root will be displayed at the bottom of the screen. Repeat steps 2 and 3 to find another root.

A.16 Finding the Inflection Points of a Function

1. Display the graph of the function. (See the procedure for graphing a single function, Section A.1.)

2. Select the **INFLC** operation as follows:
Press MORE once to display the function **MATH**. Then select ⟨**MATH**⟩. Press MORE again to obtain the **INFLC** function display. Select ⟨**INFLC**⟩ to use this operation.

3. To find the coordinates of an inflection point, move the cursor close to the point of inflection in question and then press ENTER. The coordinates of the inflection point will appear at the bottom of the screen. Repeat steps 2 and 3 to find the coordinates of any other inflection points.

A.17 Finding the Absolute Extrema of a Function

Method A

1. Display the graph of the function on a closed interval $[a, b]$.

2. To find the absolute minimum of *f* proceed as follows: Press MORE to display the function **MATH** in the menu. Then select ⟨**MATH**⟩. Press MORE again to display the **FMIN** operation. Select ⟨**FMIN**⟩ to use this function. Press ENTER, and the coordinates of the absolute minimum (point) will appear at the bottom of the screen.

3. To find the absolute maximum of *f*, press GRAPH to recall the menu. Then proceed as in step 2 but select ⟨**FMAX**⟩ instead of ⟨**FMIN**⟩.

Method B (Without displaying the graph of the function)

1. Press 2nd CALC.

2. Press MORE and then select ⟨**fMin**⟩. The screen displays

$$\text{fMin (}$$

Enter the function, the variable, and the values of *a* and *b*. For example, to find the absolute minimum value of $f(x) = x^2 + x - 1$ on the interval $[-1, 2]$, we enter

$$\text{fMin (x\^2 + x} - 1, x, -1, 2)$$

and then press ENTER. The answer displayed, -0.500000600933, gives the value of *x* at which the absolute minimum of *f* occurs. To find the absolute minimum value, we use **evalF**. Thus, the required value is

$$\text{evalF(x\^2 + x} - 1, x, -0.500000600933) = -1.25$$

The absolute maximum value of *f* on $[a, b]$ is found using the operation **fMax** in a similar manner.

A.18   Evaluating a Definite Integral

Method A. Using the fnInt Function

1. Press $\boxed{\text{2nd}}$ $\boxed{\text{CALC}}$ to display the Calculus menu. Then select ⟨**fnInt**⟩. The expression

$$\text{fnInt (}$$

is displayed.

2. Enter the integrand, the variable, and the lower and upper limits of integration—that is,

$$\text{fnInt } (f(x), x, a, b)$$

For example, to evaluate $\int_0^1 x^2 \, dx$, we enter

$$\text{fnInt } (x^2, x, 0, 1)$$

Then press $\boxed{\text{ENTER}}$. The value of the integral, 0.333333333333, will appear on the screen.

Method B. Using the $\int f(x)$ function

1. Display the graph of the function.

2. Select the $\int f(x)$ function as follows: Press $\boxed{\text{MORE}}$ to display the **MATH** function. Select ⟨**MATH**⟩ to display the $\int f(x)$ function. Then select ⟨$\int f(x)$⟩ to use this function.

3. Move the cursor to the point on the graph of f corresponding to the lower limit of integration ($x = a$). Press $\boxed{\text{ENTER}}$. Then move the cursor to the point on the graph of f corresponding to the upper limit of integration ($x = b$). Press $\boxed{\text{ENTER}}$. The value of the integral appears at the bottom of the screen.

REMARK   You may not be able to obtain the exact limits of integration. Thus, the value of the integral will be an approximation at best.

APPENDIX B

Compound Amount, Present Value, and Annuity Table

APPENDIX B   Compound Amount, Present Value, and Annuity Table

	$i = \frac{1}{4}\%$			$i = \frac{1}{2}\%$		
n	$(1 + i)^n$	$s_{\overline{n}|i}$	$a_{\overline{n}|i}$	$(1 + i)^n$	$s_{\overline{n}|i}$	$a_{\overline{n}|i}$
1	1.0025 0000	1.0000 0000	0.9975 0623	1.0050 0000	1.0000 0000	0.9950 2488
2	1.0050 0625	2.0025 0000	1.9925 2492	1.0100 2500	2.0050 0000	1.9850 9938
3	1.0075 1877	3.0075 0625	2.9850 6227	1.0150 7513	3.0150 2500	2.9702 4814
4	1.0100 3756	4.0150 2502	3.9751 2446	1.0201 5050	4.0301 0013	3.9504 9566
5	1.0125 6266	5.0250 6258	4.9627 1766	1.0252 5125	5.0502 5063	4.9258 6633
6	1.0150 9406	6.0376 2523	5.9478 4804	1.0303 7751	6.0755 0188	5.8963 8441
7	1.0176 3180	7.0527 1930	6.9305 2174	1.0355 2940	7.1058 7939	6.8620 7404
8	1.0201 7588	8.0703 5110	7.9107 4487	1.0407 0704	8.1414 0879	7.8229 5924
9	1.0227 2632	9.0905 2697	8.8885 2357	1.0459 1058	9.1821 1583	8.7790 6392
10	1.0252 8313	10.1132 5329	9.8638 6391	1.0511 4013	10.2280 2641	9.7304 1186
11	1.0278 4634	11.1385 3642	10.8367 7198	1.0563 9583	11.2791 6654	10.6770 2673
12	1.0304 1596	12.1663 8277	11.8072 5384	1.0616 7781	12.3355 6237	11.6189 3207
13	1.0329 9200	13.1967 9872	12.7753 1555	1.0669 8620	13.3972 4018	12.5561 5131
14	1.0355 7448	14.2297 9072	13.7409 6314	1.0723 2113	14.4642 2639	13.4887 0777
15	1.0381 6341	15.2653 6520	14.7042 0264	1.0776 8274	15.5365 4752	14.4166 2465
16	1.0407 5882	16.3035 2861	15.6650 4004	1.0830 7115	16.6142 3026	15.3399 2502
17	1.0433 6072	17.3442 8743	16.6234 8133	1.0884 8651	17.6973 0141	16.2586 3186
18	1.0459 6912	18.3876 4815	17.5795 3250	1.0939 2894	18.7857 8791	17.1727 6802
19	1.0485 8404	19.4336 1727	18.5331 9950	1.0993 9858	19.8797 1685	18.0823 5624
20	1.0512 0550	20.4822 0131	19.4844 8828	1.1048 9558	20.9791 1544	18.9874 1915
21	1.0538 3352	21.5334 0682	20.4334 0477	1.1104 2006	22.0840 1101	19.8879 7925
22	1.0564 6810	22.5872 4033	21.3799 5488	1.1159 7216	23.1944 3107	20.7840 5896
23	1.0591 0927	23.6437 0843	22.3241 4452	1.1215 5202	24.3104 0322	21.6756 8055
24	1.0617 5704	24.7028 1770	23.2659 7957	1.1271 5978	25.4319 5524	22.5628 6622
25	1.0644 1144	25.7645 7475	24.2054 6591	1.1327 9558	26.5591 1502	23.4456 3803
26	1.0670 7247	26.8289 8619	25.1426 0939	1.1384 5955	27.6919 1059	24.3240 1794
27	1.0697 4015	27.8960 5865	26.0774 1585	1.1441 5185	28.8303 7015	25.1980 2780
28	1.0724 1450	28.9657 9880	27.0098 9112	1.1498 7261	29.9745 2200	26.0676 8936
29	1.0750 9553	30.0382 1330	27.9400 4102	1.1556 2197	31.1243 9461	26.9330 2423
30	1.0777 8327	31.1133 0883	28.8678 7134	1.1614 0008	32.2800 1658	27.7940 5397
31	1.0804 7773	32.1910 9210	29.7933 8787	1.1672 0708	33.4414 1666	28.6507 9997
32	1.0831 7892	33.2715 6983	30.7165 9638	1.1730 4312	34.6086 2375	29.5032 8355
33	1.0858 8687	34.3547 4876	31.6375 0262	1.1789 0833	35.7816 6686	30.3515 2592
34	1.0886 0159	35.4406 3563	32.5561 1234	1.1848 0288	36.9605 7520	31.1955 4818
35	1.0913 2309	36.5292 3722	33.4724 3126	1.1907 2689	38.1453 7807	32.0353 7132
36	1.0940 5140	37.6205 6031	34.3864 6510	1.1966 8052	39.3361 0496	32.8710 1624
37	1.0967 8653	38.7146 1171	35.2982 1955	1.2026 6393	40.5327 8549	33.7025 0372
38	1.0995 2850	39.8113 9824	36.2077 0030	1.2086 7725	41.7354 4942	34.5298 5445
39	1.1022 7732	40.9109 2673	37.1149 1302	1.2147 2063	42.9441 2666	35.3530 8900
40	1.1050 3301	42.0132 0405	38.0198 6336	1.2207 9424	44.1588 4730	36.1722 2786
41	1.1077 9559	43.1182 3706	38.9225 5697	1.2268 9821	45.3796 4153	36.9872 9141
42	1.1105 6508	44.2260 3265	39.8229 9947	1.2330 3270	46.6065 3974	37.7982 9991
43	1.1133 4149	45.3365 9774	40.7211 9648	1.2391 9786	47.8395 7244	38.6052 7354
44	1.1161 2485	46.4499 3923	41.6171 5359	1.2453 9385	49.0787 7030	39.4082 3238
45	1.1189 1516	47.5660 6408	42.5108 7640	1.2516 2082	50.3241 6415	40.2071 9640
46	1.1217 1245	48.6849 7924	43.4023 7047	1.2578 7892	51.5757 8497	41.0021 8547
47	1.1245 1673	49.8066 9169	44.2916 4137	1.2641 6832	52.8336 6390	41.7932 1937
48	1.1273 2802	50.9312 0842	45.1786 9463	1.2704 8916	54.0978 3222	42.5803 1778
49	1.1301 4634	52.0585 3644	46.0635 3580	1.2768 4161	55.3683 2138	43.3635 0028
50	1.1329 7171	53.1886 8278	46.9461 7037	1.2832 2581	56.6451 6299	44.1427 8635

(continued)

	$i = \frac{1}{4}\%$			$i = \frac{1}{2}\%$		
n	$(1 + i)^n$	$s_{\overline{n}|i}$	$a_{\overline{n}|i}$	$(1 + i)^n$	$s_{\overline{n}|i}$	$a_{\overline{n}|i}$
51	1.1358 0414	54.3216 5449	47.8266 0386	1.2896 4194	57.9283 8880	44.9181 9537
52	1.1386 4365	55.4574 5862	48.7048 4176	1.2960 9015	59.2180 3075	45.6897 4664
53	1.1414 9026	56.5961 0227	49.5808 8953	1.3025 7060	60.5141 2090	46.4574 5934
54	1.1443 4398	57.7375 9252	50.4547 5265	1.3090 8346	61.8166 9150	47.2213 5258
55	1.1472 0484	58.8819 3650	51.3264 3656	1.3156 2887	63.1257 7496	47.9814 4535
56	1.1500 7285	60.0291 4135	52.1959 4669	1.3222 0702	64.4414 0384	48.7377 5657
57	1.1529 4804	61.1792 1420	53.0632 8847	1.3288 1805	65.7636 1086	49.4903 0505
58	1.1558 3041	62.3321 6223	53.9284 6730	1.3354 6214	67.0924 2891	50.2391 0950
59	1.1587 1998	63.4879 9264	54.7914 8858	1.3421 3946	68.4278 9105	50.9841 8855
60	1.1616 1678	64.6467 1262	55.6523 5769	1.3488 5015	69.7700 3051	51.7255 6075
61	1.1645 2082	65.8083 2940	56.5110 7999	1.3555 9440	71.1188 8066	52.4632 4453
62	1.1674 3213	66.9728 5023	57.3676 6083	1.3623 7238	72.4744 7507	53.1972 5824
63	1.1703 5071	68.1402 8235	58.2221 0557	1.3691 8424	73.8368 4744	53.9276 2014
64	1.1732 7658	69.3106 3306	59.0744 1952	1.3760 3016	75.2060 3168	54.6543 4839
65	1.1762 0977	70.4839 0964	59.9246 0800	1.3829 1031	76.5820 6184	55.3774 6109
66	1.1791 5030	71.6601 1942	60.7726 7631	1.3898 2486	77.9649 7215	56.0969 7621
67	1.1820 9817	72.8392 6971	61.6186 2974	1.3967 7399	79.3547 9701	56.8129 1165
68	1.1850 5342	74.0213 6789	62.4624 7355	1.4037 5785	80.7515 7099	57.5252 8522
69	1.1880 1605	75.2064 2131	63.3042 1302	1.4107 7664	82.1553 2885	58.2341 1465
70	1.1909 8609	76.3944 3736	64.1438 5339	1.4178 3053	83.5661 0549	58.9394 1756
71	1.1939 6356	77.5854 2345	64.9813 9989	1.4249 1968	84.9839 3602	59.6412 1151
72	1.1969 4847	78.7793 8701	65.8168 5774	1.4320 4428	86.4088 5570	60.3395 1394
73	1.1999 4084	79.9763 3548	66.6502 3216	1.4392 0450	87.8408 9998	61.0343 4222
74	1.2029 4069	81.1762 7632	67.4815 2834	1.4464 0052	89.2801 0448	61.7257 1366
75	1.2059 4804	82.3792 1701	68.3107 5146	1.4536 3252	90.7265 0500	62.4136 4543
76	1.2089 6291	83.5851 6505	69.1379 0670	1.4609 0069	92.1801 3752	63.0981 5466
77	1.2119 8532	84.7941 2797	69.9629 9920	1.4682 0519	93.6410 3821	63.7792 5836
78	1.2150 1528	86.0061 1329	70.7860 3411	1.4755 4622	95.1092 4340	64.4569 7350
79	1.2180 5282	87.2211 2857	71.6070 1657	1.4829 2395	96.5847 8962	65.1313 1691
80	1.2210 9795	88.4391 8139	72.4259 5169	1.4903 3857	98.0677 1357	65.8023 0538
81	1.2241 5070	89.6602 7934	73.2428 4458	1.4977 9026	99.5580 5214	66.4699 5561
82	1.2272 1108	90.8844 3004	74.0577 0033	1.5052 7921	101.0558 4240	67.1342 8419
83	1.2302 7910	92.1116 4112	74.8705 2402	1.5128 0561	102.5611 2161	67.7953 0765
84	1.2333 5480	93.3419 2022	75.6813 2072	1.5203 6964	104.0739 2722	68.4530 4244
85	1.2364 3819	94.5752 7502	76.4900 9548	1.5279 7148	105.5942 9685	69.1075 0491
86	1.2395 2928	95.8117 1321	77.2968 5335	1.5356 1134	107.1222 6834	69.7587 1135
87	1.2426 2811	97.0512 4249	78.1015 9935	1.5432 8940	108.6578 7968	70.4066 7796
88	1.2457 3468	98.2938 7060	78.9043 3850	1.5510 0585	110.2011 6908	71.0514 2086
89	1.2488 4901	99.5396 0527	79.7050 7581	1.5587 6087	111.7521 7492	71.6929 5608
90	1.2519 7114	100.7884 5429	80.5038 1627	1.5665 5468	113.3109 3580	72.3312 9958
91	1.2551 0106	102.0404 2542	81.3005 6486	1.5743 8745	114.8774 9048	72.9664 6725
92	1.2582 3882	103.2955 2649	82.0953 2654	1.5822 5939	116.4518 7793	73.5984 7487
93	1.2613 8441	104.5537 6530	82.8881 0628	1.5901 7069	118.0341 3732	74.2273 3818
94	1.2645 3787	105.8151 4972	83.6789 0900	1.5981 2154	119.6243 0800	74.8530 7282
95	1.2676 9922	107.0796 8759	84.4677 3966	1.6061 1215	121.2224 2954	75.4756 9434
96	1.2708 6847	108.3473 8681	85.2546 0315	1.6141 4271	122.8285 4169	76.0952 1825
97	1.2740 4564	109.6182 5528	86.0395 0439	1.6222 1342	124.4426 8440	76.7116 5995
98	1.2772 3075	110.8923 0091	86.8224 4827	1.6303 2449	126.0648 9782	77.3250 3478
99	1.2804 2383	112.1695 3167	87.6034 3967	1.6384 7611	127.6952 2231	77.9353 5799
100	1.2836 2498	113.4499 5550	88.3824 8346	1.6466 6849	129.3336 9842	78.5426 4477

(continued)

	$i = \frac{3}{4}\%$			$i = 1\%$		
n	$(1 + i)^n$	$s_{\overline{n}|i}$	$a_{\overline{n}|i}$	$(1 + i)^n$	$s_{\overline{n}|i}$	$a_{\overline{n}|i}$
1	1.0075 0000	1.0000 0000	0.9925 5583	1.0100 0000	1.0000 0000	0.9900 9901
2	1.0150 5625	2.0075 0000	1.9777 2291	1.0201 0000	2.0100 0000	1.9703 9506
3	1.0226 6917	3.0225 5625	2.9555 5624	1.0303 0100	3.0301 0000	2.9409 8521
4	1.0303 3919	4.0452 2542	3.9261 1041	1.0406 0401	4.0604 0100	3.9019 6555
5	1.0380 6673	5.0755 6461	4.8894 3961	1.0510 1005	5.1010 0501	4.8534 3124
6	1.0458 5224	6.1136 3135	5.8455 9763	1.0615 2015	6.1520 1506	5.7954 7647
7	1.0536 9613	7.1594 8358	6.7946 3785	1.0721 3535	7.2135 3521	6.7281 9453
8	1.0615 9885	8.2131 7971	7.7366 1325	1.0828 5671	8.2856 7056	7.6516 7775
9	1.0695 6084	9.2747 7856	8.6715 7642	1.0936 8527	9.3685 2727	8.5660 1758
10	1.0775 8255	10.3443 3940	9.5995 7958	1.1046 2213	10.4622 1254	9.4713 0453
11	1.0856 6441	11.4219 2194	10.5206 7452	1.1156 6835	11.5668 3467	10.3676 2825
12	1.0938 0690	12.5075 8636	11.4349 1267	1.1268 2503	12.6825 0301	11.2550 7747
13	1.1020 1045	13.6013 9325	12.3423 4508	1.1380 9328	13.8093 2804	12.1337 4007
14	1.1102 7553	14.7034 0370	13.2430 2242	1.1494 7421	14.9474 2132	13.0037 0304
15	1.1186 0259	15.8136 7923	14.1369 9495	1.1609 6896	16.0968 9554	13.8650 5252
16	1.1269 9211	16.9322 8183	15.0243 1261	1.1725 7864	17.2578 6449	14.7178 7378
17	1.1354 4455	18.0592 7394	15.9050 2492	1.1843 0443	18.4304 4314	15.5622 5127
18	1.1439 6039	19.1947 1849	16.7791 8107	1.1961 4748	19.6147 4757	16.3982 6858
19	1.1525 4009	20.3386 7888	17.6468 2984	1.2081 0895	20.8108 9504	17.2260 0850
20	1.1611 8414	21.4912 1897	18.5080 1969	1.2201 9004	22.0190 0399	18.0455 5297
21	1.1698 9302	22.6524 0312	19.3627 9870	1.2323 9194	23.2391 9403	18.8569 8313
22	1.1786 6722	23.8222 9614	20.2112 1459	1.2447 1586	24.4715 8598	19.6603 7934
23	1.1875 0723	25.0009 6336	21.0533 1473	1.2571 6302	25.7163 0183	20.4558 2113
24	1.1964 1353	26.1884 7059	21.8891 4614	1.2697 3465	26.9734 6485	21.2433 8726
25	1.2053 8663	27.3848 8412	22.7187 5547	1.2824 3200	28.2431 9950	22.0231 5570
26	1.2144 2703	28.5902 7075	23.5421 8905	1.2952 5631	29.5256 3150	22.7952 0366
27	1.2235 3523	29.8046 9778	24.3594 9286	1.3082 0888	30.8208 8781	23.5596 0759
28	1.2327 1175	31.0282 3301	25.1707 1215	1.3212 9097	32.1290 9669	24.3164 4316
29	1.2419 5709	32.2609 4476	25.9758 9331	1.3345 0388	33.4503 8766	25.0657 8530
30	1.2512 7176	33.5029 0184	26.7750 8021	1.3478 4892	34.7848 9153	25.8077 0822
31	1.2606 5630	34.7541 7361	27.5683 1783	1.3613 2740	36.1327 4045	26.5422 8537
32	1.2701 1122	36.0148 2991	28.3556 5045	1.3749 4068	37.4940 6785	27.2695 8947
33	1.2796 3706	37.2849 4113	29.1371 2203	1.3886 9009	38.8690 0853	27.9896 9255
34	1.2892 3434	38.5645 7819	29.9127 7621	1.4025 7699	40.2576 9862	28.7026 6589
35	1.2989 0359	39.8538 1253	30.6826 5629	1.4166 0276	41.6602 7560	29.4085 8009
36	1.3086 4537	41.1527 1612	31.4468 0525	1.4307 6878	43.0768 7836	30.1075 0504
37	1.3184 6021	42.4613 6149	32.2052 6576	1.4450 7647	44.5076 4714	30.7995 0994
38	1.3283 4866	43.7798 2170	32.9580 8016	1.4595 2724	45.9527 2361	31.4846 6330
39	1.3383 1128	45.1081 7037	33.7052 9048	1.4741 2251	47.4122 5085	32.1630 3298
40	1.3483 4861	46.4464 8164	34.4469 3844	1.4888 6373	48.8863 7336	32.8346 8611
41	1.3584 6123	47.7948 3026	35.1830 6545	1.5037 5237	50.3752 3709	33.4996 8922
42	1.3686 4969	49.1532 9148	35.9137 1260	1.5187 8989	51.8789 8946	34.1581 0814
43	1.3789 1456	50.5219 4117	36.6389 2070	1.5339 7779	53.3977 7936	34.8100 0806
44	1.3892 5642	51.9008 5573	37.3587 3022	1.5493 1757	54.9317 5715	35.4554 5352
45	1.3996 7584	53.2901 1215	38.0731 8136	1.5648 1075	56.4810 7472	36.0945 0844
46	1.4101 7341	54.6897 8799	38.7823 1401	1.5804 5885	58.0458 8547	36.7272 3608
47	1.4207 4971	56 0999 6140	39.4861 6775	1.5962 6344	59.6263 4432	37.3536 9909
48	1.4314 0533	57.5207 1111	40.1847 8189	1.6122 2608	61.2226 0777	37.9739 5949
49	1.4421 4087	58.9521 1644	40.8781 9542	1.6283 4834	62.8348 3385	38.5880 7871
50	1.4529 5693	60.3942 5732	41.5664 4707	1.6446 3182	64.4631 8218	39.1961 1753

(continued)

	$i = \frac{3}{4}\%$			$i = 1\%$		
n	$(1 + i)^n$	$s_{\overline{n}|i}$	$a_{\overline{n}|i}$	$(1 + i)^n$	$s_{\overline{n}|i}$	$a_{\overline{n}|i}$
51	1.4638 5411	61.8472 1424	42.2495 7525	1.6610 7814	66.1078 1401	39.7981 3617
52	1.4748 3301	63.3110 6835	42.9276 1812	1.6776 8892	67.7688 9215	40.3941 9423
53	1.4858 9426	64.7859 0136	43.6006 1351	1.6944 6581	69.4465 8107	40.9843 5072
54	1.4970 3847	66.2717 9562	44.2685 9902	1.7114 1047	71.1410 4688	41.5686 6408
55	1.5082 6626	67.7688 3409	44.9316 1193	1.7285 2457	72.8524 5735	42.1471 9216
56	1.5195 7825	69.2771 0035	45.5896 8926	1.7458 0982	74.5809 8192	42.7199 9224
57	1.5309 7509	70.7966 7860	46.2428 6776	1.7632 6792	76.3267 9174	43.2871 2102
58	1.5424 5740	72.3276 5369	46.8911 8388	1.7809 0060	78.0900 5966	43.8486 3468
59	1.5540 2583	73.8701 1109	47.5346 7382	1.7987 0960	79.8709 6025	44.4045 8879
60	1.5656 8103	75.4241 3693	48.1733 7352	1.8166 9670	81.6696 6986	44.9550 3841
61	1.5774 2363	76.9898 1795	48.8073 1863	1.8348 6367	83.4863 6655	45.5000 3803
62	1.5892 5431	78.5672 4159	49.4365 4455	1.8532 1230	85.3212 3022	46.0396 4161
63	1.6011 7372	80.1564 9590	50.0610 8640	1.8717 4443	87.1744 4252	46.5739 0258
64	1.6131 8252	81.7576 6962	50.6809 7906	1.8904 6187	89.0461 8695	47.1028 7385
65	1.6252 8139	83.3708 5214	51.2962 5713	1.9093 6649	90.9366 4882	47.6266 0777
66	1.6374 7100	84.9961 3353	51.9069 5497	1.9284 6015	92.8460 1531	48.1451 5621
67	1.6497 5203	86.6336 0453	52.5131 0667	1.9477 4475	94.7744 7546	48.6585 7050
68	1.6621 2517	88.2833 5657	53.1147 4607	1.9672 2220	96.7222 2021	49.1669 0149
69	1.6745 9111	89.9454 8174	53.7119 0677	1.9868 9442	98.6894 4242	49.6701 9949
70	1.6871 5055	91.6200 7285	54.3046 2210	2.0067 6337	100.6763 3684	50.1685 1435
71	1.6998 0418	93.3072 2340	54.8929 2516	2.0268 3100	102.6831 0021	50.6618 9539
72	1.7125 5271	95.0070 2758	55.4768 4880	2.0470 9931	104.7099 3121	51.1503 9148
73	1.7253 9685	96.7195 8028	56.0564 2561	2.0675 7031	106.7570 3052	51.6340 5097
74	1.7383 3733	98.4449 7714	56.6316 8795	2.0882 4601	108.8246 0083	52.1129 2175
75	1.7513 7486	100.1833 1446	57.2026 6794	2.1091 2847	110.9128 4684	52.5870 5124
76	1.7645 1017	101.9346 8932	57.7693 9746	2.1302 1975	113.0219 7530	53.0564 8638
77	1.7777 4400	103.6991 9949	58.3319 0815	2.1515 2195	115.1521 9506	53.5212 7364
78	1.7910 7708	105.4769 4349	58.8902 3141	2.1730 3717	117.3037 1701	54.9814 5905
79	1.8045 1015	107.2680 2056	59.4443 9842	2.1947 6754	119.4767 5418	54.4370 8817
80	1.8180 4398	109.0725 3072	59.9944 4012	2.2167 1522	121.6715 2172	54.8882 0611
81	1.8316 7931	110.8905 7470	60.5403 8722	2.2388 8237	123.8882 3694	55.3348 5753
82	1.8454 1691	112.7222 5401	61.0822 7019	2.2612 7119	126.1271 1931	55.7770 8666
83	1.8592 5753	114.5676 7091	61.6201 1930	2.2838 8390	128.3883 9050	56.2149 3729
84	1.8732 0196	116.4269 2845	62.1539 6456	2.3067 2274	130.6722 7440	56.6484 5276
85	1.8872 5098	118.3001 3041	62.6838 3579	2.3297 8997	132.9789 9715	57.0776 7600
86	1.9014 0536	120.1873 8139	63.2097 6257	2.3530 8787	135.3087 8712	57.5026 4951
87	1.9156 6590	122.0887 8675	63.7317 7427	2.3766 1875	137.6618 7499	57.9234 1535
88	1.9300 3339	124.0044 5265	64.2499 0002	2.4003 8494	140.0384 9374	58.3400 1520
89	1.9445 0865	125.9344 8604	64.7641 6875	2.4243 8879	142.4388 7868	58.7524 9030
90	1.9590 9246	127.8789 9469	65.2746 0918	2.4486 3267	144.8632 6746	59.1608 8148
91	1.9737 8565	129.8380 8715	65.7812 4981	2.4731 1900	147.3119 0014	59.5652 2919
92	1.9885 8905	131.8118 7280	66.2841 1892	2.4978 5019	149.7850 1914	59.9655 7346
93	2.0035 0346	133.8004 6185	66.7832 4458	2.5228 2869	152.2828 6933	60.3619 5392
94	2.0185 2974	135.8039 6531	67.2786 5467	2.5480 5698	154.8056 9803	60.7544 0982
95	2.0336 6871	137.8224 9505	67.7703 7685	2.5735 3755	157.3537 5501	61.1429 8002
96	2.0489 2123	139.8561 6377	68.2584 3856	2.5992 7293	159.9272 9256	61.5277 0299
97	2.0642 8814	141.9050 8499	68.7428 6705	2.6252 6565	162.5265 6548	61.9086 1682
98	2.0797 7030	143.9693 7313	69.2236 8938	2.6515 1831	165.1518 3114	62.2857 5923
99	2.0953 6858	146.0491 4343	69.7009 3239	2.6780 3349	167.8033 4945	62.6591 6755
100	2.1110 8384	148.1445 1201	70.1746 2272	2.7048 1383	170.4813 8294	63.0288 7877

(continued)

	$i = 1\frac{1}{4}\%$			$i = 1\frac{1}{2}\%$		
n	$(1 + i)^n$	$s_{\overline{n}\rvert i}$	$a_{\overline{n}\rvert i}$	$(1 + i)^n$	$s_{\overline{n}\rvert i}$	$a_{\overline{n}\rvert i}$
1	1.0125 0000	1.0000 0000	0.9876 5432	1.0150 0000	1.0000 0000	0.9852 2167
2	1.0251 5625	2.0125 0000	1.9631 1538	1.0302 2500	2.0150 0000	1.9558 8342
3	1.0379 7070	3.0376 5625	2.9265 3371	1.0456 7838	3.0452 2500	2.9122 0042
4	1.0509 4534	4.0756 2695	3.8780 5798	1.0613 6355	4.0909 0338	3.8543 8465
5	1.0640 8215	5.1265 7229	4.8178 3504	1.0772 8400	5.1522 6693	4.7826 4497
6	1.0773 8318	6.1906 5444	5.7460 0992	1.0934 4326	6.2295 5093	5.6971 8717
7	1.0908 5047	7.2680 3762	6.6627 2585	1.1098 4491	7.3229 9419	6.5982 1396
8	1.1044 8610	8.3588 8809	7.5681 2429	1.1264 9259	8.4328 3911	7.4859 2508
9	1.1182 9218	9.4633 7420	8.4623 4498	1.1433 8998	9.5593 3169	8.3605 1732
10	1.1322 7083	10.5816 6637	9.3455 2591	1.1605 4083	10.7027 2167	9.2221 8455
11	1.1464 2422	11.7139 3720	10.2178 0337	1.1779 4894	11.8632 6249	10.0711 1779
12	1.1607 5452	12.8603 6142	11.0793 1197	1.1956 1817	13.0412 1143	10.9075 0521
13	1.1752 6395	14.0211 1594	11.9301 8466	1.2135 5244	14.2368 2960	11.7315 3222
14	1.1899 5475	15.1963 7988	12.7705 5275	1.2317 5573	15.4503 8205	12.5433 8150
15	1.2048 2918	16.3863 3463	13.6005 4592	1.2502 3207	16.6821 3778	13.3432 3301
16	1.2198 8955	17.5911 6382	14.4202 9227	1.2689 8555	17.9323 6984	14.1312 6405
17	1.2351 3817	18.8110 5336	15.2299 1829	1.2880 2033	19.2013 5539	14.9076 4931
18	1.2505 7739	20.0461 9153	16.0295 4893	1.3073 4064	20.4893 7572	15.6725 6089
19	1.2662 0961	21.2967 6893	16.8193 0759	1.3269 5075	21.7967 1636	16.4261 6837
20	1.2820 3723	22.5629 7854	17.5993 1613	1.3468 5501	23.1236 6710	17.1686 3879
21	1.2980 6270	23.8450 1577	18.3696 9495	1.3670 5783	24.4705 2211	17.9001 3673
22	1.3142 8848	25.1430 7847	19.1305 6291	1.3875 6370	25.8375 7994	18.6208 2437
23	1.3307 1709	26.4573 6695	19.8820 3744	1.4083 7715	27.2251 4364	19.3308 6145
24	1.3473 5105	27.7880 8403	20.6242 3451	1.4295 0281	28.6335 2080	20.0304 0537
25	1.3641 9294	29.1354 3508	21.3572 6865	1.4509 4535	30.0630 2361	20.7196 1120
26	1.3812 4535	30.4996 2802	22.0812 5299	1.4727 0953	31.5139 6896	21.3986 3172
27	1.3985 1092	31.8808 7337	22.7962 9925	1.4948 0018	32.9866 7850	22.0676 1746
28	1.4159 9230	33.2793 8429	23.5025 1778	1.5172 2218	34.4814 7867	22.7267 1671
29	1.4336 9221	34.6953 7659	24.2000 1756	1.5399 8051	35.9987 0085	23.3760 7558
30	1.4516 1336	36.1290 6880	24.8889 0623	1.5630 8022	37.5386 8137	24.0158 3801
31	1.4697 5853	37.5806 8216	25.5692 9010	1.5865 2642	39.1017 6159	24.6461 4582
32	1.4881 3051	39.0504 4069	26.2412 7418	1.6103 2432	40.6882 8801	25.2671 3874
33	1.5067 3214	40.5385 7120	26.9049 6215	1.6344 7918	42.2986 1233	25.8789 5442
34	1.5255 6629	42.0453 0334	27.5604 5644	1.6589 9637	43.9330 9152	26.4817 2849
35	1.5446 3587	43.5708 6963	28.2078 5822	1.6838 8132	45.5920 8789	27.0755 9458
36	1.5639 4382	45.1155 0550	28.8472 6737	1.7091 3954	47.2759 6921	27.6606 8431
37	1.5834 9312	46.6794 4932	29.4787 8259	1.7347 7663	48.9851 0874	28.2371 2740
38	1.6032 8678	48.2629 4243	30.1025 0133	1.7607 9828	50.7198 8538	28.8050 5163
39	1.6233 2787	49.8662 2921	30.7185 1983	1.7872 1025	52.4806 8366	29.3645 8288
40	1.6436 1946	51.4895 5708	31.3269 3316	1.8140 1841	54.2678 9391	29.9158 4520
41	1.6641 6471	53.1331 7654	31.9278 3522	1.8412 2868	56.0819 1232	30.4589 6079
42	1.6849 6677	54.7973 4125	32.5213 1874	1.8688 4712	57.9231 4100	30.9940 5004
43	1.7060 2885	56.4823 0801	33.1074 7530	1.8968 7982	59.7919 8812	31.5212 3157
44	1.7273 5421	58.1883 3687	33.6863 9536	1.9253 3302	61.6888 6794	32.0406 2223
45	1.7489 4614	59.9156 9108	34.2581 6825	1.9542 1301	63.6142 0096	32.5523 3718
46	1.7708 0797	61.6646 3721	34.8228 8222	1.9835 2621	65.5684 1398	33.0564 8983
47	1.7929 4306	63.4354 4518	35.3806 2442	2.0132 7910	67.5519 4018	33.5531 9195
48	1.8153 5485	65.2283 8824	35.9314 8091	2.0434 7829	69.5652 1929	34.0425 5365
49	1.8380 4679	67.0437 4310	36.4755 3670	2.0741 3046	71.6086 9758	34.5246 8339
50	1.8610 2237	68.8817 8989	37.0128 7575	2.1052 4242	73.6828 2804	34.9996 8807

(continued)

		$i = 1\frac{1}{4}\%$			$i = 1\frac{1}{2}\%$	
n	$(1 + i)^n$	$s_{\overline{n}\,i}$	$a_{\overline{n}\,i}$	$(1 + i)^n$	$s_{\overline{n}\,i}$	$a_{\overline{n}\,i}$
51	1.8842 8515	70.7428 1226	37.5435 8099	2.1368 2106	75.7880 7046	35.4676 7298
52	1.9078 3872	72.6270 9741	38.0677 3431	2.1688 7337	77.9248 9152	35.9287 4185
53	1.9316 8670	74.5349 3613	38.5854 1660	2.2014 0647	80.0937 6489	36.3829 9690
54	1.9558 3279	76.4666 2283	39.0967 0776	2.2344 2757	82.2951 7136	36.8305 3882
55	1.9802 8070	78.4224 5562	39.6016 8667	2.2679 4398	84.5295 9893	37.2714 6681
56	2.0050 3420	80.4027 3631	40.1004 3128	2.3019 6314	86.7975 4292	37.7058 7863
57	2.0300 9713	82.4077 7052	40.5930 1855	2.3364 9259	89.0995 0606	38.1338 7058
58	2.0554 7335	84.4378 6765	41.0795 2449	2.3715 3998	91.4359 9865	38.5555 3751
59	2.0811 6676	86.4933 4099	41.5600 2419	2.4071 1308	93.8075 3863	38.9709 7292
60	2.1071 8135	88.5745 0776	42.0345 9179	2.4432 1978	96.2146 5171	39.3802 6889
61	2.1335 2111	90.6816 8910	42.5033 0054	2.4798 6807	98.6578 7149	39.7835 1614
62	2.1601 9013	92.8152 1022	42.9662 2275	2.5170 6609	101.1377 3956	40.1808 0408
63	2.1871 9250	94.9754 0034	43.4234 2988	2.5548 2208	103.6548 0565	40.5722 2077
64	2.2145 3241	97.1625 9285	43.8749 9247	2.5931 4442	106.2096 2774	40.9578 5298
65	2.2422 1407	99.3771 2526	44.3209 8022	2.6320 4158	108.8027 7215	41.3377 8618
66	2.2702 4174	101.6193 3933	44.7614 6195	2.6715 2221	111.4348 1374	41.7121 0461
67	2.2986 1976	103.8895 8107	45.1965 0563	2.7115 9504	114.1063 3594	42.0808 9125
68	2.3273 5251	106.1882 0083	45.6261 7840	2.7522 6896	116.8179 3098	42.4442 2783
69	2.3564 4442	108.5155 5334	46.0505 4656	2.7935 5300	119.5701 9995	42.8021 9490
70	2.3858 9997	110.8719 9776	46.4696 7562	2.8354 5629	122.3637 5295	43.1548 7183
71	2.4157 2372	113.2578 9773	46.8836 3024	2.8779 8814	125.1992 0924	43.5023 3678
72	2.4459 2027	115.6736 2145	47.2924 7431	2.9211 5796	128.0771 9738	43.8446 6677
73	2.4764 9427	118.1195 4172	47.6962 7093	2.9649 7533	130.9983 5534	44.1819 3771
74	2.5074 5045	120.5960 3599	48.0950 8240	3.0094 4996	133.9633 3067	44.5142 2434
75	2.5387 9358	123.1034 8644	48.4889 7027	3.0545 9171	136.9727 8063	44.8416 0034
76	2.5705 2850	125.6422 8002	48.8779 9533	3.1004 1059	140.0273 7234	45.1641 3826
77	2.6026 6011	128.2128 0852	49.2662 1761	3.1469 1674	143.1277 8292	45.4819 0962
78	2.6351 9336	130.8154 6863	49.6416 9640	3.1941 2050	146.2746 9967	45.7949 8485
79	2.6681 3327	133.4506 6199	50.0164 9027	3.2420 3230	149.4688 2016	46.1034 3335
80	2.7014 8494	136.1187 9526	50.3866 5706	3.2906 6279	152.7108 5247	46.4073 2349
81	2.7352 5350	138.8202 8020	50.7522 5389	3.3400 2273	156.0015 1525	46.7067 2265
82	2.7694 4417	141.5555 3370	51.1133 3717	3.3901 2307	159.3415 3798	47.0016 9720
83	2.8040 6222	144.3249 7787	51.4699 6264	3.4409 7492	162.7316 6105	47.2923 1251
84	2.8391 1300	147.1290 4010	51.8221 8532	3.4925 8954	166.1726 3597	47.5786 3301
85	2.8746 0191	149.9681 5310	52.1700 5958	3.5449 7838	169.6652 2551	47.8607 2218
86	2.9105 3444	152.8427 5501	52.5136 3909	3.5981 5306	173.2102 0389	48.1386 4254
87	2.9469 1612	155.7532 8945	52.8529 7688	3.6521 2535	176.8083 5695	48.4124 5571
88	2.9837 5257	158.7002 0557	53.1881 2531	3.7069 0723	180.4604 8230	48.6822 2237
89	3.0210 4948	161.6839 5814	53.5191 3611	3.7625 1084	184.1673 8954	48.9480 0234
90	3.0588 1260	164.7050 0762	53.8460 6036	3.8189 4851	187.9299 0038	49.2098 5452
91	3.0970 4775	167.7638 2021	54.1689 4850	3.8762 3273	191.7488 4889	49.4678 3696
92	3.1357 6085	170.8608 6796	54.4878 5037	3.9343 7622	195.6250 8162	49.7220 0686
93	3.1749 5786	173.9966 2881	54.8028 1518	3.9933 9187	199.5594 5784	49.9724 2055
94	3.2146 4483	177.1715 8667	55.1138 9154	4.0532 9275	203.5528 4971	50.2191 3355
95	3.2548 2789	180.3862 3151	55.4211 2744	4.1140 9214	207.6061 4246	50.4622 0054
96	3.2955 1324	183.6410 5940	55.7245 7031	4.1758 0352	211.7202 3459	50.7016 7541
97	3.3367 0716	186.9365 7264	56.0242 6698	4.2384 4057	215.8960 3811	50.9376 1124
98	3.3784 1600	190.2732 7980	56.3202 6368	4.3020 1718	220.1344 7868	51.1700 6034
99	3.4206 4620	193.6516 9580	56.6126 0610	4.3665 4744	224.4364 9586	51.3990 7422
100	3.4634 0427	197.0723 4200	56.9013 3936	4.4320 4565	228.8030 4330	51.6247 0367

(continued)

	$i = 1\frac{3}{4}\%$			$i = 2\%$		
n	$(1 + i)^n$	$s_{\overline{n}|i}$	$a_{\overline{n}|i}$	$(1 + i)^n$	$s_{\overline{n}|i}$	$a_{\overline{n}|i}$
1	1.0175 0000	1.0000 0000	0.9828 0098	1.0200 0000	1.0000 0000	0.9803 9216
2	1.0353 0625	2.0175 0000	1.9486 9875	1.0404 0000	2.0200 0000	1.9415 6094
3	1.0534 2411	3.0528 0625	2.8979 8403	1.0612 0800	3.0604 0000	2.8838 8327
4	1.0718 5903	4.1062 3036	3.8309 4254	1.0824 3216	4.1216 0800	3.8077 2870
5	1.0906 1656	5.1780 8939	4.7478 5508	1.1040 8080	5.2040 4016	4.7134 5951
6	1.1097 0235	6.2687 0596	5.6489 9762	1.1261 6242	6.3081 2096	5.6014 3089
7	1.1291 2215	7.3784 0831	6.5346 4139	1.1486 8567	7.4342 8338	6.4719 9107
8	1.1488 8178	8.5075 3045	7.4050 5297	1.1716 5938	8.5829 6905	7.3254 8144
9	1.1689 8721	9.6564 1224	8.2604 9432	1.1950 9257	9.7546 2843	8.1622 3671
10	1.1894 4449	10.8253 9945	9.1012 2291	1.2189 9442	10.9497 2100	8.9825 8501
11	1.2102 5977	12.0148 4394	9.9274 9181	1.2433 7431	12.1687 1542	9.7868 4805
12	1.2314 3931	13.2251 0371	10.7395 4969	1.2682 4179	13.4120 8973	10.5753 4122
13	1.2529 8950	14.4565 4303	11.5376 4097	1.2936 0663	14.6803 3152	11.3483 7375
14	1.2749 1682	15.7095 3253	12.3220 0587	1.3194 7876	15.9739 3815	12.1062 4877
15	1.2972 2786	16.9844 4935	13.0928 8046	1.3458 6834	17.2934 1692	12.8492 6350
16	1.3199 2935	18.2816 7721	13.8504 9677	1.3727 8571	18.6392 8525	13.5777 0931
17	1.3430 2811	19.6016 0656	14.5950 8282	1.4002 4142	20.0120 7096	14.2918 7188
18	1.3665 3111	20.9446 3468	15.3268 6272	1.4282 4625	21.4123 1238	14.9920 3125
19	1.3904 4540	22.3111 6578	16.0460 5673	1.4568 1117	21.8405 5863	15.6784 6201
20	1.4147 7820	23.7016 1119	16.7528 8130	1.4859 4740	24.2973 6980	16.3514 3334
21	1.4395 3681	25.1163 8938	17.4475 4919	1.5156 6634	25.7833 1719	17.0112 0916
22	1.4647 2871	26.5559 2620	18.1302 6948	1.5459 7967	27.2989 8354	17.6580 4820
23	1.4903 6146	28.0206 5490	18.8012 4764	1.5768 9926	28.8449 6321	18.2922 0412
24	1.5164 4279	29.5110 1637	19.4606 8565	1.6084 3725	30.4218 6247	18.9139 2560
25	1.5429 8054	31.0274 5915	20.1087 8196	1.6406 0599	32.0302 9972	19.5234 5647
26	1.5699 8269	32.5704 3969	20.7457 3166	1.6734 1811	33.6709 0572	20.1210 3576
27	1.5974 5739	34.1404 2238	21.3717 2644	1.7068 8648	35.3443 2383	20.7068 9780
28	1.6254 1290	35.7378 7977	21.9869 5474	1.7410 2421	37.0512 1031	21.2812 7236
29	1.6538 5762	37.3632 9267	22.5916 0171	1.7758 4469	38.7922 3451	21.8443 8466
30	1.6828 0013	39.0171 5029	23.1858 4934	1.8113 6158	40.5680 7921	22.3964 5555
31	1.7122 4913	40.6999 5042	23.7698 7650	1.8475 8882	42.3794 4079	22.9377 0152
32	1.7422 1349	42.4121 9955	24.3438 5897	1.8845 4059	44.2270 2961	23.4683 3482
33	1.7727 0223	44.1544 1305	24.9079 6951	1.9222 3140	46.1115 7020	23.9885 6355
34	1.8037 2452	45.9271 1527	25.4623 7789	1.9606 7603	48.0338 0160	24.4985 9172
35	1.8352 8970	47.7308 3979	26.0072 5100	1.9998 8955	49.9944 7763	24.9986 1933
36	1.8674 0727	49.5661 2949	26.5427 5283	2.0398 8734	51.9943 6719	25.4888 4248
37	1.9000 8689	51.4335 3675	27.0690 4455	2.0806 8509	54.0342 5453	25.9694 5341
38	1.9333 3841	53.3336 2365	27.5862 8457	2.1222 9879	56.1149 3962	26.4406 4060
39	1.9671 7184	55.2669 6206	28.0946 2857	2.1647 4477	58.2372 3841	26.9025 8883
40	2.0015 9734	57.2341 3390	28.5942 2955	2.2080 3966	60.4019 8318	27.3554 7924
41	2.0366 2530	59.2357 3124	29.0852 3789	2.2522 0046	62.6100 2284	27.7994 8945
42	2.0722 6624	61.2723 5654	29.5678 0136	2.2972 4447	64.8622 2330	28.2347 9358
43	2.1085 3090	63.3446 2278	30.0420 6522	2.3431 8936	67.1594 6777	28.6615 6233
44	2.1454 3019	65.4531 5367	30.5081 7221	2.3900 5314	69.5026 5712	29.0799 6307
45	2.1829 7522	67.5985 8386	30.9662 6261	2.4378 5421	71.8927 1027	29.4901 5987
46	2.2211 7728	69.7815 5908	31.4164 7431	2.4866 1129	74.3305 6447	29.8923 1360
47	2.2600 4789	72.0027 3637	31.8589 4281	2.5363 4352	76.8171 7576	30.2865 8196
48	2.2995 9872	74.2627 8425	32.2938 0129	2.5870 7039	79.3535 1927	30.6731 1957
49	2.3398 4170	76.5623 8298	32.7211 8063	2.6388 1179	81.9405 8966	31.0520 7801
50	2.3807 8893	78.9022 2468	33.1412 0946	2.6915 8803	84.5794 0145	31.4236 0589

(continued)

	$i = 1\frac{3}{4}\%$			$i = 2\%$		
n	$(1 + i)^n$	$s_{\overline{n}\,i}$	$a_{\overline{n}\,i}$	$(1 + i)^n$	$s_{\overline{n}\,i}$	$a_{\overline{n}\,i}$
51	2.4224 5274	81.2830 1361	33.5540 1421	2.7454 1979	87.2709 8948	31.7878 4892
52	2.4648 4566	83.7054 6635	33.9597 1913	2.8003 2819	90.0164 0927	32.1449 4992
53	2.5079 8046	86.1703 1201	34.3584 4632	2.8563 3475	92.8167 3746	32.4950 4894
54	2.5518 7012	88.6782 9247	34.7503 1579	2.9134 6144	95.6730 7221	32.8382 8327
55	2.5965 2785	91.2301 6259	35.1354 4550	2.9717 3067	98.5865 3365	33.1747 8752
56	2.6419 6708	93.8266 9043	35.5139 5135	3.0311 6529	101.5582 6432	33.5046 9365
57	2.6882 0151	96.4686 5752	35.8859 4727	3.0917 8859	104.5894 2961	33.8281 3103
58	2.7352 4503	99.1568 5902	36.2515 4523	3.1536 2436	107.6812 1820	34.1452 2650
59	2.7831 1182	101.8921 0405	36.6108 5526	3.2166 9685	110.8348 4257	34.4561 0441
60	2.8318 1628	104.6752 1588	36.9639 8552	3.2810 3079	114.0515 3942	34.7608 8668
61	2.8813 7306	107.5070 3215	37.3110 4228	3.3466 5140	117.3325 7021	35.0596 9282
62	2.9317 9709	110.3884 0522	37.6521 3000	3.4135 8443	120.6792 2161	35.3526 4002
63	2.9831 0354	113.3202 0231	37.9873 5135	3.4818 5612	124.0928 0604	35.6398 4316
64	3.0353 0785	116.3033 0585	38.3168 0723	3.5514 9324	127.5746 6216	35.9214 1486
65	3.0884 2574	119.3386 1370	38.6405 9678	3.6225 2311	131.1261 5541	36.1974 6555
66	3.1424 7319	122.4270 3944	38.9588 1748	3.6949 7357	134.7486 7852	36.4681 0348
67	3.1974 6647	125.5695 1263	39.2715 6509	3.7688 7304	138.4436 5209	36.7334 3478
68	3.2534 2213	128.7669 7910	39.5789 3375	3.8442 5050	142.2125 2513	36.9935 6351
69	3.3103 5702	132.0204 0124	39.8810 1597	3.9211 3551	146.0567 7563	37.2485 9168
70	3.3682 8827	135.3307 5826	40.1779 0267	3.9995 5822	149.9779 1114	37.4986 1929
71	3.4272 3331	138.6990 4653	40.4696 8321	4.0795 4939	153.9774 6937	37.7437 4441
72	3.4872 0990	142.1262 7984	40.7564 4542	4.1611 4038	158.0570 1875	37.9840 6314
73	3.5482 3607	145.6134 8974	41.0382 7560	4.2443 6318	162.2181 5913	38.2196 6975
74	3.6103 3020	149.1617 2581	41.3152 5857	4.3292 5045	166.4625 2231	38.4506 5662
75	3.6735 1098	152.7720 5601	41.5874 7771	4.4158 3546	170.7917 7276	38.6771 1433
76	3.7377 9742	156.4455 6699	41.8550 1495	4.5041 5216	175.2076 0821	38.8991 3170
77	3.8032 0888	160.1833 6441	42.1179 5081	4.5942 3521	179.7117 6038	39.1167 9578
78	3.8697 6503	163.9865 7329	42.3763 6443	4.6861 1991	184.3059 9558	39.3301 9194
79	3.9374 8592	167.8563 3832	42.6303 3359	4.7798 4231	188.9921 1549	39.5394 0386
80	4.0063 9192	171.7938 2424	42.8799 3474	4.8754 3916	193.7719 5780	39.7445 1359
81	4.0765 0378	175.8002 1617	43.1252 4298	4.9729 4794	198.6473 9696	39.9456 0156
82	4.1478 4260	179.8767 1995	43.3663 3217	5.0724 0690	203.6203 4490	40.1427 4663
83	4.2204 2984	184.0245 6255	43.6032 7486	5.1738 5504	208.6927 5180	40.3360 2611
84	4.2942 8737	188.2449 9239	43.8361 4237	5.2773 3214	213.8666 0683	40.5255 1579
85	4.3694 3740	192.5392 7976	44.0650 0479	5.3828 7878	219.1439 3897	40.7112 8999
86	4.4459 0255	196.9087 1716	44.2899 3099	5.4905 3636	224.5268 1775	40.8934 2156
87	4.5237 0584	201.3546 1971	44.5109 8869	5.6003 4708	230.0173 5411	41.0719 8192
88	4.6028 7070	205.8783 2555	44.7282 4441	5.7123 5402	235.6177 0119	41.2470 4110
89	4.6834 2093	210.4811 9625	44.9417 6355	5.8266 0110	241.3300 5521	41.4186 6774
90	4.7653 8080	215.1646 1718	44.1516 1037	5.9431 3313	247.1566 5632	41.5869 2916
91	4.8487 7496	219.9299 9798	45.3578 4803	6.0619 9579	253.0997 8944	41.7518 9133
92	4.9336 2853	224.7787 7295	45.5605 3860	6.1832 3570	259.1617 8523	41.9136 1895
93	5.0199 6703	229.7124 0148	45.7597 4310	6.3069 0042	265.3450 2094	42.0721 7545
94	5.1078 1645	234.7323 6850	45.9555 2147	6.4330 3843	271.6519 2135	42.2276 2299
95	5.1972 0324	239.8401 8495	46.1479 3265	6.5616 9920	278.0849 5978	42.3800 2254
96	5.2881 5429	245.0373 8819	46.3370 3455	6.6929 3318	284.6466 5898	42.5294 3386
97	5.3806 9699	250.3255 4248	46.5228 8408	6.8267 9184	291.3395 9216	42.6759 1555
98	5.4748 5919	255.7062 3947	46.7055 3718	6.9633 2768	298.1663 8400	42.8195 2505
99	5.5706 6923	261.1810 9866	46.8850 4882	7.1025 9423	305.1297 1168	42.9603 1867
100	5.6681 5594	266.7517 6789	47.0614 7304	7.2446 4612	312.2323 0591	43.0983 5164

APPENDIX B Compound Amount, Present Value, and Annuity Table

(continued)

	$i = 2\frac{1}{4}\%$			$i = 2\frac{1}{2}\%$		
n	$(1 + i)^n$	$s_{\overline{n}|i}$	$a_{\overline{n}|i}$	$(1 + i)^n$	$s_{\overline{n}|i}$	$a_{\overline{n}|i}$
1	1.0225 0000	1.0000 0000	0.9779 9511	1.0250 0000	1.0000 0000	0.9756 0976
2	1.0455 0625	2.0225 0000	1.9344 6955	1.0506 2500	2.0250 0000	1.9274 2415
3	1.0690 3014	3.0680 0625	2.8698 9687	1.0768 9063	3.0756 2500	2.8560 2356
4	1.0930 8332	4.1370 3639	3.7847 4021	1.1038 1289	4.1525 1563	3.7619 7421
5	1.1176 7769	5.2301 1971	4.6794 5253	1.1314 0821	5.2563 2852	4.6458 2850
6	1.1428 2544	6.3477 9740	5.5544 7680	1.1596 9342	6.3877 3673	5.5081 2536
7	1.1685 3901	7.4906 2284	6.4102 4626	1.1886 8575	7.5474 3015	6.3493 9060
8	1.1948 3114	8.6591 6186	7.2471 8461	1.2184 0290	8.7361 1590	7.1701 3717
9	1.2217 1484	9.8539 9300	8.0657 0622	1.2488 6297	9.9545 1880	7.9708 6553
10	1.2492 0343	11.0757 0784	8.8662 1635	1.2800 8454	11.2033 8177	8.7520 6393
11	1.2773 1050	12.3249 1127	9.6491 1134	1.3120 8666	12.4834 6631	9.5142 0871
12	1.3060 4999	13.6022 2177	10.4147 7882	1.3448 8882	13.7955 5297	10.2577 6460
13	1.3354 3611	14.9082 7176	11.1635 9787	1.3785 1104	15.1404 4179	10.9831 8497
14	1.3654 8343	16.2437 0788	11.8959 3924	1.4129 7382	16.5189 5284	11.6909 1217
15	1.3962 0680	17.6091 9130	12.6121 6551	1.4482 9817	17.9319 2666	12.3813 7773
16	1.4276 2146	19.0053 9811	13.3126 3131	1.4845 0562	19.3802 2483	13.0550 0266
17	1.4597 4294	20.4330 1957	13.9976 8343	1.5216 1826	20.8647 3045	13.7121 9772
18	1.4925 8716	21.8927 6251	14.6676 6106	1.5596 5872	22.3863 4871	14.3533 6363
19	1.5261 7037	23.3853 4966	15.3228 9590	1.5986 5019	23.9460 0743	14.9788 9134
20	1.5605 0920	24.9115 2003	15.9637 1237	1.6386 1644	25.5446 5761	15.5891 6229
21	1.5956 2066	26.4720 2923	16.5904 2775	1.6795 8185	27.1832 7405	16.1845 4587
22	1.6315 2212	28.0676 4989	17.2033 5232	1.7215 7140	28.8628 5590	16.7654 1324
23	1.6682 3137	29.6991 7201	17.8027 8955	1.7646 1068	30.5844 2730	17.3321 1048
24	1.7057 6658	31.3674 0338	18.3890 3624	1.8087 2595	32.3490 3798	17.8849 8583
25	1.7441 4632	33.0731 6996	18.9623 8263	1.8539 4410	34.1577 6393	18.4243 7642
26	1.7833 8962	34.8173 1628	19.5231 1260	1.9002 9270	36.0117 0803	18.9506 1114
27	1.8235 1588	36.6007 0590	20.0715 0376	1.9478 0002	37.9120 0073	19.4640 1087
28	1.8645 4499	38.4242 2178	20.6078 2764	1.9964 9502	39.8598 0075	19.9648 8866
29	1.9064 9725	40.2887 6677	21.1323 4977	2.0464 0739	41.8562 9577	20.4535 4991
30	1.9493 9344	42.1952 6402	21.6453 2985	2.0975 6758	43.9027 0316	20.9302 9259
31	1.9932 5479	44.1446 5746	22.1470 2186	2.1500 0677	46.0002 7074	21.3954 0741
32	2.0381 0303	46.1379 1226	22.6376 7419	2.2037 5694	48.1502 7751	21.8491 7796
33	2.0839 6034	48.1760 1528	23.1175 2977	2.2588 5086	50.3540 3445	22.2918 8094
34	2.1308 4945	50.2599 7563	23.5868 2618	2.3153 2213	52.6128 8531	22.7237 8628
35	2.1787 9356	52.3908 2508	24.0457 9577	2.3732 0519	54.9282 0744	23.1451 5734
36	2.2278 1642	54.5696 1864	24.4946 6579	2.4325 3532	57.3014 1263	23.5562 5107
37	2.2779 4229	56.7974 3506	24.9336 5848	2.4933 4870	59.7339 4794	23.9573 1812
38	2.3291 9599	59.0753 7735	25.3629 9118	2.5556 8242	62.2272 9664	24.3486 0304
39	2.3816 0290	61.4045 7334	25.7828 7646	2.6195 7448	64.7829 7906	24.7303 4443
40	2.4351 8897	63.7861 7624	26.1935 2221	2.6850 6384	67.4025 5354	25.1027 7505
41	2.4899 8072	66.2213 6521	26.5951 3174	2.7521 9043	70.0876 1737	25.4661 2200
42	2.5460 0528	68.7113 4592	26.9879 0390	2.8209 9520	72.8398 0781	25.8206 0683
43	2.6032 9040	71.2573 5121	27.3720 3316	2.8915 2008	75.6608 0300	26.1664 4569
44	2.6183 6444	73.8606 4161	27.7477 0969	2.9638 0808	78.5523 2308	26.5038 4945
45	2.7217 5639	76.5225 0605	28.1151 1950	3.0379 0328	81.5161 3116	26.8330 2386
46	2.7829 9590	79.2442 6243	28.4744 4450	3.1138 5086	84.5540 3443	27.1541 6962
47	2.8456 1331	82.0272 5834	28.8258 6259	3.1916 9713	87.6678 8530	27.4674 8255
48	2.9096 3961	84.8728 7165	29.1695 4777	3.2714 8956	90.8595 8243	27.7731 5371
49	2.9751 0650	87.7825 1126	29.5056 7019	3.3532 7680	94.1310 7199	28.0713 6947
50	3.0420 4640	90.7576 1776	29.8343 9627	3.4371 0872	97.4843 4879	28.3623 1168

(continued)

	$i = 2\frac{1}{4}\%$			$i = 2\frac{1}{2}\%$		
n	$(1 + i)^n$	$s_{\overline{n}|i}$	$a_{\overline{n}|i}$	$(1 + i)^n$	$s_{\overline{n}|i}$	$a_{\overline{n}|i}$
51	3.1104 9244	93.7996 6416	30.1558 8877	3.5230 3644	100.9214 5751	28.6461 5774
52	3.1804 7852	96.9101 5661	30.4703 0687	3.6111 1235	104.4444 9395	28.9230 8072
53	3.2520 3929	100.0906 3513	30.7778 0623	3.7013 9016	108.0556 0629	29.1932 4948
54	3.3252 1017	100.3426 7442	31.0785 3910	3.7939 2491	111.7569 9645	29.4568 2876
55	3.4000 2740	100.6678 8460	31.3726 5438	3.8887 7303	115.5509 2136	29.7139 7928
56	3.4765 2802	110.0679 1200	31.6602 9768	3.9859 9236	119.4396 9440	29.9648 5784
57	3.5547 4990	113.5444 4002	31.9416 1142	4.0856 4217	123.4256 8676	30.2096 1740
58	3.6347 3177	117.0991 8992	32.2167 3489	4.1877 8322	127.5113 2893	30.4484 0722
59	3.7165 1324	120.7339 2169	32.4858 0429	4.2924 7780	131.6991 1215	30.6813 7290
60	3.8001 3479	124.4504 3493	32.7489 5285	4.3997 8975	135.9915 8995	30.9086 5649
61	3.8856 3782	128.2505 6972	33.0063 1086	4.5097 8449	140.3913 7970	31.1303 9657
62	3.9730 6467	132.1362 0754	33.2580 0573	4.6225 2910	144.9011 6419	31.3467 2836
63	4.0624 5862	136.1092 7221	33.5041 6208	4.7380 9233	149.5236 9330	31.5577 8377
64	4.1538 6394	140.1717 3083	33.7449 0179	4.8565 4464	154.2617 8563	31.7636 9148
65	4.2473 2588	144.3255 9477	33.9803 4405	4.9779 5826	159.1183 3027	31.9645 7705
66	4.3428 9071	148.5729 2066	34.2106 0543	5.1024 0721	164.0962 8853	32.1605 6298
67	4.4406 0576	152.9158 1137	34.4357 9993	5.2299 6739	169.1986 9574	32.3517 6876
68	4.5405 1939	157.3564 1713	34.6560 3905	5.3607 1658	174.4286 6314	32.5383 1099
69	4.6426 8107	161.8969 3651	34.8714 3183	5.4947 3449	179.7893 7971	32.7203 0340
70	4.7471 4140	166.5396 1758	35.0820 8492	5.6321 0286	185.2841 1421	32.8978 5698
71	4.8539 5208	171.2867 5898	35.2881 0261	5.7729 0543	190.9162 1706	33.0710 7998
72	4.9631 6600	176.1407 1106	35.4895 8691	5.9172 2806	196.6891 2249	33.2400 7803
73	5.0748 3723	181.1038 7705	35.6866 3756	6.0651 5876	202.6063 5055	33.4049 5417
74	5.1890 2107	186.1787 1429	35.8793 5214	6.2167 8773	208.6715 0931	33.5658 0895
75	5.3057 7405	191.3677 3536	36.0678 2605	6.3722 0743	214.8882 9705	33.7227 4044
76	5.4251 5396	196.6735 0941	36.2521 5262	6.5315 1261	221.2605 0447	33.8758 4433
77	5.5472 1993	202.0986 6337	36.4324 2310	6.6948 0043	227.7920 1709	34.0252 1398
78	5.6720 3237	207.6458 8329	36.6087 2675	6.8621 7044	234.4868 1751	34.1709 4047
79	5.7996 5310	213.3179 1567	36.7811 5085	7.0337 2470	241.3489 8795	34.3131 1265
80	5.9301 4530	219.1175 6877	36.9497 8079	7.2095 6782	248.3827 1265	34.4518 1722
81	6.0635 7357	225.0477 1407	37.1147 0004	7.3898 0701	255.5922 8047	34.5871 3875
82	6.2000 0397	231.1112 8763	37.2759 9026	7.5745 5219	262.9820 8748	34.7191 5976
83	6.3395 0406	237.3112 9160	37.4337 3130	7.7639 1599	270.5566 3966	34.8479 6074
84	6.4821 4290	243.6507 9567	37.5880 0127	7.9580 1389	278.3205 5566	34.9736 2023
85	6.6279 9112	250.1329 3857	37.7388 7655	8.1569 6424	286.2785 6955	35.0962 1486
86	6.7771 2092	256.7609 2969	37.8864 3183	8.3608 8834	294.4355 3379	35.2158 1938
87	6.9296 0614	263.5380 5060	38.0307 4018	8.5699 1055	302.7964 2213	35.3325 0671
88	7.0855 2228	270.4676 5674	38.1718 7304	8.7841 5832	311.3663 3268	35.4463 4801
89	7.2449 4653	277.5531 7902	38.3099 0028	9.0037 6228	320.1504 9100	35.5574 1269
90	7.4079 5782	284.7981 2555	38.4448 9025	9.2288 5633	329.1542 5328	35.6657 6848
91	7.5746 3688	292.2060 8337	38.5769 0978	9.4595 7774	338.3831 0961	35.7714 8144
92	7.7450 6621	299.7807 2025	38.7060 2423	9.6960 6718	347.8426 8735	35.8746 1604
93	7.9193 3020	307.5257 8645	38.8322 9754	9.9384 6886	357.5387 5453	35.9752 3516
94	8.0975 1512	315.4451 1665	38.9557 9221	10.1869 3058	367.4772 2339	36.0734 0016
95	8.2797 0921	323.5426 3177	38.0765 6940	10.4416 0385	377.6641 5398	36.1691 7089
96	8.4660 0267	331.8223 4099	39.1946 8890	10.7026 4395	388.1057 5783	36.2626 0574
97	8.6564 8773	340.2883 4366	39.3102 0920	10.9702 1004	398.8084 0177	36.3537 6170
98	8.8512 5871	348.9448 3139	39.4231 8748	11.2444 6530	409.7786 1182	36.4426 9434
99	9.0504 1203	357.7960 9010	39.5336 7968	11.5255 7693	421.0230 7711	36.5294 5790
100	9.2540 4630	366.8465 0213	39.6417 4052	11.8137 1635	432.5486 5404	36.6141 0526

(continued)

		$i = 2\frac{3}{4}\%$			$i = 3\%$	
n	$(1 + i)^n$	$s_{\overline{n}|i}$	$a_{\overline{n}|i}$	$(1 + i)^n$	$s_{\overline{n}|i}$	$a_{\overline{n}|i}$
1	1.0275 0000	1.0000 0000	0.9732 3601	1.0300 0000	1.0000 0000	0.9708 7379
2	1.0557 5625	2.0275 0000	1.9204 2434	1.0609 0000	2.0300 0000	1.9134 6970
3	1.0847 8955	3.0832 5625	2.8422 6213	1.0927 2700	3.0909 0000	2.8286 1135
4	1.1146 2126	4.1680 4580	3.7394 2787	1.1255 0881	4.1836 2700	3.7170 9840
5	1.1452 7334	5.2826 6706	4.6125 8186	1.1592 7407	5.3091 3581	4.5797 0719
6	1.1767 6836	6.4279 4040	5.4623 6678	1.1940 5230	6.4684 0988	5.4171 9144
7	1.2091 2949	7.6047 0876	6.2894 0806	1.2298 7387	7.6624 6218	6.2302 8296
8	1.2423 8055	8.8138 3825	7.0943 1441	1.2667 7008	8.8923 3605	7.0196 9219
9	1.2765 4602	10.0562 1880	7.8776 7826	1.3047 7318	10.1591 0613	7.7861 0892
10	1.3116 5103	11.3327 6482	8.6400 7616	1.3439 1638	11.4638 7931	8.5302 0284
11	1.3477 2144	12.6444 1585	9.3820 6926	1.3842 3387	12.8077 9569	9.2526 2411
12	1.3847 8378	13.9921 3729	10.1042 0366	1.4257 6089	14.1920 2956	9.9540 0399
13	1.4228 6533	15.3769 2107	10.8070 1086	1.4685 3371	15.6177 9045	10.6349 5533
14	1.4619 9413	16.7997 8639	11.4910 0814	1.5125 8972	17.0863 2416	11.2960 7314
15	1.5021 9896	18.2617 8052	12.1566 9892	1.5579 6742	18.5989 1389	11.9379 3509
16	1.5435 0944	19.7639 7948	12.8045 7315	1.6047 0644	20.1568 8130	12.5611 0203
17	1.5859 5595	21.3074 8892	13.4351 0769	1.6528 4763	21.7615 8774	13.1661 1847
18	1.6295 6973	22.8934 4487	14.0487 6661	1.7024 3306	23.4144 3537	13.7535 1308
19	1.6743 8290	24.5230 1460	14.6460 0157	1.7535 0605	25.1168 6844	14.3237 9911
20	1.7204 2843	26.1973 9750	15.2272 5213	1.8061 1123	26.8703 7449	14.8774 7486
21	1.7677 4021	27.9178 2593	15.7929 4612	1.8602 9457	28.6764 8572	15.4150 2414
22	1.8163 5307	29.6855 6615	16.3434 9987	1.9161 0341	30.5367 8030	15.9369 1664
23	1.8663 0278	31.5019 1921	16.8793 1861	1.9735 8651	32.4528 8370	16.4436 0839
24	1.9176 2610	33.3682 2199	17.4007 9670	2.0327 9411	34.4264 7022	16.9355 4212
25	1.9703 6082	35.2858 4810	17.9083 1795	2.0937 7793	36.4592 6432	17.4131 4769
26	2.0245 4575	37.2562 0892	18.4022 5592	2.1565 9127	38.5530 4225	17.8768 4242
27	2.0802 2075	39.2807 5467	18.8829 7413	2.2212 8901	40.7096 3352	18.3270 3147
28	2.1374 2682	41.3609 7542	19.3508 2640	2.2879 2768	42.9309 2252	18.7641 0823
29	2.1962 0606	43.4984 0224	19.8061 5708	2.3565 6551	45.2188 5020	19.1884 5459
30	2.2566 0173	45.6946 0831	20.2493 0130	2.4272 6247	47.5754 1571	19.6004 4135
31	2.3186 5828	47.9512 1003	20.6805 8520	2.5000 8035	50.0026 7818	20.0004 2849
32	2.3824 2138	50.2698 6831	21.1003 2623	2.5750 8276	52.5027 5852	20.3887 6553
33	2.4479 3797	52.6522 8969	21.5088 3332	2.6523 3524	55.0778 4128	20.7657 9178
34	2.5152 5626	55.1002 2765	21.9064 0712	2.7319 0530	57.7301 7652	21.1318 3668
35	2.5844 2581	57.6154 8391	22.2933 4026	2.8138 6245	60.4620 8181	21.4872 2007
36	2.6554 9752	60.1999 0972	22.6699 1753	2.8982 7833	63.2759 4427	21.8322 5250
37	2.7285 2370	62.8554 0724	23.0364 1609	2.9852 2668	66.1742 2259	22.1672 3544
38	2.8035 5810	65.5839 3094	23.3931 0568	3.0747 8348	69.1594 4927	22.4924 6159
39	2.8806 5595	68.3874 8904	23.7402 4884	3.1670 2698	72.2342 3275	22.8082 1513
40	2.9598 7399	71.2681 4499	24.0781 0106	3.2620 3779	75.4012 5973	23.1147 7197
41	3.0412 7052	74.2280 1898	24.4069 1101	3.3598 9893	78.6632 9753	23.4123 9997
42	3.1249 0546	77.2692 8950	24.7269 2069	3.4606 9589	82.0231 9645	23.7013 5920
43	3.2108 4036	80.3941 9496	25.0383 6563	3.5645 1677	85.4838 9234	23.9819 0213
44	3.2991 3847	83.6050 3532	25.3414 7507	3.6714 5227	89.0484 0911	24.2542 7392
45	3.3898 6478	86.9041 7379	25.6364 7209	3.7815 9584	92.7198 6139	24.5187 1254
46	3.4830 8606	90.2940 3857	25.9235 7381	3.8950 4372	96.5014 5723	24.7754 4907
47	3.5788 7093	93.7771 2463	26.2029 9154	4.0118 9503	100.3965 0095	25.0247 0783
48	3.6722 8988	97.3559 9556	26.4749 3094	4.1322 5188	104.4083 9598	25.2667 0664
49	3.7784 1535	101.0332 8544	26.7395 9215	4.2562 1944	108.5406 4785	25.5016 5693
50	3.8823 2177	104.8117 0079	26.9971 6998	4.3839 0602	112.7968 6729	25.7297 6401

(continued)

	$i = 2\frac{3}{4}\%$			$i = 3\%$		
n	$(1 + i)^n$	$s_{\overline{n}|i}$	$a_{\overline{n}|i}$	$(1 + i)^n$	$s_{\overline{n}|i}$	$a_{\overline{n}|i}$
51	3.9890 8562	108.6940 2256	27.2478 5400	4.5154 2320	117.1807 7331	25.9512 2719
52	4.0987 8547	112.6831 0818	27.4918 2871	4.6508 8590	121.6961 9651	26.1662 3999
53	4.2115 0208	116.7818 9365	27.7292 7368	4.7904 1247	126.3470 8240	26.3749 9028
54	4.3273 1838	120.9933 9573	27.9603 6368	4.9341 2485	131.1374 9488	26.5776 6047
55	4.4463 1964	125.3207 1411	28.1852 6879	5.0821 4859	136.0716 1972	26.7744 2764
56	4.5685 9343	129.7670 3375	28.4041 5454	5.2346 1305	141.1537 6831	26.9654 6373
57	4.6942 2975	134.3356 2718	28.6171 8203	5.3916 5144	146.3883 8136	27.1509 3566
58	4.8233 2107	139.0298 5692	28.8245 0806	5.5534 0098	151.7800 3280	27.3310 0549
59	4.9559 6239	143.8531 7799	29.0262 8522	5.7200 0301	157.3334 3379	27.5058 3058
60	5.0922 5136	148.8091 4038	29.2226 6201	5.8916 0310	163.0534 3680	27.6755 6367
61	5.2322 8827	153.9013 9174	29.4137 8298	6.0683 5120	168.9450 3991	27.8403 5307
62	5.3761 7620	159.1336 8002	29.5997 8879	6.2504 0173	175.0133 9110	28.0003 4279
63	5.5240 2105	164.5098 5622	29.7808 1634	6.4379 1379	181.2637 9284	28.1556 7261
64	5.6759 3162	170.0338 7726	29.9569 9887	6.6310 5120	187.7017 0662	28.3064 7826
65	5.8320 1974	175.7098 0889	30.1284 6605	6.8299 8273	194.3327 5782	28.4528 9152
66	5.9924 0029	181.5418 2863	30.2953 4409	7.0348 8222	201.1627 4055	28.5950 4031
67	6.1571 9130	187.5342 2892	30.4577 5581	7.2459 2868	208.1976 2277	28.7330 4884
68	6.3265 1406	193.6914 2022	30.6158 2074	7.4633 0654	215.4435 5145	28.8670 3771
69	6.5004 9319	200.0179 3427	30.7696 5522	7.6872 0574	222.9068 5800	28.9971 2399
70	6.6792 5676	206.5184 2746	30.9193 7247	7.9178 2191	230.5940 6374	29.1234 2135
71	6.8629 3632	213.1976 8422	31.0650 8270	8.1553 5657	238.5118 8565	29.2460 4015
72	7.0516 6706	220.0606 2054	31.2068 9314	8.4000 1727	246.6672 4222	29.3650 8752
73	7.2455 8791	227.1122 8760	31.3449 0816	8.6520 1778	255.0672 5949	29.4806 6750
74	7.4448 4158	234.3578 7551	31.4792 2936	8.9115 7832	263.7192 7727	29.5928 8107
75	7.6495 7472	241.8027 1709	31.6099 5558	9.1789 2567	272.6308 5559	29.7018 2628
76	7.8599 3802	249.4522 9181	31.7371 8304	9.4542 9344	281.8097 8126	29.8075 9833
77	8.0760 8632	257.3122 2983	31.8610 0540	9.7379 2224	291.2640 7469	29.9102 8964
78	8.2981 7869	265.3883 1615	31.9815 1377	10.0300 5991	301.0019 9693	30.0099 8994
79	8.5263 7861	273.6864 9485	32.0987 9685	10.3309 6171	311.0320 5684	30.1067 8635
80	8.7608 5402	282.2128 7345	32.2129 4098	10.6408 9056	321.3630 1855	30.2007 6345
81	9.0017 7751	290.9737 2747	32.3240 3015	10.9601 1727	332.0039 0910	30.2920 0335
82	9.2493 2639	299.9755 0498	32.4321 4613	11.2889 2079	342.9640 2638	30.3805 8577
83	9.5036 8286	309.2248 3137	32.5373 6850	11.6275 8842	354.2529 4717	30.4665 8813
84	9.7650 3414	318.7285 1423	32.6397 7469	11.9764 1607	365.8805 3558	30.5500 8556
85	10.0335 7258	328.4935 4837	32.7394 4009	12.3357 0855	377.8569 5165	30.6311 5103
86	10.3094 9583	338.5271 2095	32.8364 3804	12.7057 7981	390.1926 6020	30.7098 5537
87	10.5930 0696	348.8366 1678	32.9308 3994	13.0869 5320	402.8984 4001	30.7862 6735
88	10.8843 1465	359.4296 2374	33.0227 1527	13.4795 6180	415.9853 9321	30.8604 5374
89	11.1836 3331	370.3139 3839	33.1121 3165	13.8839 4865	429.4649 5500	30.9324 7936
90	11.4911 8322	381.4975 7170	33.1991 5489	14.3004 6711	443.3489 0365	31.0024 0714
91	11.8071 9076	392.9887 5492	33.2838 4905	14.7294 8112	457.6493 7076	31.0702 9820
92	12.1318 8851	404.7959 4568	33.3662 7644	15.1713 6556	472.3788 5189	31.1362 1184
93	12.4655 1544	416.9278 3418	33.4464 9776	15.6265 0652	487.5502 1744	31.2002 0567
94	12.8083 1711	429.3933 4962	33.5245 7202	16.0953 0172	503.1767 2397	31.2623 3560
95	13.1605 4584	442.2016 6674	33.6005 5671	16.5781 6077	519.2720 2568	31.3226 5592
96	13.5224 6085	455.3622 1257	33.6745 0775	17.0755 0559	535.8501 8645	31.3812 1934
97	13.8943 2852	468.8846 7342	33.7464 7956	17.5877 7076	552.9256 9205	31.4380 7703
98	14.2764 2255	482.7790 0194	33.8165 2512	18.1154 0388	570.5134 6281	31.4932 7867
99	14.6690 2417	497.0554 2449	33.8846 9598	18.6588 6600	588.6288 6669	31.5468 7250
100	15.0724 2234	511.7244 4867	33.9510 4232	19.2186 3198	607.2877 3270	31.5989 0534

(continued)

	$i = 3\frac{1}{2}\%$			$i = 4\%$		
n	$(1 + i)^n$	$s_{\overline{n}|i}$	$a_{\overline{n}|i}$	$(1 + i)^n$	$s_{\overline{n}|i}$	$a_{\overline{n}|i}$
1	1.0350 0000	1.0000 0000	0.9661 8357	1.0400 0000	1.0000 0000	0.9615 3846
2	1.0712 2500	2.0350 0000	1.8996 9428	1.0816 0000	2.0400 0000	1.8860 9467
3	1.1087 1788	3.1062 2500	2.8016 3698	1.1248 6400	3.1216 0000	2.7750 9103
4	1.1475 2300	4.2149 4288	3.6730 7921	1.1698 5856	4.2464 6400	3.6298 9522
5	1.1876 8631	5.3624 6588	4.5150 5238	1.2166 5290	5.4163 2256	4.4518 2233
6	1.2292 5533	6.5501 5218	5.3285 5302	1.2653 1902	6.6329 7546	5.2421 3686
7	1.2722 7926	7.7794 0751	6.1145 4398	1.3159 3178	7.8982 9448	6.0020 5467
8	1.3168 0904	9.0516 8677	6.8739 5554	1.3685 6905	9.2142 2626	6.7327 4487
9	1.3628 9735	10.3684 9581	7.6076 8651	1.4233 1181	10.5827 9531	7.4353 3161
10	1.4105 9876	11.7313 9316	8.3166 0532	1.4802 4428	12.0061 0712	8.1108 9578
11	1.4599 6972	13.1419 9192	9.0015 5104	1.5394 5406	13.4863 5141	8.7604 7671
12	1.5110 6866	14.6019 6164	9.6633 3433	1.6010 3222	15.0258 0546	9.3850 7376
13	1.5639 5606	16.1130 3030	10.3027 3849	1.6650 7351	16.6268 3768	9.9856 4785
14	1.6186 9452	17.6769 8636	10.9205 2028	1.7316 7645	18.2919 1119	10.5631 2293
15	1.6753 4883	19.2956 8088	11.5174 1090	1.8009 4351	20.0235 8764	11.1183 8743
16	1.7339 8604	20.9710 2971	12.0941 1681	1.8729 8125	21.8245 3114	11.6522 9561
17	1.7946 7555	22.7050 1575	12.6513 2059	1.9479 0050	23.6975 1239	12.1656 6885
18	1.8574 8920	24.4996 9130	13.1896 8173	2.0258 1652	25.6454 1288	12.6592 9697
19	1.9225 0132	26.3571 8050	13.7098 3742	2.1068 4918	27.6712 2940	13.1339 3940
20	1.9897 8886	28.2796 8181	14.2124 0330	2.1911 2314	29.7780 7858	13.5903 2634
21	2.0594 3147	30.2694 7068	14.6979 7420	2.2787 6807	31.9692 0172	14.0291 5995
22	2.1315 1158	32.3289 0215	15.1671 2484	2.3699 1879	34.2479 6979	14.4511 1533
23	2.2061 1448	34.4604 1373	15.6204 1047	2.4647 1554	36.6178 8858	14.8568 4167
24	2.2833 2849	36.6665 2821	16.0583 6760	2.5633 0416	39.0826 0412	15.2469 6314
25	2.3632 4498	38.9498 5669	16.4815 1459	2.6658 3633	41.6459 0829	15.6220 7994
26	2.4459 5856	41.3131 0168	16.8903 5226	2.7724 6978	44.3117 4462	15.9827 6918
27	2.5315 6711	43.7590 6024	17.2853 6451	2.8833 6858	47.0842 1440	16.3295 8575
28	2.6201 7196	46.2906 2734	17.6670 1885	2.9987 0332	49.9675 8298	16.6630 6322
29	2.7118 7798	48.9107 9930	18.0357 6700	3.1186 5145	52.9662 8630	16.9837 1463
30	2.8067 9370	51.6226 7728	18.3920 4541	3.2433 9751	56.0849 3775	17.2920 3330
31	2.9050 3148	54.4294 7098	18.7362 7576	3.3731 3341	59.3283 3526	17.5884 9356
32	3.0067 0759	57.3345 0247	19.0688 6547	3.5080 5875	62.7014 6867	17.8735 5150
33	3.1119 4235	60.3412 1005	19.3902 0818	3.6483 8110	66.2095 2742	18.1476 4567
34	3.2208 6033	63.4531 5240	19.7006 8423	3.7943 1634	69.8579 0851	18.4111 9776
35	3.3335 9045	66.6740 1274	20.0006 6110	3.9460 8899	73.6522 2486	18.6646 1323
36	3.4502 6611	70.0076 0318	20.2904 9381	4.1039 3255	77.5983 1385	18.9082 8195
37	3.5710 2543	73.4578 6930	20.5705 2542	4.2680 8986	81.7022 4640	19.1425 7880
38	3.6960 1132	77.0288 9472	20.8410 8736	4.4388 1345	85.9703 3626	19.3678 6423
39	3.8253 7171	80.7249 0604	21.1024 9987	4.6163 6599	90.4091 4971	19.5844 8484
40	3.9592 5972	84.5502 7775	21.3550 7234	4.8010 2063	95.0255 1570	19.7927 7388
41	4.0978 3381	88.5095 3747	21.5991 0371	4.9930 6145	99.8265 3633	19.9930 5181
42	4.2412 5799	92.6073 7128	21.8348 8281	5.1927 8391	104.8195 9778	20.1856 2674
43	4.3897 0202	96.8486 2928	22.0626 8870	5.4004 9527	110.0123 8169	20.3707 9494
44	4.5433 4160	101.2383 3130	22.2827 9102	5.6165 1508	115.4128 7696	20.5488 4129
45	4.7023 5855	105.7816 7290	22.4954 5026	5.8411 7568	121.0293 9204	20.7200 3970
46	4.8669 4110	110.4840 3145	22.7009 1813	6.0748 2271	126.8705 6772	20.8846.5356
47	5.0372 8404	115.3509 7255	22.8994 3780	6.3178 1562	132.9453 9043	21.0429 3612
48	5.2135 8898	120.3882 5659	23.0912 4425	6.5705 2824	139.2632 0604	21.1951 3088
49	5.3960 6459	125.6018 4557	23.2765 6450	6.8333 4937	145.8337 3429	21.3414 7200
50	5.5849 2686	130.9979 1016	23.4556 1787	7.1066 8335	152.6670 8366	21.4821 8462

(continued)

	$i = 3\frac{1}{2}\%$			$i = 4\%$		
n	$(1 + i)^n$	$s_{\overline{n}|i}$	$a_{\overline{n}|i}$	$(1 + i)^n$	$s_{\overline{n}|i}$	$a_{\overline{n}|i}$
51	5.7803 9930	136.5828 3702	23.6286 1630	7.3909 5068	159.7737 6700	21.6174 8521
52	5.9827 1327	142.3632 3631	23.7957 6454	7.6865 8871	167.1647 1768	21.7475 8193
53	6.1921 0824	148.3459 4958	23.9572 6043	7.9940 5226	174.8513 0639	21.8726 7493
54	6.4088 3202	154.5380 5782	24.1132 9510	8.3138 1435	182.8453 5865	21.9929 5667
55	6.6331 4114	160.9468 8984	24.2640 5323	8.6463 6692	191.1591 7299	22.1086 1218
56	6.8653 0108	167.5800 3099	24.4097 1327	8.9922 2160	199.8055 3991	22.2198 1940
57	7.1055 8662	174.4453 3207	24.5504 4760	9.3519 1046	208.7977 6151	22.3267 4943
58	7.3542 8215	181.5509 1869	24.6864 2281	9.7259 8688	218.1496 7197	22.4295 6676
59	7.6116 8203	188.9052 0085	24.8177 9981	10.1150 2635	227.8756 5885	22.5284 2957
60	7.8780 9090	196.5168 8288	24.9447 3412	10.5196 2741	237.9906 8520	22.6234 8997
61	8.1538 2408	204.3949 7378	25.0673 7596	10.9404 1250	248.5103 1261	22.7148 9421
62	8.4392 0793	212.5487 9786	25.1858 7409	11.3780 2900	259.4507 2511	22.8027 8289
63	8.7345 8020	220.9880 0579	25.3003 5796	11.8331 5016	270.8287 5412	22.8872 9124
64	9.0402 9051	229.7225 8599	25.4109 7388	12.3064 7617	282.6619 0428	22.9685 4927
65	9.3567 0068	238.7628 7650	25.5178 4916	12.7987 3522	294.9683 8045	23.0466 8199
66	9.6841 8520	248.1195 7718	25.6211 1030	13.3106 8463	307.7671 1567	23.1218 0961
67	10.0231 3168	257.8037 6238	25.7208 7951	13.8431 1201	321.0778 0030	23.1940 4770
68	10.3739 4129	267.8268 9406	25.8172 7489	14.3968 3649	334.9209 1231	23.2635 0740
69	10.7370 2924	278.2008 3535	25.9104 1052	14.9727 0995	349.3177 4880	23.3302 9558
70	11.1128 2526	288.9378 6459	26.0003 9664	15.5716 1835	364.2904 5876	23.3945 1498
71	11.5017 7414	300.0506 8985	26.0873 3975	16.1944 8308	379.8620 7711	23.4562 6440
72	11.9043 3624	311.5524 6400	26.1713 4275	16.8422 6241	396.0565 6019	23.5156 3885
73	12.3209 8801	323.4568 0024	26.2525 0508	17.5159 5290	412.8988 2260	23.5727 2966
74	12.7522 2259	335.7777 8824	26.3309 2278	18.2165 9102	430.4147 7550	23.6276 2468
75	13.1985 5038	348.5300 1083	26.4066 8868	18.9452 5466	448.6313 6652	23.6804 0834
76	13.6604 9964	361.7285 6121	26.4798 9244	19.7030 6485	467.5766 2118	23.7311 6187
77	14.1386 1713	375.3890 6085	26.5506 2072	20.4911 8744	487.2796 8603	23.7799 6333
78	14.6334 6873	389.5276 7798	26.6189 5721	21.3108 3494	507.7708 7347	23.8268 8782
79	15.1456 4013	404.1611 4671	26.6849 8281	22.1632 6834	529.0817 0841	23.8720 0752
80	15.6757 3754	419.3067 8685	26.7487 7567	23.0497 9907	551.2449 7675	23.9153 9185
81	16.2243 8835	434.9825 2439	26.8104 1127	23.9717 9103	574.2947 7582	23.9571 0755
82	16.7922 4195	451.2069 1274	26.8699 6258	24.9306 6267	598.2665 6685	23.9972 1879
83	17.3799 7041	467.9991 5469	26.9275 0008	25.9278 8918	623.1972 2952	24.0357 8730
84	17.9882 6938	485.3791 2510	26.9830 9186	26.9650 0475	649.1251 1870	24.0728 7241
85	18.6178 5881	503.3673 9448	27.0368 0373	28.0436 0494	676.0901 2345	24.1085 3116
86	19.2694 8387	521.9852 5329	27.0886 9926	29.1653 4914	704.1337 2839	24.1428 1842
87	19.9439 1580	541.2547 3715	27.1388 3986	30.3319 6310	733.2990 7753	24.1757 8694
88	20.6419 5285	561.1986 5295	27.1872 8489	31.5452 4163	763.6310 4063	24.2074 8745
89	21.3644 2120	581.8406 0581	27.2340 9168	32.8070 5129	795.1762 8225	24.2379 6870
90	22.1121 7595	603.2050 2701	27.2793 1564	34.1193 3334	827.9833 3354	24.2672 7759
91	22.8861 0210	625.3172 0295	27.3230 1028	35.4841 0668	862.1026 6688	24.2954 5923
92	23.6871 1568	648.2033 0506	27.3652 2732	36.9034 7094	897.5867 7356	24.3225 5695
93	24.5161 6473	671.8904 2074	27.4060 1673	38.3796 0978	934.4902 4450	24.3486 1245
94	25.3742 3049	696.4065 8546	27.4454 2680	39.9147 9417	972.8698 5428	24.3736 6582
95	26.2623 2856	721.7808 1595	27.4835 0415	41.5113 8594	1012.7846 4845	24.3977 5559
96	27.1815 1006	748.0431 4451	27.5202 9387	43.1718 4138	1054.2960 3439	24.4209 1884
97	28.1328 6291	775.2246 5457	27.5558 3948	44.8987 1503	1097.4678 7577	24.4431 9119
98	29.1175 1311	803.3575 1748	27.5901 8308	46.6946 6363	1142.3665 9080	24.4646 0692
99	30.1366 2607	832.4750 3059	27.6233 6529	48.5624 5018	1189.0612 5443	24.4851 9896
100	31.1914 0798	862.6116 5666	27.6554 2540	50.5049 4818	1237.6237 0461	24.5049 9900

(continued)

		$i = 5\%$			$i = 6\%$	
n	$(1 + i)^n$	$s_{\overline{n}|i}$	$a_{\overline{n}|i}$	$(1 + i)^n$	$s_{\overline{n}|i}$	$a_{\overline{n}|i}$
1	1.0500 0000	1.0000 0000	0.9523 8095	1.0600 0000	1.0000 0000	0.9433 9623
2	1.1025 0000	2.0500 0000	1.8594 1043	1.1236 0000	2.0600 0000	1.8333 9267
3	1.1576 2500	3.1525 0000	2.7232 4803	1.1910 1600	3.1836 0000	2.6730 1195
4	1.2155 0625	4.3101 2500	3.5459 5050	1.2624 7696	4.3746 1600	3.4651 0561
5	1.2762 8156	5.5256 3125	4.3294 7667	1.3382 2558	5.6370 9296	4.2123 6379
6	1.3400 9564	6.8019 1281	5.0756 9207	1.4185 1911	6.9753 1854	4.9173 2433
7	1.4071 0042	8.1420 0845	5.7863 7340	1.5036 3026	8.3938 3765	5.5823 8144
8	1.4774 5544	9.5491 0888	6.4632 1276	1.5938 4807	9.8974 6791	6.2097 9381
9	1.5513 2822	11.0265 6432	7.1078 2168	1.6894 7896	11.4913 1598	6.8016 9227
10	1.6288 9463	12.5778 9254	7.7217 3493	1.7908 4770	13.1807 9494	7.3600 8705
11	1.7103 3936	14.2067 8716	8.3064 1422	1.8982 9856	14.9716 4264	7.8868 7458
12	1.7958 5633	15.9171 2652	8.8632 5164	2.0121 9647	16.8699 4120	8.3838 4394
13	1.8856 4914	15.7129 8285	9.3935 7299	2.1329 2826	18.8821 3767	8.8526 8296
14	1.9799 3160	19.5986 3199	9.8986 4094	2.2609 0396	21.0150 6593	9.2949 8393
15	2.0789 2818	21.5785 6359	10.3796 5804	2.3965 5819	23.2759 6988	9.7122 4899
16	2.1828 7459	23.6574 9177	10.8377 6956	2.5403 5168	25.6725 2808	10.1058 9527
17	2.2920 1832	25.8403 6636	11.2740 6625	2.6927 7279	28.2128 7976	10.4772 5969
18	2.4066 1923	28.1323 8467	11.6895 8690	2.8543 3915	30.9056 5255	10.8276 0348
19	2.5269 5020	30.5390 0391	12.0853 2086	3.0255 9950	33.7599 9170	11.1581 1649
20	2.6532 9771	33.0659 5410	12.4622 1034	3.2071 3547	36.7855 9120	11.4699 2122
21	2.7859 6259	35.7192 5181	12.8211 5271	3.3995 6360	39.9927 2668	11.7640 7662
22	2.9252 6072	38.5052 1440	13.1630 0258	3.6035 3742	43.3922 9028	12.0415 8172
23	3.0715 2376	41.4304 7512	13.4885 7388	3.8197 4966	46.9958 2769	12.3033 7898
24	3.2250 9994	44.5019 9887	13.7986 4179	4.0489 3464	50.8155 7735	12.5503 5753
25	3.3863 5494	47.7270 9882	14.0939 4457	4.2918 7072	54.8645 1200	12.7833 5616
26	3.5556 7269	51.1134 5376	14.3751 8530	4.5493 8296	59.1563 8272	13.0031 6619
27	3.7334 5632	54.6691 2645	14.6430 3362	4.8223 4594	63.7057 6568	13.2105 3414
28	3.9201 2914	58.4025 8277	14.8981 2726	5.1116 8670	68.5281 1162	13.4061 6428
29	4.1161 3560	62.3227 1191	15.1410 7358	5.4183 8790	73.6397 9832	13.5907 2102
30	4.3219 4238	66.4388 4750	15.3724 5103	5.7434 9117	79.0581 8622	13.7648 3115
31	4.5380 3949	70.7607 8988	15.5928 1050	6.0881 0064	84.8016 7739	13.9290 8599
32	4.7649 4147	75.2988 2937	15.8026 7667	6.4533 8668	90.8897 7803	14.0840 4339
33	5.0031 8854	80.0637 7084	16.0025 4921	6.8405 8988	97.3431 6471	14.2302 2961
34	5.2533 4797	85.0669 5938	16.1929 0401	7.2510 2528	104.1837 5460	14.3681 4114
35	5.5160 1537	90.3203 0735	16.3741 9429	7.6860 8679	111.4347 7987	14.4982 4636
36	5.7918 1614	95.8363 2272	16.5468 5171	8.1472 5200	119.1208 6666	14.6209 8713
37	6.0814 0694	101.6281 3886	16.7112 8734	8.6360 8712	127.2681 1866	14.7367 8031
38	6.3854 7729	107.7095 4580	16.8678 9271	9.1542 5235	135.9042 0578	14.8460 1916
39	6.7047 5115	114.0950 2309	17.0170 4067	9.7035 0749	145.0584 5813	14.9490 7468
40	7.0399 8871	120.7997 7424	17.1590 8635	10.2857 1794	154.7619 6562	15.0462 9687
41	7.3919 8815	127.8397 6295	17.2943 6796	10.9028 6101	165.0476 8356	15.1380 1592
42	7.7615 8756	135.2317 5110	17.4232 0758	11.5570 3267	175.9505 4457	15.2245 4332
43	8.1496 6693	142.9933 3866	17.5459 1198	12.2504 5463	187.5075 7724	15.3061 7294
44	8.5571 5028	151.1430 0559	17.6627 7331	12.9854 8191	199.7580 3188	15.3831 8202
45	8.9850 0779	159.7001 5587	17.7740 6982	13.7646 1083	212.7435 1379	15.4558 3209
46	9.4342 5818	168.6851 6366	17.8800 6650	14.5904 8748	226.5081 2462	15.5243 6990
47	9.9059 7109	178.1194 2185	17.9810 1571	15.4659 1673	241.0986 1210	15.5890 2821
48	10.4012 6965	188.0253 9294	18.0771 5782	16.3938 7173	256.5645 2882	15.6500 2661
49	10.9213 3313	198.4266 6259	18.1687 2173	17.3775 0403	272.9584 0055	15.7075 7227
50	11.4673 9979	209.3479 9572	18.2559 2546	18.4201 5427	290.3359 0458	15.7618 6064

(continued)

	$i = 7\%$			$i = 8\%$		
n	$(1 + i)^n$	$s_{\overline{n}|i}$	$a_{\overline{n}|i}$	$(1 + i)^n$	$s_{\overline{n}|i}$	$a_{\overline{n}|i}$
1	1.0700 0000	1.0000 0000	0.9345 7944	1.0800 0000	1.0000 0000	0.9259 2593
2	1.1449 0000	2.0700 0000	1.8080 1817	1.1664 0000	2.0800 0000	1.7832 6475
3	1.2250 4300	3.2149 0000	2.6243 1604	1.2597 1200	3.2464 0000	2.5770 9699
4	1.3107 9601	4.4399 4300	3.3872 1126	1.3604 8896	4.5061 1200	3.3121 2684
5	1.4025 5173	5.7507 3901	4.1001 9744	1.4693 2808	5.8666 0096	3.9927 1004
6	1.5007 3035	7.1532 9074	4.7665 3966	1.5868 7432	7.3359 2904	4.6228 7966
7	1.6057 8148	8.6540 2109	5.3892 8940	1.7138 2427	8.9228 0336	5.2063 7006
8	1.7181 8618	10.2598 0257	5.9712 9851	1.8509 3021	10.6366 2763	5.7466 3894
9	1.8384 5921	11.9779 8875	6.5152 3225	1.9990 0463	12.4875 5784	6.2468 8791
10	1.9671 5136	13.8164 4796	7.0235 8154	2.1589 2500	14.4865 6247	6.7100 8140
11	2.1048 5195	15.7835 9932	7.4986 7434	2.3316 3900	16.6454 8746	7.1389 6426
12	2.2521 9159	17.8884 5127	7.9426 8630	2.5181 7012	18.9771 2646	7.5360 7802
13	2.4098 4500	20.1406 4286	8.3576 5074	2.7196 2373	21.4952 9658	7.9037 7594
14	2.5785 3415	22.5504 8786	8.7454 6799	2.9371 9362	24.2149 2030	8.2442 3698
15	2.7590 3154	25.1290 2201	9.1079 1401	3.1721 6911	27.1521 1393	8.5594 7869
16	2.9521 6375	27.8880 5355	9.4466 4860	3.4259 4264	30.3242 8304	8.8513 6916
17	3.1588 1521	30.8402 1730	9.7632 2299	3.7000 1805	33.7502 2569	9.1216 3811
18	3.3799 3228	33.9990 3251	10.0590 8691	3.9960 1950	37.4502 4374	9.3718 8714
19	3.6165 2754	37.3789 6479	10.3355 9524	4.3157 0106	41.4462 6324	9.6035 9920
20	3.8696 8446	40.9954 9232	10.5940 1425	4.6609 5714	45.7619 6430	9.8181 4741
21	4.1405 6237	44.8651 7678	10.8355 2733	5.0338 3372	50.4229 2144	10.0168 0316
22	4.4304 0174	49.0057 3916	11.0612 4050	5.4365 4041	55.4567 5516	10.2007 4366
23	4.7405 2986	53.4361 4090	11.2721 8738	5.8714 6365	60.8932 9557	10.3710 5895
24	5.0723 6695	58.1766 7076	11.4693 3400	6.3411 8074	66.7647 5922	10.5287 5828
25	5.4274 3264	63.2490 3772	11.6535 8318	6.8484 7520	73.1059 3995	10.6747 7619
26	5.8073 5292	68.6764 7036	11.8257 7867	7.3963 5321	79.9544 1515	10.8099 7795
27	6.2138 6763	74.4838 2328	11.9867 0904	7.9880 6147	87.3507 6836	10.9351 6477
28	6.6488 3836	80.6976 9091	12.1371 1125	8.6271 0639	95.3388 2983	11.0510 7849
29	7.1142 5705	87.3465 2927	12.2776 7407	9.3172 7490	103.9659 3622	11.1584 0601
30	7.6122 5504	94.4607 8632	12.4090 4118	10.0626 5689	113.2832 1111	11.2577 8334
31	8.1451 1290	102.0730 4137	12.5318 1419	10.8676 6944	123.3458 6800	11.3497 9939
32	8.7152 7080	110.2181 5426	12.6465 5532	11.7370 8300	134.2135 3744	11.4349 9944
33	9.3253 3975	118.9334 2506	12.7537 9002	12.6760 4964	145.9506 2044	11.5138 8837
34	9.9781 1354	128.2587 6481	12.8540 0936	13.6901 3361	158.6266 7007	11.5869 3367
35	10.6765 8148	138.2368 7835	12.9476 7230	14.7853 4429	172.3168 0368	11.6545 6822
36	11.4239 4219	148.9134 5984	13.0352 0776	15.9681 7184	187.1021 4797	11.7171 9279
37	12.2236 1814	160.3374 0202	13.1170 1660	17.2456 2558	203.0703 1981	11.7751 7851
38	13.0792 7141	172.5610 2017	13.1934 7345	18.6252 7563	220.3159 4540	11.8288 6899
39	13.9948 2041	185.6402 9158	13.2649 2846	20.1152 9768	238.9412 2103	11.8785 8240
40	14.9744 5784	199.6351 1199	13.3317 0884	21.7245 2150	259.0565 1871	11.9246 1333
41	16.0226 6989	214.6095 6983	13.3941 2041	23.4624 8322	280.7810 4021	11.9672 3457
42	17.1442 5678	230.6322 3972	13.4524 4898	25.3394 8187	304.2435 2342	12.0066 9867
43	18.3443 5475	247.7764 9650	13.5069 6167	27.3666 4042	329.5830 0530	12.0432 3951
44	19.6284 5959	266.1208 5125	13.5579 0810	29.5559 7166	356.9496 4572	12.0770 7362
45	21.0024 5176	285.7493 1084	13.6055 2159	31.9204 4939	386.5056 1738	12.1084 0150
46	22.4726 2338	306.7517 6260	13.6500 2018	34.4740 8534	418.4260 6677	12.1374 0880
47	24.0457 0702	329.2243 8598	13.6916 0764	37.2320 1217	452.9001 5211	12.1642 6741
48	25.7289 0651	353.2700 9300	13.7304 7443	40.2105 7314	490.1321 6428	12.1891 3649
49	27.5299 2997	378.9989 9951	13.7667 9853	43.4274 1899	530.3427 3742	12.2121 6341
50	29.4570 2506	406.5289 2947	13.8007 4629	46.9016 1251	573.7701 5642	12.2334 8464

ANSWERS TO SELECTED EXERCISES

Exercises 1.1, page 7

1. Integer, rational, and real   3. Rational and real
5. Irrational and real   7. Irrational and real
9. Rational and real   11. False   13. True
15. False   17. Commutative law of addition
19. Commutative law of multiplication
21. Distributive law   23. Associative law of addition
25. Property 1 of negatives
27. Property 1 of zero properties
29. Property 2 of zero properties
31. Property 2 of quotients
33. Properties 2 and 5 of quotients
35. Property 6 of quotients and distributive law
37. False   39. False   41. False

Exercises 1.2, page 14

1. 81   3. $\frac{8}{27}$ 5. -64 7. $-\frac{54}{125}$ 9. 256
11. $243y^5$ 13. $6x - 3$ 15. $9x^2 + 3x + 1$
17. $4y^2 + y + 8$ 19. $1.2x^3 - 4.2x^2 + 2.5x - 8.2$
21. $7m^2 - 9m$ 23. $14a - 7b$ 25. $6x^2 + 5x - 6$
27. $6x^2 - 5xy - 6y^2$ 29. $12r^2 - rs - 6s^2$
31. $0.06x^2 - 0.06xy - 2.52y^2$
33. $6x^3 - 3x^2y + 4xy - 2y^2$
35. $4x^2 + 12xy + 9y^2$ 37. $4u^2 - v^2$
39. $2t^4 - 4t^3 + 9t^2 - 2t + 4$ 41. $-2x + 1$
43. $-x - 1$ 45. $x^2 - 10x + 50$ 47. $22x^3 - 20x^2 - 6x$
49. False   51. m 53. $-0.0005x^2 + 6x - 400$
55. $1.1t^2 + 1.7t + 155$; \$282; \$542

Exercises 1.3, page 20

1. $2m(3m - 1)$ 3. $3ab(3b - 2a)$
5. $5mn(2m - 3n + 4)$ 7. $(3x - 5)(2x + 1)$
9. $(2c - d)(3a + b + 4ac - 2ad)$
11. $(2m + 1)(m - 6)$ 13. $(x - 3y)(x + 2y)$
15. Prime   17. $(2a - b)(2a + b)$
19. $(uv - w)(uv + w)$ 21. Prime   23. Prime
25. $(x + 4)(x - 1)$ 27. $2y(3x + 2)(2x - 3)$
29. $(7r - 4)(5r + 3)$ 31. $xy(3x - y)(3x + y)$
33. $(x^2 - 4y)(x^2 + 4y)$ 35. $-8ab$
37. $(2m + 1)(4m^2 - 2m + 1)$
39. $(2r - 3s)(4r^2 + 6rs + 9s^2)$
41. $u^2(v^2 - 2)(v^4 + 2v^2 + 4)$
43. $(2x + 1)(x^2 + 3)$ 45. $(3a + b)(x + 2y)$
47. $(u - v)(u + v)(u^2 + v^2)$
49. $(x + y)(2x - 3y)(2x + 3y)$ 51. $(x^3 - 2)(x + 3)$
53. $(au + c)(u + 1)$

Exercises 1.4, page 26

1. $\frac{4}{3}$ 3. $\frac{2x - 1}{2x}$ 5. $\frac{x - 1}{x + 1}$ 7. $\frac{x + 2}{2x + 3}$ 9. $x + y$
11. $\frac{5x}{2}$ 13. $\frac{4}{3}$ 15. $\frac{3(3r - 2)}{2}$ 17. $\frac{k + 1}{k - 2}$
19. $\frac{10x + 7}{(2x + 3)(2x - 1)}$ 21. $\frac{5x - 9}{(x - 3)(x + 2)(x - 1)}$
23. $\frac{4m^3 - 3}{(2m^2 - 2m - 1)(2m^2 - 3m + 3)}$
25. $-\frac{x^2 - x - 3}{(x + 1)(x - 1)}$ 27. $\frac{2(x^2 - x + 2)}{(x + 2)(x - 2)}$
29. $\frac{x^3 - 2x^2 + 3x + 24}{(x + 3)(x + 2)(x - 2)(x + 1)}$ 31. $\frac{bx - ay}{ab(x - y)}$
33. $\frac{x + 1}{x - 1}$ 35. $\frac{x + y}{xy - 1}$ 37. $\frac{y - x}{x^2y^2}$ 39. $-\frac{1}{2x(x + h)}$

Exercises 1.5, page 31

1. -8 **3.** $\frac{1}{49}$ **5.** -16 **7.** $\frac{7}{12}$ **9.** 0.0004 **11.** 1

13. 1 **15.** $\frac{1}{32}$ **17.** 1 **19.** $\frac{1}{27}$ **21.** $\frac{1}{2}x^5$ **23.** $\dfrac{3}{2x}$

25. $\dfrac{1}{a^6}$ **27.** $\dfrac{8y^6}{x^6}$ **29.** $\dfrac{8}{xy}$ **31.** $-2x^2y^5$ **33.** $\dfrac{3}{2uv^2}$

35. $864x^4$ **37.** $\dfrac{1}{4x^8}$ **39.** $\dfrac{4u^4}{9v^5}$ **41.** $\dfrac{1}{1728x^2y^3z^2}$

43. $\dfrac{a^{10}}{b^{12}}$ **45.** $9a^2b^8$ **47.** $\dfrac{u^8}{16}$ **49.** $\dfrac{1-x}{1+x}$ **51.** $\dfrac{1}{uv}$

53. $\dfrac{b+a}{b-a}$ **55.** False   **57.** False

Exercises 1.6, page 39

1. 9 **3.** 4 **5.** 3 **7.** 4 **9.** -5 **11.** 4

13. $\frac{2}{3}$ **15.** $\frac{9}{4}$ **17.** $\frac{1}{4}$ **19.** $-\frac{2}{3}$ **21.** 9 **23.** $\frac{1}{16}$

25. $\frac{3}{4}$ **27.** 64 **29.** $x^{1/5}$ **31.** x **33.** $\dfrac{9}{x^6}$ **35.** $\dfrac{y^{5/2}}{x^3}$

37. $x^{12/5} - 2x^{17/5}$ **39.** $4p^2 - 2p$ **41.** $4\sqrt{2}$

43. $-3\sqrt[3]{2}$ **45.** $4xy\sqrt{y}$ **47.** m^2np^4 **49.** $\sqrt[3]{3}$

51. $\sqrt[6]{x}$ **53.** $\dfrac{2\sqrt{3}}{3}$ **55.** $\dfrac{3\sqrt{x}}{2x}$ **57.** $\dfrac{2\sqrt{3y}}{3}$ **59.** $\dfrac{\sqrt[3]{x^2}}{x}$

61. $\sqrt{3} - 1$ **63.** $-(1+\sqrt{2})^2$ **65.** $\dfrac{q(\sqrt{q}+1)}{q-1}$

67. $\dfrac{y\sqrt[3]{xz^2}}{xz}$ **69.** $\dfrac{4\sqrt{3}}{3}$ **71.** $\dfrac{\sqrt[3]{18}}{3}$ **73.** $\dfrac{\sqrt{6}}{2x}$

75. $\dfrac{\sqrt[3]{18y^2}}{3}$ **77.** $\dfrac{\sqrt{a}(1+a)}{a}$ **79.** $\dfrac{x+y}{x-y}$ **81.** $\dfrac{3x+2}{2\sqrt{x+1}}$

83. $\dfrac{3+x^{1/3}}{6x^{1/2}(1+x^{1/3})^2}$

Exercises 1.7, page 46

1. $x = -3, 2$ **3.** $x = -2, 2$ **5.** $x = -4, 3$

7. $t = -1, \frac{1}{2}$ **9.** $x = 2$ **11.** $m = -2, \frac{3}{2}$

13. $x = -\frac{3}{2}, \frac{3}{2}$ **15.** $z = -2, \frac{3}{2}$ **17.** $x = -4, 2$

19. $x = 1 - \dfrac{\sqrt{6}}{2}, 1 + \dfrac{\sqrt{6}}{2}$

21. $m = -\frac{1}{2} - \frac{1}{2}\sqrt{13}, -\frac{1}{2} + \frac{1}{2}\sqrt{13}$

23. $x = -\dfrac{3}{4} - \dfrac{\sqrt{41}}{4}, -\dfrac{3}{4} + \dfrac{\sqrt{41}}{4}$ **25.** $x = \pm\dfrac{\sqrt{13}}{2}$

27. $x = -\frac{3}{2}, 2$ **29.** $m = 2 \pm \sqrt{3}$ **31.** $x = \frac{1}{2} \pm \frac{1}{4}\sqrt{10}$

33. $x = -1 \pm \frac{1}{2}\sqrt{10}$ **35.** $x = -0.93, 3.17$

37. $x = \pm\sqrt{2}, \pm\sqrt{3}$ **39.** $y = \pm\sqrt{2}, \pm\sqrt{3}$

41. $x = -\frac{7}{2}, -\frac{5}{3}$ **43.** $y = -\frac{3}{2}, 4$

45. Two real solutions   **47.** No real solutions

49. One real solution   **51.** Two real solutions

Exercise 1.8, page 54

1. False   **3.** False

5.

7.

9.

11. $(-\infty, 2)$ **13.** $(-\infty, -5]$ **15.** $(-4, 6)$

17. $(-\infty, -3) \cup (3, \infty)$ **19.** $(-2, 3)$ **21.** $[-3, 5]$

23. $(-\infty, 1] \cup [\frac{3}{2}, \infty)$ **25.** $(-\infty, -3] \cup (2, \infty)$

27. $(-\infty, 0] \cup (1, \infty)$ **29.** 4 **31.** 2 **33.** $5\sqrt{3}$

35. $\pi + 1$ **37.** 2 **39.** False   **41.** False   **43.** True

45. $a - b < x < a + b$ **47.** False   **49.** True

51. False   **53.** \$50.70   **55.** $[362, 488.7]$

57. \$18,666.67   **59.** Between 1000 and 4000 units

61. 4 sec   **63.** $|x - 0.5| < 0.01$

Chapter 1 Review Exercises, page 57

1. Rational and real   **2.** Irrational and real

3. Irrational and real

4. Whole, integer, rational, and real

5. Rational and real   **6.** Irrational and real

7. $\frac{27}{8}$ **8.** 25 **9.** $\frac{1}{144}$ **10.** -32 **11.** $\frac{1}{4}$

12. $3\sqrt[3]{3}$ **13.** $4(x^2 + y)^2$ **14.** $\dfrac{a^{15}}{b^{11}}$ **15.** $\dfrac{2x}{3z}$

16. $-x^{1/2}$ **17.** $6xy^7$ **18.** $\frac{9}{2}a^5b^8$ **19.** $9x^2y^4$

20. $\dfrac{x}{y^{1/2}}$ **21.** $9y^2 + 12xy - 2x^2 + 7x + 3$

22. $3a - b$ **23.** $\dfrac{180}{(t+6)^2}$ **24.** $\dfrac{15x^2 + 24x + 2}{4(3x^2 + 2)(x + 2)}$

25. $\dfrac{78x^2 - 8x - 27}{3(2x^2 - 1)(3x - 1)}$ **26.** $\dfrac{2(x + 2)}{\sqrt{x + 1}}$

27. $-2\pi r^2(\pi r - 50)$ **28.** $2vw(v^2 + w^2 + u^2)$

29. $(4 - x)(4 + x)$ **30.** $6t(2t - 3)(t + 1)$

31. $-2(x - 1)(x + 3)$ **32.** $4(3x - 5)(x - 6)$

33. $(3a - 5b)(3a + 5b)$

34. $u^3(2uv + 3)(4u^2v^2 - 6uv + 9)$

35. $x = -\frac{3}{4}, \frac{1}{2}$ **36.** $x = -2, \frac{1}{3}$ **37.** $\dfrac{3 \pm \sqrt{41}}{4}$

38. $\dfrac{-5 \pm \sqrt{13}}{2}$ **39.** $y = \frac{1}{2}, 1$

40. $m \approx -1.2871, 8.2871$ **41.** $x = 0, -3, 1$

42. $x = \pm \dfrac{\sqrt{2}}{2}$ **43.** $[-2, \infty)$ **44.** $[-1, 2]$

45. $(-\infty, -4) \cup (5, \infty)$ **46.** $(-\infty, -5) \cup (5, \infty)$

47. 4   **48.** 1   **49.** $\pi - 6$ **50.** $8 - 3\sqrt{3}$

51. $[-2, \frac{1}{2}]$ **52.** $(-2, -\frac{3}{2})$ **53.** $(-1, 4)$

54. $x = \frac{2}{3}, \frac{3}{2}$ **55.** $\dfrac{1}{\sqrt{x} + 1}$ **56.** $\dfrac{x - \sqrt{x}}{2x}$

57. $1 \pm \sqrt{6}$ **58.** $-2 \pm \dfrac{\sqrt{2}}{2}$ **59.** \$100   **60.** \$400

CHAPTER 2

Exercises 2.1, page 65

1. $(3, 3)$; Quadrant I   **3.** $(2, -2)$; Quadrant IV

5. $(-4, -6)$; Quadrant III   **7.** A **9.** $E, F,$ and G

11. F **13–19.** See the following figure.

13. $(-2, 5)$

15. $(3, -1)$

17. $(8, -\frac{7}{2})$

19. $(4.5, -4.5)$

21. $\frac{1}{2}$ **23.** Not defined   **25.** 5   **27.** $\frac{5}{6}$

29. $\dfrac{d - b}{c - a}$ $(a \neq c)$ **31. a.** 4   **b.** -8 **33.** Parallel

35. $a = -5$

Exercises 2.2, page 72

1. a. v   **b.** iii   **c.** i   **d.** iv   **e.** vi   **f.** ii

3. Not perpendicular   **5.** $x = 0$ **7.** $y = -x + 6$

9. $y = -\frac{1}{2}x + \frac{5}{2}$ **11.** $x = 2$ **13.** $y = -\frac{1}{2}x - \frac{5}{2}$

15. $y = -2x - 1$ **17.** $y = -\frac{1}{2}x + \frac{3}{4}$

19. $y = 2, m = 0, b = 2$

21. $y = \frac{3}{4}x + 2, m = \frac{3}{4}, b = 2$

23. $y = -\frac{5}{8}x + 3, m = -\frac{5}{8}, b = 3$

25. $y = \frac{4}{3}x + \frac{4}{3}$ **27.** $y = \frac{3}{2}x$ **29.** $y = 4$

31. $x = a$ **33.** $k = 2$

35.

37.

39.

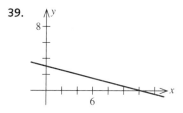

41. $y = -\frac{4}{3}x + 4$ **43.** $y = \frac{3}{2}x + \frac{3}{4}$ **45.** No

47. a. $y = 0.0765x$ **b.** \$0.0765   **c.** \$2677.50

49. a. 88.8 British tons

b.

51. a–b

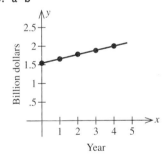

c. $y = \frac{11}{3}x - 112$ **d.** $126\frac{1}{3}$ lb

53. a–b

c. \$2.234 billion

Exercises 2.3, page 84

1. $21, -9, 5a + 6, -5a + 6, 5a + 21$

3. $-3, 6, 3a^2 - 6a - 3, 3a^2 + 6a - 3, 3x^2 - 6$

5. $\dfrac{8}{15}, 0, \dfrac{2a}{a^2 - 1}, \dfrac{2(2 + a)}{a^2 + 4a + 3}, \dfrac{2(t + 1)}{t(t + 2)}$

7. $8, \dfrac{2a^2}{\sqrt{a - 1}}, \dfrac{2(x + 1)^2}{\sqrt{x}}, \dfrac{2(x - 1)^2}{\sqrt{x - 2}}$

9. $5, 1, 1$ **11.** $\frac{5}{2}, 3, 3, 9$ **13.** Yes   **15.** Yes

17. $(-\infty, \infty)$ **19.** $(-\infty, 0) \cup (0, \infty)$ **21.** $(-\infty, \infty)$

23. $(-\infty, 5]$ **25.** $(-\infty, -1) \cup (-1, 1) \cup (1, \infty)$

27. $[-3, \infty)$ **29.** $(-\infty, -2) \cup (-2, 1]$

31. a. $(-\infty, \infty)$ **b.** $6, 0, -4, -6, -\frac{25}{4}, -6, -4, 0$

c.

33.

35.

37.

39.

41.

43.

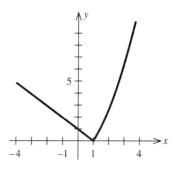

45. Yes **47.** No **49.** Yes **51.** Yes

53. 10π in. **55.** $6 billion; $43.5 billion; $81 billion

57. $S(r) = 4\pi r^2$ **59. a.** $I(x) = 1.053x$ **b.** $652.86

61. The domain of P is $(0, \infty)$.

63. 990; 2240

65. 37.5 words per minute; 42 words per minute

67. $0.1 billion; $4.5 billion

Using Technology Exercises 2.3, page 90

1. **3.**

5. **7.**

9. a. **b.**

11. a. **b.**

13. a. **b.**

15. a. **b.**

17. a. **b.**

19. a. **b.**

21.   **23.**

25.   **27.**

29.

31. 18   **33.** 2   **35.** 18.5505   **37.** 17.3850

39. 4.1616   **41.** 1.7214

43. a. 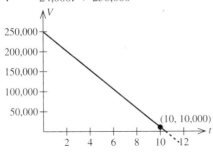   **b.** 2.1762%; 1.9095%

Exercises 2.4, page 98

1. Yes; $y = -\frac{2}{3}x + 2$ **3.** Yes, $y = \frac{1}{2}x + 2$

5. Yes; $y = \frac{1}{2}x + \frac{9}{4}$ **7.** No   **9.** No

11. a. $C(x) = 8x + 40,000$ **b.** $R(x) = 12x$
   c. $P(x) = 4x - 40,000$ **d.** ($8000); $8000

13. (2, 10)   **15.** $(4, \frac{2}{3})$ **17.** (−4, −6)

19. 1000 units; $15,000   **21.** 600 units; $240

23. $900,000; $800,000

25. a. $C(x) = 8x + 48,000$ **b.** $R(x) = 14x$
   c. $P(x) = 6x - 48,000$
   d. ($24,000); ($12,000); $12,000

27. a. $V = -24,000t + 250,000$
   b.

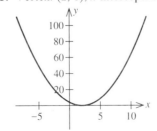

c. $154,000
d. $24,000/yr

31. $5600   **33. b.** 104.2 mg

35. a. 5000 units, $40,000   **b.** 6319 diaries

37. a. $C_1(x) = 18,000 + 15x$
   $C_2(x) = 15,000 + 20x$
   b.

In units of a hundred

c. Machine II; machine II; machine I
d. ($1500); $1500; $4750

Using Technology Exercises 2.4, page 100

1. (−3.0414, 0.1503); (3.0414, 7.4497)

3. (−2.3371, 2.4117); (6.0514, −2.5015)

5. (−1.0219, −6.3461); (1.2414, −1.5931);
   (5.7805, 7.9391)

Exercises 2.5, page 107

1. Vertex: $(-\frac{1}{2}, -\frac{25}{4})$; x-intercepts: −3, 2

3. Vertex: (2, 0); x-intercept: 2

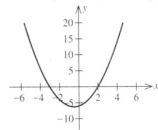

5. Vertex: $(\frac{5}{2}, \frac{1}{4})$; *x*-intercepts: 2, 3

7. Vertex: $(\frac{5}{6}, -\frac{13}{12})$; *x*-intercepts: 0.2324, 1.4343

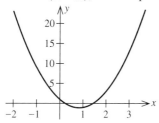

9. Vertex: $(\frac{3}{4}, \frac{15}{8})$; no *x*-intercepts

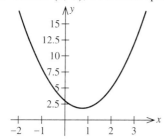

11. Vertex: $(0, -4)$; *x*-intercepts: ± 2

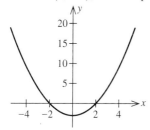

13. Vertex: $(0, 16)$; *x*-intercepts: ± 4

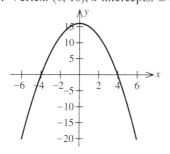

15. Vertex: $(\frac{8}{3}, -\frac{2}{3})$; *x*-intercepts: $\frac{4}{3}$, 4

17. Vertex: $(-\frac{4}{3}, -\frac{10}{3})$; *x*-intercepts: $\frac{1}{3}$, -3

19. $(-2, 0)$; $(1, 3)$ **21.** $(-2, -2)$; $(3, 3)$

23. $(-1.1205, 0.1133)$, $(2.3205, -8.8332)$

25. a.   **27. a.**

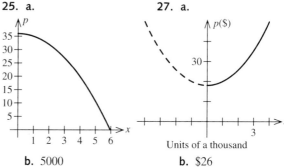

       b. 5000   **b.** \$26

29. 2500; \$67.50   **31.** 11,000; \$3

33. a.

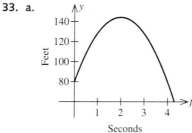

       b. $t = 2$; 144 ft

35. 3000

37. a.   **b.** \$30

39. 500; $32.50

41. a.

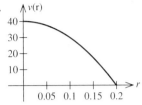

b. 0, 0.2. The velocity of blood is greatest along the central artery ($r = 0$) and smallest along the wall of the artery ($r = 0.2$). The maximum velocity is $v(0) = 40$ cm/sec and the minimum velocity is $v(0.2) = 0$ cm/sec.

Chapter 2 Review Exercises, page 110

1. $x = -2$ **2.** $y = 4$ **3.** $y = -\frac{1}{10}x + \frac{19}{5}$

4. $y = -\frac{4}{5}x + \frac{12}{5}$ **5.** $y = \frac{5}{2}x + 9$ **6.** $y = \frac{3}{4}x + \frac{11}{2}$

7. $y = -\frac{1}{2}x - 3$ **8.** $\frac{3}{5}$; $-\frac{6}{5}$ **9.** $y = -\frac{3}{4}x + \frac{9}{2}$

10. $y = -\frac{3}{5}x + \frac{12}{5}$ **11.** $y = -\frac{3}{2}x - 7$

12.

13.

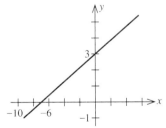

14. $(-\infty, 9]$ **15.** $(-\infty, -1) \cup (-1, \frac{3}{2}) \cup (\frac{3}{2}, \infty)$

16. a. 0 **b.** $3a^2 + 17a + 20$ **c.** $12a^2 + 10a - 2$
d. $3a^2 + 3h^2 + 6ah + 5a + 5h - 2$

17. a.   **b.** No **c.** Yes

18.

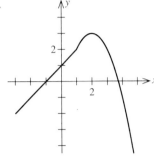

19. Vertex: $(\frac{11}{12}, -\frac{361}{24})$; x-intercepts: $-\frac{2}{3}, \frac{5}{2}$

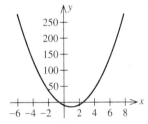

20. Vertex: $(\frac{1}{2}, 4)$; x-intercepts: $-\frac{1}{2}, \frac{3}{2}$

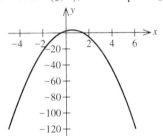

21. $(2, -3)$ **22.** $(6, \frac{21}{2})$ **23.** 60,000

24. a. $f(t) = t + 2.4$ **b.** $5.4 million **25. b.** 117 mg

26. a. $200,000/yr **b.** $4 million

27. a. $22,500/yr **b.** $V = -22,500t + 300,000$

28. a. $C(x) = 6x + 30,000$ **b.** $R(x) = 10x$
c. $P(x) = 4x - 30,000$ **d.** ($6,000); $2,000; $18,000

29. $p = -0.05x + 200$ **30.** $p = \frac{1}{36}x + \frac{400}{9}$

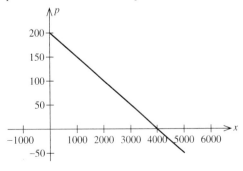

31. 2500 units; $50,000 **32.** 6000 units; $22

33. a.

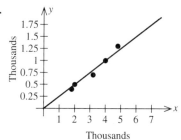

b. $y = 0.25x$ **c.** 1600

34. a. $V = -12,000n + 120,000$
b.

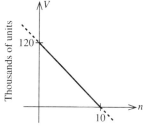

c. \$48,000 **d.** \$12,000/yr

35. a. \$620,000 **b.** \$540,000

CHAPTER 3

Exercises 3.1, page 118

1. a. 16 **b.** 27 **3. a.** 3 **b.** $\sqrt{5}$ **5. a.** -3 **b.** 8

7. a. $4x^3$ **b.** $5xy^2\sqrt{x}$ **9. a.** $\dfrac{2}{a^2}$ **b.** $\frac{1}{3}b^2$

11. a. $8x^9y^6$ **b.** $16x^4y^4z^6$ **13.** 2 **15.** 3 **17.** 3

19. $\frac{5}{4}$

21.

23.

25.

27.

29.

Using Technology Exercises 3.1, page 120

1.   **3.**

5.   **7.**

9.

Exercises 3.2, page 127

1. $\log_2 64 = 6$ **3.** $\log_3 \frac{1}{9} = -2$ **5.** $\log_{1/3} \frac{1}{3} = 1$

7. $\log_{32} 8 = \frac{3}{5}$ **9.** $\log_{10} 0.001 = -3$ **11.** 1.0792

13. 1.2042 **15.** 1.6813 **17.** $\log x + 4 \log(x + 1)$

19. $\frac{1}{2} \log(x + 1) - \log(x^2 + 1)$ **21.** $\ln x - x^2$

23. $-\frac{3}{2} \ln x - \frac{1}{2} \ln(1 + x^2)$ **25.** $x \ln x$ **27.** $x = 8$

29. $x = 3$ **31.** $x = 10$ **33.** $x = \frac{3}{2}$ **35.** $x = \frac{16}{7}$

37. $x = \frac{11}{3}$ **39.** $x = 3$

41.

43.

45.

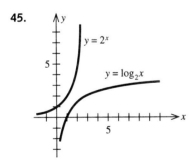

47. 5.1986 **49.** -0.0912 **51.** -8.0472 **53.** -4.9041

55. $-2 \ln \left(\dfrac{A}{B} \right)$ **57.** 105.7 mm

59. a. $10^3 I_0$ **b.** 100,000 times greater
c. 10,000,000 times greater

61. $34\frac{1}{2}$ hr earlier, at 1:30 P.M.

Exercises 3.3, page 135

1. a. 0.05 **b.** 400
c.

t	0	10	20	100	1000
Q	400	659	1087	59,365	2.07×10^{24}

3. a. $Q(t) = 100e^{0.035t}$ **b.** 266 minutes
c. $Q(t) = 1000e^{0.035t}$

5. a. 55.5 years **b.** 14.3 billion

7. 8.7 lb/sq in. **9.** $Q(t) = 100e^{-0.049t}$; 70.6 grams

11. 13,412 years ago

13. a. 60 words per minute **b.** 107 words per minute
c. 136 words per minute

15. a. 573 computers; 1177 computers; 1548 computers;
1925 computers
b. 2000 computers
c. 46 computers/month

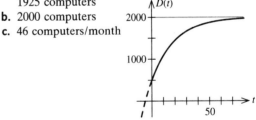

17. a. 11 **b.** 937 **c.** 1000

19. $\approx 3\%$; $\approx 65\%$

Chapter 3 Review Exercises, page 138

1.

2.

3.   **4.**

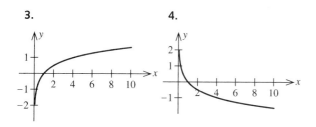

5. $\log_3 81 = 4$ **6.** $\log_9 3 = \frac{1}{2}$ **7.** $\log_{2/3} \frac{27}{8} = -3$

8. $\log_{16} 0.125 = -\frac{3}{4}$ **9.** 3.4011   **10.** 2.1972

11. 1.2809   **12.** 4.3174   **13.** $x = 3$ **14.** $x = -2, 1$

15. $x = -5$ **16.** $x = -2, 3$ **17.** $x = \frac{15}{2}$ **18.** $x = 2$

19. $x \approx 1.1610$ **20.** $x \approx -0.9464$ **21.** $x \approx -2.5025$

22. $x \approx -1.2528$ **23.** $x \approx 2.8332$ **24.** $x \approx 1.8195$

25. $x \approx 0.2409$ **26.** $x \approx 33.8672$ **27.** $x \approx \pm1.8934$

28. $x \approx 2.5903$ **29.** $x = -9.1629$ **30.** $x \approx 8.9588$

31. $x \approx 3.4657$ **32.** $x \approx -9.1629$

33.

$y = \log_2 (x + 3)$

34.

$y = \log_3 (x+1)$

35. a. $Q(t) = 2000e^{0.01831t}$ **b.** 161,992

36. $k \approx 0.0004$

37.   **a.** 1175; 2540; 3289   **b.** 4000

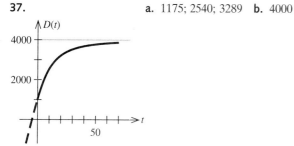

38. $\approx$ 970 students

Exercises 4.1, page 150

1. $80; $580   **3.** $836   **5.** $1000   **7.** 146 days

9. 10% per year   **11.** $1718.19   **13.** $4974.47

15. $27,566.93   **17.** $261,751.04   **19.** $214,986.69

21. $10\frac{1}{4}$% per year   **23.** 8.3% per year   **25.** $29,277.61

27. $30,255.95   **29.** $6885.64   **31.** 24% per year

33. $493   **35.** $131,080   **37.** $3.8 million

39. $26,267.49

41. a. $20,776.13   **b.** $20,331.70   **c.** $20,104.05

43. $23,329.48   **45.** 5.1289%

47. 8.6806%   **49.** Investment A

51. a. $33,885.14   **b.** $33,565.38

53. a. $16,262.79   **b.** $12,047.77   **c.** $6611.96

Technology Exercises 4.1, page 153

1.

Exercises 4.2, page 161

1. $15,937.42   **3.** $54,759.35   **5.** $37,965.57

7. $28,733.19   **9.** $15,558.61   **11.** $15,011.29

13. $109,658.91   **15.** $455.70   **17.** $44,526.45

19. $9850.12   **21.** $608.54

23. Between $120,141 and $143,927

25. Between $103,875 and $123,593

Exercises 4.3, page 168

1. $14,902.95   **3.** $444.24   **5.** $622.13   **7.** $731.79

9. $1491.19   **11.** $516.76   **13.** $172.95

15. $16,274.54   **17. a.** $212.27   **b.** $1316.36; $438.79

19. a. $387.21; $304.35   **b.** $1939.56; $2608.80

21. $587.01; $23,943.99; $29,819.93; $51,616.65

23. $3135.48   **25.** $1449.74   **27.** $2090.41; $4280.21

29. $33,835.20

Exercises 4.4, page 177

1. 30 **3.** -4.5 **5.** $-3, 8, 19, 30, 41$ **7.** $x + 6y$

9. 795 **11.** 792 **13.** 550 **15. a.** 275 **b.** -280

17. 37 weeks **19.** \$7.90 **21. b.** \$800

23. G.P.; 256; 508 **25.** Not a G.P. **27.** G.P.; 1/3; $364\frac{1}{3}$

29. 3; 0 **31.** 293,866 **33.** \$33,667.46

35. Annual raise of 8% per year

37. a. \$20,113.57 **b.** \$87,537.38 **39.** \$25,165.82

41. \$39,321.60; \$110,678.40

Using Technology Exercises 4.4, page 180

Chapter 4 Review Exercises, page 181

1. a. \$7320.50 **b.** \$7387.28 **c.** \$7422.53
   d. \$7446.77

2. a. \$19,859.95 **b.** \$20,018.07 **c.** \$20,100.14
   d. \$20,156.03

3. a. 12% **b.** 12.36% **c.** 12.5509% **d.** 12.6825%

4. a. 11.5% **b.** 11.8306% **c.** 12.0055% **d.** 12.1259%

5. \$30,000.29 **6.** \$39,999.95 **7.** \$5557.68

8. \$23,221.71 **9.** \$7861.70 **10.** \$173,804.43

11. \$694.49 **12.** \$318.93 **13.** \$332.73 **14.** \$208.44

15. 7.179% **16.** 9.563% **17.** \$2,592,702; \$8,612,002

18. \$5,491,922 **19.** \$15,000 **20.** 7.6% **21.** \$218.64

22. \$73,178.41 **23.** \$13,026.89 **24.** \$2000

25. a. \$965.55 **b.** \$227,598 **c.** \$42,684

26. a. \$1217.12 **b.** \$99,081.60 **c.** \$91,367

27. \$19,573.56 **28.** \$4727.67 **29.** \$205.09

30. 20.27% per year **31.** 12% per year **32.** \$80,000

CHAPTER 5

Exercises 5.1, page 190

1. Unique solution; $(2, 1)$ **3.** No solution

5. Unique solution; $(3, 2)$

7. Infinitely many solutions; $(t, \frac{2}{5}t - 2)$, t a parameter

9. Unique solution; $(1, -2)$ **11.** No solution

13. $k = -2$ **15.** $\begin{aligned} x + \ y &= \ \ \ 500 \\ 42x + 30y &= 18{,}600 \end{aligned}$

17. $\begin{aligned} x + \ y &= 100 \\ 2.5x + 3y &= 280 \end{aligned}$ **19.** $\begin{aligned} x + \ y &= 1000 \\ .25x + .75y &= 650 \end{aligned}$

21. $\begin{aligned} .06x + .08y + .12z &= 21{,}600 \\ z &= 2x \\ .12z &= .08y \end{aligned}$

23. $\begin{aligned} 8000x + 12{,}000y + 16{,}000z &= 1{,}000{,}000 \\ x &= 2y \\ x + \ y + \ z &= 100 \end{aligned}$

25. $\begin{aligned} 10x + \ 6y + \ 8z &= 100 \\ 10x + 12y + \ 6z &= 100 \\ 5x + \ 4y + 12z &= 100 \end{aligned}$

Exercises 5.2, page 204

1. $\begin{bmatrix} 2 & -3 & | & 7 \\ 3 & 1 & | & 4 \end{bmatrix}$ **3.** $\begin{bmatrix} 0 & -1 & 2 & | & 6 \\ 2 & 2 & -8 & | & 7 \\ 0 & 3 & 4 & | & 0 \end{bmatrix}$

5. $\begin{aligned} 3x + 2y &= -4 \\ x - \ y &= \ \ 5 \end{aligned}$ **7.** $\begin{aligned} x + 3y + 2z &= 4 \\ 2x \ \ \ \ \ \ \ \ \ &= 5 \\ 3x - 3y + 2z &= 6 \end{aligned}$

9. Yes **11.** No **13.** Yes **15.** No **17.** No

19. $\begin{bmatrix} 1 & 2 & | & 4 \\ 0 & -5 & | & -10 \end{bmatrix}$ **21.** $\begin{bmatrix} 1 & -2 & | & -3 \\ 0 & 16 & | & 20 \end{bmatrix}$

23. $\begin{bmatrix} 1 & 2 & 3 & | & 6 \\ 0 & -1 & -5 & | & -7 \\ 0 & -7 & -7 & | & -14 \end{bmatrix}$

25. $\begin{bmatrix} -6 & -11 & 0 & | & -5 \\ 2 & 4 & 1 & | & 3 \\ 1 & -2 & 0 & | & -10 \end{bmatrix}$

27. $\begin{bmatrix} 3 & 9 & | & 6 \\ 2 & 1 & | & 4 \end{bmatrix} \xrightarrow{\frac{1}{3}R_1} \begin{bmatrix} 1 & 3 & | & 2 \\ 2 & 1 & | & 4 \end{bmatrix}$

$\xrightarrow{R_2 - 2R_1} \begin{bmatrix} 1 & 3 & | & 2 \\ 0 & -5 & | & 0 \end{bmatrix} \xrightarrow{-\frac{1}{5}R_2}$

$\begin{bmatrix} 1 & 3 & | & 2 \\ 0 & 1 & | & 0 \end{bmatrix} \xrightarrow{R_1 - 3R_2} \begin{bmatrix} 1 & 0 & | & 2 \\ 0 & 1 & | & 0 \end{bmatrix}$

29.
$$\begin{bmatrix} 1 & 3 & 1 & | & 3 \\ 3 & 8 & 3 & | & 7 \\ 2 & -3 & 1 & | & -10 \end{bmatrix} \xrightarrow[R_3 - 2R_1]{R_2 - 3R_1}$$

$$\begin{bmatrix} 1 & 3 & 1 & | & 3 \\ 0 & -1 & 0 & | & -2 \\ 0 & -9 & -1 & | & -16 \end{bmatrix} \xrightarrow{-R_2}$$

$$\begin{bmatrix} 1 & 3 & 1 & | & 3 \\ 0 & 1 & 0 & | & 2 \\ 0 & -9 & -1 & | & -16 \end{bmatrix}$$

$$\xrightarrow[R_3 + 9R_2]{R_1 - 3R_2} \begin{bmatrix} 1 & 0 & 1 & | & -3 \\ 0 & 1 & 0 & | & 2 \\ 0 & 0 & -1 & | & 2 \end{bmatrix}$$

$$\xrightarrow[-R_3]{R_1 + R_3} \begin{bmatrix} 1 & 0 & 0 & | & -1 \\ 0 & 1 & 0 & | & 2 \\ 0 & 0 & 1 & | & -2 \end{bmatrix}$$

31. $(2, 0)$ **33.** $(-1, 2, -2)$ **35.** $(4, -2)$ **37.** $(-1, 2)$

39. $(\frac{7}{9}, -\frac{1}{9}, -\frac{2}{3})$ **41.** $(19, -7, -15)$ **43.** $(3, 0, 2)$

45. $(1, -2, 1)$ **47.** $(-20, -28, 13)$ **49.** $(4, -1, 3)$

51. 300 acres of corn, 200 acres of wheat

53. In 100 lb of blended coffee, use 40 lb of the $2.50/lb coffee and 60 lb of the $3/lb coffee.

55. 200 children and 800 adults

57. $40,000 in a savings account, $120,000 in mutual funds, $80,000 in money market certificates

59. 60 compact, 30 intermediate, and 10 full-size cars

61. $66,666.67 in high-risk stocks, $0 in medium-risk stocks, and $133,333.33 in low-risk stocks

63. 240 front orchestra seats, 560 rear orchestra seats, and 200 front balcony seats

Using Technology Exercises 5.2, page 210

1. $(3, 1, -1, 2)$ **3.** $(5, 4, -3, -4)$ **5.** $(1, -1, 2, 0, 3)$

Exercises 5.3, page 217

1. a. One solution   **b.** $(3, -1, 2)$

3. a. One solution   **b.** $(2, 4)$

5. a. Infinitely many solutions
b. $(4 - t, -2, t)$; t, a parameter

7. a. No solution

9. a. Infinitely many solutions
b. $(2, -1, 2 - t, t)$; t, a parameter

11. a. Infinitely many solutions
b. $(2 - 3s, 1 + s, s, t)$; s, t, parameters

13. $(2, 1)$ **15.** No solution   **17.** $(1, -1)$

19. $(2 + 2t, t)$; t, a parameter

21. $(\frac{4}{3} - \frac{2}{3}t, t)$; t, a parameter

23. $(-2 + \frac{1}{2}s - \frac{1}{2}t, s, t)$; s, t, parameters

25. $(-1, \frac{17}{7}, \frac{23}{7})$

27. $(1 - \frac{1}{4}s + \frac{1}{4}t, s, t)$; s, t, parameters

29. No solution   **31.** $(2, -1, 4)$

33. $x = 20 + z$, $y = 40 - 2z$; 25 compact cars, 30 mid-sized cars, and 5 full-sized cars; 30 compact cars, 20 mid-sized cars, and 10 full-sized cars

37. a.
$$\begin{array}{rcl} x_1 - x_2 & & = 200 \\ x_1 & - x_5 & = 100 \\ - x_2 + x_3 & + x_6 & = 600 \\ - x_3 + x_4 & & = 200 \\ x_4 - x_5 + x_6 & & = 700 \end{array}$$

b. $x_1 = s + 100$
$x_2 = s - 100$
$x_3 = s - t + 500$
$x_4 = s - t + 700$
$x_5 = s$
$x_6 = t$
$(250, 50, 600, 800, 150, 50)$
$(300, 100, 600, 800, 200, 100)$
c. $(300, 100, 700, 900, 200, 0)$

39. $k = 6$; $(-2, 2)$

Using Technology Exercises 5.3, page 221

1. $(1 + t, 2 + t, t)$; t, a parameter

3. $(-\frac{17}{7} + \frac{6}{7}t, 3 - t, -\frac{18}{7} + \frac{1}{7}t, t)$ **5.** No solution

Exercises 5.4, page 228

1. 4×4; 4×3; 1×5; 4×1 **3.** 2; 3; 8

5. D; $D^T = \begin{bmatrix} 1 & 3 & -2 & 0 \end{bmatrix}$

7. 3×2; 3×2; 3×3; 3×3

9. $\begin{bmatrix} 1 & 6 \\ 6 & -1 \\ 2 & 2 \end{bmatrix}$ **11.** $\begin{bmatrix} 1 & 1 & -4 \\ -1 & -8 & 1 \\ 6 & 3 & 1 \end{bmatrix}$

13. $\begin{bmatrix} 3 & 5 & 9 \\ 4 & 10 & 13 \end{bmatrix}$ **15.** $\begin{bmatrix} 3 & -4 & -16 \\ 17 & -4 & 16 \end{bmatrix}$

17. $\begin{bmatrix} -1.9 & 3.0 & -0.6 \\ 6.0 & 9.6 & 1.2 \end{bmatrix}$

19. $\begin{bmatrix} \frac{7}{2} & 3 & -1 & \frac{10}{3} \\ -\frac{19}{6} & \frac{2}{3} & -\frac{17}{2} & \frac{23}{3} \\ \frac{29}{3} & \frac{17}{6} & -1 & -2 \end{bmatrix}$

21. $u = 3$, $x = \frac{5}{2}$, $y = 7$, and $z = 2$

23. $u = 15$, $x = 2$, $y = 2$, and $z = -\frac{7}{3}$

31. $\begin{bmatrix} 3 \\ 2 \\ -1 \\ 5 \end{bmatrix}$　**33.** $\begin{bmatrix} 1 & 3 & 0 \\ -1 & 4 & 1 \\ 2 & 2 & 0 \end{bmatrix}$

35. $\begin{bmatrix} 220 & 215 & 210 & 205 \\ 220 & 210 & 200 & 195 \\ 215 & 205 & 195 & 190 \end{bmatrix}$

37. a. $D = \begin{bmatrix} 2960 & 1510 & 1150 \\ 1100 & 550 & 490 \\ 1230 & 590 & 470 \end{bmatrix}$

b. $E = \begin{bmatrix} 3256 & 1661 & 1265 \\ 1210 & 605 & 539 \\ 1353 & 649 & 517 \end{bmatrix}$

Using Technology Exercises 5.4, page 231

1. $\begin{bmatrix} 15 & 38.75 & -67.5 & 33.75 \\ 51.25 & 40 & 52.5 & -38.75 \\ 21.25 & 35 & -65 & 105 \end{bmatrix}$

3. $\begin{bmatrix} -5 & 6.3 & -6.8 & 3.9 \\ 1 & 0.5 & 5.4 & -4.8 \\ 0.5 & 4.2 & -3.5 & 5.6 \end{bmatrix}$

5. $\begin{bmatrix} 16.44 & -3.65 & -3.66 & 0.63 \\ 12.77 & 10.64 & 2.58 & 0.05 \\ 5.09 & 0.28 & -10.84 & 17.64 \end{bmatrix}$

7. $\begin{bmatrix} 7.4 & 7.2 & 2.9 \\ -0.1 & 5.9 & 1.4 \\ -4 & 3 & -6.9 \\ 1.5 & -1.4 & 11.2 \end{bmatrix}$

Exercises 5.5, page 240

1. 2×5; not defined　**3.** 1×1; 7×7

5. $n = s$ and $m = t$　**7.** $\begin{bmatrix} -1 \\ 3 \end{bmatrix}$　**9.** $\begin{bmatrix} 9 \\ -10 \end{bmatrix}$

11. $\begin{bmatrix} 4 & -2 \\ 9 & 13 \end{bmatrix}$　**13.** $\begin{bmatrix} 2 & 9 \\ 5 & 16 \end{bmatrix}$　**15.** $\begin{bmatrix} 0.57 & 1.93 \\ 0.64 & 1.76 \end{bmatrix}$

17. $\begin{bmatrix} 6 & -3 & 0 \\ -2 & 1 & -8 \\ 4 & -4 & 9 \end{bmatrix}$　**19.** $\begin{bmatrix} 5 & 1 & -3 \\ 1 & 7 & -3 \end{bmatrix}$

21. $\begin{bmatrix} -4 & -20 & 4 \\ 4 & 12 & 0 \\ 12 & 32 & 20 \end{bmatrix}$　**23.** $\begin{bmatrix} 4 & -3 & 2 \\ 7 & 1 & -5 \end{bmatrix}$

27. $AB = \begin{bmatrix} 10 & 7 \\ 22 & 15 \end{bmatrix}$, $BA = \begin{bmatrix} 5 & 8 \\ 13 & 20 \end{bmatrix}$

31. $A = \begin{bmatrix} -2 & -1 \\ 5 & 2 \end{bmatrix}$　**33. a.** $A^T = \begin{bmatrix} 2 & 5 \\ 4 & -6 \end{bmatrix}$

35. $AX = B$, where $A = \begin{bmatrix} 2 & -3 \\ 3 & -4 \end{bmatrix}$,

$X = \begin{bmatrix} x \\ y \end{bmatrix}$, $B = \begin{bmatrix} 7 \\ 8 \end{bmatrix}$

37. $AX = B$, where $A = \begin{bmatrix} 2 & -3 & 4 \\ 0 & 2 & -3 \\ 1 & -1 & 2 \end{bmatrix}$,

$X = \begin{bmatrix} x \\ y \\ z \end{bmatrix}$, and $B = \begin{bmatrix} 6 \\ 7 \\ 4 \end{bmatrix}$

39. $AX = B$, where $A = \begin{bmatrix} -1 & 1 & 1 \\ 2 & -1 & -1 \\ -3 & 2 & 4 \end{bmatrix}$,

$X = \begin{bmatrix} x_1 \\ x_2 \\ x_3 \end{bmatrix}$, and $B = \begin{bmatrix} 0 \\ 2 \\ 4 \end{bmatrix}$

41. $\begin{bmatrix} 2 \\ 3 \\ 4 \end{bmatrix}$; $\begin{bmatrix} 980 \\ 1590 \\ 1255 \\ 1650 \end{bmatrix}$; $5475

43.

    Dem.   Rep.   Indep.

[41,000   35,000   14,000]

45. a.

              Plush   S. Mat.   Trim

L.A. $\begin{bmatrix} 4{,}410 & 64{,}000 & 25{,}400 \\ 3{,}350 & 49{,}000 & 19{,}000 \end{bmatrix}$
Seattle

b. L.A. $\begin{bmatrix} 32{,}595 \\ 24{,}725 \end{bmatrix}$ **c.** $57,320
Seattle

Using Technology Exercises 5.5, page 244

1. $\begin{bmatrix} 18.66 & 15.2 & -12 \\ 24.48 & 41.88 & 89.92 \\ 15.39 & 7.16 & -1.25 \end{bmatrix}$

3. $\begin{bmatrix} 20.09 & 20.61 & -1.3 \\ 44.42 & 71.6 & 64.89 \\ 20.97 & 7.17 & -60.65 \end{bmatrix}$

5. $\begin{bmatrix} 32.89 & 13.63 & -57.17 \\ -12.85 & -8.37 & 256.92 \\ 13.48 & 14.29 & 181.64 \end{bmatrix}$

7. $\begin{bmatrix} 18.66 & 24.48 & 15.39 \\ 15.2 & 41.88 & 7.16 \\ -12 & 89.82 & -1.25 \end{bmatrix}$

9. $\begin{bmatrix} 87 & 68 & 110 & 82 \\ 119 & 176 & 221 & 143 \\ 51 & 128 & 142 & 94 \\ 28 & 174 & 174 & 112 \end{bmatrix}$;

$\begin{bmatrix} 113 & 117 & 72 & 101 & 90 \\ 72 & 85 & 36 & 72 & 76 \\ 81 & 69 & 76 & 87 & 30 \\ 133 & 157 & 56 & 121 & 146 \\ 154 & 157 & 94 & 127 & 122 \end{bmatrix}$

11. $\begin{bmatrix} 170 & 18.1 & 133.1 & -106.3 & 341.3 \\ 349 & 226.5 & 324.1 & 164 & 506.4 \\ 245.2 & 157.7 & 231.5 & 125.5 & 312.9 \\ 310 & 245.2 & 291 & 274.3 & 354.2 \end{bmatrix}$

Exercises 5.6, page 253

5. $\begin{bmatrix} 3 & -5 \\ -1 & 2 \end{bmatrix}$ **7.** Does not exist

9. $\begin{bmatrix} 2 & -11 & -3 \\ 1 & -6 & -2 \\ 0 & -1 & 0 \end{bmatrix}$ **11.** Does not exist

13. $\begin{bmatrix} -\frac{13}{10} & \frac{7}{5} & \frac{1}{2} \\ \frac{2}{5} & -\frac{1}{5} & 0 \\ -\frac{7}{10} & \frac{3}{5} & \frac{1}{2} \end{bmatrix}$ **15.** $\begin{bmatrix} 3 & 4 & -6 & 1 \\ -2 & -3 & 5 & -1 \\ -4 & -4 & 7 & -1 \\ -4 & -5 & 8 & -1 \end{bmatrix}$

17. a. $A = \begin{bmatrix} 2 & 5 \\ 1 & 3 \end{bmatrix}$, $X = \begin{bmatrix} x \\ y \end{bmatrix}$, $B = \begin{bmatrix} 3 \\ 2 \end{bmatrix}$

    b. $x = -1, y = 1$

19. a. $A = \begin{bmatrix} 2 & -3 & -4 \\ 0 & 0 & -1 \\ 1 & -2 & 1 \end{bmatrix}$, $X = \begin{bmatrix} x \\ y \\ z \end{bmatrix}$, $B = \begin{bmatrix} 4 \\ 3 \\ -8 \end{bmatrix}$

    b. $x = -1, y = 2$, and $z = -3$

21. a. $A = \begin{bmatrix} 1 & 4 & -1 \\ 2 & 3 & -2 \\ -1 & 2 & 3 \end{bmatrix}$, $X = \begin{bmatrix} x \\ y \\ z \end{bmatrix}$, $B = \begin{bmatrix} 3 \\ 1 \\ 7 \end{bmatrix}$

    b. $x = 1, y = 1$, and $z = 2$

23. a. $A = \begin{bmatrix} 1 & 1 & -1 & 1 \\ 2 & 1 & 1 & 0 \\ 2 & 1 & 0 & 1 \\ 2 & -1 & -1 & 3 \end{bmatrix}$,

$X = \begin{bmatrix} x_1 \\ x_2 \\ x_3 \\ x_4 \end{bmatrix}$, $B = \begin{bmatrix} 6 \\ 4 \\ 7 \\ 9 \end{bmatrix}$

    b. $x_1 = 1, x_2 = 2, x_3 = 0$, and $x_4 = 3$

25. a. $x = \frac{24}{5}, y = \frac{23}{5}$
   b. $x = \frac{2}{5}, y = \frac{9}{5}$

27. a. $x = -1, y = 3, z = 2$
   b. $x = 1, y = 8, z = -12$

29. a. $x = -\frac{2}{17}, y = -\frac{10}{17}, z = -\frac{60}{17}$
   b. $x = 1, y = 0, z = -5$

31. a. $x_1 = 1, x_2 = -4, x_3 = 5, x_4 = -1$
   b. $x_1 = 12, x_2 = -24, x_3 = 21, x_4 = -7$

33. a. $A^{-1} = \begin{bmatrix} -\frac{5}{2} & -\frac{3}{2} \\ 2 & 1 \end{bmatrix}$

35. a. $ABC = \begin{bmatrix} 4 & 10 \\ 2 & 3 \end{bmatrix}, \quad A^{-1} = \begin{bmatrix} 3 & -5 \\ 1 & -2 \end{bmatrix},$

$B^{-1} = \begin{bmatrix} 1 & -3 \\ -1 & 4 \end{bmatrix}, \quad C^{-1} = \begin{bmatrix} \frac{1}{8} & -\frac{3}{8} \\ \frac{1}{4} & \frac{1}{4} \end{bmatrix}$

37. a. 3214; 3929   **b.** 4286; 3571   **c.** 3929; 5357

39. a. \$6 million to Organization I, \$10 million to Organization II, and \$8 million to Organization III
   b. \$8 million to Organization I, \$6 million to Organization II, and \$5 million to Organization III

Using Technology Exercises 5.6, page 257

1. $\begin{bmatrix} -0.36 & 0.04 & -0.36 \\ 0.06 & 0.05 & 0.20 \\ -0.19 & 0.10 & 0.09 \end{bmatrix}$

3. $\begin{bmatrix} 0.01 & -0.09 & 0.31 & -0.11 \\ -0.25 & 0.58 & -0.15 & -0.02 \\ 0.86 & -0.42 & 0.07 & -0.37 \\ -0.27 & 0.01 & -0.05 & 0.31 \end{bmatrix}$

Chapter 5 Review Exercises, page 258

1. $\begin{bmatrix} 2 & 2 \\ -1 & 4 \\ 3 & 3 \end{bmatrix}$ **2.** $\begin{bmatrix} -2 & 0 \\ -2 & 6 \end{bmatrix}$ **3.** $\begin{bmatrix} -6 & -2 \end{bmatrix}$ **4.** $\begin{bmatrix} 17 \\ 13 \end{bmatrix}$

5. $x = 2, y = 3, z = 1$, and $w = 3$ **6.** $x = 2$ and $y = -2$

7. $a = 3, b = 4, c = -2, d = 2$, and $e = -3$

8. $x = -1, y = -2$, and $z = 1$

9. $\begin{bmatrix} 8 & 9 & 11 \\ -10 & -1 & 3 \\ 11 & 12 & 10 \end{bmatrix}$ **10.** $\begin{bmatrix} -1 & 7 & -3 \\ -2 & 5 & 11 \\ 10 & -8 & 2 \end{bmatrix}$

11. $\begin{bmatrix} 6 & 18 & 6 \\ -12 & 6 & 18 \\ 24 & 0 & 12 \end{bmatrix}$ **12.** $\begin{bmatrix} -10 & 10 & -18 \\ 4 & 14 & 26 \\ 16 & -32 & -4 \end{bmatrix}$

13. $\begin{bmatrix} -11 & -16 & -15 \\ -4 & -2 & -10 \\ -6 & 14 & 2 \end{bmatrix}$ **14.** $\begin{bmatrix} 5 & 20 & 19 \\ -2 & 20 & 8 \\ 26 & 10 & 30 \end{bmatrix}$

15. $\begin{bmatrix} -3 & 17 & 8 \\ -2 & 56 & 27 \\ 74 & 78 & 116 \end{bmatrix}$ **16.** $\begin{bmatrix} \frac{3}{2} & -2 & -5 \\ \frac{11}{2} & -1 & 11 \\ \frac{7}{2} & -3 & 0 \end{bmatrix}$

17. $x = 1, y = -1$ **18.** $x = -1$ and $y = 3$

19. $x = 1, y = 2$, and $z = 3$

20. $(2, 2t - 5, t)$; t, a parameter   **21.** No solution

22. $x = 1, y = -1, z = 2$, and $w = 2$

23. $x = 1, y = 0$, and $z = 1$

24. $\begin{bmatrix} \frac{3}{4} & -\frac{1}{2} \\ -\frac{1}{8} & \frac{1}{4} \end{bmatrix}$ **25.** $\begin{bmatrix} \frac{2}{5} & -\frac{1}{5} \\ -\frac{1}{5} & \frac{3}{5} \end{bmatrix}$

26. $\begin{bmatrix} \frac{1}{4} & \frac{1}{2} \\ \frac{1}{8} & -\frac{1}{4} \end{bmatrix}$ **27.** $\begin{bmatrix} -1 & 2 \\ 1 & -\frac{3}{2} \end{bmatrix}$

28. $\begin{bmatrix} -\frac{1}{4} & \frac{1}{2} & -\frac{1}{4} \\ \frac{7}{8} & -\frac{3}{4} & -\frac{5}{8} \\ -\frac{1}{8} & \frac{1}{4} & \frac{3}{8} \end{bmatrix}$ **29.** $\begin{bmatrix} \frac{5}{4} & \frac{1}{4} & -\frac{7}{4} \\ -\frac{1}{4} & -\frac{1}{4} & \frac{3}{4} \\ -\frac{3}{4} & \frac{1}{4} & \frac{5}{4} \end{bmatrix}$

30. $\begin{bmatrix} 0 & -\frac{1}{5} & \frac{2}{5} \\ -2 & 1 & 1 \\ -1 & \frac{1}{5} & \frac{3}{5} \end{bmatrix}$ **31.** $\begin{bmatrix} -\frac{1}{5} & \frac{2}{5} & 0 \\ \frac{2}{3} & -\frac{1}{3} & \frac{1}{3} \\ -\frac{1}{30} & \frac{1}{15} & -\frac{1}{6} \end{bmatrix}$

32. $\begin{bmatrix} \frac{11}{24} & -\frac{7}{8} \\ -\frac{1}{12} & \frac{1}{4} \end{bmatrix}$ **33.** $\begin{bmatrix} \frac{3}{2} & 1 \\ -\frac{7}{2} & -1 \end{bmatrix}$

34. $\begin{bmatrix} \frac{4}{7} & -\frac{3}{7} \\ -\frac{3}{7} & \frac{4}{7} \end{bmatrix}$ **35.** $\begin{bmatrix} \frac{2}{5} & -\frac{3}{5} \\ \frac{1}{5} & \frac{1}{5} \end{bmatrix}$

36. $A^{-1} = \begin{bmatrix} \frac{2}{5} & \frac{3}{10} \\ -\frac{1}{5} & \frac{1}{10} \end{bmatrix}$; $x = 2, y = 1$

37. $A^{-1} = \begin{bmatrix} \frac{2}{7} & \frac{3}{7} \\ \frac{1}{7} & -\frac{2}{7} \end{bmatrix}$; $x = -1, y = -2$

38. $A^{-1} = \begin{bmatrix} 0 & \frac{1}{7} & \frac{2}{7} \\ -1 & -\frac{4}{7} & \frac{6}{7} \\ -\frac{1}{2} & -\frac{1}{2} & \frac{1}{2} \end{bmatrix}$; $x = 3, y = -1, z = 2$

39. $A^{-1} = \begin{bmatrix} 1 & -\frac{2}{5} & \frac{4}{5} \\ -1 & 1 & -1 \\ -\frac{1}{2} & \frac{3}{5} & -\frac{7}{10} \end{bmatrix}$; $x = 1, y = 2,$ and $z = 4$

40. $2,300,000$; $2,450,000$; an increase of $150,000

41. $5180, $4540, and $5550

42. Houston: 100,000 gallons; Tulsa: 600,000 gallons

43. 30 of each type

CHAPTER 6

Exercises 6.1, page 267

1.

3.

5.

7.

9.
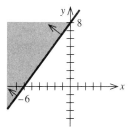

11. $x \geq 1, x \leq 5, y \geq 2,$ and $y \leq 4$

13. $2x - y \geq 2, 5x + 7y \geq 35,$ and $x \leq 4$

15. $x - y \geq -10, 7x + 4y \leq 140,$ and $x + 3y \geq 30$

17. $x + y \geq 7, x \geq 2, y \geq 3, y \leq 7$

19.

Unbounded

21.

Unbounded

23.

Bounded   No solution

25.

Bounded

27.

Unbounded

29.

Unbounded

31.

Bounded

33.

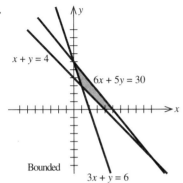

$x + y = 4$

$6x + 5y = 30$

Bounded

$3x + y = 6$

35.

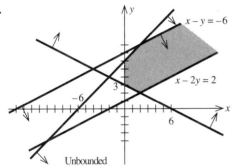

$x - y = -6$

$x - 2y = 2$

Unbounded

Exercises 6.2, page 276

1. Maximize $P = 3x + 4y$ subject to
$$6x + 9y \le 300$$
$$5x + 4y \le 180$$
$$x \ge 0, y \ge 0$$

3. Maximize $P = 2x + (3/2)y$ subject to
$$3x + 4y \le 1000$$
$$6x + 3y \le 1200$$
$$x \ge 150, y \ge 0$$

5. Minimize $C = 2x + 5y$ subject to
$$30x + 25y \ge 400$$
$$x + 0.5y \ge 10$$
$$2x + 5y \ge 40$$
$$x \ge 0, y \ge 0$$

7. Minimize $C = 32{,}000 - x - 3y$ subject to
$$x + y \le 6000$$
$$x + y \ge 2000$$
$$x \le 3000$$
$$y \le 4000$$
$$x \ge 0, y \ge 0$$

9. Maximize $P = 18x + 12y + 15z$ subject to
$$2x + y + 2z \le 900$$
$$3x + y + 2z \le 1080$$
$$2x + 2y + z \le 840$$
$$x \ge 0, y \ge 0, z \ge 0$$

11. Maximize $P = 26x + 28y + 24z$ subject to
$$\tfrac{5}{4}x + \tfrac{3}{2}y + \tfrac{3}{2}z \le 310$$
$$x + y + \tfrac{3}{4}z \le 205$$
$$x + y + \tfrac{1}{2}z \le 190$$
$$x \ge 0, y \ge 0, z \ge 0$$

13. Minimize $C = 60x_1 + 60x_2 + 80x_3 + 80x_4$
$+ 70x_5 + 50x_6$ subject to
$$x_1 + x_2 + x_3 \le 300$$
$$x_4 + x_5 + x_6 \le 250$$
$$x_1 + x_4 \ge 200$$
$$x_2 + x_5 \ge 150$$
$$x_3 + x_6 \ge 200$$
$$x_1 \ge 0, x_2 \ge 0, \ldots, x_6 \ge 0$$

15. Maximize $P = 180x + 200y + 300z$ subject to
$$2.5x + 3y + 4z \le 70$$
$$x \le 9$$
$$y \le 12$$
$$z \le 6$$
$$x \ge 0, y \ge 0, z \ge 0$$

Exercises 6.3, page 286

1. Max: 35; min: 5   **3.** No max. value; min: 27

5. Max: 44; min: 15   **7.** $x = 0$, $y = 6$, $P = 18$

9. $x = 14$, $y = 3$, $C = 58$

11. Any point (x, y) lying on the line segment joining $(5, 20)$ to $(12, 6)$, $C = 90$

13. $x = 3$, $y = 3$, $C = 75$

15. $x = 15$, $y = 17.5$, $P = 115$

17. $x = 10$, $y = 38$, $P = 134$

19. Max: $x = 6$, $y = 33/2$, $P = 258$
Min: $x = 15$, $y = 3$, $P = 186$

21. Max: $x = 5$, $y = 15$, $P = 70$
Min: $x = 0$, $y = 5$, $P = 20$

23. No model A, 2000 model B; $P = \$80{,}000$

25. 120 model A, 160 model B; $P = \$480$

27. $16 million in homeowner loans, $4 million in auto loans; $P = \$2.08$ million

29. 65 acres of crop A, 80 acres of crop B, $P = \$25,750$

31. 2000 from I to A, 4000 from I to B, 1000 from II to A, 0 from II to B; $C = \$18,000$

33. $1 million to country A; $1.5 million to country B

35. 3 oz brand A; 4 oz brand B; $0.25 per serving

Exercises 6.4, page 306

1. In final form; $x = \frac{30}{7}$, $y = \frac{20}{7}$, $u = 0$, $v = 0$, $P = \frac{220}{7}$

3. Not in final form; pivot element is $\frac{1}{2}$, lying in the first row, second column

5. In final form: $x = \frac{1}{3}$, $y = 0$, $z = \frac{13}{3}$, $u = 0$, $v = 6$, $w = 0$, $P = 17$

7. Not in final form; pivot element is 1, lying in the third row, second column

9. In final form; $x = 30$, $y = 0$, $z = 0$, $u = 10$, $v = 0$, $P = 60$ and $x = 0$, $y = 30$, $z = 0$, $u = 10$, $v = 0$, $P = 60$

11. $x = 6$, $y = 3$, $u = 0$, $v = 0$, $P = 96$

13. $x = 6$, $y = 6$, $u = 0$, $v = 0$, $w = 0$, $P = 60$

15. $x = 0$, $y = 4$, $z = 4$, $u = 0$, $v = 0$, $P = 36$

17. $x = 0$, $y = 3$, $z = 0$, $u = 90$, $v = 0$, $w = 75$, $P = 12$

19. $x = 15$, $y = 3$, $z = 0$, $u = 2$, $v = 0$, $w = 0$, $P = 78$

21. $x = \frac{5}{4}$, $y = \frac{15}{2}$, $z = 0$, $u = 0$, $v = \frac{15}{4}$, $w = 0$, $P = 90$

23. $x = \frac{15}{7}$, $y = 0$, $z = \frac{39}{7}$, $s = 0$, $t = \frac{1}{7}$, $u = 0$, $v = \frac{2}{7}$, $P = 36\frac{3}{7}$

27. No model A, 2000 model B; $P = \$80,000$

29. 65 acres of crop A, 80 acres of crop B; $P = \$25,750$

31. 180 units of product A; 140 units of product B; 200 units of product C; $P = \$7920$

33. $50,000 in growth funds; $50,000 in balanced funds, and $100,000 in income funds; maximum return: $17,000

35. 300 standard models; 300 deluxe units; 400 luxury models; $P = \$4,220,000$

Using Technology Exercises 6.4, page 311

1. $x = 1.2$, $y = 0$, $z = 1.6$, $w = 0$, and $P = 8.8$

3. $x = 1.6$, $y = 0$, $z = 0$, $w = 3.6$, and $P = 12.4$

Exercises 6.5, page 323

1. $x = 4$, $y = 0$, $C = -8$

3. $x = 4$, $y = 3$, $C = -18$

5. $x = 0$, $y = 13$, $z = 18$, $w = 14$, $C = -111$

7. $x = \frac{5}{4}$, $y = \frac{1}{4}$, $u = 2$, $v = 3$, and $C = P = 13$

9. $x = 5$, $y = 10$, $z = 0$, $u = 1$, $v = 2$, and $C = P = 80$

11. Maximize $P = 4u + 6v$ subject to
$$u + 3v \le 2$$
$$2u + 2v \le 5; \quad x = 4, y = 0, C = 8$$
$$u \ge 0, v \ge 0$$

13. Maximize $P = 60u + 40v + 30w$ subject to
$$6u + 2v + w \le 6$$
$$u + v + w \le 4; \quad x = 10, y = 20, C = 140$$
$$u \ge 0, v \ge 0, w \ge 0$$

15. Maximize $P = 10u + 20v$ subject to
$$20u + v \le 200$$
$$10u + v \le 150; \quad x = 0, y = 0, z = 10, C = 1200$$
$$u + 2v \le 120$$
$$u \ge 0, v \ge 0$$

17. Maximize $P = 10u + 24v + 16w$ subject to
$$u + 2v + w \le 6$$
$$2u + v + w \le 8; \quad x = 8, y = 0, z = 8, C = 80$$
$$2u + v + w \le 4$$
$$u \ge 0, v \ge 0, w \ge 0$$

19. Maximize $P = 6u + 2v + 4w$ subject to
$$2u + 6v \le 30$$
$$4u + 6w \le 12; \quad x = \frac{1}{3}, y = \frac{4}{3}, z = 0, C = 26$$
$$3u + v + 2w \le 20$$
$$u \ge 0, v \ge 0, w \ge 0$$

21. Loc. I: 500 to warehouse A, 200 to warehouse B; Loc. II: 200 to warehouse B, 400 to warehouse C; $C = \$20,800$

23. 8 oz orange juice; 6 oz pink grapefruit juice; 178 calories

Using Technology Exercises 6.5, page 327

1. $x = \frac{4}{3}$, $y = \frac{10}{3}$, $z = 0$, and $C = \frac{14}{3}$

3. $x = 0.9524$, $y = 4.2857$, $z = 0$, and $C = 6.0952$

Chapter 6 Review Exercises, page 328

1. Max: 18—any point (x, y) lying on the line segment joining $(0, 6)$ to $(3, 4)$; Min: 0

2. Max: 27; Min: 7   **3.** $x = 0$, $y = 4$, $P = 20$

4. $x = 0$, $y = 12$, $P = 36$

5. $x = 0, y = 0, C = 0$ **6.** $x = 1.25, y = 1.5, C = 9.75$

7. $x = 3, y = 10, P = 29$ **8.** $x = 8, y = 0, P = 48$

9. $x = 20, y = 0, C = 40$ **10.** $x = 2, y = 6$, and $C = 14$

11. Max: $x = 22, y = 0, Q = 22$; Min: $x = 3, y = \frac{5}{2}$, $Q = \frac{11}{2}$

12. Max: $x = 12, y = 6, Q = 54$; Min: $x = 4, y = 0$, $Q = 8$

13. $x = 3, y = 4, u = 0, v = 0, P = 25$

14. $x = 3, y = 6, u = 4, v = 0, w = 0, P = 36$

15. $x = \frac{56}{5}, y = \frac{2}{5}, z = 0, u = 0, v = 0, P = 23\frac{3}{5}$

16. $x = 0, y = \frac{11}{3}, z = \frac{25}{6}, u = \frac{37}{6}, v = 0, w = 0, P = \frac{119}{6}$

17. $x = \frac{3}{2}; y = 1, C = \frac{13}{2}$

18. $x = \frac{32}{11}, y = \frac{36}{11}, c = \frac{104}{11}$

19. $x = \frac{3}{4}, y = 0, z = \frac{7}{4}, C = 60$

20. $x = 0, y = 2, z = 0, C = 4$

21. $x = 45, y = 0, P = 135$

22. $x = 4, y = 4, C = 20$

23. $x = 5, y = 2, P = 16$

24. $x = 20, y = 25, C = -160$

25. $40,000 in each company; $P = \$13,600$

26. 60 model A clock-radios; 60 model B clock-radios; $P = \$1320$

27. 93 model A, 180 model B; $P = \$456$

28. 0 units product A, 30 units product B, 0 units product C; $P = \$180$

29. $70,000 in blue-chip stocks; $0 in growth stocks; $30,000 in speculative stocks; maximum return: $13,000

30. $50,000 in stocks, $100,000 in bonds, $50,000 in money market funds; $P = \$21,500$

CHAPTER 7

Exercises 7.1, page 338

1. $\{x \mid x$ is a gold medalist in the 1992 Summer Olympic Games$\}$

3. $\{x \mid x$ is an integer greater than 2 and less than 8$\}$

5. $\{2, 3, 4, 5, 6\}$ **7.** $\{-2\}$ **9. a.** T **b.** F

11. a. F **b.** F **13.** T **15. a.** T **b.** F **17. a** and **b**

19. a. $\varnothing, \{1\}, \{2\}, \{1, 2\}$
 b. $\varnothing, \{1\}, \{2\}, \{3\}, \{1, 2\}, \{1, 3\}, \{2, 3\}, \{1, 2, 3\}$
 c. $\varnothing, \{1\}, \{2\}, \{3\}, \{4\}, \{1, 2\}, \{1, 3\}, \{1, 4\}, \{2, 3\}$, $\{2, 4\}, \{3, 4\}, \{1, 2, 3\}, \{1, 2, 4\}, \{1, 3, 4\}, \{2, 3, 4\}$, $\{1, 2, 3, 4\}$

21. $\{1, 2, 3, 4, 6, 8, 10\}$

23. $\{$Jill, John, Jack, Susan, Sharon$\}$

25. a.

b.

c.

27. a.

b.

29. a.

b.

31. a.

b.

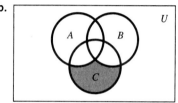

33. a. {2, 4, 6, 8, 10} **b.** {1, 2, 4, 5, 6, 8, 9, 10} **c.** U

35. a. $C = \{1, 2, 4, 5, 8, 9\}$ **b.** $\varnothing$ **c.** U

37. a. Not disjoint **b.** Disjoint

39. a. The set of all employees at the Universal Life Insurance Company who do not drink tea
   b. The set of all employees at the Universal Life Insurance Company who do not drink coffee

41. a. The set of all employees at the Universal Life Insurance Company who drink tea but not coffee
   b. The set of all employees at the Universal Life Insurance Company who drink coffee but not tea

43. a. The set of all employees in a hospital who are not doctors
   b. The set of all employees in a hospital who are not nurses

45. a. The set of all employees in a hospital who are female doctors
   b. The set of all employees in a hospital who are both doctors and administrators

47. a. $D \cap F$ **b.** $R \cap F^C \cap L^C$

49. a. B^C **b.** $A \cap B$ **c.** $A \cap B \cap C^C$

51.

53.

  =

55.

  =

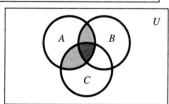

59. a. x, y, v, r, w, u **b.** v, r

61. a. s, t, y **b.** t, z, w, x, s **63.** $A \subset C$

Exercises 7.2, page 345

3. a. 4 **b.** 5 **c.** 7 **d.** 2 **7.** 3

9. a. 40 **b.** 120 **c.** 60 **11.** 11

13. 2 **15.** 600; 100 **17.** 61

19. a. 106 **b.** 64 **c.** 38 **d.** 14

21. a. 182 **b.** 118 **c.** 56 **d.** 18

23. a. 13 **b.** 15 **25. a.** 38 **b.** 64 **27.** 5

29. a. 62 **b.** 33 **c.** 25 **d.** 38

31. a. 108 **b.** 15 **c.** 45 **d.** 12

Exercises 7.3, page 352

1. 12 **3.** 64 **5.** 24 **7.** 24 **9.** 60 **11.** 5^{50} **13.** 30

15. 9990

17. a. 36
b.

19. 1024; 59,049 **21.** 2730 **23.** 217

Exercises 7.4, page 364

1. 360 **3.** 10 **5.** 120 **7.** 20

9. n **11.** 1 **13.** 35 **15.** 1 **17.** 84

19. $\dfrac{n(n-1)}{2!}$ **21.** $\dfrac{n!}{2}$ **23.** Permutation

25. Combination **27.** Permutation **29.** Combination

31. $P(4, 4) = 24$ **33.** $P(4, 4) = 24$

35. $P(9, 9) = 362,880$ **37.** $C(12, 3) = 220$

39. 151,200 **41.** $C(12, 3) = 220$

43. $C(100, 3) = 161,700$ **45.** $P(6, 6) = 720$

47. $P(12, 6) = 665,280$

49. a. $P(10, 10) = 3,628,800$
b. $P(3, 3) \cdot P(4, 4) \cdot P(3, 3) \cdot P(3, 3) = 5184$

51. a. $P(20, 20) = 20!$
b. $P(5, 5) \cdot P[(4, 4)]^5 = 5!(4!)^5 = 955,514,880$

53. $P(2, 1) \cdot P(3, 1) = 6$

55. $C(3, 3) \cdot [C(8, 6) + C(8, 7) + C(8, 8)] = 37$

57. a. $C(12, 3) = 220$ **b.** $C(11, 2) = 55$
c. $C(5, 1) \cdot C(7, 2) + C(5, 2) \cdot C(7, 1) + C(5, 3) = 185$

59. $P(7, 3) + C(7, 2) \cdot P(3, 2) = 336$

61. $[C(5, 1) \cdot C(3, 1) \cdot C(6, 2)][C(4, 1) + C(3, 1)] = 1575$

63. $C(10, 8) + C(10, 9) + C(10, 10) = 56$

65. $10 \cdot C(4, 1) = 40$

67. $4 \cdot C(13, 5) - 40 = 5108$

69. $13C(4, 3) \cdot 12C(4, 2) = 3744$

71. $C(6, 2) = 15$

73. $C(12, 6) + C(12, 7) + C(12, 8) + C(12, 9) +$ $C(12, 10) + C(12, 11) + C(12, 12) = 2510$

75. $4! = 24$

Using Technology Exercises 7.4, page 368

1. $1.307674368 \times 10^{12}$ **3.** $2.56094948229 \times 10^{16}$

5. 674,274,182,400 **7.** 133,784,560 **9.** 4,656,960

11. 658,337,004,000

Exercises 7.5, page 374

1. $\{a, b, d, f\}$; $\{a\}$ **3.** $\{b, c, e\}$; $\{a\}$ **5.** No **7.** S

9. $\varnothing$ **11.** Yes **13.** $E \cup F$ **15.** G^c

17. $(E \cup F \cup G)^c$

19. $\varnothing, \{a\}, \{b\}, \{c\}, \{a, b\}, \{a, c\}, \{b, c\}, S$

21. a. $S = \{B, R\}$ **b.** $\varnothing, \{B\}, \{R\}, S$

23. a. $S = \{(H, 1), (H, 2), (H, 3), (H, 4), (H, 5), (H, 6),$ $(T, 1), (T, 2), (T, 3), (T, 4), (T, 5), (T, 6)\}$
b. $\{(H, 2), (H, 4), (H, 6)\}$

25. $S = \{(d, d, d), (d, d, n), (d, n, d), (n, d, d), (d, n, n),$ $(n, d, n), (n, n, d), (n, n, n)\}$

27. a. $\{ABC, ABD, ABE, ACD, ACE, ADE, BCD,$ $BCE, BDE, CDE\}$
b. 6 **c.** 3 **d.** 6

29. a. E^c **b.** $E^c \cap F^c$ **c.** $E \cup F$
d. $(E \cap F^c) \cup (E^c \cap F)$

31. a. $\{x | x > 0\}$ **b.** $\{x | 0 < x \leq 2\}$ **c.** $\{x | x > 2\}$

33. a. $S = \{0, 1, 2, 3, \ldots, 10\}$ **b.** $E = \{0, 1, 2, 3\}$
c. $F = \{5, 6, 7, 8, 9, 10\}$

35. a. $S = \{0, 1, 2, \ldots, 20\}$
b. $E = \{0, 1, 2, \ldots, 9\}$ **c.** $F = \{20\}$

39. 2^n

Exercises 7.6, page 384

1. {(H, H)}, {(H, T)}, {(T, H)}, {(T, T)}

3. {(D, m)}, {(D, f)}, {(R, m)}, {(R, f)}, {(I, m)}, {(I, f)}

5. {(1, i)}, {(1, d)}, {(1, s)}, {(2, i)}, {(2, d)}, {(2, s)}, . . . , {(5, i)}, {(5, d)}, {(5, s)}

7. {(A, Rh⁺)}, {(A, Rh⁻)}, {(B, Rh⁺)}, {(B, Rh⁻)}, {(AB, Rh⁺)}, {(AB, Rh⁻)}, {(O, Rh⁺)}, {(O, Rh⁻)}

9.

Grade	A	B	C	D	F
Probability	0.10	0.25	0.45	0.15	0.05

11. a. $S = \{(0 < x \le 200), (200 < x \le 400), (400 < x \le 600), (600 < x \le 800), (800 < x \le 1000), (x > 1000)\}$

b.

Number of Cars (x)	Probability
$0 < x \le 200$	0.075
$200 < x \le 400$	0.1
$400 < x \le 600$	0.175
$600 < x \le 800$	0.35
$800 < x \le 1000$	0.225
$x > 1000$	0.075

13.

Event	A	B	C	D	E
Probability of an Event	0.026	0.199	0.570	0.193	0.012

15.

Number of Figures Produced (in dozens)	30	31	32
Probability	0.125	0	0.1875

Number of Figures Produced (in dozens)	33	34	35	36
Probability	0.25	0.1875	0.125	0.125

17. 0.469 19. a. 0.856 b. 0.144 21. 0.46

23. a. $\frac{1}{4}$ b. $\frac{1}{2}$ c. $\frac{1}{13}$ 25. $\frac{3}{8}$ 27. 0.95

29. a. 0.633 b. 0.276 31. No 33. Yes

35. a. $\frac{3}{7}$ b. $\frac{3}{14}$ c. 1 37. a. 0.35 b. 0.33

39. a. 0.17 b. 0.16

Exercises 7.7, page 393

1. $\frac{1}{2}$ 3. $\frac{1}{36}$ 5. $\frac{1}{9}$ 7. $\frac{1}{52}$ 9. $\frac{3}{13}$ 11. $\frac{12}{13}$

13. 0.002; 0.998 15. $P(a) + P(b) + P(c) \ne 1$

17. Since the five events are not mutually exclusive, Property 3 cannot be used; that is, he could win more than one purse.

19. The two events are not mutually exclusive; hence, the probability of the given event is $\frac{1}{6} + \frac{1}{6} - \frac{1}{36} = \frac{11}{36}$.

21. $E^C \cap F^C = \{e\} \ne \varnothing$

23. $P(G \cup C)^C \ne 1 - P(G) - P(C)$; he has not considered the case in which a customer buys both glasses and contact lenses.

25. a. 0 b. 0.7 c. 0.8 d. 0.3

27. a. $\frac{1}{2}, \frac{3}{8}$ b. $\frac{1}{2}, \frac{5}{8}$ c. $\frac{1}{8}$ d. $\frac{3}{4}$ 29. 0.33

31. a. 0.16 b. 0.38 c. 0.22

33. a. 0.90 b. 0.40 c. 0.40

35. a. 0.6 b. 0.332 c. 0.232 d. 0.6

Chapter 7, Review Exercises, page 398

1. {3} 2. {A, E, H, L, S, T}

3. {4, 6, 8, 10} 4. {−4} 5. Yes 6. Yes 7. Yes

8. No

9.

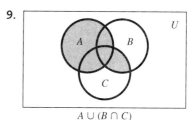

$A \cup (B \cap C)$

10.

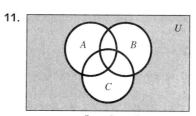

$(A \cap B \cap C)^C$

11.

$A^C \cap B^C \cap C^C$

12.

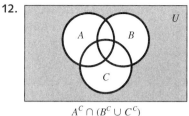

$A^C \cap (B^C \cup C^C)$

17. The set of all participants in a consumer behavior survey who both avoided buying a product because it is not recyclable and boycotted a company's products because of its record on the environment.

18. The set of all participants in a consumer behavior survey who both avoided buying a product because it is not recyclable and/or voluntarily recycled their garbage.

19. The set of all participants in a consumer behavior survey who both did not use cloth diapers rather than disposable diapers and voluntarily recycled their garbage.

20. The set of all participants in a consumer behavior survey who do not boycott a company's products because of the company's record on the environment and/or who do not voluntarily recycle their garbage.

21. 150 **22.** 230 **23.** 270 **24.** 30 **25.** 70 **26.** 200

27. 190 **28.** 181,440 **29.** 120 **30.** 8400

31. a. 0 **b.** 0.6 **c.** 0.6 **d.** 0.4 **e.** 1

32. a. 0.35 **b.** 0.65 **c.** 0.05

33. a. 0.49 **b.** 0.39 **c.** 0.48 **34.** None

35. a. 446 **b.** 377 **c.** 477 **36.** 720 **37.** 20

38. 720 **39. a.** 60 **b.** 125 **40. a.** 50,400 **b.** 5040

41. a. 1287 **b.** 288 **42. a.** 5040 **b.** 3600 **43.** 1050

44. a. $C(15, 4) = 1365$ **b.** $C(15, 4) - C(10, 4) = 1155$

45. a. 487,365 **b.** 550 **c.** 341,055

46. a. 0.019 **b.** 0.981

CHAPTER 8

Exercises 8.1, page 406

1. $\frac{1}{32}$ **3.** $\frac{31}{32}$ **5.** $P(a) = 13 \cdot C(4, 2)/C(52, 2) = 0.0588$

7. $C(26, 2)/C(52, 2) = 0.245$

9. $[C(3, 2) \cdot C(5, 2)]/C(8, 4) = \frac{3}{7}$

11. $[C(5, 3) \cdot C(3, 1)]/C(8, 4) = \frac{3}{7}$

13. $[C(3, 2) \cdot C(1, 1)]/8 = \frac{3}{8}$ **15.** $\frac{1}{8}$

17. $C(10, 6)/2^{10} = 0.205$

19. a. $C(4, 2)/C(24, 2) \approx 0.022$
b. $1 - C(20, 2)/C(24, 2) \approx 0.312$

21. a. $C(6, 2)/C(80, 2) \approx 0.005$
b. $1 - C(74, 2)/C(80, 2) \approx 0.145$

23. a. $0.12; C(98, 10)/C(100, 12) \approx 0.013$
b. 0.15; 0.015

25. $[C(12, 8) \cdot C(8, 2) + C(12, 9) \cdot C(8, 1) + C(12, 10)]/C(20, 10) \approx 0.085$

27. a. $\frac{3}{5}$ **b.** $C(3, 1)/C(5, 3) = 0.3$
c. $1 - C(3, 3)/C(5, 3) = 0.9$

29. $\frac{1}{729}$ **31.** 0.0001 **33.** 0.10

35. $40/C(52, 5) \approx 0.0000154$

37. $[4C(13, 5) - 40]/C(52, 5) \approx 0.00197$

39. $[13C(4, 3) \cdot 12C(4, 2)]/C(52, 5) \approx 0.00144$

41. a. 0.618 **b.** 0.059 **43.** 0.03

Exercises 8.2, page 421

1. a. 0.4 **b.** 0.33 **3.** 0.3 **5.** Independent

7. Independent **9. a.** 0.24 **b.** 0.76

11. a. 0.5 **b.** 0.4 **c.** 0.2 **d.** 0.35
e. No **f.** No

13. a. 0.4 **b.** 0.3 **c.** 0.12 **d.** 0.30
e. Yes **f.** Yes

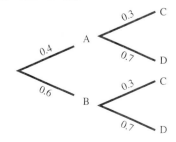

15. $\frac{4}{11}$ **17. a.** $\frac{1}{4}$ **b.** $\frac{13}{51}$ **c.** $\frac{12}{51}$ **19.** 0.012

21. a. 0.064 **b.** 0.414 **c.** 0.57 **d.** 0.43

23. a. $\frac{1}{2}$ **b.** $\frac{1}{2}$ **c.** $\frac{1}{2}$ **25.** 0.014

27. a.

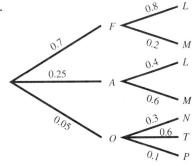

L: 30-year loans
M: 15-year loans
N: 20-year loans
T: 10-year loans
P: loans of 5 years or less

b. 0.15   **c.** 0.29

29. $\frac{1}{15}$

31. a. 0.757; 0.569; 0.393; 0.520; 0.720   **b.** No

33. $P(A \cap B) \neq P(A) \cdot P(B)$

35. a. 0.81   **b.** 0.729

37. a. 0.0000001   **b.** 0.927   **c.** 0.071

Exercises 8.3, page 430

1.

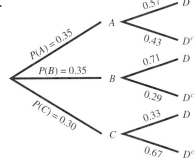

3. a. 0.45   **b.** 0.22   **5. a.** 0.48   **b.** 0.33

7. a. 0.08   **b.** 0.15   **c.** 0.35

9. a. $\frac{1}{12}$ **b.** $\frac{1}{4}$ **c.** $\frac{1}{18}$ **d.** $\frac{3}{14}$

11. $\frac{4}{17}$ **13.** 0.0784

15.

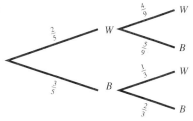

17. 0.53   **19.** 0.422   **21. a.** $\frac{3}{4}$ **b.** $\frac{2}{9}$

23. 0.856   **25. a.** 0.03   **b.** 0.29   **27.** 0.35

29. a. 0.30   **b.** 0.10   **31.** 0.62   **33.** 0.65   **35.** 0.93

Exercises 8.4, page 441

1. a. See (b)

b.

Outcome	GGG	GGR	GRG	RGG
Value	3	2	2	2

Outcome	GRR	RGR	RRG	RRR
Value	1	1	1	0

c. $\{GGG\}$

3. Any positive integer   **5.** $\frac{1}{6}$

7. Any positive integer; infinite discrete

9. $0 \leq x < \infty$; continuous

11. Any positive integer; infinite discrete

13. a. 0.20   **b.** 0.60   **c.** 0.30   **d.** 1

15.

17. a.

x	1	2	3	4	5	6
P(X = x)	$\frac{1}{6}$	$\frac{1}{6}$	$\frac{1}{6}$	$\frac{1}{6}$	$\frac{1}{6}$	$\frac{1}{6}$

y	1	2	3	4	5	6
P(Y = y)	$\frac{1}{6}$	$\frac{1}{6}$	$\frac{1}{6}$	$\frac{1}{6}$	$\frac{1}{6}$	$\frac{1}{6}$

b.

x + y	2	3	4	5	6	7
P(X + Y = x + y)	$\frac{1}{36}$	$\frac{2}{36}$	$\frac{3}{36}$	$\frac{4}{36}$	$\frac{5}{36}$	$\frac{6}{36}$

x + y	8	9	10	11	12
P(X + Y = x + y)	$\frac{5}{36}$	$\frac{4}{36}$	$\frac{3}{36}$	$\frac{2}{36}$	$\frac{1}{36}$

19. a.

x	0	1	2	3	4
P(X = x)	0.017	0.067	0.033	0.117	0.233

x	5	6	7	8	9	10
P(X = x)	0.133	0.167	0.1	0.05	0.067	0.017

b.

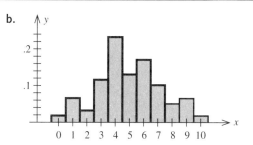

21.

x	1	2	3	4	5
P(X = x)	0.007	0.029	0.021	0.079	0.164

x	6	7	8	9	10
P(X = x)	0.15	0.20	0.207	0.114	0.029

Exercises 8.5, page 452

1. a. 2.6
  b.

x	0	1	2	3	4
P(X = x)	0	0.1	0.4	0.3	0.2

; 2.6

3. 0.86 **5.** $78.50 **7.** 0.91 **9.** 0.12

11. 1.73 **13.** $800 **15.** $162.50 **17.** First project

19. a. Dahl: 8.52; Farthington: 7.25
  b. Farthington Auto Sales

21. a. 3 to 7 **b.** 7 to 3 **23.** -10.5¢ **25.** 4 to 1; 1 to 4

27. 0.5625 **29.** 0.6154 **31.** 0.5714 **33. b.** $-$15.79

35. Mean: $5.94; mode: $6.00; median: $5.95

Exercises 8.6, page 463

1. $\mu = 2$, $V(X) = 1$, $\sigma = 1$

3. $\mu = 0$, $V(X) = 1$, $\sigma = 1$

5. $\mu = 518$, $V(X) = 1891$, $\sigma = 43.5$ **7.** a **9.** 1.56

11. $\mu = 4.5$, $V(X) = 5.25$

13. a. Let X = The annual birth rate during the years 1981–1990.
  b.

x	15.5	15.6	15.7	15.9	16.2	16.7
P(X = x)	0.2	0.1	0.3	0.2	0.1	0.1

  c. $\mu = 15.84$, $V(X) = 0.1224$, $\sigma = 0.350$

15. a. Mutual fund A: $\mu = \$620$, $V(X) = \$267{,}600$; Mutual fund B: $\mu = \$520$, $V(X) = \$137{,}600$
  b. Mutual fund A **c.** Mutual fund B

17. 1

19. $\mu = \$139{,}600$; $V(X) = \$1{,}443{,}840{,}000$; $\sigma = \$37{,}998$

21. a. At least 0.75 **b.** At least 0.96

23. At least 75% **25.** At least $\frac{7}{16}$ **27.** $c \geq 0.1$

Chapter 8 Review Exercises, page 468

1. 0.18 **2.** 0.25 **3.** 0.06 **4.** 0.367 **5.** 0.49

6. 0.364 **7. a.** $\frac{7}{8}$ **b.** $\frac{7}{8}$ **c.** No

8. a. 0.284 **b.** 0.984 **9.** 0.150 **10.** $\frac{2}{15}$ **11.** 0.00018

12. 0.00995 **13.** 0.2451 **14.** 0.510 **15.** 0.2451

16. a. {WWW, BWW, WBW, WWB, BBW, BWB, WBB, BBB}
  b.

Outcome	WWW	BWW	WBW	WWB
Value of X	0	1	1	1

Outcome	BBW	BWB	WBB	BBB
Value of X	2	2	2	3

  c.

x	0	1	2	3
P(X = x)	$\frac{1}{35}$	$\frac{12}{35}$	$\frac{18}{35}$	$\frac{4}{35}$

  d.

17. $100 **18. a.** 0.8 **b.** $\mu = 2.7$; $\sigma = 1.42$ **19.** 0.75

20. 0.457 **21.** 0.368 **22. a.** 0.68 **b.** 0.053

CHAPTER 9

Exercises 9.1, page 483

1. $\lim\limits_{x \to -2} f(x) = 3$ **3.** $\lim\limits_{x \to 3} f(x) = 3$ **5.** $\lim\limits_{x \to -2} f(x) = 3$

7. The limit does not exist.

9.

x	1.9	1.99	1.999
f(x)	4.61	4.9601	4.9960

x	2.001	2.01	2.1
f(x)	5.004	5.0401	5.41

$\lim\limits_{x \to 2} (x^2 + 1) = 5$

11.

x	−0.1	−0.01	−0.001
f(x)	−1	−1	−1

x	0.001	0.01	0.1
f(x)	1	1	1

The limit does not exist.

13.

x	0.9	0.99	0.999
f(x)	100	10,000	1,000,000

x	1.001	1.01	1.1
f(x)	1,000,000	10,000	100

The limit does not exist.

15.

x	0.9	0.99	0.999	1.001	1.01	1.1
f(x)	2.9	2.99	2.999	3.001	3.01	3.1

$\lim\limits_{x \to 1} \dfrac{x^2 + x - 2}{x - 1} = 3$

17. 3   **19.** 3   **21.** −1   **23.** 2   **25.** −4   **27.** $\frac{5}{4}$ **29.** 2

31. $\sqrt{171} = 3\sqrt{19}$ **33.** $\frac{3}{2}$ **35.** −1   **37.** −6   **39.** 2

41. $\frac{1}{6}$ **43.** 2   **45.** −1   **47.** −10

49. The limit does not exist.   **51.** $\frac{5}{3}$ **53.** $\frac{1}{2}$ **55.** $\frac{1}{3}$

57. $\lim\limits_{x \to \infty} f(x) = \infty$; $\lim\limits_{x \to -\infty} f(x) = \infty$ **59.** 0; 0

61. $\lim\limits_{x \to \infty} f(x) = -\infty$; $\lim\limits_{x \to -\infty} f(x) = -\infty$

63.

x	1	10	100	1000
f(x)	0.5	0.009901	0.0001	0.000001

x	−1	−10	−100	−1000
f(x)	0.5	0.009901	0.0001	0.000001

$\lim\limits_{x \to \infty} f(x) = 0$ and $\lim\limits_{x \to -\infty} f(x) = 0$

65.

x	1	5	10	100
f(x)	12	360	2910	2.99×10^6

x	1000	−1	−5
f(x)	2.999×10^9	6	−390

x	−10	−100	−1000
f(x)	−3090	$−3.01 \times 10^6$	$−3.0 \times 10^9$

$\lim\limits_{x \to \infty} f(x) = \infty$ and $\lim\limits_{x \to -\infty} f(x) = -\infty$

67. 3   **69.** 3   **71.** $\lim\limits_{x \to -\infty} f(x) = -\infty$ **73.** 0

75. **a.** \$0.5 million; \$0.75 million; \$1,166,667; \$2 million; \$4.5 million; \$9.5 million
    b. The limit does not exist; as the percentage of pollutant to be removed approaches 100, the cost becomes astronomical.

77. $\lim\limits_{t \to \infty} C(t) = 0$; in the long run, the concentration of drug in the bloodstream decreases to zero.

79. **a.** 5000   **b.** 25,000

81. $\lim\limits_{x \to \infty} V(x) = a$; as the amount of substrate increases indefinitely, the initial speed approaches the constant a moles per liter per second.

83. Bank B

Using Technology Exercises 9.1, page 488

1. 5   **3.** 3   **5.** $\frac{2}{3}$ **7.** $\frac{1}{2}$ **9.** e^2

Exercises 9.2, page 498

1. 3; 2; the limit does not exist.

3. The limit does not exist; 2; the limit does not exist.

5. 0; 2; the limit does not exist.

7. -2; 2; the limit does not exist. **9.** 6 **11.** $-\frac{1}{4}$

13. The limit does not exist. **15.** -1 **17.** 0 **19.** -4

21. The limit does not exist. **23.** 4 **25.** 0

27. $x = 0$; conditions 2 and 3

29. Continuous everywhere

31. $x = 0$; condition 3 **33.** $x = 0$; condition 3

35. $(-\infty, \infty)$ **37.** $(-\infty, \infty)$

39. $(-\infty, \frac{1}{2}) \cup (\frac{1}{2}, \infty)$ **41.** $(-\infty, -2) \cup (-2, 1) \cup (1, \infty)$

43. $(-\infty, \infty)$ **45.** -1 and 1 **47.** 1 and 2

49. f is discontinuous at $t = 20, 40,$ and 60. When $t = 0$, the inventory stands at 750 reams. The level drops to about 200 reams by the 20th day at which time a new order of 500 reams arrives to replenish the supply.

51. The total deposits of Franklin make a jump at each of these points as the deposits of the ailing institutions become a part of the total deposits of the parent company.

53.

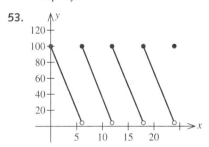

55. a. f is a polynomial of degree 2.
b. $f(1) = 3$ and $f(3) = -1$

57. a. f is a polynomial of degree 3.
b. $f(0) = 14$ and $f(1) = -23$

Using Technology Exercises 9.2, page 503

1. $x = 0, 1$ **3.** $x = -1, 2$ **5.** $x = 0, \frac{1}{2}$ **7.** $x = -\frac{1}{2}, 2$

9. $x = -2, 1$

Exercises 9.3, page 518

1. 1.5; 0.5833; 1.25 **3.** 3.075; -21.15 **5.** 0 **7.** 2

9. $6x$ **11.** $-2x + 3$ **13.** 2; $y = 2x + 7$

15. 6; $y = 6x - 3$ **17.** $\frac{1}{9}$; $y = \frac{1}{9}x - \frac{2}{3}$

19. a. $4x$ **b.** $y = 4x - 1$
c.

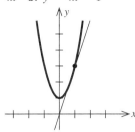

21. a. $2x - 2$ **b.** $(1, 0)$
c.

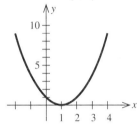

d. 0

23. a. 6; 5.5; 5.1 **b.** 5
c. The computations in (a) show that as h approaches zero, the average velocity approaches the instantaneous velocity.

25. a. 130 ft/sec; 128.2 ft/sec; 128.02 ft/sec
b. 128 ft/sec
c. The computations in (a) show that as the time intervals over which the average velocity are computed become smaller and smaller, the average velocity approaches the instantaneous velocity of the car at $t = 20$.

27. a. $C'(x) = -20x + 300$ **b.** \$100/surfboard
c. \$200

29. \$6 billion/yr; \$10 billion yr

31. Average rate of change of the seal population over the time interval $[a, a + h]$; instantaneous rate of change of the seal population at $x = a$

33. Average rate of change of the country's industrial production over the time interval $[a, a + h]$; instantaneous rate of change of the country's industrial production at $x = a$

35. a. Yes **b.** Yes **c.** Yes

37. a. No **b.** No **c.** No

39. a. No **b.** No **c.** No

41.

Using Technology Exercises 9.3, page 521

1. a. $y = 4x - 3$
b.

3. a. $y = -7x - 8$
b.

5. a. $y = 9x - 11$
b.

7. a. $y = 2$
b.

9. a. $y = \frac{1}{4}x + 1$
b.

11. a. 4
b. $y = 4x - 1$
c.

13. a. 20
b. $y = 20x - 35$
c.

15. a. $\frac{3}{4}$
b. $y = \frac{3}{4}x - 1$
c.

17. a. $-\frac{1}{4}$
b. $y = -\frac{1}{4}x + \frac{3}{4}$
c.

19. a. 4.02
b. $y = 4.02x - 3.57$
c.

21. 4   **23.** 20   **25.** $\frac{3}{4}$

27. $-\frac{1}{4}$ **29.** 4.0249

Exercises 9.4, page 528

1. 0   **3.** $5x^4$ **5.** $2.1x^{1.1}$ **7.** $6x$ **9.** $2\pi r$ **11.** $\frac{3}{x^{2/3}}$

13. $\frac{3}{2\sqrt{x}}$ **15.** $-84x^{-13}$ **17.** $10x - 3$

19. $-3x^2 + 4x$ **21.** $0.06x - 0.4$ **23.** $2x - 4 - \frac{3}{x^2}$

25. $16x^3 - 7.5x^{3/2}$ **27.** $-\frac{3}{x^2} - \frac{8}{x^3}$ **29.** $-\frac{16}{t^5} + \frac{9}{t^4} - \frac{2}{t^2}$

31. $2 - \frac{5}{2\sqrt{x}}$ **33.** $-\frac{4}{x^3} + \frac{1}{x^{4/3}}$

35. a. 20   **b.** -4 **c.** 20

37. $m = 5$; $y = 5x - 4$ **39.** $m = -2$; $y = -2x + 2$

41. a. $(0, 0)$ **b.**

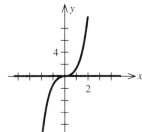

43. a. $(-2, -7), (2, 9)$
b. $y = 12x + 17$ and $y = 12x - 15$
c.

45. a. $(0, 0)$; $\left(1, -\frac{13}{12}\right)$
b. $(0, 0)$; $\left(2, -\frac{8}{3}\right)$; $\left(-1, -\frac{5}{12}\right)$
c. $(0, 0)$; $\left(4, \frac{80}{3}\right)$; $\left(-3, \frac{81}{4}\right)$

47. 30 cm/sec; -200 cm/sec; the velocity is decreasing at the rate of 200 cm/sec.

49. a. 15 pts/yr; 12.6 pts/yr; 0 pts/yr   **b.** 10 pts/yr

51. 155/month; 200/month

53. 32 turtles/yr; 428 turtles/hr; 3260 turtles

55. a. $120 - 30t$ **b.** 120 ft/sec   **c.** 240 ft

57. a. $20\left(1 - \frac{1}{\sqrt{t}}\right)$
b. 50 mph; 30 mph; 33.43 mph; the average velocity at 6, 7, and 8 A.M. is 50 mph, 30 mph, and 33.43 mph, respectively

c. -8.28 mph/hr; 0; 5.86 mph/hr; at 6:30 A.M. the average velocity is decreasing at the rate of 8.28 mph/hr, at 7 A.M. it is unchanged, and at 8 A.M. it is increasing at the rate of 5.86 mph/hr

Using Technology Exercises 9.4, page 533

1. 1 **3.** 0.4226 **5.** 0.1613

7. a.

b. 3.4295 parts/million; 105.4332 parts/million

9. a.

b. decreasing at the rate of 9 days/yr; increasing at the rate of 13 days/yr

Exercises 9.5, page 547

1. $6x^2 + 2$ **3.** $4t - 1$ **5.** $9x^2 + 2x - 6$

7. $4x^3 + 3x^2 - 1$ **9.** $5w^4 - 4w^3 + 9w^2 - 6w + 2$

11. $\dfrac{25x^2 - 10x^{3/2} + 1}{x^{1/2}}$ **13.** $\dfrac{3x^4 - 10x^3 + 4}{x^2}$ **15.** $\dfrac{-1}{(x-2)^2}$

17. $\dfrac{3}{(2x+1)^2}$ **19.** $-\dfrac{2x}{(x^2+1)^2}$ **21.** $\dfrac{s^2 + 2s + 4}{(s+1)^2}$

23. $\dfrac{1 - 3x^2}{2\sqrt{x}(x^2+1)^2}$ **25.** $\dfrac{x^2 - 2x - 2}{(x^2+x+1)^2}$

27. $\dfrac{2x^3 - 5x^2 - 4x - 3}{(x-2)^2}$

29. $\dfrac{-2(x^5 + 12x^4 - 8x^3 + 16x + 64)}{(x^4 - 16)^2}$

31. $2(3x^2 - x + 3)$; 10 **33.** $\dfrac{-3x^4 + 2x^2 - 1}{(x^4 - 2x^2 - 1)^2}$; $-\dfrac{1}{2}$

35. 60; $y = 60x - 102$ **37.** $-\frac{1}{2}$; $y = -\frac{1}{2}x + \frac{3}{2}$

39. $8x - 2$; 8 **41.** $6x^2 - 6x$; $6(2x - 1)$

43. $4t^3 - 6t^2 + 12t - 3$; $12(t^2 - t + 1)$

45. $72x - 24$ **47.** $-\dfrac{6}{x^4}$ **49.** $y = 7x - 5$

51. $(\frac{1}{3}, \frac{50}{27})$; $(1, 2)$ **53.** $(\frac{4}{3}, -\frac{770}{27})$; $(2, -30)$

55. a. $\dfrac{0.2(1 - t^2)}{(t^2 + 1)^2}$

b. 0.096 percent/hr; 0 percent/hr; -0.024 percent/hr

57. $\dfrac{6000}{(t + 12)^2}$; 18.5 mg/yr; 12.4 mg/yr

59. a. $-\dfrac{x}{(0.01x^2 + 1)^2}$ **b.** -3.2; -2.5; -1.4

61. \$38.4 million/yr; \$17.04 million/yr; \$5.71 million/yr

63. $16 - 6t$

65. a. 0, 4.2, 7.2, 9, 9.6, 9, 7.2, 4.2, 0
b. 4.8, 3.6, 2.4, 1.2, 0, -1.2, -2.4, -3.6, -4.8

67. -7.157; -2.5245; at the beginning of the year 2000, the number of Americans aged 45 to 54 will be decreasing at the rate of 7 million people/yr and the number will decrease at a rate of approximately 2.5 million people/yr/yr

Using Technology Exercises 9.5, page 548

1. 0.8750 **3.** 0.0774 **5.** -0.5000 **7.** 87,322 per year

9. 425.25 **11.** 128.7540 **13.** -13.9463 **15.** -0.0163

Exercises 9.6, page 557

1. $f(g(x)) = x^4 + x^2 + 1$; $g(f(x)) = (x^2 + x + 1)^2$

3. $f(g(x)) = \sqrt{x^2 - 1} + 1$; $g(f(x)) = x + 2\sqrt{x}$

5. $f(g(x)) = \dfrac{x}{x^2 + 1}$; $g(f(x)) = \dfrac{x^2 + 1}{x}$ **7.** 49 **9.** $\dfrac{\sqrt{5}}{5}$

11. $f(x) = 2x^3 + x^2 + 1$ and $g(x) = x^5$

13. $f(x) = x^2 - 1$ and $g(x) = \sqrt{x}$

15. $f(x) = x^2 - 1$ and $g(x) = \dfrac{1}{x}$ **17.** $8(2x - 1)^3$

19. $10x(x^2 + 2)^4$ **21.** $6x^2(1 - x)(2 - x)^2$

23. $\dfrac{-4}{(2x + 1)^3}$ **25.** $3x\sqrt{x^2 - 4}$ **27.** $\dfrac{3}{2\sqrt{3x - 2}}$

29. $\dfrac{-2x}{3(1 - x^2)^{2/3}}$ **31.** $-\dfrac{6}{(2x + 3)^4}$ **33.** $\dfrac{-1}{(2t - 3)^{3/2}}$

35. $-\dfrac{3(16x^3 + 1)}{2(4x^4 + x)^{5/2}}$ **37.** $\dfrac{-4(3x + 1)}{(3x^2 + 2x + 1)^3}$

39. $6x(2x^2 - x + 1)$ **41.** $3(t^{-1} - t^{-2})^2(-t^{-2} + 2t^{-3})$

43. $\dfrac{1}{2\sqrt{x - 1}} + \dfrac{1}{2\sqrt{x + 1}}$ **45.** $-12x(4x - 1)(3 - 4x)^3$

47. $6(x - 1)(2x - 1)(2x + 1)^3$ **49.** $-\dfrac{15(x + 3)^2}{(x - 2)^4}$

51. $\dfrac{3\sqrt{t}}{2(2t + 1)^{5/2}}$ **53.** $\dfrac{-1}{2\sqrt{u + 1}\,(3u + 2)^{3/2}}$

55. $-\dfrac{2x(3x^2+1)}{(x^2-1)^5}$ **57.** $-\dfrac{2x(3x^2+1)^2(3x^2+13)}{(x^2-1)^5}$

59. $\frac{4}{3}u^{1/3}$; $6x$; $8x(3x^2-1)^{1/3}$

61. $-\dfrac{2}{3u^{5/3}}$; $6x^2-1$; $-\dfrac{2(6x^2-1)}{3(2x^3-x+1)^{5/3}}$

63. $10x(x^2+2)^4$; $10(x^2+2)^3(9x^2+2)$

65. $14x(2x^2+2)^{5/2}$; $28(2x^2+2)^{3/2}(6x^2+1)$

67. $\dfrac{1}{(2x+1)^2}$; $-\dfrac{4}{(2x+1)^3}$ **69.** $y=-33x+57$

71. 0.333 million/wk; 0.305 million/wk; 16 million; 22.7 million

73. a. $0.027(0.2t^2+4t+64)^{-1/3}(0.1t+1)$
  b. 0.0090 parts per million

75. a. $0.21t^2(t-3)(t-7)^3$
  b. 90.72; 0; -90.72; at 8 A.M. the level of nitrogen dioxide is increasing, at 10 A.M. the level stops increasing, and at 11 A.M. the level is decreasing

77. $\dfrac{3450t}{(t+25)^2\sqrt{\frac{1}{2}t^2+2t+25}}$; 2.9 beats/min; 0.7 beats/min; 0.2 beats/min; 179 beats/min

79. 160π ft²/sec   **81.** 18.7 PCs/month

Using Technology Exercises 9.6, page 560

1. 0.5774   **3.** 0.9390   **5.** -4.9498

7. a. 20,790   **b.** 554/yr

Exercises 9.7, page 567

1. $3e^{3x}$ **3.** $-e^{-t}$ **5.** e^x+1 **7.** $x^2e^x(x+3)$

9. $\dfrac{2e^x(x-1)}{x^2}$ **11.** $3(e^x-e^{-x})$ **13.** $-\dfrac{1}{e^w}$ **15.** $6e^{3x-1}$

17. $-2xe^{-x^2}$ **19.** $\dfrac{3e^{-1/x}}{x^2}$ **21.** $25e^x(e^x+1)^{24}$ **23.** $\dfrac{e^{\sqrt{x}}}{2\sqrt{x}}$

25. $e^{3x+2}(3x-2)$ **27.** $\dfrac{2e^x}{(e^x+1)^2}$ **29.** $2(8e^{-4x}+9e^{3x})$

31. $6e^{3x}(3x+2)$ **33.** $y=2x-2$ **35.** $\dfrac{5}{x}$ **37.** $\dfrac{1}{x+1}$

39. $\dfrac{8}{x}$ **41.** $\dfrac{1}{2x}$ **43.** $-\dfrac{2}{x}$ **45.** $\dfrac{2(4x-3)}{4x^2-6x+3}$

47. $\dfrac{1}{x(x+1)}$ **49.** $x(1+2\ln x)$ **51.** $\dfrac{2(1-\ln x)}{x^2}$

53. $\dfrac{3}{u-2}$ **55.** $\dfrac{1}{2x\sqrt{\ln x}}$ **57.** $\dfrac{3(\ln x)^2}{x}$ **59.** $\dfrac{3x^2}{x^3+1}$

61. $\dfrac{(x\ln x+1)e^x}{x}$ **63.** $-\dfrac{1}{x^2}$ **65.** $\dfrac{2(2-x^2)}{(x^2+2)^2}$

67. $y=x-1$

69. $-\$6065/$day; $-\$3679/$day; $-\$2231/$day; $-\$1353/$day

71. 4505 per year; 273 cases per year

73. a. -1.68 cents/case   **b.** $\$40.36$ per case

75. -0.1694; -0.1549; -0.1415; the percentage of the total population relocating was decreasing at the rate of 16.94% in 1970, 15.49% in 1980, and 14.15% in 1990

Using Technology Exercises 9.7, page 570

1. 5.4366   **3.** 12.3929   **5.** 0.1861

7. a.

  b. 57,972/$1000; $-\$23,418/\$1000$

9. a. 69.63%   **b.** 5.09361%/decade

Exercises 9.8, page 577

1. a. $\$1.80/$record; $\$1.60/$record
  b. $\$1.80/$record; $\$1.60/$record

3. a. $100+\dfrac{200{,}000}{x}$ **b.** $-\dfrac{200{,}000}{x^2}$

  c. $\displaystyle\lim_{x\to\infty}\left(100+\dfrac{200{,}000}{x}\right)=100$

5. $\dfrac{2000}{x}+2-0.0001x$; $-\dfrac{2000}{x^2}-0.0001$

7. a. $8000-200x$
  b. 200; 0; -200; revenue seems to be maximized when the fare is approximately $\$40/$passenger

9. a. $\$750$
  b. $\$760$; the marginal profit approximates the actual profit

11. a. $600x-0.05x^2$; $-0.000002x^3-0.02x^2+200x-80{,}000$
  b. $0.000006x^2-0.06x+400$; $600-0.1x$; $-0.000006x^2-0.04x+200$
  c. 304; 400; 96

d.

Units of a thousand

Units of a thousand

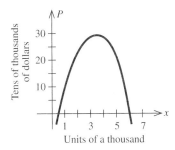

Units of a thousand

13. $0.000002x^2 - 0.03x + 400 + \dfrac{80,000}{x}$

a. $0.000004x - 0.03 - \dfrac{80,000}{x^2}$

b. -0.0132; 0.0092; the marginal average cost is negative (average cost is decreasing) when 5000 units are produced and positive (average cost is increasing) when 10,000 units are produced

c.

Units of a thousand

15. a. $\dfrac{50x}{0.01x^2 + 1}$ **b.** $\dfrac{50 - 0.5x^2}{(0.01x^2 + 1)^2}$

c. \$44,380; when the level of production is 2000 units, the revenue increases at the rate of \$44,380 per additional 1000 units produced

17. \$1.21 billion per billion dollars

19. \$0.288 billion per billion dollars

Chapter 9 Review Exercises, page 582

1. -3 **2.** 2   **3.** -21 **4.** 0   **5.** -1

6. The limit does not exist.   **7.** 7   **8.** $\frac{9}{2}$

9. 1   **10.** $\frac{1}{2}$ **11.** 1   **12.** 1   **13.** $\frac{3}{2}$

14. The limit does not exist.

15. 1; 1; 1

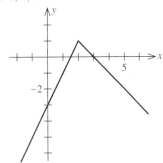

16. 4; 2; the limit does not exist

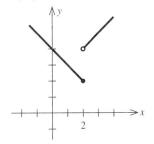

17. 2   **18.** $-\frac{1}{2}$; 1   **19.** -1 **20.** 0

21. a. 3, 2.5, 2.1   **b.** 2   **22.** 3

23. -4; $y = -4x + 4$ **24. a.** Yes   **b.** No

25. $15x^4 - 8x^3 + 6x - 2$ **26.** $24x^5 + 8x^3 + 6x$

27. $\dfrac{6}{x^4} - \dfrac{3}{x^2}$ **28.** $4t - 9t^2 + \frac{1}{2}t^{-3/2}$ **29.** $-\dfrac{1}{t^{3/2}} - \dfrac{6}{t^{5/2}}$

30. $2x - \dfrac{2}{x^2}$ **31.** $1 - \dfrac{2}{t^2} - \dfrac{6}{t^3}$ **32.** $4s + \dfrac{4}{s^2} - \dfrac{1}{s^{3/2}}$

33. $2x + \dfrac{3}{x^{5/2}}$ **34.** $-\dfrac{3}{(2x - 1)^2}$ **35.** $\dfrac{2t}{(2t^2 + 1)^2}$

36. $\dfrac{1}{2\sqrt{t}\,(\sqrt{t} + 1)^2}$ **37.** $\dfrac{1}{\sqrt{x}\,(\sqrt{x} + 1)^2}$ **38.** $\dfrac{1 - 2t^2}{(2t^2 + 1)^2}$

39. $\dfrac{2x(x^4 - 2x^2 - 1)}{(x^2 - 1)^2}$ **40.** $3(4x + 1)(2x^2 + x)^2$

41. $72x^2(3x^3 - 2)^7$ **42.** $\dfrac{5(\sqrt{x} + 2)^4}{2\sqrt{x}}$ **43.** $\dfrac{2t}{\sqrt{2t^2 + 1}}$

44. $-2t^2(1 - 2t^3)^{-2/3}$ **45.** $-\dfrac{4(3t - 1)}{(3t^2 - 2t + 5)^3}$

46. $-9x(x - 1)(2x^3 - 3x^2 + 1)^{-5/2}$ **47.** $(1 + 2x)e^{2x}$

48. $\dfrac{e^t}{2\sqrt{t}} + \sqrt{t}e^t + 1$ **49.** $\dfrac{1 - 4t}{2\sqrt{t}e^{2t}}$ **50.** $\dfrac{e^x(x^2 + x + 1)}{\sqrt{1 + x^2}}$

51. $\dfrac{2(e^{2x} + 2)}{(1 + e^{-2x})^2}$ **52.** $4xe^{2x^2 - 1}$ **53.** $(1 - 2x^2)e^{-x^2}$

54. $3e^{2x}(1 + e^{2x})^{1/2}$ **55.** $(x + 1)^2e^x$ **56.** $\ln t + 1$

57. $\dfrac{2xe^{x^2}}{e^{x^2} + 1}$ **58.** $\dfrac{\ln x - 1}{(\ln x)^2}$ **59.** $\dfrac{x - x\ln x + 1}{x(x + 1)^2}$

60. $(x + 2)e^x$ **61.** $\dfrac{4e^{4x}}{e^{4x} + 3}$ **62.** $\dfrac{(r^3 - r^2 + r + 1)e^r}{(1 + r^2)^2}$

63. $\dfrac{1 + e^x(1 - x\ln x)}{x(1 + e^x)^2}$ **64.** $\dfrac{(2x^2 + 2x^2\ln x - 1)e^x}{x(1 + \ln x)^2}$

65. $\dfrac{2(x^2 + 1)(x^2 - 1)}{x^3}$ **66.** $-\dfrac{6x^2 + 8x - 1}{(2x^2 + 1)^3}$

67. $4t^2(5t + 3)(t^2 + t)^3$ or $4t^5(5t + 3)(t + 1)^3$

68. $2x(x + 1)(2x + 1)^2(7x^2 + 7x + 1)$

69. $\dfrac{1}{2\sqrt{x}}(x^2 - 1)^2(13x^2 - 1)$ **70.** $\dfrac{4 - x^3}{2(x^3 + 2)^{3/2}}$

71. $-\dfrac{12x + 25}{2\sqrt{3x + 2}(4x - 3)^2}$ **72.** $-\dfrac{5t + 2}{\sqrt{2t + 1}(t + 1)^4}$

73. $2(12x^2 - 9x + 2)$ **74.** $-\dfrac{1}{4x^{3/2}} + \dfrac{3}{4x^{5/2}}$

75. $\dfrac{2t(t^2 - 12)}{(t^2 + 4)^3}$ **76.** $4e^{-2x}(x - 1)$ **77.** $\dfrac{e^x(1 - e^x)}{(1 + e^x)^3}$

78. $\dfrac{1}{x}$ **79.** $-\dfrac{9}{(3x + 1)^2}$ **80.** $2(15x^4 + 12x^2 + 6x + 1)$

81. $\dfrac{2}{(2x^2 + 1)^{3/2}}$ **82.** $6t(t^2 + 1)(7t^2 + 3)$ **83.** 0

84. -2

85. a. $(2, -25)$ and $(-1, 14)$
    b. $y = -4x - 17;\ y = -4x + 10$

86. a. $(-2, \frac{25}{3})$ and $(1, -\frac{13}{6})$
    b. $y = -2x + \frac{13}{3};\ y = -2x - \frac{1}{6}$

87. $y = -\dfrac{\sqrt{3}}{3}x + \dfrac{4}{3}\sqrt{3}$ **88.** $y = 112x - 80$

89. $y = -(2x - 3)e^{-2}$ **90.** $y = \dfrac{1}{e}$

91. $-\dfrac{48}{(2x - 1)^4};\ (-\infty; \frac{1}{2}) \cup (\frac{1}{2}, \infty)$

92. 20; as the level of production increases without bound, the average cost of producing the commodity steadily decreases and approaches \$20 per unit

93. 200/wk

94. a. \$2.20; \$2.20   **b.** $\dfrac{2500}{x} + 2.2;\ -\dfrac{2500}{x^2}$

    c. $\displaystyle\lim_{x\to\infty}\left(\dfrac{2500}{x} + 2.2\right) = 2.2$

95. a. $-0.02x^2 + 600x$ **b.** $-0.04x + 600$
    c. 200; the sale of the 10,001st phone will bring a revenue of \$200

96. a. $2000x - 0.04x^2;\ -0.000002x^3 - 0.02x^2 + 1000x - 120,000;\ 0.000002x^2 - 0.02x + 1000 + \dfrac{120,000}{x}$

    b. $0.000006x^2 - 0.04x + 1000;\ 2000 - 0.08x;$
    $-0.000006x^2 - 0.04x + 1000;\ 0.000004x - 0.02 - \dfrac{120,000}{x^2}$

    c. 934; 1760; 826
    d. -0.0048; 0.010125; at a production level of 5000, the average cost is decreasing by 0.48 cents/unit. At a production level of 8000, the average cost is increasing by 1.0125 cents/unit.

CHAPTER 10

Exercises 10.1, page 597

1. Decreasing on $(-\infty, 0)$ and increasing on $(0, \infty)$

3. Increasing on $(-\infty, -1) \cup (1, \infty)$ and decreasing on $(-1, 1)$

5. Decreasing on $(-\infty, 0) \cup (2, \infty)$ and increasing on $(0, 2)$

7. Decreasing on $(-\infty, -1) \cup (1, \infty)$ and increasing on $(-1, 1)$

9. Increasing on $(-\infty, \infty)$

11. Decreasing on $(-\infty, \frac{3}{2})$ and increasing on $(\frac{3}{2}, \infty)$

13. Decreasing on $\left(-\infty, -\dfrac{\sqrt{3}}{3}\right) \cup \left(\dfrac{\sqrt{3}}{3}, \infty\right)$ and increasing on $\left(-\dfrac{\sqrt{3}}{3}, \dfrac{\sqrt{3}}{3}\right)$

15. Increasing on $(-\infty, -2) \cup (0, \infty)$ and decreasing on $(-2, 0)$

17. Increasing on $(-\infty, 3) \cup (3, \infty)$

19. Decreasing on $(-\infty, 0) \cup (0, 3)$ and increasing on $(3, \infty)$

21. Decreasing on $(-\infty, 2) \cup (2, \infty)$

23. Decreasing on $(-\infty, 1) \cup (1, \infty)$

25. Increasing on $(-\infty, 0) \cup (0, \infty)$

27. Increasing on $(-1, \infty)$

29. Increasing on $(-4, 0)$; decreasing on $(0, 4)$

31. Increasing on $(-\infty, 0)$ and decreasing on $(0, \infty)$

33. Decreasing on $(-\infty, 0)$ and increasing on $(0, \infty)$

35. Increasing on $(-\infty, 0) \cup (0, \infty)$

37. Increasing on $(-\infty, 1)$; decreasing on $(1, \infty)$

39. Relative maximum: $f(0) = 1$; relative minimum: $f(-1) = 0$ and $f(1) = 0$

41. Relative maximum: $f(0) = 0$; relative minimum: $f(4) = -32$

43. Relative minimum: $f(-1) = 0$

45. Relative maximum: $f(-3) = -\frac{9}{2}$; relative minimum: $f(3) = \frac{9}{2}$

47. Relative minimum: $f(2) = -4$

49. Relative minimum: $f(2) = 2$

51. Relative minimum: $f(0) = 2$

53. Relative maximum: $g(0) = 4$; relative minimum: $g(2) = 0$

55. Relative minimum: $F(3) = -5$; relative maximum: $F(-1) = \frac{17}{3}$

57. Relative minimum: $g(3) = -19$

59. Relative minimum: $f(\frac{1}{2}) = \frac{63}{16}$

61. None

63. Relative maximum: $f(-3) = -4$; relative minimum: $f(3) = 8$

65. Relative maximum: $f(1) = \frac{1}{2}$; relative minimum: $f(-1) = -\frac{1}{2}$

67. Relative maximum: $f(0) = 0$

69. Relative minimum: $f(1) = 0$

71.

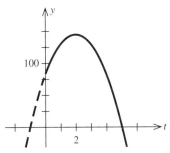

Rising in the time interval $(0, 2)$; falling in the time interval $(2, 5)$; when $t = 5$ sec

73. Rising on $(0, 33)$; descending on $(33, T)$

75. Decreasing on $(0, 1)$; increasing on $(1, 4)$; the average speed of a vehicle on Route 134 was decreasing from 6 A.M. to 7 A.M. and increasing from 7 A.M. to 10 A.M.

77. Increasing from 7 A.M. to 10 A.M.; decreasing from 10 A.M. to 2 P.M.

79. $N'(t) = \dfrac{180}{(t + 6)^2}$ is always positive.

81. Increasing on $(0, 4.5)$; decreasing on $(4.5, 11)$; the air pollution was increasing from 7 A.M. to 11:30 A.M. and decreasing from 11:30 A.M. to 6 P.M.

Using Technology Exercises 10.1, page 603

1. a. f is decreasing on $(-\infty, -0.2934)$ and increasing on $(-0.2934, \infty)$
   b. Relative minimum: $f(-0.2934) = -2.5435$

3. a. f is increasing on $(-\infty, -1.6144) \cup (0.2390, \infty)$ and decreasing on $(-1.6144, 0.2390)$
   b. Relative maximum: $f(-1.6144) = 26.7991$; relative minimum: $f(0.2390) = 1.6733$

5. a. f is decreasing on $(-\infty, -1) \cup (0.33, \infty)$ and increasing on $(-1, 0.33)$
   b. Relative maximum: $f(0.33) = 1.11$; relative minimum: $f(-1) = -0.63$

7. a. f is decreasing on $(-\infty, 0.40)$ and increasing on $(0.40, \infty)$
   b. Relative minimum: $f(0.40) = 0.79$

9. a.

   b. f is decreasing on $(0, 0.2398) \cup (6.8758, 12)$ and increasing on $(0.2398, 6.8758)$
   c. $(6.8758, 200.14)$

11. a.

b. f is decreasing on $(0, 0.8343) \cup (7.6726, 12)$ and increasing on $(0.8343, 7.6726)$

Exercises 10.2, page 613

1. Concave downward on $(-\infty, 0)$ and concave upward on $(0, \infty)$; inflection point: $(0, 0)$

3. Concave downward on $(-\infty, 0) \cup (0, \infty)$

5. Concave upward on $(-\infty, 0) \cup (1, \infty)$ and concave downward on $(0, 1)$; inflection points: $(0, 0)$ and $(1, -1)$

7. Concave downward on $(-\infty, -2) \cup (-2, 2) \cup (2, \infty)$

9. a　**11.** b

19. Concave upward on $(-\infty, \infty)$

21. Concave downward on $(-\infty, 0)$; concave upward on $(0, \infty)$

23. Concave upward on $(-\infty, 0) \cup (3, \infty)$; concave downward on $(0, 3)$

25. Concave downward on $(-\infty, 0) \cup (0, \infty)$

27. Concave downward on $(-\infty, 4)$

29. Concave downward on $(-\infty, 2)$; concave upward on $(2, \infty)$

31. Concave upward on $\left(-\infty, -\dfrac{\sqrt{6}}{3}\right) \cup \left(\dfrac{\sqrt{6}}{3}, \infty\right)$; concave downward on $\left(-\dfrac{\sqrt{6}}{3}, \dfrac{\sqrt{6}}{3}\right)$

33. Concave downward on $(-\infty, 1)$; concave upward on $(1, \infty)$

35. Concave upward on $(-\infty, 0) \cup (0, \infty)$

37. Concave upward on $(-\infty, 2)$; concave downward on $(2, \infty)$

39. Concave downward on $(-\infty, -2)$; concave upward on $(-2, \infty)$

41. Concave downward on $(0, e^{3/2})$; concave upward on $(e^{3/2}, \infty)$

43. $(0, -2)$　**45.** $(1, -15)$　**47.** $(0, 1)$ and $\left(\frac{2}{3}, \frac{11}{27}\right)$

49. $(0, 0)$　**51.** $(1, 2)$　**53.** $(1, e^{-2})$

55. $(-1, \ln 2)$ and $(1, \ln 2)$

57. Relative maximum: $f(1) = 5$

59. None

61. Relative maximum: $f(-1) = -\frac{22}{3}$; relative minimum: $f(5) = -\frac{130}{3}$

63. Relative maximum: $f(-3) = -6$; relative minimum: $f(3) = 6$

65. None

67. Relative minimum: $f(-2) = 12$

69. Relative maximum: $g(1) = \frac{1}{2}$; relative minimum: $g(-1) = -\frac{1}{2}$

71. Relative minimum: $(0, 0)$; relative maximum: $(-2, 4e^{-2})$

73. Relative minimum: $(1, 1)$

75. a. $N_1(t)$ and $N_2(t)$ are increasing on $(0, 12)$.
b. $N_1''(t) < 0$; $N_2''(t) > 0$
c. If the budget is cut, the rate of growth of the number of drug-related crimes will increase; without the budget cuts the rate will decrease.

77. $(100, 4600)$; the sales increase rapidly until $100,000 is spent on advertising; after that any additional expenditure results in increased sales but at a slower rate of increase

79. 10 A.M.

81. $(18, 20,310)$; 1452 ft/sec

Using Technology Exercises 10.2, page 618

1. a. f is concave upward on $(-\infty, 0) \cup (1.1667, \infty)$ and concave downward on $(0, 1.1667)$
b. $(1.1667, 1.1153)$; $(0, 2)$

3. a. f is concave downward on $(-\infty, 0)$ and concave upward on $(0, \infty)$
b. $(0, 2)$

5. a. f is concave downward on $(-\infty, 0)$ and concave upward on $(0, \infty)$
b. $(0, 0)$

7. a. f is concave downward on $(-\infty, 0)$ and concave upward on $(0, \infty)$
b. None

9. a.

11. a.

b. $(5.5318, 35.9483)$
c. $t = 5.5318$

b. $(3.9024, 77.0919)$

Exercises 10.3, page 628

1. $y = 0$ is a horizontal asymptote

3. $y = 0$ is a horizontal asymptote and $x = 0$ is a vertical asymptote

5. $y = 0$ is a horizontal asymptote; $x = -1$ and $x = 1$ are vertical asymptotes

7. $y = 3$ is a horizontal asymptote and $x = 0$ is a vertical asymptote

9. $y = 0$ is a horizontal asymptote

11. $y = 0$ is a horizontal asymptote and $x = 0$ is a vertical asymptote

13. $y = 0$ is a horizontal asymptote and $x = 0$ is a vertical asymptote

15. $y = 1$ is a horizontal asymptote and $x = -1$ is a vertical asymptote

17. None

19. Horizontal asymptote: $y = 1$; vertical asymptotes: $t = -3$ and $t = 3$

21. Horizontal asymptote: $y = 0$; vertical asymptotes: $x = -2$ and $x = 3$

23. Horizontal asymptote: $y = 2$; vertical asymptote: $t = 2$

25. Horizontal asymptote: $y = 1$; vertical asymptotes: $x = -2$ and $x = 2$

27. Vertical asymptote: $x = -1$

29.

31.

33.

35.

37.

39.

41.

43.

45.

47.

49.

51.

53.

55.

57. a. $x = 100$ **b.** No

59. a. $y = 0$

    b. As time passes, the concentration of the drug decreases and approaches zero.

61.

63.

65.

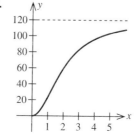

Using Technology Exercises 10.3, page 635

5. -0.9733; 2.3165, 4.6569 **7.** -1.1301; 2.9267

9. 1.5142 **11.** -0.7680, 1.6783

Exercises 10.4, page 642

1. None

3. Absolute minimum value: 0

5. Absolute maximum value: 3; absolute minimum value: -2

7. Absolute maximum value: 3; absolute minimum value: $-\frac{27}{16}$

9. Absolute minimum value: $-\frac{41}{8}$

11. No absolute extremum

13. Absolute maximum value: 1

15. Absolute maximum value: 5; absolute minimum value: -4

17. Absolute maximum value: 10; absolute minimum value: 1

19. Absolute maximum value: 19; absolute minimum value: -1

21. Absolute maximum value: 16; absolute minimum value: -1

23. No absolute extremum

25. Absolute maximum value: $\frac{37}{3}$; absolute minimum value: 5

27. Absolute maximum value ≈ 1.04; absolute minimum value: -1.5

29. Absolute maximum value: 1; absolute minimum value: 0

31. Absolute maximum value: 0; absolute minimum value: -3

33. Absolute maximum value: $\frac{\sqrt{2}}{4} \approx 0.35$; absolute minimum value: $-\frac{1}{3}$

35. Absolute maximum value: 1; absolute minimum value: e^{-4}

37. Absolute minimum value: 0; absolute maximum value: $\frac{\sqrt{2}}{2} e^{-1/2} \approx 0.4289$

39. Absolute minimum value: e; absolute maximum value: 3.1067

41. Absolute maximum value: $\frac{\sqrt{2}}{2}$; absolute minimum value: $-\frac{\sqrt{2}}{2}$

43. 144 ft   **45.** 3000   **49.** \$3600   **51.** 6000

53. 110 units

55. a. $0.000002x^2 + 5 + \frac{400}{x}$ **b.** 464 cases
c. 464 cases   **d.** Same

57. $t = 7.72$ yrs; \$160,208   **59.** 11:30 A.M.   **61.** 10,000

Using Technology Exercises 10.4, page 647

1. Absolute maximum value: 145.8985; absolute minimum value: -4.3834

3. Absolute maximum value: 16; absolute minimum value: -0.1257

5. Absolute maximum value: 11.8922; absolute minimum value: 0

7. a.

b. 200.1410 banks per year

9. a.

b. Absolute maximum value: 108.8756; absolute minimum value: 49.7773

Exercises 10.5, page 654

1. 750 yd $\times$ 1500 yd; 1,125,000 sq yd

3. $10\sqrt{2}$ ft $\times$ $40\sqrt{2}$ ft

5. $\frac{16}{3}$ inches $\times$ $\frac{16}{3}$ inches $\times$ $\frac{4}{3}$ inches

7. 5.04 inches $\times$ 5.04 inches $\times$ 5.04 inches

9. 18 inches $\times$ 18 inches $\times$ 36 inches; 11,664 cu in.

11. $r = \frac{36}{\pi}$ inches, $l = 36$ inches; $\frac{46,656}{\pi}$ cu in.

13. $x = \frac{28}{4+\pi}, y = \frac{28}{4+\pi}$ **15.** 20 trees/acre

17. $x = 750\sqrt{2}$ **19.** 56.57 mph   **21.** 45; 44,721

Chapter 10 Review Exercises, page 658

1. **a.** f is increasing on $(-\infty, 1) \cup (1, \infty)$
   b. No relative extrema
   c. Concave down on $(-\infty, 1)$; concave up on $(1, \infty)$
   d. $(1, -\frac{17}{3})$

2. **a.** f is increasing on $(-\infty, 2) \cup (2, \infty)$
   b. No relative extrema
   c. Concave down on $(-\infty, 2)$; concave up on $(2, \infty)$
   d. $(2, 0)$

3. **a.** f is increasing on $(-1, 0) \cup (1, \infty)$ and decreasing on $(-\infty, -1) \cup (0, 1)$
   b. Relative maximum value: 0; relative minimum value: -1
   c. Concave up on $\left(-\infty, -\frac{\sqrt{3}}{3}\right) \cup \left(\frac{\sqrt{3}}{3}, \infty\right)$; concave down on $\left(-\frac{\sqrt{3}}{3}, \frac{\sqrt{3}}{3}\right)$
   d. $\left(-\frac{\sqrt{3}}{3}, -\frac{5}{9}\right)$; $\left(\frac{\sqrt{3}}{3}, -\frac{5}{9}\right)$

4. **a.** f is increasing on $(-\infty, -2) \cup (2, \infty)$ and decreasing on $(-2, 0) \cup (0, 2)$
   b. Relative maximum value: -4; relative minimum value: 4
   c. Concave down on $(-\infty, 0)$; concave up on $(0, \infty)$
   d. None

5. **a.** f is increasing on $(-\infty, 0) \cup (2, \infty)$; decreasing on $(0, 1) \cup (1, 2)$
   b. Relative maximum value: 0; relative minimum value: 4
   c. Concave up on $(1, \infty)$; concave down on $(-\infty, 1)$
   d. None

6. **a.** f is increasing on $(1, \infty)$
   b. No relative extrema
   c. Concave down on $(1, \infty)$
   d. None

7. **a.** f is decreasing on $(-\infty, 1) \cup (1, \infty)$
   b. No relative extrema
   c. Concave down on $(-\infty, 1)$; concave up on $(1, \infty)$
   d. $(1, 0)$

8. **a.** f is increasing on $(1, \infty)$
   b. No relative extrema
   c. Concave down on $(1, \frac{4}{3})$; concave up on $(\frac{4}{3}, \infty)$
   d. $\left(\frac{4}{3}, \frac{4\sqrt{3}}{9}\right)$

9. **a.** f is increasing on $(-\infty, -1) \cup (-1, \infty)$
   b. No relative extrema
   c. Concave down on $(-1, \infty)$; concave up on $(-\infty, -1)$
   d. None

10. **a.** f is decreasing on $(-\infty, 0)$ and increasing on $(0, \infty)$
    b. Relative minimum value: -1
    c. Concave down on $\left(-\infty, -\frac{1}{\sqrt{3}}\right) \cup \left(\frac{1}{\sqrt{3}}, \infty\right)$; concave up on $\left(-\frac{1}{\sqrt{3}}, \frac{1}{\sqrt{3}}\right)$
    d. $\left(-\frac{1}{\sqrt{3}}, -\frac{3}{4}\right)$; $\left(\frac{1}{\sqrt{3}}, -\frac{3}{4}\right)$

11. **a.** f is increasing on $(-\infty, 3)$ and decreasing on $(3, \infty)$
    b. Relative maximum value: e^3
    c. Concave up on $(-\infty, 2)$; concave down on $(2, \infty)$
    d. $(2, 2e^2)$

12. **a.** f is decreasing on $(0, 0.61)$ and increasing on $(0.61, \infty)$
    b. Relative minimum value: $(0.61, -0.18)$
    c. Concave down on $(0, 0.22)$; concave up on $(0.22, \infty)$
    d. $(0.22, -0.07)$

13.

14.

15.

16.

17.

18.

19.

20.

21.

22.

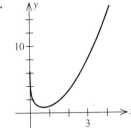

23. Vertical asymptote: $x = -\frac{3}{2}$; horizontal asymptote: $y = 0$

24. Horizontal asymptote: $y = 2$; vertical asymptote: $x = -1$

25. Vertical asymptotes: $x = -2, x = 4$; horizontal asymptote: $y = 0$

26. Horizontal asymptote: $y = 1$; vertical asymptote: $x = 1$

27. Absolute minimum value: $-\frac{25}{8}$

28. Absolute maximum value: 5; absolute minimum value: 0

29. Absolute maximum value: $\frac{5}{3}$; absolute minimum value: 1

30. Absolute maximum value: -16; absolute minimum value: -32

31. Absolute maximum value: $\frac{1}{2}$; absolute minimum value: 0

32. Absolute maximum value: $\frac{8}{3}$; absolute minimum value: 0

33. Absolute maximum value: $\frac{215}{9}$; absolute minimum value: 7

34. Absolute maximum value: $1/e$; absolute minimum value: $-2e^2$

35. Absolute maximum value: $(\ln 2)/2$; absolute minimum value: 0

36. Absolute maximum value: $\frac{1}{2}$; absolute minimum value: $-\frac{1}{2}$

37. No absolute extrema

38. $4000

39. a. $I'(t) = -\dfrac{200t}{(t^2 + 10)^2}$

   b. $I''(t) = \dfrac{-200(10 - 3t^2)}{(t^2 + 10)^3}$; concave up on $(\sqrt{10/3}, \infty)$; concave down on $(0, \sqrt{10/3})$

c.

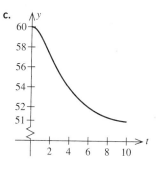

d. The rate of decline in the environmental quality of the wildlife was increasing the first 1.8 years. After that time the rate of declined decreased.

40. 168   **41. a.** $0.001x + 100 + \dfrac{4000}{x}$ **b.** 2000

42. 10 A.M.   **43.** 1 ft × 2 ft × 2 ft   **44.** 20,000 cases

CHAPTER 11

Exercises 11.1, page 671

5. b. $y = 2x + C$
c.

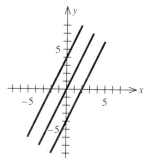

7. b. $y = \frac{1}{3}x^3 + C$
c.

9. $6x + C$ **11.** $\frac{1}{4}x^4 + C$ **13.** $-\dfrac{1}{3x^3} + C$

15. $\frac{3}{5}x^{5/3} + C$ **17.** $-\dfrac{4}{x^{1/4}} + C$ **19.** $-\dfrac{2}{x} + C$

21. $\frac{2}{3}\pi t^{3/2} + C$ **23.** $3x - x^2 + C$

25. $\frac{1}{3}x^3 + \frac{1}{2}x^2 - \dfrac{1}{2x^2} + C$ **27.** $4e^x + C$

29. $x + \frac{1}{2}x^2 + e^x + C$ **31.** $x^4 + \dfrac{2}{x} - x + C$

33. $\frac{2}{7}x^{7/2} + \frac{4}{5}x^{5/2} - \frac{1}{2}x^2 + C$ **35.** $\frac{2}{3}x^{3/2} + 6\sqrt{x} + C$

37. $\frac{1}{9}u^3 + \frac{1}{3}u^2 - \frac{1}{3}u + C$ **39.** $\frac{2}{3}t^3 - \frac{3}{2}t^2 - 2t + C$

41. $\frac{1}{3}x^3 - 2x - \dfrac{1}{x} + C$ **43.** $\frac{1}{3}s^3 + s^2 + s + C$

45. $e^t + \dfrac{t^{e+1}}{e+1} + C$ **47.** $\frac{1}{2}x^2 + x - \ln|x| - \dfrac{1}{x} + C$

49. $\ln|x| + \dfrac{4}{\sqrt{x}} - \dfrac{1}{x} + C$ **51.** $x^3 + 2x^2 - x - 5$

53. $x + \ln|x|$ **55.** $\sqrt{x}$

57. $e^x + \frac{1}{2}x^2 + 2$ **59.** $f(t) = \frac{4}{3}t^{3/2}$

61. a. $R(x) = -0.0045x^2 + 12x$ **b.** $p = -0.0045x + 12$

63. $0.001x^2 + 100x + 4000$

65. a. $-16t^2 + 400$ **b.** 5 sec   **c.** -160 ft/sec

67. 3.375 parts per million   **69.** $1.0974t^3 - 0.0915t^4$

71. $-t^3 + 96t^2 + 120t$; 63,000 ft

Exercises 11.2, page 683

1. $\frac{1}{5}(4x + 3)^5 + C$ **3.** $\frac{1}{3}(x^3 - 2x)^3 + C$

5. $-\dfrac{1}{2(2x^2 + 3)^2} + C$ **7.** $\frac{2}{3}(t^3 + 2)^{3/2} + C$

9. $\frac{1}{20}(x^2 - 1)^{10} + C$ **11.** $-\frac{1}{5}\ln|1 - x^5| + C$

13. $\ln(x - 2)^2 + C$ **15.** $\frac{1}{2}\ln|0.3x^2 - 0.4x + 2| + C$

17. $\frac{1}{6}\ln|3x^2 - 1| + C$ **19.** $-\frac{1}{2}e^{-2x} + C$ **21.** $-e^{2-x} + C$

23. $-\frac{1}{2}e^{-x^2} + C$ **25.** $e^x + e^{-x} + C$

27. $\ln(1 + e^x) + C$ **29.** $2e^{\sqrt{x}} + C$

31. $-\dfrac{1}{6(e^{3x} + x^3)^2} + C$ **33.** $\frac{1}{8}(e^{2x} + 1)^4 + C$

35. $\frac{1}{2}(\ln 5x)^2 + C$ **37.** $\ln|\ln x| + C$ **39.** $\frac{2}{3}(\ln x)^{3/2} + C$

41. $\frac{1}{2}e^{x^2} - \frac{1}{2}\ln(x^2 + 2) + C$

43. $\frac{2}{3}(\sqrt{x} - 1)^3 + 3(\sqrt{x} - 1)^2 + 8(\sqrt{x} - 1)$
$\qquad + 4\ln|\sqrt{x} - 1| + C$

45. $\dfrac{(6x + 1)(x - 1)^6}{42} + C$

47. $5 + 4\sqrt{x} - x - 4\ln(1 + \sqrt{x}) + C$

49. $-\frac{1}{252}(1 - v)^7(28v^2 + 7v + 1) + C$

51. $\frac{1}{2}[(2x - 1)^5 + 5]$ **53.** $e^{-x^2+1} - 1$

55. $21{,}000 - \dfrac{20{,}000}{\sqrt{1 + 0.2t}}; 6858$ **57.** $p(x) = \dfrac{250}{\sqrt{16 + x^2}}$

59. $30(\sqrt{2t + 4} - 2); 14{,}400\pi$ sq ft

61. $\dfrac{71.86887}{1 + 2.449e^{-0.3277t}} - 1.43760; 59.6$ inches

63. 24,555 pairs

Exercises 11.3, page 694

1. 4.27 sq units

3. **a.** 6 sq units   **b.** 4.5 sq units
   c. 5.25 sq units   **d.** Yes

5. **a.** 4 sq units   **b.** 4.8 sq units
   c. 4.4 sq units   **d.** Yes

7. **a.** 18.5 sq units   **b.** 18.64 sq units
   c. 18.66 sq units   **d.** $\approx$ 18.7 sq units

9. **a.** 25 sq units   **b.** 21.12 sq units
   c. 19.88 sq units   **d.** $\approx$ 19.9 sq units

11. **a.** 0.0625 sq units   **b.** 0.16 sq units
   c. 0.2025 sq units   **d.** $\approx$ 0.2 sq units

13. 4.64 sq units   **15.** 0.95 sq units   **17.** 9400 sq ft

Exercises 11.4, page 704

1. 6 sq units   **3.** 8 sq units   **5.** 12 sq units

7. 9 sq units   **9.** ln 2 sq units   **11.** $17\frac{1}{3}$ sq units

13. $18\frac{1}{4}$ sq units   **15.** $(e^2 - 1)$ sq units

17. 6   **19.** 14   **21.** $18\frac{2}{3}$ **23.** $\frac{4}{3}$ **25.** 45   **27.** $\frac{7}{12}$

29. ln 2   **31.** 56   **33.** $\frac{256}{15}$ **35.** $\frac{2}{3}$ **37.** $2\frac{2}{3}$ **39.** $19\frac{1}{2}$

41. **a.** \$4100   **b.** \$900   **43.** **a.** \$2800   **b.** \$219.20

45. $10{,}133\frac{1}{3}$ ft   **47.** 46% of the smoke; 24% of the smoke

Using Technology Exercises 11.4, page 706

1. 6.1787   **3.** 0.7873   **5.** -0.5888 **7.** 2.7044

9. 3.9973   **11.** 37.7 million

Exercises 11.5, page 713

1. 10   **3.** $\frac{19}{15}$ **5.** $32\frac{4}{15}$ **7.** $\sqrt{3} - 1$ **9.** $24\frac{1}{5}$ **11.** $\frac{32}{15}$

13. $18\frac{2}{15}$ **15.** $\frac{1}{2}(e^4 - 1)$ **17.** $\frac{1}{2}e^2 + \frac{5}{6}$ **19.** 0

21. 2 ln 4   **23.** $\frac{1}{3}(\ln 19 - \ln 3)$ **25.** $2e^4 - 2e^2 - \ln 2$

27. $\frac{1}{2}(e^{-4} - e^{-8} - 1)$ **29.** 5   **31.** $\frac{17}{3}$ **33.** -1 **35.** $\frac{13}{6}$

37. $\frac{1}{4}(e^4 - 1)$ **39.** 120.3 billion metric tons

41. $\approx$\$2.24 million   **43.** \$40,339.50   **45.** 26°F

47. 16,863   **49.** 0.071 mg/cu cm/hr

51. $\dfrac{2k}{3} R^2$ cm/sec

Exercises 11.6, page 723

1. 108 sq units   **3.** $\frac{2}{3}$ sq units   **5.** $2\frac{2}{3}$ sq units

7. $1\frac{1}{2}$ sq units   **9.** 3   **11.** $3\frac{1}{3}$ **13.** 27

15. $2(e^2 - e^{-1})$ **17.** $12\frac{2}{3}$ **19.** $3\frac{1}{3}$ **21.** $4\frac{3}{4}$

23. $12 - \ln 4$ **25.** $e^2 - e - \ln 2$ **27.** $2\frac{1}{2}$ **29.** $7\frac{1}{3}$ **31.** $\frac{3}{2}$

33. $e^3 - 4 + \dfrac{1}{e}$ **35.** $20\frac{5}{6}$ **37.** $\frac{1}{12}$ **39.** $\frac{1}{3}$

41. S is the additional revenue that Odyssey Travel could realize by switching to the new agency;
$$S = \int_0^b [g(x) - f(x)]\, dx$$

43. S is the additional amount of smoke that brand B will remove over brand A in the time interval $[a, b]$;
$$S = \int_a^b [f(t) - g(t)]\, dt$$

45. 42.79 million metric tons

47. 57,179 people

Using Technology Exercises 11.6, page 728

1. a.   **3. a.**

b. 1074.2857   **b.** 0.9961

5. a.   **7. a.**

b. 5.4603   **b.** 25.8549

9. a.

b. 10.5144

11. a.

b. 3.5799

Exercises 11.7, page 739

1. $11,667　**3.** $6,667　**5.** $11,667

7. Consumers' surplus: $13,333;
producers' surplus: $11,667

9. $824,200　**11.** $148,239　**13.** $52,203

15. $76,615　**17.** $111,869　**19.** $20,964

21. a.

b. 0.175; 0.816

23. a.

b. 0.104; 0.504

Using Technology Exercises 11.7, page 742

1. Consumers' surplus: $18,000,000;
producers' surplus: $11,700,000

3. Consumers' surplus: $33,120;
producers' surplus: $2,880

Chapter 11 Review Exercises, page 744

1. $\frac{1}{4}x^4 + \frac{2}{3}x^3 - \frac{1}{2}x^2 + C$　**2.** $\frac{1}{12}x^4 - \frac{2}{3}x^3 + 8x + C$

3. $\frac{1}{5}x^5 - \frac{1}{2}x^4 - \frac{1}{x} + C$　**4.** $\frac{3}{4}x^{4/3} - \frac{2}{3}x^{3/2} + 4x + C$

5. $\frac{1}{2}x^4 + \frac{2}{5}x^{5/2} + C$　**6.** $\frac{2}{7}x^{7/2} - \frac{1}{3}x^3 + \frac{2}{3}x^{3/2} - x + C$

7. $\frac{1}{3}x^3 - \frac{1}{2}x^2 + 2\ln|x| + 5x + C$

8. $\frac{1}{3}(2x + 1)^{3/2} + C$　**9.** $\frac{3}{8}(3x^2 - 2x + 1)^{4/3} + C$

10. $\dfrac{(x^3 + 2)^{11}}{33} + C$　**11.** $\frac{1}{2}\ln(x^2 - 2x + 5) + C$

12. $-e^{-2x} + C$　**13.** $\frac{1}{2}e^{x^2 + x + 1} + C$　**14.** $\dfrac{1}{e^{-x} + x} + C$

15. $\frac{1}{6}(\ln x)^6 + C$　**16.** $(\ln x)^2 + C$

17. $\dfrac{(11x^2 - 1)(x^2 + 1)^{11}}{264} + C$

18. $\frac{2}{15}(3x - 2)(x + 1)^{3/2} + C$　**19.** $\frac{2}{3}(x + 4)\sqrt{x - 2} + C$

20. $2(x - 2)\sqrt{x + 1} + C$　**21.** $\frac{1}{2}$　**22.** -6　**23.** $5\frac{2}{3}$

24. 242　**25.** -80　**26.** $\frac{132}{5}$　**27.** $\frac{1}{2}\ln 5$　**28.** $\frac{1}{15}$　**29.** 4

30. $1 - \dfrac{1}{e^2}$　**31.** $\dfrac{e - 1}{2(1 + e)}$　**32.** $\frac{1}{2}$

33. $f(x) = x^3 - 2x^2 + x + 1$　**34.** $f(x) = \sqrt{x^2 + 1}$

35. $f(x) = x + e^{-x} + 1$　**36.** $f(x) = \frac{1}{2}(\ln x)^2 - 2$

37. -4.28　**38.** $6740

39. a. $-0.015x^2 + 60x$;　**b.** $p = -0.015x + 60$

40. $V(t) = 1900(t - 10)^2 + 10,000$; $40,400

41. $3000t - 50,000(1 - e^{-0.04t})$; 16,939

42. $N(t) = 15,000\sqrt{1 + 0.4t} + 85,000$; 112,659

43. $3100　**44.** 19.4 billion metric tons　**45.** 15 sq units

46. $\frac{1}{2}(e^4 - 1)$ sq units　**47.** $\frac{2}{3}$ sq units　**48.** $\frac{9}{2}$ sq units

49. $(e^2 - 3)$ sq units　**50.** $\frac{3}{10}$ sq units　**51.** $\frac{1}{2}$ sq units

52. 234,500 barrels　**53.** $\frac{1}{3}$　**54.** $270,000

55. Consumers' surplus: $2083;
producers' surplus: $3333

56. $197,652　**57.** $98,973　**58.** $505,696

59. a.

b. 0.1017; 0.3733　**c.** 0.315

60. 70,784

CHAPTER 12

Exercises 12.1, page 755

1. $f(0, 0) = -4$; $f(1, 0) = -2$; $f(0, 1) = -1$; $f(1, 2) = 4$; $f(2, -1) = -3$

3. $f(1, 2) = 7$; $f(2, 1) = 9$; $f(-1, 2) = 1$; $f(2, -1) = 1$

5. $g(1, 2) = 4 + 3\sqrt{2}$; $g(2, 1) = 8 + \sqrt{2}$; $g(0, 4) = 2$; $g(4, 9) = 56$

7. $h(1, e) = 1$; $h(e, 1) = -1$; $h(e, e) = 0$

9. $g(1, 1, 1) = e$; $g(1, 0, 1) = 1$; $g(-1, -1, -1) = -e$

11. All real values of x and y

13. All real values of u and v except those satisfying the equation $u = v$

15. All real values of r and s satisfying $rs \geq 0$

17. All real values of x and y satisfying $x + y > 5$

19.

21.

23.

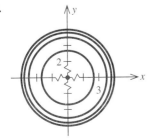

25. 9π cu ft **27.** $40,000k$ dynes

29. $25,500$; $23,580$ **31.** 8310; 7910

33. a. The set of all nonnegative values of W and H
b. 1.871 sq meters

35. $13,498.59$ **37.** $76,704.11$; $50,814.62$ **39.** 18

Exercises 12.2, page 768

1. 2; 3 **3.** $4x$; 4 **5.** $-\dfrac{4y}{x^3}$; $\dfrac{2}{x^2}$ **7.** $\dfrac{2v}{(u+v)^2}$; $-\dfrac{2u}{(u+v)^2}$

9. $3(2s - t)(s^2 - st + t^2)^2$; $3(2t - s)(s^2 - st + t^2)^2$

11. $\dfrac{4x}{3(x^2 + y^2)^{1/3}}$; $\dfrac{4y}{3(x^2 + y^2)^{1/3}}$ **13.** ye^{xy+1}; xe^{xy+1}

15. $\ln y + \dfrac{y}{x}$, $\dfrac{x}{y} + \ln x$ **17.** $e^u \ln v$; $\dfrac{e^u}{v}$

19. $yz + y^2 + 2xz$, $xz + 2xy + z^2$; $xy + 2yz + x^2$

21. ste^{rst}; rte^{rst}; rse^{rst} **23.** $f_x(1, 2) = 8$; $f_y(1, 2) = 5$

25. $f_x(2, 1) = 1$; $f_y(2, 1) = 3$

27. $f_x(1, 2) = \frac{1}{2}$; $f_y(1, 2) = -\frac{1}{4}$

29. $f_x(1, 1) = e$; $f_y(1, 1) = e$

31. $f_x(1, 0, 2) = 0$; $f_y(1, 0, 2) = 8$; $f_z(1, 0, 2) = 0$

33. $f_{xx} = 2y$, $f_{xy} = 2x + 3y^2 = f_{yx}$; $f_{yy} = 6xy$

35. $f_{xx} = 2$; $f_{xy} = f_{yx} = -2$; $f_{yy} = 4$

37. $f_{xx} = \dfrac{y^2}{(x^2 + y^2)^{3/2}}$; $f_{xy} = f_{yx} = -\dfrac{xy}{(x^2 + y^2)^{3/2}}$;

$f_{yy} = \dfrac{x^2}{(x^2 + y^2)^{3/2}}$

39. $f_{xx} = \dfrac{1}{y^2} e^{-x/y}$; $f_{xy} = \dfrac{y - x}{y^3} e^{-x/y} = f_{yx}$;

$f_{yy} = \dfrac{x}{y^3}\left(\dfrac{x}{y} - 2\right) e^{-x/y}$

41. a. $f_x = 7.5$; $f_y = 40$ **b.** Yes

43. $P_x = 10$; at $(0, 1)$, the price of land is changing at the rate of $10 per sq ft per mile change to the right. $P_y = 0$; at $(0, 1)$, the price of land is constant per mile change upward.

45. Complementary commodities

47. $30/unit change in finished desks; $-$25/unit change in unfinished desks. The weekly revenue increases by $30/unit for each additional finished desk produced (beyond 300) when the level of production of unfinished desks remains fixed at 250; the revenue decreases by $25/unit when each additional unfinished desk (beyond 250) is produced and the level of production of finished desks remains fixed at 300.

49. 0.039 liter/degree; -0.015 liter/mm of mercury. The volume increases by 0.039 liter when the temperature increases by 1 degree (beyond 300 K) and the pressure is fixed at 800 mm of mercury. The volume decreases by 0.15 liter when the pressure increases by 1 mm of mercury (beyond 800 mm) and the temperature is fixed at 300 K.

Using Technology Exercises 12.2, page 771

1. 1.3124; 0.4038   **3.** -1.8889; 0.7778

5. -0.3863; -0.8497

Exercises 12.3, page 780

1. $(0, 0)$; relative maximum value: $f(0, 0) = 1$

3. $(1, 2)$; saddle point: $f(1, 2) = 4$

5. $(8, -6)$; relative minimum value: $f(8, -6) = -41$

7. $(1, 2)$ and $(2, 2)$; saddle point: $f(1, 2) = -1$; relative minimum value: $f(2, 2), = -2$

9. $(-\frac{1}{3}, \frac{11}{3})$ and $(1, 5)$; saddle point: $f(-\frac{1}{3}, \frac{11}{3}) = -\frac{319}{27}$; relative minimum value: $f(1, 5) = -13$

11. $(0, 0)$ and $(1, 1)$; saddle point: $f(0, 0) = -2$; relative minimum value: $f(1, 1) = -3$

13. $(2, 1)$; relative minimum value: $f(2, 1) = 6$

15. $(0, 0)$; saddle point: $f(0, 0) = -1$

17. $(0, 0)$; relative minimum value: $f(0, 0) = 1$

19. $(0, 0)$; relative minimum value: $f(0, 0) = 0$

21. 200 finished units and 100 unfinished units; $10,500

23. Price of land ($200/sq ft) is highest at $(\frac{1}{2}, 1)$.

25. $(0, 1)$ gives desired location

27. 18 inches $\times$ 18 inches $\times$ 36 inches

29. 6 inches $\times$ 4 inches $\times$ 2 inches

Chapter 12 Review Exercises, page 783

1. $0, 0, \frac{1}{2}$; no   **2.** $e, \dfrac{e^2}{1 + \ln 2}, \dfrac{2e}{1 + \ln 2}$; no

3. $2, -(e + 1), -(e + 1)$

4. All ordered pairs (u, v) of real numbers such that $u \neq v$

5. All ordered pairs (x, y) such that $y \neq -x$

6. All ordered pairs (x, y) such that $x \leq 1$ and $y \geq 0$

7. The set of all triplets (x, y, z) such that $z \geq 0$ and $x \neq 1$, $y \neq 1$, and $z \neq 1$.

8. $2x + 3y = z$

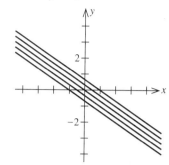

9. $z = y - x^2$

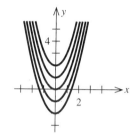

10. $z = \sqrt{x^2 + y^2}$

11. $z = e^{xy}$

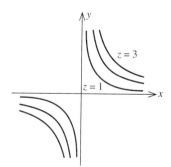

12. $f_x = 2xy^3 + 3y^2 + \dfrac{1}{y}$; $f_y = 3x^2y^2 + 6xy - \dfrac{x}{y^2}$

13. $f_x = \sqrt{y} + \dfrac{y}{2\sqrt{x}}$; $f_y = \dfrac{x}{2\sqrt{y}} + \sqrt{x}$

14. $f_u = \dfrac{v^2 - 2}{2\sqrt{uv^2 - 2u}}; f_v = \dfrac{uv}{\sqrt{uv^2 - 2u}}$

15. $f_x = \dfrac{3y}{(y + 2x)^2}; f_y = -\dfrac{3x}{(y + 2x)^2}$

16. $g_x = \dfrac{y(y^2 - x^2)}{(x^2 + y^2)^2}; g_y = \dfrac{x(x^2 - y^2)}{(x^2 + y^2)^2}$

17. $h_x = 10y(2xy + 3y^2)^4; h_y = 10(x + 3y)(2xy + 3y^2)^4$

18. $f_x = \dfrac{e^y}{2(xe^y + 1)^{1/2}}; f_y = \dfrac{xe^y}{2(xe^y + 1)^{1/2}}$

19. $f_x = 2x(1 + x^2 + y^2)e^{x^2+y^2}; f_y = 2y(1 + x^2 + y^2)e^{x^2+y^2}$

20. $f_x = \dfrac{4x}{1 + 2x^2 + 4y^4}; f_y = \dfrac{16y^3}{1 + 2x^2 + 4y^4}$

21. $f_x = \dfrac{2x}{x^2 + y^2}; f_y = -\dfrac{2x^2}{y(x^2 + y^2)}$

22. $f_{xx} = 6x - 4y; f_{xy} = -4x = f_{yx}; f_{yy} = 2$

23. $f_{xx} = 12x^2 + 4y^2; f_{xy} = 8xy = f_{yx}; f_{yy} = 4x^2 - 12y^2$

24. $f_{xx} = 12(2x^2 + 3y^2)(10x^2 + 3y^2);$
$f_{xy} = 144xy(2x^2 + 3y^2) = f_{yx};$
$f_{yy} = 18(2x^2 + 3y^2)(2x^2 + 15y^2)$

25. $g_{xx} = \dfrac{-2y^2}{(x + y^2)^3}; g_{xy} = \dfrac{2y(x - y^2)}{(x + y^2)^3} = g_{yx};$

$g_{yy} = \dfrac{2x(3y^2 - x)}{(x + y^2)^3}$

26. $g_{xx} = 2(1 + 2x^2)e^{x^2+y^2}; g_{xy} = 4xye^{x^2+y^2} = g_{yx};$
$g_{yy} = 2(1 + 2y^2)e^{x^2+y^2}$

27. $h_{ss} = -\dfrac{1}{s^2}; h_{st} = h_{ts} = 0; h_{tt} = \dfrac{1}{t^2}$

28. $f_x(1, 1, 0) = 3; f_y(1, 1, 0) = 3; f_z(1, 1, 0) = -2$

29. $(2, 3)$; relative minimum value: $f(2, 3) = -13$

30. $(8, -2)$; saddle point at $f(8, -2) = -8$

31. $(0, 0)$ and $(\frac{3}{2}, \frac{9}{4})$; saddle point at $f(0, 0) = 0$; relative minimum value: $f(\frac{3}{2}, \frac{9}{4}) = -\frac{27}{16}$

32. $(-\frac{1}{3}, \frac{13}{3})$, $(3, 11)$; saddle point at $f(-\frac{1}{3}, \frac{13}{3}) = (-\frac{445}{27})$; relative minimum value: $f(3, 11) = -35$

33. $(0, 0)$; relative minimum value: $f(0, 0) = 1$

34. $(1, 1)$; relative minimum value: $f(1, 1) = \ln 2$

35. a. $R(x, y) = -0.02x^2 - 0.2xy - 0.05y^2 + 80x + 60y$
b. The set of all points satisfying $0.02x + 0.1y \le 80$, $0.1x + 0.05y \le 60$, $x \ge 0$, $y \ge 0$.
c. 15,300; the revenue realized from the sale of 100 16-speed and 300 10-speed electric blenders is $15,300

36. Complementary

37. The company should spend $11,000 on advertising and employ 14 agents in order to maximize its revenue.

38. 337.5 yd $\times$ 900 yd

INDEX

Properties of Exponents

1. $a^m \cdot a^n = a^{m+n}$

2. $\dfrac{a^m}{a^n} = a^{m-n}$

3. $(a^m)^n = a^{mn}$

4. $(ab)^n = a^n \cdot b^n$

5. $\left(\dfrac{a}{b}\right)^n = \dfrac{a^n}{b^n}$ $(b \neq 0)$

Properties of Radicals

1. $(\sqrt[n]{a})^n = a$

2. $\sqrt[n]{ab} = \sqrt[n]{a} \cdot \sqrt[n]{b}$

3. $\sqrt[n]{\dfrac{a}{b}} = \dfrac{\sqrt[n]{a}}{\sqrt[n]{b}}$ $(b \neq 0)$

4. $\sqrt[n]{\sqrt[m]{a}} = \sqrt[mn]{a}$

5. If n is even: $\sqrt[n]{x^n} = |x|$
   If n is odd: $\sqrt[n]{x^n} = x$

Factoring Formulas

Difference of two squares:
$a^2 - b^2 = (a + b)(a - b)$

Perfect square trinomial:
$a^2 + 2ab + b^2 = (a + b)^2$
$a^2 - 2ab + b^2 = (a - b)^2$

Sum of two cubes:
$a^3 + b^3 = (a + b)(a^2 - ab + b^2)$

Difference of two cubes:
$a^3 - b^3 = (a - b)(a^2 + ab + b^2)$

Equation of a Straight Line

a. point-slope form $y - y_1 = m(x - x_1)$
b. slope-intercept form $y = mx + b$
c. general form $Ax + By + C = 0$

Compound Interest

$A = P(1 + i)^n$ $(i = r/m, \; n = mt)$
where A is the accumulated amount at the end of n conversion periods,
P is the principal, r is the interest rate per year, m is the number of conversion periods per year,
and t is the number of years.

Effective Rate of Interest

$r_{\text{eff}} = \left(1 + \dfrac{r}{m}\right)^m - 1$

where r_{eff} is the effective rate of interest, r is the nominal interest rate per year,
and m is the number of conversion periods per year.

Future Value of an Annuity $S = R\left[\dfrac{(1 + i)^n - 1}{i}\right]$

Present Value of an Annuity $P = R\left[\dfrac{1 - (1 + i)^{-n}}{i}\right]$

Amortization Formula $R = \dfrac{Pi}{1 - (1 + i)^{-n}}$

Sinking Fund Payment $R = \dfrac{iS}{(1 + i)^n - 1}$

FORMULAS

The Number of Permutations of n Distinct Objects

$$P(n, r) = \frac{n!}{(n-r)!}$$

The Number of Permutations of n Objects, Not All Distinct

$$\frac{n!}{n_1! n_2! \ldots n_r!}, \quad \text{where } n_1 + n_2 + \cdots + n_r = n$$

The Number of Combinations of n Distinct Objects

$$C(n, r) = \frac{n!}{r!(n-r)!}$$

The Product Rule for Probability

$$P(A \cap B) = P(A) \cdot P(B \mid A)$$

Bayes' Formula

$$P(A_i \mid E) = \frac{P(A_i) \cdot P(E \mid A_i)}{P(A_1) \cdot P(E \mid A_1) + P(A_2) \cdot P(E \mid A_2) + \cdots + P(A_n) \cdot P(E \mid A_n)}$$

Expected Value of a Random Variable $E(X) = x_1 p_1 + x_2 p_2 + \cdots + x_n p_n$

Basic Rules of Differentiation

1. $\dfrac{d}{dx}(c) = 0, \quad c$ a constant

2. $\dfrac{d}{dx}(u^n) = nu^{n-1}\dfrac{du}{dx}$

6. $\dfrac{d}{dx}\left(\dfrac{u}{v}\right) = \dfrac{v\dfrac{du}{dx} - u\dfrac{dv}{dx}}{v^2}$

3. $\dfrac{d}{dx}(u \pm v) = \dfrac{du}{dx} \pm \dfrac{dv}{dx}$

7. $\dfrac{d}{dx}(e^u) = e^u\dfrac{du}{dx}$

4. $\dfrac{d}{dx}(cu) = c\dfrac{du}{dx}, \quad c$ a constant

8. $\dfrac{d}{dx}(\ln u) = \dfrac{1}{u} \cdot \dfrac{du}{dx}$

5. $\dfrac{d}{dx}(uv) = u\dfrac{dv}{dx} + v\dfrac{du}{dx}$

Basic Rules of Integration

1. $\displaystyle\int du = u + C$

2. $\displaystyle\int kf(u)\, du = k\int f(u)\, du, \quad k$ a constant

3. $\displaystyle\int [f(u) \pm g(u)]\, du = \int f(u)\, du \pm \int g(u)\, du$

4. $\displaystyle\int u^n\, du = \frac{u^{n+1}}{n+1} + C, \quad n \neq -1$

5. $\displaystyle\int e^u\, du = e^u + C$

6. $\displaystyle\int \frac{du}{u} = \ln |u| + C$